Jensen · Rozenberg (Eds.) High-level Petri Nets

K. Jensen G. Rozenberg (Eds.)

High-level Petri Nets

Theory and Application

Springer-Verlag
Berlin Heidelberg New York
London Paris Tokyo
Hong Kong Barcelona
Budapest

Editors

Kurt Jensen

Computer Science Department
Aarhus University, Ny Munkegade, Bldg. 540
DK-8000 Aarhus C, Denmark

Grzegorz Rozenberg

Institute of Applied Mathematics and Computer Science
University of Leiden, Niels-Bohr-Weg 1
2333 CA Leiden, The Netherlands

ISBN-13:978-3-540-54125-7 e-ISBN-13:978-3-642-84524-6
DOI: 10.1007/978-3-642-84524-6

Library of Congress Cataloging-in-Publication Data
High-level Petri nets: theory and application/K. Jensen, G. Rozenberg (eds.). p. cm.
Includes bibliographical references.
ISBN-13:978-3-540-54125-7
1. Petri nets. I. Jensen, K. (Kurt), 1950-. II. Rozenberg, Grzegorz.
QA267.H48 1991 511.3-dc20 91-17121 CIP

45 / 3140-543210

Preface

High-level Petri nets are now widely used in both theoretical analysis and practical modelling of concurrent systems. The main reason for the success of this class of net models is that they make it possible to obtain much more succinct and manageable descriptions than can be obtained by means of low-level Petri nets—while, on the other hand, they still offer a wide range of analysis methods and tools. The step from low-level nets to high-level nets can be compared to the step from assembly languages to modern programming languages with an elaborated type concept. In low-level nets there is only one kind of token and this means that the state of a place is described by an integer (and in many cases even by a boolean value). In high-level nets each token can carry complex information which, e.g., may describe the entire state of a process or a data base.

Today most practical applications of Petri nets use one of the different kinds of high-level nets. A considerable body of knowledge exists about high-level Petri nets—this includes theoretical foundations, analysis methods and many applications. Unfortunately, the papers on high-level Petri nets have been scattered throughout various journals and collections. As a result, much of this knowledge is not readily available to people who may be interested in using high-level nets.

Within the Petri net community this problem has been discussed many times, and as an outcome this book has been compiled. The book contains reprints of some of the most important papers on the application and theory of high-level Petri nets. In this way it makes the relevant literature more available. It is our hope that the book will be a useful source of information and that, e.g., it can be used in the organization of Petri net courses. To make the book as useful as possible, the selected papers represent the current "state of the art" of high-level nets. This means that we have been forced to leave out a number of older papers which have had a profound influence on the development of high-level Petri nets—but by now have been superseded by other papers. Thus, e.g., none of the original papers introducing the first versions of high-level Petri nets have been included. The introductions to the individual sections mention a number of researchers who have contributed to the development of high-level Petri nets. Detailed references to their work can be found in the individual papers and in the available bibliographies of Petri nets, e.g., S. Dress et al.: *Bibliography of Petri Nets.* In: G. Rozenberg (ed.): *Advances in Petri Nets 1987,*

Lecture Notes in Computer Science, Vol. 266, Springer-Verlag 1987, pp. 309-451. Updated versions of this bibliography will appear in *Advances in Petri Nets* volumes at regular intervals.

In the preparation of the book the editors have been assisted by an advisory board consisting of the following well-known Petri net researchers: M. Ajmone-Marsan, H.J. Genrich, C. Girault, W. Reisig, M. Silva and P.S. Thiagarajan. The members of the advisory board have been very active and discussions with them have had a profound influence on the selection criteria, the actual selection of papers, and the entire organization of the book. The members of the advisory board have also been involved in the preparation of the introductions to the individual sections. We are very grateful for the assistance provided by these colleagues. This book would not have been possible without their help. Moreover, we are indebted to the different publishing houses and the authors for their kind permission to reprint the papers. Dr. H. Wössner and Mrs. I. Mayer from Springer-Verlag have been very helpful and cooperative in the difficult task of publishing this book.

April 1991

Kurt Jensen
Grzegorz Rozenberg

Table of Contents

Section A: Predicate / Transition Nets and Coloured Petri Nets

Section B: High-level Nets and Abstract Data Types

Section C: Hierarchical High-level Nets

Section D: Analysis by Means of Invariants

Section E: Analysis by Means of Reachability Graphs

Section F: Analysis by Means of Transformations

Section G: Analysis of Stochastic Nets

Section H: Applications of High-level Nets

Section I: Computer Tools for High-level Nets

Appendix

Section A
Predicate / Transition Nets and Coloured Petri Nets

The papers in this section describe two basic kinds of high-level Petri nets: Predicate/Transition Nets (PrT-nets) and Coloured Petri Nets (CP-nets). Today most of the practical and theoretical work in the area of high-level nets use either PrT-nets or CP-nets. The two models are very similar. The main differences concern the methods to calculate and interpret place and transition invariants. As a matter of fact, rather than viewing PrT-nets and CP-nets as two different modelling languages, it is much more adequate to view them as two slightly different dialects of the same language.

Although the two net models are closely related they are defined and presented in two rather different ways. PrT-nets are defined using the notation and concepts of many-sorted algebras—as known from abstract data types. CP-nets are defined using types, variables and expressions—as known from (functional) programming languages and from lambda calculus. The two papers can be read in any order, and it is recommend to start with the paper building on the set of concepts with which the reader is more familiar.

In addition to the basic model, the PrT-net paper clarifies the relationship between high-level nets and low-level nets, and it introduces projections, facts and synchronic distances. The paper also discusses the use of linear algebra for calculation of invariants. In addition to the basic model, the CP-net paper defines hierarchical Petri nets in which a number of separate nets together form a large complex model, and it gives a brief description of different formal analysis methods. The paper also discusses how computer tools can support the use of high-level nets, and it gives a brief description of four examples of industrial applications.

The history of high-level nets begins with the class of PrT-nets developed by H.J. Genrich and K. Lautenbach. This class of nets was the first one, among various extensions of low-level nets, to have a simple but yet powerful set of primitives covering a large range of applications areas and having a well-defined relationship to low-level nets. The ideas behind PrT-nets build upon earlier net models proposed by H.J. Genrich, K. Lautenbach, M. Schiffers, R.M. Shapiro, G. Thieler-Mevissen and H.

Wedde. Shortly after, CP-nets were defined by K. Jensen. The main ideas of CP-nets are taken from PrT-nets and the new net class was introduced to improve the method of invariant analysis. Finally a third net model, called Relation Nets, was defined by W. Reisig. This net model is closely related to CP-nets. It uses relations where the CP-nets use functions. Parallel to the development of the above models Numerical Petri Nets were developed by F.J.W Symons. This net model is in many respects similar to the others, but the primitives are more complex and less general.

1. Predicate / Transition Nets

H.J. Genrich

W. Brauer, W. Reisig and G. Rozenberg (eds.): Petri nets: central models and their properties.
Advances in Petri nets 1986, part I. Lecture Notes in
Computer Science, vol. 254. Springer, Berlin Heidelberg New York 1987, pp. 207-247

Abstract: The paper deals with conceptual, mathematical and practical aspects of developing a net theoretic system model. The model presented is based on common techniques of modelling static systems as structured sets of individuals (relational structures). These structures are 'dynamised' by allowing some relations between individuals to be changed by the processes of the modelled system.

Keywords: Predicate/transition nets (PrT-nets); higher-level Petri nets; variable relational structures; logical and linear-algebraic system invariants.

Contents

1 Introduction

Net theory is a systems theory that aims at an understanding of systems whose structure and behaviour are determined by the distributedness and combinatorial nature of their states and changes. It studies such systems at different conceptual levels, in various degrees of detail, and in many areas of application. One important branch of research in net theory is concerned with the conceptual and mathematical foundation of an adequate notion of *dynamic system* and its different ways of presentation (its *models*). The *basic* net theoretical system model is that of *condition/event systems (CE-systems)*. It was proposed by Petri as the common reference model of net theory [16]. Other models are considered theoretical in a strict sense if they are derived from or, can be translated into the basic model.

If a dynamic system has an adequate representation in a net theoretical model, we call it a *Petri system*. In this paper we give an example of introducing a net theoretical system model which is called *predicate/transition nets (PrT-nets)* and was first formulated by *Genrich&Lautenbach* [2]. There are various aspects concerned: syntactical and semantical, conceptual and analytical, theoretical and practical ones. The close relationships between these issues, the mutual effects they have on each other during the process of developing the model, constitutes one major difficulty of this presentation.

The central aim of developing PrT-nets has been to introduce, in a formal manner, the concept of *individuals with changing properties and relations* into net theory. The main ideas underlying the PrT-net model are so simple that we wish to demonstrate them at an introductory example before they are buried under all the formalism needed later.

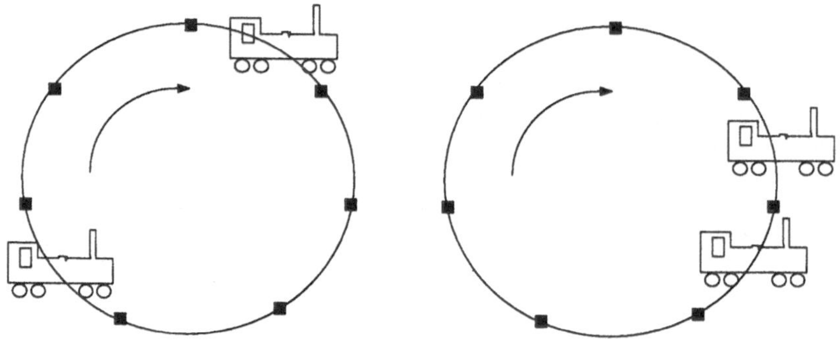

General Constellation　　　　Illegal/Unsafe Constellation (Contact)

Figure 1: Contact-free Movements of Trains on a Circular Track

To this end we consider a small dynamic system sketched in figure 1. It consists of the one-directional movements of two trains on a circular track which is divided into seven sections. Safe operation requires that the movements are *contact-free*: two adjacent sections are never occupied by more than one train at a time.

Figure 2 shows a marked annotated net. It represents a CE-system that by virtue of the interpretation given in the legend is a model of the railway system shown in figure 1. Hence we have decided to view the railway system as a Petri system rather than, for example, a mechanical system.

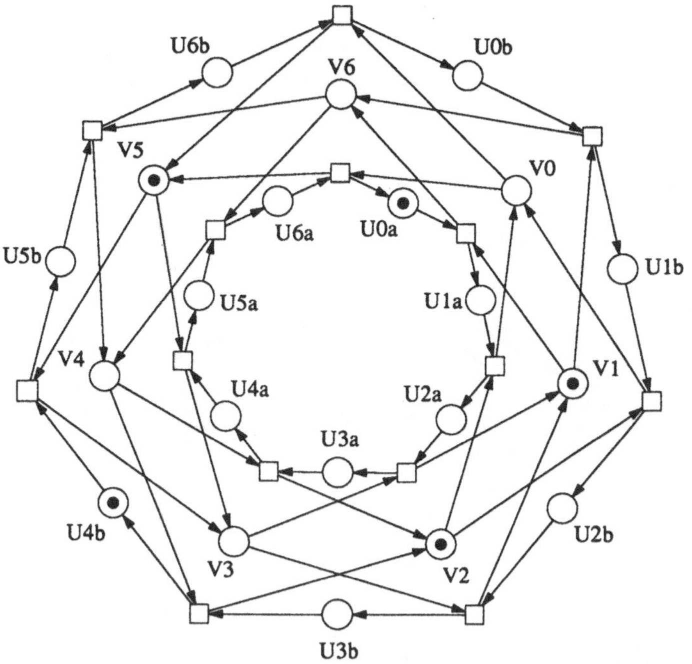

Legend: For $i = 0, \ldots, 6$ and $x = a, b$, Uix means that section i is occupied by train x, and Vi means that the sections i and $(i+1) \, mod \, 7$ are vacant.

Figure 2: Railway System as a CE-system

The annotations of the conditions are expressions that serve as shorthands for affirmative sentences in natural language. They are built from two kinds of symbols. Letters a and b and digits 0 through 6 denote individuals, namely trains and sections, respectively. Letters U and V denote relations between individuals. (We treat properties as special, unary, relations.) These relations are *variable*, ie they are changed by the movements of the trains. In contrast, the cyclic order of the sections on the track is a *static* relation. It is depicted by the flow relation of the net.

The syntactical structure of the condition annotations is that of elementary propositions of first-order predicate logic. This is not by coincidence. Rather, we shall follow logic as a model for developing a net theoretic formalism dealing with individuals and their properties and relations.

As in logic, we call the symbols that are denoting individuals *(individual) constants* and we call the symbols that are denoting relations *predicates*. Using another class of symbols as (individual) *variables*, schemes of propositions can be formulated. Replacing in a proposition one or more occurrences of a constant by variables yields a propositional scheme called a *formula*. If we take, for example, proposition $U1a$ and replace the constant a by the variable x, we get the formula $U1x$ from which both $U1a$ and $U1b$ can be derived by means of the *substitutions* $\{x \leftarrow a\}$ and $\{x \leftarrow b\}$.

The same kind of abstraction is now used to simplify the annotated net representing the railway system considerably. Together with deriving a set of propositions from a formula we introduce schemes for the corresponding sets of conditions; they are called *places*. Figure 3 shows the result of merging, for each section i, conditions (annotated by) Uia and Uib into the place (annotated by) Uix. The events of the original net are now identified by annotating the arcs by constants. These

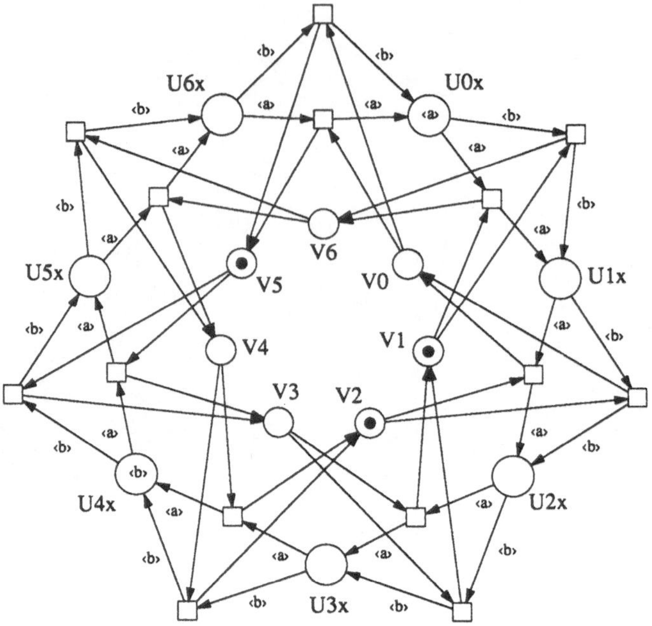

Legend: See figure 2.

Figure 3: Railway System with Sections as Places

are substituted for the variables at the places to get the corresponding condition. In the same way, some of the tokens indicating the current case of the system now carry individual constants.

Now observe how the token game translates. In the original net in figure 2, the event of train a moving from section 0 to section 1 takes a token from conditions $U0a$ and $V1$ and puts a token on conditions $U1a$ and $V6$. In the transformed net in figure 3, the same event takes an $\langle a \rangle$ from place $U0x$ and a plain token from place $V1$, and it puts an $\langle a \rangle$ on place $U1x$ and a plain token on place $V6$.

Looking closer at the net shown in figure 3 we recognise that there are pairs of events which are so similar that they can be viewed as *instances* of a single *event scheme*. For example, the event of train a moving from section 2 to section 3 is quite similar to the event of b moving from 2 to 3. Both events have the same set of input places and the same set of output places. They only differ in the annotations of the corresponding arcs. They can be represented by a single *transition* denoting *some* train moving from section 2 to 3. The result of merging pairs of similar events into transitions is shown in figure 4. The places and their markings are not affected by this transformation. For a transition to occur there must be a substitution of the train variable x at all surrounding arcs such that the resulting event has concession.

Note that this representation comes rather close to the informal presentation in figure 1. The sections are represented by the U places that may be occupied by (symbols for) trains. The borders between sections are represented by the transitions, and the direction of movements is represented by directed arcs connecting places and transitions. The control signals that avoid collisions and dangerous contacts are carried by the V places.

It now seems quite natural to try the same kind of abstraction again with respect to sections. The result of first merging places annotated by similar formulae and then merging similar transitions

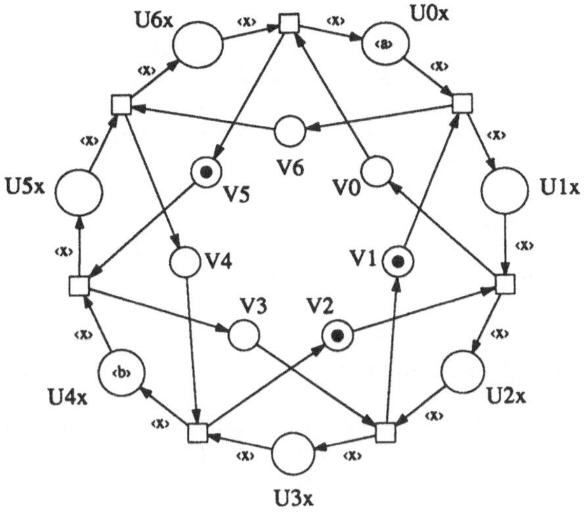

Legend: See figure 2.

Figure 4: Railway System with Places and Transitions

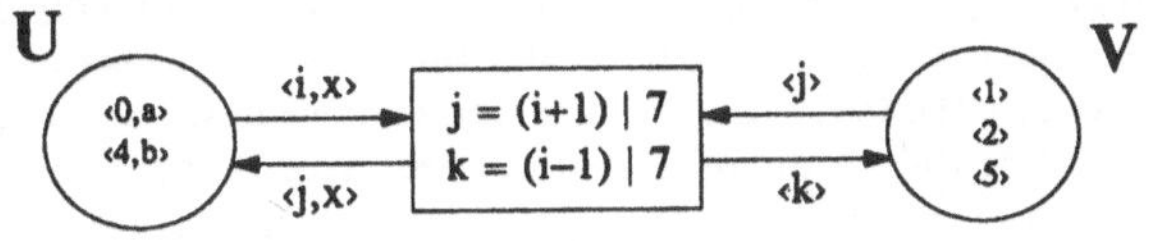

Legend: See figure 2.

Figure 5: Railway System with Variable Relations as Places

is shown in figure 5. The set of events denoted by the single transition in figure 5 cannot be determined any longer by just fixing the range of the variables. Rather, the set of substitutions generating these events is determined by the relation that exists between the sections annotating the arcs of similar transitions. Note that this relation is static; it is not dependent on the states of the system. It is expressed by the (first-order logical) formula annotating the new transition. Usually there are many ways of expressing such relations. We have expressed them in terms of the successor relation between sections in the form of addition of $1 \, modulo \, 7$.

Note again how the token game is affected by the last transformation. For the transition to occur in some 'mode', the variables x, i, j, k must be replaced by constants in such a way that the formula at the transition holds and the tuples generated by that substitution at the arcs can be removed from respectively put on the corresponding places.

The two places of figure 5 represent the two variable relations of the railway system in a one-to-one fashion. To stress this fact, they are annotated by relation symbols, ie predicates, only. This is the reason why the annotated nets of which figures 2 through 5 are examples are called *predicate/transition nets (PrT-nets).*

The rest of this paper is now organised as follows.

First we introduce in chapter 2 the basic form of predicate/transition nets as formal objects. This includes the exact definition of the language for annotating the nets, the semantics of PrT-nets as condition/event systems and the symbolic transition rule for the token game on PrT-nets that is compatible with the semantics.

In chapter 3 the basic form is refined and generalised in several respects to obtain the full modelling power and flexibility of PrT-nets.

The close affinity of Prt-nets to predicate logic that goes further than using logical formulae as annotations is demonstrated in 4.

Finally, we show in chapter 5 how linear algebraic analysis techniques known from the theory of ordinary Petri nets can be transferred to PrT-nets.

2 Predicate/Transition Nets

In the introduction we have sketched the presentation of a Petri system as an annotated net belonging to a class called *predicate/transition nets* (PrT-nets). In this chapter, PrT-nets will be defined as formal objects that can be interpreted and manipulated in a mathematical way that is comparable to working with logical formulae or algebraic expressions.

Our approach is to model a dynamic system in terms of a set of individuals that is structured by functions and relations. The structure is partially static and partially dynamic. (In the railway system of the introduction, for example, the cyclic order of the sections is static and the relation *is_occupied_by* between sections and trains is variable.

Intuitively we are going to dynamise the common notion of a relational structure. A structure is a tuple of objects, $\mathcal{R} = (D; f_1, \ldots, f_k; R_1, \ldots, R_n)$, where D is a non-empty set of *individuals* called the *domain* of $\mathcal{R}$, the f_i are *functions* in D and the R_j are *relations* in D. A *dynamic structure* will be characterised by the fact some relations are *variable* in the sense that their extensions may vary from state to state due to the occurrence of processes in the modelled system.

We shall separate the static and the variable part from each other. The static part will remain an ordinary relational structure. It is the *support* of the dynamic system. The dynamics will be represented as an annotated net, called *predicate/transition net (PrT-net)*, of the kind we have already met in the introduction. Here the variable relations appear as the places of the net. The functions and relations of the static part or more precisely, their names, ie operators respectively predicates, will appear in the annotations. In other words, the annotations of the net are interpreted in terms of a given static relational structure, the support. This will allow to derive from the annotated net the condition/event system that is the net theoretical expression for the dynamic structure. (For our introductory example this would mean to go from the PrT-net of figure 5 back to the CE-net in figure 2.)

2.1 The Language for Structures

We have chosen structured sets of individuals to support the modelling of dynamic systems. Operators (function symbols) and predicates (relation symbols) form the vocabulary of the language in which we will talk formally about structures, ie about properties and relations of individuals.

The language we use is that of *first-order predicate logic* plus a class of simple algebraic expressions for denoting linear combinations.

Definition 2.1 Let for each index $n \geq 0$, $\Omega^{(n)}$ be a set of n-ary operators and $\Pi^{(n)}$ a set of n-ary predicates and let $\Omega = \bigcup_{n \in \mathbf{N}} \Omega^{(n)}$ and $\Pi = \bigcup_{n \in \mathbf{N}} \Pi^{(n)}$. These operators and predicates form the *vocabulary* of the *first-order* language $\mathbf{L}$ that consists of two kinds of expressions, *terms* and *formulae.* In addition, there is a set of symbols, V, disjoint from Ω and Π whose elements serve as (individual) *variables*. Terms and formulae are built in the following way.

1. *Terms*:
   (a) A variable is a term.
   (b) If $f^{(n)}$ is a n-ary operator and $v_1, \ldots, v_n$ are terms then $f(v_1, \ldots, v_n)$ is a term. (Note that 0-ary operators are terms; they are used as proper names of distinct individuals.)
   (c) No other expression is a term.
2. *Formulae*:
   (a) If v_1 and v_2 are terms then $v_1 = v_2$ is an *atomic* formula.
   (b) If $P^{(n)}$ is a n-ary predicate and $v_1, \ldots, v_n$ are terms then $Pv_1 \ldots v_n$ is an atomic formula. (Note that 0-ary predicates are atomic formulae; they are unstructured propositions, the propositional variables of propositional logic.)
   (c) If p_1 and p_2 are formulae then $\neg p_1$ and $(p_1 \vee p_2)$ are formulae.
   (d) If x is a variable and p a formula then $(\exists x)p$ is a formula.
   (e) No other expression is a formula.

   Remark: The connectors $\wedge$, $\rightarrow$, $\leftrightarrow$ and $\forall$ are derived from $\neg$, $\vee$ and $\exists$ in the usual way.

An occurrence of a variable x in a formula E is called a *free occurrence* if it is not in the range of a $(\exists x)$ or $(\forall x)$. The occurrences of variables in a single term are free. The set of variables that occur freely in an expression (term or formula) is called the *index* of that expression. An expression is called *closed* iff its index is empty. Closed terms are compound names of individuals; closed formulae are propositions.

Free occurrences of a variable in an expression may be replaced by terms:

Definition 2.2 Let E be an expression (term or formula), $x_1, \ldots, x_n$ be variables and $v_1, \ldots, v_n$ terms. Then $\alpha = \{x_1 \leftarrow v_1, \ldots, x_n \leftarrow v_n\}$ is called a *substitution*, and $E{:}\alpha = E{:}\{x_1 \leftarrow v_1, \ldots, x_n \leftarrow v_n\}$ designates the result of simultaneously substituting v_i for each free occurrence of x_i for $1 \leq i \leq n$. $E{:}\alpha$ is called the *α-instance* of E. Note that $E{:}\alpha$ is a term or formula iff E is a term or formula, respectively.

We can use a first-order language $\mathbf{L}$ for talking about a relational structure $\mathcal{R}$ if we associate with each operator and each predicate in the vocabulary of $\mathbf{L}$ a function respectively a relation of $\mathcal{R}$.

Definition 2.3 Given a first-order language $\mathbf{L}$, we call a structure $\mathcal{R}$ a *structure for* $\mathbf{L}$, *or* $\mathbf{L}$-*structure,* if every operator $f^{(m)}$ of $\mathbf{L}$ denotes a m-ary function of $\mathcal{R}$ designated by $f_{\mathcal{R}}$ and every predicate $P^{(n)}$ of $\mathbf{L}$ denotes a n-ary relation of $\mathcal{R}$ designated by $P_{\mathcal{R}}$. (A 0-ary relation is either $\emptyset$ or $\{\langle\rangle\} = \{\emptyset\}$.)

To ensure that each individual in the domain of $\mathcal{R}$ can be named in a sentence, we now add to the vocabulary of $\mathbf{L}$ a new set, $U_{\mathcal{R}}$, of *constants* denoting the individuals of $\mathcal{R}$ in a one-to-one fashion. The individual denoted by a constant d is designated as $d_{\mathcal{R}}$. Every constant is defined to be a term. The language derived from this augmented vocabulary is designated as $\mathbf{L}_{\mathcal{R}}$.

The structure $\mathcal{R}$ assigns to each closed term, v, of $\mathbf{L}_{\mathcal{R}}$ an individual of $\mathcal{R}$, designated by $\mathcal{R}(v)$, and to each closed formula (proposition), p, of $\mathbf{L}_{\mathcal{R}}$ the truthvalue $\top$ (*true*) or $\perp$ (*false*), designated by $\mathcal{R}(p)$.

Definition 2.4 Let v be a closed term and p a closed formula of the language $\mathbf{L}_{\mathcal{R}}$. Then $\mathcal{R}(v)$ and $\mathcal{R}(p)$ are defined recursively on their respective syntactic structure.

1. (a) If v is a constant d, $\mathcal{R}(v)$ is the individual denoted by d, $d_{\mathcal{R}}$.

(b) If v is $f^{(n)}(v_1,\ldots,v_n)$ then $\mathcal{R}(v) = f_{\mathcal{R}}(\mathcal{R}(v_1),\ldots,\mathcal{R}(v_n))$.

2. (a) If p is $v_1 = v_2$ then $\mathcal{R}(p) = \top$ iff $\mathcal{R}(v_1)$ and $\mathcal{R}(v_2)$ are the same individual.
   (b) If p is $P^{(n)}v_1\ldots v_n$ then $\mathcal{R}(p) = \top$ iff $\langle\mathcal{R}(v_1),\ldots,\mathcal{R}(v_n)\rangle \in P_{\mathcal{R}}$.
   (c) If p is $\neg q$ then $\mathcal{R}(p) = \top$ iff $\mathcal{R}(q) = \bot$.
   (d) If p is $(p_1 \vee p_2)$ then $\mathcal{R}(p) = \top$ iff $\mathcal{R}(p_1) = \top$ or $\mathcal{R}(p_2) = \top$.
   (e) If p is $(\exists x)q$ then $\mathcal{R}(p) = \top$ iff there is a constant d such that $\mathcal{R}(q\{x = d\}) = \top$.

Definition 2.5 Let **L** be a first-order language, **A** a finite set of formulae of **L**, p, q formulae of **L**, $\mathcal{R}$ a **L**-structure and r a formula of $\mathbf{L}_{\mathcal{R}}$. Then

1. r is said to be *valid* in $\mathcal{R}$ iff each closed instance of r is true in $\mathcal{R}$. Notation: $\mathcal{R} \models r$.
2. p is said to be *valid* iff it is valid in every **L**-structure. Notation: $\models p$.
3. $\mathcal{R}$ is called a *model* of **A** iff each formula of **A** is valid in $\mathcal{R}$. Notation: $\mathcal{R} \models \mathbf{A}$
4. p implies q iff $(p \rightarrow q)$ is valid. Notation: $p \Rightarrow q$.
5. p is called a *logical consequence* of **A** iff p is valid in every model of **A**. Notation: $\mathbf{A} \models p$

In an ordinary first-order structure $\mathcal{R}$, all functions and relations are static, as opposed to dynamic structures where some relations are variable. The presentation of dynamic structures requires that we distinguish between predicates denoting static relations and predicates denoting variable relations (we do not consider variable functions). Hence we divide the set of predicates, Π, into a set of static predicates, Π_s, and a set of variable predicates that will be designated by Π_v.

In the introduction we have demonstrated the principal ideas of simplifying the net representation of dynamic structures by means of merging conditions and events into places and transitions, respectively. Because of the simplicity of the example one particular effect did not show up during the abstraction process.

Look at the example of an annotated CE-net shown in figure 6(a). When merging conditions annotated by propositions built from the same predicate into a single place, multiple arcs will be created (see figure 6(b)). Arcs of a net representation, however, represent ordered pairs of net elements rather than being elements in their own right. Hence we want a notation for combining the annotations of multiple arcs into a single expression.

We propose to use symbolic sums indicating a linear combination of the constituent tuples (see figure 6(c)). What may look here as a mere notational convention will later prove very useful for exploiting the linear algebraic properties of dynamic structures.

In order to make the use of symbolic sums precise we introduce linear integer combinations and multi-sets.

Definition 2.6 Let D be a set. A *linear combination in D with integer coefficients* is a mapping $\lambda: D \rightarrow \mathbf{Z}$. The set of all linear combinations in D is denoted by $\mathcal{L}(D)$, $\mathcal{L}(D) = [D \rightarrow \mathbf{Z}]$.
Very often we shall work with linear combinations with *finite support*, ie with a finite number of coefficients only being non-zero. Their class will be designated by $\mathcal{L}_{fin}(D)$.
For $D = D_1 \times \cdots \times D_n$ we write $\mathcal{L}(D_1,\ldots,D_n)$. (Note that $\mathcal{L}() = \mathbf{Z}$.)
The set of non-negative linear combinations in D is denoted by $\mathcal{L}^+(D)$; its elements are *multi-sets* over D. (Note that for every subset of D, its characteristic function belongs to $\mathcal{L}^+(D)$).
The combinations whose coefficients are all 0 or all 1 are denoted by **0** and **1**, respectively. (**0** corresponds to the empty set and **1** corresponds to D.)
Our notation for a single linear combination is such that for $D = \{a,b,c,d\}$, $2\langle a\rangle - 3\langle b\rangle + \langle d\rangle$ denotes $\{a \mapsto 2,\ b \mapsto -3,\ c \mapsto 0,\ d \mapsto 1\}$.

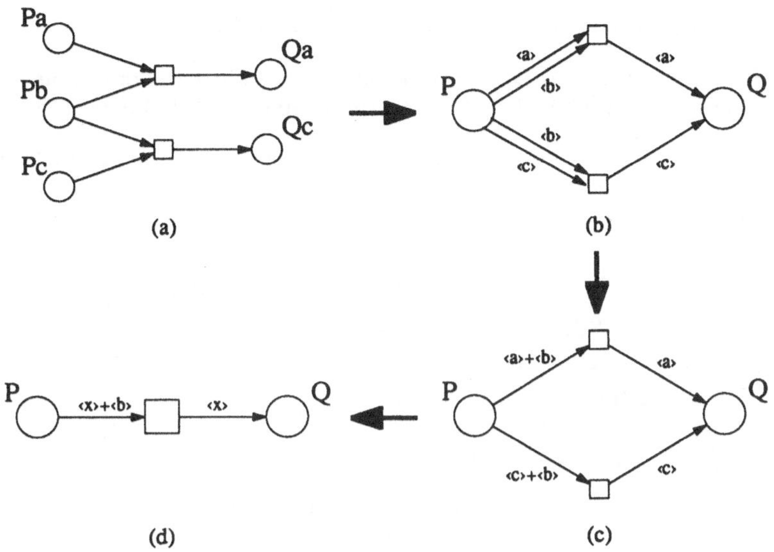

Figure 6: Notation for Multiple Arcs

In $\mathcal{L}(D)$, addition of two elements and multiplication of an element with an integer are defined in the straightforward way; additionally we define componentwise multiplication, as a generalisation of set intersection, and componentwise comparison.

Definition 2.7 Let $\lambda, \lambda_1, \lambda_2$ be in $\mathcal{L}(D)$ and z an integer. Then

1. $(\lambda_1 + \lambda_2) : x \mapsto \lambda_1(x) + \lambda_2(x) \quad (x \in D)$
2. $(z\lambda) : x \mapsto z\lambda(x) \quad (x \in D)$
3. $(\lambda_1 \sqcap \lambda_2) : x \mapsto \lambda_1(x) \cdot \lambda_2(x) \quad (x \in D)$
4. $\lambda_1 \leq \lambda_2 \iff \forall x \in X : \lambda_1(x) \leq \lambda_2(x)$

In chapter 5 where we look in detail at the linear algebraic properties of dynamic structures we shall see that by virtue of addition, multiplication and multiplication with a scalar, $\mathcal{L}(D)$ is a **Z**-linear algebra.

For annotating the arcs of PrT-nets we use symbolic sums denoting non-negative linear combinations.

Definition 2.8 Given a first-order language **L**, for each $n \geq 0$ a class $LC^{(n)}$ of *symbolic sums* of tuples of length n is defined in the following way.

1. The constant 0 is in $LC^{(n)}$.
2. If $v_1, \ldots, v_n$ are terms, then the n-tuple $\langle v_1, \ldots, v_n \rangle$ is in $LC^{(n)}$.
3. If l_1, l_2 are in $LC^{(n)}$, then $(l_1 + l_2)$ is in $LC^{(n)}$.
4. If l is in $LC^{(n)}$ and z is a non-negative integer, then zl is in $LC^{(n)}$.
5. No other expression is in $LC^{(n)}$.

The union of all classes $LC^{(n)}$ for $n \geq 0$ is designated by LC.

Each structure $\mathcal{R}$ assigns to a symbolic sum in $LC^{(n)}$ whose constituent tuples do not contain individual variables (that is closed) a linear combination in $\mathcal{L}^+(D^n)$ in the following way.

Definition 2.9 Given a structure $\mathcal{R}$ and a variable-free symbolic sum l in $LC^{(n)}$ for some $n \leq 0$, the value of l in $\mathcal{R}$ is an element of $\mathcal{L}(D^n)$, ie a linear combination of n-tuples of individuals, designated as $\mathcal{R}(l)$.

1. $\mathcal{R}(\mathbf{0})$ is $\mathbf{0}$.
2. If l is $\langle v_1, \ldots, v_n \rangle$ then for every $\langle d_1, \ldots, d_n \rangle \in D^n$, $\mathcal{R}(l)(d_1, \ldots, d_n) = 1$ iff $d_i = \mathcal{R}(v_i)$ for $1 \leq i \leq n$, and $\mathcal{R}(l)(d_1, \ldots, d_n) = 0$ otherwise.
3. If l is $(l_1 + l_2)$, $\mathcal{R}(l)$ is $\mathcal{R}(l_1) + \mathcal{R}(l_2)$.
4. If l is zl_1, $\mathcal{R}(l)$ is $z\mathcal{R}(l_1)$.

2.2 The Basic Form of PrT-Nets

We are now prepared to define, for a given first-order language $\mathbf{L}$, the class of *strict* predicate/transition nets with annotations in $\mathbf{L}$. It is that class of annotated nets that was used in the introduction for representing the railway system in its most abstract form (see figure 5). *Strict* means that we are not going to allow multiple occurrences of tuples on the places. The places represent variable relations, not multi-relations. Later we shall show how to weaken that restriction.

Definition 2.10 Let $\mathbf{L}$ be a first-order language and let $\mathbf{L}_s$ designate the sublanguage using only Π_s, the predicates denoting static relations. The class $PRT_{\mathbf{L}}$ consists of marked annotated nets, $MN = (N, A, M^0)$ where N is the underlying directed net, A is its *annotation* in $\mathbf{L}$, and M^0 is its *representative marking*.

1. N is a directed net, $N = (S, T; F)$.
2. A is the annotation of N, $A = (A_N, A_S, A_T, A_F)$ where
   (a) $A_N = \mathcal{R}$ is a first-order structure for $\mathbf{L}_s$ called the support of MN (it is the kind of legend that annotates the whole net rather than a particular element);
   (b) A_S is a bijection between the set of places, S, and the set of variable predicates, Π_v;
   (c) A_T is a mapping of the set of transitions, T, into the set of formulae (called transition *selectors*) that use only operators and static predicates (ie are in $\mathbf{L}_s$);
   (d) A_F is a mapping of the set of arcs, F, into the set of symbolic sums of tuples of terms of $\mathbf{L}$, LC, such that for an arc $(x, y) \in F$ leading into or out of a place s (ie $x = s$ or $y = s$) and n being the index of the predicate annotating s, $A_F(x, y)$ is in $LC^{(n)}$.
3. M^0 is a (*consistent*) marking of the places: it is a mapping that assigns to each place s in S a symbolic sum of tuples of constants such that if n is the index of the predicate annotating s, then $M^0(s)$ is in $LC^{(n)}$ and the value of $M^0(s)$, $\mathcal{R}(M^0(s))$, is a linear combination with coefficients being either 0 or 1, ie it is the characteristic function of a set.

Example 2.11 Figure 7 shows the representation of a simple resource management scheme for a group of $N \geq 2$ *agents* regulating the access to a common commodity. There are two access *modes*, either $\mathbf{s}$ = *shared* in which up to L agents may have access simultaneously, or $\mathbf{e}$ = *exclusive* in which only one agent at a time may have access.

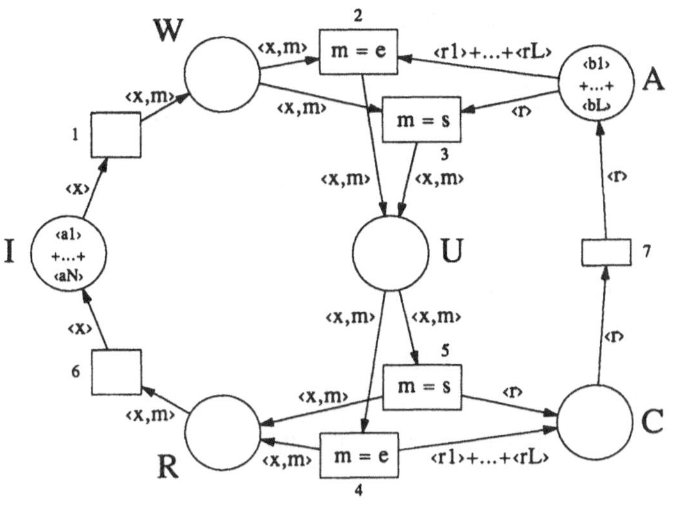

Legend: The support is $\mathcal{R} = (D; \mathbf{e}, \mathbf{s}; Ag, Tc, Md)$ where $D = Ag \cup Tc \cup Md$ with $Ag = \{a_1, \ldots, a_N\}$ being the agents, $Tc = \{b_1, \ldots, b_L\}$ being access tickets, and and $Md = \{\mathbf{e}, \mathbf{s}\}$ being the access modes.
The variable relations are:
Ix = agent x being idle, Wxm = agent x waiting for access in mode m, Uxm = agent x having access in mode m, Rxm = agent x having returned from access in mode m, Cr = access right r being closed, Ar = access right r being available.

Figure 7: Simple Resource Management Scheme as a PrT-net

2.3 The CE-Semantics of PrT-Nets

We now determine the denotation of a PrT-net $PN = (N, A, M^0)$ with $N = (S, T; F)$ and D designating the domain of the structure $A_N = \mathcal{R}$. It is a condition/event system $\Sigma(PN) = (B, E; F', [c^0])$ that is derived from PN in two steps. First, PN is unfolded into an *elementary net system* (see [21]), $\widehat{PN} = (\widehat{S}, \widehat{T}; \widehat{F}, \widehat{c}^0)$ (which corresponds to a place/transition system with all arc multiplicities and all place capacities being 1). The S-elements (conditions) of $\widehat{PN}$ are closed atomic formulae derived from the variable predicates annotating the places of PN. The T-elements (events) of $\widehat{PN}$ are 'feasible' instances of the transitions of PN. In the second step, $\widehat{PN}$ is first restricted to the subset of those events that have an occurrence in the full reachability class of $\widehat{PN}$; then it is simplified by abstracting from different elements with the same presets and postsets. The result is the CE-system $\Sigma(PN)$.

Definition 2.12 Let $t \in T$ be a transition of the PrT-net PN and $p = A_T(t)$ the formula annotating t (the selector of t). The set of all variables that occur in the tuples annotating the incident arcs, is called the *index* of t. Note that the set of variables that occur freely in p is not necessarily contained in the index of t. Those free variables of p that do not belong to the index of t will be called *dangling* variables in the sequel. They can be bound by existential quantifiers without changing the meaning of the annotation.
The index of a place s is defined to be the index of the annotating predicate, $A_S(s)$.

Definition 2.13 Let t, p be as above and let α be a substitution that replaces all variables in the index of t by constants. α is called *feasible* iff

1. α satisfies p, ie there is a substitution β replacing the dangling variables of p by constants such that $\mathcal{R}((p{:}\alpha){:}\beta) = \top$
2. α creates a set on every incident arc; for every place s and arc annotation $l = A_F(s,t)$ or $l = A_F(t,s)$, $\mathbf{0} \leq \mathcal{R}(l{:}\alpha) \leq \mathbf{1}$, ie $\mathcal{R}(l{:}\alpha)$ is a linear combination with coefficients being either 0 or 1, it is the characteristic function of a subset of D^n where n is the index of s.
3. α does not generate an impurity; for no s with both $(s,t) \in F$ and $(t,s) \in F$, there is a resulting tuple that occurs at both arcs: $\mathcal{R}(A_F(s,t){:}\alpha) \sqcap \mathcal{R}(A_F(t,s){:}\alpha) = \mathbf{0}$.

Definition 2.14 Let t,p be as above, and let α be a feasible substitution for t. Then the α-*instance* of t, designated by $t{:}\alpha$ is a pair of sets $({}^\bullet(t{:}\alpha), (t{:}\alpha)^\bullet)$ of closed atomic formulae.

A formula $Pd_1 \ldots d_n$ is in ${}^\bullet(t{:}\alpha)$ iff there is a place s of PN such that $(s,t) \in F$ and

- s is annotated by P, $P = A_S(s)$;
- The coefficient of $\mathcal{R}(d_1 \ldots d_n)$ in $\mathcal{R}(A_F(s,t){:}\alpha)$ equals 1.

A formula $Pd_1 \ldots d_n$ is in $(t{:}\alpha)^\bullet$ iff there is a place s of PN such that $(t,s) \in F$ and

- s is annotated by P, $P = A_S(s)$;
- The coefficient of $\mathcal{R}(d_1 \ldots d_n)$ in $\mathcal{R}(A_F(t,s){:}\alpha)$ equals 1.

The feasible instances of the transitions of PN are now put together to form an elementary net system $\widehat{PN} = (\widehat{S}, \widehat{T}; \widehat{F}, \widehat{c}^0)$ such that

- $\widehat{T}$ is the set of feasible instances of transitions of PN;
- $\widehat{S}$ is the set of all formulae belonging to the pre-set or post-set of a feasible instance, $\widehat{S} = \bigcup_{\widehat{t}\in\widehat{T}}({}^\bullet\widehat{t} \cup \widehat{t}^\bullet)$;
- $\widehat{F}$ is derived from the pre/post-relation, $\widehat{F} = \bigcup_{\widehat{t}\in\widehat{T}}({}^\bullet\widehat{t}\times\{\widehat{t}\} \cup \{\widehat{t}\}\times\widehat{t}^\bullet)$
- the representative case of $\widehat{PN}$, $\widehat{c}^0 \subseteq \widehat{S}$ is induced by the representative marking of PN, M^0: a condition of $\widehat{PN}$, $\widehat{s} = Pd_1 \ldots d_n$, belongs to $\widehat{c}^0$ iff there is a place $s \in S$ annotated by $A_S(s) = P^{(n)}$ and the coefficient of $\mathcal{R}(\langle d_1, \ldots, d_n\rangle)$ in $\mathcal{R}(M^0(s))$ equals 1.

The elementary net system $\widehat{PN}$ we have derived so far is almost a CE-system; however, there may be dead transitions that have no occurrence in any reachable case, and there may be several elements with same presets and postsets. The CE-system $\Sigma(PN)$ that we define as the denotation of the PrT-net PN is now the result of simplifying the restriction of $\widehat{PN}$ to the set of transitions that have an occurrence.

Definition 2.15 The denotation of the strict predicate/transition net PN is the condition/event system $\Sigma(PN) = (B, E; F', [c^0])$ which is derived from the EN system $\widehat{PN} = (\widehat{S}, \widehat{T}; \widehat{F}, \widehat{c}^0)$ in the following way.

1. Let $\widehat{T}'$ be set of transitions that have concession at some case in the full reachability class $[\widehat{c}^0]$.
2. Let $\widehat{PN}' = (\widehat{S}', \widehat{T}'; \widehat{F}', \widehat{c}'^0)$ be the restriction of $\widehat{PN}$ to $\widehat{T}'$.
3. Let two elements $\widehat{x}, \widehat{y}$ in $\widehat{X}' = \widehat{S}' \cup \widehat{T}'$ be equivalent – $\widehat{x} \sim \widehat{y}$ – iff ${}^\bullet\widehat{x} = {}^\bullet\widehat{y}$ and $\widehat{x}^\bullet = \widehat{y}^\bullet$. Note that $\sim$ is a congruence relation with respect to $\widehat{F}'$ and $\widehat{c}'^0$; for $\widehat{x}_1 \sim \widehat{x}_2$ and $\widehat{y}_1 \sim \widehat{y}_2$ we have $(\widehat{x}_1, \widehat{y}_1) \in \widehat{F}' \iff (\widehat{x}_2, \widehat{y}_2) \in \widehat{F}'$ (by definition) and $\widehat{x}_1 \in \widehat{c}'^0 \iff \widehat{x}_2 \in \widehat{c}'^0$ (because dead transitions have been removed).
4. Let $\Sigma(PN) = (B, E; F', [c^0])$ be the quotient of $(\widehat{S}', \widehat{T}'; \widehat{F}', [\widehat{c}'^0])$ by $\sim$.

2.4 The Symbolic Transition Rule

The unfolding of PrT-nets defines their semantics as CE-systems. The purpose of PrT-nets, however, is not alone the more concise representation of CE-systems. Rather, structural and behavioural properties shall be studied, as much as possible, based on the PrT-net representation. A first step to this end is to introduce the 'symbolic token game' for PrT-nets such that unfolding procedure and token game commute.

Definition 2.16 Let $PN = (N, A, M^0) \in PRT_{\mathbf{L}}$ be a strict PrT-net with $A_N = \mathcal{R}$ being the supporting structure. Let M and M' be markings of PN (consistent with the arities of the annotating predicates and such that for all places s, $\mathcal{R}(M(s)) \leq \mathbf{1}$ and $\mathcal{R}(M'(s)) \leq \mathbf{1}$), let t be a transition and α a substitution replacing the variables of the index of t by constants. Then the *α-occurrence* of t at M leading to M' is designated as $M[t{:}\alpha\rangle_{\mathcal{R}}M'$ and defined by the following requirements.

1. α is a feasible substitution for t.
2. For all arcs (s,t) entering t, $\mathcal{R}(A_F(s,t){:}\alpha) \leq \mathcal{R}(M(s))$.
3. For all arcs (t,s) leaving t, $\mathcal{R}(A_F(t,s){:}\alpha) \sqcap \mathcal{R}(M(s)) = \mathbf{0}$.
4. For all places s, $\mathcal{R}(M'(s)) = (\mathcal{R}(M(s)) - \mathcal{R}(A_F(s,t){:}\alpha)) + \mathcal{R}(A_F(t,s){:}\alpha)$.

Two transitions t and t' may occur concurrently, in one *step*, for substitutions α respectively α' ($t = t'$ or $\alpha = \alpha'$ included) iff they both may occur and the instances $t{:}\alpha$ and $t'{:}\alpha'$ are independent (have disjoint pre-sets and post-sets).

The symbolic transition rule is consistent with the CE-semantics, ie the unfolding procedure and the occurrences of transitions commute. More precisely,

Lemma 2.17 Let $M[t{:}\alpha\rangle_{\mathcal{R}}M'$ as defined above. Then the unfolding of PN at M can be followed by an occurrence of an event with the same pre-set and same post-set as $t{:}\alpha$, and the result is the same as unfolding PN at M'.
Proof: Follows immediately from the definition of the unfolding procedure and the symbolic transition rule.

The formulation of the symbolic transition rule gives us an opportunity to end this chapter with a remark that might have come already at other occasions. One may view the variable relations as the magnitudes of a system whose current values determine the actual state of the system. Not only is the state of a Petri system a *distributed* state whose magnitudes may change independently; the magnitudes are 'distributed' as well. The current extension of a variable relation may be changed *concurrently* by several events.

3 Extensions of the Basic Form of PrT-nets

The basic form of predicate/transition nets was defined as a direct formal expression for the intuitive notion of dynamic structures. We will now provide PrT-nets with several generalisations to increase their modelling flexibility, power and conciseness in modelling systems as dynamic structures. The extensions are based upon different ideas which we try to keep clearly separated. The user of PrT-nets must decide very carefully which of the proposed generalisations he wants to adopt.

3.1 Many-sorted Structures

A first, obvious generalisation is the use of *many-sorted* structures. If we distinguish for a structure different sorts of individuals, the signature has to assign as indices not just numbers but strings of sort symbols to predicates and strings paired with a single sort symbol to operators indicating the distribution of domains. If A, B, C, D are sort symbols, for example, then $P^{(A,B,D)}$ denotes a relation in $(A{\times}B){\times}D$, and $F^{(A,C:B)}$ denotes a function from $A{\times}C$ into B.

In order not to overburden the formalism we continue using single-sorted structures. When desirable, static unary predicates will classify the individuals as we did in the example of the resource management scheme 2.11. In the next section, we shall see how to do this also for variable predicates.

3.2 More General Places

A considerable gain in flexibility of the PrT-net model can be achieved if we allow places to be merged or split in small portions. When folding the railway system in the introduction, there were intermediate stages at which the places represented only parts of the variable relation *is_occupied_by*. And the resource management scheme, figure 7, is much better to analyse by the linear algebraic technics developed in chapter 5 if the place U is split into two places $Ux\mathbf{s}$ and $Ux\mathbf{e}$ according to the mode of usage.

Splitting and merging places requires that any partition of a variable relation into disjoint places should be possible. To this end we allow a place to be annotated in a slightly more complicated way than in the basic form.

Notation 3.1 The generic annotation of a place s is $\pi|p$ where

- π is an atomic formula built from a variable predicate and constants and variables only (no compound terms); it is called the *predication* of the place s.
- p is a first-order formula in which no variable predicates occur; it is called the *selector* of s.

A place s annotated by $Pv_1 \dots v_n|p$ denotes that subset of the variable relation P that is contained in $\{\mathcal{R}(\langle v_1, \dots, v_n\rangle{:}\alpha) \mid \mathcal{R}(p{:}\alpha) = \top\}$.

In this way a variable relation may be partitioned arbitrarily into subsets each of which is represented by an extra place. For example, the relation U in the resource management scheme 2.11 may be split into two places $Uxm|m{=}\mathbf{s}$ and $Uxm|m{=}\mathbf{e}$. If the selector is trivial, ie valid in all structures, it may be dropped (see, for example, figure 4).

The consistency of the model requires that different places represent *disjoint* subsets of variable relations. This may be violated if the predications of two different places are derived from the same predicate. For example, $Pay|y{\neq}a$ and $Pxb|x{\neq}b$ are not disjoint but $Pay|\top$ and $Pby|\top$ are (a, b being constants).

Some care is needed concerning the arity (the index) of a place as in contrast with the index of the annotating variable predicate. The index of a place is defined as the set of variables occurring in the annotating predication, *ordered* by their first occurrences. For example, the index of $Pxayx$ is $\langle x, y\rangle$. The tuples in the marking of a place and at the incident arcs refer to this ordered index. Note the difference between $Pxayx|\top$ and $Pxzyx|z{=}a$ whose index is $\langle x, z, y\rangle$. The index is empty iff the place is annotated by a closed atomic formula (an elementary proposition).

The original form of annotating places by predicates only is considered as a shorthand for the common case where the whole variable relation is represented by a single place. For example, $P^{(3)}$ is a shorthand for $Px_1x_2x_3|\top$.

Tuples marking a place may not satisfy the selector. A pair $\langle a, \mathbf{e}\rangle$ put on a place annotated by $Uxm|m{=}\mathbf{s}$ is void. Putting it on a place annotated by $Ux\mathbf{s}|\top$, however, would be syntactically incorrect since the index is $\langle x\rangle$.

The more general way of annotating the places allows an alternative form of specifying the representative marking marking. Rather than writing down the set tuples explicitly as a symbolic sum, a formula in $\mathbf{L}_s$ (containing no variable predicates) may be used instead. Then the problem of void or syntactically incorrect tuples cannot occur. Let the marking of a place s be a formula m. If $\langle x_1, \ldots, x_n \rangle$ is the (ordered) index of place s, the actual marking of s is the set of all $\langle x_1, \ldots, x_n \rangle{:}\alpha$ where $\mathcal{R}((p \wedge m){:}\alpha) = \top$. The explicit listing of tuples is then a notational substitute for a corresponding formula.

The use of more general places requires a slight revision of the procedure that determines the CE-system being the denotation of a PrT-net. It concerns the way the pre- and post-sets of a feasible instance $t{:}\alpha$ are derived from the annotation of arcs. Some of the tuples given by the sum at an arc may be removed by the selector of the incident place.

Assume that l is annotating an arc (s,t) and the annotation of s is $P^{(n)}v_1 \ldots v_n | p$ with $\langle x_1, \ldots, x_k \rangle$ being the index. Then for some substitution β, $Pv_1 \ldots v_n{:}\beta$ belongs to ${}^\bullet t{:}\alpha$ iff both $p{:}\beta$ holds in $\mathcal{R}$ and the coefficient of $\mathcal{R}(\langle x_1, \ldots, x_k \rangle{:}\beta)$ in $\mathcal{R}(l{:}\alpha)$ equals 1.

One consequence of this is that the transitions of the PrT-net may no longer be *uniform* in the sense that all feasible instances of a transition have for every predicate the same number of state elements in their pre-sets and their post-sets. For example, some instances of an arc annotation $\langle x, y \rangle$ are void if the adjacent place is annotated by $Pxy|x{\neq}y$.

3.3 More General Arcs

The next extension is to increase the possibilities for merging events into transitions. In figure 7, the transitions 2 and 3 on top of place U have the same presets and same postsets of places. Furthermore they are intentionally closely related since they both represent ways of getting access to the commodity. They cannot be merged, however, because not all pairs of corresponding arcs are annotated by the same number of tuples.

The property that for a given arc of the PrT-net all instances of the respective transition yield the *same number* of conditions is called *uniformity*. It should not be given up easily. Rather there are good reasons to keep it. In particular, ordinary P/T-nets can be viewed as a special kind of uniform PrT-nets where the transitions ignore all differences between individuals.

To allow greater flexibility in merging transitions, we introduce arc selectors that work similarly to transition and place selectors. In the symbolic sums they appear as symbolic scalars, ie coefficients whose values depend on the truth value of a formula (1 for $\top$, 0 for $\bot$). Transition selectors can then be viewed as arc selectors that are common to all arcs around the transition.

Using such *conditional sums* we can equivalently transform the annotation of the arcs of transition 2 and 3 in figure 8(a) in a way that corresponding arcs are annotated identically; the transitions are *unified* (see figure 8(b)). The unified transitions can be merged into a single one whose selector is just the adjunction of the two original selectors (see figure 8(c)).

First we extend the class of expressions denoting linear combinations (definitions 2.8 and 2.9) by

- If l is in $LC^{(n)}$ and p is a formula, then $[p]l$ is in $LC^{(n)}$.
- If l is $[p]l_1$, $\mathcal{R}(l)$ is $\mathcal{R}(l_1)$ if $\mathcal{R}(p) = \top$ and $\mathbf{0}$ otherwise.

The symbolic coefficients derived from formulae are a kind of generalised *Kronecker* symbols: $\delta_{ij} \simeq [i{=}j]$. To stress their role as scalars, we treat them independently of the symbolic sums.

Definition 3.2 Let p be a closed formula. Then the value of $[p]$ in $\mathcal{R}$ is designated by $\mathcal{R}[p]$; it is defined such that $\mathcal{R}[p] = 1$ iff $\mathcal{R}(p){=}\top$ and $\mathcal{R}[p] = 0$ otherwise.
For two arbitrary formulae p, q, we call two symbolic scalars $[p]$ and $[q]$ equivalent – $[p] = [q]$ – if for all structures $\mathcal{R}$ and for all substitutions α such that $p{:}\alpha$ and $q{:}\alpha$ are closed, $\mathcal{R}[p{:}\alpha] = \mathcal{R}[q{:}\alpha]$.

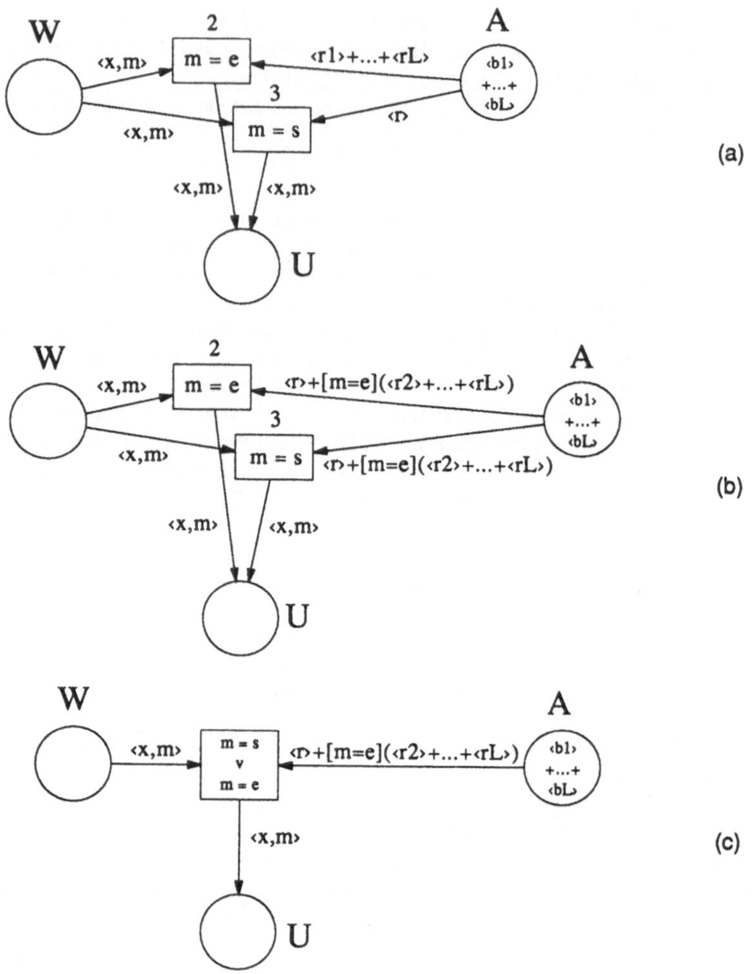

Figure 8: Unifying and Merging Non-uniform Transitions

From these definitions we get immediately some obvious rules for symbolic scalars and conditional sums.

Lemma 3.3 Let p, q be formulae. Then

1. $[\top] = 1$, $[\bot] = 0$
2. $[p] = [q]$ iff $p \Leftrightarrow q$
3. $[\neg p] = 1 - [p]$
4. $[p \wedge q] = [p][q]$
5. $[p \vee q] = [p] + [q] - [p][q]$
6. $[p]l_1 + [p]l_2 = [p](l_1 + l_2)$

7. $[p]l + [q]l = ([p] + [q])l$

Proof: The verification of these laws is straightforward.

3.4 Multi-Sets and the Weak Transition Rule

An object may belong to several sets at the same time but the notion *set* excludes that it occurs more than once in the same set. ($\{a,b\} \triangleq \{a\} \cup \{b\}$, hence $\{a,a\} = \{a\}$.) Therefore the feasibility constraint 2.13 for transition occurrences had to exclude those substitutions from being applicable that would put or take the same tuple more than once to respectively from a place.

The usual way of allowing multiple occurrences of an object at a place is to generalise the marking of places from sets to *multi-sets* (also called *bags*). Given a set D, a multi-set in D (or, more precisely, a multi-subset of D) is a function $B : D \rightarrow \mathbf{N}$, ie an element of $\mathcal{L}^+(D)$ as defined in 2.6.

The whole formalism of PrT-nets can be generalised easily from places representing sets to places representing multi-sets. The items 2 and 3 of 2.13 (feasibility) may be dropped, and the *strict* symbolic transition rule 2.16 is changed into the *weak* transition rule by dropping requirement 3 thus allowing arbitrarily many copies of a tuple on the same place.

Later we shall make extensive use of multi-sets when studying linear algebraic techniques for analysing PrT-nets. At this stage of the development, however, we take an alternative approach. We shall sketch an approach that treats PrT-nets marked by multi-sets as a *special* case, and not as a generalisation, of *strict* PrT-nets.

There are two reasons for doing so. Firstly, we want to demonstrate that *conceptually* there is no need for using multi-sets and the complication of the formalism that follows from it. Secondly, we have to show that multiple occurrences of an object on a place and the weak transition rule still allow a reduction to CE-systems and hence, do not lead us out of the realm of Petri systems.

Rather than counting the number of occurrences of an object in a multi-set we distinguish different copies of an object by attaching to it a *tag* that functions like the serial number of a bank-note. Tags are a special sort of individuals that have no structure and do not appear in formulae. We take the tags from the set of natural numbers, $\mathbf{N}$, and call, for example, the tagged object $\langle a,b\rangle.3$ the third copy of the object $\langle a,b\rangle$. ($\langle a,b\rangle.0$ may be called the original.)

Formally there is no difference between $\langle a,b\rangle.3$ and $\langle a,b,3\rangle$. Hence the use of tags means that a sort *tag* is appended to the index of each variable predicate.

In the same way, to each tuple occurring at an arc another variable has to be appended whose occurrence there is unique for the adjacent transition. Hence the transition occurrences will totally ignore the value of the tag that is assigned to it.

To stress this *don't-care* property of the tag variables, and to avoid using too many variables, their occurrences are indicated by a special *don't-care symbol* $\sim$. Each occurrence of $\sim$ represents an occurrence of a variable that occurs nowhere else at the respective transition. (This usage of $\sim$ is not restricted to tag variables.)

By appending tags, an arc annotation like $\langle x\rangle + \langle x\rangle$, or $2\langle x\rangle$ for short, that would not allow any feasible substitution becomes $\langle x,\sim\rangle + \langle x,\sim\rangle$, or equivalently, $2\langle x,\sim\rangle$. Any substitution that assigns different tags to the two different variables represented by the two occurrences of $\sim$ could be feasible, eg $\langle a,0\rangle + \langle a,7\rangle$.

In most cases of actual system modelling, the set of tags doesn't have to be $\mathbf{N}$ but some rather small number $n = \{0,\ldots,n-1\}$. One may even assign to each variable predicate a different number as tag sort indicating the upper bound for the number of copies of the same object that may occur on places annotated by that predicate. In this way we get a generalisation of the strict transition rule allowing more than one copy but restricting the number of copies to an upper bound.

The use of tags and don't-care symbols reduces PrT-nets with multi-sets as markings to strict PrT-nets. We don't have to leave ordinary set theory and logic. What is more important, PrT-nets with multi-sets have a denotation as CE-systems. They are standard models of net theory.

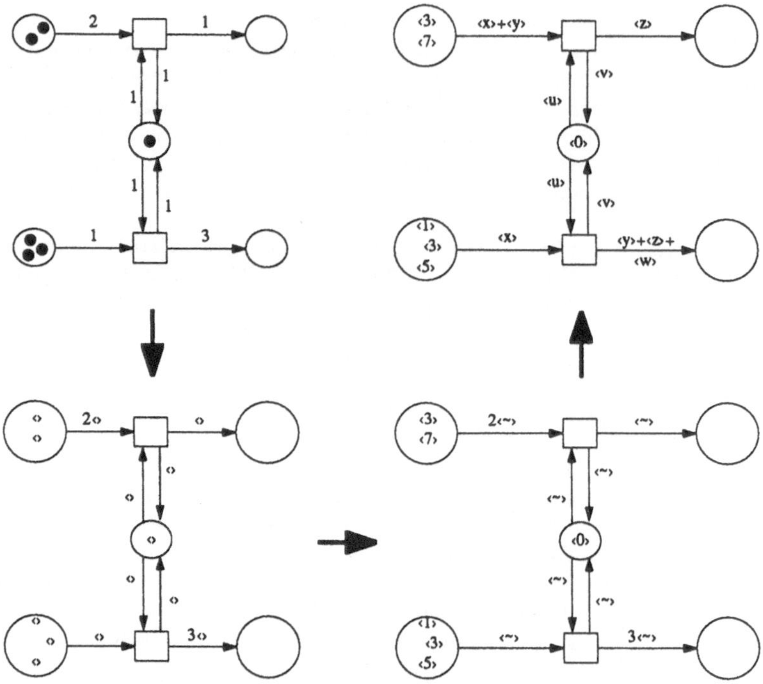

Figure 9: Semantics of PT-nets as CE-systems

Ordinary PT-nets in particular can be seen as a special class of PrT-nets with multi-sets where all places are 0-ary, ie the annotations of arcs and the marking the places are multiples of the zero-tuple $\langle\rangle$. By using tags, there are multiples of $\langle\sim\rangle$ at the arcs and sets of tags on the places. A place capacity n turns into the sort $n = \{0, \ldots, n-1\}$ of tags for that place.

An example of this transformation is shown in figure 9. In the first step, tokens are turned into 0-tuples; places become 0-ary multi-relations. In the next step, the 0-tuples at the places are distinguished by appending arbitrary but different tags and at the arcs, don't-care-symbols are introduced. Finally, the don't-care-symbols are replaced consistently by unique occurrences of variables. As in the original PT-net, the transitions of the PrT-net depend only the presence of tokens but not on their identity.

3.5 Place Projections

In this section we wish to reverse the operation of introducing tags and don't-care-symbols. Let $R = \{\langle a,b\rangle, \langle a,c\rangle, \langle b,c\rangle\}$ be a binary relation in the set $D = \{a,b,c\}$. If we are not interested in knowing which individuals occur at the second position of R, we may describe R as follows: R contains two pairs of the form $\langle a, \textit{something}\rangle$ and one pair of the form $\langle b, \textit{something}\rangle$. Using the don't-care symbol $\sim$ we write $R \in \{2\langle a,\sim\rangle, \langle b,\sim\rangle\}$ indicating that $\{2\langle a,\sim\rangle, \langle b,\sim\rangle\}$ denotes a whole family of relations of that form. More formally we have $R \in \{2\langle a,\sim\rangle, \langle b,\sim\rangle\}$ iff there are x_1, x_2, y such that $x_1 \neq x_2$ and $R=\{\langle a,x_1\rangle, \langle a,x_2\rangle, \langle b,y\rangle\}$. The affinity to cardinal numbers is obvious: $2 = |\{x \mid \langle a,x\rangle \in R\}|$ and $R \in \{|R|\langle\sim,\sim\rangle\}$.

If we are not interested in the details of the second position of R, we may eliminate this position as well. The idea is to introduce a kind of partial cardinal numbers that do not abstract totally from the identity of the elements of a set but only partially. We do so by means of an operation called *projection along the i-th position* and denoted by $|\ |_i$. As a result, we shall get $|R|_2 = \{2\langle a\rangle, \langle b\rangle\}$ and $||R|_2|_1 = \{3\langle\rangle\} \cong 3$.

Multi-sets and multi-relations may always be viewed as the result of projecting a relation along some position. For our purposes we define projections formally for linear combinations as defined in 2.6. It is important, however, to have finite supports only in order to avoid infinite multiplicities.

Definition 3.4 For $\mathcal{L}_{fin}(D_1,\dots,D_n)$ $(n \geq 1)$, the *projection* of its elements *along the i-th position* $(1 \leq i \leq n)$ is defined such that for $\lambda \in \mathcal{L}_{fin}(D_1,\dots,D_n)$,

$$|\lambda|_i: (x_1,\dots,x_{i-1},x_{i+1},\dots,x_n) \mapsto \sum_{y\in D_i} \lambda(x_1,\dots,x_{i-1},y,x_{i+1},\dots,x_n)$$

The *total projection* is denoted by $|\lambda|$: $|\lambda| = \sum_{D_1\times\cdots\times D_n} \lambda(x_1,\dots,x_n)$

It is easy to verify that $|\lambda|_i \in \mathcal{L}_{fin}(D_1,\dots,D_{i-1},D_{i+1},\dots,D_n)$, that $|\ |_i$, $|\ |$ are linear, and $||\lambda|_j|_i = ||\lambda|_i|_{j-1}$ for $n \geq 2$ and $1 \leq i < j \leq n$.

Example 3.5 For $\lambda = 2\langle a,b\rangle - 3\langle a,c\rangle + \langle b,c\rangle$,

$$\begin{aligned}
|\lambda|_1 &= 2\langle b\rangle - 3\langle c\rangle + \langle c\rangle = 2\langle b\rangle - 2\langle c\rangle \\
|\lambda|_2 &= 2\langle a\rangle - 3\langle a\rangle + \langle b\rangle = -\langle a\rangle + \langle c\rangle \\
|\lambda| &= ||\lambda|_2|_1 = ||\lambda|_1|_1 = \mathbf{0}
\end{aligned}$$

We will use the projection of linear combinations as a general means for abstracting from unnecessary or inconvenient details in a PrT-net. The idea is to project a place in order to introduce a partially quantitative view of the system. Tuples on a place that differ only at their $i-th$ position are no longer distinguished but their occurrences are counted if $|\ |_i$ is applied to the place.

Projecting a place s means to eliminate the $i-th$ position from its index. It is done in the following way.

1. Delete the $i-th$ position in all tuples of the marking of s.
2. Delete the $i-th$ position in all tuples annotating the arcs adjacent to s.
3. Prefix $|\ |_i$ to the annotation of s to indicate that the projection was applied.

As example, project the places A and C of the resource management scheme 7 along its first (and only) position. The tickets turn into tokens, the arc annotations become arc multiplicities, 1 or L. If projections are applied to a PrT-net it is assumed that the weak transition rule will be used such that projections commute with the occurrence of transitions. The result of totally projecting all places yields the merely quantitative presentation of a dynamic structure. If the PrT-net is uniform, its total projection is an ordinary PT-net. We have seen in the previous section that conversely, every PT-net can be viewed as the total projection of a strict PrT-net.

3.6 Structures as Parameters

The extensions to the PrT-net model introduced so far did not leave the concept of presenting a single dynamic structure as a PrT-net supported by a specific first-order structure $\mathcal{R}$. The last generalisation of the PrT-net model introduces a new layer of the presentation of dynamic structures. The idea is to use the structure $\mathcal{R}$ as a parameter of a family of similar dynamic structures.

To this end, we change the annotation A_N. Instead of a particular structure $\mathcal{R}$, it may carry a finite set of formulae, $\mathbf{A} \subseteq \mathbf{L}$, that specifies a whole class of **L**-structures. For a given **L**-structure $\mathcal{R}$ to be a feasible support, it must be a model of **A** (see 2.5).

Replacing a particular structure $\mathcal{R}$ by a set of formulae, however, does not yet accomplish the full job of parameterising a PrT-net. While the generic annotation of places introduced in section 3.2 in connection with marking formulae takes care of parameterised markings, the parameter L and the ellipsis "..." in figure 7 show that we may have to generate variables and certain expressions dependent on the domain of the actual structure $\mathcal{R}$. For example, the sum $\langle r_1\rangle + \cdots + \langle r_L\rangle$ of ticket variables in figure 7 must be generated dependent on the number of tickets.

The use of indices for generating variables and of "..." for generating symbolic sums can be made formal by using the generalised +-operator $\sum_i^p E$ such that i is the variable bound by $\sum$, p is a formula specifying the range of i, and E is the argument expression. Then $\langle r_1\rangle + \cdots + \langle r_L\rangle$ becomes $\sum_r^{\top}\langle r\rangle$. Furthermore, $\sum_r^{\top}([r \neq s]\langle s,r\rangle)$ is a *macro* operator (not an operator in the language **L**) that assigns to every individual s the sum of all pairs $\langle s,r\rangle$ where $r \neq s$. Note that when this macro appears at the arc of a PrT-net (as in figure 16), it can be expanded once the domain of individuals has been fixed (at 'compile time') while $\sum_r^{r\neq s}\langle s,r\rangle$ could be expanded only after s has been fixed as well for the occurrence of the corresponding transition (at 'run time').

3.7 Other Formalisms for Structures

Rather than the formalism of first-order logical formulae and their structures, related ways of specifying the supporting structure and formulating annotations may be used. Examples could be a formalism for abstract data types or a data base description language.

4 Logical Invariants of PrT-Nets

When we chose first-order predicate logic as a model for introducing structured sets of individuals into net theoretical system modelling, it was not only a matter of taste. Rather, there is a kind of canonical relationship between net theory and logic which suggests to look for something of the kind of PrT-nets. It was first discovered by *Petri* when he set out to study net theoretical systems in terms of their *invariants* [16].

Petri defines the *enlogic structure* of a CE-system as the classification of all conceivable event-like elements in relation to the set of cases and the *synchronic structure* as the classification of all conceivable condition-like elements in relation to the set of processes. One such class of event-like elements is called *facts* and defined as those elements that at no reachable case would have concession (that are *dead*). Facts correspond to *invariant* propositions that hold at all cases although their constituents (the atoms are the conditions) may vary from case to case. And conversely, every such proposition corresponds to a set of facts.

The net representation of propositions yields a graphical calculus for propositional logic which was lifted to first-order predicate logic before the PrT-net model was found.[1,4,22] We shall now see that PrT-nets are those transition nets whose dead transitions have the power to express all first-order logical invariant assertions. The language of PrT-nets allows to formulate both the dynamic and the static properties of dynamical relational structures.

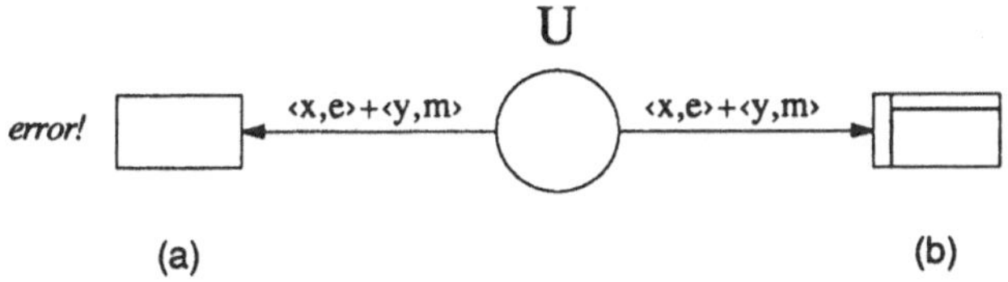

Figure 10: Mutual Exclusion Property Specified as a Dead Transition

4.1 Dead Transitions and Universal Invariants

Let us take the resource management scheme 2.11 as an example. The main restriction to the un-coordinated behaviour of the set of agents is the requirement that when one agent has exclusive access to the common commodity, no other agent may have access in either mode at the same time. We can formulate this requirement as a logical formula that must hold at all reachable markings.

$$⊟\,(\forall x)[U x \mathbf{e} \rightarrow \neg(\exists y)(\exists m)((y{\neq}x \vee m{\neq}\mathbf{e}) \wedge U y\, m)]$$

By prefixing the ⊟ symbol we indicate that the formula is to hold at all reachable markings. It is the symbol for facts introduced by *Petri* and an instance of the necessity operator of modal logic (see eg [8]) which is often written as $\Box$. Here it is a shorthand for the universal quantifier $(\forall M{\in}[M^0])$ (which is *not* part of the first-order language **L**). An expression of the form ⊟p where p is a formula of the first-order language **L** is called a *logical invariant* or, in the context of this chapter where no other invariants will be considered, just *invariant*.

Now assume that we want to add to the resource management scheme an error check that would issue a message whenever it detects a violation of the mutual exclusion requirement. In the PrT-net model of the scheme we could represent this check by a transition labelled by *error!* as shown in part (a) of figure 10. Obviously, this transition is not supposed to have concession at any reachable marking. Rather, if, and only if, the error transition is dead, the design of the system will be called correct with respect to the requirement.

In the sequel we shall indicate dead transitions by the ⊟ symbol as in part (b) of figure 10. Normally we will not indicate, however, whether the ⊟ transition is actually dead with respect to the class of reachable markings or not. All elements of a net theoretical model are to specify certain properties, and we do not demand that the elements of a specification are independent of each other.

If we add part (b) of figure 10 to the net in figure 2.11, we can easily derive from the control mechanism using access tickets that the transition is dead. As a specification, the ⊟ transition would be redundant. If the ⊟ transition replaces the control mechanism, however, it represents the mutual exclusion property without specifying a particular control mechanism. It tells that the class of system states is the class of those markings reachable by the transition rule at which the ⊟ transition has no concession. In this sense, the ⊟ transition and the ⊟ formula have exactly the same meaning. They both represent the mutual exclusion property.

In chapter 5 we will study one particular technique of verifying such specifications. In this chapter, however, we are concerned with the expressive power of dead transitions.

To start with, we give a precise formulation to our observation that dead transitions are invariants.

Theorem 4.1 Every dead transition of a PrT-net *MN* represents a formula that holds at all reachable markings of *MN*; it is a logical invariant of *MN*.

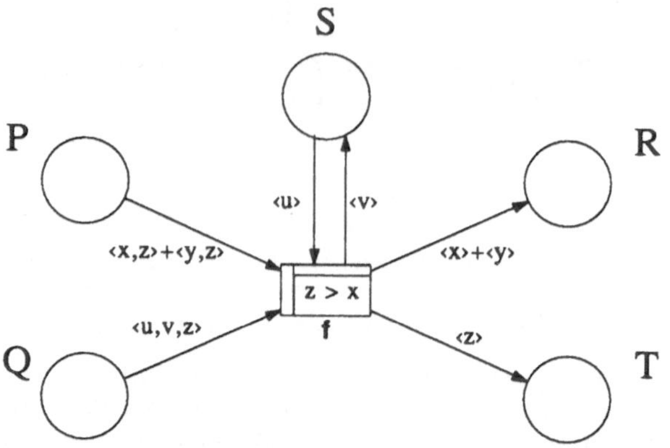

Figure 11: Universal Invariant Expressed by a Dead Transition

Proof: To avoid too complicated a formalism, let us consider a (totally meaningless) example that is general enough to show us all relevant aspects. It is shown in figure 11. The places are assumed to belong to a PrT-net MN with the marking class $[M^0]$. From the (strict) transition rule for PrT-nets (see 2.16) follows immediately that transition f is dead iff at all $M \in [M^0]$ and for all substitutions $\alpha = \{u \leftarrow u_0, v \leftarrow v_0, x \leftarrow x_0, y \leftarrow y_0, z \leftarrow z_0\}$, at least one of the following criteria applies.

1. α is not feasible (see 2.13);
2. there is at some incoming arc (s, f) a resulting tuple that does not belong to $M(s)$;
3. there is at some outgoing arc (s, f) a resulting tuple that belongs already to $M(s)$.

Consequently, the dead transition f represents the invariant

$$\Box(\forall u, v, x, y, z)[\underbrace{\neg(z{>}x \wedge x{\neq}y \wedge u{\neq}v)}_{1.} \vee \underbrace{(\neg Pxz \vee \neg Pyz \vee \neg Quvz \vee \neg Su)}_{2.} \vee \underbrace{(Sv \vee Rx \vee Ry \vee Tz)}_{3.}]$$

To study the consequences of this result we first collect some material known from logic.

A formula is said to be *open* if it contains no quantifiers. A formula p is said to be in *prenex form* if it has the form $(Q_1x_1)\ldots(Q_nx_n)q$ where the (Q_ix_i) are quantifiers and q is open. The part $(Q_1x_1)\ldots(Q_nx_n)$ is called the *prefix* and q the *matrix* of p.

Proposition 4.2 For every formula p exists a logically equivalent formula p' in prenex form.

For example, the formula $(\forall x)(\forall y)(\forall m)(\neg Ux\mathbf{e} \vee \neg Uym \vee (y{=}x \wedge m{=}\mathbf{e}))$ is a prenex form of the mutual exclusion property. Although the original formula contains an existential quantifier, the prefix of the prenex form contains only universal quantifiers.

If a formula in prenex form contains only universal quantifiers, it is called *universal.* We call an invariant $\Box p$ universal if p is universal.

An (open) formula q is said to be in *conjunctive form* if it has the form $q_1 \wedge \ldots \wedge q_m$ where each q_j is a disjunction of *literals.* (A literal is an atomic or negated atomic formula.)

Proposition 4.3 For every open formula q exists a logically equivalent formula q' in conjunctive form.

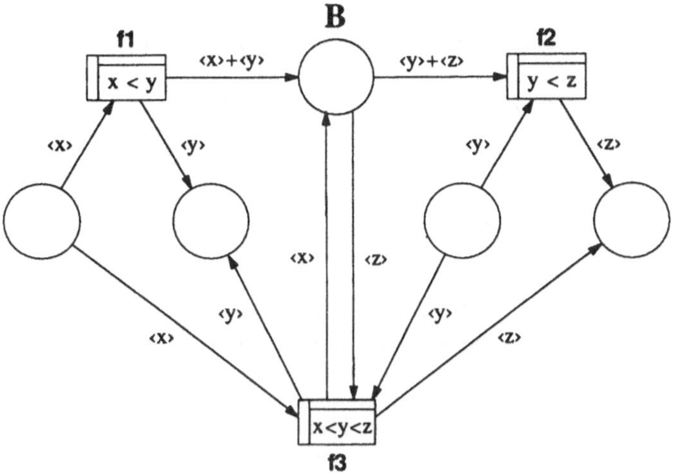

Figure 12: Mechanical Derivation of a Dead Transition

To begin our study of invariants, we state two obvious laws that show that the rules of first-order logic for inferring formulae from given ones can be also applied to invariants.

Proposition 4.4 For every PrT-net *MN* and two formulae p, q we have

1. If $\Box p$ holds for *MN* and p implies q, $\Box q$ holds for *MN*.
2. If both $\Box p$ and $\Box q$ hold for *MN*, $\Box(p \wedge q)$ holds for *MN*. (The converse follows from 1.)

This simple fact allows to derive new dead transitions from given ones in a mechanical manner. An example of applying the *cut* rule (resolution) is shown in figure 12. The places are assumed to belong to some PrT-net. Whenever transitions f_1 and f_2 are dead, transition f_3 will be dead as well. f_3 is the *resolvent* of f_1 and f_2. It is the result of merging f_1 and f_2 where the 'bridge' By is eliminated. The selector of f_3 is the conjunction of the selectors of f_1 and f_2 (in contrast with the merging of similar transitions where the selectors have to be $\vee$–ed; see figure 8).

It is now not difficult to see that universal invariants are of particular importance for us.

Theorem 4.5 Every universal invariant $\Box p$ of a PrT-net *MN* can be expressed by a finite set of dead transitions in *MN*.
Proof: Since p is universal, $\Box p$ has a form $\Box Q(q_1 \wedge \ldots \wedge q_m)$ where Q is the prefix containing only universal quantifiers and the q_j are disjunctive clauses. Hence it follows from proposition 4.4 that $\Box p$ is logically equivalent to the set of invariants $\{\Box Qq_1, \ldots, \Box Qq_m\}$.

For each $\Box Qq_j$ we add to the net *MN* a transition f_j in the following way: q_j is divided into two parts, $q_j \Leftrightarrow (c_j \vee v_j)$ where c_j is the disjunction of all literals in q_j that are built from static predicates and v_j is the disjunction of all literals in q_j that are built from variable predicates. The negation of the constant part, $\neg(c_j)$, becomes the selector annotating f_j. For each negated literal in v_j of the form $P^{(n)}u_1 \ldots u_n$, an arc is inserted leading from the place P to f_j and annotated by $\langle u_1, \ldots, u_n \rangle$. For each non-negated literal in v_j of the form $P^{(n)}u_1 \ldots u_n$, an arc is inserted leading from f_j to the place P and annotated by the tuple $\langle u_1, \ldots, u_n \rangle$. Multiple arcs are merged using symbolic sums.

From theorem 4.1 we know that f_j is dead in *MN* iff q_j holds at all reachable markings. Hence $\Box p$ and the set of dead transitions $f_1, \ldots, f_m$ are equivalent.

With this result at hand, what remains to do is to study the general case where for an invariant $\Box p$, the prefix of p (in prenex form) contains an existential quantifier.

4.2 Existential Invariants and Skolem Places

In ordinary predicate logic where we don't have variable predicates, every formula p in prenex form can be translated into a universal formula by eliminating, step by step, the existential quantifiers in the following way.

Let p be $(\forall x_1)\dots(\forall x_n)(\exists y)Q'q$. Let $\tilde{y}^{(n)}$ be a new n-ary operator (called *Skolem* operator) that is added to the given vocabulary of **L**. Then p is valid in some **L**-structure $\mathcal{R}$ iff there is an extension of $\mathcal{R}$, $\mathcal{R}'$ by a function interpreting the *Skolem* operator $\tilde{y}$ such that the formula $(\forall x_1)\dots(\forall x_n)Q'q{:}\{y{\leftarrow}\tilde{y}(x_1,\dots,x_n)\}$ is valid there. (Rather than eliminating $(\exists y)$, it would be more correct to say that the explicit occurrence of $(\exists y)$ is removed and replaced by an implicit occurrence within the interpretation of the operator $\tilde{y}$).

Unfortunately, this procedure no longer works once we prefix a formula containing variable predicates by the modal operator $\Box$. Since $\Box$ stands for $(\forall M{\in}[M^0])$, the Skolem operator $\tilde{y}$ had to be $(n+1)$-ary depending at its first position on markings. In other words, it would denote a *variable* function. However, we do not consider dynamic structures with variable functions. The global constraint to the occurrence of transitions needed for maintaining variable functions would be inconsistent with the whole PrT-net approach.

There may be systems where some variable relation has the invariant property of being a function. However, then we are back again at the same question. How to express in net theoretical terms that, for example, Z denotes the graph of a variable binary function, ie how to express the invariant $\Box(\forall x)(\forall y)(\exists z)(\forall z')(Zxyz \wedge (Zxyz' \rightarrow z{=}z'))$.

The second part of this requirement, the uniqueness of the value for every pair of arguments, is a universal invariant $\Box(\forall x)(\forall y)(\forall z)(\forall z')((Zxyz \wedge Zxyz') \rightarrow z{=}z')$. It can be expressed by a dead transition.

For the first, existential, part we remember that the existential quantifier $\exists$ is a generalisation of the logical $\vee$. For a finite domain of individuals, $D = \{d_1,\dots,d_n\}$, the formula $(\forall x)(\forall y)(\exists z)Zxyz$ is equivalent to the universal formula $(\forall x)(\forall y)(Zxyd_1 \vee \dots \vee Zxyd_n)$ which may be written as $(\forall x)(\forall y)(\bigvee_{i=1,\dots,n} Zxyd_i)$. It can be expressed by the dead transition shown in figure 13(a). In the same way as $\vee$ is generalised to $\exists$ (another notation is $\bigvee$), the symbolic $+$ can be generalised to $\sum$ as we suggested already in section 3.6. This is shown in parts (b) and (c) of figure 13.

This way of expressing $\exists$ by $\sum$, however, does still not enable us to express existential invariants in general. The *Skolem* operators allow one to separate the different disjunctive clauses of the conjunctive form of the matrix. While for example, $(\forall x)(\exists y)(Pxy \wedge Qxy)$ is not equivalent to $((\forall x)(\exists y)Pxy) \wedge (\forall x)(\exists y)Qxy)$, the formula $(\forall x)(Px\tilde{y}(x) \wedge Qx\tilde{y}(x))$ can be equivalently split into $(\forall x)(Px\tilde{y}(x)$ and $(\forall x)(Qx\tilde{y}(x)$.

However, there is a way of revising the procedure of eliminating $\exists$ quantifiers by using additional predicates rather than operators. Again, the $\exists$'s are not really eliminated; rather, they are brought into a special form such that we can express them by $\sum$ and separate several disjunctive subformulae.

Proposition 4.6 Let p be $(\forall x_1)\dots(\forall x_n)(\exists y)Q'q$. Let $Y^{(n+1)}$ be a new $(n+1)$-ary predicate that we call *Skolem* predicate and add to the given vocabulary. Then p is valid in some **L**-structure $\mathcal{R}$ iff there is an extension of $\mathcal{R}$, $\mathcal{R}'$, by a relation interpreting the *Skolem* predicate Y such that the formulae $(\forall x_1)\dots(\forall x_n)(\exists y)Yx_1\dots x_n y$ and $(\forall x_1)\dots(\forall x_n)(\forall y)Q'(Yx_1\dots x_n y \rightarrow q)$ are both valid in $\mathcal{R}'$.

If the matrix q has the form $q_1 \wedge\dots\wedge q_m$, the transformed matrix, $(Yx_1\dots x_n y \rightarrow q)$, is logically equivalent to $(\neg Yx_1\dots x_n y \vee q_1) \wedge\dots\wedge(\neg Yx_1\dots x_n y \vee q_m)$.

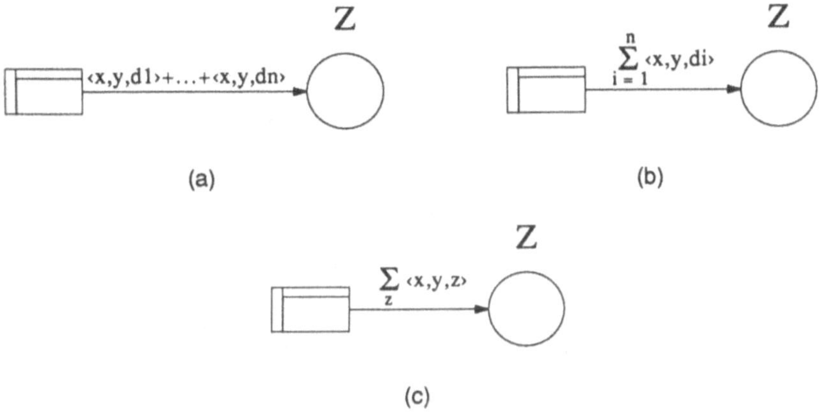

Figure 13: Generalised + for Existential Invariants

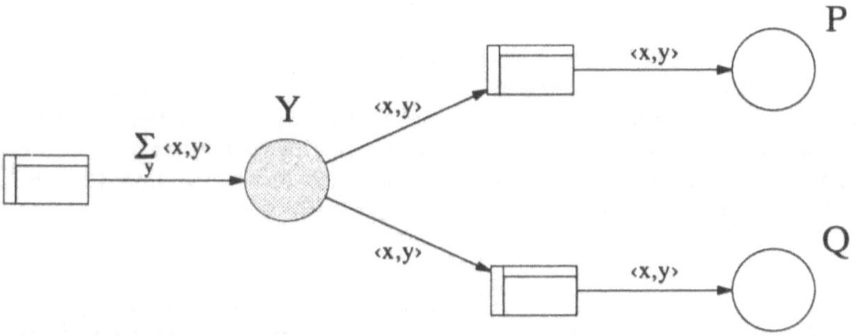

Figure 14: Existential Invariant Expressed by a Set of Dead Transitions

The revised procedure solves both our problems. It can be applied to the invariant $\Box p$ in like manner if we only require that the *Skolem* predicate Y denotes a *variable* relation. And it yields the two invariants $\Box(\forall x_1)\dots(\forall x_n)(\exists y)Y x_1 \dots x_n\, y$ and $\Box(\forall x_1)\dots(\forall x_n)(\forall y)Q'(Y x_1 \dots x_n\, y \rightarrow q)$ such that the existential invariant concerning Y can be expressed by a dead transition.

For example, the invariant $\Box(\forall x)(\exists y)(P x\, y \wedge Q x\, y)$ is translated into the set of three invariants, $\Box(\forall x)(\exists y)Y x\, y$, $\Box(\forall x)(\forall y)(\neg Y x\, y \vee P x\, y)$, and $\Box(\forall x)(\forall y)(\neg Y x\, y \vee Q x\, y)$. They are shown as dead transitions in figure 14 where the places P and Q are assumed to belong to a PrT-net but the place Y is added.

To summarise, we have arrived at the main result of this chapter.

Theorem 4.7 Every first-order logical invariant $\Box p$ can be expressed by a finite set of dead transitions using a finite number of *Skolem* places.

There is still one open question, however. We have to introduce new predicates Y that have no interpretation as variable relations in the original dynamic structure and do not appear as places in the PrT-net *MN*. They are added to *MN* together with the dead transitions expressing the invariant. What is their marking?

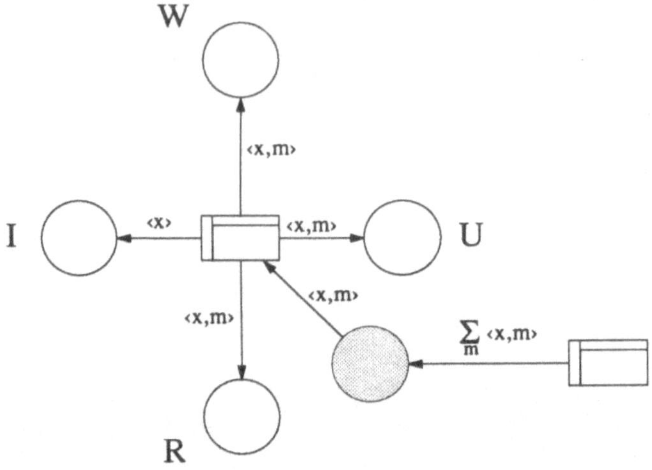

Figure 15: Example of an Existential Invariant

The answer is that the *Skolem* predicates are not *independent* variable predicates like the original ones. Rather, they are derived, virtual, predicates that are defined in terms of the given (independent) predicates. The kind of dependency can be left unspecified in the same way as the identity of individuals whose existence is stated by an $\exists$ quantifier is left unspecified. However, we may also specify it using the device of a place selector introduced in section 3.2 – but with a twist. The selector may contain not only operators and static predicates but variable (independent!) predicates as well.

As an example, take the obvious property of the resource management scheme 2.11 that every agent is always in some state. It can be expressed by the invariant $\Box(\forall x)(Ix \vee (\exists m)(Wx\,m \vee Ux\,m \vee Rx\,m))$ or, equivalently, by the subnet of dead transitions shown in figure 15. To verify the claim that the transitions are actually dead, define the *Skolem* place by annotating it with the selector $((Ix \wedge m{=}\mathbf{e}) \vee Wx\,m \vee Ux\,m \vee Rx\,m)$.

To end this chapter with, let us discuss what we have achieved. We have shown that the expressive power of dead transitions of a PrT-net is as strong as it could be. We are able to express every first-order logical invariant of a dynamic structure by a finite set of dead transitions. (Note that this power does not depend on annotating the transitions with first-order logical selectors. Rather, by representing the static relations of the supporting structure $\mathcal{R}$ of a PrT-net by places with constant markings, the selectors can be translated into net structure as well. This was done in [4]. It then follows that an invariant is contradictory iff from its representation as a set of selector-free dead transitions, the isolated transition can be derived.)

We do not suggest, however, to always translate invariants into dead transitions. Our main reason for assigning a net theoretic interpretation to every formula prefixed by the modal symbol $\Box$ was to make sure that it can be used safely, beyond any doubt about its meaning, for expressing an invariant property of a PrT-net. In addition, the translation into dead transition yields a decomposition of an arbitrary logical invariant into a set of elementary net theoretic system invariants.

5 Linear Algebraic Analysis of PrT-nets

The transition rules for ordinary PT-nets as well as for PrT-nets show that the occurrences of transitions have the following linearity property.

- The effects of the occurrences of a transition on markings are *independent* of the markings at which the transition occurs:
  $M_1[t\rangle M_2$ and $M_1'[t\rangle M_2'$ imply $M_2 - M_1 = M_2' - M_1'$ (where t is a (feasible instance of a) transition and $-$ denotes componentwise subtraction).

Let for the moment C^t denote this effect of a transition t. Then we get immediately:

- The total effect of several transition occurrences in a process is the *sum* of the individual effects; $M_1[t_1\rangle M_2[t_2\rangle M_3$ implies $M_3 - M_1 = (M_2 - M_1) + (M_3 - M_2) = C^{t_1} + C^{t_2}$.

In this chapter we want to exploit this linearity property for the analysis of PrT-nets. We will try to unify several approaches taken by *Genrich&Lautenbach* [2,3], *Jensen* [9], *Reisig* [18], and *Lautenbach&Pagnoni* [14].

5.1 Some Basic Mathematics

Before we start developing linear algebraic methods for analysing Petri systems modelled by PrT-nets we collect some basic notions and results we shall use throughout this chapter. Details can be found, for example, in [12].

Notation 5.1 For two sets A, B we designate the set of all functions of A into B as $[A \rightarrow B]$. For $B = \mathbf{N}$ or $B = \mathbf{Z}$, $[A \rightarrow B]_{fin}$ designates the set of all functions of A into B with *finite support*, ie with only finitely many elements of A having an image different from 0. (See also definition 2.6.)

There are two kinds of mathematical structures that are central to this chapter: rings and modules.

Definition 5.2 A *ring* is a set R provided with *addition* $(x, y) \mapsto x+y$, *multiplication* $(x, y) \mapsto x\,y$, and two constants 0,1 satisfying the following conditions:

1. Under addition, R is an abelian group with unit element 0. For all $x, y, z \in R$ we have
   - $(x + y) + z = x + (y + z)$ (*associativity*)
   - $x + y = y + x$ (*commutativity*)
   - $x + 0 = 0 + x = x$ (*unit element*)
   - There is an element x', usually designated as $-x$ such that $x + x' = 0$ (*inverse*)
2. Under multiplication, R is a momoid with unit element 1. For all $x, y, z \in R$ we have
   - $(xy)z = x(yz)$ (*associativity*)
   - $x1 = 1x = x$ (*unit element*)
3. For all $x, y, z \in R$ we have
   - $(x + y)z = xz + yz$ and $x(y + z) = xy + xz$ (*distributivity*)

R is called a *commutative* if multiplication is commutative.

Examples of rings are:

1. The integers $\mathbf{Z}$ (commutative).

2. For $n \geq 1$, the integer n, n-matrices.

3. The set of polynomials in several formal parameters $x_1, \ldots, x_n$ with integer coefficients (commutative).

Definition 5.3 Let R be a ring. A *(left) module over* R, or R-module, is a set M with addition $(v, w) \mapsto v{+}w$ and (left) multiplication by elements of R (called *scalars*), $(i, v) \mapsto i\,v$, satisfying the following conditions ($i, j \in R$ and $v, w \in M$).

1. Under addition, M is an abelian group.
2. $(ij)v = i(jv)$
3. For 1 being the unit element of multiplication in R, $1\,v = v$.
4. $(i + j)v = iv + jv$ and $i(v + w) = iv + iw$.

Definition 5.4 Let R be a commutative ring. A *R-linear algebra*, also called just R-algebra or algebra over R, is a R-module A which is provided with *multiplication* $(v, w) \mapsto v\,w$ satisfying the following conditions:

1. Under addition and multiplication, A is a ring.
2. For all $v, w \in A$ and $i \in R$, $i\,(v\,w) = (i\,v)\,w = v\,(i\,w)$.

Note that a ring R can be viewed as a R-module and a commutative ring can be viewed as a R-algebra. Another example of a R-algebra, for $R = \mathbf{Z}$, is $\mathcal{L}(D)$ with multiplication $\sqcap$ (see definition 2.7).

The modules in this chapter are all built according to the following scheme. We start with the ring of integers, $\mathbf{Z}$, and construct 'higher-order' $\mathbf{Z}$-modules by using the elements of a given $\mathbf{Z}$-module as entries for *vectors* and *matrices*.

Definition 5.5 Let R be ring, let M be a R-module, and let X, Y be arbitrary sets.

- A mapping $v{:}X \to M$ is called a *X-vector* over M.
- A mapping $F{:}X{\times}Y \to M$ is called a *X, Y-matrix* over M.

For vectors and matrices we define addition and multiplication by scalars as componentwise addition and multiplication by scalars in M. For $v, w{:}X \to M$, $F, G{:}X{\times}Y \to M$ and $i \in R$ we have

- $(v{+}w){:}x \mapsto v(x){+}w(x)$ and $(i\,v){:}x \mapsto i\,v(x)$ for $x \in X$,
- $(F{+}G){:}(x, y) \mapsto F(x, y){+}G(x, y)$ and $(i\,F){:}(x, y) \mapsto i\,F(x, y)$ for $x \in X, y \in Y$.

Note that in the special case that $M = R = \mathbf{Z}$, the set of X-vectors over M is the same as $\mathcal{L}(X)$ and the set of X, Y-matrices is $\mathcal{L}(X, Y)$ (linear combinations with integer coefficients as defined in 2.6).

Proposition 5.6 Let R be ring, let M be a R-module, and let X, Y be arbitrary sets. Under addition and multiplication by scalars, the set of X-vectors over M and the set of X, Y-matrices over M are R-modules.

Notation 5.7 For X-vectors and X, Y-matrices over a R-module M we use the following notations (vector v, matrix F, $x \in X, y \in Y$).

- 0 denotes the unit element of addition in M.

- $\mathbf{0}$ is the constant vector, $\mathbf{0}{:}X \to \{0\}$, respectively the constant matrix, $\mathbf{0}{:}X \times Y \to \{0\}$.
- $v_x = v(x)$ denotes the x-th entry of the vector v.
- F_x denotes the x-th *row* of the matrix F (when considered in the usual way as an rectangular array); it is a single-row matrix.
- F^y denotes the y-th *column* of F; it is a single-column matrix. A vector may always be viewed as a single-column matrix, and vice versa. Note that the X, Y-matrices can also be viewed as Y-vectors over the R-module of X-vectors, $[X \times Y \to M] \simeq [Y \to [X \to M]]$.
- $F_x^y = (F^y)_x$ denotes x, y-th entry of F.
- $F^\top{:}Y \times X \to M$ with $(F^\top)_y^x = F_x^y$ is the *transpose* of F.

If M is also a R-algebra with 1 denoting its unit under multiplication, we have:

- $\mathbf{1}$ denotes the constant vector, $\mathbf{1}{:}X \to \{1\}$.
- $\mathbf{e}^{(u)}$ is the u-th unit vector, $\mathbf{e}^{(u)}{:}x \mapsto [x{=}u]$ $(x \in X)$. (Hence $v = \sum_x v_x \mathbf{e}^{(x)}$.)

Normally the use of vectors and matrices implies that M is a R-algebra and X, Y and Z are *finite* sets. We then have:

- For two X-vectors v and w, the *scalar product* of v and w is defined as $v{\cdot}w = \sum_{x \in X} v_x w_x$.
- For two matrices $F{:}X \times Y \to M$ and $G{:}Y \times Z \to M$, the *(matrix) product* of F and G is the matrix $F\,G{:}X \times Z \to M$ defined by $(F\,G)_x^z = \sum_y F_x^y G_y^z$ (the scalar product of $(F_x)^\top$ and G^z). Note that $v{\cdot}w = v^\top w$ (vector v being viewed as a single-column matrix).

The main role of matrices is to serve as representations of *linear mappings* between modules.

Definition 5.8 Let R be a ring and M, M' be two R-modules. A function $f{:}M \to M'$ is called *R-linear* – designated as $f{:}M \overset{lin}{\to} M'$ – iff for all $i \in R$ and $v, w \in M$, $f(v{+}w) = f(v){+}f(w)$ and $f(iv) = if(v)$.
If in particular $M' = R$, f is called a *linear form*.

Proposition 5.9 Let R be a ring, M, M' be two R-modules, and $f{:}M \overset{lin}{\to} M'$. The set of elements $v \in M$ for which $f(v) = 0$ (the *kernel* of f) is a submodule of M.

Proposition 5.10 If M is a the module of X-vectors over a ring R, a linear mapping $f{:}M \overset{lin}{\to} M'$ is uniquely determined by the values it takes on the unit vectors of M. For every vector $v \in M$, $f(v) = f(\sum_x v_x \mathbf{e}^{(x)}) = \sum_x v_x f(\mathbf{e}^{(x)})$.

Definition 5.11 For a mapping $g{:}X \to M'$, the R-linear mapping $f{:}[X \to R] \overset{lin}{\to} M'$ with $f(\mathbf{e}^{(x)}) = g(x)$ is called the *linear extension* of g.

Proposition 5.12 Let R be a ring and X, Y be finite sets. A R-linear mapping f of the module of X-vectors into the module of Y-vectors, $f{:}[X \to R] \overset{lin}{\to} [Y \to R]$, can be represented, in a unique manner, by a matrix $F{:}Y \times X \to R$ such that for all $v \in [X \to R]$, $f(v) = F\,v$.
Conversely, every matrix $F{:}Y \times X \to R$ represents a R-linear mapping $f{:}[X \to R] \overset{lin}{\to} [Y \to R]$ by virtue of $f(v) = F\,v$.
Proof: Take F such that for all $x \in X$, $F^x = f(\mathbf{e}^{(x)})$. Then we get from the linearity of f, $f(v) = f(\sum_x v_x \mathbf{e}^{(x)}) = \sum_x v_x f(\mathbf{e}^{(x)}) = \sum_x v_x F^x = F\,v$. Uniqueness follows from proposition 5.10. The converse is obvious.

Corollary 5.13 For a linear form $f:[X\to R] \overset{lin}{\to} R$, the matrix representation of f reduces to a vector (single-column matrix) $\tilde{f}:X\to R$ such that for all $v \in [X\to R]$, $f(v) = v\cdot\tilde{f}$.

Proposition 5.14 For a finite set X, the set of square matrices $[X\times X\to R]$ under addition and (matrix) multiplication is a ring and, if R is commutative, a R-algebra. The unit element under multiplication is the diagonal matrix $\mathbf{D}:x,y \mapsto [x=y]$ $(x,y \in X)$.

Let for $X = \{a,b,c\}$, $v = 2a - b$ and $w = -3a + 2b - c$ be two elements of $\mathcal{L}(X)$. While the scalar product $v\cdot w$ is an integer, $v\cdot w = -6 - 2 + 0 = -8$, we now wish to extend $\mathcal{L}(X)$ to the ring of formal polynomials in X such that the *ring product* of v and w is the polynomial $v\,w = -6a^2 + 7ab - 2ac - 2b^2 + bc$.

Definition 5.15 Let X be a finite set, $X = \{x_1,\dots,x_n\}$. The set of polynomials in X over $\mathbf{Z}$ is designated as $\mathcal{A}(X)$ ($\mathcal{A}$ because we get a linear algebra over X). It is defined as $\mathcal{A}(X) = [[X\to\mathbf{N}]\to\mathbf{Z}]_{fin}$.
The elements of $[X\to\mathbf{N}]$ are called the *monomials*. The monomials form a commutative monoid which we write with multiplication. A monomial $v:X\to\mathbf{N}$ is written as $v = x_1^{v_{x_1}}\dots x_n^{v_{x_n}}$. Hence the product of two monomials is $vw = x_1^{(v+w)_{x_1}}\dots x_n^{(v+w)_{x_n}}$.
$\mathcal{A}(X)$ is provided with addition $(p,q) \mapsto p + q$ and (ring) multiplication $(p,q) \mapsto p\,q$ defined by $(p+q)_v = p_v + q_v$ and $(p\,q)_u = \sum_{u=vw} p_v q_w$ for $u,v,w \in [X\to\mathbf{N}]$.

Proposition 5.16 Let X be a finite set and $\mathcal{A}(X)$ the set of formal polynomials in X over $\mathbf{Z}$. Under addition, multiplication and multiplication by scalars, $\mathcal{A}(X)$ is a $\mathbf{Z}$-algebra.
Proof: All properties (see defintiton 5.4) are very easy to verify. The unit element of multiplication is $x_1^0\dots x_n^0$. The finite support property is preserved by addition and multiplication. It allows to define for each polynomial p its *degree* $\delta(p)$ which is the largest sum of powers in a monomial $v \in [X\to\mathbf{N}]$ with non-zero coefficient in p. Hence for each number $d \geq 0$, a polynomial p of degree d can be written as $\sum_{i_1+\dots+i_n\leq d} c_{(i_1,\dots,i_n)} x_1^{i_1}\dots x_n^{i_n}$.

Note that $\mathcal{A}(X)$ is an extension of $\mathcal{L}(X)$; $\mathcal{L}(X)$ is the set of all elements of $\mathcal{A}(X)$ of degree 1 where the unit monomial $x_1^0\dots x_n^0$ has the coefficient 0.

5.2 The Representation of Linear Transformations

At the beginning of this chapter we have mentioned that the important role that linear algebra plays in net theory is due to fact that the transition rule of ordinary Petri nets has a linear-algebraic representation.

Definition 5.17 Let $N = (S,T;F)$ is a net of places S and transitions T. The *incidence matrix* of N is a matrix $C:S\times T\to\mathbf{Z}$ such that for $s \in S$, $t \in T$, $C(s,t) = \begin{cases} -1 & \text{if } (s,t) \in F \setminus F^{-1} \\ 1 & \text{if } (s,t) \in F^{-1} \setminus F \\ 0 & \text{otherwise} \end{cases}$.

Proposition 5.18 Let $N = (S,T;F)$ is a net and C the incidence matrix of N. Then for two markings $M, M':S\to\mathbf{N}$ the following holds.

· If M' is reachable from M, there exists a vector $p:T\to\mathbf{Z}$ such that $M' = M + Cp$

To concentrate on the linear algebraic aspects we abstract from Petri nets for a while and consider what we call $\mathbf{Z}$-linear systems.

Definition 5.19 A $\mathbf{Z}$-linear system is a tuple $\mathcal{S} = (S,T;C,[m^0])$ where

- S is a finite, non-empty set of integer variables called *places*; a vector $m{:}S \to \mathbf{Z}$ is called a *state* of S.
- T is a finite, non-empty set, disjoint from S, of integer variables called *transitions*; a vector $p{:}T \to \mathbf{Z}$ is called an *process* of S.
- C is a matrix $C{:}S{\times}T \to \mathbf{Z}$ called *the transformation matrix* of S.
- m^0 is the *representative* state of S. $[m^0]$ is the set of all states that can be reached from m^0: $m \in [m^0]$ iff there is a process $p{:}T \to \mathbf{Z}$ such that $m = m^0 + C\,p$.

The power of linear algebra for the analysis of Petri systems is based above all on the possibility to compute an important class of system invariants.

Definition 5.20 Let $\mathcal{S} = (S, T; C, [m^0])$ be a $\mathbf{Z}$-linear system. A linear form on the integer S-vectors,, $l{:}\mathcal{L}(S) \overset{lin}{\to} \mathbf{Z}$, is called a (linear) S*-invariant* of $\mathcal{S}$ iff for every state $m \in [m^0]$, $l(m) = l(m^0)$.

Linear algebra tells us that the S-invariants of $\mathcal{S}$ can be found by solving a homogeneous system of linear equations whose coefficient matrix is transpose of the transformation matrix C.

Theorem 5.21 Let $\mathcal{S} = (S, T; C, [m^0])$ be a $\mathbf{Z}$-linear system. A linear form $l{:}\mathcal{L}(S) \to \mathbf{Z}$ is a S-invariant of $\mathcal{S}$ iff there is an integer S-vector $i \in \mathcal{L}(S)$ such that

1. i is a solution of the homogeneous system of linear equations $C^{\top} i = \mathbf{0}$,
2. for all $m \in \mathcal{L}(S)$, $l(m) = m{\cdot}i$.

Proof: (a) Let l be a S-invariant. Let i be the vector representing l according to corollary 5.13. Then $l(m) = m{\cdot}i$ for all S-vectors m. For every transition t there is a state m such that $m = m^0 + C\mathbf{e}^{(t)}$. Since $l(m) = l(m^0)$, $0 = l(m) - l(m^0) = l(C^t) = C^t{\cdot}i = (C^{\top} i)_t$.
(b) Let i be a solution of $C^{\top} i = \mathbf{0}$. Let $l{:}\mathcal{L}(S) \to \mathbf{Z}$ be defined by $l(m) = m{\cdot}i$ for all S-vectors $m \in \mathcal{L}(S)$. For all states $m \in [m^0]$ there is a process $p \in \mathcal{L}(T)$ such that $m = m^0 + C\,p$. Hence $l(m) = l(m^0 + C\,p) = l(m) + l(C\,p)$, and $l(Cp) = (Cp){\cdot}i = (Cp)^{\top} i = (p^{\top} C^{\top}) i = p^{\top}(C^{\top} i) = \mathbf{0}$.

Corollary 5.22 The S-invariants do not depend on the representative state $[m^0]$.

Corollary 5.23 The S-invariants form a submodule of (S).
Proof: $C^{\top}$ represents a linear mapping of $\mathcal{L}(S)$ into $\mathcal{L}(T)$ (see 5.12). The kernel of a linear mapping between modules is a submodul of the domain (see 5.9).

Because of their close relationship to S-invariants, the solutions of $C^{\top} i = \mathbf{0}$ will be called S-invariants, too.

The rest of this chapter will be concerned with three questions: How to compress the matrix representation of the system $\mathcal{S}$ in a way that is comparable to folding a CE-net into a PrT-net; how to compute S-invariants of $\mathcal{S}$ based on the compressed representation; how to apply the results to the analysis of PrT-nets.

5.3 The Compression of Matrices

Given a $\mathbf{Z}$-linear system $\mathcal{S} = (S, T; C, [m^0])$ we wish to compress its transformation matrix C in such a way that

- the number of rows and columns can be decreased;
- the original representation can be reproduced;

· there are techniques for computing S-invariants of $\mathcal{S}$ using the compressed representation.

The basic idea is to partition the rows and columns of C and to merge the elements of every block to form a *coarse* matrix $\mathbf{C}$. To be able to identify the original constituents of a new row or column, they are 'painted' in different colours.

Definition 5.24 Let $C{:}S{\times}T{\rightarrow}\mathbf{Z}$ be an integer matrix, and let A and B be two sets of identifiers, not necessarily disjoint, serving as place colours and transition colours, respectively. Let $\mathbf{S} \subseteq \mathcal{P}(S)$ and $\mathbf{T} \subseteq \mathcal{P}(T)$ be partitions (into disjoint non-empty *blocks*) of S respectively T. A *colouring* of C is a pair of mappings $p_S{:}S{\rightarrow}A$ and $p_T{:}T{\rightarrow}B$. It is said to be *consistent* with the partitions of S and T if any two different elements of the same block are coloured differently.

Given a matrix C with partitions $\mathbf{S}$ and $\mathbf{T}$ and a consistent colouring (p_S, p_T) in colours A and B, we now construct the quotient matrix $\mathbf{C}$ by first merging rows and then merging columns.

The result of merging for each block $\mathbf{s} \in \mathbf{S}$ all rows of $\mathbf{s}$ into a single row associated with $\mathbf{s}$ is a matrix $\tilde{C}{:}\mathbf{S}{\times}T{\rightarrow}\mathcal{L}(A)$. Every entry $\tilde{C}^t_{\mathbf{s}}$ is the linear combination of the colours used for $\mathbf{s}$ with the corresponding entries in C^t as coefficients. $\tilde{C}^t_{\mathbf{s}}{:}A{\rightarrow}\mathbf{Z}$ is defined such that for every $a \in A$, $\tilde{C}^t_{\mathbf{s}}(a) = C^t_s$ if $s \in \mathbf{s}$ and $\tilde{C}^t_{\mathbf{s}}(a) = 0$ otherwise.
(Note that $\tilde{C}^t_{\mathbf{s}}$ is really a function since p_S is consistent with $\mathbf{S}$.)

In the second step, the columns of $\tilde{C}$ are merged according to the partition $\mathbf{T}$ in very much the same way the rows were merged. The difference is that the entries of $\tilde{C}$ are no longer integers but linear combinations of place colours. Hence the entries of $\mathbf{C}$ are functions that assign to transition colours linear combinations of place colours. We get $\mathbf{C}{:}\mathbf{S}{\times}\mathbf{T}{\rightarrow}[B{\rightarrow}\mathcal{L}(A)]$ defined such that for $b \in B$, $\mathbf{C}^{\mathbf{t}}_{\mathbf{s}}(b) = \tilde{C}^t_{\mathbf{s}}$ if $t \in \mathbf{t}$ and $\mathbf{C}^{\mathbf{t}}_{\mathbf{s}}(b) = 0$ otherwise.

Example 5.25 An example of compressing an integer matrix is shown in table 1. Places $s_1, \ldots, s_6$ and transitions $t_1, \ldots, t_4$ are given with their partitioning into $\mathbf{S} = \{L, R\}$ and $\mathbf{T} = \{O, E\}$ and their painting with colours $A = \{r, b, g, y\}$ and $B = \{l, h\}$. First the rows of the upper left matrix are compressed. The result is the lower left matrix. Its columns are then compressed to form the 2x2-matrix in the lower right part of the scheme. Its entries are functions that are defined separately below the scheme.

The method of compressing an integer matrix just described is, except for some technical details, the one used by *Jensen* in his model of *Coloured Petri nets (CP-nets)* [9,10]. As arc annotations and hence as entries of the incidence matrix of a coloured Petri net, he uses simple linear combinations of names for functions that map transition colours into integer vectors of place colours. An he develops a technique for finding S-invariants based on this representation.

5.4 Structured Colours and Individual Variables

So far we have used plain colours for compressing the transformation matrices of linear systems. Next we present a method that is strongly influenced by the PrT-nets. In the first step of compression, it uses tuples of individual constants as place colours. This leads to quite a different way of proceeding in the second step.

Definition 5.26 Let as before $C{:}S{\times}T{\rightarrow}\mathbf{Z}$ be the transformation matrix of a $\mathbf{Z}$-linear system $\mathcal{S} = (S, T; C, [m^0])$. Let U be a finite set of identifiers (individual constants), and let for a given $n \geq 0$, the colour set A be the set of tuples of elements of U of length $\leq n$: $A = \bigcup_{0 \leq m \leq n} U^m$. Let $\mathbf{S} \subseteq \mathcal{P}(S)$ be a partition of the places, and let $p_S{:}S{\rightarrow}A$ be a place colouring. p_S is called *consistent* with $\mathbf{S}$ if for each block $\mathbf{s}$,

		t_1 O,l	t_2 E,l	t_3 O,h	t_4 E,h	O	E
s_1	L,r	-1			1		
s_2	R,r	1	-1				
s_3	L,b		1	-1			
s_4	R,b			1	-1		
s_5	L,y	1			-1		
s_6	L,g		-1	1			
	L	$-\langle r\rangle+\langle y\rangle$	$\langle b\rangle-\langle g\rangle$	$-\langle b\rangle+\langle g\rangle$	$\langle r\rangle-\langle y\rangle$	$-G$	$\tilde{G}$
	R	$\langle r\rangle$	$-\langle r\rangle$	$\langle b\rangle$	$-\langle b\rangle$	F	$-F$

$F = \{l \mapsto \langle r\rangle, h \mapsto \langle b\rangle\}$
$G = \{l \mapsto \langle r\rangle-\langle y\rangle, h \mapsto \langle b\rangle-\langle g\rangle\}$
$\tilde{G} = \{h \mapsto \langle r\rangle-\langle y\rangle, l \mapsto \langle b\rangle-\langle g\rangle\}$

Table 1: Compression of an Integer Matrix

1. different places in $\mathbf{s}$ are coloured differently,
2. all places of $\mathbf{s}$ are coloured by tuples of the same length (the *arity* of $\mathbf{s}$).

Let $\tilde{C}{:}\mathbf{S}{\times}T{\to}\mathcal{L}(A)$ be the result of merging the rows of each block of $\mathbf{S}$ into a single coarse place $\mathbf{s}$, as described above. The second step of merging columns of $\tilde{C}$ will now differ essentially from the previous method (the 'Jensen method'). We no longer take an arbitrary partition $\mathbf{T}$ of T with some consistent colouring to compress $\tilde{C}$. Rather, we shall employ the general technique of abstracting from a set of similar expressions to a *scheme* from which the expressions can be derived by means of substitutions.

Definition 5.27 Two elements $v, w \in \mathcal{L}(A)$ are called *similar* if

1. all tuples with non-zero coefficients in v or w have the same length,
2. the sum of coefficients is the same for v and w, $v{\cdot}\mathbf{1} = w{\cdot}\mathbf{1}$ (*uniformity*).

Definition 5.28 Let $\tilde{C}{:}\mathbf{S}{\times}T{\to}\mathcal{L}(A)$ be the result of merging the rows of C as described above. Let $\mathbf{T}$ be a partition of T. $\tilde{C}$ is called *uniform* with respect to $\mathbf{T}$ if for each block $\mathbf{t} \in \mathbf{T}$ and any two transitions $t, t' \in \mathbf{t}$, the columns $\tilde{C}^t$ and $\tilde{C}^{t'}$ are similar, ie $\tilde{C}^t_{\mathbf{s}}$ and $\tilde{C}^{t'}_{\mathbf{s}}$ are similar for all $\mathbf{s}$.

The uniformity of $\tilde{C}$ is the key to merging its columns according to $\mathbf{T}$.

Theorem 5.29 Let V be a set of individual variables disjoint from U (and large enough as specified in the proof) and let $W = U \cup V$. Let $L \subseteq \mathcal{L}(A)$ be a set of linear combinations of tuples that are pairwise similar. Then there exists a scheme λ for L being a linear combination of tuples of individual variables such that for each element $v \in L$ there is a substitution $\beta{:}V{\to}U$ such that $v = \lambda{:}\beta$.

Proof: Let m be the length of tuples and k the sum of coefficients that are common to L due to similarity. Let l be the maximal 'length' of the elements of L, namely $l = max_{v\in L} \sum_x |v_x|$. Note that $l \geq |k|$. We split l into $l^+ = max_v \sum_{v_x>0} v_x$ and $l^- = max_v \sum_{v_x<0} -v_x$. Then $l = l^+ + l^-$ and $k = l^+ - l^-$.

To form the scheme λ, we take a set of l m-tuples of individual variables such that all variables are different. (Hence V must contain at least $l\,m$ variables.) We assign to l^+ tuples the coefficient $+1$ and to the remaining l^- tuples the coefficient -1. Then each member of L can be derived from λ by means of a substitution. If not all tuples of λ are needed for some element of L because its length is less than l, the same number of tuples with positive coefficients and with negative coefficients are left since the sum of coefficients $k = l^+ - l^-$ is the same for all elements of L. These remaining tuples can be made to sum up to zero if all their variables are replaced, for example, by the same constant.

This result now helps to construct from $\tilde{C}$:$\mathbf{S}\times T\rightarrow\mathcal{L}(A)$ the compressed representation $\mathbf{C}$:$\mathbf{S}\times \mathbf{T}\rightarrow\mathcal{L}(\tilde{A})$ where $\tilde{A}$ is built from W in the same way as A was built from U. Formally, every element λ of $\mathcal{L}(\tilde{A})$ denotes a function λ:$B\rightarrow[A\rightarrow\mathbf{Z}]]$ where the set of transition colours, B is the set of all substitutions of variables by constants, $B = [V\rightarrow U]$. (The representation of these functions forming the entries of the compressed matrix $\mathbf{C}$, however, differs essentially from those used by the other methods described previously (which are all tabular, ie in some matrix form). The advantage, as we shall see a little later, of this method is that algorithms for computing S-invariants will exploit the information about the function present in its representation as a scheme.

For every coarse transition $\mathbf{t}$, we have a subset $B_{\mathbf{t}} \subseteq B$ of transition colours (substitutions) that is one-to-one related to the elements of $\mathbf{t}$; for every $t \in \mathbf{t}$ we take exactly one element $\beta_t \in B_{\mathbf{t}}$ with $\tilde{C}^t_{\mathbf{s}} = \lambda$:$\beta_t$. In the PrT-net model, these 'firing modes' of the transitions are determined by the formulae annotating the transitions.

Note that like place colours, also the transition colours may be viewed as tuples of individual constants. Assume that the variables are alphabetically ordered and list for a transition $\mathbf{t}$, the variables that occur in its entries. Then a substitution can be represented by the tuple resulting from applying it to the tuple of variables.

To form the entries of a column $\mathbf{C}^{\mathbf{t}}$, the construction of theorem 5.29 requires disjoint sets of variables for different coarse places $\mathbf{s}$. What may look here as a terrible waste of variables leading to unnecessarily large substitution vectors, isn't that bad in practice where constants and shared variables are used.

5.5 Computing S-invariants

Let for the rest of this chapter, $\mathcal{S}/_{\mathbf{S},\mathbf{T}} = (\mathbf{S}, \mathbf{T}, \mathbf{C})$ be the compressed representation of the linear system $\mathcal{S} = (S, T; C, [m^0])$ constructed according to the 'PrT-method' described in the previous section. The ingredients of $\mathcal{S}/_{\mathbf{S},\mathbf{T}}$ are as follows:

- $\mathbf{S}$: the set of (coarse) places
- $\mathbf{T}$: the set of (coarse) transitions
- U: the set of (individual) constants
- V: the set of (individual) variables
- $W = U \cup V$: the set of (individual) names
- A: the set of place colours; for a given maximal arity of places, n, $A = \bigcup_{0\leq m\leq n} U^m$ with $U^0 = \{\langle\rangle\}$
- $\tilde{A}$: the set of place colour schemes; $\tilde{A}$ is built from W as A is built from U
- B: the set of transition colours; B is the set of substitutions, $B = [V\rightarrow U]$
- $B_{\mathbf{t}}$: the colour set of transition $\mathbf{t}$, ie the set of substitutions ('firing modes') of $\mathbf{t}$
- m: the generic state vector, m:$\mathbf{S}\rightarrow\mathcal{L}(A)$
- p: the generic process, p:$\mathbf{T}\rightarrow\mathcal{L}(B)$, with $p_{\mathbf{t}}(\beta) = 0$ if $\beta \notin B_{\mathbf{t}}$

· $\mathbf{C}$:$\mathbf{S}\times\mathbf{T}\rightarrow\mathcal{L}(\tilde{A})$: the compressed transformation matrix

· $m = m^0 + \mathbf{C}{*}p$: the system equation in terms of $\mathbf{C}$ where $\mathbf{C}{*}p = \sum_{\mathbf{t},\beta} p_{\mathbf{t},\beta}\mathbf{C}^{\mathbf{t}}{:}\beta$

· l: a linear form on states, $l{:}[\mathbf{S}\rightarrow\mathcal{L}(A)]\rightarrow\mathbf{Z}$; l is a S-invariant if $l(m) = l(m^0)$ for all m such that $m = m^0 + \mathbf{C}{*}p$ for some process p

We are now going to present a method for deriving S-invariants for the linear system S from its PrT-matrix $\mathbf{C}$. It is based on the ring extension of $\mathcal{L}(\tilde{A})$. For a second method that is based on a generalisation of the scalar product in $\mathcal{L}(A)$ to schemes, the reader is referred to [3,14].

Let $\mathcal{A}(\tilde{A})$ be the linear algebra based on the ring extension of $\mathcal{L}(\tilde{A})$, as defined in definition 5.15. In $\mathcal{A}(\tilde{A})$ we can solve the homogeneous equation system $\mathbf{C}^{\mathsf{T}}i = \mathbf{0}$, for example by a generalisation of the Gaussian algorithm (see [11,15]). If such a solution is variable-free, ie if its coefficients are in $\mathcal{A}(A)$ (the subring of $\mathcal{A}(\tilde{A})$ generated by $A \subseteq \tilde{A}$), it can be used as a S-invariant in the same way as the integer solutions of the ordinary integer equation system $C^{\mathsf{T}}i = \mathbf{0}$ for the uncompressed system (see theorem 5.21).

Theorem 5.30 [5] Let $i{:}\mathbf{S}\rightarrow\mathcal{A}(A)$ be a *variable-free* solution of the equation system $\mathbf{C}^{\mathsf{T}}i = \mathbf{0}$, ie for all $\mathbf{t}$, $\sum_{\mathbf{s}} \mathbf{C}^{\mathbf{t}}_{\mathbf{s}} i_{\mathbf{s}} = 0$. Then for all processes $p{:}\mathbf{T}\rightarrow\mathcal{L}(B)$, the scalar product of $\mathbf{C}{*}p$ and i (in the $\mathcal{A}(A)$-module $[\mathbf{S}\rightarrow\mathcal{A}(A)]$ is $\mathbf{0}$.
Proof:

$$\begin{aligned}
(\mathbf{C}{*}p){\cdot}i &= \textstyle\sum_{\mathbf{s}}(\mathbf{C}{*}p)_{\mathbf{s}} i_{\mathbf{s}} \\
&= \textstyle\sum_{\mathbf{s}}(\sum_{\mathbf{t},\beta} p_{\mathbf{t},\beta}\mathbf{C}^{\mathbf{t}}{:}\beta)_{\mathbf{s}} i_{\mathbf{s}} \\
&= \textstyle\sum_{\mathbf{t},\beta} p_{\mathbf{t},\beta} \sum_{\mathbf{s}}(\mathbf{C}^{\mathbf{t}}_{\mathbf{s}}{:}\beta) i_{\mathbf{s}} && \text{(linearity)} \\
&= \textstyle\sum_{\mathbf{t},\beta} p_{\mathbf{t},\beta} \sum_{\mathbf{s}}(\mathbf{C}^{\mathbf{t}}_{\mathbf{s}} i_{\mathbf{s}}){:}\beta && (i_{\mathbf{s}} \text{ is variable-free}) \\
&= \textstyle\sum_{\mathbf{t},\beta} p_{\mathbf{t},\beta} \underbrace{(\sum_{\mathbf{s}}\mathbf{C}^{\mathbf{t}}_{\mathbf{s}} i_{\mathbf{s}})}{:}\beta && (\mathbf{C}^{\mathsf{T}}i) \\
&= \textstyle\sum_{\mathbf{t},\beta} p_{\mathbf{t},\beta}(0{:}\beta) \\
&= \mathbf{0}
\end{aligned}$$

As a consequence every solution of $\mathbf{C}^{\mathsf{T}}\, i = \mathbf{0}$ whose entries do not contain individual variables determines a family of S-invariants.

Lemma 5.31 Let $i{:}\mathbf{S}\rightarrow\mathcal{A}(A)$ be a variable-free solution of $\mathbf{C}^{\mathsf{T}}\, i = \mathbf{0}$. Then for all monomials $v \in [A\rightarrow\mathbf{N}]$ the linear form $l_v{:}[\mathbf{S}\rightarrow\mathcal{L}(A)]\rightarrow\mathbf{Z}$, defined by $l_v(m) = (m^{\mathsf{T}}i){\cdot}\mathbf{e}^{(v)}$ is a S-invariant. ($l_v(m)$ is the coefficient of v in $m^{\mathsf{T}}i \in \mathcal{A}(A)$. Of course, l_v is trivial (constantly 0) for most monomials v.)
Proof: Let m,m', p be such that $m' = m + \mathbf{C}{*}p$. By definition of l_v we have

$$\begin{aligned}
l_v(m') &= (m'^{\mathsf{T}}i){\cdot}\mathbf{e}^{(v)} \\
&= ((m + \mathbf{C}{*}p)^{\mathsf{T}}i){\cdot}\mathbf{e}^{(v)} \\
&= (m^{\mathsf{T}}i + \underbrace{(\mathbf{C}{*}p)^{\mathsf{T}}i}){\cdot}\mathbf{e}^{(v)} \\
&= (m^{\mathsf{T}}i + \mathbf{0}){\cdot}\mathbf{e}^{(v)} \\
&= l_v(m)
\end{aligned}$$

Very often, not enough variable-free solutions of $\mathbf{C}^{\mathsf{T}}i = \mathbf{0}$ will exist. The partial projection of places as defined in section 3.5 may then help to transform $\mathbf{C}$ in such a way that the variables are eliminated.

Theorem 5.32 Let $\mathbf{C}$ be a PrT-matrix and let $\mathbf{s}$ be a place whose arity is $m \geq 1$ (each entry of $\mathbf{C}^{\mathbf{t}}_{\mathbf{s}}$ is an element of $\mathcal{L}(U^m)$). Let for some k with $1 \leq k \leq m$, $\bar{\mathbf{C}} = |\mathbf{C}|^{(\mathbf{s})}_k$ designate the result of projecting in $\mathbf{C}$ all entries of row $\mathbf{C}_{\mathbf{s}}$ along the k-th position. Let $\bar{i}{:}\mathbf{S}\rightarrow\mathcal{L}(A)$ be a variable-free solution of $\bar{\mathbf{C}}^{\mathsf{T}}\bar{i} = \mathbf{0}$. Then for every monomial $v{:}A\rightarrow\mathbf{N}$, the linear form $\bar{l}_v$ defined by

$\tilde{l}_v(m) = |m^\top|_k^{(s)}\,\tilde{i}$ is a S-invariant.
Proof: The result follows easily from the fact that projections are **Z**-linear operations on $\mathcal{L}_{fin}(A)$ and commute with substitution. Due to uniformity we have
$|(m + \mathbf{C}{*}p)|_k^{(s)} = |m|_k^{(s)} + |\mathbf{C}{*}p|_k^{(s)} = |m|_k^{(s)} + |\mathbf{C}|_k^{(s)}{*}p = |m|_k^{(s)} + \tilde{\mathbf{C}}{*}p$
Hence the above proofs work also for $\tilde{\mathbf{C}}$ instead of $\mathbf{C}$.

Lemma 5.33 The total projection of $\mathbf{C}$, $|\mathbf{C}|$, is the transformation matrix of an ordinary linear system $|\mathcal{S}/_{\mathbf{S},\mathbf{T}}|$ that represents the mere quantitative aspect of $\mathcal{S}/_{\mathbf{S},\mathbf{T}}$. The total projection $|i|$ of every solution of $\mathbf{C}^\top i = \mathbf{0}$ is an $\mathcal{S}$-invariant of $|\mathcal{S}/_{\mathbf{S},\mathbf{T}}|$.
Proof: From the theorem above follows immediately that $\mathbf{C}^\top i = \mathbf{0}$ implies $|\mathbf{C}|^\top |i| = \mathbf{0}$.

5.6 Two Examples

We are now going to demonstrate the S-invariant method for PrT-nets at two examples. The first one is to get us a little familiar with the practical aspects of finding invariants. We shall see how to combine solving linear symbolic equations with projections and other transformations of the net representation. The other example is to show how to use S-invariants in a mathematical proof of system propertries.

For our first example we return to the simple resource management scheme 2.11. The goal is to derive from its net representation, in a strictly formal manner, the validity of the two logical invariants (1) and (2) that we have already looked at in chapter 4 from a different point of view.

$$\Box(\forall x)[Uxe \rightarrow \neg(\exists y)(\exists m)((y{\neq}x \vee m{\neq}\mathbf{e}) \wedge Uy\,m)] \tag{1}$$

$$\Box(\forall x)(Ix \vee (\exists m)(Wx\,m \vee Ux\,m \vee Rx\,m)) \tag{2}$$

Table 2 shows the matrix form of the net representation where the tickets of the control mechanism have been replaced by plain tokens (places A and C are totally projected). L is an integer parameter denoting the number of tickets (access rights).

The rows belong to the places. The columns, as place vectors, correspond to the transitions 1 through 7, the representative marking M^0, two 'quasi' invariants q_1, q_2 and a proper S-invariant i_1. 0 entries are omitted.

	1	2 $m{=}e$	3 $m{=}s$	4 $m{=}e$	5 $m{=}s$	6	7	M^0	q_1	i_1	q_2 $L{=}1$	
I	$-\langle x\rangle$					$\langle x\rangle$		$\sum_x^{A\ni x}\langle x\rangle$	$\langle x,m\rangle$	1		I
W	$\langle x,m\rangle$	$-\langle x,m\rangle$	$-\langle x,m\rangle$						$\langle x\rangle$	$1\|\ \|_2$		W
U		$\langle x,m\rangle$	$\langle x,m\rangle$	$-\langle x,m\rangle$	$-\langle x,m\rangle$				$\langle x\rangle$	$1\|\ \|_2$	L	U
R				$\langle x,m\rangle$	$\langle x,m\rangle$	$-\langle x,m\rangle$			$\langle x\rangle$	$1\|\ \|_2$		R
$\|A\|$		$-L$	-1				1	L			$\langle x,m\rangle$	$\|A\|$
$\|C\|$				L	1		-1				$\langle x,m\rangle$	$\|C\|$

Table 2: The Resource Management Scheme in Matrix Form

The left part of the scheme is the incidence matrix $\mathbf{C}$ of the PrT-net. The vector q_1 is a solution of the symbolic equation system $\mathbf{C}^\top i = \mathbf{0}$ in the ring $\mathcal{R}(\tilde{A})$. This is easy to check by multiplying the rows of $\mathbf{C}$ by the corresponding entries of q_1 and adding them up; the result is the 0-row.

Not so easy to verify is that q_1 is the only relevant solution provided that L is different from 1. All solutions are multiples of it. The programs, however, that compute a basis for the module of solutions (cf [11,15]) deliver only one element. Another system [20] (developed for other purposes than net theory) gives also q_2 for the special case $L = 1$.

q_1 is not variable-free, hence it does not satify the premises of theorem 5.30. However, if we could eliminate the second position of the I entry, $\langle x, m \rangle$, it would become a multiple of the vector that has 1's as entries for I, W, U, and R. The origin of the I entry, $\langle x, m \rangle$, is in the rows W, U, and R. If we project these rows along the second position, the four rows add up to the 0-row.

This result is given in the column i_1 which represents a *proper* S-invariant of the system. It states the following invariant equation that we split into two inequalities.

$$I + |W|_2 + |U|_2 + |R|_2 = I^0 = \sum_x^{Agx} \langle x \rangle \quad (3)$$

$$I + |W|_2 + |U|_2 + |R|_2 \geq I^0 = \sum_x^{Agx} \langle x \rangle \quad (4)$$

$$I + |W|_2 + |U|_2 + |R|_2 \leq I^0 = \sum_x^{Agx} \langle x \rangle \quad (5)$$

From inequality (4) follows immediately the validity of the logical invariant (1). In fact, (1) and (4) are equivalent. Inequality (5) states that no agent is ever in more than one state (knowing that the magnitudes I, W, U, R – the markings of the places I, W, U, R – are non-negative).

Since the case $L = 1$ is not of interest to us, the other solution q_2 does not allow us to come to conclusions about the control mechanism. However, it tells us what to do. If we use the values for m given on top of columns 2 through 5 (by the transition selectors) for splitting the place U into two places, $Ux\mathbf{e}$ and $Ux\mathbf{s}$, (cf section 3.2) we get the slightly more detailed yet equivalent representation of the system shown in table 3.

	1	2 ⊤	3 ⊤	4 ⊤	5 ⊤	6	7	M^0	q_1	i_1	q_2	i_2	
I	$-\langle x \rangle$					$\langle x \rangle$		$\sum_x^{Agx} \langle x \rangle$	$\langle x, m \rangle$	1			I
W	$\langle x, m \rangle$	$-\langle x, \mathbf{e} \rangle$	$-\langle x, \mathbf{s} \rangle$						$\langle x \rangle$	$1\| \|_2$			W
$Ux\mathbf{e}$		$\langle x \rangle$		$-\langle x \rangle$					$\langle x, \mathbf{e} \rangle$	1	L	$L\| \|$	$Ux\mathbf{e}$
$Ux\mathbf{s}$			$\langle x \rangle$		$-\langle x \rangle$				$\langle x, \mathbf{s} \rangle$	1	1	$1\| \|$	$Ux\mathbf{s}$
R				$\langle x, \mathbf{e} \rangle$	$\langle x, \mathbf{s} \rangle$	$-\langle x, m \rangle$			$\langle x \rangle$	$1\| \|_2$			R
$\|A\|$		$-L$	-1				1	L			$\langle x \rangle$	1	$\|A\|$
$\|C\|$				L	1		-1				$\langle x \rangle$	1	$\|C\|$

Table 3: The Transformed Resource Management Scheme

Here the solution q_2 exists in general. It yields, without difficulties, the (proper) S-invariant i_2 which represents the invariant integer equation

$$L|Ux\mathbf{e}| + |Ux\mathbf{s}| + |A| + |C| = |A^0| = L \quad (6)$$

Since all magnitudes are non-negative, this equation implies the logical invariant (2). (Assume indirectly that there is an agent x_0 using the resource in exclusive mode and simultaneously, another agent x_1 using it in any mode. Then $|Ux\mathbf{e}| \geq 1$ and $(|Ux\mathbf{e} \geq 2 \vee |Ux\mathbf{s}| \geq 1)$. Hence $L|Ux\mathbf{e}| + |Ux\mathbf{s}| + |A| + |C| > L$, violating (6).)

The other example we look at is the (toy) scheme for maintaining multiple copies of a database taken from Reisig's book [17] (p.117). Figure 16 shows the PrT-net and table 4 shows its incidence matrix $\mathbf{C}$, the initial marking M^0, and eight solutions of the symbolic equation system $\mathbf{C}^\top i = \mathbf{0}$.

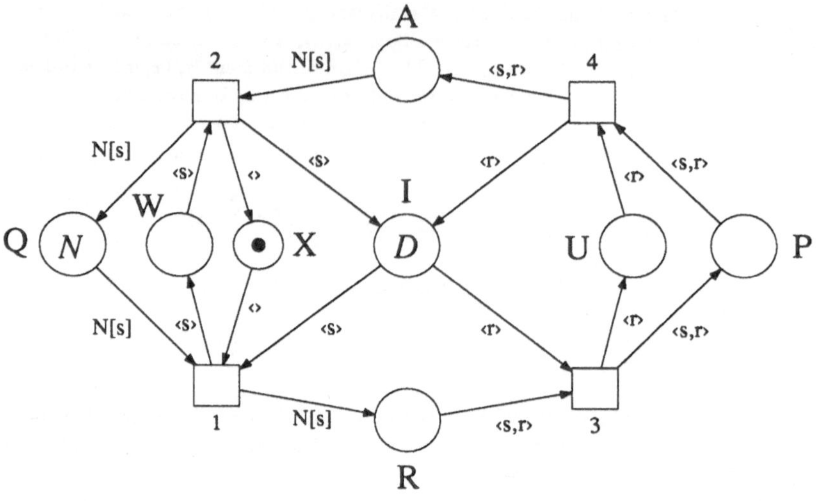

$$D = a_1 + \ldots + a_n; \quad N = \sum_{r,s}([r{\neq}s]\langle s,r\rangle); \quad N[s] = \sum_{r}([r{\neq}s]\langle s,r\rangle)$$

Figure 16: Protocol for Maintaining Multiple Copies of a Database

	1	2	3	4	M^0	i_1	i_2	q_3	q_4	q_5	q_6	q_7	q_8	
X	$-\langle\rangle$	$\langle\rangle$			$\langle\rangle$			$\langle s\rangle$			$-N[s]$	$\langle r\rangle\, N[s]$		X
I	$-\langle s\rangle$	$\langle s\rangle$	$-\langle r\rangle$	$\langle r\rangle$	D	1								I
W	$\langle s\rangle$	$-\langle s\rangle$				1		$\langle\rangle$		$N[s]$			$N[s]$	W
U			$\langle r\rangle$	$-\langle r\rangle$		1			$\langle s,r\rangle$			$\langle s,r\rangle$		U
Q	$-N[s]$	$N[s]$			N		1			$\langle s\rangle$	$\langle\rangle$			Q
R	$N[s]$		$-\langle s,r\rangle$				1					$\langle r\rangle$	$-\langle s\rangle$	R
P			$\langle s,r\rangle$	$-\langle s,r\rangle$			1		$-\langle r\rangle$				$-\langle s\rangle$	P
A		$-N[s]$		$\langle s,r\rangle$			1					$\langle r\rangle$	$-\langle s\rangle$	A

Table 4: The Database Example in Matrix Form

At most five solutions are linear independent; i_1, i_2, q_3, q_4, q_5, for example, form a basis. Only two solutions are free of individual variables, however, namely i_1 and i_2. They are *proper* S-invariants for which the equation $M{\cdot}i = M^0{\cdot}i$ holds for all markings M reachable from M^0 (see lemma 5.31).

The vectors q_3 through q_8 are 'quasi invariants'; they contain individual variables in some coefficients. To get rid of these variables we project those places from which they originate. For example, the coefficient $\langle s,r\rangle$ of row U in solution q_4 has its origin in row P. So we project row P such that the s disappears.

The result is shown in table 5. In some cases we cannot do better than projecting totally. In other cases, however, we can save a rest of the qualitative information given by the model. The vectors i_1 through i_8 yield the following eight system equations.

$$I + W + U = D \tag{7}$$
$$Q + R + P + A = N \tag{8}$$

	i_1	i_2	i_3	i_4	i_5	i_6	i_7	i_8	
X			1			$n-1$	$n-1$		X
I	1								I
W	1		$1\|\,\|$		$n-1$			$n-1$	W
U	1			1			$1\|\,\|$		U
Q		1			$1\|\,\|_2$	$-1\|\,\|$			Q
R		1					$1\|\,\|$	$-1\|\,\|_2$	R
P		1		$-1\|\,\|_1$				$-1\|\,\|_2$	P
A		1					$1\|\,\|$	$-1\|\,\|_2$	A

Table 5: Proper S-Invariants of the Database Scheme

$$X + |W| = 1 \quad (9)$$
$$U - |P|_1 = 0 \quad (10)$$
$$(n-1)W + |Q|_2 = (n-1)D \quad (11)$$
$$|Q| - (n-1)X = (n-1)^2 \quad (12)$$
$$|U| + |R| + |A| = (n-1)(1-X) \quad (13)$$
$$|R|_2 + |P|_2 + |A|_2 = (n-1)W \quad (14)$$

These equations can now be used to verify system properties in an ordinary mathematical way.

Theorem 5.34 The scheme shown in figure 16 is deadlock-free (live) and contact-free (safe).

Proof: Contact-freeness means that even under the weak transition rule allowing multi-sets on places there is no marking reachable that puts the same tuple on a place more than once. This follows immediately from equations (7) and (8) since D and N are characteristic vectors of sets and markings are non-negative. Hence in the following, we only have to look at the input places of a transition to determine whether it is enabled.

Deadlock-freeness means that there is no forward reachable marking at which no transition is enabled. Let M be any marking that is reachable from M^0. Assume that neither t_4 nor t_3 nor t_1 is enabled. *Claim:* Then t_2 is enabled.

(a) t_4 not enabled implies that there is no s, r such that $\langle r\rangle$ is on U and $\langle s, r\rangle$ is on P. Hence because of equation (10), $U = P = \mathbf{0}$.

(b) t_3 not enabled implies that there is no s, r such that $\langle r\rangle$ is on I and $\langle s, r\rangle$ on R. Claim: $R = \mathbf{0}$. Assume that some $\langle s, r\rangle$ is on R. Then $\langle r\rangle$ is not on I, hence $\langle r\rangle$ on W (equ. (7) and $U = \mathbf{0}$), and $W = \langle r\rangle$ due to (9). So (14) and (8) give $R \le N_r = \sum_{t\neq r}\langle r, t\rangle$ which contradicts $\langle s, r\rangle$ on R. Hence $R = \mathbf{0}$.

(c) t_1 not enabled implies that there is no s such that $\langle s\rangle$ is on I, $\langle\rangle$ is on X, and for all $r \neq s$, $\langle s, r\rangle$ is on Q. *Claim:* $W \neq \mathbf{0}$. Assume that $W = \mathbf{0}$. Then $X = \langle\rangle$ (9), $R = P = A = \mathbf{0}$ (14), so $Q = N$ (8). So t_1 would be enabled. Hence $W \neq \mathbf{0}$, and in fact $W = \langle s\rangle$ for some s (9). Then $A = N[s]$ (14) because $R = P = \mathbf{0}$ and $A \le N$ (8).
Consequently, t_2 is enabled since $X = \mathbf{0}$ (9).

Conclusion

We have sketched the process of introducing a net theoretic system model called predicate/transition nets. Although the considerations leading to the model seem quite simple, there is a large amount of formalism needed to allow mathematical reasoning. Any improvement or simplification of the presentation would be most welcome.

Acknowledgement: Part of the material presented in this paper is based on course given by the author at the University of Nijmegen, The Netherlands. Many thanks to students and colleagues, in particular to Wil Dekkers for fruitful discussions and many improvements. Thanks also to the unknown referee of the preliminary version of this report. His criticisms and suggestions were of great help when preparing the final version.

References

[1] Darlington, J.L.: *A Net Based Theorem Prover for Program Verification and Synthesis.* Gesellschaft für Mathematik und Datenverarbeitung, GMD-IST Internal Report 3/79 (1979)

[2] Genrich, H.J.; Lautenbach, K.: *System Modelling with High-Level Petri Nets.* Theor. Comp. Science 13 (1981) 109-136

[3] Genrich, H.J.; Lautenbach, K.: *S-Invariance in Predicate/Transition Nets.* Informatik-Fachberichte 66: Application and Theory of Petri Nets. — Selected Papers from the Third European Workshop on Application and Theory of Petri Nets, Varenna, Italy, September 27–30, 1982 / Pagnoni, A.; Rozenberg, G. (eds.) — Springer-Verlag, pp. 98–111 (1983)

[4] Genrich, H.J.; Thieler-Mevissen, G.: *The Calculus of Facts.* Mathematical Foundations of Computer Science 1976 / Mazurkiewicz, A. (ed.) — Berlin, Heidelberg, New York: Springer-Verlag, pp. 588–595 (1976)

[5] Gerhards, B.: *S-Invarianten in Prädikat/Transitionsnetzen.* Diplomarbeit, Universität Bonn (1982) (in German)

[6] Halmos, P.R.: *Naive Set Theory.* Springer-Verlag (1974)

[7] Holt, A.W.; Commoner, F.; Even, S.; Pnueli, A.: *Marked Directed Graphs.* J. Comp. Sys. Sc. 5 (1971) 511-523

[8] Hughes, G.E.; Cresswell, M.J.: *An Introduction to Modal Logic.* Methuen (1982)

[9] Jensen, K.: *Coloured Petri Nets and the Invariant Method.* Theor. Comp. Science 14 (1981) 317-336

[10] Jensen, K.: *Coloured Petri Nets.* In this volume.

[11] Kujansuu, R.; Lindqvist, M.: *Efficient Algorithms for Computing S-invariants for Predicate/Transition Nets.* Proceedings of the 5th European Workshop on Applications and Theory of Petri Nets. — Aarhus University (1984) pp. 156–173

[12] Lang, S.: *Algebra.* Addison-Wesley Publ. Comp. (1965)

[13] Lautenbach, K.: *Linear Algebraic Techniques for Place/Transition Nets.* In this volume.

[14] Lautenbach, K.; Pagnoni, A.: *Invariance and Duality in Predicate/Transition Nets and in Coloured Nets.* Gesellschaft für Math. und Datenverarbeitung mbH Bonn, Arbeitspapiere der GMD Nr. 132 (Feb., 1985)

[15] Mevissen, H.: *Algebraische Bestimmung von S-Invarianten in Prädikat/Transitions-Netzen.* Gesellschaft für Math. und Datenverarbeitung mbH Bonn, ISF-Report 81.02 (März, 1985) (In German)

[16] Petri, C.A.: *Interpretations of Net Theory.* Gesellschaft für Math. und Datenverarbeitung mbH Bonn, Technical Report ISF 75-07, 2nd ed. (Dec., 1976)

[17] Reisig, W.: *Petri Nets.* Springer-Verlag (1985)

[18] Reisig, W.: *Petri Nets with Individual Tokens.* Theor. Comp. Science 41 (1985) 185-213

[19] Schoenfield, J.R.: *Mathematical Logic.* Addison-Wesley Publ. Comp. (1967)

[20] Schwarz, F.: *A* REDUCE *Package for Determining First Integrals of Autonomous Systems of Ordinary Differential Equations.* Computer Physics Communications 39,2 (1986) 285-296

[21] Thiagarajan, P.S.: *Elementary Net Systems.* In this volume.

[22] Thieler-Mevissen, G.: *The Petri Net Calculus of Predicate Logic.* Gesellschaft für Math. und Datenverarbeitung mbH Bonn, Technical Report ISF 76-09, 2nd ed. (May, 1977)

2.
Coloured Petri Nets: A High Level Language for System Design and Analysis

K. Jensen

G. Rozenberg (ed.): Advances in Petri nets 1990. Lecture Notes in Computer Science, vol. 483. Springer, Berlin Heidelberg New York 1990, pp. 342-416

Abstract

This paper describes how Coloured Petri Nets (CP-nets) have been developed – from being a promising theoretical model to being a full-fledged language for the design, specification, simulation, validation and implementation of large software systems (and other systems in which human beings and/or computers communicate by means of some more or less formal rules).

First CP-nets are introduced by means of a small example and a formal definition of their structure and behaviour is presented. Then we describe how to extend CP-nets by a set of hierarchy constructs (allowing a hierarchical CP-net to consist of many different subnets, which are related to each other in a formal way). Next we describe how to analyse CP-nets, how to support them by various computer tools, and we also describe some typical applications. Finally, a number of future extensions are discussed (of the net model and the supporting software).

The non-hierarchical CP-nets in the present paper are analogous to the CP-nets defined in [35] and the High-level Petri Nets defined in [33]. In all three papers CP-nets (and HL-nets) have two different representations: The *expression representation* uses arc expressions and guards, while the *function representation* uses linear functions between multi-sets. Moreover, there are formal translations between the two representations (in both directions). In [33] and [35] we used the expression representation to describe systems, while we used the function representation for all the different kinds of analysis. It has, however, turned out that it only is necessary to turn to functions when we deal with invariant analysis, and this means that we now use the expression representation for all purposes – except for the calculation of invariants. This change is important for the practical use of CP-nets – because it means that the function representation and the translations (which are a bit mathematically complex) no longer are parts of the basic definition of CP-nets. Instead they are parts of the invariant method (which anyway demands considerable mathematical skills).

The development of CP-nets has been supported by several grants from the Danish National Science Research Council.

Contents

To find a given page please add the page number in this table to the page number that precedes this article.

1. Informal Introduction to Non-Hierarchical CP-nets

High-level nets, such as Coloured Petri Nets (CP-nets) and Predicate/Transition Nets are now in widespread use for many different practical purposes.[1] The main reason for the large success of these kinds of net models is that they – *without loosing the possibility of formal analysis* – allow the modeller to make *much more succinct and manageable descriptions* than can be produced by means of low-level nets (such as Place/Transition Nets and Elementary Nets). In high-level nets the complexity of a model can be divided between the net structure, the net inscriptions and the declarations. This means that it is possible to handle the description of much larger and more complex systems. It also means that we can describe simple data manipulation (such as the addition of two integers) by means of arc expressions (such as x+y) – instead of having to describe this by a complex set of places, transitions and arcs. The step from low-level nets to high-level nets can be compared to the step from assembly languages to modern programming languages with an elaborated type concept: In low-level nets there is only one kind of token and this means that the state of a place is described by an integer (and in many cases even by a boolean). In high-level nets each token can carry a complex information (which e.g. may describe the entire state of a process or a data base).[2]

However, looking at the history of high-level programming languages, it is obvious that their success also to a very large degree depends upon issues that do not concern typing. In particular, the development of subroutines and modules has played a key role, because they have made it possible to divide a large description into smaller units which can be investigated more or less independently of each other. In fact, the absence of compositionality has been one of the main critiques raised against Petri net models. To meet this critique hierarchical CP-nets have been developed. In this net model it is possible to create a number of individual CP-nets, which then can be related to each other in a formal way – i.e. in a way which has a well-defined behaviour and thus allows formal analysis.

The remaining parts of this chapter contains an informal introduction to non-hierarchical CP-nets and their behaviour.

1.1 A simple example of a non-hierarchical CP-net

The non-hierarchical CP-net in Fig. 1 describes a system in which a number of processes compete for some shared resources. As in all other kinds of Petri nets there is a set of places (drawn as circles/ellipses) and a set of transitions (drawn as rectangles). The places and their tokens represent states, while the transitions represent state changes. However, each place may contain several tokens and each of these carries a data value – which may be of arbitrarily complex type (e.g. a record where the first

---

1 A selection of references can be found in section 6.5.

2 We shall in this paper not use any more space to compare high-level and low-level Petri nets. The reason is that we primarily are interested in the practical applications of Petri nets – and in this field the superiority of high-level nets is now generally accepted. A more detailed comparison of high-level and low-level nets can be found in [20] and [22].

field is a real, the second a text string, while the third is a list of integer pairs). The data value which is attached to a given token is referred to as the **token colour**.

In Fig. 1 there are two kinds of processes: three q-processes start in state A and cycle through five different states (A, B, C, D and E), while two p-processes start in state B and cycle through four different states (B, C, D and E). Each of these five processes is represented by a token – where the token colour is a pair such that the first element tells whether the token represents a p-process or a q-process while the second element is an integer telling how many full cycles that process has completed. In the initial marking, there are three (q,0)-tokens at place A and two (p,0)-tokens at place B.

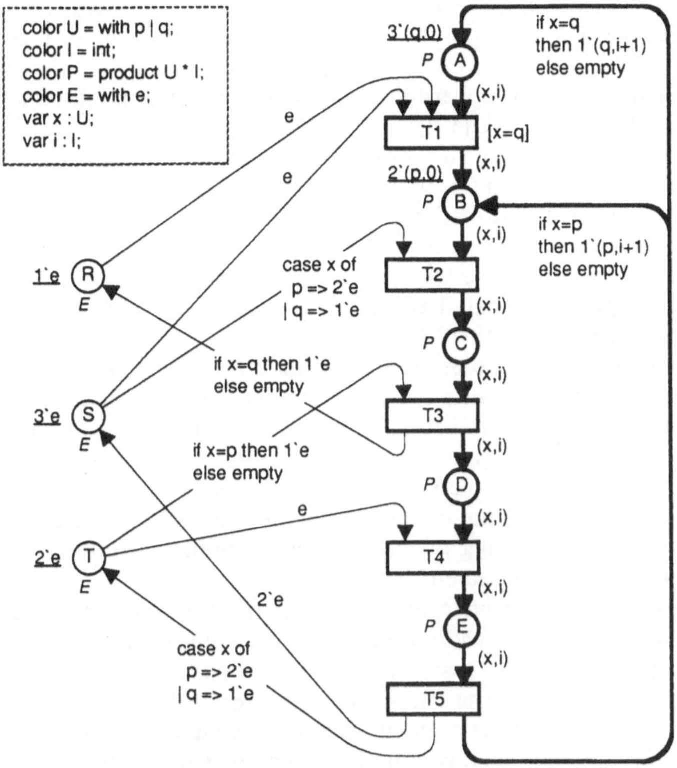

Figure 1. Non-hierarchical CP-net describing a simple resource allocation

There are three different kinds of resources: one r resource, three s resources and two t resources (each resource is represented by an e-token, on R, S or T). The arc expressions tell how many resources the different kinds of processes reserve/release. As an example, "case x of p=>2`e | q=>1`e" (at the arc from S to T2) tells that a p-process needs two s resources in order to go from state B to C, while a q-process only needs one.[3] Analogously, "if x=q then 1`e else empty" (at the arc from T3 to R) tells that each

3 The operator takes an integer n and a colour c and it returns the multi-set that contains n appearances of c (and nothing else).

q-process releases an r resource when it goes from state C to D, while a p-process releases none.[4] It should be noticed that the processes in this system neither consume nor create tokens (during a full cycle the number of releases matches the number of reservations). Now let us take a closer view of the CP-net in Fig. 1. It consists of three different parts: the **net structure**, the **declarations** and the **net inscriptions**.

The net structure is a directed graph with two kinds of nodes, **places** and **transitions**, interconnected by **arcs** – in such a way that each arc connects two different kinds of nodes (i.e. a place and a transition).[5] In Fig. 1 the right hand part of the net (describing how processes change between different states) is drawn with thick lines. This distinguishes it from the rest of the net (describing how resources are reserved and released). It should, however, be stressed that such graphical conventions have no formal meaning. The only purpose is to make the CP-net more readable for human beings.

The declarations in the upper left corner tell us that we in this simple example have four **colour sets**, (U, I, P, and E) and two **variables** (x and i). The use of colour sets in CP-nets is analogous to the use of types in programming languages: Each place has a colour set attached to it and this means that each token residing on that place must have a colour (i.e. attached information) which is a member of the colour set. Analogously to types, the colour sets not only define the actual colours (which are members of the colours sets), they also define the operations and functions which can be applied to the colours. In this paper we shall define the colour sets using a syntax that is similar to the way in which types are defined in most programming languages. It should be noticed that a colour set definition often implicitly introduces new operators and functions (as an example the declaration of a colour set of type integer introduces the ordinary addition, subtraction, and multiplication operators). In our present example, the colour set U contains two elements (p and q) while the colour set I contains all integers.[6] The colour set P is the set of all pairs, where the first component is of type U while the second is of type I. Finally, the colour set E only contains a single element – and this means that the corresponding tokens carry no information (often we think of them as being "ordinary" or "uncoloured" tokens).

Each net inscription is attached to a place, transition or arc. In Fig. 1 places have three different kinds of inscriptions: **names**, **colour sets** and **initialization expressions**, transitions have two kinds of inscriptions: **names** and **guards**, while arcs only have one kind of inscription: **arc expressions**. All net inscriptions are positioned next to the corresponding net element – and to make it easy to distinguish between them we write names in plain text, colour sets in italics, while initialization expressions are underlined and guards are contained in square brackets.

Names have no formal meaning. They only serve as a mean of identification that makes it possible for human beings and a computer systems to refer to the individual places and transitions. Names can be omitted and one can use the same name for several nodes (although this may create confusion). As explained above each place must have a

4 *empty* denotes the empty multi-set.

5 Such a graph is called a bipartite directed graph.

6 To be more precise, I only contains the integers in the interval MinInt..MaxInt – where MinInt and MaxInt are determined by the implementation of the Integer data type. In general, each colour set is demanded to be finite, although it (as I) may have very many elements.

colour set and this determines the kind of tokens which may reside on that place. The initialization expression of a place must evaluate to a multi-set over the corresponding colour set. Multi-sets are analogous to sets except that they may contain multiple appearances of the same element. In the case of CP-nets, this implies that two tokens on the same place may have identical colours. By convention we omit initialization expressions which evaluate to the empty multi-set.

The guard of a transition is a boolean expression which must be fulfilled before the transition can occur. By convention we omit guards which always evaluate to true. The arc expression of an arc is an expression, and it may (as the guard) contain variables, constants, functions and operations that are defined in the declarations (explicitly or implicitly). When the variables of an arc expression are bound (i.e. replaced by colours from the corresponding colour sets) the arc expression must evaluate to a colour (or a multi-set of colours) that belong to the colour set attached to the place of the arc. When the same variable appears more than once, in the guard/arc expressions of a *single* transition, all these appearances must be bound to the same colour. In contrast to this appearances, in the guard/arc expressions of *different* transitions, are totally independent, and this means that they may be bound to different colours. As explained in the sequel, a CP-net may have several other kinds of inscriptions (e.g. used to describe hierarchical relationships and time delays).

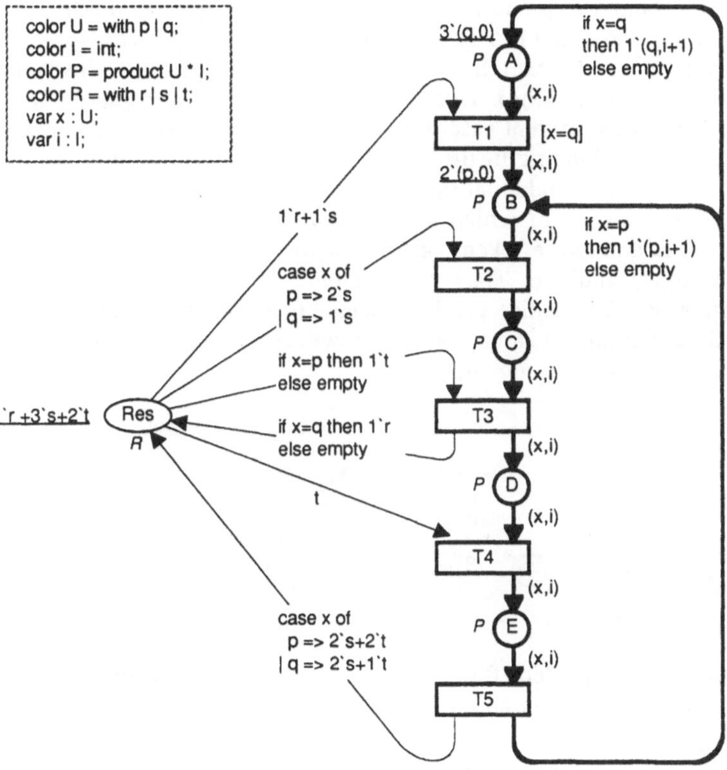

Figure 2. A slightly different CP-net describing the same resource allocation

As mentioned above a CP-net consists of three different parts: the net structure, the declarations and the net inscriptions. The complexity of a description is distributed among these three parts and this can be done in many different ways. As an example, each arc to or from a resource place could have had a very simple arc expression of the form f(x), where the function f was defined in the declaration part. As another, and perhaps more sensible example, we could have represented all resources by means of a single place RES, as shown in Fig. 2. The + operator in the arc expressions denotes addition of multi-sets. As an example, 2`s+1`t is the multi-set that contains two appearances of s and one appearance of t:

1.2 Dynamic behaviour of non-hierarchical CP-nets

One of the most important properties of CP-nets (and other kinds of Petri nets) is that they – in contrast to many other graphical description languages – have a well-defined semantics which in an unambiguous way defines the behaviour of the system. The ideas behind the semantics are very simple, as we shall demonstrate by means of Fig. 3 - which contains one of the transitions from Fig. 1.[7]

The transition has two variables (x and i) and before we can consider an occurrence of the transition these variables have to be bound to colours of the corresponding types (i.e. elements of the colour sets U and I). This can be done in many different ways: One possibility is to bind x to p and i to zero. Then we get: b_1 = <x=p,i=0>. Another possibility is to bind x to q and i to 37. Then we get: b_2 = <x=q,i=37>.

For each **binding** we can check whether the transition, with this binding, is **enabled** (in the current marking). This is done by evaluating the guard and all the input arc expressions: In the present case the guard is trivial (a missing guard always evaluates to true). For the binding b_1 the two arc expressions evaluate to (p,0) and 2`e, respectively. Thus we conclude that $b_1$ is enabled – because each of the input places contains at least the tokens to which the corresponding arc expression evaluates (one (p,0)-token on B and two e-tokens on S). For the binding $b_2$ the two arc expressions evaluate to (q,37) and 1`e. Thus we conclude that b_2 is *not* enabled (there is no (q,37)-token on B). A transition can be executed in as many ways as the variables (in its arc expressions and guard) can be bound. However, for a given state, it is usually only a few of these bindings that are enabled.

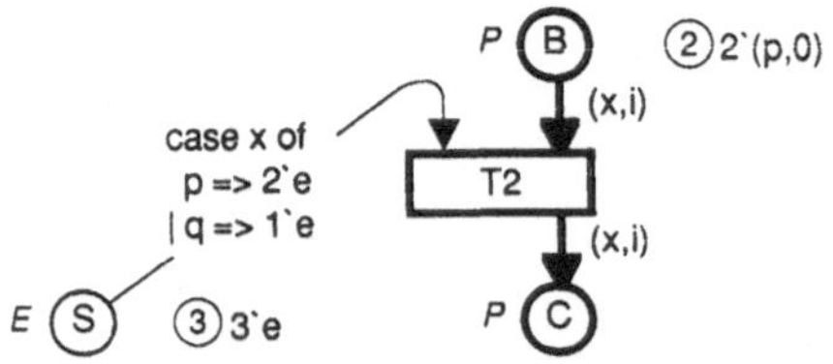

Figure 3. A transition from the resource allocation system

When a transition is enabled (for a certain binding) it may **occur** and it then removes tokens from its input places and adds tokens to its output places. The number of re-

7 The inscriptions at the right hand side of the places indicate the current marking. The number of tokens are indicated in the small circle while the colours are described by the multi-set next to the circle.

moved/added tokens and the colours of these tokens are determined by the value of the corresponding arc expressions (evaluated with respect to the binding in question). An occurrence of the binding b_1 removes a (p,0)-token from B, removes two e-tokens from S and adds a (p,0)-token to C.[8] The binding b_2 is not enabled and thus it cannot occur.

A distribution of tokens (on the places) is called a **marking**. The **initial marking** is the marking determined by evaluating the initialization expressions. A pair, where the first element is a transition and the second element a binding of that transition, is called an **occurrence element**. Now we can ask whether an occurrence element O is enabled in a given marking M_1 – and when this is the case we can speak about the marking M_2 which is **reached** by the occurrence of O in M_1. It should be noticed that several occurrence elements may be enabled in the same marking. In that case there are two different possibilities: Either there are enough tokens (so that each occurrence element can get its own share) or there too few tokens (so that several occurrence elements have to compete for the same input tokens). In the first case the occurrence elements are said to be **concurrently enabled**. They can occur in the same **step** and they each remove their own input tokens and produce their own output tokens. In the second case the occurrence elements are said to be in **conflict** with each other and they cannot occur in the same step.

In the initial marking of Fig. 1 we observe that the occurrence element O_1 = (T1,<x=q,i=0>) is concurrently enabled with O_2 = (T2,<x=p,i=0>). This means that we can have a step where both O_1 and O_2 occur. Such a step is denoted by the multi-set 1\`$O_1$ + 1\`O_2 and when it occurs a (q,0)-token is moved from A to B and a (p,0)-token from B to C. Moreover, an e-token is removed from R and three e-tokens from S. It should be noticed that the effect of the step 1\`$O_1$ + 1\`O_2 is the same as when the two occurrence elements occur after each other in arbitrary order. This is an example of a general property: Whenever an enabled step contains more than one occurrence element, it can (in any thinkable way) be divided into two or more steps, which then are known to be able to occur after each other (in any thinkable order) and together have the same total effect as the original step.[9]

The above informal explanation of the occurrence rule, tells us how to understand the behaviour of a CP-net – and it explains the intuition on which CP-nets build. It is, however, very difficult (probably impossible) to make an informal explanation which is complete and unambiguous, and thus it is extremely important that the intuition is complemented by a more formal definition (which we shall present in chapter 2). It is the formal definition that has formed the basis for the implementation of a CP-net simulator, and it is also the formal definition that has made it possible to develop the analysis methods by which it can be *proved* whether a given CP-net has certain properties (e.g. absence of dead-locks).

---

8 We often think of the (p,0)-token as being moved from B to C. However, in the formal definition of CP-nets, the (p,0)-token added to C has no closer relationship to the (p,0)-token removed from B than it has to the two e-tokens removed from S.

9 Without this property it is very difficult to construct occurrence graphs because it no longer is sufficient to consider steps that correspond to single occurrence elements. It is, however, easy to violate this property and this is in fact done by many of the ad hoc extensions which are presented in the Petri net literature (e.g. the use of inhibitor arcs and some definitions of capacity).

Consider again the resource allocation system. It can easily be proved that this system has no dead-lock.[10] Now let us, in the initial marking, add an extra s resource (i.e. an extra e-token on S). Obviously, this cannot lead to a dead-lock (dead-locks appear when we have too few resource tokens and thus an extra resource token cannot cause a dead-lock). Is this argumentation convincing? At a first glance: yes! However, the argument is *wrong*, and the extra s resource actually means that we can reach a dead-lock (by letting the two p-processes advance to state D while the q-processes remain in state A). Hopefully, this small exercise demonstrates that informal arguments about behavioural properties are dangerous – and this is our motivation for the development of more formal analysis methods. We shall return to such methods in chapter 4.

2. Formal Definition of Non-Hierarchical CP-nets

The non-hierarchical CP-nets in the present paper are analogous to the CP-nets defined in [35] – but not identical to them (for more details see the abstract).

2.1 Multi-sets and expressions

In this section we define multi-sets and introduce the notation which we use to talk about expressions:

Definition 2.1: A **multi-set** m, over a non-empty and *finite* set S, is a function $m \in [S \rightarrow \mathbb{N}]$.[11] The non-negative integer $m(s) \in \mathbb{N}$ is the **number of appearances** of the element s in the multi-set m.

We usually represent the multi-set m by a formal sum:

$$\sum_{s \in S} m(s)\, s.$$

By S_{MS} we denote the **set of all multi-sets** over S. The non-negative integers $\{m(s) \mid s \in S\}$ are called the **coefficients** of the multi-set m, and m(s) is called the **coefficient** of s. An element $s \in S$ is said to **belong** to the multi-set m iff $m(s) \neq 0$ and we then write $s \in m$. The **empty multi-set** is the multi-set in which all coefficients are zero, and it is denoted by Ø (or by *empty*).[12]

As an example, consider the set S = {a,b,c,d,e} and the two multi-sets m_1 = a+2c+e, and m_2 = a+2b+3c+e which both are members of S_{MS}. As it can be seen, we usually omit S-values which have a zero coefficient and we also omit coefficients which are equal to one.[13] For multi-sets we define the following operations:

---

10 This can e.g. be done by means of occurrence graphs or by means of place invariants.

11 $\mathbb{N}$ denotes the set of all non-negative integers while $[A \rightarrow B]$ denotes the set of all functions from A to B.

12 To be precise there is an empty multi-set for each element set S. We shall, however, ignore this and allow ourselves to speak about *the* empty multi-set – in a similar way as we speak about *the* empty set and *the* empty list.

13 When the CPN tools described in chapter 5 was designed, it turned out to be convenient to insert an explicit operator between the coefficients and the S-values and include coefficients which are equal to

Definition 2.2: Summation, scalar-multiplication, comparison, and multiplicity of multi-sets are defined in the following way, for all m, m_1, $m_2 \in S_{MS}$ and $n \in \mathbb{N}$:

(i) $m_1 + m_2 = \sum_{s \in S} (m_1(s) + m_2(s))\ s$ (summation).

(ii) $n * m = \sum_{s \in S} (n \cdot m(s))\ s$ (scalar-multiplication).

(iii) $m_1 \neq m_2 = \exists s \in S: [m_1(s) \neq m_2(s)]$ (comparison).
$m_1 \leq m_2 = \forall s \in S: [m_1(s) \leq m_2(s)]$ (the relations $<$, $\geq$, $>$, and $=$ are defined analogously to $\leq$).

(iv) $|m| = \sum_{s \in S} m(s)$ (multiplicity).

When $m_1 \leq m_2$ we also define **subtraction**:

(v) $m_2 - m_1 = \sum_{s \in S} (m_2(s) - m_1(s))\ s$ (subtraction).

It can be shown that the multi-set operations above have a number of nice properties. As an example $(S_{MS}, +)$ is a commutative monoid.

For CP-nets, we use the terms *variables* and *expressions* in the same way as in typed lambda-calculus and functional programming languages. This means that expressions do *not* have side-effects and variables are *bound* to values (instead of being assigned to). It also means that complex expression are built, from variables and simpler subexpressions, by means of functions and operations. To give the abstract definition of CP-nets it is not necessary to fix the concrete syntax in which the modeller writes expressions, and thus we shall only assume that such a syntax exists (together with a well-defined semantics) – making it possible in an unambiguous way to talk about:

- The *type of a variable* v – denoted by Type(v).
- The *type of an expression* expr – denoted by Type(expr).
- The *set of variables in an expression* expr – denoted by Var(expr). The set of variables only includes the free variables – i.e. those which are *not* bound internally in the expression (e.g. by a local definition).
- A *binding of a set of variables* $V = \{v_1, v_2, \ldots, v_n\}$ – denoted by $<v_1=c_1, v_2=c_2, \ldots, v_n=c_n>$. It is demanded that $c_i \in \text{Type}(v_i)$ for each variable $v_i \in V$.[14]
- The *value obtained by evaluating an expression* expr *in a binding* b – denoted by expr<b>. It is demanded that Var(expr) is a subset of the variables of b, and the evaluation is performed by substituting each variable $v_i \in \text{Var(expr)}$ with the value $c_i \in \text{Type}(v_i)$ determined by the binding.

---

one. In this case we write m_1=1\`a+2\`c+1\`e and $m_2$=1\`a+2\`b+3\`c+1\`e. This makes it easier to perform type checking (and it makes it easier to deal with multi-sets over integers, e.g. 3\`1+2\`35+1\`59).

14 For a type A we also use A to denote the *set of elements* in A, and we use $c \in A$ to denote that the value c is an element of A.

As an example, illustrating this notation, we may have:

Type(x) = Type(y) = S.
Var(2 * (x + 3y)) = {x,y}.
Type(2 * (x + 3y)) = S_{MS}.
(2 * (x + 3y))<x=b,y=d> = 2`b+6`d.

2.2 Definition of non-hierarchical CP-nets

In this section we define non-hierarchical CP-nets as a many-tuple. It should, however be understood, that the only purpose of this is to give a mathematically sound and unambiguous definition of CP-nets and their semantics. Any concrete net – created by a modeller - will always be specified in terms of a CP-graph (i.e. a diagram similar to Fig. 1). In the following Bool is the boolean type (containing the elements Bool = {False,True} and having the standard logic operations). Some motivation and explanation of the individual parts of the definition is given immediately below the definition:

Definition 2.3: A **non-hierarchical** CP-net is a tuple CPN = (Σ, P, T, A, N, C, G, E, IN) satisfying the requirements below:

(i) Σ is a finite set of types, called **colour sets**. Each colour set must be finite and non-empty.
(ii) P is a finite set of **places**.
(iii) T is a finite set of **transitions**.
(iv) A is a finite set of **arcs** such that:
- $P \cap T = P \cap A = T \cap A = \emptyset$.

(v) N is a **node** function. It is defined from A into $P \times T \cup T \times P$.
(vi) C is a **colour** function. It is defined from P into Σ.
(vii) G is a **guard** function. It is defined from T into expressions such that:
- $\forall t \in T$: [Type(G(t)) = Bool $\wedge$ Type(Var(G(t))) $\subseteq \Sigma$].

(viii) E is an **arc expression** function. It is defined from A into expressions such that:
- $\forall a \in A$: [Type(E(a)) = $C(p(a))_{MS}$ $\wedge$ Type(Var(E(a))) $\subseteq \Sigma$]

where p(a) is the place of N(a).
(ix) IN is an **initialization** function. It is defined from P into expressions such that:
- $\forall p \in P$: [Type(IN(p)) = $C(p)_{MS}$ $\wedge$ Var(IN(p)) = $\emptyset$].

(i) The set of **colour sets** determines the types, operations and functions that can be used in the net inscriptions (i.e. arc expressions, guards, initialization expressions, colour sets, etc.). If desired, the colour sets (and the corresponding operations and functions) can be defined by means of a many-sorted sigma algebra (in the same way as known from the theory of abstract data types). We demand all colour sets to be finite - although they may have a very large cardinality (e.g. be equivalent to all the real numbers which can be represented on a given computer). This restriction means that the linear extension of a function $F \in [A \rightarrow B_{MS}]$ to a function $\hat{F} \in [A_{MS} \rightarrow B_{MS}]$ always is known to be convergent. Such functions are used in the theory of place invariants and transition invariants.

(ii) + (iii) + (iv) The **places, transitions** and **arcs** are described by three sets P, T and A which are demanded to be finite and pairwise disjoint. In contrast to classical Petri nets, we allow the net structure to be empty (i.e. $P \cup T = \emptyset$). The reason is pragmatic: It allows the user to define and syntax check a set of colour sets without having to invent a dummy net structure.

(v) The **node** function maps each arc into a pair where the first element is the source node and the second the destination node. The two nodes have to be of different kind (i.e. one must be a place while the other is a transition). In contrast to classical Petri nets, we allow a CP-net to have several arcs between the same ordered pair of nodes (and thus we define A as a separate set and not as a subset of $P \times T \cup T \times P$). The reason is pragmatic: We often have nets where each occurrence element moves exactly one token along each of the surrounding arcs, and it is then awkward to be forced to violate this convention in the cases where an occurrence element removes/adds two or more tokens to/from the same place. It is of course easy to combine such multiple arcs to a single arc by adding the corresponding arc expressions (which must be of the same multi-set type). We also allow nodes to be isolated. Again the reason is pragmatic: When we build computer tools for CP-nets we want to be able to check whether a diagram is a CP-net (i.e. fulfils the definition above). There is, however, no conceptual difference between an isolated node and a node where all the arc expressions of the surrounding arcs always evaluate to the empty multi-set (and the latter is difficult to detect in general, since arc expressions may be arbitrarily complex). It is of course easy to exclude such degenerate nets when this is convenient for theoretical purposes.

(vi) The **colour** function C maps each place p into a set of possible **token colours** C(p). Each token on p must have a colour that belongs to the type C(p).

(vii) The **guard** function G maps each transition t into an expression of type boolean, i.e. a predicate. Moreover, all variables in G(t) must have types that belong to Σ.[15]

(viii) The **arc expression** function E maps each arc *a* into an expression which must be of type $C(p(a))_{MS}$. This means that each evaluation of the arc expression must yield a multi-set over the colour set that is attached to the corresponding place. We shall, as a shorthand, also allow an arc expression to be of type C(p(a)). In this case the arc expression evaluates to a colour in C(p(a)) which we then consider to be a multi-set with only one element.

(ix) The **initialization** function IN maps each place p into an expression which must be of type $C(p)_{MS}$ – i.e. a multi-set over C(p). The expression is not allowed to contain any variables. Analogously to (viii), we shall, as a shorthand, also allow an initial expression to be of type C(p).

As mentioned in the abstract, the "modern version" of CP-nets (presented in this paper) uses the expression representation (defined above) – not only when a system is being described, but also when it is being analysed. It is only during invariant analysis that it may be adequate/necessary to translate the expression representation into a function representation.

---

15 For a set of variables Vars we use Type(Vars) to denote the set $\{Type(v) \mid v \in Vars\}$.

In addition to the concepts introduced in Def. 2.3, we use $X = P \cup T$ to denote the set of all **nodes**, and we define the following functions:[16]

- $s \in [A \to X]$ maps each arc *a* into the **source** of *a*, i.e. the first component of N(a).
- $d \in [A \to X]$ maps each arc *a* into the **destination** of *a*, i.e. the second component of N(a).
- $p \in [A \to P]$ maps each arc *a* into the **place** of N(a), i.e. that component of N(a) which is a place.
- $t \in [A \to T]$ maps each arc *a* into the **transition** of N(a), i.e. that component of N(a) which is a transition.
- $A \in [(P \times T \cup T \times P) \to A_S]$[17] maps each ordered pair of nodes (x_1,x_2) into the set of **connecting arcs**, i.e. the arcs that have the first node as source and the second as destination:
  $A(x_1,x_2) = \{a \in A \mid N(a) = (x_1,x_2)\}$.
- $A \in [X \to A_S]$[18] maps each node x into the set of **surrounding arcs**, i.e. the arcs that have x as source or destination:
  $A(x) = \{a \in A \mid \exists x' \in X: [N(a) = (x,x') \vee N(a) = (x',x)]\}$.
- $X \in [X \to X_S]$ maps each node x into the set of **surrounding nodes**, i.e. the nodes that are connected with x by an arc:
  $X(x) = \{x' \in X \mid \exists a \in A: [N(a) = (x,x') \vee N(a) = (x',x)]\}$.

All the functions above can, in the usual way, be extended to take sets as input (then they all return sets and thus all the function names are written with a capital letter).

2.3 Dynamic behaviour of non-hierarchical CP-nets

Having defined the static structure of CP-nets we are now ready to consider their behaviour – but first we introduce the following notation where Var(t) is called the set of **variables** of t while $E(x_1,x_2)$ is called the **expression** of (x_1,x_2):

- $\forall t \in T: [Var(t) = \{v \mid v \in Var(G(t)) \vee \exists a \in A(t): v \in Var(E(a))\}]$.
- $\forall (x_1,x_2) \in (P \times T \cup T \times P): [E(x_1,x_2) = \sum_{a \in A(x_1,x_2)} E(a)]$.[19]

Next we define what we mean by a binding. Intuitively, a binding, of a transition t, is a substitution that replaces each variable of t with a colour. It is demanded that each colour is of the correct type and that the guard evaluates to true:

---

[16] Each function name indicates the range of the function – as an example p maps into places, while A maps into *sets* of arcs.

[17] A_S denotes the set of all subsets of A.

[18] From the argument(s) it will always be clear whether we deal with the function $A \in [X \to A_S]$, the function $A \in [(P \times T \cup T \times P) \to A_S]$ or the set A.

[19] The summation indicates addition of expressions (and it is well-defined because all the participating expressions have a common multi-set type). From the arguments(s) it will always be clear whether we deal with the function $E \in [A \to Exp]$ or the function $E \in [(P \times T \cup T \times P) \to Exp]$.

Definition 2.4: For a transition $t \in T$ with variables $Var(t) = \{v_1, v_2, \ldots, v_n\}$ we define the **binding type** BT(t) as follows:

$$BT(t) = Type(v_1) \times Type(v_2) \times \ldots \times Type(v_n).$$[20]

Moreover, we define the set of all **bindings** B(t) as follows:

$$B(t) = \{(c_1, c_2, \ldots, c_n) \in BT(t) \mid G(t)\langle v_1{=}c_1, v_2{=}c_2, \ldots, v_n{=}c_n\rangle\}.$$[21]

For convenience we denote bindings in two different ways: Either in the form $\langle v_1{=}c_1, v_2{=}c_2, \ldots, v_n{=}c_n\rangle$ or in the form $(c_1, c_2, \ldots, c_n)$. In both cases this denotes an element of BT(t). Next we define token distributions, binding distributions, markings and steps:[22]

Definition 2.5: A **token distribution** is a function M, defined on P such that $M(p) \in C(p)_{MS}$ for all $p \in P$. The set of all token distributions (for a given CP-net CPN) is denoted by TD_{CPN}, and for all $M_1, M_2 \in TD_{CPN}$ we define the relations $\neq$ and $\leq$ in the following way:

(i) $M_1 \neq M_2 \;\Leftrightarrow\; \exists p \in P: [M_1(p) \neq M_2(p)]$.

(ii) $M_1 \leq M_2 \;\Leftrightarrow\; \forall p \in P: [M_1(p) \leq M_2(p)]$.

The relations $<$, $\geq$, $>$, and $=$ are defined analogously to $\leq$. When $c \in M(p)$ for some $c \in C(p)$, we say that the pair (p,c) is an **element** of M, and we write $(p,c) \in M$. Moreover, we say that M is **non-empty** iff it has at least one element.

A **binding distribution** is a function Y, defined on T such that $Y(t) \in B(t)_{MS}$ for all $t \in T$.[23] The set of all binding distributions (for a given CP-net CPN) is denoted by BD_{CPN}, and the relations $\neq$, $\leq$, $<$, $\geq$, $>$, and $=$ are defined analogously to the way they were defined for token distributions. When $b \in Y(t)$ for some $b \in B(t)$, we say that the pair (t,b) is an **element** of Y, and we write $(t,b) \in Y$. Moreover, we say that Y is **non-empty** iff it has at least one element.

A **marking** of a CP-net is a token distribution and a **step** is a *non-empty* binding distribution. The set of all markings (for a given CP-net CPN) is denoted by M_{CPN}, and the set of all steps is denoted by Y_{CPN}. The **initial marking** M_0 is the marking which is obtained by evaluating the initialization expressions, i.e. the marking where $M_0(p) = IN(p)\langle\rangle$ for all $p \in P$.[24]

---

20 We assume that the set of variables Var(t) is ordered – in some arbitrary way.

21 As defined in section 2.1, $G(t)\langle v_1{=}c_1, v_2{=}c_2, \ldots, v_n{=}c_n\rangle$ denotes the evaluation of the guard expression G(t) in the binding $\langle v_1{=}c_1, v_2{=}c_2, \ldots, v_n{=}c_n\rangle$.

22 There is no difference between the set of token distributions and the set of markings, and there is very little difference between the set of binding distributions and the set of steps. In this paper we only use token/binding distributions to define markings/steps and thus it may seem unnecessary to introduce all four sets. Token/binding distributions are however, general concepts, which are useful in a number of other contexts (in which it would be misleading to talk about markings/steps).

23 It should be noticed that all bindings of a binding distribution, according to Definition 2.4, automatically satisfy the corresponding guard.

24 $IN(p)\langle\rangle$ denotes the evaluation of IN in the empty binding $\langle\rangle$ (which is used because IN(p) has an empty set of variables).

Definition 2.6: A step Y is **enabled** in a marking M iff the following property is satisfied:

$$\forall p \in P: \left[\sum_{(t,b)\in Y} E(p,t)\langle b\rangle \le M(p) \right].$$

Let Y be an enabled step, with respect to a given marking M. When $(t,b)\in Y$, we say that t is **enabled** in M for the **binding** b. We also say that the pair (t,b) is enabled in M, or simply that t is enabled in M. When two different transitions $t_1, t_2 \in T$ satisfy $Y(t_1) \neq \emptyset \neq Y(t_2)$, we say that t_1 and t_2 are **concurrently enabled.** When a transition $t\in T$ satisfies $|Y(t)| \geq 2$, we say that t is **concurrently enabled with itself** and when it for a binding $b\in B(t)$ satisfy $Y(t) \geq 2`b$, we say that (t,b) is **concurrently enabled with itself.**

When a step is enabled it may occur and this means that tokens are removed from the input places and added to the output places of the occurring transitions. The number and colours of the tokens are determined by the arc expressions, evaluated for the occurring bindings:

Definition 2.7: When a step Y is enabled in a marking M_1 it may **occur**, changing the marking M_1 to another marking M_2, defined by:

$$\forall p \in P: \left[M_2(p) = \left(M_1(p) - \sum_{(t,b)\in Y} E(p,t)\langle b\rangle\right) + \sum_{(t,b)\in Y} E(t,p)\langle b\rangle\right].$$

The first sum is called the **removed** tokens while the second is called the **added** tokens. Moreover we say that M_2 is **directly reachable** from M_1 by the occurrence of the step Y, which we also denote:

$$M_1\,[Y\rangle\,M_2.$$

Definition 2.8: A **finite occurrence sequence** is a sequence of markings and steps:

$$M_1\,[Y_1\rangle\,M_2\,[Y_2\rangle\,M_3 \ldots\ldots M_n\,[Y_n\rangle\,M_{n+1}$$

such that $n\in\mathbb{N}$, and $M_i\,[Y_i\rangle\,M_{i+1}$ for all $i\in 1..n$.[25] The marking M_1 is called the **start marking** of the occurrence sequence, while the marking M_{n+1} is called the **end marking.** The non-negative integer n is called the **number of steps** in the occurrence sequence, or the **length** of it.

Analogously, an **infinite occurrence sequence** is a sequence of markings and steps:

$$M_1\,[Y_1\rangle\,M_2\,[Y_2\rangle\,M_3 \ldots\ldots$$

such that $M_i\,[Y_i\rangle\,M_{i+1}$ for all $i\in\mathbb{N}_+$.[26] The marking M_1 is called the **start marking** of the occurrence sequence, which is said to have **infinite length.**

[25] By 1..n we denote the set of all integers i that satisfy $1 \leq i \leq n$.

[26] $\mathbb{N}_+$ denotes the set of all positive integers.

The start marking of an occurrence sequence will often, but not always, be identical to the initial marking of the CP-net. We allow the user to omit some parts of an occurrence sequence and e.g. write:

$$M_1 [Y_1 Y_2 \ldots Y_n \rangle M_{n+1}.$$

Definition 2.9: A marking M" is **reachable from** a marking M' iff there exists a finite occurrence sequence having M' as start marking and M" as end marking – i.e. iff there, for some $n \in \mathbb{N}$, exists a finite sequence of steps such that:

$$M' [Y_1 Y_2 \ldots Y_n \rangle M''.$$

We then also say that M" is reachable from M' in **n steps**. The *set* of markings which are reachable from M' is denoted by $[M'\rangle$. As a shorthand, we say that a marking is **reachable** iff it is reachable from the initial marking M_0 – i.e. contained in $[M_0\rangle$.

It should be obvious that behavioural properties, such as dead-lock, liveness, home markings, boundedness and fairness, can be defined for CP-nets in a similar way as for Place/Transition Nets (PT-nets). It is well-known that each CP-net has an equivalent PT-net, and each behavioural property is defined in such a way that a given CP-net has the property iff the equivalent PT-net has. The definitions of the behavioural properties are outside the scope of this paper.

2.4 Some historical remarks about the development of CP-nets

The foundation of Petri nets was presented by Carl Adam Petri in his doctoral-thesis [48]. The first nets were called Condition/Event Nets (CE-nets). This net model allows each place to contain at most one token – because the place is considered to represent a boolean condition, which can be either true or false. In the following years a large number of persons contributed to the development of new net models, basic concepts, and analysis methods. One of the most notable results was the development of Place/Transition Nets (PT-nets). This net model allows a place to contain several tokens. The first coherent presentation of the theory and application of Petri nets was given in the course material developed for the First Advanced Course on Petri Nets [5] and later this was supplemented by the course material for the Second Advanced Course on Petri Nets [6] and [7].

For theoretical considerations CE-nets turned out to be more tractable than PT-nets and much of the theoretical work concerning the definition of basic concepts and analysis methods has been performed on CE-nets. Later, a new net model called Elementary Nets (EN-nets) has been proposed in [51] and [57]. The basic ideas of this net model are very close to CE-nets – but EN-nets avoid some of the technical problems which have turned out to be present in the original definition of CE-nets.

For practical applications, PT-nets were used. However, it often turned out that this net model was too low-level to cope with the real-world applications in a manageable way, and different researchers started to develop their own extensions of PT-nets - adding concepts such as: priority between transitions, time delays, global variables to be tested and updated by transitions, zero testing of places, etc. In this way a large number of different net models were defined. However, most of these net models were designed with a single – and often very narrow – application area in mind. This created a serious problem: Although some of the net models could be used to give adequate descriptions of certain systems, most of the net models possessed nearly no analytic power. The main reason for this was the large variety of different net models. It often turned out to be a difficult task to translate an analysis method developed for one net model to another – and in this way the efforts to develop suitable analysis methods were widely scattered.

The breakthrough with respect to this problem came when Predicate/Transition Nets (PrT-nets) were presented in [20]. PrT-nets were the first kind of high-level nets which was constructed without any particular application area in mind. PrT-nets form a nice generalization of PT-nets and CE-nets (exploiting the same kind of reasoning that leads from propositional logic to predicate logic). PrT-nets can be related to PT-nets and CE-nets in a formal way – and this makes it possible to generalize most of the basic concepts and analysis methods that have been developed for these net models – so that they also become applicable to PrT-nets. Later, an improved definition of PrT-nets has been presented in [22]. This definition draws heavily on sigma algebras (as known from the theory of abstract data types).

However, it soon turned out that PrT-nets present some technical problems when the analysis methods of place invariants and transition invariants are generalized. It is possible to calculate invariants for PrT-nets, but the interpretation of the invariants is difficult and must be done with great care to avoid erroneous results. The problem arises because of the variables which appear in the arc expressions of PrT-nets. These variables also appear in the invariants, and to interpret the invariants it is necessary to bind the variables, via a complex set of substitution rules. To overcome this problem the first version of Coloured Petri Nets (CP^{81}-nets) was defined in [32]. The main ideas of this net model are directly inspired by PrT-nets, but the relation between an occurrence element and the token colours involved in the occurrence is now defined by functions and not by expressions as in PrT-nets. This removes the variables, and invariants can now be interpreted without problems.

However, it often turns out that the functions attached to arcs in CP^{81}-nets are more difficult to read and understand than the expressions attached to arcs in PrT-nets. Moreover, as indicated above, there is a strong relation between PrT-nets and CP^{81}-nets and from the very beginning it was clear that most descriptions in one of the net models could be informally translated to the other net model and vice versa. This lead to the idea of an improved net model – combining the qualities of PrT-nets and CP^{81}-nets. This net model was defined in [33] where it was called High-level Petri Nets (HL-nets). Unfortunately, this name has given rise to a lot of confusion since the term "high-level nets" at that time started to become used as a generic name for PrT-nets, CP^{81}-nets, HL-nets, and several other kinds of net models. To avoid this confusion it was necessary to rename HL-nets to Coloured Petri Nets (CP^{87}-nets). CP^{87}-nets have two different representations (and formal translations between them). The expression representation is nearly identical to PrT-nets (as presented in [20]), while the function representation is nearly identical to CP^{81}-nets. The first coherent presentation of CP^{87}-nets and their analysis methods was given in [35].

Today most of the practical applications of Petri nets (reported in the literature) use either PrT-nets or CP-nets – although several other kinds of high-level nets have been proposed. There is very little difference between PrT-nets and CP-nets (and many modellers do not make a clear distinction between the two kinds of net models). The main differences between the two net models are today hidden inside the methods to calculate and interpret place and transition invariants (and this is of course not surprising when you think about the original motivation behind the development of CP^{81}-nets). Instead of viewing PrT-nets and CP-nets as two different modelling languages it is, in our opinion, much more adequate to view them as two slightly different dialects of the same language.

3. Hierarchical CP-nets

Hierarchical CP-nets were first presented in [31] and it should be understood that this (as far as we know) was the very first successful attempt to create a set of hierarchy concepts for a class of high-level Petri nets. This means that the proposed concepts are likely to undergo many improvements and refinements (in the same way as the first very simple concept of subroutines has undergone dramatical changes to become the procedure concept of modern programming languages). In other words: We do not claim that our current proposal will be the "final solution". However, we do think that it constitutes a good starting point for further research and practical experiences in this area. In chapter 6 we describe a number of industrial applications of hierarchical CP-nets and more information about some of these can be found in [49], [54] and [55].

In [31] individual CP-nets, called pages, are related in five different ways, known as the five **hierarchy constructs**: substitution of transitions, substitution of places, invocation of transitions, fusion of places and fusion of transitions. In the present paper we shall, however, only deal with the first and fourth of these hierarchy constructs.[27] For an explanation of the other three hierarchy constructs the reader is referred to [31].

The intention has been to make a set of hierarchy constructs, which is general enough to be used with many different development methods and with many different analysis techniques. This means that we present the hierarchy constructs *without* prescribing specific methods for their use. Such methods have to be developed and written down – but this can only be done as we get more experiences with the practical use of the hierarchy constructs. Eventually the new development methods and analysis techniques will influence the definition of the hierarchy constructs – in the same way as modern programming languages have been influenced by the progress in the areas of programming methodology and verification techniques.[28]

3.1 Substitution of transitions

The intuitive idea behind substitution transitions is to allow the user to replace a transition (and its surrounding arcs) by a more complex CP-net – which usually gives a more precise and detailed description of the activity represented by the substituted transition.

---

27 These are the two hierarchy constructs that are supported by the current version of the CPN tools described in chapter 5 – and they are the easiest to define, understand and use.

28 During the design of the hierarchy constructs we have, of course, been influenced by the constructs and methods used with other graphical description languages and with modern programming languages.

The idea is analogous to the hierarchy constructs found in many graphical description languages (e.g. IDEF/SADT diagrams [43] and Yourdon diagrams [64]) – and it is also, in some respects, analogous to the module concepts found in many modern programming languages: At one level we want to give a simple description of the activity (without having to consider internal details about how it is carried out). At another level we want to specify the more detailed behaviour. Moreover, we want to be able to integrate the detailed specification with the more crude description and this integration must be done in such a way that it becomes meaningful to speak about the behaviour of the *combined* system.

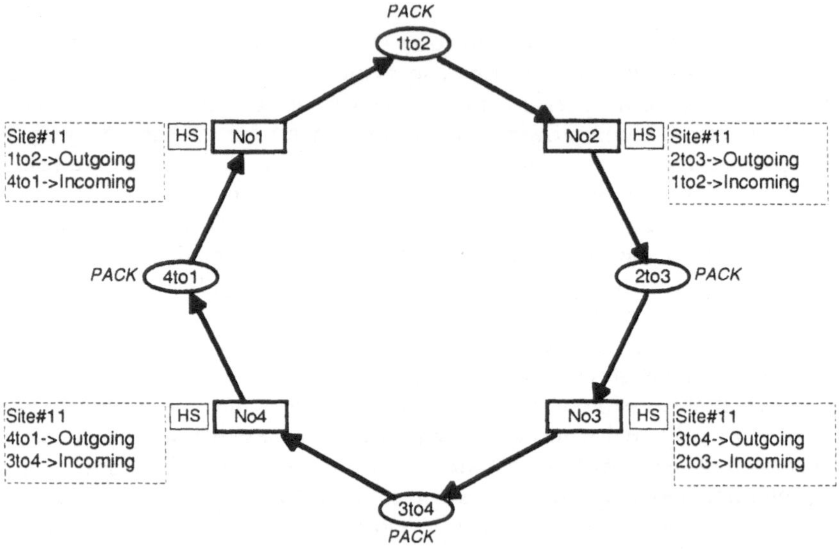

Figure 4. NetWork#10 describes a ring network with four different sites

As mentioned above, we want to relate individual CP-nets to nodes, which are members of other CP-nets, and this means that our description will contain a *set* of (non-hierarchical) CP-nets – which we shall call **pages**. Now let us consider a small example.[29] Imagine that we have a ring network with four different sites. This can be described by the page NetWork#10 in Fig. 4.[30] The four sites are represented by the four substitution transitions – NO1, NO2, NO3 and NO4 – each of which has an HS-tag adjacent (HS ≈ Hierarchy + Substitution). The inscription next to the HS-tag is called a **hierarchy inscription** and it defines the details of the actual substitution. We shall return to the hierarchy inscriptions in a moment, but let us first consider Site#11 in Fig. 5. This page describes an individual site.

[29] The purpose of the example is to explain the semantics of substitution transitions. The described ring network is far too simple to be realistic.

[30] To be able to refer to the individual pages we give each of them a page name (e.g. NetWork) and a page number (e.g. 10).

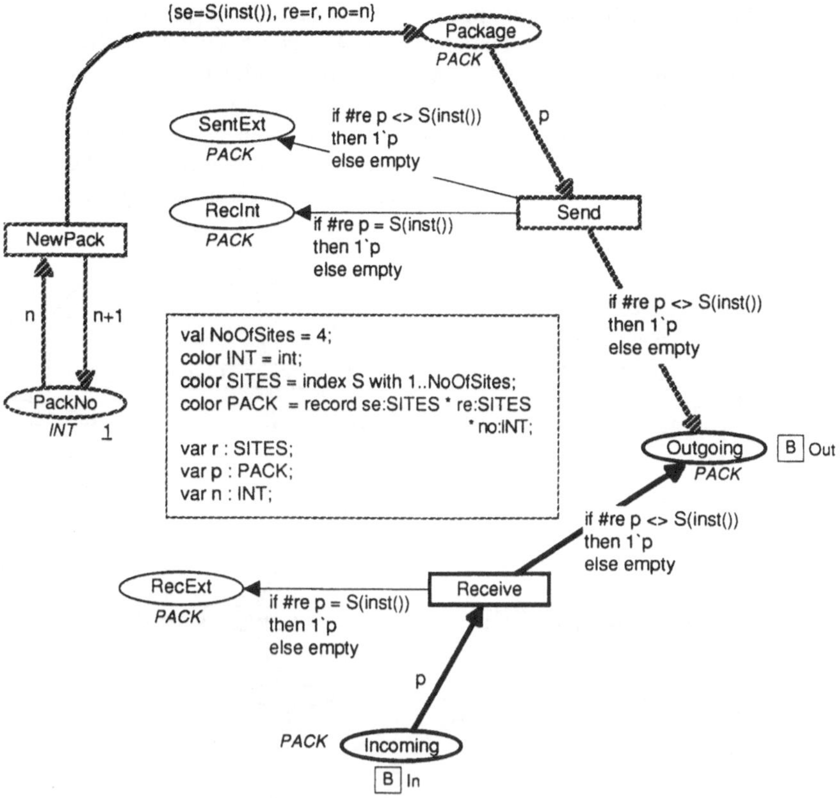

Figure 5. Site#11 describes an individual site of the ring network

Some of the declarations in the middle of Fig. 5 need a little explanation: The first line declares a constant NoOfSites. It is used in one of the other declarations – and it could also have been used e.g. in the arc expressions of NetWork#10 and Site#11, if desired. The colour set SITES contains four different elements which are denoted by S(1), S(2), S(3) and S(4).[31] The colour set PACK contains all records which have an se-field (identifying the sender), an re-field (identifying the receiver) and a no-field (containing a package number).

Site#11 has three different transitions: Each occurrence of NEWPACK creates a new package. The se-field of the new package becomes identical to S(inst()) where the pre-declared function inst() returns the identity number of the page instance on which the transition occurs[32] while the no-field is read from PACKNO. The re-field is determined

31 The idea behind index colour sets is to make it easy for the user to define colour sets which are of the form $\{S_1, S_2, \ldots S_n\}$.

32 Allowing net inscriptions (such as arc expressions, guards and initialization expressions) to be dependent on the page instance is a generalization – with respect to the class of hierarchical CP-nets formally defined in section 3.4. The extension, which has turned out to be extremely useful, will be

by the variable r – which does not appear anywhere else, and this means that r can take an arbitrary value (from SITES). The created packages are handled by SEND, which inspects the re-field of the package.[33] When the re-field indicates that the receiver is different from the present site, the package is transferred to the network via OUTGOING (and a copy is put on SENTEXT). Otherwise the package is sent directly to RECINT. Finally, RECEIVE inspects all packages which are transferred from the network via INCOMING. Again the re-field is inspected, and based on this inspection the package is routed, either to OUTGOING or to RECEXT.

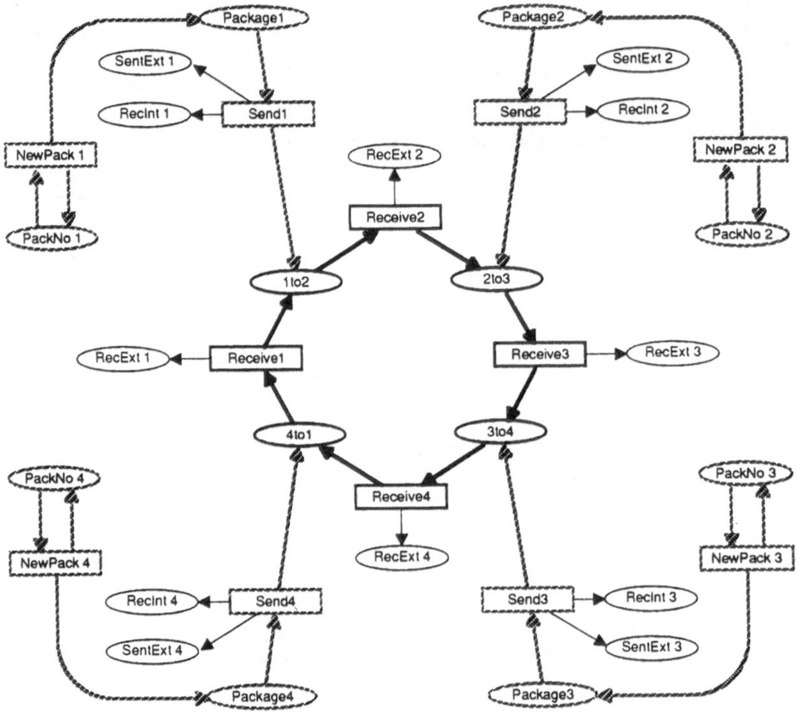

Figure 6. Non-hierarchical CP-net with the same behaviour as the hierarchical CP-net that contains the pages NetWork#10 and Site#11

Now let us return to the hierarchy inscriptions of the four substitution transitions on NetWork#10: The first line of each hierarchy inscription identifies the **subpage**, i.e. the page that is going to replace the substitution transition. In our present example, the four substitution transitions are being replaced by the same subpage Site#11. Each substitution transition gets, however, its own "private copy" of Site#11.

supported by one of the next versions of the CPN tool described in chapter 5. The page instance numbers are consecutive positive numbers, starting from 1 (i.e. in this case: 1, 2, 3, and 4).

33 We use *#re p* to denote the re-field of a package p.

The remaining lines of the hierarchy inscription contain the **port assignment** which tells how the subpage (Site#11) is going to be inserted into the **superpage** which contains the substitution transition (NetWork#10). Each line of the port assignment relates a **socket node** on the superpage (i.e. one of the nodes surrounding the substitution transition) to a **port node** on the subpage (i.e. one of the nodes which have a B-tag next to it (B ≈ Border)). In our example, let us now consider the hierarchy inscription next to NO2. The first line of the port assignment tells us that the socket node 2TO3 is assigned to (i.e. "glued" together with) the port node OUTGOING. Analogously, the second line tells us that 1TO2 is assigned to INCOMING. The remaining three hierarchy inscriptions (of NO1, NO3 and NO4) are interpreted in a similar way – and this tells us that the hierarchical CP-net with the two pages NetWork#10 and Site#11 is equivalent – i.e. has the same behaviour – as the non-hierarchical CP-net in Fig. 6 (where we for clarity have omitted the net inscriptions).

When we consider the behaviour of a hierarchical CP-net each page has its own marking. We allow a single page to replace several substitution transitions, and then the page has several **page instances**, each having its own marking. In the example above, their are four instances of Site#11 – and thus four different markings.[34]

3.2 Fusion of places

The intuitive idea behind fusion of places is to allow the user to specify that a set of places are considered to be identical – i.e. they all represent a single place even though they are drawn as individual places. This means that when a token is added/removed at one of the places, an identical token has to be added/removed at all the others. The places of a fusion set may belong to a single page or to several different pages.

When all members of a fusion set belong to a single page and that page only has one instance, place fusion is nothing other than a drawing convenience that allows the user to avoid too many crossing arcs. However, things become much more interesting when the members of a fusion set belong to several different pages *or* to a page that has many different page instances. In that case fusion sets allow the user to specify a behaviour which it might be cumbersome to describe without fusion. To allow modular analysis of hierarchical CP-nets, global fusion sets should be used with care.

There are three different kinds of fusion sets: Global fusion sets are allowed to have members from many different pages, while page and instance fusion sets only have members from a single page. The difference between the last two is the following: A page fusion unifies all the instances of its places (independently of the page instance at which they appear), and this means that the fusion set only has one "resulting place" which is "shared" by all instances of the corresponding page. In contrast to this, an instance fusion set only identifies place instances that belong to the *same* page instance, and this means that the fusion set has a "resulting place" for each page instance. A global fusion set is analogous to a page fusion set, in the sense that it only has one "resulting place" (which is common for all instances of all the participating pages).

---

[34] When a CP-net is simulated by means of the CPN tools described in chapter 5, we have a window for each page. The window shows the marking of one instance at a time, and it is possible for the user to switch from one instance to another.

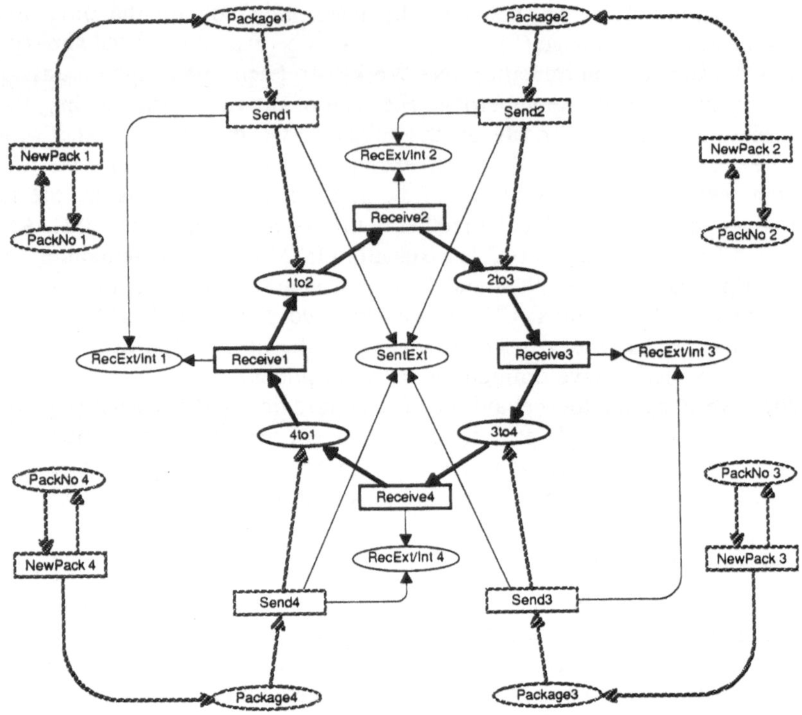

Figure 7. Non-hierarchical equivalent of a CP-net with two fusion sets

The difference between page and instance fusion sets can be illustrated by the ring network. In Fig. 7 we show how the non-hierarchical CP-net in Fig. 6 is modified when we on Site#11 define two fusion sets: An instance fusion set containing {RECEXT, RECINT} and a page fusion set containing {SENTEXT}.

Above we have illustrated the difference between page and instance fusion sets by drawing a non-hierarchical CP-net which is behaviourally equivalent to our hierarchical CP-net. It should, however, be understood that hierarchical CP-nets is a modelling language in its own right. This means that it is possible (and desirable) to model and analyse a complex system by a hierarchical CP-net – without ever constructing the equivalent non-hierarchical CP-net.

3.3 Partitions

To give a formal definition of hierarchical CP-nets we need the concept of a partition. Intuitively, a partition is a division of a set into a number of subsets, which are non-empty and pairwise disjoint:

Definition 3.1: Let a finite set Z be given. A **partition** of Z is a family of sets $X = \{X_i\}_{i \in I}$ such that:

(i) The **index set** I is a finite set.

(ii) Each **component** X_i is a non-empty subset of Z:
- $\forall i \in I: [\emptyset \subset X_i \subseteq Z]$.

(iii) The components are pairwise disjoint:
- $\forall (i,k) \in I: [i \neq k \Rightarrow X_i \cap X_k = \emptyset]$.

The **range** of the partition is the set:

$$X_R = \bigcup_{i \in I} X_i \subseteq Z.$$

The partition is said to be **total** iff $X_R = Z$. Otherwise it is **partial**.

It should be obvious that there is a very close relationship between partitions and equivalence relations: On the one hand, each partition determines an equivalence relation for its range (two elements are equivalent iff they belong to the same component - and each component is an equivalence class). On the other hand, each equivalence relation determines a total partition (two elements belong to the same component if they are equivalent – and each equivalence class is a component).

3.4 Formal definition of hierarchical CP-nets

This section contains the formal definition of hierarchical CP-nets. Some motivation and explanation of the individual parts of the definition is given immediately below the definition:

Definition 3.2: A **hierarchical CP-net** is a tuple HCPN = (S, SN, SA, PN, PA, FS, FT, PP) satisfying the requirements below:

(i) $S = \{S_i \mid i \in I\}$ is a finite set of **pages** such that:
- Each page is a non-hierarchical CP-net: $S_i = (\Sigma_i, P_i, T_i, A_i, N_i, C_i, G_i, E_i, IN_i)$.
- The sets of net elements are pairwise disjoint: $\forall (i,k) \in I: [i \neq k \Rightarrow (P_i \cup T_i \cup A_i) \cap (P_k \cup T_k \cup A_k) = \emptyset]$.

When XX_i is a set, relation or a function, defined for all $i \in I$, we use XX to denote the union.[35] When YY is a set, relation or function, defined for HCPN, we use YY_i (and YY_{S_i}) to define the restriction to S_i.

(ii) $SN \subseteq T$ is a set of **substitution nodes**.

(iii) SA is a **page assignment** function. It is defined from SN into S such that:
- No page is a subpage of itself:[36]
$\{i_0\, i_1 \dots i_n \in I^* \mid n \in \mathbb{N}_+ \wedge i_0 = i_n \wedge \forall k \in 1..n: S_{i_k} \in SA(SN_{i_{k-1}})\} = \emptyset$. *(cont.)*

35 The union of a set of functions is the union of the corresponding set of relations and this is known to become a function (because the set of domains are pair-wise disjoint).

36 I* denotes all finite sequences with elements from I, and we extend SA so that it can be used on a *set* of substitution nodes..

(iv) $PN \subseteq P$ is a set of **port nodes**.

(v) PA is a **port assignment** function. It is defined from SN into binary relations such that:

- Socket nodes are related to port nodes:
  $PA(x) \subseteq X(x) \times PN_{SA(x)}$.
- Related nodes have identical colour sets and equivalent initialization expressions:
  $\forall x \in SN\ \forall (x_1,x_2) \in PA(x)$: $[C(x_1) = C(x_2) \ \wedge\ IN(x_1)\langle\rangle = IN(x_2)\langle\rangle]$.

(vi) $FS = \{FS_r\}_{r \in R}$ is a finite set of **fusion sets** such that:

- FS is a partition of P.
- Members of a fusion set have identical colour sets and equivalent initialization expressions:

$\forall r \in R\ \forall x_1,x_2 \in FS_r$: $[C(x_1) = C(x_2) \ \wedge\ IN(x_1)\langle\rangle = IN(x_2)\langle\rangle]$.

(vii) FT is a **fusion type** function. It is defined from fusion sets such that:

- Each fusion set is of type: global, page or instance.
- Page and instance fusion sets belong to a single page:
  $\forall r \in R$: $[FT(FS_r) \neq \text{global} \ \Rightarrow\ \exists i \in I: FS_r \subseteq P_i]$.

(viii) $PP \in S_{MS}$ is a multi-set of **prime pages**.

(i) Each **page** is a non-hierarchical CP-net. We use Σ to denote the union of all the colour sets Σ_i of the individual pages (these colour sets do *not* need to be disjoint). The pages have pairwise disjoint sets of nodes and arcs, and this means that for functions and relations, defined on places, transitions and arcs, we can omit the page index without any ambiguity. As an example we can write C(p), G(t) and E(a) instead of $C_i(p)$, $G_i(t)$ and $E_i(a)$. Analogously, we use P, T and A to denote the set of all places, the set of all transitions and the set of all arcs in HCPN. The notational conventions described above allows us to move our point of focus from a given page to the entire CP-net by omitting the page index. It is, however, also possible to do the opposite and this means that we restrict a set, relation or function, defined for elements of the entire CP-net, to elements of a particular page. As an example, we use SN_i (and SN_{S_i}) to denote the subset of substitution nodes that belong to page S_i.

(ii) Each **substitution node** is a transition.[37]

(iii) The **page assignment** relates substitution transitions to pages. When a transition $t \in SN_i$ is related to a page S_k, we say that S_k is a *direct subpage* of the page S_i which is a *direct superpage* of S_k. Analogously, we say that S_k is a *direct subpage* of the node x which is a *direct supernode* of S_k. These four relations are in the usual way extended by taking the transitive closure and we then omit the word "direct" and talk about subpages, superpages and supernodes. It is demanded that no page is a subpage of itself. Otherwise, the process of substituting supernodes with their direct subpages will be infinite and it would be impossible to construct an equivalent non-hierarchical CP-net (without allowing P, T and A to be infinite).

(iv) Each **port node** is a place. It should be noticed that we allow a page to have port nodes even when it is not a subpage. Such port nodes have no semantic meaning

[37] As described in [31] it is also possible to allow places to be substitution nodes. The semantics of such a model is, however, slightly more complex.

(and thus they can be turned into non-ports without changing the behaviour of the CP-net).

(v) The **port assignment** relates *socket nodes* (i.e. the places surrounding a substitution transition) with port nodes (on the corresponding direct subpage). Each related pair of socket/port nodes must have identical colour sets and equivalent initialization expressions. It should be noticed that it is possible to relate several sockets to the same port and vice versa. It is also possible to have sockets and ports which are totally unrelated. Usually, most port assignments are bijective functions and in that case there is a one to one correspondence between sockets and ports.

(vi) The **fusion sets** are the components in a partition of P and this means that a place can belong to at most one fusion set. All members of a fusion set must have identical colour sets and equivalent initialization expressions. Usually, it is only a few places that belong to fusion sets and thus the partition is partial.

(vii) The **fusion type** divides the set of fusion sets into global, page and instance fusion sets. For the last two kinds of fusion sets all members must belong to the same page.

(viii) The **prime pages** is a multi-set over the set of all pages and they determine, together with the page assignment, how many instances the individual pages have. Often the multi-set contains only a single page (with coefficient one).

It should be obvious that each non-hierarchical CP-net is a hierarchical CP-net with a single page. There are no substitution, port and fusion nodes – and thus the page assignment, port assignment and fusion type functions become trivial. The single page belongs to the multi-set of prime pages, with coefficient one.

3.5 Page, place, transition and arc instances

A page may have several instances: There is a page instance for each time the page appears in the multi-set PP and, moreover there is a page instance for each way in which the page is a subpage of an element of PP.

In the following definition s and n identify the element of PP from which the page instance is constructed, while $x_1 x_2 \ldots x_m$ identifies the sequence of substitution nodes that leads to the page instance (in this sequence each node x_{k+1} belongs to the direct subpage of x_k). It should be noticed that the sequence may be empty:

Definition 3.3: The **page instances** of a page $S_i \in S$ is the set SI_i of all triples $(s, n, x_1 x_2 \ldots x_m)$ that satisfy the following requirements:

(i) $s \in PP \wedge n \in 1..PP(s)$.

(ii) $x_1 x_2 \ldots x_m$ is a sequence of substitution nodes, with $m \in N$, such that:
$m > 0 \Rightarrow (x_1 \in SN_s \wedge [k \in 1..(m-1) \Rightarrow x_{k+1} \in SN_{SA(x_k)}] \wedge SA(x_m) = S_i)$.

When a page has several page instances each of these have their own instances of the corresponding places, transitions and arcs. It should, however, be noticed that substitution nodes and their surrounding arcs do not have instances (because they are replaced by instances of the corresponding direct subpages):

Definition 3.4: The **place instances** of a page $S_i \in S$ is the set PI_i of all pairs (p,id) that satisfy the following requirements:
(i) $p \in P_i$.
(ii) $id \in SI_i$.

The **transition instances** of a page $S_i \in S$ is the set TI_i of all pairs (t,id) that satisfy the following requirements:
(iii) $t \in T_i - SN_i$.
(iv) $id \in SI_i$.

The **arc instances** of a page $S_i \in S$ is the set AI_i of all pairs (a,id) that satisfy the following requirements:
(v) $a \in A_i - A(SN_i)$.
(vi) $id \in SI_i$.

Place instances may be related to each other, either by means of fusion sets or by means of port assignments and this leads to the following concepts:

Definition 3.5: The **place instance relation** is the smallest equivalence relation on PI[38] containing all those pairs $((p_1,(s_1,n_1,xx_1)),(p_2,(s_2,n_2,xx_2))) \in PI$ that satisfy one of the following conditions:
(i) $\exists r \in R$: $[p_1,p_2 \in FS_r \wedge (FT(FS_r) = \text{instance} \Rightarrow (s_1,n_1,xx_1) = (s_2,n_2,xx_2))]$.
(ii) $\exists t \in SN$: $[(p_1,p_2) \in PA(t) \wedge (s_1,n_1) = (s_2,n_2) \wedge xx_1\text{^}t = xx_2]$.[39]

An equivalence class of the place instance relation is called a **place instance group** and the set of all such equivalence classes is denoted by PIG.

3.6 Equivalent non-hierarchical CP-net

In sections 3.1 and 3.2 we have sketched how to define the behaviour of a hierarchical CP-net – by constructing a non-hierarchical CP-net that is behaviourally equivalent. In this section we define the non-hierarchical equivalent in a much more formal way, but before doing this we again want to stress that the construction of the non-hierarchical equivalent plays a similar role as the unfolding of a CP-net to a behaviourally equivalent PT-net: The construction is only performed in order to define and understand the semantics. When we describe a system we directly use hierarchical CP-nets – and we never construct the non-hierarchical equivalent. Analogously, we directly analyse a hierarchical CP-net – without having to construct the non-hierarchical equivalent. The existence of the non-hierarchical equivalent is, however, very important – because it tells us how to generalize the basic concepts and the analysis methods of non-hierarchical CP-nets to hierarchical CP-nets.

---

38 Following our notational conventions we use PI to denote the set of all place instances in the entire CP-net (i.e. the union of PI_i over $i \in I$).

39 The ^ operator denotes concatenation of two sequences.

Definition 3.6: Let a hierarchical CP-net HCPN = (S, SN, SA, PN, PA, FS, FT, PP) be given. Then we define the **equivalent non-hierarchical** CP-net to be CPN = (Σ*, P*, T*, A*, N*, C*, G*, E*, IN*) where:
(i) $\Sigma^* = \Sigma$.
(ii) $P^* = PIG$.
(iii) $T^* = TI$.
(iv) $A^* = AI$.
(v) $\forall a^*=(a,id) \in AI\ \forall (p,t) \in P \times T$:
$$[\ N(a) = (p,t) \Rightarrow N^*(a^*) = ([(p,id)],(t,id)) \wedge$$
$$N(a) = (t,p) \Rightarrow N^*(a^*) = ((t,id),[(p,id)])].$$ [40]
(vi) $\forall p^*=[(p,id)] \in PIG$: $[C^*(p^*) = C(p)]$.
(vii) $\forall t^*=(t,id) \in TI$: $[G^*(t^*) = G(t)]$.
(viii) $\forall a^*=(a,id) \in AI$: $[E^*(a^*) = E(a)]$.
(ix) $\forall p^*=[(p,id)] \in PIG$: $[IN^*(p^*) = IN(p)]$.

(i) The non-hierarchical CP-net has the same set of colour sets as the hierarchical CP-net.

(ii) The non-hierarchical CP-net has a place for each place instance group of the hierarchical CP-net. This means that there is place for each place instance – unless that place instance either belongs to a fusion set (in which case the place instance is merged with the other members of the fusion set) or it is an assigned socket/port node (in which case it is merged with the place instance to which it is assigned).

(iii) + (iv) The non-hierarchical CP-net has a transition for each transition instance of the hierarchical CP-net. Analogously, it has an arc for each arc instance of the hierarchical CP-net.

(v) The basic idea behind the definition of the node function is that each page instance has the same arcs as the original page. This means that a place instance and a transition instance only can have connecting arcs if they belong to the same page instance – and in that case they have connecting arcs iff the original place and transition have. It should, however, be noticed that the node function (due to place fusion and socket/port assignment) maps into place instance groups (and not into individual place instances). This is done in such a way that each place instance group gets a set of surrounding arcs that is the union of those arcs that the corresponding place instances would have got (if they had not participated in any fusion or socket/port assignment).

(vi) The colour set of a place instance group is determined by the colour set of the participating places. From Def. 3.2 (v) + (vi) it follows that all these places must have identical colour sets.

(vii) The guard of a transition instance is determined by the guard of the corresponding transition.

(viii) The arc expression of an arc instance is determined by the arc expression of the corresponding arc.

(ix) The initialization expression of a place instance group is determined by the initialization expression of one of the participating places. From Def. 3.2 (v) + (vi) it

40 We use [(p,id)] to denote the equivalence class to which (p,id) belongs.

follows that all these places must have initialization expressions which evaluate to the same value, and thus it does not matter which one we choose.

3.7 Dynamic behaviour of hierarchical CP-nets

Having defined the static structure of CP-nets we are now ready to consider their behaviour – but first we introduce the following notation, where E'(p',t') and E'(t',p') are called the **expressions** of (p',t') and (t',p'):[41]

- $\forall p'=(p,id_p)\in PI\ \forall t'=(t,id_t)\in TI$:
$$[\ id_p = id_t \Rightarrow (E'(p',t') = E(p,t)\ \wedge\ E'(t',p') = E(t,p))\ \wedge$$
$$id_p \neq id_t \Rightarrow (E'(p',t') = E'(t',p') = 0)\ ].$$

Next we define token distributions, binding distributions, markings and steps:

Definition 3.7: A **token distribution** is a function M, defined on PIG such that $M(p^*)\in C(p)_{MS}$ for all $p^* = [(p,id)] \in PIG$ and a **binding distribution** is a function Y, defined on TI such that $Y(t^*)\in B(t)_{MS}$ for all $t^* = (t,id)\in TI$. We define TD_{HCPN}, BD_{HCPN}, ≠, ≤, <, ≥, >, =, **element** and **non-empty** in exactly the same way as for non-hierarchical CP-nets.

A **marking** is a token distribution and a **step** is a *non-empty* binding distribution. The set of all markings is denoted by M_{HCPN}, and the set of all steps is denoted by Y_{HCPN}. The **initial marking** M_0 is the marking where $M_0(p^*) = M_0(p)$[42] for all $p^* = [(p,id)] \in PIG$.

Finally we define enabling and occurrence:

Definition 3.8: A step Y is **enabled** in a marking M iff the following property is satisfied:

$$\forall p^*\in PIG: [\ \sum_{\substack{(t',b)\in Y \\ p'\in p^*}} E'(p',t')\langle b\rangle \leq M(p^*)].$$

We define **enabled** transition instances and **concurrently enabled** transition instances/bindings analogously to the corresponding concepts in a non-hierarchical CP-net.

(continues)

41 We use p' and t' to denote a place instance and a transition instance, respectively.

42 We use M_0 for two different purposes: On the left-hand side of the equation it denotes a marking of HCPN (i.e. a function defined on PIG). On the right-hand side it denotes the union constructed from the initial markings M_{0i} of the individual pages (i.e. a function defined on P). From the argument it will always be clear which of the two functions we deal with.

When a step is enabled in a marking M_1 it may **occur**, changing the marking M_1 to another marking M_2, defined by:

$$\forall p^* \in PIG: [M_2(p) = \Big(M_1(p) - \sum_{\substack{(t',b)\in Y \\ p'\in p^*}} E(p,'t)'\langle b\rangle\Big) + \sum_{\substack{(t',b)\in Y \\ p'\in p^*}} E(t',p')\langle b\rangle].$$

The first sum is called the **removed** tokens while the second is called the **added** tokens. Moreover we say that M_2 is **directly reachable** from M_1 by the occurrence of the step Y, which we also denote:

$$M_1\,[Y\rangle\, M_2.$$

We define **occurrence sequences** and **reachability** analogously to the corresponding concepts for a non-hierarchical CP-net.

The following theorem shows that there is a one to one correspondence between the behaviour of a hierarchical CP-net and the corresponding non-hierarchical equivalent:

Theorem 3.9: Let HCPN be a hierarchical CP-net and CPN the non-hierarchical equivalent. Then we have the following properties:

(i) $M_{HCPN} = M_{CPN}$.

(ii) $Y_{HCPN} = Y_{CPN}$.

(iii) $\forall M_1, M_2 \in M_{HCPN}\ \forall Y \in Y_{HCPN}$: $[\, M_1\,[Y\rangle_{HCPN}\, M_2 \Leftrightarrow M_1\,[Y\rangle_{CPN}\, M_2 \,]$.

Proof: Property (i) is an immediate consequence of Def. 2.5, Def. 3.6 (ii) and Def. 3.7, while property (ii) is an immediate consequence of Def. 2.5, Def. 3.6 (iii) and Def. 3.7. Property (iii) follows from Def. 2.6, Def. 2.7, Def. 3.6 and Def. 3.8. The proof is omitted. It is straightforward but tedious – due to the large number of details which have to be considered.

4. Analysis of CP-nets

This chapter describes how CP-nets can be analysed. The most straightforward kind of analysis is simulation – which is very useful for the understanding and debugging of a system, in particular in the design phase and the early validation phases. There are, however, also more formal kinds of analysis – by which it is possible to *prove* that a given system has a set of desired properties (e.g. absence of dead-lock, the possibility to return to the initial state, and an upper bound on the number of tokens). This chapter contains a brief introduction to the main ideas behind the most important analysis methods and it contains references to papers in which the technical details of these methods can be found.

4.1 Simulation

Simulation can be supported by a computer tool or it can be totally manually (e.g. performed on a blackboard or in the head of the modeller). Simulation is similar to the debugging of a program, in the sense that it can reveal errors, but in practice never be sufficient to prove the correctness of a system. Some people argue that this makes simulation uninteresting and that the user instead should concentrate on the more formal analysis methods. We do not agree with this conclusion but consider simulation to be just as important and necessary as the more formal analysis methods.

In our opinion, all users of CP-nets (and other kinds of Petri nets) are forced to make simulations – because it is impossible to construct a CP-net without thinking about the effects of the individual transitions. Thus the proper question is not whether the modeller should make simulations or not, but whether he wants computer support for this activity. With this rephrasing the answer becomes trivial: Of course, we want computer support. This means that the simulation can be done much faster and with no errors. Moreover, it means that the modeller can use all his mental capabilities to interpret the simulation results (instead of using most of his efforts to calculate the possible occurrence sequences). Simulation is often used in the design phases and the early investigation of a system design (while the more formal analysis methods are used for the final validation of the design). In section 5.5 we give a detailed description of an existing CPN simulator.

4.2 Occurrence graphs

The basic idea behind occurrence graphs is to construct a graph which contains a node for each reachable state and an arc for each possible change of state. Obviously such a graph may, even for small CP-nets, become very large (and perhaps infinite). Thus we want to construct and inspect the graph by means of a computer – and we want to develop techniques by which we can construct a reduced occurrence graph without loosing too much information. The reduction can be done in many different ways:[43]

One possibility is to reduce by means of covering markings. This method looks for occurrence sequences leading from a system state to a larger system state (one with additional tokens) and the method guarantees that the reduced occurrence graph always becomes finite. The method has, however, some drawbacks. First of all it only gives a reduction for unbounded systems (and most practical systems are bounded). Secondly, so much information is lost by the reduction that several important properties (e.g. liveness and reachability) no longer are decidable. For more information see [18] and [40].

A second possibility is to reduce by ignoring some of the occurrence sequences which are identical, except for the order in which the elements occur. This method often gives a very significant reduction, in particular when the modelled system contains a large number of relatively independent processes. Unfortunately, it is with this method necessary to construct several different occurrence graphs (because the construction method depends upon the property which we want to investigate). For more information see [59].

A third possibility is to reduce by means of the symmetries which often are present in the systems which we model by CP-nets. To do this the modeller defines, for each colour set, an algebraic group of allowed bijections (each bijection defines a possible way in which the elements of the colour set can be interchanged with each other) – and

---

[43] For all the methods described below, it is possible to construct the reduced occurrence graph without first constructing the full occurrence graph.

this induces an equivalence relation on the set of all system states. The reduced occurrence graph only contains a node for each equivalence class and this means that it often is much smaller than the full occurrence graph. The reduced graph contains, however, exactly the same information as the full graph – and this means that the reduced graph can be used to investigate all the system properties which can be investigated by means of the full graph.[44] For more information see [30] and [35].

A fourth possibility is to construct an occurrence graph where each state is denoted by a symbolic expression (which describes a number of system states, in a similar way as the equivalence classes in method three). For more information see [9] and [42].

Finally, it is possible to construct occurrence graphs in a modular way. The model is divided into a number of submodels, an occurrence graph is constructed for each submodel, and these subgraphs are combined to form an occurrence graph for the entire model. For more information see [60].

When an occurrence graph has been constructed it can be used to prove properties about the modelled system. For bounded systems a large number of questions can be answered: Dead-locks, reachability and marking bounds[45] can be decided by a simple search through the nodes of the occurrence graph, while liveness and home markings can be decided by constructing and inspecting the strongly connected components. One problem with occurrence graph analysis is the fact that it, usually, is necessary to fix all system parameters (e.g. the number of sites in a ring protocol) before an occurrence graph can be constructed – and this means that the found properties always are specific to the chosen values of the system parameters. In practice the problem isn't that big: If we e.g. understand how a ring protocol behaves for a few sites we also know a lot about how it behaves when it has more sites.[46]

As described above, the occurrence graph method can be totally automated – and this means that the modeller can use the method, and interpret the results, without having much knowledge about the underlying mathematics. For the moment it is, however, only possible to construct occurrence graphs for relatively small systems and for selected parts of large systems. This doesn't mean that the method is uninteresting. On the contrary, the method seems to be a very effective way to debug new subsystems (because trivial errors such as the omission of an arc or a wrong arc expression often means that some of the system properties are dramatically changed). In the future, it may also be possible to use occurrence graph analysis for larger systems. This can be done by combining some of the reduction techniques described above – and by using the increased computing power of the next generations of hardware. In section 7.2 we

---

44 The reduced occurrence graph (called an OE-graph) has more complex node and arc inscriptions than the full occurrence graph (called an O-graph). The OE-graph is a folded version of the O-graph, in the same way as a CP-net is a folding of the equivalent PT-net. The O-graph can be constructed from the OE-graph, but this is never necessary since the analysis can be done directly on the OE-graph.

45 There are two kinds of marking bounds. Integer bounds only deal with the number of tokens while multi-set bounds also deal with the token colours. It can be proved that a place is integer bounded if and only if it is multi-set bounded. There are, however, situations in which the integer bound gives more information than the multi-set bound (and vice versa) – and thus it is useful to calculate both kinds of bounds.

46 This is of course only true when we talk about the correctness of the protocol, and not when we speak about the performance.

describe the plans to implement a CPN tool to support the calculation and analysis of occurrence graphs.

4.3 Place and transition invariants

The basic idea behind place invariants is to find a set of equations which characterize all reachable markings, and then use these equations to prove properties of the modelled system (in a way which is analogous to the use of invariants in program verification). To illustrate the idea, let us consider the resource allocation system from Fig. 1. This system has the five place invariants shown below.[47] A place invariant is a linear sum of the markings of the individual places: Each place marking is by a weight function (attached to the place) mapped into a new multi-set. All the new multi-sets are over the same colour set and thus they can be added together – to give a **weighted sum** (determined from the given marking by the given set of weight functions).

The invariants use the three functions *P*, *Q* and *PQ* as weight functions. Each of them maps P-colours into multi-sets of E-colours. Intuitively, *P* "counts" the number of p-tokens (it maps (p,i) into 1`e and (q,i) into the empty multi-set). Analogously *Q* counts the number of q-tokens and *PQ* counts the number of p/q-tokens (i.e. the total number of tokens).[48] The invariants also use identity functions and zero functions as weights. The five invariants are satisfied for all reachable markings M (later we shall discuss how this can be proved). The right hand side of the invariants are found by evaluating the left hand side in the initial marking.

Intuitively PI_P and PI_Q tell what happens to the two different kinds of processes, while PI_R, PI_S and PI_T tell what happens to the three different kinds of resources. Each invariant can be seen as a way of extracting specific information – from the general information provided by the entire marking.

PI_P	$P(M(B) + M(C) + M(D) + M(E)) = 2`e$
PI_Q	$Q(M(A) + M(B) + M(C) + M(D) + M(E)) = 3`e$
PI_R	$M(R) + Q(M(B) + M(C)) = 1`e$
PI_S	$M(S) + Q(M(B)) + 2 * PQ(M(C) + M(D) + M(E)) = 3`e$
PI_T	$M(T) + P(M(D)) + (PQ + P)M(E) = 2`e$

The five invariants above can be used to prove that the resource allocation system doesn't have a dead-lock. *The proof is by contradiction:* Let us assume that we have a reachable state with no enabled transitions. From PI_P we know that there are two p-tokens distributed on the places B-E and from PI_Q that three are three q-tokens distributed on A-E. Now let us investigate in more detail where these tokens can be positioned. *First, assume that there are tokens on E:* Then T5 is enabled (and we have a contradiction with the assumption of no enabled transitions). *Secondly, assume that*

[47] There are many other place invariants for the system – but these are the most simple and useful.

[48] A weight function is usually specified as a function $f \in [C(p) \rightarrow A_{MS}]$ (i.e. a function from the colour set C(p) of the place into multi-sets over a colour set A). We always extend f to a function $f_{ext} \in [C(p)_{MS} \rightarrow A_{MS}]$ (for each multi-set $m \in C(p)_{MS}$ we calculate $f_{ext}(m)$ by adding the results of applying f to all the individual elements of m). Usually we do not distinguish between f and f_{ext} (and we use f to denote both functions).

there are tokens on C and/or D (and no tokens on E): From PI_S it follows that there can be at most one such token and then PI_T tells that there is at least one e-token on T (because $P(M(D)) \leq 1`e$ and $(PQ + P)M(E) = empty$). Thus T3 or T4 can occur. *Thirdly, assume that there are tokens on B (and no tokens on C, D and E):* From $PI_R$ it follows that there can be at most one q-token on B and then $PI_S$ tells us that there is at least two e-tokens on S (because $Q(M(B)) \leq 1`e$ and $2 * PQ(M(C) + M(D) + M(E)) = empty$). Thus T2 can occur. *Now we have shown that it is impossible to position the two p-tokens (without violating the dead-lock assumption) – and thus we conclude that all reachable states have at least one enabled transition.* From the fact there are no dead-locks and the cyclic structure of the net, it is easy to prove other system properties e.g. that the initial marking is a home marking, thai the system is live and that all reachable markings are reachable from each other.

Next let us discuss how we can find place invariants: As mentioned earlier, each CP-net has a function representation – which is a matrix where each element is a function (mapping multi-sets of bindings into multi-sets of token colours).[49] The matrix determines a homogeneous matrix equation and the place invariants are the solutions to this matrix equation (each solution is a vector of weight functions).[50] The matrix equation can be solved in different ways: One possibility is to translate the matrix of functions into a matrix of integers[51] for which the homogeneous matrix equation can be solved by standard Gauss elimination. Another, and more attractive, possibility is to work directly on the matrix of functions (this is, however, more complicated e.g. because some functions do not have an inverse). With both methods we do not explicitly find all solutions (there are usually infinitely many). Instead we find a basis from which all invariants can be constructed (as linear combinations). This leaves us with a second problem: How do we from the basis find the interesting place invariants – i.e. those from which it is easy to prove system properties? In our opinion, the best solution is to allow the user to tell the analysis program where to look for invariants – and thus calculate invariants in an interactive way. For more details about the calculation of invariants, see [12], [35], [44] and section 7.2.

Above, we have discussed how to calculate invariants by solving a homogeneous matrix equation. The problem is, however, often of a different nature – because we (instead of starting from scratch) already have a set of weight functions and just want to verify that these are invariants. This task is much easier and it can, without any problems, be done totally automatically. The potential invariants, to be checked, can be derived from the system specification and the modellers knowledge of the expected system properties. The potential invariants may be specified after the system design has been finished. It is, however, much more useful (and easier) to use CP-nets during the design and construct the invariants as an integrated part of the design (in the same way as a good programmer specifies a loop invariant at the moment he creates the loop). For this use of invariants it is important to notice that the check of invariants are constructive – in the sense that it, in the case of failure, is told where in the CP-net the

---

49 The translation into the function representation can easily be defined by means of the lambda calculus. For more details see [35].

50 Each solution to the matrix equation is a place invariant. The other direction is, however, only true when it is known that each occurrence element is enabled in at least one reachable marking.

51 This is exactly the same as unfolding the CP-net to the behavioural equivalent PT-net.

problems are. Thus it is often relatively easy to see how the CP-net (or the invariant) should be modified.

Transition invariants are the duals of place invariants and the basic idea behind them is to find occurrence sequences with no effects (i.e. with the same start and end marking). Transition invariants can be calculated in a similar way as place invariants[52] - but, analogously to place invariants, it is more useful to construct them during the system design. Transition invariants are used for similar purposes as place invariants (i.e. to investigate the behavioural properties of CP-nets).

Place/transition invariants have several very attractive properties: First of all invariant analysis can be used for large systems – because it can be performed in a modular way[53] and does not involve the same kind of complexity problems as occurrence graph analysis. Secondly, invariant analysis can be done without fixing system parameters (e.g. the number of sites in a ring protocol). Thirdly, the the use of invariants during the design of a system will (as described above) usually lead to a better design. The main drawback of invariant analysis is that the skills, required to perform it, are considerably higher than for the other analysis methods. In section 7.2 we describe the plans to implement a CPN tool to support the interactive calculation and use of place/transition invariants.

4.4 Other analysis methods

CP-nets can also be analysed by means of reduction. The basic idea behind this method is to select one or more behavioural properties (e.g. liveness and dead-locks), define a set of transformation rules, prove that the rules do not change the selected set of properties, and finally apply the rules to obtain a reduced CP-net – which usually is so small that it is trivially to see whether the desired properties are fulfilled or not. Reduction methods are well-known for PT-nets and they have in [25] been generalised to CP-nets. A serious problem with reduction methods is that they often are non-constructive (because the absence of a property in the reduced net, usually, do not tell much about why the original net doesn't have the property).[54]

Most applications of CP-nets are used to design and validate the correctness of a system (e.g. whether the system executes the desired functions and whether it is deadlock free). CP-nets can, however, also be used to investigate the performance of a system (i.e. how fast it executes). To perform this kind of analysis it is necessary to specify the time consumption in the modelled system, and this can be done in many different ways: As a delay between the enabling and occurrence of a transition, a delay between the removal of input tokens and the creation of output tokens, or as a delay between the creation of a token and the time at which that token can be used. In all three cases, the delay may be a fixed value, a value inside a given interval, or a value deter-

52 Transition invariants are found by solving a homogeneous matrix equation (obtained by transposing the matrix used to find place invariants). Each transition invariant is a solution to the matrix equation. The opposite is, however, not always true (even for "nice" CP-nets).

53 As shown in [45] invariants can be obtained by the composition of existing invariants and this means that we can construct invariants of a hierarchical CP-net – from invariants of the individual pages.

54 An exception is the reduction method to calculate place/transition invariants, mentioned in section 7.2. In this case it is, from the reduced net, possible to determine a set of the invariants for the original net – and this means that the analysis results can be interpreted in terms of the original net.

mined by a probability distribution. Performance analysis is often made by simulation, and we shall in section 7.1 briefly describe how this can be done. For some kinds of delays, it is also possible to translate the net model into a Markovian chain – from which analytic solutions of the performance values can be calculated. For more information about performance analysis see [47].

For ordinary Petri nets at least two other kinds of analysis methods are known. One method translates the net structure into a set of logical equations, transforms the equations by a general theorem prover, and obtains results above the behaviour of the system. For more information see [10]. The other method uses structural properties[55] of a Petri net to deduce behavioural properties. For more information see [3]. Unfortunately, neither of these methods have yet been generalized to CP-nets (or other kinds of high-level Petri nets).

5. Computer Tools for CP-nets

The practical use of Petri nets is, just as all other description techniques, highly dependent upon the existence of adequate computer tools – helping the user to handle all the details of a large description. For CP-nets we need an editor (supporting construction, syntax check and modification of CP-nets) and we also need a number of analysis programs (supporting a wide range of different analysis methods). The recent development of fast and cheap raster graphics gives us the opportunity to work directly with the graphical representations of CP-nets (and occurrence graphs). This chapter describes some existing CPN tools (the CPN editor and CPN simulator from [1]). In chapter 7 we discuss other kinds of CPN tools that are needed, but have not yet been fully developed.

5.1 Why do we need computer tools for CP-nets?

The most important advantage of using computerized CPN tools is the possibility to create *better results*. As an example, the CPN editor provides the user with a precision and drawing quality, which by far exceeds the normal manual capabilities of humans beings. Analogously, computer support for complex analysis methods (e.g. occurrence graphs) makes it possible to obtain results, which could not have been achieved manually (since the calculations would have been too error-prone).

A second advantage is the possibility to create *faster results*. As an example, the CPN editor multiplies the speed by which minor modifications can be made: It is easy to change the size, form, position and text of the individual net elements without having to redraw the entire net. It is also possible to construct new parts of a net by copying and modifying existing subnets. Analogously, analysis methods may be fully or partially automated. As an example, the manual construction of an occurrence graph is an extremely slow process – while it can be done on a computer in a few minutes/hours (even when there are several hundred thousand nodes).

---

55 Structural properties are properties which can be formulated *without* considering the behaviour (i.e. occurrence sequences). In a CP-net structural properties may involve properties of the net structure, but also properties of the net inscriptions and the declarations.

A third advantage is the possibility to make *interactive presentations* of the analysis results. The CPN simulator makes it easy to trace the different occurrence sequences in a CP-net. Between each occurrence step, the user can (on the graphical representation of the CP-net) see the transitions which are enabled, and choose between them in order to investigate different occurrence sequences. Analogously, it is possible to make an interactive investigation of a complex occurrence graph – using an elaborated search system.

A fourth advantage is the possibility of *hiding technical aspects* of the CP-net theory inside the tools. This allows the users to apply complicated analysis methods without having a detailed knowledge of the underlying mathematics. Often the analysis is performed in an interactive way: The user proposes the operations to be done. Then the computer checks the validity of the proposals, performs the necessary calculations (which often are very complex) and displays the results.

For industrial applications the possibility of producing fast results of good quality - without requiring too deep knowledge of Petri net theory – is a necessary prerequisite for the entire use of CP-nets. Furthermore it is important to be able to use CP-nets together with other specification/implementation languages (we shall return to this question in chapters 6 and 7).

The remaining sections of this chapter describe the basic design criteria behind the CPN editor and the CPN simulator. For a more complete and detailed description the user is referred to [36]. The sections can also be seen as a list of design criteria which is relevant for all high-quality Petri net editors and simulators. There are a large number of different groups which work with the development of Petri net tools. Many of the tools are, however, still research prototypes – and for the moment it is only few of them which are able to deal with large high-level nets and are sufficiently robust to be used in an industrial environment. A list of available Petri net tools can be found in [17].

5.2 CPN editor

The CPN editor allows the user to construct, modify and syntax check hierarchical CP-nets. It is also easy to construct and modify many other kinds of graphs (but they can of course not be syntax checked).[56] All figures in this paper has been produced by means of the CPN editor.

A CP-net constructed by means of the CPN editor is called a **CPN diagram** and it contains a large number of different types of graphical **objects**. Each object is either a **node**, a **connector** (between two nodes) or a **region** (i.e. a subordinate of another object). Places and transitions are nodes, arcs are connectors, while all the net inscriptions are regions. As examples, colour sets and initialization expressions are regions of the corresponding places, guards of the corresponding transitions and arc expressions of the corresponding arcs.

The division of objects into nodes, connectors and regions reflects the fact that the CPN editor works with the **graph** (and not just an unstructured set of objects, as it is the case for most general purpose drawing tools, such as MacDraw™ or MacDraft™).

56 In this paper, the word graph denotes the mathematical concept of a graph (i.e. a structure which consists of a set of nodes interconnected by a set of edges).

This is important because it means that the construction and modification of the CPN diagrams become much faster (and with more accurate results): When the user constructs a connector he identifies the source and destination nodes (and perhaps some intermediate points). Then the editor automatically draws the connector in such a way that the two endpoints are positioned at the border of the two nodes. When the user changes the position or size of a node the regions and surrounding arcs are automatically redrawn by the editor. A repositioning implies that the regions keep their relative position (with respect to the node). A resizing implies that the relative positions of the regions are scaled while their sizes are either unchanged or scaled (depending upon an attribute of each region). When a node is deleted the regions and arcs are deleted too. This is illustrated by Fig. 8 where the node X is first repositioned, then resized and finally deleted. Similar rules apply for the repositioning, resizing and deletion of arcs and regions.

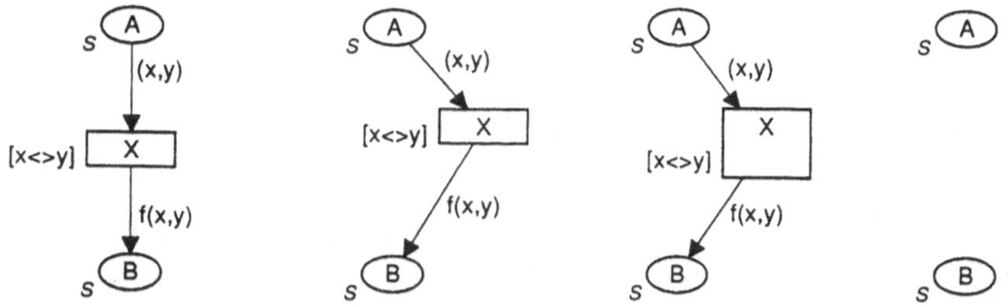

Figure 8. When a node is repositioned, resized or deleted, the regions and surrounding arcs are automatically updated.

In addition to the **CPN objects** (e.g. places, transitions, arcs and net inscriptions), which are formal parts of the model there may also be **auxiliary objects** which have no formal meaning but play a similar role as the comment facilities in programming languages. Finally, there are **system objects** which are special objects created and manipulated by the CPN editor itself. Each object has an **object type** and it should be noticed that it is the object type which determines the formal meaning of the object - independently of the object position and object form. The CPN editor distinguishes between nearly 50 different object types.

It is possible for the user to determine, in great detail, how he wants the CPN diagram to look. One of the most attracting features of CP-nets (and Petri nets in general) is the very appealing graphical representation, and it would be a pity to put narrow restrictions on how this representation can look (e.g. by making an editor in which the user cannot give two transitions different forms and/or sizes). In our opinion a good editor must allow the user to draw nearly all kinds of CP-nets which can be constructed by a pen and a typewriter. In the CPN editor each object has its own set of **attributes** which determine e.g. the position, shape, size, line thickness, line and fill patterns, line and fill colours and text type (including font, size, style, alignment and colour). There are 10-30 attributes for each object (depending upon the object type). When a new object is constructed the attributes are determined by a set of **defaults** (each object type has its own set of defaults). At any time the user can change one or more attributes for each individual objects.[57] Moreover, it is easy to change the defaults and it can be specified whether such changes apply to the current diagram or to future diagrams (or both).

In addition to the attributes the CPN editor (and in particular the CPN simulator) has a large set of **options** – which determines how the detailed functions in the editor are performed (e.g. the scroll speed, the treatment of duplicate arcs when two nodes are merged, and details about how the syntax check is performed). The difference between attributes and options is that the former relate to an individual object while the latter do not. Also options have defaults and these can be changed by the user.[58]

The CPN editor supports hierarchical CP-nets[59] and this means that each CPN diagram contains a number of pages. Each page is displayed in its own window (which in the usual way can be opened, closed, resized and repositioned). The relation between the individual pages is shown by the **page hierarchy** (which is positioned on a separate page called the **hierarchy page** and automatically maintained by the CPN editor). The page hierarchy is a graph in which each node represents a page and each connector a (direct) superpage/subpage relationship. The nodes are **page nodes** and each of them contains the corresponding page name and page number. The connectors are **page connectors** and each of them has a set of **page regions** containing the names of the involved supernodes.[60] The page objects can be moved and modified in exactly the same way as all other types of objects, and this means that the user can determine how the page hierarchy looks. The editor uses the line pattern of a page node to indicate whether the corresponding page window is active, open or closed. As an example, the ring network from Fig. 4-5 has a hierarchy page with three page nodes and one page arc, and it may look as shown in Fig. 9, where NetWork#10 is open but not active, Site#11 is closed, while the hierarchy page Hierarchy#10010 is open and active. In general, the hierarchy pages are much more complex.

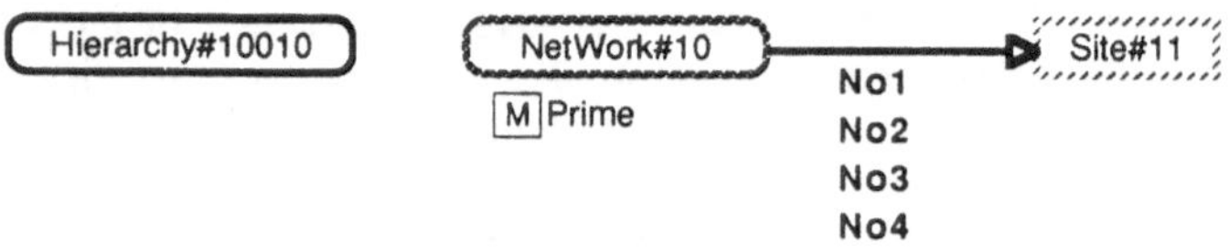

Figure 9. Hierarchy page for the ring network

The hierarchies in a CPN diagram can be constructed in many different ways – ranging from a pure top-down approach to a pure bottom-up: Part of a page can by a single editor operation be *moved to a new subpage:* The user selects the nodes to be moved and invokes the operation, then the editor checks the legality of the selection,[61] creates

[57] This is done by specifying an explicit value, selecting another object (from which the attribute is copied) or by resetting the attribute to the current default.

[58] For options a change in the default value only effects future diagrams (while a change in the option value itself, of course, effects the current diagram).

[59] For the moment the CPN editor supports substitution transitions and place fusion. The other hierarchy constructs from [31] will be added later (some of them perhaps in an improved form).

[60] Page nodes, page connectors and page regions are system objects.

[61] All perimeter nodes (i.e. nodes with external arcs) must be transitions – in order to guarantee that the selection forms a closed subnet.

the new page, moves the subnet, creates the port nodes (by copying those nodes which were next to the selection), creates the border regions for the port nodes, constructs the necessary arcs between the the port nodes and the subnet, asks the user to create a new transition (which becomes the supernode for the new subpage), draws the arcs surrounding the new transition, creates a hierarchy inscription for it, and updates the hierarchy page. As it can be seen, a lot of rather complex checks, calculations and manipulations are involved. However, nearly all of these are automatically performed by the CPN editor. The user only selects the subnet and creates the new supernode.

There is also an editor operation to *turn an existing transition into a supernode* (by relating it to an existing page). Again most of the work is done by the editor: The user selects the transition and invokes the operation, then the editor makes the hierarchy page active and enters a mode in which the user by means of the mouse can select the desired subpage,[62] the editor creates the hierarchy inscription,[63] and updates the hierarchy page. To destroy the hierarchical relationship between a supernode and a subpage the user simply deletes the corresponding hierarchy inscription (or the corresponding page connector/region). It is also possible to *replace the supernode by the contents of the subpage:* This involves a lot of complex calculations and manipulations, but again all of them are carried out by the CPN editor. The user simply selects the supernode, invokes the operation and uses a simple dialogue box to specify how the operation shall be performed – e.g. he tells whether the page shall be deleted (in the case where no other supernodes exist).

The user works with a high-resolution raster graphical screen and a mouse.[64] The CPN diagram under construction can be seen in a number of windows (where it looks as close as possible to the final output obtained by a matrix or laser printer). The editor is menu driven and have self-explanatory dialogue boxes (as known e.g. from many Macintosh programs). The user moves and resizes the objects by direct manipulation - i.e. by means of the mouse (instead of typing coordinates and object identification numbers on the keyboard). This also applies to the pages which can be opened, closed, scrolled and scaled by means of the corresponding page node. When the user deletes a page node the corresponding page is deleted (after a confirmation). Analogously, the deletion of a page connector or a page region means that the corresponding hierarchical relationship is destroyed (and thus the corresponding supernodes become ordinary transitions).

One important difference between the CPN editor and many other drawing programs is the possibility to work with **groups** of objects. This means that the user is able to select a *set* of objects and *simultaneously* change the attributes, delete the objects, copy them, move them or reposition them (e.g. vertical to each other). The user can select groups in many different ways (e.g. by dragging the mouse over a rectangular area or by pressing a key while he points to a sequence of objects). The CPN editor

---

62 When the mouse is moved over a page node it blinks – unless it is illegal (because selection of it would make the page hierarchy cyclic). Only blinking page nodes can be selected.

63 The user can ask the editor to try to deduce the port assignment by means of a set of rules (which looks at the node names, the port types and the arcs between the transition and the sockets).

64 For the moment the CPN tools are implemented on Macintosh, SUN and HP machines – and they can easily be moved to other machines running UNIX and X-Windows. It is recommended, but not necessary, to have a large colour screen.

allows the user to perform operations on groups in exactly the same way as they can be performed on individual objects[65] – and this has the same effect as when the corresponding operation is performed on each group member one at a time. All members of a group has to belong to the same page and be of the same kind – i.e. all be nodes, all be connectors or all be regions.[66] Otherwise there are no restrictions on the way in which groups can be formed. The group facility has a very positive impact upon the speed and ease by which editing operations are performed. By selecting a group of page nodes it is possible to work on several pages at the same time.

In the design of the CPN editor it has been important for us to make it as flexible as possible. As described above, this means that it is possible to construct CPN diagrams which *look* very different. However, it also means that each diagram can be *created* in many different ways. One example of this principle is the many different ways in which the page hierarchy can be constructed. Another example is the fact that the CPN editor allows the user to construct the various objects in many different orders: Some users prefer first to construct the net structure (i.e. the places, transitions and arcs). Later they add the net inscriptions (i.e. the CPN regions) – and doing this they either finish one node at a time or one kind of CPN regions at a time, and they either type from scratch or copy from existing regions. Other users prefer to create templates - e.g. a place with a colour set region and an initialization region. Then they create the diagram by copying the appropriate templates to the desired positions and modifying the texts (if necessary).[67] Finally, most users work in a way which is a mixture of the possibilities described above. We think that this kind of flexibility – where the user controls the detailed planning of the editing process – is extremely important for a good tool. Thus the CPN editor has been designed to allow most operations to be performed in several different ways.

A CPN diagram contains many different kinds of information and this means that the individual pages very easy become cluttered. To avoid this the user is allowed to make objects invisible (without changing the semantics of the objects). As an example the user may hide all colour set regions and instead indicate the colour sets by giving the corresponding places different attributes (e.g. different line patterns/colours). In this case it is still the invisible regions that determine the formal behaviour, and it is the responsibility of the user to keep the pattern/colour coding of the places correctly updated (there are several facilities in the CPN editor which helps him in this task). Another facility, which also helps avoiding cluttered diagrams, is the concept of key and popup regions, which are used for a number of different object types (both in the editor and the simulator). The idea is very simple: Instead of having a single region (containing a lot of information) we have both a **key region** (which is a region of the object to which we want to attach the information) and a **popup region** (which is a region of the key). The key region is small (it usually only contains one or two characters) and its main purpose is to give access to the popup region which contains the ac-

65 There are only very few operations which do not make sense for groups.

66 In a later version of the CPN editor, we may allow a group to have members from different pages. This is easy to implement and it creates no conceptual problems. It is, however, unlikely that we will allow mixed groups. The reason is that the semantics of many operations then become a bit obscure.

67 When the user copies a node, the editor automatically copies the regions. Analogously, when a group of nodes is copied, the internal connectors (between two members of the group) are copied too.

tual information. A double click on the key region makes the popup region visible/invisible and in this way it is extremely easy to hide and show large amounts of information. For examples of key/popup regions see the hierarchy regions in Fig. 4 (with the HS-keys), the border regions in Fig. 5 (with the B-keys) and the marking regions in Fig. 3 (containing the current marking). It should be noticed that the use of key/popup regions is more general than the use of popup windows (in which information can be displayed on demand). The difference is that the popup regions are objects in the diagram itself and thus the user can leave all of them or some of them permanently visible. Actually, it is an attribute of each key region that determines whether the corresponding popup region is visible or not.[68]

It should be noticed that the generality of the CPN editor means that the user can create very confusing CPN diagrams. As examples, it may be impossible to distinguish between auxiliary objects and CPN objects (because they have been given identical attributes), transitions may be drawn as ellipses while places are boxes, and some or all of the objects may be invisible – just to mention a few possibilities. We do *not* believe it is sensible to try to construct a tool which makes it *impossible* to produce bad nets. Such a tool will, in our opinion, inevitably be far too rigid and inflexible. However, we do of course believe that the tool should make it easy for the user to make good nets.

There are many other facilities in the CPN editor: Operations to open, close, save and print diagrams.[69] An operation which allows the editor to import diagrams created by other tools (e.g. SADT diagrams created by the IDEF/CPN tool described in section 6.3). The standard Undo,[70] Cut, Copy, Paste and Clear operations known e.g. from the Macintosh concept. Operations to define fusion sets, specify port nodes and perform port assignments. Operations to create many different types of auxiliary objects (e.g. connectors, boxes, rounded boxes, ellipses, polygons, wedges and pictures[71]). Operations to turn auxiliary objects into CPN objects (and vice versa). An operation to syntax check the CPN diagram (and other operations to start/stop the ML compiler, see section 5.3). A large set of operations to change attributes and options – and their defaults. Operations which assist the user to select the correct object (when many are close to each other or on top of each other), move objects to another position (on the same page or on another page), change object size (e.g. to fit the size of the text in the object), change object shape (e.g. from ellipse to box),[72] merge a group of nodes into a

---

68 For efficiency reasons the popup region can also be missing. In this case a double click on the key implies that the popup is generated (with the correct information) and becomes visible.

69 It is also the intention to allow the user to save part of a diagram and later load it into another diagram. In this way it will be possible to create libraries of reusable submodels. This facility is, however, not yet implemented.

70 For the moment, Undo only works for a limited set of operations.

71 A picture is a bit map which is obtained from a CPN diagram (by copying part of a page) or from another program (via the clipboard). Pictures makes it easy to work with icons.

72 All objects can take many different shapes. Nodes and regions can e.g. be boxes, rounded boxes, ellipses, polygons, wedges and pictures. Connectors can be single headed, double headed and without heads. As an example, of a creative use of this generality, it is possible to let a substitution transition be a picture which is a diminished version of the corresponding subpage.

single node, duplicate a node[73], hide and show regions and change the graphical layering of the objects. Operations to redraw the page hierarchy – when this has become too cluttered (e.g. because the user has made a number of manual changes to the automatic layout proposed by the CPN editor). Operations to select groups (e.g. by means of fusion sets, text searches and object types).[74] Operations to search for specified text strings and replace them by others (either in the entire diagram, on a single page, or in one or more selected objects). Operations to search for matching brackets, create hyper text structures,[75] and copy the contents of external text files into nodes (and vice versa). A large number of alignment operations. Some of these make it easy to position nodes and regions relative to each other (e.g. vertically below each other, with equal distances, on a circle, with the same center, etc.). Others make it easy to create arcs with right angles and vertical/horizontal segments.

The CPN editor can be used at many different skill levels. Casual and novice users only have to learn and apply a rather small subset of the total facilities. The more frequent and experienced users gradually learn how to use the editor more efficiently: All the more commonly used commands can be invoked by means of key shortcuts, and these can be changed by the users. Many commands have one or more modifier keys, allowing the user, in one operation, to do things which otherwise would require several operations. The user can create a set of templates (e.g. a set of nodes with different attributes and object types). These nodes can then be positioned on special palette pages, from where they, in one operation, can be copied to the different pages of a diagram. In this way it is easy to make company standards for the graphics of CPN diagrams.

To make it easier to use the CPN editor we have tried to make the user interface as consistent and self-explanatory as possible. To do this, we have defined a set of concepts allowing us to give a precise description of the different parts of the interface: As an example, a list box with a scroll bar can behave in many slightly different ways: It may be possible to select only a single line at a time, a contiguous set of lines, an arbitrary set of lines, or no lines at all – and when the dialogue box is opened, the list box may have the same selection as last time, have the first line selected, have no lines selected, or have a selection which depends upon the current selection in the diagram. Hopefully, this simple example demonstrates that it is important to identify the possibilities – and use them in a consistent way.

When the user creates a CPN diagram, the editor stores all the semantic information in an abstract data base – from which it easily can be retrieved by the CPN simulator (and other analysis programs). The abstract data base was designed as a relational data base but for efficiency implemented by means of a set of list structures (making the most commonly used data base operations as efficient as possible). The existence of the abstract data base makes it much easier to integrate new/existing editors and analysis programs with the CPN tools – and for this purpose there are three sets of predeclared functions: The first set makes it possible to read the information which is present in the abstract data base (e.g. get information about the colour set of place). The second set

---

73 The new node get a set of regions and connectors which are similar to the original node. By using the command on a group of nodes, it is possible to get a subnet which is identical to an existing subnet (and has the same connectors to/from the environment).

74 Some of the group selection facilities are not yet implemented.

75 This facility is not yet fully implemented.

makes it possible to create auxiliary objects (which have a graphical representation but no representation in the abstract data base). Finally, the third set makes it possible to convert auxiliary objects to CPN objects (which means that the abstract data base is updated accordingly). Using these three sets of predeclared functions it is a relatively straightforward task to write programs which translates textual/graphical representations of a class of Petri nets (or another formalism with a well-defined semantics) into CPN diagrams – and vice versa.

Finally, it should be mentioned that the CPN editor is designed to work with large CPN diagrams – i.e. diagrams which typically have 50-100 pages, each with 5-25 nodes (and 10-50 connectors plus 10-200 regions).

5.3 Inscription language for CP-nets

When the user creates a CPN diagram he simultaneously creates a *drawing* and a *formal model*. The behaviour of the formal model is determined by the objects, their object types, the relationships between the objects[76], and the text strings inside the objects. Obviously these text strings need to have a well-defined syntax and semantics, and this is achieved by using a programming language called Standard ML (SML). It is by means of this language we declare colour sets, functions, operations and specify arc expressions and guards. SML has been developed by a group at Edinburgh University and it is one of the most well-known functional languages. For details about SML and functional languages, see [26], [27], [50] and [63].

By choosing an existing programming language we obtained a number of advantages. First of all we got a much better, more general and better tested language than we could have hoped to develop ourselves.[77] Secondly, we only had to port the compiler to the relevant machines and integrate it with our editor (instead of developing it from scratch).[78] Thirdly, we can use the considerable amount of documentation and tutorial material which already exists for SML (and for functional languages in general).

Why did we choose SML? First of all, we need a functional language: Arc expressions and guards are not allowed to have side effects and when a CP-net is translated into matrix form (e.g. for invariant analysis) the arc expressions and guards are, via lambda expressions, translated into functions. Secondly, we need a strongly typed language: Because CP-nets use colour sets in a way which is analogous to types in programming languages. Thirdly, we need a language with a flexible and extendible syntax: This makes it possible to allow the user to write arc expressions and guards in a form which is very close to standard mathematics (as an example, multi-set plus is de-

76 There are many different kinds of relationships – e.g. the relationship between connectors and their source/destination nodes, between nodes and their regions, and between substitution transitions and their subpages.

77 The development of a new programming language is a very slow and expensive process that requires resources comparable with the entire CPN tool project.

78 The CPN tools use two different SML compilers. On the Macintosh we use the original compiler developed at Edinburgh University. On the Unix machines we use a more modern compiler developed by AT&T. It is also possible to run the graphics on one machine and the SML compiler on another (connected to the first by a local area network).

noted by "+").[79] SML is only one out of a number of languages which fulfil the three requirements above. SML was chosen because it was one of the best known, it had commercially available compilers, and some of us already had good experiences with the language.

We have many times been amazed by the high quality of SML, the generality of the language, and the ease by which complex programs can be written.[80] Thus we consider the choice of SML as one of the most successful design decisions in the CPN tool project. This choice has given us a very powerful and general inscription language and it has saved a lot of implementation time. As we shall see in section 5.5, the use of SML also makes it easy to make a smooth integration between the net inscriptions of a CP-net and code segments (which are sequential pieces of code attached to the individual transitions and executed each time a binding of the transition occurs).

To make it easier for the user we have made three small extensions of SML – and this yields a language called CPN ML: As the first extension, syntactical sugar has been added for the declaration of colour sets. This makes it easy to declare the most common kinds of colour sets, and it also means that a large number of predeclared functions and operations can be made accessible, just by including their names in the colour set declaration.[81] As examples, each enumeration type has a function mapping colours into ordinal numbers, each product type has a function mapping a set of multi-sets into their product multi-set, and each union type has a set of functions performing membership tests. SML allows the user to declare integers, reals, strings, enumerations, products, records, discrete unions and lists – and nest the type constructors arbitrarily inside each other. As an example we may declare the following colour sets (which should be rather self-explanatory):

```
color Name = string;
color NameList = list Name;

color Year = int;
color Month = with  Jan | Feb | Mar | Apr | May | Jun |
                    Jul | Aug | Sep | Okt | Nov | Dec;
color Day = int with 1..31;
color Date = product Year * Month * Day;

color Person = record name : Name * BirthDay : Date * Children : NameList;
```

79 The "+" operator is infixed (i.e. written between the two arguments). It is polymorphic (i.e. it works for multi-sets over all different types) and it is overloaded (i.e. it uses the same operator symbol as integer plus and real plus).

80 Much of the more intrinsic code of the CPN simulator is written in SML. In particular, all the code that calculates the set of enabled bindings. This code is rather complex: It defines a function which maps an arbitrary set of arc expressions (plus a guard) into a function mapping a set a multi-sets into a set of enabled bindings.

81 This convention saves a lot of space in the ML heap, because it turns out that most CPN diagrams only use few of the predeclared functions. A later version of the CPN editor will automatically detect the predeclared functions applied by the user (and then it will no longer be necessary to list their names).

Via the syntactic sugar, it is in CPN ML easy to declare colour sets from all the SML types mentioned above (and from subranges, substypes, and indexed types, which do not exist as standard SML types). In SML it is also possible to declare function types and abstract data types. However, such types do not have an equality operation and thus it does not immediately make sense to use them as colour sets (because you cannot talk about multi-sets without being able to talk about equality).[82]

As the second extension, we have added syntax which allows the user to declare the CPN variables – i.e. the typed variables used in arc expressions and guards. This extension is necessary because SML do not have variable declarations (in SML a value is bound to a name and this determines the current type of the name; later the name may get a new value and a new type).

As the third extension, we have added syntax which allows the user to declare three different kinds of reference variables. This is a non-functional part of SML and we only allow reference variables to be used in code segments. We distinguish between global, page and instance reference variables – in the same way as we distinguish between global, page and instance fusion sets: A global reference variable can be used by all code segments in the entire CPN diagram, while a page and instance reference variable only can be used by the code segments on a single page. A page reference variable is shared by all instances of the page, while an instance reference variable has a separate value for each page instance.

SML (and thus CPN ML) can be viewed as being a syntactical sugared version of typed lambda calculus, and this means that it is possible to declare arbitrary mathematical functions (as long as they are computable). It should be noticed that the use of SML gives an immense generality: The user can declare arbitrarily complex functions[83] and, if he wants, he can turn them into operations (i.e. use infix notation). This generality has been heavily used in the implementation of the CPN tools. Multi-sets are implemented as a polymorphic type constructor "ms" which maps an arbitrary type *A* into a new type, denoted by *A ms* and containing all multi-sets over *A*. Then we have declared a large number of polymorphic and sometimes overloaded operations/functions – by which multi-sets can be manipulated (e.g. operations to add and subtract multi-sets and functions to calculate the coefficients and the size of multi-sets).

The generality of the CPN ML language means that some legal CPN diagrams cannot be handled by the CPN simulator. As an illustration consider the transition in Fig. 10, where x is a CPN variable of type X, while $f \in [X \rightarrow A]$ and $g \in [X \rightarrow B]$ are two functions. To calculate the set of all enabled bindings for such a transition it is either necessary to try all possible values of X or use the inverse relations of f and g (and neither is possible, in general – because X may have two may values and the inverse relations may be unknown).

---

[82] The user can, with some extra work, use an arbitrary ML type as a colour set – as long as the standard equality operator "=" exists (and the type is non-polymorphic). In this way it is possible to declare abstract data types and turn them into colour sets. Details are outside the scope of this paper.

[83] Many CPN diagrams use recursive functions defined on list structures.

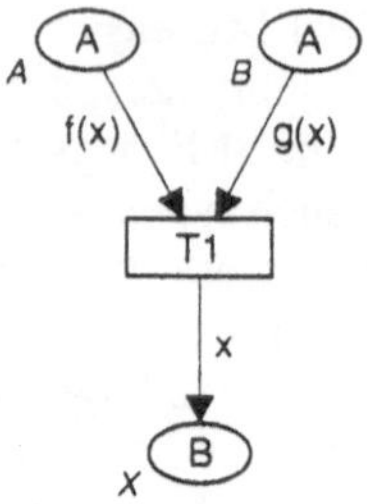

Figure 10. A syntactically legal CPN transition which cannot be handled by the CPN simulator

To avoid such problems the CPN simulator demands that each CPN variable, used around a transition, must appear either in an input arc expression without functions or operations[84] (then the possible values can be determined from the marking of the corresponding input place), be determinable from the guard, have a small colour set (in which case all possibilities can be tried),[85] or only appear on output arcs (in which case all possible values can be used). It is very seldom that these restrictions present any practical problems. Most net inscriptions, written by a typical user, fulfil the restrictions – and those which do not, can usually be rewritten by the user, without changing the semantics. As an example, consider the three transitions in Fig. 11. None of these can be directly handled by the CPN simulator. The first transition is identical to the transition in Fig. 10. The second transition has a guard which is a list of boolean expressions, and this means that each of the expressions must be fulfilled. The third transition uses the function exp(x,y) which takes two non-negative integers as arguments and returns x^y.

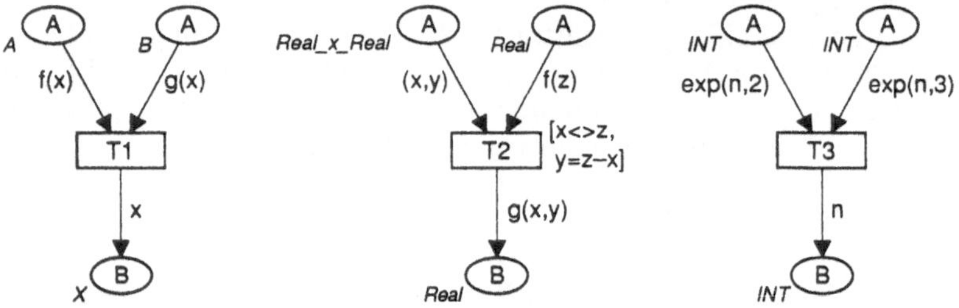

Figure 11. Three transitions which cannot be handled by the CPN simulator

Now let us assume that f has an inverse function f1 ∈ [A→X]. Then we can, as shown in Fig. 12, rewrite the three transitions – so that their semantics is unchanged and they can be handled by the CPN simulator. In the first transition z is a variable of type A. In the second transition z can now be determined from the guard – because there is an equality in which z appears on one side (alone or in a matchable pattern) while the value of

84 The arc expression is allowed to contain matchable operations such as the tuple constructor (,,) in (x,y,z), the list constructor :: in head::tail, and the record constructor {,,} in {se=S(inst()), re=r, no=n}. It is also allowed to contain multi-set "+" and "`".

85 Intuitively a small colour set is a type with few values. A precise definition can be found in [36].

the other side is known (x and y are bound by one of the input arc expressions). In the third transition the function sq(x) takes a non-negative integer as argument and it returns the integer which is closest to √x.

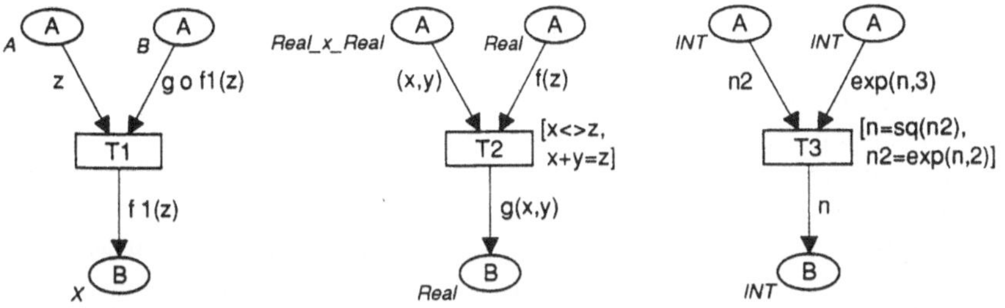

Figure 12. Three transitions that are behaviourally equivalent to those in Fig. 11 and which can be handled by the CPN simulator

It is important to understand that the general definition of CP-nets talks about expressions and colour sets – without specifying a syntax for these. It is only when we want to implement a CPN editor and a CPN simulator (and other kinds of CPN tools) that we need a concrete syntax. Thus it is for the CPN tools, and not for CP-nets in general, that CPN ML has been developed. Other implementations of CP-nets may use different inscription languages – and still they deal with CP-nets.

5.4 Syntax check

The CPN editor is syntax directed – in the sense that it recognizes the structure of CP-nets and prevents the user to make many kinds of syntax errors. This is done by means of a large number of **built-in** syntax restrictions. All the built-in restrictions deal with the net structure and the hierarchical relationships. As examples, it is impossible to make an arc between two transitions (or between two places), to give a place two colour set regions (or give a transition a colour set region), to create cycles in the substitution hierarchy, and to make an illegal port assignment (involving nodes which aren't sockets/ports or are positioned on a wrong page).

The CPN editor also operates with **compulsory** syntax restrictions. These restrictions are necessary in order to guarantee that the CPN diagram has a well-defined semantics – and thus they must be fulfilled before a simulation (and other kinds of behavioural analysis) is performed. Many of the compulsory restrictions deal with the net inscriptions and thus with CPN ML. As examples, it is checked that each colour set region contains the name of a declared colour set *A* (and that all surrounding arc expressions have a type which is identical to either *A* or *A ms)*, that all members of a fusion set have the same colour set and equivalent initialization expressions, and that all identifiers in arc expressions and guards are declared (e.g. as CPN variables or functions). Many of the compulsory syntax restrictions could have been implemented as built-in restrictions. This would, however, have put severe limits on the way in which a user can construct an edit a CPN diagram. As examples, we could have demanded that each place always has a colour set (and this would mean that the colour set has to be specified at the moment the place is created) and we could have demanded that each arc expression always is of the correct type (and this would mean that a colour set cannot be changed without simultaneously changing all the surrounding arc expressions).

Finally, the CPN editor operates with **optional** syntax restrictions.[86] These are restrictions which the user imposes upon himself – e.g. because he knows that he usually does not use certain facilities of the editor and wants to be warned when he does (in order to check whether this was on purpose or due to an error). As examples, it can be checked whether port assignments are injective, surjective and total, whether all arcs have an explicit arc expression (otherwise they by default evaluate to the empty multi-set) and whether the place/transition names are unique (on each page).

All the type checking is done by the SML compiler and it is the error messages of this compiler which is presented to the user (together with a short heading produced by the CPN editor). The fact that these messages are easy to understand and uses CP-net terminology tells a lot about the generality and quality of SML. To illustrate this, let us imagine that we, in Fig. 1, change the arc expression between A and T1 from (x,i) to x. This will result in an error message which looks as follows:[87]

```
C.11 Arc Expression must be legal
Type clash in: x:((P)ms)
Looking for a: P ms
I have found a: U
««135»»
```

To speed up the syntax check we avoid duplicate tests: As an example, the same arc expression may appear at several arcs and it is then only checked once (provided that the places have identical colour sets). We also apply incremental tests: When the user changes part of a CPN diagram as little as possible is rechecked. Changing an arc expression or a guard means that the use of variables in the code segment must be rechecked. Changing a colour set means that the initialization expression and all surrounding arc expressions have to be rechecked.[88] Changing the global declaration node (which contains the declarations of colour sets, functions, operations, and CPN variables), unfortunately means that the entire CPN diagram has to be rechecked. To avoid using too much time for such total rechecks, the CPN editor allows the user to add a temporal declaration node which extends the declarations of the global declaration node.[89]

The CPN editor allows the user to give each page, transition and place a name (i.e. a text string) and a number (which must be non-negative).[90] It should, however, be un-

---

86 Optional syntax checks are not implemented in the current version of the CPN editor.

87 This is how the error message looks when the SML compiler runs on a Macintosh (on a Unix system another SML compiler is used, and thus the error messages looks a little bit different). C11 means that it is the 11th kind of compulsory restriction, while ««135»» is a hyper text pointer which allows the user to jump to the error position (i.e. to the arc with the erroneous arc expression).

88 If the place belongs to a fusion set or is an assigned port/socket it also has to be checked whether the restrictions in Def. 3.2 (v) and (vi) still are satisfied.

89 It is also possible, but not recommended, to use the temporal declaration node to overwrite existing declarations.

90 In the current version it is not possible to give transitions and places a number.

derstood that these names have no semantic meaning.[91] Names are used in the feedback information from the editor to the user (e.g. in the page hierarchy and in the hierarchy inscriptions). To make this information unambiguous it is recommended to keep names unique,[92] but this is not enforced (unless the user activates an optional syntax restriction). Many users have a large number of transitions and places with an empty name (and this is no problem, as long as these nodes are not used in a way which generates system feedback).

The possibility of performing an automatic syntax check means that the user has a much better chance of getting a consistent and error-free CPN diagram. This is very useful – also in situations where the user isn't interested in making a simulation (or other kinds of machine assisted behavioural analysis).

5.5 CPN simulator

The CPN editor and CPN simulator are two different parts of the same program and they are closely integrated with each other: In the editor it is possible to prepare a simulation (e.g. change the many options which determine how the simulation is performed). In the simulator it is possible to perform simple editing operations (those which change the attributes of objects without changing the semantics of the model).[93]

The CPN simulator is able to work with large CP-nets, i.e. CP-nets with 50-500 page instances, each with 5-25 nodes. Fortunately, it turns out that a CP-net with 100 page instances, typically, simulates nearly as fast as a CP-net with only a single page instance (measured in terms of the number of occurring transitions). This surprising result is due to the fact that the CPN simulator, during the execution of a step, goes through three different phases: First it makes a random selection between enabled transitions, then it removes and adds tokens at the input/output places of the occurring transitions, and finally it calculates the new enabling. The first of these phases is fast (compared to the others), the second is independent of the model size and the third only depends upon the model size to a very limited degree. This is due to the fact that the enabling and occurrence rule of CP-nets are strictly local – and this means that it only is the transitions in the immediate neighbourhood of the occurring transitions that need to have their enabling recalculated.[94] Without a local rule the calculation of the new enabling would grow linearly with the model size and that would make it very cumbersome to deal with large systems. We have not yet tried to work with very large systems (e.g. containing 10.000 page instances) but our present experiences tell us that the upper limit is more likely to be set by the available memory than by the processor speed.

The user must be able to follow the on-going simulation – and it is obvious that no screen (or set of screens) will be able simultaneously to display all page instances of a large model. Like the editor, the CPN simulator uses a window for each page and in

91 In the current version of the CPN editor the names of fusion sets play a semantic role, and thus they have to be unambiguous. This will be changed in a later version.

92 For places and transitions it is sufficient to demand the names to be unique on each individual page.

93 In one of the next versions of the CPN simulator we will also allow the user to make changes that modify the behaviour of the model – as long as these changes cannot make the current marking illegal.

94 When the neighbourhood of an occurring transition is defined, fusion sets and port/socket assignments must be taken into consideration.

this window the simulator displays the marking of one of the corresponding page instances. The user can see the names of the other page instances and switch to any of these. When a transition occurs the simulator automatically opens the corresponding page window (if necessary), brings it on top of all other windows, switches to the correct page instance, and scrolls the window so that the transition becomes visible. The user can, however, tell that he doesn't want to observe all page instances. In that case the simulator still executes the transitions of the non-observed page instances but this cannot be seen by the user (unless the relevant part of corresponding page instance happens to be visible on the screen without any rearrangements). The user can set breakpoints and in this way ask the simulator to pause before, during, and/or after each simulation step. Breakpoints can be preset or added on the fly, i.e. at any point during a simulation. At each breakpoint the user can investigate the system state (and decide whether he wants to continue or cancel the remaining part of the step).

It is possible to simulate a selected part of a large CPN diagram (without having to copy this part to a separate file, which would give all the usual inconsistency problems). This is achieved by allowing the user to change the multi-set of prime pages and tell that certain page instances should be temporarily ignored. When a page instance is ignored it is no longer generated, and this means that the corresponding direct supernode becomes an ordinary transition with enabling and occurrence calculated in the usual way (i.e. by means of the surrounding arc expressions and guard). As a short hand, it is also possible to ignore a page and this means that all instances of the page are ignored.

When we simulate a CP-net it is sometimes convenient to be able to equip some of the transitions with a **code segment** – i.e. a sequential piece of code which is executed each time a binding of the transition occurs. Each code segment has a code guard, an input pattern, an output pattern and a code action. The code guard replaces the corresponding guard (in a simulation with code segments). A missing code guard means that the ordinary guard is used. The input pattern contains some of the CPN variables of the transition, and this indicates that the code action is allowed to use (but not update) these variables. Analogously, the output pattern contains some of the CPN variables (but only those which do not appear in the input arc expressions and the code guard) and this indicates that the binding of these variables is determined by the code segment. Finally, the action part is an SML expression (with the same type as the output pattern).[95] The action part may declare local variables, share reference variables with other code segments, use the CPN variables from the input pattern and manipulate input/output files. When the transition occurs the action part is evaluated and the resulting value determines the binding of the CPN variables in the output pattern. It should be noticed that the code segment is executed once for each occurring binding, and this means that it may be executed several times in the same step.[96]

---

95 In a later version of the CPN simulator it will also be possible to use other programming languages in the code action, e.g. C++, Pascal and Prolog.

96 The order of these executions is non-deterministic (but it is guaranteed that each execution is indivisible, in the sense that it is finished before the next is started).

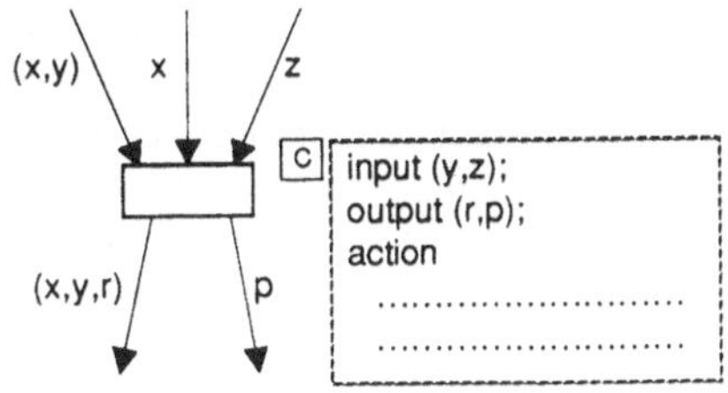

Figure 13. A simple example of a code segment

Code segments can be used for many different purposes: They can be used to gather statistical information about the simulation: It is easy to dump the value of all occurring bindings on a file (which then later can be analysed e.g. by means of a spread sheet program). It is also possible to use the graphic routines of the CPN tools (which can be invoked via predeclared SML functions) and in this way make a visual representation of the simulation results (as an example, it is easy to make a window which has an node for each site in a communication network and a connector for each pair of sites which are engaged in a communication).[97] Code segments also allow interactive user input, and they can be used to communicate with other programs (as an example, it is possible to run different parts of a very large CPN model on separate computers and let the different submodels communicate via input/output statements).

Although code segments are extremely useful for many purposes, they also have severe limitations. This is due to the fact that they allow the occurrence of transitions to have side effects and allow bindings to be determined by input files (and other kinds of user input). This means that it doesn't make sense to talk about occurrence graphs for CP-nets with code segments, and it also becomes more difficult to use the invariant method for such nets (because the relation between the CPN variables surrounding a transition may be determined by the code action, instead of the arc expressions). For this reason it is important to have a well-understood relationship between a CPN diagram executed with code segments and the same CPN diagram executed without code segments. This is one of the main reasons for the introduction of the input and output patterns.

It is possible to perform both **manual** and **automatic** simulations. In a manual simulation the simulator calculates and displays the enabling, the user chooses the occurrence elements (i.e. the transitions and bindings) to be executed and finally the simulator calculates the effect of the chosen step. During the construction of a step, the simulator assists the user in many different ways: First of all, the simulator always shows the current enabling (and updates it each time a new occurrence element is added/removed at the step). Secondly, the user can ask the simulator to find all bindings for a given enabled transition – or he can specify a partial binding and ask the simulator to finish it, if possible. In an automatic simulation the simulator chooses among the enabled occurrence elements by means of a random number generator. It is possible to specify how large each step should be: It may contain a single occurrence element or as

[97] We do not allow code segments to create or delete CPN objects (but the attributes can be changed).

many as possible (and between these two extremes there is a continuum of other possibilities).

It is possible to vary the amount of graphical feedback provided by the CPN simulator. In the most detailed mode the user sees the enabled transitions, the occurring transitions, the tokens which are being moved and the current markings. Each of these feedback mechanisms is, however, controlled by one or more options, and thus they can be fully or partially omitted. In this way it is possible to speed up the simulation. As an extreme a special **super-automatic** mode has been provided. In this mode there is no user interaction (for the selection of bindings) and there is no feedback during the simulation (on the CPN diagram) – and this means that the simulation runs much faster than usual, because the simulation is performed by a SML program alone (while an ordinary simulation is performed by a SML program and a C program, with a heavy intercommunication).[98] At the end of a super-automatic simulation it is possible to inspect the effect of the simulation. This can be done either by means of the usual page windows (in which the marking is updated when the super-automatic simulation finishes) or by means of files manipulated by code segments. Finally, the code segments may, as described above, use the graphic routines of the CPN tools to create a visual representation of the simulation results – and this can be inspected while a super-automatic simulation is going on.

The user can, at any time during a simulation, change between manual, automatic and super-automatic simulation (and there are many other possibilities in between these three extremes).[99] It is usual to apply the more manual simulation modes early in a project (e.g. when a design is being created and investigated) while the more automatic modes are used in the later phases (e.g. when the design is being validated). There are no restrictions on the way in which the different simulation options can be mixed and this means that each of them can be chosen totally independently of the others (as an example manual/automatic/semi-automatic simulation can be with/without code (and with/without time, see section 7.1)).

There are many other facilities in the CPN simulator: An operation that proposes a step (which can be inspected and modified by the user before it is executed). Operations to return to the initial marking of the CPN diagram and to change the current marking of an arbitrary place instance (this means that it often is possible to continue a simulation in the case where a minor modelling error is encountered). Operations to save and load system states. Operations to activate/deactivate a large number of warning and stop options (i.e. different criteria under which a manual simulation issues a warning while an automatic simulation stops).[100] An operation to determine the order in which the different occurrence elements in a step is executed. Moreover, the earlier comments about different skill levels and a consistent and self-explanatory user interface also apply to the CPN simulator.

98 Super-automatic simulation is not available in the released version of the CPN simulator, but a prototype version has been used in several projects (e.g. the one described in section 6.1). One of the next versions of the CPN simulator will contain a super-automatic mode which is fully integrated with the rest of the simulator.

99 In the current version of the CPN simulator it is, during a simulation, not possible to change to/from super-automatic mode. This will, however, be possible in one of the next versions.

100 The load/save operations and the warning/stop options are not yet implemented.

Finally, it should be mentioned that many modellers use simulation during the construction of CPN diagrams – in a similar way as a programmer tests selected parts of the program which he is writing. It is thus very important that it is reasonably fast to shift between the editor and the simulator (and that it is possible to simulate selected parts of a large model).

6. Applications of CP-nets

This chapter describes a number of projects which have used hierarchical CP-nets and the CPN tools. All the described projects have worked with reasonably large models and this have been done in an industrial environment – where parameters such as turn-around time and use of man-hours have been important.

6.1 Communication protocol

This project was carried out in cooperation with a large telecommunications company and it involved the modelling and simulation of selected parts of an existing ISDN protocol for digital telephone exchanges.[101] The modelling started from an SDL diagram, and it was straightforward to make a manual translation of the SDL diagram to a hierarchical CP-net.[102] The translation and simulation of the basic part of the protocol was finished in 16 days by a single modeller (which had large experience with the CPN tools, but no prior knowledge of communication protocols). The model was presented to engineers at the participating company. This was done by making a manual simulation of selected occurrence sequences – and by a super-automatic simulation, where code segments were used to update a page containing a visual representation of the travelling messages and the status of the user sites.[103] According to the engineers, who all had large experience with telephone systems, the CPN diagram provided the most detailed behavioural model which they had seen for this kind of system.

Later the modelling of a hold-feature was included in the CPN diagram. This was done in a single day, and it tuned out that it could be done by adding two extra pages, and making a simple modification of the existing pages (a colour set was changed from a triple to quadruple). In SDL the inclusion of the hold-feature made it necessary to duplicate the entire model, i.e. include many new pages – and so did three other features (which were not modelled in the project, but could have been handled in a similar way). Obviously, this makes it easier to maintain the CPN diagram (because it is sufficient to make modifications to one page instead of five).

---

[101] ISDN stands for Integrated Services Digital Network. The protocol is a BRI protocol (Basic Rate Interface) and it is the network layer which has been modelled.

[102] SDL is one of the standard graphical specification languages used by telecommunications companies. For information about SDL and how it can be translated into high-level Petri nets, see [13], [41] and [53].

[103] The simulation traces a call from the originating user to the terminating user, and to do this it was necessary to include a page which models the underlying protocol layers.

Fig. 14 shows the page hierarchy for the CPN diagram.[104] The subpages of UserTop#2 describe the actions of the user part while the subpages of NetTop#19 describe the actions of the network part. Most of these pages have a supernode which is called Ui (or Ni) and this indicates that the page describes the activity which can happen when the user part is in state Ui (the network part is in state Ni). The bracket in front of the pages U_PROG#41...U_REL_CO#40 indicates that they are subpages of all the pages in Null#3...Release_#17. The five pages describe activities which are carried out in the same way in all user states. If one of these activities is to be changed it is sufficient to modify one page of the CPN diagram (while it in the SDL diagram would be necessary to modify a large number of pages). The hold-feature is modelled by U_HOLD#45 and N_HOLD#44, while ROUTING#24 models the underlying protocol layers.

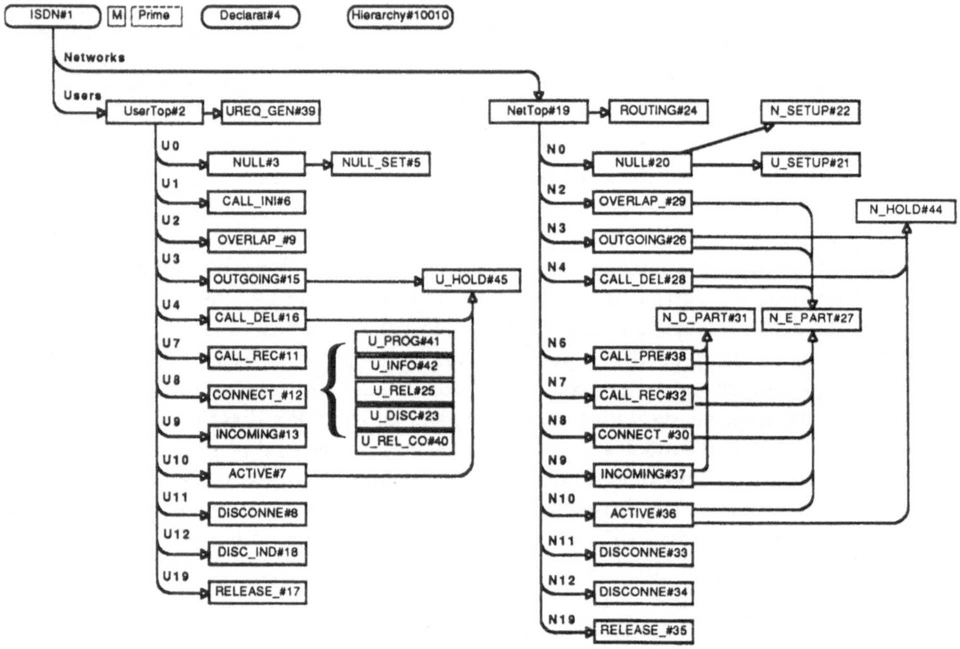

Figure 14. Page hierarchy for the ISDN protocol

A typical representative of the Ui/Ni pages is shown in Fig. 15.[105] It shows that, in the state U8, there are six different possibilities. When there is an internal user request the

---

104 Names are truncated to the first eight characters, unless one of these is a format character (such as space, TAB, RETURN, etc). This convention keeps the feedback readable (also in diagrams with very long text strings). One of the next versions of the CPN tools will have a set of name options allowing the user to specify how names are truncated.

105 The vertical lines and triangular figures inside the transitions are carried over from the SDL diagram, where they have a formal meaning. In the CPN diagram they have no formal meaning but they are retained, because they make the diagram more accessible for people who have experience with SDL.

first transition can occur. It creates a message to the network and the new user state becomes U11. When there is a message from the network one of the last five transitions can occur (the guards determine which one).[106] Two of the transitions create a message to the network, and the new user state becomes either U10, U12, U0, or U8. Three of the transitions are drawn with thick borders, this indicates that they are substitution transitions (having the pages U_DISC#23, U_REL#25 and U_REL_CO#40 as subpages). It should also be noticed that a global fusion set is used to glue all the U0-places together (and analogously for all the other 23 kinds of Ui/Ni-places).[107]

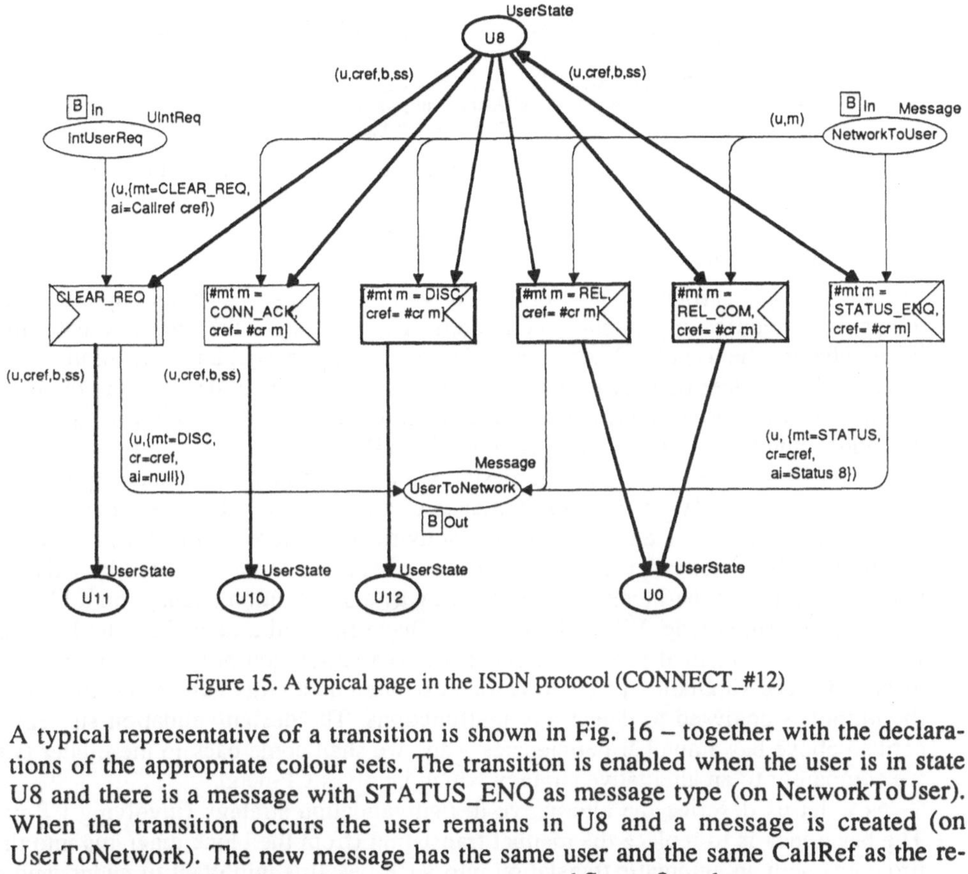

Figure 15. A typical page in the ISDN protocol (CONNECT_#12)

A typical representative of a transition is shown in Fig. 16 – together with the declarations of the appropriate colour sets. The transition is enabled when the user is in state U8 and there is a message with STATUS_ENQ as message type (on NetworkToUser). When the transition occurs the user remains in U8 and a message is created (on UserToNetwork). The new message has the same user and the same CallRef as the received message, it has STATUS as message type and Status 8 as data.

[106] To improve the readability the modeller has made some of the arc expressions invisible. All output arcs of NetworkToUser have identical arc expressions – and only one of these is visible.

[107] To improve the readability the modeller has made all the fusion regions invisible.

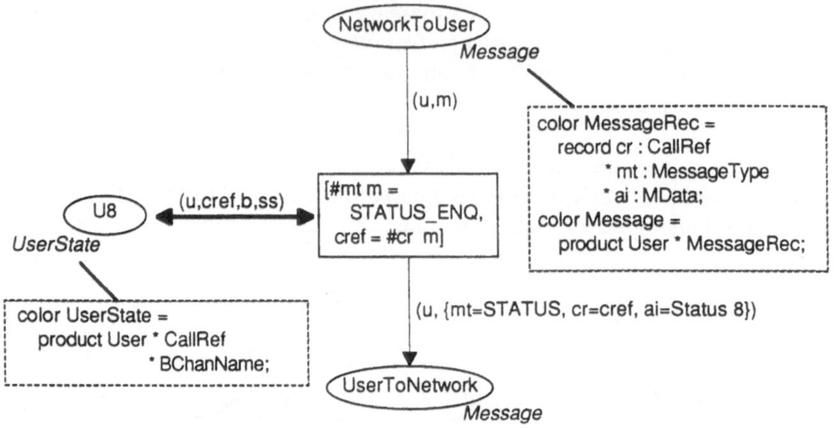

Figure 16. A typical transition in the ISDN protocol (rightmost on CONNECT_#12)

6.2 Hardware chip

This project was carried out in cooperation with a company which, among many other things, is a manufacturer of super-computers. The purpose of the project was to investigate whether the use of CP-nets is able to speed up the design and validation of VLSI chips (at the register transfer level). Below we sketch the main ideas behind the project and the most important conclusions. A much more detailed description of the project, the model, and the conclusions can be found in [54].

Let us first describe the existing design/validation strategy (without CP-nets): The chip designers specify a new chip by means of a set of block diagrams. Each diagram contains an interconnected set of blocks (activities), where each block has a specified input/output behaviour. A complex block may be specified in a separate block diagram, which is related to the block in a similar way as a substitution subpage in a CP-net is related to its supernode. When the designers have finished a new chip, the block diagrams are (by a manual process) translated into a simulation program written in a dialect of C. The simulation program is then executed on a large number of test data and the output is analysed to detect any malfunctions. The design/validation strategy described above has a number deficiencies – and we shall come back to these later (when we compare it to an alternative strategy which involves CP-nets).

Now let us describe the alternative design/validation strategy (involving CP-nets): The basic idea is to replace the manual translation (from the block diagrams into the C program) with an automatic translation into a CP-net. It is important to understand that it is *not* the intention to stop using block diagrams. The designers will still specify the designs by means of block diagrams, and they will during a simulation of the CP-net see the simulation results on the block diagrams. To support the new strategy three things are needed: The existing drawing tool for the block diagrams must be modified (to have a formal syntax and semantics). The set of block diagrams must be translated into CP-nets. Finally, it must be investigated whether the CPN simulator is powerful enough to handle complex VLSI designs.

The project only dealt with the last two issues (which were considered to be the most difficult). It was shown that the block diagrams could be translated into hierarchical CP-nets. This was done manually, but the translation process is rather straightforward and we see no problems in implementing an automatic translation. The obtained CP-net only contained 15 pages, but during a simulation there is nearly 150 page instances (due to the repeated use of substitution subpages representing adders and multipliers). The CP-net was simulated on the CPN simulator.[108]

Fig. 17 shows a subpage from which it can be seen that the VLSI chip has a pipelined design with six different stages. Each stage is modelled on a separate subpage and two of the more complex stages are shown in Fig. 18.[109] The eight transitions in the leftmost part of stage 1 are all substitution transitions (and they have the same subpage). In stage 2 the four transitions SUM1L, SUM1R, SUM2L and SUM2R represent registers. These registers establish the border to stage 3, and the transitions can only occur when they receive a clock pulse from stage 3 (via the two c-transitions in the rightmost part of stage 2). All the remaining transitions in stage 2 are substitution transitions (OR3 and OR4 denote or-gates while "+" denotes 16 bit adders).

Now let us compare the new design/validation strategy with the old: First of all, it is easier to translate the block diagrams into a CP-net than it is to translate them into a C program (the latter takes often several man-months while the construction of the CP-net only took a few man-weeks). The translation is also more transparent – in the sense that it is much easier to recognize those parts of the CP-net which models a given block than it is to find the corresponding parts in the C program (each page in the CP-net has nearly the same graphical layout as the corresponding block diagram). This means that it is relatively easy to change the CP-net to reflect any changes in the design, while this (according to the chip manufacturer) often is rather difficult for the C program. As stated above we think that it will be easy to automate the translation.

Secondly, the new strategy (when it is fully implemented) allows the designer to make simulations during the design process. This means that the knowledge and understanding which is acquired during the simulation of the model can be used to improve the design itself (in a much more direct way than in the old strategy where the validation is performed after the design has been finished).

Thirdly, the validation techniques of the old strategy concentrates on the logic correctness (tested by an inspection of the output data from the C program) and very little concern is given to those design decisions which deal with timing issues (e.g. the division into stages and the clock rate).[110] Using CP-nets it is possible to validate both the logic correctness and the timing issues – inside the same basic model.[111]

---

108 When maximal graphical feedback was used the simulation was slow (due to the many graphical objects which had to be updated in each step). However, when a more selective feedback was used, the speed became reasonable.

109 It is our intention to give the reader an idea about the complexity of the model (without explaining it in any detail).

110 This is surprising, because the timing issues are crucial for the correct behaviour and the effectiveness of the chip (too fast clocking means malfunctioning while too slow means loss of speed).

111 The timing issues were not modelled in the project – but with the time extensions of the CPN simulator (described in section 7.1) this can easily be done.

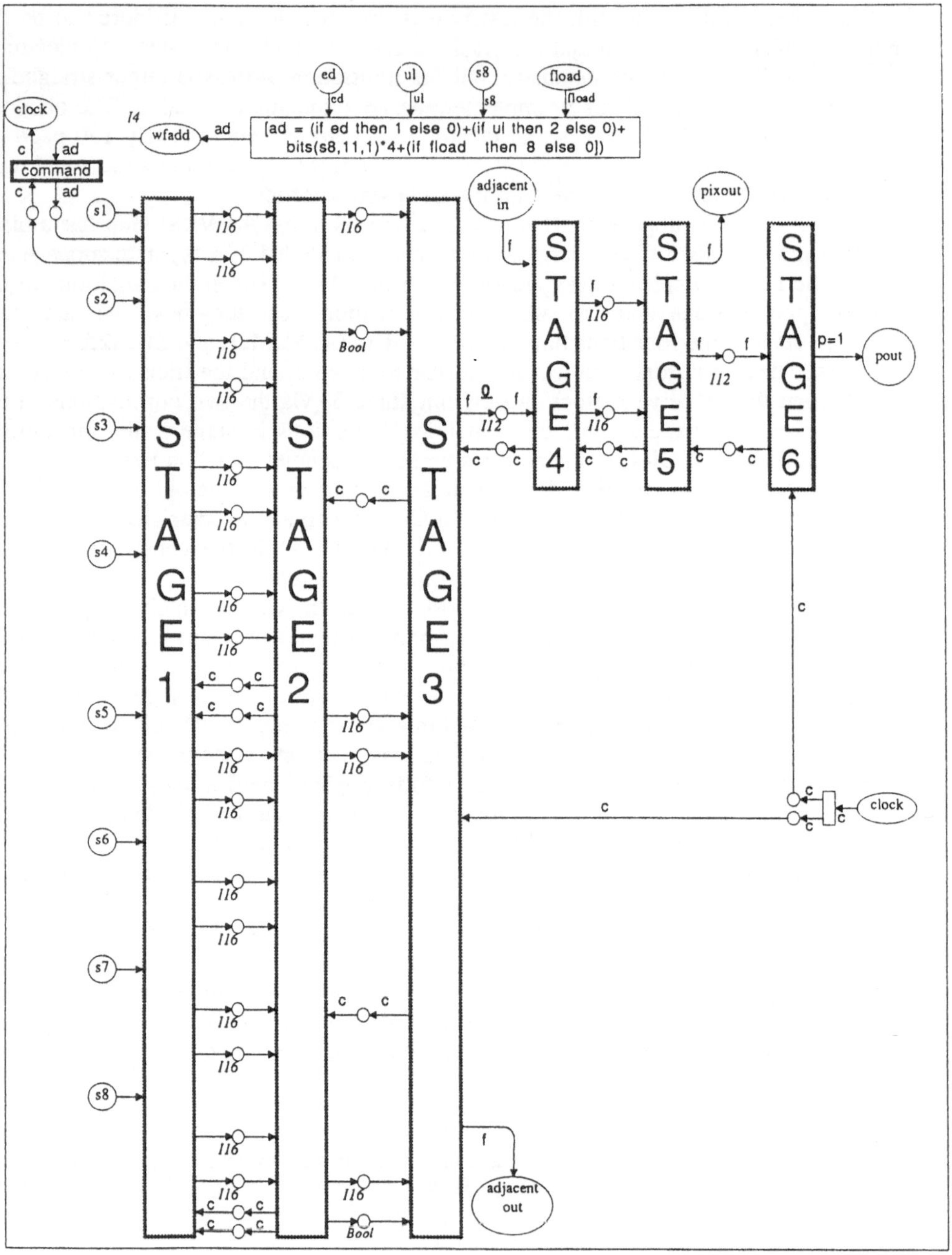

Figure 17. A page from the VLSI chip (showing the division into six pipe-lined stages)

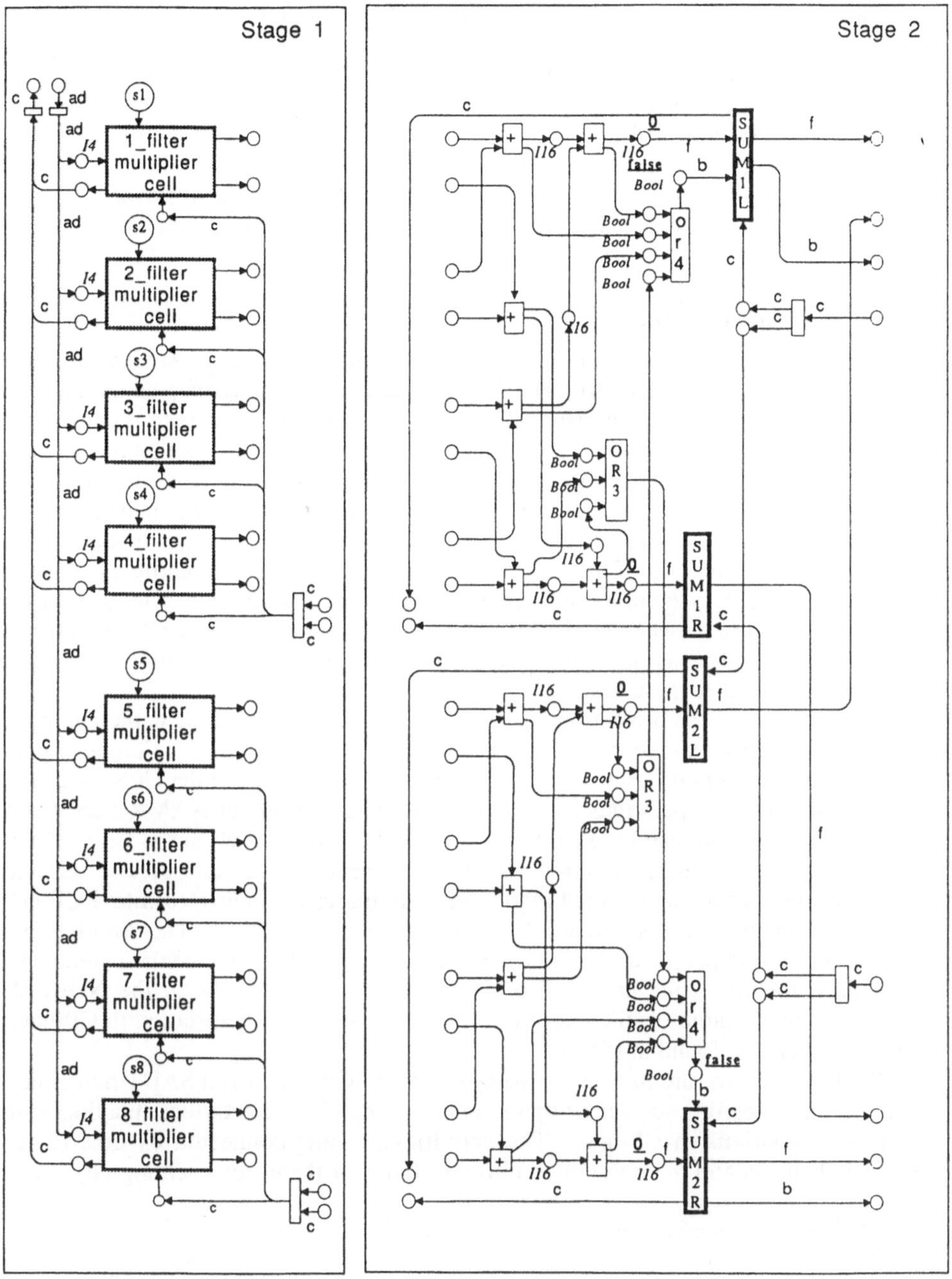

Figure 18. Two subpages of the page in Fig. 17 (modelling stage 1 and 2)

Finally, it was noticed that the execution of the C program was much faster than the CPN simulator – and that it with the latter would be impossible to make the usual amount of test runs (which typically include 10-20.000 sets of test data). It should, however, be noticed that the project was carried out immediately after the first version of the CPN simulator had been released – and that we (based on the experience with this and other large models) now have improved the speed of the CPN simulator with more than a factor 10. Moreover, super-automatic mode has been provided – and this means that we now are in a situation where it makes sense also to deal with large sets of test data.

6.3 Radar surveillance

This project was carried out in cooperation with Armstrong Aerospace Medical Research Laboratory (AAMRL) and it involved the modelling of a command post in the NORAD system.[112] The responsibility of the command post is to recommend different actions – based upon an assessment of the (rapidly changing) status of surveillance networks, defensive weapons and air traffic information. To do this the individual crew member communicates with many different types of equipment, other control posts and other members of the crew, and there is a complex set of detailed rules telling what he must do in the different types of situations. The entire system can be compared to a very complex communication protocol (although a large part of the communication is between human beings and not between computers). The proper design of command posts, including procedures, equipment and staffing, is an on-going problem – typical of the *Command and Control* area.

The purpose of the project was to get an executable model of the command post and use this model to get a better understanding of the command post – in order to improve its effectiveness and robustness. It was never the intention to use the CPN tools directly in the surveillance operations. A team of modellers working at AAMRL created a description of the command post, by means of SADT [43] (which in the United States is known as IDEF). This description was then augmented with more precise behavioural information, and the augmented model was automatically translated into a CP-net and simulated on the CPN simulator (for more details see below). The simulation gave (according to the people at AAMRL) an improved understanding of the command post, and they are now continuing the project modelling other parts of the NORAD system.[113] A much more detailed description of the project, the translation to CP-nets, and the model can be found in [55].

SADT diagrams are in many respects similar to CP-nets: Each SADT model consists of a set of pages,[114] and each of these contains a number of activities (playing a similar role as transitions in CP-nets). The activities are interconnected by arcs (these are called channels and there are three different kinds of them: representing physical flow,

112 NORAD is the North American Radar Defense system.

113 It is the plan to model a number of command posts – and run the submodel for each of these on a separate machine (using a separate copy of the CPN simulator). The submodels will then communicate via input/output statements in code segments (and this will be similar to the way in which the real control posts communicate with each other via electronic networks).

114 In the SADT terminology each page is called a diagram. In this paper we shall, however, use the term diagram for the *set of pages* which constitutes a model.

control flow and availability of resources). SADT has no counterpart to places, but each channel has an attached date type (playing a similar role as the colour sets in CP-nets). Each SADT page (except for the top page) is a refinement of an activity of its parent page (and this works in a way which is totally analogous to transition substitution in CP-nets).

SADT diagrams are often ambiguous. As an example, a branching output channel may mean that the corresponding information/material sometimes is sent in one direction and sometimes in another. It may, however, also mean that the information/material is split in two parts, or that it is copied (and sent in both directions). Although some ambiguity may be tolerable as long as SADT is used to describe the structure of a system,[115] it is obvious that all ambiguity must be removed before the behaviour of a SADT model can be defined (i.e. before simulations can be made) – and this means that SADT must be augmented with better facilities to describe behaviour (e.g. to tell what a branching output channel means).

There are many different ways in which this can be done. One possibility (proposed in several SADT papers) is to attach a table to each activity. Each line in the table describes a possible set of acceptable input values and it specifies the corresponding set of output values. Another, and in our opinion much more attractive possibility, is to describe the input/output relation by a set of channel expressions and a guard – in exactly the same way as the behaviour of a CP-net transition is described by means of a set of arc expressions and a guard. Thus we introduce a new SADT dialect – called IDEF/CPN. In addition to the added channel expressions and guards there is a global declaration node (containing the declarations of types, functions, operations and IDEF variables). Finally, it is possible to use place fusion sets in a similar way as in CPN diagrams.

It is easy to translate an IDEF/CPN diagram into a behavioural equivalent CPN diagram, and this means that the CPN simulator can be used to investigate the behaviour of IDEF/CPN models. For the moment there is a separate IDEF/CPN tool which allows the user to construct, syntax check and modify IDEF/CPN diagrams. This tool works in a similar way as the CPN editor (and many parts of the two user interfaces are identical or very similar). The IDEF/CPN tool can create a file containing a textual representation of the IDEF/CPN diagram, and this file can then be read into the CPN simulator (where it is interpreted as a CPN diagram). The translation from IDEF/CPN to CPN diagrams is thus totally automatic. Later it is the plan to integrate a copy of the CPN simulator into the IDEF/CPN tool itself, and this will mean that the turn-around time will be faster (because it then is possible to edit and simulate in the same tool). Such an integration will also mean that the user will see the simulation results directly on the IDEF/CPN diagram. For the moment he sees the results on the CPN diagram – but this is not a big problem because the two diagrams look nearly identical (except that the former does not have places). Fig. 19 shows an IDEF/CPN page (from the radar surveillance system) and Fig. 20 shows the corresponding CPN page (as it is obtained by the automatic translation).

---

[115] The designers of SADT argue that it is fine to allow such ambiguities – because SADT should be used to "design" the information/material flow, without having to worry about the detailed behaviour (which in their opinion is an "implementation detail").

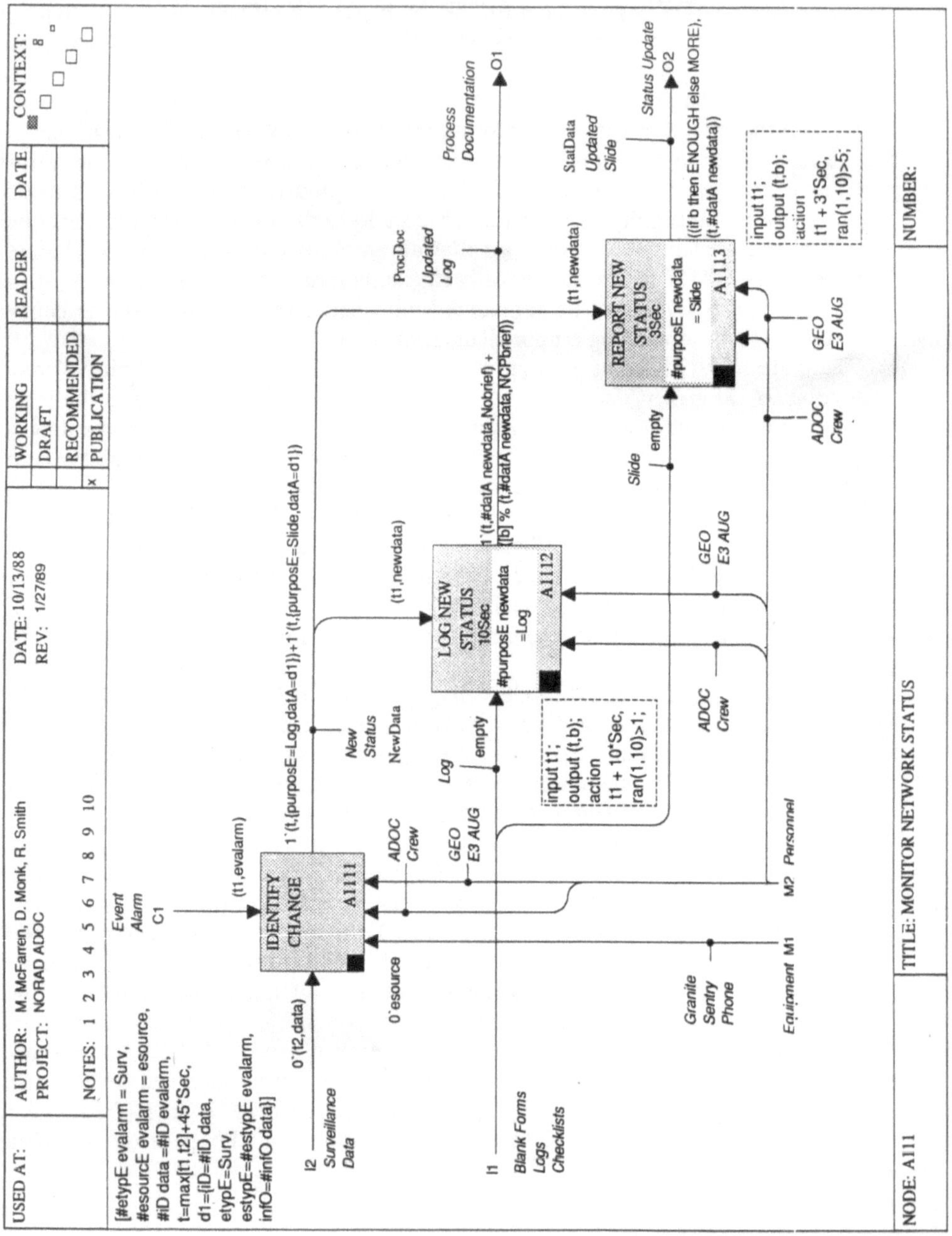

Figure 19. An IDEF/CPN page from the radar surveillance model

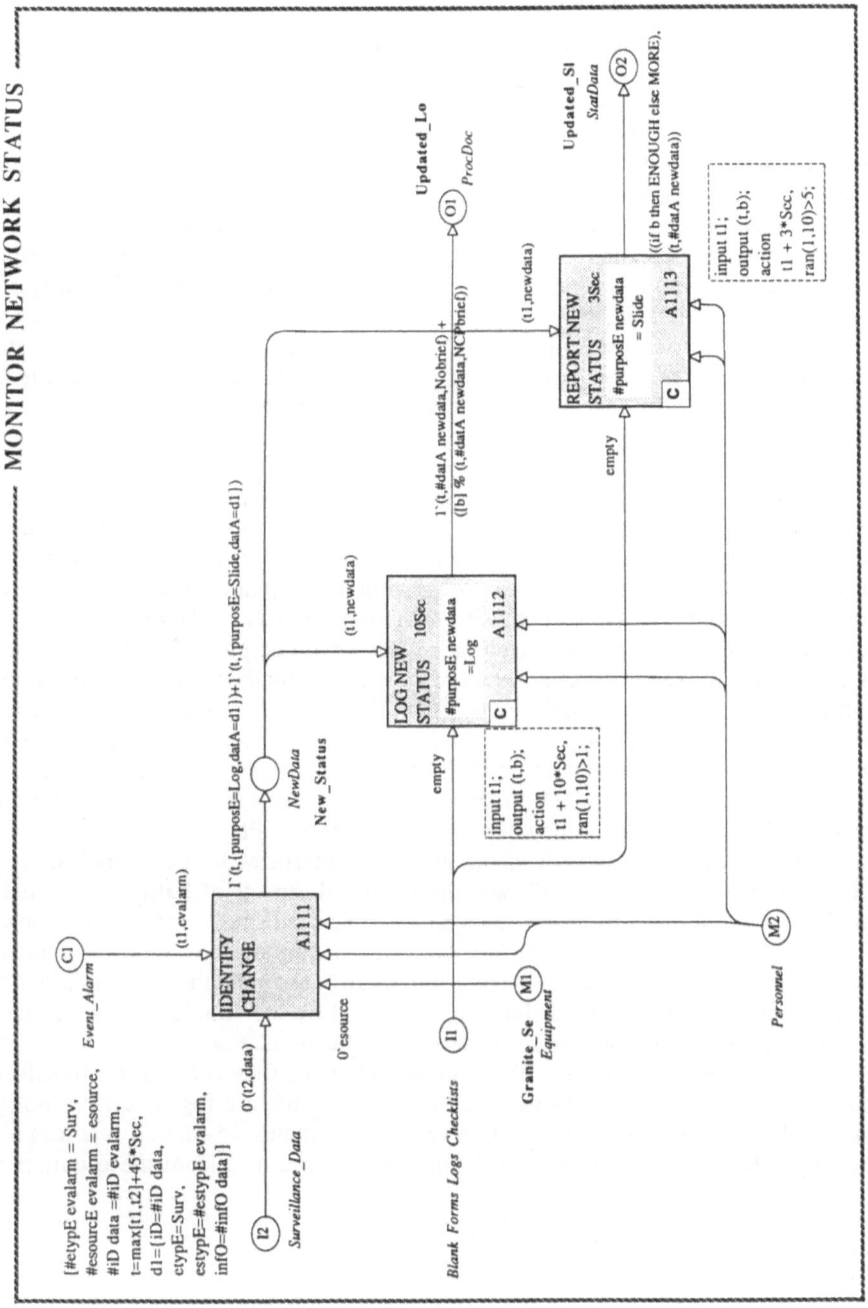

Figure 20. The CPN page obtained from the IDEF/CPN page in Fig. 19

6.4 Electronic funds transfer

This project was carried out in cooperation with two banks (Societé Générale and Marine Midland Bank of New York) and it involved the design and implementation of software to *control* the electronic transfer of money between banks. The speed of modern bank operations means that banks often make commitments which are based on money which they do not have (but expect to receive inside the next few minutes). What happens if these money are delayed – or never arrive? Two managers (at the involved banks) had an idea for a new control strategy – allowing the responsible staff to use computer support to control the electronic funds transfer.[116] The two managers concretized their idea in terms of a relatively small SADT diagram which was created by means of the IDEF/CPN tool (see section 6.3) and contained a rather informal description of the proposed algorithm. The IDEF/CPN diagram was translated to a CP-net and more accurate behavioural information was added, by a CPN modeller.[117] The translation was done in close cooperation with the two bank managers and they participated in the debugging (which also resulted in improvements of the original proposal).

During the project there were several different versions of the CPN model. The first of these was obtained more or less directly from the IDEF/CPN diagram, and it was rather crude (with simple arc expressions and very simple types). This model was primarily used to describe the data flow (while the actual data manipulations were ignored). Later the arc expressions were made more precise, a large number of complex data types were declared (and used as colour sets), complex CPN ML functions were declared (e.g. to search, sort and merge files), and finally most of the behavioural information were moved to code segments. In the final CPN model most transitions have arc expressions which consist of a single variable, and complex code segments determining the values of the output variables from the values of the input variables. It took 5 man-weeks to create the IDEF/CPN diagram, 1 man-week to get the first CPN diagram, and 16 man-weeks to develop this into the final CPN model.

In the first part of the project the graphical interface (in the editor/simulator) was of very large importance– and it was the graphical aspects of IDEF and CP-nets which made it possible for the bank managers to concretize their ideas. Later, however, it turned out that the graphical interface became less important while the output files produced by the simulation became more important – and thus the project started to use a stand-alone SML program (which was generated by the simulator in a similar way as the internal SML code needed for super-automatic simulation).

Now a simulation works with a number of input files (describing transfers which have already been made that day, and transfers which are registered but not yet executed). From these input files (which typically contains 15-50.000 records) a number of output files are produced (in 5-10 minutes) – and it is from these output files the

---

[116] Today the control of the transfer (i.e. the decisions about acceptance/rejection of the individual transactions) is made totally manual – although the transactions themselves are performed via special computer networks.

[117] The additional behavioural information could just as well have been added before the translation (i.e. by means of the IDEF/CPN tool instead of the CPN editor).

staff determine the transfer strategy to be used for the next 15-20 minutes (at which time a new set of simulation results is ready).

In this project the CPN tools (together with the IDEF/CPN tool) was used as a case tool. When the new strategy had been specified (by means of IDEF/CPN and the CPN editor) and validated (by means of the CPN simulator) the resulting SML code was automatically produced (by the CPN simulator).[118] The new control strategy, proposed by the two bank managers, seems to be working as expected – and it is for the moment being tested on historical bank data (using the SML code produced by the CPN simulator). When these tests are finished it will be determined whether the project will continue. If the project is continued the CPN model (and the IDEF/CPN model) will be extended to reflect additional aspects of funds transfer – and a graphical user interface will be added allowing the staff to interact with the model in a more natural way. The user interface will be created by letting the code segments use the graphical routines of the CPN tools. These routines are also available in the stand-alone ML environment, and thus it will still be possible to obtain the final SML code automatically from the CPN simulator (including the added graphical interface). A much more detailed description of the project, the models, and the conclusions can be found in [49].

6.5 Other application areas

CP-nets and other kinds of high-level Petri nets are used in many other application areas. For more information see e.g. [56] (flexible manufacturing systems); [28] and [52] (distributed algorithms); [58] (computer organization); [61] (data bases); [62] (office automation); [2] (computer architecture); [46] (human-machine interaction); [34] (semantics of programming languages); [15] and [29] (software development methods); [4], [8], [11], [14], [16], [19] and [24] (protocols).

From the applications reported in sections 6.1-6.4 (and some of the applications mentioned above) two interesting observations can be made: First of all, it is often adequate to use CP-net models in connection with different front-end languages (e.g. SDL, SADT and block diagrams). The reason may be that there already exist descriptions in these languages, or that the projects involve people who are familiar with some of the languages and thus prefer to use them (instead of learning a totally new formalism). It will also sometimes be sensible to make a tailored language (with a semantics based on CP-nets, but a syntax adopted to the problem area). This is for instance done by the designers of the Vista language [38], who have defined the semantics of their graphical specification language in terms of CP-nets.

Secondly, it is often the case that the graphical representation (which is very important in the early phases) later becomes less interesting. In this case the modellers may turn to super-automatic simulation and this yields a prototype implementation – or (for certain applications areas) even a final implementation. In this way the CPN tools are used as a case tool – and this will be even more attractive when it becomes possible to write code segments in different languages (such as C++, Pascal and Prolog).

---

118 It was necessary to make a few manual operations to create the stand-alone SML code. These operations were trivial, and with the full support of super-automatic simulation they will disappear.

7. Future Plans for CP-nets

This chapter describes our plans for the further development of CP-nets. First we describe a number of extensions which is being made to the existing CPN tools (i.e. the CPN editor and the CPN simulator). Then we describe a number of new CPN tools which are being developed (e.g. to support occurrence graph analysis and invariant analysis). Finally we describe a book project which will provide the necessary introduction and documentation for CP-nets, their analysis methods, and selected examples of industrial applications.

7.1 Extensions of the CPN editor and CPN simulator

The CPN editor/simulator are being extended to handle timed CP-nets, which is an extension of ordinary CP-nets making it easy to describe systems which are time-driven. It will then be possible to use the same net model to analyse both the logic correctness and the time performance of a system. A timed CP-net has a global clock and the value of this is called the current *model time.*[119] The user can specify that certain colour sets are *with time* and this means that the corresponding tokens carry a time stamp (in addition to the ordinary colour information). Intuitively, the time stamp tells when the token is ready to be used (i.e. consumed by a transition). An occurrence element is said to be *colour enabled* if it satisfies the usual enabling criteria (defined by the arc expressions and the guard) and it is then said to have an *enabling time* which is the maximum of all the time stamps in the input tokens and the current model time. A colour enabled occurrence element is *time enabled* iff no other colour enabled occurrence elements have a smaller enabling time.[120] Only time enabled occurrence elements are allowed to occur (and this means that the transitions are executed in the order in which their tokens become ready). The occurrence rule is the same as for CP-nets without time – except that the time stamps of timed output tokens are determined by adding a delay to the current model time. The delays are specified by SML expressions and they may depend upon the colours of the input and output tokens (and via code segments also depend on reference variables and input files).[121] Each time a step has been executed the model time is advanced to match the minimal enabling time in the new system state[122] and this works very much like an event queue in a traditional simulation language. For more information about timed CP-net and the corresponding editor/simulation extensions see [37].

The CPN simulator is also being extended with a set of reporting facilities which will allow much easier visualization of the simulation results (e.g. during a super-automatic simulation). By means of code segments the user will be able to manipulate a

[119] The values of model time may be discrete (integers) or continuous (reals). In both cases each system state exists at a given model time – and the model time is monotonically increased throughout the simulation.

[120] A set of occurrence elements can be concurrently time enabled, but this requires that they all have the same enabling time.

[121] It is also possible to specify that different output tokens get different delays.

[122] The new model time may be identical to the old. This can e.g. happen when some (non conflicting) time enabled occurrence elements do not participate in the step or when some output tokens are created with a time stamp identical to the old model time (or without a time stamp).

large number of different charts (e.g. bar charts, function charts, pie charts and matrix charts). For each chart the code segments update an SML structure (with a predeclared type) while it is the CPN simulator which automatically updates the graphical representation of the chart (based on the value of the SML structure). The frequency by which the chart is updated is specified by the user (either in terms of the number of steps or in terms of model time). The charts are constructed by a special command in the CPN editor and they each consist of a number of auxiliary objects (which can be modified, e.g. resized, recoloured and repatterned, by the same editor operations as the other objects in the CPN diagram). For more information about the reporting facilities and their implementation see [37] and [39].

The implementation of timed CP-nets and the reporting facilities will be finished during the first half of 1991. Later we will also extend the CPN editor to allow the user to construct and modify CP-nets by means of a set of behaviour preserving transformation rules (for more information see [23]). We will also extend the CPN simulator to handle code segments written in other languages[123] and we will extend the CPN editor/simulator to handle the remaining hierarchy constructs and different extensions of CP-nets (e.g. capacities, inhibitor arcs and FIFO places). These projects have, however, lower priority than the creation of the occurrence graph and invariant tools described in section 7.2.

7.2 Additional CPN tools

A CPN tool will be created to support occurrence graph analysis. The tool will construct occurrence graphs for CP-nets (with/without equivalence classes) and it will also assist the user in the analysis of the constructed graphs. As described in section 4.2, a large number of system properties can be automatically determined from the occurrence graph (by an inspection of the individual markings and from the strongly connected components). There is, however, also a need to develop more complex search systems by which the user can perform an interactive inspection of a large occurrence graph. The CPN occurrence graph tool will be able to handle hierarchical CP-nets[124] and it will be tightly integrated into the existing CPN tools. It will e.g. be possible to ask the CPN simulator to execute an occurrence sequence which is found in the occurrence graph – or ask the occurrence graph analyser to search for markings which are identical to or larger than the current marking of the CPN simulator.

To keep the size of occurrence graphs manageable it will be necessary to create occurrence graphs for selected parts of a large model (and this will be done in exactly the same way as in the simulator – i.e. by defining prime pages and being able to ignore specified page instances). It will, moreover, be possible to simplify a model by means of *colour set restrictions*. The basic idea behind this concept is to be able to ignore parts of complex token colours – e.g. one or more components of a record type.[125] As an example, it may during the analysis of a communication protocol be adequate to ig-

---

123 With the SML compiler running under Unix it is already today possible to use object code produced by other compilers.

124 It is straightforward to extend the theory of occurrence graphs with equivalence classes to hierarchical CP-nets with transition substitution and place fusion (and this has already been done).

125 This is analogous to (and inspired by) the concept of projections defined in [21].

nore the data contents of the messages. The restrictions are specified together with the colour set declarations, and this means that it is unnecessary to change the arc expressions or other net inscriptions. Colour set restrictions are also useful for simulation and it will in the future be possible to simulate a model with/without restrictions. For more information about colour set restrictions see [37].

Occurrence graphs can, for a given model, be constructed with/without time and with/without colour set restrictions. It makes, however, no sense to create occurrence graphs with code segments (at least not when these have side effects). The first version of the occurrence graph tool will be available during 1991. Later we will try to integrate our occurrence graph technique (building upon equivalence classes) with the techniques of other groups (see section 4.2).[126]

Analogously, a CPN tool will be created to support invariant analysis. The tool will calculate and check invariants for CP-nets and it will also assist the user when he applies the invariants to prove properties of the modelled system. The calculation of invariants are done in two steps: The first step is automatic and performs a reduction of the CP-net by a set of transformation rules which are proved to preserve the set of invariants.[127] The second step is interactive and it is performed directly on the CPN diagram (i.e. upon the graphical representation of the CP-net): The user proposes weight functions for a number of places. Typically he will define a small number of non-zero weight functions for places he is interested in (but also tell that certain places have zero weight). Then the invariant tool calculates those weight functions which can be uniquely determined from the weights proposed by the user. In this process the tool may also determine that some weights are inconsistent and high-light those transitions that create problems.[128] To calculate new weights and detect inconsistencies the invariant tool uses the reduced matrix obtained in the first step – but it shows the weights and the inconsistencies on the CPN diagram (i.e. in terms of the original CP-net). The user inspects the calculated weights and the high-lighted transitions – and based on this he may add new weights, modify existing weights, or change the behaviour of transitions (e.g. by modifying arc expressions and guards). The process continues, with a number of iterations, and at the end an invariant will be constructed (with some weights specified by the user and the remaining calculated by the invariant tool). The method described above may seem primitive and cumbersome – but this is *not* the case. On the contrary, it is often possible for the user to obtain useful invariants by defining a few weights.[129] It should, moreover, be remembered that the user often have a good idea about what the invariants will be (and thus e.g. knows that certain weights should be zero).

---

126 In particular the technique described in [59] is interesting, because it seems to be orthogonal to our equivalence class technique (in the sense that the former exploits concurrency while the latter exploits symmetries).

127 There are two different sets of reduction rules. One of them preserves place invariants while the other preserves transition invariants.

128 To have an invariant each transition must be neutral, in the sense that the input tokens balance the output tokens (when the weights are taken into account).

129 Each of the five place invariants PI_X from section 4.3 can be determined by specifying the weight of the single place X (and telling that some other places have weight zero).

To check a proposed invariant is even simpler: The user specifies all the weights and the invariant tool checks their consistency. When a set of invariants have been found they can be used to prove system properties, and this is also supported by the tool: As an example, the user may specify the marking of some places. Then the invariant tool calculates upper and lower bounds for other places (by means of the invariants) and in this process the tool may also determine that the specified set of place markings is inconsistent (i.e. impossible in all reachable markings).[130] The invariant tool will be able to handle hierarchical CP-nets[131] and it will, as described above, be tightly integrated into the existing CPN tools. The first version of the invariant tool is planned to be available during 1992. It is, however, obvious that this, among other things, will depend upon the priority given to the improvement of the new occurrence graph tool (and other extensions of existing CPN tools).

Finally we want to develop CPN tools to support reduction methods and the analysis of special subclasses of CP-net – e.g. as described in [9], [12] and [25]. Such tools have, however, lower priority than those described above.

7.3 CPN book

It is our plan to develop a coherent course material for those who want to study the theoretical and practical aspects of CP-nets. This material will be published as a three volume book in EATCS Monographs on Theoretical Computer Science. The book will contain the formal definition of CP-nets and the mathematical theory behind their analysis methods. It is, however, the intention to write the material in such a way that it also becomes attractive to people who are more interested in applications than the underlying mathematics. This means that a large part of the book will be written in a way which is closer to an engineering text book (or a users manual) than it is to a typical textbook in theoretical computer science.

The first volume of the book will introduce and define the net model (i.e. hierarchical CP-nets) and the basic concepts (e.g. the different behavioural properties such as dead-locks, fairness and home markings). It will in detail present a number of small examples and have brief overviews of some industrial applications. It will also contain a description of the CPN editor and the CPN simulator. Most of the material in this volume will be application oriented. The purpose of the volume is to teach the readers how to construct CPN models and how to analyse these by means of simulation.

The second volume will describe the theory behind the formal analysis methods – in particular occurrence graphs with equivalence classes, place/transition invariants and reductions. It will also describe how these analysis methods can be supported by CPN tools, and illustrate this by means of a number of examples. Part of this volume will be rather theoretical while other parts will be application oriented. The purpose of the volume is to teach the readers how to use the formal analysis methods (and this will not necessarily require a deep understanding of the underlying mathematical theory - although such knowledge of course will be a help).

---

130 Performed in this way, the non dead-lock proof in section 4.3 becomes much easier, faster and more reliable.

131 It is straightforward to extend the theory of invariants to hierarchical CP-nets with transition substitution and place fusion (and this has already been done).

The third volume will contain a detailed description of approximately ten different industrial applications. The purpose is to document the most important ideas and experiences from the projects – in a way which is useful for people who do not yet have personal experiences with the construction and analysis of large CPN diagrams. Another purpose is, of course, to document the feasibility of using CP-nets and the CPN tools for such projects.

For the moment approximately 400 pages have been written. Volume 1 will be available at the end of 1991 and we hope that volume 2 will be available during 1992/93. This depends, among other things, upon the speed by which the additional CPN tools are implemented.

8. Conclusions

This paper has presented the theory behind CP-nets, the supporting CPN tools and some of the practical experiences with them. In our opinion it is extremely important to develop these three research ares simultaneously. The three areas influence each other and none of them can be adequately developed without the other two. As an example we think it would have been totally impossible to develop the hierarchy concepts of CP-nets without simultaneously having a solid background in the theory of CP-nets, a good idea about a tool to support the hierarchy concepts and a thorough knowledge of the typical application areas.

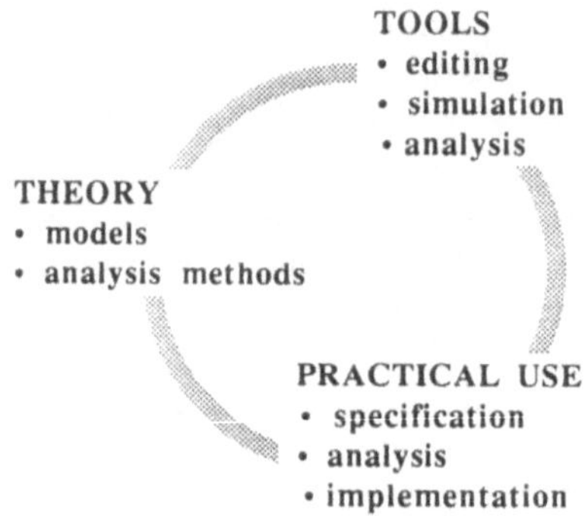

Acknowledgments

Many different persons have contributed to the development of CP-nets and the CPN tools. Below some of the most important contributions are listed:

- CP-nets were derived from Predicate/Transition Nets which were developed by *Hartmann Genrich & Kurt Lautenbach.*
- The first version of occurrence graphs with equivalence classes was developed together with *Peter Huber, Arne Møller Jensen & Leif Obel Jepsen.*
- Many students and colleagues – in particular at *Aarhus University* – have influenced the development of CP-nets.
- *Grzegorz Rozenberg* has been a great support and inspiration for my book project (and for many other of my Petri net activities).
- The hierarchy constructs and the basic structure of the CPN tools were developed together with *Peter Huber & Robert M. Shapiro.*
- The idea to use an extension of Standard ML for the inscriptions of CP-nets is due to *Jawahar Malhotra.*
- The idea of a super-automatic simulation mode and a stand-alone SML program is due to *Valerio Pinci & Robert M. Shapiro..*
- The user interface of the CPN tools was designed together with *Søren Christensen* and it was implemented by *Ole Bach Andersen.* Valuable critique and suggestions were provided by *Michel Beaudouin-Lafon.*
- The abstract date base of the CPN tools was designed by *Peter Huber* and it was implemented by *Vino Gupta.*
- The ML functions to calculate enabling and bindings were designed and implemented together with *Søren Christensen & Peter Huber.*
- IDEF/CPN was designed and implemented by *Robert M. Shapiro.*
- The Unix + X-Windows version of the CPN tools was implemented by *Jane Eisenstein, Ivan Hajadi & Greg Alonso.*
- The reporting facilities was implemented by *Alain Karsenty.*
- Some of the first hierarchical CPN models were made by *Vino Gupta, Peter Huber, Robert Mameli, Valerio Pinci & Robert M. Shapiro.*
- *Hartmann Genrich* has participated in many parts of the development of the CPN tools.
- *Bob Seltzer* has been a continuous supporter of the CPN tool project (and the daily chats with him have been of great importance for the mood of the project group).
- *Meta Software* has provided the financial support for the CPN tool project. So far more than 25 man years have been used. The project is also supported by the Danish National Science Research Council, the Human Engineering Division of the Armstrong Aerospace Medical Research Laboratory at Wright-Patterson Air Force Base, and the Basic Research Group of the Technical Panel C3 of the US Department of Defense Joint Directors of Laboratories at the Navel Ocean Systems Center.

Finally I thank the anonymous referees for their contributions to this paper.

References

[1] K. Albert, K. Jensen and R.M. Shapiro: **Design/CPN. A tool package supporting the use of Coloured Petri Nets.** Petri Net Newsletter 32 (April 1989), 22-36.

[2] J.L. Baer: **Modelling architectural features with Petri nets.** In: W. Brauer, W. Reisig and G. Rozenberg (eds.): Petri Nets: Applications and Relationships to Other Models of Concurrency, Advances in Petri Nets 1986 Part II, Lecture Notes in Computer Science vol. 255, Springer-Verlag 1987, 258-277.

[3] E. Best: **Structure theory of Petri nets: the free choice hiatus.** In: W. Brauer, W. Reisig and G. Rozenberg (eds.): Petri Nets: Central Models and Their Properties, Advances in Petri Nets 1986 Part I, Lecture Notes in Computer Science vol. 254, Springer-Verlag 1987, 168-205.

[4] J. Billington, G. Wheeler and M. Wilbur-Ham: **Protean: a high-level Petri net tool for the specification and verification of communication protocols.** IEEE Transactions on.

Software Engineering, Special Issue on Tools for Computer Communication Systems, SE-14(3), 1988, 301-316.

[5] W. Brauer (ed.): **Net theory and applications**. Proceedings of the Advanced Course on General Net Theory of Processes and Systems, Hamburg 1979, Lecture Notes in Computer Science vol. 84, Springer-Verlag 1980, 213-223.

[6] W. Brauer, W. Reisig and G. Rozenberg (eds.): **Petri nets: Central models and their properties.** Advances in Petri Nets 1986 Part I, Lecture Notes in Computer Science vol. 254, Springer-Verlag 1987

[7] W. Brauer, W. Reisig and G. Rozenberg (eds.): **Petri nets: Applications and relationships to other models of concurrency**. Advances in Petri Nets 1986 Part II, Lecture Notes in Computer Science vol. 255, Springer-Verlag 1987

[8] G. Chehaibar: **Validation of phase-executed protocols modelled with coloured Petri nets.** Proceedings of the 11th International Conference on Application and Theory of Petri Nets, Paris 1990, 84-103.

[9] G. Chiola, C. Dutheillet, G. Franceschinis and S. Haddad: **On well-formed coloured nets and their symbolic reachability graph.** Proceedings of the 11th International Conference on Application and Theory of Petri Nets, Paris 1990, 387-411.

[10] C. Choppy and C. Johnen: **Petrireve: proving Petri net properties with rewriting systems.** J.P. Jouannaud (ed.): Rewriting Techniques and Applications, Lecture Notes in Computer Science vol. 202, Springer-Verlag 1985, 271-286.

[11] B. Cousin et. al.: **Validation of a protocol managing a multi-token ring architecture.** Proceedings of the 9th European Workshop on Applications and Theory of Petri Nets, Vol. II, Venice 1988.

[12] J.M. Couvreur: **The general computation of flows for coloured Petri nets.** Proceedings of the 11th International Conference on Application and Theory of Petri Nets, Paris 1990, 204-223.

[13] F. De Cindio, G. Lanzarone and A. Torgano: **A Petri net model of SDL.** Proceedings of the 5th European Workshop on Applications and Theory of Petri Nets, Aarhus 1984, 272-289.

[14] M. Diaz: **Petri net based models in the specification and verification of protocols.** In: W. Brauer, W. Reisig and G. Rozenberg (eds.): Petri Nets: Applications and Relationships to Other Models of Concurrency, Advances in Petri Nets 1986 Part II, Lecture Notes in Computer Science vol. 255, Springer-Verlag 1987, 135-170.

[15] R. Di Giovanni: **Putting Petri nets into use: the Columbus programme**. Proceedings of the 11th International Conference on Application and Theory of Petri Nets, Paris 1990, 123-138.

[16] P. Estraillier and C. Girault: **Petri nets specification of virtual ring protocols.** In: A. Pagnoni and G. Rozenberg (eds.): Applications and Theory of Petri Nets, Informatik-Fachberichte vol. 66, Springer-Verlag 1983, 74-85.

[17] F. Feldbrugge: **Petri net tool overview 1989.** In: G. Rozenberg (ed.): Advances in Petri Nets 1989. Lecture Notes in Computer Science vol. 424, Springer-Verlag 1990, 151-178.

[18] A. Finkel: **A minimal coverability graph for Petri nets.** Proceedings of the 11th International Conference on Application and Theory of Petri Nets, Paris 1990, 1-21.

[19] G. Florin, C. Kaiser, S. Natkin: **Petri net models of a distributed election protocol on undirectional ring**. Proceedings of the 10th International Conference on Application and Theory of Petri Nets, Bonn 1989, 154-173.

[20] H.J. Genrich and K. Lautenbach: **System modelling with high-level Petri nets.** Theoretical Computer Science 13 (1981), 109-136.

[21] H.J. Genrich: **Projections of C/E-systems.** In: G. Rozenberg (ed.): Advances in Petri Nets 1985. Lecture Notes in Computer Science vol. 222, Springer-Verlag 1986, 224-232.

[22] H.J. Genrich: **Predicate/Transition nets.** In: W. Brauer, W. Reisig and G. Rozenberg (eds.): Petri Nets: Central Models and Their Properties, Advances in Petri Nets 1986 Part I, Lecture Notes in Computer Science vol. 254, Springer-Verlag 1987, 207-247.

[23] H.J. Genrich: **Equivalence transformations of PrT-nets.** In: G. Rozenberg (ed.): Advances in Petri Nets 1989, Lecture Notes in Computer Science, vol. 424, Springer-Verlag 1990, 179-208.

[24] C. Girault, C. Chatelain and S. Haddad: **Specification and properties of a cache coherence protocol model.** In: G. Rozenberg (ed.): Advances in Petri Nets 1987, Lecture Notes in Computer Science, vol. 266, Springer-Verlag 1987, 1-20.

[25] S. Haddad: **A reduction theory for coloured nets.** In: G. Rozenberg (ed.): Advances in Petri Nets 1989, Lecture Notes in Computer Science, vol. 424, Springer-Verlag 1990, 209-235.

[26] R. Harper: **Introduction to Standard ML.** University of Edinburgh, Department of Computer Science, The King's Buildings, Edinburgh EH9 3JZ, Technical Report ECS-LFCS-86-14, 1986.

[27] R. Harper, D. MacQueen and R. Milner: **Standard ML.** University of Edinburgh, Department of Computer Science, The King's Buildings, Edinburgh EH9 3JZ, Technical Report ECS-LFCS-86-2, 1986.

[28] G. Hartung: **Programming a closely coupled multiprocessor system with high level Petri nets.** In: G. Rozenberg (ed.): Advances in Petri Nets 1988, Lecture Notes in Computer Science vol. 340, Springer-Verlag 1988, 154-174.

[29] T. Hildebrand, H. Nieters, and N Trèves: **The suitability of net-based Graspin tools for monetics applications.** Proceedings of the 11th International Conference on Application and Theory of Petri Nets, Paris 1990,139-160.

[30] P. Huber, A.M. Jensen, L.O. Jepsen and K. Jensen: **Reachability trees for high-level Petri nets.** Theoretical Computer Science 45 (1986), 261-292.

[31] P. Huber, K. Jensen and R.M. Shapiro: **Hierarchies in coloured Petri nets.** In: G. Rozenberg (ed.): Advances in Petri Nets 1990, Lecture Notes in Computer Science, Springer-Verlag.

[32] K. Jensen: **Coloured Petri nets and the invariant method.** Theoretical Computer Science 14 (1981), 317-336.

[33] K. Jensen: **High-level Petri nets.** In: A. Pagnoni and G. Rozenberg (eds.): Applications and Theory of Petri Nets, Informatik-Fachberichte vol. 66, Springer-Verlag 1983, 166-180.

[34] K. Jensen and E.M. Schmidt: **Pascal semantics by a combination of denotational semantics and high-level Petri nets.** In: G. Rozenberg (ed.): Advances in Petri Nets 1985. Lecture Notes in Computer Science vol. 222, Springer-Verlag 1986, 297-329.

[35] K. Jensen: **Coloured Petri nets.** In: W. Brauer, W. Reisig and G. Rozenberg (eds.): Petri Nets: Central Models and Their Properties, Advances in Petri Nets 1986 Part I, Lecture Notes in Computer Science vol. 254, Springer-Verlag 1987, 248-299.

[36] K. Jensen et. al.: **Design/CPN: A tool supporting coloured Petri nets.** User's manual, vol 1-2. Meta Software Corporation, 150 Cambridge Park Drive, Cambridge MA 02140, USA, 1988.

[37] K. Jensen et. al.: **Design/CPN extensions.** Meta Software Corporation, 150 Cambridge Park Drive, Cambridge MA 02140, USA, 1990.

[38] E. de Jong and M.R. van Steen: **Vista: a specification language for parallel software design.** Proceedings of the 3rd International Workshop on Software Engineering and its Applications, Toulouse, 1990.

[39] A. Karsenty: **Interactive graphical reporting facilities for Design/CPN.** Master Thesis, University of Paris Sud, Computer Science Department, 1990.

[40] R.M. Karp and R.E. Miller: **Parallel program schemata.** Journal of Computer and System Sciences, vol. 3, 1969, 147-195.

[41] M. Lindqvist: **Translation of the specification language SDL into predicate/transition nets.** Licentiate's Thesis, Helsinki University of Technology, Digital Systems Laboratory, 1987.

[42] M. Lindqvist: **Parameterized reachability trees for predicate/transition nets.** Proceedings of the 11th International Conference on Application and Theory of Petri Nets, Paris 1990, 22-42.

[43] D.A. Marca and C.L. McGowan: **SADT.** McGraw-Hill, New York, 1988.

[44] G. Memmi and J. Vautherin: **Analysing nets by the invariant method.** In: W. Brauer, W. Reisig and G. Rozenberg (eds.): Petri Nets: Central Models and Their Properties, Advances in Petri Nets 1986 Part I, Lecture Notes in Computer Science vol. 254, Springer-Verlag 1987, 300-336.

[45] Y. Narahari: **On the invariants of coloured Petri nets.** In: G. Rozenberg (ed.): Advances in Petri Nets 1985. Lecture Notes in Computer Science vol. 222, Springer-Verlag 1986, 330-345.

[46] H. Oberquelle: **Human-machine interaction and role/function/action-nets.** In: W. Brauer, W. Reisig and G. Rozenberg (eds.): Petri Nets: Applications and Relationships to Other Models of Concurrency, Advances in Petri Nets 1986 Part II, Lecture Notes in Computer Science vol. 255, Springer-Verlag 1987, 171-190.

[47] **Petri nets and performance models.** Proceedings of the third international workshop, Kyoto Japan 1989, IEEE computer society press, order number 2001, ISBN 0-8186-20001-3.

[48] C.A. Petri: **Kommunikation mit automaten.** Schriften des IIM Nr. 2, Institut für Instrumentelle Mathematik, Bonn, 1962. *English translation:* Technical Report RADC-TR-65-377, Griffiss Air Force Bas, New York, Vol. 1, Suppl. 1, 1966.

[49] V.O. Pinci and R.M. Shapiro: **Development and implementation of a strategy for electronic funds transfer by means of hierarchical coloured Petri nets.** Proceedings of the 11th International Conference on Application and Theory of Petri Nets, Paris 1990, 161-180.

[50] C. Reade: **Elements of functional programming.** Addison Wesly, International Computer Science Series, ISBN 0-201-12915-9, 1989.

[51] G. Rozenberg: **Behaviour of elementary net systems.** In: W. Brauer, W. Reisig and G. Rozenberg (eds.): Petri Nets: Central Models and Their Properties, Advances in Petri Nets 1986 Part I, Lecture Notes in Computer Science vol. 254, Springer-Verlag 1987, 60-94.

[52] M. Rukoz and R. Sandoval.: **Specification and correctness of distributed algorithms by coloured Petri nets.** Proceedings of the 9th European Workshop on Applications and Theory of Petri Nets, Vol. II, Venice 1988.

[53] **Functional specification and description language SDL.** In: CCITT Yellow Book, Vol. VI, recommendations Z.101 - Z.104, CCITT, Geneva, 1981.

[54] R.M. Shapiro: **Validation of a VLSI chip using hierarchical coloured Petri nets.** Proceedings of the 11th International Conference on Application and Theory of Petri Nets, Paris 1990, 224-243.

[55] R.M. Shapiro, V.O. Pinci and R. Mameli: **Modelling a NORAD command post using SADT and coloured Petri nets.** Proceedings of the IDEF Users Group, Washington DC, May 1990.

[56] M. Silva and R. Valette: **Petri nets and flexible manufacturing.** In: G. Rozenberg (ed.): Advances in Petri Nets 1989, Lecture Notes in Computer Science, vol. 424, Springer-Verlag 1990, 374-417.

[57] P.S. Thiagarajan: **Elementary net systems.** In: W. Brauer, W. Reisig and G. Rozenberg (eds.): Petri Nets: Central Models and Their Properties, Advances in Petri Nets 1986 Part I, Lecture Notes in Computer Science vol. 254, Springer-Verlag 1987, 26-59.

[58] R. Valk: **Nets in computer organization.** In: W. Brauer, W. Reisig and G. Rozenberg (eds.): Petri Nets: Applications and Relationships to Other Models of Concurrency, Advances in Petri Nets 1986 Part II, Lecture Notes in Computer Science vol. 255, Springer-Verlag 1987, 218-233.

[59] A. Valmari: **Stubborn sets for reduced state space generation.** Proceedings of the 10th International Conference on Application and Theory of Petri Nets, Bonn 1989, Vol II.

[60] A. Valmari: **Compositional state space generation.** Proceedings of the 11th International Conference on Application and Theory of Petri Nets, Paris 1990, 43-62.

[61] K. Voss: **Nets in data bases.** In: W. Brauer, W. Reisig and G. Rozenberg (eds.): Petri Nets: Applications and Relationships to Other Models of Concurrency, Advances in Petri Nets 1986 Part II, Lecture Notes in Computer Science vol. 255, Springer-Verlag 1987, 97-134.

[62] K. Voss: **Nets in office automation.** In: W. Brauer, W. Reisig and G. Rozenberg (eds.): Petri Nets: Applications and Relationships to Other Models of Concurrency, Advances in Petri Nets 1986 Part II, Lecture Notes in Computer Science vol. 255, Springer-Verlag 1987, 234-257.

[63] Å. Wikström: **Functional programming using Standard ML.** Prentice Hall International Series in Computer Science, ISBN 0-13-331968-7, ISBN 0-13-331661-0 Pbk, 1987

[64] E. Yourdon: **Managing the system life cycle.** Yourdon Press, 1982.

CORRECTIONS

Section 3.2: The last sentence in the second paragraph *To allow modular......with care* should be moved to the end of the third paragraph.

Defs. 3.1 and 3.2: "$\forall(i,k) \in I$" should be "$\forall i,k \in I$".

Comment (iii) after Def. 3.2: "node x" should be "node t".

Def. 3.3: Add an extra line, "$m > 0 \Rightarrow s = S_i$." to the end of (ii).

Section 4.3: In the table "2`p" and "3`q", in PI_P and PI_Q, should be changed to "2`e" and "3`e", respectively.

Fig. 16: Add " * Substate" to the end of the dashed box in the lower left part (immediately before the semi-colon).

Section B
High-level Nets and Abstract Data Types

Petri Nets are usually combined with abstract data types, not merely to obtain something principally new, but to find a mathematically adequate style for formulating ideas that have been present in high-level nets from the very beginning: The net formalism should be extended by data structures, and existing Petri net analysis techniques should be generalized to cover those concepts. Furthermore, and particularly for implementation purposes, a syntactic rather than a semantic representation is strived at.

Algebraic specifications provide a formalism for data representations, which is independent of concrete programming languages. They are therefore a natural candidate for a formalism to integrate Petri nets and data representations. Research in this area started in the early 1980s. The SEGRAS project belongs to the earliest efforts, and the first paper in this section describes recent results of this project, including concepts for abstract specifications of strict nets and symbols for partial operations. Based on the specification language OBJ2, the second paper abstractly specifies a particular class of nets, consisting of superposed automata with special requirements to the terms occurring as arc inscriptions. Tokens have a constant individuality and a variable data part.

An important concern of all combinations of nets and algebraic specifications is the border between the algebraically specified concepts, and concepts formulated in usual mathematical style. In the first two papers, everything under consideration is abstractly specified, including the involved nets and the occurrence rule. In the third paper the algebraically specified parts are quite limited: arcs are inscribed by multi-sets of terms, and the occurrence rule is formulated with respect to a concrete semantical interpretation of the involved function symbols. This paper adds threshold inhibitor inscriptions and capacities. The fourth paper suggests an algebraic specification of markings (the token load of a place being given by a term) and of arc inscriptions. Each such term evaluates semantically to a multi-set. Terms in arc inscriptions may include variables (in contrast to markings). A central concern of this paper is a formalism for place invariants, as powerful as (and inspired by) that of CP-nets, but entirely in a symbolic setting. Additionally, a formalism for transition invariants is suggested.

Additional credit for the combination of high-level nets and abstract data types goes to the groups of B. Berthomieu and A. Choquet, formulating (in 1986) the basics of terms being used as tokens. The surprisingly simple and elegant idea of formulating the product of terms, needed in the invariant formalism, just by substitution of terms, is due to J. Vautherin. Recently, attempts have been made by U. Hummert, L. Petrucci and others to employ category theory in a more abstract, schematic presentation of the considered area.

The four papers in this section can be read in any order. Some basic knowledge of algebraic specifications will be useful, but all papers give a short introduction to this technique. Algebraic specifications provide an adequate framework for formally presenting and proving concepts in the area of high-level concepts. A transparent presentation of the results, without explicit reference to general algebra, remains a future challenge.

3.
Many-sorted High-level Nets

J. Billington*

Proc. 3rd Int. Workshop on Petri Nets and Performance Models. IEEE Computer Society Press, Kyoto 1989, pp. 166-179

Abstract

Many-sorted high-level nets (MHLNs) combine abstract data types and Petri nets within the same algebraic framework, and include inhibitor arcs and place capacities. Many-sorted signatures are used to define inscriptions. MHLNs are defined at two different levels of abstraction. At an abstract level markings and capacities are defined by terms. This is suitable for specifying classes of systems. At the concrete level, a many-sorted algebra satisfying the signature, is used for markings and capacities. Both abstract and concrete MHLNs can be given an interpretation in terms of Coloured Petri Nets extended by place capacities and inhibitors, known as P-nets. A hierarchy of high-level nets, including many-sorted versions of Predicate-Transition (PrT) nets and Algebraic nets, is developed and differences with their single-sorted versions are discussed.

1 Introduction

Since the early 1970's Petri nets have been used for the modelling and analysis of systems that involve communication, synchronization, co-operation and concurrency. Some of the reasons for this are their foundation in concurrency, their ability to be analysed and executed by machine and their graphical appeal allowing the dynamics of a system to be visualised by playing the token game.

Although Petri nets have sufficient modelling power for most (if not all) practical systems, they suffer from a lack of modelling convenience or elegance particularly for the representation of data. This led to the development of a number of hybrid net/data models [18,17,13,21], that associated a set of variables with the net and/or attributes with tokens, that could be modified on the occurrence of transitions. Thus the 1970's witnessed a number of useful experiments with adding a more convenient data representation to nets, driven by the needs of practical applications. These models could be analysed using reachability techniques and simulation, but other techniques used to analyse nets (structure theory, invariants, reductions and synchronic distance) were no longer applicable as these extended nets did not come with an underlying Petri net semantics.

This problem was tackled in the next decade, where we have seen the development of *high-level* nets where tokens are data items and arcs and transitions are inscribed with symbolic expressions. The earliest of these were Predicate/Transition nets (PrT nets) [7,9] and Coloured Petri nets (CPNs) [11] which included methods for calculating invariants. Since then, Predicate/Event (P/E) nets, Relation nets [20] and Algebraic nets [19] have been developed and PrT nets and CPNs have been reformulated [8,10]. The links between abstract data types (ADTs) [6] and high-level nets appear to have been discovered in the mid 1980's [15,22,2,14,1,5]. Here the approach has been to combine the strengths of ADTs for data representation with the strengths of Petri nets (synchronisation, concurrency, graphics) within the same algebraic framework. Both [22] and [1] go further and provide some interesting analysis techniques. Vautherin [22] enables classes of systems to be analysed by considering a single representative that turns out to be an ordinary Petri net, while in [1] invariants analysis and composition are provided.

The main purpose of this paper is to define a class of high-level nets that combines abstract data types and Petri nets by including a many-sorted signature in the high-level Petri net structure. This class is called many-sorted high-level nets (MHLNs) to distinguish them from PrT-nets, P/E-nets, Relation nets and Algebraic nets which are all single-sorted. The approach is similar to that of [22] but differs in a number of aspects. Firstly we consider nets with inhibitors and capacities and more general arc inscriptions. (In this presentation we do not consider the axioms of ADTs but they can easily be added). Secondly the nets can be defined at two levels of abstraction. At the abstract level, places are many-sorted (i.e. a sort is associated with each place) and inscriptions, capacities and markings are defined on the level of terms. This is similar to [22,2,14,1] and is appropriate for specifying classes of systems. At the concrete level, places are typed by sets chosen from the carriers of a many-sorted algebra that satisfies the ADT signature, and markings and capacities are multisets over these sets. Inscriptions are

*A major part of this work was performed while the author was at the University of Cambridge Computer Laboratory supported by a Telecom Australia Postgraduate Scholarship and an ORS Award.

again at the level of terms. This has similarities with [19] and is appropriate for the specification of concrete systems.

The abstract MHLN can be interpreted by a *class* of Coloured Petri nets, extended by capacities and inhibitors, known as P-nets and defined elsewhere in these proceedings [3]. The relationship between the concrete form and a P-net is also given. The concrete MHLN is called a P-Graph, because it provides a definition for the graphical form of a P-net. Similarly, the abstract MHLN is known as a P-Graph schema or abstract P-Graph.

The paper concentrates on the (concrete) P-Graph and shows how subclasses such as CP-Graphs, Many-sorted PrT-nets (MPrT-nets) and Many-sorted Algebraic nets can be derived. Why do we need many-sorted versions? The many-sorted versions overcome some difficulties experienced with their singled-sorted predecessors. To allow for the many-sorted nature of applications, the carrier of the single-sorted HLN has to be a union of more basic sets, necessitating the use of partial functions. All variables are typed by the single carrier. This leads to a number of difficulties in typing and interpretation. These problems are investigated and it is shown how they are elegantly solved by allowing the terms to be built from a many-sorted signature with variables.

The paper is organised as follows. Concepts from algebraic specification are recalled in section 2 to provide the necessary background for the definition of P-Graphs in section 3 and for their interpretation as a P-net in section 4. Section 5 discusses the graphical representation of the P-Graph and in section 6, some elementary examples provide a vehicle for introducing some conventions adopted for the graphical form. By placing restrictions on the structure of the P-Graph, Many-sorted Algebraic nets, CP-Graphs and Many-sorted PrT nets are obtained. These classes are defined and discussed in sections 7 and 8 with the aid of examples which also allow them to be compared with the single-sorted versions. The work on the P-Graph is concluded in section 9 by providing an example to illustrate the use of the inhibitor and capacity extensions. The abstract P-Graph is introduced in section 10 and illustrated by an example of a generic queue. Finally some conclusions are drawn in the closing section.

2 Concepts from Algebraic Specification

In the P-Graph, we shall inscribe arcs with multisets of terms involving variables, and transitions with Boolean expressions. Many-sorted signatures provide an appropriate mathematical framework for this representation. Signatures provide a convenient way to characterise many-sorted algebras at a syntactic level. This section introduces the concepts of signatures, terms and many-sorted algebras that will be required for the definition of the P-Graph and abstract P-Graph. We make use of the ideas found in [6,16] for example.

2.1 Signatures

A *many-sorted* (or *R-sorted*) signature, Σ, is a pair:

$$\Sigma = (R, \Omega)$$

where

- R is a set of sorts (the **names** of sets, e.g. *Int* for the integers); and
- Ω is a set of operators (the **names** of functions) together with their *arity* in R which specifies the names of the domain and co-domain of each of the operators.

The arity is a function from the set of operator names to $R^* \times R$, where R^* is the set of finite sequences, including the empty string, ε, over R. Thus every operator in Ω is indexed by a pair (σ, r), $\sigma \in R^*$ and $r \in R$ denoted by $w_{(\sigma,r)}$. $\sigma \in R^*$ is known as the *input* or *argument* sorts, and r as the *output* or *range* sort of operator w. (The sequence of input sorts will define a cartesian product as the domain of the function corresponding to the operator and the output sort will define its co-domain - but this is jumping ahead to the many-sorted algebra.)

For example, if $R = \{Int, Bool\}$, then $w_{(Int.Int,Bool)}$ would represent a binary predicate symbol such as *equality* ($=$) or *less than* ($<$). Using a standard convention, the type of a constant may be declared by letting $\sigma = \varepsilon$. For example an integer constant would be denoted by $cons_{(\varepsilon,Int)}$ or simply $cons_{Int}$.

Types of variables may also be declared in the same way. This leads to the consideration of signatures with variables.

2.2 Signatures with Variables

A many-sorted signature with variables is the triple:

$$\Sigma = (R, \Omega, V)$$

where R is a set of sorts, Ω a set of operators with associated arity as before and V is a set of typed variables, known as an *R-sorted set of variables*. It is assumed that R, Ω and V are disjoint. The type of the variable is defined by the arity function, in a similar way to that of constants, from the set of variable names to $\{\varepsilon\} \times R$. A variable in V of sort $r \in R$ would be denoted by $v_{(\varepsilon,r)}$ or more simply by v_r. For example, if $Int \in R$, then an integer variable would be $v_{(\varepsilon,Int)}$ or v_{Int}.

V may be partitioned according to sorts, where V_r denotes the set of variables of type (sort) r (i.e. $v_a \in V_r$ *iff* $a = r$).

Including the variables in the signature is a convenient way of ensuring that they are appropriately typed.

2.3 Natural and Boolean Signatures

The term *Boolean Signature* is used to mean a many-sorted signature where one of the sorts is Boolean. Similarly, the term *Natural Signature* is used when one of the sorts corresponds to the Naturals (N).

2.4 Terms of a Signature with Variables

Terms of sort $r \in R$ may be built from a signature $\Sigma = (R, \Omega, V)$ in the normal way. We denote a term, e, of sort r by $e : r$ and the set of terms of sort r by $TERM(\Omega \cup V)_r$, and generate them inductively as follows. For $r, r_1, \ldots, r_n \in R$ $(n > 0)$

1. $V_r \subseteq TERM(\Omega \cup V)_r$;
2. For all $w_{(\epsilon,r)} \in \Omega$, $w_{(\epsilon,r)} \in TERM(\Omega \cup V)_r$; and
3. If $e_1 : r_1, \ldots, e_n : r_n$ are terms and $w_{(r_1 \ldots r_n, r)} \in \Omega$, is an operator, then $w_{(r_1 \ldots r_n, r)}(e_1, \ldots, e_n) \in TERM(\Omega \cup V)_r$

Thus if *Int* is a sort, integer constants and variables, and operators (with appropriate arguments) of output sort *Int* are terms of sort *Int*.

We denote the set of all terms of a signature with variables by $TERM(\Omega \cup V)$, the set of all *closed* terms (those not containing variables, also known as *ground* terms) by $TERM(\Omega)$. Thus

$$TERM(\Omega \cup V) = \bigcup_{r \in R} TERM(\Omega \cup V)_r$$

2.5 Multisets of Terms

Multisets or *bags* of terms can also be built inductively from the signature if we assume that we have a Natural signature. We define multisets of terms this way to allow the multiplicities to be terms of sort *Nat*, rather than just the Naturals themselves. (This allows, for example, the introduction of conditions into arc expressions - see sections 3.2.2 and 8.4.)

Let $BTERM(\Omega \cup V)$ denote the set of multisets of terms, defined inductively as follows.

- $TERM(\Omega \cup V) \subset BTERM(\Omega \cup V)$;
- if $b1, b2 \in BTERM(\Omega \cup V)$, then $(b1 + b2) \in BTERM(\Omega \cup V)$; and
- if $i \in TERM(\Omega \cup V)_{Nat}$ and $b \in BTERM(\Omega \cup V)$, then $i \times b \in BTERM(\Omega \cup V)$ where '×' represents scalar multiplication.

Where there is no confusion the '×' will be dropped and juxtaposition will be used for scalar multiplication (e.g. '$3 \times x$ can be replaced by $3x$ and $4 \times 3 \times x$ by $4 \times 3x$ which is distinctly different from $43x$.)

This may be extended to the set of bags with infinite multiplicities, $B_\infty TERM(\Omega \cup V)$, as follows

- $BTERM(\Omega \cup V) \subset B_\infty TERM(\Omega \cup V)$; and
- if $b \in BTERM(\Omega \cup V)$, then $\infty \times b \in B_\infty TERM(\Omega \cup V)$.

where multiplication by ∞ is defined in the appendix of [3].

2.6 Many-sorted Algebras

A many-sorted algebra, (or Σ-Algebra), H, provides an interpretation (meaning) for the signature Σ. For every sort, $r \in R$, there is a corresponding set, H_r, known as a *carrier* and for every operator $w_{(r_1 \ldots r_n, r)} \in \Omega$, there is a corresponding function

$$w_H : H_{r_1} \times \ldots \times H_{r_n} \rightarrow H_r.$$

In case an operator is a constant, w_r, then there is a corresponding element $w_H \in H_r$. They may be considered as functions of arity zero.

Definition: A many-sorted Algebra, H, is a pair

$$H = (R_H, \Omega_H)$$

where $R_H = \{H_r | r \in R\}$ is the set of carriers and
$\Omega_H = \{w_H | w_{\sigma,r} \in \Omega, \sigma \in R^* \text{and } r \in R\}$ the set of corresponding functions.

For example, if $\Sigma = (\{Int, Bool\}, \{<_{(Int.Int,Bool)}\})$ then a corresponding many-sorted algebra would be

$$H = (Z, Boolean; lessthan)$$

where Z is the set of integers: $\{\ldots, -1, 0, 1, \ldots\}$
$Boolean = \{true, false\}$
and $lessthan : Z \times Z \rightarrow Boolean$ is the usual integer comparison function.

It could also be

$$B = (N, Boolean; lessthan)$$

where N is the set of non-negative integers: $\{0, 1, \ldots\}$
$Boolean = \{true, false\}$
and $lessthan : N \times N \rightarrow Boolean$.

(The power of the signature is that it allows a class of algebras to be categorised.)

For signatures with variables, variables are R-sorted. In the algebra, the variable is typed by the carrier corresponding to the sort.

2.7 Assignment and Evaluation

Given an R-sorted algebra, H, with variables in V, an *assignment* [1] for H and V is a set of functions α, comprising an assignment function for each sort $r \in R$,

$$\alpha_r : V_r \rightarrow H_r.$$

[1]The terms *binding* and *valuation* are also used in this context.

This function may be extended to terms by considering the family of functions ass comprising

$$ass_r : TERM(\Omega \cup V)_r \rightarrow H_r$$

for each sort $r \in R$. The values are determined inductively as follows. For $\sigma \in R^*\backslash \varepsilon, \sigma = r_1 r_2 \ldots r_n$, with $r, r_1, \ldots, r_n \in R$ and $e, e_1, \ldots, e_n \in TERM(\Omega \cup V)$,

- If $e \in V_r$ is a variable, then $ass_r(e) = \alpha_r(e)$
- For a constant, $w_r \in \Omega$, $ass_r(w_r) = w_H \in H_r$.
- If $e = w_{(\sigma,r)}(e_1, \ldots, e_n)$, then $ass_r(e) = w_H(ass_{r_1}(e_1), \ldots, ass_{r_n}(e_n)) \in H_r$, where $e_1 : r_1 \ldots e_n : r_n$.

Knowing the values of terms we can determine the value of multisets of terms by expanding the multiset into a sum of scaled terms and evaluating each scalar and term for a particular assignment to variables. This is defined inductively as follows for $a \in TERM(\Omega \cup V)$, $i \in TERM(\Omega \cup V)_{Nat}$ and $b1, b2 \in BTERM(\Omega \cup V)$

- $Val_H(i \times a) = ass(i) \times ass(a)$
- $Val_H(b1 + b2) = Val_H(b1) + Val_H(b2)$

3 P-Graphs

In this section a definition of a graphical form of P-nets is given by defining a **P-Graph**. P-nets are Coloured Petri Nets extended by place capacities and inhibitors as defined in another paper in these proceedings [3]. Due to space limitations, they cannot be discussed further here, and to understand the terminology and notation of P-nets used in this paper the reader is referred to [3].

A P-Graph consists of an inhibitor net where the arcs are annotated by multisets of terms. The multiplicities of the multisets are non-negative integer terms. Transitions are annotated by Boolean terms. The terms are built from a Natural-Boolean signature which has an associated many-sorted algebra. The colour function restricted to places is included. It associates a carrier of the many-sorted algebra with a place. The capacity and initial marking are multisets over the place colour set as usual.

3.1 Definition

A **P-Graph** is a structure

$$\mathbf{PG} = (IN, \Sigma, C, AN, K, M_0)$$

where

- $IN = (S, T; F, IF)$ is an inhibitor net, with
  - S a finite set of places;
  - T a finite set of transitions disjoint from S;
  - $F \subseteq (S \times T) \cup (T \times S)$ a set of arcs; and
  - $IF \subseteq S \times T$ a set of inhibitor arcs.
- $\Sigma = (R, \Omega, V)$ is a Natural-Boolean signature with variables. It has a corresponding Σ-Algebra, $H = (R_H, \Omega_H)$.
- $C : S \rightarrow R_H$ is the colour function restricted to places, such that $\forall s \in S, C(s) \neq \emptyset$.
- $AN = (A, IA, TC)$ is a triple of net annotations.
  - $A : F \rightarrow BTERM(\Omega \cup V)$ such that for $C(s) = H_r$, then for all $(s,t),(u,s) \in F$, $A(s,t), A(u,s) \in BTERM(\Omega \cup V)_r$. It is a function that annotates arcs with a multiset of terms of the same sort as the carrier associated with the arc's place.
  - $IA : IF \rightarrow B_\infty TERM(\Omega \cup V)$ such that for $C(s) = H_r$, then for all $(s,t) \in IF$, $IA(s,t) \in B_\infty TERM(\Omega \cup V)_r$. It is a function that annotates inhibitor arcs with a multiset of terms of the same sort as the carrier associated with the arc's place.
  - $TC : T \rightarrow TERM(\Omega \cup V)_{Bool}$ where for all $t \in T$, $TC(t) \in TERM(\Omega \cup V(t))_{Bool}$ and $V(t)$ is the set of variables occurring in the arc inscriptions associated with t. TC annotates transitions with Boolean expressions.
- $K : S \rightarrow \bigcup_{s \in S} \mu^+_\infty C(s)$ where $K(s) \in \mu^+_\infty C(s)$ is the capacity function.
- $M_0 : S \rightarrow \bigcup_{s \in S} \mu C(s)$ such that $\forall s \in S$, $M_0(s) \leq K(s)$, is the initial marking.

3.2 Discussion

3.2.1 Concrete Colour Sets

In defining P-Graphs, we have intentionally associated a concrete colour set with each place. This colour set is a carrier of the chosen many-sorted algebra, H. This allows us to specify concrete systems where the sets and functions have already been determined.

There is also a need for a more abstract or syntactic form that allows classes of systems to be specified. In this case the places become R-sorted. This leads us to the notion of a P-Graph schema which is defined later in section 10.

3.2.2 Arc Annotations

When generating multisets of terms for the arc inscriptions, we allow the multiplicities to be natural number terms, so that the value can depend on the values of variables and operators of other types. In particular this includes as a special case, the *generalised Kronecker delta* extension to PrT-nets [8].

3.2.3 Strong Typing vs Weak Typing

The inclusion of the colour function, C, may be considered unnecessary. This is because the co-domain of the capacity function and initial marking function could be represented as the set of multisets of terms in $TERM(\Omega)$ evaluated in the Σ-Algebra, H. The colour set of a place would be determined by the types of the terms in the annotations of the surrounding arcs (evaluated in H) and the capacity and initial marking functions.

The inclusion of the colour function has a number of advantages. Firstly it encourages good design, as the typing of places needs to be considered early in the specification of a system. Secondly, it ensures that the initial marking, capacity function and arc annotations are all consistently typed. This can be used to great advantage for type checking specifications with automated tools. Finally, it allows a straightforward interpretation in terms of a P-net.

We shall use the term *strongly-typed* for P-Graphs in which the colour function is included and *weakly-typed* when it is not included.

4 Interpretation of the P-Graph as a P-net

The P-Graph may be given an interpretation as a P-net in the following way.

1. Places: S is the set of places in the P-net.
2. Transitions: T is the set of transitions in the P-net.
3. Colour Sets: The colour set for a transition is determined by the types of the variables occurring in the surrounding arc annotations restricted by its transition condition.

   Let there be n_t free variables associated with the arcs surrounding a transition $t \in T$. Let these have names $v_{r_1}(t), \ldots, v_{r_{n_t}}(t) \in V$. In the Σ-Algebra, H, for all $i \in \{1, 2, \ldots, n_t\}$, let the carrier corresponding to r_i, H_{r_i}, be denoted by G_i with typed variables $v_i(t) : G_i$. Following [10], for all i, let $g_i \in G_i$, then

   $$C(t) = \{(g_1, \ldots, g_{n_t}) \mid TC'(t)\} \text{ where}$$

   $TC'(t) = \lambda(v_1(t), \ldots, v_{n_t}(t)).TC(t))(g_1, \ldots, g_{n_t})$.
   Tuples which satisfy $TC(t)$ are included in $C(t)$.

   The colour sets for the places are obtained from the colour function. Thus the structuring set (of colour sets) is given by $C = \{C(x) | x \in S \cup T\}$.
4. The Colour Function: The colour function restricted to places is defined in the P-Graph and $C(t)$ is given above.
5. Pre and Post Maps.

   The pre and post maps are given, for all $(s,t), (t,s) \in F$, by the following mappings from $C(t)$ into $\mu C(s)$

   $$Pre(s,t) = \lambda(v_1(t), \ldots, v_{n_t}(t)).A(s,t)$$

   $$Post(s,t) = \lambda(v_1(t), \ldots, v_{n_t}(t)).A(t,s)$$

   For $(s,t) \notin F$ and $\forall m \in C(t)$, $Pre(s,t;m) = \emptyset$ and for $(t,s) \notin F$ and $\forall m \in C(t)$, $Post(s,t;m) = \emptyset$.
6. Inhibitor Map

   The inhibitor map is a function from $C(t)$ into $\mu_\infty C(s)$ where for all $(s,t) \in IF$

   $$I(s,t) = \lambda(v_1(t), \ldots, v_{n_t}(t)).IA(s,t)$$

   and for $(s,t) \notin IF$,

   $$\forall g \in C(s), \forall m \in C(t), mult(g, I(s,t;m)) = \infty$$
7. Capacity Function.

   $K(s)$ is as defined in the P-Graph.
8. Initial Marking.

   $M_0(s)$ is as defined in the P-Graph.

With this translation from the P-Graph to P-nets in place, we may now use the definitions of marking, enabling and transition rule for P-nets (see [3]) to allow the P-Graph to be executed. Alternatively, we could define enabling and the transition rule directly for the P-Graph, by considering assignments for terms in a similar way to [19].

5 Graphical Form of P-Graph

5.1 General

The graphical form comprises two parts: a *Graph* which represents the net elements graphically and carries textual inscriptions; and a *Declaration*, defining all the sets, variables, constants and functions that will be used to annotate the Graph part. The declaration may also include the initial marking, the capacity and the colour function if these cannot be inscribed on the graph part due to lack of space.

5.2 Places

In the usual way we shall represent places by circles (or ellipses). A place s may carry four inscriptions.

- the place name;
- the colour set associated with the place, $C(s)$;
- the place capacity, $K(s)$; and
- the initial marking, $M_0(s)$.

The first three would be inscribed close to the place, whereas the initial marking would be inscribed inside the circle representing the place. $C(s)$, $K(s)$ and $M_0(s)$ can be defined in the Declaration if there is insufficient space in the Graph part. We shall adopt the convention that if a place $s \in S$ is not annotated by a capacity multiset, then it will have infinite capacity for all tokens in $C(s)$, unless specified otherwise in the Declaration.

5.3 Transitions

Transitions are represented by rectangles, annotated by a name and may be inscribed by a boolean expression, known as the *Transition Condition.* The Transition Condition may only involve the variables of the inscriptions of its surrounding arcs. If a transition, t, is left blank, then the Transition Condition is true ($TC(t) = true$).

5.4 Arcs

As usual arcs are represented by arrows. For $(s,t) \in F$, an arrow is drawn from place s to transition t and vice versa for $(t,s) \in F$. If (s,t) and (t,s) have the same inscriptions (s is a side place of t), $A(s,t) = A(t,s)$, then this may be shown by a single arc with an arrowhead at both ends and annotated by a single inscription.

An inhibitor arc, $(s,t) \in IF$, is represented by an edge from place s to transition t with a small circle instead of an arrow head at its destination.

The arcs will be annotated with multisets of terms of the same type of their associated place. We therefore need a convenient representation for multisets. We use the symbolic sum or vector representation described in the appendix of [4]. In order to distinguish multiplicities from terms, the convention is adopted that terms may be enclosed in angular brackets.

5.5 Markings and Tokens

A token is a member of $\bigcup_{s \in S} C(s)$. A Marking of the net may be shown graphically by annotating a place with its multiset of tokens $M(s)$. We again use the symbolic sum representation and distinguish multiplicities from tokens, by enclosing tokens in angular brackets. Thus if $g \in M(s)$, g or $< g >$ could appear written in the circle representing place s. We use the natural numbers greater than one, to represent the multiplicity of the token in $M(s)$. Thus if $mult(g, M(s)) = m_g$ we would represent this by juxtaposition: $m_g < g >$ and this would be written inside the circle representing s. If $m_g = 1$, it would be omitted from the inscription. If g is an n-tuple (for example $g = (a,b,c)$), then we adopt the convention of dropping the parentheses (e.g. (a,b,c) would be represented by $< a,b,c >$ and not $< (a,b,c) >$.)

6 Simple Examples

This section provides an introduction to the graphical form of the P-Graph via some simple examples which illustrate some of the conventions adopted in the graph. An interpretation in terms of a P-net is also presented.

Linear P-Graph

$S = \{p1\}, T = \{t1\}, F = \{(p1,t1)\}, IF = \emptyset$
$\Sigma = (\{\mathbf{A}, Bool\}, \{true_{Bool}\}, \{x_{\mathbf{A}}\})$
$H = (\{A, Boolean\}, \{true\})$
$C(p1) = A$
$A(p1,t1) = x, IA = \emptyset, TC(t1) = true$
$K(p1) = \{(a,\infty) | a \in A\}$
$M_0(p1) = A$

Figure 1: Subset Consumption

P-Graph

Declarations

$A = \{a_1, \ldots, a_n\}$
$M_0(p1) = A$

Graph

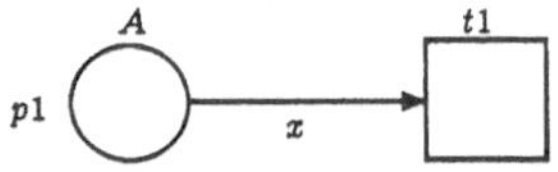

P-Net

$S = \{p1\}$
$T = \{t1\}$
$C = \{A\}$
$C(p1) = C(t1) = A$
$\forall a \in A, Pre(t1,a) = \{(p1,a)\}$
$\forall a \in A, Post(t1,a) = \emptyset$
$K = \{((p1,a),\infty) | a \in A\}$
$M_0 = \{(p1,a) \mid a \in A\}$

Figure 2: Folding Input Place/Transition Pairs

6.1 Consume any subset

Consider the following example which represents the consumption of any non-empty subset of a set of $n \in N^+$ elements, $A = \{a_1, \ldots, a_n\}$, followed by further consumption of a non-empty subset of the remainder and so on until all elements are consumed. The linear form of the P-Graph is given in figure 1, while its graphical form and corresponding P-Net are shown in figure 2.

For each value of $x : A$, there is an occurrence mode of $t1$. All modes of the transition are enabled and any (non-empty) subset could occur simultaneously, consuming the corresponding subset of A. The P-Net represents a set of $|A|$

independent underlying input place/transition pairs, which are concurrently enabled.

This example illustrates a number of conventions that are adopted in the graphical form.

- Inhibitors: It is quite often the case that inhibitor arcs are not present so that $IF = \emptyset$ and hence IA is also empty.
- Omission of a capacity annotation or declaration indicates infinite capacity.
- Quite often it is not necessary to state the signature explicitly and we can operate at the level of the algebra. Thus we can just declare the sets and operators and type variables. In this case we have adopted the convention that the type *Bool* can be considered primitive and that there is no need to explicitly declare the constant ***true***. This is consistent with the use of the default transition condition as discussed below.
- Implicit typing of variables. When the colour set of a place is a simple product of carriers (or a union of products of different degree), then the type of a variable in an arc annotation is determined from its position in the tuple, the degree of the tuple and the colour set definition. (If the variable occurs in the argument of a function, then it is typed by the domain of the function.) In this example, $x : A$.

  If the variable is used in a number of arc insciptions, then it is possible for mistakes to be made with implicit typing, so that the typing of a specific variable is inconsistent. Considerable care is required with implicit typing and ambiguity will be avoided if all variables are declared in the Declaration.
- Default Transition Condition. If for $t \in T, TC(t) =$ ***true***, t is left blank rather than annotating it with the constant ***true***. This is the convention adopted for $t1$ in this example.
- Multisets as sets. We adopt the convention that when a multiset is a set (i.e. its multiplicities are chosen from $\{0,1\}$), then it can be represented as a set. This has been followed for the initial marking and the images of the pre and post maps.

In the rest of the paper we shall only give the graphical form of the P-Graph.

6.2 Transition Condition

This example demonstrates the use of a simple transition condition. The P-Graph is given in figure 3. The less than operator, <, must be defined in the Declaration. Infix notation is used as it is customary. In the corresponding P-net the occurrence modes are limited by the condition $x < y$ so that $C(t1) = \{(a,b) | a \in A, b \in B, \text{and } a < b\}$. The other elements of the P-net are easily derived.

Declarations

A, B: Non-empty sets
$<: A \times B \rightarrow Boolean$
$M_0(p1) = A$, $M_0(p2) = \emptyset$

Graph

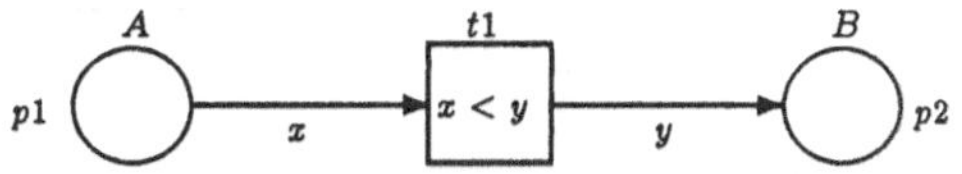

Figure 3: P-Graph with Transition Condition

7 Many-sorted Algebraic Nets

Algebraic nets were proposed as a reformulation of PrT-nets with an improved invariants calculus in [19], where a *partial* algebra over a single carrier was employed. The many-sorted nature of applications was captured by allowing the carrier to be the union of a number of sets. This then lead to the definition of partial functions and their associated operators to be used in the multiset of terms for arc inscriptions.

A colour function is not included in [19] and the net is therefore *weakly-typed*. We shall consider two *many-sorted* algebraic nets: one weakly-typed and the other strongly-typed.

7.1 Weakly-typed many-sorted algebraic nets

A weakly-typed many-sorted algebraic net, MAN, is one of the simplest special cases of a P-Graph, where the inhibitor arcs and annotations, the Transition Condition, and the colour and capacity functions are removed (i.e. $IF = \emptyset$; $(\forall t \in T) TC(t) =$ ***true***; all places have infinite capacity; and the colour function is not included). The arc annotations are also restricted to multisets where the multiplicities are constants rather than natural number terms.

7.1.1 Definition

A weakly-typed MAN is a structure

$$(N, \Sigma, A, M_0)$$

where

- $N = (S, T; F)$ is a net.
- $\Sigma = (R, \Omega, V)$ is a an R-sorted signature with variables. It has a corresponding R-sorted algebra, D.

$D = A \cup B$
$A = \{a_1, \ldots, a_n\}, n \in N^+$
$B = \{b_1, \ldots, b_n\}$
$f : D \rightarrow D$ where
$f(a_i) = b_i$ for $i = 1, \ldots, n$
$f(b_i)$ is undefined for $i = 1, \ldots, n$
$M_0(p1) = B$, $M_0(p2) = \emptyset$

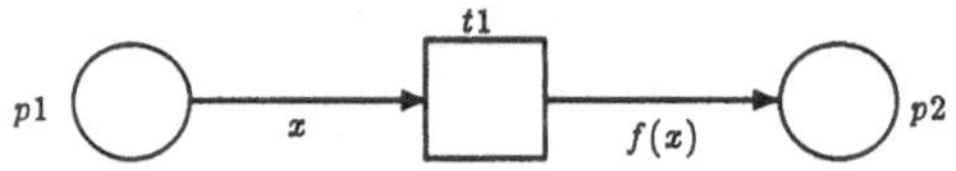

Figure 4: Algebraic Net with an undefined follower marking

- $A : F \rightarrow \mu TERM(\Omega \cup V)$ is the arc annotation function.
- $M_0 : S \rightarrow \mu\{Val_D(\tau) \mid \tau \in TERM(\Omega)\}$ is the initial marking.

I believe that this net captures the spirit of Algebraic nets in terms of a specification language and it has the following advantages:

1. Functions are total.

   Because functions are partial in [19], it is possible to annotate arcs with terms that are not defined in the algebra. This leads to difficulties in interpreting the behaviour of such nets. An example of an Algebraic net illustrating the difficulty is shown in figure 4.

   Firstly, consider the situation when $M_0(p1) = A$. Using the terminology of Algebraic nets, a *valuation* (assignment) for x, $\beta(x) = a_1$ for example, will enable $t1$ in mode β. When $t1$ occurs in mode β, a_1 is removed from place $p1$ and $f(a_1) = b_1$ is added to $p2$, i.e. $M(p1) = A \setminus \{a_1\}$ and $M(p2) = \{b_1\}$. A similar situation occurs for any valuation, $\beta(x) \in A$. Any valuation, $\beta(x) \in B$, will not enable $t1$, due to the initial marking of $p1$.

   Now consider when $M_0(p1) = B$, (a perfectly legal initial marking as $M_0 : S \rightarrow \mu D$, where $D = A \cup B$). A valuation, $\beta(x) \in B$, will now enable $t1$. When $t1$ occurs in mode b_1, the follower marking for $p1$ is clear, $M_0(p1) = B \setminus \{b_1\}$, **but** the follower marking for $p2$ is undefined as the value $f(b_1)$ is not defined.

   This problem does not occur with many-sorted algebraic nets as defined above because functions are total. The intention of the designer of the above algebraic net is unclear. A possible interpretation would be that transition, $t1$, is only enabled when x is bound to an element of A. This interpretation is easily han-

Declarations

Sorts: $R = \{r1, r2\}$
Carriers: $D_{r1} = A$ and $D_{r2} = B$
Operators: All constants from A and B and unary operator $f = w_{(r1,r2)}$
Variable: $x = v_{r1}$ thus $x : A$
$A = \{a_1, \ldots, a_n\}, n \in N^+$
$B = \{b_1, \ldots, b_n\}$
$f_D : A \rightarrow B$ where
$f_D(a_i) = b_i$ for $i = 1, \ldots, n$
$M_0(p1) = A \cup B$, $M_0(p2) = \emptyset$

Graph

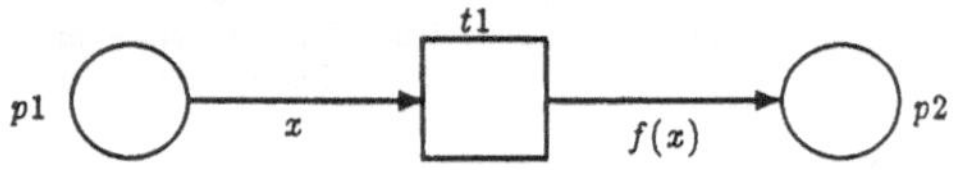

Figure 5: Weakly-typed MAN interpretation of above Algebraic Net

   dled with a many-sorted algebraic net (MAN) (see figure 5).

   The MAN has essentially the same graphical form. The graph part and initial marking are identical. The main difference is that a signature with variables is explicitly included. The sorts $R = \{r1, r2\}$ have corresponding carriers $D_{r1} = A$ and $D_{r2} = B$. The set of operators includes a unary operator $f = w_{(r1,r2)}$ and enough constants of type $A \cup B$ to define the initial marking. The set of variables, V, is a singleton $x = v_{r1}$ and thus $x : A$. The function corresponding to the operator f is a bijection $f_D : A \rightarrow B$, where for $A = \{a_1, \ldots, a_n\}$ and $B = \{b_1, \ldots, b_n\}$, $f_D(a_i) = b_i$ for $i = 1, \ldots, n$. To make the example more interesting we have set the initial marking to $M_0(p1) = A \cup B$ and $M_0(p2) = \emptyset$.

   Transition, $t1$, is enabled in all modes, $m \in A$, and once all the $a's$ in $p1$ have been transformed into $b's$ in $p2$, $t1$ is dead. There is no possibility of binding x to an element of B, as it is of type $x : A$ as defined in the signature. Thus there are no difficulties of interpretation.

2. Sets can be simple.

   The sets of the many-sorted algebra are simple (as opposed to complex unions of other component sets) and correspond to the sets of the physical world that is being modelled. This contrasts with Algebraic nets where there is only one carrier which needs to contain the union of all the simple sets. This is more than an aesthetic problem when developing automated tools, as valuations for each variable will be over the rather large set D, instead of a much smaller domain corresponding to a carrier of the many-sorted algebra.

7.2 Strongly-typed many-sorted algebraic nets

7.2.1 Definition

A strongly-typed many-sorted algebraic net, includes a colour function and is given by

$$(N, \Sigma, C, A, M_0)$$

where

- $N = (S, T; F)$ is a net.
- $\Sigma = (R, \Omega, V)$ is an R-sorted signature with variables. It has a corresponding R-sorted algebra, $H = (R_H, \Omega_H)$.
- $C : S \rightarrow R_H$ is the colour function restricted to places.
- $A : F \rightarrow \mu TERM(\Omega \cup V)$ is the arc annotation function, restricted so that for $C(s) = H_r$, and for all $(s,t),(u,s) \in F$, $A(s,t), A(u,s) \in \mu TERM(\Omega \cup V)_r$.
- $M_0 : S \rightarrow \bigcup_{s \in S} \mu C(s)$ such that $\forall s \in S$, $M_0(s) \in \mu C(s)$ is the initial marking.

The strongly-typed many-sorted algebraic net has the advantage that static type checking can be done to eliminate errors as discussed before. In the above example, it may have been that place p1 should never be marked with tokens from B and that the initial marking was just a mistake. In this case it would be appropriate to set $C(p1) = A$ and, depending on the application, $C(p2) = B$. In this case, setting $M_0(p1) = B$, would violate the typing rules and be detected in a static check. This would not be the case in a weakly-typed MAN, where the error would be detected at run time when an attempt to execute the net would reveal that $t1$ was dead.

The P-Graph of figure 2, is a strongly-typed MAN but figure 3 is not a MAN as it has a transition condition different from *true*. In the next section we define CP-Graphs which allow for transition conditions. It will be seen that a strongly-typed *MAN* is a special class of CP-Graph where $\forall t \in T, TC(t) = true$ and the multiplicities of terms in arc expressions are natural numbers rather than natural number terms.

8 CP-Graphs and Many-sorted PrT-Nets

8.1 CP-Graphs

On removing the inhibitor arcs and the place capacities from the P-Graph, we obtain a subclass that is very similar to Jensen's 'CP-**graph**' [10]. We shall distinguish our class, called CP-Graphs, from that of Jensen by using an upper case 'G' in 'Graph'. The CP-**graph** differs from the CP-Graph defined here in two respects:

- it is a multigraph (i.e. multiple arcs are allowed between places and transitions); and
- the arc inscriptions and transition conditions ('**guards**') are not explicitly defined.

Jensen [10] states that the expressions and guards may be defined by means of a many-sorted algebra (but excludes this from his scope of concern) and that has provided part of the stimulus for the definition of P-Graphs. For CP-Graphs to be a subclass of P-Graphs it includes a signature rather than the algebra. This seems to be the simplest approach, since a signature is always required to build terms and it can also be used to type variables.

Definition

A CP-Graph (**CPG**) is a P-Graph

$$(IN, \Sigma, C, AN, K, M_0)$$

with the following restrictions

- $IN = (S, T; F, \emptyset)$ i.e. no inhibitor arcs.
- $AN = (A, \emptyset, TC)$ i.e. no inhibitor arc annotations.
- For all $s \in S$, $K(s) = \{(g, \infty) | g \in C(s)\}$ i.e. the capacities of the places are infinite.

8.2 Many-sorted PrT-nets

Predicate/Transition Nets (PrT-nets) have been developed over the last decade with the latest definition appearing in [8]. PrT-nets are defined on a syntactic level accompanied by a *relational structure* that provides an interpretation at the concrete level of sets, functions and relations. In this section we shall consider a many-sorted PrT-net as a form of P-Graph, and return to PrT-nets defined at the syntactic level in a later section.

In [8], Genrich mentions the use of many-sorted structures and a 'formalism for abstract data types' (many-sorted algebras) but does not pursue these ideas. PrT-nets are single-sorted, do not include the inhibitor extension nor the colour function, all variables range over a single carrier, and in [8] a capacity function is not defined. A predicate associated with a place has a fixed index, so terms annotating arcs associated with the place can only be multisets of tuples of the same length as the index.

Declarations

Set of Trains:T = {a,b}
Set of track sections:I = $\{0, 1, \ldots, n-1 \mid n > 4\}$
n: number of sections
Variables x:T; i:I
Function $\oplus$:I×I→I is modulo n addition
Place $p1$: Sections occupied by trains
Place $p2$: Vacant sections
$M_0(p1) = \{<0,a>,<2,b>\}$
$M_0(p2) = I \setminus \{0, 2\}$

Graph

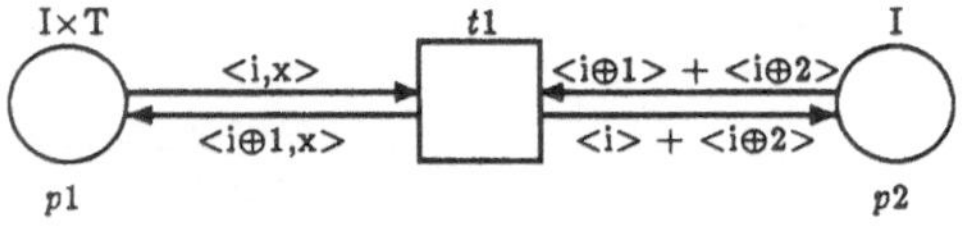

Figure 6: MPrT-Net of Safe Train Operation

In the following we define a subclass of the CP-Graph, known as a many-sorted PrT-net which includes the colour function (i.e. it is strongly typed), and types variables via the signature, but retains the PrT-net flavour of restricting the colour sets to products. Hence tuple lengths for terms annotating arcs are constant.

Definition

A many-sorted PrT-net (MPrT-net) is a CP-Graph with the restriction that for all $s \in S$, $C(s)$ is a simple set or a cartesian product of simple sets, where a simple set comprises elements that are singletons. This ensures that the tuples annotating arcs are of the same length.

The advantages of many-sorted PrT-nets over the PrT-nets of [8] are the same as those for many-sorted algebraic nets over Algebraic nets (see section 7.1). These problems may be overcome with PrT-nets by typing variables in transition conditions, but as this is an **option** of the specifier, mistakes can easily arise. The need to type variables in the transition condition may unduly clutter the graph with information that is best handled in a declaration to allow for static checks.

8.3 Train Example

In [8], Genrich describes the operation of two trains travelling in the same direction on a circular track of seven sections. For safe operation, the trains must never be on the same section or even on adjacent sections. A MPrT-net is given in figure 6 where any number of sections greater than 4 is allowed.

The model is a little different from that in [8]. Apart from the minor difference of generalising the number of track sections, the marking of place $p2$ represents which track sections are vacant. In the original model, the same place represented the predicate that sections i and i⊕1 were vacant. As a minor modelling point, the simpler meaning for a place is preferred. The less intuitive predicate also necessitates the definition of two functions, the modulo 7 successor and predecessor functions, whereas only one (modulo n addition) is required in the MPrT-net. There is also no need for the transition condition and extra variables.

The drawback of the PrT-net is that the successor functions are partial, whereas the variables all range over $I \cup T$. Thus there are legal substitutions for the variables for which the transition condition is undefined. This situation does not arise with the MPrT-net.

It can be seen that this net is also a strongly-typed MAN.

8.4 Example of Conditionals in arc expressions

In this example we use a variant of the readers/writers problem to illustrate the use of conditionals in arc expressions. It is essentially the same as the resource management scheme example of [8], but the model is considerably simplified by removing unnecessary states and colours. The identities of the agents wishing to access the common resource have been retained, but the access 'tickets' are not distinguished.

A number (N) of agents (processes) wish to access a shared resource (such as a file). Access can be in one of two modes: shared (s), where up to L agents may have access at the same time (e.g. reading) and exclusive (e), where only one agent may have access (e.g. writing). No assumptions are made regarding scheduling. An MPrT-net model is given in figure 7.

It has been assumed that the initial state is when all the agents are idle or waiting to gain access to the shared resource (with no queueing discipline assumed). Place *Wait* is marked with all agents; *Access* is empty and the *Control* place contains L ordinary tokens. An agent can obtain access in one of two modes: if shared (m=s), then a single token is removed from *Control* (as m=e is false) when *enter* occurs in a single mode; if exclusive (m=e), then all L tokens are removed preventing further access until the resource is released (transition *Leave*). Shared access is limited to a maximum of L agents as transition *enter* is disabled when *Control* is empty.

Following [8] outfix notation has been used for the function $Bool \to \{0, 1\}$ and this will be used as a standard convention. It is assumed that integer addition and subtraction and the equality predicate are primitive and do not need to be defined in the Declaration.

Remark: The net of figure 7 is very like a PrT-net. If the places were annotated by predicates rather than colour sets and the domain was formed as the union $D = A \cup M \cup C$,

Declarations

Set of Agents:A = $\{a_1, \ldots, a_N\}$
Set of Access Modes:M = {s,e}
Control: C = {•}
Positive integer constants: N,L
Variables x:A ; m:M
Function [ ]:*Bool* → {0,1} where
[*true*] = 1 and [*false*] = 0
M_0(Wait) =A
M_0(Control) =L<•>
M_0(Accessing) = ∅

Graph

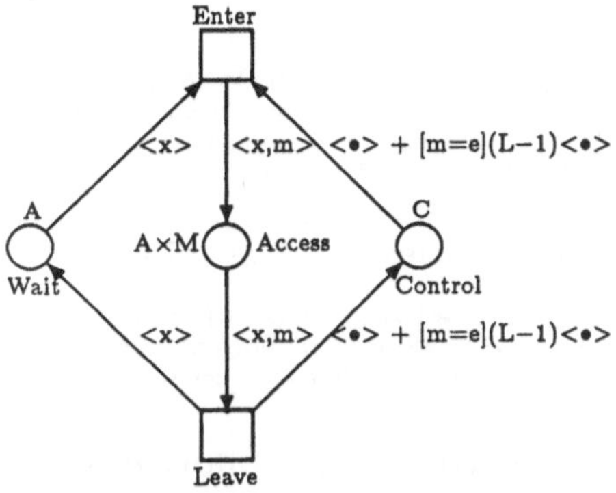

Figure 7: MPrT-Net of Resource Management

Declarations

Set of Trains:T = {a,b}
Set of track sections:I = $\{0, 1, \ldots, n-1 \mid n > 4\}$
$n \in N$: number of sections
Variables x:T; i:I
Function ⊕:I×I→I is modulo n addition
Place *p*1: Sections occupied by trains
$K(p1)$ = I×T
$M_0(p1)$ = {<0,a>,<2,b>}

Graph

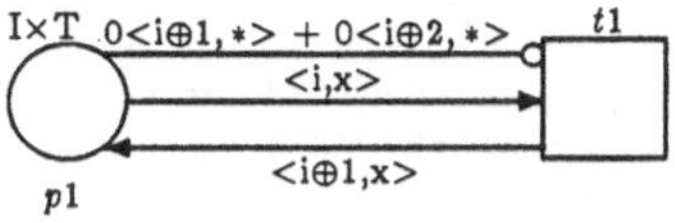

Figure 8: P-Graph of Safe Train Operation

with variables x and m of type D, it would be a PrT-net. The indices of the predicates annotating *Wait* and *Control* would each be one, and that of *Access*, two. It is important to note that the behaviour of the two nets is not the same. In the PrT-net m can be bound to any element of D. Hence an agent could gain access to the resource in a meaningless mode • or a_i for example. The meaning of this is unclear and contrary to the intention of the specification.

9 P-Graph Example: Genrich's Train revisited

The train example above provides us with a very simple illustration of the use of the inhibitor arc to provide a more compact, and I think more intuitive, model of the trains travelling on a circular track.

The graphical form of the P-Graph for the operation of the train is given in figure 8. As usual we include only the information about the algebra in the Declaration and type variables with the appropriate carrier. The tupling operator and function are considered primitive without any need to define them each time in a Declaration. I have also been less formal with the use of operator names and functions in not distinguishing between them (i.e. ⊕ has been used as an operator and also as a function).

For inhibitor arcs we use the convention that zero multiplicities are shown explicitly, whereas infinite multiplicities are assumed for any term that is not shown explicitly (c.f. pre map arcs which assume that zero multiplicities are not shown in the sum). We have also used '∗' notation to represent sums of tuples. It is defined as follows:

Let $(x, y) : A \times B$, then $(x, *) = \sum_{b \in B}(x, b)$.

This can be generalised to tuples of any length, by allowing the sum to be over the domains of all the variables replaced by stars.

The graphical form provides an intuitively appealing specification of the behaviour of the trains on the track. The occurrence of $t1$ again indicates the movement of a train from section i to section i⊕1. This is possible if there is a train on section i, (pre condition) and there are no trains on sections i⊕1 and i⊕2 (inhibitor condition). Of course the concurrent moving of trains is allowed, so long as the conditions are met for different trains on different sections of track. For example, on a 10 section track ($n = 10$) if train 'a' is on section 4 and train 'b' on section 9, then the bindings of i=4 and x=a **and** i=9 and x=b, both satisfy the enabling condition when taken together.

10 Abstract P-Graphs

The P-Graph defined previously included concrete colour sets, markings and capacities. This is often the level at which telecommunication and other systems are specified. However it is very useful to have a more abstract specification that allows classes of systems to be specified. For example the range of sequence numbers or window sizes in protocols may be left open. The hope is that it will be possible to prove properties about systems for a whole range of parameter values by just considering the more abstract specification.

This is the approach adopted by Vautherin [22] where he defines a Petri net-like schema, Σ-schema, and provides an interpretation for it with a class of CP-nets. Vautherin does not include capacity or inhibitor functions, only allows equations to be associated with transitions, does not allow conditionals in arc expressions and does not type variables in his definition (although he does in examples). The following defines a schema addressing these points. The term *Abstract P-Graph* used for the schema, was suggested by Jensen in [12].

10.1 Definition

An **Abstract P-Graph, (APG)** or P-Graph Schema is a structure

$$(IN, \Sigma, \tau, AN, \mathcal{K}, \mathcal{M}_0)$$

where

- $IN = (S,T;F,IF)$ is an inhibitor net, with
  - S a finite set of places;
  - T a finite set of transitions disjoint from S;
  - $F \subseteq (S \times T) \cup (T \times S)$ a set of arcs; and
  - $IF \subseteq S \times T$ a set of inhibitor arcs.
- $\Sigma = (R, \Omega, V)$ is a Natural-Boolean signature with variables.
- $\tau : S \to R$ is a function that types places. (Places are R-sorted.)
- $AN = (A, IA, TC)$ is a triple of net annotations.
  - $A : F \to BTERM(\Omega \cup V)$ such that for $s \in S$, $(s,y),(x,s) \in F$, $A(s,y), A(x,s) \in BTERM(\Omega \cup V)_{\tau(s)}$. Arcs are annotated with a multiset of terms that have the same type as the associated place.
  - $IA : IF \to B_\infty TERM(\Omega \cup V)$ such that for every $(s,t) \in IF$, $IA(s,t) \in B_\infty TERM(\Omega \cup V)_{\tau(s)}$. Inhibitor arcs are annotated with a multiset of terms that have the same type as the associated place.
  - $TC : T \to TERM(\Omega \cup V)_{Bool}$ where for all $t \in T$, $TC(t) \in TERM(\Omega \cup V(t))_{Bool}$ and $V(t)$ is the set of variables occurring in the arc inscriptions associated with t. TC annotates transitions with Boolean expressions.
- $\mathcal{K} : S \to \mu_\infty^+ TERM(\Omega)$ where $\forall s \in S$, $\mathcal{K}(s) \in \mu_\infty^+ TERM(\Omega)_{\tau(s)}$ is the capacity function associating a multiset of closed terms with each place.
- $\mathcal{M}_0 : S \to \mu TERM(\Omega)$ such that $\forall s \in S$, $\mathcal{M}_0(s) \le \mathcal{K}(s)$, is the initial marking at a syntactic level which respects the capacity.

The definition mirrors that of the P-Graph, where the colour function is replaced by the typing function and the capacity and initial markings are defined at the syntactic level of terms rather than at the concrete level of sets.

10.2 Interpretation as a P-net

For a many-sorted algebra, H, satisfying the R-sorted signature, (i.e. $H = (R_H, \Omega_H)$) the interpretation of the abstract P-Graph as a P-net is given in section 4, with the following exceptions.

1. Place Colour Sets. $\forall s \in S$, $C(s) = H_{\tau(s)}$.
2. Capacity Function. $\forall s \in S$, $K(s) = Val_H(\mathcal{K}(s))$.
3. Initial Marking. $\forall s \in S$, $M_0(s) = Val_H(\mathcal{M}_0(s))$.

10.3 Abstract CP-Graphs and other classes of Abstract P-Graphs

Like the CP-Graph, the Abstract CP-Graph does not include inhibitor arcs and has infinite capacities for places.

An Abstract CP-Graph (**ACPG**), or CP-Graph schema is a structure

$$(IN, \Sigma, \tau, AN, \mathcal{K}, \mathcal{M}_0)$$

with the following restrictions

- $IN = (S,T;F,\emptyset)$ i.e. no inhibitor arcs.
- $AN = (A,\emptyset,TC)$ i.e. no inhibitor arc annotations.

Declarations

$R = \{item, queue\}$
$\Omega = \{empty, enq, deq\}$
$empty : \rightarrow queue$
$enq, deq : item \times queue \rightarrow queue$
Variables $x : item; q : queue$
$\tau(Queue) = queue$
$M_0(Queue) = \{empty\}$

Graph Schema

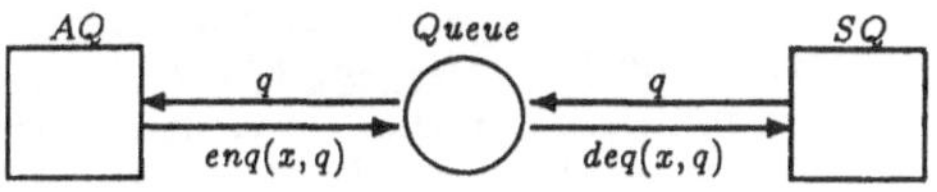

Figure 9: Generic Queue Specification with a CP-Graph Schema

- $\forall s \in S,\ K(s) = \{(term, \infty) | term \in TERM(\Omega)_{\tau(s)}\}$ i.e. the capacities of the places are infinite.

We may also have classes such as a many-sorted PrT-net or algebraic net schema by restricting the above structure in a similar way to that explored for P-Graphs.

10.4 CP-Graph Schema Example: A Generic Queue

Queues with different service disciplines (e.g. first-in-first-out (FIFO), last-in-first-out (LIFO), arbitrary) are important components of computer and communication systems. There may be times in an early part of a design when the service discipline and the items to be queued are undecided. It is at this stage when a class of queues can be specified by using a P-Graph Schema.

A CP-Graph schema of a generic queue is shown in figure 9. There are two sorts: *item*: the name of the set of items to be queued and serviced; and *queue*: the name of the queue structure (such as strings of items, or sets of items). We define a constant *empty* and enqueueing and dequeueing operations which compose items and queues to form queues. Variables are typed (x of sort *item* and q of sort *queue*) as is the place *Queue*. The capacity of place *Queue* is infinite and it is marked by the constant *empty*.

In the graph, the transition *AQ* models the arrival of items and *SQ* their servicing. On the arrival of an item, the current queue is removed, the item is added, and the new queue placed in *Queue*. If there is an item in the queue, then on servicing, this item is removed. The way in which items are added to and removed from the queue is not specified. This is the role of the algebra. For an arbitrary queueing discipline, the enqueueing and dequeueing operations correspond to multiset union and for FIFO and LIFO queues they correspond to concatenation.

11 Conclusions

Many-sorted high-level nets have been defined as inhibitor nets that include a many-sorted signature. Variables can now be appropriately typed and functions are total removing any difficulties in interpretation that can arise with single-sorted high-level nets.

The full power of abstract data types has been incorporated into the same algebraic framework used to define high-level nets, significantly increasing their expressive power. (Although not covered here, the axioms of ADTs can be included in the structure very simply as has been done in [22].) MHLNs allow synchronisation and concurrency to be shown explicitly and provide a visually appealing graphical form for ADTs as pointed out in [2].

A hierarchy of high-level nets can be defined by restricting the structure of the MHLN, to include CP-Graphs, many-sorted PrT and Algebraic nets. If partial functions are allowed, then PrT-nets and Algebraic nets can also be included in the hierarchy, but there appears to be little motivation to do so.

The abstract P-Graph provides a vehicle for the specification of classes of systems and the possibility of their analysis via a single member of the class as has been demonstrated by Vautherin [22]. This opens up exciting possibilities which need to be investigated in various application domains.

Acknowledgements

I gratefully acknowledge the valuable discussions and comments of Prof. Glynn Winskel and Prof. Kurt Jensen on earlier drafts of this work. The paper has also benefitted from the comments of my colleague Geoff Wheeler.

The permission of the Executive General Manager, Telecom Australia Research Laboratories, to present this paper is hereby acknowledged.

References

[1] E. Battiston, F. De Cindio, and G. Mauri. OBJSA nets: a class of high-level nets having objects as domains. In G. Rozenberg, editor, *Advances in Petri Nets 1988, Lecture Notes in Computer Science 340*, pages 20 – 43, Springer-Verlag, Berlin, 1988.

[2] B. Berthomieu, N. Choquet, C. Colin, B. Loyer, J.M. Martin, and A. Mauboussin. Abstract Data Nets:

combining Petri nets and abstract data types for high level specifications of distributed systems. In *Proceedings of the Seventh European Workshop on Application and Theory of Petri Nets*, pages 25 – 48, Oxford, England, 30 June - 2 July 1986.

[3] J. Billington. Extensions to Coloured Petri Nets. In *Proceedings of the Third International Workshop on Petri Nets and Performance Models, Kyoto, Japan, 11-13 December 1989*, IEEE CS Press, Washington, D.C., USA, 1989.

[4] J. Billington. Formal specifications of protocols: protocol engineering. in the Encyclopedia of Microcomputers to be published by Marcel Dekker, Inc., New York, 1989.

[5] Jonathan Billington. *Extending Coloured Petri Nets.* Technical Report 148, University of Cambridge Computer Laboratory, New Museums Site, Pembroke Street, Cambridge, England, October 1988.

[6] H. Ehrig and B. Mahr. *Fundamentals of Algebraic Specification 1, Equations and Initial Semantics.* Volume 6 of *EATCS Monographs on Theoretical Computer Science*, Springer-Verlag, Berlin, 1985.

[7] H. J. Genrich and K. Lautenbach. The analysis of distributed systems by means of predicate/transition-nets. *Lecture Notes in Computer Science*, 70:123–146, 1979.

[8] Hartmann J. Genrich. Predicate/Transition Nets. In W. Brauer, W. Reisig, and G. Rozenberg, editors, *Petri Nets: Central Models and their Properties. Advances in Petri Nets 1986, Part 1: Proceedings of an Advanced Course, Bad Honnef, September 1986*, pages 207 – 247, Springer-Verlag, Berlin, February 1987. Lecture Notes in Computer Science, Volume 254.

[9] Hartmann J. Genrich and Kurt Lautenbach. System modelling with high-level Petri nets. *Theoretical Computer Science*, 13:109–136, 1981.

[10] Kurt Jensen. Coloured Petri Nets. In W. Brauer, W. Reisig, and G. Rozenberg, editors, *Petri Nets: Central Models and Their Properties. Advances in Petri Nets 1986, Part 1: Proceedings of an Advanced Course, Bad Honnef, September 1986*, pages 248 – 299, Springer-Verlag, Berlin, February 1987. Lecture Notes in Computer Science, Vol. 254.

[11] Kurt Jensen. Coloured Petri Nets and the invariant-method. *Theoretical Computer Science*, 14:317–336, 1981.

[12] Kurt Jensen. Private communication. July 1988.

[13] R.M. Keller. Formal verification of parallel programs. *Communications of the ACM*, 19(7):371 – 384, July 1976.

[14] Bernd Krämer. SEGRAS - A Formal and Semigraphical Language combining Petri Nets and Abstract Data Types for the Specification of Distributed Systems. In *Proceedings of the Ninth International Conference on Software Engineering*, pages 116 – 125, CS Press, Los Alamitos, California, March 1987.

[15] Bernd Krämer. Stepwise construction of non-sequential software systems using a net-based specification language. In G. Rozenberg, editor, *Advances in Petri Nets 1984*, pages 307 – 330, Springer-Verlag, Berlin, 1985. Lecture Notes in Computer Science, Vol. 188.

[16] J. Loeckx. Algorithmic specifications: a constructive specification method for abstract data types. *ACM Transactions on Programming Languages and Systems*, 9(4):646 – 685, October 1987.

[17] J.D. Noe and G.J. Nutt. Macro E-Nets for Representation of Parallel Systems. *IEEE Trans. Comput.*, C-22(8):718 – 727, August 1973.

[18] G.J. Nutt. Evaluation nets for computer system performance analysis. In *Proc FJCC, AFIPS*, pages 279 – 286, AFIPS Press, Montvale, N.J., 1972.

[19] W. Reisig and J. Vautherin. An algebraic approach to high level Petri nets. In *Proceedings of the Eighth European Workshop on Application and Theory of Petri Nets*, pages 51–72, Zaragoza, Spain, 24-26 June 1987.

[20] Wolfgang Reisig. *Petri Nets, An Introduction.* Volume 4 of *EATCS Monographs on Theoretical Computer Science*, Springer-Verlag, Berlin, 1985.

[21] F.J.W. Symons. *Modelling and Analysis of Communication Protocols using Numerical Petri Nets.* PhD thesis, University of Essex, 1978. Dept. of Elec. Eng. Sci. Telecommunication Systems Group Report No. 152, May 1978.

[22] J. Vautherin. Parallel systems specifications with Coloured Petri Nets and algebraic specifications. In G. Rozenberg, editor, *Advances in Petri Nets 1987*, pages 293 – 308, Springer-Verlag, Berlin, April 1987. Lecture Notes in Computer. Science, Vol. 266.

CORRECTIONS

Reference [4] should be deleted. The citing of reference [4] in section 5.4, third last line, is in error. The correct citation is reference [3].

4.
Petri Nets and Algebraic Specifications*

W. Reisig

Theoretical Computer Science *80* (1991) 1-34

Communicated by G. Rozenberg
Received March 1989
Revised February 1990

Abstract

Reisig, W., Petri nets and algebraic specifications, Theoretical Computer Science 80 (1991) 1–34.

Petri nets gain a great deal of modelling power by representing dynamically changing items as structured tokens (instead of "black dots"). Algebraic specifications turned out adequate for dealing with structured items. We will use this formalism to construct Petri nets with structured tokens. Place- and transition-invariants are useful analysis techniques for conventional Petri nets. We derive corresponding formalisms for nets with structured tokens, based on term substitution.

Contents

* In part supported by the ESPRIT basic research action DEMON, and by the subproject A3 (SEMAFOR) of the Sonderforschungsbereich 342 at TU Munich.

Introduction

Conventional Petri nets (place/transition nets) use "black dots" (*tokens*) for modelling dynamically changing items. A system state is then determined by the number of tokens on each place of a net. Other Petri net models use structured tokens. Data structures and any kind of structured items can be represented in this way. Their dynamic change is described by particular rules of change. The term *high level Petri net* usually refers to (the several versions of) such net models.

It is almost obvious and confirmed by experience that, for practical applications, high level Petri nets are much more useful than ordinary Petri nets. High level nets support the construction of concise, but nevertheless comprehensible and transparent models of real-world systems.

These advantages must be payed for by a more involved formalism. A formalism is needed coping with data structures and particularly with multisets over any kind of domains because multisets are a fundamental feature for high level Petri nets. This formalism furthermore should include a proper handling of *terms* (used as net inscriptions) and should clarify the step from nets and terms as syntactical objects to their interpretation (meaning) in concrete domains. Finally, the formalism should support analysis techniques for high level Petri nets, particularly place- and transition-invariants.

Algebraic specifications are a promising candidate for this purpose: They turned out to be an adequate and flexible instrument for handling structured items. Multisets over any domain can be specified by additional sorts, operations and equations. The concept of initial algebra semantics, relating (syntactical) terms to (semantical) interpretations, will appear to be directly applicable to high level Petri nets. Last but not least, the calculi of place- and transition-invariants can be based on term substitution.

The central concern of this paper is not the presentation of entirely new results. It is rather intended to present and to integrate known ideas in a—possibly adequate—new setting. A couple of concepts to cope with structured tokens, developed in various papers more or less completely from the scratch, will turn out to be representable by well established concepts of algebraic specifications.

After an introductory example in Section 1, in Section 2, we recall the—elementary—fundamentals of algebraic specifications to the extent needed in this paper. The particular case of abstractly specifying multisets, in fact a well-known standard construction, is discussed in Section 3. On this basis it is a simple step to define nets with structured tokens in Section 4: Markings and arc inscriptions are given by multiset ground terms and arbitrary multiset terms, respectively. Occurrences of transitions can then easily be formulated (depending on the equations of the underlying specification). Section 5 then deals with "place invariants" as a fundamental analysis tool. This technique is based on solutions of linear equations in the domain of multiset terms. Term substitution serves as product in this formalism. The central invariant theorem (5.6) is based on a well-known lemma of general algebra. In Section 6, we discuss some examples to derive system properties, using the technique of place invariants. An analysis technique dual to "place invariants" are the "transition invariants" introduced and investigated in Section 7. Systematic transformations of nets with structured tokens are considered in Sections 8 and 9. Some useful extensions of the formalism are discussed in Section 10. The conclusion finally relates the formalism to predicate/transition nets, coloured nets, and algebraic specifications.

1. An introductory example

As an example to explain the essential ideas of this paper, we consider a basic version of the well-known system of dining philosophers [6]. It is based on an algebra consisting of a set $P = \{p_0, \ldots, p_4\}$ of philosophers and a set $G = \{g_0, \ldots, g_4\}$ of forks. There are two operations *lf* and *rf*, assigning to each philosopher p_i his left and right forks $lf(p_i) = g_i$ and $rf(p_i) = g_{i+1}$ $(0 \leq i \leq 4,\ g_5 = g_0)$, respectively. The corresponding term algebra, generated by a set X of variables, includes terms such as $RF(p_1)$ and $LF(x)$. The embedding of this algebra into a multiset environment yields terms such as $RF(x) + LF(x)$ which occur as arc inscriptions in Fig. 1.

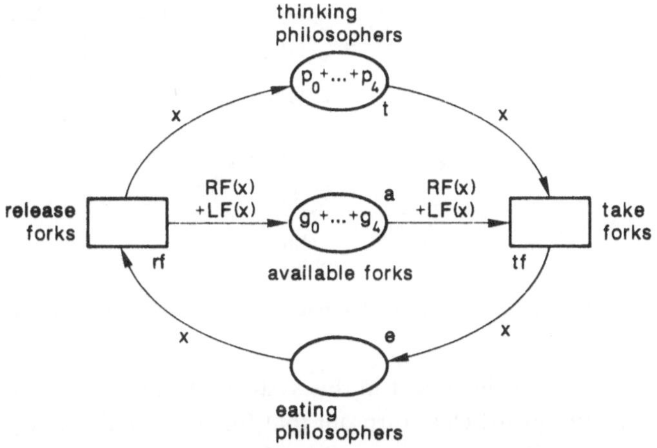

Fig. 1. The system of dining philosophers in its initial situation.

This figure shows a net which specifies the dynamic aspects of the philosophers system. It consists of three *places* (drawn as ellipses) called "thinking philosophers" (**t**), "available forks" (**a**) and "eating philosophers" (**e**), respectively. They can be inscribed by terms, representing objects such as philosophers or forks. Figure 1 shows the initial situation where all philosophers are thinking and all forks are available. The *transitions* (drawn as rectangles) "take forks" (**tf**) and "release forks" (**rf**) with their surrounding arcs and the arc inscriptions indicate the dynamics of the system.

System dynamics is based on the *occurrence of transitions in certain modes.* In Fig. 1, a mode is given by a substitution β of a philosopher p_i for the variable x, that is, $\beta(x) = p_0$. The arc inscription $RF(x) + LF(x)$ in this mode yields the term $RF(p_0) + LF(p_0)$. Assuming the equations $g_0 = LF(p_0)$ and $g_1 = RF(p_0)$, this term represents the set $\{g_0, g_1\}$, viz. the set of forks p_0 is to use.

The arc inscriptions around the transition "take forks" indicate that each philosopher p_i can start his meals only if his left and right forks g_i and g_{i+1} are available. The occurrence of "take forks" in mode $\beta(x) = p_i$ then removes these forks from "available forks", thus they are no longer available for the other philosophers during the meal of p_i. As an example, Fig. 2 shows the situation after p_0 and p_2 both having taken their forks. The occurrence of "release forks" in the mode $\beta(x) = p_0$ as well as in the mode $\beta(x) = p_2$ turns the situation shown in Fig. 2 again to the situation shown in Fig. 1.

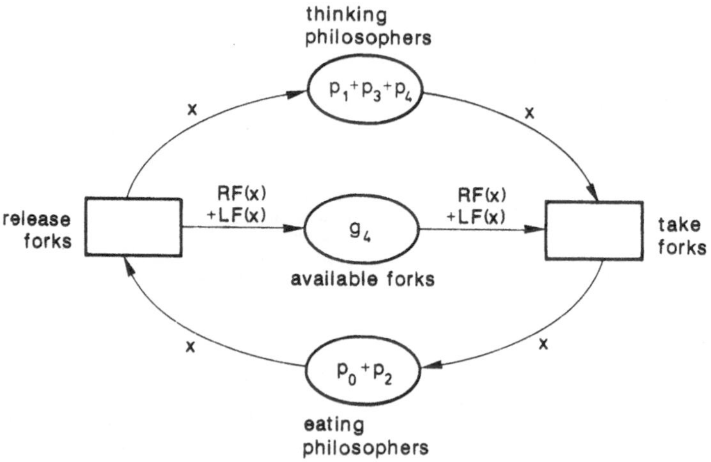

Fig. 2. The system of dining philosophers after p_0 and p_2 having taken their forks.

Notice that no multisets occur in this system. Ordinary sets suffice to describe its behaviour. It will nevertheless turn out that for the proof of system properties (e.g. to show that neighboured philosophers never eat concurrently), multisets are required.

2. Algebraic preliminaries

We recall here some fundamentals of algebraic specifications according to [7]. This serves fixing the—elementary—scope of algebraic notions used in this paper.

2.1. A *signature* $\Sigma = (S, OP)$ consists of a set S of *sorts* and of a famlly $OP = (OP_{w,s})_{w \in S^*, s \in S}$ of *operation symbols.* We particularly distinguish the sets $K_s := OP_{\lambda,s}$ of *constant symbols* (λ denotes the empty word over S).

2.2. A *Σ-algebra* $A=(S_A, OP_A)$ consists of a family $S_A=(A_s)_{s\in S}$ of *domains* and a family $OP_A=(N_A)_{N\in OP}$ of *operations* $N_A: A_{s_1}\times\cdots\times A_{s_n}\to A_s$ for all $N\in OP_{s_1\ldots s_n,s}$. Clearly, $N_A\in A_s$ iff $N\in K_s$.

2.3. A set X of *Σ-variables* is a family $X=(X_s)_{s\in S}$ of *variables*, disjoint to OP.

2.4. The set $T_{OP,s}(X)$ of *(OP, X)-terms of sort* s is inductively defined by

(i) $X_s\cup K_s\subseteq T_{OP,s}(X)$, and

(ii) $N(u_1,\ldots,u_n)\in T_{OP,s}(X)$ for $N\in OP_{s_1\ldots s_n,s}$ and $n\geq 1$, in case $u_1\in T_{OP,s_1}(X),\ldots,u_n\in T_{OP,s_n}(X)$.

2.5. The set $T_{OP,s}:=T_{OP,s}(\emptyset)$ contains the *ground terms of sort s*, $T_{OP}(X):=\bigcup_{s\in S}T_{OP,s}(X)$ is the set of *Σ-terms over X*, and $T_{OP}:=T_{OP}(\emptyset)$ is the set of *Σ-ground terms.*

2.6. An *evaluation* is a mapping *eval*: $T_{OP}\to A$ of Σ-ground terms into a Σ-algebra A, inductively defined by

(i) $eval(N)=N_A$ for all constant symbols N, and

(ii) $eval(N(u_1,\ldots,u_N))=N_A(eval(u_1),\ldots,eval(u_n))$ for all $N(u_1,\ldots,u_n)\in T_{OP}$.

2.7. *An assignment of Σ-variables X to a Σ-algebra A* is a mapping $ass: X\to A$ with $ass(x)\in A_s$ iff $x\in X_s$. *ass* is canonically extended to $\overline{ass}: T_{OP}(X)\to A$, inductively defined by $\overline{ass}(x)=ass(x)$ for $x\in X$, $\overline{ass}(N)=N_A$ for $N\in K_s$ and $\overline{ass}(N(u_1,\ldots,u_n))=N_A(\overline{ass}(u_1),\ldots,\overline{ass}(u_n))$ for $N(u_1,\ldots,u_n)\in T_{OP}(X)$.

2.8. A *Σ-equation over X of sort s* is a pair (L, R) of terms $L, R\in T_{OP,s}(X)$.

2.9. A Σ-equation (L, R) over X is *valid in a Σ-algebra A* iff for all $ass: X\to A$, $\overline{ass}(L)=\overline{ass}(R)$.

2.10. A *specification* $SPEC=(S, OP, E)$ consists of a signature $\Sigma=(S, OP)$ and a set E of Σ-equations.

2.11. For $SPEC=(S, OP, E)$, a *SPEC-algebra* $A=(S_A, OP_A)$ is a (S, OP)-algebra in which all equations in E are valid.

2.12. Two ground terms $u, v\in T_{OP}$ are *congruent* in $SPEC=(S, OP, E)$ (written $u\equiv_E v$) iff $eval_A(u)=eval_A(v)$ for all *SPEC*-algebras A. $\equiv_E$ is an equivalence on T_{OP}.

2.13. $\equiv_E$ is a congruence on T_{OP}, i.e. substitution of congruent terms retains congruence: $u\equiv_E v$ implies $N(\ldots u\ldots)\equiv_E N(\ldots v\ldots)$, cf. [7, Fact 3.11]. When E is clear from the context, $[u]$ denotes the congruence class of term u under $\equiv_E$.

2.14. The *quotient term algebra* $T_{SPEC}=((Q_s)_{s\in S}, (N_Q)_{N\in OP})$ of a specification $SPEC=(S, OP, E)$ has the equivalence classes $Q_s=\{[u]\mid u\in T_{OP,s}\}$ of $\equiv_E$ as carrier sets and the operations N_Q defined by $N_Q([u_1],\ldots,[u_n]):=[N(u_1,\ldots,u_n)]$.

The semantics (meaning) of a specification *SPEC* is any algebra which is isomorphic to T_{SPEC}.

2.15. Congruence on ground terms is extended to terms $u, v\in T_{OP}(X)$: $u\equiv_E v$ iff for all $ass: X \to T_{OP}$, $\overline{ass}(u) \equiv_E \overline{ass}(v)$. Whenever it is clear from the context, the index E in $\equiv_E$ may be skipped.

2.16. A specification $SPEC1$ may consist of a given specification $SPEC = (S, OP, E)$ and additional sets of sorts $S1$, operation symbols $OP1$ and equations $E1$. Then the notation $SPEC1 = SPEC + (S1, OP1, E1)$ means $SPEC1 = (S+S1, OP+OP1, E+E1)$ where + stands for disjoint union of sets. For a signature $\Sigma = (S, OP)$, $\Sigma + (S1, OP1, E1)$ stands of course for the specification $(S+S1, OP+OP1, E1)$.

2.17. Compatibility of evaluation is preserved by extended evaluations: Commutativity of the following diagram (1) implies commutativity of diagram (2), cf. [7, Fact 1.12]:

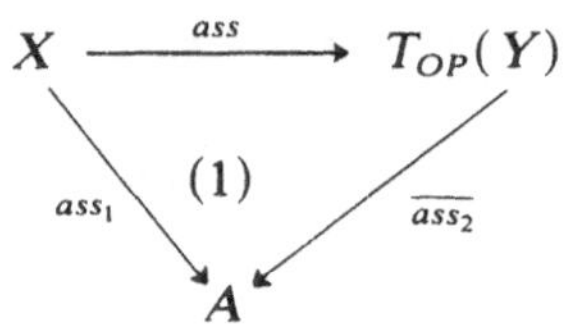

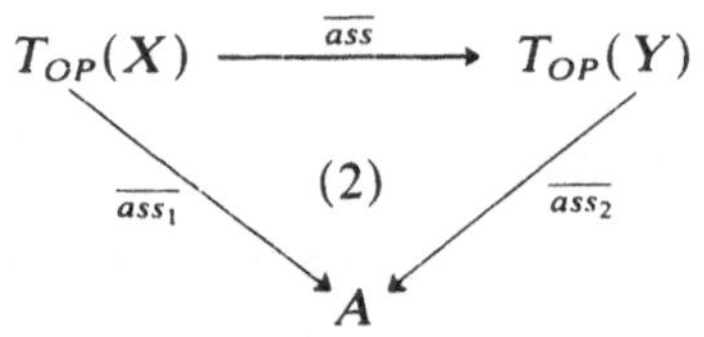

Returning to the introductory example of Section 1, the system of dining philosophers can be constructed from the following signature:

2.18. **phils-base** =
sorts: *phils*
forks

opns: $p_0, \ldots, p_4: \to phils$
$g_0, \ldots, g_4: \to forks$
$LF, RF: phils \to forks$

The arc- and place-inscriptions of the net in Fig. 1 include terms of this signature over the variable x of sort philosopher. (The "+"symbol denotes union of multisets.) The intended meaning of the inscriptions is based on the following specification, extending the above signature by the following equations:

2.19. **phils** = **phils-base** +
eqns: $\left.\begin{array}{l} LF(p_i) = g_i \\ RF(p_i) = g_{i+1} \end{array}\right\} i = 0, \ldots, 4, \text{ with } g_5 := g_0$

$RF(p)$ and $LF(p)$ denote the right and left forks of philosophers p.

3. Multiset specifications

System modelling often requires different copies (items) of data or items which should not be distinguished in the model. This is reflected by *nonnegative multisets* (or *bags*), i.e. collections of elements, some of which may be undistinguishable.

Formally, a nonnegative multiset M over a given set D is a mapping $M: D \to \mathbb{N}$. The *empty multiset* ϑ_D over D is given by $\vartheta_D(d) = 0$ for all $d \in D$. Single elements $d \in D$ can be considered as one-elementary multisets m_d, defined by $m_d(x) = 1$ if $x = d$ and $m_d(x) = 0$, otherwise.

For nets with structured tokens we shall later on study analysis and representation techniques based on general multisets $M: D \to \mathbb{Z}$. The case of nonnegative multisets is thus extended to negative numbers $M(d)$. For general multisets M_1 and M_2, addition is defined component wise, by $(M_1 + M_2)(d) = M_1(d) + M_2(d)$. The inverse $-M$ of a multiset M is defined by $(-M)(d) = -(M(d))$.

Any specification *SPEC* can be extended to its corresponding multiset specification **m**_*SPEC*.

To each constant symbol K_s of *SPEC* a term $MAKE_s(K_s)$ is associated. If K_s is evaluated to the element d, $MAKE_s(K_s)$ is evaluated to the set $\{d\}$. A term ϑ_s denotes the empty multiset; addition and subtraction symbols are defined in the obvious way:

3.1. Definition. Given a specification $SPEC = (S, OP, E)$, let

m_*SPEC* = *SPEC* +
sorts: m_s

$$\left.\begin{array}{ll} \text{opns:} & \vartheta_s : \to m_s \\ & MAKE_s : s \to m_s \\ & +_s : m_s m_s \to m_s \\ & -_s : m_s \to m_s \\ \text{eqns:} & a \in s;\ p, q, r \in m_s \\ & +_s(p, \vartheta_s) = p \\ & +_s(p, q) = +_s(q, p) \\ & +_s(p, +_s(q, r)) = +_s(+_s(p, q), r) \\ & +_s((p, -_s(p)) = \vartheta \end{array}\right\} \quad \text{for all } s \in S$$

A couple of notations and shorthands supports handling the formalism:

3.2. Notations

(i) For a signature Σ, the specification **m**_Σ is defined in the obvious way, cf. 2.16.

(ii) Given a specification $SPEC = (S, OP, E)$ we denote the specification **m**_*SPEC* by $(\hat{S}, \widehat{OP}, \hat{E})$, respectively.

(iii) Within multiset terms we often skip the sort indices s of operation symbols, and write ϑ instead of ϑ_s.

(iv) We use infix notations $u+v$ and $u-v$ for $+(u, v)$ and $-(u, v)$, respectively.
(v) We furthermore write a for $MAKE(a)$.
(vi) Whenever ambiguities are excluded, brackets may be skipped.

As an example, with constant symbols a and b of some sort s, $a-b$ stands for the term $+_s(MAKE_s(a), -_s(MAKE_s(b)))$.

Nonnegative multisets can be specified using (besides the operation symbols of the underlying specification) only the operation symbols ϑ_s, $MAKE_s$ and $+_s$. This motivates the following concepts:

3.3. Definition. Let $SPEC=(S, OP, E)$ be a specification.
(i) $NNS := \widehat{OP}\{-_s \mid s \in S\}$ is the set of *nonnegative operation symbols in m_SPEC.*
(ii) Let X be a family of *SPEC*-variables.

$$T_{OP^+}(X) := \{u \in T_{\widehat{OP}}(X) \mid u \equiv_{\hat{E}} v \text{ for some } v \in T_{NNS}(X)\}$$

is the set of *nonnegative terms of SPEC.*
(iii) Corresponding to 2.5, let $T_{OP^+} := T_{OP^+}(\emptyset)$, and for all $s \in S$, let $T_{OP^+,m_s}(X) := T_{OP^+}(X) \cap T_{\widehat{OP},m_s}(X)$ and $T_{OP^+,s} := T_{OP^+,m_s}(\emptyset)$.
(iv) Given two multiset terms $u, v \in T_{\widehat{OP}}(X)$, u is said to be *smaller or equal to v in SPEC*, written $u \leq_{\hat{E}} v$ if $v-u$ is nonnegative in *SPEC*.

Nonnegativity of terms depends in fact on the assumed equations. As an example, a term $a+a-b$ (with a and b constant symbols of some sort s) is nonnegative iff the equation $a=b$ is assumed.

The net in Fig. 1 can now entirely be explained: based on the specification **phils** $=(S, OP, E)$ of 2.19 and the set $X=X_{phils}=\{x\}$, the arc inscriptions of Fig. 1 are taken from $T_{OP^+}(X)$.

4. Nets with structured tokens

Nets with structured tokens can now be described in an algebraic framework. We start with the conventional definition of (uninscribed) nets:

4.1. Definition. A triple $N=(P, T, F)$ is called a *net* iff
(i) P and T are nonempty, finite, disjoint sets (the *places* and *transitions* of N, respectively), and
(ii) $F \subseteq (P \times T) \cup (T \times P)$ is a relation (the *arcs* of N).

Places, transitions, and arcs will graphically be represented as usual by circles, boxes and arrows, respectively.

Structured tokens are now introduced by inscribing a net w.r.t. a specification and a corresponding set of variables: Each place p of the net is assigned its sort $\varphi(p)$ and its initial marking $M_0(p)$ which is a nonnegative multiset ground term, and each arc f is inscribed by a nonnegative multiset term $\lambda(f)$. Both M_0 and λ should be sort-preserving, i.e. $M_0(p)$ is of sort $m_{\varphi(p)}$, and the multiset sort of $\lambda(f)$ corresponds to the sort of the place adjacent to f.

Based on the notations for multiset sorts of Definition 3.3 we have the following definition.

4.2. Definition. Let $N=(P, T, F)$ be a net, let $SPEC=(S, OP, E)$ be a specification, and let X be a family of Σ-variables

(i) A mapping $\varphi: P \to S$ is called a *sort assignment* of N. Assuming φ, for places $p \in P$ let $\tilde{p}$ denote the multiset sort $m_{\varphi(p)}$.

(ii) A mapping $M_0: P \to T_{OP^+}$ with $M_0(p) \in T_{OP^+,\tilde{p}}$ for each $p \in P$ is called a *φ-respecting initial marking of N.*

(iii) A mapping $\lambda: F \to T_{OP^+}(X)$ with $\lambda(f) \in T_{OP^+,\tilde{p}}(X)$ for each $f=(t, p)$ or $f=(p, t)$ is called a *φ-respecting arc inscription of N.*

(iv) A triple $ins=(\varphi, M_0, \lambda)$ of a sort assignment φ of N, a φ-respecting initial marking M_0 of N, and a φ-sorted arc inscription λ of N, is called a *SPEC-inscription of N*, and (N, ins, E) is a *SPEC-inscribed net.* As a shorthand, N is said to be *inscribed* assuming that *ins* and E can be understood from the context.

Within the **phils**-inscribed net of Fig. 1, the sort assignment φ is given by φ(thinking philosophers) $=\varphi$(eating philosophers) $=$ *phils*, and φ(available forks) $=$ *forks*. M_0 and λ are obvious in Fig. 1. (Entries $M(s)=\vartheta$ are skipped.)

The following notations will be useful when dealing with dynamics of inscribed nets:

4.3. Notations. Let (φ, M_0, λ) be a Σ-inscription of a net $N=(P, T, F)$ over X.

(i) For all $(x, y) \in (T \times P) \cup (P \times T)$ let

$$\xrightarrow{x,y} = \begin{cases} \lambda(x, y) & \text{iff } (x, y) \in F, \\ \vartheta & \text{otherwise.} \end{cases}$$

(ii) For each $t \in T$ we define the vector $\underline{t}: P \to T_{\widehat{OP}}(X)$ by

$$\underline{t}(p) = \xrightarrow{t,p} - \xrightarrow{p,t}.$$

Dynamics of inscribed nets is now defined as follows:

4.4. Definition. Let the net $N=(P, T, F)$ be inscribed over a specification $SPEC=(S, OP, E)$ and variables X.

(i) *Markings* of N are mappings $M: P \to T_{OP^+}$ with $M(p) \in T_{OP^+,\tilde{p}}$ for each $p \in P$.

(ii) An *occurrence mode* of N is an assignment $\beta: X \to T_{OP}$.

(iii) Given a marking M, a transition $t \in T$ and an occurrence mode β, t is *β-enabled at M* (or *enabled at M in mode β*) iff, for all $p \in P$,

$$\bar{\beta}(\xrightarrow{p,t}) \leq_E M(p).$$

(iv) If t is β-enabled at M, t may *occur in mode* β. This returns the marking M' which is defined for each $p \in P$ by

$$M'(p) = M(p) - \bar{\beta}(\xrightarrow{p,t}) + \bar{\beta}(\xrightarrow{t,p})$$

We write $M \xrightarrow{t,\beta} M'$ in this case.

(v) For a marking M of N, the set $[M\rangle$ of *markings reachable from* M is the smallest set of markings such that $M \in [M\rangle$ and if $M' \in [M\rangle$ and $M' \xrightarrow{t,\beta} M''$ then $M'' \in [M\rangle$.

With Notation 4.3(ii) and 2.7 we get immediately the following corollary.

4.5. Corollary. *If* $M \xrightarrow{t,\beta} M'$ *then* $M'(p) \equiv_{\hat{E}} M(p) + \bar{\beta}(\underline{t}(p))$ *for all places* p.

Considering markings M and mappings $\underline{t}$ as P-indexed vectors and extending sum and extended assignments $\bar{\beta}$ component-wise to vectors, the above corollary reads as follows.

4.6. Corollary. *If* $M \xrightarrow{t,\beta} M'$ *then* $M' \equiv_{\hat{E}} M + \bar{\beta}(\underline{t})$.

As an example, the transition **tf** of Fig. 1 is enabled at M_0 in all modes $\beta : \{x\} \to \{p_0, \ldots, p_4\}$. There is however no mode β to enable **rf** at M_0. Assuming $\beta_0(x) = p_0$, let M_1 be reached by $M_0 \xrightarrow{tf,\beta_0} M_1$. Now in M_1, **tf** is enabled in both modes $\beta(x) = p_2$ and $\beta(x) = p_3$. The assignment β_0 additionally enables **rf** under M_1. With $M_1 \xrightarrow{rf,\beta_0} M_2$ we return to $M_0 \equiv_{\hat{E}} M_2$. Notice that without the equations of **phils**, each philosopher is assumed to have two forks of his own.

This completes the notion of *SPEC*-inscribed nets. Other versions of nets with structured tokens can be regarded as being based on particular specifications *SPEC*: *PrT*-nets with tuples of variables as arc inscriptions use the specification of tuples; place/transition nets do with one sort and one constant of this sort. Details of such classes will be discussed later.

Transitions may additionally be inscribed by logical formulae. Then a transition is enabled in a mode β only if in addition to the requirements of Definition 4.4(iii), its formula evaluates with β to *TRUE*. We skip this feature here, since it does not contribute to this paper's topic.

As a further example, let the specification **phils**′ be given by

phils′ =
sorts: *phils*
forks
opns: $g_0, \ldots, g_4 : \to phils$
$p_0, \ldots, p_4 : \to forks$
$RU : forks \to phils$
$RSF : forks \to forks$
eqns: $\left.\begin{array}{l} RU(g_i) \;= p_i \\ RSF(g_i) = g_{i+1}. \end{array}\right\} i = 0, \ldots, 4$, with $g_5 := g_0$

The operations *RU* and *RSF* are to return for each fork its right user and its right successor fork, respectively.

Figure 3 shows a **phils'**-inscribed net. Intuitively its behaviour is identical to the behaviour of the net in Fig. 1. We will discuss in Section 8 how algebraic specifications provide means to prove this formally.

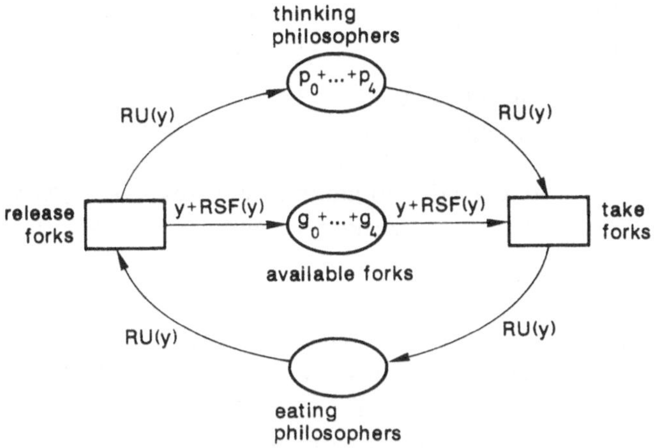

Fig. 3. A variant of the philosophers system.

5. Place invariants

Place invariants are one of the most important analysis tools for several versions of Petri nets. A place invariant provides a weight $W(M)$ to markings M in such a way that in each set $[M\rangle$ of reachable markings, the weighted markings $W(M')$ are constant for all $M' \in [M\rangle$. We will show that such weights can be represented by a place vector of multiset terms. The application of a weight function W to a marking M will be defined as a scalar product of the vectors W and M, with the product of components being defined as a term substitution. Products $W \cdot M$ thus amount to multiset ground terms. Safety properties can be derived from knowing the product $W \cdot M$ to remain constant.

A fundamental property of place invariants is their characterizability as solutions of homogeneous systems of linear equations: Each net N with places P and transitions T canonically defines a $P \times T$-matrix $\underline{N}$ with entries $\underline{N}(p, t) = \underline{t}(p)$ (cf. Notations 4.3). The product of matrix entries with invariant entries is again defined as term substitution, just as the above-mentioned product of markings and invariant entries. Place invariants will then be the solutions of $\underline{N}^\tau \cdot i \equiv_{\hat{E}} \gamma_s$ ($\underline{N}^\tau$ denotes the transpose of matrix $\underline{N}$).

The term product for place invariants is based on a distinguished, quite simple kind of "constant" assignments, ass_u, of Σ-variables X. The range of ass_u is the set $T_{OP}(X)$ of Σ-terms over X. Given a sort s, ass_u maps all $x \in X_s$ to $u \in T_{OP,s}(X)$, leaving all other variables untouched.

5.1. Definition. Let $\Sigma = (S, OP)$ be a signature, let X be a set of Σ-variables with X_s the variables of sort $s \in S$, and let $u \in T_{OP,s}(X)$. Then the *constant u-assignment*, $ass_u : X \to T_{OP}(X)$ is defined as

$$ass_u(x) = \begin{cases} u & \text{iff } x \in X_s, \\ x & \text{iff } x \in X \setminus X_s. \end{cases}$$

Assignments ass_u are extended to multiset terms by $\overline{ass_u} : T_{\widehat{OP}}(X) \to T_{\widehat{OP}}(X)$ in the usual way (cf. 2.7). Based on constant assignment we define the following product for multiset terms:

5.2. Definition. Let (S, OP) be a signature and let X be a set of (S, OP)-variables. A *product* $u \cdot u' \in T_{\widehat{OP}}(X)$ for multiset terms $u, u' \in T_{\widehat{OP}}(X)$ is defined by induction over the structure of u: $MAKE(u) \cdot u' = \overline{ass_u}(u')$ iff $u \in T_{OP}(X)$, $\vartheta \cdot u' = \vartheta$, $(u_1 + u_2) \cdot u' = u_1 \cdot u' + u_2 \cdot u'$, and $(-u) \cdot u' = -(u \cdot u')$.

For vectors of multiset terms we extend this product furthermore to the usual inner product of linear algebra:

5.3. Definition. Let P be a finite set, let (S, OP) be a signature and let X be a set of (S, OP)-variables. For vectors $\underline{u}, \underline{u}' : P \to T_{\widehat{OP}}(X)$, let the *product* $\underline{u} \cdot \underline{u}' \in T_{\widehat{OP}}(X)$ be defined

$$\underline{u} \cdot \underline{u}' = \sum_{p \in P} \underline{u}(p) \cdot \underline{u}'(p).$$

This definition is (semantically) unique, due to associativity of addition (cf. Definition 3.1).

The central notion of *place invariants* is based on the above product:

5.4. Definition. Let a net $N = (P, T, F)$ be inscribed over a specification $SPEC = (S, OP, E)$ and a set X of variables. Let furthermore Y be a family of (S, OP)-variables, disjoint from X, and let $s \in S$ be freely chosen.

A vector $i : P \to T_{\widehat{OP},m_s}(Y)$ is a *place invariant of sort s* of N iff for all markings M and M' of N with $M' \in [M\rangle$:

$$M \cdot i \equiv_{\hat{E}} M' \cdot i.$$

We shall show that place invariants can be characterized as solutions of an equational system derived from N. The proof is based on the following lemma. It shows how the product of Definition 5.2 interacts with assignments:

5.5. Lemma. *Let $\Sigma = (S, OP)$ be a signature, let X and Y be disjoint sets of Σ-variables, let $u \in T_{\widehat{OP}}(X)$, $u' \in T_{\widehat{OP}}(Y)$ and let $\gamma: X \cup Y \to T_{OP}(Y)$ be an assignment with $\gamma(y) = y$ for all $y \in Y$. Then*

$$\bar{\gamma}(u \cdot u') = \bar{\gamma}(u) \cdot u'.$$

Proof. We proceed by induction over the structure of u.

(1) For $u \in T_{\widehat{OP}}(X)$ and all $u' \in T_{\widehat{OP}}(Y)$, we have to show: $\bar{\gamma}(\overline{ass_u}(u')) = \overline{ass_{\bar{\gamma}(u)}}(u')$. With 2.17 it suffices to show for all $y \in Y$: $\bar{\gamma}(ass_u(y)) = ass_{\bar{\gamma}(u)}(y)$. In a graphical representation, we have to show that the following diagram commutes:

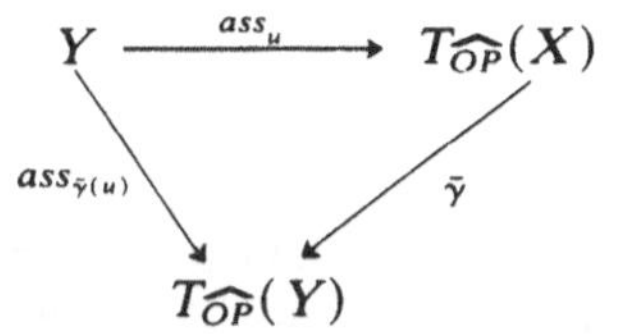

Let now s be the sort of u, i.e. $u \in T_{OP,s}(X)$. $\bar{\gamma}(u)$ has of course the same sort, i.e. $\bar{\gamma}(u) \in T_{OP,s}(Y)$.

In case $y \in Y_s$, we get $ass_{\bar{\gamma}(u)}(y) = \bar{\gamma}(u)$ (by Definition 5.1) $= \bar{\gamma}(ass_u(y))$ (by 5.1). Otherwise, for all $y \in Y \setminus Y_s$, $ass_{\bar{\gamma}(u)}(y) = y$ (by 5.1) $= \bar{\gamma}(y)$ (by construction of γ) $= \bar{\gamma}(ass_u(y))$ (by 5.1).

This completes the proof of $\bar{\gamma}(u) \cdot u' = \bar{\gamma}(u \cdot u')$ for $u \in T_{OP}(X)$. Thus the basis for the induction over the structure of u is given.

(2) To show the induction step over the structure of u, we distinguish three cases.

Case (a) For $u = \vartheta$ we get

$$\begin{aligned} \bar{\gamma}(\vartheta \cdot u') &= \bar{\gamma}(\vartheta) && \text{(by Definition 5.2)} \\ &= \vartheta && \text{(by 2.7)} \\ &= \vartheta \cdot u' && \text{(by Definition 5.2)} \\ &= \bar{\gamma}(\vartheta) \cdot u' && \text{(by 2.7).} \end{aligned}$$

Case (b) For $u = u_1 + u_2$ we show $\bar{\gamma}((u_1 + u_2) \cdot u') = \bar{\gamma}(u_1 + u_2) \cdot u'$ as follows:

$$\begin{aligned} \bar{\gamma}((u_1 + u_2) \cdot u') &= \bar{\gamma}(u_1 \cdot u' + u_2 \cdot u') && \text{(by Definition 5.2)} \\ &= \bar{\gamma}(u_1 \cdot u') + \bar{\gamma}(u_2 \cdot u') && \text{(by 2.7)} \\ &= \bar{\gamma}(u_1) \cdot u' + \bar{\gamma}(u_2) \cdot u' && \text{(by induction hypothesis)} \\ &= (\bar{\gamma}(u_1) + \bar{\gamma}(u_2)) \cdot u' && \text{(by 5.2)} \\ &= \bar{\gamma}(u_1 + u_2) \cdot u' && \text{(by 2.7).} \end{aligned}$$

Case (c) For $u=-u_1$ we show $\bar{\gamma}((-u_1)\cdot u')=\bar{\gamma}(-u_1)\cdot u'$ as follows:

$$\begin{aligned}\bar{\gamma}((-u_1)\cdot u') &= \bar{\gamma}(-(u_1\cdot u')) \quad \text{(by 5.2)}\\ &= -\bar{\gamma}(u_1\cdot u') \quad \text{(by 2.7)}\\ &= -(\bar{\gamma}(u_1)\cdot u') \quad \text{(by induction hypothesis)}\\ &= (-\bar{\gamma}(u_1)\cdot u') \quad \text{(by 5.2)}\\ &= \bar{\gamma}(-u_1)\cdot u' \quad \text{(by 2.7)}\end{aligned}$$

This completes the proof of Lemma 5.5. □

The following theorem states the central property of place invariants:

5.6. Theorem. *Let a net $N=(P,T,F)$ be inscribed over a specification $SPEC=(S,OP,E)$ and a set X of variables. Let furthermore Y be a family of (S,OP)-variables disjoint from X, and let $i:P\to T_{\widehat{OP},m_s}(Y)$ for some $s\in S$. If $\underline{t}\cdot i\equiv_{\hat{E}}\vartheta_s$ for all $t\in T$, then i is a place invariant.*

Proof. Assume $\underline{t}\cdot i\equiv_{\hat{E}}\vartheta_s$ for all $t\in T$. (a) We first show for all $t\in T$ and all assignments $\beta:X\to T_{OP}(Y)$: $\sum_{p\in P}\bar{\beta}(\underline{t}(p))\cdot i(p)\equiv_{\hat{E}}0_s$. To do so, we extend β to $\gamma:X\cup Y\to T_{OP}(Y)$ by $\gamma(x)=\beta(x)$ for $x\in X$ and $\gamma(y)=y$ for $y\in Y$. Now,

$$\begin{aligned}\sum_{p\in P}\bar{\beta}(\underline{t}(p))\cdot i(p) &= \sum_{p\in P}\bar{\gamma}(\underline{t}(p))\cdot i(p) \quad \text{(by definition of } \gamma)\\ &= \sum_{p\in P}\bar{\gamma}(\underline{t}(p)\cdot i(p)) \quad \text{(by Lemma 5.5)}\\ &= \bar{\gamma}\Big(\sum_{p\in P}\underline{t}(p)\cdot i(p)\Big) \quad \text{(by 2.7)}\\ &\equiv_{\hat{E}} \bar{\gamma}(\underline{t}\cdot i) \quad \text{(by 5.3)}\\ &\equiv_{\hat{E}} \bar{\gamma}(\vartheta_s) \quad \text{(by 2.13 and the assumption on } i)\\ &= \vartheta_s \quad \text{(by 2.7).}\end{aligned}$$

(b) To show the theorem, it is sufficient to show $M\cdot i=M'\cdot i$ for all $M\xrightarrow{t,\beta}M'$. So we get

$$\begin{aligned}M'\cdot i &= \sum_{p\in P}M'(p)\cdot i(p) \quad \text{(by Definition 5.3)}\\ &= \sum_{p\in P}(M(p)+\bar{\beta}(\underline{t}(p)))\cdot i(p) \quad \text{(by Corollary 4.6)}\\ &= \sum_{p\in P}(M(p)\cdot i(p)+\bar{\beta}(\underline{t}(p))\cdot i(p)) \quad \text{(by Definition 5.2)}\\ &\equiv_{\hat{E}} \sum_{p\in P}(M(p)\cdot i(p)+\vartheta_s) \quad \text{(by part (a) of this proof and 2.13)}\\ &\equiv_{\hat{E}} \sum_{p\in P}M(p)\cdot i(p) \quad \text{(by 3.1)}\\ &= M\cdot i \quad \text{(by Definition 5.3).} \qquad \square\end{aligned}$$

The inverse of Theorem 5.6 is also valid:

5.7. Theorem. *Let N, SPEC, X, Y and i be as in the assumption of Theorem 5.6. If i is a place invariant of N, then* $\underline{t} \cdot i \equiv_{\hat{E}} \vartheta$ *for all* $t \in T$.

Proof. Let $t \in T$, $\beta : X \to T_{OP}(Y)$ and let M, M' be markings of N with $M \xrightarrow{t,\beta} M'$. (Choose e.g. $M(p) = \bar{\beta}(\xrightarrow{p,t})$.) Then, we get

$$\begin{aligned} M' \cdot i &= (M + \bar{\beta}(\underline{t})) \cdot i \quad \text{(by Corollary 4.6)} \\ &= M \cdot i + \bar{\beta}(\underline{t}) \cdot i \quad \text{(by Definitions 5.3 and 5.2).} \end{aligned}$$

The assumption $M' \cdot i \equiv_{\hat{E}} M \cdot i$ now implies $\bar{\beta}(\underline{t}) \cdot i \equiv_{\hat{E}} \vartheta$.

Now we extend β to $\gamma : X \cup Y \to T_{OP}(Y)$ by $\gamma(x) = \beta(x)$ for $x \in X$ and $\gamma(y) = y$ for $y \in Y$. Clearly, $\bar{\gamma}(\underline{t}) \cdot i \equiv_{\hat{E}} \vartheta$, and with Lemma 5.5 we get $\bar{\gamma}(\underline{t} \cdot i) \equiv_{\hat{E}} \vartheta$. As this holds for all assignments β, Definition 2.15 implies $\underline{t} \cdot i \equiv_{\hat{E}} \vartheta$. □

From Definition 5.2 it follows directly that place invariants are additive:

5.8. Corollary. *If* i_1 *and* i_2 *are place invariants of some inscribed net N, then* $i_1 + i_2$ *and* $-i_1$ *are also place invariants of N.*

6. Application examples for place invariants

6.1. Properties of the dining philosophers system

Figure 4 shows the matrix, the initial marking and some invariants for the dining philosophers system in Fig. 1. (Entries ϑ are skipped.) The invariants are useful for proving some properties of this system, avoiding the consideration of its runs. These properties include:

(1) *Each philosopher always is either eating or thinking*: For each reachable marking $M \in [M_0\rangle$ we get with invariant i_1: $M(\mathbf{t}) + M(\mathbf{e}) = M(\mathbf{t}) \cdot y + M(\mathbf{e}) \cdot y = M \cdot i_1 = M_0 \cdot i_1 = p_0 + \cdots + p_4$. Hence, each philosopher occurs exactly once in $M(\mathbf{t}) + M(\mathbf{e})$.

	tf	**rf**	M_0	i_1	i_2	i_3
t	$-x$	x	$p_0 + \ldots + p_4$	y	$RF(y)$ $+LF(y)$	
a	$-(RF(x)$ $+LF(x))$	$RF(x)$ $+LF(x)$	$g_0 + \ldots + g_4$		$-z$	z
e	x	$-x$		y		$RF(y)+$ $LF(y)$

$$M_0 \cdot i_1 = p_0 + \ldots + p_4$$
$$M_0 \cdot i_2 = M_0 \cdot i_3 = g_0 + \ldots + g_4$$

Fig. 4. Matrix, initial marking and three place invariants to Fig. 1 with $y \in Y_{phils}$ and $z \in Y_{forks}$.

(2) *If a fork is available, then both of its potential users are thinking*: for each reachable marking $M \in [M_0\rangle$, we get from i_2:

$$-M(\mathbf{a}) + RF(M(\mathbf{t})) + LF(M(\mathbf{t}))$$

$$= M(\mathbf{a}) \cdot (-z) + M(\mathbf{t}) \cdot (RF(y) + LF(y))$$

$$= M \cdot i_2 = M_0 \cdot i_2$$

$$= (g_0 + \cdots + g_4) \cdot (-z) + (p_0 + \cdots + p_4) \cdot (RF(y) + LF(y))$$

$$= (-g_0 - \cdots - g_4) + (g_0 + \cdots + g_4) + (g_0 + \cdots + g_4) = g_0 + \cdots + g_4.$$

So we get $M(\mathbf{a}) = RF(M(\mathbf{t})) + LF(M(\mathbf{t})) - (g_0 + \cdots + g_4)$. Now, if $g_i \in M(\mathbf{a})$, it follows $g_i \in RF(M(\mathbf{t}))$ as well as $g_i \in LF(M(\mathbf{t}))$. This yields p_i and p_{i+1} in $M(t)$.

(3) *Neighboured philosophers never eat at the same time*: Neighbours p_i, p_{i+1} both eating at the same time are represented by a marking $\bar{M}$ with $\bar{M}(\mathbf{e}) \geq p_i + p_{i+1}$. Then, $\bar{M} \cdot i_3 \geq \bar{M}(\mathbf{e}) \cdot i_3(\mathbf{e}) = RF(\bar{M}(\mathbf{e})) + LF(\bar{M}(\mathbf{e})) \geq RF(p_i) + LF(p_{i+1}) = g_i + g_i$. But for each reachable marking $M \in [M_0\rangle$ we get with i_3: $M(\mathbf{a}) + RF(M(\mathbf{e})) + LF(M(\mathbf{e})) = M(\mathbf{a}) \cdot z + M(\mathbf{e}) \cdot (RF(y) + LF(y)) = M \cdot i_3 = M_0 \cdot i_3 = (g_0 + \cdots + g_4) \cdot z = g_0 + \cdots + g_4$.

Notice that the invariants of Fig. 4 rely on the multiset equations only. They make no use of the particular equations of **phils** and therefore might be denoted *Σ-invariants.*

6.2. N-Tuples as net inscriptions

Nets with structured tokens often include pairs or generally *n*-tuples of constants and variables as markings and arc inscriptions, respectively. Pairs and generally *n*-tuples can easily be specified in the algebraic framework discussed in this paper, providing all universal properties and constructs of cartesian products. A most elementary specification is the following one:

pair =
sorts: s
$\quad$ *pair*

opns: $a_1, a_2 : \to s$
$\quad PAIR : s\, s \to pair$
$\quad PR1 : pair \to s$
$\quad PR2 : pair \to s$

eqns: $x_1, x_2 \in s$
$\quad PR1(PAIR(x_1, x_2)) = x_1$
$\quad PR2(PAIR(x_1, x_2)) = x_2$

It is obvious how to construct pairs with components of different sorts or general n-tuples, or how to use predefined sorts as component sorts of such tuples.

Figure 5 shows a **pair**-inscribed net, and one of its invariants is discussed in Fig. 6. With $\beta(x_1)=a_1$ and $\beta(x_2)=a_2$, we get $M \xrightarrow{t,\beta} M'$. As usual, we write (x_1, x_2) for $PAIR(x_1, x_2)$.

Notice that the invariant i of Fig. 6 makes use of the particular equations of **pair** (whereas the invariants of Fig. 4 rely on the multiset equations only). i of Fig. 6 might therefore be called a (proper) *SPEC*- (or **pair**-)*invariant.*

6.3. A database maintaining scheme

As a last example, we refer to the often considered scheme for maintaining multiple copies of a database [11, 14].

A set D of sites is assumed, each of which being able to send update requests to all other sites. Upon receiving an update request, a site performs the required update of its database and returns an acknowledgement to the sender.

We assume a successor function SUC on the set $D=\{a_1, \ldots, a_n\}$ of sites, by $a_{i+1}=SUC(a_i)$ and $a_1=SUC(a_n)$. We furthermore assume the specification **pair** considered above. This leads to the following specification, assuming a given natural number n:

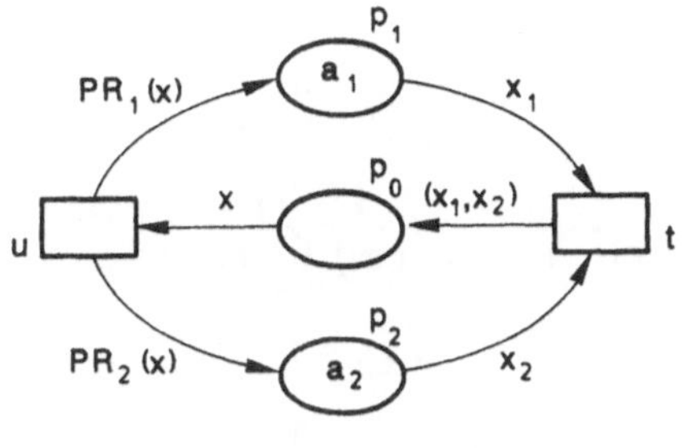

Fig. 5. A **pair**-inscribed net with $x_1, x_2 \in X_s$ and $x \in X_{pair}$.

	t	u	i	$t \cdot i$	$u \cdot i$	M	$M \cdot i$	M'	$M' \cdot i$
p_0	(x_1, x_2)	$-x$	$PR_1(y)$	x_1	$-PR_1(x)$			(a_1, a_2)	a_1
p_1	$-x_1$	$PR_1(x)$	z	$-x_1$	$PR_1(x)$	a_1	a_1		
p_2	$-x_2$	$PR_2(x)$				a_2			
				θ	θ		a_1		a_1

Fig. 6. The matrix to Fig. 5, a place invariant i with $y \in Y_{pair}$ and $z \in Y_s$, and two markings M, M' with $M' \in [M\rangle$.

number n:

database$^{(n)}$ = **pair**+

sorts: *const*

opns: $a_1, \ldots, a_n : \to s$

$dot : \to const$

$SUC : s \to s$

$d_1 : s \to const$

$d_2 : pair \to const$

eqns: $x, y \in s$;

$SUC(a_i) = a_{i+1}\ (i = 1, \ldots, n-1)$

$SUC(a_n) = a_1$

$d_1(x) = d_2(x, y) = dot$

Figure 7 shows the database-maintaining scheme as a **data-base**$^{(n)}$-inscribed net over $X_s = \{q, r\}$. The constant "*dot*" is as usual represented as a black dot. As shorthands for the initial marking, let $D = a_1 + \cdots + a_n$ and let N be the sum of all pairs (a_i, a_j) with $i \neq j$. In Fig. 7 as well as in Fig. 8, for variables x in $X_s \cup Y_s$ we use the shorthand N_x for $(x, SUC(x)) + \cdots + (x, SUC^{n-1}(x))$.

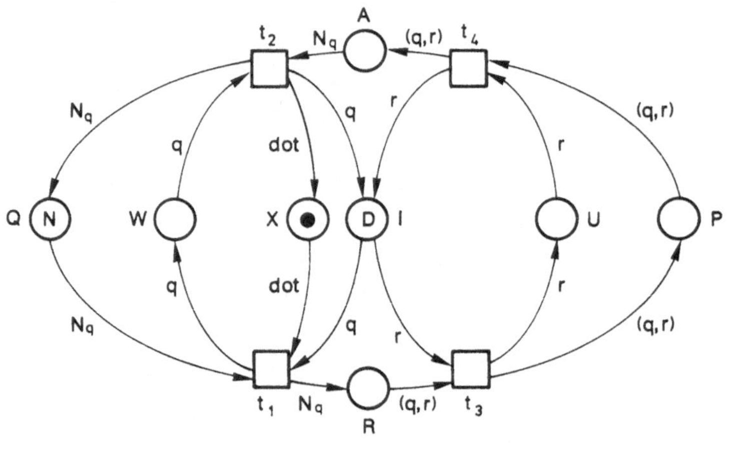

Fig. 7. Scheme for maintaining multiple copies of a database.

Initially, all sites are idle (marking D on I), all "envelopes" for messages are empty (marking N on Q) and no update requests are under way (dot on X). Transition t_1 models the dispatch of update requests by some site q, t_3 the reception of such requests by each single recipient, t_4 the dispatch of acknowledgements, and t_2 the reception of all acknowledgements by q.

Figure 8 gives the matrix, the initial marking and some place invariants over $Y = Y_s \cup Y_{pair} \cup Y_{const}$ with $Y_s = \{x\}$, $Y_{pair} = \{z\}$ and $Y_{const} = \{c\}$. For terms u and

	$\mathbf{t}_1$	$\mathbf{t}_2$	$\mathbf{t}_3$	$\mathbf{t}_4$	M_o	i_1	i_2	i_3	i_4	i_5	i_6	i_7	i_8	i_9
X	$-dot$	dot			dot			c			$(n-1)c$	$(n-1)(ABS(c))$	$(n-1)(ABS(c))$	
I	$-q$	q	$-r$	r	D	x								
W	q	$-q$				x		$d_1(x)$		N_z				N_z
U			r	$-r$		x			$-x$			$ABS(x)$		
Q	$-N_q$	N_q			N		z			z	$-d_2(z)$			
R	N_q		$-(q,r)$				z					$ABS(z)$	$ABS(z)$	$-z$
P			(q,r)	$-(q,r)$			z		$pr_2(z)$				$ABS(z)$	$-z$
A		$-N_q$		(q,r)			z					$ABS(z)$	$ABS(z)$	$-z$

Fig. 8. Matrix, initial marking and nine invariants for the net in Fig. 7.

naturals n, we write nu as a shorthand for $u+\cdots+u$ (n times). A new operation symbol *ABS* is used in the invariants of Fig. 8. $ABS(u)$ is to give the cardinality of the multiset represented by the term u. To cover this formally, the multiset specification of 3.1 is extended to

$$\begin{aligned}
&\mathbf{m_}SPEC' = \mathbf{m_}SPEC + \mathbf{int} + \\
&\quad \text{opns: } ABS_s : m_s \rightarrow int \\
&\quad \text{eqns: } \left.\begin{aligned} & a \in s;\ p, q \in m_s \\ & ABS_s(\vartheta_s) = 0 \\ & ABS_s(MAKE(a)) = 1 \\ & ABS_s(+_s(p,q)) = ABS_s(p) + ABS_s(q) \\ & ABS_s(-_s(p)) = -ABS_s(p) \end{aligned}\right\}, \text{ for all } s \in S
\end{aligned}$$

assuming any reasonable specification **int** of the natural numbers.

7. Transition invariants

The previous section showed solutions of $\underline{N}^T \cdot i \equiv_{\hat{E}} \vartheta$ to represent properties of inscribed nets N. Symmetry and duality of the net calculus suggest to study also equations formed $\underline{N} \cdot j \equiv_{\hat{E}} \vartheta$. A solution j then assigns an object—of whatever kind—to each transition. As in the case of place invariants we have to specify which kinds of solutions we are interested in and how to derive system properties from them.

Transition invariants of place/transition nets return a natural number j_t for each transition t. Each occurrence sequence $M_0[t_1\rangle M_1 \ldots [t_n\rangle M_n$ with each t occurring j_t times, reproduces the initial marking (i.e. $M_0 = M_n$). We will here obtain a similar result, but we have of course to take into consideration the modes of transition occurrences.

Formally we must be capable of summing up several modes in which a transition may occur. This is captured by the following notion of *multi-assignments*:

7.1. Definition. Let $\Sigma = (S, OP)$ be a signature and let X, Y be sets of Σ-variables. The set $MA_\Sigma(X, Y)$ of *multiassignments over* Σ, X *and* Y is the smallest set of mappings $\Gamma : T_{\widehat{OP}}(X) \to T_{\widehat{OP}}(Y)$ such that

(i) for each assignment $ass : X \to T_{OP}(Y)$, $\overline{ass} \in MA_\Sigma(X, Y)$;

(ii) if $\Gamma_1, \Gamma_2 \in MA_\Sigma(X, Y)$ then $(\Gamma_1 + \Gamma_2) \in MA_\Sigma(X, Y)$ and $(-\Gamma_1) \in MA_\Sigma(X\ Y)$, inductively given by $(\Gamma_1 + \Gamma_2)(u) = \Gamma_1(u) + \Gamma_2(u)$ and $(-\Gamma_1)(u) = -(\Gamma_1(u))$, respectively, for each $u \in T_{\widehat{OP}}(X)$.

A multiassignment Γ is *constant* iff for all u, $\Gamma(u) \in T_{\widehat{OP}}$.

There is a particular multiassignment ϑ_{MA}, definable as $\Gamma + (-\Gamma)$ for any Γ. Obviously, $\vartheta_{MA}(u) = \vartheta$ for each $u \in T_{\widehat{OP}}(X)$.

The product of multiset terms defined in Definition 5.2 can be considered as a special case of the above definition, based on constant (instead of arbitrary) assignments: $(u_1 + u_2) \cdot u'$ and $(-u_1) \cdot u'$ read now $(\overline{ass_{u_1}} + \overline{ass_{u_2}})u'$ and $(-\overline{ass_{u_1}})(u')$, respectively.

The occurrence count of each transition in a sequence of transition occurrences can now be defined as a constant multiassignment:

7.2. Definition. Let $N = (P, T, F)$ be an inscribed net and let $\sigma = M_0 \xrightarrow{t_1, \beta_1} \cdots \xrightarrow{t_n, \beta_n} M_n$ be an occurrence sequence in N. For each $t \in T$, the *occurrence count* Γ_t of t in σ is defined by $\Gamma_t = \Sigma\{\overline{\beta_i} \mid t_i = t\}$. ($\Sigma$ of course denotes the sum of multiset assignments.)

Initial and final markings of occurrence sequences can now be related by occurrence counts:

7.3. Theorem. *Let* $N = (P, T, F)$ *be an inscribed net and for each* $t \in T$, *let* Γ_t *be the occurrence count of* t *in* $\sigma = M_0 \xrightarrow{t_1, \beta_1} \cdots \xrightarrow{t_n, \beta_n} M_n$. *Then for all* $p \in P$ *it holds*:

$$M_n(p) = M_0(p) + \sum_{t \in T} \Gamma_t(\underline{t}(p)).$$

Proof. By induction over the length n of σ.

If $n = 0$, for all $t \in T$, the occurrence count of t in σ is $\Gamma_t = \vartheta_{MA}$. Then we get for all $p \in P$: $M_n(p) = M_0(p) = M_0(p) + \vartheta = M_0(p) + \sum_{t \in T} \vartheta_{MA}(\underline{t}(p)) = M_0(p) + \sum_{t \in T} \Gamma_t(\underline{t}(p))$.

To show the induction step, for each $t \in T$ let Γ'_t be the occurrence count of $M_0 \xrightarrow{t_1, \beta_1} \cdots \xrightarrow{t_{n-1}, \beta_{n-1}} M_{n-1}$. The definition of occurrence counts in Definition 7.2 implies

(*) $\Gamma_{t_n} = \Gamma'_{t_n} + \overline{\beta_n}$ and

(**) $\Gamma_t = \Gamma'_t$ for all $t \neq t_n$.

For each $p \in P$, let $u'_p := \sum_{t \in T\setminus\{t_n\}} \Gamma'_t(\underline{t}(p))$ and $u_p := \sum_{t \in T\setminus\{t_n\}} \Gamma_t(\underline{t}(p))$. From (**) it follows that for each $p \in P$: $u'_p = u_p$.

Now we get for each $p \in P$:

$$\begin{aligned} M_n(p) &= M_{n-1}(p) + \overline{\beta_n}(\underline{t_n}(p)) \quad \text{(by Corollary 4.6)} \\ &= M_0(p) + u'_p + \Gamma'_{t_n}(\underline{t_n}(p)) \\ &\quad + \overline{\beta_n}(\underline{t_n}(p)) \quad \text{(by the induction assumption)} \\ &= M_0(p) + u'_p + \Gamma_{t_n}(\underline{t_n}(p)) \\ &= M_0(p) + u_p + \Gamma_{t_n}(\underline{t_n}(p)) \\ &= M_0(p) + \sum_{t \in T} \Gamma_t(\underline{t}(p)). \quad \square \end{aligned}$$

We now turn to the notion of "transition invariants" and obtain their essential properties as a corollary to the above Theorem 7.3:

7.4. Definition. Let a net $N = (P, T, F)$ be inscribed over (S, OP, E) and X. Let $\sigma = M_0 \xrightarrow{t_1, \beta_1} \cdots \xrightarrow{t_n, \beta_n} M_n$ be an occurrence sequence of N with $M_0 \equiv_{\hat{E}} M_n$. For each $t \in T$, let j_t be the occurrence count of t in σ. Then the vector $(j_t)_{t \in T}$ is a *transition invariant* of N.

7.5. Corollary. *Let a net $N = (P, T, F)$ be inscribed over (S, OP, E), and let $(j_t)_{t \in T}$ be a transition invariant of N. Then for each $p \in P$,*

$$\sum_{t \in T} j_t(\underline{t}(p)) \equiv_{\hat{E}} \vartheta.$$

Proof. This follows from Theorem 7.3 and the above definition. □

To achieve in Corollary 7.5 a product notation comparable to the product of Definition 5.3, we have to define a product $j \cdot v$ for $MA_\Sigma(X, Y)$-vectors j with $T_{\widehat{OP}}(X)$-vectors v:

7.6. Definition. Let T be a finite set, let $\Sigma = (S, OP)$ be a signature and let X be a set of Σ-variables. For $j: T \to MA_\Sigma(X, Y)$ and $v: T \to T_{\widehat{OP}}(X)$ we define a product $j \cdot v \in T_{\widehat{OP}}(Y)$ by

$$j \cdot v = \sum_{t \in T} j(t)(v(t)).$$

Based on Definitions 7.4 and 7.6 we then get the following corollary.

7.7. Corollary. *Let $N = (P, T, F)$ be an inscribed net. For each $p \in P$, let $\underline{p}: T \to T_{OP}(X)$ be defined by $\underline{p}(t) := \underline{t}(p)$. If a vector $j: T \to MA_\Sigma(X, Y)$ is a transition invariant of N, then $j \cdot \underline{p} \equiv_{\hat{E}} \vartheta$.*

We close this section with some examples:

Transition invariants of the system of dining philosophers (Fig. 1) are quite simple: For each assignment *ass* of the only variable x, $j_{tf} = j_{rf} = \overline{ass}$ yields a transition invariant. This shows that initial markings are retained upon firing of both transitions equally often in the same modes.

For Fig. 5, we get a transition invariant (j_t, j_u) with $j_t(x_1) = a_1$, $j_t(x_2) = a_2$ and $j_u(x) = (a_1, a_2)$. Thus the initial marking is retained by t occurring in mode $\beta(x_i) = a_i$ $(i = 1, 2)$ and u occurring in mode $\beta(x) = (a_1, a_2)$.

Transition invariants for Fig. 7 are somewhat more involved: For $i = 1, \ldots, n$, let $\beta_i : \{q, r\} \to \{a_1, \ldots, a_n\}$ be defined by $\beta_i(q) = a_1$ and $\beta_i(r) = a_i$. Then we define a transition invariant $(j_{t_i})_{i=1,\ldots,4}$ by $j_{t_1} = j_{t_2} = \beta_1$ and $j_{t_3} = j_{t_4} = \beta_2 + \cdots + \beta_n$. This invariant describes an update cycle of the data base, initiated by a_1: Firing t_1 in mode β_1 describes a_1 sending messages to $a_2, \ldots, a_n$. Each of $a_2, \ldots, a_n$ then performs its local update (occurrences of t_3 and t_4 in modes $\beta_2, \ldots, \beta_n$). a_1 finally collects all commitments (occurrence of t_2 in mode β_1) and releases a dot to X, allowing for a further update cycle.

8. Homomorphic transformations of arc inscriptions

Here we investigate the effect of transforming (by extended assignments) arc inscriptions. It turns out that the overall behaviour will in general be restricted but never be extended under this kind of transformation. The behaviour is retained by the special case of bijectively renaming variables in the environment of a transition. Place invariants are retained and transition invariants are transformed by extended assignment transformations.

8.1. Definition. Let $ins = (\varphi, M_0, \lambda)$ be an inscription of a net $N = (P, T, F)$.

(i) For $t \in T$ and $ass : X \to T_{OP}(X)$, let $\lambda_{t,ass}$ be defined by

$$\lambda_{t,ass}(x, y) = \begin{cases} \overline{ass} \circ \lambda(x, y) & \text{iff } x = t \text{ or } y = t, \\ \lambda(x, y) & \text{otherwise.} \end{cases}$$

(ii) Let the inscription $ins_{t,ass}$ of N be defined as $ins_{t,ass} = (\varphi, M_0, \lambda_{t,ass})$.

As an application example for homomorphic transformation we consider the relationship among the two versions of the dining philosophers system in Fig. 1 and Fig. 3: To this end we combine both underlying specifications and augment two obvious equations: Let

phils″ = **phils** + **phils′** +
  eqns: $y \in fork$
    $LF(RU(y)) = y$
    $RF(RU(y)) = RSF(y)$

Then the assignment $ass(x) = RU(y)$, applied to both transitions of Fig. 1, yields the inscriptions of Fig. 3. One likewise transforms Fig. 3 to Fig. 1 by e.g. an assignment ass' with $ass'(y) = LF(x)$, assuming the equations $RU(LF(x)) = x$ and $RSF(LF(x)) = RF(x)$.

Next we investigate the behaviour of transformed nets:

8.2. Lemma. *Let ins be an inscription of a net N over (S, OP, E) and X, and let M, M' be markings of N. Let $ass : X \to T_{OP}(X)$ be an assignment, and let $\beta : X \to T_{OP}$ be an occurrence mode. Then $M \xrightarrow{t,\beta\circ ass} M'$ in (N, ins, E) iff $M \xrightarrow{t,\beta} M'$ in $(N, ins_{t,ass}, E)$.*

Proof. $M \xrightarrow{t,\beta\circ ass} M'$ for *ins* iff for each $p \in P$,

$$\begin{aligned} M'(p) &= M(p) - \bar{\beta} \circ \overline{ass}(\lambda(p, t)) + \bar{\beta} \circ \overline{ass}(\lambda(t, p)) \\ &= M(p) - \bar{\beta}(\overline{ass} \circ \lambda(p, t)) + \bar{\beta}(\overline{ass} \circ \lambda(t, p)). \end{aligned}$$

This holds iff $M \xrightarrow{t,\beta} M'$ for $ins_{t,ass}$. □

Hence each step in the transformed net corresponds to a step in the original net. Vice versa, only $\overline{ass}$-prefixed assignments in the original net correspond to steps in the transformed net. Consequently, reachability sets $[M\rangle$ under $ins_{t,ass}$ are subsets of $[M\rangle$ under *ins*.

Above we have shown that both systems in Fig. 1 and Fig. 3 can mutually be transformed by the assignments $ass(x) = RF(y)$ and $ass'(y) = LF(x)$, respectively. So Lemma 8.2 implies that both nets in fact behave equally, i.e. a step $M \xrightarrow{t,\beta} M'$ can occur in Fig. 1 if and only if a step $M \xrightarrow{t,\tilde{\beta}} M'$ can occur for some assignment $\tilde{\beta}$ in Fig. 3.

Next we show that place invariants are retained by homomorphic transformations of arc inscriptions:

8.3. Theorem. *Let ins be an inscription of a net N, let t be a transition of N and let ass be an assignment of the involved variables. Then each place invariant i of (N, ins, E) is also a place invariant of $(N, ins_{t,ass}, E)$.*

Proof. For all $u \in T$, if $u \neq t$, the vectors $\underline{u}$ are equal for both nets. So, according to Definitions 5.4 and 5.3, we have to show: If $\sum_{p\in P} \underline{t}(p) \cdot i(p) \equiv_E \vartheta$ for *ins*, then $\sum_{p\in P} \underline{t}(p) \cdot i(p) \equiv_E \vartheta$ for $ins_{t,ass}$. With Notation 4.3, $\underline{t}(p) = \lambda(t, p) - \lambda(p, t)$ for *ins* and $\underline{t}(p) = \overline{ass} \circ \lambda(t, p) - \overline{ass} \circ \lambda(p, t)$ for $ins_{t,ass}$. Now we get for $ins_{t,ass}$:

$$\begin{aligned} \sum_{p\in P} \underline{t}(p) \cdot i(p) &= \sum_{p\in P} (\overline{ass} \circ \lambda(t, p) - \overline{ass} \circ \lambda(p, t)) \cdot i(p) \\ &= \sum_{p\in P} (\overline{ass}(\lambda(t, p)) - \overline{ass}(\lambda(p, t))) \cdot i(p) \\ &= \sum_{p\in P} \overline{ass}(\lambda(t, p) - \lambda(p, t)) \cdot i(p) \quad \text{(by 2.7)} \\ &= \sum_{p\in P} \overline{ass}((\lambda(t, p) - \lambda(p, t)) \cdot i(p)) \end{aligned}$$

(by Lemma 5.5, extending *ass* to Y by identity)

$$= \overline{ass}\left(\sum_{p \in P} (\lambda(t, p) - \lambda(p, t)) \cdot i(p) \right) \quad \text{(by 2.7)}$$

$$= \overline{ass}(\vartheta) \quad \text{(by } i \text{ being a place invariant of } N \text{ for } ins\text{)}$$

$$= \vartheta. \quad \square$$

As Fig. 1 and Fig. 3 can be mutually transformed, the above theorem implies that the place invariants of both systems coincide.

Next we consider the effect of transformations to transition invariants. In contrast to place invariants, they are not retained but transformed by the assignment applied to the underlying net:

8.4. Theorem. *Let ins be an inscription of a net $N = (P, T, F)$, let t be a transition of N and let ass be an assignment of the involved variables. If $(j_{t'})_{t' \in T}$ is a transition invariant of (N, ins, E) with $j_t = j \circ \overline{ass}$ for some assignment j, the vector $(j'_{t'})_{t' \in T}$ is a transition invariant of $(N, ins_{t,ass}, E)$, with $j'_{t'} = j_{t'}$ for $t' \neq t$ and $j_t = \bar{j}$.*

Proof. It is sufficient to show

$$\overline{j \circ ass}(\underline{t}(p)) \text{ in } (N, ins, E)$$

$$= \overline{j \circ ass}(\lambda(t, p) - \lambda(p, t))$$

$$= \bar{j}(\overline{ass} \circ \lambda(t, p) - \overline{ass} \circ \lambda(p, t))$$

$$= \bar{j}(\underline{t}(p)) \text{ in } (N, ins_{t,ass}, E). \quad \square$$

In case two inscribed nets can be mutually transformed, the above lemma and theorems imply entirely identical behaviour:

8.5. Corollary. *Let ins be an inscription of a net $N = (P, T, F)$ over (S, OP, E) and X and let ass_1, $ass_2 : X \to T_{OP}(X)$ be two assignments such that $\overline{ass}_2 \circ ass_1 \equiv_{\hat{E}} \overline{ass}_1 \circ ass_2 \equiv_{\hat{E}} id$.*

(i) *For each marking $M : P \to T_{OP^+}$, the reachability sets $[M\rangle$ are identical for both (N, ins, E) and (N, ins_{t,ass_1}, E).*

(ii) *Both ins and ins_{t,ass_1} yield identical sets of place invariants for N.*

(iii) *If $(j_t)_{t \in T}$ is a transition invariant for (N, ins, E) then $(j_t \circ ass_2)_{t \in T}$ is a transition invariant for (N, ins_{t,ass_1}, E), and if $(j_t)_{t \in T}$ is a transition invariant for (N, ins_{t,ass_1}, E) then $(j_t \circ ass_1)_{t \in T}$ is a transition invariant for (N, ins, E).*

Proof. (i) If $M \xrightarrow{t,\beta} M'$ for *ins*, then $M \xrightarrow{t,\beta \circ ass_2 \circ ass_1} M'$ for *ins* (by assumptions of the theorem). Then $M \xrightarrow{t,\beta \circ ass_2} M'$ for ins_{t,ass_1} (by Lemma 8.2).

Vice versa, if $M \xrightarrow{t,\beta} M'$ for $ins_{t,ass1}$, then $M \xrightarrow{t,\beta \circ ass_1 \circ ass_2} M'$ for ins_{t,ass_1} (by assumptions of the theorem). Then, by Lemma 8.2, $M \xrightarrow{t,\beta \circ ass_1} M'$ for $ins_{t,ass_1 \circ ass_2} =$ *ins*.

(ii) follows from Theorem 8.3, as $ins_{t,ass_1 \circ ass_2} = ins$.

(iii) $(j_t)_{t \in T}$ is a transition invariant of N for *ins* iff $(j_t \circ ass_2 \circ ass_1)_{t \in T}$ is one. Then by Theorem 8.4, $(j_t \circ ass_2)_{t \in T}$ is a transition invariant for ins_{t,ass_1}. Vice versa, $(j_t)_{t \in T}$ is a transition invariant of N for ins_{t,ass_1} iff $(j_t \circ ass_1 \circ ass_2)_{t \in T}$ is one. Then by Theorem 8.4, $(j_t \circ ass_1)_{t \in T}$ is a transition invariant of N for $ins_{t,ass_1 \circ ass_2} = ins$. □

The bijective renaming of variables in the environment of transitions is a special case of this theorem.

Homomorphic transformations of arc inscriptions preserve dead transitions. If a net N has a dead transition t_0 under an inscription *ins*, then t_0 remains dead under each $ins_{t,ass}$. This follows directly from Lemma 8.2. Homomorphic transformations may produce additional dead transitions as Fig. 9 shows.

The situation is slightly different if we consider dead markings. Call a marking M of a net N *dead* iff M has no successor marking M', i.e. for no t and no β,

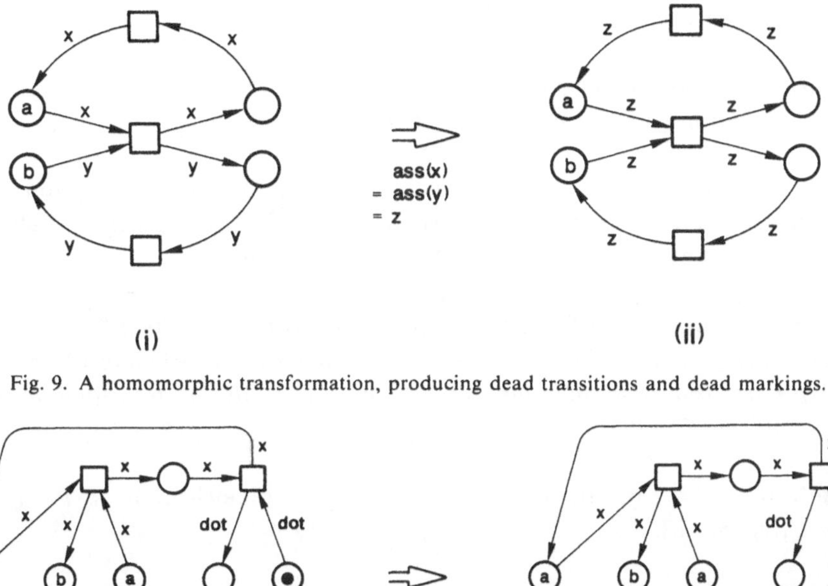

Fig. 9. A homomorphic transformation, producing dead transitions and dead markings.

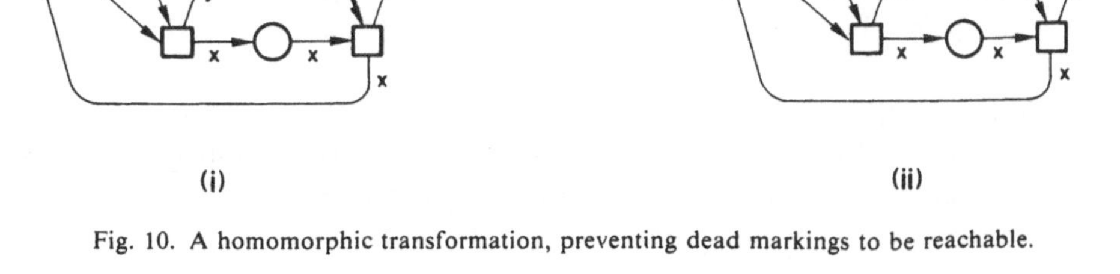

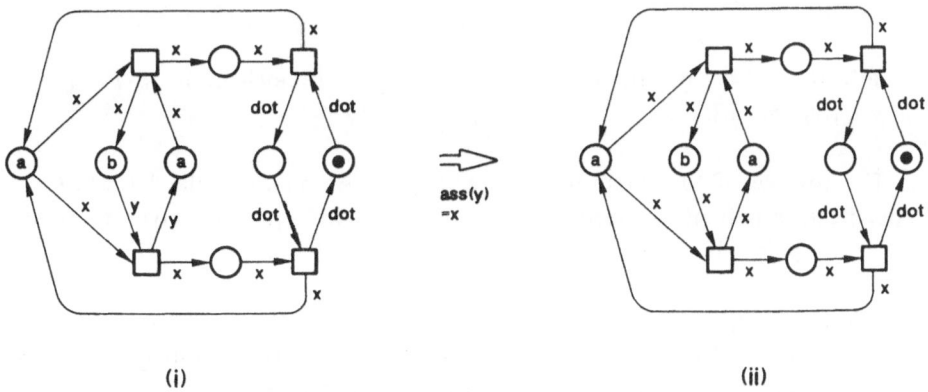

Fig. 10. A homomorphic transformation, preventing dead markings to be reachable.

$M \xrightarrow{t,\beta} M'$. It is almost obvious that homomorphic transformations may produce additional reachable dead markings, and Fig. 9 shows an example. But homomorphic transformations additionally may prevent dead markings to be reachable any more; Fig. 10 shows an example for this fact (assuming no equations).

9. Enhancing equations

Let a net N be inscribed over a specification $SPEC = (S, OP, E)$, and let E_1 be an additional set of (S, OP)-equations. Here we consider properties retained or lost by interpreting N over $SPEC1 = (S, OP, E + E1)$.

Transition occurrences, place and transition invariants, and the absence of dead transitions turn out to be retained under additional equations:

9.1. Proposition. *Let $SPEC = (S, OP, E)$ and $SPEC1 = (S, OP, E + E1)$ be two specifications and let ins be a SPEC-inscription of a net N.*

(i) *If $M \xrightarrow{t,\beta} M'$ in (N, ins, E), then $M \xrightarrow{t,\beta} M'$ also in $(N, ins, E + E_1)$.*

(ii) *For each marking M of N, the set $[M\rangle$ in (N, ins, E) is a subset of $[M\rangle$ in $(N, ins, E + E_1)$.*

Proof. (i) follows directly from Corollary 4.5.
(ii) is an immediate consequence of (i). □

The reverse of this proposition is not valid under the assumption of initial semantics. As an example, in Fig. 9 (ii) no step $M \xrightarrow{t,\beta} M'$ is possible at all, whereas the additional equation "$a = b$" leads to a live system.

9.2. Proposition. *Let SPEC, SPEC1 and ins be as in Proposition* 9.1.

(i) *Each place invariant of (N, ins, E) is also a place invariant of $(N, ins, E + E_1)$.*

(ii) *Each transition invariant of (N, ins, E) is also a transition invariant of $(N, ins, E + E_1)$.*

Proof. *If* $\underline{t} \cdot i \equiv_{\hat{E}} \vartheta$, then also $\underline{t} \cdot i \equiv_{\widehat{E+E_1}} \vartheta$. Likewise, if $j \cdot \underline{p} \equiv_{\hat{E}} \vartheta$, then $j \cdot \underline{p} \equiv_{\widehat{E+E_1}} \vartheta$. □

The reverse of this theorem does not hold. As an example, let *SPEC* be the specification **pair** of Subsection 6.2 without the two equations given there, and let *SPEC*1 = **pair**. Then the vector i of Fig. 6 is no invariant under *SPEC*, but under *SPEC*1.

Additional equations preserve the absence of dead transitions: If a net N has no dead transition under a specification *SPEC*, it has also no dead transition if additional equations are assumed to be valid. Additional equations may turn dead transitions into nondead ones. As an example, in Fig. 9(ii) all transitions are dead, whereas the additional equation "$a = b$" makes the net entirely live. This step shows that dead markings may turn into nondead ones, but additional equations also can lead to dead markings. As an example, Fig. 10(ii) has no reachable dead markings, whereas, with the equation "$a = b$", dead markings become reachable.

10. Extending the formalism

This paper strives at the basic notions and techniques for construction and analysis of Petri nets with structured tokens. A number of extensions make the formalism more handy for practical applications. We start with short look at using ordinary sets instead multisets. Then five generalizations of the formalism are glanced over; place capacities, transition inscriptions, extended arc inscriptions, schematic markings and more general equivalence transformations.

10.1. Place capacities

Capacity functions K_p may be assigned to places p, indicating for each item a maximum number of copies in allowable markings. Transition occurrences $M \xrightarrow{t,\beta} M'$ are discharged if in M', the multiplicity of some item d in some place $p \in t^{\bullet}$ exceeds the capacity $K_p(d)$.

In the setting of the above term calculus, an item may be represented by different ground terms u, $u' \in T_{OP}$. A capacity function for place p, $K_p : T_{OP,\varphi(p)} \rightarrow \mathbb{N} \cup \{\omega\}$ must therefore assign equal multiplicities $K_p(u) = K_p(v)$ in case $u \equiv_E v$.

In case of finitely many $\equiv_E$-equivalence classes and finite capacities for all items, the well-known construct of complements as outlined in Fig. 11 renders capacity functions superfluous. More general capacity functions cannot be implemented this way as infinite multisets cannot be represented in the term calculus.

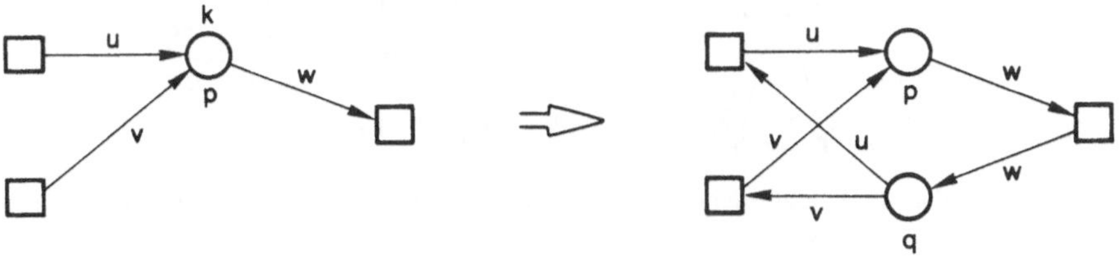

Fig. 11. Place complementation: Let $M(q)$ be such that the number of items in $M(p)+M(q)$ is just the item's capacity.

The calculi of place and transition invariants remain valid under the introduction of capacities. This holds likewise for the theorems on homomorphic transformations and additional equations.

10.2. Strict nets

A further variant of nets with structured tokens assumes markings to represent ordinary sets (instead of multisets). Transitions are prevented from occurrence in case an item to be put to a place does already belong to the place's marking. Places p in this setting represent predicates $\tilde{p}$ with variable extensions. The overall maximal extension of $\tilde{p}$ is the set of all items of sort $\varphi(p)$. The respective actual extension of $\tilde{p}$ is given by the actual markings $M(p)$. The denotation of "Predicate/Transition Nets" is due to those predicates.

This model can be considered as being based on the capacity function $K_p(d)=1$ for all places p and all d of sort $\varphi(p)$. In our formalism a specification **set_***SPEC* could describe this model (**set_***SPEC* should of course provide the usual operations on sets).

10.3. Transition inscriptions

Coming back to the remark following Corollary 4.6, additional predicates can be assigned to transitions. This provides a means for formulating additional requirements to the enabling of transitions. In our setting, terms of sort *bool* will do this job. Assuming in Definition 4.4 a further component $\eta: T \to T_{OP,bool}(X)$, we define in Definition 4.4(iii) a transition to be enabled in a mode β if additionally $\bar{\beta}(\eta(t)) \equiv_E TRUE$.

Given a transition inscription u, the set $U := \{\bar{\beta}(u) \mid \beta: X \to T_{OP}\}$ may decompose into finitely many $\equiv_{\hat{E}}$-equivalence classes $U = [u_1] \cup \cdots \cup [u_n]$ for some $u_1, \ldots, u_n \in T_{OP^+}$. In this case, transition inscriptions may equivalently be replaced by the construction of a loop as Fig. 12 outlines.

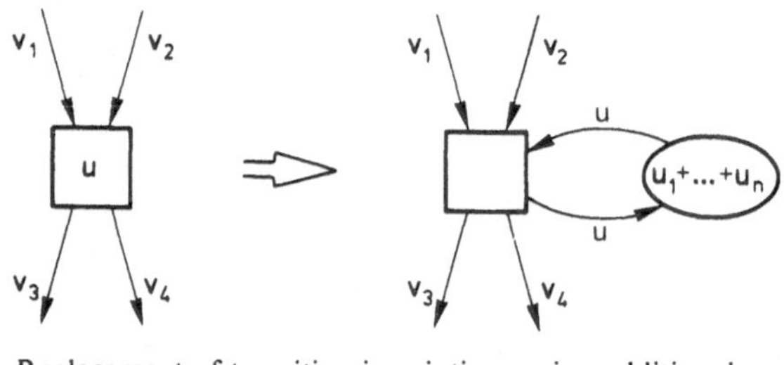

Fig. 12. Replacement of transition inscriptions using additional predicates.

The general case of *fool*-typed transition inscriptions can not be replaced this way. Below we shall discuss a further transformation for this case.

Like the above introduction of place capacities, additional transition inscriptions at most limit the overall behaviour of a net. Therefore, even in this case, the calculi of invariants as well as the theorems on homomorphisms and additional equations remain valid.

10.4. Flexible arc inscriptions

In this paper the operations on multisets are limited to the operations introduced by **m**_*SPEC*, viz. addition, negation and the constant empty multisets. Of course, one could think of more general operations on multisets, and also of using variables of multiset sorts m_s.

Introducing multiplication of terms with natural numbers is a nearby extension: As long as n represents a constant natural, with $u \in T_{\widehat{OP}}(X)$, the term $n * u$ may be considered just as a shorthand for the n-fold sum $u + \cdots + u$. A corresponding algebraic specification should then specify the items of sort s to yield a (left written) module over the integers as discussed in [21]. This extension does not principally exceed the formalism and could be considered must a "syntactic sugaring".

A more general formalism is obtained with terms $v \in T_{OP,nat}(X)$ where *nat* denotes the natural numbers. Then an arc inscription $v * u$ yields a "flexible throughput": At event occurrences $M \xrightarrow{t,\beta} M'$, with an arc $f = (p, t)$ or $f = (t, p)$ inscribed $v * u$, the "number of tokens flowing through f" essentially depends on $\bar{\beta}(v)$ (and hence on the chosen occurrence mode β), whereas in the nets of this paper the throughput of each arc is constant for all occurrence modes.

In a formal setting, for a given specification *SPEC*, one may consider some extended multiset specification including any kind of operations over multisets (and other sorts), particularly the product with integers. Such specifications may be formed by

extm_*SPEC* = **m**_*SPEC* + **int** +
opns: $*: int\ m_s \rightarrow m_s$
⋮
$F_i : m_s \ldots m_s \rightarrow m_s$
⋮

With **extm**_*SPEC* = $(S, \widehat{\widehat{OP}})$ one can then inscribe arcs by terms in $T_{\widehat{\widehat{OP}}}(X)$ with X including variables of sort m_s.

The extension from **m**_*SPEC* to **extm**_*SPEC* influences of course the invariant calculi and the theorems on homomorphic transformations. Place invariants can easily be generalized if the product with terms of sort integer is the only additional operator. In this case, the theory resembles invariants for self-modifying nets [23]. Details are beyond the scope of this paper.

Products with integer terms can be used for moving the above considered transition inscriptions to arcs: With the additional operation [ ]: *bool* → *nat*, defined for $u \in T_{OP,bool}$ by $[u] = 1$ iff $u \equiv_{\hat{E}}$ *TRUE* and $[u] = 0$ iff $u \equiv_{\hat{E}}$ *FALSE*, the scheme of Fig. 13 outlines a meaning-preserving transformation. This extension has been suggested in [9].

10.5. Marking schemes

The intended scheme of using nets implies markings to represent distributions of items in systems. In our formalism such items are adequately represented as ground

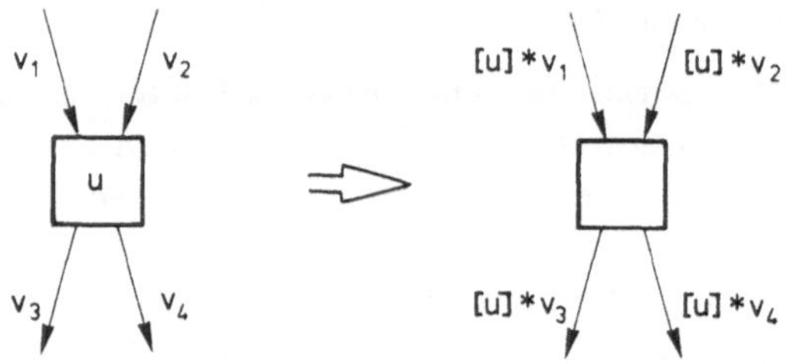

Fig. 13. Replacement of transition inscriptions using flexible arc inscriptions.

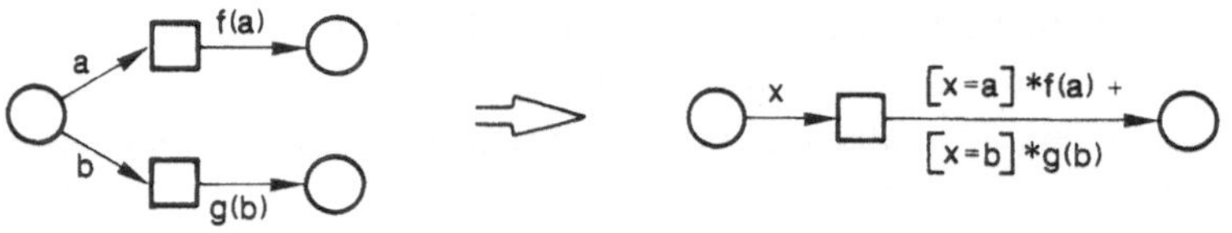

Fig. 14. Equivalent reduction of the net structure.

terms. If only partial knowledge of markings is available or if one is interested in relationships or properties of structured sets of markings, it may nevertheless be useful to consider terms with variables as markings. A mapping $M: P \to T_{OP^+}(Z)$ then can be considered as a *scheme* for all markings gained from M by assignments $ass: Z \to T_{OP}$. (Marking schemes have been suggested in [20] in a different context.) The invariant calculi (particularly Theorems 5.6, 5.7 and 7.5) remain valid, provided the set Z of variables is disjoint from the set $X \cup Y$ of variables appearing in arc inscriptions and invariant components, respectively. Under this assumption, also the theorems on homomorphic transformations and on additional equations remain.

10.6. Equivalence transformations

In the context of transition inscriptions (Subsection 10.3) and flexible arc inscriptions (Subsection 10.4), we discussed already some equivalence transformations, keeping the underlying net structure untouched. Equivalence transformations may change the net structure itself. Figure 14 outlines an example. Genrich [10] introduces a complete list of such transformations for inscriptions based on n-tuples (cf. Subsection 6.2).

A systematic approach to such transformations in the style of this paper might include an algebraic specification of the underlying net structure itself (as e.g., in [18]). Net transformations can then be formulated as algebra homomorphisms.

11. Conclusion

11.1. The formalisms of Predicate/Transition nets and coloured nets

The several versions and variants of Petri nets with structured tokens can roughly be divided into two groups: The first kind of formalisms is based on the idea of "dynamizing" predicate logic, using predicates with changing extensions. We denote them in the following "Predicate/Transition nets" (PrT-nets) in accordance with their introduction in [11, 12]. PrT-nets are schemes of system models with n-tuples of expressions (including variables) as arc inscriptions and as invariant entries. The product of such expressions is assumed to be commutative. Due to the use of variables, formal expressions, products and sums, PrT-nets are essentially a syntax-based formalism.

A second, more semantically oriented line of models is based on "coloured tokens". We call such models in the following "coloured nets", according to their first introduction in [14]. Coloured nets have been motivated as shorthands for conventional Petri nets. Sums of functions serve as arc inscriptions and as invariant entries. Their product is based on the composition of functions.

A greater number of papers relate or reformulate various versions and aspects of Petri nets with structured tokens: Two different types of place invariants have been suggested in [13, 19] for PrT-nets; techniques to easier construct place invariants for coloured nets are discussed in [15] and for special PrT-nets in [25]. Different versions of high level nets are compared and interrelated in [16, 19, 21]. Recent reformulations of PrT-nets and coloured nets include [8, 17].

In [21] we aimed at a transparent mathematical treatment of Petri nets with structured tokens, suggesting the set of multirelations over some given set to be taken as the underlying domain of the formalism. Multirelations form a homogeneous semantical domain, in fact an integer module. This approach in the above classification is totally semantics-oriented, and close to coloured Petri nets.

In this paper we follow the opposite way, suggesting a heterogeneous, syntax-oriented approach which resembles PrT-nets. The formalism for (place) invariants differs however substantially from PrT-nets: We replaced the formal, commutative product by—in general—noncommutative term substitution. As term substitution is essentially the syntactical analogon to the composition of functions, our (place) invariants resemble those of coloured nets. In [24], term substitution was first suggested to base the invariant calculus upon; cf. also [22].

Syntax oriented approaches are useful because each system model needs a syntactical representation for being communicated or implemented. The explicit use of terms including variables is common mathematics. We use them by means of a lot of concepts which are standard notions in general algebra, hence algebraic specifications are a natural basis for a formalism to deal with structured tokens. This idea has been suggested several times and will be considered next.

11.2. Combining Petri nets and algebraic specifications

A couple of papers combine Petri nets and algebraic specifications. The specification language SEGRAS [18] includes an abstract specification of strict nets (also called Predicate/event nets) in the sense of Subsection 10.2 and uses partial operations. Based on the specification language OBJ2, Battistion et al. [2] abstractly specify a particular class of nets, consisting of superposed automata with special requirements to the terms occurring as arc inscriptions. Tokens have a constant individuality and a variable data part. Place invariants are based on a product which essentially tests equality of terms, in the style of [13]. A special product is defined for transition invariants according to [19]. Berthomien et al. [3] and Vautherin [24] suggest multisets of variables and of terms, respectively (whereas we use particular terms to handle multisets). Vautherin [26] additionally introduces a formalism of place invariants, called "type1-semiflows", based on multiple occurrences of terms. The tokens of PROT nets [1], are essentially Pascal records to be transformed according to transition inscriptions.

An important aspect in all those papers concerns the border between the abstract data type formalism and what is formulated in usual mathematics. In SEGRAS and OBJSA [18, 30, 2], everything under consideration is abstractly specified, including the involved nets and the occurrence rule. On the other hand, Berthomien et al. [3] and Vautherin [26] keep the formally specified parts quite limited: arcs are inscribed by multisets of terms, and the occurrence rule is formulated with respect to interpreted net schemes. Billington [4] presents a similar approach, adding threshold inhibitor inscriptions and capacities.

Reference [27] suggests net inscriptions over specifications similar to our formalism. For analysis purposes they derive place/transition nets from given high-level nets, and they consider reachability trees. A couple of papers besides algebraic techniques also employ category theory. Reference [29] defines parameterized net schemes and structuring concepts, [30] introduces a notion of "implementation", and [28] gives several versions of semantics of the same schematic inscription.

We saw in Definition 4.4 that enabledness of a transition is already a semantical notion in the sense that enabledness depends on the validity of equations: On the purely syntactical, uninterpreted level, a transition may appear not enabled, whereas equations may cause enabledness. In [22], this was covered by a mixture of syntactical and semantical concepts: Arc inscriptions were multiset terms, i.e. syntactical constructs, whereas multisets over concrete algebras were taken as markings.

In this paper we consequently apply corresponding concepts of initial algebra semantics: Ground terms are to represent markings. They are to be considered equivalent if and only if they belong to one equivalence class induced by the involved equations.

We aimed at obtaining at a particularly adequate border and integration of general algebra and nets. In fact, it turned out that the essential aspects of nets with structured tokens can concisely be represented this way and, vice versa, that a lot of elementary concepts of abstract data types have been applied. This includes ground terms for markings, general terms as arc inscriptions, assignments and initiality for the occurrence rule and term substitution for place and transition invariants.

On this basis, one might hope that more involved problems of nets with structured tokens will be adequately solvable by more involved but well-known algebraic techniques. This includes particularly systematic net transformations and equivalence notions.

Acknowledgment

Jacques Vautherin showed me how place invariants can be defined by term substitution. Walter Dosch and Hartmut Ehrig advised me in technicalities of algebraic specification. Walter Dosch, Ekkart Kindler and anonymous referees gave valuable hints for a proper presentation of the paper. Thanks to all of them.

References

[1] M. Baldassari and G. Bruno, An environment for object oriented conceptual programming based on PROT nets, in: G. Rozenberg, ed., *Advances in Petri Nets 1988*, Lecture Notes in Computer Science **340** (Springer, Berlin, 1988) 1-19.

[2] E. Battistion, F. de Cindio and G. Mauri, OBJSA nets: a class of high-level nets having objects as domains, in: G. Rozenberg, ed., *Advances in Petri Nets 1988*, Lecture Notes in Computer Science **340** (Springer, Berlin, 1988) 20-43.

[3] B. Berthomieu et al., Abstract data nets: combining Petri nets and abstract data types for high level specifications of distributed systems, in: *Proc. 7th European Workshop on Applications and Theory of Petri nets*, Oxford (1986).

[4] Jonathan Billington, Extending coloured Petri nets, Tech. Report No. 148 University of Cambridge Computer Laboratory, England, 1988.

[5] CIP Language group, *The Munich Project CIP. Vol. 1: The Wide Spectrum Language CIP-L*, Lecture Notes in Computer Science **183** (Springer, Berlin, 1985).

[6] E.W. Dijkstra, Hierarchical ordering of sequential processes, *Acta Inform.* **1** (1971) 115-138.

[7] H. Ehrig and B. Mahr, *Fundamentals of Algebraic Specifications 1*, EATCS Monographs on Theoretical Computer Science **6** (Springer, Berlin, 1985).

[8] H.J. Genrich, Predicate/Transition nets, in: Lecture Notes in Computer Science **254** (Springer, Berlin, 1987) 207-247.

[9] H.J. Genrich, Equivalence tranformations of PrT-nets, Arbeitspapiere der GMD 284, 1988.

[10] H.J. Genrich, Re: *S*-invariants of PrT-nets. On feasible transformations of the incidence matrix, unpublished manuscript, GMD-F1, 1988.

[11] H.J. Genrich and K. Lautenbach, The analysis of distributed systems by means of Predicate/Transition nets, in: Lecture Notes in Computer Science **70** (Springer, Berlin, 1979) 123-146.

[12] H.J. Genrich and K. Lautenbach, System modelling with high level Petri nets, *Theoret. Comput. Sci.* **13** (1981) 109-136.

[13] H.J. Genrich and K. Lautenbach, S-invariance in Predicate/Transition nets, Informatik-Fachberichte **66** (Springer, Berlin, 1983) 98-111.

[14] K. Jensen, Coloured Petri nets and the invariant method, *Theoret. Comput. Sci.* **14** (1981) 317-336.

[15] K. Jensen, How to find invariants for coloured Petri nets, in: Lecture Notes in Computer Science **118** (Springer, Berlin, 1981) 327-338.

[16] K. Jensen, High level Petri nets, in: Informatik-Fachberichte **66** (Springer, Berlin, 1983) 166-180.

[17] K. Jensen, Coloured Petri nets, in: Lecture Notes in Computer Science **254** (Springer, Berlin, 1987) 248-299.

[18] B. Krämer and H.-W. Schmidt, Types and modules for net specifications, in: K. Voss, H. Genrich and G. Rozenberg, eds., *Concurrency and Nets* (Springer, Berlin, 1987) 269-286.

[19] K. Lautenbach and A. Pagnoni, Invariance and duality in Predicate/Transition nets and in coloured nets, Arbeitspapiere der GMD 132, 1985.

[20] M. Lindqvist, Parameterized reachability trees for Predicate/Transition nets, Acta Polytechnica Scandinavica, Mathematics and Computer Science, No. 54, 1989.

[21] W. Reisig, Petri nets with individual tokens, *Theoret. Comput. Sci.* **41** (1985) 185-213.

[22] W. Reisig and J. Vautherin, An algebraic approach to high level Petri nets, in: *Proc. Eighth Workshop on Applications and Theory of Petri Nets*, Zaragoza (Spain) (1987) 51-72.

[23] R. Valk, Generalizations of Petri nets, in: Lecture Notes in Computer Science **118** (Springer, Berlin, 1981) 140–155.
[24] J. Vautherin, Un modèle algébrique, basé sur les réseaux de Petri, pour l'étude des systèmes parallèles, Thèse de Doctorat d'Ingénieur, Univ. de Paris-Sud, Centre d'Orsay, 1985.
[25] J. Vautherin and G. Memmi, Computation of flows for unary Predicate/Transition nets, in: Lecture Notes in Computer Science **188** (Springer, Berlin, 1985) 455–467.
[26] J. Vautherin, Parallel systems specifications with coloured Petri nets and algebraic specifications, in: G. Rozenberg, ed., *Advances in Petri Nets 87*, Lecture Notes in Computer Science **266** (Springer, Berlin, 1987) 293–308.
[27] G. Berthelot and L. Pétrucci, Putting algebraic nets into practice, CEDRIC-11E, Internal Report, Conservat. Nat. des Arts et Métiers, Evry, France (1989).
[20] C. Dimitrovici, U. Hummert and L. Pétrucci, The properties of algebraic net schemes in some semantics, L.R.I. Rap. de Recherche No. 539, Univ. de Paris-Sud (1990).
[29] U. Hummert, Algebraische Theorie von high-level Netzen, Dissertation, F.U. Berlin (1989).
[30] H. W. Schmidt, Specification and Correct Implementation of Non-sequential Systems Combining Abstract Data Types and Petri Nets (Oldenburg Verlag, München, 1989).

5.
Types and Modules for Net Specifications

B. Krämer and H.W. Schmidt

K. Voss, H. J. Genrich and G. Rozenberg (eds.): Concurrency and nets.
Advances in Petri nets. Springer, Berlin Heidelberg New York 1987, pp. 269-286

Abstract

A specification language for nonsequential systems that unifies algebraic specifications of abstract data types with high-level Petri net specifications of dynamic behavior is presented. The data structure of a system, the information content of local states, and static constraints to state changing operations are specified by sorted Horn clause rules with equality. Behavior is specified by schemes of Predicate-Event nets together with an initial (distributed) state. Many-sorted algebras provide a standard interpretation of such specifications in terms of the initial models satisfying the rules, the flow, and the initial state given. One concern of this paper is to sketch the mathematical semantics of the core language. The other is to define a notion of abstract system that supports modularity and reusability of specifications similar to abstract data types.

1 Introduction

Algebraic specifications of abstract data types (ADT's) have been studied for many years as a powerful tool for the development of software systems (cf. e.g., [10], [6], [9], [5]). Algebraic ADT specifications provide a formal semantics for many useful programming concepts and are therefore widely accepted for specifying basically sequential systems. But their value proved to be limited for defining the behavior of abstract machines in a distributed environment (cf. [8]).

To tackle behavioral issues raised by nonsequential and distributed systems such as concurrency, nondeterminism, communication, and synchronization, a number of dedicated design and specification methods evolved. Among these we mention Milner's Calculus of Communicating Systems (CCS) [19], temporal logic methods [21,17], and various Petri net based methods [4]. Petri nets provide a notion of concurrency and nonsequential processes that is based on partial orders. This makes Petri nets particularly valuable for distributed application domains in which a clear distinction between concurrency, nondeterminism, and sequentiality is required [16].

The specification language *SEGRAS* [12,11], whose core concepts are presented here, combines ADT's and Petri nets in a uniform syntactic and semantic framework. Specifications of nonsequential systems given in *SEGRAS* generally consist of a) abstract data types describing data structures on which the system operates, b) Predicate-Event nets [23] describing the dynamic behavior of the system, and c) a description of the initial (distributed) system state. Such specifications can be considered as a variant of Predicate-Transition nets (PrT-net) [7]. In our variant a many-sorted partial algebra is used in place of the set theoretic and logic approach proposed in [7]. A partial algebra provides different data domains and partial operations on these domains. Their effects are recursively specified by conditional equations which are equivalent to many-sorted Horn clause logic with equality. This logic was shown to have yet "standard", i.e., initial models [18].

SEGRAS is the specification language of the ESPRIT project GRASPIN[1]. In this project we are developing techniques and prototypes of tools which are to aid in the construction and verification of nonsequential systems specifications and their systematic transformation into executable programs. For such a software engineering approach it is essential that the specification formalism used has computable models. *SEGRAS*

[1]GRASPIN is supported in part by the Commission of the European Communities within the ESPRIT program.

seems to be a good compromise between expressive power and semantic strength as there is a strong relationship between initiality and computability.

SEGRAS was designed with the goal to provide an appropriate high-level Petri net language with *data types* and *modularization* capabilities to carry the methodological use of ADT's in programming over to net specifications. This idea was first presented informally in [14] and [13]. A formal treatment of PrE nets combined with many-sorted partial algebras was then given in [24]. Later on it was also proposed to integrate colored nets [26] or Predicate-Transitions nets [1] and ADT's.

Modularity is essential for writing large specifications. Modules comprising only small specifications can be verified and implemented once and for all and can be reused and combined freely to make large specifications. We are going to define a notion of *abstract system* that comprises data abstraction with an interface of named operations and which additionally has a hidden local, possibly distributed state and state changing operations that can actually see and transform parts of the state.

In this paper emphasis is put on the semantics of the core language. As the formalization of the full language (see [12]) is overly complex, a simplified abstract syntax is defined in the next section. This syntax is used in Section 3 to present the mathematical semantics of core concepts of the language. Section 4 introduces concepts for structuring specifications and sketches restrictions on preserved properties of combined specifications. The last section gives a small example of a structured specification.

2 Abstract Syntax of the Language

The abstract syntax of our specification language is based on conditional algebraic specifications of abstract data types, as defined in [15], and on Predicate-Event nets labelled by expressions over a given signature (see [24]). Most of our notions and definitions are kept close to these sources to take advantage of theorems and proofs given there.

This section partly recalls fundamental notions about algebraic specifications adapted to our specific needs. The reader not familiar with the concepts may consult [10,15,5] for detailed information on the topic.

2.1 Algebraic Language

Classically, ADT's consist of sorts, operation symbols, and equations. The sorts serve for naming data domains. The operation symbols name operations on these domains. The operation symbols generate an algebra of terms on which the equations induce a congruence relation. Thus the effect of each operation is abstractly defined in the sense that neither a particular implementation of the operations nor a particular representation of data is referred to.

Signatures

Signatures formalize the notion of a sorted collection of operations available to the user of an abstraction.

Definition 1 *Let S be a countable set, whose elements are called* sorts, *and let S^* be the set of all finite strings over S, including the empty string λ. Then an S-*sorted algebra signature *Σ is an $S^* \times S$-indexed family of sets $(\Sigma_{u,s})_{u \in S^*, s \in S}$ of* operation symbols *of* arity *u,* coarity *s, and of* type *$u \rightarrow s$.*

Notation. In examples, we write $\sigma : u \longrightarrow s$ to denote $\sigma \in \Sigma_{u,s}$; an S-sorted signature is given as a sequence of operation symbols headed by the keyword **fu**; the sorts indexing a signature are listed after the keyword **sorts**.

As a matter of notational convenience, let S be an arbitrary but fixed set of sorts and Σ an arbitrary but fixed S-sorted signature for the rest of this paper.

Terms, Equations, and Rules

A signature determines a language of terms. Terms are used to formulate equations and inequations, from which Horn clause like conditional equations, called rules, are build up.

Definition 2 *Let X be an S-indexed family of sets $(X_s)_{s\in S}$ of variable symbols disjoint*[2] *from $(\Sigma_{\lambda,s})_{s\in S}$. Then the elements of the S-indexed family of sets $T_{\Sigma(X)} = (T^s_{\Sigma(X)})_{s\in S}$ that is minimally defined by the following list of points are called* Σ-terms *(with variables in X).*

1. $X_s \subseteq T^s_{\Sigma(X)}$ *for* $s \in S$
2. $\Sigma_{\lambda,s} \subseteq T^s_{\Sigma(X)}$ *for* $s \in S$
3. $\sigma(t_1,\ldots,t_n) \in T^s_{\Sigma(X)}$ *if* $t_i \in T^{s_i}_{\Sigma(X)}$ *and* $\sigma \in \Sigma_{s_1\ldots s_n,s}$ *for* $s_i, s \in S, s_i \neq \lambda (1 \leq i \leq n)$

A Σ-term is called *ground* if it is without variables. T_Σ denotes the subfamily of ground Σ-terms.

Definition 3 *A* Σ-equation (Σ-inequation) *is a pair of Σ-terms (t, t') of sort $s \in S$ written $t =_s t'$ (and $t \neq_s t'$ resp.). An equation without variables is called* ground.

Notation. For the remaining part of this text, let X be a fixed family of variable symbols and let $vars(t)$ denote the smallest family $Y \subseteq X$ of variables occurring in term t such that $t \in T_{\Sigma(Y)}$.
We will drop the sort index in equations and inequations, i.e., we write $x = y$ for $x =_s y$.

Definition 4 *A* Σ-rule, *written $e_0 : e_1, \ldots, e_n$. (for $n \geq 0$), consists of an equation (or inequation) e_0, called* conclusion, *and a possibly empty sequence of equations e_i, called* premises. *A rule is called* positive *if e_0 is an equation and* negative *if e_0 is an inequation* $(1 \leq i \leq n)$.

Partial Specifications of Data Types

Now we show how the ideas of total specification of data types by equations [5] may be modified so as to yield partial specifications. They differ from total specifications by the use of partial operations, i.e., operations that are meaningful only on part of their domain. For example, *tail* is only defined for non-empty lists, or *top* and *pop* are only meaningful for non-empty stacks. In the case of partial specifications, such exceptional situations are simply made undefined (here be means of inequations, so that the problem of error recovery and exception handling is left to the implementor of the data type.

We have chosen partial specifications and partial algebras as their models because we wanted to adequately handle the partiality of operations and relations used in PrE nets in the algebraic framework. Albeit the mathematical theory involved with this approach is more complex than for total specifications, partial specifications have been advocated by other authors, too (e.g., [3]), so that we can adopt some of their results.

Definition 5 *A* conditional specification *SPEC is a triple (S, Σ, R) where R is a finite set of Σ-rules.*

The example in fig. 1 gives the signature and the rules for the ADT "bintree". We use a mixfix notation for some of the operation symbols. We assume that the sort **Nat** of natural numbers is built-in together with operations on **Nat**. The binary trees have leaves labelled by natural numbers.

In this example, the rules only serve to specify the definedness of terms. Actually, here all terms are defined or, in other words, all "bintree " operations are total. Since this situation occurs so often in partial specifications, in *SEGRAS* we distinguish between total and partial operations on the syntactic level of signatures. For simplicity of the present text we avoid the technicalities required for this distinction. The results however remain the same [15].

Notation. To ease the reading of the examples, equations like $t = t$ are simply written as *literals* of the form "t" in subsequent examples.

[2]For A, B S-indexed families of sets, $A \cap B$ means $A^s \cap B^s$ for all $s \in S$.

```
sorts  bintree, Nat

fu   leaf:   Nat  -> bintree.
     tree:   bintree bintree -> bintree.

forall T, T1, T2: bintree; n: Nat.
leaf n = leaf n:  n.
tree(T,tree(T1,T2) = tree(tree(T,T1),T2): T = T, T1 = T1, T2 = T2.
```

Figure 1: Abstract Data Type bintree

2.2 Net Language

Petri nets which are labelled by symbols and terms over a given signature and which are accompanied by a set of rules will be used to specify the nonsequential behavior of an abstract system. The rules specify a) data structures the abstract system is operating on, b) the definedness of net elements, c) constraints to event occurrences, and d) the initial state of the system.

This reflects the idea that the various sorts of data involved in the dynamic behavior are considerd integral parts of a system specification and that static constraints to high-level events are formulated by the same axiomatic means as data structures are specified.

PrE Signatures

Now, the notion of algebra signature is extended to a notion of PrE signature in which operation symbols naming net elements are distinguished and two distinct symbols are assumed that capture the flow relation and the initial case in the associated algebra.

Definition 6 *A* signature for a Predicate-Event system *(*PrE signature*) over Σ is a triple* $\Pi = (\Sigma, \Pi^C, \Pi^E)$ *such that*

1. *Σ an algebra signature*
2. *$\Pi^C, \Pi^E \subseteq \Sigma$ are disjoint subsignatures containing the* condition *and* event symbols, *resp.*
3. *There are distinctive symbols*
   flow $\in \Sigma_{s_1 s_2,\lambda}$ *for $s_1, s_2 \in Sys$, the* flow symbol, *and*
   case $\in \Sigma_{s,\lambda}$ *for $s \in Sys$, the* initial case symbol,

where the set $Sys \stackrel{\text{def}}{=} \{s \mid \sigma \in \Pi^C_{u,s} \cup \Pi^E_{u,s}, u \in S^*, s \in S\}$ of system sorts *is formed by collecting the coarities of all condition and event symbols.*

Remark. The terms of a sort in *Sys* are used to represent markings and occurrences of events, i.e., they model information related to states and their changes.
Obviously an algebra signature is the degenerate case of a PrE signature with $\Pi^C \cup \Pi^E = \emptyset$.

Notation. We write **ac** $\sigma : u \longrightarrow s$ to denote $\sigma \in \Pi^E_{u,s}$ and **st** $\sigma : u \longrightarrow s$ to denote $\sigma \in \Pi^C_{u,s}$.

Π-labelled Nets

We omit fundamental net theoretic definitions here and simply use the terminology and definitions given in [2]. We shall take the symbols of Π^C as the *state predicates* and the symbols of Π^E as the *event schemes* of a *Petri net* $\mathbf{N} = (\Pi^C, \Pi^E; F)$. The set of net *elements* is denoted by $N = \Pi^C \cup \Pi^E$. A labeling of the event schemes and of the *flow* F of a net serves for specifying an appropriate transition rule in the following section. In contrast to [2] we admit the empty net because we want an algebraic specification to be a special case of a PrE net specification.

Definition 7 *Let* Π *be a PrE signature. A* Π-labelled net **N** *is a structure* $(\Pi^C, \Pi^E; F, l)$*, where* $(\Pi^C, \Pi^E; F)$ *is a finite (possibly empty) net with* $F \subseteq (\Pi^C \times \Pi^E) \cup (\Pi^E \times \Pi^C)$ *and* $l = (l_E, l_F)$ *is a* labeling *of event schemes and of the flow F such that*

1. $l_F : F \longrightarrow \mathcal{F}(T_{\Sigma(X)})$ *(where* $\mathcal{F}(A)$ *denotes the finite subsets of A) such that for* $(\sigma_1, \sigma_2) \in F$ *and forall* $t \in l_F(\sigma_1, \sigma_2)$*, we have that* $t \in T^u_{\Sigma(X)}$ *if* $\sigma_1 \in \Pi^C_{u,s}$ *or* $\sigma_2 \in \Pi^C_{u,s}$ *(with* $s \in S, u \in S^*$*).*
2. $l_E : E \longrightarrow T_{\Sigma(X)}$ *such that for* $\sigma \in \Pi^E_{u,s}$*:* $l_E(\sigma) = \sigma(t)$ *for some* $t \in T^u_{\Sigma(X)}$ *(with* $s \in S, u \in S^*$*).*

PrE Net Specification

A PrE net specification is now introduced as a Π-labelled net together with a set of Σ-rules.

Definition 8 *A* PrE specification *SPEC is a quadruple* $(S, \Pi, R, \mathbf{N})$ *where* Π *is a PrE signature, R is a finite set of* Σ*-rules, and* **N** *is a* Π*-labelled net.*

Example

Fig. 2 gives an example of a PrE net specification. We assume the specification bintree of our first example to be included. We also assume that some operation "f" on pairs of natural numbers is specified elsewhere. The task now is to apply this operation recursively to determine the value of a tree based on the values of its left and right subtrees.

We specify an evaluation behaviour in which f is applied concurrently to the subtrees of a given tree t. The tree t is completely split up, and then f is recursively applied to the values computed for sibling subtrees starting from the values on the leaves of t. The splitting and computation processes can be partly concurrent. To preserve the knowledge about the tree structure in the subtrees split up, a data abstraction "treepath" is included. It is used for uniquely referencing the subtrees of a binary tree. According to this specification, a path and hence a subtree can be uniquely determined by a term such as "l(r(r(l(r(l(0))))))". The operation "0" denotes the empty path, i.e., it refers to the main tree; "l" denotes the left and "r" the right subtree of a given tree.

Fig. 3 gives a particular tree "mytree" as a term and its corresponding graphical representation. The result expected from evaluating mytree according to the specified strategy is indicated by the Nat term "myval" on which the state predicate _value_ will be defined when the computation has finished.

3 Semantics

The semantic domain for our specifications is based on many-sorted algebras and a many-sorted variant of PrE nets. We call the latter Predicate-Event systems (PrE systems) in analogy to Condition-Event systems. In [24] it is shown that each PrE system abstracts from a possibly infinite Condition-Event system. The definition of partial algebras satisfying a conditional specification is widely standard. Details can be found in [15].

3.1 Many-Sorted Partial Algebras

An algebra signed by a signature Σ, a Σ-algebra for short, provides a set of data elements for each sort in S and (possibly partial) operations for the symbols in Σ.

Definition 9 *A* Σ-algebra **A** *is a structure* $(A; (\sigma_{A,u,s})_{s \in S, u \in S^*, \sigma \in \Sigma_{u,s}})$ *providing a family of nonempty sets of* carriers $(A^s)_{s \in S}$ *and a partial operation* $\sigma_{A,u,s} : A^u \longrightarrow A^s$ *for each* σ *in* $\Sigma_{u,s}$ *where* $A^{us} \stackrel{\text{def}}{=} A^u \times A^s$ *and* $A^\lambda \stackrel{\text{def}}{=} \{\emptyset\}$*, for* $s \in S$ *and* $u \in S^*$*.*

```
sorts  path, e-state

fu  0:         ->  path.
    l: path    ->  path.
    r: path    ->  path.

st  _tree_:             path tree      -> e-state.
    _value_:            path Nat       -> e-state.
ac  split up tree__:    tree path      -> e-state.
    get_at_:            Nat path       -> e-state.
    eval__at_:          Nat Nat path   -> e-state.

forall T, L, R: tree; p: path; n, n1, n2: Nat.
0:  .                                   l:  .
r:  .                                   p tree T:  p, T.
p value n:  p, n.                       split up tree T p:  T = tree(L,R), p.
get n at p:  n, p.                      eval n1 n2 at p:  n1, n2, p.
case ((0) tree T):  T.
```

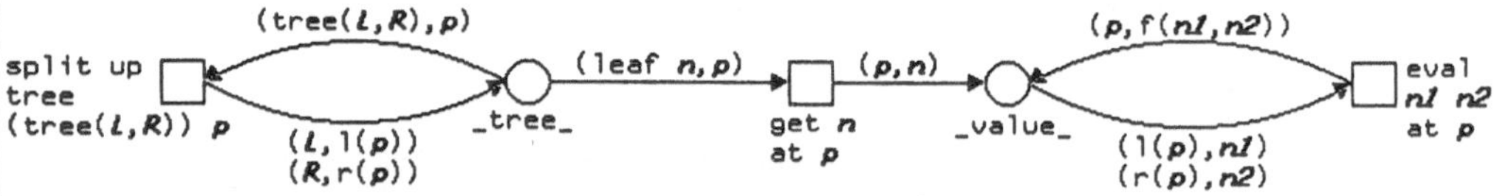

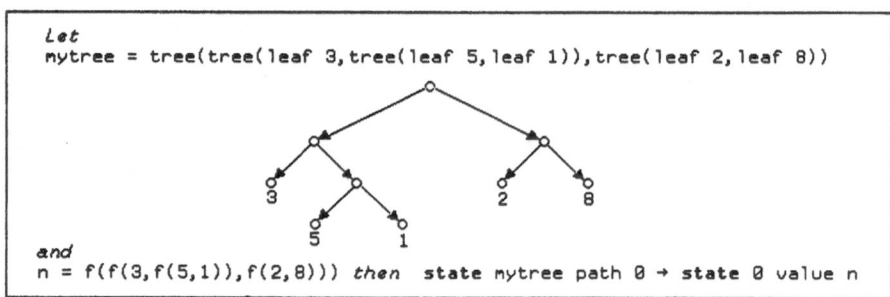

Figure 2: Concurrent evaluation of binary trees

```
Let
mytree = tree(tree(leaf 3,tree(leaf 5,leaf 1)),tree(leaf 2,leaf 8))
```

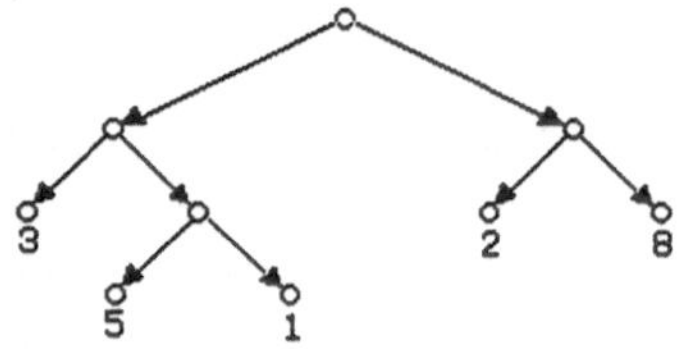


```
and
n = f(f(3,f(5,1)),f(2,8))) then  state mytree path 0 → state 0 value n
```

Figure 3: A binary tree and the exepected result of evaluation

Notation. We denote elements of the cartesian product $a \in A^{s_1} \times \cdots \times A^{s_n}$ by tuples $(a_1 \ldots a_n)$ where a_i is the i-th projection of a for $1 \leq i \leq n$.

Σ-algebras form a category with morphism being weak homomorphism, i.e., homomorphisms which preserve definedness but need not preserve undefinedness. We omit the details of such morphism here, although in the theory of ADT's and PrE-systems, the consistency of modular compositions and the correctness of implementation steps consist in showing injectivity and surjectivity of the homomorphisms between certain algebras.

3.2 Models of Algebraic Specifications

A conditional specification $SPEC$ has many-sorted partial Σ-algebras as models. Such models satisfy the given specification in a sense defined below. In particular, we use initial algebras as unique standard models. For defining the notion of satisfaction, we first introduce the notion of assignment of values in the model to variables in the specification. Satisfaction then leads to a class of models, called $SPEC$-algebras, and to a standard model in that class, denoted $\mathbf{P}_{SPEC}$. $\mathbf{P}_{SPEC}$ has as its elements equivalence classes of ground terms under a congruence relation generated by the conclusions of positive rules in R.

We cannot go into the details of these nötions here, but refer the interested reader to [10] and the textbook on equational specifications by Ehrig and Mahr [5].

Definition 10 *Let* **A** *be a* Σ*-algebra. Then an* assignment *is a operation* $\varepsilon : X \longrightarrow A$ *which assigns values in* **A** *to variables in* X. *The following extension of* ε *to a partial operation* $\bar{\varepsilon} : T_{\Sigma(X)} \not\longrightarrow A$ *is called* evaluation of Σ-terms *in* **A**. *It is recursively defined by*

1. $\bar{\varepsilon}(x) = \varepsilon(x)$ *for all variables* $x \in X$
2. $\bar{\varepsilon}(\sigma) = \sigma_A$ *iff* $\sigma \in \Sigma_{\lambda,s}, s \in S$
3. $\bar{\varepsilon}(\sigma(t_1, \ldots, t_n))$ *is defined and equals* $\sigma_A(\bar{\varepsilon}(t_1), \ldots, \bar{\varepsilon}(t_n))$ *iff* $\bar{\varepsilon}(t_1), \ldots, \bar{\varepsilon}(t_n)$, *and* $\sigma_A(\bar{\varepsilon}(t_1), \ldots, \bar{\varepsilon}(t_n))$ *are all defined.*

Remark. In the limit case, where X is the empty set, $\bar{\varepsilon}$ defines a unique *ground term evaluation* in **A** [15], which we also denote ε_A.

For a subfamily $Y \subseteq X$, $T_{\Sigma(Y)}$ is a Σ-algebra and a $\Sigma(Y)$-algebra. As a special case of the above definition, we can therefore consider assignments $\varepsilon : Y \longrightarrow T_{\Sigma(Z)}$ with $Z \subseteq X$. They assign terms with variables to variables and are therefore called *substitutions*.

Definition 11 *Let* $SPEC = (S, \Sigma, R)$ *be a* Σ*-specification and* **A** *a* Σ*-algebra. We say*

1. $\mathbf{A} \models t = t'$ — *(read "***A** *satisfies* $t = t'$*") for* $t, t' \in T^u_\Sigma$ *with* $u \in S^*$ *iff* $\varepsilon_A(t)$ *and* $\varepsilon_A(t')$ *are defined and equal in* **A**
2. $\mathbf{A} \models t \neq t'$ — *for* $t, t' \in T^u_\Sigma$ *iff not* $\mathbf{A} \models t = t'$ *with* $u \in S^*$
3. $\mathbf{A} \models e_0 : e_1, \ldots, e_n.$ — *iff forall ground substitutions* $\varepsilon : X \longrightarrow T_\Sigma$ *we have that* $\mathbf{A} \models \bar{\varepsilon}(e_0)$ *whenever* $\mathbf{A} \models \bar{\varepsilon}(e_i)$ *for all* $1 \leq i \leq n$
4. $\mathbf{A} \models R$ — *iff* $\mathbf{A} \models r_j$ *for all* $r_j \in R$

If $\mathbf{A} \models R$ *we also say that* **A** satisfies *SPEC and that* **A** *is a model of SPEC.*

Remark. The equality we use in the above definition is also called *weak equality*. Weak equality can be used to specify (conditionally) the definedness of a term t under all assignments of variables in $vars(t)$ by the rule "$t = t : e_1, \ldots .e_n.$". We call it a *definedness rule*.
Hence, the definedness rules given in the previous examples assert the definedness of the bintree and path operations and of the net elements in each Σ-algebra which satisfies the specification.

A key result of the initial algebra approach to data structure specifications is that a particular initial[3] algebra, the quotient term algebra, can be constructed for equational specifications (cf. [10] Theorem 6), and it was shown (by Theorem 8) that it always exists. These results have been adapted in [15] to partial algebras that are specified by rules of our kind and that include multi-valued operations and distinguish between total and partial operations. It was also shown that an initial model exists for a consistent specification of that kind and that it is the initial model for the subset of positive rules of R. Hence we can formulate the following theorem:

Theorem 1 *Let* $SPEC = (S, \Sigma, R)$ *be a conditional specification and let* $\mathbf{PALG}_{SPEC}$ *be the category of all partial* Σ*-algebras satisfying the rules in* R*, together with* Σ*-homomorphisms between them. Then* $\mathbf{PALG}_{SPEC}$ *has an initial* Σ*-algebra.*

The initial algebra provides a framework in which those and only those equations between ground terms are valid which are satisfied by all models of a specification. The idea underlying the proof of this theorem (cf. [15]) is that the carriers are shown to consist of R-induced equivalence classes of Σ-terms. These terms must be consistently defined by the rules in R.

[3] Based on the notion of Σ-homomorphisms one can define an *initial* Σ-algebra in a category **PALG** of Σ-algebras to be one which belongs to **PALG** and for which there is a unique Σ-homomorphisms to any other Σ-algebra in **PALG**.

3.3 Dynamic Interpretation of Π-labelled Nets

In [24] the semantics of PrE net specifications was defined by referring explicitly to the Condition-Event system underlying a PrE net specification. It was shown that a net morphism exists from this underlying system to the PrE-net specification. This formalization has the advantage that any concept on the level of Condition-Event systems carries over from the level of the underlying system to the level of PrE-systems. Its disadvantage is that some notions cannot be expressed any more as structural properties on the level of PrE specifications. In the present text, we use many-sorted algebras to model the data associated with a net and provide a dynamic interpretation of PrE net specifications directly.

In Reisig's definition of PrE nets [23], the (one-sorted) value domain and the operations which are labeling the net are given directly by enumeration. In our method, these domains are specified and the specification can be used as a basis for symbolic computations. Their interpretation then is given by a suitable model algebra similar to the pure algebraic case. Moreover, in our PrE nets both conditions and events are labelled as we want the events being the operations that are applicable from the environment of use.

Remark. In the following definitions we speak about markings m instead of cases c to avoid notational confusion.

Definition 12 *Let* $\Pi = (\Sigma, \Pi^C, \Pi^E)$ *be a PrE signature,* $\mathbf{N}$ *be a* Π*-labelled net, and* $\mathbf{A}$ *be a* Σ*-algebra. Then*

1. *A* marking of $\mathbf{N}$ under $\mathbf{A}$ *is a family of operations* $m_{u,s} : \Pi^C_{u,s} \longrightarrow \mathcal{F}(A^u)$ *where* $s \in S, u \in S^*$.
2. *The* initial marking of $\mathbf{N}$ under $\mathbf{A}$, m_A, *is the marking defined by*

$$\{(\sigma, a) \mid case_A(\sigma_A(a_1, \ldots, a_n)) \text{ defined where } \sigma \in \Sigma_{u,s}, a \in A^u, s \in S, u \in S^*\}.$$

3. *An* event of $\mathbf{N}$ under $\mathbf{A}$ *is a pair* (σ, a) *where* $\sigma \in \Pi^E_{u,s}$ *and* $\sigma_A(a_1, \ldots, a_n)$ *defined.*
4. *Let* $e = (\sigma, a)$ *be an event of* $\mathbf{N}$ *under* $\mathbf{A}$ *with* $\sigma \in \Sigma_{u,s}, a \in A^u, s \in S, u \in S^*$. *Moreover, let* ${}^\bullet\sigma$ *be the preset and* $\sigma^\bullet$ *be the postset of* σ *in* $\mathbf{N}$, *and let* $l_E(\sigma) = \sigma(t)$. *Then*
$${}^\bullet e \stackrel{\text{def}}{=} \{(\sigma', a') \mid \sigma' \in {}^\bullet\sigma, \exists t' \in l_F(\sigma', \sigma) \text{ and } \exists \varepsilon : X \longrightarrow A \text{ an assignment such that } \bar{\varepsilon}(t') = a' \wedge \bar{\varepsilon}(t) = a\}$$
$$e^\bullet \stackrel{\text{def}}{=} \{(\sigma', a') \mid \sigma' \in \sigma^\bullet, \exists t' \in l_F(\sigma, \sigma') \text{ and } \exists \varepsilon : X \longrightarrow A \text{ an assignment such that } \bar{\varepsilon}(t') = a' \wedge \bar{\varepsilon}(t) = a\}$$
For a set of events Π^E *this notation is extended to* ${}^\bullet\Pi^E$ *and* $\Pi^{E\bullet}$ *in the usual way.*
5. *Given two markings* m_1, m_2 *and a set* Π^E *of events of* $\mathbf{N}$ *under* $\mathbf{A}$. *Then we say that* m_2 is reachable from m_1 by Π^E in one step, *written* $m_1[\Pi^E\rangle m_2$, *if the following conditions hold for* $e, e_1, e_2 \in \Pi^E$:
   (a) conflict free: ${}^\bullet e_1 \cap {}^\bullet e_2 \neq \emptyset \Rightarrow e_1 = e_2$ *and* $e_1^\bullet \cap e^\bullet \neq \emptyset \Rightarrow e_1 = e_2$
   (b) concession: ${}^\bullet e \subseteq m_1$ *and* $e^\bullet \cap m_1 = \emptyset$
   (c) change: $m_2 = (m_1 \setminus \bigcup_{e \in \Pi^E} {}^\bullet e) \cup \bigcup_{e \in \Pi^E} e^\bullet$

Notation. We also use the equivalent set representation of a marking m and write $(\sigma, a) \in m$ to express that $a \in m(\sigma)$, for $\sigma \in \Sigma_{u,s}$.

Note that different state predicates and event schemes give rise to different conditions (elements of a marking) and events. Congruences of Σ-terms lead to congruences of the corresponding conditions and events.

An example for a graphical representation of a marking is depicted in fig. 4.

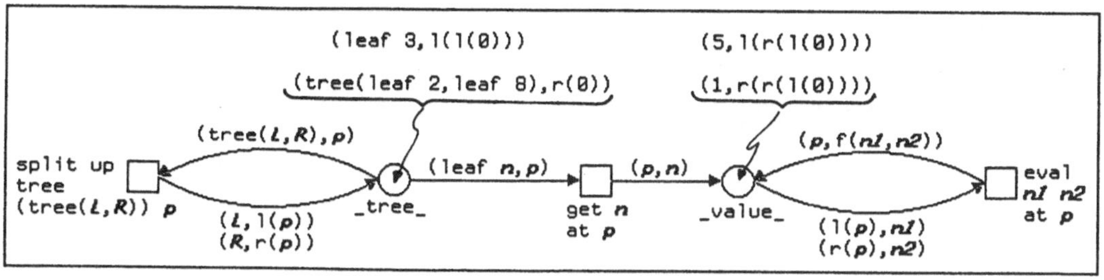

Figure 4: A The Π-labelled net of fig. 2 in some state of computation

Processes of PrE Nets

We now come to defining the behavior of a Predicate-net specification in terms of the processes that may occur on its Π-labelled net. As usual, occurrence nets are taken as domains of nonsequential processes of PrE nets.

Definition 13 *Let* $\Pi = (\Sigma, \Pi^C, \Pi^E)$ *be a PrE-signature,* $(S, \Pi, R, \mathbf{N})$ *a PrE-net specification,* $\mathbf{A}$ *a* Σ-*algebra,* $N_A \stackrel{\text{def}}{=} \{(\sigma, a) \mid \sigma \in \Pi^C_{u,s} \cup \Pi^E_{u,s}, a \in A^u, s \in S, u \in S^*\}$ *and let* **ON** *be an occurrence net (cf. [2]). Then the operation* $p : ON \longrightarrow N_A$ *is called a* process of **N** *iff it satisfies the following conditions:*

1. *for each event* $e \in ON$, $p(e)$ *is an event of* **N** *under* **A**;
2. *for each* Π^C-*cut* B *of* ON : $p\mid_B$ *is injective;*
3. $p({}^\bullet e) = {}^\bullet p(e)$ *and* $p(e^\bullet) = p(e)^\bullet$ *for all events* e *of* ON;
4. $p({}^\bullet e) \cap p(e^\bullet) = \emptyset$.

ON *is called the* process domain *of* **N** *under* p.

Notation. In the graphical representation of a process p we draw the process domain as a net and label its elements n by $p(n)$.

Example

Let s_t denote some event (split up tree_,(t,p)) of **N**, the net in fig. 2, under some algebra **A** which has t as an element of the carrier $A^{tree}$and $p \in A^{path}$. Similarly, let g_n denote an event (get_at_,(n,p)) and $e_f(n1,n2)$ denote some event (eval_at_,$(n1,n2,p)$) of **N** under **A**. Then an example of a process of **N** is shown in fig. 5.

3.4 Predicate-Event Systems

A model of a PrE net specification is a many-sorted algebra that satisfies elementary net theoretic conditions.

Definition 14 *Let* $\Pi = (\Sigma, \Pi^C, \Pi^E)$ *be a PrE-signature. A* Predicate-Event System signed by Π *(*Π-system, *for short) is a* Σ-*algebra* **A** *satisfying the following two conditions:*

1. *for all* $\sigma \in \Pi^C_{u,s}, \sigma' \in \Pi^E_{v,s}, a \in A^u, a' \in A^v$ *with* $s, s' \in S, u, v \in S^*$: $\sigma_A(a) \neq \sigma'_A(a')$, *i.e., the sets of condition and the set of event instances of the same sort must be disjoint;*
2. *for all* $\sigma \in \Pi^C_{u,s}(\sigma \in \Pi^E_{u,s}), a \in A^u$ *with* $\sigma_A(a)$ *defined* $(s, s' \in S, u \in S^*)$: *there is a* $\sigma' \in \Pi^E_{v,s'}(\sigma \in \Pi^C_{v,s'}), a' \in A^v, v \in S^*$ *such that* $flow_A(\sigma_A(a), \sigma'_A(a'))$ *defined. That is there are no "isolated" condition and event instances.*

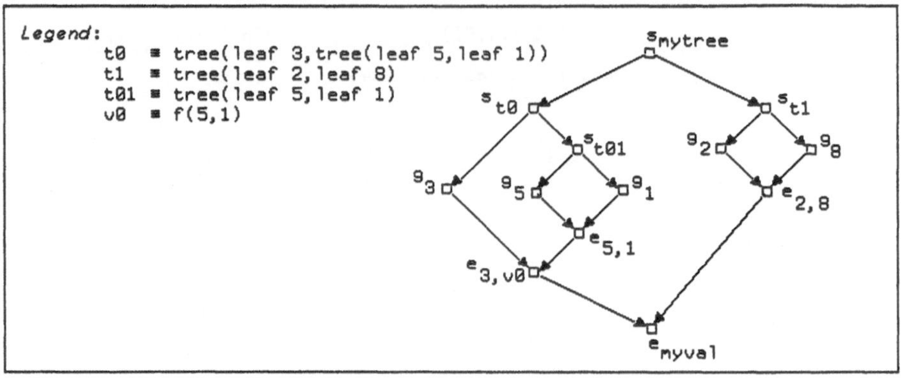

Figure 5: A process of fig. 2

3.5 Models of PrE-Specifications

Now we can explain how the syntactic domain of PrE-net specifications relates to the semantic domain of many-sorted algebras.

Definition 15 *Let* $\Pi = (\Sigma, \Pi^C, \Pi^E)$ *be a PrE signature and* $SPEC = (S, \Pi, R, \mathbf{N})$ *be a PrE net specification. The a PrE-system* $\mathbf{A}$ *satisfies* $SPEC$ *iff the following conditions hold:*

1. $\mathbf{A} \models R_{SPEC}$
2. *Given an event* $e = (\sigma, a)$ *of* $\mathbf{N}$ *under* $\mathbf{A}$. *Then we have that* $(\sigma', a') \in {}^\bullet e$ *iff* $flow_A((\sigma'_A(a'), \sigma_A(a)))$ *defined and* $(\sigma', a') \in e^\bullet$ *iff* $flow_A((\sigma_A(a), \sigma'_A(a')))$ *defined.*

Theorem 2 *Let* $SPEC = (S, \Pi, R, \mathbf{N})$ *be a consistent PrE net specification. Then there is an initial model of SPEC in the category of PrE-systems satisfying SPEC.*

In analogy to the pure algebraic case, we denote this initial model by $\mathbf{P}_{SPEC}$.
The following list of points sketches the proof (cf. [24]):

1. First assume that the preconditions of the theorem are fulfilled, i.e. there exists at least one model of SPEC.
2. The net N of SPEC is then translated into a set of rules.
3. Combined with R, the resulting set of rules defines an inital algebra according to theorem 1.
4. This algebra is shown to satisfy the points required for PrE-system by showing that these are implications of the rules.
5. Finally, this algebra is shown to be initial. That is it is demonstrated that any model of SPEC satisfies the rules resulting from the translation.

As an implication of this proof, the following corollary can be formulated.

Corollary 1 *Let* $SPEC = (S, \Pi, R, \mathbf{N})$ *be a PrE net specification. Then SPEC can be translated into an equivalent algebraic specification in the sense that the induced congruences uniquely define the class of PrE systems satisfying SPEC.*

The interesting point about this fact is that PrE systems can be viewed as algebras with some operators distinguished via the PrE signature. Some of the operators are used to identify a variable state of affairs, some are used to identify the changes in states, and the rest describes invariant (structural) properties of the system. Nevertheless all of them can be used in a pure algebraic way to analyze consistency or implementation correctness and to study the use of algebraic verification techniques for concurrent systems.

4 Modularization

For large systems and data structures we may not want to give their entire specification at once. Rather we may find it easier to extend given specifications or to combine them systematically to form a larger system which possibly introduces additional synchronization or extends the freedom of behavioral choice. Synchronization is increased when high level events in the interface of different components combined can be identified, while freedom of choice is increased when high level conditions of different component interfaces can be identified.

SEGRAS offers various features that allow structuring specifications: combination, parameterization, abstract implementation, and scoping. In the framework of ADT's, these concepts have been given an algebraic semantics. As for *SEGRAS* we adopted these concepts and extended them by restrictions on behavioral properties of combined specifications. These behavioral properties are based on the net theoretic notion of processes of PrE-systems and ensure a kind of behavioral consistency and completeness.

Here we confine ourselves to the notion of modularization of PrE-specifications by means of combination and extension.

4.1 Combination and Extension of Specifications

Combinations provide a way of building specifications on existing ones by adding new sorts, operations, rules, and nets to a given specification or by putting existing specifications together. The user of our specification language can view this union as a "glueing" of the labelled nets in the graphical representation. Extension is a specific kind of combination by which the semantics of the extended specifications is preserved. Extendable specifications can be verified once and for all and can be implemented independently from their various extensions as the carrier sets of their models are protected against modifications.

Definition 16 *Let $SPEC = (S, \Pi, R, \mathbf{N})$ and $SPEC' = (S', \Pi', R', \mathbf{N}')$ be two PrE-specifications.*

1. *We call $SPEC$ a* subspecification *of $SPEC'$ if $S \subseteq S', \Pi \subseteq \Pi', R \subseteq R'$, and $\mathbf{N}$ is a subnet of $\mathbf{N}'$ defined by:*
   (a) $\Pi^C \subseteq \Pi^{C'}$ and $\Pi^E \subseteq \Pi^{E'}$
   (b) $l_E(e) = l'_E(e)$ forall $e \in \Pi^E$ and $l_F(f) \subseteq l'_F(f)$ forall $f \in F$
2. *Let $SPEC$ be a subspecification of $SPEC'$. Then $SPEC'$ is called a* combination *if $SPEC' = (S \cup S'', \Pi \cup \Pi'', R \cup R'', \mathbf{N} \cup \mathbf{N}'')$ such that $S \cap S'' = \Sigma \cap \Sigma'' = \emptyset$ and $\mathbf{N} \cup \mathbf{N}'' = (\Pi^C \cup \Pi^{C''}, \Pi^E \cup \Pi^{E''}; F \cup F'', l \cup l'')$ where $l_F \cup l''_F(f) \stackrel{def}{=} l_F(f) \cup l''_F(f)$ forall $f \in F'$. If $SPEC'$ is* consistent, *there exists a unique morphism $f_{SPEC'} : \mathbf{P}_{SPEC} \longrightarrow \mathbf{P}_{SPEC'}$.*
3. *If $SPEC'$ is a combination and if the initial $SPEC$-algebra $\mathbf{P}_{SPEC}$ is congruent to the Σ-reduct of $\mathbf{P}_{SPEC'}$, written $\mathbf{P}_{SPEC} \cong \mathbf{T}_{SPEC'|\Sigma}$, then $SPEC'$ is called an* extension *of $SPEC$.*

Remark. In the Σ-reduct of a Σ'-algebra with $\Sigma' = \Sigma \cup \Sigma''$ we forget the sorts in S'', the operation symbols in Σ'', and all data elements which cannot be expressed without these symbols in Σ''.
Note that, if $SPEC'$ is an extension, then the morphism $f_{SPEC'}$ is injective and surjective on the carriers of S.
This definition degrades to the purely algebraic case if $\mathbf{N}' = \emptyset$.

In the categorical treatment of PrE systems and their underlying Condition-Event systems (CE systems), combination corresponds to the disjoint union of possibly infinite CE nets modulo some congruence of net-elements. In an extension, the CE-net underlying $\mathbf{P}_{SPEC}$ is isomorphic to a subnet of the CE-net underlying $\mathbf{P}_{SPEC'}$. For a behavioural extension moreover, the CE-system (i.e. the CE-net plus a selected behaviour in terms of markings and reachability) underlying $\mathbf{P}_{SPEC'}$ additionally "carries" all the processes of $\mathbf{P}_{SPEC}$.

With the above requirements we may ensure consistency and completeness of extensions w.r.t. the terms denoting data and net elements. However, we have not yet ensured similar properties for the processes of $SPEC$ and $SPEC'$. It is still possible that $SPEC$ has processes which are not mapped on processes of $SPEC'$. It is also possible that processes of $SPEC'$ change the marking of the subnet $\mathbf{N}$ in $\mathbf{N}'$ such that there is no process in $SPEC$ which affects the same change on $\mathbf{N}$.

Therefore we refine our previous definition by two further requirements.

Definition 17 *Let $SPEC' = SPEC + (S'', \Pi'', R'', \mathbf{N}'')$ be an extension of $SPEC$. Then $SPEC'$ is called a* behavioural extension

1. *if for every process p with occurrence net ON of $\mathbf{P}_{SPEC}$ can be embedded into some process p' with occurrence net ON' of $\mathbf{P}_{SPEC'}$ by an injective operation $inj : ON \longrightarrow ON'$, such that,*

$$f_{SPEC'}(p(a)) = p(inj(a)).$$

   This means, each process in the behaviour specified by $SPEC$ is the Σ-reduct of some process in the behaviour specified by $SPEC'$.

2. *forall markings m_1, m_2 of $\mathbf{N}$ and forall processes p of $\mathbf{N}$ sucht that $m_1[p\rangle m_2 \Rightarrow m_{1|\Sigma}[p_{|\Sigma}\rangle m_{2|\Sigma}$.*

Notation. $m_1[p\rangle m_2$ denotes the extension from steps to processes.

5 Example: Extension and Combination of PrE-net Specifications

In this section we give some small examples to illustrate the concepts defined.

Storage Cell

Assume an algebraic specification of strings, called string, which provides the sort string and appropriate operations and rules for a monoid over an alphabet. Assume further a specification "length" giving a constant operation "length" of sort **Nat**. Then objects of sort cell each of which may keep arbitrary string values are specified by the PrE net specification cell in fig. 6.

The above specification tells us that an arbitrary system C of sort cell exchanges data of sort string with its environment by means of two actions. This is defined by the terms "put S in C" and "get S from C" where C is a term denoting the data exchanged. The net defines the (distributed) state of C before and after communication has taken place. It also tells us that in case of an occurrence of action "put_in_" S is an input to C because it is undefined in the state before such an occurrence. For an occurrence of "get_from_", however, S is an output of C because it is determined by the state before such an occurrence.

Concurrent Array

A group of cells can be put together to form a kind of array which can be accessed concurrently when different indeces are used at a time (cf. fig. 7).

To illustrate the fact that this specification is a behavioral extension of the first, a typical process is given in fig. 8.

```
spec string-cell extends string by

   sorts  cell, c-ac, c-st

   fu     cell:       Nat -> cell.
   ac     put_in_:    string cell -> cell-ac.
          get_from_:  string cell -> cell-ac.
   st     _empty:     cell -> cell-st.
          _in_:       string cell -> cell-st.

   forall C: cell; S: string.
   case (C empty): C.
   S in C:  S, C.                 C empty: C.
   put S in C:  S, C.             get S from C:  S, C.
```

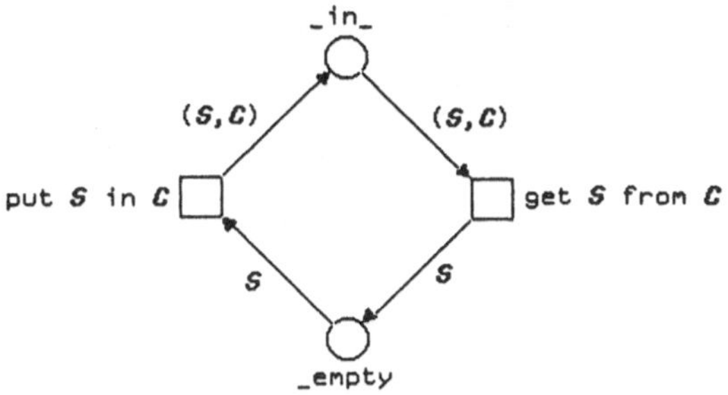

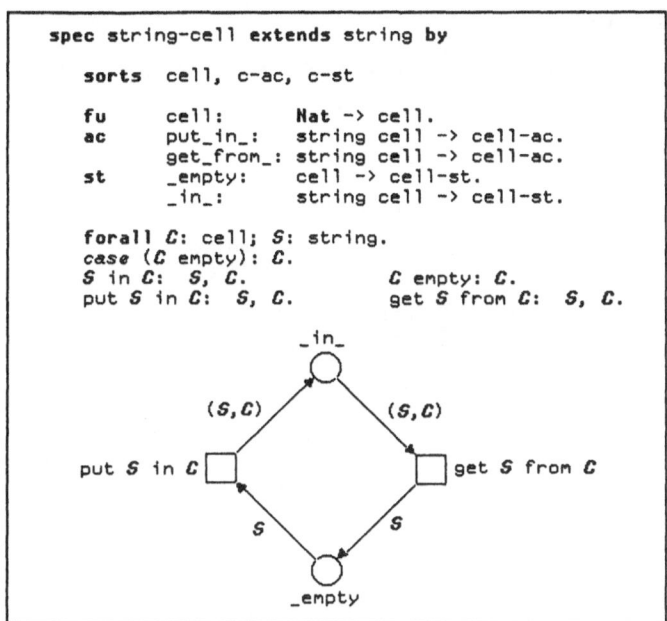

Figure 6: Specification of a cell able to carry strings

```
spec concurrent-array extends string-cell, length by

   sorts  array

   fu     _sub_:          array Nat -> cell.
   fu     make-array_:    cell -> array Nat.

   forall A: array; i: Nat.
   make-array (cell i) = (A,i):   i < length.
   A sub i = cell i : A.
```

Figure 7: Specification of a concurrent array

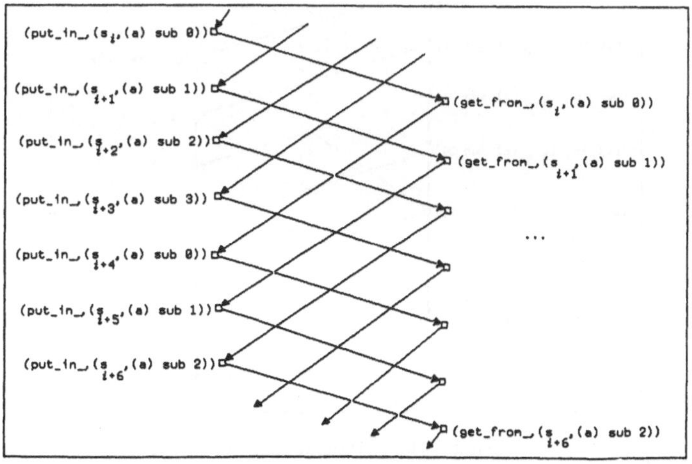

Figure 8: A typical array process

Bounded Ring Buffer

When combining the previous specification with additional behavioral constraints as shown in fig. 9, we arrive at the specification of a ring buffer.

```
spec ring-buffer combines concurrent-array with

   sorts  head-ptr, tail-ptr

   st     _head:  cell -> head-ptr.
   fu     _tail:  cell -> tail-ptr.

   case ((cell 0) head): .
   case ((cell 0) tail): .
   forall A: array; i: Nat.
   (A sub i) tail: A.
   (A sub i) head: A.
   forall array; i: Nat; S: string.
```

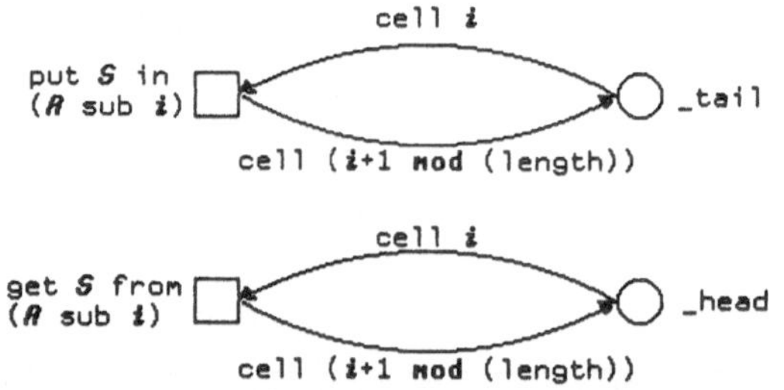

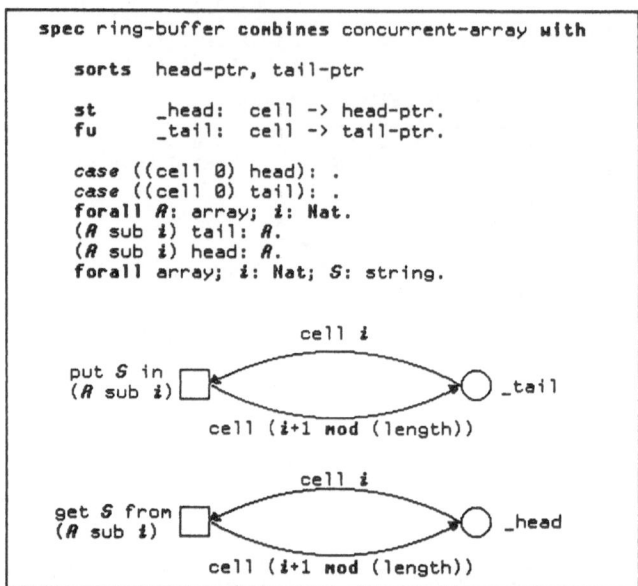

Figure 9: Specification of a concurrent ring buffer

It still allows concurrent put and get operations but all put and all get operations are put in a certain order to achieve FIFO properties. As we also identify the first and the last cell of or buffer by means of the modulo (mod) operation on **Nat**, we have a combination but no extension of concurrent array. Again, a typical process is shown in fig. 10.

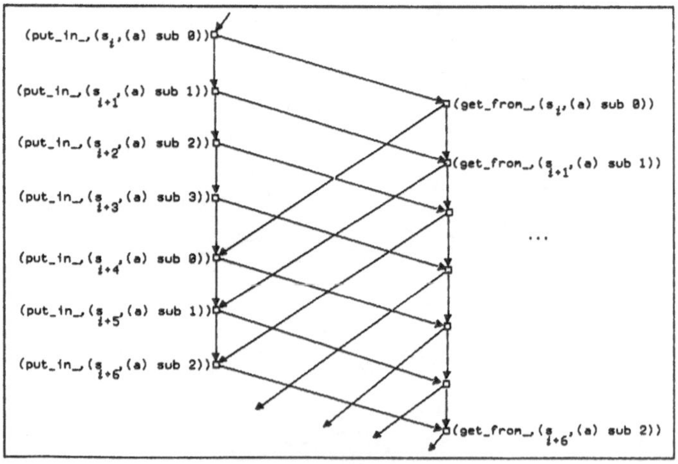

Figure 10: A typical ring buffer process

Note that in the full language this example can be simplified by parameterization and it can be made more readable by taking advantage of renaming concepts.

Discrete Pipeline

Finally, we show another example of how FIFO properties can be imposed on a concurrent array. In the shift buffer in fig.11 the put and get operation of any two subsequent cells are identified.

```
spec shift-buffer combines concurrent-array with

   forall A: array; i: Nat; S: string.

   put S in (A sub i) = get S from (A sub (i+1)):
                              put S in (A sub i),
                                get S from (A sub (i+1)),
                                i < ((length) - 1).
```

Figure 11: Specification of a concurrent shift buffer

The idea underlying this specification is that a string produced is put in the last cell of the buffer and passes along the cell sequence to be gotten from a consumer when it arrived in the first cell (see also fig. 12).

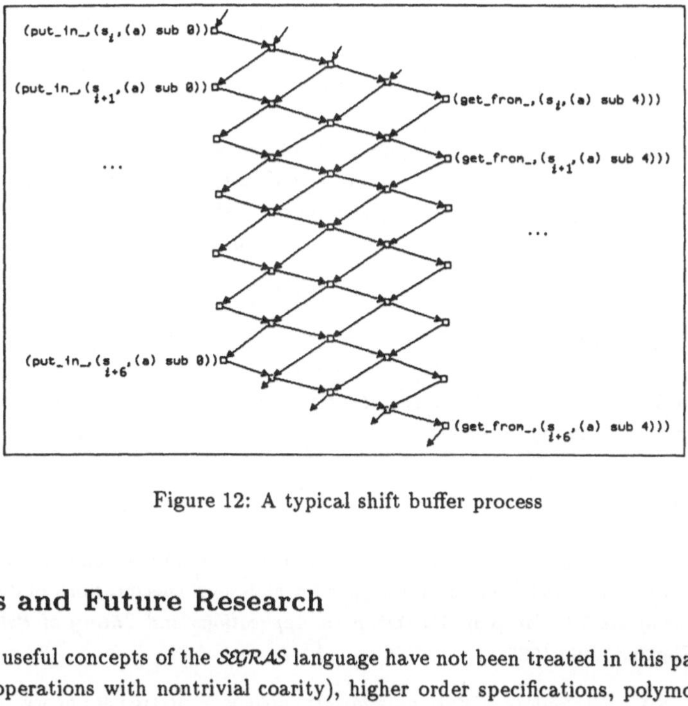

Figure 12: A typical shift buffer process

Conclusions and Future Research

Many practically useful concepts of the SEGRAS language have not been treated in this paper: multi-valued operations (i.e., operations with nontrivial coarity), higher order specifications, polymorphism, subsorts, parameterized specifications, and abstract implementation. The intention was to study our combination of partially specified abstract data types and Predicate-Event nets in the smallest possible framework. However, there are various efforts extending the algebraic theory of abstract data types to handle such concepts. For example, the definition of partial algebras given by Kreowski and Schmidt in [15] already deals with nontrivial coarity of operations. Möller provided a definition of higher order specifications and algebras [20]. Poigné gave a formalization of two-level specifications [22] which allows him to handle polymorphism and higher order types. Our notion of polymorphism relies on multi-leveled specifications which can be considered a generalization of the two-level case. The formal treatment of subsorts, parameterization, and abstract implementation is nearly classics in the algebraic specification framework.

The small examples given in this paper might have shown the potential of the unified framework of many-sorted algebraic and net theoretic specifications. But they might also have raised the demand for a less restrictive methodology and notation to be of practical use. The concrete syntax of the *SEGRAS* specification language [12] is an attempt to provide as much programming power as possible within the framework. The idea is to achieve this by displacing much of the burden of incrementally constructing and manipulating specifications into support tools. Such tools can also supplement omissions by default and may survey the observance of context conditions in a tolerant way.

SEGRAS is designed for interactive use. It is supported by a programming environment, called the *SEGRAS* Lab, which currently consists of a *syntax-directed editor* including the graphical representation of PrE nets[4] a grammar driven database, and a polymorphic typechecker.

Rewrite rules which can be derived from conditional algebraic specifications and the transition rule defined for PrE-systems provide an operational semantics for the language. This semantics can be used to test specifications for adequacy and consistency with their intended initial semantics. In the GRASPIN project we are adapting and integrating a rewrite rule engine into the Lab to support extension checking, consistency and completeness checks for algebraic specifications. An interactive net simulator is under development to provide an interpreter for PrE system specifications. It will permit to observe the token flow through a net on the graphical representation. As tokens are denoted by terms, the rewrite engine will supply its capabilities to rewrite the tokens according to the flow labeling and the behavioral constraints defined.

It is still an open problem how the requirements of behavioral extension can be checked, in general. One solution might be to translate processes into algebraic terms and then to apply the usual algebraic machinery. This was proposed in [25]. Another solution might be to find equivalent definitions of these requirements such that net-theoretic machinery can be applied for checking. This is one of the research topics of the GRASPIN project.

Acknowledgments

We are grateful to Harald Fonio and the 'Advances' referee for their careful reading of this manuscript and valuable suggestions for its improvement.

References

[1] B. Berthomieu, N. Choquet, C. Colin, B. Loyer, J. Martin, and A. Mauboussin. Abstract data nets combining Petri nets and abstract data types for high level specifications of distributed systems. In *Proceedings of the 7th European Workshop on Applications and Theory of Petri Nets*, pages 25–48, Oxford, England, July 1986.

[2] E. Best and C. Fernàndez. Notations and terminology on Petri net theory. *Petri Net Newsletter*, 23:21–46, April 1986.

[3] M. Broy and M. Wirsing. Partial abstract types. *Acta Informatica*, 18:47–64, 1982.

[4] M. Diaz. Modelling and analysis of communication and cooperation protocols using Petri net based models. In C. A. Sunshine, editor, *Protocol Specification, Testing and Verification*, pages 465–510, North-Holland Publishing Company, 1982.

[4] All examples in this paper have been produced with this component of the *SEGRAS* Lab.

[5] H. Ehrig and B. Mahr. *Fundamentals of Algebraic Specification 1.* Volume 6 of *EATCS Monographs on Theoretical Computer Science*, Springer-Verlag, Berlin, Heidelberg, New York, Tokyo, 1985.

[6] K. Futatsugi, J. A. Goguen, J. Jouannaud, and J. Meseguer. Principles of OBJ2. In *Conference Record of the Twelfth Annual ACM Symposium on Principles of Programming Languages*, pages 52–66, Louisiana, New Orleans, 1985.

[7] H. J. Genrich and K. Lautenbach. System modelling with high-level Petri nets. *Theoretical Computer Science*, 13(1):109–136, 1981.

[8] S. L. Gerhart, D. Musser, D. Thompson, D. Baker, R. Bates, R. Erickson, R. London, D. Taylor, and D. S. While. An overview of Affirm: a specification and verification system. In S. Lavington, editor, *INFORMATION PROCESSING 80*, pages 343–347, IFIP, North-Holland Publishing Company, 1980.

[9] J. A. Goguen and J. Meseguer. EQLOG: equality, types and generic modules for logic programming. In D. DeGroot and G. Lindstrom, editors, *Functional and Logic Programming*, pages 295–363, Prentice-Hall, 1986.

[10] J. A. Goguen, J. W. Thatcher, and E. G. Wagner. An initial algebra approach to the specification, correctness, and implementation of abstract data types. In R. T. Yeh, editor, *Current Trends in Programming Methodology*, pages 80–149, Prentice-Hall, Englewood Cliffs, New Jersey, 1978.

[11] B. Krämer. *SEGRAS* – a formal language combining Petri nets and Abstract Data Types for specifying distributed systems. In *Proceedings of the 9th Annual International Conference on Software Engineering*, pages 116–125, Monterey, California, March 1987.

[12] B. Krämer. *SEGRAS – The GRASPIN Specification Language. GRASPIN Technical Paper*, Sankt Augustin, July 1986.

[13] B. Krämer. Stepwise construction of non-sequential software systems using a net based specification language. In G. Rozenberg, editor, *Advances in Petri nets 1984*, pages 307–327, Springer-Verlag, Berlin, Heidelberg, New York, Tokyo, 1985.

[14] B. Krämer and H. Schmidt. An approach to algebraic specification and stepwise implementation of non-sequential systems. In *Poster Session Proceedings of the 6th International Conference on Software Engineering*, pages 63–64, Information Processing Society of Japan, Tokyo, Japan, September 1982.

[15] H. Kreowski and H. Schmidt. *Some Algebraic Concepts of the Specification Language SEGRAS and their Initial Semantics. GMD-Studien*, Gesellschaft für Mathematik und Datenverarbeitung, October 1984.

[16] L. Lamport. On interprocess communication, part I. *Distributed Computing*, 1:77–85, 1986.

[17] L. Lamport. Specifying concurrent program modules. *ACM Transactions on Programming Languages*, 5(2):190–222, April 1983.

[18] B. Mahr and J. Makowsky. Characterizing specification languages which admit initial semantics. *Theoretical Computer Science*, 31:49–60, 1984.

[19] R. A. Milner. *A Calculus of Communicating Systems. Lecture Notes in Computer Science*, Springer-Verlag, Berlin, Heidelberg, New York, 1980.

[20] B. Möller. Algebraic specification with higher-order operators. In L. Meertens, editor, *Proceedings of the IFIP TC 2 Working Conference on Program Specification and Transformation*, North-Holland, Amsterdam, to appear.

[21] A. Pnueli. The temporal logic of programs. In *Proceedings of the 18th Annual Symposium on Foundations of Computer Science*, pages 46–57, Providence, October 1986.

[22] A. Poigné. On specifications, theories, and models with higher types. *Information and Control*, 68(1–3), January, February, March 1986.

[23] W. Reisig. *Petri Nets.* Volume 4 of *EATCS Monographs on Theoretical Computer Science*, Springer-Verlag, Berlin, Heidelberg, New York, Tokyo, 1985.

[24] H. Schmidt. *Towards a Net-Theoretic Notion of Type based on Predicate-Transition Nets.* Arbeitspapiere der GMD 117, Gesellschaft für Mathematik und Datenverarbeitung, November 1984.

[25] H. Schmidt and M. Papazoglou. *Abstract Implementation of Predicate-Event Systems.* Sankt Augustin, August 1985.

[26] J. Vautherin. Parallel systems specifications with colored Petri nets and algebraic abstract data types. In *Proceedings of the 7th European Workshop on Applications and Theory of Petri Nets*, pages 5–23, Oxford, England, July 1986.

6.
OBJSA Nets: A Class of High-level Nets Having Objects as Domains

E. Battiston, F. De Cindio and G. Mauri

G. Rozenberg (ed.): Advances in Petri nets 1988. Lecture Notes in Computer Science, vol. 340.
Springer, Berlin Heidelberg New York 1988, pp. 20-43

ABSTRACT

To define classes of high level nets having structured (individual) tokens is a very fundamental goal for making nets actually usable in real concurrent system modelling. A promising approach is that of combining nets with algebraic specification techniques. This results in a formal specification language which supports both aspects of system modelling, namely data structure and control structure modelling, with suitable abstraction notions.

Some different formalisms combining nets and abstract data types have been proposed. In this paper, we define a class of high-level Petri nets, namely OBJSA net systems (or OBJSA nets for short), in which: 1) the net can be decomposed into state-machine components, i.e. it preserves the main characteristics of Superposed Automata (SA) nets; 2) the domains to which individual tokens belong are defined as abstract data types by using the language OBJ2. For this class of nets two products (namely an S-product ⊗ and a T-product ⊙) are then provided for defining, respectively, the S- and T-invariants as the first step for preserving in the resulting specification language the possibility, typical of nets, of deriving properties of the modelled system by using algebraic techniques.

1. Combining process abstraction and data abstraction for real system modelling

In the area of complex system modelling the need for combining techniques supporting process (control) abstraction and techniques supporting data abstraction in a well-founded and usable formalism is by now widely recognized (see for instance |CHJ|). This need stems from the evidence that: a) the complexity in real system modelling originated neither only from the number of concurrent components the system consists of, nor from the complexity of the data structures they modify, but by the combination of both; b) the most popular formalisms supporting process abstraction, such as (high-level) Petri nets (|Gen|, |Jen|), Milner's CCS |Mil| or specialized specification languages such as SDL for telecommunication applications, are weak in supporting data abstraction; c) analogously, the most popular formalisms for data abstraction, in particular the algebraic ones (|LZ|, |Zil|, |GTW|, |EM|), have points of weakness, mainly a limited ability to support process abstraction, which becomes serious for modelling highly concurrent systems.

This awareness has given rise in recent years to some attempts to integrate both aspects, such as: the new definition of SDL standard, where data are no longer described with a Pascal-like formalism, but through an algebraic initial model |CCITT|; the LOTOS specification language |ISO| combining the algebraic language Act-One |EM| with CCS; the SMoLCS methodology |AMRW|, bringing together in a unique framework denotational, algebraic and operational techniques (based on transition systems) in order to specify large concurrent systems.

As for the combination of nets with algebraic specification techniques, let us mention the SEmiGRAphical Specification language SEGRAS (|Kra|), where nets are used to introduce some degree of concurrency among the operations on abstract objects, i.e. to optimize the rewriting process; the abstract data type representation of nets with individual tokens given in |Rei1|, where algebraic techniques are used for giving the semantics of a class of high-level nets; the Coloured Petri nets with abstract data types proposed in |Vau|; and finally the Abstract Data Nets of |Ber|.

|Vau| redefines the domains of coloured nets in algebraic terms, but he does not accomplish the task of defining a new class of high-level nets, with graphical and matrix representation, a firing rule, products for invariant calculus and so on. Nevertheless, he gives a number of interesting results for deriving properties of the modelled system through the application of standard analysis techniques to the underlying Place/Transition net.

To some extent Abstract Data Nets go the direction of defining a new class of high-level nets, but no technique is provided for analyzing the resulting nets. Rather, the proposed way for property verification consists in transforming the composite model into a full-algebraic specification where the usual techniques can be applied.

Hence, although all the proposals combining nets and algebraic specification published so far refer to high-level nets, none of them yields the definition of a new class of high-level nets using algebraic techniques instead of (multi)set theory for specifying the individual tokens flowing into the net. This is the main goal of our work. In this paper we will define, both in graphic and matrix form, a new class of high-level nets where individual tokens are specified as algebraic terms. Two products (S-product and T-product) are then provided for defining, respectively, S- and T-invariants, so preserving in the resulting language the possibility, typical of nets, of deriving properties of the modelled system by using algebraic techniques.

1.1 Systems (models) through components (models) composition

In defining this new class of nets we aimed to overcome a further weakness which is often ascribed to nets, i.e. the difficulty of structuring them in conformity with the system to be modelled. As is well-known to computer scientists, engineers, biologists and organization analysts, the behaviour of whatever real system results in the combination of the behaviours of its autonomous, i.e.containing local non-deterministic choices, components, as constrained by their mutual interactions. In the field of embedded systems specification and of concurrency models, in order to reflect and preserve this characteristic, Hoare's CSP |Hoa|, Milner's CCS, or COSY language |LTS|, among others, build the overall system model through the composition of its component models. All the approaches mentioned essentially perform the composition by distinguishing the actions local to a component from the interactions among components. In terms of nets, where states also must be considered, the preserving of components identity (as discussed in |Obe|) requires also the partition of places among the different components, so that the global system state (the case) results from the composition of their local states.

State-Machine Decomposable nets introduced by Hack |Hac|, and recalled in |Bert|, preserve the components in the overall model, but, since they focus attention on decomposition, allow a place to belong to more than one component. On the contrary, Superposed Automata (SA) nets |DDPS| focus attention on the possibility of building the net system model through composition of its (sequential non-deterministic) components, as CSP, CCS, and COSY do. An SA net therefore results from the combination, through transition superposition, of a set of *state-machine (sm) components*, each one representing a different sequential component of the system to be modelled. Each transition has therefore the same number of incoming and outgoing places. If this number is equal to one, then the transition models an action local to a sm-component (a state-machine net, in the Hack' terms); otherwise, the transition represents an interaction among two or more sm-components.

Ever since their definition, SA nets have been tested in real system modelling. In particular, the application of 1-safe SA nets, where just one (unstructured) token flows in each sm-component (called state net decomposable in |BFP|), appeared effective and satisfactory |DDS|. Furthermore, the definition of two orthogonal notions of equivalence allows the nice possibility of building the overall net system model through a network of partial, therefore simpler, models connected by formally defined relationships. Nevertheless, SA nets which inherits Predicate/Transition nets approach to data specification often becomes cumbersome and unreadable when data must be taken into account (see for instance lDLTl). The goal in defining a new class of high-level nets is then to reach a synergy between SA nets and algebraic specification techniques.

In order to achieve this goal, the class of nets this paper introduces retains the assumption, typical of a lot of languages already mentioned, that each sm-component has autonomy and is owner of its data structure whose state is modified both by local actions and by interactions with other components. Furthermore, it retains structural characteristics of SA nets, i.e.:

1) the sm-components are well distinguished, i.e. their states (represented by places) are partitioned into disjoint sets and the overall system state results from the composition of the component states;
2) transitions can represent actions local to a sm-component or interactions among sm-components; in the first case they are extensionally characterized by the sm-component input and output states; the same holds in the second case when considering the state of all the involved sm-components.

1.2 Modularity and reusability of specifications

The class of nets this paper introduces retains another characteristics of SA nets:

3) the individual flowing in each sm-component models a sequential component of the system. If several individuals flow in the same sm-component they represent different instances of the same sequential component; therefore, in the line of lObel, their individuality must be preserved through transitions firing.

Algebraic specification is the technique chosen for specifying such individual tokens, together with the related arc labels and transition inscriptions which are used, as in Predicate/Transition and Coloured nets, for imposing further constraints on transition firings. Furthermore, we have preferred to make use of an existing well-known tool supported specification language, such as OBJ2 lFGJMl, instead of redefining all the algebraic machinery.

The specification languages in the OBJ family are based on the initial algebra semantics lGTWl. OBJ2, in particular, extends the previous versions of OBJ mainly by generalizing the standard notion of abstract object as a triple $<S,\Sigma,E>$ (where S is a set of sort names, Σ a set of operator names of given arity, E a family of equations defining the operators) to the notion of (possibly parameterized) module (lGogl, lFGJMl). Parameterization is a powerful mechanism which supports modularity in the development of algebraic specifications and hence facilitates specification reusability, which is in our opinion a major issue (see also lBCGl and lGogl). A further OBJ2 facility supporting module reusability consists in the possibility of defining a new module starting from existing ones, by means of the clauses *protecting, extending* and *using*, which guarantee different protection levels of the basic modules.

1.3 Intuition behind OBJSA nets

As a consequence of the choices presented above, the intuition behind the merging of SA nets and OBJ2, obtaining thus the class of OBJSA net systems, or OBJSA nets for short, can be summarized as follows:

a) the net structure is the same as Hack's State-Machine Decomposable nets, with the further condition that places are partitioned into disjoint classes; as a consequence, transitions are balanced, i.e. they have the same number of input and output places, where couples of input/output places belong to the same sm-component;
b) the individuals flowing in the net consist in a name part, which models instances individuality and is not modified by transition firing, and a data part, which represents the data structure of the sm-component and can be modified by transition firing;
c) the overall net system can be obtained through composition of the net models of its components.

The main idea for the merging of SA nets and OBJ2 is that of defining the domains to which the individual tokens flowing in the net belong as instantiations of a predefined *parameterized object* called COMP-DOM (for COMPonent

DOMain). By using suitable *theories*, this parameterized object characterizes the generic individual flowing in whatever component as consisting in a *name* part and in a *data* part (see point b here above). Furthermore, the parameterized object TRANSITION is introduced, and its instantiations associated with transitions. These actual objects contain, really consist in, the definitions of the operators which capture how the transition firing modifies the individuals flowing from the input to the output places.

Let us point out that the integration of Petri nets and OBJ should be carried out without requiring that the underlying net is an SA net. Nevertheless, the choice made, on one hand, guarantees that the previous discussed needs for modularity, compositionality and reusability are quite easily satisfied; on the other, it is a guideline for associating objects to places and transitions embedded in the built-in parameterized objects COMP-DOM and TRANSITION.

The definitions of the net structure and of the mentioned parameterized objects are given in Section 2, and are used in Section 3 for defining OBJSA net systems, their firing rule and the associated matrix. In the class of OBJSA net systems two products (namely an S-product ⊗ and a T-product ⊙) are then provided (Section 4), for defining, respectively, the S- and T-invariants. All the definitions are illustrated on a simple and uninterpreted example. Section 5 shows on this uninterpreted example how an OBJSA net system can be obtained through composition, starting from its sequential components, each of which is parametric with respect to its interaction transitions. The formal definitions of OBJSA components and of the composition function, which are omitted here for lack of space, can be found in |BDM1| where an interpreted example is also presented. Other examples in the field of the specification of telecommunication systems are in |BDM2|.

2. Preliminary definitions

The definition of SA nets, first given in |DDPS|, is recalled, with minor changes, and a toy example is given which will be used throughout the paper. Afterward, some definitions of parameterized abstract objects are introduced. They will be used in the next section for the definition of OBJSA net systems.

Def.2.1- A Superposed Automata (SA) net is a quadruple $N=\langle S, T, F, \Pi\rangle$ where:

- S is a finite non empty set of places;
- T is a finite non empty set of transitions, such that: $S \cup T \neq \emptyset$ and $S \cap T = \emptyset$;
- $F \subset S\times T \cup T\times S$ is the flow relation, such that: $dom(F) \cup ran(F) = S \cup T$;
- Π is a partition of S into classes $\Pi_1,\ldots,\Pi_m$ such that $\forall i\ (1\leq i\leq m)\ \forall t\in T$:
  (i) $0 \leq |\Pi_i \cap {}^\bullet t| \leq 1$
  (ii) $|\Pi_i \cap {}^\bullet t| = |\Pi_i \cap t^\bullet|$

The nets generated by the classes $\Pi_1,\ldots,\Pi_m$ of the partition Π are called state-machine components of N (sm-components in the following). Furthermore, by π we will denote the natural mapping, which associates with every place s the class to which it belongs: $\pi(s) = \Pi_i$ iff $s \in \Pi_i$ □

Note that conditions (i) and (ii) imply condition (iii) here below:
(iii) $\forall t\in T\ (|{}^\bullet t| = |t^\bullet|)$

Example 1

The net shown in Fig.1 is a SA net, where $\Pi_1=\{s_1,s_2\}$ and $\Pi_2=\{s_3,s_4\}$ are the two sm-components. Following the intuition presented in 1.4 above, we want to associate a domain with each sm-component. The individuals (structured tokens) "belong" to a sm-component and take values in its domain. Each domain consists of two parts: the first one represents the possible colours (names) of the individuals for distinguishing the various instances of the same sm-component; the second part represents the data structure of each instance.

In the example, we want to express that n individuals flow within Π_1 and n within Π_2. The names of the individuals of Π_1 are $\{a_1,\dots,a_n\}$ and $\{b_1,\dots,b_n\}$ for Π_2. This means that there are n instances of Π_1, each one identified by a different a_i, and n instances of Π_2, each one identified by a different b_j; all of them have an empty data structure. In the initial marking all the individuals of Π_1 are in s_1 and all the individuals of Π_2 are in s_3. The individuals of Π_1 and Π_2 are in a bijective correspondence and transitions are enabled to fire only if mutually corresponding individuals (for instance, individuals whose names have the same index) are in the input places. □

Then for defining OBJSA net systems we have to define: a function for associating with each sm-component the appropriate domain for its individuals (called *interpretation function*, since it associates a meaning to the individuals flowing in net structure); a function which associates labels with arcs and transitions (therefore called *labelling function*) in order to express further conditions to transition firing, as other classes of high-level nets do; a function which attributes individuals to places so giving the initial marking (the *initial marking* function).

Following the motivations discussed in the introduction, we aim at giving these definitions in the formalism of OBJ2. In order to do this, some objects and theories have to be introduced. We recall that, in the OBJ2 vocabulary, an object is a (possibly parameterized) module, encapsulating executable code, which generally introduces new sorts of data and new operations on them; a theory is a specification of (non executable) properties of modules and interfaces. In other words, a theory imposes some constraints that have to be satisfied by objects in order to use them as actual parameters in the instantiation of a parameterized object (for more details, see |Gog| and |FGJM|). First of all we introduce a trivial theory (called STH for SomeTHeory) containing a unique constant operator, #.

```
th    STH is
   sorts SRT
   op     # : ---> SRT
endt
```

Let us now introduce the parameterized object RECORD, which uses the above defined theory STH as its parameter.

```
obj  RECORD [ X,Y :: STH ] is
   sorts  RECSORT
   op     REC : SRT.X  SRT.Y  ---> RECSORT
   op     _.L :  RECSORT  --->  SRT.X
   op     _.R  :  RECSORT  --->  SRT.Y
   op     TUPLE :  RECSORT  --->  RECSORT
   var    r : RECSORT
   var    s1 : SRT.X
   var    s2 : SRT.Y
   eq     REC(s1,s2).L = s1
   eq     REC(s1,s2).R = s2
   eq     TUPLE(r) = REC(r.L,r.R)
endo
```

Here, the operator TUPLE allows one to look at a record r as obtained by coupling its left and right projections, r.L and r.R, using the constructor operator REC; obviously, it can be shown that TUPLE(r)=r. For the sake of simplicity, in the following we will indifferently look at a record r as a "global" object or as a pair, following our convenience, without explicitly using, in the latter case, the form TUPLE(r) .

The parameterized object RECORD is then used in order to define a key object in the OBJSA net systems definition, namely the parameterized object which represents the domain associated to each sm-component, therefore named COMP-DOM, i.e. the abstract object defining the characteristics of the individuals flowing in the sm-components. It catches that each individual has a *name part* for distinguishing the various instances, and a *data part*, representing the data structure of each particular instance. This is done by using the structure of RECORD.

obj COMP-DOM [N :: STH , D :: STH] *is*
  dfn DOM := RECORD [N,D] * (
  op ((REC) to (<_;_>)),
  op ((_.L) to (name)),
  op ((_.R) to (data))).
endo

Let us point out the role played by the theory STH. It formalizes the requirements that an object must satisfy in order to be an acceptable actual parameter for the instantiation of the parametrized object COMP-DOM. Then we could say that in the specification of a parametrized object, the theory specifies the "type" of the parameter, which can obviously be different for each parameter. In the above case STH requires that both the name part and the data part of the domain associated to a sm-component must contain at least a constant, representing: in the first case, the name of the possibly unique instance of the component; in the second case, an empty data structure.

Example 2

In order to build up, step by step, a OBJSA net system corresponding to the requirements given in Example 1, we first define the two objects, say A and B, that will play the role of "name" part for the domains of Π_1 and Π_2, respectively, and an object NOTHING, for the "data" part of both domains.

obj A *is*
  sort Asort
  op a_1 : ---> Asort
  ....
  op a_n : ---> Asort
endo

obj B *is*
  sort Bsort
  op b_1 : ---> Bsort
  ....
  op b_n : ---> Bsort
endo

obj NOTHING *is*
  sort NTHsort
  op λ : ---> NTHsort
endo

The instantiations of the parameterized object COMP-DOM are therefore defined as follows.

obj P1 *is*
  dfn P1sort :=
    COMP-DOM [A,NOTHING]
  op p_1 : ---> P1sort
  ....
  op p_n : ---> P1sort
  eq $p_1 = <a_1,\lambda>$
  ....

obj P2 *is*
  dfn P2sort :=
    COMP-DOM [B,NOTHING]
  op q_1 : ---> P2sort
  ....
  op q_n : ---> P2sort
  eq $q_1 = <b_1,\lambda>$
  ....

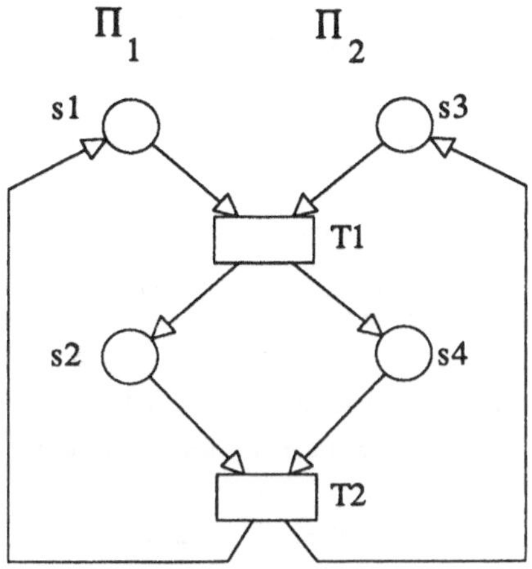

Fig. 1

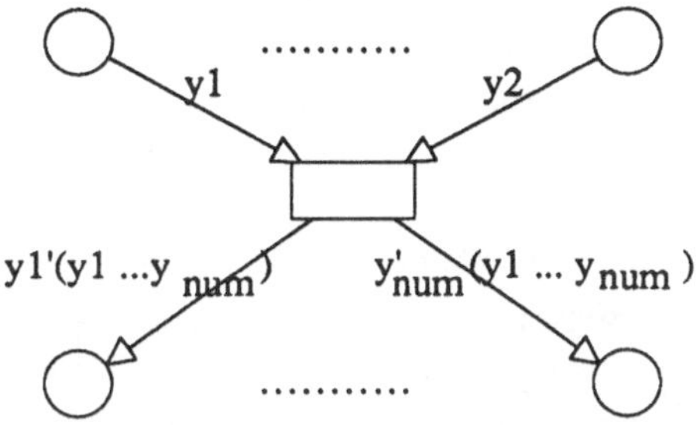

Fig. 2

Fig. 3

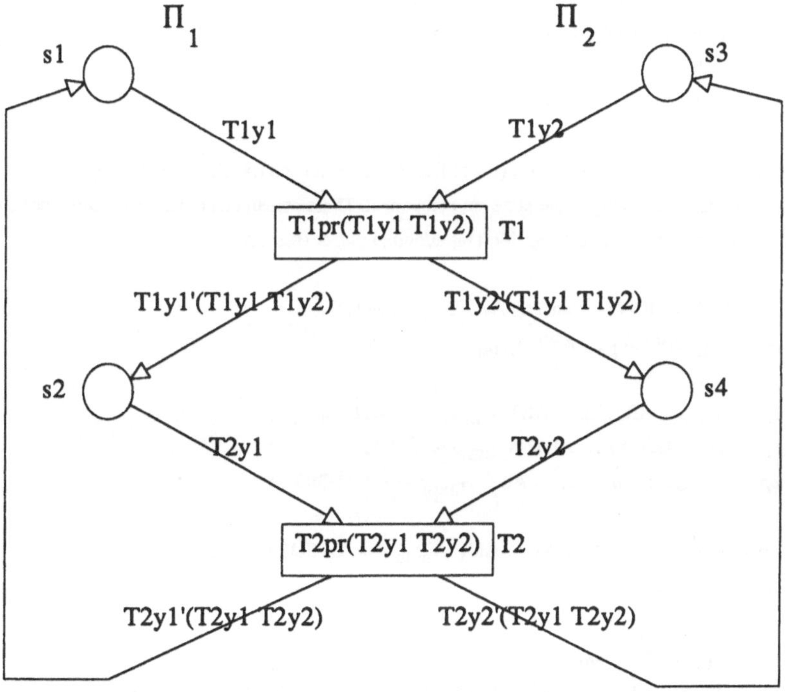

```
    eq    p_n = <a_n,λ>                    eq  q_n = <b_n,λ>
endo                                   endo
```
□

The next step is to introduce the objects that will be associated with each transition. The purpose of these objects is to allow a formal definition, in terms of OBJ2, of:

- predicates (relations into the sort BOOL) which introduce constraints for the individuals enabling the transition firing; and of:
- functions which capture how the transition firings modify the (data part of the) individuals flowing from input to output places.

Therefore, these objects TRANSITION must operate on the sorts of the individuals flowing in their input and output places, i.e. on the main sorts of some objects COMP-DOM. Since these sorts are in general different for each sm-component, they must be taken as parameters.

In OBJ2 objects cannot be formal parameters: only theories can play this role. Therefore, it is now necessary to introduce a theory ATH (for Automata THeory) which retains the main characteristics of the object COMP-DOM, i.e. a theory by definition satisfied by all the instantiation of parameterized object COMP-DOM. This theory is defined as follows, where NMSORT stands for NaMeSORT, DTSORT for DaTaSORT and ART for AutomatasoRT:

```
th ATH [NMSORT,DTSORT  :: STH ] is
   dfn ART := RECORD [NMSORT,DTSORT] * (
   op ( (REC)  to  (<_;_>) ),
   op ( (_.L)  to  (name) ),
   op ( (_.R)  to  (data) )              ).
endt
```

Furthermore, conditions (i), (ii), (iii) of Def.2.1 say that each transition has the same number of input and output places, coupled in pairs belonging to the same sm-component. These constraints require us to introduce a parameterized object which depends on the number of the transition input/output places (see also Fig.2).

```
obj    TRANSITION [I::NumTh ; A_1,..., A_num::ATH] is
   op     y'_1 : ART.A_1 ... ART.A_num ---> ART.A_1
   ........................
   op     y'_num : ART.A_1 ... ART. A_num ---> ART. A_num
   op     pr  :  ART.A_1 ... ART. A_num  ---> BOOL
   op     change_1:   ART.A_1 ... ART. A_num ---> DTSORT.A_1
   ........................
   op     change_num:   ART.A_1 ... ART. A_num ---> DTSORT.A_num
   var    y_1 : ART.A_1
   ........................
   var    y_num : ART.A_num
   ceq    y'_1(y_1,...,y_num) = <name(y_1);change_1(y_1,...,y_num)>  if  pr (y_1,...,y_num)
   ........................
   ceq    y'_num (y_1,...,y_num) = <name(y_num);change_num (y_1,...,y_num)>  if  pr (y_1,...,y_num)
endo
```

Let us point out that the first parameter of the parameterized object TRANSITION is constrained by the theory NumTh (see here below) which, through the *protecting* clause, inherits the standard object NAT (and its main sort Nat) and contains a generic natural number represented by the constant operator num. The instantiation of TRANSITION corresponding to a $t \in T$ will have as its first actual parameter an object DELTA defined as a *view* of Nat, where the generic num is mapped into the specific $\partial = |{\bullet}t|$. In the following we will assume that for each integer n, e.g. 2, a corresponding object DELTA, e.g. TWO, is defined to be used as first parameter in the instantiation of TRANSITION.

th NumTh *is*	*view* DELTA *of* NAT *as*	*view* TWO *of* NAT *as*
protecting NAT	NumTh *is*	NumTh *is*
op num : ---> Nat	*op* num *to* ∂	*op* num *to* 2
endt	*endv*	*endv*

Let us note that in the parametrized object TRANSITION the form of the conditional equations guarantees that the firing of a transition does not modify the name part of the individuals while it can change their data part.

3. OBJSA net systems

We can now give the definition of OBJSA net systems, then introduce for this class of high-level nets the standard notions of marking, concession and firing of a transition.

Def.3.1 - A OBJSA net system (or OBJSA net for short) is a quadruple $\mathfrak{R}$ = <N, I, L, M_0> where:

1) N is a SA net;
2) I is the interpretation function which associates with each $\Pi_i \in \Pi$ an object obtained as instantiation of the parameterized object COMP-DOM [N::STH , D::STH] by means of two actual parameters, respectively NM_i and DT_i, under the condition that in each object $I(\Pi_i)$ each constant operator defined in NM_i appears only once at the right side of an equation (this restriction is related with condition 3 at the beginning of section 1.2);
3) L is the labelling function which associates with each transition $t \in T$ an object obtained as instantiation of the parameterized object TRANSITION with the object DELTA corresponding to the integer $\partial = |{\bullet}t| = |t{\bullet}|$ as first actual parameter. The actual parameters $U_1,\ldots, U_\partial$ corresponding to the formal parameters $A_1,\ldots, A_\partial$ are determined as follows:

   let ${\bullet}t = \{s_1, \ldots, s_\partial\}$; then $U_i = I(\pi(s_i))$ for $1 \le i \le \partial$.

   The instantiation of the parametrized object TRANSITION:
   - labels each arc (s,t) such that $s \in \Pi_i$ with a variable ty_i of sort ART.U_i;
   - labels each arc (t,s) such that $s \in \Pi_i$ with the actual operator $ty_i'(ty_1,\ldots,ty_\partial)$ associated through the mapping with one of the ∂ primed operators y_k' of TRANSITION and introduces ∂ actual operators $chng_i(ty_1,\ldots,ty_\partial)$ associated through the mapping to the ∂ operators $change_k$;
   - maps the boolean valued operator pr into an actual operator tpr defined by a suitable equation.

   Furthermore, if a place s_i is both an input place and an output place for transition t, in order to preserve the purity of the net, the actual operator $chng_i$ must satisfy the following condition:

   $$chng_i(ty_1 \ldots ty_\partial) \neq data(ty_i).$$
4) M_0 is the initial marking function which associates with each $s \in S$ a (possibly empty) set $M_0(s) = \{ind^s{}_1, \ldots, ind^s{}_{sk}\}$, where each $ind^s{}_j$ $(1 \le j \le s_k)$ is a different constant operator belonging to the instantiation of the object $I(\pi(s))$, under the condition that each constant operator defined in each $I(\Pi_i)$ belongs to a unique $M_0(s)$ (so that each individual token marks a well-identified place). □

Example 3

Now we want to complete the definition of an OBJSA net system corresponding to the requirements given in Example 1, shown in Fig.3. First of all, in order to define the interpretation function (Def.3.1, point 2) we use the objects P1 and P2 defined in Example 2 as instantiation of the parameterized object COMP-DOM. It is sufficient to formally state $P1 = I(\Pi_1)$ and $P2 = I(\Pi_2)$.

In order to define the labelling function we must now define the transition objects TR1 and TR2 such that TR1=L(T1) and TR2=L(T2). TR1 and TR2 are defined by instantiating twice the parameterized object TRANSITION, with TWO, P1 and P2, which are, respectively, the object corresponding to the number of input/output places and the domains associated with the two interacting sm-components, as actual parameters. By Def.3.1, point 3, the instantiation concerning T1:

- introduces the variables $T1y_1$ and $T1y_2$ for labelling the incoming arcs of T1;
- maps the operator y_i' (for i=1,2) into the operator $T1y_i'$ for labelling the outcoming arcs of T1;
- maps the operator pr into the actual operator f, whose equations define, through the hidden operator g, the bijective correspondence between the constants of Asort and Bsort;
- maps the operator $change_i$ (i=1,2) into the operator $IDOP_i$, whose corresponding equations say that no actual change occurs in the data part of the individuals when the transition T1 fires.

Since TR2 is close to TR1, the last is used in the TR2 definition.

```
obj TR1 is
  using TRANSITION[TWO,P1,P2] *   (
    (op (y1') to (T1y1') )
    (op (y2') to (T1y2') )
    (op (pr) to (T1pr) )
    (op (change1) to (IDOP1) )
    (op (change2) to (IDOP2) )        )
  op   f : Asort ---> Bsort  (hidden)
  var  T1y1: P1sort
  var  T1y2: P2sort
  ceq  T1pr(T1y1,T1y2) = true if name(T1y2)==f(name(T1y1))
  eq   IDOP1(T1y1,T1y2) = data(T1y1)
  eq   IDOP2(T1y1,T1y2) = data(T1y2)
  eq   f(a1) = b1
  ................................
  eq   f(an) = bn
endo
```

```
obj TR2 is
  extending   TR1 *          (
    (op (T1y1') to ( T2y1') )
    (op (T1y2') to ( T2y2') )     )
  var  T2y1 : P1sort
  var  T2y2 : P2sort
endo
```

Let us note that in both instantiations we could avoid defining a mapping for y_i', since it is implicitly induced by the mapping of $change_i$. The choice done allows us to associate a "standard" label with each outcoming arc and aims to simplify the definition of the incidence matrix corresponding to the net (see Def.3.4). Nevertheless, through suitable rewriting steps, the actual meaning of the labellings becomes evident. Let us consider for instance the label

$$T1y_1'(T1y_1\ T1y_2)$$

of the arc $(T1,s_2)$. First, by using the conditional equations inherited by the object TR1 from the parameterized object TRANSITION (assuming the condition holds) we can rewrite the label as

<name(Tly$_1$) ; IDOP$_1$(Tly$_1$ Tly$_2$)>

Now, by using the equations in TR1, we can rewrite the last as

<name(Tly$_1$) ; data (Tly$_1$)>

and hence, using the equations of the object RECORD, simply as: Tly$_1$.

Therefore, the labelling of the OBJSA net system says that the firing of T1 (T2) does not change the individual flowing in Π_1 (Π_2).

Let us finally note that the instantiation of TRANSITION into TR1 (and TR2) really consists only in defining the mapping for pr, change$_1$ and change$_2$, i.e. in defining the actual form of the predicate associated with the transition and the two functions which capture how the transition firing modifies the data part of the individuals.

The initial marking function distributes into the places of each sm-component the constants (the individuals) defined in the associated domain (in our case in the objects P1 and P2):

$M_0(s_1)=\{p_1,...,p_n\}$ $\quad$ $M_0(s_2)=M_0(s_4)=\emptyset$ $\quad$ $M_0(s_3) = \{q_1,...,q_n\}$ □

We want now to formalize for OBJSA net systems the notions of concession and firing of a transition under a given marking.

Def.3.2 - A marking M of a OBJSA net system $\mathfrak{R}$ is a function which associates with each s∈S a (possibly empty) set $M(s) = \{ind^{s}_1, ... , ind^{s}_{sk}\}$, where each ind^{s}_j ($1 \le j \le s_k$) is a different constant operator belonging to the object $I(\pi(s))$, under the condition that each constant operator defined in each $I(\Pi_i)$ belongs to a unique M(s) □

Def.3.3 - Let $\mathfrak{R}$ be an OBJSA net system, M a marking of $\mathfrak{R}$, t∈T such that $|{\bullet}t| = |t{\bullet}| = \partial$ and ${\bullet}t = \{s_1, ..., s_\partial\}$. Then:
t has concession in M iff every $M(s_i)$ (for $i=1,...,\partial$) contains an ind_i such that $tpr(ind_1,...,ind_\partial)$, where:

- ind_i has value in the main sort of the object $I(\pi(s_i))$;
- tpr is the boolean valued operator associated by the mapping defined in the object L(t) with the operator pr defined in the parameterized object TRANSITION.

If t has concession in M for the objects ind_i ($i=1,...,\partial$), then M [t> M' is defined as follows:

$$M'(p) = \begin{cases} M(p) - \{ind_i\} & \text{if } p = s_i \text{ and } p \notin t{\bullet} \quad \text{for } i=1,...,\partial \\ M(p) \cup \{ty_i'(ind_1,...,ind_\partial)\} & \text{if } p \in t{\bullet} \cap \pi(s_i) \text{ and } p \notin {\bullet}t \quad \text{for } i=1,...,\partial \\ M(p) - \{ind_i\} \cup \{ty_i'(ind_1,...,ind_\partial)\} & \text{if } p = s_i \text{ and } p \in t{\bullet} \cap {\bullet}t \quad \text{for } i=1,...,\partial \\ M(p) & \text{otherwise} \end{cases}$$

with ty_i' as defined in Def.3.1, point 2. □

Example 4

Let us consider again our previous example. It is easy to see, following Def.3.3, that under the initial marking transition T1 has concession for each pair of individuals $<p_i,q_i>$ (with $1 \le i \le n$). The marking M reached through the firing of T1 for a particular pair $<p_k,q_k>$ (say $<p_1,q_1>$) is the following:

$M(s_1) = \{p_2,...,p_n\}$ $\quad$ $M(s_2) = \{p_1\}$ $\quad$ $M(s_3) = \{q_2,...,q_n\}$ $\quad$ $M(s_4) = \{q_1\}$

where the markings of s_2 and s_4 are determined, by following Def.3.3, through the standard rewriting, respectively, of the two following OBJ2 expressions:

$Tly_1'(p_1\ q_1)$ and $Tly_2'(p_1\ q_1)$.

Let us sketch this rewriting process which can be performed by an OBJ2 interpret by considering, for instance, the first one. It is first rewritten by using the equations inherited by the objects TR1 and TR2 from the parameterized object TRANSITION, since for the substitution: p_1 for Tly_1 and q_1 for Tly_2, the condition $T1pr(Tly_1 Tly_2)$ holds, as turns out from the following rewriting sequence:

T1pr $(p_1\, q_1)$ ---> name(q_1)==f(name(p_1)) ---> name(<b_1,λ>)==f(name(<a_1,λ>)) ---> b_1==f(a_1) ---> true

Therefore, $T1y_1'(p_1\, q_1)$ is rewritten as:

$$<name(p_1);IDOP_1\,(p_1\, q_1)>$$

Now we can apply the equation characterizing in the objects TR1 the operator $IDOP_1$ obtaining:

$$<name(p_1);data(p_1)>$$

Finally, the equations of the object RECORD allow us to rewrite the last expression simply as: p_1.

More details about the rewriting procedure and the possibility of conflict and loops will be given in the following (see Example 6). □

Def.3.4 - Let $\mathfrak{R}$ be an OBJSA net system. For each transition t ∈T a vector **t**, having as many elements as there are places in the net, is defined as follows:

$$t(s) = \begin{cases} -ty_i & \text{if } s \in {}^\bullet t - t^\bullet \ \wedge\ s \in \Pi_i \\ +ty_i'(ty_1,\ldots,ty_\partial) & \text{if } s \in t^\bullet - {}^\bullet t \ \wedge\ s \in \Pi_i \\ -ty_i + ty_i'(ty_1,\ldots,ty_\partial) & \text{if } s \in {}^\bullet t \cap t^\bullet \ \wedge\ s \in \Pi_i \\ 0 & \text{otherwise} \end{cases}$$

where the vector elements are formal series (on the ring of integers) in the variables and operators defined in the object **L**(t) and associated to the arc (s,t) or (t,s).

The elements of the <u>incidence matrix</u> W, having as many rows as there are places and as many columns as transitions, are defined as:

$$W(s,t) = \mathbf{t}(s)$$ □

Consistent with this definition, as usual:

- the marking of a place is represented as a formal sum of the individuals belonging to it; furthermore, in order to allow the correct application of the operator ⊗ (see section 4.1), the empty marking is represented by the constant operator NUL;
- the firing of a transition t, leading from marking M to the marking M' can be expressed as : M + **t** = M'.

Example 5

The incidence matrix of our previous example, is shown in Fig.4, where:

- each $i_1,\ldots,i_4$ is a variable having value in the main sort of the corresponding object $\mathbf{I}(\pi(s_1)),\ldots, \mathbf{I}(\pi(s_4))$;
- each $T1j_1$, $T1j_2$, $T2j_1$, $T2j_2$ is a variable having value in the main sort of the corresponding object $\mathbf{I}(\pi(\bullet T1))$, $\mathbf{I}(\pi(\bullet T2))$.

	T1	T2	S-vector I
s_1	$-T1y_1$	$T2y_1'(T2y_1,T2y_2)$	i_1
s_2	$T1y_1'(T1y_1,T1y_2)$	$-T2y_1$	i_2
s_3	$-T1y_2$	$T2y_2'(T2y_1,T2y_2)$	i_3
s_4	$T1y_2'(T1y_1,T1y_2)$	$-T2y_2$	i_4
T-vector J	$\mid T1j_1\ \ T1j_2 \mid$	$\mid T2j_1\ \ T2j_2 \mid$	

Fig. 4

4. OBJSA nets invariants

In order to make available for OBJSA nets a calculus for S- and T-invariants, first of all we must introduce two product operators: ⊗, to be used in the calculus of S-invariants, and ⊙, to be used in the calculus of T-invariants.

The definitions which follow preserve the usual intuition behind S- and T- invariants: as in other classes of (high-level) Petri nets, S-invariants are sets of places marked with a constant set of tokens; T-invariants capture markings which are reproducible through sequences of transition firings. Example 6 below will support this intuition by showing for our example how to interpret the resulting invariants.

4.1. The ⊗-product and the properties of S-invariants

The ⊗-product is an operator for comparing individuals, in the line of the type-2 product of |GL2|, called the object-S-product in the terms proposed in |LP|. It is introduced in the parameterized object COMP-DOM and is therefore inherited by all the instantiation of such parameterized object. It acts on a pair in the main sort DOM, representing the two individuals to be compared: if they are equal, their ⊗-product gives 1, otherwise 0. For completing the set of equations defining ⊗ in COMP-DOM the constant operator NUL and the operations of sum of individuals (+) and product between an integer and an individual (*) are also introduced. The extension of COMP-DOM to such operators is performed through the OBJ2 clause *extending*:

```
obj  COMP-DOM-⊗ [ N :: STH , D :: STH ] is
     protecting  INT
     extending   COMP-DOM [N,D]
     op     _ ⊗ _ : DOM DOM ---> INT    (COMM)
     op     NUL : ---> DOM
     op     _ + _ : DOM DOM ---> DOM    (COMM , ASS)
     op     _ * _ : INT DOM   ---> DOM
     vars   p, q : DOM
     eq     (q1+ q2) ⊗ p = q1 ⊗ p + q2 ⊗ p
     eq     (z * q) ⊗ p = z * (q ⊗ p) = z * (p ⊗ q)
     ceq    p ⊗ q = if p == q and p≠NUL then 1 else 0
endo
```

Let now I, M be S-vectors and W an incidence matrix. We can extend the ⊗-product as follows, ambiguously using the same symbol ⊗ for all the extensions:

a) $I \otimes M = \Sigma_{i=1,\dots,|S|} I(s_i) \otimes M(s_i)$;

b) $I \otimes W(t_j) = \Sigma_{i=1,\dots,|S|} I(s_i) \otimes W(s_i,t_j)$, where $W(t_j)$ denotes the S-vector corresponding to the t_j column in W;

c) $I \otimes W = < I \otimes W(t_1), \dots , I \otimes W(t_{|T|})>$.

Def.4.1 - Let W(s,t) the incidence matrix representing an OBJSA net system $\mathfrak{R}$.

An <u>S-vector</u> x consistent with W(s,t) is a tuple $x = <x_1,\dots, x_{|S|}>$, where x_k is a term in the main sort of the object $I(\pi(s_k))$. □

An S-vector I consistent with the incidence matrix of our example is shown in Fig.4.

Def.4.2 - An object-S-invariant of the OBJSA net system $\mathfrak{R}$ is an S-vector I of constants satisfying the system of equations $I \otimes W = \mathbf{0}$, where $\mathbf{0}$ is the T-vector having all its element equal to 0 of sort INT. □

The above definitions guarantee the usual results on S-invariants linearity, and the main theorem for S-invariants ($\forall M \in [M_0>, \forall$ S-invariant x: $x \otimes M = x \otimes M_0$) holds.

Th.4.1 - Given $z \in INT$ and two S-invariants I_1 and I_2, even $I_1 + I_2$ and $z * I_1$ are S-invariants.

Proof -
$$\begin{aligned}(I_1+I_2) \otimes W &= <\Sigma_{i=1,\ldots,|S|} [I_1(s_i)+I_2(s_i)] \otimes W(s_i,t_1), \ldots, \Sigma_{i=1,\ldots,|S|} [I_1(s_i)+I_2(s_i)] \otimes W(s_i,t_{|T|})> = \\ &= <\Sigma_{i=1,\ldots,|S|} [I_1(s_i) \otimes W(s_i,t_1) + I_2(s_i) \otimes W(s_i,t_1)], \ldots > = \\ &= <\Sigma_{i=1,\ldots,|S|} I_1(s_i) \otimes W(s_i,t_1) + \Sigma_{i=1,\ldots,|S|} I_2(s_i) \otimes W(s_i,t_1), \ldots > = \\ &= <\Sigma_{i=1,\ldots,|S|} I_1(s_i) \otimes W(s_i,t_1), \ldots > + <\Sigma_{i=1,\ldots,|S|} I_2(s_i) \otimes W(s_i,t_1), \ldots > = \\ &= I_1 \otimes W + I_2 \otimes W = \mathbf{0} + \mathbf{0} = \mathbf{0}\\ (z * I_1) \otimes W &= <\Sigma_{i=1,\ldots,|S|} [z * I_1(s)] \otimes W(s_i,t_1), \ldots > = \\ &= <z * \Sigma_{i=1,\ldots,|S|} (I_1(s) \otimes W(s,t)), \ldots > = z * [I_1 \otimes W] = z * \mathbf{0} = \mathbf{0}\end{aligned}$$

Th.4.2 - Let I be an S-invariant of the OBJSA net system $\mathfrak{R}$ and let $M \in [M_0>$ be any reachable marking of $\mathfrak{R}$. Then, it holds: $I \otimes M = I \otimes M_0$.

Proof - Let $M, M' \in [M_0>$ and let $t \in T$ such that $M [t> M'$. Then
$$\begin{aligned}I \otimes M' = I \otimes [M + t] &= \Sigma_{i=1,\ldots,|S|} I(s_i) \otimes [M(s_i) + t(s_i)] = \Sigma_{i=1,\ldots,|S|} [I(s_i) \otimes M(s_i) + I(s_i) \otimes t(s_i)] = \\ &= I \otimes M + I \otimes t = I \otimes M + \mathbf{0} = I \otimes M.\end{aligned}$$

4.2. The ⊙-product and the properties of T-invariants

The ⊙-product is introduced for the calculus of T-invariants. Since T-invariants capture invariances in the transitions firing (namely allowing the identification of firing sequences which reproduce the initial marking), the ⊙-product is defined in the object TRANSITION. Its definition draws its inspiration from the definition of the object-T-product introduced in |LP|, where the axiom equations are taken from, but for the fact that we can omit some axioms which would never been used.

The ⊙-product performs substitutions on tuples having as many elements as there are transition input places, each element having value in the main sort of the corresponding parameter. This fact is captured by introducing in the object TRANSITION a sort IN, having the main parameter sorts as subsorts, and the operator | |. Furthermore, for completing the set of equations defining ⊙ in TRANSITION, the constant operator NIL and the operations of sum (+) and product times an integer (*) are also introduced, for representing multiple firings of a transitions.

The extension of TRANSITION to such operators is performed through the OBJ2 clause *extending* , as follows:

obj TRANSITION -⊙ [I::NumTh ; $A_1,\ldots, A_{num}$::ATH] *is*

extending TRANSITION [I, $A_1,\ldots, A_{num}$]

sort IN

subsort ART.A_1 < IN

....

subsort ART.A_{num}< IN

op _⊙_ : IN IN ---> IN

op |_| : ART.A_1 ---> IN

....

op $|\_ \ldots \_|$: $ART.A_1 \ldots ART.A_{num}$ ---> IN

op NIL : ---> IN

op _+_ : IN IN ---> IN (COMM , ASS)

op _*_ : INT IN ---> IN

var $j_1 : ART.A_1$

....

var $j_{num} : ART.A_{num}$

vars $J_A, J_B, y_1, \ldots, y_{num}$: IN

var z : INT

eq $y_1 \odot (J_A + J_B) = y_1 \odot J_A + y_1 \odot J_B$

eq $y_1 \odot (z * J_A) = z * (y_1 \odot J_A)$

eq $y_1 \odot |j_1 \ldots j_{num}| = j_1$

....

eq $y_{num} \odot |j_1 \ldots j_{num}| = j_{num}$

eq $y_1'(y_1, \ldots, y_{num}) \odot |j_1 \ldots j_{num}| = y_1'(j_1, \ldots, j_{num})$

....

eq $y'_{num}(y_1, \ldots, y_{num}) \odot |j_1 \ldots j_{num}| = y'_{num}(j_1, \ldots, j_{num})$

endo

As for the ⊗-product and S-invariants, given T-vectors I, J and an incidence matrix W, we can extend the ⊙-product as follows:

a) $I \odot J^T = \Sigma_{i=1,\ldots,|T|} I(t_i) \odot J(t_i)$;

b) $W(s_j) \odot J^T = \Sigma_{i=1,\ldots,|T|} W(s_j,t_i) \odot J(t_i)$, where $W(s_j)$ denotes the T-vector corresponding to the s_j row in W;

c) $W \odot J^T = < W(s_1) \odot J^T, \ldots, W(s_{|S|}) \odot J^T >$

and then define T-invariants and prove the usual results on their linearity and the main theorem for T-invariants.

Def.4.3 - Let W(s,t) the incidence matrix representing an OBJSA net system $\mathfrak{R}$.

A T-vector v consistent with W(s,t) is a tuple $v = <v_1, \ldots, v_{|T|}>$ where v_k is a term in the sort IN built by means of the operators | | having as many arguments as $|{\bullet}t|$. □

A T-vector I consistent with the incidence matrix of our example is shown in Fig.4.

Def.4.4 - An object-T-invariant of the OBJSA net system $\mathfrak{R}$ is a T-vector J of constants satisfying the system of equations $W \odot J^T = \emptyset^T$, where $\emptyset^T$ is the transpose of an S-vector having each element equal to the constant operator NUL defined in COMP-DOM. □

Th.4.3 - Given $z \in$ INT and two T-invariants J_1 e J_2, even $J_1 + J_2$ and $z * J_1$ are T-invariants.

Proof - $W \odot (J_1+J_2)^T = <\Sigma_{i=1,\ldots,|T|} W(s_1,t_i) \odot [J_1(t_i)+J_2(t_i)], \ldots, \Sigma_{i=1,\ldots,|T|} W(s_{|S|},t_i) \odot [J_1(t_i)+J_2(t_i)] > =$

$= <\Sigma_{i=1,\ldots,|T|}[W(s_1,t_i) \odot J_1(t_i) + W(s_1,t_i) \odot J_2(t_i)], \ldots > =$

$= W \odot J_1{}^T + W \odot J_2{}^T = \emptyset^T + \emptyset^T = \emptyset^T$

$W \odot (z * J_1)^T = <\Sigma_{i=1,\ldots,|T|} W(s_1,t_i) \odot [z * J_1(t_i)], \ldots >=$

$= z * <\Sigma_{i=1,\ldots,|T|} W(s_1,t_i) \odot [z * J_1(t_i)], \ldots >= z * [W \odot J_1{}^T] = z * \emptyset^T = \emptyset^T$

Th.4.4 - Let $v_n = M1[t1> M2[t2> \ldots Mn[tn>Mn+1$ be a firing sequence and $Jvn: T \to IN$ a vector defined by:

$$\forall \quad t \in T\ (Jv_n(t) = \Sigma_{0 \le i \le c(t)}\, j_t^{\,i})$$

where c(t) is the cardinality of the set $\{k \,/\, 1 \le k \le n \ \wedge\ t_k = t\}$, $j_t^0 = NIL$ and $j_t^{\,i} = |\ j_t^1 \ \ldots\ j_t^{|\bullet t|}\ |$ if $i > 0$. Then it holds that :

$$M_1 = M_{n+1} \quad \Longleftrightarrow \quad Jv_n \text{ is a T-invariant.}$$

Proof. Let $s \in S$ and for each $1 \le k \le n$ let $v_k = M_1[t_1> M_2 \ldots M_k[t_k>M_{k+1}$. By induction over k we will show :

$$(*) \quad \Sigma_{0 \le i \le k}\, t_i(s) = \Sigma_{s \in S}\, \Sigma_{t \in T}\ W(s,t) \odot Jv_k(t)$$

The expressions on the left- and on the right-hand side of the equation (*) will be denoted a_k and b_k, respectively. In case $k = 0$, the proposition is obvious because $a_0 = NIL = b_0$.

Assume $a_{k-1} = b_{k-1}$ and let $v' = M_k[t_k>M_{k+1}$. Obviously, it holds: $Jv_k(t) = Jv_{k-1}(t) + Jv'(t)$.

Furthermore, $Jv'(t) =$ if $t_k = t$ then j_t else NIL .

With $a = t_k(s)$ and $b = \Sigma_{s \in S}\, \Sigma_{t \in T}\ W(s,t) \odot Jv'(t)$ it holds that $a = b$, and we get

$$a_k = a_{k-1} + a = b_{k-1} + b = b_k.$$

The theorem is now proved as follows :

$M_1 = M_{n+1} \iff \Sigma_{0 \le i \le n}\, t_i(s) = \emptyset \iff \Sigma_{s \in S}\, \Sigma_{t \in T}\ W(s,t) \odot Jv_n(t) = \emptyset$

$\iff W \odot Jv_n^T = \emptyset \iff Jv_n$ is a T-invariant. □

4.3. An example of rewriting

Let us point out that if we aims to calculate the S-invariants (respectively the T-invariants) of a OBJSA net system, we must define the interpretation (respectively the labelling) function by instantiating the parameterized object COMP-DOM-⊗ instead of COMP-DOM (respectively TRANSITION-⊙ instead of TRANSITION).

In the system of equations for the calculus of the S-invariants $I \otimes W = \mathbf{0}$ (respectively $W \odot J^T = \emptyset^T$ for the calculus of the T-invariants), the left side of each single equation is an expression in the main sort of the object INT inherited in the parameterized object COMP-DOM-⊗ (respectively in the sort IN defined in the parameterized objects TRANSITION-⊙), to which the rewriting rules defined by the equations in the objects introduced for the definition of $\Re$ can be applied. The rewriting steps necessary for calculating the solutions of the system of equations in the unknown vectors I and J respectively are sketched below for the particular case already considered.

Example 6

The extended form of the system of equations $I \otimes W = \mathbf{0}$, consisting of as many equations as there are transitions (in our case two), is rewritten, in subsequent steps, by using:

1) the equations inherited by the objects TR1 and TR2 from the parameterized object TRANSITION

   $T1y_1'(T1y_1, T1y_2) = \langle name(T1y_1); IDOP_1\,(T1y_1, T1y_2)\rangle$ if $f(T1y_1, T1y_2)$

   $T1y_2'(T1y_1, T1y_2) = \langle name(T1y_2); IDOP_1\,(T1y_1, T1y_2)\rangle$ if $f(T1y_1, T1y_2)$

   and the like for T2. The resulting system consists therefore of 2 conditional equations.

2) the equations characterizing in the objects TR1 and TR2 the operators f and $IDOP_i$ which rename the operators pr and $change_i$ of the parameterized object TRANSITION:

   $T1pr(T1y_1, T1y_2) =$ true if $name(T1y_2) == f(name(T1y_1))$

   $IDOP_1(T1y_1, T1y_2) = data\ (T1y_1)$

   $IDOP_2(T1y_1, T1y_2) = data\ (T1y_2)$

   and the like for T2. The resulting system still consists of 2 conditional equations.

3) the equations capturing that each individual flowing in the net consists in a name part and a data part; these equations are inherited by all the instantiations of the object COMP-DOM from the object RECORD:

Tly_1 = <name(Tly_1) ; data(Tly)>

Tly_2 = <name(Tly_2) ; data(Tly_2)>

and the like for T2. The resulting system still consists of 2 conditional equations.

At this point there are no more possibilities for rewriting the system. Since it must be satisfied for all the possible values that the variables Tly_1, $T2y_1$ (resp., Tly_2, $T2y_2$) can assume in P1sort (resp., in P2sort), we can now proceed to the substitution of these variables with all their possible values. In our particular case the system consists of 2 conditional equations with 4 variables, each one with n possible values (as defined in the objects A, B). The form of the condition (see objects TR1, TR2) is the following: $name(ty_2)=f(name(ty_1))$. By substitution we obtain a system of $2n^4$ conditional equations, which can be reduced by canceling the equations which have the associated condition false (only the pairs (a_i,b_i), with a_i defined in object A and b_i defined in object B, satisfy the condition).

The resulting system, consisting of n^2 couples of equations, is the following, with $1 \le i,k \le n$:

$$\begin{cases} -\langle a_i,\lambda\rangle \otimes i_1 + \langle a_i,\lambda\rangle \otimes i_2 - \langle b_i,\lambda\rangle \otimes i_3 + \langle b_i,\lambda\rangle \otimes i_4 = 0 \\ \langle a_k,\lambda\rangle \otimes i_1 - \langle a_k,\lambda\rangle \otimes i_2 + \langle b_k,\lambda\rangle \otimes i_3 - \langle b_k,\lambda\rangle \otimes i_4 = 0 \end{cases}$$

The following 3n linearly independent S-invariants ($1 \le k \le n$) are solutions of the system:

$I_{1k} = (\langle a_k,\lambda\rangle\ , \langle a_k,\lambda\rangle\ , 0\ , 0\ )$

$I_{2k} = (\ 0\ , 0\ , \langle b_k,\lambda\rangle\ , \langle b_k,\lambda\rangle\ )$

$I_{3k} = (\langle a_k,\lambda\rangle\ ,0\ , 0\ , \langle b_k,\lambda\rangle\ )$

By using the S-invariant main theorem, with the initial marking defined in Example 4, these invariants can be interpreted as follows. For whatever reachable marking it holds that:

- each individual $\langle a_k,\lambda\rangle$ is either in place s_1 or in place s_2 (from I_{1k});
- each individual $\langle b_k,\lambda\rangle$ is either in place s_3 or in place s_4 (from I_{2k});
- if the individual $\langle a_k,\lambda\rangle$ is in place s_1 the corresponding (under the predicate f) individual $\langle b_k,\lambda\rangle$ cannot be in place s_4 (from I_{3k}), i.e., for I_{2k}, is in s_3.

In IBDM1I, it is shown how the extended form of the system concerning the T-invariants $W \odot J^T = \emptyset^T$, consisting of as many equations as there are places (in our case four) can be analogously rewritten . Its solutions are the following n linearly independent T-invariants ($1 \le k \le n$):

$j_k = (\ |\langle a_k,\lambda\rangle\langle b_k,\lambda\rangle|\ \ |\langle a_k,\lambda\rangle\langle b_k,\lambda\rangle|\ )$

j_k says that the firing sequence:

T1 fires moving the individual $\langle a_k,\lambda\rangle$ from s_1 to s_2 and moving $\langle b_k,\lambda\rangle$ from s_3 to s_4;

T2 fires moving the individual $\langle a_k,\lambda\rangle$ from s_2 to s_1 and moving $\langle b_k,\lambda\rangle$ from s_4 to s_3;

reproduce the initial marking of the net. □

Remarks

Let us point out that the steps shown for the calculus of S- and T- invariants do not depend on the particular example, but are the general ones. In particular, they do not change if in some instantiation of TRANSITION the renaming of pr is a relation. Therefore, the confluence property of the rewriting system associated to the definition of an OBJSA net system is guaranteed under the following conditions:

- in the suitable modules a priority is associated to the different operators in such a way that the following ordering relation is satisfied: $\odot > y_i' > change_i = pr > REC > \otimes$
- the confluence property is satisfied by the objects (A, B, NOTHING in the example) used in the instantiations of the parameterized objects COMP-DOM. Nevertheless their finiteness is required in order to carry out the assignment (of constant values to the variables y_i) performed as the last step in the calculus of S-invariants. □

5. OBJSA nets obtained by components composition

We have said in Section 1.2 that SA nets allow one to build the net system model through composition of the models of its components. We want OBJSA nets to retain this characteristics so that the specification of complex systems is performed by composing components in turn resulting from the previous composition of more simple ones. Specifications reusability is enhanced by this possibility. This goal is achieved in |BDM1| by defining OBJSA components and a composition operation based on transition superposition. The definition of OBJSA components is closed under the composition operation. For reason of space we give here only hints of the definition of OBJSA component and exemplify the composition operation on two very simple components, yielding the OBJSA net of our previous example.

An OBJSA component is an OBJSA net system (as in Def.3.1) where a transition can be already fully extensionally characterized, i.e. all its input/output places are already identified, and in this case is said to be <u>closed</u>; or a transition can be only partially extensionally characterized, i.e. some of its input/output places have to be identified through superposition with some other transition, and it is said to be <u>open</u>. The transition labelling function **L** associates:

- with each closed transition an object obtained by instantiating the parameterized object TRANSITION (as defined in Section 2.);
- with each open transition a theory obtained by instantiating the parameterized theory TTH; the theory TTH (for Transition THeory) retains the characteristics of the object TRANSITION, i.e it is a theory by definition satisfied by all the instantiations of the parameterized object TRANSITION.

th TTH [I::NumTh ; A_1T,..., $A_{num}T$::ATH] *is*
  op y_1' : ART.A_1T ... ART.$A_{num}T$ ---> ART.A_1T
  ....
  op y_{num}' : ART.A_1T ... ART. $A_{num}T$ ---> ART. $A_{num}T$
  op pr : ART.A_1T ... ART. $A_{num}T$ ---> BOOL
  op $change_1$: ART.A_1T ... ART. $A_{num}T$ ---> DTSORT.A_1T
  ....
  op $change_{num}$: ART.A_1T ... ART. $A_{num}T$ ---> DTSORT.$A_{num}T$
  var y_1 : ART.A_1T
  ....
  var y_{num} : ART.$A_{num}T$
  ceq $y'_1(y_1,\ldots,y_{num})$ = <name(y_1);$change_1(y_1,\ldots,y_{num})$> *if* pr $(y_1,\ldots,y_{num})$
  ....
  ceq $y'_{num}(y_1,\ldots,y_{num})$ = <name(y_2);$change_{num}(y_1,\ldots,y_{num})$> *if* pr $(y_1,\ldots,y_{num})$
endt

Fig.5 shows two OBJSA components. In this case the underlying nets are in particular two state-machine nets. All the four transitions are open and therefore the labelling function **L** associates with them the following theories.

th $T_{11}TH$ *is*
  using
    TTH [TWO,P1, A_2T_{11}:: $A_2T_{11}H$] * (
      (*op* ($change_1$) *to* ($IDOP_1$))

th $T_{21}TH$ *is*
  using
    TTH [TWO, A_1T_{21}:: $A_1T_{21}H$, P2] * (
      (*op* ($change_2$) *to* ($IDOP_2$))

```
      (op (y'1) to (T11y'1))            )
   var  T11y1: P1sort
   var  z1 : A2T11sort
   eq   IDOP1(T11y1,z1) = data(T11y1)
endt
```

```
      (op (y'2) to ( T12y'2))           )
   var  T12y2: P2sort
   var  x1 : A1T21sort
   eq   IDOP2(x1, T12y2) = data(T12y2)
endt
```

```
th T12TH is
   using
     TTH [TWO,P1, A2T12:: A2T12H] * (
        (op (change1) to (IDOP1))
        (op (y'1) to (T12y'1))          )
   var  T12y1: P1sort
   var  z2 : A2T12sort
   eq   IDOP1(T12y1,z2) = data(T12y1)
endt
```

```
th T22TH is
   using
     TTH [TWO, A1T22:: A1T22H, P2] * (
        (op (change2) to (IDOP2))
        (op (y'2) to ( T22y'2))         )
   var  T22y2: P2sort
   var  x2 : A1T22sort
   eq   IDOP2(x2,T22y2) = data(T22y2)
endt
```

The first parameter of the instantiation states the number of input/output places the transition must have for becoming closed. In regards to the other parameters, one consists of the domain associated with the component: respectively the object P1 for component C1 and the object P2 for component C2, both defined as in Example 1.

The theories $A_2T_{11}H$, $A_2T_{12}H$, $A_1T_{21}H$, $A_1T_{22}H$ are restrictions of the theory ATH , and formalize the constraints which must be satisfied in order to make the transition superposition possible.In this case all the four are actually the same, except for the mapping of the sort , and requires (theory NelemTH) that the name part must contain n different constant operators, while the data part just one (theory STH, definend in Section 2.) standing for "empty". For instance, $A_2T_{11}H$ is defined as follows:

```
th A2T11H is
   extending
     ATH [ A2T11Nm :: NelemTH,
           A2T11Dt :: STH] * (
       (sort (Art) to ( A2T11sort))  )
endt
```

```
th NelemTH is
   sort NelemSort
   op  el1 : ---> NelemSort
   ..............................
   op  eln : ---> NelemSort
endt
```

The open transitions, or more precisely the corresponding theories, therefore represent the interface exhibited by a component to the other components and the constraints imposed on them for making the composition operation possible. That is, they are the parameters of a parameterized object which defines the component interface.

```
obj COMPONENT [K:: NumTh; I1, .. , Ik :: TTH] is
  endo
```

where k is the number of open transitions of the component. The instantiation of this object together with the objects defining the component domains and the objects associated with the closed transitions constitute the object associated with the component. In our case these two objects are:

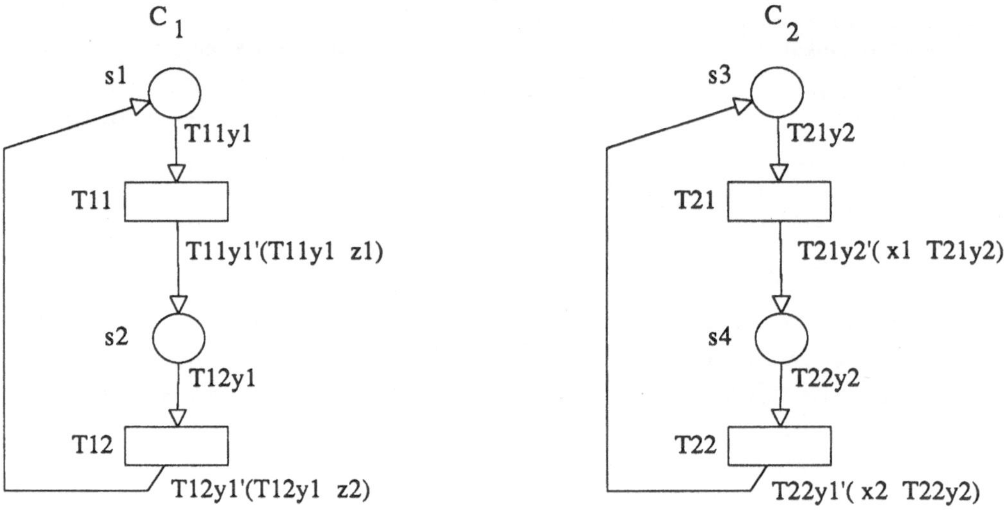
C1
C2
s1
T11y1
T11
T11y1'(T11y1 z1)
s2
T12y1
T12
T12y1'(T12y1 z2)
s3
T21y2
T21
T21y2'(x1 T21y2)
s4
T22y2
T22
T22y1'(x2 T22y2)

Fig. 5

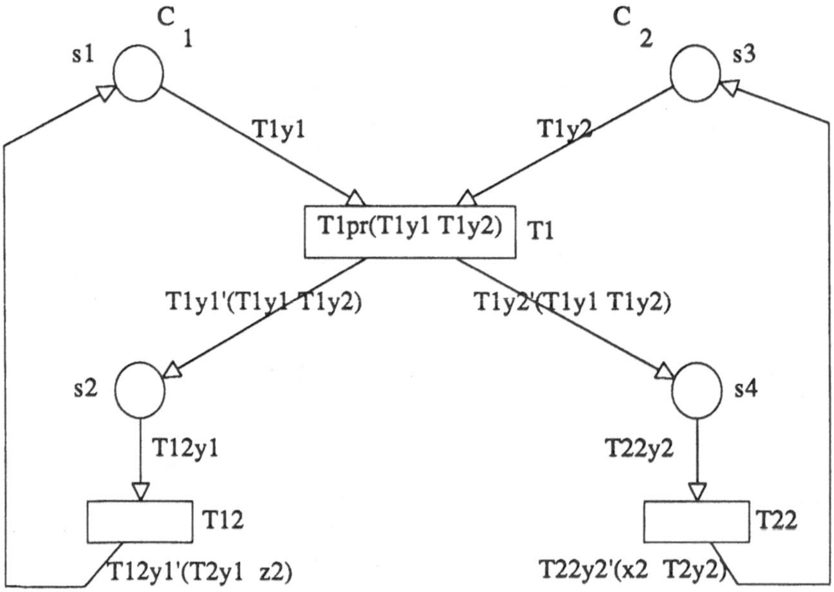
C1
C2
s1
s3
T1y1
T1y2
T1pr(T1y1 T1y2)
T1
T1y1'(T1y1 T1y2)
T1y2'(T1y1 T1y2)
s2
s4
T12y1
T22y2
T12
T22
T12y1'(T2y1 z2)
T22y2'(x2 T2y2)

Fig. 6

obj C1 *is*
extending P1
protecting COMPONENT
[TWO, $t_{11}h :: T_{11}TH$, $t_{12}h :: T_{12}TH$]
endo

obj C2 *is*
extending P2
protecting COMPONENT
[TWO, $t_{21}h :: T_{21}TH$, $t_{22}h :: T_{22}TH$]
endo

The operation of superposition of a tuple of n open transitions is possible if, considered the corresponding instantiations of TTH, the objects and theories which are the instantiation's actual parameters mutually correspond. For instance, transitions t_{11} and t_{21} can be superposed since, considered $T_{11}TH$ and $T_{21}TH$, P1 (of $T_{11}TH$) satisfies the theory $A_1T_{21}H$ (of $T_{21}TH$) and P2 (of $T_{21}TH$) satisfies $A_2T_{11}H$ (of $T_{11}TH$). The same holds for t_{12} and t_{22}.

The superposition gives rise to a new component having, as interface, all the open transitions of the sources minus the transitions of the superposed tuple which generate a new closed transition associated with a corresponding object obtained by "joining" the various modules (objects and theories) associated with the source open transitions in the various components. This object can introduce a new boolean operator defining further constraints to the firing of the new closed transition, and is inherited by the object associated with the new component.

For instance, the superposition of t_{11} and t_{21} gives rise to a new OBJSA component characterized: by the net of Fig.6 and by the object C3 which combines the objects C1 and C2. In particular, it extends the object TR1 (as defined in Example 3) obtained by composing $T_{11}TH$ and $T_{21}TH$ and associated to the closed transition T1, and the instantiation of COMPONENT having as parameter the theories associated to the two remaining open transitions.

obj C3 *is*
extending P1
extending P2
extending TR1
extending COMPONENT [TWO, $t_{12}h :: T_{12}TH$, $t_{22}h :: T_{22}TH$]
endo

At this point the component C3, which is still open since it contains open transitions, could be combined with two different components satisfying the constraints defined by $T_{12}TH$ and $T_{22}TH$. In particular, since we already know that t_{12} and t_{22} can be superposed, we can perform this superposition yielding a new OBJSA component characterized: by the net of Fig.1 and by the object C4 obtained by combining C3 with itself. In particular, it extends the object TR2 (as defined in Example 3) obtained composing $T_{12}TH$ and $T_{22}TH$ and associated to the closed transition T2, and the object CLOSEDCOMP indicating that the component has an empty interaace, and is then called closed.

obj C4 *is*
extending P1
extending P2
extending TR1
extending TR2
extending CLOSEDCOMP
endo

obj CLOSEDCOMP *is*
endo

The resulting OBJSA component coincides with the OBJSA net system defined in Section 2, the object C4 representing its rewriting environment.

6. Conclusions

This paper introduces a new class of high-level Petri nets with the purpose of supplying a language for the specification of concurrent systems which takes full advantage of the best features of Petri nets and of algebraic specification techniques and which provides suitable linguistic features supporting compositionality, modularity and reusability. The comparison with the most popular classes of high-level nets - namely Predicate/Transition and Coloured nets - should take into account this goal, but cannot be founded on this paper only. On one hand, in regard to the Predicate/Transition nets, the weakness of set theory for data specification has been pointed out by Liskov and Zilles among the others lLZl. On the other, in regard to the approach embedded in Coloured nets, its weakness is in the use of concrete functions (i.e. concrete algebras) instead of founding data representation on an abstract (initial) model, as OBJ does.

These considerations of course do not imply that OBJSA nets will become as popular as Predicate/Transition and Coloured nets are. With respect to the overall goal, the results achieved until now and summarized in this paper satisfies the main initial requirements, but also shows that major improvements to deal with language usability and effectiveness are required.

The first improvement consists in the development of a real specification language having OBJSA nets as semantics. The point of view of users engaged in complex systems modelling has to be considered in designing both syntactic and pragmatic features of the target specification language. We are committed to achieve this definition in narrow connection with a couple of potential users lPFIl.

In this framework we also plan to carry on:

- the development of automatic tools for supporting the language, first of all an editor, driving the instantiation of the predefined parameterized objects, and then a real OBJSA nets interpreter, containing for instance, among other facilities, a net simulator which, by automatically performing term rewriting, shows to the user how the individual tokens, namely their data part, are changed through transition firing (as shown above in Example 4). A first sketch of the main requirements for such an interpreter are given in lBatl, while the experience gained in developing an interpreter, written in C, for OBJ lCDMl, together with the literature on ADT computer-based environments such as Reve lLesl or RAP lGHl represents a good basis for accomplishing this task;
- the development of methods for deriving from the OBJSA net model properties of the modelled system. First candidate is a calculus for deriving S- and T- invariants, i.e. the development of algorithms and heuristics for calculating the solutions of the systems of equations $I \otimes W = \mathbf{0}$ and $W \odot J^T = \emptyset$; also techniques such as those presented in lVaul will be considered.
- the extension to them of two orthogonal equivalence notions, supporting dual notions of abstractions. The need of two abstraction/refinement mechanisms to be used in a combined way in the stepwise development of complex systems specifications is recognized, with differences of accents and proposals, both in the ADT community (see for instance lSTl, lBVl) and in the net community (see lRei2l and lDDSl). The goal is to extend to OBJSA nets the orthogonal notions of equivalence defined for 1-safe SA net systems, first by considering behavioural equivalence from both the algebraic (lGGMl, lReicl) and the net lPoml) perspectives.

7. Acknowledgements

We want to thank K.Nygaard for his stimulus in accomplishing this work; J.Goguen, W.Reisig, K.Jensen and a number of anonymous referees for their suggestions on preliminary versions of this paper; G. Degli Antoni, G. De Michelis, A.Giacovelli, L.Pomello and C.Simone for their encouragement and fruitful discussions.

8. References

|AMRW| E. Astesiano, G.F. Mascari, G. Reggio, M. Wirsing, On the parameterized algebraic specification of concurrent systems, Proc. CAAP '85, *LNCS* 185, Springer Verlag, 1985

|Bat| E. Battiston, Definizione di una classe di reti di alto livello aventi per domini tipi di dati astratti, Thesis, Dept. Information Sciences, University of Milano, A.A. 85/86 (in italian)

|Ber| B. Berthomieu et al., Abstract data nets: combining Petri nets and abstract data types for high level specifications of distributed systems, Proc. 7th European Workshop on Applications and Theory of Petri Nets, Oxford, UK, 1986

|Bert| G. Berthelot, Transformations and decompositions of nets, Proc. Advanced Course on Petri nets, Bad Honnef, September 1986

|BCG| R. Balzer, T. Cheatham, C. Green, Software Technology in the 1990's Using a new Paradigm, in *Computer* , 11.85, pp. 39-45

|BDM1| E. Battiston, F. De Cindio, G. Mauri, OBJSA Net Systems, Int. Rep., Dip. Scienze dell'Informazione, Milano, 1987

|BDM2| E. Battiston, F. De Cindio, G. Mauri, OBJSA nets : OBJ2 and Petri Nets for specifying concurrent systems, accepted for pubblication in "Experiences with OBJ" (R. Gallimore ed.), to appear

|BFP| E. Best, C. Fernandez, H. Plünnecke, Concurrent systems and processes, GMD-Studien Nr.104, 1985

|BV| C. Beierle, A. Voss, On implementation of loose abstract data type specifications and their vertical composition, Proc. STACS '87, LNCS 247, Springer Verlag, 1987

|CCITT| CCITT, Recommendation Z 100, 1988 (preliminary version)

|CDM| C. Cavenaghi, M. De Zanet, G.Mauri, MC-Obj: a C interpreter for OBJ, accepted for pubblication in "Experiences with OBJ" (R. Gallimore ed.), to appear

|CHJ| B. Cohen, W.T. Harwood, M.I. Jackson, The specification of complex systems, Addison Wesley, 1986

|DDPS| F.De Cindio, G. De Michelis, L. Pomello, C. Simone, Superposed Automata Nets, in "Application and Theory of Petri Nets" (C. Girault and W. Reisig eds.), IFB 52, Springer Verlag, 1982

|DDS| F.De Cindio, G. De Michelis, C. Simone, Gameru: a language for the analysis and design of human communication pragmatics, in "Advances in Petri Nets 87", (G. Rozemberg ed.), LNCS 266, Springer Verlag, 1987

|DLT| F. De Cindio, G.A.Lanzarone, A. Torgano, A Petri Net Model of SDL, Proc. 5th European Workshop on Petri nets, Aarhus (Dk), 1984

|EM| H. Ehrig, B. Mahr, Fundamentals of algebraic specification 1, Springer Verlag, 1985

|FGJM| K.Futatsugi, J.A. Goguen, J.P. Jouannaud, J. Meseguer, Principles of OBJ2, Proc. ACM Symp. on Principles of Programming Languages, 1985

|Gen| H. Genrich, Predicate/Transition nets, in "Petri Nets: Central Models and Their Properties", (W. Brauer, W. Resig, G. Rozemberg eds.), LNCS 254, Springer Verlag, 1987

|GGM| V. Giarratana, F. Gimona, U. Montanari, Observability Concepts in Abstract Data Type Specifications, Proc. 5th Symp. Math. Found. of Comp. Science 1976, LNCS 45, Springer Varlag 1976, 576-587

|GH| A. Geser, H. Hussmann, Experiences with the RAP system - a specification interpreter combining term rewriting and resolution, Proc. ESOP 86, LNCS 213, 1986, 339-350

|GL1| H. Genrich, K. Lautenbach, System Modelling with High-level Petri nets, TCS 13, North-Holland 1981

|GL2| H. Genrich, K. Lautenbach, S-invariance in Predicate-Transition Nets, in "Application and Theory of Petri Nets" (A. Pagnoni and G. Rozenberg eds.), IFB 66, Springer Verlag, 1983

|Gog| J.A.Goguen, Parameterized programming, IEEE Trans. on Soft. Eng., SE-10(5), 528-543, 1984

|GTW| J.A. Goguen, J.W. Thatcher, E.G. Wagner, An initial algebra approach to the specification, correctness and implementation of abstract data types, in "Current trends in programming methodology IV: Data structuring, (R. Yeh, Ed.), Prentice Hall, 1978, 80-144

|Hac| M. Hack, Extended State-Machine Allocatable Nets, an extension of Free Choice Petri Nets results, MIT Project MAC, MAC-TR 78-1, Cambridge (Ma, Usa), 1974

|Hoa| C.A.R. Hoare, Communicating sequential processes, CACM 21, 666-677, 1978

|ISO| ISO, Information Processing Systems - Open Systems Interconnection - The definition of the specification language LOTOS. Draft proposal ISO/TC 97/SC 16/WG1N157, August 1983

|Jen| K. Jensen, Coloured Petri nets and the invariant method, TCS 14, 1981, 317-336

|Kra| B. Kraemer, Stepwise construction of non-sequential software systems using a net-based specification language, in "Advances in Petri nets 1984" (G. Rozenberg ed.), LNCS 188, Springer Verlag, 1985

|Les| P. Lescanne, Computer experiments with the REVE term rewriting system generator, Proc. 10th ACM Symp. on Principles of Programming Languages, 1983, 99-108

|LP| K. Lautenbach, A. Pagnoni, Invariance and Duality in Predicate/Transition Nets and Coloured Nets, GMD Report n.132, 1985

|LTS| P.E. Lauer, P.R. Torrigiani, M.W. Shields, COSY - A System Specification Language Based on Paths and Processes, Acta Informatica, 12, 1979, 109-158

|LZ| B. Liskov, S. Zilles, An introduction to Formal Specifications of Data Abstractions, in 'Current Trends in Programming Methodology' (R. Yeh ed.), Prentice-Hall, 1978

|Mil| R. Milner, A calculus for communicating systems, LNCS 92, Springer Verlag, 1980

|Obe| H. Oberquelle, Some concepts for studiyng flow and modification of actors and objects in high level nets, Proc. 3rd European Workshop on Petri nets, Varenna (Italy), 1982

|PFI| Languages and Tools for Concurrent and Distributed System, proposal submitted to Italian National Research Council, Progetto Finallizzato Informatica, area 4.2.1, nov. 1987

|Pom| L. Pomello, Some equivalence notions for concurrent systems. An overview, in "Advances in Petri Nets 1985" (G. Rozenberg ed.), LNCS 222, 1986, 381-400

|Rei1| W. Reisig, Petri Nets with Individual Tokens, TCS 41, North Holland, 1985

|Rei2| W. Reisig, Petri Nets in Software Engineering, in 'Petri Nets: Applications and Relationships to Other Models of Concurrency", (W. Brauer, W. Resig, G. Rozemberg eds.), LNCS 254, Springer Verlag, 1987

|Reic| H. Reichel, Behavioural Equivalence - A unifying concept for initial and final specification methods, Proc. 3rd Hungarian Comp. Sci. Conf., Budapest, 1981, 27-39

|ST| D. Sannella, A. Tarlecki, Toward formal development of programs from algebraic specifications: implementations revisited (extended abstract), Proc. TAPSOFT '87, LNCS 249, Springer Verlag,1987, 96-110

|Vau| J. Vautherin, Parallel systems specifications with colored Petri nets and algebraic abstract data types, in "Advances in Petri Nets 87", (G. Rozemberg ed.), LNCS 266, Springer Verlag, 1987

|Zil| S.N. Zilles, Algebraic specification of data types, Project MAC Progress Report 11, MIT, Cambridge, Mass., 1974, 28-52

This research has been developed with the financial support of the Italian Ministero della Pubblica Istruzione.

Section C
Hierarchical High-level Nets

Looking at the history of high-level programming languages, it is obvious that their success, to a large degree, is due to the introduction of subroutines and modules—by which it is possible to construct a large description from smaller units which can be investigated independently of each other. The absence of compositionality has been one of the main criticisms raised against Petri net models. The paper in this section addresses this by defining hierarchical high-level nets—the use of which makes it possible to relate a number of individual nets to each other in a formal way (i.e. in a way which has a well-defined semantics and thus allows formal analysis). The last two papers in Section H describe industrial applications of hierarchical high-level nets.

The basic idea behind hierarchical nets is to allow the modeller to construct a large model by combining a number of small nets into a larger net. This is similar to the situation in which a programmer constructs a large program from a set of modules and subroutines The idea is different from the many approaches which relate two or more separate subnets to each other, in order to compare their behaviour, but *without* combining them into a single net. Such approaches are analogous to program transformations.

It is possible to translate a hierarchical high-level Petri net into a non-hierarchical high-level net—which in turn can be translated into a low-level Petri net. This means that the theoretical modelling powers of these three classes of nets are the same. However, from a practical point of view, the three net classes have very different properties. To cope with large systems we need to develop strong structuring and abstraction concepts. The first substantial step along this path was to replace low-level Petri nets with high-level nets. The second step is to introduce hierarchical nets. In terms of programming languages, the first step can be compared to the introduction of types—allowing the programmer to work with structured data elements rather than with single bits. The second step may then be compared to the development of programming languages with subroutines—allowing the programmer to work with reusable patterns. From a theoretical point of view machine languages (or even Turing machines) are equivalent to the most powerful modern programming languages. From a practical point of view, this is of course not the case. One of the most important limitations that system developers face today is the inability to cope with many details at the same time. In order to develop and analyse complex systems one needs structuring and abstraction concepts that allow to work with selected parts of the model—without being distracted by low-level details of other parts. Hierarchical nets provide such abstraction mechanisms.

The ideas behind hierarchical nets build upon earlier work by H.J. Genrich, H. Oberquelle and R.M. Shapiro. References to related work can be found in section 1 of the hierarchy paper.

7.
Hierarchies in Coloured Petri Nets

P. Huber, K. Jensen and R.M. Shapiro

G. Rozenberg (ed.): Advances in Petri nets 1990. Lecture Notes in Computer Science, vol. 483. Springer, Berlin Heidelberg New York 1990, pp. 313-341

ABSTRACT The paper shows how to extend Coloured Petri Nets with a hierarchy concept. The paper proposes five different hierarchy constructs, which allow the analyst to structure large CP-nets as a set of interrelated subnets (called pages). The paper discusses the properties of the proposed hierarchy constructs, and it illustrates them by means of two examples. The hierarchy constructs can be used for theoretical considerations, but their main use is to describe and analyse large real-world systems. All of the hierarchy constructs are supported by the editing and analysis facilities in the CPN Palette tool package (see [1-5]).

Keywords high-level nets, Coloured Petri Nets, structuring mechanisms, hierarchies, re-usable components, subnets.

CONTENTS

0. Introduction

This paper is an introduction to hierarchical Coloured Petri Nets (CP-nets). It shows how a set of subnets, called pages, can be related to each other – in such a way that they together constitute a single model. The basic idea is to allow the system modeller to describe a set of submodels which all contribute to a much larger model – in which the submodels interact with each other in a well-defined way. This idea is well known from other kinds of artificial languages, e.g. submodels in SADT and subroutines in programming languages. The purpose is to break down the complexity of the large model, by dividing it into a number of submodels.

It is important not to confuse this idea of hierarchical nets with the many approaches which relate two or more separate subnets to each other *without* combining them into a single model. In these approaches each page defines its own model – and the goal of the approach is to compare these individual models – e.g. showing that they have an equivalent behaviour or that they describe the system from different viewpoints [6,7]. In the domain of programming languages, these latter approaches are equivalent to program transformations. The models are different descriptions of the same system – and they are not used to obtain a larger and more complex description.

Although our main objective is to synthesize large models from smaller submodels, our approach also has some elements which allow the user to create a number of related models. However, each of these models are usually synthesized from several submodels. The idea is most easily explained by considering our notion of a substitution node, which is a place or a transition related to a submodel. Usually, the submodel totally replaces the substitution node and the surrounding arcs, and hence it doesn't make sense to say that a substitution transition occurs or a substitution place is marked. The substitution node is not itself part of the final model – its role is merely to describe how the related submodel is inserted in the synthesized model (in the same way as an in-line subroutine call describes where to insert the subroutine code). During the analysis of a complex system it is often convenient temporarily to be able to ignore parts of the model – or replace them with simpler components. In our methodology we achieve this by allowing the user to specify, for each substitution node, whether the related subpage should be included in the current synthesized model or not. When the submodel isn't included, the substitution node becomes an ordinary node. This means that the node can occur (if it is a transition) or become marked (if it is a place). In order for this to be useful it is of course necessary that there be some formal or informal correspondence between the behaviour described by the model in which the submodel is inserted and the model in which the substitution node is an ordinary node. Probably the most obvious relationship is to make the two models equivalent. This means that the substitution node with its surrounding arcs is behaviourally equivalent to the related submodel. Such a strict equivalence is, however, by no means the only interesting kind of relationship. Another possibility is to let a substitution transition describe the normal behaviour of an activity while the related submodel in addition describes several kinds of abnormal behaviour (e.g. time-outs and loss of messages or signals). This allows the user to have two related models: a crude one in which he can investigate the normal behaviour and a more complex one also coping with abnormal behaviour. The user can switch between the two models, e.g. first investigate the simple one and then go on with the more complex one. The two models share most of the submodels, and this means that changes in one of them automatically apply to the other.

Our hierarchy constructs do not extend the theoretical modelling power of CP-nets, and some readers might at a first glance be tempted to characterize at least some of them merely as graphical conveniences. This may be justifiable from a theoretical point of view. However, from a practical point of view, we do not think that this is a fair characterization. To cope with large systems in practice we need to develop strong structuring tools. The first very substantial step on this path was to replace ordinary Petri Nets with high-level Petri nets, such as Predicate/Transition-nets [8,9] and Coloured Petri Nets. The second step is, in our opinion, to introduce hierarchical models. In terms of programming languages, the first step can be compared to the introduction of types – allowing the programmer to work with structured data elements instead of single bits. The second step may then be compared to the development of programming languages with subroutines – allowing the programmer to work with reusable patterns. From a theoretical point of view machine languages (or even Turing machines) are equivalent to the most powerful modern programming languages. From a practical point of view, this is of course not the case. One of the most important limitations that system developers face to day, is their own inability to cope with many details at the same time. In order to develop and analyse complex systems they need structuring tools which allow them to work with a selected part of the model – without being distracted by the low-level details in the remaining parts. Hierarchical CP-nets is an attempt to provide the Petri Net modeller with such abstraction mechanisms. From a theoretical point of view it can be judged whether our framework has a sound mathematical basis. Its real success or failure can, however, only be judged by Petri Net modellers using the concepts to develop large models.

This paper assumes that the reader is familiar with non-hierarchical CP-nets – as defined in [10,11,12]. The paper gives a rather informal description of hierarchical CP-nets and concentrates on the explanation of the intuition behind the different hierarchy constructs. It is, however, not particularly difficult to give a formal definition of hierarchical CP-nets, and that will be done in a future paper.

The paper describes five different ways to relate submodels to each other. These five hierarchy constructs are not at all independent, and it is often possible to choose between them – in order to obtain a certain modelling goal (in the same way as many programming tasks can be achieved either by means of a while loop or a recursive procedure). The current formulation of the hierarchy constructs is very general – and gives the user many modelling choices. Practical experiences will probably lead to a number of more restricted hierarchy constructs – perhaps aimed at different application areas. The hierarchy constructs are in this paper formulated in terms of CP-nets. We are, however, convinced that it will be nearly trivial to reformulate them to apply to most other kinds of Petri Nets.

The remaining part of the paper is organized as follows: First we take a closer look at the idea of hierarchies in behavioural models in general. Then we present our five hierarchy constructs: substitution transitions, substitution places, invocation transitions, place fusion and transition fusion. We introduce our constructs by means of a small, original example. At the end of the paper we discuss two examples known from the Petri Net literature.

1. Hierarchies in Behavioural Models

In the area of system modelling and design the inadequacies of single level system models are well known:

- missing overview,
- too many details at one time,
- the structure of the system in question is not mirrored adequately.

Hierarchical modelling languages have been introduced to overcome these problems and have been in practical use for quite some time. Among others we have SADT and IDEF [13], Yourdon's data flow diagrams [14], and Statecharts [15]. For the majority of such hierarchical modelling languages we can list a set of nice features:

- hiding of details in a consistent way,
- separation into well defined components,
- reusable components,
- support of both top down and bottom up development strategies,
- strong graphical expressive capabilities.

However, at present – on the threshold of the development of powerful execution and simulation tools – we have to look for yet another model quality: executability. This implies that the modelling language must:

- support the notion of behaviour for its components *in a precise and consistent way,*
- make it possible to observe the execution of large, complex system models at different levels of detail.

Most of the modelling languages in the group above do not possess these executability properties. In contrast, high-level Petri Nets have in recent years been acclaimed as excellent modelling languages for expressing concurrent behaviour in a natural and sound way. Part of the reason is the mathematical (formal) basis of high-level Petri Nets, see [8,9,10,11,12].

Recently the practical use of CP-nets (and Predicate/Transition-nets) has been accelerating in such diverse areas as protocol verification, design of computer integrated flexible manufacturing systems, air traffic control problems, hardware and operating system design, and real time banking applications. However, up to recently the CP-nets and most other kinds of Petri Nets have been considered as flat models. Our rationale here is to start out from this well-proven framework for behavioural modelling and add hierarchical structure. We hope this will be an important step towards a more easy and widespread practical use of CP-nets for many different kinds of complex systems. Although the hierarchy constructs are general, their design is closely related to the CPN Palette project [1-5]. This means that the hierarchy constructs are designed in a way which makes it easy to support them by computer tools.

In the literature there is almost no work on hierarchies in Petri Nets, which take a behavioural point of view. The concept of net morphisms [16] is in its present formulation too general to be used as a basis for computer supported modelling and execution. Thus our work is more related to hierarchy concepts found in other kinds of system description languages. In both literature and practice, Petri Nets models with a consistent use of hierarchical structuring are sparse and isolated. Some ideas can be found in papers like [17,18,19,20,21,22,23]. Recently several articles showing rather complex high-level Petri Net models have been presented, e.g. [24,25]. However, these models only use multi-level structuring in an informal and rather ad-hoc way. Especially for computer supported editing, execution and formal analysis of such models a more consistent framework is necessary. In [3] we illustrate our framework by means of nine medium size examples – some known from the literature [11,24,25] and some original ones.

We have tried to look at hierarchies in an as general way as possible and impose as few restrictions on the hierarchy constructs as possible. From discussions, experiments, application of metaphors and comparison to related models, we have then developed five hierarchy constructs. Each of these gives CP-nets a more useful and more flexible expressive power. The aim of the hierarchy constructs is to guide the analyst to produce structured models by supplying a set of sound and consistent structuring concepts.

2. Hierarchies in CP-nets

Our point of departure is the ordinary non-hierarchical CP-nets as presented in [10,11,12]. A brief review of the terminology is given below:

A **place** is a node, where **tokens** from a specified **colour set** may reside. The distribution of tokens in the CP-net is called a **marking**. A place has an **initial marking**, which specifies the initial load of tokens. Places represent states and are normally drawn as ellipses. A **transition** is a node representing an action and it is normally drawn as a box. For each transition a **guard** can be specified. This is a boolean expression restricting the conditions under which the transition can occur. Places and transitions are called **nodes**. An **arc** represents an input or output relationship between a place and a transition. The actual amount and the colours of tokens moved are specified by the corresponding **arc expression**. Arc-expressions may be non-simple and may evaluate to a multi-set with 0 to n tokens.

From the arcs, the current marking, the arc expressions, and the guards one can calculate which transitions are **enabled** with respect to which **bindings** (of the **variables** in the arc expressions). If not in conflict with other transitions, an enabled transition may **occur**, whereby tokens are removed from the **input places** of the transition, and tokens are added to the **output places** of the transition, as specified by the arc expressions. The transitions can be seen as *schemes* for behaviour, in the sense that the actual binding determines the details of the behaviour. The number of tokens moved along an arc may depend upon the actual binding and it may even for some bindings be zero. A set of transitions and bindings occurring concurrently is called a **step**. A **diagram** is a set of related non-hierarchical CP-nets, called **pages**.

We define the semantics of the new hierarchy constructs by showing how each use of them can be translated into an **equivalent non-hierarchical CP-net** which has exactly the same reachable system states and enabled steps. This approach of introducing new language constructs by specifying a translation to well-known old constructs is traditional. Exactly the same thing was done, when CP-nets (and Predicate/Transition Nets) were defined by means of ordinary Petri Nets (PT-nets). It should be stressed that the only purpose of this translation is to define and present the hierarchy constructs in a precise way. The analyst works directly with the hierarchy constructs – without constructing an intermediate flat CP-net. Our extension of CP-nets implies that the existing analysis methods must be extended to cope directly with hierarchical CP-nets. For place invariants, occurrence graphs and reductions [11,26,27,28] we are convinced that this extension will present no serious conceptual problems. There are, however, a very large number of technical details which have to be worked out.

3. Substitution Transitions

We now take a closer look at the first metaphor used to develop our CP-net hierarchies: the hardware plug-in, e.g. a silicon chip. Imagine a component with a set of interface posts. We can connect such a component to a given environment by means of eyelet connectors, which are attached to the posts. Whenever a specific component of this kind is used, the only thing to be done is to specify the eyelet/post correspondence. Going back to the CP-nets, we will consider a transition as such a component and its surrounding places as the interface to the environment.

Example 1: Simple assembly line

Let us explain the idea by means of a small example. We consider a simple assembly line in a factory consisting of three machines and two intermediate buffers. The machines are identical, hence we only want to model them once. The same is the case for the buffers. The page in the left part of Figure 1 represents each machine (Mach1-Mach3) and each buffer (Buf1-Buf2) as a **substitution transition**. The details of the machines and buffers are described on two other pages in the right part of Figure 1. The result is a hierarchical CP-net where five substitution transitions at page AssemblyLine#1 are related to two **subpages** Machine#2 and Buffer#3 (we denote each page by a page name, followed by "#", followed by a page number). The interfaces of Machine#2 and Buffer#3 are defined by the B-tags and the inscriptions next to them (B≈Border). The relationship between each of the substitution transitions and the corresponding subpage is defined by the inscription next to the HS tag (HS≈Hierarchy+Substitution). This inscription tells the name and the number of the subpage and it describes how each of the places surrounding the compound transition is assigned to one of the border nodes of the subpage. We have omitted most of the other net inscriptions, e.g. initial marking, since we are focussing more on the net structure than the details of colour sets, arc expressions, etc.

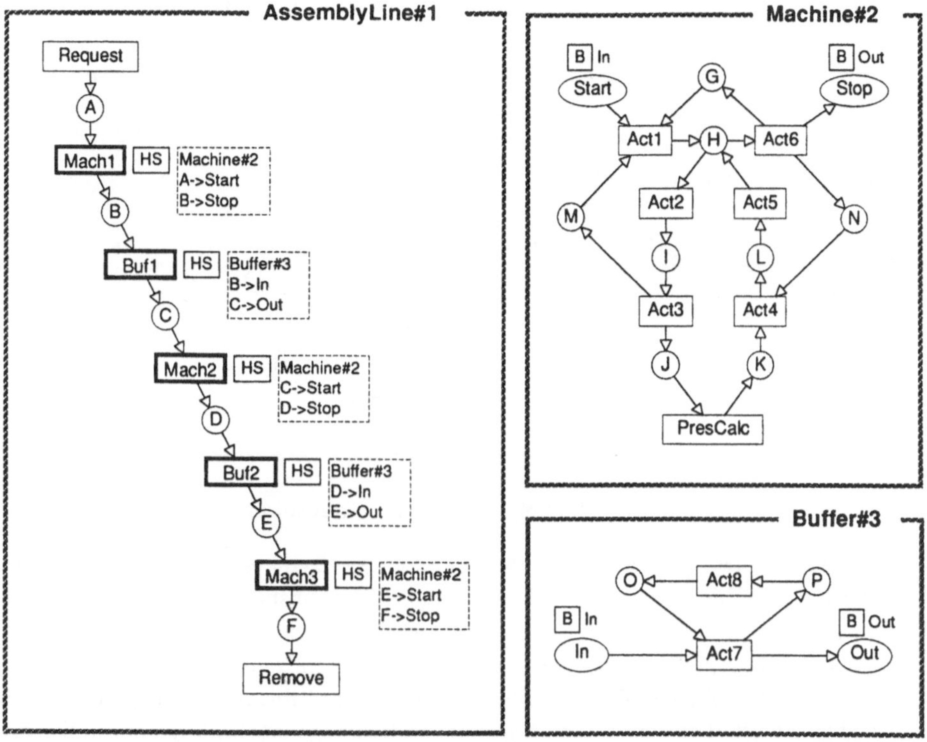

Figure 1: Assembly line with three machines and two buffers

From page AssemblyLine#1 the reader gets an overview of the assembly line. To feed the line and remove the produced items we have added two ordinary transitions: Request and Remove. The behaviour of the machines is described at Machine#2 by means of seven lower level activities (transitions). The buffers are described at Buffer#3. They are very simple and thus we might instead have inserted them at AssemblyLine#1. However, by describing them on a separate page we have abstracted them, in a way which later allows for confined experimentation with the details of the buffers without altering any other part of the diagram.

Let us pause for a moment and compare this approach with the earlier abstraction mechanisms in CP-nets. In order to represent three instances of the same machine in a compact way, we could have folded them into the CP-net of Machine#2 by introducing an additional machine-id component of the colour sets for all the places representing the machine states. We could then have done a similar thing with the buffers. The problem would however arise when we afterwards tried to relate the machines and buffers to each other. For example the fact that Mach1 is the only machine that delivers input to Buf1 would then have had to be encoded in the guard of transition Act7 in Buffer#3. To continue in that way would soon become cumbersome. Folding is well suited, when we have exactly identical objects, but less suited for asymmetrical arrangements.

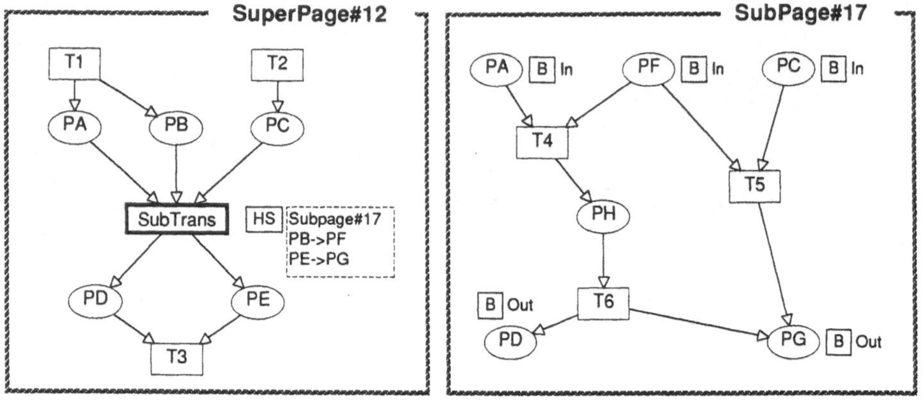

Figure 2: Substitution transition

We will now be more precise on the terminology: Each **substitution transition** designates a page. This page is said to be a **subpage**. The substitution transition represents the component seen as a black box, whereas the subpage contains the details of how this component actually performs the activity. The substitution transition is a short hand for the CP-net on the subpage. Figure 2 illustrates the idea of this node-to-page relationship for an abstract net. The substitution transition and its page are, with respect to the subpage, said to be a **supernode** and a **superpage**. For simplicity, most of the CP-net inscriptions are again omitted. The substitution transition, named SubTrans, can be recognized by the HS-tag. SubTrans has five surrounding places, which are called **socket places**: three input socket places and two output socket places. Five places on the subpage are defined to be **port places** and hence marked by B-tags. The port places represent the posts from the hardware metaphor. They are the interface to the upper level, at which the subpage is plugged in and used. The relationship between socket places and port places is called the **port assignment**. It is a function mapping sockets into ports. The port assignment is shown in the inscription next to the HS-tag of the substitution transition: The first line tells the name and number of the subpage. Each of the remaining lines describe the assignment between a socket and a port. By convention we only mention sockets which are unassigned or assigned to a port with a different name. In this case, we have omitted the lines PA->PA, PC->PC and PD->PD. We require that all socket nodes must be assigned, and that a socket node must be assigned to a port node with an identical colour set. The port assignment function is, however, allowed to be non-injective and non-surjective. The inscriptions next to the B-tags of port nodes tell whether the assigned socket node has to be an input, output or input/output node for the substitution transition. The modeller can also define a port node to be general and this means that all three kinds of socket nodes can be assigned.

It has been important for us to design concepts, which allow system components to be reusable. Once a given building block has been designed and verified it should be possible to use it at several locations. Hence, an important feature of our framework is that the same page may be used as a subpage for several substitution transitions, even on different pages. The assembly line example uses both Machine#2 and Buffer#3 as multiple plug-ins.

In the CPN Palette there are facilities for manual and automatic port assignment, optional syntax restrictions and system-aided editing. All these facilities are set up to help the user produce a consistent diagram in an easy way: It is possible to move part of a page to a subpage, whereby a supernode is created and the port assignment automatically performed. It is also possible to replace a supernode by the contents of its subpage. Finally there are facilities for connecting compound nodes to already constructed pages. For more details, see [4].

Semantics of substitution transitions

To define the semantics of a substitution transition we show how to translate it into an equivalent non-hierarchical CP-net. This is done in two steps:

a) Delete the substitution transition (together with the surrounding arcs).

b) Insert a copy of the subpage.

   The left-hand side of Figure 3 illustrates steps a + b for the net from Figure 2.

c) **Merge** each socket place with the assigned port node. The result of a merge of two nodes A and B is a node C with a set of arcs which is the union of the arcs of A and B. In the case where two or more socket places are assigned to the same port, we only merge one of them with the port place and then "glue" all the socket places together by means of an instance fusion set (which we introduce later in this paper). Unassigned ports are left alone, i.e. they are not merged into another node.

   The right-hand side of Figure 3 shows the result of step c for the net from Figure 2.

Each subpage is a template. From that *copies* can be made to replace the corresponding substitution transitions. We refer to such copies as (**substitution**) **instances** of the actual page. The initial marking of the subpage is copied together with the net structure. Each port place, however, inherits (and shares) the initial marking of the assigned socket (if any). The page instances form the **instance hierarchy** and we speak about **superinstances** and **subinstances**, in a similar way as we talk about superpages and subpages.

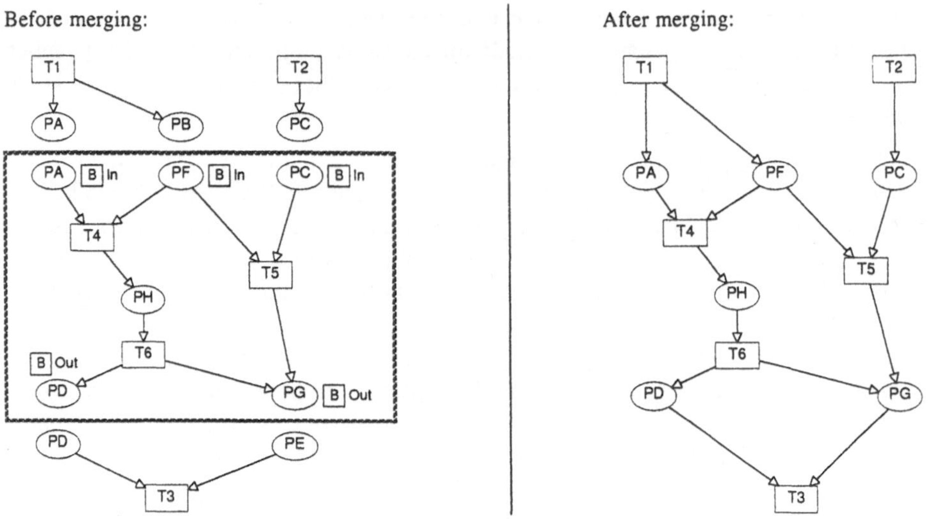

Figure 3: Semantics of substitution transitions:

The subpage replaces the substitution transition:

1. Each socket is merged into the corresponding port.

Given the semantics above, it should be clear why it makes no sense to talk about enabled or occurring substitution transitions. They are not an executable part of the model, because they are substituted by the subpage.

At this point we would like to stress, that a substitution transition should *not* be considered as a normal transition.– it has a special tag signifying this fact. The semantics is defined as a substitution by its subpage as shown above. In order to make a model more readable, it is up to the modeller to write some meaningful arc-expressions surrounding the substitution transition. Typically, the modeller would leave inscriptions empty for some of the surroundings arcs. In some nets it would make sense to exhibit a strong equivalence, but more often the modeller would choose to exhibit *some* of the possible behaviour at the upper level. The other extreme would be to have no equivalence what so ever. More work will go into defining a notion of equivalence in between these extremes. In Section 8, we explore the relationship between upper and lower level inscriptions by means of an example.

4. Substitution Places

Analogous to the substitution transitions we use the plug-in metaphor for places. We let a substitution place be a short hand for a more detailed subnet, which the analyst want to hide at the upper level of abstraction. Each substitution place has an interface which consists of transitions and it behaves in a way which resembles an abstract data type. Figure 4 shows a **substitution place** SubPlace together with its subpage Queue#4.

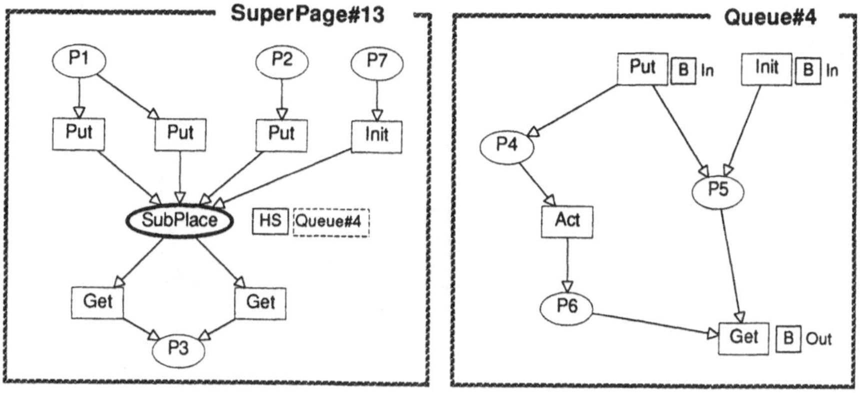

Figure 4: Substitution place

At the superpage, the substitution place outlines a simple data type: a queue with three operation handles: Init, Put, and Get. In this case Init is called once, Put three times and Get twice. These calls are represented by the six **socket transitions** surrounding SubPlace. As for substitution transitions we allow that some port nodes are unassigned. Moreover, as illustrated above, we often assign several socket transitions to the same port. At the subpage the three operations are described in more detail and the interface to them is represented by the three **port transitions**.

Semantics of substitution places

The semantics of substitution places is similar to the semantics of substitution transitions. The roles of places and transitions are, however, reversed and in addition to this there are a number of minor differences:

a) Delete the substitution place (together with the surrounding arcs).

b) Create a copy of the subpage and **duplicate** each port transition to obtain a copy for each of the assigned socket transitions. The result of a duplication of a node A is a new node B which have exactly the same set of arcs as A. When a port has only one assigned socket node, no duplication is needed. Un-assigned ports are deleted. The left-hand side of Figure 5 illustrates steps a + b for the net in Figure 4.

c) Merge each socket transition with a copy of the port transition to which it is assigned. The guard in the resulting transition is the conjunction of the guards for the socket node and and the port node. The merge implies that a socket transition has to occur together with its corresponding port transition, as a single indivisible state change.
The right-hand side of Figure 5 illustrates step c for the net in Figure 4.

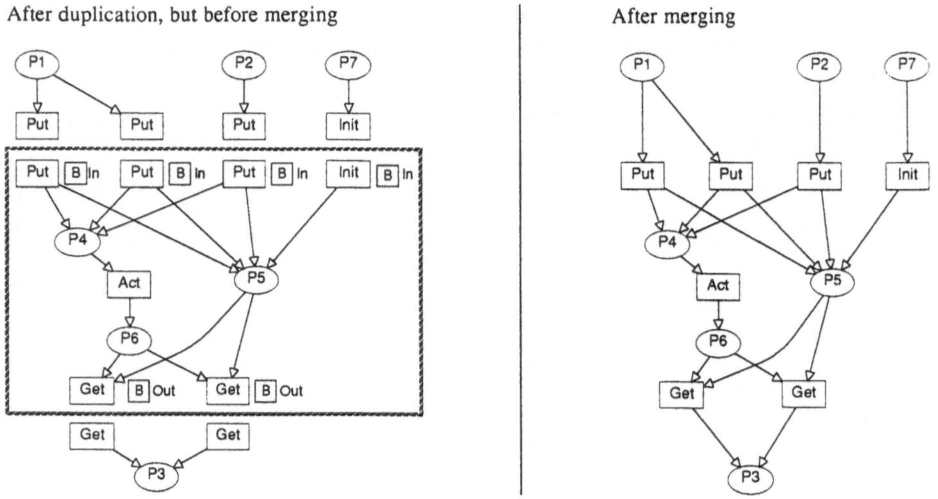

Figure 5: Semantics of substitution places:

The subpage replaces the substitution place:

1. Each port is duplicated to match the number of attached sockets.

2. Each socket is merged into its own copy of the corresponding port.

Example 2: Small factory unit

Let us now return to the assembly line. We want to use it as a building block in a larger system. Hence, we define Request and Remove to be port transitions (by placing a B-tag next to them). We then use AssemblyLine#1 as a subpage for two substitution places Assembly1 and Assembly2, as shown in Figure 6. In addition we use Queue#4 from Figure 4 as a subpage for a front-end queue named InQueue and a back-end queue named OutQueue. This small example illustrates how it is easy to extend the scope of a model by inserting it as a submodel in a larger model.

We do not allow a substitution place SP to be neighbour to a substitution transition ST. The reason is that it then would be impossible to construct an equivalent non-hierarchical CP-net by the method defined above – because SP is socket for ST and vice versa. It is possible to extend our concepts to cover such a case. We have, however, not been able to do this in a natural and simple way, and we do not think that the restriction presents a serious problem. In the models we have been working with there has been a tendency to use either compound places or compound transitions in a given part of the diagram.

Moreover, should two substitution nodes come too close together, they can always be separated by inserting an extra place and an extra transition between them.

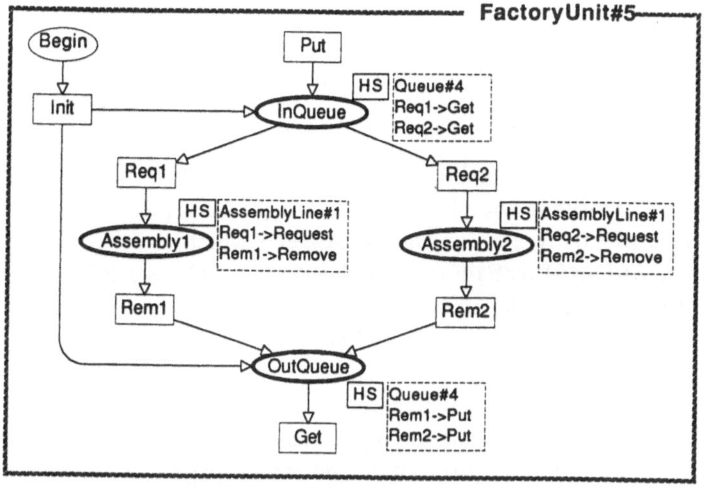

Figure 6: Factory unit with two assembly lines

Abstract data types is one of the most well-known and useful abstraction mechanisms in computer science, and it has been important for us to allow the user to apply a similar concept for CP-nets. As shown above, this can be done by means of substitution places where the port transitions represent the exported operations and the socket transitions the operation calls. There is, however, an additional set of problems which we have to solve: Suppose that the three calls of Put in Figure 4 use different expressions to describe the tokens which are given to the queue. Such a difference should not be specified at the subpage, but at the socket transitions. Thus we allow the arc expressions of the port transitions to refer to the arc expressions of the socket transitions by means of special identifiers (details are described in [3]). However, we think this method is too primitive, and it will be refined as we get more experience with the use of our framework. One solution might be to allow the supernodes to specify instance dependent properties of their subpages. This would also make it possible for arbitrary places to have different initial markings in different page instances. For the moment this is only possible for port places, since they inherit their marking from the assigned socket nodes.

As for substitution transitions, substitution places should not be considered as real places, since they are substituted by a subpage. Accordingly, the substitution places carry a tag.

5. Invocation Transitions

In our search for useful constructs we have been looking for other ways to relate nodes and pages. For transitions it was rather obvious to apply the metaphor of subroutines, as known from many programming languages. A subroutine is declared with a set of formal parameters and it can be invoked (i.e. called) from different locations by supplying a set of actual parameters. Each call implies a temporary instantiation of the subroutine.

Example 3: Pressure calculation

Let us again return to the assembly line and let us assume that the machines use a complicated recursive algorithm for calculating the correct work pressure for Act4. To specify this we define in Figure 7 a subpage PresCalc#6 which is invoked by three different invocation transitions: PresCalc (at Machine#2) and PC1 and PC2 (at PresCalc#6). In contrast to substitution nodes, the invocation transitions are *not* substituted by their subpage. This means that they can occur and each of their occurrences triggers the creation of a new instance of the subpage. These subpage instances are executed concurrently with the other page instance in the model, until some specified exit condition is reached (more details below). When an invocation page instance is created or terminated, tokens are passed between the invocation transition and the subpage instance in a similar way as parameters are passed between a subroutine call and the subroutine execution (more details below).

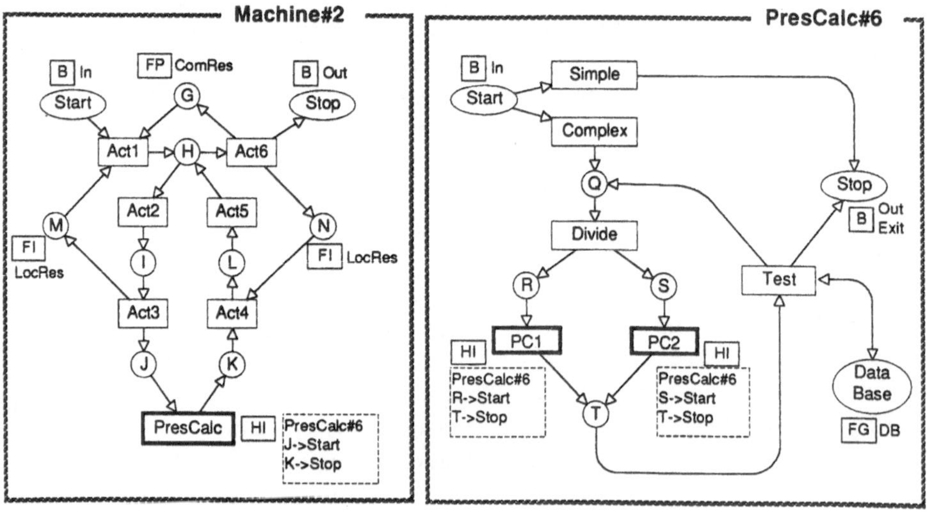

Figure 7: Recursive pressure calculation

Each invocation of PresCalc#6 receives a token from the input node of the calling invocation transition, and then classifies the task as either simple or complex. In the first case the result is immediately calculated and passed back to the invocation transition via Stop – and the execution of the subpage is destroyed, because an exit place received a token. In the second case the task is divided into two recursive subtasks (represented by PC1 and PC2), the results of the two subtasks are tested against a database, and finally a token is put either on Stop (if the test was positive) or on Q (if the test was negative) – remember, that in CP-nets am arc-expression may evaluate to the empty multi-set for certain bindings. For the moment ignore the FG/FP/FI-tags – we will return to them later.

In terms of the subroutine metaphor, the invocation subpage represents the subroutine description, while the three invocation transitions represent the subroutine calls. All the ports must be places and they represent the formal parameters. The places surrounding the invocation transitions are called **parameter places** and they represent the actual parameters. The invocation transitions are distinguishable by the HI-tags (HI≈Hierarchy+Invocation) and the inscription next to them specifies the **subpage** and the **port assignment**, relating parameter nodes to port nodes. The rule for port assignment is identical to that of substitution nodes and this means that there may be more formal than actual parameters.

The termination of a subroutine execution is usually triggered by execution of the last statement or by an explicit exit statement. In our framework it is not always possible to talk about the last node and thus we allow the analyst to define **exit nodes**. The execution is terminated the first time an exit transition occurs or an exit place receives a token.

Semantics of invocation transitions

The enabling rule for invocation transitions is identical to that of ordinary transitions, but an occurrence of the invocation transition implies a temporary *extension* of the CP-net:

a) A new instance of the invocation subpage is created. This subpage may contain substitution nodes, and when this is the case it is necessary also to create new page instances of the corresponding substitution subpages. These page instances become subinstances of the invocation page instance and they may themselves have subinstances (if they contain substitution nodes).
The arc expressions on the input arcs of the invocation transition are evaluated and the corresponding tokens are subtracted from the input parameter places and added to the assigned port places (in the invocation instance).
Multiple assignment of parameter nodes to the same port implies, as for substitution transitions, that the parameter nodes are "glued" together by means of an instance fusion set (which we introduce later in this paper).

b) The invocation instance and its subinstances are executed as a normal part of the diagram. This execution is concurrent to the rest of the diagram and it continues until an exit transition occurs or an exit place receives a token.

c) All the tokens on output port places are copied to the assigned parameter places.The invocation instance and its subinstances are removed from the CP-net, and any token information which might be left in them are lost.
Multiple assignment of parameter nodes to the same port implies, as for substitution transitions, that the parameter nodes are "glued" together by means of an instance fusion set (which we introduce later in this paper), so that tokens are copied back into that fusion set.

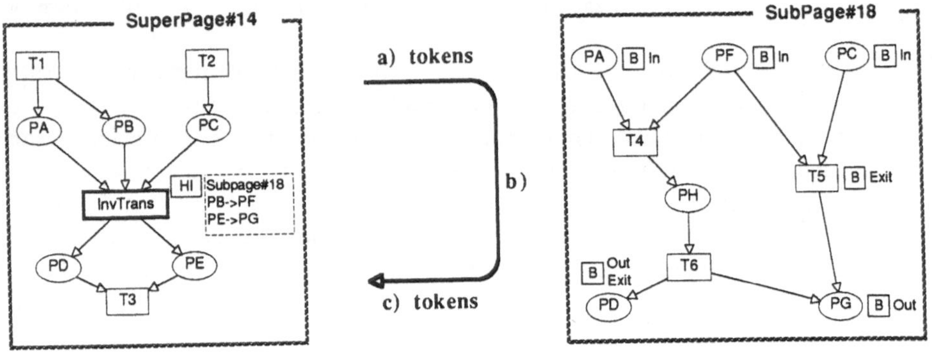

Figure 8: Semantics of invocation transitions:

When the invocation transition occurs a new instance of the subpage is created:

1. Tokens are transferred from the input places of the invocation transition to the input ports of the subpage
2. The subpage instance is executed concurrent to the other page instances until an exit is reached.
3. Tokens are transferred from the output ports of the subpage to the output places of the invocation transition.

For substitution nodes we could (although we in practice never want to do it) statically calculate the equivalent non-hierarchical CP-net. For invocation transitions this is not possible and it is necessary dynamically to extend and shrink the equivalent non-hierarchical CP-net. Without invocation each page has a constant number of instances, but with invocation the number of page instances may change (even for pages which only are substitution subpages). Moreover, each invocation transition can have any page as a subpage (as long as the ports are places). This means that the invocation hierarchy is allowed to contain circular (i.e. recursive) dependencies, while the substitution hierarchy is demanded to be acyclic (to avoid infinite substitution).

The token passing in step a) and c) is analogous to the use of in, out, and in+out parameters in subroutines. This is also sometimes known as call-by-value and call-by-result. Some programming languages, e.g. Pascal, allow subroutines to be passed as parameters to other subroutines. Analogously, we might allow token colours to represent CP-nets – but for the moment we don't allow this.

6. Fusion Sets

The main idea behind fusion is to allow the system modeller conceptually to fold a set of nodes into a single node – without graphically having to represent them as a single object. A fusion is obtained by defining a **fusion set** containing an arbitrary number of places or an arbitrary number of transitions. The nodes of a fusion set are called **fusion set members**. This idea is illustrated in Figure 9, where the leftmost CP-net has a fusion set called FusA. This fusion set contains the fusion set members A1 and A2, which are distinguishable by the FP-tags (FP≈Fusion+Page). FusA is a page fusion set and this means that it only is allowed to have fusion set members from a single page in the diagram. For the moment let us assume that this page only has one instance. Then the equivalent non-hierarchical CP-net is shown in the right part of Figure 9. It is obtained by merging A1 into A2 (or vice versa). Intuitively, this semantics means that the places A1 and A2 share the same marking.

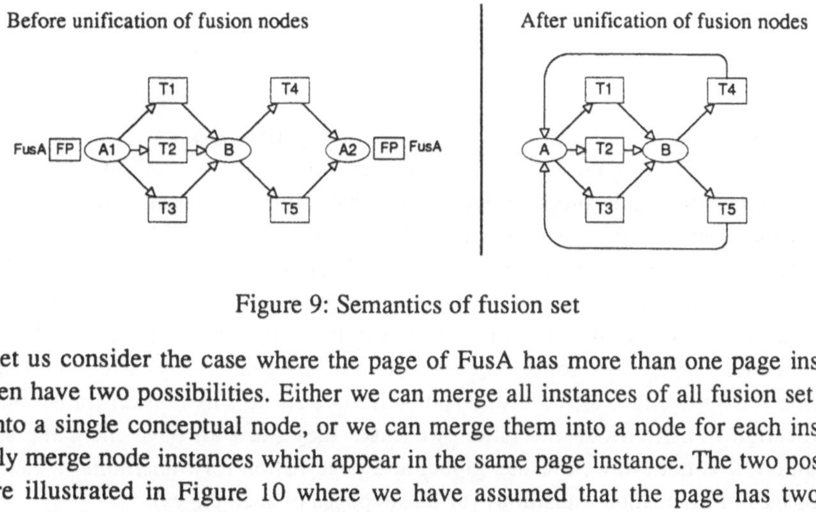

Figure 9: Semantics of fusion set

Now let us consider the case where the page of FusA has more than one page instance. We then have two possibilities. Either we can merge all instances of all fusion set members into a single conceptual node, or we can merge them into a node for each instance, i.e. only merge node instances which appear in the same page instance. The two possibilities are illustrated in Figure 10 where we have assumed that the page has two page instances. Both possibilities are useful and thus we allow the user to specify which of them he wants. The leftmost is obtained by making FusA a page fusion set while the rightmost is obtained by making it an instance fusion set (with FI-tags).

Finally we allow global fusion sets (with FG-tags). This allows fusion set members from all pages in the diagram and all instances of these nodes are merged into a single conceptual node. This means that a page fusion set is a special case of global fusion set, and seen from a theoretical point of view we could have omitted the concept of page fusion sets. However, when CP-nets are used to model large systems, it is important to be able to distinguish between global fusion sets and page fusion sets. In a computer supported environment, such as the CPN Palette, this gives the analyst an easy way to avoid unintended fusion of two fusion sets which by coincidence have the same name.

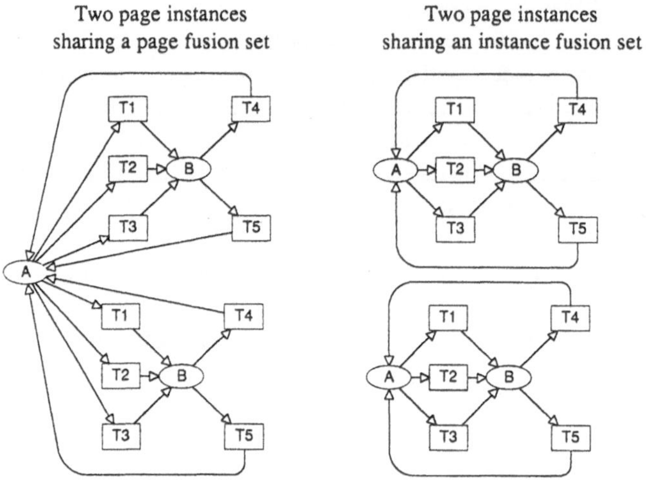

Figure 10: Page fusion set versus instance fusion set

The members of a fusion set must be comparable to each other. For places this means that they must have the same colour set and the same initial marking. It also means that they either all must be ordinary places or all be substitution places, and in the latter case they must all have the same subpage. Fusion of substitution places is useful, when we want to apply the same instance – e.g. an abstract data type – at several locations in the diagram. The cache coherence protocol modeled in [3] illustrates the use of this facility.

Above we have concentrated on fusion sets with places. However, exactly the same set of concepts applies to transition fusion. For transitions we do not demand that the guards are identical. Instead we form the conjunction of the guards. The members of a transition fusion set must either all be ordinary transitions, all be substitution transitions or all be invocation transitions. In the latter two cases they must all have the same subpage. In addition, it is not allowed to use global and page fusion for transitions, which appear on subpages of invocation transitions (or on subpages of such pages).

When applied to nodes at a single page with only one page instance, fusion is mainly a drawing convenience. However, when applied to nodes at different pages, or pages which have several instances, fusion becomes a strong description primitive of its own right, and it supplements the notions of substitution and invocation in a very fruitful way.

Example 4: Resources in the assembly line

In Figure 7 we have illustrated the use of all three kinds of fusion sets. Resources shared by all machines are modeled by a page fusion set ComRes, while resources local to one machine are modeled by an instance fusion set LocRes. Finally, the Data Base is modeled by a global fusion set DB, since another page (not shown) has the responsibility of updating the database. If we want to model the data base as an abstract datatype, the corresponding places become both substitution and fusion places.

7. Page Hierarchy

To get an overview of a given hierarchical CP-net we use a graph called the **page hierarchy.** Each node represents a page, and the shape of each such **page node** tells what kinds of supernodes the page can have. Ellipse shape indicates that all supernodes must be places, box shape that they must be transitions and rounded box shape that there is no restriction. Each arc represents a hierarchical relationship between two pages, and the graphics tells whether the arc represents a substitution relationship, an invocation relationship or a global fusion set. Page and instance fusion sets are not represented in the page hierarchy, because they involve only a single page. Figure 11 contains the page hierarchy for the small factory unit described earlier in this paper. The page hierarchy graph is generated automatically by the CPN Palette. The user can, however, change the layout and graphics in any way he might want. The page hierarchy is an integrated part of the user interface. As an example, the user deletes a page by deleting the corresponding page node in the page hierarchy.

To specify the initial state for an execution of a hierarchical CP-net, the user must define a set of starting pages, called **prime pages**. Declaring FactoryUnit#5 to be a prime page means that the execution will start with one instance of FactoryUnit#5, two instances of Queue#4, two instances of AssemblyLine#1, six instances of Machine#2 and four instances of Buffer#3. Instead we could have declared AssemblyLine#1 to be a prime page and then we would only have executed a single assembly line containing one instance of AssemblyLine#1, three instances of Machine#2 and two instances of Buffer#3. In general we allow the user to have more than one prime page and he can even let the same page be a multiple prime. Intuitively, the prime pages tell what should be included in the execution. It is, however, also possible, explicitly, to exclude certain pages. If we in the factory unit exclude Buffer#3, no instances will be created for this page, and Buf1 and Buf2 will be treated as if they were ordinary places. It is important to be able to include and exclude parts of a model – without having to change the model itself.

The CPN Palette applies an elaborated naming scheme for page instances: As an example consider "(5:Machine#2) Mach2@(2:AssemblyLine#1) Assembly2@(1:FactoryUnit#5)" which denotes the fifth instance of Machine#2, which is a subinstance of the transition Mach2 at the second instance of AssemblyLine#1, which in turn is a subinstance of the transition Assembly2 at the first instance of FactoryUnit#5. More details about naming schemes, prime pages and exclusion of pages can be found in [4-5].

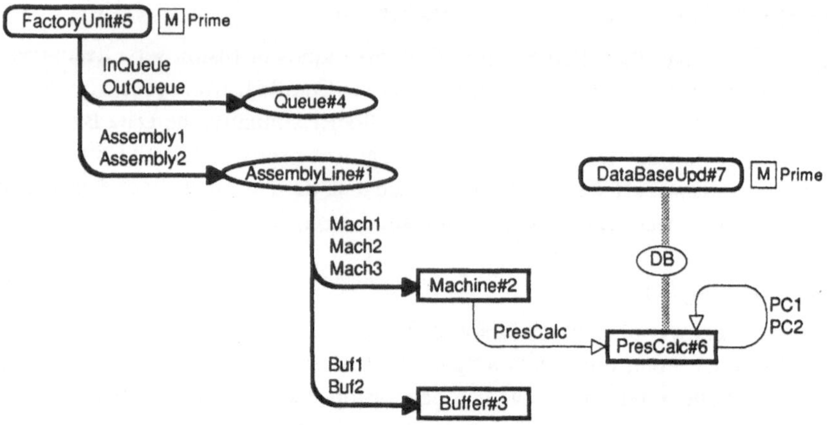

Figure 11: Page hierarchy for the factory unit

8. Example 5: Telephone System

This section presents a hierarchical version of the telephone system from [11]. It contains a single prime page Phone#51 (see Figure 13) and five substitution subpages (see Figure 13-14). In the original non-hierarchical model the place Engaged is defined to be complementary to the place Inactive (and all arcs surrounding Engaged are omitted). In the hierarchical version we update Engaged explicitly and to do this we use a global fusion set, called Engaged. The arc expression "1`x+1`y" denotes the multi-set which contains one x-token and one y-token.

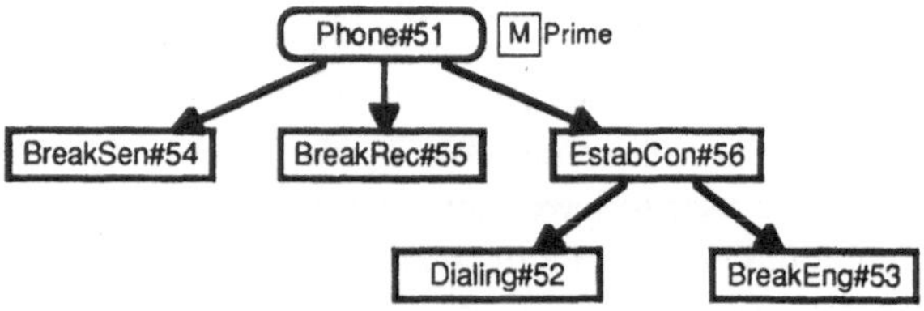

Figure 12: Page hierarchy for the telephone system

The arc expressions of the substitution transition BreakSen at Phone#51 gives a slightly less detailed description of the corresponding activity than the subpage BreakSen#54. The substitution transition doesn't show that the activity has two subactivities (which are executed after each other) and it doesn't show that Engaged is updated. However, BreakSen still gives the reader a very good idea about what the activity does. It describes the combined effect of the two subactivities on the markings of the socket places.

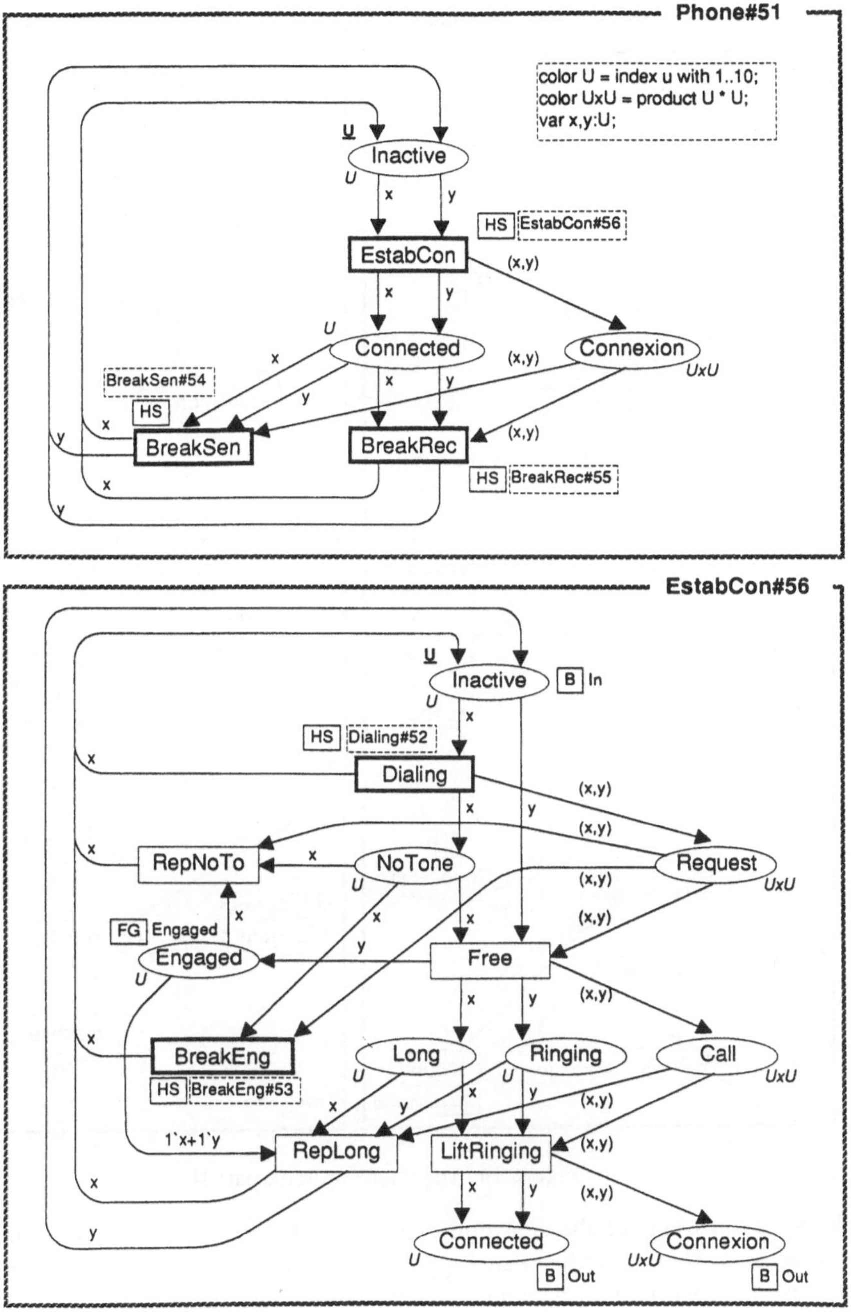

Figure 13: Telephone system, part I

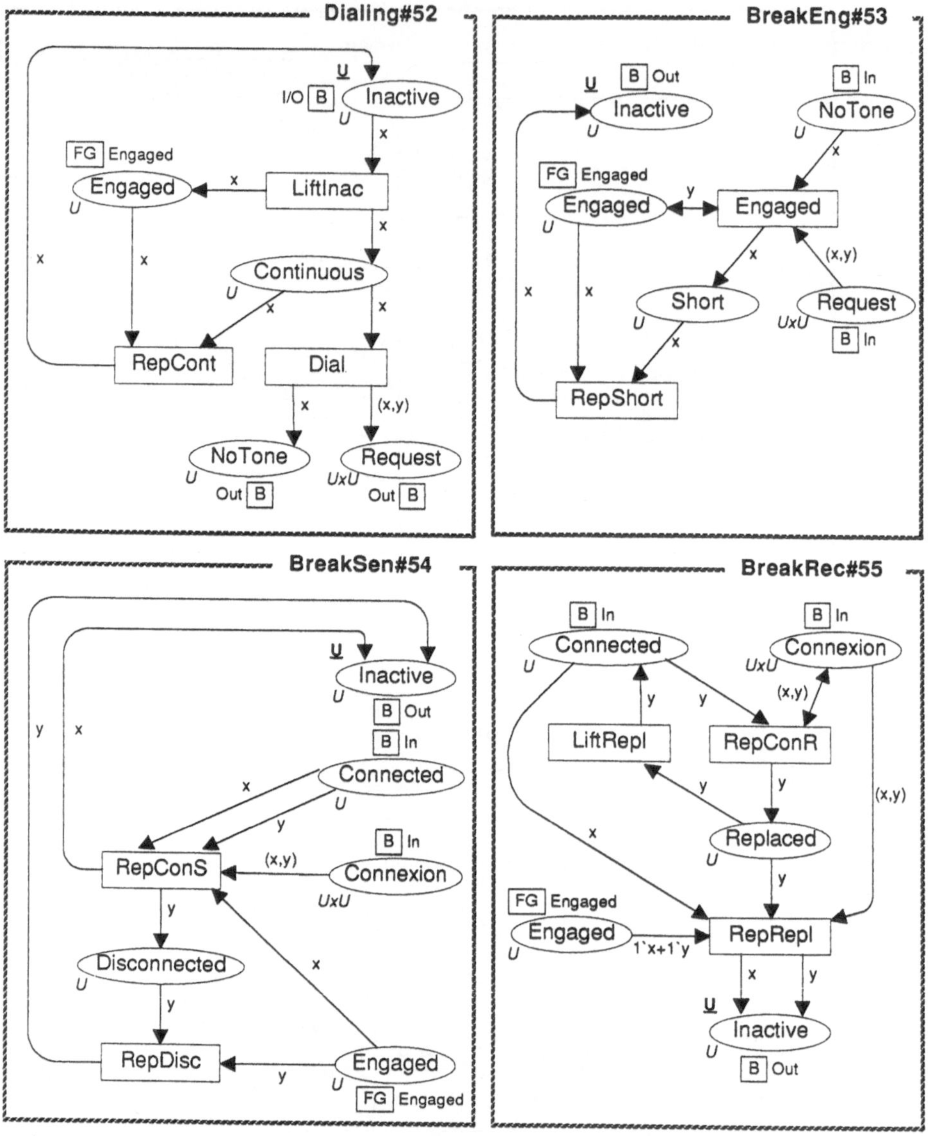

Figure 14: Telephone system, part II

The arc expressions of the substitution transition EstabCon at Phone#51 are used in another way. They describe what happens in the normal case, where a connexion is established. However, they do not say that the result of the subpage activities may be that no connexion is established (routing both the x and y token to Inactive instead of

Connected). It would not have been particularly difficult to describe this possibility in the arc expressions, but the analyst has – at Phone#51 – chosen to concentrate on the normal case.

For substitution transitions the arc expressions at the upper level never influence the execution – unless the corresponding subpage is excluded from the execution. However, for substitution places and invocation transitions this is not always true (cf. section 4 and 5).

9. Example 6: Multi-Token Ring Protocol

This section presents a hierarchical version of the net from [24], where a protocol called PLASMA is proposed. The protocol uses several tokens rings, switching and re-transmission to ensure the network service. The protocol fits between the LLC and MAC sublayers in the ISO standard for local area networks.

Our model contains a single prime page Prot#61 (see Figure 15) and seven subpages (reproduced in Figures 16-18). At execution time, we have eight page instances – one for each page in the net.

Each of the seven main components are here formally made into substitution nodes at different levels in the hierarchy. Due to their nature five of them are represented by substitution places with shared socket transitions – subpages #62–66. The last two are represented as substitution transitions, which share information (places) with the environment – subpages #67–68.

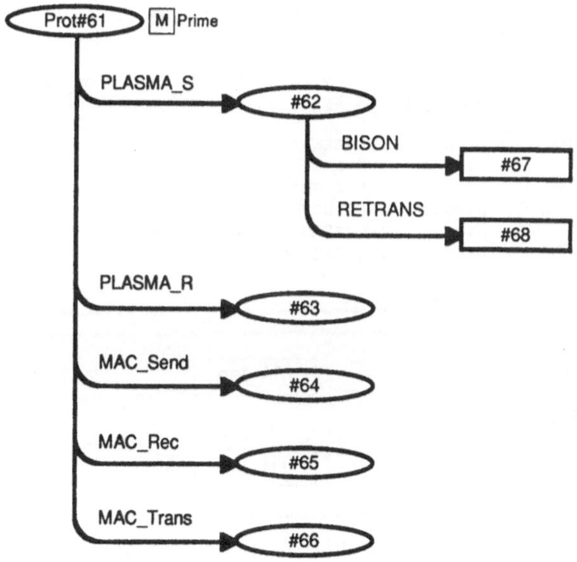

Figure 15: Page hierarchy for the multi-token ring protocol

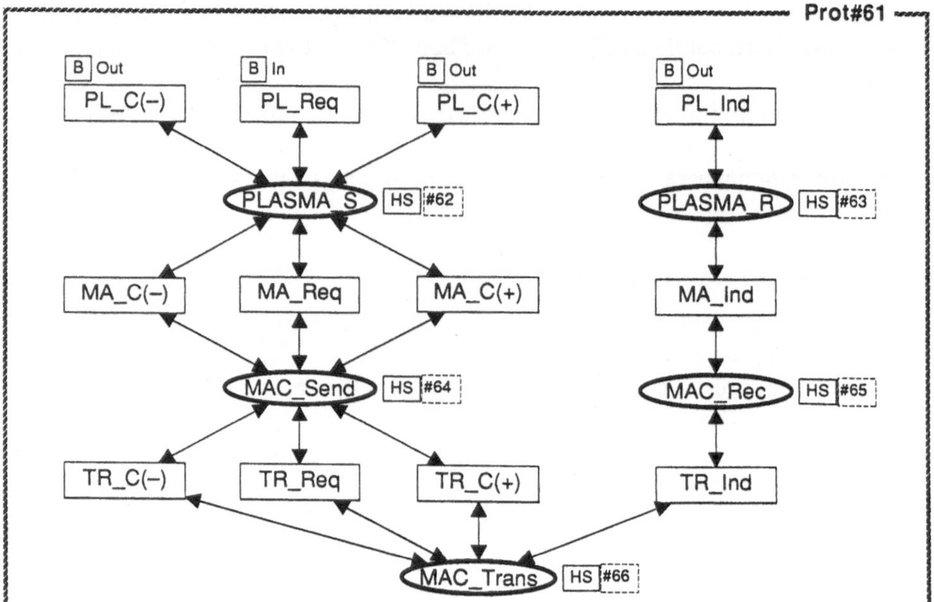


```
val    N     = 5;  (* Window size *)
val    ND    = 3;  (* No of recievers *)
val    NS    = 4;  (* No of senders *)
val    NR    = 2;  (* No of token rings *)

color  REC   = index D with 0..ND-1 declare ms;   var d     : REC;
color  SEND  = index S with 0..NS-1 declare ms;   var s     : SEND;
color  RING  = index R with 0..NR-1 declare ms;   var r,r'  : RING;
color  FRAME = int with 0..1024;                  var x     : FRAME;

color  LOWER = subset FRAME by fn (x:FRAME) => x<N declare ms;
color  UPPER = subset FRAME by fn (x:FRAME) => x>=N declare ms;

color  MESS  = product RING*SEND*REC*FRAME;
color  PAIR  = product SEND*REC;
color  RPAIR = product RING*SEND*REC declare mult;
color  FPAIR = product SEND*REC*FRAME declare mult;
color  RSEND = product RING*SEND declare mult;
color  RREC  = product RING*REC declare mult;
```

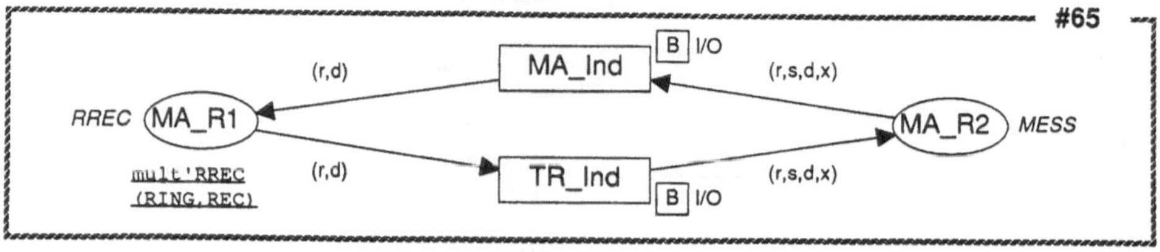

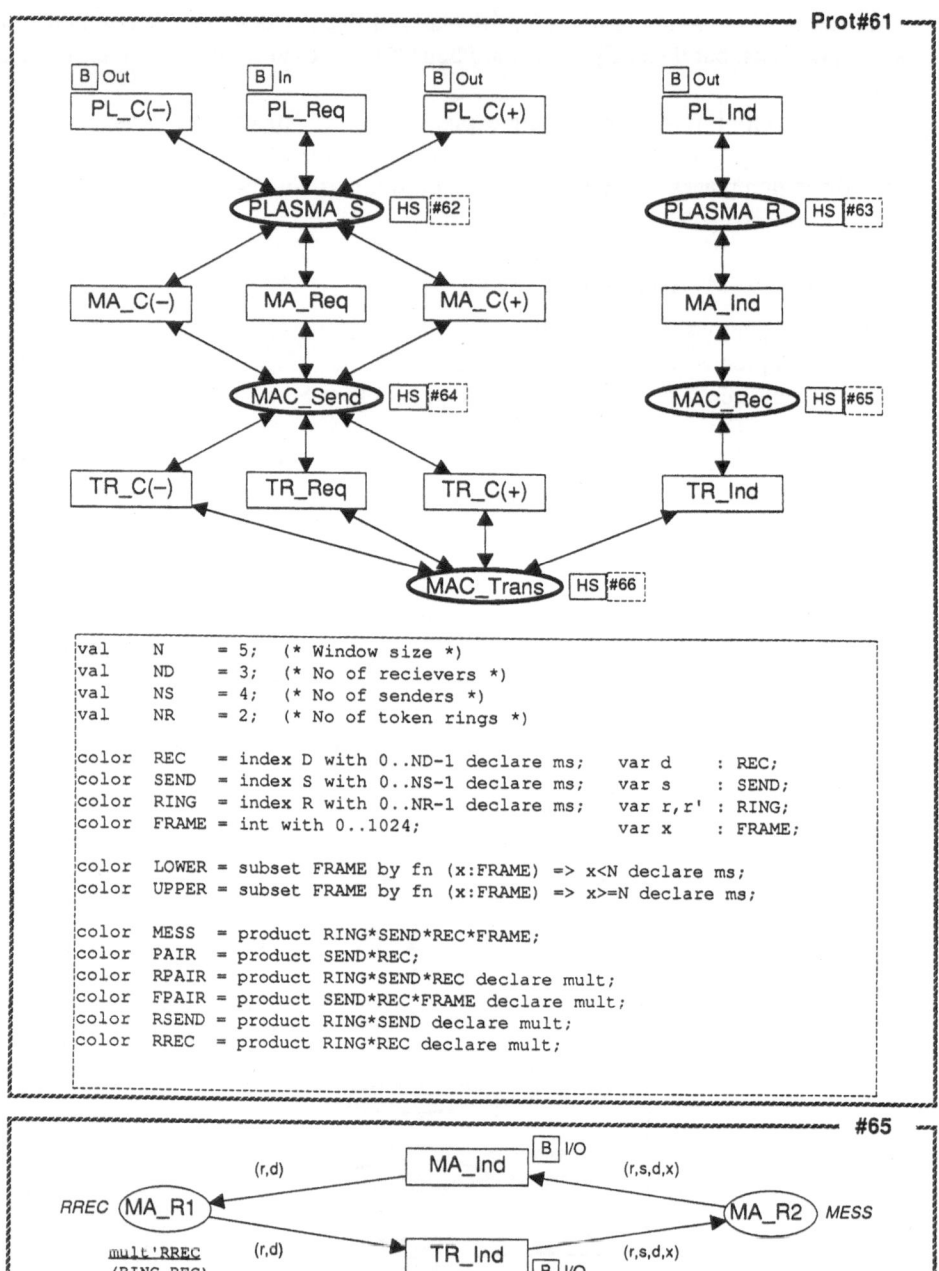

Figure 16: Multi-Token Ring Protocol, part I

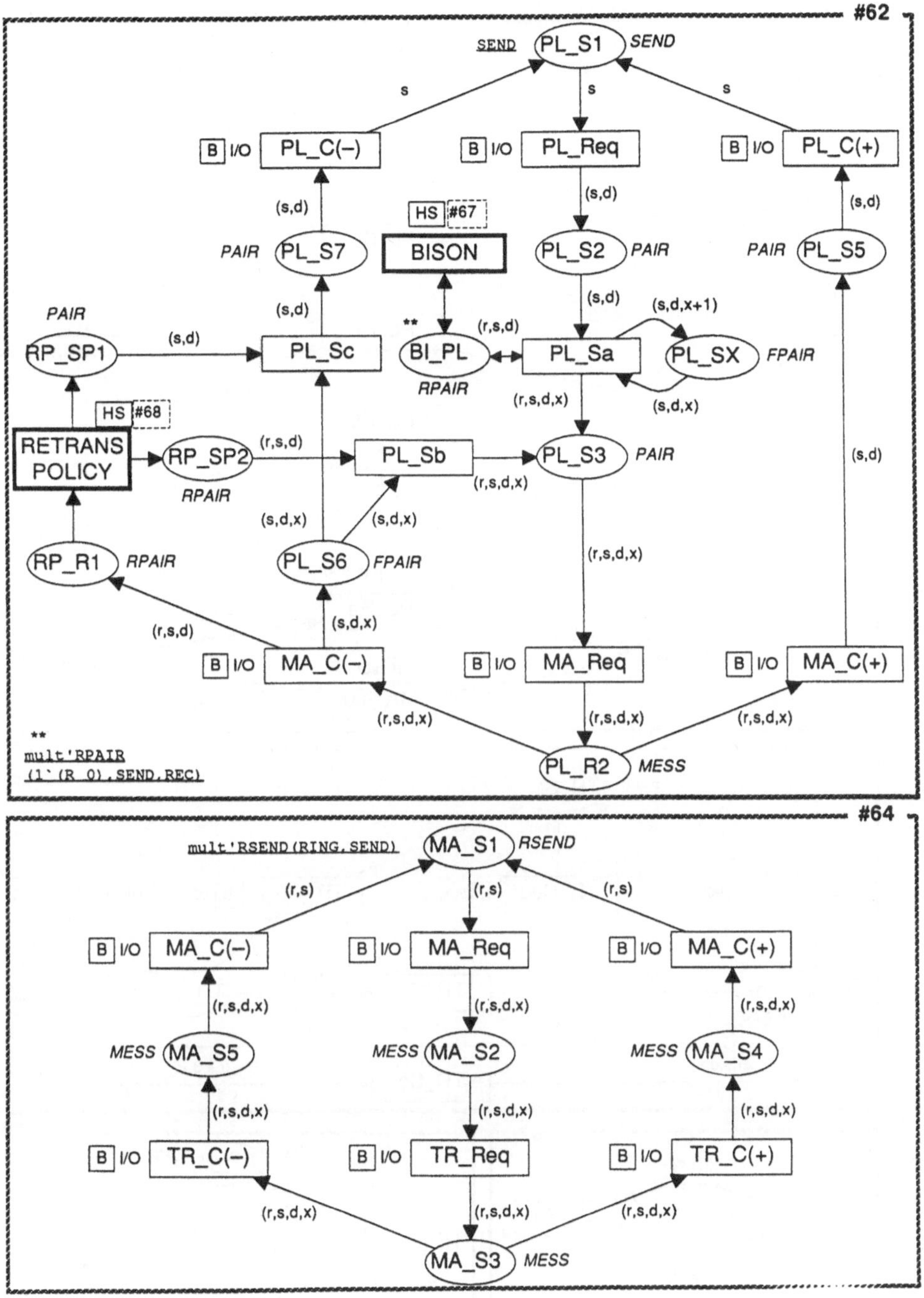

Figure 17: Multi-Token Ring Protocol, part II

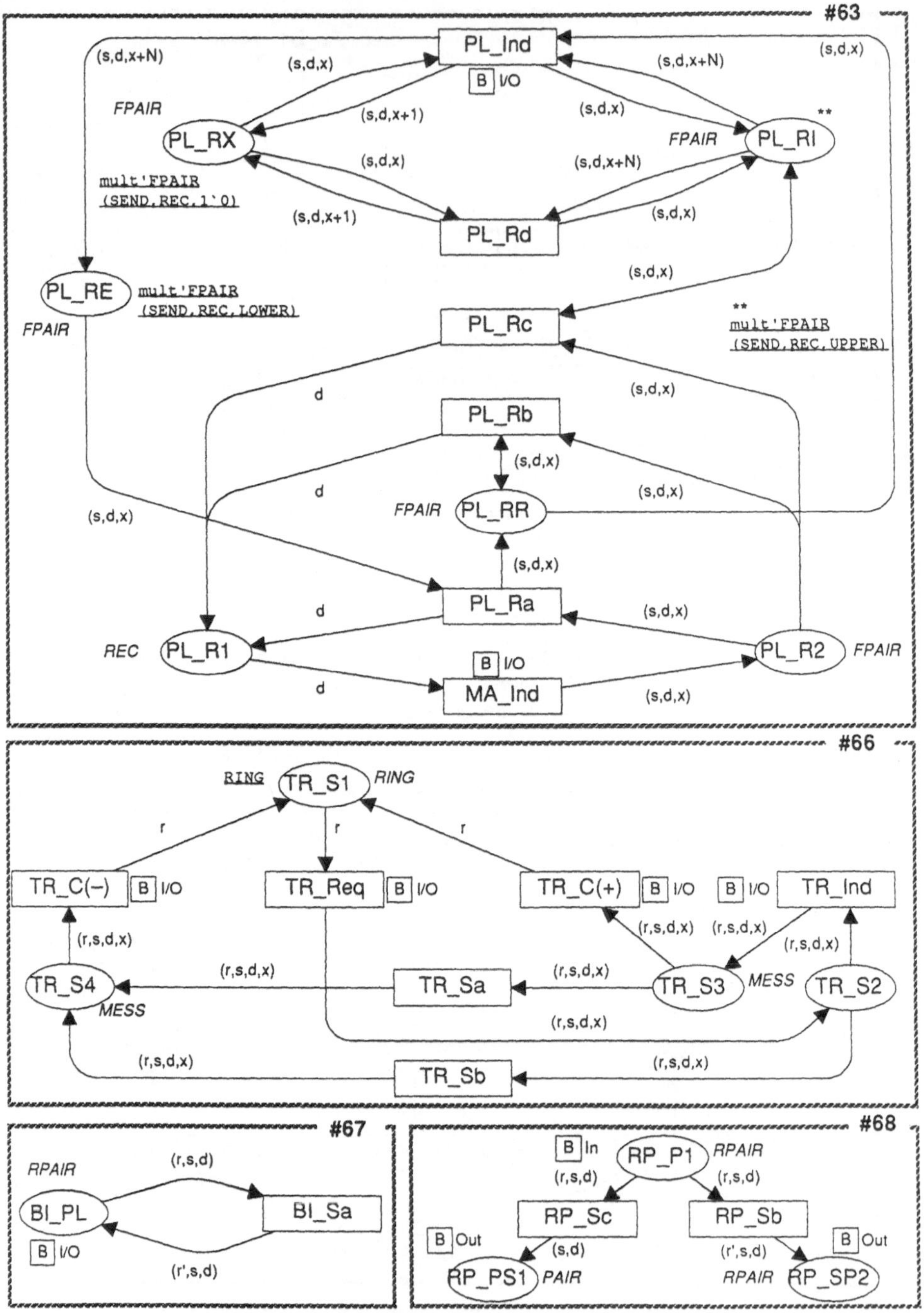

Figure 18: Multi-Token Ring Protocol, part III

We have subordinated the BISON and the Retransmission Policy process to the PLASMA_Send process, since it is the only one that uses them. Even though they are small, it is nice to have them on separate pages. This allows us to hide and experiment with the strategies for these error detection and ring switching processes separately.

We have specified port nodes for the top page Prot#61, even though it is not used as a subpage for the moment. However, this indicates, that we later might use it as a subpage for a substitution place in an even larger network system.

The definition of constants and colour-sets is given on page Prot#61 as well. Note, that we in the net only used functions, which are *predefined* in the CPN Editor. We have used the multi-set multiplication operator – mult'<Color set name> - to formulate the initial markings for product color sets in a compact form.

10. Conclusion

We have explored hierarchies in a specific behavioural modelling language: the CP-nets. This led us to the formulation of hierarchical CP-nets, which contains ordinary CP-nets as a subclass. We have introduced the notions of pages and page instances together with five elaborated concepts for hierarchical relationships:

- substitution transitions,
- substitution places,
- invocation transitions,
- fusion sets for places,
- fusion sets for transitions.

We do not claim that our five hierarchy constructs are the only sensible ones for CP-nets. As we get more experience with the modelling of large systems, the hierarchy constructs will probably be modified and perhaps augmented by new kinds. However, based on our present experience with a number of very different and rather complex examples, we believe that the current hierarchy constructs will form the base for a useful and theoretically sound extension of CP-nets.

Acknowledgements

Some of the ideas in our hierarchy concepts originates from the coarsen operation described in [19].

The participation by Kurt Jensen in the CPN Palette project is partially supported by a grant from the Danish National Science Research Council.

Reference List

[1] K. Albrect, K. Jensen & R.M. Shapiro: **CPN Palette. A Tool Package Supporting the Use of Coloured Petri Nets.** The Petri Net Newsletter, April 1989.

[2] K. Jensen: **CPN ML.** Specification paper for the CPN Palette – Part 1. Meta Software Corporation, Cambridge, Massachusetts, USA, 1989.

[3] P. Huber: **Hierarchies in Coloured Petri Nets.** Specification paper for the CPN Palette – Part 2. Meta Software Corporation, Cambridge, Massachusetts, USA, 1989.

[4] K. Jensen and S. Christensen: **CPN Editor.** Specification paper for the CPN Palette – Part 3. Meta Software Corporation, Cambridge, Massachusetts, USA, 1989.

[5] K. Jensen and S. Christensen: **CPN Simulator.** Specification paper for the CPN Palette – Part 4. Meta Software Corporation, Cambridge, Massachusetts, USA, 1989.

[6] W. Reisig: **Petri Nets in Software Engineering.** In: W. Brauer, W. Reisig and G. Rozenberg (eds.): Petri Nets: Applications and Relationships to Other Models of Concurrency,
Advances in Petri Nets 1986-Part II, Lecture Notes in Computer Science, vol. 255, Springer-Verlag 1987, 207-247.

[7] H. Oberquelle: **Human-machine Interaction and Role/Function/Action Nets.** In: W. Brauer, W. Reisig and G. Rozenberg (eds.): Petri Nets: Applications and Relationships to Other Models of Concurrency, Advances in Petri Nets 1986-Part II, Lecture Notes in Computer Science, vol. 255, Springer-Verlag 1987, 207-247.

[8] H.J. Genrich and K. Lautenbach: **System Modelling with High-level Petri Nets** Theoretical Computer Science 13. 1981, 109-136.

[9] H.J. Genrich: **Predicate/Transition Nets** In: W. Brauer, W. Reisig and G. Rozenberg (eds.): Petri Nets: Central Models and Their Properties, Advances in Petri Nets 1986-Part I, Lecture Notes in Computer Science, vol. 254, Springer-Verlag 1987, 207-247.

[10] K. Jensen: **Coloured Petri Nets. A Way to Describe and Analyse Real World Systems – Without Drowning in Unnecessary Details.** Proceedings of the 5'th International Conference on Systems Engineering, Dayton 1987, New York: IEEE, 395-401.

[11] K. Jensen: **Coloured Petri Nets.** In: W. Brauer, W. Reisig and G. Rozenberg (eds.): Petri Nets: Central Models and Their Properties, Advances in Petri Nets 1986-Part I, Lecture Notes in Computer Science, vol. 254, Springer-Verlag 1987, 248-299.

[12] K. Jensen: **Informal Introduction to Coloured Petri Nets.** Chapter 1 of a three-volume book on CP-nets. The book will be published by Springer-Verlag in the series: EATCS Monographs on Theoretical Computer Science.

[13] D.A. Marca and C.L. McGowan: **SADT.** McGraw-Hill, New York, 1988.

[14] E. Yourdon: **Managing the System Life Cycle.** Yourdon Press, 1982.

[15] D. Harel: **Statecharts: A Visual Formalism for Complex Systems.** In: Science of Computer Programming, Vol. 8, North-Holland 1987, 231-274.

[16] H.J. Genrich, K. Lautenbach and P.S. Thiagarajan: **Elements of General Net Theory.** In: G. Goos and J. Hartmanis (eds.): Net Theory and Applications, Lecture Notes in Computer Science, vol. 84, Springer-Verlag 1980, 248-299.

[17] R.M. Shapiro and P. Hardt: **The Impact of Computer Technology. A Case Study: The Dairy Industry.** GMD Internal Report, ISF-76-11, 1976.

[18] R.R Razouk and M.T. Rose: **Verifying Partial Correctness of Concurrent Software using Contour/Transition Nets.** In: Proceedings of the Hawaii International Conference on System Sciences, 1986.

[19] H.J. Genrich and R.M. Shapiro: **A Diagram Editor for Line Drawing with Inscriptions.** Proceedings of the 3'rd European Workshop on Applications and Theory of Petri Nets, Varenna, Italy, 1982, 193-212.

[20] **Network Tool Net: System Analysis and Simulation with Petri-Nets.** PSI Gesellschaft für Prozesssteuerungs- und Informationssysteme, Berlin, undated, 23 pages.

[21] H. Oberquelle: **Some Concepts for Studying Flow and Modification of Actors and Objects in High-level Nets.** Proceedings of the 3'rd European Workshop on Applications and Theory of Petri Nets, Varenna, Italy, 1982, 343-363.

[22] A. Kiehn: **A Structuring Mechanism for Petri Nets.** Institut für Informatik der Technischen Universität München, 1988, 127 pages.

[23] K.M. van Hee, L.J. Somers, and M. Voorhoeve: **Executable Specifications for Distributed Information Systems.** In: E.D. Falkenberg and P. Lindgreen (eds.): Information System Concepts: An In-depth Analysis, North Holland, 1989, 139-156

[24] B. Cousin et. al.: **Validation of a Protocol Managing a Multi-token Ring Architecture.** Proceedings of the 9'th European Workshop on Applications and Theory of Petri Nets, Vol. II, Venice 1988.

[25] C. Girault, C. Chatelain and S. Haddad: **Specification and Properties of a Cache Coherence Protocol Model.** In: G. Rozenberg (ed.): Advanced in Petri Nets 1987, Lecture Notes of Computer Science, vol. 266, Springer-Verlag, 1987, 1-20.

[26] S. Haddad.: **Generalization of Reduction Theory to Coloured Nets.** Proceedings of the 9'th European Workshop on Applications and Theory of Petri Nets, Vol. II, Venice 1988.

[27] P. Huber, A.M. Jensen, L.O. Jepsen and K. Jensen: **Reachability Trees for High-level Petri Nets.** Theoretical Computer Science 45 (1986), 261-292.

[28] K. Jensen: **How to Find Invariants for Coloured Petri Nets.** In: J. Gruska, M. Chytill (eds.): Mathematical Foundations of Computer Science 1981, Lecture Notes in Computer Science vol. 118, Springer-Verlag 1981, 327-338.

Section D
Analysis by Means of Invariants

Invariant analysis allows logical properties of Petri nets to be investigated in a formal way—which is similar to the use of invariants in program verification. There are two dual classes of invariants. A place invariant characterises the conservation of a weighted set of tokens, while a transition invariant characterises a set of occurrence sequences having no effect, i.e. with identical start and end markings.

The main advantages of invariant analysis are the low computational complexity (in particular, compared to the method of reachability graphs described in Section E) and the easy parametrization with respect to system parameters (through the definition of different initial markings). The main drawbacks is the difficulty to automate the interpretation of invariants and the incompleteness (in the sense that it is usually only possible to calculate either necessary or sufficient conditions, for a given property).

It is desirable to be able to make an automatic computation of invariants. This can be done by solving a matrix equation, and this provides a bridge between convex geometry, linear programming and Petri net theory. The result is usually given in terms of a generating family of invariants. However, it is also important to have suitable proof techniques, and this is closely related to the interpretation—i.e. to the extraction of information (about the behaviour of the modelled system). By means of invariants it is possible to investigate many different kinds of system properties, e.g. deadlocks, mutual exclusion and marking bounds. At this point it should be remembered that the main differences between PrT-nets and CP-nets are in the methods to compute and interpret invariants.

The first paper in this section is a tutorial, addressing aspects of both the computation and the use of invariants—for low and high-level nets. The other two papers deal with the computation of invariants. The second paper defines a number of subclasses of high-level nets, by constraining the functions which may be used in the arc inscriptions. In this way it becomes possible to obtain efficient algorithms for computation of invariants. The third paper discusses the computation of symbolic invariants by means of generalized inverses (of the linear functions in the arc inscriptions). This is done by a generalization of the well known Gauss-elimination technique.

Invariant analysis has been considered since the early days of high-level nets. Algorithms for computing invariants of PrT-nets were designed and implemented by H. Mevissen, R. Kujansuu, J. Lindquist, G. Memmi and J. Vautherin. The first method to find invariants for CP-nets was developed by K. Jensen. It uses a set of heuristic transformation rules to make an interactive simplification of the equation system, without changing the set of invariants. The automatic computation of invariants has followed two different tracks, represented by the second and third paper of this section. The former approach has primarily been developed by the group of C. Girault, S. Haddad and J.M. Couvreur, while the latter, primarily, has been developed by J. Martínez and M. Silva.

8.
Analysing Nets by the Invariant Method[1]

G. Memmi and J. Vautherin

W. Brauer, W. Reisig and G. Rozenberg (eds.): Petri nets: central models and their properties.
Advances in Petri nets 1986, part I.
Lecture Notes in Computer Science, vol. 254. Springer, Berlin Heidelberg New York 1987, pp. 300-336

CONTENTS

I. INTRODUCTION

Methods for analysing P/T-systems can be roughly divided into several categories : study of the reachability set, transformation by homomorphism, and invariants. Each of these methods have advantages and disadvantages.

Here we will focus on the methods based upon the computation and the analysis of invariants. The invariant method [Lautenbach... 74], [Jensen 81], [Memmi 83] is well known for at least two advantages : on the one hand, analysis can be performed on local subnets while ignoring how the whole system behaves ; on the other hand, this method still holds, even when the P/T-system is enriched with parameters - i.e. even when considering Pr/T-systems or other high level nets.

In this paper, three theoretical results about linear invariants and semi-flows computation are presented (we call *semi-flows* the weighting vectors f - solutions of the linear system $f^T.C = 0$ - which generate the linear invariants) :

- The first one deals with the computation of a smallest set of generators of the positive semi-flows of a P/T-system. An algorithm - the Farkas algorithm - is presented and proved.

- The second result is about the computation of semi-flows of unary Pr/T-systems : it states that the set of semi-flows of such a system can be generated by a finite number of integers vectors which are computable

[1] A part of this work was granted by the Esprit project FOR-ME-TOO.

directly from the incidence matrix of the system ; thus without unfolding and independently of the colors which are not effectively present in the incidence matrix.

- The third one shows how to characterize all the minimal supports of the set of positive semi-flows while the net contains some parameters but with special connection constraints.

These results are illustrated by the modelling and analysis of a parallel system - namely the "general channel" of the L language. Other important issues of this study are

- to show how different kinds of nets (P/T-systems, unary Pr/T-systems, fifo nets) can be used at different steps of an analysis, depending on the aspects of the system that we want to study.

- to outline an analysis methodology of nets based on the proof of invariants. For that purpose, we introduce the notion of home space (a set of markings that we are always sure to reach after any evolution of the system).

- to introduce and show the interest of a combined use of nets and algebraic specifications.

The paper is organized as follows :

The general channel is described in section II and modelled by a P/T-system in section III, omitting some details voluntarily. In this first model, we describe the Farkas algorithm (III.1) and the notion of home space (III.2). We show how to use these results in order to obtain informations on the length of some data and the behaviour of the system. We show that the P/T-system can reach a deadlock.

This is due to a lack of precision in the description of the data structures ; thus we make the model evolve from a P/T-system to a unary Pr/T-system [Vautherin... 85] where some aspects of the system can be detailed. Then the invariants also bring more information (IV.2). In this model, we describe the computation of a set of generators of semi-flows, and we characterize all the minimal supports of the set of positive semi-flows (IV.3 and IV.4).

Then we change the model again and choose the fifo nets model [Memmi... 85] (see also the lecture of G. Roucairol in this course) for taking into account the fifo mechanism of the main data structure of the system (V.I). Here, by stepwise refinements of home spaces, also with the help of non linear invariants, we prove that the initial marking is a home state, and then that the system is live (V.2).

Next we try to find the most general data structure such that the system remains live. at the same time, we want this data structure to impose the least synchronization constraints, so that the whole system runs as fast as possible. This structure is specified by an algebraic abstract data type associated with a Pr/T-system [Vautherin 85] (VI.1). Here again the invariant method is transposed : the liveness question is studied for different data structure with the same sketch of proof (VI.2).

We must warn the reader that we had to homogenize some notations between the different nets models. However, we tried to remain unchanged the most standard ones though it may induce some slight ambiguities (such as C for the incidence matrix and for a set of colors). Our belief is that they are context-sensitive.

II. THE GENERAL CHANNEL EXAMPLE

The general channel is a central element in the L language [Boussinot 83] for processes to communicate. It has been analysed first in [Memmi 81]. Mainly it is a bounded fifo ; one can put into or get messages from it. Its functioning is formally described in the Boussinot-Kahn model [Boussinot 81] ; here, we simplify the original description a little and consider the schema of figure 1.

Sequential processes (represented by rectangles) communicate via infinite fifo channels :

- *put (c, e)* put into the channel c the value e,

- *get (c, v)* get from the channel c some value assigned to the variable v. If c is empty, the process is locked until some process puts a value into c. Only one process can get from a given channel c.

The general channel is a bounded fifo Fileval managed by a sequential process G. Its length is max. ip producers can put information into it ; ic consumers can get information from it.

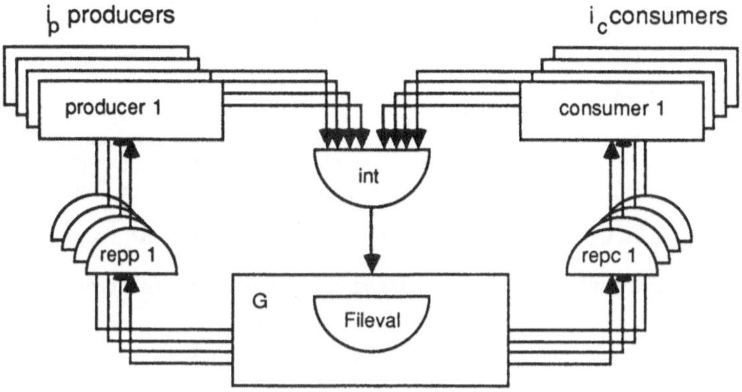

Figure 1

Before putting some value e into Fileval, a producer pr_k must send its identity in order to get from the channel $repp_k$ an acknowledgement from G. Then its code is :

loop put (int, pr_k) ; get ($repp_k$, v) ; put (int, e) end loop

Let us point out that the variable v is used in order to store the acknowledgment from G which is only a way to synchronize the two processes. A consumer c_k will ask G some value and will store it in z by :

loop put (int, c_k) ; get ($repc_k$, z) end loop

G uses auxiliary queues :

- Depos, of length ip, contains the identities of producers which sent a request while Fileval was full.

- Prise, of length ic, contains the identities of consumers which sent a request while Fileval was empty.

Moreover, G controls Fileval via a variable Per which indicates the number of acknowlegments G can deliver. In its loop, G begins to get a message in its input channel int :

In case of a request from producer pr_k, G executes

If Per = 0 then put (Depos, pr_k) else {Per ← Per - 1; put ($repp_k$, ok)} end if.

In case of a request from consumer c_k, G executes :

If Fileval is empty then put (Prise, c_k)
else {get (Fileval, x) ; put ($repp_k$, x) ;
if Depos is empty then Per ← Per + 1
else {get (Depos, pr_h) ; put ($repp_h$, ok)}
end if
end if.

In case of a value e sent by a producer, G executes :

If Prise is empty then put (Fileval, e)

else {get (Prise, c_h) ; put ($repc_h$, e) ; Per ← Per + 1}

endif.

At the initial state, processes are idle ; all the fifos are empty ; the value of Per is max.

III. THE P/T-SYSTEM MODEL

With this first net, int is unstructured and modelled by three places intp, intc, inte receiving the three different types of messages which can be sent to int. The channels $repp_i$ are folded in one place repp, the channels $repc_i$ in the place repc. All the messages are merged and modelled by a mere token. The initial marking m° is given by :

$$m°(p1) = ip,\ \ m°(p8) = ic,\ \ m°(p3) = 1,\ \ m°(Per) = max,\ \ m°(p) = 0 \ \text{ otherwise.}$$

In the figure 2, we have duplicated the place p3, and used inhibitor arcs (between Fileval and t7 for example) for readability.

The places p1 and p2 represents the states of producers ; p8 and p9 the states of consumers ; p3 and p7 the states of G :

When m(p3) = 1, G is idle and waits for a message in int.
When m(p4) = 1, G treats a request from a producer and tests Per.
When m(p5) = 1, G treats a request from a consumer and tests Fileval.
When m(p6) = 1, G treats a value and tests Prise.
When m(p7) = 1, G sent a value into repc and tests Depos.

A first approach to analyze the net should be to construct the reachability set for ip = ic = max = 1 and to invoke some induction step or to developp a kind of high-level reachability tree like in [Hubert 85] ; but it is still open to investigation. Thus, let us apply the invariant method. First, we have to find a family of semi-flows to generate our invariants.

It is well known that the set of semi-flows, over Z or Q, of a given P/T-system admits a basis which can be computed by using Hermite's or Gauss algorithms for instance. In the following we are interested in the computation of of *positive* semi-flows. A reason for doing that is given below ; others can be found in [Memmi 83].

III.1. The Farkas algorithm

This algorithm which we delicate to Farkas has been rediscovered at the same time by several authors [Martinez 82], [Alaiwan... 85], [Memmi 83], with slight variations. In fact, a first attempt to solve a system of linear equations with solutions in the natural numbers is found in the Farkas works [Farkas 02]. In [Alaiwan 85] a program of this algorithm is presented. Here, we give only the proof that this algorithm provides a smallest set of generator of the set of the semi-flows of a given system over Q^+, i.e. a set of generators of minimal support (see [Memmi... 80]).

Let us recall that the support of a vector v, denoted by ||v||, is defined by :

$$||v|| = \{i \mid v_i \neq 0\}.$$

As usual, S and T denote the sets of places and transitions of the P/T-system. We take n = | S | ; then I_n denotes the (n×n) identity matrix.

For each k, e_i in Z^k is defined by $e_i(i) = 1$, $e_i(j) = 0$ otherwise.

C(.,i) denotes the column indexed by i of the incidence matrix C.

Figure 1

The P/T-system model

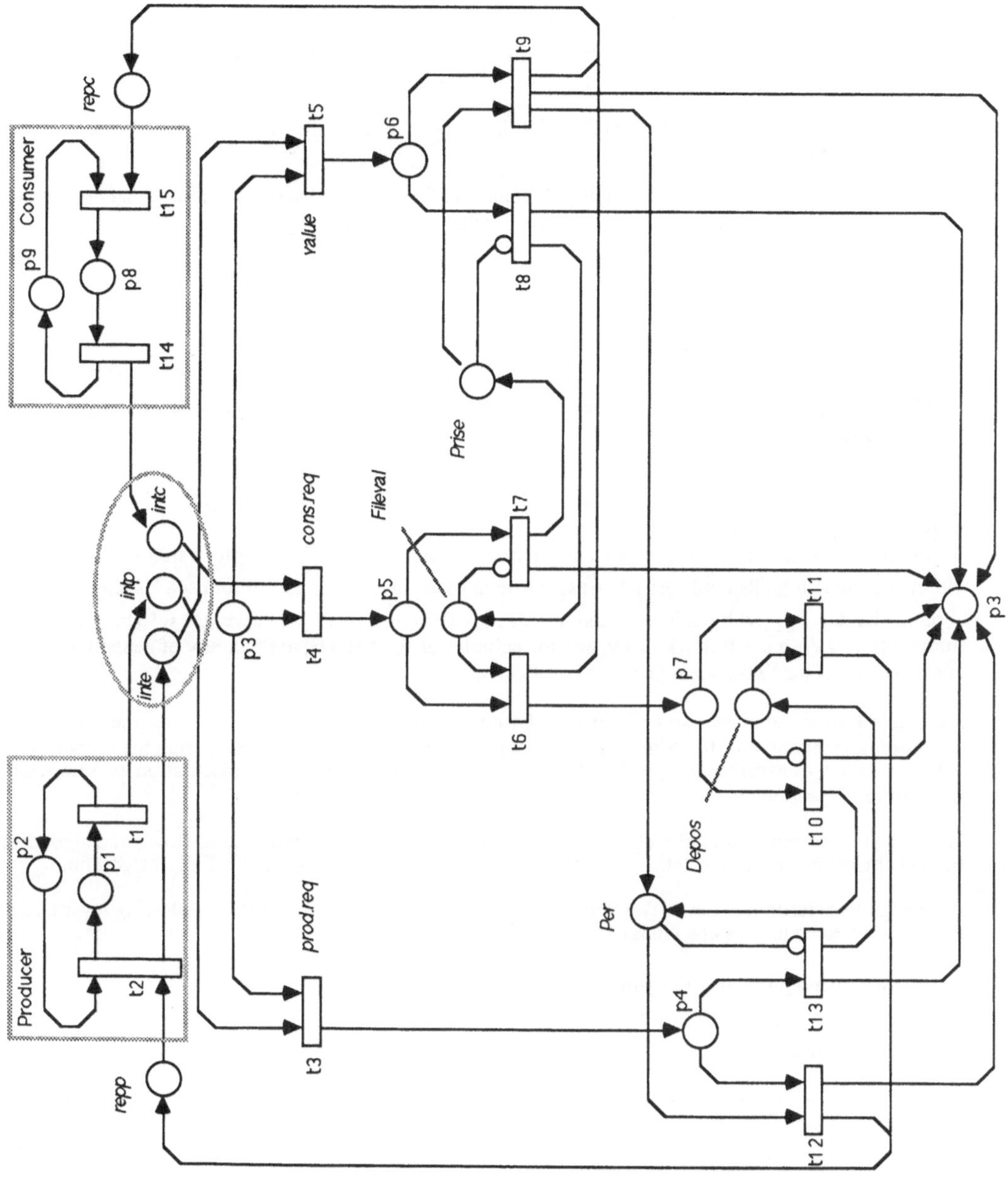

Algorithm :

Initialisation : Let $F_0^* = I_n$; $m_0 = |T|$, $T_0 = \emptyset$

Step i : choose a transition t_i not in T_{i-1}

Let $x = (F_{i-1}^*)^T.C(.,i)$

$E = \{e_k \in Z^{m_{i-1}} \mid x_k = 0\}$

$P = \{i \mid x_i > 0\}$, $Q = \{j \mid x_j < 0\}$

$F = \{f_{ij} \in N^{m_{i-1}} \mid \forall i \in P, \forall j \in Q, f_{ij}(i) = -x_j, f_{ij}(j) = x_i, f_{ij}(k) = 0 \text{ otherwise}\}$

$F_i = F_{i-1}^*.[F \mid E]$,

where $[F \mid E]$ is the matrix obtained by the justaposition of the vectors of F with the vectors of E considered as columns.

F_i^* is obtained from F_i by canceling all columns of non minimal support and such that there is not two columns with the same support. m_i is the number of columns in F_i.

$T_i = T_{i-1} \cup \{t_i\}$.

Termination : when $(F_k^*)^T.C = 0$.

Theorem 1. *This algorithm does terminate at some step $k < n_0$ where F_k is a smallest set of generators of the set of semi-flows over Q^+.*

Proof. By induction on step i: at this step, F_i^* is a smallest set of generators of the set of semi-flows of the system generated by the treated transitions.
i) At step 0, each place p_j is isolated, so e_j is a semi-flow, and the result is straigtforward.
ii) Let us assume it holds until step i-1. At step i we have to solve $u^T.C(.,k) = 0$ for each k such that t_k is in T_i ; in particular, $u = F_{i-1}^*.w$ by induction and, $u^T.C(.,i) = 0$. With $u = F_{i-1}^*.w$, we get $w^T.(F_{i-1}^*)^T.C(.,i) = 0$. But $w = [F \mid E].v$, so $u = F_{i-1}^*.[F \mid E].v$, and the columns of $F_{i-1}^*.[F \mid E]$ are a family of generators for the set of semi-flows of the P/T-system generated by T_i. []

This algorithm is complex ; some heuristics based upon the choice one can make on the transition t_i are described in [Memmi 83]. In [Martinez... 82], a proposition is made for not comparing the supports of every columns when transforming F_i in F_i^*. In [Graubmann 85] an interesting calculation of semi-flows by composition is presented.

When the P/T-system is a marked graph, a circuit can be associated with any semi-flow ; and algorithms for finding circuits in a graph [Mateti... 76] are in many cases more efficient than the Farkas algorithm.

Applying this algorithm to our P/T-system we get six semi-flows generating the six following invariants : three of them are related to the three kinds of processes :

IPP : $\forall m \in [m^\circ\rangle$, $m(p1) + m(p2) = ip$

IPC : $\forall m \in [m^\circ\rangle$, $m(p8) + m(p9) = ic$

IPG : $\forall m \in [m^\circ\rangle$, $\Sigma_{i=3,\dots,7}\, m(p_i) = 1$

while the three others are related to the three kinds of messages :

IP : $\forall m\in [m°>$, m(p1) + m(intp) + m(p4) + m(Depos) + m(repp) = ip

IC : $\forall m\in [m°>$, m(p8) + m(intc) + m(p5) + m(Prise) + m(repc) = ic

IV : $\forall m\in [m°>$, m(repp) + m(inte) + m(p6) + m(Fileval) + m(p7) + m(Per) = max

We know ([Memmi... 80]) that all linear invariants with positive coefficients can be generated with these six ones: they contain all the information that any linear invariant can bring.

With IP, IC, IV, we already can conclude that all the channels are bounded with :

$$m(repp) \leq ip \; ; \; m(repc) \leq ic \; ; \; m(int) \leq max + ic + ip.$$

Moreover, from IV we directly have : m(Fileval) ≤ max for any behaviour of the net. Thus, when a producer sends a value for Fileval we are sure that there exists a place for it in Fileval : Per controls Fileval efficiently.

In fact, we could not conclude such properties so straightforwardly if coefficients of the semi-flows were not positive. This justifies to investigating the set of positive semi-flows.

To go further in our study, we need to introduce the notion of home space.

III.2. The notion of home space

Home states are well known in protocol analysis ; the notion of home space generalizes this idea. A home space is a set of markings such that after any sequence of transistions there always exists a possibility to reach one element of this set. This notion is closely related to the notion of liveness.

Definition 1. *$E \in N^S$ is a <u>home space</u> if and only if ,*

$\forall m\in [m°>, \exists e\in E \cap [m>$

If $E = \{m\}$, m is said to be a home state.

The equivalence of the three following properties can easily be stated:

i) The initial marking m° is a home state.
ii) Each marking m of [m°> is a home state.
iii) The reachability graph is strongly connectèd.

The set of markings at which a given transition t is enabled is a home space if and only if t is live. But, if there exists a home space for a given P/T-system, this one is not necessarily live. Let us point out that N^S is a home space for any marked P/T-system.

If m is a home state then each transition enabled at m is live. Our method of analysis will consist in finding some home space and trying to reach a home state by successive refinements.

In our example, let us prove that $E3 = \{m\in N^S \mid m(p3) = 1\}$ is a home space. Let M be a marking of [m°>. From IPG, there exists a token among p3 to p7 :

- if M(p4) = 1, either t12 or t13 is enabled, then we reach a marking of E3
- if M(p7) = 1, either t10 or t11 is enabled, then we reach E3
- if M(p5) = 1, either t6 is enabled then we reach m' with m'(p7) = 1 or t7 is enabled and we reach E3.
- if M(p6) = 1, either t8 or t9 is enabled and allows to reach E3.

Thus E3 is a home space. It means that the process G can always return to its idle state (p3) after any occurrence sequence of the system.

Moreover, we have proved that we always can reach E3 without t1, t2 or t14 occurring (and these transitions are the only ones that put messages in int). This means that we can easily refine E3 by :

$$E3int = \{m \in E3 \mid m(intp) = m(intc) = m(inte) = 0\}.$$

E3int is a home space, meaning that G can always return to its idle state while int is empty.

Even though E3int exists, there exists a deadlock for this P/T-system : the producers send max values in inte, then they are all put in Depos ; the ic consumers are then put in Prise. At this state, the sequence : t5, t9, t15, t14, t4, t5 may occur max times (Fileval is always empty) ; and then the P/T-system is dead.

Let us point out that the dead markings are necessarily in E3int.

We can refine E3int again by :

$$E3R = \{m \in E3int \mid m(repp) = m(repc) = 0\}.$$

Indeed, let us consider m∈E3int with m(repp) = xp, m(repc) = xc. From IP and IPP, we have

$$m(p2) = m(repp) + m(Depos) + m(p4) + m(intp).$$

So m(p2) > xp and t2 may occur xp times yielding m' with m'(inte) = xp. Then the sequences t5, t8 or t5, t9 may occur xp times reaching m'' of E3int m''(repp) = 0 and m''(repc) > xc. From IC and IPC, we have :

$$m''(p9) = m''(repc) + m''(Prise) + m''(intc) + m''(p5).$$

So t15 may occur as many times as necessary to empty repc. Hence we reach a marking of E3R.

Thus a message sent by G can always be consumed by some process but we are not at all sure that a message sent for pr_i has been be consumed by itself.

IV. COLOURED P/T-SYSTEMS AND UNARY Pr/T-SYSTEMS

This analysis can be made more precise by distinguishing the processes from each other. For this purpose, we use a P/T-system with individual tokens. More precisely, we use a system in a particular class of coloured P/T-systems (or coloured Petri nets [Jensen 81]) : unary Predicate/Transition-systems ([Vautherin... 85]).

IV.1. Definitions

In a P/T-system with individual tokens, the marking of a place is a multiset instead of a positive integer. Before recalling the definitions of coloured P/T-systems, we precise some notations about multisets and neighbouring concepts.

In the following, K denotes either N or Z or Q^+ or Q. Let A be any set, we denote by K^A the set of functions from A into K provided with the usual "pointwise" operations :

$$(w + w')(x) = w(x) + w'(x), \qquad (n.w)(x) = n.w(x) \qquad \forall x \in A,$$

for every w and w' in K^A and every n in K. We write δ for the null function of K^A (i.e. $\delta(x) = 0$ for every x in A). The set K^A is ordered by defining $w \geq w'$ iff

$$\forall x \in A, \qquad w(x) \geq w'(x).$$

For each x in A, we write e_x the element of K^A such that $e_x(x) = 1$ and $e_x(y) = 0$ when $y \neq x$.

We denote by $K^{(A)}$ the subset of K^A containing the functions whose support is finite (remember that the support of w is defined by $\| w \| = \{x \mid w(x) \neq 0\}$). An element of $N^{(A)}$ is called a *multiset* over A. Obviously we have:

$$w = \Sigma_{x \in A}\, w(x).e_x$$

for every w in $K^{(A)}$. Notice that this summation is well defined because there is only a finite number of non null terms.

When no confusion is possible, we only write <x> for e_x. So, for instance, with A = {a,b}, 2<a> + <b> denotes the multiset w over A such that w(a) = 2 and w(b) = 1.

Now let us consider a finite set X of variables. We can represent by multisets over $A \cup X$ some functions into $N^{(A)}$: let w be a multiset over $A \cup X$ and let A^X be the set of all variables assignment $\sigma : X \rightarrow A$; w determines a function [[w]] from A^X into $N^{(A)}$ by :

$$[[w]](\sigma) = \Sigma_{a \in A}\, w(a).<a> + \Sigma_{x \in X}\, w(x).<\sigma(x)>$$

For instance, with X = {x} and $\sigma : x \rightarrow a$, [[a + b + x]](σ) = 2<a> + <b>.

The set of functions from A^X into $N^{(A)}$ that can be represented in this way by elements of $N^{(A \cup X)}$ is denoted by $[[N^{(A \cup X)}]]$.

Finally, for every w in $K^{(A)}$, we write | w | for $\Sigma_{a \in A}\, w(x)$. When w is a multiset, | w | is the length of w. Notice that for all σ in A^X, $| [[w]](\sigma) | = | w |$.

Definition 2. *A* <u>*coloured P/T-system*</u> *is a 5-tuple* $R = <S, T, C, W, M°>$ *where :*
i) $S = \{p1, \ldots, pn\}$ *and T are the finite sets of places and transitions.*
ii) C is a $(S \cup T)$*-indexed family of non empty sets (possibly infinite) C(x). For each place p, elements of C(p) are called colours.*
iii) W is a $(S \times T \cup T \times S)$*-indexed family of functions, such that, for each p in S and each t in T, W(p,t) and W(t,p) have C(t) for domain and* $N^{(C(p))}$ *for codomain.*
iv) M° is a marking of R, i.e. an element of the product set $N^{(C(p1))} \times \ldots \times N^{(C(pn))}$ *; M° is called the initial marking of R.*

The behaviour of a coloured P/T-system is defined just as for a P/T-system : a transition t may occur at the marking M, yielding to the marking M', if and only if :

$$\exists \sigma \in C(t),\ \forall p \in S, \qquad M(p) \geq W(p,t)(\sigma) \text{ and,}$$
$$M'(p) + W(p,t)(\sigma) = M(p) + W(t,p)(\sigma).$$

This is denoted by : M[t > M'.

A unary Predicate/Transition-system is a coloured P/T-system whose functions valuating the arcs can be represented by multisets over $A \cup X$, where A is a set of colours and X a set of variables.

Definition 3. *A (free) unary Predicate/Transition-system (UPr/T-system for short), is a coloured P/T-system R = < S, T, C, W, M°> such that :*

i) all places are associated with the same set of colours A : $\forall p \in S, C(p) = A$

ii) there is a set X such that for every t in T, $C(t) = A^X$

iii) for every p and t, the functions W(p,t) and W(t,p) can be represented by multisets over $A \cup X$ *:*

$W(p,t), W(t,p) \in [[N^{(A \cup X)}]]$.

Notice that, since all the places have the same set of colors, a marking of an UPr/T-system can be seen as a function from S into $N^{(A)}$.

Remark. The original definition of UPr/T-systems ([Vautherin... 85]) differs slightly from this one on the sets C(t) which may be proper subsets of A^X. This is the reason why we call these systems "free" UPr/T-systems.

According to the definition of $[[N^{(A \cup X)}]]$, a UPr/T-system is totally described by a 6-tuple <S, T, A, X, w, M°> where :

i) S, T, A, X and M° are defined as previously,

ii) w is an $(S \times T \cup T \times S)$-indexed family of multisets over $A \cup X$.

Notice that we use the lower case form w when the valuations of the arcs are multisets and we use the capital form W when these valuations are functions. The relation between the two notations is W(x,y) = [[w(x,y)]] for every arc (x,y).

It is well known that each coloured P/T-system R whose sets C(x) (for $x \in S \cup T$) are finite, admits an equivalent P/T-system which is obtained by unfolding R (one unfolds p in | C(p) | places, t in | C(t) | transitions). Then, we define the semi-flows of a coloured P/T-system R in such a way that they coincide with those of the equivalent P/T-system, when it is defined.

Definition 4. *Let R = <S, T, C, W, M°> be a coloured P/T-system with S = {p1,...,pn}, a K-semi-flow of R is an element* $f = (f_{p1},\dots,f_{pn})$ *of* $K^{C(p1)} \times \dots \times K^{C(pn)}$ *such that*

$$\forall t \in T,\ \forall \sigma \in C(t),\ \sum_{p \in S, a \in C(p)} f_p(a).W(p,t)(\sigma)(a) = \sum_{p \in S, a \in C(p)} f_p(a).W(t,p)(\sigma)(a)$$

The set of K-semi-flows of R is denoted by $F_K(R)$.

Notice that f_p may have an infinite support. However, since $W(p,t)(\sigma)$ and $W(t,p)(\sigma)$ have a finite support, the summations in this definition are always defined. Let us also mention that in [Jensen 81] and [Silva... 85], any set Z(U) can be used instead of K ; but in that case, each semi-flow with respect to U thus defined, can be viewed as a family of Z-semi-flows as we defined them here.

Finally notice that for a UPr/T-system, since all the places have the same set of colors A, a semi-flow is a vector $f = (f_p)_{p \in S}$ with $f_p \in K^A$.

Proposition 1. (Lautenbach) *If f is a K-semi-flow of R, then*

$$\forall M \in [M°>,\ \sum_{p \in S, a \in C(p)} f_p(a).M(p)(a) = \sum_{p \in S, a \in C(p)} f_p(a).M°(p)(a)$$

When satisfied, this formula is called a linear invariant of R.

With the usual componentwise operations $(f_1,\dots, f_n) + (g_1,\dots, g_n) = (f_1 + g_1,\dots, f_n + g_n)$ and $k.(f_1,\dots, f_n) = (k.f_1,\dots, k.f_n)$, it is easy to see that $F_K(R)$ is closed under addition and product by a scalar ; hence $F_Z(R)$ has a Z-module structure.

IV.2. The channel modelling

A unary Pr/T-system representing the channel is drawn in figure 3. Here we have : $A = Cp \cup Cc \cup \{*\}$ where

$Cp = \{pr_i \mid i \in [1,..., ip]\}$ represents the set of producers,

$Cc = \{c_i \mid i \in [1,..., ic]\}$ the set of consumers,

and * is used to denote a produced value in inte and Fileval, as well as the usual token marking Per, p3, p6 and p7. We have $X = \{x,y\}$.

The initial marking M° is such that :

$M°(p1) = \Sigma_i <pr_i>, M°(p8) = \Sigma_i <c_i>, M°(Per) = \max <*>,$

$M°(p3) = <*>$ and $M°(p) = \delta$ otherwise

This coloured system has the same deadlock as the P/T-system of figure 2, but its behaviour is more precise : for each marking, we know in which state the different processe are and the position of their request, if any.

By considering the invariants of the P/T-system, one can foresee that the coloured system admits the following ones :

$\forall i \in [1,..., ip],$

IPPi : $\forall M \in [M°>, M(p1)(pr_i) + M(p2)(pr_i) = 1$

IPi : $\forall M \in [M°>, M(p1)(pr_i) + M(intp)(pr_i) + M(p4)(pr_i) + M(Depos)(pr_i) + M(repp)(pr_i) = 1$

$\forall i \in [1,..., ic],$

IPCi : $\forall M \in [M°>, M(p8)(c_i) + M(p9)(c_i) = 1$

ICi : $\forall M \in [M°>, M(p8)(c_i) + M(intc)(c_i) + M(p5)(c_i) + M(Prise)(c_i) + M(repc)(c_i) = 1$

IPG : $\forall M \in [M°>, \Sigma_{i=3,...,7} \mid M(p_i) \mid = 1$

IV : $\forall M \in [M°>, \mid M(repp) \mid + \mid M(inte) \mid + \mid M(p6) \mid + \mid M(Fileval) \mid + \mid M(p7) \mid + \mid M(Per) \mid = \max$

From the two families of invariants IPi and ICi, we can conclude that each channel $repp_i$ and each channel $repc_i$ has length 1. Also from IPi and IPPi we can deduce :

$M(p2)(pr_i) = M(intp)(pr_i) + M(p4)(pr_i) + M(Depos)(pr_i) + M(repp)(pr_i),$
and $M(p2)(pr_i) \leq 1.$

Hence any message sent by G to a producer pr_i is well consumed by pr_i. That could not be deduced from the ordinary P/T-system.

Now, some questions arise :

- How are these invariants really connected to those of the P/T-system of figure 1 ?
- Do they constitute an exhaustive list of "basic" invariants of the Pr/T-system ?
- Was it possible to generate them by a computation over the Pr/T-system ?

The following two paragraphs answer these questions.

Figure 3

The Unary Pr/T-system model

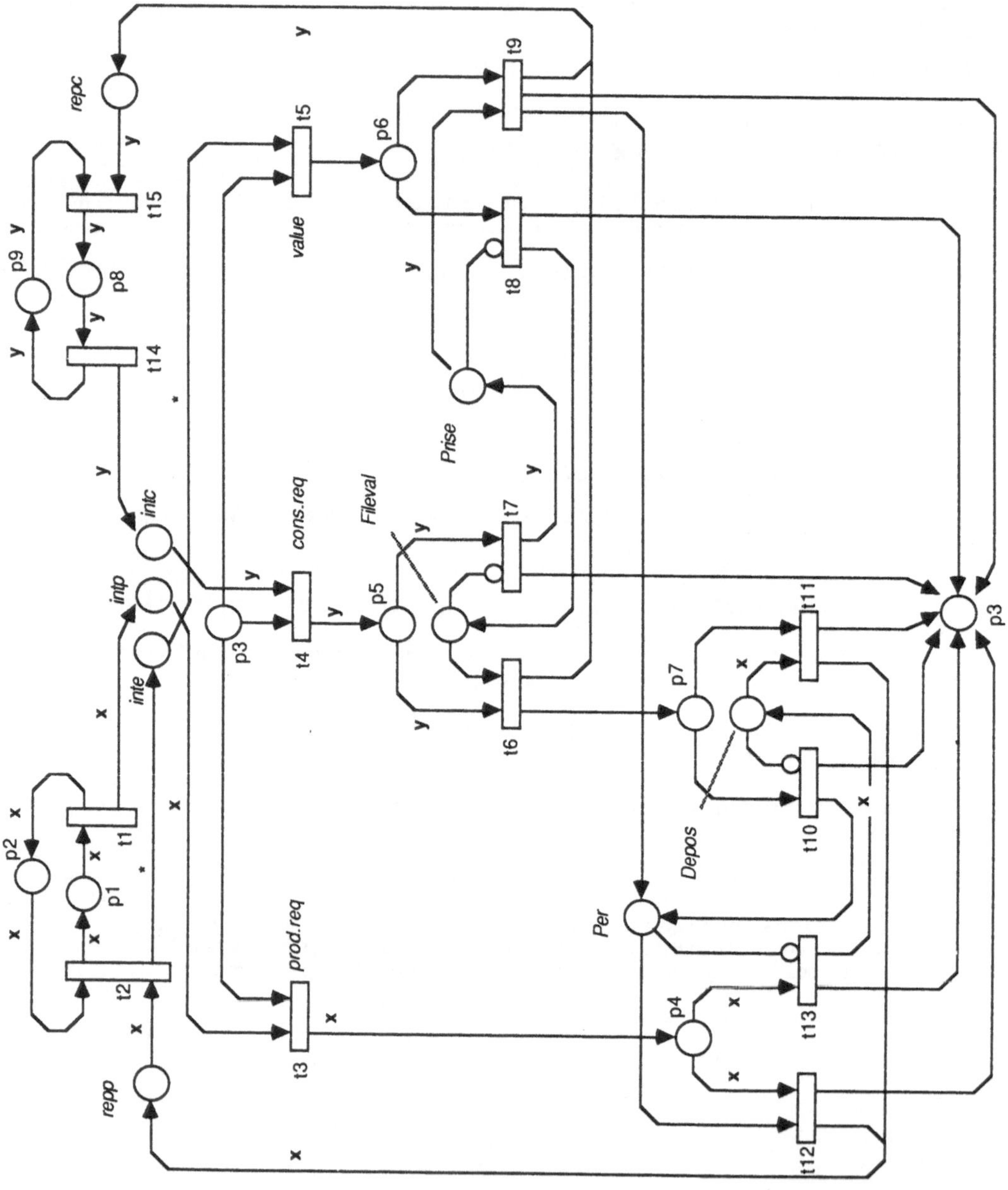

IV.3. Computation of invariants for unary Pr/T-systems

At first we describe more precisely the structure of $F_Z(R)$ when R is a unary Pr/T-system.

Let R = < S, T, A, X, w, M°> be a unary Pr/T-system ; we denote by $(c_{p,t})_{p\in S, t\in T}$ the incidence matrix of R; so for every p and t, $c_{p,t} = w(p,t) - w(t,p)$. We write E for the subset of colors which are effectively present in the incidence matrix (i.e., $a\in E \iff \exists p,t \,/\, c_{p,t}(a) \neq 0$). The complementary of E in A is denoted by NE. Notice that E is always finite, but NE is infinite as soon as A is infinite.

Now we choose an arbitrary element a° in A. We denote by $\mathbb{1}$ the element of Z^A such that $\mathbb{1}(a) = 1$ for all a in A ; remember that e_a is such that $e_a(a) = 1$ and $e_a(a') = 0$ for $a' \neq a$.

Proposition 2. *Let* $u = ((X_p)_{p\in S}, (Y_{p,a})_{p\in S, a\in E\setminus\{a°\}})$ *be a solution of*

$$(I)\begin{cases} \forall t\in T, \ \sum_{p\in S}\left(X_p \,/\, c_{p,t} \,/ + \sum_{a\in E\setminus\{a°\}} Y_{p,a}\, c_{p,t}(a)\right) = 0 \\ \forall a\in E\setminus\{a°\},\ \forall x\in X,\ \sum_{p\in S} Y_{p,a}\, c_{p,t}(x) = 0 \end{cases}$$

then the vector $f = (f_p)_{p\in S}$ *such that*

$$\forall p\in S,\quad f_p = X_p.\mathbb{1} + \sum_{a\in E\setminus\{a°\}} Y_{p,a}.e_a$$

is a semi-flow of R. The corresponding invariant is

$$\forall M\in[M°>,\ \sum_{p\in S}\left(X_p \,/\, M_p \,/ + \sum_{a\in E\setminus\{a°\}} Y_{p,a}\, M_p(a)\right) = C^{te}$$

Proof. Cf. [Vautherin 86b].

Proposition 3. *Let* $w = ((Z_p)_{p\in S})$ *be a solution of*

$$(II)\ \left\{\forall x\in X,\ \forall t\in T,\ \sum_{p\in S} Z_p\, c_{p,t}(x) = 0\right.$$

then for each color α *in* $NE\setminus\{a°\}$, *the vector* $f^\alpha = (f^\alpha_p)_{p\in S}$ *such that*

$$\forall p\in S,\quad f^\alpha_p = Z_p.e_\alpha$$

is a semi-flow of R. The corresponding invariant is

$$\forall \alpha\in NE\setminus\{a°\},\ \forall M\in[M°>,\ \sum_{p\in S} Z_p\, M_p(\alpha) = C^{te}$$

Proof. Cf. [Vautherin 86b].

Theorem 2. *Let* $B_I = (u^1, \ldots, u^i, \ldots, u^n)$ *be a basis of solutions to (I) and* $B_{II} = (w^1, \ldots, w^j, \ldots, w^m)$ *be a basis of solutions to (II). Denoting by* f^i *the semi-flow constructed from* u^i *as indicated by proposition 2, and by* $f^{j\alpha}$, *for* α *in* $E\setminus\{a°\}$, *the semi-flows constructed from* w^j *as indicated by proposition 3, then the two following propositions are equivalent:*

i) *f is a semi-flow of R*

ii) *there is a family of integers* $((\lambda_i)_{i=1..n}, (\mu_{j\alpha})_{j=1..m,\alpha\in NE\setminus\{a°\}})$ *such that*

$$f = \sum_{i=1}^{n} \lambda_i f^i + \sum_{\alpha\in NE\setminus\{a°\}} \sum_{j=1}^{m} \mu_{j\alpha} f^{j\alpha}$$

When ii) is satisfied, the family $((\lambda_i)_{i=1..n}, (\mu_{j\alpha})_{j=1..m,\alpha\in E\setminus\{a°\}})$ *is unique.*

Proof. Cf. [Vautherin 86b].

Thus, even in the case where A is infinite, one can compute a finite number of integer vectors which generates all the Z-semi-flows of R. The computation of these vectors does not depend on the set NE of colors which are not effectively present in the incidence matrix of the system. Especially that allows parametrization of the set of colors.

For instance, in the channel example, the computation is done independently of the parameters ip and ic :

Let us take a° = * ; so NE = $\{pr_1,..., pr_{ip}\} \cup \{c_1,..., c_{ic}\}$. From proposition 2, we deduce the following invariants :

IPG : $\forall M\in [M°>, \Sigma_{i=3,...,7} \mid M(p_i) \mid = 1$

IV : $\forall M\in [M°>, \mid M(repp) \mid + \mid M(inte) \mid + \mid M(p6) \mid + \mid M(Fileval) \mid + \mid M(p7) \mid + \mid M(Per) \mid = max$

I1 : $\forall M\in [M°>, \mid M(p1) \mid + \mid M(p2) \mid = 1$

I2 : $\forall M\in [M°>, \mid M(p8) \mid + \mid M(p9) \mid = 1$

I3 : $\forall M\in [M°>, \mid M(p1) \mid + \mid M(intp) \mid + \mid M(p4) \mid + \mid M(Depos) \mid + \mid M(repp) \mid = ip$

I4 : $\forall M\in [M°>, \mid M(p8) \mid + \mid M(intc) \mid + \mid M(p5) \mid + \mid M(Prise) \mid + \mid M(repc) \mid = ic$

and from proposition 3 we deduce

$\forall i\in [1,..., ip]$,

IPPi : $\forall M\in [M°>, M(p1)(pr_i) + M(p2)(pr_i) = 1$

IPi : $\forall M\in [M°>, M(p1)(pr_i) + M(intp)(pr_i) + M(p4)(pr_i) + M(Depos)(pr_i) + M(repp)(pr_i) = 1$

I5i : $\forall p\in \{inte, p3, p6, Fileval, p7, Per\}, \forall M\in [M°>, M(p)(pr_i) = 0$

I6i : $\forall M\in [M°>, M(p8)(pr_i) + M(p9)(pr_i) = 0$

I7i : $\forall M\in [M°>, M(p8)(pr_i) + M(intc)(pr_i) + M(p5)(pr_i) + M(Prise)(pr_i) + M(repc)(pr_i) = 0$

$\forall i\in [1,..., ic]$,

IPCi : $\forall M\in [M°>, M(p8)(c_i) + M(p9)(c_i) = 1$

ICi : $\forall M\in [M°>, M(p8)(c_i) + M(intc)(c_i) + M(p5)(c_i) + M(Prise)(c_i) + M(repc)(c_i) = 1$

I8i : $\forall p\in \{inte, p3, p6, Fileval, p7, Per\}, \forall M\in [M°>, M(p)(c_i) = 0$

I9i : $\forall M\in [M°>, M(p1)(c_i) + M(p2)(c_i) = 0$

I10i : $\forall M\in [M°>, M(p1)(c_i) + M(intp)(c_i) + M(p4)(c_i) + M(Depos)(c_i) + M(repp)(c_i) = 0$

Following theorem 2 all linear invariants can be generated with these ones.

Notice that the additional invariants (I1-I10i) are trivial ones : invariants I5i-I7i means that there is never any token marked by pr_i in the places inte, intc, Fileval, Prise, Per, repc, p3, p5, p6, p7 ; invariants I8i-I10i means that there is never any token marked by c_i in the places inte, intp, Fileval, Depos, Per, repp, p3, p4, p6, p7 ; invariants I1-I4, according to IPPi, IPi, IPCi, ICi can be rewritten

$$I'1 : \forall M \in [M^\circ>, M(p1)(*) + M(p2)(*) = 0$$

$$I'2 : \forall M \in [M^\circ>, M(p8)(*) + M(p9)(*) = 0$$

$$I'3 : \forall M \in [M^\circ>, M(p1)(*) + M(intp)(*) + M(p4)(*) + M(Depos)(*) + M(repp)(*) = 0$$

$$I'4 : \forall M \in [M^\circ>, M(p8)(*) + M(intc)(*) + M(p5)(*) + M(Prise)(*) + M(repc)(*) = 0$$

which means that there is never any token marked by * in the places intp, intc, p1, p2, p8, p9, Prise, Depos, repp, repc, p4, p5.

Also notice that in this example, system (I) is equivalent to the system of equations which gives the semi-flows of the P/T-system considered in section III.

More generally, when R = <S, T, A, X, w, M°> is a UPr/T-system, we define an ordinary P/T-system | R |, called the *skeleton* of R, by | R | = <S,T, | w |, | M° |> with :

$$| w |(x,y) = | w(x,y) | \quad \forall (x,y) \in S\times T \cup T\times S$$

$$| M^\circ |(p) = | M^\circ(p) | \quad \forall p \in S$$

Roughly speaking, | R | is obtained from R by "forgetting the colours of the tokens". The following proposition is a corollary of proposition 2 ; it expresses a connection between the semi-flows of R and those of | R |.

Proposition 4. *Let* $g \in K^S$ *be a semi-flow of | R |, then the vector f of* $K^{S\times A}$ *defined by :*

$$f(p,a) = g(p) \quad \forall p \in S,\ \forall a \in A$$

is a semi-flow of R (such a vector g is called a type-1 semi-flow of R in [Vautherin... 85]).

Now, notice that all the semi-flows that we have found here are positive. Thus one can ask whether they constitue also a smallest set of generators of positive linear invariants. We answer this question in the following paragraph.

IV.4. Parameter reduction

Often, subsets of transitions are folded in the same manner for a subset of colours ; unfolding them gives birth to isomorphic incidence submatrices. A general study about this fact can be found in [Memmi 83] for fifo nets.

Here, we consider a family of unary Pr/T-systems defined with the help of a parameter $d \in N$.

We set R_d = <P∪Q, TP∪TQ, A∪Ad, X, w, M°> with $| A_d | = d$

R_d is a unary Pr/T-system with the following constraints :

i) $\forall p \in P, \forall t \in TP, \forall x \in X$, $w(p,t) \in [[N^{(\{x\})}]]$ and $w(t,p) \in [[N^{(\{x\})}]]$

ii) $\forall p \in P, \forall t \in TQ$, $w(p,t) = w(t,p) = \delta$

iii) $\forall t \in TP, \forall p \in Q$, $w(p,t) \in [[N^{(A)}]]$ and $w(t,p) \in [[N^{(A)}]]$

iv) $\forall t \in TP, C(t) = A_d{}^X$; $\forall t \in TQ, C(t) = A^X$

We want to characterize $F(R_d)$ from $F(R_1)$ and to be able to compute a set of generators for R_d from a set of generators for R_1. Let us point out that R_1 is defined with $| A_1 | = 1$, it is to say that R_1 behaves like a P/T-system over P and TP.

Let us set $A_1 = \{a_1\}$ and $A_d = \{a_1,..., a_d\}$, and let us define a set of projections :

$$proj_i : N^{P\times A_d \cup Q\times A} \rightarrow N^{P\times A_1 \cup Q\times A}$$

$$g \rightarrow proj_i(g)$$

such that

$$proj_i(g)(p,a_1) = g(p,a_i) \text{ if } p\in P$$

$$proj_i(g)(q,a) = g(q,a) \text{ if } q\in Q \text{ and } a\in A$$

We need the followling equivalence relation $\equiv^Q$ over $N^{P\times A_1 \cup Q\times A}$:

$$g \equiv^Q g' \text{ if and only if } \forall a\in A, \forall g\in Q, g(q,a) = g'(q,a)$$

i.e., g and g' have the same projection on Q. Then we define a fonction ext from $(N^{(P\times A_1 \cup Q\times A)})^d$ into $N^{P\times A_d \cup Q\times A}$ with

$$dom(ext) = \{v = (v_1,..., v_d) \in (N^{(P\times A_1 \cup Q\times A)})^d \mid \forall i, j\in [1,..., d], v_i \equiv^Q v_j\}$$

and

$$ext(v)(p,a_i) = v_i(p,a_1) \text{ if } p\in P$$

$$ext(v)(q,a) = v_1(q,a) \text{ if } g\in Q \text{ and } a\in A$$

When $| A | = 1$, R_1 is clearly the skeleton of R_d ; proposition 4 says that if g is a semi-flow of R_1 then ext(g,..., g) is a semi-flow of R_d. With the constraints set on R_d, we can be more precise :

Theorem 3.
i) If $g\in F(R_d)$ then $\forall i\in [1,..., d]$, $proj_i(g)\in F(R_1)$
ii) If $(g_1,..., g_d)\in (F(R_1))^d \cap dom(ext)$ then $ext(g_1,..., g_d)\in F(R_d)$
Moreover, $F(R_d)$ is isomorphic to $(F(R_1))^d \cap dom(ext)$.

The proof is somewhat tedious and is given in [Memmi 83]. Now, we still need to characterize when a given semi-flow v of $F(R_d)$ has minimal support.

First, let us point out that if $e\in F(R_1)$ and $\| e \|$ is minimal then :

If $\| e \| \cap Q\times A \neq \emptyset$ then $\| ext(e,..., e) \|$ is minimal else $\| ext(0,..., e,..., 0) \|$ is minimal (where e appears one times in any position).

If $\| e_1 \|$ and $\| e_2 \|$ are minimal in R_1, with $e_1 \equiv^Q e_2$ then $\| ext(e_{Y(1)},..., e_{Y(d)}) \|$ where $Y\in \{1, 2\}^{[1,..., d]}$, is minimal.

But these two remarks are not sufficient to characterize any minimal support of $F(R_d)$, and examples of minimal supports of $F(R_d)$ not captured by these two cases can be constructed : consider for instance the following system.

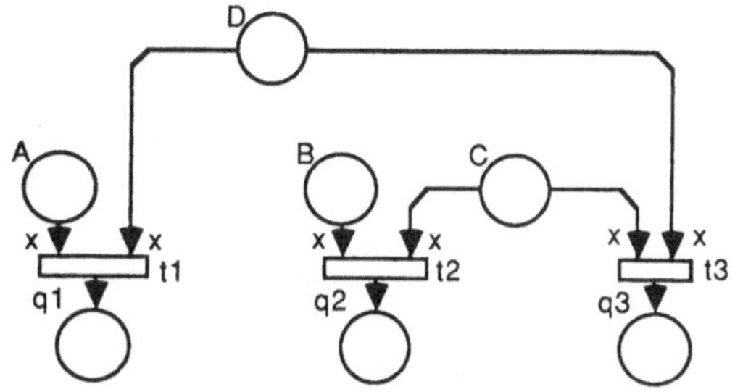

Figure 4

with $A = \{*\}$, $A2 = \{al, a2\}$, $X = \{x\}$, $P = \{A, B, C, D\}$, $Q = \{q1, q2, q3\}$, $TP = \{t1, t2, t3\}$, $TQ = \varnothing$.

Neither v1 = (0,1,0,1,1,1,1) nor v2 = (1,0,1,0,1,1,1) have a minimal support in R_1. However ext((v1,v2)) = ((0,1,0,1), (1,0,1,0), (1,1,1)) is such that || ext ((v1,v2)) || = {(B,a1), (D,a1), (A,a2), (C,a2), (q1,*), (q2,*), (q3,*)} is minimal.

Now, let v be a semi-flow of R_1. We denote by Comp(v) = $\{e \in F(R_1) \mid \| e \| \text{ is minimal and } \| e \| \subseteq \| v \| \}$. Let $B = \{b_1, \ldots, b_k\}$ be a family of semi-flows,

$$[B] = \{v \mid \exists\, \mu_1, \ldots, \mu_k \in Q^+,\ v = \Sigma_{i=1,\ldots,k}\, \mu_i . b_i\}$$

Then, the following result characterizes when v has minimal support in R_d by considering its projections $proj_i(v)$ in R_1. In many cases it will allow to compute a set of generators in a unary Pr/T-system by partial unfoldings.

Theorem 4. *Let v be a semi-flow in R_d, // v // is minimal if and only if :*

i) either there exists a unique i such that $\forall j \neq i$, $proj_j(v) = 0$, // $proj_i(v)$ // is minimal in R_1 and is included in $P \times A_l$.

ii) or
i) $\forall i \in [1, \ldots, d]$, $\forall u \in [Comp(proj_i(v))]$,
if $\exists k \in N$, $u \equiv^{Q} k.proj_i(v)$ then $u = k.proj_i(v)$
ii) if $u_1, \ldots, u_d$ are such that $u_i \in [Comp(proj_i(v))] \setminus [\{proj_i(v)\}]$
then there exist k and j such that $u_k \equiv^{Q} u_j$.

The proof uses theorem 3 and the decomposition theorem given in [Memmi... 80] for the set of positive semi-flows. Let us point out that this result only depends on the definition of $proj_i$, ext and $\equiv^Q$; it gives us an important insight on $F(R_d)$.

Now, we can go back to our example. We define P with the help of all the arcs valuated by x and all the arcs valuated by y. Then R_1 corresponds to the P/T-system defined in III ; and $F(R_1)$ contains the six semi-flows of minimal support associated with the three kinds of processes :

f_{IPP} is such that $f_{IPP}(pl) = f_{IPP}(p2) = 1$, $f_{IPP}(x) = 0$ otherwise
f_{IPC} is such that $f_{IPC}(p8) = f_{IPC}(p9) = 1$, $f_{IPC}(x) = 0$ otherwise
f_{IPG} is such that $f_{IPG}(p3) = f_{IPG}(p4) = f_{IPG}(p5) = f_{IPG}(p6) = f_{IPG}(p7) = 1$, $f_{IPG}(x) = 0$ otherwise

and with the kinds of data :

$f_{IP}(pl) = f_{IP}(p4) = f_{IP}(intp) = f_{IP}(p4) = f_{IP}(Depos) = f_{IP}(repp) = 1, f_{IP}(x) = 0$ otherwise
$f_{IC}(p8) = f_{IC}(intc) = f_{IC}(p5) = f_{IC}(Prise) = f_{IC}(repc) = 1, f_{IC}(x) = 0$ otherwise
$f_{IV}(repp) = f_{IV}(intc) = f_{IV}(p6) = f_{IV}(Fileval) = f_{IV}(p7) = f_{IV}(Per) = 1, f_{IV}(x) = 0$ otherwise

Now let v be a semi-flow of $F(R_d)$ (here d = ip + ic + 1) such that || v || is minimal. The application of theorem 4 is straightforward. Either there exists a unique i such that :

$$proj_i(v) \in \{f_{IPP}, f_{IPC}, f_{IP}, f_{IC}\} \text{ or } Comp(proj_i(v)) \subseteq \{f_{IPG}, f_{IV}\}.$$

We then point out that p3 of Q appears uniquely in f_G and Fileval of Q uniquely in f_{IV}. So

$$\text{either } v = ext(f_{IPG},..., f_{IPG}) \text{ or } v = ext(f_{IV},..., f_{IV}).$$

In conclusion, a set of generators for $F(R_d)$ is given by :

$ext(f_{IPG},..., f_{IPG})$ and $ext(f_{IV},..., f_{IV})$;
$ext(0,..., 0, f, 0,..., 0)$ where $f \in \{f_{IPP}, f_{IPC}, f_{IP}, f_{IC}\}$ is in the ith position, for i in [1,..., d].

And the associated invariants are

IPG, IV
IPPi, I9i and I'1 ; IPCi, I6i and I'2 ; IPi, I10i and I'3 ; ICi, I7i and I'4.

Notice that we do not find the invariants I5 and I8, because here, we have not suppose that the places inte, p3, p6... , which constitute the set Q, are colored.

V. THE FIFO NET MODEL

For our P/T-system model as for our Pr/T-system model, we have not structured the channel int. In fact, in the L-language, int has been designed as an infinite queue. Fifo nets are particularly fitted for that king of modelling. Examples and some theoretical results about this kind of nets are presented in the lecture of G. Roucairol in this course.

On a historical point of view, the code of the general channel has been automatically translated into a fifo net by a software tool named Rafael. The six linear invariants were also computed via Rafael [Behm... 84], [Memmi 83].

Here, we just recall the basic definition of fifo nets, then extend their descriptive power by changing the transition rule again so that one can pass from coloured P/T-systems to fifo nets in a continuous manner. At last, we prove the liveness of the general channel.

V.1. Definitions

Let A be a set, we denote by A* the set of finite words over A. We write λ for the empty word.

Definition 5. *A fifo net is a 5-tuple $R = <F, T, A, W, M°>$ where :*
i) F is the finite set of fifos, T the finite set of transitions ($F \cap T = \emptyset$) ;
ii) A is a finite alphabet. Each letter is called a message.
iii) W is a function from $(F \times T) \cup (T \times F)$ into A.*
iv) M° is a marking of R, i.e. a function from F into A ; M° is called the initial marking.*

A transition t is enabled at the marking M if and only if :

$$\forall f \in F, W(f,t) \leq M(f) \text{ (i.e. } \exists x \in A^*, M(f) = W(f,t).x \text{ ; thus } W(f,t) \text{ is a prefix of } M(f)).$$

Then, we reach M' from M by removing the prefix W(f,t) and appending W(f,t) as a suffix to the marking of each fifo f :

$$\forall f\in F,\ W(f,t)\,.M'(f) = M(f)\,.W(t,f).$$

As for coloured P/T-systems, we can generalize fifo nets by folding sets of transitions (i.e. extending the transition rule). Then we can add a set X of variables to A and introduce the unary fifo nets.

Definition 6. *A* *general fifo net* *is a 6-tuple* $R = \langle F, T, A, C, W, M^\circ \rangle$ *where :*
i) F and T are the finite sets of fifos and transitions respectively.
ii) A is a finite alphabet, C is a T-indexed family of non empty sets C(t) (possibly infinite).
iii) W is a $(F\times T \cup T\times F)$*-indexed family of functions such that, for each f in F and each t in T, W(f,t) and W(t,f) are from C(t) into* A^*.
iv) M° *is a marking of R, i.e. a function from F to* A^* *;* M° *is called the initial marking.*

A transition t is enabled at the marking M and yields to the marking M' if and only if :

$$\exists\sigma\in C(t),\ \forall p\in F,\qquad W(f,t)(\sigma) \le M(f) \text{ and,}$$
$$W(f,t)(\sigma)\,.\,M'(f) = M(f)\,.W(t,f)(\sigma)$$

We can see a word w of A^* as a function from N into $A\cup\{\lambda\}$ with :

$w(0) = \lambda$
$w(i)$ is the ith letter in w if $i \le |w|$
$w(i) = \lambda$ if $i > |w|$

thus, $w = w(1)w(2)...w(|w|)$. Then let X be a set of variables, a word w over $(A\cup X)^*$ determines a function $[[w]]$ from A^X into A^* with

$[[w]](\sigma)(i) = w(i)$ if $w(i)\in A$
$[[w]](\sigma)(i) = \sigma(w(i))$ if $w(i)\in X$

for each $i \le |w|$. For $L \subseteq (A\cup X)^*$, $[[L]]$ will denote $\{[[w]] \mid w\in L\}$.

Definition 7. *A* *unary fifo net* *is a general fifo net* $R = \langle F, T, A, A^X, [[w]], M^\circ \rangle$ *where :*
i) X is a set of variables,
ii) for every transition t, $C(t) = A^X$ *,*
iii) w is a $(F\times T \cup T\times F)$*-indexed family of words over* $A\cup X$.

It is easy to deduce a multiset w~ from a word w over A with w~(a) = #(a,w), where #(a,w) denotes the number of occurrences of a in w. Then, each fifo net R can be associated with a coloured P/T-system R~.

The semi-flows of a fifo net are defined as being exactly those of R~. This definition implies that for a fifo net, linear invariants bring information on the number of occurrences of messages in fifos but never on their order.

V.2. The general channel is live

For our example, we obtain the unary fifo net of the figure 5 by merging the places intp, intc, inte into a fifo int. Notice that p1, p2, p8, p9, repp and repc remain "coloured" : i.e. messages have to be considered merged in these elements. This functioning can easily be simulated with fifos (see [Memmi 83]).

It is easy to see that all the invariants and the home spaces established for the coloured system of figure 2 remain true (the initial marking is the same as for the UPr/T-system).

Figure 5

The fifo net model

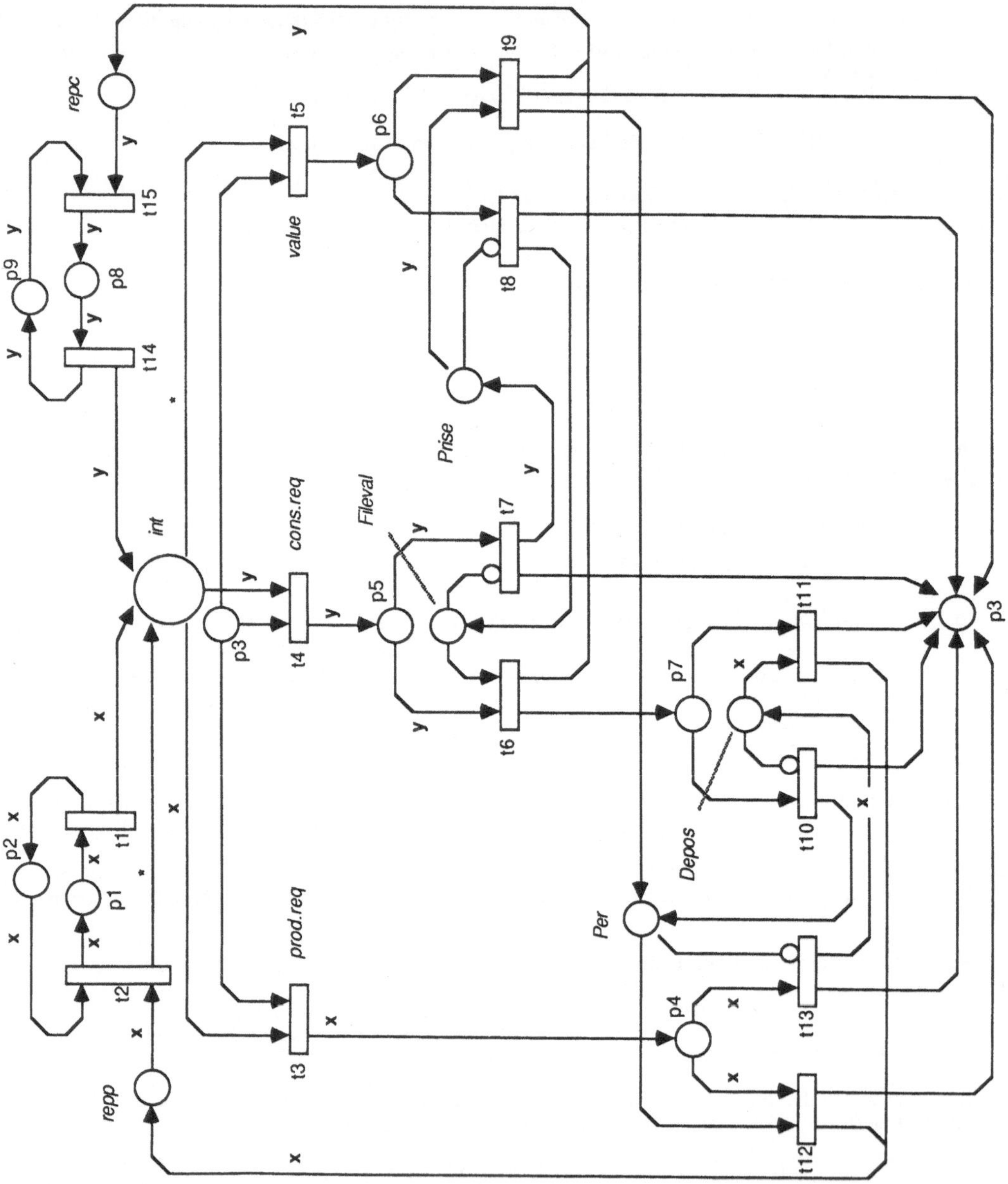

Now, two additional (non linear) invariants can be proved :

IPF : $\forall M \in [M°>, | M(Fileval) | . | M(Prise) | = 0$

This is true at M°. We can put a message in Fileval only by t8, i.e., if and only if |M(Prise) | = 0. Similarly, we can put a message in Prise only by t7, i.e., if and only if | M(Fileval) | = 0. Let us point out that this first non linear invariant can be proved in the initial P/T-system.

IDP : $\forall M \in [M°>, | M(Per) | + | M(Depos) | < max + ip$.

For proving IDP, we need to consider the fifo mechanism in int : a value produced for Fileval can always be followed by a production request. Hence t9 cannot be enabled with | M(Depos) | = ip. This is the main difficulty for proving IDP.

Now, we can show that the initial marking M° is a home state :

Let M be a reachable marking of E3R (it is not empty since M° belongs to this set). According to IV, M verifies | M(Fileval) | + | M(Per) | = max.

a) If M(Fileval) = λ, then M(Per) = $*^{max}$ and | M(Depos) | < ip (from IDP) ; then from IP, M(p1) $\neq \lambda$. If |M(Prise)| = k with k > 0, then we empty Prise with the sequence $(t1, t3, t12, t2, t5, t9, t15)^k$. Then we reach a marking M' such that (according to IC), M(p8) $\neq \lambda$. Now if | M'(Depos) | = k' with k' > 0, we empty Depos with the sequence t1, t3, t12, t2, t5, t8, t14, t4, $(t6, t15, t11, t2, t5, t8)^{k'}$. Finally we empty Fileval by t14, t4, t6, t15, t10 as many times as needed, and we reach M°.

b) If M(Fileval) $\neq \lambda$, then from IPF, M(Prise) = λ and | M(p8) | = ic according to IC. Then we empty Depos and Fileval successively as in the previous case, and we reach M°.

From that, we easily conclude that

The general channel is live.

Now, one can ask whether there exists some buffer structures which are both less compelling than the fifo queue structure and such that the system remains live. It is the purpose of the following section.

VI. ALGEBRAIC ABSTRACT DATA TYPES AND COLOURED SYSTEMS

In order to answer, this question, we need to be able to define what a buffer is in an abstract way ; that is to say, without precisely specifying its implementation. Then we need a specification method for the whole system which respects this abstraction.

VI.1. The model

Such a method is described in [Vautherin 86a] and, applied to the channel example, gives the specification presented in figures 6 and 7.

In this method a specification has two parts. In the first one (fig. 6), the data structure of the system that we consider is specified by an *algebraic specification of abstract data type* ([ADJ 78], [Gaudel 79],[Ehrig... 85]) ; that is to say, a triplet $<\underline{S}, \Sigma, E>$ where $\underline{S}$ is a set of sorts names, Σ a set of symbols with their corresponding arity in $\underline{S}$ (the names of domains and codomains) and E a set of axioms. These axioms are equations between terms composed of operations symbols and free variables. The variables are implicitly universally quantified over their corresponding sort.

We denote by $F : s_1 \times .. \times s_n \rightarrow s$ the arity of F. A function without any specified domain ($F : \rightarrow s$) is a constant.

Here we consider four non-constant functions : repp(-) and repc(-) associate each process with its corresponding request, unit(-) associates a message with the buffer containing only this message and app(-,-) (for "append") constructs a buffer from two others ; this function is used in the following with the function unit(-) to express the fact that messages go into or out of the buffer.

($\underline{S}°$) <u>Sorts</u> : Producer, Consumer, Message, Buffer ;
($\Sigma°$) <u>Operations</u> : $pr_1,\ldots, pr_{ip}$: $\rightarrow$ Producer,
$c_1,\ldots, c_{ic}$: $\rightarrow$ Consumer,
* : $\rightarrow$ Message
repp(-) : Producer $\rightarrow$ Message
repc(-) : Consumer $\rightarrow$ Message
unit(-) : Message $\rightarrow$ Buffer
empty : $\rightarrow$ Buffer
app(-,-) : Buffer, Buffer $\rightarrow$ Buffer
(E°) <u>Equations</u> : app(empty, f) == f
app(f, empty) == f
app(app(f, f'), f'') == app(f, app(f', f''))

Figure 6

The couple <$\underline{S}$,Σ> is called a *signature*. It characterizes some many-sorted algebras called Σ-algebras : more precisely, a Σ-algebra is a many-sorted algebra A such that for each name of sort s there corresponds a carrier A_s ; and for each operation symbol F there corresponds an operation F_A of A.

The algebras which satisfy the axioms of E are called the *models* of the specification <$\underline{S}$,Σ,E>. Each model is intended to represent a possible implementation of the specified data structure.

For instance, we give below two models of the specification given in figure 6. The first algebra, A°, represents the implementation where the buffer is structured as a fifo queue ; while the second algebra, B°, represents the implementation where the buffer is managed as a multiset.

$A°_{Producer} = B°_{Producer} = \{pr_1,\ldots, pr_{ip}\}$
$A°_{Consumer} = B°_{Consumer} = \{c_1,\ldots, c_{ic}\}$
$A°_{Message} = B°_{Message} = \{*\} \cup \{pr_1,\ldots, pr_{ip}\} \cup \{c_1,\ldots, c_{ic})$

$repp_{A°}(x) = repp_{B°}(x) = x$	(we identify a request from a
$repc_{A°}(y) = repc_{B°}(y) = y$	process and the process itself).

$A°_{Buffer} = (A°_{Message})^*$	(the set of words over $A°_{Message}$)
$Unit_{A°}(m) = m$	($Unit_{A°}(m)$ is the word of containing only the letter m)
$empty_{A°} = \lambda$	(the empty word)
$app_{A°}(w, w') = w.w'$	(the concatenation of w and w')

$B°_{Buffer} = N^{(B°_{Message})}$	(the set of multisets over B° Message)
$Unit_{B°}(m) = <m>$	(unitB(m) is the multiset which contains only m)
$empty_{B°} = \delta$	(the empty multiset)
$app_{B°}(w, w') = w + w'$	(the union of multisets)

The second part of the channel specification (figure 7) is a Pr/T-system like schema $\Omega°$, which is intended to specify how the data structure can evolve in the system.

Figure 7

The algebraic specification

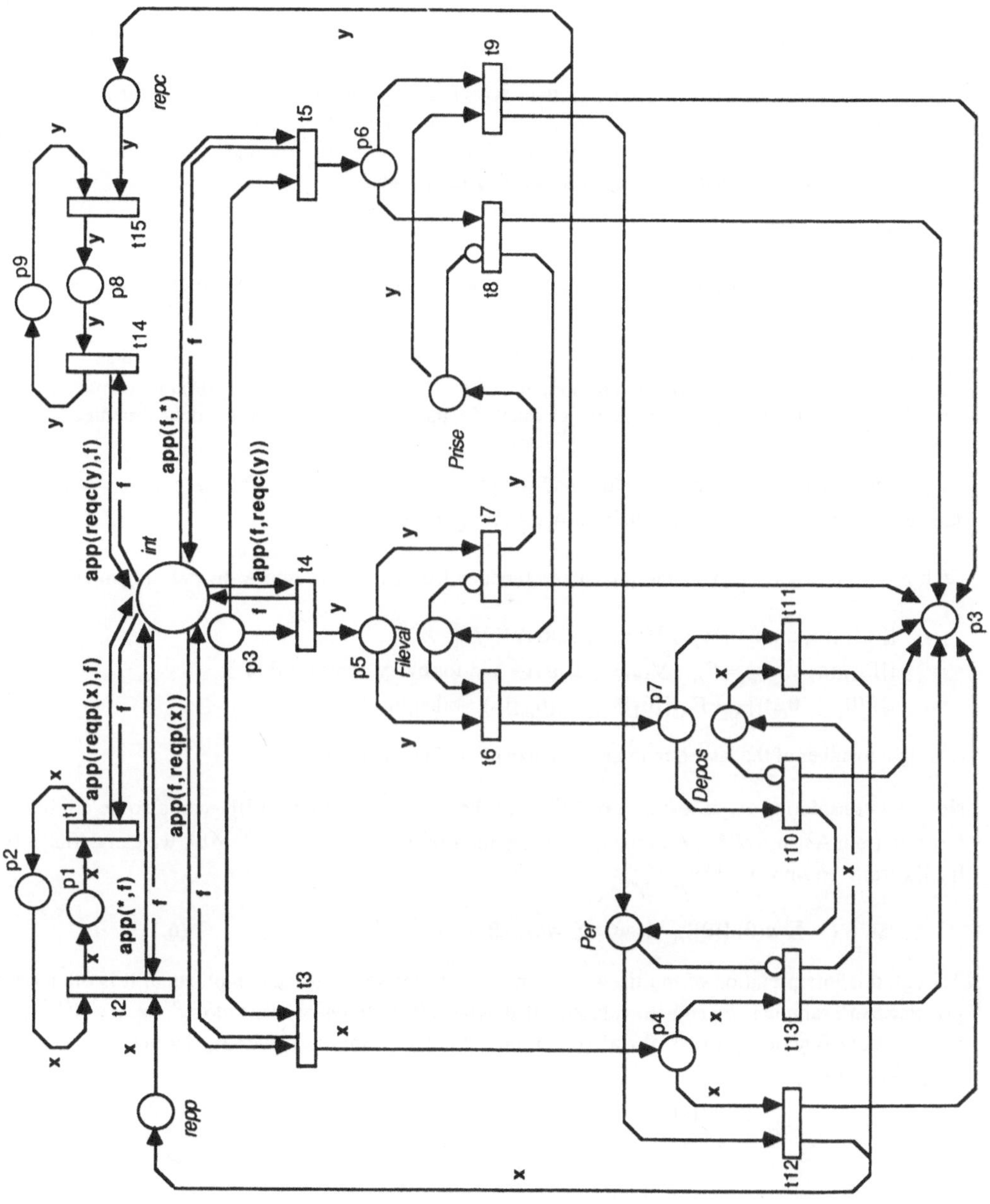

The arcs of $\Omega°$ are valuated by terms composed of variables and symbols of $\Sigma°$: let X be a set of $\underline{S}$-sorted variables ; we write X_s the subset of X which contains the variables whose sort is s. We denote by $T(\Sigma \cup X)$ the set of well-formed terms that one can write using symbols of Σ and variables of X. These terms are called Σ-terms. The subset of Σ-terms without variables (closed Σ-terms) is denoted by $T(\Sigma)$ and $T(\Sigma \cup X)_s$ denotes the subset of Σ-terms whose sort is s.

Definition 8. *A (free)*[1] *Σ-schema is a 5-tuple $\Omega = <S, T, X, w, M°>$ where :*
i) S and T are two finite sets of places and transitions respectively. The places are $\underline{S}$-sorted ; i.e. we also suppose an application $z : S \rightarrow \underline{S}$.
ii) X is a finite set of variables.
iii) w is a $(S{\times}T \cup T{\times}S)$-indexed family of multisets of Σ-terms which is coherent with the places sorts ; i.e. such that for every (p,t) in $S{\times}T$, w(p,t) and w(t,p) are in $N^{(T(\Sigma \cup X)_{z(p)})}$.
iv) $M°$ is a terms-marhing, i.e., an S-indexed family of multisets of closed terms, coherent with the places sorts ; i.e. such that for every place p, $M°(p) \in N^{((T(\Sigma)_{z(p)}))}$.

[1] Cf. [Vautherin 85a]

Notice that, in order to simplify the notations, the operation symbol unit is omitted in the valuations of $\Omega°$ (figure 7).

For each algebra A specified by the first part of the specification, the schema is interpreted as a coloured P/T-system $[[\Omega]]_A$ whose sets of colours are domains of A. The behaviour of this system is intended to represent the behaviour of the real system (here, the channel) for the implementation of the data structure coresponding to A.

The interpretation of the valuation w(p,t) and w(t,p) as functions is done as follows : let $X = \{x_1,..., x_n\}$ and denote by s_i the sort of x_i ; the product set $A_{s1} \times ... \times A_{sn}$ is denoted by A^X.

Each Σ-term θ of sort s can be interpreted as a function $[[\theta]]_A$ from A^X into A_s by defining inductively :

$$[[x_i]]_A(a_1,..., a_n) = a_i, \quad \forall (a_1,..., a_n) \in A^X, \forall x_i \in X$$
$$[[F]]_A(a_1,..., a_n) = F_A, \quad \forall (a_1,..., a_n) \in A^X \text{ and for every constant F of } \Sigma.$$
$$[[F(\theta_1,..., \theta_m)]]_A = F_A([[\theta_1]]_A,..., [[\theta_m]]_A) \quad \text{otherwise}$$

(the right member of the last equation is a composition of operations).

Next, by considering A_s as a subset of $N^{(A_s)}$ (by the injection $a \rightarrow e_a$), $[[\theta]]_A$ can also be considered as a function from A^X into $N^{(A_s)}$. And then, for every multiset of terms $w \in N^{(T(\Sigma \cup X)_s)}$, we can define a function $[[w]]_A$ from A^X into $N^{(A_s)}$ by :

$$[[w]]_A = \Sigma\, w(\theta).[[\theta]]_A \quad \text{if } w = \Sigma\, w(\theta).<\theta> \qquad (6.1)$$

Though this interpretation of multisets of terms a functions seems to be complicated, it is often obvious in practice. An example is given below. Notice that, when all the terms of w are closed, i.e. $w \in N^{(T(\Sigma)_s)}$, $[[w]]_A$ is a constant function into $N^{(A_s)}$ and so can be viewed as a multiset over A_s. Notice also that for every assignment σ A^X,

$$|\,[[w]]_A(\sigma)\,| = |\,w\,|. \qquad (6.2)$$

Definition 9. *Let Ω = <S, T, X, w, m°> be a Σ-schema and A be a Σ-algebra, the* <u>*interpretation of Ω*</u> *in A is the coloured P/T-system $[[\Omega]]_A$ = <S, T, C_A, w_A, $M°_A$> where*
i) for every transition t, $C_A(t) = A^X$
ii) for every place p, $C_A(p) = A_{z(p)}$
iii) for every (x,y) in (S×T∪T×S), $W_A(x,y) = [[w(x,y)]]_A$
iv) for every p in S, $M°_A(p) = [[m°(p)]]$.

Then we call *system specification* a 4-tuple <S, Σ, E, Ω > where <S, Σ, E> is an algebraic specification of an abstract data type and Ω a Σ-schema. The *models* of such a specification are the interpretations of Ω in the models of <S,Σ,E>.

Hence, from now on, we are working on a class of coloured systems instead of a single one.

Let us by example consider the following schema Ω^1 :

<u>Place</u> p : Buffer
<u>Var</u> f : Message

m°(p) = <app(*, repc(c_1))>

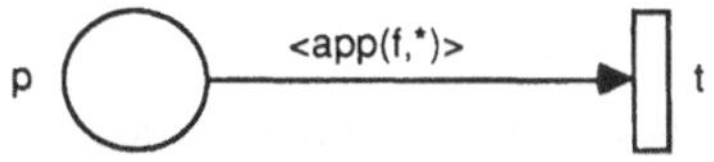

Figure 8

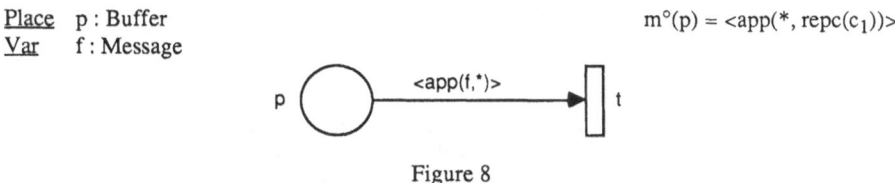

(here again, we omit the symbol "unit")

When it is interpreted in the algebra A = A° defines above, this schema gives a coloured system $[[\Omega^1]]_A$ such that $M°_A(p)$ = <*.c_1> and $W_A(p,t)$ is the function from $A^X \approx A_{Message}$ which associates each message m with <m.*>.

When it is interpreted in B = B°, it gives a coloured system $[[\Omega^1]]_B$ such that $M°_B(p)$ = <<*> + <c_1>> and $W_B(p,t)$ is the function from $B^X \approx B_{Message}$ into B_{Buffer} which associates each message m with <<m> + <*>>.

Notice that in both interpretations, there is only *one* token in p at the initial marking. In the first system, this token is "colored" by the word *.c_1 and in the second one, it is "colored" by the multiset <*> + <c_1> which is *one* value of B_{Buffer}.

Likewise, we show below (IB) that for every interpretation of the schema $\Omega°$ (figure 7) there is always one and only one token in the place int ; but this token can be "colored" by a word, or a multiset or something else, depending on the choosen interpretation.

Notice also that the transition t of Ω^1 may occur at the initial marking in the system $[[\Omega^1]]_B$, but not in the system $[[\Omega^1]]_A$, because the addition of multiset is commutative, but the concatenation of words is not. So, different interpretations of a schema may have different behaviours. The following proposition gives more details about the relationship between two interpretations of a schema.

Let A and B be two Σ-algebras ; a *Σ-morphism* from A into B is an $\underline{S}$-indexed family of functions h = $(h_s)_{s\in \underline{S}}$ which preserves the operations of Σ (i.e. such that $h(F_A(x)) = F_B(h(x))$).

Proposition 5. *If A and B are two Σ-algebras such that there is a Σ-morphism h from A into B, then the behaviour of $[[\Omega]]_A$ is included in the behaviour of $[[\Omega]]_B$. More precisely, $M°_B = h(M°_A)$ and $h(M)$ $(t > h(M')$ in $[[\Omega]]_B$ whenever $M(t >M'$ in $[[\Omega]]_A$ (here $h(M)$ denotes the marking of $[[\Omega]]_B$ such that for every place p, $h(M)(p) = <h(c_1)> + \ldots + <h(c_k)>$ when $M(p) = <c_1> + \ldots + <c_k>$).*

Proof. Cf. [Vautherin 86a]. Intuitively, B satisfies more properties than A and thus, the synchronization imposed by Ω is less compelling in the system $[[\Omega]]_B$ than in $[[\Omega]]_A$.

Each operation F of Σ naturally defines an operation between terms which associates the term $F(\theta_1,\ldots,\theta_n)$ to $\theta_1,\ldots,\theta_n$. Then, $T(\Sigma \cup X)$ is naturally provided with a Σ-algebra structure. We denote by $T(\Sigma,E)$ the quotient algebra of $T(\Sigma)$ by the smallest congruence $\equiv_E$ which contains the equations of E. Roughly speaking, this congruence $\equiv_E$ is defined by $\theta \equiv_E \theta'$ if and only if the equation $\theta == \theta'$ is a consequence of E by substitutions and replacements of equal terms. The algebra $T(\Sigma,E)$ is initial in the class $Alg(\Sigma,E)$ of all Σ-algebras satisfying E ; i.e. there is a unique Σ-morphism from $T(\Sigma,E)$ into every algebra A of $Alg(\Sigma,E)$.

An other particular Σ-algebra is the trivial algebra $\perp(\Sigma)$ where all carriers are reduced to singletons : $\perp(\Sigma)_s = \{¢_s\}$. This algebra is terminal in the class of all Σ-algebra ; i.e. there is a unique Σ-morphism from every Σ-algebras (and especially those of $Alg(\Sigma,E)$) into $\perp(\Sigma)$.

Thus, we represent $Alg(\Sigma,E)$ by the diagram of figure 9.a, denoting Σ-morphisms by arrows. And then, from the previous proposition, we deduce the diagram of figure 9.b which represents the class of all models of a specification $<S, \Sigma, E, \Omega>$, where arrows denote behaviour inclusion.

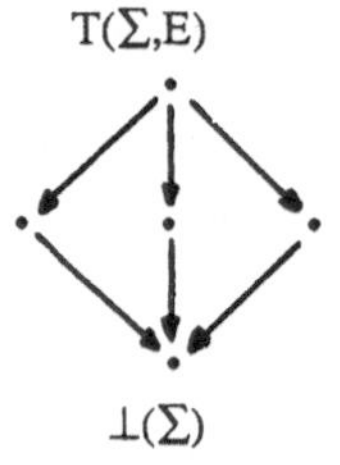

Models of $<S,\Sigma,E>$

Figure 9.a

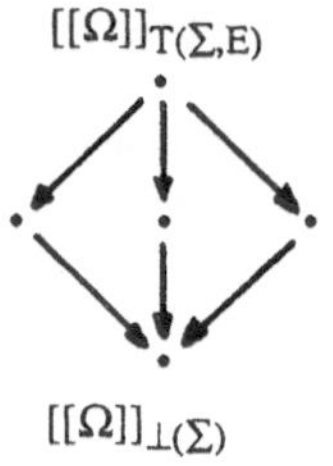

Models of $<S,\Sigma,E,\Omega>$

Figure 9.b

As far as the channel is concerned, it is not very difficult to see that the algebra A° is isomorphic to the initial algebra of $<S°, \Sigma°, E°>$ (figure 6). Also, we claim that the coloured system $[[\Omega°]]_{A°}$ has the same behaviour as our fifo-net (figure 5), that the coloured system $[[\Omega°]]_{B°}$ has the same behaviour as the UPr/T-system of figure 3 and that the P/T-system of figure 2 also is equivalent to a model of the specification $<S°, \Sigma°, E°, \Omega°>$. Hence, we have the diagram of figure 10.

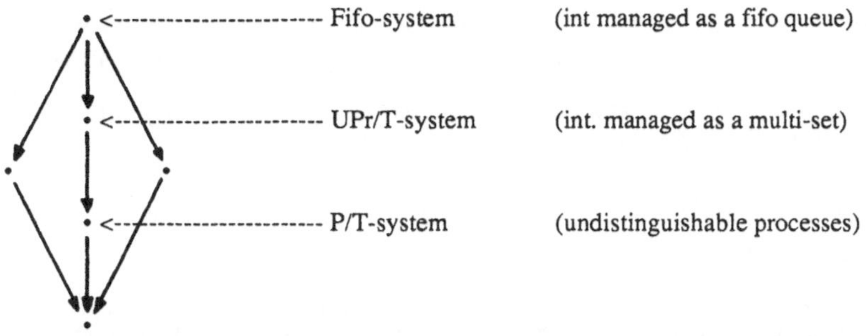

Figure 10

The following proposition gives light on the terminal model. At first, we define the skeleton of a schema, just as for UPr/T-systems : let $\Omega = <S, T, X, W, M°>$; the *skeleton* of Ω is the (ordinary) P/T-system $|\ \Omega\ | = <S, T, |\ w\ |, |\ M°\ |>$ with :

$$|\ w\ |(x,y) = |\ w(x,y)\ | \quad \forall (x,y) \in S \times T \cup T \times S \tag{6.3}$$

$$|\ M°\ |(p) = |\ M°(p)\ | \quad \forall p \in S \tag{6.4}$$

Remember that w(x,y) and M°(p) are multisets over $T(\Sigma \cup X)$ and that $|\ w\ |$ denotes the length of the multiset w.

Proposition 6.*The interpretation of a Σ-schema in the terminal algebra $\perp\!\!\perp(\Sigma)$ is (equivalent to) the skeleton of Ω ; and for every Σ-algebra A,*

$$|\ M°_A\ | = M°_{\perp\!\!\perp(\Sigma)} \text{ and, if } M(t > M' \text{ in } [[\Omega]]_A \text{, then } |\ M\ |\ (t > |\ M'\ | \text{ in } |\ \Omega\ |.$$

Proof. The equivalence between $[[\Omega]]_{\perp(\Sigma)}$ and $|\ \Omega\ |$ results from the isomorphism |.| from $N^{(\{\mathrm{¢}\})}$ into N which associates w with $|\ w\ |$. The second part of the proposition results from (6.2). []

Thus the terminal model of $<S°, \Sigma°, E°, \Omega°>$ is the P/T-system obtained by erasing the arcs valuations of the figure 7.

VI.2. Proof of liveness

Now let us study the liveness of the different models of the channel specification.

At first, we consider the set of equations

$$E^1 = \{ \ \text{app(repp(x),*)} == \text{app(*, repp(x))}$$
$$\text{app(repc(y),*)} == \text{app(*, repc(y))} \ \}$$

The sequence which leads to a deadlock in the P/T-system of figure 2 also can occur in $R = [[\Omega°]]_{T(\Sigma°,E°\cup E1)}$. Let M be the marking of R reached by this sequence. It satisfies

$$|\ M(\text{Depos})\ | = ip, \ \ |\ M(\text{Prise})\ | = ic, \ \ |\ M(p3)\ | = 1 \ \text{ and } \ |\ M(p)\ | = 0 \ \text{ otherwise.}$$

Hence $|\ M\ |$ is a deadlock in $|\ \Omega°\ | = [[\Omega\ °]]_{\perp(\Sigma°)}$. Now, let A be an algebra of $Alg(\Sigma°,E°)$ which also satisfies E^1. There is a morphism $h : T(\Sigma°,E°\cup E^1) \to A$ and thus, according to the proposition 5, h(M) can be reached in $[[\Omega°]]_A$. Moreover, by the definition of h(M), | h(M) | = | M |. Then, according to the proposition 6, since |M| is a deadlock in | Ω° | , also h(M) is a deadlock in $[[\Omega°]]_A$. we can summarize that by :

If the buffer structure allows the commutation of values and production requests, and the commutation of values and consummation requests, then the channel can reach a deadlock.

Next, we try to see for which conditions over the algebra A the liveness proof given above for the fifo net can also be applied to $[[\Omega°]]_A$.

The first step of this proof consists in establishing that it is always possible to empty the buffer int and to put the manager back in its initial state. In both the P/T-system and the fifo-net this property can be deduced from the fact that there is always one and only one token in the set of places {p3,..., p7}. The following proposition shows that for the interpretations of a schema such invariants over the number of tokens can also be computed by way of the skeleton.

Proposition 7. *Let f be a semi-flow of | Ω |, then for every interpretation $[[\Omega]]_A$,*

$$\forall M \in [M°_A>,\ \sum_{p\in S} f_p\,|\,M(p)\,| = \sum_{p\in S} f_p\,|\,M°(p)\,| \qquad (6.5)$$

Proof. It is a consequence of the proposition 6 by also using the fact that $|\,M°_A\,| = |\,M_\circ(p)\,|$ (cf (6.2)).

Applying that to the schema Ω°, we deduce that for every Σ°-algebra A, the interpretation $[[\Omega°]]_A$ satisfies

$$\text{IPG}: \forall M \in [M°_A>,\ \Sigma_{i=3..7}\,|\,M(p_i)\,| = 1$$

$$\text{IB}: \forall M \in [M°_A>,\ |\,M(int)\,| = 1$$

Then we deduce that for every algebra A of Alg(Σ°,E°), the set

$$\text{E3int} = \{M \mid M(int) = [[empty]]_A \text{ and } |\,M(p3)\,| = 1\}$$

is a home space of $[[\Omega°]]_A$.

Indeed, let A be an algebra of Alg(Σ°,E°).

a) At first, we notice that since for each transition of Ω°, every variable on an incoming arc is also on an incoming arc, then :

$$\forall M \in [M°_A>,\ \forall p \in S,\ M(p) \in N^{([[T(\Sigma°)]]_A)} \qquad (6.6)$$

where $[[T(\Sigma°)]]_A$ is the subset of A whose elements can be obtained by interpreting a closed Σ°-term.

b) Then, according to (IB),

$$\forall M \in [M°_A>,\ \exists b \in T(\Sigma°)_{Buffer},\ M(int) = [[b]]$$

Let suppose that b ≠ empty and denote by n the size of b. Since n ≠ 1, either b $\equiv_{E°}$ app(b', repp(x)) or b $\equiv_{E°}$ app(b, repc(y)) or b $\equiv_{E°}$ app(b',*). In any case, there is a transition t (t = t3, t4 or t5) such that M(t >M' with $M'(int) = [[b']]_A$. The size of b' is then smaller than n. Thus, by induction on n, we show that $\{M \mid M(int) = [[empty]]_A\}$ is a home space of $[[\Omega°]]_A$.

c) Finally, by IPG, this home space can be refined in E3int, just as for the P/T-system of figure 2.

According to (6.6), in any $\Sigma°$-algebra A , the only useful part for $[[\Omega°]]_A$ is the finitely generated part of A, $[[T(\Sigma°)]]_A$. Thus, from now on, *we consider only finitely generated algebras* ; i.e., algebras such that $[[T(\Sigma°)]]_A = A$. The class of finitely generated algebras satisfying E is denoted by Gen(Σ,E).

The remainder of the liveness proof given for the fifo-net requires invariants about the number of messages of each kind in the buffer int (IPi, ICi, IV). Of course, such invariants may exist only for implementations such that the number of messages in a buffer can be defined. For instance, it is not the case when the algebra satisfies an equation like : app(repp(x), repp(x)) == repp(x). Indeed, in such an implementation, a buffer can create or destroy some messages. In order to formalize that, we enrich the specification <S°, Σ°, E°> with observers which count the number of messages of each kind in the buffer :

(S^i) Sorts : Integer, Boolean.

(Σ^i) Operations :
- Occ_* : Buffer → Integer
- Occ_p : Producer, Buffer → Integer
- Occ_c : Consumer, Buffer → Integer

(E^i) Equations :

$Occ_*(app(f,f')) == Occ_*(f) + Occ_*(f')$
$Occ_*(unit(*)) == 1$
$Occ_*(unit(repp(x))) == 0$
$Occ_*(unit(repc(y))) == 0$
$Occ_p(x,app(f,f')) == Occ_p(x,f) + Occ_p(x,f')$
$Occ_p(x,unit(*)) == 0$
$Occ_p(x,unit(repp(x'))) ==$ if $x = x'$ then 1 else 0
$Occ_p(x,unit(repc(y)) == 0$
$Occ_c(y,app(f,f')) == Occ_c(y,f) + Occ_c(y,f')$
$Occ_c(y,unit(*)) == 0$
$Occ_c(y,unit(repp(x))) == 0$
$Occ_c(y,unit(repc(y'))) ==$ if $y = y'$ then 1 else 0

Figure 11

Remark. We assume that the sort integer is provided with the operations 0, 1, + and =. For simplicity, these operations are not specified here. Similarly, the specification of booleans and equalities in the sorts Producer and Consumer are omitted.

Now, let us denote by $\mathfrak{I}(A)$ the free algebra over A generated by this enrichment. A formal construction of $\mathfrak{I}(A)$ is given in the appendix. Roughly speaking, denoting by Σ^i and E^i the sets of new operations symbols and equations that we are considering, $\mathfrak{I}(A)$ is the $\Sigma° \cup \Sigma^i$ - algebra which just satisfies the same equations as A and the equations of E^i. Then we have :

(R1) : for each algebra A of Gen($\Sigma°$,E°) such that $\mathfrak{I}(A)$ is consistent, i.e. does not satisfy true == false, the interpretation $[[\Omega°]]_A$ satisfies :

For every reachable marking M and every terms marking m ($\forall p$, $m(p) \in N^{(T(\Sigma))}$) which represents M, i.e. such that $M(p) = [[m(p)]]_A$, $\forall p$,

$\forall i \in [1,..., ip]$,
IPP'i : $\#(pr_i, m(repp)) + \#(pr_i, m(int)) + \#(pr_i, m(p4)) + \#(pr_i, m(Depos)) = \#(pr_i, m(p2))$

$\forall j \in [1,..., ic]$,
$IPC'j : \#(c_j, m(repc)) + \#(c_j, m(int)) + \#(c_j, m(p5)) + \#(c_j, m(Prise)) = \#(c_j, m(p9))$

$IP' : | m(repp) | + | m(p1) | + \#(repp, m(int)) + | m(p4) | + | m(Depos) | = ip$

$IC' : |m(repc) | + | m(p8) | + \#(repc, m(int)) + | m(p5) | + | m(Prise) | = ic$

$IV' : | m(repp) | + \#(*, m(int)) + | m(p6) | + | m(Fileval) | + | m(p7) | + | m(Per) | = max$

where #(F,w), with $F \in \Sigma$ and $w \in N^{(T(\Sigma))}$, denotes the number of occurrences of F in the multiset of terms w.

The proof of this result is given in the appendix. From IPP'i and IPC'j, we deduce that

For every algebra A in Gen($\Omega°$,E°) such that $\Im$(A) is consistent, E3R = {$M \in$ E3int / |M(repp) | = |M(repc)| = 0} is a home space of $[[\Omega°]]_A$.

Indeed, let M be a marking in E3int and let m be a marking with terms which represents M. Assume that $|M(repp)| \neq 0$; then there is a multiset of terms, $w \in N^{(T(\Sigma°))}$ and $i \in [1,..., ip]$ such that m(repp) = $<pr_i>$ + w (otherwise, 0 = | m(repp) | = | M(repp) | by (6.2)). Then, according to IPP'i, m(p2) = $<pr_i>$ + w' and thus, m(t2 >m' in the initial model $[[\Omega°]]_{T(\Sigma°,E°)}$ with m'(repp) = w. Then, M(t2 >M' in $[[\Omega°]]_A$ with M'(repp) = $[[w]]_A$; and so | M'(repp) | = | m'(repp) | < | m(repp) | = | M(repp) |. Hence, by induction over | M(repp) |, we conclude that E3int $\cap$ {M | | M(repp) | = 0} is a home space of $[[\Omega°]]_A$; and, by similar arguments with repc, that E3R is a home space.

Intuitively, as far as $\Im$(A) is consistent, the buffer int cannot generate arbitratry messages and thus, no unexpected requests are sent by the manager.

The remainder of the liveness proof of the fifo net, is based on the fact that each time there is a value in the buffer int, then either a producer is in its initial state, or has sent a request in the buffer that will *necessarily* go out of the buffer *after* the considered value.

As above, in order to formalize this intuition, we enrich the abstract data type specification with some observers :

(Σ^{ii}) <u>Operations</u> : in?p(-), in ?*(-) : Buffer → Boolean
P?(-) : Buffer → Boolean

(E^{ii}) <u>Equations</u> :

```
in?p(empty) == false
in?p(app(f,*)) == in?p(f)
in?p(app(f, repp(x))) == true
in?p(app(f, repc(x))) == in?p(f)
in?*(empty) == false
in?*(app(f,*)) == true
in?*(app(f, repp(x))) == in?*(f)
in?*(app(f, repc(y))) == in?*(f)
P?(empty) == false
P?(app(f,*)) == (in?p(f) and non in?*(f)) or P?(f)
P?(app(f, repp(x))) == P?(f)
P?(app(f, repc(y))) == P?(f)
```

Figure 12

The predicate P? indicates if there is in the considered buffer a production request which came in after the last value entered. Let us denote by $\mathfrak{I}'(A)$ the free algebra over A generated by the enrichment $<\Sigma^i \cup \Sigma^{ii}, E^i \cup E^{ii}>$. We have

(R2) : for every algebra A in $Alg(\Sigma°,E°)$ and every reachable marking M in $[[\Omega]]_A$, $M(int) = a$ with $a \in A_{Buffer}$ and :

if $in?^*(a) \equiv true$ in $\mathfrak{I}'(A)$, then either $|M(p_1)| > 1$ or $P?(a) \equiv true$ in $\mathfrak{I}'(A)$.

And from that, we can deduce :

(R3) : for every algebra A in $Gen(\Sigma°,E°)$ such that $\mathfrak{I}'(A)$ is consistent, $[[\Omega°]]_A$ satisfies :

$$IDP : \forall M \in [M°_A>, \ |M(Per)| + |M(Depos)| < max + ip$$

The proof of these two results is given in the appendix. Finally, in order to establish that the initial marking of $[[\Omega]]_A$ is a home state, we need the following invariant :

$$IPF : \forall M \in [M°_A>, \ |M(Prise)| \,.\, |M(Fileval)| = 0$$

which is satisfied by every interpretation $[[\Omega°]]_A$ since it is satisfied by the skeleton $|\Omega°|$ (cf proposition 6). Then, from IP', IC', IV', IDP and IPF, we conclude, just like for the fifo net that for every algebra A in $Gen(\Sigma,E°)$ such that $\mathfrak{I}'(A)$ is consistent, $M°_A$ is a home state ; and thus $[[\Omega°]]_A$ is deadlock free.

Moreover, we already know that the initial model $[[\Omega°]]_{T(\Sigma°,E°)}$, which is equivalent to the fifo net of figure 5, is live. Thus for this model, for every transition t, there is a reachable marking M' such that M'[t >. According to proposition 5, this property is also satisfied by any model $[[\Omega°]]_A$ ($A \in Alg(\Sigma°,E°)$) ; and thus , from the previous result, we conclude that

For every algebra A in $Gen(\Sigma°,E°)$, such that $\mathfrak{I}'(A)$ is consistent, $[[\Omega°]]_A$ is live.

Now, let us take an example. We consider the set of equations

PERM = { app(repc(y),*) == app(*,repc(y))
app(repc(y), repp(x)) == app(repp(x), repc(y))
app(repc(y), repc(y')) == app(repc(y'), repc(y))
app(repp(x), repp(x')) == app(repp(x'), repp(x)) }

The algebra $C° = T(\Sigma°,E° \cup PERM)$ is such that $\mathfrak{I}'(C°)$ is consistent. Thus $[[\Omega°]]_{C°}$ is live, and so one can allow the commutation of consummation requests and values, and the commutation of consummation and production requests among themselves.

The following implementation is equivalent to C°, but more concrete :

- A buffer is implemented by three (multi-) sets S_p, S_c, S_v and a fifo queue over the two symbols p and v.
- The entry of a production request (resp. a value) in the buffer is done by adding this request to S_p (resp. this value to S_v) and pushing a symbol p (resp. v) in q.
- The entry of a consummation request is done by adding this request to S_c.
- Any consummation request in S_c can always be taken out of the buffer.
- Any production request in S_p (resp. any value in S_v) can be taken out of the buffer provided that the head of q is a sysmbol p (resp. a symbol v). When a message is taken out, the corresponding symbol is pulled out of q.

The results of this section are summarized in the following diagram, where the models corresponding to A°, B° and C° are respectively denoted by FIFO, SET and FIFO/PERM.

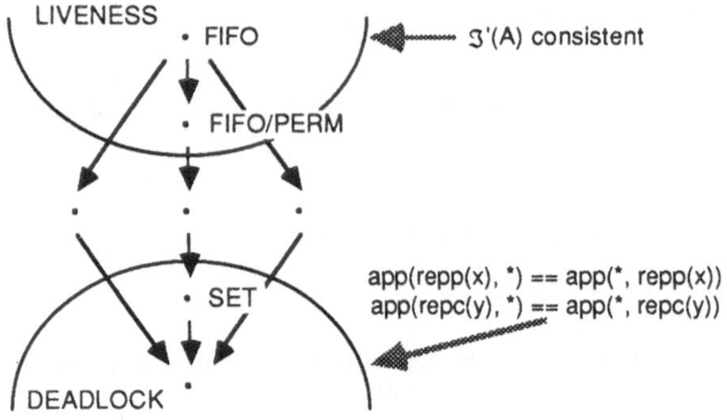

Figure 10

Let $E^2 = \{app(repp(x),*) == app\ (*, repp(x))\}$. The algebra $D° = T(\Sigma°, E° \cup E^2)$ is such that $\mathfrak{I}'(D°)$ is not consistent. So, when we just assume the commutation of production requests and values we cannot conclude anything from the previous analysis, about the liveness of the corresponding model. This model is just between the two areas "LIVENESS" and "DEALOCK" of the previous diagram. In fact, *for this model*, the liveness depends on the parameters i_c and max of the channel : we claim that

The system corresponding to D° is live if and only if $i_c < max$.

On the other hand, for the models which are on the areas "LIVENESS" or "DEADLOCK", the liveness or the presence of deadlock does not depend on the parameter values of the system.

VII. CONCLUSION

At each step of the analysis we have changed the model in order to describe elements of our example more and more precisely. But in no case, we had to make a translation from one model into another one: the changes have been done rather continuously, adding each time some descriptive power, leading to more and more acute results, verifying how the invariant method holds and showing how complementary the models are.

The study of our example can be pursued by making precise again which data structures of int lead to a deadlock, which ones do not, and for which values of max, ip and ic.

We think we have shown how the invariant notion is fundamental in programming. Though the calculus of semi-flows defined for the P/T-systems has been somewhat extended to high level systems, it should be extended again towards other models of parallel computation and completed. Also, we think it would be interesting to unify the results of section IV.3 (computation of invariants for UPr/T-systems) with the results in [Haddad 86] where the invariants of another subclass of coloured P/T-systems (namely the regular coloured nets) are investigated.

From a theoretical point of view, the Farkas algorithm has to be optilized. Some transformations on a P/T-system preserving the set of invariants are under study. We are also studying a polynomial algorithm giving one linear invariant if it exists.

The notion of home space is a nice extension of the home state notion. Nevertheless, we do not know whether it is decidable if a given set is a home space or not. The stepwise refinement of home spaces method seems to be a helpful guideline for reasoning about systems. Today, we do not know of another way to prove the liveness of the general channel.

ACKNOWLEGMENT

We would like to thank L. Meima for many improvements of the english version of this paper. Also many thanks to the two referees of this paper for their careful reading of a rough preliminary version.

VIII. BIBLIOGRAPHY

[ADJ 78] : Groupe ADJ : J.A. Goguen, J.W. Thatcher and E.G. Wagner
"An initial algebra approach to the specification, correctness and implementation of abstract data types". Current Trends in Programming Methodology, Vol. IV, R.T. Yeh (Ed.), Prentice Hall, New Jersey (1978).

[Alaiwan... 85] : H. Alaiwan and J.M. Toudic
"Recherche des semi-flows, des verrous et des trappes dans les réseaux de Petri". T.S.I., Vol.4, N°1 - Numero spécial réseaux de Petri, G. Memmi (Ed.), pp 103 - 112 (1985).

[Behm... 84] : P. Behm and G. Memmi
"Rafael : Un outil d'analyse de systèmes temps réel". 2ème Colloque de Génie Logiciel AFCET, Nice, pp 13-32 (1984).

[Boussinot 81] : F. Boussinot
"Réseaux de processus avec mélange équitable : une approche du temps réél". Thèse d'état, Université Paris VII (1981).

[Boussinot... 83] : F. Boussinot, R. Martin, G. Memmi, G. Ruggiu and J. Vapné
"A language for formal descriptions of real time systems". Proc. of SAFECOMP'83 - 3rd IFAC/IFIP Workshop, J.A. Baylis (Ed.), Pergamon Press, Cambridge UK. (1983).

[Brams 82] : G. W. Brams
"Réseaux de Petri : théorie et pratique". Tome 1, Edition Masson, Paris (1982).

[Ehrig... 85] : H. Ehrig and B. Mahr
"Fundamentals of Algebraic Specification 1 : Equations and Initial Semantics". EATCS Monographs on Theoretical Computer Science, Vol. 6, W. Brauer, G. Rozenberg, A. Salomaa (Eds.), Springer Verlag (1985)

[Farkas 02] : J. Farkas
"Theorie der einfachen Ungleichungen". Journal für reine und angew. Mathematik 124, pp 1-27 (1902).

[Gaudel 79] : M.C. Gaudel
"Algebraic Specification of Abstract Data Type". R.R. N° 360, INRIA, Le Chesnay (1979).

[Graubmann 85] : P. Graubmann
"Composition of place transition nets using additional places (or transitions) and the calculation of their invariants". Internal Report of the Esprit project n° 283 FO-ME-TOO, september 1985 (1985).

[Haddad... 86] : S. Haddad and C. Girault
"Algebraic structure of flows of a regular coloured net". 7th European Workshop on Application and Theory of Petri Nets, Oxford, June 1986 (1986).

[Huber... 85] : P. Huber, A.M. Jensen, L.O. Jepsen and K. Jensen
"Towards reachability trees for high-level Petri nets". in "Advances in Petri Nets 1984", L.N.C.S. 188, G.Rozenberg (Ed.), Springer Verlag, pp 215-233. (1985).

[Lautenbach... 74] : K. Lautenbach and H. Schmid
"Use of Petri nets for proving correctness of concurrent process systems". Information Processing 1974 - North Holland Pub. Co., pp 187-191 (1974).

[Jensen 81] : K. Jensen
"Coloured Petri nets and the invariant method". T.C.S. 14, pp 317-336 (1981).

[Martinez... 82] : J. Martinez and M. Silva
"A simple and fast algorithm to abtain all invariants of a generalized Petri Net". Informatik - Fachbrichte 52, C. Girault and W. Reisig (Eds.), Springer Verlag, pp 301-310, (1982).

[Mateti... 76] : P. Mateti and D. Nardingh
"On algorithms for enumerating all circuits of a graph". SIAM J. Comput., Vol.5, N°1, pp 90-99 (1976).

[Memmi... 80] : G. Memmi and G. Roucairol
"Linear algebra in net theory". Proc. of "Advanced Course on genral Net Theory of Processes and Systems" Hambourg 1979, L.N.C.S. 84, W. Brauer (Ed.), Springer Verlag (1980).

[Memmi 81] : G. Memmi
"Contrôle du parallèlisme et détection des blocages". Journée de Synthèse "Quelques outils d'aideà la conception et à la réalisation de systèmes informfatiques". AFCET - Informatique Gif/s/Yvette, pp 35-69, (1981).

[Memmi 83] : G. Memmi
"Methode d'analyse de réseaux de Petri, réseaux à files, et applications aux systèmes temps réel". Thèse de Doctorat d'Etat, Université Pierre et Marie Curie, Juin 1983 (1983)

[Memmi... 85] : G. Memmi and A. Finkel
"An introduction to fifo nets - monogeneous nets : a subclass of fifo nets". T.C.S. 35, pp 191-214, (1985).

[Silva... 85] : M. Silva, J. Martinez, P. Ladet and H. Alla
"Generalized inverses and the calculation of symbolic invariants for coloured Petri nets". T.S.I., Vol.4, N°1 - Numéro spécial Réseaux de Petri, G. Memmi (Ed.), pp 113-126, (1985).

[Toudic 81] : J.M. Toudic
"Algorithmes d'analyse structurelle des réseaux Petri". Thèse de 3ème cycle, Université Pierre et Marie Curie, Octobre 1981 (1981).

[Vautherin 85] : J. Vautherin
"Un modèle algebrique, basé sur les réseaux de Petri, pour l'étude des systèmes parallèles". Thèse de Docteur-Ingénieur, Université Paris-Sud, Juin 1985 (1985).

[Vautherin 85] : J. Vautherin and G. Memmi
"Computation of flows for unary Predicates/Transitions nets". in "Advances in Petri Nets 1984", L.N.C.S. 188, G.Rozenberg (Ed.), Springer Verlag, pp 307-327 (1985).

[Vautherin 86a] : J. Vautherin
"Parallel systems specifications with coloured Petri nets and algebraic abstract data types". 7th European Workshop on Application and Theory of Petri Nets, Oxford, June 1986 (1986).

[Vautherin 86b] : J. Vautherin
"Calculation of semi-flows of Pr/T-systems". Research Report L.R.I., N° 130, Université Paris Sud, Octobre 1986 (1986).

IX. APPENDIX

IX.1. Construction of $\mathfrak{I}(A)$

Let $\langle S', \Sigma', E' \rangle = \langle S \cup S^i, \Sigma \cup \Sigma^i, E \cup E^i \rangle$. The free algebra over A, $\mathfrak{I}(A)$, generated by the enrichment $\langle S^i, \Sigma^i, E^i \rangle$ is constructed as follows:

Let us denote by $T(\Sigma'\cup A)$ (resp. $T(\Sigma\cup A)$) the algebra of terms constructed with operations symbols of Σ' (resp. Σ) and values of A considered as constants. $T(\Sigma\cup A)$ is a Σ-algebra and there is a unique Σ-morphism $eval_A$ from $T(\Sigma\cup A)$ into A which extends the identity of A.

We define in $T(\Sigma'\cup A)$ a relation |--| by θ|--|θ' iff

either there is an equation g == d or d == g in E', an occurrence u and a substitution σ such that $\theta|_u = g\sigma$ and $\theta' = \theta[u \leftarrow d\sigma]$

or there is an occurrence u and a term θ_1 in $T(\Sigma\cup A)$ such that $\theta|_u \in T(\Sigma\cup A)$, $eval_A(\theta|_u) = eval_A(\theta_1)$ and $\theta' = \theta[u \leftarrow \theta_1]$

Now, let $\equiv$ be the reflexive and transitive closure of |--|. It is a Σ'-congruence and $\mathfrak{Z}(A)$ is the quotient algebra $T(\Sigma'\cup A)/\equiv$.

Notice that $\mathfrak{Z}(A)$ satisfies E' and, if A is a finitely generated algebra, $\mathfrak{Z}(A)$ is also finitely generated. Indeed, if A is finitely generated, then for every a in A, there is a term $\underline{a}$ in $T(\Sigma)$ such that $eval_A(\underline{a}) = a$. Then, let θ be a term of $T(\Sigma'\cup A)$; by substituting each value $a\in A$ occuring in θ by the term $\underline{a}$, we get a term θ' of $T(\Sigma')$ such that $\theta' \equiv \theta$. So the unique Σ'-morphism from $T(\Sigma')$ into $\mathfrak{Z}(A)$ which associates a term with its $\equiv$-class ($T(\Sigma') \subseteq T(\Sigma'\cup A)$) is surjective and thus $\mathfrak{Z}(A) \in Gen(\Sigma',E')$.

IX.2. Proof of (R1), a notion of semi-flows for schemas

Let $X = \{x^1,..., x^n\}$; we associate with each variable x^i and each symbol F a new variable x^i_F. Then we denote by $Z[\{x^i_F\}]$ the set of integer polynoms over these new variables.

For every F in Σ, we define a function Q_F from $T(\Sigma\cup X)$ into $Z[\{x^i_F\}]$ by

$$Q_F(\theta) = \#(F, \theta) + \sum_{i=1..n} \#(x^i, \theta).\ x^i_F.$$

(remember that $\#(F, \theta)$ denotes the number of occurrence of F in θ). Then this function is extended to $N^{(\Sigma\cup X)}$ by

$$Q_F(\sum w(\theta).\theta) = \sum w(\theta).Q_F(\theta)$$

Thus Q_F associates with each multiset of terms an integer polynom, and, for instance, with X = {f, x},

$$Q_*(app(f,*) + app(repp(x), f)) = 1 + 2\ f_* + x_*$$

When all the terms of w are closed, then $Q_F(w)$ is constant and equals to the number of occurrence of F in w. When w contains non closed terms, for each substitution σ: $X \rightarrow T(\Sigma)$, $Q_F(w\sigma)$ can be obtained by substituting in the polynom $Q_F(w)$ each variable x^i_F by the number of occurrences of F in $\sigma(x^i)$.

Now, let Ω = <S, T, X, w, m°> be a Σ-schema ; f be a vector of integers in $Z^S \times Z^{\Sigma\times S}$: $f = ((f_p), (f_{F,p}))$; and let A be a finitely generated algebra. Assume that

(A1) for every F in Σ such that $f_{F,p} \neq 0$ for one p at least,

$$\forall w, w' \in T(\Sigma), \quad ([[w]]_A = [[w']]_A) \Rightarrow (\#(F,w) = \#(F,w'))$$

(A2) for every transition t,

$$\sum_p \left[f_p . | c_{p,t} | + \sum_F f_{F,p} . Q_F(c_{p,t}) \right] = 0$$

where $c_{p,t} = w(p,t) - w(t,p)$ and $Q_F(c_{p,t}) = Q_F(w(p,t)) - Q_F(w(t,p))$.

Then for every reachable marking M of $[[\Omega]]_A$ and every terms-marking m which represents M (i.e., for each place p, $m(p) \in N^{(T(\Sigma))}$ and $[[m(p)]]_A = M(p)$),

$$\sum_p \left[f_p . | m(p) | + \sum_F f_{F,p} . \#(F, m(p)) \right] = \sum_p \left[f_p . | m^\circ(p) | + \sum_F f_{F,p} . \#(F, m^\circ(p)) \right] \quad (B1)$$

Indeed, let suppose that M(t >M' in $[[\Omega]]_A$; since A is finitely generated, there is a substitution $\sigma : X \rightarrow T(\Sigma)$ and, for each place p, there is w(p) in $N^{(T(\Sigma))}$ such that $M(p) = [[w(p,t) + w(p)]]_A$ and $M'(p) = [[w(t,p) + w(p)]]_A$. Let $m(p) = w(p,t) + w(p)$ and $m'(p) = w(t,p) + w(p)$;

$$\sum_p \left[f_p . | m(p) | + \sum_F f_{F,p} . \#(F, m(p)) \right] - \sum_p \left[f_p . | m'(p) | + \sum_F f_{F,p} . \#(F, m'(p)) \right]$$

$$= \sum_p \left[f_p . | c_{p,t} | + \sum_F f_{F,p} . Q_F(c_{p,t}) \right] = 0$$

according to (A2). According to (A1), this is right also for every terms-markings representing M and M'. Then, by induction on the set of reachable markings, we deduce the expected result.

Vectors f satisfying (A2) are called *general semi-flows* of the schema Ω. Hence, the general semi-flows of a schema give invariants assertions (B1) over the markings *for every algebra satisfying (A1).*

Notice that a basis of general semi-flows can be computed from (A2).

Now, let us come back to the channel example. One can easily prove that the vector f such that :

$f_{repp} = f_{p6} = f_{Fileval} = f_{p7} = f_{Per} = 1$, $f_p = 0$ otherwise,
$f_{*,int} = 1$ and $f_{p,F} = 0$ otherwise,

is a general semi-flow of Ω°. In order to conclude from (B1) that the invariant IV' is satisfied, we need to establish that $[[w]]_A = [[w']]_A$ implies $\#(*,w) = \#(*,w')$ whenever $\mathfrak{I}(A)$ is consistent:

Let suppose that $[[w]]_A = [[w']]_A$ and $\#(*,w) \neq \#(*,w')$. Since (x = x) == true is a theorem of $T(\Sigma',E')$, it is satisfied by $\mathfrak{I}(A)$. Hence true $\equiv$ ($Occ_*(w) = Occ_*(w)$). Moreover, since $[[w]]_A = [[w']]_A$, ($Occ_*(w) = Occ_*(w)$) $\equiv$ ($Occ_*(w) = Occ_*(w')$) in $\mathfrak{I}(A)$; thus true $\equiv$ ($Occ_*(w) = Occ_*(w')$). But one can prove by induction on the size of w and w' that $\#(*,w) \neq \#(*,w')$ implies that ($Occ_*(w) = Occ_*(w')$) $\equiv_{E'}$ false. Then ($Occ_*(w) = Occ_*(w')$) $\equiv$ false in $\mathfrak{I}(A)$. Thus true $\equiv$ false in $\mathfrak{I}(A)$.

The other invariants of (R1) are proved in the same way.

IX.3. Proof of (R2)

At first the proof requires to establish the following theorems in $T(\Sigma', E')$:

In?*(app(repp(x), f) == In?*(f)	(th1)
P?(app(repp(x), f)) == In?*(f)	(th2)
In?*(app(*, f)) == true	(th3)
P?(app(repc(y), f)) == P?(f)	(th4)
[(In?*(f) and P?(app(f, *)) => P?(f)] == true	(th5)

which can be proved by structural induction over f. Since $\mathfrak{S}'(A)$ is finitely generated, then these theorems are also satisfied by $\mathfrak{S}'(A)$. Then we show (R2) by induction over the sequence leading to M :

a) For M = M°, a = $[[\text{empty}]]_A$ thus In?*(a) ≡ In?(empty) ≡ false in $\mathfrak{S}'(A)$. Then either false ≢ true in $\mathfrak{S}'(A)$ and thus In?*(a) ≢ true or false ≡ true in $\mathfrak{S}'(A)$ and then P?(a) ≡ true anyway.

b) Assume that M(t >M' and that the property is satisfied by M. Then M'(int) = a' with $a' \in A_{\text{Buffer}}$.

1) If t = t1, then a' = $\text{app}_A(\text{repp}_A(x), a)$ with $x \in A_{\text{Producer}}$. If In?*(a) ≢ true, then In?*(a') ≡ In?*(app(repp(x), a)) $\equiv_{(\text{th1})}$ In?*(a) ≢ true. Similarly, if In?*(a) ≡ true, then In?*(a') ≡ true and then, P?(a') ≡ P?(app(repp(x), a)) $\equiv_{(\text{th2})}$ In?*(a) ≡ true.

2) If t = t2, then a' = $\text{app}_A(*_A, a)$ and thus In?(a') ≡ In?*(app(*, a)) $\equiv_{(\text{th3})}$ true. Moreover | M'(p1) | ≥ 1.

3) If t = t14, then a' = $\text{app}_A(\text{repc}_A(y), a)$ with y in A_{Consumer} and so, In?*(a') ≡ In?*(a) (by the same theorem as th2 for repc(y)) and P?(a') ≡ P?(a) by th5.

4) If t = t3, then a' = $\text{app}_A(\text{repp}_A(x), a)$ with x in A_{Producer} and so, In?*(a') ≡ In?*(a) and P?(a') ≡ P?(a).

5) If t = t4, idem.

6) If t = t5, then a = $\text{app}_A(a', *_A)$ and thus In?*(a) ≡ true. Now, assume that In?*(a') ≡ true. If | M(p1) | ≥ 1 then also | M'(p1) | ≥ 1. Otherwise, P?(a) ≡ true and then P?(a') ≡ (true => P?(a')) ≡ [(In?*(a') and P?(a)) => P?(a')] ≡ [(In?*(app(a', *)) and P?(a)) => P?(a')] $\equiv_{(\text{th5})}$ true.

7) For every other transition the property is obvious.

IX.4. Proof of (R3)

From IV' and IP', we already know that | M(Per) | = | m(Per) | ≤ max and | M(Depos) | ≤ ip.

If | M(per) | = max, then no transition can occur and increase M(Depos).

If | M(Depos) | = ip, then a marking M' has necessarily been reached before M such that | M'(Depos) | = ip, and M'(t5 >. Let us put M'(int) = a with $a \in A_{\text{Buffer}}$; a = $\text{app}_A(a', *_A)$ and thus In?*(a) ≡ true in $\mathfrak{S}'(A)$. But, according to IP', since | M'(Depos) | = ip, then | M(p1) | = 0 ; and so, according to (R2), P?(a) ≡ true in $\mathfrak{S}'(A)$. In addition, by structural induction, one can establish the following theorem in $T(\Sigma', E')$: [P?(f) => In?p(f)] == true. Thus In?p(a) ≡ [true => In?p(a)] ≡ [P?(a) => In?p(a)] ≡ true in $\mathfrak{S}'(A)$.

On the other hand, let m be a terms marking which represents M'. According to IP', #(repp, m(int)) = 0. Let m(int) = x with x in $T(\Sigma)$; a = $[[x]]_A$. By induction over the size of x, one can prove that #(repp, x) = 0 implies In?p(x) $\equiv_{E'}$ false and thus In?p(x) ≡ false in $\mathfrak{S}'(A)$. Hence we get a contradiction with the consistency of $\mathfrak{S}'(A)$.

9.

Linear Invariants in Commutative High Level Nets

J. M. Couvreur and J. Martínez

G. Rozenberg (ed.): Advances in Petri nets 1990. Lecture Notes in Computer Science, vol. 483.
Springer, Berlin Heidelberg New York 1990, pp. 146-165

ABSTRACT. Commutative nets are a subclass of colored nets whose color functions belong to a ring of commutative diagonalizable endomorphisms. Although their ability to describe models is smaller than that of colored nets, they can handle a broad range of concurrent systems. Commutative nets include net subclasses such as regular homogeneous nets and ordered nets, whose practical importance has already been shown.

Mathematical properties of the color functions of commutative nets allow a symbolic computation of a family of generators of flows. The method proposed decreases the number of non-null elements in a given color function matrix, without adding new columns. By iteration, the entire matrix is annulled and a generative family of flows is obtained. The interpretation of the invariants associated with each flow is straightforward.

Keywords. Linear invariants, flow computation, structural analysis methods, subclasses of Petri nets, colored nets

CONTENTS

1. INTRODUCTION

Petri nets are a family of tools with which concurrent systems can be modeled. One of their main advantages is that the constructed models can be analyzed. The best-known of these nets are the place/transition nets. In them, the tokens contained in any place are identical, and the arcs are labeled by natural numbers.

Colored nets can be considered as high level nets in relation to place/transition nets. They permit the construction of more concise and abstract models. Their tokens are distinguished by a differentiating attribute denoted color. In a colored net, a place can contain a non-negative number of colored tokens, or colors. A transition admits various firing modes, parametrized by colors. An arc from a transition to a place (or from a place to a transition) is labeled by a linear function (color function) which determines the colors to be added to (or to be removed from) the place upon firing the transition.

Many results on place/transition net analysis have been elucidated (see, for example (Brams 83), (Silva 85) or (Reisig 85)). Little has been done to generalize these results to high level nets. This paper concentrates on the problem of calculating linear invariants in colored nets. These invariants are the bases for the structural analysis of a net (Jensen 81) (Genrich,Lautenbach 83). The linear invariants can be directly deduced from the left annihilators or flows of the net's incidence matrix.

(Silva et al. 85) presents a method for calculating flows in colored nets, based on a generalization of the Gauss elimination method applied to the net's incidence matrix. In the general case, the process is controlled by heuristic rules which do not guarantee that a generative family of flows will be obtained.

Other works have studies the problem of calculating flows in subclasses of colored nets. Thus there are papers on unary nets (Vautherin,Memmi 85); factorizable nets with commutative functions (Alla et al. 85); regular nets (Haddad 87); associative nets and ordered nets (Haddad, Couvreur 88); etc.

This paper presents commutative nets (§2), a subclass of colored nets which contains regular homogeneous nets, ordered nets, and projection nets (which are defined in §2.3), among others. The fact that we consider a subclass of nets limits the capacity for modeling. However, commutative nets can describe a broad range of practical problems (communication protocols; queue, ring and counter management systems; filters; etc..).

One compensation is that the possibilities of analyzing the models are greatly enhanced.

The color functions of a commutative net belong to a ring of commutative diagonalizable endomorphisms. Section §3 studies these properties and their effect on calculating the flows of a net.

Section §4 deals with the problem of calculating a generative family of flows for a commutative net. The key result is the algorithm presented in §4.3 which, when applied to a color function matrix, decreases the number of non-null functions in a given column. By iteration, the entire color function matrix is annulled, giving a generative family of flows.

Finally, §5 gives some examples of modeling with commutative nets in order to illustrate the ideas presented before. In each case, a generative family of flows is computed, and the corresponding linear invariants are obtained and interpreted.

2. COMMUTATIVE HIGH LEVEL NETS

2.1 Definition of a colored net

Definition 2.1.1: A ***colored net*** is a 6-tuple $CPN=\langle P,T,C,W^+,W^-,M_0\rangle$ where:

- P is a non-empty set of places
- T is a non-empty set of transitions
- $C: P \cup T \rightarrow \Omega$, where Ω is a set of finite non-empty sets
- W^+ (W^-) is the pre-(post-) incidence matrix of $P\times T$, where $W^+(p,t)$ ($W^-(p,t)$) is a function of $C(p)\times C(t)$ in **N** (the set of natural numbers).
- M_0, the initial marking, is a vector indexed by the elements of P, where $M_0(p)$ is a function of C(p) in **N**.

Notes:

- The color functions $W^+(p,t)$ and $W^-(p,t)$ are matrices $C(p)\times C(t)$ with coefficients in **N** which, consequently, define respective linear applications of Bags(C(t)) in Bags(C(p)). (Bags(A) is the set of multisets over A).
- The initial marking $M_0(p)$ of place p takes its values in Bags(C(p)).

Definition 2.1.2: The ***transition firing rule*** is given by:

- A transition t is firable under a marking M with respect to a color c_t of C(t) if and only if: $\forall p \in P,\ \forall c \in C(p),\ M(p)(c) \geq W^+(p,t)(c,c_t)$
- Firing a transition t with respect to the color c_t of C(t) leads to a new marking M' defined by:

$$\forall p \in P,\ \forall c \in C(p),\ M'(p)(c) = M(p)(c) + W^+(p,t)(c,c_t) - W^-(p,t)(c,c_t)$$

Definition 2.1.3: The ***incidence matrix*** of a colored net is defined by:$W=W^+-W^-$, where W(p,t) is a linear mapping whose associated matrix $C(p)\times C(t)$ takes values in **Z**. When a transition t is fired with respect to a color $c_t \in C(t)$ then, for every color

$c_p \in C(p)$, $W(c_p,c_t)$ gives the number of colors c_p to be added to (if the number is positive) or to be removed from (if it is negative) place p.

2.2 Definition of a commutative net

Definition 2.2.1: A ***homogeneous net*** is a colored net in which the domains of all its places and transitions are identical.

Notes:
- In a homogeneous net, the color functions are endomorphisms of Bags(C), where C is the common domain of the colors of places and transitions.
- We denote **E** as the vector space Bags(C) over the field **Q** (rational numbers) and End(E) as the ring of endomorphisms on **E**.

Definition 2.2.2: Two endomorphisms f and g are ***commutative*** iff $f.g = g.f$

Definition 2.2.3: Let f be an endomorphism of End(E). The function f is said to be ***diagonalizable*** on the field of complex numbers if there exists an invertible matrix H such that $H.f.H^{-1}$ is a complex diagonal matrix.

Proposition 2.2.1 (Chambalad 72) (Blyth,Robertson 86): ***Characterization of diagonalizable functions.*** Let f be an endomorphism of End(E). Then the following two statements are equivalent:

(i) There exists a polynomial, P(x), without multiple complex roots, which annuls the function f:
$P(f)=0$
P(x) has no multiple factors

(ii) The function f is diagonalizable on the field of complex numbers.

Definition 2.2.4: Let $\{u_1,\ldots,u_p\}$ be a finite family of endomorphisms of End(E),we denote $Q[u_1,\ldots,u_p]$ as the subring of polynomials generated by the family.

Definition 2.2.5: A ***commutative net*** is a homogeneous net such that the color functions are generated by a finite family of diagonalizable endomorphisms $\{u_1,\ldots,u_p\}$ which commute: $\forall u_i,u_k \in \{u_1,\ldots,u_p\}\ \ u_i.u_k = u_k.u_i$

2.3 Subclasses of commutative colored nets

Previous definition allows to define subclasses of commutative nets. In effect, every finite family of commutative and diagonalizable endomorphisms $\{u_1,\ldots,u_p\}$ defines a subclass. Functions $u_1,...,u_p$ will be named elementary functions of the subclass.

Below are the basic characteristics of three subclasses of colored nets which are particular cases of commutative nets.

Regular homogeneous nets (Haddad 87)

Color domain: $C=C_1\times ... \times C_k$, where $C_1, ..., C_k$ are color sets.
Family of elementary functions: $\{1, S_1, ..., S_k\}$ where 1 is the identity function and S_i is the **diffusion** function on C_i:

$S_i(X_1, ..., X_k) = \Sigma_{y_i\in C_i}<X_1, ..., X_{i-1}, y_i, X_{i+1}, ..., X_k>$, where X_j $(1\leq j\leq k)$ is a variable taking values in $Bag(C_j)$.
There exist polynomials $P_i(x)$, without multiple roots, which annul the functions S_i :

$P_i(X) = X^2 - n_i.X = 0$; $P_i(S_i) = S_i^2 - n_i.S_i = 0$, with $n_i = Card(C_i)$
The diffusion functions commute :

$S_i.S_j(X_1, ..., X_k) = S_j.S_i(X_1, ..., X_k) = \Sigma_{y_i\in C_i, y_j\in C_j}<X_1, ..., y_i, ..., y_j, ..., X_k>$

Ordered nets (Haddad,Couvreur 88)

Color domain: $C=C_1\times ... \times C_k$, where $C_1, ..., C_k$ are ordered color sets:

$C_i= \{c_{i,0}, c_{i,1}, ..., c_{i,n_i-1}\}$ $(1\leq i\leq k)$.
Family of elementary functions: $\{1, s_1, ..., s_k\}$ where 1 is the identity function and s_i is the **successor** function on color set C_i:

$s_i(X_1, ..., X_k) = <X_1, ..., X_{i-1}, X_i\oplus 1, X_{i+1}, ..., X_k>$, where X_j $(1\leq j\leq k)$ is a variable taking values in $Bag(C_j)$ and $\oplus 1$ is the unary operator **successor** defined for each color set C_j $(c_{j,0}\oplus 1=c_{j,1}; c_{j,1}\oplus 1=c_{j,2}; ... ; c_{j,n_j-1}\oplus 1=c_{j,0})$
There exist polynomials $P_i(x)$, without multiple roots, which annul the functions s_i:

$P_i(X) = X^{n_i} - 1$; $P_i(s_i) = s_i^{n_i} - 1 = 0$, where $n_i = Card(C_i)$
The successor functions commute :

$s_i.s_j(X_1, ..., X_k) = s_j.s_i(X_1, ..., X_k) = <X_1, ..., X_i\oplus 1, ..., X_j\oplus 1, ..., X_k>$
Net in figure 6 belongs to ordered net subclass.

Projection nets

Color domain: A
Family of elementary functions: $\{1, \Pi_{A1}, ..., \Pi_{Ak}\}$, where 1 is the identity function and the function Π_{Ai} is the **projection** of A on the subset A_i defined as follows:

if $c\in A_i$ then $\Pi_{Ai}(c) = c$ else $\Pi_{Ai}(c) = 0$
There exist polynomials $P_i(x)$, without multiple roots, which annul the functions Π_{Ai}:

$P_i(X) = X^2 - X$; $P_i(\Pi_{Ai}) = \Pi_{Ai}^2 - \Pi_{Ai} = 0$

The projection functions commute :

$\Pi_{Ai}.\Pi_{Aj} = \Pi_{Aj}.\Pi_{Ai} = \Pi_{Ai \cap Aj}$

Nets in figures 1, 2, 3, 4 and 5 belong to projection net subclass.

Projection net shown in figure 1 is a color filter. Let E be the domain of colors of places, P and Q, and of transition t. Let A and B be two subsets of E. Functions Π_A and Π_B are defined as follows.

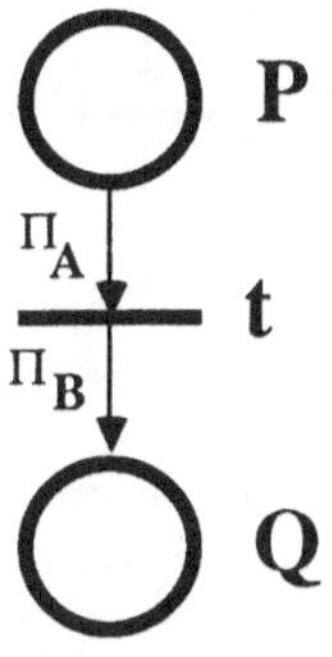

Fig.1: Color filter

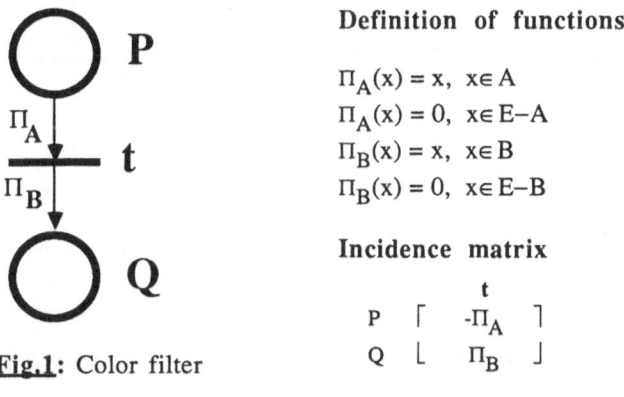

Definition of functions

$\Pi_A(x) = x,\ x \in A$
$\Pi_A(x) = 0,\ x \in E-A$
$\Pi_B(x) = x,\ x \in B$
$\Pi_B(x) = 0,\ x \in E-B$

Incidence matrix

$$\begin{array}{c} \\ P \\ Q \end{array} \begin{array}{c} t \\ \begin{bmatrix} -\Pi_A \\ \Pi_B \end{bmatrix} \end{array}$$

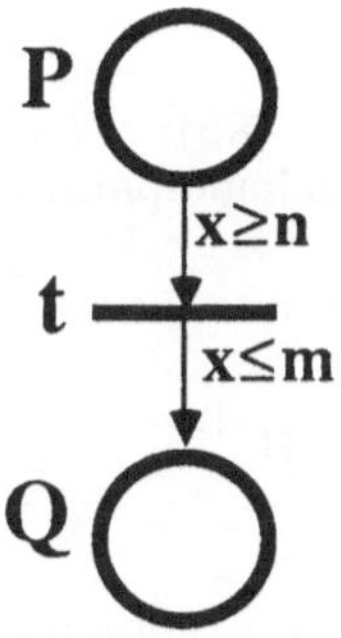

Fig.2: Particular case of a color filter

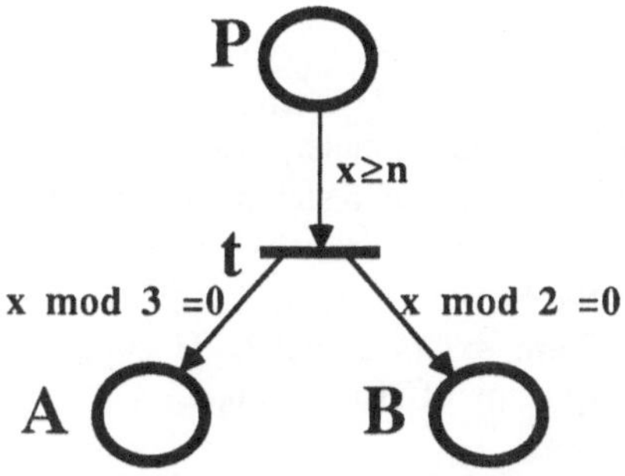

Fig.3: Selective filter of colors greater than or equal to a threshold n which are multiples of 2 (in B) and of 3 (in A).

Figures 2, 3 and 4 give specific examples of projection nets. In them the color functions labeling arcs appear as predicates, pred(x), instead of using a notation of type Π_A. The correspondence between the two is straightforward:

$x \in A \Leftrightarrow$ pred(x)

Note, finally, that the net in figure 4 can be rewritten in figure 5 by associating predicates with transitions. The transformation rule applied has been introduced in (Genrich 88). This allows us to use predicates, under certain conditions, in the process of computing the flows of a net.

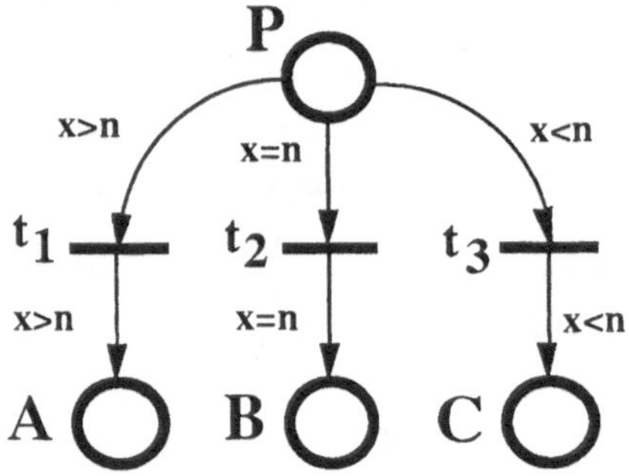

Fig.4: Color selector

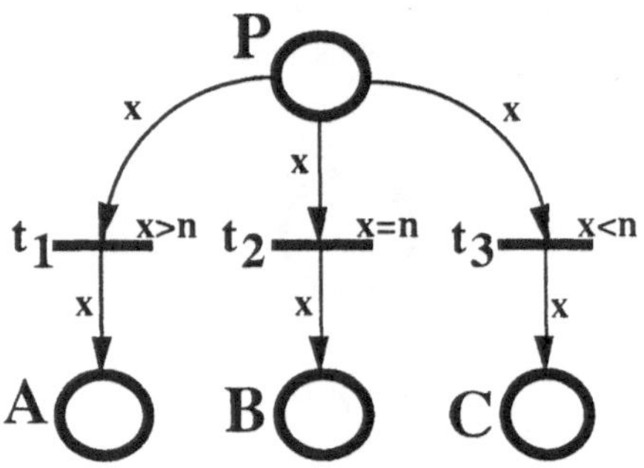

Fig.5: Alternative form of the net in figure 4

3. PROPERTIES OF THE FUNCTIONS OF A COMMUTATIVE NET

3.1 Stability theorem

Proposition 3.1.1: (Chambalad 72) (Blyth,Robertson 86): ***Simultaneous diagonalization***. Let $\{u_1, \dots, u_p\}$ be a finite family of endomorphisms of End(E). Then the following two statements are equivalent:

(i) The family $\{u_1, \dots, u_p\}$ is composed of commutative diagonalizable endomorphisms.

(ii) There exists a matrix H such that the matrix $H.u_i.H^{-1}$ is diagonal for the endomorphisms of $\{u_1, \dots, u_p\}$.

Proposition 3.1.2: ***Diagonalizable color function.*** Every color function of a commutative net is a diagonalizable function.

Proof:

Let $\{u_1, \dots, u_p\}$ be a family of elementary functions of a commutative net.

Every color function f can be expressed as a polynomial of elementary functions:

$$f = P(u_1, \dots, u_p)$$

The family $\{u_1, \dots, u_p\}$ is composed of diagonalizable endomorphisms which commute. According to propositions 2.2.1 and 3.1.1, there exists a matrix H such that the matrix $H.u_i.H^{-1}$ is diagonal for the endomorphisms of $\{u_1, \ldots, u_p\}$.

Then we have:

$$H.f.H^{-1} = H.P(u_1, \ldots, u_p).H^{-1} = P(H.u_1.H^{-1}, \ldots, H.u_p.H^{-1})$$

Since the sum and product of diagonal matrices are diagonal matrices, then $H.f.H^{-1}$ is a diagonal matrix. Therefore, the function f is diagonalizable. ♦

3.2 Properties of diagonalizable functions

The aim of this section is to characterise the solutions h of the equation **h.f=0**, where f is a diagonalizable function.

Definition 3.2.1: ***Annihilator of a function.*** Let f be an endomorphism of End(E). The set of solutions h of h.f=0 is a left ideal. This ideal is called a left annuller and is denoted by Ann(f).

Proposition 3.2.1: ***Annihilator of a diagonalizable function.*** Given a diagonalizable function f, there exists a polynomial R(x) such that R(f) is a generator of Ann(f).

Proof:

Let f be a diagonalizable function and P(x) a polynomial annihilator of f without multiple roots: P(f)=0.

If P(0)≠0, Let us put R(x)=P(x). Otherwise (P(0)=0) the polynomial P(x) is divisible by x and we put R(x)=P(x)/x.

We immediately find that R(f) is an element of Ann(f): R(f).f=0

Let us prove that R(f) is a generator of Ann(f):

Since P(x) has not got multiple roots, R(x) and x are relatively prime. Applying Bezout's theorem (Bourbaki 81), there are two polynomials a(x) and b(x) such that:

$$a(x).x + b(x).R(x) = 1$$

Putting x=f we get the following function equation:

$$a(f).f + b(f).R(f) = 1$$

Let h be an element of Ann(f). From the above èquation we get:

$$h.a(f).f + h.b(f).R(f) = h.b(f).R(f) = h$$

Therefore, f was generated by the function R(f). ♦

Proposition 3.2.2: ***Annihilator of the annihilator of a diagonalizable function.*** Given a diagonalizable function f and a polynomial S(x) which is a generator of Ann(f), then the left annihilator of S(f), Ann(S(f)), can be generated by f.

Proof:

1) f is an element of Ann(S(f)) since f . S(f) = 0

2) Let h be an element of Ann(S(f)), i.e. h.S(f)=0. We will prove that h can be expressed as h=g.f, where g belongs to End(E).

We consider two possible cases: $S(0) \neq 0$ and $S(0) = 0$

Case $S(0) \neq 0$: x and S(x) are necessarily relatively prime. Applying Bezout's theorem (Bourbaki 81), there exist two polynomials a(x) and b(x) such that:

$$a(x).x + b(x).S(x) = 1$$

Putting x=f we get the function equation:

$$a(f).f + b(f).S(f) = 1$$

Let h be an element of Ann(S(f)). From the above equation we get:

$$h.a(f).f + h.b(f).S(f) = h.a(f).f = h$$

Therefore, h is of the form g.f with g=h.a(f)

Case $S(0) = 0$: Let us prove that necessarily S(f)=0. This implies that Ann(f)= {0} and, therefore, that f is invertible and that Ann(S(f))=End(E) since f is a generator of Ann(S(f)).

Let P(x) be a non-null polynomial without multiple roots which satisfies P(f)=0 and let R(x) be a polynomial defined by:

If $P(0) \neq 0$ then R(x)=P(x)
else R(x)=P(x)/x (P(0)=0 means that P(x) is divisible by x)

Since P(x) has no multiple roots, then $R(0) \neq 0$. let us prove that R(x) is a generator of Ann(f). Let c(x)be the highest common denominator of R(x) and S(x).

$c(0) \neq 0$ since c(x) divides R(x) and $R(0) \neq 0$

Since c(x) divides S(x) and $c(0) \neq 0$, there exists a polynomial d(x) which satisfies:

$$S(x) = x.d(x).c(x)$$

Applying Bezout's theorem, there exist two polynomials a(x) and b(x) such that:

$$a(x).S(x) + b(x).R(x) = c(x)$$

Putting x=f we get the function equation:

$$a(f).S(f) + b(f).R(f) = c(f)$$

Multiplying both sides of the equation by f, and taking into account that R(f)=0 and that f.S(f)=0 then we get: f.c(f) = 0.

Therefore: S(f)= f.d(f).c(f) = 0 ♦

Proposition 3.2.3: ***Annihilator of the product of two functions.*** Given two endomorphism of End(E), f and g, which are diagonalizable and commute with each other, the following holds

$$Ann(f.g) = Ann(f) + Ann(g)$$

Proof:

Let R(x) and S(x) be two polynomials which satisfy:
- R(f) is a generator of Ann(f) and S(g) is a generator of Ann(g)
- $R(0)\neq 0$ and $S(0)\neq 0$

We will prove that every annihilator h of f.g has the form h1.R(f) + h2.S(g), where h1 and h2 are endomorphisms of End(E).

Applying Bezout's theorem, there are four polynomials a(x), b(x), c(x) and d(x) which satisfy:

$$a(x).x + b(x).R(x) = 1$$
$$c(x).x + d(x).S(x) = 1$$

Putting x=f in the first equation and x=g in the second we get the following equations:

$$a(f).f + b(f).R(f) = 1$$
$$c(g).g + d(g).S(g) = 1$$

Multiplying the two equations element by element and recalling that f and g commute with each other, we get:

$$a(f).c(g).f.g + a(f).d(g).f.S(g) + b(f).c(g).g.R(f) + b(f).d(g).R(f).S(g) = 1$$

Multiplying both elements by function h and taking into account that h.g.f=h.f.g=0 we get:

$$(h.a(f).d(g).f).S(g) + (h.b(f).c(g).g + h.b(f).d(g).S(g)).R(f) = h$$

Therefore, h has the form h1.R(f) + h2.S(g) with:

$$h1 = h.b(f).c(g).g + h.b(f).d(g).S(g) \quad \text{and}$$
$$h2 = h.a(f).d(g).f$$ ♦

Corollary 3.2.1: ***Annihilator of the product of k-functions.*** Given k endomorphisms of End(E), $\{f_1, ..., f_k\}$, which are diagonalizable and commutative, the following holds:

$$Ann(f_1.f_2. \ldots .f_k) = Ann(f_1) + Ann(f_2) + \cdots + Ann(f_k)$$

4. FLOWS OF A COMMUTATIVE NET

4.1 Symbolic flows of a homogeneous net

Definition 4.1.1: A ***symbolic flow*** of a homogeneous net is a vector F on $(End(E))^{card(P)}$ which satisifies the equation: F . W = 0

Note: The problem of computing the symbolic flows of a homogeneous commutative net is transformed into that of solving a system of linear equations on the ring End(E).

4.2 Elementary operations

Three types of elementary operations on the commutative net's incidence matrix allow us to get a generative family of flows:

- Adding a new column to the matrix, which is the result of multiplying one of its columns by a given color function.
- Adding to a row of the matrix the result of the multiplication of another row by a given color function.
- Substituting in the matrix the solutions associated with the annullment of a column.

Proposition 4.2.1: Let W be the incidence matrix of a commutative net, h a color function and W' the matrix obtained by adding to W the column resulting from the multiplication of one of its columns by the color function h. Then the flows associated with matrices W and W' are identical.

Proposition 4.2.2: Let W be the incidence matrix of a commutative net, h a color function and W' the matrix obtained by adding to row (b) of W the result of multiplying another of its rows, row (a), by the color function h.

Then $F' = (f_1, \dots, f_a, \dots, f_b, \dots, f_k)$ is a flow of W'
iff $F = (f_1, \dots, f_a+f_b.h, \dots, f_b, \dots, f_k)$ is a flow of W.

Proposition 4.2.3: Let W be the incidence matrix of a commutative net, (t) a column of W and $\{G_1, \dots, G_r\}$ a family which generates the flow associated with column (t). Let us define the matrix G composed of the linear vectors G_i and the matrix $W' = G.W$. The flows of matrices W and W' are related by the following properties:

(1) F' is a flow of W' $\Leftrightarrow$ F'.G is a flow of W

(2) F is a flow of W $\Leftrightarrow$ there exist F' of W' such that $F=F'.G$

Proof:

(1) F' is a flow of W' $\Leftrightarrow$ $F'.G.W = 0$ $\Leftrightarrow$ F'.G is a flow of W

(2) If there exist flows F' of W' such that $F=F'.G$, according to property (1) F is a flow of W.

If F is a flow of W, F is also a flow of column (t) and can, therefore, be expressed as a linear combination of the family $\{G_1, \dots, G_r\}$ of the form F'.G. According to property (1) F' is a flow of W'. ♦

4.3 Solution algorithm

With the sequence of elementary operations which will be presented here, we can decrease the number of non-null functions in a given column of the color function matrix without increasing the number of columns. Thus, by iterating this sequence we get a generative family of flows.

Let W be an incidence matrix of the form:

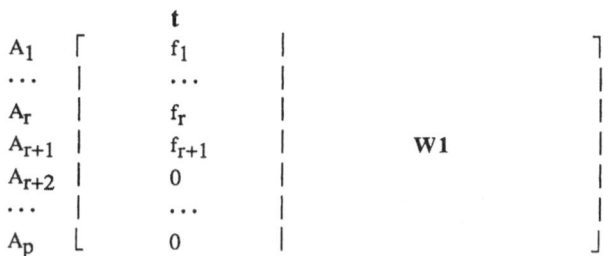

$$
\begin{array}{l}
\\
A_1 \\ \cdots \\ A_r \\ A_{r+1} \\ A_{r+2} \\ \cdots \\ A_p
\end{array}
\begin{array}{c}
t \\
\left[\begin{array}{c|c}
f_1 & \\
\cdots & \\
f_r & \\
f_{r+1} & \mathbf{W1} \\
0 & \\
\cdots & \\
0 &
\end{array}\right]
\end{array}
$$

Step 1: Add a column

Let $K_i(f_i)$ be a polynomial in f_i which generates $Ann(f_i)$ for $i \in [1,r]$. Let us add to matrix W the column $K_1(f_1). \ldots . K_r(f_r).t$

$$
\begin{array}{l}
\\
A_1 \\ \cdots \\ A_r \\ A_{r+1} \\ A_{r+2} \\ \cdots \\ A_p
\end{array}
\begin{array}{c}
t \qquad\qquad \mathbf{K_1(f_1). \ldots . K_r(f_r).t} \qquad\qquad \\
\left[\begin{array}{c|c|c}
f_1 & 0 & \\
\cdots & \cdots & \\
f_r & 0 & \\
f_{r+1} & K_1(f_1). \ldots . K_r(f_r).f_{r+1} & \mathbf{W1} \\
0 & 0 & \\
\cdots & \cdots & \\
0 & 0 &
\end{array}\right]
\end{array}
$$

Step 2: Substitute the flows of the added column

According to the properties of color functions $\{f_1, \ldots, f_r, K_{r+1}(f_{r+1})\}$ is a family which generates the annihilators of the added column $K_1(f_1). \cdots . K_r(f_r).t$. Substituting we get:

$$
\begin{array}{r}
\\
A_1 \\ \cdots \\ A_r \\ f_1.A_{r+1} \\ \cdots \\ f_r.A_{r+1} \\ K_{r+1}(f_{r+1}).A_{r+1} \\ A_{r+2} \\ \cdots \\ A_p
\end{array}
\begin{array}{c}
t \qquad\qquad \mathbf{K_1(f_1). \ldots .K_r(f_r).t} \qquad\qquad \\
\left[\begin{array}{c|c|c}
f_1 & 0 & \\
\cdots & \cdots & \\
f_r & 0 & \\
f_1.f_{r+1} & 0 & \\
\cdots & \cdots & \mathbf{W3} \\
f_r.f_{r+1} & 0 & \\
0 & 0 & \\
0 & 0 & \\
\cdots & \cdots & \\
0 & 0 &
\end{array}\right]
\end{array}
$$

Step 3: Subtract from each row $f_i.A_{r+1}$ the row A_i premultiplied by f_{r+1}

With this we eliminate from column (t) all the added non-null functions. In addition, we can eliminate the added column, since it has been completely annulled.

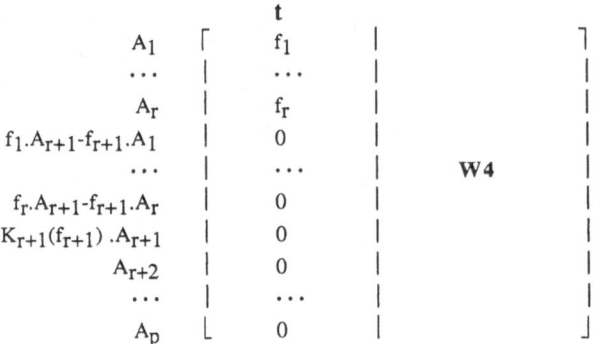

$$\begin{array}{r} \\ A_1 \\ \cdots \\ A_r \\ f_1.A_{r+1}-f_{r+1}.A_1 \\ \cdots \\ f_r.A_{r+1}-f_{r+1}.A_r \\ K_{r+1}(f_{r+1}).A_{r+1} \\ A_{r+2} \\ \cdots \\ A_p \end{array} \begin{array}{c} t \\ \left[\begin{array}{c|c} f_1 & \\ \cdots & \\ f_r & \\ 0 & \\ \cdots & W4 \\ 0 & \\ 0 & \\ 0 & \\ \cdots & \\ 0 & \end{array}\right] \end{array}$$

5. APPLICATION OF THE COMPUTATION ALGORITHM

5.1 Filter models

We will study the computation of a generative family of flows of the projection net shown in figure 1, which is a color filter.

Incidence matrix

$$\begin{array}{c} \\ P \\ Q \end{array} \begin{array}{c} t \\ \left[\begin{array}{c} -\Pi_A \\ \Pi_B \end{array}\right] \end{array}$$

Step 1: Add a column

The annihilator of Π_A is $(1 - \Pi_A)$.

$$\begin{array}{c} \\ P \\ Q \end{array} \begin{array}{c} \begin{array}{cc} t & (1 - \Pi_A).t \end{array} \\ \left[\begin{array}{c|c} -\Pi_A & 0 \\ \Pi_B & \Pi_B.(1 - \Pi_A) \end{array}\right] \end{array}$$

Step 2: Sustitute the flows of the added column

$$\begin{array}{c} \\ P \\ \Pi_A.Q \\ (1 - \Pi_B).Q \end{array} \begin{array}{c} \begin{array}{cc} t & (1 - \Pi_A).t \end{array} \\ \left[\begin{array}{c|c} -\Pi_A & 0 \\ \Pi_A.\Pi_B & 0 \\ 0 & 0 \end{array}\right] \end{array}$$

Step 3: Add to the row $\Pi_A.Q$ the row P premultiplied by Π_B

$$\begin{array}{r} P \\ \Pi_A.Q+\Pi_B.P \\ (1-\Pi_B).Q \end{array} \begin{bmatrix} t \\ -\Pi_A \\ 0 \\ 0 \end{bmatrix}$$

Row P can be eliminated by premultiplying by the function $(1 - \Pi_A)$. The computed generative family of flows is:

	P	Q
(F1)	[$(1-\Pi_A)$	0]
(F2)	[0	$(1-\Pi_B)$]
(F3)	[Π_B	Π_A]

Interpreting the invariants associated with the flows:

We denote P(x) the number of colors x in the marking of place P and $P_o(x)$ the number of colors x in the initial marking of place P.

Given that $\Pi_C(x) = x$ if $x \in C$ and $\Pi_C(x) = 0$ if $x \notin C$, we can deduce the following invariant relations:

(F1) $\forall x \notin A \quad (1-\Pi_A)(x) = x - 0 = x \Rightarrow P(x) = P_o(x) = const$
Color x belonging to E-A cannot leave place P. P(x) is an invariant.

(F2) $\forall x \notin B \quad (1-\Pi_B)(x) = x - 0 = x \Rightarrow Q(x) = Q_o(x) = const$
Color x belonging to E-B cannot reach place Q. Q(x) is an invariant

(F3) $\forall x \in A \cap B \quad \Pi_A(x) = \Pi_B(x) = x \Rightarrow P(x)+Q(x) = P_o(x)+Q_o(x) = const$
Number of colors $x \in A \cap B$ in places P and Q is invariant.

In a similar way we can compute flows for nets in figures 2, 3 and 4:

Incidence matrix of net in figure 2:

$$\begin{array}{r} P \\ Q \end{array} \begin{bmatrix} t1 \\ -[x \geq n] \\ [x \leq m] \end{bmatrix}$$

Generative family of flows:

	P	Q
(F1)	[$[x < n]$	0]
(F2)	[0	$[x > m]$]
(F3)	[$[x \leq m]$	$[x \geq n]$]

Incidence matrix of net in figure 3:

	t1
P	$-[x \geq n]$
A	$[x \bmod 3 = 0]$
B	$[x \bmod 2 = 0]$

Generative family of flows:

	P	A	B
(F1)	$[x < n]$	0	0
(F2)	0	$[x \bmod 3 <> 0]$	0
(F3)	0	0	$[x \bmod 2 <> 0]$
(F4)	$[x \bmod 3 = 0]$	$[x \geq n]$	0
(F5)	$[x \bmod 2 = 0]$	0	$[x \geq n]$
(F6)	0	$[x \bmod 2 = 0]$	$[x \bmod 3 = 0]$

Incidence matrix of net in figure 4:

	t1	t2	t3
P	$-[x > n]$	$-[x = n]$	$-[x < n]$
A	$[x > n]$	0	0
B	0	$[x = n]$	0
C	0	0	$[x < n]$

Generative family of flows:

	P	A	B	C
(F1)	0	$[x \leq n]$	0	0
(F2)	0	0	$[x <> n]$	0
(F3)	0	0	0	$[x \geq n]$
(F4)	1	1	1	1

5.2 Synchronized clocks arranged on a virtual ring

Figure 6 shows a commutative net which models the synchronization of the local clocks of M stations connected to a virtual ring. The clock of each station, H, has an associated value X. The values or dates of the clocks are synchronized by sending messages along the net. Each station, H, sends a message to the next one, H⊕1 (firing transition t2). When a station H has sent its synchronization message to the next station (the color <H,X> marks place B) and has received the corresponding message from the previous station (the color <H,X> marks place C) it can proceed to update its local clock with the value (date) X⊕1 (firing transition t1).

The domain of the colors of the places and transitions in the net is C1 × C2, where:

C1 = { Stations } = {0 , ··· , M-1}
C2 = { Dates of the local clocks } = {0 , ··· , N-1}
M and N are parameters

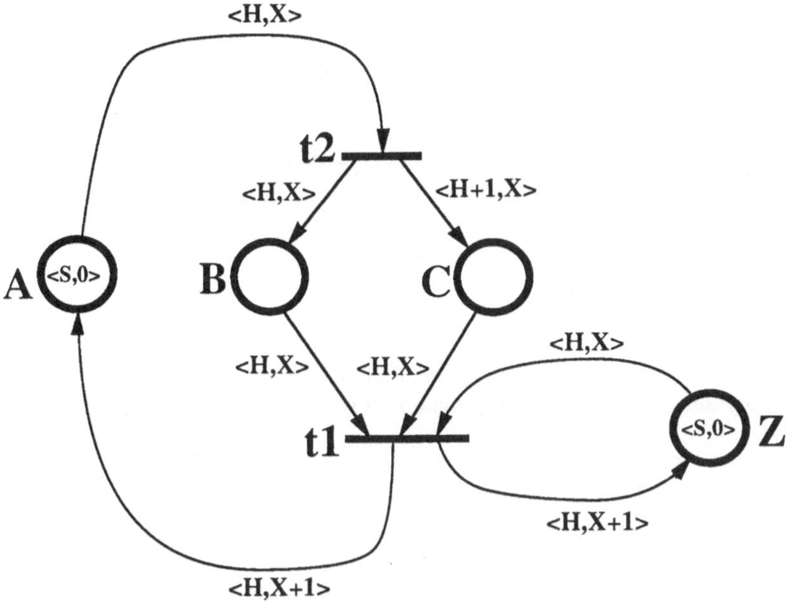

Fig.6: Commutative net which models a protocol for synchronizing the local clocks of stations on a virtual ring. (Note: <S,0> = <0,0> + <1,0> + ... + <M-1,0>)

Incidence matrix

	t1	t2
Z	<H,X⊕1> - <H,X>	0
A	<H,X⊕1>	-<H,X>
B	–<H,X>	<H,X>
C	–<H,X>	<H⊕1,X>

Writing the incidence matrix in terms of functions

To make easier the notation, we have associated names (identifiers) with the color functions of the model:

1 for the "identity" function { i.e.: **1**(<H,X>) = <H,X> }

λ for the "next_value_of_clock" function { i.e.: $\lambda(\langle H,X\rangle) = \langle H,X\oplus 1\rangle$ }

μ for the "next_station" function { i.e.: $\mu(\langle H,X\rangle) = \langle H\oplus 1,X\rangle$ }

The incidence matrix of the commutative net can be rewritten as:

	t1	t2
Z	$\lambda - 1$	0
A	λ	-1
B	−1	1
C	−1	μ

By elementary transformations of the incidence matrix we get:

	t1	t2
Z	$\lambda - 1$	0
Z - A - B	0	0
B	−1	1
C - μ.B	μ-1	0

Let us first consider column t2. Only one element remains to be eliminated: function 1 whose polynomial annihilator is polynomial 0: Ann(1)=0. Therefore, row B can be eliminated.

	t1	t2
Z	$\lambda - 1$	0
Z - A - B	0	0
C - μ.B	μ-1	0

Applying the elimination algorithm given in §4.3 to column (t1) we immediately get a generative family of flows:

	Z	Z-A-B	C - μ . B
(F1)	S_λ	0	0
(F2)	0	0	S_μ
(F3)	0	1	0
(F4)	$-(\mu - 1)$	0	$(\lambda - 1)$

Where: $S_\lambda = 1 + \lambda + ... + \lambda^{N-1}$ is a generator of Ann(λ - 1) and

$S_\mu = 1 + \mu + ... + \mu^{M-1}$ is a generator of Ann(μ - 1)

Interpreting the invariants associated with the flows:

(F1) $Z(H,0) + \cdots + Z(H,N-1) = 1$

For each station clock, H, there is only one token (H,X) marking place Z. Then X is the current date of clock H.

(F2) $C(0,X) + \cdots + C(M-1,X) = B(0,X) + \cdots + B(M-1,X)$

For each date, X, the number of messages (-,X) waiting for updating stations clocks (colors (-,X) marking place C) is equal to the number the station with current date X waiting for an update message (colors (-,X) marking place B).

(F3) Z(H,X) = A(H,X) + B(H,X)
A station H, independently of its state (place A or B marked), could send or receive only its clock date, X.

(F4) C(H,X) - C(H,X⊖1) = Z(H,X) - Z(H⊖1,X) + B(H⊖1,X) - B(H⊖1,X⊖1) =
= Z(H,X) - A(H⊖1,X) - B(H⊖1,X⊖1)
The difference between the number of consecutive update messages sent by a station H and not yet received only depends on the state of the stations H⊖1 and H (markings Z(H,X), A(H⊖1,X) and B(H⊖1,X⊖1)).

6. CONCLUSION

Commutative nets, a particular case of colored nets, have been introduced. Commutative nets include, as subclasses, regular homogenous nets, ordered nets and projection nets, among others. These nets can model a broad range of concurrent systems of practical importance, such as communication protocols; queue, ring and counter management systems; filters; etc.. Furthermore, other subclasses of nets can be transformed into commutative nets by applying homogenization and decomposition rules.

The color functions which appear both in the model (incidence matrix) and in the process of computing flows belong to a ring of commutative functions which are diagonalizable with respect to a single base. This enables us to compute flows symbolically.

We have proposed an algorithm for computing a flow generator family. The algorithm is based on an elimination method which, in each step, decreases the number of non-null functions in a given column of the color function matrix.

These results are a generalization of those already known for computing flows in regular nets (Haddad 87) and ordered nets (Haddad, Couvreur 88). Furthermore, they complement the method of computation by elimination proposed by (Silva et al. 85) since they give partial solutions to the problem of controlling the elimination algorithm.

ACKNOWLEDGEMENTS

The authors are indebted to the referees for their helpful comments. This work was partially supported by the DEMON Esprit Research Action 3148 and the Plan Nacional de Investigación, Grant TIC-0358/89.

REFERENCES

(Alla et al. 85) H.Alla, P.Ladet, J. Martínez, M. Silva. Modelling and validation of complex systems by coloured Petri nets. *Advances in Petri nets 1984. L.N.C.S. 188*, Springer-Verlag, pp.15-31

(Blyth,Robertson 86) T.S.Blyth, E.F.Robertson. *Linear Algebra (Vol. 4)*. Chapman and Hall. London.

(Bourbaki 81) N.Bourbaki . *Algèbre (Chapitres 4 à 7)*. Masson. Paris.

(Brams 83) G.W.Brams. *Réseaux de Petri:théorie et pratique*. Masson. Paris

(Chambalad 72) L.Chambalad. *Algèbre multilinaire*. Dunod. Paris

(Genrich 88) H.J.Genrich.Equivalence transformations of PrT-Nets. *9th European Workshop on Application and Theory of Petri Nets. Vol. II.* Venice (Italy). June. pp. 229-248

(Genrich,Lautenbach 83) H.J.Genrich, K.Lautenbach.S-invariance in predicate transition nets.*Informatik Fachberichte 66: Application and Theory of Petri Nets.* A.Pagnoni,G.Rozenberg (eds.). Springer-Verlag. pp. 98-111

(Jensen 81) K.Jensen. Coloured Petri nets and the invariant method. *Theoretical Computer Science 14*. North Holland Publ. Co. pp.317-336

(Haddad 87) S.Haddad. *Une catégorie régulière de réseau de Petri de haut niveau: définition, propietés et reductions. Application à la validation des systèmes distribués.* Ph.D. University Paris VI. June.

(Haddad, Couvreur 88) S.Haddad, J.M.Couvreur. Towards a general and powerful computation of flows for parametrized coloured nets. *9th European Workshop on Application and Theory of Petri Nets. Vol. II.* Venice (Italy). June.

(Reisig 85) W.Reisig. *Petri nets.* EATCS Monographs on Theoretical Computer Science, Vol. 4. Springer Publ. Co.

(Silva 85) M.Silva. *Las redes de Petri en la automática y la informática.* Ed. AC. Madrid.

(Silva et al. 85) M.Silva, J.Martínez, P.Ladet, H.Alla. Generalized inverses and the calculation of invariants for coloured Petri nets. *Technique et science informatique.* Vol.4 nº1, pp. 113-126

(Vautherin,Memmi 85) J.Vautherin, G.Memmi. Computation of flows for unary predicates transition nets. *Advances in Petri nets 1984. L.N.C.S. 188*, Springer-Verlag. pp.455-467

10.

Generalized Inverses and the Calculation of Symbolic Invariants for Coloured Petri Nets

M. Silva, J. Martinez, P. Ladet and H. Alla

Technique et Science Informatiques *4,1* (1985) 113-126

SUMMARY Coloured Petri Nets permit construction of more compact models than do generalized Petri Nets. In coloured Petri Nets the real parallelism can be considerable which leads to large sets of reachable markings. To analyse coloured Nets, the method of linear invariants of markings has been used successfully. In this paper, we present a systematic method for calculating symbolic invariants of a coloured Petri Net that is to say, invariant relations in terms of place markings, of functions associated to the arcs of the coloured net and, eventually, of generalised inverses of these functions.

Table of Contents

1. Introduction

Coloured Petri nets permit the construction of more compact models than do generalized Petri nets ([1]). A coloured net can be seen as a folding of a generalized Petri net : (1) to each place we associate a set of places of the generalized net; (2) to each transition we associate a set of transitions of the generalized net and (3) to each arc we associate a linear function which relates the places and transitions of the underlying net.

In coloured Petri nets the real parallelism can be considerable, which leads to large sets of reachable markings. To analyse coloured nets, the method of linear invariants of the marking has been used successfully. It has been possible to show that certain coloured nets are bounded or without deadlocks [Jensen 81a], [Jensen 81b], [Alla 84]. These invariants are of the form $x.W = 0$, in which W is the incidence matrix of the net under consideration.

The linear invariants were introduced in [Lautenbach 74] for generalized Petri nets. For these, the automatic search for a basis of invariants or for the generating set of non-negative invariants can be carried out with no difficulty [Silva 80], [Martinez 82]. In this paper we present a systematic method for calculating symbolic invariants of a coloured Petri net. That is to say, invariant relations expressed directly in terms of place markings (or transition firings), of functions associated to the arcs of the coloured net and, eventually, of generalized inverses of these functions.

The set of four rules set out in [Jensen 81b] permits the transformation of an incidence matrix of a coloured net into a simpler matrix (that is, with a lower number of columns or/and rows) but with the same set of invariants.

The above mentioned rules are complementary, in every way, with the results shown in this paper.

([1]) The references [Brams 83] et [Silva 85], for example, contain the basic results concerning the theory of generalized Petri nets.

2. Basic concepts

2.1. Definitions

Let S be a non-empty set and let DS be $\mathbb{N}$ or $\mathbb{Z}$. By $[S \rightarrow DS]_f$ we denote the set of functions $g : S \rightarrow DS$, where the support $\{ s \in S \mid g(s) \neq 0 \}$ is finite. In the rest of this paper, supports will always be finite and then for simplicity we will only write $[S \rightarrow DS]$ in future. Let CS be a non-empty set of elements named colours and $\mathcal{P}(CS)$ be the set of CS parts.

DEFINITION 1 [Jensen 81a] : *A coloured Petri Net is a 5-tuple* $CPN = \langle P, T, C, W, m_0 \rangle$ *where :*

1) P *is a set of places ;*

2) T *is a set of transitions ;*

3) $P \cap T = \emptyset$ *and* $P \cup T \neq \emptyset$ *;*

4) $C : P \cup T \rightarrow \mathcal{P}(CS) - \{ \emptyset \}$ *is the colour function ;*

5) $W : P \times T \rightarrow [C(t) \rightarrow [C(p) \rightarrow \mathbb{Z}]]$ *is the incidence function ;*

6) $m_0 : P \rightarrow [C(p) \rightarrow \mathbb{N}]$ *is the initial marking.* □

Each function $h \in [C(t) \to [C(p) \to \mathbb{Z}]]$ has a unique linear extension in $[[C(t) \to \mathbb{Z}] \to [C(p) \to \mathbb{Z}]]$, which will also be denoted by h.

To each linear function h we can associate a matrix $h \in \mathbb{Z}^{cp \times ct}$ which represents the linear transformation under consideration, where cp and ct are the cardinals of the supports of $[C(p) \to \mathbb{Z}]$ and $[C(t) \to \mathbb{Z}]$, respectively. In the future we will denote with the same name the linear functions and their associated matrices.

The incidence function W is in principle a matrix of functions. In virtue of the proposed linear extension, we can consider W as a matrix defined by boxes, such that its box $W(p, t)$ is the matrix associated with $[C(t) \to [C(p) \to \mathbb{Z}]]$.

A *weighted set of transitions* is a function y defined on T such that $y(t) \in [C(t) \to \mathbb{Z}]$ for all $t \in T$. It is non-negative if $y(t)(c) \geq 0$ for all pairs $t \in T$ and $c \in C(t)$.

Let m be a marking reachable from the initial marking m_0. There exists a non-negative weighted set of transitions, y, such that :

$$m = m_0 + W.y. \tag{1}$$

Let U be a non-empty set. A *weighted set of places*, with respect to U, is a function v defined on P, such that :

$$v(p) \in [C(p) \to [U \to \mathbb{Z}]] \quad \text{for all} \quad p \in P.$$

Let x be a weighted set of places (with respect to U). If $x.W = 0$ then $x.m = x.m_0$ for all markings m reachable from the initial marking m_0.

In fact, if we premultiply (1) by x :

$$x.m = x.m_0 + x.W.y = x.m_0 \quad \text{given that} \quad x.W = 0.$$

We say that x is a *p-flow* or *flow* of W and $x.m$ is a *marking invariant* of the net or a *p-invariant*. In the rest of this paper, except where the contrary is clearly stated, a flow denotes a p-flow and an invariant denotes a p-invariant.

Let us consider a classical example of modeling with *CPN* before the presentation of the definitions and basic properties related with generalized inverses. This example has been handled by several authors (see for example : [Jensen 81b] [Reisig 82]).

A set of database managers communicate with each other via a fixed set of message buffers. Each manager can make an update to his own database. At the same time, he must send a message to each of the other managers, thereby informing them about the update. After that, the sending manager waits until all other managers have received his message, performed an update and sent an acknowledgment. When all acknowledgments are present, the sending manager returns to be inactive. At that time, but not before, another manager may perform an update and send messages.

The system can be described by the coloured Petri net in figure 1.

Let :

— $h \in \mathbb{Q}^{n \times m}$ be a linear function;
— $R(h)$ be the range of h, that is the linear span of the h rows. That is $R(h) = \{ y \in \mathbb{Q}^m \mid \exists x \in \mathbb{Q}^n \text{ such that } x^T.h = y^T \}$;
— Ker (h) be the kernel of h, that is

$$\mathrm{Ker}(h) = \{ x \in \mathbb{Q}^n \mid x^T.h = 0 \};$$

— dim (h) be the $R(h)$ dimension, where

$$\dim(h) = \mathrm{rank}(h) = m';$$

— CK and CR be complementary subspaces of Ker (h) and $R(h)$, respectively.

DEFINITION 2 : ([Campbell 79], *algebraic definition*) : *h^* is called the (CK-CR)-generalized inverse of h if :*

1) $R(h^*) = CK$
2) Ker $(h^*) = CR$
3) $h.h^*.h = h$
4) $h^*.h.h^* = h^*$. □

For a given CK and CR, h^* is a uniquely defined linear transformation from $\mathbb{Q}^{m \times n}$.

Let $\mathrm{id}_n \in \mathbb{Q}^{n \times n}$ be a identity function (matrix) and let h^* be a generalized inverse of $h \in \mathbb{Q}^{n \times m}$.

PROPERTY 1 : *The rows of the matrix $(\mathrm{id}_n - h.h^*)$ generate* Ker (h). *That is to say : $R(\mathrm{id}_n - h.h^*) = \mathrm{Ker}\,(h)$.* □

Let $l \in \mathbb{Q}^{m' \times n}$ and $r \in \mathbb{Q}^{m \times m'}$ be linear transformations which satisfy : rank $(l.h)$ = rank $(h.r)$ = rank (h) = m'.

PROPOSITION 1 : *The function $h^* = r.(l.h.r)^{-1}.l$ is a generalized inverse of h.*

Proof : Following definition 2, conditions 1 and 2 are deduced directly on considering the equivalence of the algebraic and functional definitions of a $(CK\text{-}CR)$-generalized inverse (see [Campbell 79]).

Condition 3 : $\mathrm{rank}(h) = \mathrm{rank}(l.h) \Rightarrow h = v.l.h$ $\{ v \in \mathbb{Q}^{n \times m'} \}$

$$h = v.l.h = v.\mathrm{id}_{m'}.l.h = v.(l.h.r).(l.h.r)^{-1}.l.h$$
$$= h.r.(l.h.r)^{-1}.l.h = h.h^*.h.$$

Condition 4 : $h^*.h.h^* = r.(l.h.r)^{-1}.l = h^*$. □

The choice of « r » and « l » implies the definition of subspaces CK and CR.

To illustrate the construction of a generalized inverse, let us consider the function « mine », which appears in the *CPN* of figure 1 and let us write its corresponding matrix for $n = 3$ managers :

$$\text{mine} = \begin{bmatrix} 1 & 0 & 0 \\ 1 & 0 & 0 \\ 0 & 1 & 0 \\ 0 & 1 & 0 \\ 0 & 0 & 1 \\ 0 & 0 & 1 \end{bmatrix}$$

rank (mine) = 3, therefore the functions « r » and « l » can be chosen as selecting matrices of three linearly independent rows of « mine » (e.g. : 1, 3 and 5) and of three linearly independent columns (1, 2 and 3), respectively. That implies :

$$l = \begin{bmatrix} 1 & 0 & 0 & 0 & 0 & 0 \\ 0 & 0 & 1 & 0 & 0 & 0 \\ 0 & 0 & 0 & 0 & 1 & 0 \end{bmatrix} \qquad r = \begin{bmatrix} 1 & 0 & 0 \\ 0 & 1 & 0 \\ 0 & 0 & 1 \end{bmatrix} = \mathrm{id}.$$

PLACES

I : inactive manager
W : waiting manager
P : performing manager
E : exclusion
U : unused buffer
S : sent message
R : received message
A : sent acknowledge

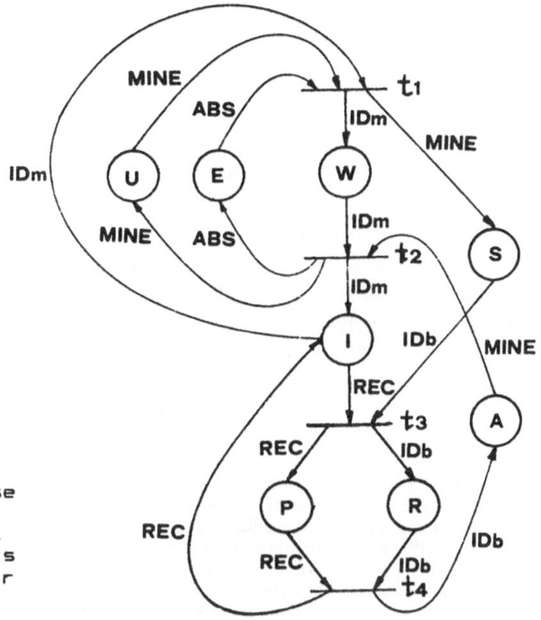

TRANSITIONS

t1 : update and send message
t2 : receive acknowledgments
t3 : receive message
t4 : send acknowledgments

COLOURS

DBM = $\{m_1 \ldots m_n\}$ set of database managers
MB = $\{(s,r) \mid \forall s,r \in DBM \wedge s \neq r\}$ set of message buffers, where s represents the sender and r the receiver
ε token without colour

FUNCTIONS

id_m : $\forall m \in DBM$, $id_m(m)=m$
id_b : $\forall b \in MB$, $id_b(b)=b$
id_ε : $id_\varepsilon(\varepsilon)=\varepsilon$
abs : $\forall m \in DBM$, $abs(m)=\varepsilon$
rec : $\forall (s,r) \in MB$, $rec(s,r)=r$
mine : $\forall s \in DBM$, $mine(s)=\sum_{r \neq s}(s,r)$

INITIAL MARKING

$m(I) = \Sigma\, DBM$
$m(E) = \varepsilon$
$m(U) = \Sigma\, MB$

INCIDENCE MATRIX W

		t1 DBM	t2 DBM	t3 MB	t4 MB
I	DBM	$-id_m$	id_m	$-rec$	rec
W	DBM	id_m	$-id_m$		
P	DBM			rec	$-rec$
E	ε	$-abs$	abs		
U	MB	$-mine$	$mine$		
S	MB	$mine$		$-id_b$	
R	MB			id_b	$-id_b$
A	MB		$-mine$		id_b

Figure 1. — Coloured Petri net which describes a net of databases.

From the previous choice of « l » and « r », the generalized inverse of « mine », called « mine* », is :

$$mine^* = r.(l.mine.r)^{-1}.l = \begin{bmatrix} 1 & 0 & 0 & 0 & 0 & 0 \\ 0 & 0 & 1 & 0 & 0 & 0 \\ 0 & 0 & 0 & 0 & 1 & 0 \end{bmatrix}$$

In general, the elements of a generalized inverse matrix are rationals as may be seen by considering the following definition of « func » :

$$func = \begin{bmatrix} 2 & 0 & 0 \\ 3 & 0 & 0 \\ 0 & 3 & 0 \\ 0 & 2 & 0 \\ 0 & 0 & 5 \\ 0 & 0 & 1 \end{bmatrix}$$

Maintaining the same matrices « r » and « l » as for « mine », we get :

$$\text{func}^* = r.(l.\text{func}.r)^{-1}.l = \begin{bmatrix} 1/2 & 0 & 0 & 0 & 0 & 0 \\ 0 & 0 & 1/3 & 0 & 0 & 0 \\ 0 & 0 & 0 & 0 & 1/5 & 0 \end{bmatrix}$$

Given that in « func* » all the elements are rationals, we can get a matrix whose elements are integers by multiplying it by the lowest common multiple, in this case $2.3.5 = 30$.

2.2. Invariants of a *CPN* with identity functions

Let the $CPNa = \langle P, T, C, A, m_0 \rangle$ where

1) $P = \{ p_1, ..., p_n \}$;
2) $T = \{ t_1, ..., t_m \}$;
3) $C(p_i) = C(t_j) = D \; \forall p_i \in P$ and $\forall t_j \in T$ and where D is a finite set non-empty of colours;
4) $A(p_i, t_j) = a_{ij}.\text{id}$, where $a_{ij} \in \mathbb{Z}$ and id is the identity function of $[[D \to [D \to \mathbb{Z}]]$, that is $\text{id}(c) = c$, $\forall c \in D$;
5) m_0 is any initial marking of *CPNa*.

The coloured Petri net thus defined is characterised by having the same set of colours associated with all the places and transitions, and the function id with all its arcs. We call this type of coloured Petri net a *CPN with identity functions.* If we proceed to unfold *CPNa*, associating a place with each $C(p_i)$ and a transition with each $C(t_j)$, we obtain as many generalized Petri nets (*GPN*) as D has colours. Associated with each of them will be the same incidence matrix $A' = [a_{ij}]_{1 \leq i \leq n,\ 1 \leq j \leq m}$.

Definition 3 : *Given CPNa, a coloured Petri net with identity functions, the Generalized Petri net associated with CPNa is* $GPNa' = \langle P, T, A', m'_0 \rangle$ *where :*

1) $P = \{ p_1, ..., p_n \}$ *;*
2) $T = \{ t_1, ..., t_m \}$ *;*
3) $A'(p_i, t_j) = a_{ij}$ *;*
4) m'_0 *is any initial marking of GPNa'.*

Definition 4 : *Let* $a' = (a_1 \ldots a_n)$ *a flow of A' and let the application* h *:* $\text{Ker}(A') \to \text{Ker}(A)$ *be such that* $h(a') = \text{id}.a'$*, that is* $h[(a_1 \ldots a_n)] = (a_1.\text{id} \ldots a_n.\text{id})$*,* $\forall (a_1 \ldots a_n) \in \text{Ker}(A')$*, where* $a_1, ..., a_n \in \mathbb{Q}$. □

$\text{Ker}(A')$ *is a subspace of* $\mathbb{Q}^n$ *and* $\text{Ker}(A)$ *is a submodule of E, where E is the module of linear functions* $\mathbb{Q}^{d \times d}$ *and* $d = \text{card}(D)$.

Proposition 2 : *The images, under h, of any pair of linearly independent elements* $a, a' \in \text{Ker}(A')$*, are linearly independent.*

Proof : We are going to prove that there do not exist two non-null functions $f, f' \in [D \to [D \to \mathbb{Z}]]$ such that :

$$f.h(a) + f'.h(a') = 0 \Leftrightarrow$$

$$\Leftrightarrow (a_1.f \ldots a_n.f) + (a'_1.f' \ldots a'_n.f') = 0 .$$

We can write the vectorial equation in terms of a system with « n » equations :

$$\begin{cases} a_1.f + a_1.f' = 0 \\ \ldots\ldots\ldots\ldots\ldots\ldots \\ a_n.f + a'_n.f' = 0 . \end{cases}$$

Given that a and a' are, by assumption, linearly independent, then $f = f' = 0$ from which we deduce that $h(a)$ and $h(a')$ are linearly independent. □

Proposition 3 : *The images under h of the elements of a basis of* $\text{Ker}(A')$ *constitute a basis of* $\text{Ker}(A)$.

Proof : Let $Ba' = \{ v_1, ..., v_q \}$ a basis of $\text{Ker}(A')$. From proposition 1 we deduce that $\{ h(v_1), ..., h(v_q) \}$ are linearly independent and as a result :

$$\dim [R[h(v_1), ..., h(v_q)]^T] = \sum_{i=1}^{q} \dim [R[h(v_i)]] =$$

$$= \sum_{i=1}^{q} \dim [R[v_i.\text{id}]] = q.d .$$

It will be sufficient to demonstrate that $\dim [\text{Ker}(A)] = q.d$.

In effect :

$$\dim [\text{Ker}(A)] = d.n - \text{rank}(A) = d.n - d.\text{rank}(A')$$
$$= d.[n - \text{rank}(A')]$$
$$= d.\dim [\text{Ker}(A')] = d.q . \quad \square$$

This proposition indicates that we can obtain a basis of flows of A immediately if we have a basis of flows of A'. Bearing in mind that A and A' are the incidence functions of *CPNa* and *GPNa'* we can state that the calculation of invariants of a coloured Petri net with identity functions can be reduced to the calculation of the invariants of a Generalized Petri net. To resolve this problem there are algorithms to calculate : 1) the basis of invariants of a *GPN* by means of the triangulation of the incidence matrix A' and 2) all the non-negative invariants with minimal support (minimal support invariant) (see [Silva 84] and [Martinez 82]).

3. Calculation of the symbolic invariants of a *CPN* (I) : special case when *W* is factorizable

The calculation of the invariants of a *CPN* is expressed in terms of the resolution of the matrix equation : $x.W = 0$ where :

$$x : P \to [C(p) \to [U \to \mathbb{Z}]]$$

and U is a non-empty set of colours.

If the set U consisted of only one colour, x would be a row matrix and the problem would be equivalent to calculating the flows of the incidence matrix of a *GPN*. As was noted in § 2.2, there exist algorithms to perform this task efficiently.

Given that the invariants calculated on a *CPN* will be used in a subsequent validation process, in general, it is not of interest to have the invariants with respect to only one colour, but to a set of colours which would have an overall significance in the model represented by the *CPN*. These sets will normally be associated with places or transitions of the net due to the colour function C, or subsets of these with special characteristics.

A method for calculating the invariants with respect to a set U of colours could be based on the calculation of the invariants associated with each colour of U and their subsequent combination. These invariants would be defined numerically, and as such their use in a validation process would be very cumbersome.

As a result it would be more useful if invariants were defined by a set of functions $x_i \in [C(P_i) \to [U \to \mathbb{Z}]]$ such that the expressions $x_i . m(p_i)$ have a clear meaning for the modeller. This can be fully achieved if all the non-null functions x_i, can be written as a composition of known linear functions (matrix sums and products) : identity functions, functions of W and generalized inverse functions of any of these above. In future, these invariants will be named « symbolic invariants ». This title can be applied also to the invariants of a *CPN* with identity functions (*CPNa* of § 2.2) obtained from the images, under h, of the flows of the matrix A' of the *GPNa'*.

The method for calculating symbolic invariants of a *CPNw*, which is proposed in this paragraph, is based on the factorization of the incidence matrix, W. That means $W = F.A$, where A is the incidence matrix of a *CPNa* with identity functions (§ 2.2). The symbolic invariants of *CPNw* will be calculated by previously obtaining the symbolic invariants of *CPNa* (§ 2.2). The process for obtaining the set of symbolic invariants which make up a basis or a generator of invariants will be considered in (§ 3.3) and applied to a problem which is very popular in the existing literature on the subject [Jensen 81b] [Reisig 82].

3.1. Factorization of W

Let $CPNw = \langle P, T, C, W, m_0 \rangle$ where :

1) $P = \{ p_1, ..., p_n \}$
2) $T = \{ t_1, ..., t_m \}$
3) $C(p_i) = D_i$, $\forall p_i \in P$ and $C(t_j) = D$, $\forall t_j \in T$
4) $W(p_i, t_j) = a_{ij} . f_i$, $\forall (p_i, t_j) \in P \times T$

$$a_{ij} \in \mathbb{Z}, \quad f_i \in [C(t_j) \to [C(p_i) \to \mathbb{Z}]] .$$

The definition of *CPNw* assumes that the colour function applied to any transition gives, as a result, the same set of colours, D, and that all the arcs which arrive at or leave a place, p_i, have the same associated function f_i, weighted by different integer coefficients a_{ij}, it being possible that $a_{ij} = 0$.

Under these circumstances, the incidence matrix, W, can be factorised as follows : $W = F.A$, where F is a diagonal matrix with $\text{diag}(F) = (f_1, f_2, ..., f_n)$ and $A = [a_{ij} . \text{id}]_{1 \leq i \leq n,\ 1 \leq j \leq m}$ is the incidence matrix of a net *CPN* with identity functions : $\text{id} \in [D \to [D \to \mathbb{Z}]]$ (§ 2.2).

The problem of calculating the flows of W can be expressed as follows :

$$x . W = x . F . A = z . A = 0 , \quad \text{where} \quad z = x . F .$$

To obtain symbolic flows, x, from W we proceed in two steps :

1) Calculating the symbolic flows y of A, which will be of the form $y = (y_1 . \text{id} ... y_n . \text{id})$.

2) Calculating the symbolic x's which will satisfy $x . F = z$ where $z = f . y = (y_1 . f ... y_n . f)$, with $f \in [D \to [U \to \mathbb{Z}]]$.

The calculation of flows, y, of A was treated in § 2.2. The symbolic solution of $x . F = f . y$ is given below (§ 3.2).

3.2. Calculation of symbolic invariants of *CPNw* based on symbolic invariants of *CPNa*

Let $y = (y^1 . \text{id} ... y^n . \text{id})$, be a symbolic flow of A, where $\forall i \in [1, n]$, $y^i \in \mathbb{Z}$.

Definition 5 : *Let FSy be the set of functions $f_i \in \text{diag}(F)$ associated with the support of y. That is :*

$$FSy = \{ f_i \in \text{diag}(F) \mid y^i \neq 0 \} .$$

Definition 6 : *Let $R(y)$ be the range of the symbolic flow y, that is :*

$$R(y) = \{ (y^1 . g ... y^n . g) \mid g \in [D \to [U \to \mathbb{Z}]] \}$$

where U is a non-empty set of colours.

Proposition 4 : *If $x . F \in R(y)$ then $x . W = 0$ and $\exists g \in [D \to [U \to \mathbb{Z}]]$ such that $x_i . f_i = y^i . g, \forall f_i \in \text{diag}(F)$.*

Proof : $x . F \in R(y) \Leftrightarrow \exists g$ such that $x . F = g . y \Leftrightarrow (x_1 . f_1 ... x_n . f_n) = (y^1 . g ... y^n . g)$.

On the other hand

$x . W = x . F . A = (x_1 . f_1 ... x_n . f_n) . A$

$= (y^1 . g ... y^n . g) . A = g . (y^1 . \text{id} ... y^n . \text{id}) . A$

$= g . y . A = 0 .$ □

Function « g » plays a capital role in the flow calculation of the matrix W. It is possible to characterize this function by considering proposition 4 :

$$x_i . f_i = y^i . g , \quad \forall f_i \in \text{diag}(F) \Rightarrow R(g) \subseteq R(f_i) , \quad \forall f_i \in FSy .$$

Among all the functions, g, which can verify the last relation, those with $R(g)$ maximum will have a special interest. Let f be a function with maximum range defined by the row space intersection of functions f_i of FSy :

$$R(f) = R(g)_{\max} = \bigcap_{f_i \in FSy} R(f_i) .$$

Let $\{ h_i \}$ be a function set, where $\forall i$ such that $y^i \neq 0$ h_i is defined as $h_i \in [D_i \to [U \to \mathbb{Z}]]$, and verifies :

1) $h_i . f_i = f$.

2) $R(f) = \bigcap_{\forall f_i \in FSy} R(f_i)$.

Proposition 5 : *$x = (y^1 . h_1 ... y^n . h_n)$ is a flow of W and consequently $x . m$ is a symbolic invariant.*

Proof :

$x . W = x . F . A = (y^1 . h_1 . f_1 ... y^1 . h_n . f_n) . A$

$= f . (y^1 . \text{id} ... y^n . \text{id}) . A = f . y . A = 0 .$ □

Proposition 6 let us to express h_i functions in terms of known functions. If a given condition is satisfied, corollary 2 let us to express h_i functions in terms of functions of W and their generalized inverses.

Proposition 6 : *If we take $h_i = f . f_i^*$, then $x = (y_1 . f . f_1^* ... y^n . f . f_n^*)$, and x is a flow of W, where f_i^* is a (CK-CR)-generalized inverse function of f_i, $\forall i \in [1, n]$.*

Proof : $x . W = x . F . A = (y^1 . f . f_1^* . f_1 ... y^n . f . f_n^* . f_n) . A$.

By considering that :

1) $f = h_i . f_i$ then $f . f_i^* . f_i = h_i . f_i . f_i^* . f_i = h_i . f_i = f$, $\forall f_i \in FSy$, and

2) $y^i = 0, \forall f \notin FSy$.

We can write :

$$x.W = (y^1.f \dots y^n.f).A = f.(y^1.\mathrm{id} \dots y^n.\mathrm{id}).A$$
$$= f.y.A = 0 . \qquad \square$$

COROLLARY 1 : *The following flows are symbolic flows of W :*

a) $x_1 = (y^1.h_m.f_m.f_1^* \dots y^m.h_m.f_m.f_m^* \dots y^n.h_m.f_m.f_n^*)$.

b) $x_2 = (0 \quad \dots \quad 0 \quad \mathrm{id}_m - f_m.f_m^* \quad 0 \quad \dots 0)$.
$\quad 1) \qquad m-1) \qquad m) \qquad m+1) \quad n)$.

c) $x_3 = (y^1.h_m.f_m.f_1^* \dots y^m.h_m \dots y^n.h_m.f_m.f_n^*)$.

Proof : $x = (y^1.f.f_1^* \dots y^n.f.f_n^*)$ is a flow of W (Prop. 6).

a) For $f = h_m.f_m$ then $x_1 = x \Rightarrow x_1.W = 0$.

b) $(\mathrm{id}_m - f_m.f_m^*).f_m = 0$ (property 1 of § 2.1) $\Rightarrow x_2.F = 0 \Rightarrow x_2.W = 0$.

c) $x_3 = x_1 + y^m.h_m.x_2 \Rightarrow x_3.W = 0$. $\square$

COROLLARY 2 : *If $\exists f_m \in FSy$ such that $R(f_m) \subseteq R(f_i)$, $\forall f_i \in FSy$ then $x = (y^1.f_m.f_1^* \dots y^m.\mathrm{id}_m \dots y^n.f_m.f_n^*)$ is a flow of W.*

Proof : It is directly deduced from corollary 1*c*) by considering that $f = f_m$, and then $h_m = \mathrm{id}_m$. $\square$

The results presented in this paragraph constitute a method for calculate symbolic flows of W. Let us suppose that :

$$y = (y^1.\mathrm{id} \quad y^2.\mathrm{id} \quad y^3.\mathrm{id} \quad 0 \dots 0).$$

Figure 2 shows graphically three possible particular cases which we can find when we calculate the row space intersection of FSy functions.

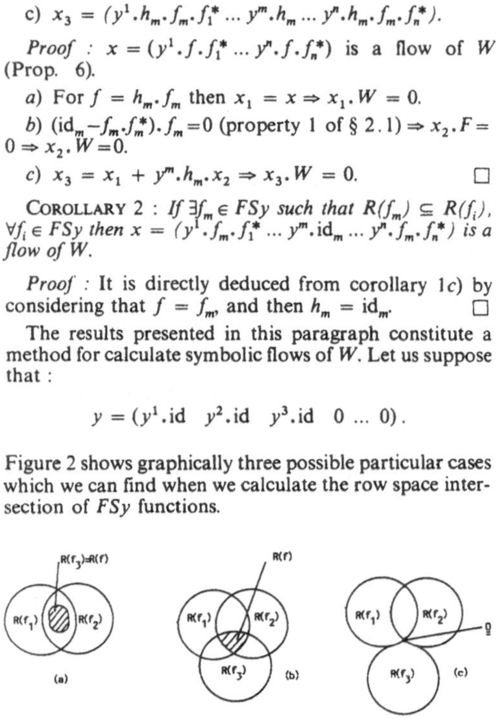

Figure 2. — Row space intersection of *FSy* functions of *CPNa p*-flow $y = (y^1.\mathrm{id}, y^2.\mathrm{id}, y^3.\mathrm{id}\ 0 \dots 0)$: three possible particular cases.

Case *a* : $R(f) = \bigcap_{f_i \in FSy} R(f_i) = R(f_3)$, that is, f_3 range is included in ranges or FSy remaining functions. By applying corollary 2 and considering that $f = h_3.f_3 = f_3$, that is, $h_3 = \mathrm{id}$, we can write the flow :

$$x = (y^1.f_3.f_1^* \quad y^2.f_3.f_2^* \quad y^3.\mathrm{id} \quad 0 \dots 0).$$

Case b : $R(f) = \bigcap_{f_i \in FSy} R(f_i) \neq R(f_k)$, $\forall f_k \in FSy$, that is, $R(f)$ is not equal to any FSy function range. The flow calculated will be :

$$x = (y^1.h_3.f_3.f_1^* \quad y^2.h_3.f_3.f_2^* \quad y^3.h_3 \quad 0 \dots 0)$$

and h_3 must verify $f = h_3.f_3$.

Case *c* : $R(f) = 0$, that is, range intersection is the null subspace. In this case $h_3 = 0$ and consequently only trivial flow exists :

$$x = (0 \quad 0 \quad 0 \quad 0 \dots 0).$$

Finally, let us suppose that the « r » functions of FSy can be ordered as $FSy = \{ g_1, g_2, \dots, g_r \}$, where the g_i function verifies the following commutativity property :

$$\prod_{j=i}^{r} g_j = \left(\prod_{j=i+1}^{r} g_j\right).g_i, \quad \forall i \in [1, r].$$

That means, g_i commutes with the product $g_{i+1}.\dots.g_r$.

COROLLARY 3 : *If such ordinance exists then $x = (y^1.h_1 \dots y^n.h_n)$ is a symbolic flow of W, where :*

$$h_i = \prod_{\substack{j \in [1,r] \\ \wedge g_j \neq f_i}} g_j .$$

Proof : It is enough to consider that $\forall f_i \in FSy$, $h_i.f_i = \prod_{j=1}^{r} g_j = f$ and then the present corollary is a direct result of proposition 5. $\square$

3.3. CALCULATION OF A SET OF SYMBOLIC INVARIANTS OF *CPNw*

The validation of a *CPN* basically consists in the verification of a series of design specifications. This verification can be done, in part, by considering a set of linear invariant relations of the net marking. It is easy to understand how having an adequate set of symbolic invariants is important and indeed vital to be able to make a good job of validation.

In § 3.2 we have presented a series of results which make up a method for calculating symbolic flows of $W(W = F.A)$. The goal pursued by the modeller will be centred on obtaining a certain set of symbolic invariants of *CPNw*, as function of the necessities of the validation. For example, a set capable of generating all the non-negative linear invariants of *CPNw*, a set which constitutes a basis of linear invariants, etc. To achieve this goal it will be necessary to carry out an adequate selection of flows of A on which to apply the above-mentioned method. We will shortly present a proposition which permits us to considerably reduce the field of selection.

Let $y = (y^1.\mathrm{id} \dots y^n.\mathrm{id})$ be a flow of A with non-minimal support. It is always possible to find a basis of flows of A, where each y_i is a minimal support flow. Vector y can be expressed as follows :

$$\lambda.y = \sum_{i=1}^{q} \lambda_i.y_i ,$$

where

$$\begin{cases} \forall i \in [1, q]\ \lambda_i \in \mathbb{Z} \quad \text{and} \quad y = (y_i^1.\mathrm{id} \dots y_i^n.\mathrm{id}) \\ \lambda \in \mathbb{Z} . \end{cases}$$

Let $x = (y^1.f.f_1^* \dots y^n.f.f_n^*)$ and $x_i = (y_i^1.g_i.f_1^* \dots y_i^n.g_i.f_n^*)$ be the flows of W calculated by applying proposition 6 to y and y_i, respectively. It is easy to verify that $R(f) \subseteq R(g_i)$, $\forall i \in [1, q]$, and consequently $\exists h_i$ such that $f = h_i.g_i$.

Proposition 7 : *Any symbolic flow, x, of a factorizable matrix, can be expressed as linear combination of minimal support flows, x_i :*

$$\lambda . x = \sum_{i=1}^{q} \lambda_i . h_i . x_i .$$

Proof :

$$\sum_{i=1}^{q} \lambda_i . h_i . x_i = \left(\sum_{i=1}^{q} \lambda_i . h_i . y_i^1 . g_i . f_1^* \ldots \sum_{i=1}^{q} \lambda_i . h_i . y_i^n . g_i . f_n^* \right)$$

$$= \left(\sum_{i=1}^{q} \lambda_i . y_i^1 . f . f_1^* \ldots \sum_{i=1}^{q} \lambda_i . y_i^n . f . f_n^* \right)$$

$$= \lambda . (y^1 . f . f_1^* \ldots y^n . f . f_n^*) = \lambda . x . \qquad \square$$

From this result we can conclude that the application of the method of calculating symbolic invariants of *CPNw* with minimal support will give not inferior results to those obtained by applying it to another of higher support. With this, the process of calculation of symbolic invariants of *CPNw* can be restricted to those of minimal support.

Let us consider again the *CPN* in figure 1. The incidence matrix of this net is not factorizable and thus, in principle, the proposed method of calculating invariants cannot be applied. Let's consider, for the moment, the sub *CPN* which contains all its places, but only the transitions t_1 and t_2. Let's call its incidence matrix, W_1. Matrix W_1 is factorizable, $W_1 = F_1 . A_1$:

$$W = \begin{array}{c} \\ I \\ W \\ P \\ E \\ U \\ S \\ R \\ A \end{array} \begin{array}{c} \begin{array}{cc} t_1 & t_2 \end{array} \\ \begin{bmatrix} -\mathrm{id}_m & \mathrm{id}_m \\ \mathrm{id}_m & -\mathrm{id}_m \\ & \\ -\mathrm{abs} & \mathrm{abs} \\ -\mathrm{mine} & \mathrm{mine} \\ \mathrm{mine} & \\ & \\ & -\mathrm{mine} \end{bmatrix} \end{array} = \begin{bmatrix} \mathrm{id}_m & & & & & & & \\ & \mathrm{id}_m & & & & & & \\ & & 0 & & & & & \\ & & & \mathrm{abs} & & & & \\ & & & & \mathrm{mine} & & & \\ & & & & & \mathrm{mine} & & \\ & & & & & & 0 & \\ & & & & & & & \mathrm{mine} \end{bmatrix} \cdot \begin{array}{c} \\ I \\ W \\ P \\ E \\ U \\ S \\ R \\ A \end{array} \begin{array}{c} \begin{array}{cc} t_1 & t_2 \end{array} \\ \begin{bmatrix} -\mathrm{id}_m & \mathrm{id}_m \\ \mathrm{id}_m & -\mathrm{id}_m \\ & \\ -\mathrm{id}_m & \mathrm{id}_m \\ -\mathrm{id}_m & \mathrm{id}_m \\ \mathrm{id}_m & \\ & \\ & -\mathrm{id}_m \end{bmatrix} \end{array}$$

Table 1 shows symbolic flows with minimal support of matrix A_1. Taking into account that $R(\mathrm{id}_m) = R(\mathrm{mine})$ and $R(\mathrm{abs}) \subset R(\mathrm{id}_m)$ we can proceed to calculate the twelve symbolic flows of table 2. Each one results from the application of corollary 2 to a flow of A_1 (table 1).

To the set of symbolic flows of W_1 of table 2, we must add those which are directly flows of F_1. If k_f denotes any linear function belonging to Ker (f), that is $k_f . f = 0$, table 3 presents the mentioned flows.

Observation : $k_f = \mathrm{id}_n - f . f^*$ is a solution of the equation in question. (Note : « n » is the number of rows of the matrix f associated with the linear function f).

As an illustration of this process let us show in detail the method for obtaining two flows of table 2 : v_6 and v_4.

v_6) We start by considering the flow a_6, of A_1. The support of a_6 and the set of functions of F which are

TABLE 1

Symbolic flows of A_1 with minimal support

	I	W	P	E	U	S	R	A
a1			id_m					
a2							id_m	
a3	id_m	id_m						
a4	id_m					id_m		id_m
a5		id_m		id_m				
a6				id_m		id_m		id_m
a7		id_m			id_m			
a8					id_m	id_m		id_m
a9		$-\mathrm{id}_m$				id_m		id_m
a10	$-\mathrm{id}_m$			id_m				
a11	$-\mathrm{id}_m$				id_m			
a12				id_m	$-\mathrm{id}_m$			

TABLE 2

Symbolic flows of $W_1 = F_1 . A_1$ with minimal support

	I	W	P	E	U	S	R	A
v1			id_m					
v2							id_b	
v3	id_m	id_m						
v4	mine					id_b		id_b
v5		abs		id_ε				
v6				id_ε		abs.mine*		abs.mine*
v7		mine			id_b			
v8					id_b	id_b		id_b
v9		-mine				id_b		id_b
v10	-abs			id_ε				
v11	-mine				id_b			
v12				id_ε	-abs.mine*			

TABLE 3

Flows of W_1 which are also flows of F_1

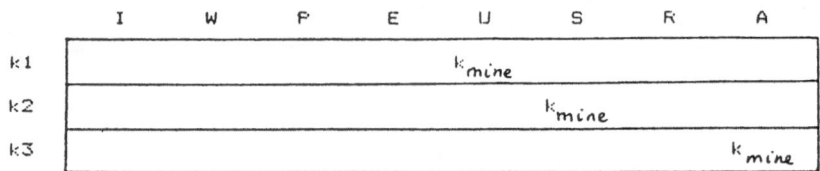

	I	W	P	E	U	S	R	A
k1					k_{mine}			
k2						k_{mine}		
k3								k_{mine}

associated with that support are :

$$\text{support}\,(a_6) = \{\ E, S, A\ \}$$

$$FSa_6 = \{\ f_E, f_S, f_A\ \},$$

where

$$f_E = \text{abs} \quad \text{and} \quad f_S = f_A = \text{mine}\,.$$

Applying corollary 2, and taking $f_m = \text{abs}$, because $R(\text{abs}) \subset R(\text{mine})$, we can write the flow of W :

$$v_6 = (0 \quad 0 \quad 0\ x_E\ 0\ x_S\ 0\ x_A),$$

$$\text{where :} \quad \begin{cases} x_E = \text{id}_\varepsilon \\ x_S = f_E . f_S^* = \text{abs.mine}^* \\ x_A = f_E . f_A^* = \text{abs.mine}^* . \end{cases}$$

v_4) Considering the flow a_4, of A_1 :

$$\text{support}\,(a_4) = \{\ I, S, A\ \}$$

$$FSa_4 = \{\ f_I, f_S, f_A\ \} \quad \text{where} \quad f_I = \text{id}_m$$

and

$$f_S = f_A = \text{mine}\,.$$

Any one of these three functions can be taken as f_m in corollary 2, because $R(\text{id}) = R(\text{mine})$. Taking f_I, and applying corollary 2 we can write the symbolic flow of W :

$$v_4 = (x_I \quad 0 \quad 0 \quad 0 \quad 0 \quad x_S \quad 0 \quad x_A),$$

$$\text{where :} \quad \begin{cases} x_I = \text{id}_m \\ x_S = f_I . f_S = \text{id}_m.\text{mine}^* = \text{mine}^* \\ x_A = f_I . f_A = \text{id}_m.\text{mine}^* = \text{mine}^* . \end{cases}$$

On the other hand, if we choose f_S as f_m, the flow we obtain will be,

$$v'_4 = (x_I \quad 0 \quad 0 \quad 0 \quad 0 \quad x_S \quad 0 \quad x_A),$$

$$\text{where :} \quad \begin{cases} x_I = f_S . f_I = \text{mine.id}_m^* = \text{mine} \\ x_S = \text{id}_b \\ x_A = f_S . f_A = \text{mine.mine}^* . \end{cases}$$

We should not be surprised by the different forms of the two symbolic invariants, which is due to the fact that the invariant k_2 (table 2) has been integrated into v'_4. Thus, letting

$$k_2 = (0 \quad 0 \quad 0 \quad 0 \quad 0 \quad 0 \quad \text{id}_b\text{-mine.mine}^* \quad 0 \quad 0).$$

It is easy to show that : $v'_4 = \text{mine}.v_4 + k_2$.
Given that $f_A = f_S$, we can rewrite v'_4 (corollary 1)

$$x'_I = \text{mine}$$
$$x'_S = \text{id}_b$$
$$x'_A = \text{id}_b$$

which is just as it appears in table 2, under the name v_4.

It is, finally, worth noting that k_2 can be obtained as a linear combination of v_4 and k_3 : $k_2 = k_{\text{mine}}.v_4 - k_3$.

Considering together table 2 and 3, they contain redundant information since there exist flows which are linear combinations of other flows.

As far as the modeller, who must validate the net, is concerned, it is not usually of interest to have an over-populated set of symbolic flows of W, from which to obtain the marking invariants. A more restricted set is preferable. Thus, for example, we can imagine that it would be of interest to have a set of symbolic flows which is in turn a generator of flows. This set could eventually constitute a basis. Another possible point of interest could be to obtain a set of symbolic flows which would generate all the non-negative symbolic flows or symbolic semi-flows. This interest is due to the special physical significance which the non-negative invariants of a net usually have.

The process of obtaining non-negative symbolic flows of W is based on results of § 3.2, and on the algorithms presented in [Martinez 82].

Returning to table 2, the flows $v_1, v_2, \ldots, v_8$ are non-negative, and as we will see later, make up a set which generates all the flows of W_1, including those which are not non-negative.

To calculate a basis of symbolic flows of W_1 ($W_1 = F_1.A_1$), we cannot proceed from just any basis of flows of A_1. As an example we will consider the database *CPN* (*fig.* 1).

Let B_{A_1} be a basis of flows of A_1 :

$$B_{A_1} = \{ a_1, a_2, a_5, a_6, a_{10}, a_{12} \},$$

where the a_i are the same presented in table 1.

Using these elements we can calculate the following set of flows of W_1 :

$$S_{W_1} = \{ v_1, v_2, v_5, v_6, v_{10}, v_{12} \}.$$

Let B_{F_1} be a set of flows of F_1 which constitutes a basis :

$$B_{F_1} = \{ k_1, k_2, k_3 \}.$$

It is easy to show that in this case $S_{W_1} \cup B_{F_1}$ does not constitute a set which generates all the flows of W_1. In fact, it is sufficient to show that flows such as v_3 or v_4 cannot be expressed as linear combinations of the flows of $S_{W_1} \cup B_{F_1}$.

On the contrary, if we take $B'_{A_1} = \{ a_1, a_2, a_3, a_4, a_5, a_7 \}$ and its induced $S'_{W_1} = \{ v_1, v_2, v_3, v_4, v_5, v_7 \}$, it is possible to show that $G'_{W_1} = S'_{W_1} \cup B_{F_1}$ constitutes a set which generates all flows of A_1 and

$$B_{W_1} = G'_{W_1} - \{ k_1, k_2 \} = \{ v_1, v_2, v_3, v_4, v_5, v_7, k_3 \}$$

is a basis.

We can summarise what has gone before by saying that, to calculate a basis of flows of W, we cannot proceed from just any basis of flows of A. Let us consider now a set of conditions that if verified permits to calculate directly a basis of W symbolic flows.

Let $B_A = \{ a_i \}_{1 \leq i \leq q}$ be a basis of symbolic flows of A such that the two following conditions are verified :

(C_1) B_A has the following form :

$$B_A^T = \begin{array}{c} \\ a_1 \\ a_2 \\ -- \\ a_q \end{array} \begin{bmatrix} \pi_1 & \pi_2 & \cdots & \pi_q & \cdots & \pi_n \\ a_{11}.\text{id} & \cdots & & & & \\ & a_{22}.\text{id} & \cdots & & \boxed{a_{ij}.\text{id}} & \\ & & \ddots & & \cdots & \\ 0 & & & a_{qq}.\text{id} & \cdots & \end{bmatrix}$$

where π represents a permutation of the coloured places $p_i \in P$.

(C_2) $\forall a_i \in B_A$, $\forall f_j \in FS_{a_i}$, $R(f_i) \subseteq R(f_j)$ and consequently it is possible to apply corollary 2 to a_i.

We can suppose, to make easier to write, that π is the identity permutation.

PROPOSITION 8 : *The image, under F, of any flow, x, of W expressed in the basis* B_A *has the form :*

$$y = x.F = (u_1.f_1 \ldots u_q.f_q).B_A^T,$$

$$\textit{where} : \begin{cases} x_j \in [C(p_j) \rightarrow [U \rightarrow \mathbb{Z}]], & \forall j \in [1, n] \quad \textit{and} \\ u_i \in [C(p_i) \rightarrow [U \rightarrow \mathbb{Q}]], & \forall i \in [1, q] \end{cases} \quad \square$$

We are going to give a constructive proof by presenting a method for the calculation of u_i functions ($i \in [1, q]$).

Proof :

$$y = x.F = (x_1.f_1 \ldots x_n.f_n) \qquad (1)$$

$$x.W = 0 \Leftrightarrow x.F.A = y.A = 0 \Rightarrow y \in \text{Ker}(A)$$

and consequently y can be expressed in basis B_A as :

$$y = (y_1 \ldots y_q).B_A^T. \qquad (2)$$

From (1) and (2) we can write the system :

$$\left[\begin{array}{l} (1)\ a_{11}.y_1 = x_1.f_1 \\ (2)\ a_{12}.y_1 + a_{22}.y_2 = x_2.f_2 \\ \ldots\ldots\ldots\ldots\ldots\ldots\ldots\ldots \\ (j)\ a_{1j}.y_1 + \cdots + a_{jj}.y_j = x_j.f_j \\ \ldots\ldots\ldots\ldots\ldots\ldots\ldots\ldots \\ (q)\ a_{1q}.y_1 + \cdots + a_{qq}.y_q = x_q.f_q. \end{array} \right.$$

We can calculate y_j if we have previously calculated $y_1, \ldots, y_{j-1}$.

From (1) : $y_1 = u_1 . f_1$, where $u_1 = \frac{1}{a_{11}} x_1$.

. . . .

From (j) : $y_j = u_j . f_j$, where :

$$\begin{cases} u_j = \frac{1}{a_{jj}}\left(x_j - \sum_{\substack{i \in [1, j-1] \\ \wedge a_{ij} \neq 0}} a_{ij} . u_i . h_{ij}\right) . f_j, \text{ and} \\ h_{ij} \text{ is a function such that } f_i = h_{ij} . f_j . \end{cases}$$ □

Let $G_W = \{ g_i \}_{1 \leq i \leq q}$ be the set of W flows calculated by applying corollary 2 to the elements of basis B_A :

$$\begin{aligned} g_i . F &= (0 \ldots 0 \, a_{ii} . \mathrm{id} \ldots \boxed{a_{ij} . f_i . f_j^*} \ldots) . F \\ &= (0 \ldots 0 \, a_{ii} . f_i \ldots \boxed{a_{ij} . f_i . f_j^* . f_j} \ldots) . \end{aligned}$$

As $R(f_i) \subseteq R(f_j) \Rightarrow f_i . f_j^* . f_j = f_i$ then we can write :

$$g_i . F = f_i . (0 \ldots 0 \, a_{ii} . \mathrm{id} \ldots \boxed{a_{ij} . \mathrm{id}} \ldots) = f_i . a_i .$$

And consequently

$$G_W^T . F = \begin{bmatrix} f_1 & & \\ & \ldots & \\ & & f_q \end{bmatrix} . B_A^T .$$

Let FQ and FC be diagonal matrices of linear functions defined as :

$$FQ = \begin{bmatrix} f_1 & & & & \\ & \ldots & & & \\ & & f_q & & \\ & & & \mathrm{id} & \\ & & & & \ldots \\ & & & & & \mathrm{id} \end{bmatrix}$$

$$FC = \begin{bmatrix} \mathrm{id} & & & & \\ & \ldots & & & \\ & & \mathrm{id} & & \\ & & & f_{q+1} & \\ & & & & \ldots \\ & & & & & f_n \end{bmatrix}$$

It is easy to verify that $B_{KF} = B_{KFQ} \cup B_{KFC}$, where B_{KF}, B_{KFQ} and B_{KFC} are basis of Ker (F), Ker (FQ) and Ker (FC) respectively.

Proposition 9 : $B_{KW} = G_W \cup B_{KFC}$ *is a symbolic basis of W flows.*

Proof : $B_{KW} = B_{CKF} \cup B_{KF} = B_{CKF} \cup B_{KFQ} \cup B_{KFC}$, where B_{CKF} is a basis of a complementary subspace, CKF, of Ker (F) respect to Ker (W). We must prove :

1) G_W generates CKF. That is equivalent to prove that :

$$\forall y \in R(F) \cap \mathrm{Ker}(A) \begin{cases} \exists x \in \mathrm{Ker}(W) \text{ such that } x . F = y \text{ and} \\ x \text{ can be written as a linear combination} \\ \text{of } G_W \text{ elements} : x = (x_1 \ldots x_q) . G_W^T . \end{cases}$$

In effect, from proposition 8 :

$$y = (u_1 . f_1 \ldots u_q . f_q) . B_A^T .$$

On the other hand :

$$y = x . F = (x_1 \ldots x_q) . G_W^T . F = (x_1 . f_1 \ldots x_q . f_q) . B_A^T ,$$

and we can conclude that x exists because $x_i = u_i$, $\forall i \in [1, q]$.

2) G_W generates Ker (FQ). Let k_i a linear function such that $R(k_i) = \mathrm{Ker}(f_i)$. We can define a set of vectors :

$$\begin{aligned} v_i &= k_i . g_i = k_i . (0 \ldots 0 \, a_{ii} . \mathrm{id} \ldots a_{ij} . f_i . f_j^* \ldots) \\ &= (0 \ldots 0 \, a_{ii} . k_i \, 0 \ldots 0) . \end{aligned}$$

It is clear that $\{ v_i \}_{i = [1, q]}$ is a generator of Ker (FQ).

3) Elements of G_W are independents because matrix G_W is triangular. □

Usually, conditions C1 and C2 are fullfilled in *CPN* with factorizable matrix. Otherwise, is is possible to transform the *CPN* in other equivalent wich satisfies both conditions through a folding process ([Martinez 84]).

4. Calculation of the symbolic invariants of a *CPN* (II) : general case, when *W* is not factorisable

Considering the *CPN* of figure 1, it can be noted that its incidence matrix W cannot be factorised in the form $W = F . A$.

This example draws our attention to a very common problem.

The technique which we propose to solve the problem is based on partitioning W into columns :

$$W = [W_1 \; W_2 \ldots W_k]$$

where W_i is the incidence matrix of *CPNi*, a subset of *CPNw* defined $\forall i \in [1, k]$ as $CPNi = (P, T_i, C_i, W_i, m_0)$ such that :

1) $\bigcup_{i \in [1, k]} T_i = T$ and $T_i \cap T_j = \varnothing$, $\forall j \neq i$ (partition in T_i)
2) $C_i(p) = C(p)$, $\forall p \in P$ and $C_i(t) = C(t)$, $\forall t \in T_i$
3) $W_i(p, t) = W(p, t)$, $\forall (p, t) \in P \times T$
4) $W_i = F_i . A_i$ (factorisable following § 3.1).

In the general case there exist different partitions of W which satisfy the above mentioned conditions. The condition which restricts their number is condition (4) which contains implicity, the following incompatibility relation between transitions of *CPN*.

Definition 7 : *Two transitions t_j, $t_k \in T$ are incompatible as elements of the set of transitions T_i of CPNi if $\exists p \in P$ such that $W(p, t_j) \neq 0$ and $W(p, t_k) \neq 0$ and $\nexists \lambda, \mu \in \mathbb{Z}$ such that $\lambda . W(p, t_j) = \mu . W(p, t_k)$.* □

In other words, each element of the partition is *a compatible* (all transitions are compatible two-by-two).

A trivial solution to the problem of partitioning T in such a way that the condition of compatibility is satisfied consists of taking one unique transition per element, that is : $W_i = [t_i]$ .For the *CPN* of figure 1, the table of incompatible transitions can be constructed from figure 3. On inspection, we can deduce immediately that the solution for minimum cardinality is :

$W_1 = \{ t_1, t_2 \}$ $W_2 = \{ t_3, t_4 \}$ and consequently

$$W = [W_1\ W_2].$$

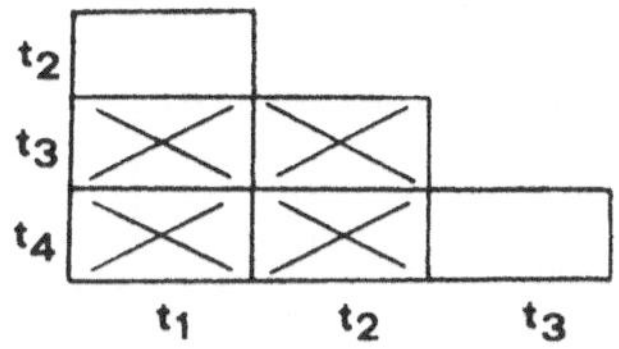

Figure 3. — Table of incompatible transition in the *CPN* of figure 1.

In § 3.3 we have calculated a basis, B_{W_1}, of W_1 flows. Now $\mathrm{Ker}(W) \subseteq \mathrm{Ker}(W_1)$ given that $\mathrm{Ker}[W_1\ W_2] \subseteq \mathrm{Ker}[W_1]$ and thus we must proceed to restrict the subspace of flows of W_1 to those which are also flows of W_2. We present below an algorithm which will allow the calculation of a basis of flows of W, which may consist of symbolic elements.

Let :

(1) I_0 be an identity matrix of such dimension that :

$$I_0 . W = W .$$

(2) $B = \{ b_1, ..., b_k \}$ be a basis of flows of a matrix H of linear functions to which we associate the matrix B, defined as $B = (b_1, ..., b_k)$, then $B^T . H = 0$.

Algorithm for obtaining a basis of flows of W.

(1) $Q := I_0 ; i := 1 ; H^i := W ;$
(2) ***while*** $H^i \langle \rangle [\,]$ *DO* { [ ] *is a empty matrix* }

2.1 Partition the matrix H^i by satisfying transition incompatibility relation :

$$H^i = [H_1^i \dots H_q^i] \text{ \{ where « } q \text{ » is variable \}}$$

2.2 Calculate the matrix B_i^T associated with the basis B_i of H_1^i flows.

2.3 $Q := B_i^T . Q;\ \ H^{i+1} := [B_i^T . H_2^i \dots B_i^T . H_q^i];$
$i := i + 1$ { the assigning sentences should be taken as a simple symbolic notation }

end of while

Let : (1) T_i be the transition set associated with H_1^i in the ith iteration.

(2) $W = [W_1 \dots W_k]$ be an ordinance of matrix W such that W_i columns have associated T_i transition set.

PROPOSITION 10 : *The matrix Q obtained by using the algorithm is the matrix associated with a basis of symbolic flows of W.*

Proof :

$$\mathrm{Ker}[W_1 \dots W_k] \subseteq \mathrm{Ker}[W_1 \dots W_{k-1}] \subseteq \subseteq \dots \subseteq \mathrm{Ker}[W_1] .$$

Proceeding by induction, for $i = 1$ we have : $Q = B_1^T . I_0 = B_1^T$ which is the matrix associated with a basis of $\mathrm{Ker}[W_1]$, because $W_1 = H_1^1$.

Now taking the general case, $i = i$, the following conditions are fulfilled :

a) $Q = B_i^T . B_{i-1}^T . \dots . B_1^T$ is the matrix associated with a basis of $\mathrm{Ker}[W_1 \dots W_{i-1}\ W_i]$, and

b) B_{i+1}^T is the matrix associated with a basis of H_1^{i+1},

where $$H_1^{i+1} = Q . W_{i+1}$$

and we can conclude immediately that :

$$Q := B_{i+1}^T . Q = B_{i+1}^T . B_i^T . B_{i-1}^T . \dots . B_1^T$$

is the matrix associated with a basis of

$$\mathrm{Ker}[W_1 \dots W_i\ W_{i+1}] .$$

In fact, it is sufficient to show that :

$$B_j^T . B_{j-1}^T . \dots . B_1^T . W_j = 0 , \quad \forall j \in [1, i + 1] .$$

By continuing the inductive process until H^i is a empty matrix the proposition follows. □

The application of the algorithm to the example in figure 1 can be followed by observing tables 4 to 7. The symbolic invariants of the marking can be written, based on the symbolic flows of table 7 :

i1) $\mathrm{abs}.m(W) + \mathrm{id}_\varepsilon . m(E) = \varepsilon$
i2) $\mathrm{mine}.m(W) + \mathrm{id}_b . m(U) = \sum MB$
i3) $\mathrm{id}_m . m(P) = \mathrm{rec}.m(R)$
i4) $\mathrm{mine}.m(I) + \mathrm{id}_b . m(S) + (\mathrm{mine}.\mathrm{rec} + \mathrm{id}_b) . m(R) + \mathrm{id}_b . m(A) = \sum MB$
i5) $\mathrm{id}_m . m(I) + \mathrm{id}_m . m(W) + \mathrm{id}_m . m(P) = \sum DBM .$

In conclusion, it is important to emphasize that the algorithm given for calculating a basis of symbolic flows of W, based on its partition into columns, is applicable if we wish to calculate any other set which generates symbolic flows. Thus, if we wish to obtain a set which generates the non-negative flows, it is sufficient to substitute the matrix B_i, in the algorithm, by the matrix G_i associated with a set which generates non-negative flows of H_1^i.

Proceeding in this way, for the coloured Petri net of figure 1 we get the non-negative symbolic flows of table 8.

5. Conclusion

The method given for calculating the invariants of a *CPN* allows us to obtain symbolic invariants constructed from symbolic expressions which contain identity functions, functions of the incidence matrix and some generalized inverse functions of subexpression of the latters.

The invariant calculation of a factorizable matrix is accomplished in two steps :

1) Calculation of the flows of minimal support of the incidence matrix of a generalized Petri net.

2) Calculation of symbolic flows of W, based on the precedings, by using the results presented in § 3.2.

For the calculation of symbolic flows of $W = F.A$ associated to a_j, symbolic flow of A, a function f must

TABLE 4
Matrix $[B_1^T . W_1 \ B_1^T . W_2] = [0 \ R]$

		t1	t2	t3	t4
		DBM	DBM	MB	MB
v1	DBM	0	0	rec	−rec
v2	MB	0	0	id_b	$-id_b$
v3	DBM	0	0	−rec	rec
v4	MB	0	0	$-mine.rec-id_b$	$mine.rec+id_b$
v5	ε	0	0	0	0
v7	MB	0	0	0	0
k3	−	0	0	0	k_{mine}

TABLE 5
Basis of flows of A_2, *where* $R = F_2 . A_2$

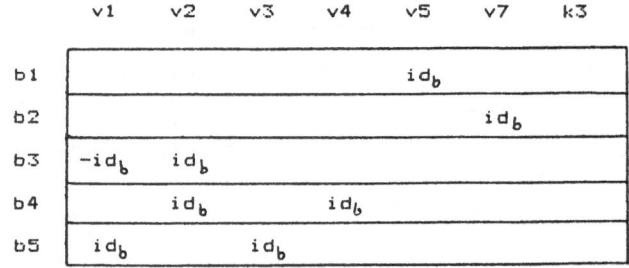

	v1	v2	v3	v4	v5	v7	k3
b1					id_b		
b2						id_b	
b3	$-id_b$	id_b					
b4		id_b		id_b			
b5	id_b		id_b				

TABLE 6
Basis of flows of $[0 \ R]$

	v1	v2	v3	v4	v5	v7	k5
w1					id_ε		
w2						id_b	
w3	$-id_m$	rec					
w4		$mine.rec+id_b$		id_b			
w5	id_m		id_m				

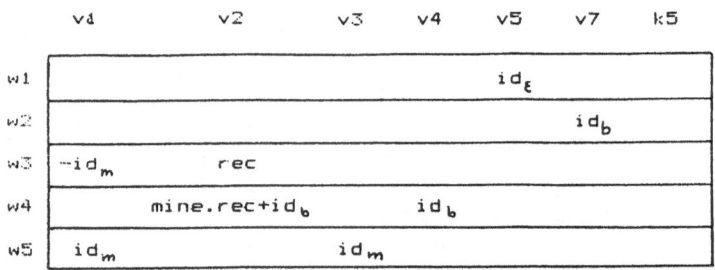

TABLE 7
Basis of flows of $W = [W_1 \ W_2]$

	I	W	P	E	U	S	R	A
w1		abs		id_ε				
w2		mine			id_b			
w3			$-id_m$				rec	
w4	mine					id_b	$mine.rec+id_b$	id_b
w5	id_m	id_m	id_m					

TABLE 8

Non-negative symbolic left flows obtained by applying the modified algorithm (substituting B_i for G_i)

	I	W	P	E	U	S	R	A
c1	id_m	id_m	id_m					
c2			id_m		rec	rec		rec
c3	rec.mine		rec.mine+id_m			rec		rec
c4			h	id_ε		h.rec		h.rec
c5	id_m	id_m					rec	
c6	mine					id_b	mine.rec+id_b	id_b
c7				id_ε		abs.mine*	abs.mine*	abs.mine*
c8					id_b	id_b	id_b	id_b
c9		abs		id_ε				
c10		mine			id_b			

(*) h = abs.(rec.mine)*

be characterized such that :

$$R(f) \subseteq \bigcap_{\forall f_i \in FSa_j} R(f_i).$$

If there is $f_k \in FSa_j$ such that $R(f_k) = R(f)$, the symbolic flow can be written by using W functions and their generalized inverses.

When W is not a factorizable matrix, reapplying the preceding method to factorizable submatrices, it is possible to obtain symbolic invariants in which can appear composed functions like $(f_1 + f_2 . f_3^*)$ and their generalized inverses.

An algorithm for calculating a symbolic basis of flows of a factorisable submatrix has been given. Its iterated application leads to the calculation of a basis of flows of the *CPN*, when premises C_1 and C_2 of proposition 8 are fulfilled. In the other case, the method presented in § 4 can be directly generalized, but in the calculated flows will appear functions f that « a priori » are unknown and which must be calculated.

The method has been applied successfully to the validation of complex *CPNs* such as in the model of a flexible manufacturing system [Alla 84].

The application of the method to the incidence matrix W gives linear invariants of the marking (p-invariants). If we applied it to its transpose, W^T, we would get linear symbolic invariants of the firing (t-invariants),

BIBLIOGRAPHIE

[Alla 84] H. ALLA, P. LADET, J. MARTINEZ and SILVA, M. : *Modelling and Validation of Complex Systems by Coloured Petri Nets : Application to a Flexible Manufacturing System ;* 5nd European Workshop on Petri Net Applications and Theory, Aarhus, June 1984, 122-140.

[Brams 83] G. W. BRAMS : *Réseaux de Petri : Théorie et Pratique.* Masson, Paris, 1983.

[Campbell 79] S. L. CAMPBELL and C. D. MEYER : *Generalized Inverses of Linear Transformations ;* Pitman, London, 1979.

[Jensen 81a] K. JENSEN : *Coloured Petri Nets and the Invariant Method ;* Theoretical Computer Science **14**, North Holland. Publ. Co., 1981, 317-336.

[Jensen 81b] K. JENSEN : *How to Find Invariants for Coloured Petri Nets ;* Mathematical Foundations of Computer Science, Lecture Notes in Computer Science **118**, Springer Verlag, Berlin, 1981.

[Lautenbach 74] K. LAUTENBACH and H. A. SCHMID : *Use of Petri Nets for Proving Correctness of Concurrent Process Systems ;* IFIP 74, North Holland Pub. Co., 1974, 187-191.

[Martinez 82] J. MARTINEZ and M. SILVA : *A Simple and Fast Algorithm to Obtain all Invariants of a Generalized Petri Net ;* 2nd European Workshop on Petri Nets Theory and Applications, Informatik Fachberichte **52**, Springer Verlag, Berlin, 1982 301-310.

[Martinez 84] J. MARTINEZ : *Contribucion al anàlisis y modelado de sistemas concurrentes mediante redes de Petri ;* Ph. D. Thesis, Universidad de Zaragoza. Spain. October 1984.

[Reisig 82] W. REISIG : *Petri Nets with Individual Tokens ;* 3th European Workshop on Applications and Theory of Petri Nets, Varenna, September 1982, 386-406.

[Silva 80] M. SILVA : *Simplification des réseaux de Petri par élimination de places implicites ;* Digital Processes **6** (4), 1980, 245-256.

[Silva 85] M. SILVA : *Las redes de Petri en la Automàtica y la Informàtica ;* Ed. AC, Madrid, 1985.

Section E
Analysis by Means of Reachability Graphs

The basic idea behind reachability graphs is to construct a graph which contains a node for each reachable state and an arc for each possible change of state. However, even for small nets such a graph may become very large—and sometimes infinite. Thus it is necessary to construct and analyse the graph using automated methods—and it is desirable to develop techniques which make it possible to work with reduced reachability graphs without losing too much information.

When a reachability graph has been constructed it can be used to prove properties about the modelled system. For bounded systems a large number of questions can be answered. Deadlocks, mutual exclusion, reachability and marking bounds can be decided by a simple search through the nodes of the reachability graph, while liveness and home markings can be decided by constructing and inspecting the strongly connected components.

The reachability graph method can be totally automated. This means that the modeller can use the method, and interpret the results, without having much knowledge about the underlying mathematics. However, at present it is only possible to construct reachability graphs for relatively small systems and for selected parts of large systems. In spite of this, reachability graphs form a very effective way to debug new subsystems (because trivial errors such as the omission of an arc or a wrong arc expression often means that some of the system properties are dramatically changed).

The three papers in this section describe different ways to obtain reduced reachability graphs (without building the full reachability graph). The first paper deals with equivalence classes of markings, which correspond to symmetry properties of the given high-level net. The two other papers deal with symbolic representation of markings. The main ideas of the three approaches are closely related to each other, and the papers can be read in any order.

11.
Reachability Trees for High-level Petri Nets*

P. Huber, A.M. Jensen, L.O. Jepsen and K. Jensen

Theoretical Computer Science *45* (1986) 261-292

Communicated by H. Genrich
Received August 1985
Revised March 1986

Abstract. High-level Petri nets have been introduced as a powerful net type by which it is possible to handle rather complex systems in a succinct and manageable way. The success of high-level Petri nets is undebatable when we speak about description, but there is still much work to be done to establish the necessary analysis methods. In other papers it is shown how to generalize the concept of place- and transition invariants from place/transition nets to high-level Petri nets. Our present paper contributes to this with a generalization of reachability trees, which is one of the other important analysis methods known for place/transition nets.

Contents

1. Introduction

High-level Petri nets [1, 4, 5, 6, 9] have been introduced as a powerful net type by which it is possible to handle rather complex systems in a succinct and manageable way. The success of high-level Petri nets is undebatable when we speak about description, but there is still much work to be done to establish the necessary analysis methods. In [1, 4, 5] it is shown how to generalize the concept of place invariants (S-invariants) from place/transition nets (PT-nets) to high-level Petri nets (HL-nets). Analogously, [9] shows how to generalize transition invariants (T-invariants). Our

* This paper is a revised and enlarged version of the paper by P. Huber, A.M. Jensen, L.O. Jepsen, and K. Jensen, Towards reachability trees for high-level Petri nets, in: G. Rozenberg, ed., *Advances in Petri Nets 1984*, Lecture Notes in Computer Science **188** (Springer, Berlin, 1985) pp. 215–233.

present paper contributes with a generalization of reachability trees, which is one of the other important analysis methods known for PT-nets [2, 7, 8].

The central idea in our paper is the observation, that HL-nets often possess classes of equivalent markings. As an example the HL-net describing the five dining philosophers in [4] has an equivalence class consisting of those five markings in which exactly one philosopher is eating. These five markings are interchangeable, in the sense that their subtrees represent equivalent behaviours, where the only difference is the identity of the involved philosophers and forks. If we analyse one of these subtrees, we will also understand the behaviour of the others.

This paper shows how to define reachability trees for HL-nets (HL-trees). For PT-nets the reachability trees in [2, 7, 8] are kept finite by means of *covering* markings (introducing ω-symbols) and by means of *duplicate* markings (cutting away their subtrees). For HL-trees we reduce by means of *covering* markings and by means of *equivalent* markings (for each equivalence class we only develop the subtree of one node, while the other equivalent nodes become leaves of the tree). Reduction by equivalent markings is a generalization of reduction by duplicate markings. We describe an algorithm which constructs the HL-tree. The algorithm can easily be automated and we will soon start working on an implementation. The constructed HL-trees turn out to be considerably smaller than the corresponding PT-trees (the reachability trees for the equivalent PT-nets obtained from the HL-nets by the method described in [4]).

The rest of the paper is organized as follows. Section 2 reviews the formal definition of HL-nets and ω-bags. In Section 3, HL-trees are introduced by means of an example. Section 4 contains the formal definition of HL-trees and the algorithm to construct them. Section 5 discusses how to establish proof rules by which properties of HL-nets can be derived from properties of the corresponding HL-trees. Section 6 contains two examples where HL-trees are constructed and compared with the corresponding PT-trees. Appendix A contains some proofs that are omitted in Section 5.

2. A brief review of HL-nets and definition of ω-bags

In this section we will review the basic concepts of HL-nets [6] and we generalize bags (allowing their elements to have multiplicity ω, representing an unlimited number of occurrences). Bags (multisets) are represented as formal sums as shown in [6]. By BAG(S) we denote the set of all finite bags over a nonempty set S. By $[A \to B]_L$ we denote the set of all linear functions with domain A and range B.

2.1. Definition. An HL-net is a 6-tuple $H = (P, T, C, I_-, I_+, m_0)$, where

(1) P is a set of *places*,

(2) T is a set of *transitions*,

(3) $P \cap T = \emptyset$, $P \cup T \neq \emptyset$,

(4) C is the *colour-function* defined from $P \cup T$ into nonempty sets,

(5) I_- and I_+ are the *negative* and *positive incidence function* defined on $P \times T$ such that $I_-(p, t), I_+(p, t) \in [\text{BAG}(C(t)) \to \text{BAG}(C(p))]_L$ for all $(p, t) \in P \times T$,

(6) m_0, the *initial marking*, is a function defined on P such that $m_0(p) \in \text{BAG}(C(p))$ for all $p \in P$.

Throughout this paper we assume P, T, $C(p)$, and $C(t)$ to be finite for all $p \in P$ and $t \in T$. A *marking* of H is a function m defined on P such that $m(p) \in \text{BAG}(C(p))$ for all $p \in P$. A *step* of H is a function x defined on T, such that $x(t) \in \text{BAG}(C(t))$ for all $t \in T$. The step x has *concession* at the marking m iff $\forall p \in P$: $\sum_{t \in T} I_-(p, t)(x(t)) \leq m(p)$. A marking is *dead* iff only the empty step has concession at it.

When x has concession at m, it may *occur* yielding a new *directly reachable marking* given by the equation

$$\forall p \in P\colon m'(p) = m(p) - \sum_{t \in T} I_-(p, t)(x(t)) + \sum_{t \in T} I_+(p, t)(x(t)).$$

We indicate this by the notation $m[x\rangle m'$. In this paper we will only consider steps which map a single transition $t \in T$ into a single colour $c \in C(t)$, while all other transitions are mapped into the empty bag. Such a step is denoted by (t, c), where we sometimes omit the parentheses. When, for $n \geq 0$, $m[t_1, c_1\rangle m_1[t_2, c_2\rangle m_2 \ldots m_{n-1}[t_n, c_n\rangle m'$, the sequence $\sigma = (t_1, c_1)(t_2, c_2) \ldots (t_n, c_n)$ is called a *transition sequence* at m, and m' is (forward) *reachable* from m, which we shall denote by $m[\sigma\rangle m'$. By $R(m)$ we denote the set of all markings which are reachable from m. An HL-net is *bounded at place* $p \in P$ *and colour* $c \in C(p)$ iff $\exists k \in \mathbb{N}\, \forall m \in R(m_0)$: $m(p)(c) \leq k$, and it is *bounded* iff it is bounded at all places and all colours.

2.2. Definition. An ω-bag over a nonempty set S is a function $b: S \to \mathbb{N} \cup \{\omega\}$ and it is represented as a formal sum $\sum_{s \in S} b(s)s$, where $b(s) \in \mathbb{N} \cup \{\omega\}$.

$b(s)$ represents the number of occurrences of the element s. If $b(s) = \omega$ the exact value is unknown and may be arbitrarily large. An ω-bag b over the set S is *finite* iff its support $\{s \in S \mid b(s) \neq 0\}$ is finite. The set of all finite ω-bags over the nonempty set S will be denoted by ω-BAG(S). Summation, scalar multiplication, comparison, and multiplicity of ω-bags are defined in the following way, where $b_1, b_2, b \in \omega\text{-BAG}(S)$, $n \in \mathbb{N}$ and $m \in \mathbb{N} \cup \{\omega\}$:

$$\omega + m = \omega, \qquad \omega > n, \qquad m\omega = \begin{cases} \omega & \text{if } m \neq 0, \\ 0 & \text{if } m = 0, \end{cases}$$

$$\omega - m = \omega, \qquad \omega \geq m,$$

$$b_1 + b_2 = \sum_{s \in S} (b_1(s) + b_2(s))s, \qquad m \times b = \sum_{s \in S} (mb(s))s,$$

$$b_1 \geq b_2 \Leftrightarrow \forall s \in S\colon b_1(s) \geq b_2(s),$$

$$b_1 > b_2 \Leftrightarrow (b_1 \geq b_2 \wedge b_1 \neq b_2).$$

When $b_1 \geq b_2$, we also define subtraction: $b_1 - b_2 = \sum_{s \in S} (b_1(s) - b_2(s))s$.

A function $F \in [S \to \text{BAG}(R)]$, where S and R are nonempty sets, can be uniquely extended to a linear function $\hat{F} \in [\text{BAG}(S) \to \text{BAG}(R)]$, called the *bag-extension* of F: $\forall b \in \text{BAG}(S)$: $\hat{F}(b) = \sum_{s \in S} b(s) \times F(s)$.

Analogously, we define the *ω-bag-extension* of $F \in [S \to \omega\text{-BAG}(R)]$ to be $\bar{F} \in [\omega\text{-BAG}(S) \to \omega\text{-BAG}(R)]$, where $\forall b \in \omega\text{-BAG}(S)$: $\bar{F}(b) = \sum_{s \in S} b(s) \times F(s)$.

An *ω-marking of H* is a function m defined on P such that $m(p) \in \omega\text{-BAG}(C(p))$ for all $p \in P$. The concepts of step, concession, and reachability are generalized from markings to ω-markings by replacing the word 'marking' by 'ω-marking'. An ω-marking m_1 *covers* another ω-marking m_2 ($m_1 \geq m_2$) iff $\forall p \in P$: $m_1(p) \geq m_2(p)$, and it *strictly covers* m_2 ($m_1 > m_2$) iff $m_1 \geq m_2 \wedge m_1 \neq m_2$.

3. Informal introduction to reachability trees for HL-nets

In this section we will give, by means of an example, an informal introduction to our notion of reachability trees for HL-nets. The basic idea of a reachability tree is to organize all reachable markings in a tree structure where, to each node, a reachable marking is attached, while to each arc a transition and a colour are attached (which transforms the marking of its source node into the marking of its destination node). Such a tree contains all reachable markings and all possible transition sequences. By inspection of the tree it is possible to answer a large number of questions about the system. However, in general the reachability tree described above will be infinite. For practical use it is necessary to reduce it to a finite size. This is done by *covering* markings and by *equivalent* markings which is a generalization of *duplicate* markings. Reduction by covering markings and duplicate markings are well known from PT-trees. Reduction by equivalent markings is, however, a new concept suitable for HL-trees and this idea is the primary result of our paper.

Covering markings

When a node has a marking m_2 which strictly covers the marking m_1 of a predecessor, the transition sequence transforming m_1 into m_2 can be repeated several times starting from m_2 [1]. Thus it is possible to get an arbitrarily large value for each coefficient which has increased from m_1 to m_2. In the tree, we indicate this by substituting, in m_2, the ω-symbol for each such coefficient. The situation is analogous to the idea behind the 'pumping lemma' of automata theory, and it means that some of the places can obtain an arbitrarily large number of tokens of certain colours.

This kind of reduction results in a loss of information. In [8], it is shown that if ω occurs in a PT-tree, it is not always possible to determine from the tree whether the net has a dead marking or not.

[1] If m_2 already contains ω, the situation is more complicated and it may be necessary to involve some extra occurrences, cf. the proof of Lemma 5.7 in Appendix A.

Duplicate markings

If there are several nodes with identical markings, only one of them is developed further, while the others are marked as 'duplicate'. This reduction will not result in a loss of information, because we can construct the missing subtrees from the one developed. Due to reduction by covering markings, two such subtrees may not be completely identical, but they will represent the same set of markings and transition sequences.

Equivalent markings

In order to introduce our notion of equivalent markings, we will now look at the HL-net for the five dining philosophers in [4], see Fig. 1 and Table 1.

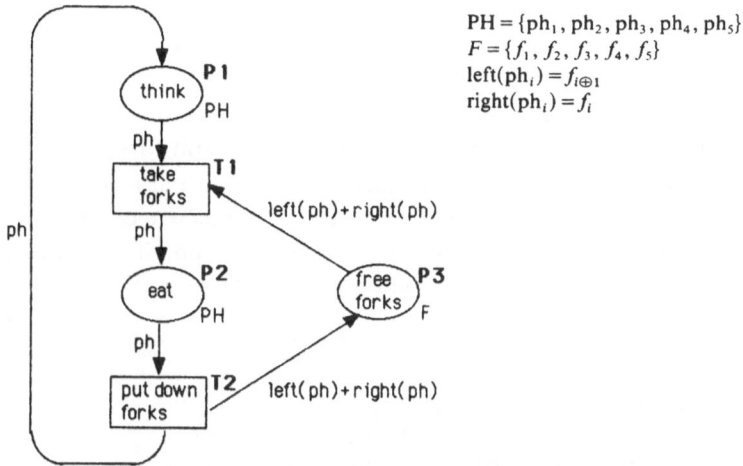

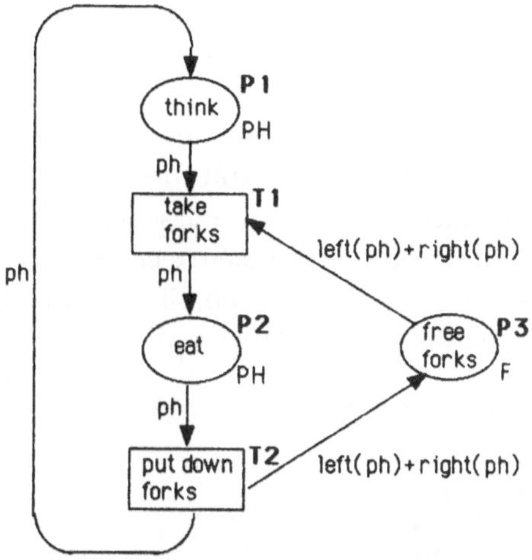

$\mathrm{PH} = \{\mathrm{ph}_1, \mathrm{ph}_2, \mathrm{ph}_3, \mathrm{ph}_4, \mathrm{ph}_5\}$
$F = \{f_1, f_2, f_3, f_4, f_5\}$
$\mathrm{left}(\mathrm{ph}_i) = f_{i\oplus 1}$
$\mathrm{right}(\mathrm{ph}_i) = f_i$

Fig. 1. HL-net for the dining philosophers problem.

We will now analyse the following markings:

$$\begin{aligned}
m_1 &= (\mathrm{ph}_2+\mathrm{ph}_3+\mathrm{ph}_4+\mathrm{ph}_5, \mathrm{ph}_1, f_3+f_4+f_5),\\
m_2 &= (\mathrm{ph}_1+\mathrm{ph}_3+\mathrm{ph}_4+\mathrm{ph}_5, \mathrm{ph}_2, f_1+f_4+f_5),\\
m_3 &= (\mathrm{ph}_2+\mathrm{ph}_4+\mathrm{ph}_5, \mathrm{ph}_1+\mathrm{ph}_3, f_5),\\
m_4 &= (\mathrm{ph}_2+\mathrm{ph}_3+\mathrm{ph}_4+\mathrm{ph}_5, \mathrm{ph}_1, f_2+f_4+f_5),\\
m_5 &= (\mathrm{ph}_3+\mathrm{ph}_4+\mathrm{ph}_5, \mathrm{ph}_1+\mathrm{ph}_2, f_5).
\end{aligned}$$

By intuition, we want m_1 and m_2 to be *equivalent*. The point is that we do not need to know the identity of eating philosophers, because all philosophers 'behave in the same way'. The marking m_3 contains a different number of eating philosophers and thus it is not equivalent to m_1 or m_2. However, two markings may be

Table 1

		$T1$	$T2$	m_0
		PH	PH	
$P1$	PH	$-$id	id	$\sum$ PH
$P2$	PH	id	$-$id	
$P3$	F	$-$left$-$right	left+right	$\sum F$

nonequivalent even though they have the same number of eating philosophers and the same number of free forks. In m_1 and m_2, the non-free forks are those belonging to the eating philosopher. This is not the case in m_4, and thus, m_4 is not equivalent to m_1 or m_2. In m_5, the two eating philosophers are neighbours. This is not the case in m_3, and so these markings are not equivalent either. To obtain equivalent markings we must demand that the identity of all philosophers and forks are changed by the same *rotation*. As an example, m_1 is obtained from m_2 by the rotation that adds 4 (in a cyclic way) to the index of each philosopher and fork.

To formalize the notion of equivalent markings we associate, to the colour set PH, the symmetry type 'rotation' and we define a bijective correspondence between F and PH by a function $r \in [F \to \text{PH}]$, where $r(f_i) = \text{ph}_i$. Two markings m' and m'' are equivalent iff there exists a rotation φ_{PH} of PH such that

$$\begin{aligned} m'(p) &= \overline{\varphi_{\text{PH}}}(m''(p)) \quad \text{for } p = P1, P2 \\ m'(P3) &= \overline{r^{-1} \circ \varphi_{\text{PH}} \circ r}(m''(P3)). \end{aligned} \tag{1}$$

In our example the markings m_1 and m_2 are equivalent because the rotation $\varphi_{\text{PH}} \in [\text{PH} \to \text{PH}]$, defined by $\varphi_{\text{PH}}(\text{ph}_i) = \text{ph}_{i \oplus 4}$, satisfies (1). On the other hand, m_2 and m_4 are not equivalent. From the place $P2$ it is demanded that $\text{ph}_2 = \varphi_{\text{PH}}(\text{ph}_1)$, i.e., $\varphi_{\text{PH}}(\text{ph}_i) = \text{ph}_{i \oplus 1}$, but this does not work at $P3$:

$$m_2(P3) = f_1 + f_4 + f_5 \neq f_1 + f_3 + f_5 = \overline{r^{-1} \circ \varphi_{\text{PH}} \circ r}(m_4(P3)).$$

As a generalization of reduction by duplicate markings we will now reduce the reachability tree by equivalent markings: only one element of each class of equivalent markings is developed further, and when a marking has several direct successors which are equivalent, only one of them is included in the tree.

Figure 2 shows an HL-tree obtained for the philosopher system. In the initial marking, transition $T1$ can occur in all colours of PH producing five equivalent markings of which only one is included in the tree, while the existence of the others are indicated by the label attached to the corresponding arc. If we only reduced by covering markings and duplicate markings, the tree would have had 31 nodes (and exactly the same tree structure as the PT-tree corresponding to the equivalent PT-net).

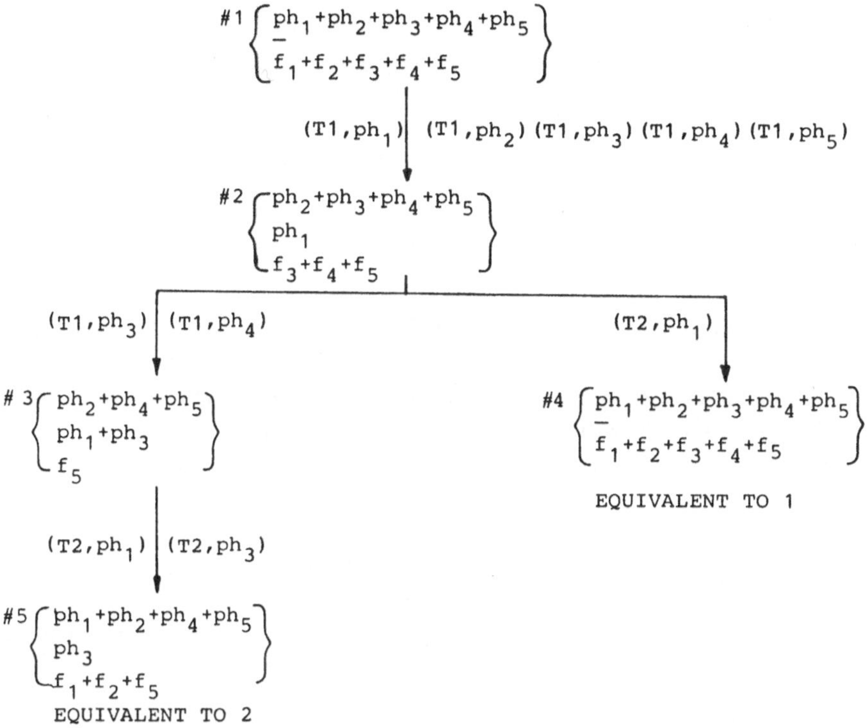

Fig. 2. HL-tree of the dining philosophers problem reduced by covering marking (in this tree, none) and equivalent markings.

The relation of equivalent markings is determined by the persons who analyse the system, and it must respect the inherent nature of the system. In the philosopher system, rotation is the suitable symmetry type. But in the telephone system of [6], arbitrary permutation would be the suitable symmetry type (since there is no special relation between a phone number and its nearest neighbours). In general, several symmetry types (rotation, permutation or identity-function) may be involved in the same system (for different colour sets).

When the relation of equivalent markings is defined in a sound way (to be formalized in Section 5), the reduction by means of covering markings and equivalent markings does not result in a greater loss of information than reduction by covering markings and duplicate markings only. This means, that all net properties which can be proved by means of the PT-tree of the equivalent PT-net can also be proved by means of our (much smaller) HL-tree.

4. Definition of reachability trees for HL-nets

In this section we consider a fixed HL-net $H = (P, T, C, I_-, I_+, m_0)$.

4.1. Definition. The set of colour sets $\{C(x) | x \in P \cup T\}$ is *partitioned* into three pairwise disjoint classes:

(1) A is the set of *atomic* colour sets, where, to each $\mathrm{Ca} \in A$, is attached a symmetry type: $\mathrm{sym(Ca)} \in$ {permutation, rotation[2], identity}.

(2) R is the set of *related* colour sets, where each $\mathrm{Cr} \in R$ is related to an atomic colour set Ca by a bijective function $r \in [\mathrm{Cr} \to \mathrm{Ca}]$.

(3) Π is the set of *product* colour sets, where each $\mathrm{C\pi} \in \Pi$ is the cartesian product of atomic and related colour sets.

4.2. Definition. A *symmetry* (allowed by the given partition) is a set of bijective functions $\varphi = \{\varphi_C\}_{C \in A \cup R \cup \Pi}$ where $\varphi_C \in [C \to C]$ for all C, and

(1) for all $\mathrm{Ca} \in A$, φ_{Ca} is a function of the kind specified by sym(Ca);

(2) for all $\mathrm{Cr} \in R$, with $r \in [\mathrm{Cr} \to \mathrm{Ca}]$, $\varphi_{\mathrm{Cr}} = r^{-1} \circ \varphi_{\mathrm{Ca}} \circ r$;

(3) for all $\mathrm{C\pi} \in \Pi$, with $\mathrm{C\pi} = C1 \times C2 \times \cdots \times Cn$, we have $\varphi_{\mathrm{C\pi}} = \varphi_{C1} \times \varphi_{C2} \times \cdots \times \varphi_{Cn}$.

The *set of symmetries* (allowed by the given partition) is denoted by Φ. It is finite since P, T, $C(p)$, and $C(t)$ are assumed to be finite for all $p \in P$ and $t \in T$.

The definition of φ_{Cr} can be visualized by the following commutative diagram:

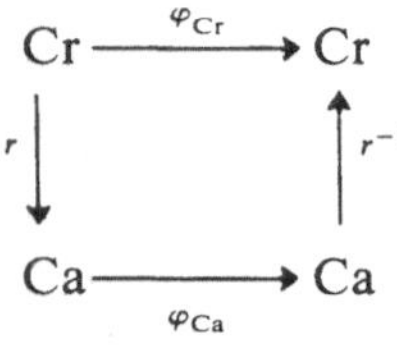

Since r is a bijection, it follows that φ_{Cr} is a function of the kind specified by sym(Ca).

Technical remark: The definition of partition is here presented in its simplest form. In some cases (cf. the database example in Section 6), it may be convenient or necessary to allow Π to contain *subsets* of cartesian products. If $\mathrm{C\pi} = C^n \backslash \{(a, a, \ldots, a) | a \in C\}$, we define $\varphi_{\mathrm{C\pi}} = (\varphi_C \times \varphi_C \times \cdots \times \varphi_C)|_{\mathrm{C\pi}}$, yielding a bijection on $\mathrm{C\pi}$ as requested. Secondly, in special cases, there can be sets in use to construct products in Π which are not themselves ordinary colour sets in the HL-net. These sets have to be included as atomic or related sets. (*end of remark.*)

Given an ω-marking m, a transition sequence $\sigma = (t_1, c_1)(t_2, c_2) \ldots (t_n, c_n)$, and a symmetry $\varphi \in \Phi$, we define an equivalent ω-marking $\varphi(m)$ by

$$\varphi(m)(p) = \bar{\varphi}_{C(p)}(m(p)) \quad \text{for all } p \in P$$

and an equivalent transition sequence $\varphi(\sigma)$ by

$$\varphi(\sigma) = (t_1, \varphi_{C(t_1)}(c_1))(t_2, \varphi_{C(t_2)}(c_2)) \ldots (t_n, \varphi_{C(t_n)}(c_n)).$$

[2] When an atomic colour set has rotation as symmetry type, it must be a finite set indexed by 1, 2, ..., n, where n is the cardinality.

4.3. Definition. Two ω-markings m_1 and m_2 of H are *equivalent*, denoted $m_1 \sim m_2$, iff there exists a symmetry $\varphi \in \Phi$ such that $m_1 = \varphi(m_2)$. It is easy to show that $\sim$ is an equivalence relation.

We would like to draw the reader's attention to the fact that, given a net, there are often several meaningful ways to define a partition. It is the user who chooses the partition, and this choice determines the possible symmetries and thus, the relation of equivalent markings. In Section 5 we will define two soundness criteria for partitions and we will establish four proof rules which, for sound partitions, allow us to deduce properties of HL-nets from properties of the corresponding HL-trees.

Given the notions above, we are now able to formalize the definition of reachability trees for HL-nets.

4.4. Definition. A reachability tree (*HL-tree*) for an HL-net with an equivalence relation $\sim$ (specified by a partition) is the full reachability tree[3] reduced with respect to covering markings and equivalent markings as follows:

(1) If a node y strictly covers a predecessor z, then we assign $m_y(p)(c) := \omega$ for all $p \in P$ and $c \in C(p)$ satisfying $m_y(p)(c) > m_z(p)(c)$.

(2) Only one node in each (reachable) equivalence class of $\sim$ is developed further. Only one node from a set of equivalent brothers is included in the tree; the other nodes are removed, but the arc to the included brother node contains information of their existence.

(3) Associated to each node is an *ω-marking* and a *node label.* The node label is a (possibly empty) sequence of status information, which may indicate that the marking is either *equivalent* to the marking of an earlier processed node, *covering* the marking of a predecessor node, or *dead.*

(4) Associated to each arc from node n_1 to n_2 is an *arc label* which is a list of occurrence information. Each element is a pair (t, c) where $t \in T$ and $c \in C(t)$. Each pair in the list has concession at the marking of n_1. An occurrence of the first pair in the list results in the marking of n_2, whereas occurrence of the other pairs results in markings which are equivalent to the marking of n_2.

Now we will describe our algorithm to produce the HL-trees. To create a new node we use the operation "NEWNODE(m, l)", where m and l are the ω-marking and node label of the node. A new arc is created by "NEWARC(n_1, n_2, l)" where n_1, n_2, and l are the source node, destination node, and arc label respectively. It is possible to append new information to an existing label l by the operation "APPEND(l, new-inf)". The ω-marking and the node label of a node x is denoted by m_x and l_x respectively. The arc label of the arc from node x to node y is denoted by l_{xy}. By "NEXT(m, t, c)" we denote the ω-marking obtained by the occurrence of transition t with colour $c \in C(t)$ in the ω-marking m.

[3] The full reachability tree contains all reachable markings and all transition sequences starting from the initial marking.

Algorithm to produce HL-trees

```
UNPROCESSED := {NEWNODE(m_0, empty)}; PROCESSED := ∅
repeat
  SELECT some node x ∈ UNPROCESSED
  if m_x ~ m_y for some node y ∈ PROCESSED
  then APPEND(l_x, "equivalent to y")
  else if no pair (t, c) has concession at m_x
    then APPEND(l_x, "dead")
    else
    begin {x is nonequivalent and non-dead}
      for all (t, c) having concession in m_x do
      begin
        m := NEXT(m_x, t, c); l := empty
        for all ancestors z with m > m_z do
        begin
          for all p ∈ P, c ∈ C(p) where m(p)(c) > m_z(p)(c) do
            m(p)(c) := ω
          APPEND(l, "covering of z")
        end
        if m ~ m_u for some node u being a son of x
        then APPEND(l_xu, "(t, c)")
        else
        begin
          v := NEWNODE(m, l)
          UNPROCESSED := UNPROCESSED ∪ {v}
          NEWARC(x, v, "(t, c)")
        end
      end
    end
  UNPROCESSED := UNPROCESSED \ {x}; PROCESSED := PROCESSED ∪ {x}
until UNPROCESSED = ∅.
```

The algorithm works in the following way: as long as there still are unprocessed nodes, one is selected and processed. The processing of a node starts with a check for equivalence with an already processed node, i.e., only the first processed node in each equivalence class of $\sim$ is developed further. If no equivalent node has been found, the node is checked for being dead. If it is not dead, for each pair (t, c) with concession, a son is produced and included in the tree (unless it is an equivalent brother). Each HL-tree is a subtree of a PT-tree for the equivalent PT-net, obtained from the HL-net by the method described in [4]. In [2, 7, 8], it is shown that each PT-tree is finite. Thus, each HL-tree is finite and our algorithm always halts.

Technical remark. The constructed HL-tree normally depends on the order in which the nodes are processed. This means that each HL-net may have several corresponding HL-trees. Normally, an implementation enforces an ordering rule for the processing of nodes and this rule then determines the actual HL-tree that is constructed for the HL-net by that implementation.

Technical remark: In an implementation of the algorithm it is crucial to minimize the time spent on testing for equivalence. In [3, Appendix 3] we describe a fairly effective algorithm to test two ω-markings for equivalence. Moreover, our implementation will use hash coding to divide markings into subclasses in such a way that equivalent markings always belong to the same subclass. This hash coding drastically decreases the number of pairs to be tested for equivalence.

5. What can be proved by means of HL-trees?

In this section we discuss how HL-trees can be used to prove properties of the corresponding HL-nets.

A *proof rule* is a theorem by which properties of HL-nets can be deduced from properties of HL-trees (or vice versa). For PT-trees, Hack [2] and Peterson [8] describe a number of such proof rules, from which it is possible to deduce information concerning boundedness, coverability, reachability, liveness, etc. Some of the proof rules are total, in the sense that the question concerning presence or absence of the particular net property can always be answered by means of the proof rule. Other proof rules are partial, in the sense that the question can only sometimes be answered.

For HL-trees, the situation is a bit more complicated since the observed tree properties may, in a crucial way, depend on the chosen partition which determines the relation of equivalent markings. Hence, it is necessary to introduce the notion of a *sound* partition, which intuitively means that the partition respects the inherent symmetry properties of the HL-net. If, for the philosopher system, we allowed arbitrary permutation instead of just rotation, this would be a typical example of a non-sound partition since it neglects the fact that in this system there is another relationship between neighbours than between non-neighbours. Analogously, it would be non-sound to have both PH and F as atomic colour sets since this would neglect the fact that there is another relationship between a philosopher and the two nearest forks than between the philosopher and the three remote forks.

5.1. Definition. A partition is *sound* iff it satisfies the following criteria:

(SC1) $\forall p \in P\, \forall t \in T\, \forall \varphi \in \Phi$: $\hat{\varphi}_{C(p)} \circ I_{\pm}(p, t) = I_{\pm}(p, t) \circ \hat{\varphi}_{C(t)}$,

(SC2) $\forall \varphi \in \Phi$: $m_0 = \varphi(m_0)$.

(SC1) can be visualized by the following commutative diagram:

$$\begin{array}{ccc}
\mathrm{BAG}(C(t)) & \xrightarrow{\hat{\varphi}_{C(t)}} & \mathrm{BAG}(C(t)) \\
{\scriptstyle I_{\pm}(p,t)}\downarrow & & \downarrow{\scriptstyle I_{\pm}(p,t)} \\
\mathrm{BAG}(C(p)) & \xrightarrow{\hat{\varphi}_{C(p)}} & \mathrm{BAG}(C(p))
\end{array}$$

(SC1) demands that the chosen partition for the HL-net, and hence the set of allowed symmetries, agree with the occurrence of transitions in the sense that equivalent colours have to be treated in the 'same' way. (SC2) demands that the initial marking be symmetric. In practice, it is often almost trivial to verify the soundness criteria by means of the following rules:

(R1): Due to the linearity of the functions, (SC1) can be verified by checking only steps of the form (t, c).

(R2): If $I_{\pm}(p, t)$ is an identity function or a zero function, then (SC1) is always satisfied.

(R3): When $I_{\pm}(p, t)$ is a sum of several functions, (SC1) can be verified for each of them separately.

(R4): When a function appears in $I_{\pm}(p, t)$ for several places or transitions, it only needs to be considered once to verify (SC1).

(R5): When the symmetry types of $C(t)$ and $C(p)$ both are identity, (SC1) is always satisfied.

(R6): When the symmetry type of $C(t)$ is rotation, it is enough to consider the 'one-step-forward' rotation to verify (SC1).

(R7): When the symmetry type of $C(t)$ is permutation, it is enough to consider transpositions (interchanging of two elements) to verify (SC1).

(R8): (SC2) is satisfied iff

$$\forall p \in P: [\text{sym}(C(p)) \neq \text{identity} \Rightarrow \exists k \in \mathbb{N}_0: m_0(p) = k \times \sum C(p))],$$

where $\sum C(p)$ denotes the bag which contains exactly one occurrence of each colour in $C(p)$.

As an example, the soundness of the partition chosen for the philosopher system in Section 3 can easily be verified. We only have to prove the following properties (where r is the function relating F to PH, while φ_{PH} is the 'one-step-forward' rotation on PH):

$$r^{-1} \circ \varphi_{\text{PH}} \circ r \circ \text{left} = \text{left} \circ \varphi_{\text{PH}}, \qquad r^{-1} \circ \varphi_{\text{PH}} \circ r \circ \text{right} = \text{right} \circ \varphi_{\text{PH}}.$$

To formulate our proof rules we need some notation. $R(m_0)$ is the set of markings which are reachable from m_0. $R(m_0)(p) = \{m(p)(c) \mid m \in R(m_0) \wedge c \in C(p)\}$ is the set of coefficients appearing at place p, while $R(m_0)(p)(c) = \{m(p)(c) \mid m \in R(m_0)\}$ is the set of coefficients appearing at place p for colour c. $T(m_0)$ is the set of nodes in the HL-tree having m_0 as root. $T(m_0)(p)$ and $T(m_0)(p)(c)$ are defined analogously to $R(m_0)(p)$ and $R(m_0)(p)(c)$, respectively. Furthermore, we define the function $\text{map}_{C(p)}$ from $C(p)$ into subsets of $C(p)$ as follows:

$$\text{map}_{C(p)}(c) = \{c' \in C(p) \mid \exists \varphi \in \Phi: \varphi_{C(p)}(c') = c\}.$$

Observation:

(O1) $$\text{map}_{C(p)}(c) = \begin{cases} \{c\} & \text{if sym}(C(p)) = \text{identity}, \\ C(p) & \text{if sym}(C(p)) \in \{\text{rotation, permutation}\}. \end{cases}$$

We now formulate our four proof rules for HL-trees. They are generalizations of the proof rules for PT-trees given in [8].

Proof rules for HL-nets

(PR1): H is bounded $\Leftrightarrow \forall p \in P: \omega \notin T(m_0)(p)$,
prerequisite: (SC1);

(PR2): $\sup R(m_0)(p)(c)^4 = \max \bigcup_{c' \in \mathrm{map}_{C(p)}(c)} T(m_0)(p)(c')$,
prerequisite: (SC1), (SC2);

(PR3): $\exists \alpha \in T(m_0)$: "dead" $\in l_\alpha \Rightarrow \exists m \in R(m_0)$: m is dead,
prerequisite: none;

(PR4): $\exists m \in R(m_0)$: m is dead $\Rightarrow (\exists \alpha \in T(m_0)$: "dead" $\in l_\alpha)$
$\vee (\exists p \in P: \omega \in T(m_0)(p))$,
prerequisite: (SC1).

As an example on how the proof rules can be used, we again turn to the philosopher system with the HL-tree shown in Fig. 2. By applying (PR1) we derive that the net is bounded, and from (PR2) we see that 1 can be used as a uniform bound for all places and all colours. (PR4) tells us that no reachable marking is dead.

To prove the correctness of our proof rules we need the following four lemmas:

5.2. Lemma. *Assume* (SC1), *then* $\forall \varphi \in \Phi$: $m_1[\sigma\rangle m_2 \Rightarrow \varphi(m_1)[\varphi(\sigma)\rangle\varphi(m_2)$ *for all ω-markings and all transition sequences.*

Proof. By induction on the length of σ. □

5.3. Corollary. *Assume* (SC1) *and* (SC2), *then*

(a) $m_1 \sim m_2 \Rightarrow [m_1 \in R(m_0) \Leftrightarrow m_2 \in R(m_0)]$,

(b) $m_1 \sim m_2 \Rightarrow [m_1$ *is dead* $\Leftrightarrow m_2$ *is dead*$]$.

5.4. Definition. Given an ω-marking m_ω and a marking m, we define that m_ω *agrees* with m, denoted $m_\omega \rhd m$, iff

$$\forall p \in P\ \forall c \in C(p): m_\omega(p)(c) \neq \omega \Rightarrow m_\omega(p)(c) = m(p)(c),$$

i.e., for each pair p and c, the coefficients in m_ω and m are identical or that of m_ω is ω. It is easy to prove the following observations.

Observations:

(O2) $m_\omega \rhd m \Rightarrow \varphi(m_\omega) \rhd \varphi(m)$ for all $\varphi \in \Phi$,

(O3) $m_\omega \rhd m \wedge m[\sigma\rangle m' \Rightarrow \exists m'_\omega: m_\omega[\sigma\rangle m'_\omega \wedge m'_\omega \rhd m'$ for all transition sequences σ.

5.5. Lemma. *Assume* (SC1), *then* $\forall m \in R(m_0)\ \exists \varphi \in \Phi\ \exists \alpha \in T(m_0)$: $m_\alpha \rhd \varphi(m)$.

[4] By convention, $\sup A = \omega$ for $A \subseteq \mathbb{N}$ when $\forall k \in \mathbb{N}\ \exists a \in A: a \geq k$.

Proof. In order to deal with reduction by equivalent markings, the corresponding proof for PT-trees in [2, Lemma 3.7] can be generalized. The proof is done by induction on the length of σ, where $m_0[\sigma\rangle m$. □

5.6. Definition. Given an ω-marking m and $k \in \mathbb{N}$, we then define $m[{}^{\omega}_{k}]$, the *substitution* of ω by k, as follows:

$$m\begin{bmatrix}\omega\\ k\end{bmatrix}(p)(c) = \begin{cases} k & \text{if } m(p)(c)=\omega,\\ m(p)(c) & \text{otherwise,}\end{cases}$$

for all $p \in P$ and $c \in C(p)$.

5.7. Lemma. $\forall \alpha \in T(m_0)\ \forall k \in \mathbb{N}\ \exists m \in R(m_0)$: $m_\alpha \rhd m \geqslant m_\alpha[{}^{\omega}_{k}]$.

Proof. See Appendix A. The proof of this lemma is by far the most complicated and it involves several induction arguments. □

5.8. Corollary. (a) $\omega \in T(m_0)(p)(c) \Rightarrow \sup R(m_0)(p)(c) = \omega$.
(b) $\forall k \in \mathbb{N}$: $k \in T(m_0)(p)(c) \Rightarrow k \in R(m_0)(p)(c)$.

5.9. Lemma. *Assume* (SC1) *and* (SC2), *then*

$$\sup R(m_0)(p)(c) = \max \bigcup_{c' \in \mathrm{map}_{C(p)}(c)} T(m_0)(p)(c').$$

If only (SC1) *is assumed, we get* "$\leqslant$" *instead of* "$=$".

Proof. See Appendix A. □

5.10. Theorem. *The four proof rules* (PR1)-(PR4) *are valid, under the given prerequisites.*

Proof. (PR1): The proof is by contradiction. Assume that H is bounded, and $\exists p \in P$: $\omega \in T(m_0)(p)$. Then, $\omega \in T(m_0)(p)(c)$ for some colour $c \in C(p)$ and, by Corollary 5.8, $R(m_0)(p)(c)$ is unbounded—a contradiction with H being bounded.

Next, assume that $\forall p \in P$: $\omega \notin T(m_0)(p)$ and H unbounded, i.e.,

$$\exists p \in P\ \exists c \in C(p)\ \forall k \in \mathbb{N}\ \exists m \in R(m_0)\colon m(p)(c) > k. \tag{2}$$

For each of these m, by Lemma 5.5,

$$\exists \alpha \in T(m_0)\ \exists \varphi^m \in \Phi\colon m_\alpha \rhd \varphi^m(m). \tag{3}$$

We then get

$$m_\alpha(p)(\varphi^m_{C(p)}(c)) \geqslant \varphi^m(m)(p)(\varphi^m_{C(p)}(c)) = m(p)(c) > k \tag{4}$$

for each k in (2). "$\geqslant$" follows from (3), "$=$" is an immediate consequence of the way $\varphi^m(m)$ is defined, while "$>$" follows from (2). Since $T(m_0)$ and Φ are finite, it follows from (4) that $\exists\alpha' \in T(m_0)$: $m_{\alpha'}(p)(\varphi^m(c)) = \omega$: a contradiction with $\omega \notin T(m_0)(p)$.

(PR2): Identical to Lemma 5.9.

(PR3): Assume that $\exists\alpha \in T(m_0)$: "dead" $\in l_\alpha$. By Lemma 5.7, $\exists m \in R(m_0)$: $m_\alpha \trianglerighteq m$. The marking m_α is dead and since m is smaller, m is dead, too.

(PR4): Assume that $\exists m \in R(m_0)$: m is dead, and $\forall\alpha \in T(m_0)$: "dead" $\notin l_\alpha$. By Lemma 5.5, $\exists\varphi \in \Phi\ \exists\alpha \in T(m_0)$: $m_\alpha \trianglerighteq \varphi(m)$. The marking $\varphi(m)$ is dead, by Corollary 5.3, m_α is not dead and thus, we conclude $m_\alpha > \varphi(m)$ which, together with $m_\alpha \trianglerighteq \varphi(m)$, yields $m_\alpha(p)(c) = \omega$ for some $p \in P$ and $c \in C(p)$. □

The following two lemmas are not necessary to establish the proof rules, but they provide useful insight in the structure of the reachability tree.

5.11. Lemma. $\forall \alpha 1, \alpha 2 \in T(m_0)$ *with* $(t, c) \in l_{\alpha 1\, \alpha 2}$ $\exists m_1, m_2 \in R(m_0)$ *with* $m_1[t, c\rangle m_2$ *such that*

(i) $m_{\alpha 1} \trianglerighteq m_1$ *and*

(ii) $$m_{\alpha 2} \trianglerighteq \begin{cases} m_2 & \text{if } (t, c) = \text{head}(l_{\alpha 1\, \alpha 2}), \\ \varphi(m_2) & \text{for some } \varphi \in \Phi \text{ otherwise.} \end{cases}$$

Proof. Using Lemma 5.7 with $\alpha 1$ for α and $\max\{I_-(p, t)(c)(c') \mid p \in P \wedge c' \in C(p)\}$ for k, we get an $m_1 \in R(m_0)$ such that $m_{\alpha 1} \trianglerighteq m_1$ and (t, c) has concession at m_1 (with $m_1[t, c\rangle m_2$). If $(t, c) = \text{head}(l_{\alpha 1\, \alpha 2})$, we have $m_{\alpha 1}[t, c\rangle m_{\alpha 2}$ and (ii) directly follows from observation (O3); otherwise we can get the deleted brother node by means of a symmetry such that $\varphi^{-1}(m_{\alpha 2}) \trianglerighteq m_2$, and we finish with observation (O2). □

5.12. Lemma. *Assume* (SC1), *then* $\forall m_1, m_2 \in R(m_0)$ *with* $m_1[t, c\rangle m_2$ *there exists a* $\varphi \in \Phi$ *and* $\exists \alpha 1, \alpha 2 \in T(m_0)$ *with* $\varphi(t, c) \in l_{\alpha 1\, \alpha 2}$ *such that*

(i) $m_{\alpha 1} \trianglerighteq \varphi(m_1)$ *and*

(ii) $$m_{\alpha 2} \trianglerighteq \begin{cases} \varphi(m_2) & \text{if } \varphi(t, c) = \text{head}(l_{\alpha 1\, \alpha 2}), \\ \varphi' \circ \varphi(m_2) & \text{for some } \varphi' \in \Phi \text{ otherwise.} \end{cases}$$

Proof. The existence of φ and $\alpha 1$ follows from Lemma 5.5 and we have (i). (As in the proof of Lemma 5.5, we choose $\alpha 1$ to be in the node in the actual equivalence class which is further processed.) Lemma 5.2 yields

$$\varphi(m_1)[\varphi(t, c)\rangle \varphi(m_2). \tag{5}$$

We can then use observation (O3) on (i) and (5) and get

$$m_{\alpha 1}[\varphi(t, c)\rangle \tilde{m} \quad \text{and} \quad \tilde{m} \trianglerighteq \varphi(m_2). \tag{6}$$

Hence, we can select $\alpha 2$ to be that son of $\alpha 1$ which has $\varphi(t, c) \in l_{\alpha 1\, \alpha 2}$. If $\varphi(t, c) = \text{head}(l_{\alpha 1\, \alpha 2})$, we have $m_{\alpha 2} = \tilde{m}$ and the second part of (6) gives (ii). Otherwise, we have to apply a symmetry as in the proof of Lemma 5.11. □

6. Examples of the use of HL-trees

This section contains two examples which, together with the system of the five dining philosophers treated in Section 3, illustrate a spectrum of the problems concerning the construction and analysis of HL-trees. The first example is a system where the equivalence relation involves permutation, identity, and products. The second example illustrates covering markings.

6.1. Example (*Data base system*). This example is taken from [4], but originally it was given by Genrich and Lautenbach.

Three database managers, DBM = $\{a, b, c\}$, communicate with each other. Each manager can make an update to his own database. At the same time he must send a message to each of the other managers, thereby informing them about the update. Having sent this set of messages, the sending manager waits until all other managers have received his message, performed an update, and sent an acknowledgment. When all acknowledgments are present, the sending manager returns to be inactive. At that time (but not before) another manager may perform an update and send messages.

Each manager can be in three states: 'inactive', 'waiting' (for acknowledgments), and 'performing' (an update on request of another manager). The managers communicate via a fixed set of message buffers, MB = $\{(s, r) \mid s, r \in \text{DBM} \wedge s \neq r\}$, where s represents the sender and r represents the receiver. Each message buffer may be in four different states: 'unused', 'sent', 'received' and 'acknowledged'.

The system can be described by the HL-net in Fig. 3. The corresponding incidence matrix is shown in Table 2, and a partition is defined by

atomic: DBM: permutation; **E**: identity;

product: MB: subset of DBM × DBM.

Table 2
Incidence matrix for the data base system.

		$T1$	$T2$	$T3$	$T4$	m_0
		DBM	DBM	MB	MB	
$P1$	DBM	−ID	ID	−REC	REC	Σ DBM
$P2$	DBM	ID	−ID			
$P3$	DBM			REC	−REC	
$P4$	E	−ABS	ABS			ε
$P5$	MB	−MINE	MINE			Σ MB
$P6$	MB	MINE		−ID		
$P7$	MB			ID	−ID	
$P8$	MB		−MINE		ID	

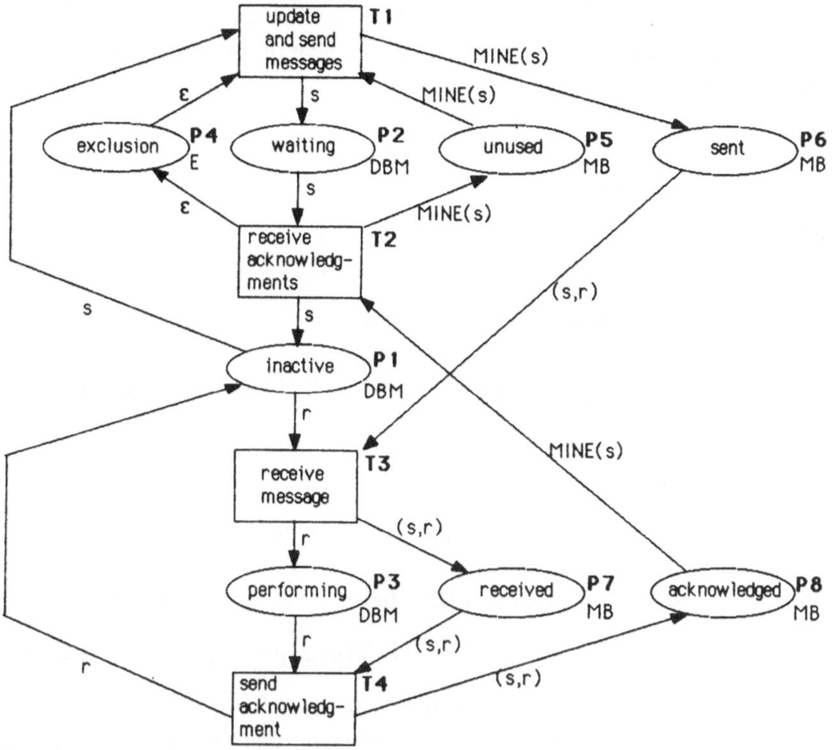

$\mathrm{DBM} = \{a, b, c\}$, $\mathrm{MB} = \mathrm{DBM} \times \mathrm{DBM} \backslash \{(u, u) \mid u \in \mathrm{DBM}\}$, $\mathbf{E} = \{\varepsilon\}$.

Fig. 3. HL-net for the data base system.

The functions

$$\mathrm{ID} \in [\mathrm{BAG}(\mathrm{DBM}) \to \mathrm{BAG}(\mathrm{DBM})]_{\mathrm{L}},$$

$$\mathrm{ABS} \in [\mathrm{BAG}(\mathrm{DBM}) \to \mathrm{BAG}(\mathbf{E})]_{\mathrm{L}},$$

$$\mathrm{MINE} \in [\mathrm{BAG}(\mathrm{DBM}) \to \mathrm{BAG}(\mathrm{MB})]_{\mathrm{L}},$$

$$\mathrm{ID} \in [\mathrm{BAG}(\mathrm{MB}) \to \mathrm{BAG}(\mathrm{MB})]_{\mathrm{L}},$$

$$\mathrm{REC} \in [\mathrm{BAG}(\mathrm{MB}) \to \mathrm{BAG}(\mathrm{DBM})]_{\mathrm{L}}$$

are defined by

$$\mathrm{ID}(s) = s \qquad \text{for all } s \in \mathrm{DBM},$$

$$\mathrm{ABS}(s) = \varepsilon \qquad \text{for all } s \in \mathrm{DBM},$$

$$\mathrm{MINE}(s) = \sum_{r \neq s} (s, r) \qquad \text{for all } s \in \mathrm{DBM},$$

$$\mathrm{ID}((s, r)) = (s, r) \qquad \text{for all } (s, r) \in \mathrm{MB},$$

$$\mathrm{REC}((s, r)) = r \qquad \text{for all } (s, r) \in \mathrm{MB}.$$

Soundness criterion (SC1) is verified by means of rules (R1)-(R7), given in Section 5. By (R1), (R2), and (R4) it is sufficient to check that the incidence functions ABS, MINE, and REC satisfy

$$\hat{\varphi}_{C(p)} \circ I_{\pm}(p, t)(c) = I_{\pm}(p, t) \circ \varphi_{C(t)}(c)$$

for each $\varphi \in \Phi$ and $c \in C(t)$.

(ABS): Let $\varphi \in \Phi$ and $s \in \mathrm{DBM}$. Then

$$\varphi_{\mathbf{E}}(\mathrm{ABS}(s)) = \varphi_{\mathbf{E}}(\varepsilon) = \varepsilon = \mathrm{ABS}(\varphi_{\mathrm{DBM}}(s)).$$

(MINE): Let $\varphi \in \Phi$ and $s \in \mathrm{DBM}$. Then

$$\hat{\varphi}_{\mathrm{MB}}(\mathrm{MINE}(s)) = \hat{\varphi}_{\mathrm{MB}}\left(\sum_{x \neq s} (s, x) \right) = \sum_{x \neq s} (\varphi_{\mathrm{DBM}}(s), \varphi_{\mathrm{DBM}}(x))$$

$$= \sum_{y \neq \varphi_{\mathrm{DBM}}(s)} (\varphi_{\mathrm{DBM}}(s), y) = \mathrm{MINE}(\varphi_{\mathrm{DBM}}(s)).$$

In this particular case we do not use rule (R7) since it is just as easy to prove the property for arbitrary permutations.

(REC): Let $\varphi \in \Phi$ and $(s, r) \in \mathrm{MB}$. Then

$$\varphi_{\mathrm{DBM}}(\mathrm{REC}((s, r))) = \varphi_{\mathrm{DBM}}(r) = \mathrm{REC}((\varphi_{\mathrm{DBM}}(s), \varphi_{\mathrm{DBM}}(r))) = \mathrm{REC}(\varphi_{\mathrm{MB}}(s, r)).$$

Soundness criterion (SC2) immediately follows from rule (R8) in Section 5.

Having verified soundness for the partition we can now apply the proof rules on the HL-tree shown in Fig. 4.

(PR1): The HL-net is bounded.

(PR2): All places and all colours in the HL-net have 1 as a uniform bound. This follows from the following observations:

$$\mathrm{map}_{\mathrm{DBM}}(s) = \mathrm{DBM} \quad \text{for all } s \in \mathrm{DBM} \qquad \text{(by (O1))},$$

$$\mathrm{map}_{\mathbf{E}}(\varepsilon) = \{\varepsilon\} = \mathbf{E} \qquad \text{(by (O1))},$$

$$\mathrm{map}_{\mathrm{MB}}((s, r)) = \mathrm{MB} \quad \text{for all } (s, r) \in \mathrm{MB} \qquad \text{(from the definition of 'map')}.$$

(PR3): Cannot be applied.

(PR4): The HL-net has no reachable marking which is dead.

The leaves of the tree are identical with #1 and #6, respectively. This is, however, a coincidence and it changes if the nodes are processed in another order. As mentioned earlier, an alternative to the HL-tree is to construct the PT-tree for the equivalent PT-net. In Table 3 we compare the size of the HL-tree with the size of the PT-tree (for different sizes of DBM).

The HL-trees are not just smaller than the corresponding PT-trees, but they also seem to grow slower when the sizes of the involved colour sets increase. It is, however, normally not necessary to consider colour sets which have more than a few elements. If you know how a system with five philosophers works, you also

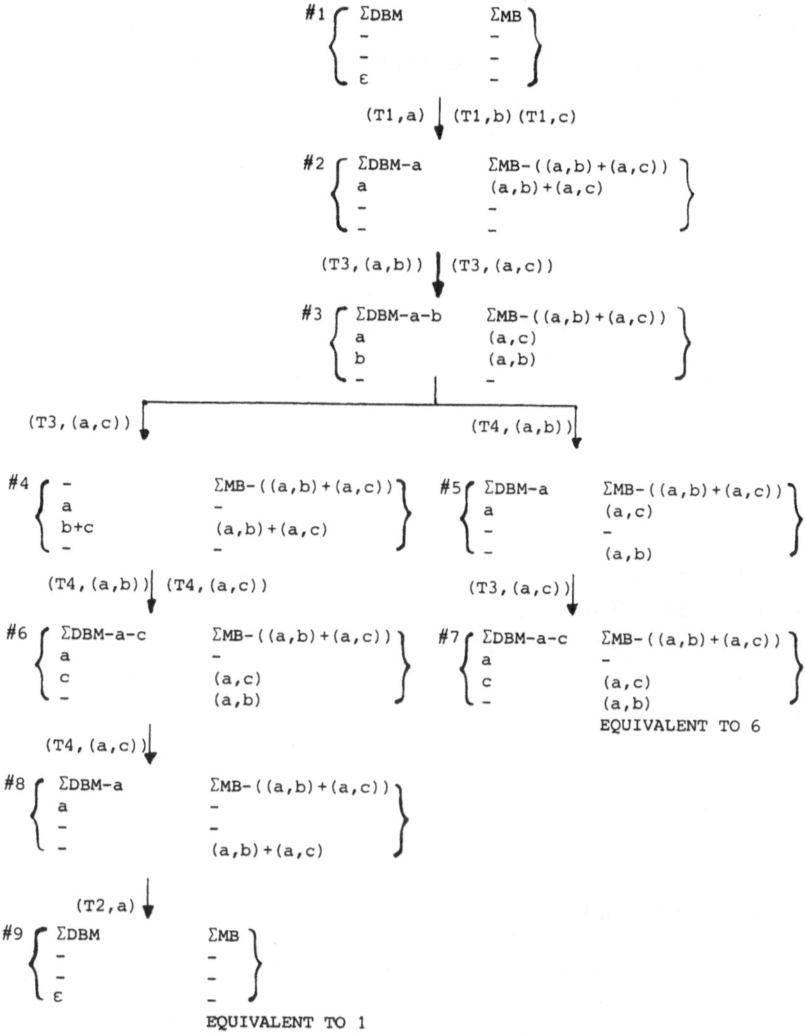

Fig. 4. HL-tree for the data base system with three data base managers.

Table 3
Data base system.

Number of data base managers	Number of nodes in the HL-tree	Number of nodes in the PT-tree
2	5	9
3	9	43
4	14	225
5	23	>1400

know how a system with six or more works. Analogously, you have to investigate a system with three data base managers in order to know how a system with arbitrarily many managers would work. But, even for small colour sets, the HL-trees are considerably smaller than the corresponding PT-trees. This is illustrated for the data base system by Table 3, and for the philosopher system by Table 4.

Table 4
Philosopher system.

Number of philosophers	Number of nodes in the HL-tree	Number of nodes in the PT-tree
3	3	7
4	5	17
5	5	31

6.2. Example (*Producer-consumer system*). Two producers, $A = \{a1, a2\}$, each produce their own kind of message which they repeatedly send to a consumer via an unbounded buffer. The consumer can only receive pairs of messages consisting of one message from each producer.

The system can be described by the HL-net in Fig. 5. The corresponding incidence matrix is shown in Table 5, where the function

$$\text{PAIR} \in [\text{BAG}(B) \rightarrow \text{BAG}(A)]_{\text{L}}$$

is defined by

$$\text{PAIR}(b) = a1 + a2.$$

We define a partition by

atomic: A: permutation; B: identity.

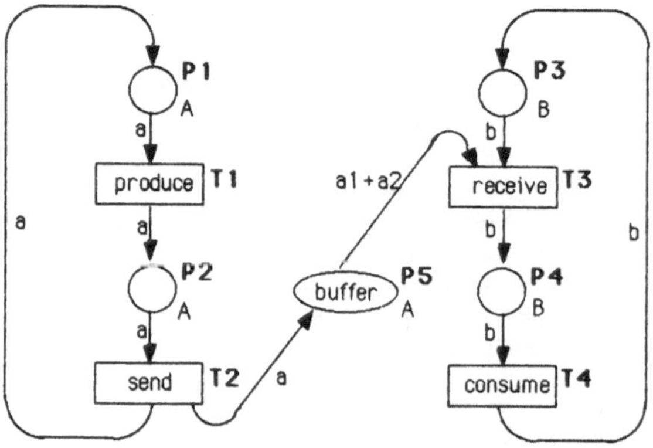

$A = \{a1, a2\}$, $B = \{b\}$.

Fig. 5. HL-net for the producer-consumer system.

Table 5
Incidence matrix for the producer-consumer system.

		$T1$	$T2$	$T3$	$T4$	m_0
		A	A	B	B	
$P1$	A	$-$ID	ID			$\sum A$
$P2$	A	ID	$-$ID			
$P3$	B			$-$ID	ID	$\sum B$
$P4$	B			ID	$-$ID	
$P5$	A		ID	$-$PAIR		

Soundness of the partition immediately follows from rules (R1)–(R8) in Section 5. One of the corresponding HL-trees is shown in Fig. 6.

We can now apply the proof rules.

(PR1): The HL-net is unbounded.

(PR2): The places $P1$–$P4$ are bounded for all colours, with 1 as a uniform bound. The place $P5$, which represents the buffer, is unbounded for both colours in its colour set.

(PR3): Cannot be applied.

(PR4): Cannot be applied.

The HL-tree has 30 nodes of which 17 are coverings (some of them even cover two other markings). As in the two other examples of this paper, the HL-tree for this system is remarkably smaller than the corresponding PT-tree (see Table 6).

Table 6
Producer-consumer system.

Number of producers	Number of nodes in the HL-tree	Number of nodes in the PT-tree
2	30	93

Appendix A

This appendix contains the proofs of Lemmas 5.7 and 5.9. Furthermore, it contains five propositions which are necessary for the proofs.

Definition A.1. Let $I = I_+ - I_-$. Then, for each transition sequence

$$\sigma = (t_1, c_1)(t_2, c_2)\ldots(t_n, c_n) \quad \text{with } n \geq 0,$$

we define $\Delta(\sigma)$ to be the change in marking caused by σ:

$$\forall p \in P: \Delta(\sigma)(p) = \sum_{i=1}^{n} I(p, t_i)(c_i).$$

Analogously we define Δ_+ and Δ_- by means of I_+ and I_- respectively.[5]

[5] In the terminology of [4, 6] $\Delta(\sigma) = I * \sigma$, where "*" is the generalized matrix product.

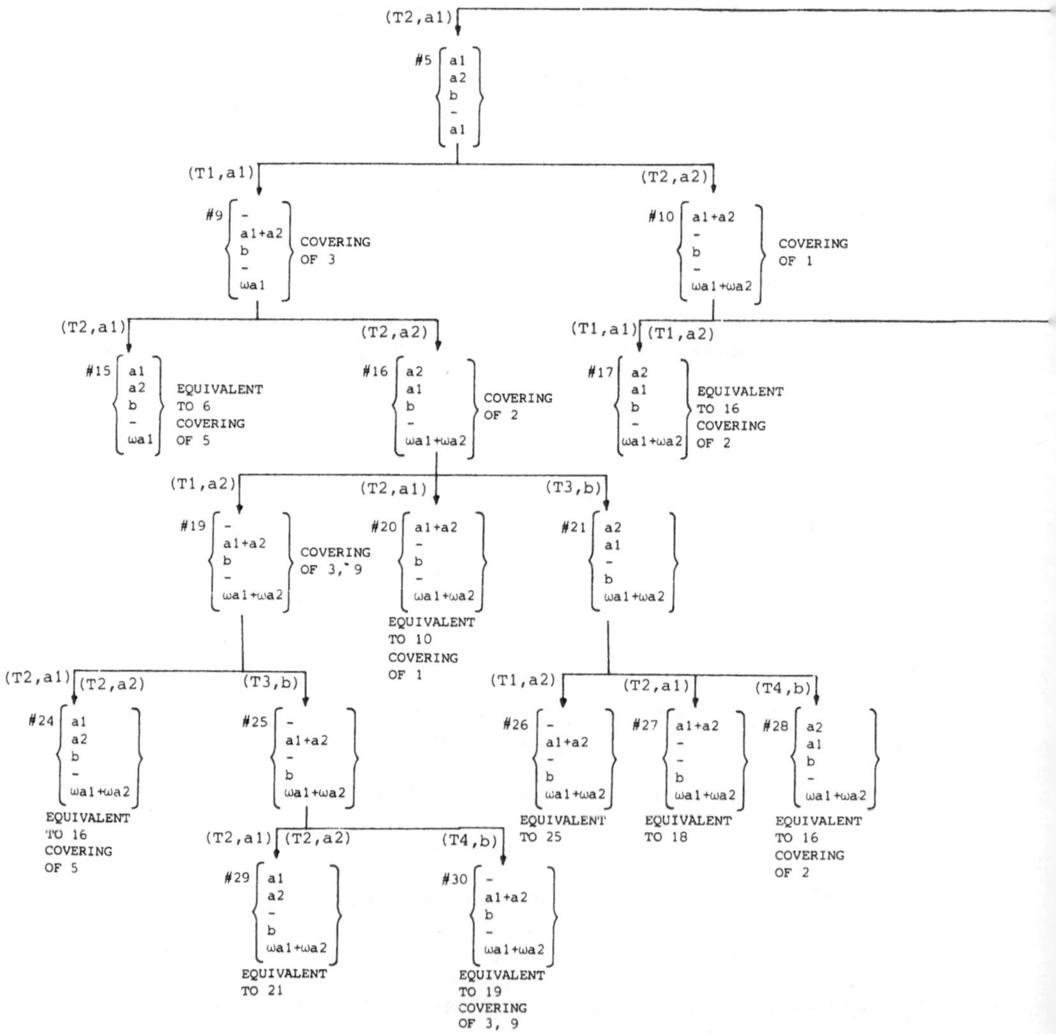

Fig. 6. HL-tree for the producer-consumer system.

A31.

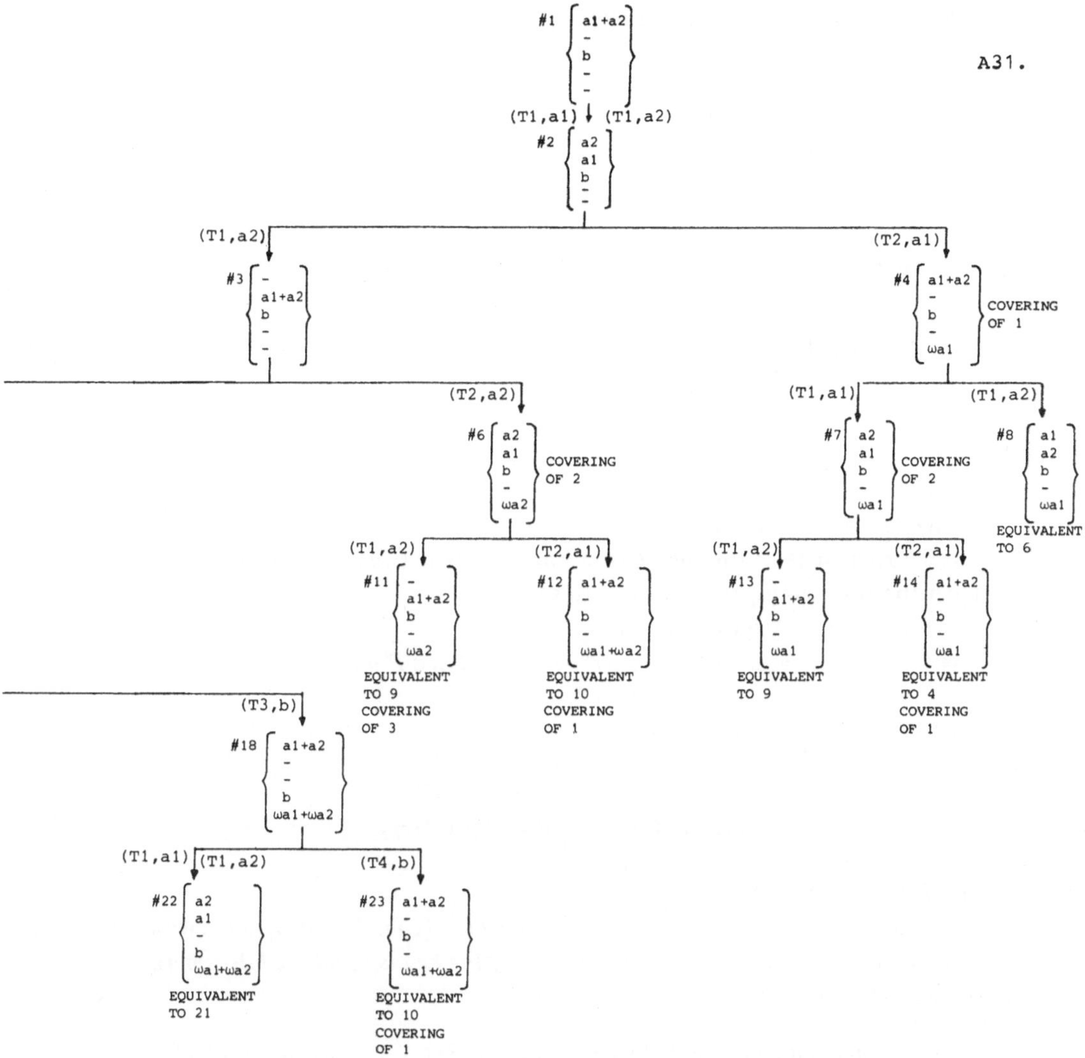
#1
a1+a2
-
b
-
-
(T1,a1)
(T1,a2)
#2
a2
a1
b
-
-
(T1,a2)
(T2,a1)
#3
-
a1+a2
b
-
-
#4
a1+a2
-
b
-
ωa1
COVERING OF 1
(T2,a2)
#6
a2
a1
b
-
ωa2
COVERING OF 2
(T1,a1)
(T1,a2)
#7
a2
a1
b
-
ωa1
COVERING OF 2
#8
a1
a2
b
-
ωa1
EQUIVALENT TO 6
(T1,a2)
(T2,a1)
#11
-
a1+a2
b
-
ωa2
EQUIVALENT TO 9 COVERING OF 3
#12
a1+a2
-
b
-
ωa1+ωa2
EQUIVALENT TO 10 COVERING OF 1
(T1,a2)
(T2,a1)
#13
-
a1+a2
b
-
ωa1
EQUIVALENT TO 9
#14
a1+a2
-
b
-
ωa1
EQUIVALENT TO 4 COVERING OF 1
(T3,b)
#18
a1+a2
-
-
b
ωa1+ωa2
(T1,a1)
(T1,a2)
(T4,b)
#22
a2
a1
-
b
ωa1+ωa2
EQUIVALENT TO 21
#23
a1+a2
-
b
-
ωa1+ωa2
EQUIVALENT TO 10 COVERING OF 1

A.1. Propositions

Assume the following situation:

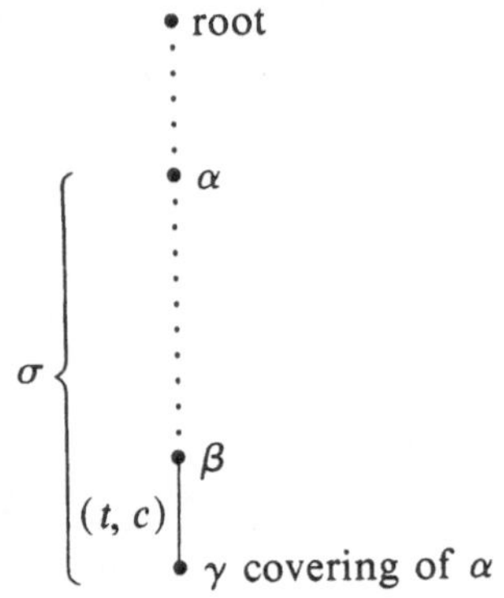

where

(i) 'covering of α' $\in l_\gamma$,

(ii) $(t, c) = \text{head}(l_{\beta\gamma})$,

(iii) σ consists of the heads of all arc labels between α and γ.

Then the following propositions are satisfied:

(a) $m_\gamma \geq m_\beta + \Delta(t, c) \geq m_\alpha$.

(b) $m_\beta(p)(c') \neq \omega \Rightarrow \Delta(\sigma)(p)(c') \geq 0$ *for all* $p \in P$ *and* $c' \in C(p)$.

(c) $m_\gamma(p)(c') \neq \omega \Rightarrow \Delta(\sigma)(p)(c') = 0$ *for all* $p \in P$ *and* $c' \in C(p)$.

(d) σ *is a transition sequence at an* ω*-marking* m *if*

$$m(p)(c') \geq \begin{cases} m_\alpha(p)(c') & \text{if } m_\beta(p)(c') \neq \omega, \\ \Delta_-(\sigma)(p)(c') & \text{if } m_\beta(p)(c') = \omega \end{cases}$$

for all $p \in P$ *and* $c' \in C(p)$.

(e) *If* $\alpha_1, \alpha_2, \ldots, \alpha_n$ *are* all *nodes for which 'covering of* α_i*'* $\in l_\gamma$ *(with transition sequences* $\sigma_1, \sigma_2, \ldots, \sigma_n$ *where* σ_i *consists of the heads of all arc labels between* α_i *and* γ*), we also get*

$$(m_\beta(p)(c') \neq \omega \wedge m_\gamma(p)(c') = \omega) \Rightarrow \exists i \in 1, \ldots, n\colon \Delta(\sigma_i)(p)(c') > 0$$

for all $p \in P$ *and* $c' \in C(p)$.

Proof. Propositions A.1(a)–(c) and (e) immediately follow from the HL-tree algorithm.

To prove Proposition A.1(d), let σ have the form

$$\sigma = (t_1, c_1)(t_2, c_2) \ldots (t_k, c_k) \quad \text{with } k \geq 1$$

and let $\delta_0, \delta_1, \ldots, \delta_k$ be the nodes between α and γ:

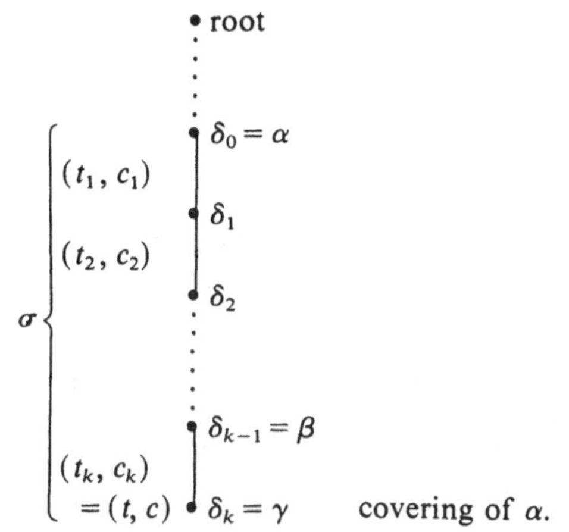

The proof is by induction on $j \in 0, \ldots, k$, using the following induction hypothesis: there exists a marking m' such that

(i) $\quad m[(t_1, c_1) \ldots (t_j, c_j)\rangle m'$,

(ii) $$m'(p)(c') \geq \begin{cases} m_{\delta_j}(p)(c') & \text{if } m_\beta(p)(c') \neq \omega, \\ \Delta_-((t_{j+1}, c_{j+1}) \ldots (t_k, c_k))(p)(c') & \\ & \text{if } m_\beta(p)(c') = \omega \end{cases}$$

for all $p \in P$ and $c' \in C(p)$ and $j < k$.

When $j = k$, part (i) of the induction hypothesis immediately yields that σ is a transition sequence at m, and the proof of Proposition A.1(d) is finished.

Basis step: $j = 0$. Then m can be used as m' in (i). Part (ii) immediately follows from the assumption of Proposition A.1(d).

Induction step: $j > 0$. By the induction hypothesis, there exists a marking m' such that

$$m[(t_1, c_1) \ldots (t_{j-1}, c_{j-1})\rangle m', \tag{A.1}$$

$$m'(p)(c') \geq \begin{cases} m_{\delta_{j-1}}(p)(c') & \text{if } m_\beta(p)(c') \neq \omega, \\ \Delta_-((t_j, c_j) \ldots (t_k, c_k))(p)(c') & \\ & \text{if } m_\beta(p)(c') = \omega \end{cases} \tag{A.2}$$

for all $p \in P$ and $c' \in C(p)$. Since $(t_j, c_j) = \text{head}(l_{\delta_{j-1}\delta_j})$, it has concession at $m_{\delta_{j-1}}$ and by (A.2) it has concession at m' too, i.e., $\exists m''$:

$$m'[t_j, c_j\rangle m''. \tag{A.3}$$

Together with (A.1) this yields $m[(t_1, c_1) \ldots (t_j, c_j)\rangle m''$, i.e., part (i) of the induction hypothesis for j. Now, let $p \in P$ and $c' \in C(p)$ and $j < k$ to check part (ii) of the induction hypothesis for j. There are two cases:

Case 1: $m_\beta(p)(c') \neq \omega$. Since $j < k$, $\delta_j \neq \gamma$, and since $m_\beta(p)(c') \neq \omega$, there is not introduced any ω at place p and colour c' at any predecessor node of β. Especially,

$m_{\delta_j}(p)(c') \neq \omega$ and then

$$\begin{aligned} m_{\delta_j}(p)(c') &= m_{\delta_{j-1}}(p)(c') + \Delta(t_j, c_j)(p)(c') \\ &\leq m'(p)(c') + \Delta(t_j, c_j)(p)(c') && \text{(by (A.2))} \\ &= m''(p)(c'). && \text{(by (A.3))} \end{aligned}$$

Case 2: $m_\beta(p)(c') = \omega$. Then

$$\begin{aligned} m''(p)(c') &= m'(p)(c') + \Delta(t_j, c_j)(p)(c') && \text{(by (A.3))} \\ &\geq \Delta_-((t_j, c_j)(t_{j+1}, c_{j+1}) \ldots (t_k, c_k))(p)(c') \\ &\quad + \Delta(t_j, c_j)(p)(c') && \text{(by (A.2))} \\ &= \Delta_-((t_j, c_j)(t_{j+1}, c_{j+1}) \ldots (t_k, c_k))(p)(c') \\ &\quad + \Delta_+(t_j, c_j)(p)(c') - \Delta_-(t_j, c_j)(p)(c') \\ &\geq \Delta_-((t_{j+1}, c_{j+1}) \ldots (t_k, c_k))(p)(c'). \qquad \square \end{aligned}$$

A.2. *Proof of Lemma 5.7*

The proof is by induction on the number of arcs from the root to α.

Basis step: *number of arcs* $= 0$. Then α is the root node and $m_\alpha = m_0$. We can use $m_0 \in R(m_0)$ for m for all $k \in \mathbb{N}$.

Induction step: *number of arcs* > 0. The general situation can be pictured as follows:

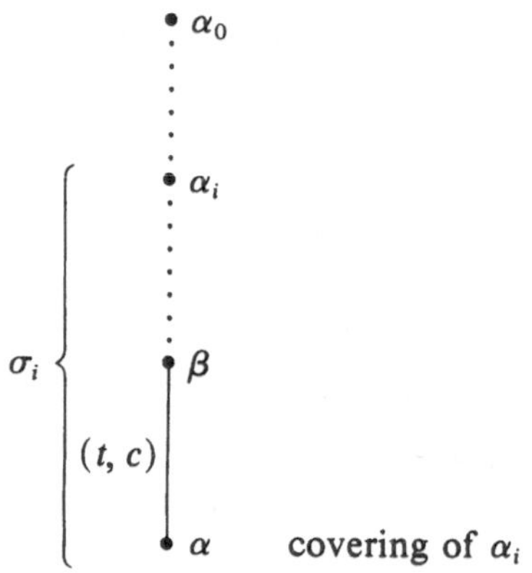

$\alpha_1, \ldots, \alpha_n$ are *all* nodes with 'covering of α_i' $\in l_\alpha$, and σ_i consists of the heads of all arc labels between α_i and α. Now, let $k \in \mathbb{N}$ be given. The induction step is rather complicated. The proof is divided into four parts, I–IV, and the idea can be visualized as follows:

$$\begin{array}{ccccccl}
m_\beta & & \rhd & m\beta & \geqslant & m_\beta\left[{\omega \atop k}\right] & \\
\Big\Downarrow & & & (t,c)\ \Big\Downarrow & & & \leftarrow \text{part I} \\
(t,c)\ \Big\Downarrow & & & m\alpha & & & \leftarrow \text{part II} \\
\Big\Downarrow & & & \sigma_1^k \ldots \sigma_n^k\ \Big\Downarrow & & & \leftarrow \text{part III} \\
m_\alpha & & \rhd & m & \geqslant & m_\alpha\left[{\omega \atop k}\right] & \leftarrow \text{part IV.}
\end{array}$$

By the induction hypothesis we get the marking $m\beta \in R(m_0)$ corresponding to m_β. From $m\beta$, (t, c) occurs to get the marking $m\alpha$. Then, for each 'covering of α_i' $\in l_\alpha$, σ_i occurs k times to get the marking m corresponding to m_α. When choosing $m\beta$ we have to ensure that (t, c) followed by $\sigma_1^k \ldots \sigma_n^k$ can occur from $m\beta$.

Now, let $K_1, K_2 \in \mathbb{N}$ satisfy

$$K_1 \geqslant \Delta_-(t, c)(p)(c') \qquad \text{for all } p \in P \text{ and } c' \in C(p), \tag{A.4}$$

$$K_2 \geqslant \Delta_-(\sigma_1^k \ldots \sigma_n^k)(p)(c') \qquad \text{for all } p \in P \text{ and } c' \in C(p). \tag{A.5}$$

By the induction hypothesis $\exists m\beta \in R(m_0)$:

$$\forall p \in P\ \forall c' \in C(p)\colon m_\beta(p)(c') \geqslant m\beta(p)(c') \geqslant m_\beta\left[\begin{matrix}\omega \\ k+K_1+K_2\end{matrix}\right](p)(c'). \tag{A.6}$$

The choice of K_1 and K_2 assures that (t, c) followed by $\sigma_1^k \ldots \sigma_n^k$ can occur from $m\beta$, which can be a problem if $m_\beta(p)(c')$ is infinite while $m\beta(p)(c')$ is finite.

Part I: We want to show that (t, c) has concession at $m\beta$; take $p \in P$ and $c' \in C(p)$. There are two cases:

Case 1: $m_\beta(p)(c') \neq \omega$. Then, by definition,

$$m_\beta\left[\begin{matrix}\omega \\ k+K_1+K_2\end{matrix}\right](p)(c') = m_\beta(p)(c')$$

and (by (A.6))

$$m\beta(p)(c') = m_\beta(p)(c') \geqslant \Delta_-(t, c)(p)(c')$$

since (t, c) has concession at m_β.

Case 2: $m_\beta(p)(c') = \omega$. Then

$$\begin{aligned} m\beta(p)(c') &\geqslant k + K_1 + K_2 && \text{(by (A.6))} \\ &\geqslant \Delta_-(t, c)(p)(c') && \text{(by (A.4)).} \end{aligned}$$

Since (t, c) has concession at $m\beta$, there exists an $m\alpha \in R(m_0)$ satisfying $m\beta[t, c\rangle m\alpha$, i.e.,

$$\forall p \in P\ \forall c' \in C(p)\colon m\alpha(p)(c') = m\beta(p)(c') + \Delta(t, c)(p)(c'). \tag{A.7}$$

Part II: The marking $m\alpha$ satisfies the following two conditions: $\forall p \in P\ \forall c' \in C(p)$:

$$(m_\alpha(p)(c') = \omega \Leftrightarrow m_\beta(p)(c') = \omega) \\ \Rightarrow m_\alpha(p)(c') \geq m\alpha(p)(c') \geq m_\alpha\begin{bmatrix} \omega \\ k+K_2 \end{bmatrix}(p)(c'), \tag{A.8}$$

$$m_\beta(p)(c') \neq \omega \Rightarrow m\alpha(p)(c') = m_\beta(p)(c') + \Delta(t, c)(p)(c'). \tag{A.9}$$

Condition (A.9) immediately follows from (A.6) and (A.7).

To show (A.8), take $p \in P$ and $c' \in C(p)$ and assume $m_\alpha(p)(c') = \omega \Leftrightarrow m_\beta(p)(c') = \omega$. Then

$$m_\alpha(p)(c') = m_\beta(p)(c') + \Delta(t, c)(p)(c'). \tag{A.10}$$

There are now two cases:

Case 1: $m_\alpha(p)(c') \neq \omega \wedge m_\beta(p)(c') \neq \omega$. Then, by (A.6), $m_\beta(p)(c') = m\beta(p)(c')$. From (A.7) and (A.10) it follows that $m_\alpha(p)(c') = m\alpha(p)(c')$ and since $m_\alpha(p)(c') \neq \omega$, we have

$$m_\alpha(p)(c') = m\alpha(p)(c') = m_\alpha\begin{bmatrix} \omega \\ k+K_2 \end{bmatrix}(p)(c').$$

Case 2: $m_\alpha(p)(c') = \omega \wedge m_\beta(p)(c') = \omega$. Then

$$\begin{aligned} m\alpha(p)(c') &= m\beta(p)(c') + \Delta(t, c)(p)(c') && \text{(by (A.7))} \\ &\geq m\beta(p)(c') - \Delta_-(t, c)(p)(c') \\ &\geq k + K_1 + K_2 - \Delta_-(t, c)(p)(c') && \text{(by (A.6))} \\ &\geq k + K_2 && \text{(by (A.4))} \end{aligned}$$

and thus we have

$$m_\alpha(p)(c') = \omega \geq m\alpha(p)(c') \geq k + K_2 = m_\alpha\begin{bmatrix} \omega \\ k+K_2 \end{bmatrix}(p)(c').$$

Part III: We will now show that $\sigma_1^k \ldots \sigma_n^k$ is a transition sequence at $m\alpha$. Let $\sigma_1^k \ldots \sigma_n^k = \xi_1\xi_2$. We will use induction on the length of ξ_1 and the induction hypothesis is as follows: $m\alpha[\xi_1\rangle m\xi_1$, where $m\xi_1$ satisfies

(1) $m_\beta(p)(c') \neq \omega \Rightarrow m\xi_1(p)(c') \geq m_{\alpha_i}(p)(c')$ for all $i \in 1, \ldots, n$,

(2) $m_\beta(p)(c') = \omega \Rightarrow m\xi_1(p)(c') \geq k + K_2 - \Delta_-(\xi_1)(p)(c')$.

Basis step of part III: $|\xi_1| = 0$. Then $\xi_1 = \lambda$ and $m\xi_1 = m\alpha$.

(1) Assume $m_\beta(p)(c') \neq \omega$. Then

$$\begin{aligned} m\alpha(p)(c') &= m_\beta(p)(c') + \Delta(t, c)(p)(c') && \text{(by (A.9))} \\ &\geq m_{\alpha_i}(p)(c') && \text{for all } i \in 1, \ldots, n. \end{aligned}$$

The inequality follows from Proposition A.1(a) used on each of the involved coverings.

(2) Assume $m_\beta(p)(c')=\omega$. Then $m_\alpha(p)(c')=\omega$, too and $m\alpha(p)(c') \geq k+K_2$ (by (A.8)).

Induction step of part III: $|\xi_1|>0$. Then let $\xi_1=\xi_0\sigma_i$. By the induction hypothesis we have:

$$m\alpha[\xi_0\rangle m\xi_0, \tag{A.11}$$

where $m\xi_0$ satisfies

$$m_\beta(p)(c')\neq\omega \Rightarrow m\xi_0(p)(c')\geq m_{\alpha_i}(p)(c') \quad \text{for all } i\in 1,\ldots,n, \tag{A.12}$$

$$m_\beta(p)(c')=\omega \Rightarrow m\xi_0(p)(c')\geq k+K_2-\Delta_-(\xi_0)(p)(c'). \tag{A.13}$$

In the case of (A.12) it immediately follows that σ_i is a transition sequence at $m\xi_0$, while in the case of (A.13) this follows since

$$\begin{aligned}
m\xi_0(p)(c') &\geq k+K_2-\Delta_-(\xi_0)(p)(c')\\
&\geq K_2-\Delta_-(\xi_0)(p)(c')\\
&\geq \Delta_-(\sigma_1^k\ldots\sigma_n^k)(p)(c')-\Delta_-(\xi_0)(p)(c') \quad \text{(by (A.5))}\\
&= \Delta_-(\xi_1\xi_2)(p)(c')-\Delta_-(\xi_0)(p)(c')\\
&\geq \Delta_-(\xi_1)(p)(c')-\Delta_-(\xi_0)(p)(c')\\
&= \Delta_-(\xi_0\sigma_i)(p)(c')-\Delta_-(\xi_0)(p)(c')\\
&= \Delta_-(\sigma_i)(p)(c').
\end{aligned}$$

Thus, let $m\xi_1\in R(m_0)$ be defined by

$$m\xi_0[\sigma_i\rangle m\xi_1. \tag{A.14}$$

Then $m\alpha[\xi_1\rangle m\xi_1$ by (A.11). Now $m\xi_1$ satisfies (1) and (2) of the induction hypothesis:

(1): Assume $m_\beta(p)(c')\neq\omega$. Then

$$\begin{aligned}
m\xi_1(p)(c') &= m\xi_0(p)(c')+\Delta(\sigma_i)(p)(c') && \text{(by (A.14))}\\
&\geq m\xi_0(p)(c') && \text{(by Proposition A.1(b))}\\
&\geq m_{\alpha_i}(p)(c') \quad \text{for all } i\in 1,\ldots,n && \text{(by (A.12))}.
\end{aligned}$$

(2): Assume $m_\beta(p)(c')=\omega$. Then

$$\begin{aligned}
m\xi_1(p)(c') &= m\xi_0(p)(c')+\Delta(\sigma_i)(p)(c') && \text{(by (A.14))}\\
&\geq k+K_2-\Delta_-(\xi_0)(p)(c')+\Delta(\sigma_i)(p)(c') && \text{(by (A.13))}\\
&\geq k+K_2-\Delta_-(\xi_1)(p)(c').
\end{aligned}$$

This ends the induction step. We have now proved that $\sigma_1^k\ldots\sigma_n^k$ is a transition sequence at $m\alpha$ and thus we let $m\in R(m_0)$ be defined by

$$m\alpha[\sigma_1^k\ldots\sigma_n^k\rangle m. \tag{A.15}$$

Part IV: To complete the induction step of Lemma 5.7, we have to prove that $m_\alpha \rhd m \geq m_\alpha[{}^\omega_k]$, i.e., the following inequalities:

$$\forall p \in P \ \forall c' \in C(p)\colon m_\alpha(p)(c') \geq m(p)(c') \geq m_\alpha\begin{bmatrix}\omega\\ k\end{bmatrix}(p)(c').$$

Take $p \in P$ and $c' \in C(p)$. There are three cases:

Case 1: $m_\beta(p)(c') \neq \omega \wedge m_\alpha(p)(c') = \omega$. Then

$$\begin{aligned} m(p)(c') &= m\alpha(p)(c') + \Delta(\sigma_1^k \ldots \sigma_n^k)(p)(c') && \text{(by (A.15))}\\ &\geq \Delta(\sigma_1^k \ldots \sigma_n^k)(p)(c')\\ &\geq k \cdot \left[\sum_{i=1}^{n} \Delta(\sigma_i)(p)(c')\right]\\ &\geq k \cdot 1 && \text{(by Proposition A.1(b) and (e)).} \end{aligned}$$

Thus, $m_\alpha(p)(c') = \omega \geq m(p)(c') \geq k = m_\alpha[{}^\omega_k](p)(c')$.

Case 2: $m_\beta(p)(c') \neq \omega \wedge m_\alpha(p)(c') \neq \omega$. Then

$$\begin{aligned} m(p)(c') &= m\alpha(p)(c') + \Delta(\sigma_1^k \ldots \sigma_n^k)(p)(c') && \text{(by (A.15))}\\ &= m\alpha(p)(c') && \text{(by Proposition A.1(c))}\\ &= m_\alpha(p)(c') \end{aligned}$$

by (A.8) since $m_\alpha(p)(c') \neq \omega$. Thus, $m_\alpha(p)(c') = m(p)(c') = m_\alpha[{}^\omega_k](p)(c')$.

Case 3: $m_\beta(p)(c') = \omega \wedge m_\alpha(p)(c') = \omega$. Then

$$\begin{aligned} m(p)(c') &= m\alpha(p)(c') + \Delta(\sigma_1^k \ldots \sigma_n^k)(p)(c') && \text{(by (A.15))}\\ &\geq k + K_2 + \Delta(\sigma_1^k \ldots \sigma_n^k)(p)(c') && \text{(by (A.8))}\\ &\geq k + K_2 - \Delta_-(\sigma_1^k \ldots \sigma_n^k)(p)(c')\\ &\geq k && \text{(by (A.5)).} \end{aligned}$$

Thus, $m_\alpha(p)(c') = \omega \geq m(p)(c') \geq k = m_\alpha[{}^\omega_k](p)(c')$. □

A.3. Proof of Lemma 5.9

Let $p \in P$ and $c \in C(p)$ be given. We will show the inequality of Lemma 5.9 in both directions.

("$\geq$"-*Direction*):

Case 1: Assume

$$\max \bigcup_{c' \in \mathrm{map}_{C(p)}(c)} T(m_0)(p)(c') = k \in \mathbb{N}.$$

Then we have, for some element $c' \in C(p)$ and some $\varphi \in \Phi$, that

$$k \in T(m_0)(p)(c'), \tag{A.16}$$

$$\varphi_{C(p)}(c') = c. \tag{A.17}$$

By Corollary 5.8 we get from (A.16) that $k \in R(m_0)(p)(c')$, i.e., $\exists m \in R(m_0)$:

$$m(p)(c') = k. \tag{A.18}$$

Then

$$\begin{aligned} \varphi(m)(p)(c) &= \varphi(m)(p)(\varphi_{C(p)}(c')) && \text{(by (A.17))} \\ &= m(p)(c') && \text{(see below)} \\ &= k && \text{(by A.18)).} \end{aligned}$$

The second equality can be seen by observing that, in the first expression, we permute the marking and then take a look at the coefficient of the image of c'. Instead, we can simply look at the coefficient of c' in the nonpermuted marking.

Since $m \in R(m_0)$, we get $\varphi(m) \in R(m_0)$ from Corollary 5.3, hence $k \in R(m_0)(p)(c)$.

Case 2: Assume

$$\max \bigcup_{c' \in \mathrm{map}_{C(p)}(c)} T(m_0)(p)(c') = \omega.$$

Since $T(m_0)$ and $\mathrm{map}_{C(p)}(c)$ are finite, there exist an element $c' \in C(p)$ and a symmetry $\varphi \in \Phi$ such that $\omega \in T(m_0)(p)(c')$ and $\varphi_{C(p)}(c') = c$. By Corollary 5.8, $R(m_0)(p)(c')$ is unbounded, hence,

$$\forall k \in \mathbb{N}\ \exists m \in R(m_0)\colon m(p)(c') \geq k. \tag{A.19}$$

Analogously to Case 1, we get

$$\varphi(m)(p)(c) \geq k \text{ and } \varphi(m) \in R(m_0) \quad \text{for each } m \text{ in (A.19).} \tag{A.20}$$

Thus, $R(m_0)(p)(c)$ is unbounded, i.e., $\sup R(m_0)(p)(c) = \omega$.

("$\leq$"-*Direction*):

Case 3: Assume $\sup R(m_0)(p)(c) = k \in \mathbb{N}$. We choose m such that

$$m(p)(c) = k. \tag{A.21}$$

By Lemma 5.5, $\exists \varphi \in \Phi\ \exists \alpha \in T(m_0)$: $m_\alpha \rhd \varphi(m)$. Then, in particular,

$$\begin{aligned} m_\alpha(p)(\varphi_{C(p)}(c)) &\geq \varphi(m)(p)(\varphi_{C(p)}(c)) && \text{(A.22)} \\ &= m(p)(c) && \text{(see Case 1)} \\ &= k && \text{(by (A.21)).} \end{aligned}$$

Let $c' = \varphi_{C(p)}(c)$. Then $c = \varphi^{-1}_{C(p)}(c')$ and (A.22) yields

$$\sup R(m_0)(p)(c) = k \leq m_\alpha(p)(c') \leq \max \bigcup_{c' \in \mathrm{map}_{C(p)}(c)} T(m_0)(p)(c').$$

Case 4: Assume $\sup R(m_0)(p)(c) = \omega$. Then p is unbounded on colour c, i.e.,

$$\forall k \in \mathbb{N}\ \exists m \in R(m_0)\colon m(p)(c) \geq k. \tag{A.23}$$

Analogously to Case 3 we have $\exists\varphi \in \Phi\ \exists\alpha \in T(m_0)$:

$$m_\alpha(p)(\varphi_{C(p)}(c)) \geq m(p)(c) \geq k \quad \text{for each } m \text{ in (A.23).} \tag{A.24}$$

Thus

$$\sup \bigcup_{c' \in \text{map}_{C(p)}(c)} T(m_0)(p)(c') = \omega$$

and since $T(m_0)$ and $\text{map}_{C(p)}(c)$ are finite, we get

$$\max \bigcup_{c' \in \text{map}_{C(p)}(c)} T(m_0)(p)(c') = \omega. \quad \square$$

Acknowledgment

Some of the ideas in this paper are founded on a student project at Aarhus University with the following participants: Arne M. Jensen, Peter A. Nielsen, Erik Schjøtt, Kasper Østerbye and Kurt Jensen (supervisor). We also thank the (unknown) referees for their comments.

References

[1] H.J. Genrich and K. Lautenbach, System modelling with high-level Petri nets, *Theoret. Comput. Sci.* **13** (1981) 109-136.

[2] M. Hack, Decidability questions for Petri Nets, Intern. Rept. TR 161, MIT, 1976.

[3] P. Huber, A.M. Jensen, L.O. Jepsen and K. Jensen, Towards reachability trees for high-level Petri nets, Intern. Rept. PB-174, Computer Science Dept., Univ. of Aarhus, 1985.

[4] K. Jensen, Coloured Petri nets and the invariant-method, *Theoret. Comput. Sci.* **14** (1981) 317-336.

[5] K. Jensen, How to find invariants for coloured Petri nets, in: J. Gruska and M. Chytill, eds., *Mathematical Foundations of Computer Science 1981*, Lecture Notes in Computer Science **118** (Springer, Berlin, 1981) 327-338.

[6] K. Jensen, High-level Petri nets, in: A. Pagnoni and G. Rozenberg, eds., *Applications and Theory of Petri Nets*, Informatik-Fachberichte **66** (Springer, Berlin, 1983) 166-180.

[7] R.M. Karp and R.E. Miller, Parallel program schemata, *J. Comput. System Sci.* **3** (1969) 147-195.

[8] J.L. Peterson, *Petri Net Theory and the Modelling of Systems* (Prentice-Hall, Englewood Cliffs, NJ, 1981).

[9] W. Reisig, Petri nets with individual tokens, in: A. Pagnoni and G. Rozenberg, eds., *Applications and Theory of Petri Nets*, Informatik-Fachberichte **66** (Springer, Berlin, 1983) 229-249.

12.
Parametrized Reachability Trees for Predicate / Transition Nets

M. Lindqvist

11th Int. Conference on Applications and Theory of Petri Nets, Paris 1990

Abstract

Elaboration of reachability analysis in the context of Predicate/Transition nets is studied. For this purpose, parameters are introduced into the markings of Predicate/Transition nets. These parameters represent any fixed individual values potentially appearing in the marking. The formalism for dealing with parameterized markings is developed and the dynamics of Predicate/Transition nets are augmented to cope with parameters. These are used to define parameterized reachability trees and an algorithm for generating them is presented. They are shown to be significantly smaller than ordinary reachability trees and to contain the same information: from one parameterized reachability tree several instances of ordinary reachability trees can be derived.

1 Introduction

The main area of application of nets is to provide adequate means for modelling the behavior of systems featuring concurrency. As a result of such modelling an opportunity to analyse the behaviour of these systems becomes available. In this analysis the machinery offered by the modelling formalism, nets, may be exploited.

Several different techniques for the analysis of nets have been suggested over the years. The most prominent of these remain the *s-invariant method* [5], [9], [11] and *the generation of a reachability tree* or exhaustive search [11]. Here we will concentrate on the latter.

Reachability tree generation consists of generating the full state space of a system which in the case of nets means generating all reachable markings. Equivalently this means that a state machine describing the global states of a system and transitions between these states is produced. At present reachability analysis is undoubtly the most versatile analysis method for nets. Its most severe drawback is probably its inability to express true concurrency. This is due to the fact that the generation of a reachability tree is based on interleaving semantics [12]. Another drawback is that this method of analysis is quite arduous. Therefore only systems expressable by a net of reasonably small size are analysable at all. To be able to analyse larger problems as well one would need to be able to reduce the size of the reachability tree. But reducing the size of a reachability tree so that no relevant information is lost is a very difficult problem and usually possible only under special circumstances.

In the literature two approaches to elaborate reachability analysis have been presented. The first one uses C/E-system like nets and reduces the reachability tree by discarding certain transition sequences which all begin from and end in the same same global state [13]. As a result some markings are not generated. This discarding preserves the information of the reachability tree when it comes to certain properties concerning the analysed system. The other approach is developed for Coloured Petri nets and does not work with net formalisms like P/T- or C/E-systems [7], [8]. With it the generation of the reachability tree is done in the ordinary way, but certain parts of the tree can be omitted without loss of information. The use of this method can sometimes lead to drastic reductions in the size of the generated tree.

The latter method mentioned above relies on the existence of internal symmetries in the net and they are a prerequisite for the method to be effective. In other words, this imposes restrictions on the nets for which the method works well. In [4] and [3] the method has been elaborated further to avoid the need of heuristics in finding out the symmetries. They also introduce a level of symbolic treatment of markings under conditions where these symmetries hold. Nevertheless, the symmetries are still crucial.

In relating these different reduction methods it should be clear that nothing is gained for free: there obviously is an unsurmountable trade off between the restrictions put on a net in order to make a reduction method work and of the expressiveness of the resulting reachability tree. If restrictions put on the net are loosened the expressiveness of the tree can be reduced either by

1. letting the size of it grow, i.e. letting the number of markings in it increase or
2. making it more of intermediate nature when it comes to proving properties of the net it was derived from.[1]

The method that will be introduced here for reducing the size of a reachability tree is based on the latter item in the above list. This trade off is made in order to allow the application of the method without having to introduce any restrictions on nets thus analyzed. In that respect it differs rather drastically from the reduction methods based on the symmetry concept.

So, our aim here is to reduce the size of the reachability tree produced by reachability analysis. We would not like to lose any information because of this reduction. We would also like to produce directly a reachability tree such that it is reduced. The means for trying to accomplish this task will be to use some kind of folding following the line of ideas applied in Predicate/Transition nets, or PrT-nets for short. In PrT-nets a folding is accomplished by introducing, among other things, individuals and variables into the net. This way the number of s- and t-elements can be decreased. But the marking of one s-element being a part of a node in a reachability tree is, however, not folded in a PrT-net in any way. To accomplish this something more is needed.

In PrT-nets variables on arcs may stand for anything since they are not bounded at all. The tuples representing markings on s-elements consist of individuals and are therefore totally bounded. Thus what is needed is something to allow some amount of freedom in the s-elements. The objects in the s-elements should represent, not just anything, but *any* marking from the *set of possible* markings. Therefore, these objects should not consist of totally unbounded variables, but rather of something which is bounded to some extent. For this purpose we will introduce *parameters* into nets – or to put it better – into markings.

The paper starts with a definition of the Predicate/Transition nets involved in Section 1.1. In Section 2 the formalism as well as the notation for dealing with parameterized markings is developed. In Section 3 the formalism is applied to parameterization of the dynamics of PrT-nets. This leads to the central results: the parameterized transition rule and parameterized reachability trees. Before conclusions are drawn potentials in the use of parameterized reachability trees are briefly discussed. Because of the limitations on the length of this paper only an introduction to the central points of the formalism can be provided. For a thorough and detailed treatment of the matter we refer to [10].

1.1 Predicate/Transition nets

To make the paper self contained we start with a brief introduction into PrT-nets following the line of [6]. Defining PrT-nets starts with considering a structure $\mathcal{R}$ with a domain D of individuals. The

[1]Actually, the net itself can be perceived as the most compact representation of a reachability tree.

set D is assumed always to be finite. Further, we have a first order language $\mathbf{L}$ whose non-logical alphabet consist of a set Π predicate symbols, a set Ω of operator symbols and a set V of individual variable symbols. By $\mathcal{M} = (D, \mathcal{W})$ we denote a model for $\mathbf{L}$ in the usual way [1], where D is the set of individuals as above and $\mathcal{W}$ the interpretation function. Used also is the set Ψ of variable predicates, the meaning of which is defined in the following definition:

Definition 1.1 *Let $\psi^n \in \Psi$, where n is the arity of Ψ, be a variable predicate annotating the S-element $s \in S$ of a net $N = (S, T; F)$. We say that ψ^n is true in marking M under the assignment $\alpha = [v_1 \leftarrow d_1, \ldots, v_n \leftarrow d_n]$, $d_i \in D, v_i \in V$ assigning individuals to the variables of ψ^n, iff $\langle d_1, \ldots, d_n \rangle \in M(s)$. A model $\mathcal{M}$ for Ψ in marking M is defined by: $\mathcal{M} \models \psi[d_1, \ldots, d_n]$ iff $\langle d_1, \ldots, d_n \rangle \in M(s)$ such that s is annotated by ψ.*

Predicate/Transition net is then defined as a directed net whose elements are annotated with expressions of the language $\mathbf{L}$ as follows [6]:

Definition 1.2 *Let D be the set of individuals, $\mathbf{L}$ a language built from Ω, Π and Ψ and $\mathcal{M} = (D, \mathcal{W})$ a model. Let V the set of variables of $\mathbf{L}$ and $X = D \cup V$.*

Predicate/Transition nets (PrT-nets) are marked annotated nets $MN = (N, A, M_0)$, where N is the underlying net, A its annotation and M_0 the initial marking.

1. *N is a net, $N = (S, T; F)$.*
2. *A is the annotation of N, $A = (A_S, A_T, A_F)$, where:*
   (a) *A_S is an injective mapping of S into the set of variable predicates Ψ.*
   (b) *A_T is a mapping of T into the set of selectors which are syntactically correct formulae of $\mathbf{L}$ not involving elements of Ψ.*
   (c) *A_F is a mapping of F into the set of non-empty sets of tuples τ of individuals and variables, $\mathcal{P}(\langle X^* \rangle) - \{\emptyset\}$, such that if $(x, y) \in F$ is an arc connected to an s-element s (i. e. $x = s$ or $y = s$) and n is the index of the variable predicate annotating s then the tuples annotating (x, y) are of length n, $\emptyset \neq A_F(x, y) \subseteq \langle X \rangle^n$.*
3. *M_0 is a marking of N, i. e. a mapping from S into $\mathcal{P}(\langle D^* \rangle)$, that assigns to each place $s \in S$ a set $M_0(s)$ of tuples τ of individuals such that if $\psi^{(n)} \in \Psi$ is the variable predicate annotating s then the model $\mathcal{M}_0$ for M_0: $\mathcal{M}_0 \models \psi^{(n)} u_1, \ldots, u_n$ iff $\tau = \langle u_1, \ldots, u_n \rangle \in M_0(s)$.*

To define the dynamics of PrT-nets we present the transition rule in which the notions of feasibility and permissibility are used. The feasibility condition is concerned with, so to speak, static requirements which are not dependent of the marking. The permissibility condition deals then with more dynamic aspects and depends on the marking in question.

Definition 1.3 *Let $MN = (N, A, M_0)$ be a PrT-net, t a transition of MN and φ the selector formula annotating t. Let α be an assignment that replaces all variables that occur in the tuples $\tau \in A_F(s, t) \cup A_F(t, s)$, $s \in S$, or in the formula φ. The assignment α is called feasible for t at marking M iff*

1. *α satisfies the selector φ of t.*
2. *α does not generate an impurity; if $(s, t) \in F$ and $(t, s) \in F$ then for any two tuples $\tau_1 \in A_F(s, t), \tau_2 \in A_F(t, s), \tau_1\alpha \neq \tau_2\alpha$.*
3. *α does not create a multiple arc; if $\tau_1, \tau_2 \in A_F(s, t)$ (or $A_F(t, s)$) and $\tau_1 \neq \tau_2$ then $\tau_1\alpha \neq \tau_2\alpha$.*

Definition 1.4 *Let $MN = (N, A, M_0)$ be a PrT-net, t a transition of MN and φ the selector formula annotating t. Let α be an assignment that replaces all variables that occur in the tuples $\tau \in A_F(s,t) \cup A_F(t,s)$, $s \in S$ and $\mathcal{M}$ the model corresponding with a marking M. The assignment α is called permissible for t at M iff*

1. *For all arcs (s,t) leading to t, $\mathcal{M} \models \psi(\tau\alpha)$ if $\tau \in A_F(s,t)$.*
2. *For all arcs (t,s) leading to s, $\mathcal{M} \not\models \psi(\tau\alpha)$ if $\tau \in A_F(t,s)$.*

Using the feasibility and permissibility conditions we can now express the transition rule rather concisely:

Definition 1.5 *Let $MN = (N, A, M_0)$ be a PrT-net. Let M and M' be markings of MN and $\mathcal{M}$ and $\mathcal{M}'$ the corresponding models, t a transition and α as in Definition 1.3. Let α' be an assignment that replaces the variables in the index of a variable predicate $\psi \in \Psi$. Then the α-occurrence of t at M leading to M' is designated as $M\ [t\alpha\rangle\ M'$ and defined by: $M\ [t\alpha\rangle\ M'$ iff:*

1. *α is a feasible assignment for t at M.*
2. *α is a permissible assignment for t at M.*
3. *The follower marking $M'(s)$, such that $A_S(s) = \psi \in \Psi$, is the following:*
   (a) *For all places $s \in S$ such that $(s,t) \notin F$ and $(t,s) \notin F$, $\mathcal{M}' \models \psi\alpha'$ iff $\mathcal{M} \models \psi\alpha'$.*
   (b) *For all places $s \in S$ such that $(s,t) \in F$ and $\tau \in A_F(s,t)$, $\mathcal{M}' \not\models \psi(\tau\alpha)$ and $\mathcal{M}' \models \psi(\tau\alpha')$ iff $\mathcal{M} \models \psi(\tau\alpha')$ for $\tau\alpha' \neq \tau\alpha$.*
   (c) *For all places $s \in S$ such that $(t,s) \in F$ and $\tau \in A_F(t,s)$, $\mathcal{M}' \models \psi(\tau\alpha)$ and $\mathcal{M}' \models \psi(\tau\alpha')$ iff $\mathcal{M} \models \psi(\tau\alpha')$ for $\tau\alpha' \neq \tau\alpha$.*

We say that t is α enabled in M iff $M[t\alpha\rangle M'$.

As the reader may notice PrT-nets have been defined here according to the strict interpretation [6] limiting the marking to being sets of tokens rather than multisets. In this paper we will stick to the strict interpretation. The weak one has been briefly discussed in [10].

Having now the transition rule for PrT-nets we may speak of reachable markings and reachability trees of Predicate/Transition nets. The following sections will introduce parameters both to PrT-nets and to their dynamics. This will allow us to define finally parameterized reachability trees that will be significantly smaller than ordinary ones. Before going into the formal introduction of the method an informal outline of the basic ideas is given in the following.

2 Parameterized markings

To start with the introduction of parameterized markings some basic notation is needed as stated in the following. The convention to be followed througout this presentation is that all expressions dealing with parameters consist of symbols with the *hat* symbol ^ over them:

Notation 2.1 *The set of individuals is denoted by D, the set of variables by V and the set of parameters by $\widehat{V}$. Further $X = D \cup V$ and $\widehat{X} = D \cup \widehat{V}$. Then we have:*

1. *$d \in D$ is an individual.*

2. *$v \in V$ is a (local) variable.*

3. *$\hat{v} \in \hat{V}$ is a (global) parameter.*

4. *$\hat{x} \in \hat{X}$ is a parameter or an individual.*

5. *$\tau \in X^*$ is a tuple of individuals and variables appearing on the arcs of a PrT-net.*

6. *$\hat{\tau} \in \hat{X}^*$ is a tuple of individuals and parameters appearing as tokens in parameterized markings of s-elements of a PrT-net, also called a parameterized token.*

7. *M and $\widehat{M}$ are a marking and a parameterized marking of a PrT-net, respectively.*

8. *$\hat{X}(\widehat{M}) = \hat{V}(\widehat{M}) \cup D(\widehat{M})$ ($\hat{X}(\hat{\tau})$) denote the set of parameters and individuals in a parameterized marking $\widehat{M}$ (parameterized token $\hat{\tau}$).*

9. *$|M(s)|$ and $|\widehat{M}(s)|$ denote the cardinality, i.e. the number of tokens of a marking $M(s)$ and a parameterized marking $\widehat{M}(s)$, respectively.*

10. *$\widehat{M}[\hat{\tau}]$ denote the number of tokens $\hat{\tau}$ in $\widehat{M}$.*

11. *$\hat{\tau}(i)$ denote the i:th component of the token $\hat{\tau}$, i.e. if $\hat{\tau} = \langle \hat{x}_1, \ldots, \hat{x}_i, \ldots, \hat{x}_n \rangle$ then $\hat{\tau}(i) = \hat{x}_i$.*

12. *$\alpha, \hat{\alpha}, \beta$ and $\hat{\beta}$ are the following mappings:*

    (a) $\alpha : V \to D$ is called an assignment,

    (b) $\hat{\alpha} : V \to \hat{X}$ is called a parameterized assignment,

    (c) $\beta : \hat{V} \to D$ is called a fixing and

    (d) $\hat{\beta} : \hat{V} \to \hat{X}$ is called a substitution, in some cases a unifier.

In a parameterized marking $\widehat{M}$ we will have, in addition to individuals, parameters from the set $\hat{V}$ as well. Only after these parameters have been *fixed* to some particular individuals do we get an actual marking M. We say that M is a marking represented by $\widehat{M}$. Let us have, for example, the following three different markings on s-element s of a PrT-net:

$$M_1(s) = \{\langle a_1 \rangle, \langle b_1 \rangle\}, M_2(s) = \{\langle a_2 \rangle, \langle b_2 \rangle\}, M_3(s) = \{\langle a_3 \rangle, \langle b_3 \rangle\}.$$

These can all be expressed by using parameters instead of individuals with the following parameterized marking:

$$\widehat{M}(s) = \{\langle \hat{x}_1 \rangle, \langle \hat{x}_2 \rangle\}.$$

Suppose a transition t occurs in such a way that one of the two tuples in M_1, M_2 or M_3 is removed. How should this be reflected in the parameterized marking ? Suppose that the other parameter, say $\hat{x}_2$, is simply removed and we get the parameterized marking $\widehat{M}' = \{\langle \hat{x}_1 \rangle\}$. This leads, however, to problems. Let us consider the marking $M(s) = M_1(s) = \{\langle a_1 \rangle, \langle b_1 \rangle\}$. This is represented by the parameterized marking $\widehat{M} = \{\langle \hat{x}_1 \rangle, \langle \hat{x}_2 \rangle\}$ from which we get M by fixing the parameters, say, $\hat{x}_1 \leftarrow a_1, \hat{x}_2 \leftarrow b_1$. Transition t now occurs so that a_1 is removed and we get $M' = \{\langle b_1 \rangle\}$ which should now be represented by $\widehat{M}' = \{\langle \hat{x}_1 \rangle\}$ as we agreed above. The problem, however, is that $\hat{x}_1$ has already been fixed to a_1 and from this it follows that $\widehat{M}'$ can no longer represent M'. We should have guessed better and removed $\hat{x}_1$ from $\widehat{M}$ instead.

The solution to the above problem is to introduce a new parameter $\hat{x}_3$ which is fixed to either $\hat{x}_1$ or $\hat{x}_2$ and removed from $\widehat{M}$. To do this the parameterized marking $\widehat{M}'$ has to be divided into two parameterized markings $\widehat{M}'^{+}$ and $\widehat{M}'^{-}$ so that

$$\widehat{M}' = \widehat{M}'^{+} - \widehat{M}'^{-} = \{\langle\hat{x}_1\rangle, \langle\hat{x}_2\rangle\} - \{\langle\hat{x}_3\rangle\}.$$

There are also some other requirements a parameterized marking should fulfill and they will be returned to shortly. For now, however, we have enough material to present the definition of a parameterized marking:

Definition 2.2 Parameterized marking:
Let $MN = (N, A, M_0)$ be a PrT-net with $N = (S, T; F)$ and k, m integers such that $k \geq 1, m < 2k$. A non-empty parameterized marking $\widehat{M}$ of MN is for all $s \in S$ of the form:

$$\widehat{M}(s) = \{\nu_1\hat{\tau}_1, \ldots, \nu_k\hat{\tau}_k\} - \{\nu_{k+1}\hat{\tau}_{k+1}, \ldots, \nu_m\hat{\tau}_m\},$$

where $\nu_i \in \mathbf{N}^+$, $\hat{\tau}_i$ is a parameterized token such that $\hat{\tau}_i \in (\widehat{V} \cup D)^n, i = 1, \ldots, m$, and n is the index of the predicate $\psi \in \Psi$ annotating s. By $\widehat{M}^+(s)$ we denote $\{\nu_1\hat{\tau}_1, \ldots, \nu_k\hat{\tau}_k\}$ and by $\widehat{M}^-(s)$ correspondingly $\{\nu_{k+1}\hat{\tau}_{k+1}, \ldots, \nu_m\hat{\tau}_m\}$.

2.1 Denotation of parameterized markings

So far we have defined a syntactical object called a parameterized marking but have not yet presented its exact meaning or semantics. It is, however, quite clear from the introduction that a parameterized marking represents a set of markings derived by replacing the parameters with individuals from the set D or by *fixing* the parameters.

Definition 2.3 *Let $\hat{o}$ be a parameterized object, marking or a token, with $\widehat{V}(\hat{o}) = \{\hat{v}_1, \ldots, \hat{v}_n\}$. Let β be a mapping $\beta : \widehat{V}(\hat{o}) \rightarrow D$ or $\beta = [\hat{v}_1 \leftarrow d_1, \ldots, \hat{v}_n \leftarrow d_n]$, where $d_i \in D, i = 1, \ldots, n$. Then β is called a fixing of the parameters of $\hat{o}$ and $\hat{o}\beta$ denotes the object obtained by substituting each occurrence of $\hat{v}_i$ in $\hat{o}$ with $d_i, i = 1, \ldots, n$ and it is called an unfolding of $\hat{o}$.*

After having defined the fixing of parameters we may now formally present what a parameterized marking denotes. In the language used to talk about parameterized objects we have adopted the convention that such objects are denoted by symbols with ^ over them. The denotation of these objects is then the set of objects they represent and it is obtained by fixing the parameters to individuals or by unfolding the parameterized object in a certain way.

Definition 2.4 *Let $\widehat{M}$ be a parameterized marking of a PrT-net $N = (S, T; F)$ with $\widehat{V}(\widehat{M}) = \{\hat{v}_1, \ldots, \hat{v}_n\}$. Let β be a fixing for the parameters in $\widehat{V}(\widehat{M})$. Then*

$$\mathcal{V}(\widehat{M}) = \{M \mid M = \widehat{M}\beta \wedge M(s) \text{ is a set } \forall s \in S\}$$

is the denotation of the parameterized marking $\widehat{M}$ according to the strict denotation.[2]

It is clear that such constraints which apply to parameterized markings are absent when parameterized tokens are considered. The denotation of a parameterized token is then the set of tokens resulting from all possible fixings replacing the parameters with individuals.

We now have the means to talk about parameterized markings and tokens, and also a meaning for expressions involving them. Therefore, we may now build the machinery needed for applying parameterized markings to constructing reachability trees. Here we will develop the parts of this machinery necessary to be able to present the parameterization of the dynamics of PrT-nets and of reachability trees. For a detailed presentation we refer to [10].

[2] As already mentioned, we deal here only with the strict interpretation, for the weak one the denotation consists of positive multisets rather than of sets.

2.2 Unification and matching

Consider two non-identical parameterized markings $\widehat{M}$ and $\widehat{M}'$. Based on this information, what can be said about $\mathcal{V}(\widehat{M})$ and $\mathcal{V}(\widehat{M}')$: Are they disjoint, are they identical, or do they just have some markings in common? To decide this we have to compare the denotations of the parameterized tokens appearing in the markings and try to decide how they relate to each other. This process of comparing and deciding is very closely related to *unification* of expressions which is one of the main tools used in mechanical theorem proving, [2, pages 74 – 80].

The intuition behind the concept of matching to be defined here is that two parameterized tokens match exactly when there exist a fixing β such that $\hat{o}_1\beta$ and $\hat{o}_2\beta$ are the same. It is not, however, usually necessary to go as far as to find such a fixing to be able to decide whether the two tokens match. It is enough to find a substitution, or a unifier, replacing parameters with other parameters and individuals in such a way that the tokens become the same. After finding a unifier it is easy to find any fixing equating the two tokens since any fixing will thereafter be such a one. Consider, for example, the following two parameterized tokens:

$$\hat{\tau}_1 = \langle \hat{v}_1, a, \hat{v}_3, b, \hat{v}_5 \rangle, \hat{\tau}_2 = \langle \hat{v}_1, \hat{v}_2, \hat{v}_3, \hat{v}_4, \hat{v}_6 \rangle.$$

Then they become the same with the following substitutions: $[\hat{v}_2 \leftarrow a, \hat{v}_4 \leftarrow b, \hat{v}_6 \leftarrow \hat{v}_5]$. Only the parameters which had to be fixed or bound in some way were considered. The rest was still left open and one can freely choose the fixing of $\hat{v}_1, \hat{v}_3$ and $\hat{v}_5$ and still keep the unfolding of the two tokens the same. Therefore, we are interested in something which is the most general way of unifying parameterized tokens.

Definition 2.5 *Let $\hat{o}_1$ and $\hat{o}_2$ be two parameterized objects, markings or tokens and $\hat{\beta} : \widehat{V}(\hat{o}_1) \cup \widehat{V}(\hat{o}_2) \rightarrow \widehat{X}(\hat{o}_1) \cup \widehat{X}(\hat{o}_2)$. Then $\hat{\beta}$ is called a unifier iff $\hat{o}_1\hat{\beta} = \hat{o}_2\hat{\beta}$.*

Further, $\hat{\beta}$ is called the most general unifier iff for each unifier $\hat{\beta}'$ there exists a substitution $\hat{\beta}''$ such that $\hat{\beta}' = \hat{\beta}\hat{\beta}''$.

Now we may formalise the concept of matching with the help of a unifier in the following way:

Definition 2.6 *Let $\hat{o}_1, \hat{o}_2$ be two parameterized objects, tokens or markings. Then we say that $\hat{o}_1$ matches $\hat{o}_2$ or $\hat{o}_1 \simeq \hat{o}_2$ iff*

$$\exists \text{ a unifier } \hat{\beta} : \widehat{V}(\hat{o}_1) \cup \widehat{V}(\hat{o}_2) \rightarrow \widehat{X}(\hat{o}_1) \cup \widehat{X}(\hat{o}_2) \text{ such that } \hat{o}_1\hat{\beta} = \hat{o}_2\hat{\beta}.$$

Corollary 2.7 *Let $\hat{o}_1, \hat{o}_2$ be two parameterized objects such that $\hat{o}_1 \simeq \hat{o}_2$. Then clearly $\hat{o}_2 \simeq \hat{o}_1$.*

Corollary 2.8 *Let $\hat{o}_1, \hat{o}_2$ be two parameterized objects such that $\hat{o}_2 \simeq \hat{o}_1$. Then*

$$\mathcal{V}(\hat{o}_1) \cap \mathcal{V}(\hat{o}_2) \neq \emptyset.$$

The concept of matching is of paramount importance in reasoning about parameterized markings. It is not, however, suitable if the denotation has to be considered in order to be able to say whether an object matches another one or not. Of course there is always the trial-and-error method fr exhaustive search as a means of trying to locate a unifying fixing[3] but its use is not that attractive. Especially if the set D of individuals is large there might be a vast number of possible fixings. It is, however, rather straightforward to find a unifier $\hat{\beta}$, especially when two tokens are considered. The same principle can also be extended to sets of tokens providing us with the possibility to compare

[3] Note that we are still dealing only with a finite D resulting in a finite number of different possible fixings β.

parameterized markings with each other. For details and a description of the unification algorithms needed we again refer to [10].

Let us now consider two parameterized markings $\widehat{M}$ and $\widehat{M}'$ and their denotations in particular. How do these denotations relate to each other under all possible global fixings of their parameters ? We can have the following four cases:

1. The denotations of the two markings are disjoint,
2. the denotations have a common intersection,
3. the denotation of $\widehat{M}$ is included in the denotation of $\widehat{M}'$ or
4. the denotation of $\widehat{M}'$ is included in the denotation of $\widehat{M}$.

The cases which are of primary interest are 3 and 4 of the above list. If we know that $\widehat{M}$ includes, so to speak, $\widehat{M}'$ then all the markings that can be reached from $\widehat{M}'$ can also be deduced from $\widehat{M}$. This immediately points to the possibility of using this idea for defining a covering relation for parameterized markings. How including markings are actually used is postponed until Section 3.1. For now we will return to formalising the concept and to telling how including markings can be detected.

Definition 2.9 *Let $\widehat{M}$ and $\widehat{M}'$ be two parameterized markings of a PrT-net. Then we say that $\widehat{M}'$ includes $\widehat{M}$ denoted by $\widehat{M} \subseteq \widehat{M}'$ iff*

$$\mathcal{V}(\widehat{M}\beta) \subseteq \mathcal{V}(\widehat{M}'\beta)$$

for all fixings $\beta : \widehat{\mathcal{V}}(\widehat{M}) \cap \widehat{\mathcal{V}}(\widehat{M}') \to D$ of their common parameters.

In the above definition we have only considered those parts of the denotations of the two markings that are obtained through global fixings, i.e. fixings where each parameter is fixed only once. The reasoning for this is that when parameterized reachability trees are considered, only global fixings make sense whereas fixings local to a marking are useless.

When a parameterized marking $\widehat{M}'$ includes another one $\widehat{M}$ it is clear that any fixing possible for $\widehat{M}$ is also possible for $\widehat{M}'$ as stated in the following:

Theorem 2.10 *Let $\widehat{M}$ and $\widehat{M}'$ be two parameterized markings such that $\widehat{M} \subseteq \widehat{M}'$ and D the set of individuals. Then for all fixings $\beta : \widehat{\mathcal{V}}(\widehat{M}) \to D$ such that $\widehat{M}\beta \in \mathcal{V}(\widehat{M})$ there exists a fixing β' such that $\widehat{M}'\beta' = \widehat{M}\beta$.*

Proof:

Follows directly from Definitions 2.4 and 2.9.

In order to detect such a pair of markings where the one includes the other we should concentrate on the possible fixings of the markings. But in this case we would, as before, not like to consider the denotation of the markings in order decide whether one includes the other. This, however, is not necessary. Again, we may use unification of sets of parameterized tokens and find a unifier which unifies the two markings [10, pages 66–67]. However, when constructing the unifier one must take care that it does not change the denotation of the marking which we hope will be included in the other one:

Theorem 2.11 *Let $\widehat{M}$ and $\widehat{M}'$ be parameterized markings of a PrT-net MN with $N = (S,T;F)$ such that they are of the same cardinality and $\forall s \in S : |\widehat{M}'^{+}(s)| \geq |\widehat{M}^{+}(s)|$. Then*

$$\widehat{M} \subseteq \widehat{M}' \Leftrightarrow \exists\hat{\beta} : \hat{V}(\widehat{M}') - \hat{V}(\widehat{M}) \rightarrow \hat{X}(\widehat{M}) \cup \hat{X}(\widehat{M}') \text{ such that } \widehat{M}'\hat{\beta} = \widehat{M}\hat{\beta} = \widehat{M}.$$

Proof:

See Theorem 3.28 of [10].

We have now rather briefly introduced the concept of parameterized markings. We have also shown the central points of how parameterized markings can be dealt with. For those interested in the details we refer to [10]. The purpose of introducing parameters into markings was to be able to fold markings in a manner similar to the one which is used to fold s- and t-elements in PrT-nets. How this helps in cutting down the size of a reachability tree (which is our ultimate goal) will be shown in the next section.

3 Parameterizing the dynamics

When constructing a reachability tree with ordinary markings a method for generating new markings from old ones by transition occurrences is needed. Therefore, to have a corresponding method for parameterized markings the transition rule has to be modified to cope with parameters. This means that parameters will also be introduced to the occurrence of a transition by possible assignment of parameters instead of individuals to the variables appearing in the tuples of A_F.

What should be the semantics of a parameterized occurrence of a transition which changes one parameterized marking into another ? What should its interpretation be ? Clearly this parameterized occurrence should stand for all those occurrences which may lead from the denotation of the one marking into the denotation of the other. Therefore we can speak of the denotation of a parameterized occurrence and present the following definition:

Definition 3.1 *Let $\widehat{M}$ and $\widehat{M}'$ be two parameterized markings of a PrT-net MN with $N = (S,T;F)$. Let $t \in T$ be a transition and $\hat{\alpha}$ a parameterized assignment replacing the variables in the index of t with parameters and individuals. Further, let β be a fixing $\beta : \hat{V}(\widehat{M}) \cup \hat{V}(\widehat{M}') \cup \hat{V}(\hat{\alpha}) \rightarrow D$. Then*

$$\mathcal{V}(\widehat{M}[t\hat{\alpha}\rangle\widehat{M}') = \{t\hat{\alpha}\beta \mid \widehat{M}\beta[t\hat{\alpha}\beta\rangle\widehat{M}'\beta \wedge \widehat{M}\beta \in \mathcal{V}(\widehat{M}) \wedge \widehat{M}'\beta \in \mathcal{V}(\widehat{M}')\},$$

is the (strict) denotation of the parameterized transition occurrence $\widehat{M}[t\hat{\alpha}\rangle\widehat{M}'$.

The "ordinary" transition rule of Definition 1.5 uses the requirement that an assignment replacing the variables with individuals has to be both *feasible* and *permissible* as defined in Definitions 1.3 and 1.4. The former involves fulfilling static requirements which do not change as processes occur in the net. The requirements set by the latter depend, on the other hand, solely on the current marking of the net and are thus of a dynamic nature. When parameterized markings are involved, this distinction is still somewhat valid. The role of these requirements will be as significant as in the ordinary case. They alone should be enough to ensure that the denotation of a parameterized occurrence is not empty.

The feasibility requirement should guarantee that the assignment is such that there exists a fixing of the parameters in $\hat{V}(\widehat{M})$ such that the requirements of Definition 1.3 are fulfilled. The feasibility can, however, be checked symbolically only in the case that the selector formula is restricted to being a conjunction of identities and their negations. This restriction, however, is not absolutely necessary and it may be dropped without hampering the use of parameterized markings as explained in [10, Section 5.3].

Definition 3.2 Feasible assignment of parameters

Let MN be a PrT-net with $N = (S, T; F)$, $\widehat{M}$ a parameterized marking of MN, t its transition and φ the selector formula annotating t being a conjuction of identities and their negations. Let $\hat{\alpha} = [v_1 \leftarrow \hat{x}_1, \ldots, v_n \leftarrow \hat{x}_n]$ be a parameterized assignment replacing variables in the index of t with individuals or parameters. Then $\hat{\alpha}$ is called feasible iff:

1. *$\hat{\alpha}$ satisfies φ*
   (a) *if $\varphi\hat{\alpha} \Rightarrow v_i\hat{\alpha} = v_j\hat{\alpha}$ then $\hat{\alpha}$ must assign the same parameter or individual to variables v_i and v_j.*
   (b) *if $\varphi\hat{\alpha} \Rightarrow v_i\hat{\alpha} \neq v_j\hat{\alpha}$ then $\hat{\alpha}$ must assign different parameters or individuals to variables v_i and v_j.*
2. *$\hat{\alpha}$ does not generate an impurity; if $(s,t) \in F$ then for any pair $\hat{\tau}_1, \hat{\tau}_2$ of tuples such that $\hat{\tau}_1 \in A_F(s,t)$ and $\hat{\tau}_2 \in A_F(t,s)$ it holds that $\hat{\tau}_1\hat{\alpha} \neq \hat{\tau}_2\hat{\alpha}$.*
3. *$\hat{\alpha}$ does not generate a multiple arc: if $\hat{\tau}_1, \hat{\tau}_2 \in A_F(s,t)$ or $A_F(t,s)$ and $\hat{\tau}_1 \neq \hat{\tau}_2$ then $\hat{\tau}_1\hat{\alpha} \neq \hat{\tau}_2\hat{\alpha}$.*

The two last items of Definition 3.2 correspond directly to the items of Definition 1.3. Item 1 ensures that all dependencies that are set by the selector formula φ are taken explicitly into account and are reflected directly in the assignment. This way no hidden constraints are left in the parameterized marking resulting from an occurrence of a transition with a selector formula.

The permissibility requirement of Definition 1.4 ensures that the assignment is such that the result of the occurrence of the transition is a new marking. The permissibility requirement should, in the parameterized case, ensure that the denotation of the new parameterized marking produced by the parameterized transition occurrence is not empty. This then requires considering the parameterized marking of each s-element not separately but as one global entity.

Definition 3.3 Permissible assignment of parameters

Let MN be a PrT-net with $N = (S, T; F)$, $\widehat{M}$ a parameterized marking of MN and t its transition. Let $\hat{\alpha} = [v_1 \leftarrow \hat{x}_1, \ldots, v_n \leftarrow \hat{x}_n]$ be a parameterized assignment replacing variables in the index of t with individuals or parameters. Then $\hat{\alpha}$ is called permissible iff:

1. *$\hat{x}_i, i = 1, \ldots, n$, is an individual $\hat{x}_i \in D$ or an old parameter $\hat{x}_i \in \widehat{V}(\widehat{M})$ iff:*
   (a) *$\exists\ \tau \in A_F(s,t)$ such that $\tau(j) = v_i$ and $\forall\ \hat{\tau} \in \widehat{M}^+(s)$ such that $\hat{\tau} \simeq \tau\hat{\alpha} : \hat{\tau}(j) = \hat{x}_i$ and*
   (b) *$\forall \tau \in A_F(s',t)$ such that $\tau(k) = v_i\ \exists\ \hat{\tau} \in \widehat{M}^+(s')$ such that $\hat{\tau} \simeq \tau\hat{\alpha}$.*
2. *$\hat{x}_i, i = 1, \ldots, n$, is a new parameter $\hat{x}_i \notin \widehat{V}(\widehat{M})$ iff $\forall\ \tau \in A_F(s,t)$ such that $\tau(j) = v_i\ \exists\ \hat{\tau}_1, \hat{\tau}_2$ such that $\hat{\tau}_1 \simeq \tau\hat{\alpha} \wedge \hat{\tau}_2 \simeq \tau\hat{\alpha}$ and $\hat{\tau}_1(j) \neq \hat{\tau}_2(j)$.*
3. *$\exists\ \hat{\beta} : \widehat{V}(\widehat{M}) \cup \widehat{V}(\hat{\alpha}) \to \widehat{X}(\widehat{M})$ such that*
   (a) *$\forall\ s \in S\ \widehat{M}^+\hat{\beta}(s) \geq \widehat{M}^-\hat{\beta}(s)$ and $\neg\exists\ \hat{\tau}$ such that $\widehat{M}^+\hat{\beta}(s)[\hat{\tau}] - \widehat{M}^-\hat{\beta}(s)[\hat{\tau}] > 1$, i.e. $\widehat{M}\hat{\beta}(s)$ is a set of parameterized tokens.*
   (b) *$\forall\ s \in S$ such that $\tau \in A_F(s,t)\ \widehat{M}^+\hat{\beta}(s)[\tau\hat{\alpha}\hat{\beta}] - \widehat{M}^-\hat{\beta}(s)[\tau\hat{\alpha}\hat{\beta}] = 1$.*
   (c) *$\forall\ s \in S$ such that $\tau \in A_F(t,s)\ \widehat{M}^+\hat{\beta}(s)[\tau\hat{\alpha}\hat{\beta}] - \widehat{M}^-\hat{\beta}(s)[\tau\hat{\alpha}\hat{\beta}] = 0$.*

We have now presented the feasibility and permissibility requirements for the parameterized case. The motivation for the feasibility requirement of Definition 3.2 should be quite clear but may seem

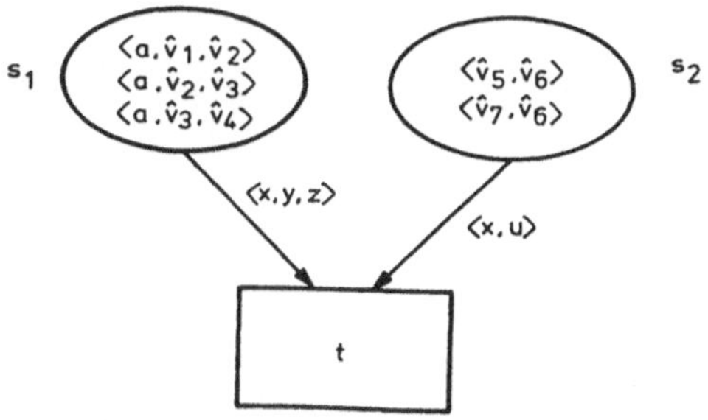

Figure 1: A transition of a PrT-net with its input s-elements and a parameterized marking $\widehat{M}$.

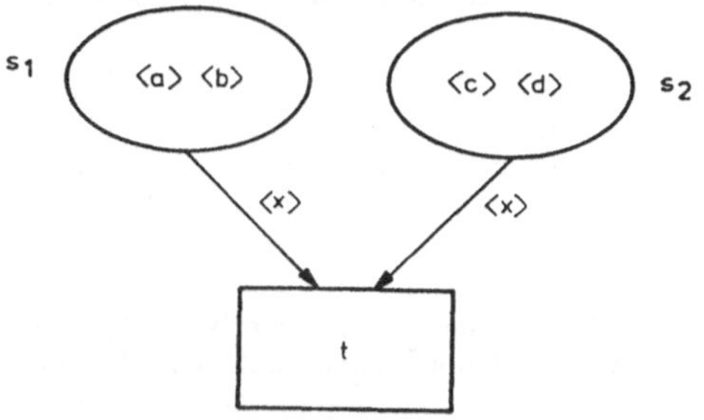

Figure 2: A situation where a permissible parameterized assignment does not exist.

rather cumbersome for the permissibility requirement of Definition 3.3. Therefore, its meaning will be studied somewhat more closely.

The meaning of items 1 and 2 of Definition 3.3 is to tell what the prerequisities are for assigning an individual, an old parameter, or a new parameter to a variable. It is easily motivated by studying the following example:

Example 3.1 In the situation of Figure 1 items 1, 2 and 3 of Definition 3.3 require that the assignment $\hat{\alpha} = [u \leftarrow \hat{v}_6, x \leftarrow a, y \leftarrow \hat{v}_8, z \leftarrow \hat{v}_9]$, where $a \in D, \hat{v}_6 \in \widehat{V}(\widehat{M})$ and $\hat{v}_8, \hat{v}_9 \notin \widehat{V}(\widehat{M})$. For u we can only assign $\hat{v}_6$ since all tokens of s_2 contain it as their second component. For y and z we must take new parameters since there is more than one possibility for choosing the assignment for them. The only thing known is that the pair $\hat{v}_8, \hat{v}_9$ has the same value as the pair $\hat{v}_1, \hat{v}_2$ or $\hat{v}_2, \hat{v}_3$ or $\hat{v}_3, \hat{v}_4$ in any unfolding. The only possible assignment for x is individual a because of the tokens on s-element s_1. Additionally, the components which may be assigned to x in the tuple from s_2 to t are parameters and thus we can assume that either $\hat{v}_5$ or $\hat{v}_7$ will have the value a. If there were only individuals different from a as first components of tokens in s_2 a permissible assignment could not be found.

The rest of the items of Definition 3.3 are needed to ensure that both the markings involved in an occurrence of a transition, the source and the target so to speak, are still feasible after the occurrence. They are somewhat complex but unfortunately also unavoidable since the feasibility constraints put on a marking of some s-element effect all those markings of other s-elements which share the common parameters with it. This will become clear when studying the following examples. They are chosen to reflect those troublesome cases which require Definition 3.3 to be enforced all the way before the permissibility of an assignment can be ensured.

Example 3.2 In the situation of Figure 3 where $\widehat{M}(s_1) = \{\langle a\rangle, \langle b\rangle\}$ and $\widehat{M}(s_2) = \{\langle c\rangle, \langle d\rangle\}$ items 1

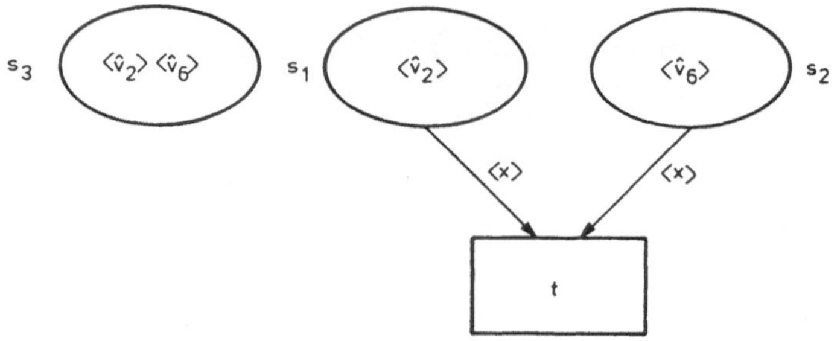

Figure 3: A situation where the occurrence of t causes $\widehat{M}(s_3)$ to become unfeasible under any possible assignment.

and 2 of Definition 3.3 would give, say, the following assignment $\hat{\alpha} = [x \leftarrow \hat{v}]$ where $\hat{v}$ is a new parameter. But there is no fixing β for $\hat{v}$ such that $\tau\hat{\alpha}\beta \in \widehat{M}\beta(s_1)$ and $\tau\hat{\alpha}\beta \in \widehat{M}\beta(s_2)$ since a, b, c and d are all different individuals. This problem arises with the new parameter introduced by the assignment $\hat{\alpha}$. It represents a set of possible parameters and individuals that could be assigned to the variables on the arcs. On different s-elements these sets may, however, be disjoint and thus in some cases no feasible fixing for the parameters exists. This also means that one cannot find a substitution $\hat{\beta}$ as required by item 3.b of Definition 3.3 and thus the assignment is not permissible.

The case in the previous example was still quite simple. However, already in this simple case we had to fall back on item 3 of Definition 3.3 in order to see that using a certain assignment in an occurrence of a transition would have produced a parameterized occurrence with an empty denotation. The next example shows that it is also necessary to consider the marking of s-elements which are not immediately connected to the transition which occurs:

Example 3.3 In Figure 3 the assignment $\hat{\alpha}$ assigns to variable x either $\hat{v}_2$ or $\hat{v}_6$. An implicit consequence of this is that for any fixing β it must be that $\hat{v}_2\beta = \hat{v}_6\beta$. Thus, any assignment causes the marking $\widehat{M}(s_3)$ to become unfeasible even if s_3 is not connected to the transition which occurs. It is also impossible to find a substitution such that items 3.a and 3.b of Definition 3.3 would be fulfilled and therefore a permissible assignment does not exist. And neither should it, since the denotation of a parameterized occurrence of t would necessarily be empty.

It is necessary to check the effect of the dependencies caused by the assignment. A variable appearing in more than one of the input tuples forces the parameters assigned for the different appearances to represent the same individual, like parameters $\hat{v}_2$ and $\hat{v}_6$ in the above example. One would very probably tend to think that it would be enough to check the feasibility of markings for only those s-elements which contain parameters involved in the assignment. But unfortunately that is not the case as shown in the next example.

Example 3.4 Consider the parameterized marking depicted in Figure 4. Parameters $\hat{v}_1$ and $\hat{v}_2$ are not involved in the assignment at all. However, since the token $\langle\hat{v}_3, \hat{v}_4\rangle$ is necessarily removed from $\widehat{M}(s_2)$ by the occurrence of t we get that $\langle\hat{v}_1, \hat{v}_2\rangle$ must be $\langle c, a\rangle$. But now we run into a problem since such a token is not present in $\widehat{M}^+(s_1)$ and the transition cannot occur without causing either $\widehat{M}(s_1)$ or $\widehat{M}(s_2)$ to become unfeasible. Therefore the assignment should not be permissible, and indeed, it is not, since it is impossible to fulfill items 3.a and 3.b of Definition 3.3.

We have now studied some examples which clearly show why the permissibility requirement of Definition 3.3 must be as exacting as it is. What it is ultimately about is that it tells when a new

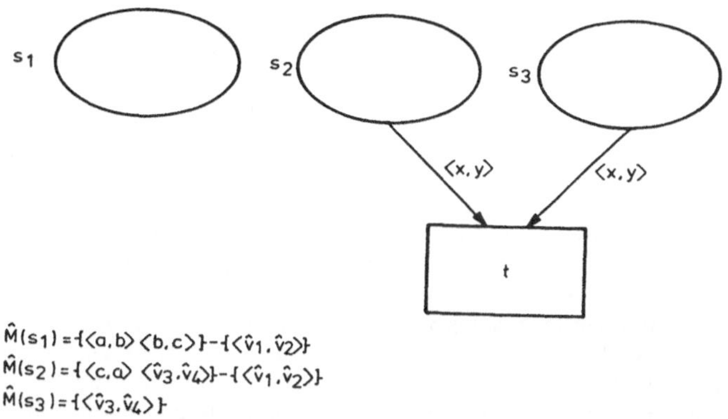

Figure 4: Transition t cannot occur because either $\widehat{M}(s_1)$ or $\widehat{M}(s_3)$ would become unfeasible.

parameter expressing a possibility of choice has to be introduced. In addition this new parameter must stand for some old one and therefore at least one feasible parameterized marking where new parameters have been replaced with old ones must exist. In fact, this is quite similar to what the permissibility requirement of ordinary markings requires of variables and the values assigned to them.

In [10, pages 77–78] an algorithm to check the permissibility of a parameterized assignment is presented. Here we content ourselves with presenting the parameterized transition rule and some of its properties.

Definition 3.4 Parameterized transition rule:

Let $\widehat{M}$ and $\widehat{M}'$ be two parameterized markings of a PrT-net with $N = (S, T; F)$, t a transition and $\hat{\alpha} = [v_1 \leftarrow \hat{x}_1, \ldots v_n \leftarrow \hat{x}_n]$ a parameterized assignment replacing the variables in the index of t with parameters or individuals. Then $\widehat{M}[t\hat{\alpha}\rangle\widehat{M}'$ iff:

1. *$\hat{\alpha}$ is a permissible parameterized assignment,*
2. *$\hat{\alpha}\hat{\beta}$ is a feasible assignment, where $\hat{\beta} : \hat{V}(\widehat{M}) \cup \hat{V}(\hat{\alpha}) \rightarrow \hat{X}(\widehat{M})$ is a substitution as defined in Definition 3.3.*
3. *For all places $s \in S$:*
   - *(a) If $(s, t) \in S$ then $\widehat{M}'^{-}(s) = \widehat{M}^{-}(s) + \{\tau\hat{\alpha} \mid \tau \in A_F(s, t)\}$.*
   - *(b) If $(t, s) \in S$ then $\widehat{M}'^{+}(s) = \widehat{M}^{+}(s) + \{\tau\hat{\alpha} \mid \tau \in A_F(t, s)\}$.*
   - *(c) $\widehat{M}'^{+}(s) = \widehat{M}^{+}(s)$ and $\widehat{M}'^{-}(s) = \widehat{M}^{-}(s)$ otherwise.*

We say that t is $\hat{\alpha}$ enabled in $\widehat{M}$ iff $\widehat{M}[t\hat{\alpha}\rangle\widehat{M}'$.

The transition rule for the parameterized case now gives the possibility of generating follower markings when an initial marking has been given. This already allows for generating a reachability tree consisting of parameterized markings. But before starting to study how this is done a theorem of central importance will be presented. It tells that using the transition rule of Definition 3.4 ensures that the denotation of the parameterized occurrence will be non empty:

Theorem 3.5 *Let $\widehat{M}$ and $\widehat{M}'$ be two parameterized markings of a PrT-net such that $\widehat{M}$ is feasible, t a transition and $\hat{\alpha}$ a parameterized assignment replacing the variables in the index of t with parameters and individuals. Suppose $|D|$ is sufficiently large[4] and the selector φ of t as in Definition 3.2.[5] Then*

[4] At most the maximum of $|\hat{V}(\widehat{M}'(s))|$ of all $s \in S$.

[5] The consequences of dropping this assumption has been delt with in Section 5.3 in [10].

$$\widehat{M}[t\hat{\alpha}\rangle\widehat{M}' \Leftrightarrow \mathcal{V}(\widehat{M}[t\hat{\alpha}\rangle\widehat{M}') \neq \emptyset.$$

Proof:

See the Appendix.

We have now defined a parameterized transition rule and shown that its use ensures that the denotation of a parameterized occurrence is non empty. This, however, holds only under the assumption that the fixing is not limited by any other means but those resulting from the feasibility requirements of the two markings involved. Therefore we have added into the premises the assumption that D has enough members to allow for a fixing which fixes each parameter to a different individual. We also assumed that the selector formula contains only identities and their negations. For one direction of the proof this assumption was not needed and even for the other it is not always necessary but it is sufficient. Before starting with the introduction of parameterized reachability trees we have yet to present a theorem about parameterized occurrences telling more of their denotation:

Theorem 3.6 *Let M be a marking such that $M[t\alpha\rangle M'$ and $M \in \mathcal{V}(\widehat{M})$ where $\widehat{M}$ is a parameterized marking. Further, let D be the set of individuals and $\widehat{M}[t\hat{\alpha}\rangle\widehat{M}'$ be a parameterized occurrence according to Definition 3.4. Then $M[t\alpha\rangle M' \in \mathcal{V}(\widehat{M}[t\hat{\alpha}\rangle\widehat{M}')$.*

Proof: See the Appendix.

We have now almost everything that is needed for constructing reachability trees with parameterized markings. One crucial concept, however, is still missing: that is the covering relation for parameterized markings. As we already mentioned at the end of previous section, the inclusion relation of parameterized markings can be used. The reason for this is shown by the next theorem:

Theorem 3.7 *Let $MN = (N, A, M_0)$ be a PrT-net with $N = (S, T; F)$ and $t \in T$ a transition of MN. Let $\widehat{M}, \widehat{M}'$ be two parameterized markings of MN and $\widehat{M} \subseteq \widehat{M}'$. Further, let $\hat{\alpha}, \hat{\alpha}'$ be two parameterized assignments such that t is $\hat{\alpha}$ enabled in $\widehat{M}$ and $\hat{\alpha}'$ enabled in $\widehat{M}'$. Then*

$$\mathcal{V}(\widehat{M}[t\hat{\alpha}\rangle\widehat{M}_1) \subseteq \mathcal{V}(\widehat{M}'[t\hat{\alpha}'\rangle\widehat{M}_1').$$

Proof:

See the Appendix.

Theorem 3.7 shows that the denotation of a marking $\widehat{M}$ included in some other marking $\widehat{M}'$ can not have more followers than the denotation of $\widehat{M}'$. One immediate result of this is that everything that can be reached from $\widehat{M}$ can also be reached from $\widehat{M}'$. Therefore the use of the inclusion relation as the covering relation of parameterized markings is justified.

3.1 Parameterized reachability trees

We will start the study of parameterized reachability trees, PR-trees for short, by looking at an example involving a well known PrT-net model. Consider the net of Figure 5 describing a system maintaining multiple copies of a database [5]. Further, consider the case where there are three different agents, a_1, a_2 and a_3, in the system. The initial marking, denoted by $\widehat{M}_0$ is then the following:

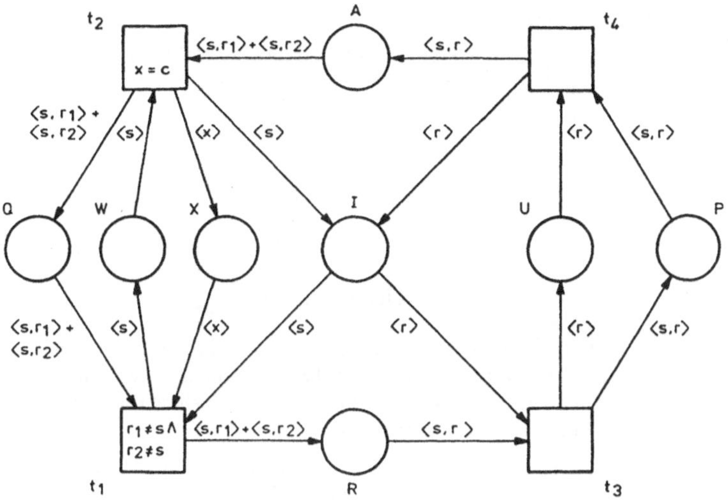

Figure 5: A system for maintaining multiple copies of a bata base.

$$\widehat{M}_0 \begin{cases} \mathbf{A} & \\ \mathbf{Q} & \{\langle a_1, a_2\rangle, \langle a_1, a_3\rangle, \langle a_2, a_1\rangle, \langle a_2, a_3\rangle, \langle a_3, a_1\rangle, \langle a_3, a_2\rangle\} = N \\ \mathbf{W} & \\ \mathbf{X} & \{\langle c\rangle\} \\ \mathbf{I} & \{\langle a_1\rangle, \langle a_2\rangle, \langle a_3\rangle\} \\ \mathbf{U} & \\ \mathbf{P} & \\ \mathbf{R} & \end{cases}$$

The set of individuals is $D = \{a_1, a_2, a_3, c\}$. From here only transition t_1 may fire but it may fire with three different values assigned to variable s. We use parameter $\hat{s}_1$ to denote the possible values assignable to s, i. e. the values, or individuals a_1, a_2 or a_3. The selector formula φ of t_1 restricts the possible values assignable to variables r_1 and r_2 to be different from the value assigned to s.[6] We will denote these values with parameters $\hat{r}_1$ and $\hat{r}_2$. Since the only possibility is to assign individual c for x the parameterized assignment becomes $\hat{\alpha} = [s \leftarrow \hat{s}_1, r_1 \leftarrow \hat{r}_1, r_2 \leftarrow \hat{r}_2, x \leftarrow c]$. In this case it is trivial to find such a $\hat{\beta} : \widehat{V}(\widehat{M}_0) \cup \widehat{V}(\hat{\alpha}) \rightarrow \widehat{X}(\widehat{M}_0)$ that $\hat{\alpha}$ is permissible, say $\hat{\beta} = [\hat{s}_1 \leftarrow a_1, \hat{r}_1 \leftarrow a_2, \hat{r}_2 \leftarrow a_3]$. Clearly $\hat{\alpha}\hat{\beta}$ is also feasible. Therefore t may occur under the parameterized assignment $\hat{\alpha}$ and a new parameterized marking $\widehat{M}_1$ is produced:

[6] We could actually do without the selector φ in this case.

$$\widehat{M}_0 \xrightarrow[\varphi:\ \hat{r}_1 \neq \hat{s}_1 \wedge \hat{r}_2 \neq \hat{s}_1]{t_1:\ \hat{\alpha} = [s \leftarrow \hat{s}_1, r_1 \leftarrow \hat{r}_1, r_2 \leftarrow \hat{r}_2, x \leftarrow c]}$$

$$\widehat{M}_1 \begin{cases} \mathbf{A} & \\ \mathbf{Q} & N - \{\langle \hat{s}_1, \hat{r}_1\rangle, \langle \hat{s}_1, \hat{r}_2\rangle\} \\ \mathbf{W} & \hat{s}_1 \\ \mathbf{X} & \{\langle c\rangle\} - \{\langle \hat{x}\rangle\} = \emptyset \\ \mathbf{I} & \{\langle a_1\rangle, \langle a_2\rangle, \langle a_3\rangle\} - \{\hat{s}_1\} \\ \mathbf{U} & \\ \mathbf{P} & \\ \mathbf{R} & \{\langle \hat{s}_1, \hat{r}_1\rangle, \langle \hat{s}_1, \hat{r}_2\rangle\} \end{cases}$$

The edge connecting $\widehat{M}_0$ to $\widehat{M}_1$ is labeled with the transition that occurred and the used parameterized assignment $\hat{\alpha}$. Also the selector formula φ of the transition is shown with its variables replaced by parameters and individuals in the way $\hat{\alpha}$ defines. This is to show the static restrictions that are put on all occurrences of t_1 regardless of the marking.

The complete parameterized reachability tree of the net of Figure 5 is depicted in Appendix 1. It is obtained by applying reasoning based on Definition 3.4 in a manner similar to that described above. This works nicely and no such complex considerations are necessary as in the examples presented in the previous section.

A reachability tree is made finite by the use of covering markings. This is also employed in PR-trees: consider the marking presented below. It is $\widehat{M}_6$ of the PR-tree depicted in Appendix 1. There it is claimed that $\widehat{M}_6$ is included in marking $\widehat{M}_5$. How this claim is justified is resolved in the following. To make it easier to follow the discourse the markings $\widehat{M}_6$ and $\widehat{M}_5$ are repeated below:

$$\widehat{M}_4 \xrightarrow{t_4:\ \hat{\alpha} = [s \leftarrow \hat{s}_1, r \leftarrow \hat{r}_6]}$$

$$\widehat{M}_6 \begin{cases} \mathbf{A} & \{\langle \hat{s}_1, \hat{r}_6\rangle\} \\ \mathbf{Q} & N - \{\langle \hat{s}_1, \hat{r}_1\rangle, \langle \hat{s}_1, \hat{r}_2\rangle\} \\ \mathbf{W} & \hat{s}_1 \\ \mathbf{X} & \\ \mathbf{I} & \{\langle \hat{r}_6\rangle\} \\ \mathbf{U} & \{\langle \hat{r}_3\rangle, \langle \hat{r}_4\rangle\} - \{\langle \hat{r}_6\rangle\} = \{\langle \hat{r}_8\rangle\} \\ \mathbf{P} & \{\langle \hat{s}_1, \hat{r}_3\rangle, \langle \hat{s}_1, \hat{r}_4\rangle\} - \{\langle \hat{s}_1, \hat{r}_6\rangle\} = \{\langle \hat{s}_1, \hat{r}_8\rangle\} \\ \mathbf{R} & \end{cases}$$

$$\widehat{M}_3 \xrightarrow{t_3:\ \hat{\alpha} = [s \leftarrow \hat{s}_1, r \leftarrow \hat{r}_3]}$$

$$\widehat{M}_5 \begin{cases} \mathbf{A} & \{\langle \hat{s}_1, \hat{r}_3\rangle\} \\ \mathbf{Q} & N - \{\langle \hat{s}_1, \hat{r}_1\rangle, \langle \hat{s}_1, \hat{r}_2\rangle\} \\ \mathbf{W} & \hat{s}_1 \\ \mathbf{X} & \\ \mathbf{I} & \{\langle a_1\rangle, \langle a_2\rangle, \langle a_3\rangle\} - \{\langle \hat{s}_1\rangle, \langle \hat{r}_5\rangle\} = \{\langle \hat{r}_7\rangle\} \\ \mathbf{U} & \{\langle \hat{r}_5\rangle\} \\ \mathbf{P} & \{\langle \hat{s}_1, \hat{r}_5\rangle\} \\ \mathbf{R} & \end{cases}$$

Looking at $\widehat{M}_5$ and $\widehat{M}_6$ we can immediately see that they are of the same cardinality and thus it might be possible that $\widehat{M}_6 \subseteq \widehat{M}_5$. If this were true then it should be possible according to Theorem 2.11 to construct a unifier $\hat{\beta} : \widehat{V}(\widehat{M}_5) \to \widehat{X}(\widehat{M}_6)$ such that $\widehat{M}_5\hat{\beta} = \widehat{M}_6$. The prerequisite for this is that for all s-elements $\widehat{M}_5^+(s) \geq \widehat{M}_6^+(s)$, which is not true for s-elements **U** and **P**. We can, however, replace these markings with other ones so that this prerequisite would be satisfied [10, pages 70–71]. We use markings $\{\langle\hat{r}_8\rangle\}$ and $\{\langle\hat{s}_1, \hat{r}_8\rangle\}$ to stand for $\widehat{M}_6(\mathbf{U})$ and $\widehat{M}_6(\mathbf{P})$, respectively, as can be seen above. But we still have problems with $\widehat{M}_5(\mathbf{I})$ since it has parameters only in $\widehat{M}_5^-(\mathbf{I})$ and thus no unifier unifying it with $\widehat{M}_6(\mathbf{I})$ can be found. But here again $\widehat{M}_5(\mathbf{I})$ can be replaced with $\{\langle\hat{r}_7\rangle\}$ since the denotation of the marking remains intact. Now it is rather straightforward to find a unifier $\hat{\beta} : \widehat{V}(\widehat{M}_5) \rightarrow \widehat{X}(\widehat{M}_6)$ such that $\widehat{M}_5\hat{\beta} = \widehat{M}_6$. It is in this case

$$\hat{\beta} = [\hat{r}_3 \leftarrow \hat{r}_6, \hat{r}_5 \leftarrow \hat{r}_8, \hat{r}_7 \leftarrow \hat{r}_6].$$

It is easily seen that indeed $\widehat{M}_5\hat{\beta} = \widehat{M}_6$. Now we can use Theorem 2.11 and conclude that $\widehat{M}_6$ is included in $\widehat{M}_5$. Therefore we can conclude that all that can be reached from any marking in the denotation of $\widehat{M}_6$ can also be reached from a marking in the denotation of $\widehat{M}_5$ as shown by Theorem 3.7. Therefore there it is no use to developing $\widehat{M}_6$ further.

Returning to our example of the data base system of Figure 5 there is still one more marking to consider. That is the marking $\widehat{M}_8$ which is shown below:

$$\widehat{M}_7$$

$$\big\downarrow\ t_2 : \quad \hat{\alpha} = [s \leftarrow \hat{s}_1, r_1 \leftarrow \hat{r}_9, r_2 \leftarrow \hat{r}_{10}]$$

$$\widehat{M}_8 \left\{ \begin{array}{ll} \mathbf{A} & \\ \mathbf{Q} & N + \{\langle\hat{s}_1, \hat{r}_9\rangle, \langle\hat{s}_1, \hat{r}_{10}\rangle\} - \{\langle\hat{s}_1, \hat{r}_1\rangle, \langle\hat{s}_1, \hat{r}_2\rangle\} = N \\ \mathbf{W} & \\ \mathbf{X} & \{\langle c\rangle\} \\ \mathbf{I} & \{\langle a_1\rangle, \langle a_2\rangle, \langle a_3\rangle\} \\ \mathbf{U} & \\ \mathbf{P} & \\ \mathbf{R} & \end{array} \right.$$

This marking is equal to $\widehat{M}_1$, the initial marking, except on s-element **Q**. This marking can, however, be reduced. Since we are dealing with strict parameterized markings we can delete all the tokens that are in excess of N since N is the complete denotation of each of these tokens. Therefore we obtain that $\widehat{M}_8 \subseteq \widehat{M}_0$ and the generation of the reachability tree has been completed.

To generate the reachability tree we have used reasoning based on several of the definitions and theorems presented earlier in this and the previous section. By pulling all this material together we can now present the algorithm for generating parameterized reachability trees. [7]

Algorithm 3.1 An algorithm to produce a parameterized reachability tree of a PrT-net with an initial marking $\widehat{M}_0$ under the strict interpretation. The $\mathcal{U}$ denotes the set of unprocessed nodes, $\mathcal{N}$ the set of nodes in the tree and $\mathcal{E}$ edges connecting these nodes.

INPUT: A PrT-net MN with $N = (S, T; F)$.
The initial marking $\widehat{M}_0$ of MN.

OUTPUT: A parameterized reachability tree $\mathcal{T}$ of MN with the root $\widehat{M}_0$.

GENERATETREE($MN, \widehat{M}_0$, **VAR** $\mathcal{T}$)

Let unprocessed nodes $\mathcal{U} := \{\widehat{M}_0\}; \mathcal{T} := \widehat{M}_0$;

[7] It is a slightly simplified version of the one presented in [10].

REPEAT
  Select some node $\widehat{M} \in \mathcal{U}$
  FORALL transitions t enabled at $\widehat{M}$ under some $\hat{\alpha}$
    according to Definition 3.4 **DO**
    Generate follower $\widehat{M}'$ of $\widehat{M}$ via $t\hat{\alpha}$ according to Definition 3.4
    Reduce the representation of $\widehat{M}'$ if possible
    Append $\widehat{M}'$ to $\mathcal{N}$ and $(\widehat{M}, \widehat{M}')$ labeled by $t\hat{\alpha}$ and $\varphi\hat{\alpha}$ into $\mathcal{E}$
    IF $\exists\ \widehat{M}'' \in \mathcal{N}$ such that $|\widehat{M}'(s)| = |\widehat{M}''(s)|\ \forall s \in S$
      THEN try to find a unifier $\hat{\beta}$ according to Theorem 2.11
      IF a unifier was found
        THEN $\widehat{M}' \subseteq \widehat{M}''$, add $\hat{\beta}$ to $\widehat{M}'$
        ELSE add $\widehat{M}'$ to $\mathcal{U}$
    ELSE add $\widehat{M}'$ to $\mathcal{U}$
  END FORALL
UNTIL $\mathcal{U} = \emptyset$.

We now have an algorithm to produce PR-trees.[8] It works in the way that was described with the example involving the net of Figure 5 and the PR-tree in Appendix 1. We are now able to produce reachability trees of PrT-nets such that the number of nodes in it is reduced. This reduction is achieved by using a folding such that one node in the tree represents not only one marking but a whole set of markings. It is clear that the number of nodes in a reachability tree of a PrT-net is strongly decreased using this method since one parameterized reachability tree may represent a whole set of instances of ordinary reachability trees.

But before concluding the presentation we should still discuss what PR-trees are good for. After all, the crucial point is how a PR-tree can be used to deduce properties of the net it was generated from. This will be briefly addressed in the following section.

4 How can PR-trees be used

In this section we will briefly state without further proof some properties of PR-trees and outline the method that can be used to deduce properties of the net they are derived from. For a detailed treatment with necessary theorems we refer to [10, pages 86–103].

First of all PR-trees are finite. One can also say a little about their size when compared to an underlying P/T-system [8]. If no selectors are involved the lower limit to the size of a PR-tree of some net is the size of the reachability tree of the underlying P/T-net. If selectors are involved it can be even lower. Not very much can be said about the upper limit without some rather strong assumptions. This concerns the order in which markings of the same cardinality are generated: Suppose that the first generated marking of a certain cardinality always includes all the markings of the same cardinality that are generated later. In this case the size of the PR-tree is at most the size of the reachability tree of the underlying P/T-system. [10, pages 86–89]

The central thing now to ask is what the relation of a PR-tree and an ordinary reachability tree of a PrT-net is. What we naturally want is for them to be in some sense equivalent in their information content. A theorem addressing this question is presented in the following.

Theorem 4.1 *Let $MN = (N, A, \widehat{M}_0)$ be a PrT-net with $N = (S, T; F)$ and $\mathcal{T} = (\mathcal{N}, \mathcal{E})$ a PR-tree produced by Algorithm 3.1 under the strict interpretation. Then*

[8]The reduction of the representation of a parameterized marking $\widehat{M}$ has not been discussed here. It has been studied in [10, pages 61–64]. However, the results concerning reduction are still somewhat initial.

$$\exists\pi = M_0[t_1\alpha_1\rangle M_1 \ldots M_{n-1}[t_n\alpha_n\rangle M_n$$
$$\Leftrightarrow$$
$$\exists\hat{\pi} = M_0[t_1\hat{\alpha}_1\rangle \widehat{M}_1 \ldots \widehat{M}_{n-1}[t_n\hat{\alpha}_n\rangle \widehat{M}_n$$
and a fixing $\beta : \widehat{V}(\hat{\pi}) \to D$ *such that*

1. $M_0, \widehat{M}_i \in \mathcal{N}, i = 1, \ldots, n,$
2. $\widehat{M}_i\beta = M_i, i = 1, \ldots, n$ *and*
3. $t_i\hat{\alpha}\beta = t_i\alpha, i = 1, \ldots, n.$

Proof:

See Theorem 5.8 in [10].

Even though the $\Leftarrow$ direction of the above theorem seems rather trivial (what is on the left side is more or less qiven on the right side) the other direction shows that all that is in an ordinary reachability tree is represented in a PR-tree. But the converse is not true: a PR-tree may in some cases represent something that does not have an origin in the ordinary case, i.e. sometimes a PR-tree may contain parts that do not have any meaningful interpretation. In such cases no legal fixing β can be found for the path.[9].

Even though PR-trees may contain unnecessary information in the context of a given initial marking they have the nice property that everything contained in an ordinary reachability tree is represented in the corresponding PR-tree. Therefore PR-trees can be used to decide the reachability between two markings.

PR-trees often describe the behavior of a system modelled as a net on a more abstract level than is desirable. Consequently a PR-tree can be seen as an intermediate point between the net and statements expressing the properties of the net. An ordinary tree takes us considerably closer than a PR-tree towards being able to state these properties. The information of a PR-tree must be elaborated further in parts which seem interesting. This allows us to focus our effort more accurately than with ordinary reachability trees and concentrate only on what seems to be of importance.

The essential information in a reachability tree is that it tells what the states of a system are and how they are reached from each other. All the other properties can be seen as consequences of this reachability relation. Therefore one should investigate how this reachability relation can be decided from PR-trees. This involves unfolding a part of the reachability tree by fixing the parameters such that the reachability relation between the particular members of the denotations of the parameterized markings is revealed. Two markings are then reachable if one can find them in the denotation of two parameterized markings which are reachable in the PR-tree. In addition the fixing of parameters involved must not make the path between the two markings unfeasible. As before, for details we refer to [10, pages 92–99].

Conclusions

In this presentation parameterized markings were introduced as a means for folding reachability trees of Predicate/Transition nets into significantly conciser parameterized reachability trees, PR-trees for short.

In the beginning parameterized markings and central points of the formalism were introduced. The core of the matter was to study the relationship between a parameterized marking as a syntactical

[9]For details see [10, pages 91–92]

object and its meaning or denotation. Including markings were introduced to allow for an ordering of parameterized markings according to their denotation. The dynamics of Predicate/Transition nets was then parameterized. The validity of the parameterized transition rule was shown by numerous examples and by a theorem concerning its denotation: the denotation of a parameterized transition occurrence is always non empty provided that the selector can be evaluated without fixing the parameters. It was also shown that the use of including markings in the coverability relation was justified. Finally, parameterized reachability trees were introduced and an algorithm for producing them was presented. An example of generating a PR-tree for a well-known net was presented in detail.

In the last section, the properties and the use of parameterized reachability trees were discussed. A PR-tree of a finite Pr/T-net is finite and the information it covers the information of the corresponding ordinary reachability tree. Therefore PR-trees can be used to decide the reachability of a marking.

The work presented here introduces a new method for generating reachability trees for Predicate/Transition nets. It does the reachability analysis on a 'high level' all the way, i.e. one does not necessarily have to consider the actual unfolded markings while generating the reachability tree. In addition it is a method that does not require any specific properties of the analyzed net to be effective in the reduction. The method can also be applied to other higher level net models once it is generalized to deal with the weak interpretation.

Although the reachability trees or PR-trees produced by this method can be significantly smaller in size than those produced by ordinary methods it is premature to make claims concerning its practical value. The results are still somewhat initial when it comes to using PR-trees in proving properties of the analyzed net. The basic framework has, however, been built here which opens a door into a new and little explored area of research. This, in turn, may offer several possibilities for speeding up reachability analysis so as to make it an attractive analysis method even with nets of larger size.

The PR-tree method should not be taken as self contained. Quite on the contrary; possibilities to profitably combine the PR-tree method with other reduction methods to improve reachability analysis should be looked for. Therefore the different reachability tree reduction methods should not be thought of as mutually exclusive but rather complementary.

Acknowledgements

I am indebted to Dr. Hartmann Genrich for his valuable advice and constructive criticism he has presented over the time the work with parameterized reachability trees was conducted.

A significant part of the work has been done under a grant from Academy of Finland.

References

[1] C. Chang, H. Keisler: *Model Theory.* North Holland, Amsterdam, (1973).

[2] C. Chang, R. Lee: *Symbolic Logic and Mechanical Theorem Proving.* Academic Press, New York, (1973).

[3] G. Chiola et al: *On Well-Formed Coloured Nets and their Symbolic Reachability Graph.* Proceedings of the Eleventh International Conference on Applications and Theory of Petri nets, Paris, June 1990.

[4] C. Dutheillet, S. Haddad: *Regular Stochastic Petri Nets.* Proceedings of the Tenth International Conference on Applications and Theory of Petri nets, Bonn, June 1989.

[5] H. Genrich: *Projections on C/E-systems.* Proceedings of Sixth European Workshop on Applications and Theory of Petri nets, Espoo, June 1985.

[6] H. Genrich: *Predicate/Transition nets.* Petri Nets: Central Models and Their Properties (eds. W. Brauer, W. Reisig, G. Rozenberg), Lecture Notes in Computer Science, Vol. 254, Springer Verlag, Berlin (1987).

[7] K. Jensen: *Coloured Petri Nets.* Petri Nets: Central Models and Their Properties (eds. W. Brauer, W. Reisig, G. Rozenberg), Lecture Notes in Computer Science, Vol. 254, Springer Verlag, Berlin (1987).

[8] P. Huber, A. Jensen, L. Jepsen, K. Jensen: *Reachability Trees for High Level nets.* Theoretical Computer Science, Nr. 45, (1986), pp. 261 - 292.

[9] R. Kujansuu, M. Lindqvist: *Efficient Algorithms for Computing S-invariants for Predicate/Transition Nets.* Proceedings of Fifth European Workshop on Applications and Theory of Petri Nets, Aarhus, Denmark, June 1984.

[10] M. Lindqvist: *Parameterized Reachability Trees for Predicate/transition Nets.* Acta Polytehcnica Scandinavica, Mathematics and Computer Science Series No. 54, Helsinki, 1989, 120 pp.

[11] W. Reisig: *Petri Nets, An Introduction.* Springer Verlag, Berlin, (1985).

[12] W. Reisig: *A Strong Part of Concurrency.* Advances in Petri Nets 1987 (ed. G. Rozenberg), Lecture Notes in Computer Science, Vol. 266, Springer Verlag, Berlin (1987).

[13] A. Valmari: Error Detection by Reduced Reachability Graph Generation. Proceedings of 9th European Workshop on Applications and Theory of Petri Nets, Venice, June 1988.

Appendix

Proof of Theorem 3.5:

Since $\hat{\alpha}$ is permissible a substitution $\hat{\beta} : \hat{V}(\widehat{M}) \cup \hat{V}(\hat{\alpha}) \rightarrow \hat{X}(\widehat{M})$ exists such that for all $s \in S$ $\widehat{M}^+(s) \geq \widehat{M}^-(s)$. Therefore $\widehat{M}^-(s)$ can be removed from the presentation of the markings and let $\widehat{M}''(s)$ be such that $\mathcal{V}(\widehat{M}''(s)) = \mathcal{V}(\widehat{M}^+(s) - \widehat{M}^-(s))$. Since $\neg\exists$ $\hat{\tau}$ such that $\widehat{M}^+(s)[\hat{\tau}] - \widehat{M}^-(s)[\hat{\tau}] > 1$ all $\widehat{M}''(s)$ are sets of parameterized tokens.

For all input places it holds that if $\tau \in A_F(s,t)$ then $\widehat{M}''(s)[\tau\hat{\alpha}\hat{\beta}] = 1$. Since $\hat{\alpha}\hat{\beta}$ is feasible there are no two $\tau_1, \tau_2 \in A_F(s,t)$ such that $\tau_1\hat{\alpha}\hat{\beta} = \tau_2\hat{\alpha}\hat{\beta}$. Therefore all $\tau_i\hat{\alpha}\hat{\beta}$ such that $\tau_i \in A_F(s,t), i = 1, \ldots n$ can be removed from $\widehat{M}''(s)$ and the resulting parameterized marking which is equivalent to $\widehat{M}'\hat{\beta}(s)$ is still a set of parameterized tokens.

For all output places it holds that if $\tau \in A_F(t,s)$ then $\widehat{M}''(s)[\tau\hat{\alpha}\hat{\beta}] = 0$. Since $\hat{\alpha}\hat{\beta}$ is feasible there are no two $\tau_1, \tau_2 \in A_F(t,s)$ such that $\tau_1\hat{\alpha}\hat{\beta} = \tau_2\hat{\alpha}\hat{\beta}$. Therefore all $\tau_i\hat{\alpha}\hat{\beta}$ such that $\tau_i \in A_F(t,s), i = 1, \ldots n$ can be added to $\widehat{M}''(s)$ and the resulting parameterized marking which is equivalent to $\widehat{M}'\hat{\beta}(s)$ is still a set of parameterized tokens.

Now we can conclude that for all $s \in S$ both $\widehat{M}\hat{\beta}(s)$ and $\widehat{M}'\hat{\beta}(s)$ are sets of parameterized tokens. Let $\beta = \hat{\beta}\beta'$ where β' is a fixing $\beta' : \hat{V}(\widehat{M}\hat{\beta}) \cup \hat{V}(\widehat{M}'\hat{\beta}) \rightarrow D$. Then $\widehat{M}\beta \in \mathcal{V}_S(\widehat{M})$ and $\widehat{M}'\beta \in \mathcal{V}_S(\widehat{M}')$. Also $\hat{\alpha}\beta$ is feasible and permissible according to Definitions 1.3 and 1.4 since we have assumed that there are no predicates or functions in the selector formula. Therefore

$$\exists \text{ a fixing } \beta \text{ such that } \widehat{M}\beta[t\hat{\alpha}\beta\rangle\widehat{M}'\beta$$

and $\widehat{M}\beta \in \mathcal{V}_S(\widehat{M})$ and $\widehat{M}'\beta \in \mathcal{V}_S(\widehat{M}')$. Consequently $\mathcal{V}(\widehat{M}[t\hat{\alpha}\rangle\widehat{M}')$ is non empty.

For the other direction: if $\mathcal{V}(\widehat{M}[t\hat{\alpha}\rangle\widehat{M}')$ is non empty then there exists a β such that $\widehat{M}\beta[t\hat{\alpha}\beta\rangle\widehat{M}'\beta$. Now, let $\hat{\beta} = \beta$ and the claim trivially follows even in the case when the selectors are not restricted in any way.

Proof of Theorem 3.6:

Since $M \in \mathcal{V}(\widehat{M})$ there must be a fixing β' such that $\widehat{M}\beta' = M$.

Suppose $\alpha = [v_1 \leftarrow d_1, \ldots, v_n \leftarrow d_n]$ and $\hat{\alpha} = [v_1 \leftarrow \hat{x}_1, \ldots, v_n \leftarrow \hat{x}_n]$. Then we can not find a fixing β'' such that $\hat{\alpha}\beta'' = \alpha$ if

1. $\hat{x}_i = d_k \in D$ and $d_k \neq d_i$ or
2. $\hat{x}_i = \hat{x}_j, i \neq j$ and $d_i \neq d_j$ where $i, j \in \{1, \ldots, n\}$.

But then from item 1.a of Definition 3.3 it would follow that either $M \notin \mathcal{V}(\widehat{M})$ or α is not permissible as defined by Definition 1.4. Since these are assumed we can conclude that $\exists \beta'' : \hat{\alpha}\beta'' = \alpha$. Since α is feasible and permissible it follows that $t\hat{\alpha}\beta''$ is enabled in $\widehat{M}\beta'$.

Next we have to show that there is no parameter $\hat{v}_i$ such that $\hat{v}_i\beta' \neq \hat{v}_i\beta''$. If $\hat{v}_i$ is a parameter such that $\hat{v}_i \notin \widehat{V}(\widehat{M})$ or $\hat{v}_i \notin \widehat{V}(\hat{\alpha})$ this naturally holds. But, if $\hat{v}_i = \hat{x}_i, i \in \{1, \ldots, n\}$, is not such a parameter then according to item 1.a of Definition 3.3

$$\begin{aligned} &\exists\, \tau \in A_F(s,t) \text{ for some } s \in S \text{ such that } \tau(j) = v_i, v_i \in V \text{ and} \\ &\forall\, \hat{\tau} \in \widehat{M}^+(s) \text{ such that } \hat{\tau} \simeq \tau\hat{\alpha} \Rightarrow \\ &\hat{\tau}(j) = \hat{x}_i. \end{aligned}$$

Suppose $\hat{x}_i\beta' \neq \hat{x}_i\beta''$. Then $t\alpha = t\hat{\alpha}\beta''$ could not be enabled in $M = \widehat{M}\beta'$ since there would be no token $\tau' \in M(s)$ such that $\tau' = \tau\alpha$: there is always a j such that $\tau'(j) \neq \tau\alpha(j)$. Since α is permissible it must be that $\hat{x}_i\beta' = \hat{x}_i\beta''$.

Now we have a $\beta = \beta'\beta''$ such that $\widehat{M}\beta = M$ and $t\hat{\alpha}\beta = t\alpha$. Since $M[t\alpha\rangle M'$ it must also be that $\widehat{M}\beta[t\hat{\alpha}\beta\rangle M'$. It still remains to show that $M' = \widehat{M}'\beta$. Since $\widehat{V}(\widehat{M}') \subseteq (\widehat{V}(\widehat{M}) \cup \widehat{V}(\alpha))$ this directly follows and thus

$$M[t\alpha\rangle M' \in \mathcal{V}(\widehat{M}[t\hat{\alpha}\rangle\widehat{M}').$$

Proof of Theorem 3.7:

According to Theorem 3.6

$$M \in \mathcal{V}(\widehat{M}) \Rightarrow M[t\alpha\rangle M_1 \in \mathcal{V}(\widehat{M}[t\hat{\alpha}\rangle\widehat{M}_1).$$

Now, $\widehat{M} \subseteq \widehat{M}'$. Therefore $M[t\alpha\rangle M_1 \in \mathcal{V}(\widehat{M}'[t\hat{\alpha}'\rangle\widehat{M}_1')$. From this we can conclude that

$$\mathcal{V}(\widehat{M}[t\hat{\alpha}\rangle\widehat{M}_1) \subseteq \mathcal{V}(\widehat{M}'[t\hat{\alpha}'\rangle\widehat{M}_1').$$

13.
On Well-Formed Coloured Nets and Their Symbolic Reachability Graph

G. Chiola, C. Dutheillet, G. Franceschinis and S. Haddad

11th Int. Conference on Applications and Theory of Petri Nets, Paris 1990

February 11, 1991

Abstract

The new class of Well Formed Coloured Nets (WN) is formally defined as an extension of Regular Nets (RN), together with an extended Symbolic Reachability Graph (SRG) construction algorithm. WNs allow the representation of any colour function in a structured form, so that they have the same modelling power as general coloured nets (CPN). In particular, with respect to RN, WNs allow the use of non-symmetric initial markings, of repeated occurrences of the same basic class in the Cartesian product defining the colours for transitions and places, and of the "constant" and "successor" functions as arc labels. The SRG allows colour symmetries to be exploited to reduce the space and time complexity of the analysis by reachability graph. The advantage of using WNs instead of unconstrained CPNs is that the detection of symmetries to construct the SRG is totally algorithmic, and requires no special heuristics.

1 Introduction

Regular Nets (RN) have been proposed in [Had87] as a restriction of Coloured Petri Nets (CPN) [Jen81]. The interest in introducing such a restriction on the colour domains and on the arc functions was that complete algorithms have been proposed for the computation of flows, reductions, and the definition of a Symbolic Reachability Graph (SRG) in this case. The computation of flows and the reduction techniques have been subsequently extended to more and more general subclasses of CPNs [HC88,CM89], and now the whole class of CPNs is covered by such extended algorithms [Had88,Cou90]. In this paper we propose a similar extension also for the computation of the SRG, until now working only for RNs.

Symbolic marking representations to exploit the symmetries of the reachability graph (RG) have already been proposed for general CPNs [HJJJ84,LM88,Car89] as well, but in all these cases heuristics were needed to decide the type of aggregations. These heuristics were based on an explicit knowledge of the symmetries present in a particular model. In [CBD88] an obvious algorithm was given to construct automatically the SRG, but it was based on the actual firing of all transition instances for each colour instantiation of the symbolic markings; hence only the space complexity problem was addressed, while the time complexity was at least as large as that of the non-symbolic RG construction algorithm, thus making its use impractical. From this point of view, the superiority of RNs was due to the availability of a general *symbolic firing rule* which allows the computation of the SRG without any actual instantiation of colours and without explicit knowledge of the symmetry of the model being studied.

Recently, stochastic models based on RNs called Regular Stochastic Petri Nets (RSPN) have been introduced for performance evaluation purposes [DH89b,DH89a]. The steady-state performance of an RSPN model can be obtained by numerically solving a Markov chain (MC) isomorphic to the SRG generated by the RSPN. The complexity in the computation of this solution is polynomial in the number of symbolic markings of the SRG, which can be much less than the number of ordinary markings generated by the Place/Transition net (P/T) resulting from the unfolding of the RN.

*This work has been done while G. Chiola was visiting researcher at the Lab. MASI of the University of Paris 6, with the financial support of a NATO-CNR annual research grant.

This use of the SRG for performance evaluation purposes is the main motivation that led us to the proposal of the extended SRG computation for more general classes of CPNs. The original RNs and their related SRG generation algorithm can be used to model many interesting systems, but unfortunately the strong restrictions imposed on the definition of the arc labelling functions, the object colour sets, and the initial marking, prevent their use in general cases. In [CF89], some semi-formal transformation rules have been introduced and exploited to transform "quasi-regular" CPN models into RN models, in order to exploit the SRG construction algorithm to be able to reduce the complexity of the performance evaluation. The possibility of using quasi-regular net models in which some of the restrictions of RNs were relaxed with respect to the original definition and yet that are able to exploit the same SRG construction algorithm proposed for RNs, indicated the possibility of extending the original definition.

In this paper we formally define the new class of well-formed coloured nets (WN) as extensions of RNs together with an extended SRG construction algorithm. The SRG defined for this class of models allows the same kind of performance evaluation presented in [DH89a] and [CF89] in the case of much more general systems. Moreover, this extended algorithm has the same advantages as the one originally proposed for RN, i.e.,

1. it uses a symbolic firing rule, so that both its time and space complexities depend only on the size of the SRG, and not on the size of the actual RG;

2. it does not require any special heuristics to explicitly define the symmetries of the model: it uses the information that is implicit in the well structured function and colour domain definitions.

From the modelling power point of view, any colour function can be broken down into the sum of the basic colour functions provided by the definition, possibly guarded by the allowed form of predicates. Thus any general CPN model can be translated into an equivalent WN model with the same underlying structure; only the expression of the colour functions and of the composition of colour classes is re-written in a more explicit (and parametric) form, in terms of the basic constructs provided by the WN formalism. Moreover, in practical modelling this formalism translation is hardly needed: most (if not all) CPN models published in the literature can be directly represented as WNs, even without exploiting the power of predicate guards on the arc labelling functions.

The balance of the paper is as follows. Section 2 summarizes the basic notation and outlines the SRG construction algorithm for RNs. Sections 3 presents the formal definition of WN, while Section 4 contains the extended algorithm for the computation of their SRG. Finally, Section 5 contains some concluding remarks and perspectives of this work. All the notation used throughout the paper and some more technical definitions are collected in the Appendix.

2 Regular Nets and their SRG

In order to understand the notation for the definition of WN and the algorithms for the construction and analysis of their SRG, we think it is essential to have a clear idea of the definition, notation and algorithms initially proposed for the more restricted case of RNs. Since most of these topics have been published only in French in the PhD thesis of one of the authors ([Had87]), we start by a revised and condensed but rigorous presentation of the state of the art in the case of RN.

First we present the underlying model (without time): the actual Regular Nets. Then we present the SRG and the original enumeration algorithm which will be modified in the following sections in order to handle the proposed extensions. Finally, we outline the definition as well as the analysis technique for the stochastic extension of the model.

2.1 RN definition and notation

A Regular Net is a CPN in which the colour domains of places and transitions are made of any Cartesian product of basic object classes, each class appearing no more than once in the product. A colour domain can be either a set of undistinguished resources if the product is null (neutral colour domain), or a set of objects if there is a single element in the product, or a set of object tuples if the product is made of several elements. We denote $C = \{C_1, \ldots, C_n\}$ the family of basic object domains. $C_J = \bigotimes_{i \in J} C_i$ is a standard colour domain, where $J \subset I = \{1, \ldots, n\}$.

As shown in [DH89a], every RN can be transformed into a *normalized* RN with equivalent behaviour, in which the domain of all places and transitions is C_I. Hence, we recall only the definition of normalized RNs. For more details, see [DH89a].

The colour functions are built from two kinds of basic colour functions, X_i and S_i. The former results in the selection of one object of C_i whose behaviour is independent of the other objects of the class when firing the transition. S_i corresponds either to a synchronization of all the objects in a class if it labels an input arc, or to a diffusion to all the objects in a class if it labels an output arc.

Definition 2.1 *The general form of a colour function is:* $\bigotimes_{i \in I}(\alpha_i.S_i + \beta_i.X_i)$
with $(\alpha_i.S_i + \beta_i.X_i) \; : \; C_i \times C_I \rightarrow I\!N, \quad (\alpha_i.S_i + \beta_i.X_i)(c'_i, \bigotimes_{i \in I} c_i) \; = \; (\textit{If } c_i = c'_i \textit{ then } \alpha_i + \beta_i \textit{ else } \alpha_i)$

Definition 2.2 *A normalized regular net* $RN = \langle P, T, C, W^-, W^+, \pi, M_0 \rangle$ *is defined by:*

P *the finite set of places,*

T *the finite set of transitions,* $P \cap T = \emptyset, P \cup T \neq \emptyset$,

C *the family of object classes :* $C = \{C_1, \ldots, C_n\}$, *with* $C_i \cap C_j = \emptyset$ *(we will denote* $I = \{1, \ldots, n\}$ *the set of indexes),*

$W^-, W^+ \; : \; W^-(p,t), W^+(p,t) \in [C_I \times C_I \rightarrow I\!N]$ *the input and output functions are colour functions as defined above,*

$\pi \; : \; T \rightarrow I\!N$ *the priority function.*

$M_0(p) \; : \; C_I \rightarrow I\!N$ *is the initial marking of the place p, with* $\forall c, c' \in C_I, M_0(p,c) = M_0(p,c')$ *(symmetry condition on the initial marking).*

Definition 2.3 *Firing rule. A transition t is enabled for a colour c in a marking M iff:*

i) $\forall p \in P, \forall c' \in C(p), W^-(p,t)(c',c) \leq M(p,c')$
which is the expression of the firing rule in a net without priorities

ii) $\forall t'$ *with* $\pi(t') > \pi(t), \forall c'' \in C(t'), \exists p \in P, \exists c' \in C(p), W^-(p,t')(c',c'') > M(p,c')$,
which means that no higher priority transition is enabled.

The firing of t for colour c leads to a new marking $M' = M[t,c\rangle$ *defined by:*
$\forall p \in P, \; \forall c' \in C(p), \quad M'(p,c') = M(p,c') + W^+(p,t)(c',c) - W^-(p,t)(c',c)$

The colour c is called the colour *instantiated* by the firing. As c is a product of basic objects, each component will also be said to be instantiated by the firing.

2.2 Symbolic reachability graph

The SRG is a graph whose nodes are classes of ordinary markings. These classes are called symbolic markings and can be built using a symbolic firing rule without developing the whole reachability graph.

2.2.1 Symbolic markings and their representation

Definition 2.4 *Let Eq be a relation between markings defined by M Eq M' if there exists a family of permutations* $(s_i)_{i \in I}$ *(where* s_i *denotes a permutation on* C_i*), such that if we apply the corresponding permutation on every colour component of any token of M, we obtain M':*
$\forall p \in P, \; \forall c_i \in C_i, \quad M\left(p, \bigotimes_{i \in I} s_i(c_i)\right) = M'\left(p, \bigotimes_{i \in I} c_i\right)$

Eq is an equivalence relation and the equivalence classes of the relation are called symbolic markings. We denote $\mathcal{M}$ the symbolic marking associated with the ordinary marking M.

The basic principle for representing a symbolic marking starting from the representation of ordinary markings consists in grouping into a subclass all the objects of a class that have the same marking (i.e., the marking is left unchanged when permuting two objects of the subclass). The identity of the objects in a subclass can be forgotten and only the number of objects is taken into account for each subclass. Thus we see that the representation of a symbolic marking requires only the above information.

Definition 2.5 *A representation of a symbolic marking $\mathcal{M}$ is a 3-tuple $\mathcal{R} = \langle m,\ card,\ mark\rangle$:*

- $m\ :\ I \rightarrow \mathbb{N}^+$, *such that $m(i)$ (which will be also denoted m_i) is the number of dynamic subclasses of C_i in $\mathcal{M}$. The set of dynamic subclasses of C_i is denoted :* $\hat{C}_i = \{Z_i^j \mid 0 < j \leq m_i\}$
- $card\ :\ \left(\bigcup_{i\in I} \hat{C}_i\right) \rightarrow \mathbb{N}$
- $mark\ :\ \left(\bigotimes_{i\in I} \hat{C}_i\right) \rightarrow [P \rightarrow \mathbb{N}]$
- $\forall M \in \mathcal{M},\ \exists \psi_i\ :\ C_i \rightarrow \hat{C}_i$ *such that*
  i) $|\psi_i^{-1}(Z_i^j)| = card(Z_i^j)$
  ii) $\forall c_i \in C_i,\ \forall p,\quad M\left(p, \bigotimes_{i\in I} c_i\right) = mark\left(\bigotimes_{i\in I} \psi_i(c_i)\right)(p)$

Notice that the definition of dynamic subclasses is local to a representation of a symbolic marking. Thus, for instance, the notation Z_i^j appearing in a representation $\mathcal{R}$ and the same notation Z_i^j appearing in another representation $\mathcal{R}'$ are not related. When a distinction is needed we use the notation $\mathcal{R}.Z_i^j$. The same kind of notation applies to m_i, *mark*, and *card* as well.

$\hat{C}_i$ can be interpreted as a representation of the set of all partitions of objects of C_i that group objects with the same marking together in some subset C_i^j that correspond to a possible instantiation of Z_i^j. Thus, $card(Z_i^j)$ represents the cardinality of any subset C_i^j that can instantiate Z_i^j, and $mark\left(\bigotimes_{i\in I} Z_i^j\right)$ represents the marking of any colour belonging to $\bigotimes_{i\in I} C_i^j$. Finally, $\psi_i^{-1}(Z_i^j)$ represents one of the possible instantiations of Z_i^j.

Using the above representation without further constraints, one can find many representations for a given symbolic marking $\mathcal{M}$. Hence we define the concepts of minimality and order to obtain a canonical representation.

Definition 2.6 *Let $\mathcal{M}$ be a symbolic marking. One of its representations $\mathcal{R}$ is minimal iff:* [1]

$$\forall i, j, k, \qquad j \neq k \implies mark_{sc}(Z_i^j) \neq mark_{sc}(Z_i^k)$$

The above definition implies that a representation is minimal when we use different dynamic subclasses only in cases where instantiation produces different markings.

The minimal representations of a symbolic marking is unique within a permutation $\bigotimes_{i\in I} s_i$ of the indexes of dynamic subclasses. However, it is possible to define and to compute a canonical representation for each symbolic marking by introducing an adequate ordering based on the definition of the marking of a n-product of subclasses [Had87].

Definition 2.7 *Let $\mathcal{M}$ be a symbolic marking. One of its representations $\mathcal{R}$ is ordered iff:* [2]
$\forall \bigotimes_{i\in I} s_i$, *where s_i is a permutation of the subclasses of $\hat{C}_i$ (we use the notation $Z_i^{s_i(u_i)}$ for $s_i(Z_i^{u_i})$)*

either $\quad \forall(u_1,\ldots,u_n),\ mark(\bigotimes_{i\in I} Z_i^{u_i}) = mark(\bigotimes_{i\in I} Z_i^{s_i(u_i)})$

or $\quad$ **both** $\exists(u_1,\ldots,u_n)$ *such that* $mark(\bigotimes_{i\in I} Z_i^{u_i}) < mark(\bigotimes_{i\in I} Z_i^{s_i(u_i)})$

$\quad$ **and** $\ \forall(v_1,\ldots,v_n) < (u_1,\ldots,u_n),\ mark(\bigotimes_{i\in I} Z_i^{v_i}) = mark(\bigotimes_{i\in I} Z_i^{s_i(v_i)})$

Proposition 2.1 *Let $\mathcal{M}$ be a symbolic marking. Then there is one and only one minimal and ordered representation of $\mathcal{M}$, which is called canonical.*

In the following, we will use $\mathcal{M}$ to denote both a symbolic marking and its canonical representation, unless an explicit distinction is needed.

[1] The formal definition of $mark_{sc}$ can be found in the Appendix. This marking is obtained from the symbolic representation by suppressing all the tuples in which the subclass Z_i^j does not appear. If Z_i^j appears in a tuple, then it is suppressed from the tuple.

[2] Note that the order relation for vectors is the lexicographic one.

2.2.2 Symbolic firing rule

In order to build the SRG directly (i.e., without building the RG and then grouping markings), we first define a symbolic firing rule on the symbolic marking representations. This rule must be sound, i.e. an ordinary marking enables a coloured transition if and only if its symbolic marking enables an equivalent symbolic firing, and the ordinary marking obtained by the firing belongs to the symbolic marking obtained by the symbolic firing.
In a symbolic firing, one dynamic subclass per class is instantiated. Thus we need an evaluation of colour functions for dynamic subclasses.

Definition 2.8 *Let $\mathcal{R}$ be a representation of a symbolic marking. Then $\bigotimes_{i\in I}(\alpha_i.S_i + \beta_i.X_i)$ is a function from $(\bigotimes_{i\in I}\hat{C}_i) \times (\bigotimes_{i\in I}\hat{C}_i) \to I\!N$ which has the same definition as the one given in 2.1 with dynamic subclasses replacing objects.*

The primary effect of the symbolic firing is to split each instantiated dynamic subclass in two subclasses, one representing the set of objects instantiated in the underlying firings and the other representing the remaining objects in the subclass.

Definition 2.9 *Splitting. Let $\mathcal{R}$ be a representation of a symbolic marking $\mathcal{M}$. Let $\mathcal{R}.Z_i^{u_i} \in \hat{C}_i$ for all $i \in I$. Then $\mathcal{R}_s = \mathcal{R}[\bigotimes_{i\in I} Z_i^{u_i}]$ is another representation of $\mathcal{M}$ defined by:*

- $\forall i \in I$, **if** $\mathcal{R}.card(\mathcal{R}.Z_i^{u_i}) > 1$
  then *a new dynamic subclass $\mathcal{R}_s.Z_i^k$ is created (where $k = \mathcal{R}_s.m(i) + 1$) with cardinality $\mathcal{R}_s.card(\mathcal{R}.Z_i^k) = \mathcal{R}.card(\mathcal{R}.Z_i^{u_i}) - 1$, and hence $\mathcal{R}_s.m(i) = \mathcal{R}.m(i) + 1$*
  else $\mathcal{R}_s.m(i) = \mathcal{R}.m(i)$.
- *$\forall i \in I$, $\mathcal{R}_s.card(\mathcal{R}_s.Z_i^{u_i})$ is set to 1.*
- *$\forall i \in I$, $\forall l$ such that $0 < l \leq \mathcal{R}.m(i)$ and $l \neq u_i$, $\mathcal{R}_s.card(\mathcal{R}_s.Z_i^l) = \mathcal{R}.card(\mathcal{R}.Z_i^l)$.*
- *$\mathcal{R}_s.mark$ is equal to $\mathcal{R}.mark$ for all common $\bigotimes_{i\in I} Z_i^j$, while in case $j = \mathcal{R}.m(i) + 1$ the value is obtained by substituting j by u_i and applying $\mathcal{R}.mark$ on this new product.*

Let $\mathcal{R}$ be a representation, t a transition, and $\bigotimes_{i\in I} Z_i^{u_i}$ a tuple of dynamic subclasses such that $Z_i^{u_i} \in \hat{C}_i$. Then $Z_i^{u_i}$ is the subclass instantiated in $\hat{C}_i$ for the (possible) firing of t.

Definition 2.10 *Transition t is enabled in $\mathcal{R}$ for $\bigotimes_{i\in I} Z_i^{u_i}$ iff:*

i) *The tokens required by the backward incidence functions of t applied to $\bigotimes_{i\in I} Z_i^{u_i}$ are all present in the split representation $\mathcal{R}[\bigotimes_{i\in I} Z_i^{u_i}]$*

ii) *There exists no t' with $\pi(t') > \pi(t)$ such that for any $\bigotimes_{i\in I} Z_i^{v_i}$ condition i) holds*

Note that condition ii) has never to be tested explicitly if we order the application of the test for condition i) by decreasing transition priority.

In the case where an instantiated dynamic subclass had cardinality greater than 1, the symbolic firing should be enabled not only for the object instantiated in the underlying firing but also for the other objects of the subclass which are not instantiated; thus the definition should be different. However, since we apply our definitions on split markings, this case never arises.

Symbolic firing algorithm : the canonical representation of the symbolic marking obtained by firing $(t, \bigotimes_{i\in I} Z_i^{u_i})$ in $\mathcal{M}$ (i.e., $\mathcal{M}' = \mathcal{M}[t, \bigotimes_{i\in I} Z_i^{u_i}\rangle$) is computed in four steps, that use different intermediate (non canonical) representations. Note that the presence of priorities affects only the enabling test, but does not add any complexity to the handling of symbolic marking representations.

1. let $\mathcal{R}_s = \mathcal{R}[\bigotimes_{i\in I} Z_i^{u_i}]$ be the split representation of $\mathcal{M}$ in which $(t, \bigotimes_{i\in I} Z_i^{u_i})$ is enabled;
2. Define $\mathcal{R}_f$ by copying from $\mathcal{R}_s$ the components m and $card$, and by computing $\mathcal{R}_f.mark$ by applying the incidence functions on $\mathcal{R}_s.mark$. $\mathcal{R}_f$ is a (possibly non canonical) representation of $\mathcal{M}'$;

3. Compute a minimal representation $\mathcal{R}_m$ of $\mathcal{M}'$ by grouping the representation $\mathcal{R}_f$ (using the information provided by $\mathcal{R}_f.mark_{sc}$); [3]

4. Compute the canonical representation of $\mathcal{M}'$ by transforming $\mathcal{R}_m$ into an ordered representation. This can be computed as follows in the case of two classes. If we have for instance $\hat{C}_1 = \{Z_1^1, \ldots, Z_1^{m_1}\}$ and $\hat{C}_2 = \{Z_2^1, \ldots, Z_2^{m_2}\}$, we build a matrix with rows indexed by C_1 and columns indexed by C_2, such that the entry (i,j) is $mark(Z_1^i, Z_2^j)$. Then we re-order rows and columns so that the matrix is lexicographically minimal. The case of more than two classes can be computed as the obvious extension to multidimensional matrices of the above schema.

2.2.3 SRG properties

The algorithm for constructing the SRG differs from the construction of the ordinary RG only in the marking representation, the firing rule, and the labels of the arcs. In the SRG, the latter are made of transitions and tuples of firing subclasses; in the RG instead we have transitions and tuples of firing objects.
Notice that the initial symbolic marking $\mathcal{M}_0$ contains only the ordinary initial marking M_0 because of the symmetry of the initial marking.

Many properties have been proven on the graph of symbolic markings, such as quasi-liveness, and the possibility of finding home states. However, here we are interested only in some properties that are useful for the proof of the algorithm that computes the symbolic marking probabilities for RSPNs. All the proofs of the propositions given here can be found in [Had87].

Property 2.1 *The reachability property is equivalent for the ordinary and the symbolic markings :*

$$M \in RG \iff \mathcal{M} \in SRG$$

(where, as usual, $\mathcal{M}$ is the symbolic marking containing M).

Property 2.2 *The strong connection is equivalent for the ordinary and the symbolic reachability graphs:*

$$RG \text{ is Strongly Connected} \iff SRG \text{ is Strongly Connected}$$

Property 2.3 *There is an exact relation between the arcs in the RG and the SRG :*
Let $\mathcal{M}$ and $\mathcal{M}'$ be two symbolic markings of the SRG.
Let A be the set of ordinary outgoing arcs from M to any $M' \in \mathcal{M}'$, and B the set of symbolic arcs connecting $\mathcal{M}$ to $\mathcal{M}'$.
Then there is an application mapping A on B such that the reciprocal image of a symbolic arc labelled by $(t, \bigotimes_{i \in I} Z_i^{j_i})$ is a set of arcs labelled by some $(t, \bigotimes_{i \in I} c_i^{k_i})$. The cardinality of this set of arcs is $\prod_{i=1}^{n} card(Z_i^{j_i})$.

2.3 Regular Stochastic Petri Nets

RSPNs are a timed extension of RNs in which transitions are either immediate, or have an exponentially distributed firing delay. To preserve the symmetry of the underlying model, we impose that all the colour instances of a transition have the same firing delay.

2.3.1 Definition

We denote $T_\pi = \{t \in T \mid \pi(t) = \pi\}$ the set of all priority-π transitions. $\overline{M}(p) = \sum_{c \in C(p)} M(p,c)$ is the total number of tokens in p, whatever the colour. Let M be a reachable marking of an RN. If M is not a dead marking, then there exists a unique π such that the set of the transitions enabled in marking M is included in $\{(t,c) \mid t \in T_\pi, c \in \bigotimes_{i \in I} C\}$. If $\pi = 0$ the stochastic behaviour of the net is entirely determined by the firing rates of the enabled transitions. If $\pi > 0$ the specification of the stochastic behaviour would require the definition of a switch table [AMBC84].

[3] In fact, only the split subclasses may be equivalent to previously existing ones in $\mathcal{R}$. This allows for an optimization of the grouping algorithm.

Here we give a definition of RSPN which is a slight simplification of the one given in [DH89a]. As proposed in [AMBCC87] for Generalized Stochastic Petri Nets, the resolution of conflicts among immediate transitions is obtained by giving a weight to each immediate transition rather than defining switch tables. In practice this restricted definition covers most of the interesting models and leads to a more efficient analysis algorithm. The probability of firing a transition of the enabled set is given by its weight divided by the sum of the weights in the enabled set.

An RSPN is defined by (RN, Λ), where RN is a regular net such that $\Lambda(t)(\overline{M})$ is the (possibly marking-dependent) firing rate (or weight) associated with the transition t. Implicit conditions on the model: all the colour instances of a transition have the same firing rate or the same weight, and the firing rates and weights depend only on $\overline{M}$ and not on M in a more general form.

2.3.2 Computation of the solution for RSPN

It has been shown in [DH89b] that the Markov chain derived from the SRG is isomorphic to a lumping of the Markov chain of the original process constructed on the RG. Moreover, all the states within a class have the same steady-state probability, so that the solution of the original process can be computed from the solution of the process derived from the SRG and from the knowledge of the number of actual markings represented by each symbolic marking.

Once the symbolic graph has been computed, we apply a method similar to the one in [AMBC84]. This method consists in solving an embedded Markov chain where the instant a transition occurs corresponds to a change of state. The solution of the process derived from the SRG is then obtained by a simple renormalization. Thus the steady-state solution of an RSPN model can be efficiently computed.

The first step of the algorithm consists in the computation of the transition probability matrix A^*. A^* is indexed on symbolic markings, and the entry $(\mathcal{M}, \mathcal{M}')$ is the probability of going from the symbolic marking $\mathcal{M}$ to the symbolic marking $\mathcal{M}'$, disregarding the notion of time. This coefficient is computed using the firing rates (or weights) of the transitions leading from $\mathcal{M}$ to $\mathcal{M}'$, and taking into account the cardinalities of the dynamic subclasses instantiated by the firing. The solution of the embedded Markov chain is obtained by solving the linear system $Y^*.A^* = 0$, Y^* being a probability vector. The state probabilites of the original process are obtained from Y^* [AMBC84]. More details on the algorithm can be found in [DH89a].

3 Well-Formed Coloured Nets

We start by defining the basic object classes, then we define the standard predicates, the colour functions, and the concept of state equivalence. Finally we give the formal definition of WNs, and we briefly discuss their modelling power as compared with other High-Level nets.

3.1 Object classes

The family of colour classes of a WN is defined by: $C = \{C_1, \ldots, C_h, C_{h+1}, \ldots, C_n\}$ where $0 \leq h \leq n$, $i \neq j \Rightarrow C_i \cap C_j = \emptyset$, and

- C_i, $0 < i \leq h$, are non ordered classes,
- C_i, $h < i \leq n$, are ordered classes, with $|C_i| > 1$. The successor of object c in class C_i is denoted $\oplus c$. The successor function associated with class C_i is defined by: $\forall c_j \in C_i$, $\bigcup_{k=1}^{|C_i|}\{\oplus^k c_j\} = C_i$

The partition of a colour class into static subclasses is defined by: $C_i = \bigcup_{q=1}^{n_i} D_{i,q}$ with $D_{i,q}$ predefined and disjoint sets of elements, i.e., $\forall 0 < i \leq n$, $\forall 0 < q \leq n_i$, $|D_{i,q}| > 0$ and $\forall 0 < r \leq n_i$, $q \neq r \Rightarrow D_{i,q} \cap D_{i,r} = \emptyset$

In an RN, all the objects belonging to the same class always behave in the same way. In a WN instead, a colour class can be partitioned into static subclasses, each static subclass being made up of all the objects that have the same behaviour; objects with different behaviour are thus allowed in the same class.

3.2 Colour domains

Let $I = \{1, \ldots, n\}$ the set of indexes of the object classes.
Let $J = \bigotimes_{i=1}^{n} e_i$ be a n-tuple of integers belonging to $Bag(I)$.
We denote $C_J = \bigotimes_{i=1}^{n} (C_i)^{e_i}$,[4] a Cartesian product of object domains. If $J = \vec{0}$ then by definition $C_{\vec{0}} = \{\varepsilon\}$, where ε is called the *neutral colour*. An object tuple $c_J \in C_J$ is denoted $\bigotimes_{i=1}^{n} \bigotimes_{j=1}^{e_i} c_i^j$.

Definition 3.1 *Let $r \in P \cup T$. The colour domain of r is defined by $C(r) = C_{J(r)}$, where $J(r)$ is a tuple of integers.*

3.3 Standard predicates

When firing a transition, the same colour class can be instantiated several times. It is then interesting to introduce some restriction on the possible firings by introducing predicates on the transitions. The predicates that we allow in our model are: $X_i^j = X_i^k$ (the same object is chosen for both instantiations), $X_i^j = \oplus X_i^k$ (the object chosen for the jth instantiation is the successor of the object chosen for the kth instantiation), $X_i^j \in D_{i,q}$ (the chosen object belongs to a particular static subclass), and combinations of those predicates using the $\wedge$, $\vee$, and $\neg$ logical operators.
The values of the predicates are given by: $\phi_J : C_J \rightarrow \{True, False\}$

- $(X_i^k = X_i^{k'})(c_J) = (c_i^k = c_i^{k'})$
- $(X_i^k = \oplus X_i^{k'})(c_J) = (c_i^k = \oplus c_i^{k'})$
- $(X_i^k \in D_{i,q})(c_J) = (c_i^k \in D_{i,q})$

and all the possible combinations with the operators $\wedge, \vee$, and $\neg$:

- $(\phi_J \vee \phi'_J)(c_J) = (\phi_J(c_J) \vee \phi'_J(c_J))$
- $(\phi_J \wedge \phi'_J)(c_J) = (\phi_J(c_J) \wedge \phi'_J(c_J))$
- $(\neg\phi_J)(c_J) = \neg(\phi_J(c_J))$

3.4 Colour functions

The definition of colour functions is given in three subsequent steps. First a set of *basic functions* is defined. Then the composition of basic colour functions and predicates is introduced in order to define *guarded functions*. Finally, *standard functions* labelling arcs are defined as a sum of guarded functions.

3.4.1 basic functions

Let J and J' be two n-tuples. In this section, we define the possible basic colour functions between a place p and a transition t, with $C(p) = C_J$ and $C(t) = C_{J'}$. The set of such functions is denoted $F_{J,J'}$. We can notice that $F_{J,J'}$ is a subset of $\{f : C_J \times C_{J'} \rightarrow I\!N\}$. We proceed by induction on the size of J.

case $J = \vec{0}$: $(C(p) = \{\varepsilon\})$

These functions are equivalent to valuated arcs in an ordinary net.

$F_{\vec{0},J'} = \{ \, \overline{\delta} \in [\{\varepsilon\} \times C_{J'} \rightarrow I\!N] \mid \forall c \in C_{J'},\ \overline{\delta}(\varepsilon, c) = \delta \, \}$.

In the rest of the paper, this will simply be denoted by δ.

case $J = \vec{1}_i$: $(C(p) = C_i)$

We first define three basic colour functions, the third one applying only to ordered classes.

[4] Meaning $C_J = \bigotimes_{i=1}^{n} \left(\bigotimes_{j=1}^{e_i} C_i \right)$. Note that if $e_i = 0$, C_i does not appear.

- The first function corresponds to a synchronization of all the objects in a static subclass if it labels an input arc, or to a diffusion to all the objects of the static subclass if it labels an output arc.
$\forall J', \forall\alpha \geq 0, \alpha.S_{i,q} \ : \ C_i \times C_{J'} \rightarrow I\!N$

$$\alpha.S_{i,q}(c_i', c) = (\text{ if } c_i' \in D_{i,q} \text{ then } \alpha \text{ else } 0\ )$$

where the subscript q denotes the selected static subclass.
- The second function allows one to select one object whose behaviour is independent of that of the other objects of the class when firing the transition.
$\forall J'$ such that $e_i \geq k$, $\forall\beta \geq 0, \beta.X_i^k \ : \ C_i \times C_{J'} \rightarrow I\!N$

$$\beta.X_i^k(c_i', \bigotimes_{j=1}^{n}\bigotimes_{l=1}^{e_j} c_j^l) \ = \ \beta.Id(c_i', c_i^k) \ = \ (\text{ if } c_i^k = c_i' \text{ then } \beta \text{ else } 0\ )$$

where the superscript k denotes that X_i^k is used to instantiate the kth occurrence of C_i in the colour domain of t. Note that, in particular, if $e_i = 0$ the identity function cannot be used.
- The third function is the successor function and can apply only to the case of ordered classes. Let $h < i \leq n$. $\forall J'$ such that $e_i \geq k$, $\forall\gamma \geq 0, \gamma.\oplus X_i^k \ : \ C_i \times C_{J'} \rightarrow I\!N$

$$\gamma.\oplus X_i^k(c_i', \bigotimes_{j=1}^{n}\bigotimes_{l=1}^{e_j} c_j^l) \ = \ \gamma.Id(c_i', \oplus c_i^k) \ = \ (\text{ if } \oplus c_i^k = c_i' \text{ then } \gamma \text{ else } 0\ )$$

Now we can define the sets of basic colour functions. When combining functions, the coefficients of the combination can be negative, provided that some conditions on $\alpha_{i,q}$, $\beta_{i,k}$, and $\gamma_{i,k}$ ensure that a non negative number of objects is selected in each static subclass. Non necessary but sufficient conditions are very easily checked:

if $0 < i \leq h$ **then**

$$F_{\vec{1}_i,J} \ = \ \left\{ \sum_{q=1}^{n_i} \alpha_{i,q}.S_{i,q} + \sum_{k=1}^{e_i} \beta_{i,k}.X_i^k \ \middle| \ \mathop{Inf}_{\substack{0<q\leq n_i \\ K\subseteq\{1,\ldots,e_i\}}} \left(\alpha_{i,q} + \sum_{k\in K}\beta_{i,k}\right) \geq 0 \right\}$$

if $h < i \leq n$ **then** $F_{\vec{1}_i,J}$ is the set of all $\sum_{q=1}^{n_i}\alpha_{i,q}.S_{i,q} + \sum_{k=1}^{e_i}\left(\beta_{i,k}.X_i^k + \gamma_{i,k}.\oplus X_i^k\right)$
such that $\mathop{Inf}_{\substack{0<q\leq n_i \\ K\subseteq\{1,\ldots,e_i\}}} (\alpha_{i,q} + \sum_{k\in K} Inf(\beta_{i,k}, \gamma_{i,k})) \geq 0$

We introduce the following simplifications in the notations.

- S_i is a shorthand for $\sum_{q=1}^{n_i} S_{i,q}$.
- X_i is a shorthand for X_i^1 when C_i is instantiated only once.
- X^k is a shorthand for X_1^k when there is only one colour class in the net.
- X is a shorthand for X_1^1 when there is a single colour class instantiated once.

case $J \neq \vec{0}$: (this case involves and generalizes case $J = \vec{1}_i$).

On each component of the product, the function behaves as an elementary one.

$$F_{J,J'} \ = \ \left\{ \bigotimes_{i=1}^{n}\bigotimes_{j=1}^{e_i} f_i^j \text{ such that } f_i^j \in F_{\vec{1}_i,J'} \right\}$$

where $\bigotimes_{i=1}^{n}\bigotimes_{j=1}^{e_i} f_i^j$ is a notation for: $\left(\bigotimes_{i=1}^{n}\bigotimes_{j=1}^{e_i} f_i^j\right)(c_J, c_{J'}) \ = \ \prod_{i=1}^{n}\prod_{j=1}^{e_i}\left(f_i^j(c_i^j, c_{J'})\right)$
Notice that neither X_i^k nor $\oplus X_i^k$ can appear if $J'(i) < k$. This due to the fact that the firing of a transition cannot instantiate k objects in C_i if this transition does not include at least k occurrences of C_i in its colour domain. In particular, $\forall i$ such that $J'(i) = 0$, only the $\alpha_{i,q}.S_{i,q}$ functions can be used.

3.4.2 guarded functions

Let J and J' be two n-tuples. In this section, we define the possible guarded colour functions between a place p and a transition t, with $C(p) = C_J$ and $C(t) = C_{J'}$.

Definition 3.2 *A guarded function g between the colour domains C_J and $C_{J'}$ can have the following forms:* $g : C_J \times C_{J'} \rightarrow \mathbb{N} \quad g = [\phi_{J'}]\, f$
where $f \in F_{J,J'}$ is a basic function, and $\phi_{J'} : C_{J'} \rightarrow \{True, False\}$ is a standard predicate.

The value of a guarded function $g = [\phi_{J'}]\, f$ is defined as follows:

$$\forall c_J \in C_J, \forall c_{J'} \in C_{J'} \quad g(c_J, c_{J'}) = (\text{ if } \phi_{J'}(c_{J'}) \text{ then } f(c_J, c_{J'}) \text{ else } 0 \text{ })$$

Note that, in particular, $\phi_{J'}$ can be the standard predicate $True$, so that a guarded function may be reduced to a basic function.

3.4.3 standard functions

Definition 3.3 *A standard function labelling an arc connecting a transition t and a place p is defined as* $W^*(p,t) = \sum_l g_l$ *where* $* \in \{+,-\}$, *and* $g_l : C(p) \times C(t) \rightarrow \mathbb{N}$ *are guarded functions.*

3.5 State equivalence

The introduction of a new basic function and of the partition of classes into static subclasses requires that we modify the state equivalence relation in order to preserve the soundness of the aggregation.

Definition 3.4 *Let $\xi = \{ s = \langle s_1, \ldots, s_h, s_{h+1}, \ldots, s_n \rangle \}$ (where, as usual, s will be also denoted $\bigotimes_{i=1}^{n} s_i$) be a subgroup of the permutations on $\bigotimes_{i=1}^{n} C_i$ such that:*

- *$\forall 0 < i \leq h$ s_i is a permutation on C_i such that $\forall D_{i,q}$, $s_i(D_{i,q}) = D_{i,q}$;*
- *$\forall h < i \leq n$ s_i is a rotation on C_i such that $\forall D_{i,q}$, $s_i(D_{i,q}) = D_{i,q}$. Note that this condition implies that if $n_i > 1$ then the only allowed rotation s_i is the identity.*

Let $c \in C_J$, $s \in \xi$. Then $s(c)$ is defined by : $s\left(\bigotimes_{i=1}^{n} \bigotimes_{j=1}^{e_i} c_i^j\right) = \bigotimes_{i=1}^{n} \bigotimes_{j=1}^{e_i} s_i\left(c_i^j\right)$

Definition 3.5 *Let M be a marking, and $s \in \xi$ a permutation. Then $s.M$ is a marking defined by:*

$$\forall p \in P,\ \forall c \in C(p), \quad s.M(p,c) = M(p, s(c))$$

Thus, if we apply the permutation s on every colour component of any token of M, we obtain M':

$$\forall p \in P,\ \forall c_i^j \in C_J, \quad M\left(p, \bigotimes_{i=1}^{n} \bigotimes_{j=1}^{e_i} s_i(c_i^j)\right) = M'\left(p, \bigotimes_{i=1}^{n} \bigotimes_{j=1}^{e_i} c_i^j\right)$$

Definition 3.6 *A symbolic marking $\mathcal{M}$ is an equivalence class of the relation Eq defined by:*

$$M \ Eq \ M' \iff \exists s \in \xi,\ M' = s.M$$

3.6 WN formal definition

Definition 3.7 *A well formed net* $WN = \langle P, T, C, J, W^-, W^+, \Phi, \pi, \mathcal{M}_0 \rangle$ *is composed of:*

P the finite set of places;

T the finite set of transitions, $P \cap T = \emptyset$, $P \cup T \neq \emptyset$;

C the family of object classes: $C = \{C_1, \ldots, C_n\}$, with $C_i \cap C_j = \emptyset$ (we denote $I = \{1, \ldots, n\}$ the ordered set of indexes), C_i is partitionned in static subclasses: $C_i = \bigcup_{q=1}^{n_i} D_{i,q}$;

$J : P \cup T \rightarrow Bag(I)$, where $Bag(I)$ is the multiset on I. $C(r) = C_{J(r)}$ denotes the domain of node r;

$W^-, W^+ : W^-(p,t), W^+(p,t) \in [C_{J(p)} \times C_{J(t)} \rightarrow \mathbb{N}]$ the input and output standard functions;

$\Phi(t) : C_{J(t)} \rightarrow \{True, False\}$ *is a standard predicate associated with the transition* t*. By default we will assume* $\forall t \in T$ *the standard predicate* $\Phi(t) = True$*;*

$\pi : T \rightarrow \mathbb{N}$ *the priority function. By default we will assume* $\forall t \in T$ *the value* $\pi(t) = 0$*;*

$\mathcal{M}_0$ *is the symbolic initial marking representing the equivalence class of initial markings.*

Definition 3.8 *Firing rule. A transition* t *is enabled for a colour* c *in a marking* M *iff:*

i) $\forall p \in P$, $\forall c' \in C(p)$, $W^-(p,t)(c',c) \leq M(p,c')$, *and* $\Phi(t)(c)$
which is the expression of the firing rule in a net without priorities,

ii) $\forall t'$ *with* $\pi(t') > \pi(t)$, $\forall c'' \in C(t')$, $\exists p \in P$ *and* $\exists c' \in C(p)$ *such that either* $W^-(p,t')(c',c'') > M(p,c')$ *or* $\neg\Phi(t')(c'')$
which means that no higher priority transition is enabled.

The firing of t *for colour* c *leads to a new marking* $M' = M[t,c\rangle$ *defined by:*

$$\forall p \in P,\ \forall c' \in C(p), \quad M'(p,c') = M(p,c') + W^+(p,t)(c',c) - W^-(p,t)(c',c)$$

The colour c is called the colour *instantiated* by the firing. As c is a product of basic objects, each component will also be said to be instantiated by the firing.

Property 3.1 *The firing property is preserved by applying a permutation both on the markings and on the transition.* $\forall M$ *ordinary marking,* $\forall t \in T$, $\forall c \in C(t)$, $\forall s \in \xi$, $\quad M[t,c\rangle M' \iff s.M[t,s(c)\rangle s.M'$

3.7 Modelling power of WN

Proposition 3.1 *Any CPN can be transformed into a WN with the same basic structure, same colour domains (possibly partitioned in static subclasses), equivalent arc labelling, and the possible addition of transition predicates.*

The above proposition, which is not proved here for the sake of brevity, implies that WNs have a modelling power at least as large as general (unconstrained) CPNs [Jen81], in the sense that they can represent the same model with the same level of conciseness. In fact CPNs were initially proposed as a sort of short-hand notation for the concise representation of large and symmetric P/T systems. WN, even though imposing restrictions on the form of arc labelling functions, achieve the same goal.

On the other hand, a short-hand notation is not the only advantage that one can exploit with a High-level Net formalism. Another (more important) aspect of the modelling power of a High-level formalism is its *parametrization power*. The idea of parametric Petri net models representing classes of similar systems instead of a single system was first proposed by Genrich [Gen87,Gen88], in the framework of Predicate/Transition nets (Pr/T) [GL81]. Indeed, the definition of arc labels and transition predicates in WN is completely independent of the cardinalities of the basic object (sub)classes, which is not the case for CPNs. In this sense, WN are more similar to Pr/T than to CPNs, and in fact have a degree of modelling parametrization that CPNs have not.

Unfortunately, this modelling parametrization is very difficult to exploit in order to parametrize the construction of the SRG as a function of the cardinalities of the basic object (sub)classes. On the other hand, this kind of parametric results can be obtained in the case of flow analysis and structural reuctions, which are beyond the scope of this paper.

4 SRG

We now present the extension to the case of WNs of the SRG construction algorithm originally proposed for RNs. To do this, we first define a canonical representation of the symbolic markings and then present the symbolic firing rule that allows one to compute the symbolic markings $\mathcal{M}'$ reachable from a given symbolic marking $\mathcal{M}$ without instantiating the dynamic subclasses of $\mathcal{M}$.

4.1 Representation of a symbolic marking

As in the case of Regular nets, we give a formal definition of a *representation* of a symbolic marking, and we introduce the concept of *canonical representation*. Then we give an outline of the algorithm for the computation of the canonical representation of a symbolic marking $\mathcal{M}$ starting from an arbitrary representation $\mathcal{R}$.

4.1.1 Canonical Representation

Definition 4.1 *A representation of $\mathcal{M}$ is a 4-tuple $\mathcal{R} = \langle m,\ card,\ d,\ mark\rangle$ verifying :*

- $m : I \to I\!N^+$, *such that $m(i)$ (which will be also denoted m_i) is the number of dynamic subclasses of C_i in $\mathcal{M}$. The set of dynamic subclasses of C_i is denoted:* $\hat{C}_i = \{Z_i^j \mid 0 < j \leq m_i\}$
  In the following we will extend to $\hat{C}_i$ the notation introduced for C_i, e.g., $\hat{C}_J$, $\hat{\xi}$, etc.
- $card : \left(\bigcup_{i\in I} \hat{C}_i\right) \to I\!N$
- $d : \left(\bigcup_{i\in I} \hat{C}_i\right) \to I\!N$ *such that $d(Z_i^j)$ is the index q of a static subclass $D_{i,q}$*
- $mark : P \to [\hat{C}_{J(P)} \to I\!N]$
- $\forall M \in \mathcal{M},\ \exists \psi_i : C_i \to \hat{C}_i$ *such that*
  i) $|\psi_i^{-1}(Z_i^j)| = card(Z_i^j)$
  ii) $\forall i \in I$, $\exists D_{i,q}$ *such that* $\psi_i^{-1}(Z_i^j) \subseteq D_{i,q}$ *and* $q = d(Z_i^j)$
  iii) $\forall c_i^j \in C_J,\ \forall p,\quad M\left(p, \bigotimes_{i=1}^{n} \bigotimes_{j=1}^{e_i} c_i^j\right) = mark(p)\left(\bigotimes_{i=1}^{n} \bigotimes_{j=1}^{e_i} \psi_i(c_i^j)\right)$
  iv) $\forall Z_i^j$ *such that* $i > h$, $\exists c \in \psi_i^{-1}(Z_i^j)$ *such that*
  $\oplus c \in \psi_i^{-1}(Z_i^{\oplus j}) \quad \wedge \quad \forall c' \in \psi_i^{-1}(Z_i^j),\ c' \neq c,\ \oplus c' \in \psi_i^{-1}(Z_i^j)$

Using the above representation without further constraints, one can find many representations for a given symbolic marking $\mathcal{M}$.

Definition 4.2 *Let $\mathcal{M}$ be a symbolic marking. One of its representations $\mathcal{R}$ is minimal iff:* [5]

$$\forall i \leq h,\ \forall j,k, \qquad j \neq k \implies \left(mark_{sp}(Z_i^j) \neq mark_{sp}(Z_i^k)\right) \vee \left(d(Z_i^j) \neq d(Z_i^k)\right)$$

$$\forall i > h,\ \forall j,k, \qquad k = \oplus j \implies \left(mark_{sp}(Z_i^j) \neq mark_{sp}(Z_i^k)\right) \vee \left(d(Z_i^j) \neq d(Z_i^k)\right)$$

The minimal representations of a symbolic marking is unique within a permutation $s \in \hat{\xi}$ of the indexes of dynamic subclasses. However, it is possible to define and to compute a canonical representation for each symbolic marking by introducing an adequate ordering, similar to the one used for RN.

Definition 4.3 *Let* $\hat{\xi} = \{\ s = \langle s_1, \ldots, s_h,\ s_{h+1}, \ldots,\ s_n\rangle\ \}$ *(where, as usual, s will be also denoted $\bigotimes_{i=1}^{n} s_i$) be the set of permutations on $\bigotimes_{i=1}^{n} \hat{C}_i$ such that:*

- $\forall 0 < i \leq h$ *s_i is a permutation on $\hat{C}_i$ such that* $\forall Z_i^j \in \hat{C}_i\ :\ d(s_i(Z_i^j)) = d(Z_i^j)$;
- $\forall h < i \leq n$ *s_i is a rotation on $\hat{C}_i$ such that* $\forall Z_i^j \in \hat{C}_i\ :\ d(s_i(Z_i^j)) = d(Z_i^j)$.

Definition 4.4 *Let $\mathcal{M}$ be a symbolic marking. One of its representations $\mathcal{R}$ is ordered iff:*[6]

i) $\forall i,\ w(i)\ <\ w'(i) \implies d(Z_i^{w(i)}) \leq d(Z_i^{w'(i)})$

ii) $\forall s \in \hat{\xi}$,

[5]The formal definition of $mark_{sp}$ can be found in the Appendix. Its meaning can be intuitively explained as follows. The application of the function to an arbitrary tuple of basic objects returns all possible projections (corresponding to different positions of the elements of the argument tuple in the Cartesian product) over all remaining dimensions (i.e., the dimensions of the Cartesian product not involved in the argument tuple) of the marking.

[6]As usual, the order relation for vectors is the lexicographic one.

either $\forall \bigotimes_{i=1}^{n} u_i,\ mark_{sp}(\bigotimes_{i=1}^{n} Z_i^{u_i}) = mark_{sp}(\bigotimes_{i=1}^{n} Z_i^{s_i(u_i)})$

or both $\exists \bigotimes_{i=1}^{n} u_i$ such that $mark_{sp}(\bigotimes_{i=1}^{n} Z_i^{u_i}) < mark_{sp}(\bigotimes_{i=1}^{n} Z_i^{s_i(u_i)})$

and $\forall \bigotimes_{i=1}^{n} v_i < \bigotimes_{i=1}^{n} u_i,\quad mark_{sp}(\bigotimes_{i=1}^{n} Z_i^{v_i}) = mark_{sp}(\bigotimes_{i=1}^{n} Z_i^{s_i(v_i)})$

Proposition 4.1 *Let $\mathcal{M}$ be a symbolic marking. Then there is one and only one minimal and ordered representation of $\mathcal{M}$, which is called canonical.*

In the following, we will use $\mathcal{M}$ to denote both a symbolic marking and its canonical representation, unless an explicit distinction is needed.

4.1.2 Computation of the canonical representation

Here we define the algorithm for the computation of the canonical representation of any symbolic marking. This computation is organized in two steps: the grouping into a minimal representation, and the ordering of the subclasses.

Algorithm M: Minimality. Given a representation $\mathcal{R}$, one can compute a minimal representation $\mathcal{R}_m$ of the same symbolic marking $\mathcal{M}$ as follows.

```
Rm := R;
for i := 1 to h do (* non-odered classes *)
   for k := R.m(i) downto 2 do
      j := 1;
      while j < k do
         if R.mark_sp(R.Z_i^j) = R.mark_sp(R.Z_i^k) ∧ R.d(R.Z_i^j) = R.d(R.Z_i^k)
         then  (* include Z_i^k into Z_i^j *)
            Rm.card(Rm.Z_i^j) := Rm.card(Rm.Z_i^j) + Rm.card(Rm.Z_i^k);
            for l := k + 1 to Rm.m(i) do (* shift down higher subclasses *)
               Rm.card(Rm.Z_i^{l-1}) := Rm.card(Rm.Z_i^l);
               Rm.d(Rm.Z_i^{l-1}) := Rm.d(Rm.Z_i^l);
               Rm.mark(Rm.Z_i^{l-1}) := Rm.mark(Rm.Z_i^l)
            od;
            Rm.m(i) := Rm.m(i) - 1;
            j := k (* exit from while *)
         fi;
         j := j + 1
      od
   od
od;
for i := h + 1 to n do (* ordered classes *)
   for j := R.m(i) downto 2 do
      if R.mark_sp(R.Z_i^{j-1}) = R.mark_sp(R.Z_i^j) ∧ R.d(R.Z_i^{j-1}) = R.d(R.Z_i^j)
      then (* include Z_i^j into Z_i^{j-1} *)
         Rm.card(Rm.Z_i^{j-1}) := Rm.card(Rm.Z_i^{j-1}) + Rm.card(Rm.Z_i^j);
         for k := j + 1 to Rm.m(i) do (* shift down higher subclasses *)
            Rm.card(Rm.Z_i^{k-1}) := Rm.card(Rm.Z_i^k);
            Rm.d(Rm.Z_i^{k-1}) := Rm.d(Rm.Z_i^k);
            Rm.mark(Rm.Z_i^{k-1}) := Rm.mark(Rm.Z_i^k)
         od;
         Rm.m(i) := Rm.m(i) - 1
      fi
   od;
   if Rm.m(i) > 2
   then (* according to ⊕, Z_i^1 and Z_i^{m_i} are adjacent as well *)
```

```
        j := R_m.m(i);
        if R_m.mark_sp(R_m.Z_i^1) = R_m.mark_sp(R_m.Z_i^j) ∧ R_m.d(R_m.Z_i^1) = R_m.d(R_m.Z_i^j)
        then (* include Z_i^{m_i} into Z_i^1 *)
            R_m.card(R_m.Z_i^1) := R_m.card(R_m.Z_i^1) + R_m.card(R_m.Z_i^j);
            R_m.m(i) := R_m.m(i) - 1
        fi
    fi
od.
```

Algorithm O_1: Ordering. Given a minimal representation $\mathcal{R}$, that satisfies the condition of Definition 4.4.i one can compute an ordered representation $\mathcal{R}_o$ of the same symbolic marking $\mathcal{M}$ by applying a permutation $s \in S$ to the dynamic subclasses of $\mathcal{R}$, where the set S is computed by the following algorithm.

```
S := ξ̂;
if |S| > 1
then  (* restrict S *)
    ⊗_{i=1}^n j_i := ⊗_{i=1}^n 1; (* initialize the loop *)
    while ⊗_{i=1}^n j_i ≠ ⊗_{i=1}^n R.m(i) do
        S := { s ∈ S | R.mark_sp(⊗_{i=1}^n Z_i^{s_i(j_i)}) is lexicographically minimal };
        if |S| > 1
        then (* lexicographic increment of ⊗_{i=1}^n j_i *)
            k := n;
            while j_k = R.m(k) do
                j_k := 1;
                k := k - 1
            od;
            j_k := j_k + 1
        else (* exit from loop *)
            ⊗_{i=1}^n j_i := ⊗_{i=1}^n R.m(i)
        fi
    od
fi.
```

Here the ordering is achieved in terms of minimality of the lexicographic order relation. Note that if the representation $\mathcal{R}$ is minimal, then the cardinality of the set S depends on the symbolic marking, but not on the chosen representation. This cardinality will be denoted $K(\mathcal{M})$.

4.2 Computation of $\mathcal{M}_0$

The canonical representation $\mathcal{M}_0$ of $\mathcal{R}_0$ is computed in two steps:

1. Compute a minimal representation $\mathcal{R}_m$ by applying Algorithm M of Section 4.1.2 to $\mathcal{R}_0$;
2. Compute the canonical representation $\mathcal{M}_0$ by applying the following Algorithm O_0 to $\mathcal{R}_0$.

Algorithm O_0: Computation of $\mathcal{M}_0$. Given a minimal representation $\mathcal{R}_m$, one can compute an ordered representation $\mathcal{R}_o$ of the same simbolic marking $\mathcal{M}$ by first sorting the dynamic subclasses in order to satisfy the condition of Definition 4.4.i and then applying Algorithm O_1 to the resulting intermediate representation $\mathcal{R}_d$.

$\mathcal{R}_d$:= sort_dyn_sc($\mathcal{R}_m$) such that $\forall i,\ w(i) < w'(i) \implies \mathcal{R}_d.d(\mathcal{R}_d.Z_i^{w(i)}) \leq \mathcal{R}_d.d(\mathcal{R}_d.Z_i^{w'(i)})$;
$\mathcal{R}_o$:= $s(\mathcal{R}_d)$, where $s \in S$ as computed by Algorithm O_1 on $\mathcal{R}_d$.

The difference between Algorithm O_0 and Algorithm O_1 is that the latter assumes that the representation $\mathcal{R}$ given as input already satisfies the condition of Definition 4.4.i. Indeed, this new version of the algorithm is needed only once to compute the canonical representation $\mathcal{M}_0$ of the initial marking $\mathcal{R}_0$ since by construction the firing rule defined in Section 4.3 preserves condition 4.4.i.

4.3 Symbolic firing rule

In order to build the SRG directly as in the case of RN (i.e., without building the RG and then grouping markings), we first define a symbolic firing rule on the symbolic marking representations. In a symbolic firing, dynamic subclasses are instantiated. Thus we need an evaluation of colour functions and predicates for dynamic subclasses.

Definition 4.5 *Let $\mathcal{R}$ be a representation of a symbolic marking.*
Then $W^*(p,t)$ is a standard function from $\hat{C}_{J(p)} \times \hat{C}_{J(t)} \to I\!N$ which has the same definition as the one given in 3.3 with dynamic subclasses replacing objects.
Similarly, $\Phi(t)$ is a standard predicate from $\hat{C}_{J(t)} \to \{True, False\}$ which has the same definition as the one given in Section 3.3 with dynamic subclasses replacing objects.

4.3.1 Splitting

Definition 4.6 *Symbolic instantiation. Let t be a transition with $C(t) = \bigotimes_{i=1}^{n} \bigotimes_{j=1}^{e_i} C_i$. Let $\mathcal{R}$ be a symbolic representation. Then $[\lambda, \mu]$ is an instantiation of $\hat{C}(t)$ for $\mathcal{R}$ defined by :*
$\lambda = \{\lambda_i : \{1, \ldots, e_i\} \to I\!N^+\}, \qquad \mu = \{\mu_i : \{1, \ldots, e_i\} \to I\!N^+\},$
such that $\forall i \in I$, $\forall 0 < x \leq e_i$,

- $\lambda_i(x) \leq \mathcal{R}.m(i)$,
- $\mu_i(x) \leq \mathcal{R}.card(\mathcal{R}.Z_i^{\lambda_i(x)})$
- *if* $i \leq h$ *then* $\forall 0 < k < \mu_i(x)$, $\exists x' < x$ *such that* $\lambda_i(x') = \lambda_i(x) \wedge \mu_i(x') = k$.

$\lambda_i(x)$ is used to choose the subclasses of $\hat{C}_i$ to be instantiated. In case of non ordered classes C_i, $\mu_i(x)$ is used to distinguish already instantiated elements from the other elements within the subclass selected by $\lambda_i(x)$; the additional conditions on $\mu_i(x)$ guarantees that the functions λ and μ define a partition of the ordinary arcs in symbolic arcs (an ordinary arc cannot satisfy two different pairs $[\lambda, \mu]$ simultaneously). We denote $\mu_i^j = \sup(\mu_i(x) \mid \lambda_i(x) = j)$ the number of different instantiations in the dynamic subclass Z_i^j. For any Z_i^j that is not instantiated (i.e.: $\nexists x$ such that $\lambda_i(x) = j$), then $\mu_i^j = 0$ by definition.

We now extend the notion of *split symbolic marking*. In each subclass, a separation is made between the objects that will be selected for the firing and the other objects. For ordered classes we always split the classes into dynamic subclasses of cardinality 1.

Definition 4.7 *Splitting. Let $\mathcal{R}$ be a representation of a symbolic marking $\mathcal{M}$.*
Then $\mathcal{R}_s = \mathcal{R}[\lambda, \mu]$ is another representation of $\mathcal{M}$ defined by the following transformations on $\mathcal{R}$:
$\mathcal{R}_s.\hat{C}_i = \{Z_i^{j,k}\}$ with $Z_i^{j,k}$ such that:

1. $Z_i^j \in \mathcal{R}.\hat{C}_i$
2. **if** $i \leq h$ *(C_i is not ordered)*, **then**
   either $\exists x$ *such that* $\langle \lambda_i(x), \mu_i(x) \rangle = \langle j, k \rangle$ *(subclasses of instantiated objects)*
   or $\quad k = 0 \wedge \mu_i^j < \mathcal{R}.card(Z_i^j)$ *(subclass of remaining objects)*
3. **if** $i > h$ *(C_i is ordered)*, **then** $0 < k \leq \mathcal{R}.card(Z_i^j)$
4. $\mathcal{R}_s.card(Z_i^{j,k}) = ($ *if* $(k = 0)$ *then* $(\mathcal{R}.card(Z_i^j) - \mu_i^j)$ *else* 1 $)$.
5. $\forall u, v, \quad \mathcal{R}_s.mark\left(p, \bigotimes_{i=1}^{n} \bigotimes_{e=1}^{e_i} Z_i^{u(i,e),v(i,e)}\right) = \mathcal{R}.mark\left(p, \bigotimes_{i=1}^{n} \bigotimes_{e=1}^{e_i} Z_i^{u(i,e)}\right)$
6. $\mathcal{R}_s.d(Z_i^{j,k}) = \mathcal{R}.d(Z_i^j)$

With such a definition of the split marking, subclasses can substitute objects in the transition firing.

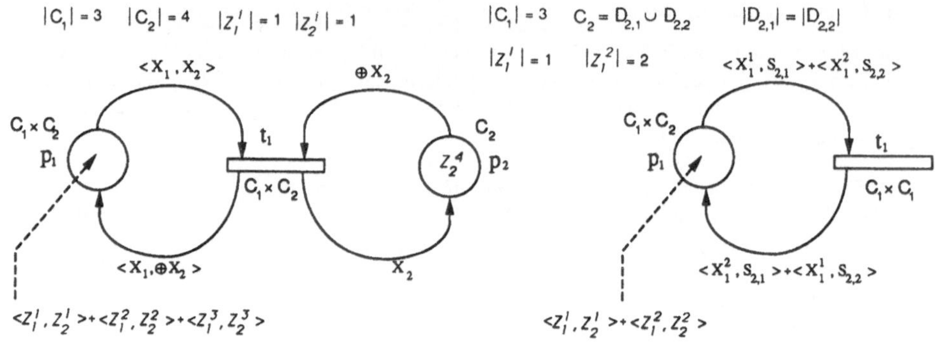

Figure 1: Two very simple WNs.

4.3.2 Enabling

Definition 4.8 *Transition t is enabled in $\mathcal{R}$ for $[\lambda, \mu]$ iff:*

i) **both** *the predicate* $\Phi(t, \bigotimes_{i=1}^{n} \bigotimes_{j=1}^{e_i} Z_i^{\lambda_i(j), \mu_i(j)})$ *is True*

and *the tokens required by the backward incidence functions of t applied to* $\bigotimes_{i=1}^{n} \bigotimes_{j=1}^{e_i} Z_i^{\lambda_i(j), \mu_i(j)}$ *are all present in the split representation* $\mathcal{R}[\lambda, \mu]$

ii) *There exists no t' with $\pi(t') > \pi(t)$ such that for any* $\bigotimes_{i=1}^{n} \bigotimes_{j=1}^{e_i} Z_i^{\lambda_i'(j), \mu_i'(j)}$ *condition i) holds*

4.3.3 Symbolic firing algorithm

The canonical representation of the symbolic marking obtained by firing $(t, \bigotimes_{i=1}^{n} \bigotimes_{j=1}^{e_i} Z_i^{\lambda_i(j), \mu_i(j)})$ in $\mathcal{M}$ (i.e., $\mathcal{M}' = \mathcal{M}[t, \bigotimes_{i=1}^{n} \bigotimes_{j=1}^{e_i} Z_i^{\lambda_i(j), \mu_i(j)}\rangle)$ is computed in four steps, that use different intermediate (non canonical) representations.

1. let $\mathcal{R}_s = \mathcal{R}[\lambda, \mu]$ be the split representation of $\mathcal{M}$ in which $(t, \bigotimes_{i=1}^{n} \bigotimes_{j=1}^{e_i} Z_i^{\lambda_i(j), \mu_i(j)})$ is enabled;
2. Define $\mathcal{R}_f$ by copying the components m, $card$, and d from $\mathcal{R}_s$, and computing $\mathcal{R}_f.mark$ by applying the incidence functions on $\mathcal{R}_s.mark$. $\mathcal{R}_f$ is a (possibly non canonical) representation of $\mathcal{M}'$;
3. Compute a minimal representation $\mathcal{R}_m$ applying Algorithm M of Section 4.1.2 to $\mathcal{R}_f$;
4. Compute the canonical representation of $\mathcal{M}'$ by transforming $\mathcal{R}_m$ into an ordered representation. The ordering is obtained by applying one of the permutations in S as computed by Algorithm O_1 of Section 4.1.2 to the dynamic subclasses of $\mathcal{R}_m$.

Let us make a final remark on the splitting of ordered classes. The complete splitting into dynamic subclasses of cardinality 1 is not strictly necessary, but it is convenient since it allows a homogeneous treatment of the different cases that arise considering all the different positions occupied by the selected element in the instantiated dynamic subclass. The reason for considering each case separately is due to the need to maintain the ordering among the dynamic subclasses of an ordered class throughout the algorithm.

4.4 Examples of SRG computation

As an example of the application of the SRG construction algorithm, we will show the details of the various computation steps on the examples of WNs depicted in Figures 1.a and 1.b.

Example a). The initial marking of the WN in Figure 1.a is given in a *symbolic non symmetric* form, in terms of three dynamic subclasses for the first class C_1, and four dynamic subclasses for the second (ordered) class C_2. The colour sets are made of Cartesian products of one ordered and one non-ordered basic classes. No repetition of basic classes nor partitioning in static subclasses is introduced here.

Example b). The initial marking of the WN in Figure 1.b is given in a symbolic non symmetric form, in terms of two dynamic subclasses for the first class C_1, and two dynamic subclasses reducing to the two static subclasses $D_{2,1}$ and $D_{2,2}$ for the second class C_2. In this case both basic classes are non ordered (so that the successor function is not used). Note the repetition of the first basic class in the colour domain of transition t_1, and the use of static subclasses and the diffusion functions $S_{2,1}$ and $S_{2,2}$ to obtain constant arc valuations.

4.4.1 Canonical representation of $\mathcal{M}_0$

The first step is the computation of the canonical representation of the initial marking $\mathcal{M}_0$ starting from this specification $\mathcal{R}_0$.

Example a). In the first example, the representation of the initial marking $\mathcal{R}_0$ given by the modeller is composed as follows.

m : $m(1) = 3$, $m(2) = 4$; $\hat{C}_1 = \{ Z_1^1, Z_1^2, Z_1^3 \}$, $\hat{C}_2 = \{ Z_2^1, Z_2^2, Z_2^3, Z_2^4 \}$

$card$: $\forall 0 < j \leq 3$, $card(Z_1^j) = 1$, $\forall 0 < j \leq 4$, $card(Z_2^j) = 1$;

d : $\forall 0 < j \leq 3$, $d(Z_1^j) = 1$, $\forall 0 < j \leq 4$, $d(Z_2^j) = 1$ (by definition, since there is no static subclass partition in this case);

$mark$: $mark(p_1)(\langle Z_1^1, Z_2^1\rangle) = mark(p_1)(\langle Z_1^2, Z_2^2\rangle) = mark(p_1)(\langle Z_1^3, Z_2^3\rangle) = 1$ and $mark(p_1) = 0$ for all other arguments. $mark(p_2)(Z_2^4) = 1$ and $\forall 0 < j < 4$, $mark(p_2)(Z_2^j) = 0$.

This representation, according to Definition 4.2, is already minimal. Indeed, $mark_{sp}(Z_1^1) = Z_2^1.p_1$, $mark_{sp}(Z_1^2) = Z_2^2.p_1$, and $mark_{sp}(Z_1^3) = Z_2^3.p_1$, are all different from each other, so that no dynamic subclass of $\hat{C}_1$ can be grouped by the application of Algorithm M. Similarly, $mark_{sp}(Z_2^1) = Z_1^1.p_1$, $mark_{sp}(Z_2^2) = Z_1^2.p_1$, $mark_{sp}(Z_2^3) = Z_1^3.p_1$, and $mark_{sp}(Z_2^4) = p_2$, are all different from each other, so that no dynamic subclass of $\hat{C}_2$ can be grouped by the application of Algorithm M either.

Since there is no static subclass partition in this example the first sorting phase of Algorithm O_0 does not apply. The values of $mark_{sp}$ for this representation $\mathcal{R}_0$ computed according to Definition 4.4 can be represented by the following 2-dimensional matrix, where each entry represents the marking in places p_1 and p_2:

	Z_2^1	Z_2^2	Z_2^3	Z_2^4
Z_1^1	1,0	0,0	0,0	0,1
Z_1^2	0,0	1,0	0,0	0,1
Z_1^3	0,0	0,0	1,0	0,1

The application of Algorithm O_1 to this case can be traced as follows.

1. Initialization of S with $\hat{\xi}$. Since no static subclass partition is defined, all 24 permutations resulting from the 4 rotations of $\hat{C}_2$ times the 6 permutations of $\hat{C}_1$ are possible. Thus $|S| = 24 > 1$, and the "while" loop must be executed starting with $j_1 = 1$ and $j_2 = 1$;

2. Restriction of S to the permutations that allow a lexicographically minimal pair of $\langle Z_1^k, Z_2^l\rangle$. By inspection of the $mark_{sp}$ matrix, 6 pairs corresponding to the lexicographically minimal string '0,0' exist. One rotation of $\hat{C}_2$ and two different permutations of $\hat{C}_1$ correspond to each of these pairs. In particular, rotations id, $\oplus$, and $\oplus^2$ can be considered for $\hat{C}_2$, respectively associated with Z_1^2 or Z_1^3, Z_1^1 or Z_1^3, and Z_1^1 or Z_1^2, as first elements of $\hat{C}_1$. The combinations result in a set S of cardinality 12;

3. The next iteration of the "while" loop is then prepared, with the indexes $j_1 = 1$ and $j_2 = 2$;

4. Restriction of S with these new indexes. By inspection of the $mark_{sp}$ matrix, one can see that only the two rotations id (with the choice Z_1^3 as first element of $\hat{C}_1$) and $\oplus$ (with the choice of Z_1^1 as first element of $\hat{C}_1$) yield the lexicographically minimal string 0,0. Thus the total number of elements of S is reduced to 4.

5. Since $|S| = 4 > 1$, the next iteration of the "while" loop is prepared, with the indexes $j_1 = 1$ and $j_2 = 3$;

6. Restriction of S with these new indexes. By inspection of the $mark_{sp}$ matrix, one can see that only the rotation $\oplus$ associated with the choice of Z_1^1 as first element of $\hat{C}_1$ yields the lexicographically minimal string 0, 1. Thus the total number of elements of S is reduced to 2: $\langle s_1^1, \oplus\rangle$, and $\langle id, \oplus\rangle$, where s_1^1 maps the indexes as follows; $1 \to 1$, $2 \to 3$, $3 \to 2$.

7. Since $|S| = 2 > 1$, some more iterations are allowed. The next choice of the indexes $j_1 = 1$ and $j_2 = 4$ cannot reduce the size of S, since $\hat{C}_2$ is already completely ordered. Thus we arrive to the following step, with indexes $j_1 = 2$ and $j_2 = 1$;

8. Restriction of S with indexes $j_1 = 2$ and $j_2 = 1$. By inspection of the $mark_{sp}$ matrix, one can see that only the permutation $\langle s_1^1, \oplus\rangle$ yields the lexicographically minimal string 0, 0. Thus the total number of elements of S is reduced to 1 and the Algorithm terminates.

Since Algorithm O_1 produces a set S that contains a single permutation which is different from the identity permutation, we conclude that the representation $\mathcal{R}_0$ given by the modeller to define the initial marking was not ordered, hence non canonical. The canonical representation $\mathcal{M}_0$ is computed by applying the permutation $\langle s_1^1, \oplus\rangle$ to the dynamic subclasses of $\mathcal{R}_0$.

Example b). In the second example, the representation of the initial marking $\mathcal{R}_0$ given by the modeller is composed as follows.

m : $m(1) = 2, m(2) = 2$; $\hat{C}_1 = \{ Z_1^1, Z_1^2 \}$, $\hat{C}_2 = \{ Z_2^1, Z_2^2 \}$

$card$: $card(Z_1^1) = 1, card(Z_1^2) = 2, card(Z_2^1) = card(Z_2^2) = 1$;

d : $d(Z_1^1) = d(Z_1^2) = 1$ (by definition, since there is no static subclass partition in the first class), $d(Z_2^1) = 1$, $d(Z_2^2) = 2$;

$mark$: $mark(p_1)(\langle Z_1^1, Z_2^1\rangle) = 1$, $mark(p_1)(\langle Z_1^2, Z_2^2\rangle) = 1$, and $mark(p_1) = 0$ for all other arguments.

This representation, according to Definition 4.2, is already minimal. Indeed, $mark_{sp}(Z_1^1) = Z_2^1.p_1$ and $mark_{sp}(Z_1^2) = Z_2^2.p_1$ are different from each other, so that no dynamic subclass of $\hat{C}_1$ can be grouped by the application of Algorithm M. Moreover, $m(2) = n_2 = 2$, so that no dynamic subclass of $\hat{C}_2$ can be grouped by the application of Algorithm M either.

In this example there are static subclasses, however $\mathcal{R}_0$ already satisfies the condition of Definition 4.4.i; thus no preliminary sorting is required, and we can apply Algorithm O_1 directly. Let us compute the values of $mark_{sp}$ according to Definition 4.4. The results are shown in the following 2-dimensional matrix:

	Z_2^1	Z_2^2
Z_1^1	1	0
Z_1^2	0	1

The application of Algorithm O_1 to this case can be traced as follows.

1. Initialization of S with $\hat{\xi}$. Since class C_2 is partitioned in two static subclasses, only the identity permutation is allowed for $\hat{C}_2$. The two permutations id and s_1^1 such that $s_1^1(1) = 2$ and $s_1^1(2) = 1$ are allowed instead for $\hat{C}_1$. Thus $|S| = 2 > 1$ after the execution of the first statement, and the "while" loop is entered with $j_1 = 1$ and $j_2 = 1$;

2. Restriction of S. By inspection of the $mark_{sp}$ matrix, we realize that permutation $\langle s_1^1, id\rangle$ (yielding the string '0') is lexicographically strictly lower than $\langle id, id\rangle$ (which yields the string '1'), so that $|S|$ is immediately reduced to 1, and the Algorithm terminates.

Also in this example Algorithm O_1 produces a set S that contains a single permutation different from the identity, so we conclude that the representation $\mathcal{R}_0$ given by the modeller was not ordered, hence non canonical. The canonical representation $\mathcal{M}_0$ is computed by applying the permutation $\langle s_1^1, id\rangle$ to the dynamic subclasses of $\mathcal{R}_0$.

4.4.2 Splitting and enabling test for $\mathcal{M}_0$

Once a canonical representation of the initial marking $\mathcal{M}_0$ has been computed, the SRG construction algorithm can start by taking into consideration one symbolic marking (in this case $\mathcal{M}_0$) and perform all possible splitting of dynamic subclasses and apply the transition enabling check on the split representation $\mathcal{R}_s$.

Example a). In the first example, the representation of the initial marking $\mathcal{M}_0$ already contains only dynamic subclasses of cardinality 1, so that no actual splitting is necessary (i.e., $\mathcal{R}_s = \mathcal{M}_0$). In this particular case, the only pair of dynamic subclasses that satisfies the enabling condition of t_1 is $\langle \mathcal{M}_0.Z_1^2, \mathcal{M}_0.Z_2^4 \rangle$, due to the fact that only $\langle \mathcal{M}_0.Z_2^1 \rangle = \langle \mathcal{M}_0.Z_2^{\oplus 4} \rangle$ is present in place p_2. According to Definitions 4.6 and 4.7, the only corresponding symbolic instantiation $[\lambda, \mu]$ is defined by: $\lambda_1(1) = 2,\ \lambda_2(1) = 4,\ \mu_1(1) = 1,\ \mu_2(1) = 1$

Example b). In the second example, the representation of the initial marking $\mathcal{M}_0$ contains the dynamic subclass Z_1^1 of cardinality 2, which must be split in order to apply the enabling test. According to Definitions 4.6 and 4.7, the only symbolic instantiation $[\lambda, \mu]$ that is enabled in $\mathcal{M}_0$ is defined by: $\lambda_1(1) = 2,\ \lambda_1(2) = 1,\ \mu_1(1) = 1,\ \mu_1(2) = 1$
The split representation $\mathcal{R}_s = \mathcal{M}_0[\lambda, \mu]$ is composed as follows:

m : $m(1) = 3, m(2) = 2;\quad \hat{C}_1 = \{ Z_1^{1,0}, Z_1^{1,1}, Z_1^2 \},\quad \hat{C}_2 = \{ Z_2^1, Z_2^2 \}$

$card$: equal to 1 for all dynamic subclasses;

d : $d(Z_1^{1,0}) = d(Z_1^{1,1}) = d(Z_1^2) = 1$ (by definition), $d(Z_2^1) = 1$, $d(Z_2^2) = 2$;

$mark$: $mark(p_1)(\langle Z_1^{1,1}, Z_2^2 \rangle) = mark(p_1)(\langle Z_1^{1,0}, Z_2^2 \rangle) = mark(p_1)(\langle Z_1^2, Z_2^1 \rangle) = 1$, and $mark(p_1) = 0$ for all other arguments.

4.4.3 Firing $\mathcal{M}_0[t_1, \lambda, \mu\rangle$

The firing of the enabled transition with the given symbolic instantiation produces a (non canonical) symbolic marking $\mathcal{R}_f$, as follows.

Example a). In the first example, the representation of the symbolic marking $\mathcal{R}_f$ is composed as follows:

m, $card$, d : same as for $\mathcal{R}_s = \mathcal{M}_0$;

$mark$: $mark(p_1)(\langle Z_1^1, Z_2^2 \rangle) = mark(p_1)(\langle Z_1^3, Z_2^3 \rangle) = mark(p_1)(\langle Z_1^2, Z_2^1 \rangle) = 1$ and $mark(p_1) = 0$ for all other arguments. $mark(p_2)(Z_2^4) = 1$ and $mark(p_2) = 0$ for all other arguments.

This representation, as in the case of $\mathcal{R}_0$, is already minimal. However it is not ordered, hence non canonical. Indeed applying algorithm O_1 we can obtain the following set S of permutations: $S = \{\langle s_1^2, \oplus \rangle\}$, with permutation s_1^2 defined by $1 \rightarrow 3$, $2 \rightarrow 1$, and $3 \rightarrow 2$. Applying this permutation s to $\mathcal{R}_m = \mathcal{R}_f$, we obtain, by sheer coincidence, the same symbolic representation of $\mathcal{M}_0$. Thus, in this particular example, $\mathcal{M}_0 = \mathcal{M}_0[t_1, \lambda, \mu\rangle$, and the SRG is composed of a single symbolic marking connected to itself by a single symbolic arc.

Example b). In the second example, the representation of the symbolic marking $\mathcal{R}_f$ contains a double index for some of the dynamic subclasses (that have been split). Of course the double index can be eliminated by renaming $Z_1^{1,0}$ as Z_1^1, $Z_1^{1,1}$ as Z_1^2, and Z_1^2 as Z_1^3. After this renaming, $\mathcal{R}_f$ is composed as follows:

m : $m(1) = 3, m(2) = 2;\quad \hat{C}_1 = \{ Z_1^1, Z_1^2, Z_1^3 \},\quad \hat{C}_2 = \{ Z_2^1, Z_2^2 \}$

$card$: equal to 1 for all dynamic subclasses;

d : $d(Z_1^1) = d(Z_1^2) = d(Z_1^3) = 1$ (by definition), $d(Z_2^1) = 1$, $d(Z_2^2) = 2$;

mark : $mark(p_1)(\langle Z_1^1, Z_2^2\rangle) = mark(p_1)(\langle Z_1^2, Z_2^1\rangle) = mark(p_1)(\langle Z_1^3, Z_2^2\rangle) = 1$, and $mark(p_1) = 0$ for all other arguments.

We can apply Algorithm M to $\mathcal{R}_f$, in order to compute a minimal representation $\mathcal{R}_m$. We can see by inspection of the $mark_{sp}$ function that the dynamic subclasses Z_1^1 and Z_1^3 can be grouped, since they both have the same projected marking $\langle p_1, Z_2^2\rangle$. Thus the following minimal representation $\mathcal{R}_m$ can be computed:

m : $m(1) = 2$, $m(2) = 2$; $\quad \hat{C}_1 = \{ Z_1^1, Z_1^2 \}$, $\quad \hat{C}_2 = \{ Z_2^1, Z_2^2 \}$

card : $card(Z_1^1) = 2$, $card(Z_1^2) = 1$, $card(Z_2^1) = card(Z_2^2) = 1$;

d : $d(Z_1^1) = d(Z_1^{2,0}) = 1$ (by definition), $d(Z_2^1) = 1$, $d(Z_2^2) = 2$;

mark : $mark(p_1)(\langle Z_1^1, Z_2^2\rangle) = mark(p_1)(\langle Z_1^2, Z_2^1\rangle) = 1$, and $mark(p_1) = 0$ for all other arguments.

We realize that the representation $\mathcal{R}_m$ is coincidentally identical to $\mathcal{M}_0$, hence

i) it is already ordered and canonical, so that the application of Algorithm O_1 is not necessary, and

ii) also in this second example, $\mathcal{M}_0 = \mathcal{M}_0[t_1, \lambda, \mu\rangle$, and the SRG is composed of a single symbolic marking connected to itself by a single symbolic arc.

4.5 Some properties of the SRG

Property 4.1 *Reachability is equivalent for symbolic markings and for the sets of corresponding of ordinary markings, i.e.:*

$$\bigcup_{M_0 \in \mathcal{M}_0} [M_0\rangle = \{ M' \text{ such that } \mathcal{M}' \in [\mathcal{M}_0\rangle \}$$

Property 4.2 *Strong connection of RG $\Longrightarrow$ Strong connection of SRG, but not vice-versa.*

The example of WN in Figure 1.a is a case in which the SRG (made of the single symbolic marking $\mathcal{M}_0$) is strongly connected while the underlying RG is not, as only rotations but not general permutations $s \in \xi$ on $C_1 \times C_2$ are implemented by the transition firing.

Property 4.3 *Strong connection of SRG* $\wedge$ $\forall 0 < i \leq h$ $\exists \mathcal{M}' \in$*SRG such that* $(\mathcal{M}'.m(i) = n_i)$ $\wedge$ $\forall h < i \leq n$ $((n_i > 1) \vee (\exists \mathcal{M}' \in$*SRG such that* $(\mathcal{M}'.m(i) = 1)\ )$ $\Longrightarrow$ *Strong connection of RG*

The above property is an extension of Property 2.2. Indeed, the additional condition on the dynamic subclass partition is always satisfied by the (symmetric) initial marking in an RN.

On the other hand, the condition is not necessary in order for the RG to be strongly connected, even in the simpler case of RNs. Indeed, the WN in Figure 1.b is a case in which the SRG (made of the single symbolic marking $\mathcal{M}_0$) is strongly connected, $\mathcal{M}_0.m(1) = 2 > n_1 = 1$, and the RG is strongly connected.

Property 4.4 *Relation between the arcs.*

$$N_a(\lambda, \mu) = \prod_{i=1}^{h} \prod_{j=1}^{m_i} \frac{card(Z_i^j)!}{(card(Z_i^j) - \mu_i^j)!}$$

Property 4.5 *Number of ordinary markings.*

$$N_m = \frac{1}{K(\mathcal{M})} \left(\prod_{i=1}^{h} \prod_{q=1}^{n_i} \frac{|D_{i,q}|!}{\prod_{d(Z_i^j)=q} card(Z_i^j)!} \right) \prod_{i=h+1}^{n} \nu(i)$$

where $\nu(i) =$ *if* $(m(i) > 1 \wedge n_i = 1)$ *then* $|C_i|$ *else* 1.

5 Conclusions

Restrictions were initially imposed to Regular nets in order to simplify the task of defining efficient analysis tools, such as structural reduction, computation of flows, and construction of a Symbolic Reachability Graph. Indeed, this simplification allowed the development of very interesting results, but they posed a real constraint to the modelling power of the model. Due to space limitations, a detailed discussion of these limitations with respect to actual modelling requirements could not be included in this paper.

In this paper we define Well Formed Coloured Nets as a useful extension of RNs, together with their associated Symbolic Reachability Graph, that allow an easier modelling of interesting systems. Moreover, with these extensions and the use of guard predicates on transitions and arcs, WN reach the same modelling power of non-constrained CPNs without losing the factorization power of RN in terms of exploitation of symmetries for the SRG computation.

We also want to point out the similarity that exists between WN and Predicate/Transition nets [GL81]. Indeed, the description of a WN can easily be put in parametric form with respect to the cardinality of the basic colour domains, which is not possible using the classical CPN formalism [Jen81]. Genrich [Gen87,Gen88] was the first to identify the idea of parametrization as a different and more important concept with respect to the shorthand notation provided by CPNs. In one way, WNs can be seen as the basis for the definition of classes of coloured nets with a given structure and behaviour for various populations of coloured tokens.

Not all combinations of the features offered by WN are useful from a practical point of view. For instance, the combination of static subclass partitioning and ordering through the use of the successor function for the same class reduces symbolic markings to ordinary markings, thus losing the factorization power of the SRG representation. On the other hand, multidimensional colour sets in which some classes are ordered and some others are partitioned into subclasses make sense, since the SRG representation is not reduced to the RG in this case. In general, one can say that in any case the SRG of a WN captures the symmetries of the model, if there are any; if the model is intrinsically non symmetric, then of course the use of the SRG has no advantage over that of the RG.

In conclusion we notice that the analysis by RG of a Petri net model is in practice hampered by the large size of the graph caused by two different reasons. First, the representation of the parallelism in terms of all possible interleavings that can be observed on the global state of the model. Second, the possible presence of intrinsic symmetries of the model, that produce many different portions of a graph representing the same behaviour. With the proposal of our SRG construction algorithm we have addressed and solved the second problem. The first problem is inherent to the state-space representation, and in our opinion can be avoided only by using non state-space based analysis techniques.

6 Acknowledgements

The authors wish to express their gratitude to the anonymous referees that made many valuable comments on the first draft of this paper. Special thanks are due to Ms. Frances Valeria Perricone, who spent one afternoon proof-reading the last version of the paper in order to improve the English style.

References

[AMBC84] M. Ajmone Marsan, G. Balbo, and G. Conte. A class of generalized stochastic Petri nets for the performance analysis of multiprocessor systems. *ACM Transactions on Computer Systems*, 2(1), May 1984.

[AMBCC87] M. Ajmone Marsan, G. Balbo, G. Chiola, and G. Conte. Generalized stochastic Petri nets revisited: random switches and priorities. In *Proc. Int. Workshop on Petri Nets and Performance Models*, pages 44–53, IEEE-CS Press, Madison, WI, USA, August 1987.

[Car89] J. A. Carrasco. Automated construction of compound Markov chains from generalized stochastic high-level Petri nets. In *Proc. 3rd Intern. Workshop on Petri Nets and Performance Models*, pages 93–102, IEEE-CS Press, Kyoto, Japan, December 1989.

[CBD88] G. Chiola, G. Bruno, and T. Demaria. Introducing a color formalism into generalized stochastic Petri nets. In *Proc.* 9th *Europ. Workshop on Application and Theory of Petri Nets*, Venezia, Italy, June 1988.

[CF89] G. Chiola and G. Franceschinis. Colored GSPN models and automatic symmetry detection. In *Proc. 3rd Intern. Workshop on Petri Nets and Performance Models*, IEEE-CS Press, Kyoto, Japan, December 1989.

[CM89] J.M. Couvreur and J. Martinez. Linear invariants in commutative high-level nets. In *Proc.* 10th *International Conference on Application and Theory of Petri Nets*, Bonn, Germany, June 1989.

[Cou90] J.M. Couvreur. The general computation of flows for coloured nets. In *Proc.* 11th *International Conference on Application and Theory of Petri Nets*, Paris, France, June 1990.

[DH89a] C. Dutheillet and S. Haddad. Aggregation and disaggregation of states in colored stochastic Petri nets: application to a multiprocessor architecture. In *Proc. 3rd Intern. Workshop on Petri Nets and Performance Models*, IEEE-CS Press, Kyoto, Japan, December 1989.

[DH89b] C. Dutheillet and S. Haddad. Regular stochastic Petri nets. In *Proc. 10th Intern. Conf. Application and Theory of Petri Nets*, Bonn, Germany, June 1989.

[Gen87] H.J. Genrich. Predicate/transition nets. In W. Brawer, W. Reisig, and G. Rozenberg, editors, *Advances on Petri Nets '86 - Part I*, pages 207–247, Springer Verlag, Bad Honnef, West Germany, February 1987.

[Gen88] H.J. Genrich. Equivalence transformations of Pr/T nets. In *Proc.* 9th *Europ. Workshop on Application and Theory of Petri Nets*, Venezia, Italy, June 1988.

[GL81] H. J. Genrich and K. Lautenbach. System modelling with high-level Petri nets. *Theoretical Computer Science*, 13:109–136, 1981.

[Had87] S. Haddad. *Une Categorie Regulier de Reseau de Petri de Haut Niveau: Definition, Proprietes et Reductions.* PhD thesis, Lab. MASI, Universite P. et M. Curie (Paris 6), Paris, France, Oct 1987. These de Doctorat, RR87/197 (in French).

[Had88] S. Haddad. Generalization of reduction theory to coloured nets. In *Proc.* 9th *Europ. Workshop on Application and Theory of Petri Nets*, Venezia, Italy, June 1988.

[HC88] S. Haddad and J.M. Couvreur. Towards a general and powerful computation of flows for parametrized coloured nets. In *Proc.* 9th *Europ. Workshop on Application and Theory of Petri Nets*, Venezia, Italy, June 1988.

[HJJJ84] P. Huber, A.M. Jensen, L.O. Jepsen, and K. Jensen. Towards reachability trees for high-level Petri nets. In G. Rozenberg, editor, *Advances on Petri Nets '84*, pages 215–233, Springer Verlag, 1984.

[Jen81] K. Jensen. Coloured Petri nets and the invariant method. *Theoretical Computer Science*, 14:317–336, 1981.

[LM88] Chuang Lin and C. Marinescu. Stochastic high level Petri nets and applications. *IEEE Transactions on Computers*, 37(7):815–825, July 1988.

A Appendix

A.1 Notation

$|C|$: If C is a finite set then $|C|$ is its cardinality.

$Bag(C)$: If C is a finite set, then $Bag(C)$ denotes the class of multisets that can be defined on C, represented by $|C|$-tuples of integers.

$\oplus$: If C is a finite ordered set, then $\oplus$ denotes the "successor function" associated with C, i.e., $\forall c_i \in C$ $c_j = \oplus c_i$ is a different element of C, which is called the successor of c_i.
$\oplus^k$ denotes the k-th successor function, i.e., $\underbrace{(\oplus \ldots \oplus)}_{k}$

$\bigotimes_{i=1}^{n} C_i$ Denotes the n-dimensional Cartesian product: $C_1 \times \ldots \times C_n$

$\{C_i\}_{i \in I}$: Let $\{C_i\}_{i \in I}$ be a family of sets indexed by $I = \{1, \ldots, n\}$.

i) $\bigotimes_{i \in I} C_i$ is the Cartesian product of the sets C_i defined by:

$$\bigotimes_{i \in I} C_i = \{\langle c_1, \ldots, c_n \rangle \mid c_i \in C_i\}$$

ii) $\langle c_1, \ldots, c_n \rangle \in \bigotimes_{i \in I} C_i$ is also denoted $\bigotimes_{i \in I} c_i$. In the examples, we use the notation $\sum_{i \in I} c_i.i$ that allows us to omit c_i if $c_i = 1$ and $c_i.i$ if $c_i = 0$.

iii) If $I = \emptyset$, then $\bigotimes_{i \in I} C_i = \{\varepsilon\}$

$\{f_i\}_{i \in I}$: Let $\{f_i : C_i \times C \to I\!N\}_{i \in I}$ be a set of functions. $\bigotimes_{i \in I} f_i$ is the product of functions:
$\bigotimes_{i \in I} f_i : \bigotimes_{i \in I} C_i \times C \to I\!N$, $\quad \bigotimes_{i \in I} f_i\,(\bigotimes_{i \in I} c_i, c) = \prod_{i \in I} f_i(c_i, c)$

f^{-1} : Let f be a function, $f : E \to F$. Then f^{-1} is a function mapping F on $\mathcal{P}(E)$ such that:

$$f^{-1}(y) = \{x \in E \mid f(x) = y\}$$

$f(x)(y)$: Let $f : X \to [\, Y \to Z \,]$ be a function. Then $f(x) : Y \to Z$ is also a function that can be applied to the argument $y \in Y$. In this case $f(x)(y)$ is also denoted $f(x, y)$.

I/i is the set I, except the element i.

$I\!N$ is the set of integers.

$Z_i^{s_i(u_i)}$ is a notation for $s_i(Z_i^{u_i})$

$\bigotimes_i \bigotimes_j Z_i^{u(i,j)}$ superscript $u(i,j)$ in this notation stands for any value in $1, \ldots, m(i)$

A.2 More technical definitions

$mark_{sc}$: In order to check the minimality of the grouping in an RN, we must define a function $mark_{sc}$ such that $mark_{sc}(Z_i^j)$ is the marking of the dynamic subclass Z_i^j.

$mark_{sc}$ is a function from $\bigcup_{i \in I} \hat{C}_i$ where
$mark_{sc}(Z_i^j) \in \left[\bigotimes_{k \in I/i} \hat{C}_k \to [P \to I\!N]\right]$ such that
$mark_{sc}(Z_i^{j(i)}) \left(\bigotimes_{k \in I/i} Z_k^{j(k)}\right) = mark \left(\bigotimes_{k \in I} Z_k^{j(k)}\right)$.

$mark_{sp}$: The function $mark_{sp}$ defines the marking of a tuple of subclasses in a WN. This marking is needed to compute the unique canonical representation of a symbolic marking.

Let p be a place such that $C(p) = \bigotimes_{i \in I} (C_i)^{e_i} \neq \{\varepsilon\}$, $e_i \geq 0$. Let $\mathcal{M}$ be a symbolic marking, and let $I' \subseteq I$ (Note that I' can be considered as a particular case of multiset $J^\star$ in $Bag(I)$). A selection operation that chooses one occurrence of C_i in $C(p)$ for all $i \in I'$ whenever possible, i.e.: $\sigma : I \to I\!N$ with

- If $e_i = 0$ or $i \notin I'$ then $\sigma(i) = 0$ else $1 \leq \sigma(i) \leq e_i$
- $\exists i, \sigma(i) > 0$

We denote $Sl(I', C(p))$ the set of selections of I' in $C(p)$.
$I'(i) =$ (If $i \in I'$ then 1 else 0) denotes the characteristic function of I'.
Then, the restriction to I' of $C(p)$ is defined by:

$$C(p)_{/I'} = \bigotimes_{i \in I} (C_i)^{e'_i}$$

with $e'_i =$ (If $e_i = 0 \vee i \notin I'$ then e_i else $(e_i - 1)$).
A marking $\mathcal{M}$ of a tuple in I' with respect to a selection $\sigma \in Sl(I', C(p))$.
$\sigma.mark_{sp}(p)$ is defined by: $\sigma.mark_{sp}(p) : C_{I'} \rightarrow \left(C(p)_{/I'} \rightarrow I\!N\right)$

$$\sigma.mark_{sp}(p) \left(\bigotimes_{i \in I'} Z_i^{w(i)} \right) \left(\bigotimes_{i \in I} \bigotimes_{j=1}^{e'_i} Z_i^{u(i,j)} \right) = mark(p) \left(\bigotimes_{i \in I} \bigotimes_{j=1}^{e_i} Z_i^{v(i,j)} \right)$$

with $v(i,j) =$ If $j > \sigma(i)$ then $u(i, j - I'(i))$ else if $j = \sigma(i)$ then $w(i)$ else $u(i,j)$.
Note that $v(i,j)$ takes into account the insertion of $Z_i^{w(i)}$ as the $\sigma(i)^{th}$ instantiation of C_i in $mark(p)$.
Let us consider a symbolic marking $\mathcal{M}$, a subset $I' \subseteq I$, and a tuple $\bigotimes_{i \in I'} Z_i^{k(i)}$.

$$mark_{sp} \left(\bigotimes_{i \in I'} Z_i^{k(i)} \right) \in \left\{ \bigotimes_{p \in P} \bigotimes_{\sigma \in Sl(I', C(p))} \left(C(p)_{/I'} \rightarrow I\!N \right) \right\}$$

is a function such that:

$$mark_{sp} \left(\bigotimes_{i \in I'} Z_i^{k(i)} \right) = \bigotimes_{p \in P} \bigotimes_{\sigma \in Sl(I', C(p))} \sigma.mark_{sp} \left(\bigotimes_{i \in I'} Z_i^{k(i)} \right)$$

In particular $mark_{sp}(Z_i^j)$ is used in Definition 4.2 and $mark_{sp}(\bigotimes_{i \in I} Z_i^{w(i)})$ is used in Definition 4.4.

Section F
Analysis by Means of Transformations

The basic idea behind transformations is to modify a Petri net—without changing a selected set of properties, e.g., liveness, marking bounds, deadlocks and reachability. The modification of the net is performed by means of a set of transformations rules and may be carried out manually, automatically or interactively. In the latter case the strategy is decided by a person, while the detailed computations and checks are made by a computer.

The two papers in this section complement each other, and they can be read in any order. The first paper extends a set of low-level net transformation rules—so that they can be used for high-level nets. The transformation rules are proved to preserve liveness and boundedness. The purpose of the transformation is to obtain a small and simple net for which it is easy to investigate the given properties. The low-level net transformation rules were developed by G. Berthelot.

The second paper discusses the great modelling flexibility introduced by high-level nets, and the arbitrariness in choosing a suitable net, from some given class, to represent a system at hand. Two high-level nets are said to be equivalent if they unfold to the same low-level net—i.e. have the same behaviour. It is not the objective of the method to find the most suitable net for a given purpose. Instead, the question is how to generate equivalent representations systematically, and how to translate two equivalent nets into each other.

14.
A Reduction Theory for Coloured Nets

S. Haddad

G. Rozenberg (ed.): Advances in Petri nets 1989. Lecture Notes in Computer Science, vol. 424.
Springer, Berlin Heidelberg New York 1990, pp. 209-235

ABSTRACT This paper presents the generalization to the coloured nets of the most efficient reductions defined by Berthelot for Petri nets. First, a generalization methodology is given that is independent from the reduction one wants to generalize. Then based on that methodology, we define extensions of the implicit place transformation and the pre and post agglomeration of transitions. For each reduction we prove that the reduced net has exactly the same properties as the original net. Finally we completely reduce an improved model of the data base management with multiple copies, thus showing its correctness.

KEYWORDS coloured Petri nets, behavioural properties, methodology, reductions

CONTENTS

1 INTRODUCTION

In any theoretical model of computation, a useful method to prove properties of an object (program, protocol, ...) of this model is to reduce this object such that the simplified object has the same properties as the original one. In his thesis, Berthelot [Ber83] has defined ten reductions of Petri nets and has shown how they can be efficiently used in various Petri net modellings (See also [Ber85]).

As abbreviations of Petri nets - coloured nets [Jen82] and Predicate-Transition nets [Gen81], [Lau85] - are now defined in order to model complex systems, an interesting contribution to high-level net theory would be the definition of similar reduction rules. The main works that have been done are [Col86] , [Kro89] and [Gen88]. In [Col86] the authors define different rules where the main "coloured" condition is the orthonormality of the colour functions (too strong a condition in our opinion) that valuate some arcs and where the structural conditions are similar but not equivalent to the conditions of Berthelot. In [Kro89] the authors study the extension of reduction rules to Predicate-Transition nets but they only require that their reductions preserve local properties. The work of Genrich is a deep study of equivalence for Predicate-Transition nets, namely it defines a set of transformation rules (sound and complete) which ensure that the original and the transformed Predicate-Transition nets have the same unfolded net in all valid interpretations. These rules are complementary to the rules we will define here, since they do not reduce the unfolded net but can extend the application field of our rules by adequate transformations. In fact we will give here a transformation (the orthonormalized reduction) that can be deduced from Genrich's rules and that has allowed us to extend the reductions presented in [Had88].

The work we present here leads to alternative definitions for reduction rules which can be added to those defined in [Col86] and provides a more powerful tool. But rather than these additional rules, the main contribution is the principle of the generalization of the reductions. Indeed the reduction theory is nothing but **an heuristic method** and then :

- It only provides **sufficient conditions** to the simplification problem,
- It will always be improvable **by more sophisticated rules**.

So, in our opinion, the presentation of a methodology to generalize reductions for coloured nets is at least as important as the definition of new reduction rules. The methodology we present here is based on two principles :

- Not define, if possible, additional structural conditions for the extended reduction rules.
- Only define the functional conditions necessary to ensure the equivalence between the reduced net and the original net.

Preserving these two principles while one generalizes a reduction makes the extended reduction as accurate as the original one. In order to respect these two principles, it is essential to take into account the unfolded Petri net of the coloured net while defining and proving the extension.

In order to illustrate this methodology, we have chosen to extend Berthelot's most frequently used rules, namely the implicit place simplification and the pre and post agglomeration of transitions. We emphasize two advantages of our reductions : on the one hand, they are strictly equivalent to the reductions defined by Berthelot and they then have the numerous properties proved by him; on the other hand, the functional conditions are not predefined but are as weak as possible to obtain this equivalence in each case, thus giving our reductions a large field of application.

The coloured reductions we have defined completely reduce an improved version [Had87b] of the data base management [Jen82] with multiple copies. In [Had87b], one can also find the complete reduction of the two-step commitment protocol [Bae81]. In that thesis, we also showed that these reductions were programmable for the regular coloured nets.

General notations

- **N** is the set of non negative integers, **Z** is the set of integers, **Q** is the set of rational numbers
- M.N , where M and N are matrices, denotes the matrices product (this notation includes the product of a vector by a matrix, since a vector is a special case of matrix)
- M^t, where M is a matrix n x p, denotes the matrix p x n such that : $M^t_{i,j} = M_{j,i}$
- Let U be a finite set. Then the set of functions from U to **N** is denoted Bag(U). An item a of Bag(U) is noted $\sum a_u.u$ where the summation is over $u \in U$.
- A partial order on Bag(U) is defined by : $a \leq b$ if and only if $\forall u \in U, a_u \leq b_u$
- The sum of two items of Bag(U) is defined by $a+b = \sum (a_u+b_u).u$ where the summation is over $u \in U$
- The difference between two items $a \geq b$ of Bag(U) is defined by : $a-b = \sum (a_u-b_u).u$ where the summation is over $u \in U$

2 COLOURED NETS

We recall the definitions of a coloured net, the firing rule in a coloured net, some particular colour functions, the equivalent unfolded Petri net and the flow definition that we need for the implicit place rule.

Definition 1 A coloured net $R = \langle P,T,C,I^+,I^-,M \rangle$ is defined by :

- P the set of places
- T the set of transitions with $P \cup T \neq \emptyset$ and $P \cap T = \emptyset$
- C the "colour function" from $P \cup T$ to W , where W is some finite set of finite and not empty sets. An item of C(s) is called a colour of s and C(s) is called the colour set of s.
- I^+ (I^-) is the forward (backward) incidence matrix of P x T , where $I^+(p,t)$ is a function from C(p) x C(t) to **N** (i.e. a linear application from Bag(C(t)) to Bag(C(p)))
- M the "initial marking" of the net is a vector of P, where M(p) is a function from C(p) to **N** (i.e. an item of Bag(C(p)))

Notation

We note $I^+,I^-(p,t)(c_t)$, where c_t belongs to C(t), the corresponding item of Bag(C(p)) for $I^+,I^-(p,t)$ considered as a linear application from Bag(C(t)) to Bag(C(p)).

Definition 2 The firing rule is defined by :

- A transition t is enabled for a marking M and a colour $c_t \in C(t)$ if and only if :

  $\forall p \in P,\ M(p) \geq I^-(p,t)(c_t)$

- The firing of t for a marking M and a colour $c_t \in C(t)$ gives a new marking M' defined by : $\forall p \in P,\ M'(p) = M(p) - I^-(p,t)(c_t) + I^+(p,t)(c_t)$

We present the particular functions we need for the definition of coloured reductions. As we have already said, a colour function can be defined either as a linear application from Bag(C(t)) to Bag(C(p)) or as a function from C(p) x C(t) to **N** . As we need the two definitions for this paper we use the same symbol for the two functions and the formula below shows how to translate one definition into the other :

$$\mathbf{f(c) = \Sigma\ f(c',c).c'\ \text{where}\ c'\ \text{ranges over}\ C(p)}$$

where f(c) denotes the mapping of c to an item of Bag(C(p)) by f as a linear application and where f(c',c) denotes the mapping of (c',c) to an integer value. Notice that no confusion can appear since the first definition implies one argument while the second definition implies two arguments. In the next definitions, all the functions are linear applications.

Definition 3 The identity function of Bag(C) "Id" is defined by Id(c) = c

Remark : The second definition gives Id(c',c) == If c=c' then 1 else 0

Definition 4 A function f from Bag(C) to Bag(C) is orthonormal if and only if there exists a substitution s of C such that f(c) = s(c)

Remark : The second definition gives f(c',c) == If s(c)=c' then 1 else 0

Definition 5 The projection from Bag(CxC') to Bag(C) "Proj" is defined by :

Proj(<c,d>) = c

Remark :

- This definition is not the usual definition of projection on vector spaces.
- The second definition gives Proj(c',<c,d>) == If c=c' then 1 else 0

Definition 6 A function f from Bag(C) to Bag(C') is quasi injective if and only if :

(Let us recall that f(c) = Σ f(c',c).c' where c' ranges over C')

$\forall c' \in C', \forall c_1 \in C, \forall c_2 \in C,\ f(c', c_1) \neq 0 \text{ and } f(c', c_2) \neq 0 \Rightarrow c_1 = c_2$

Definition 7 A function f from Bag(C) to Bag(C') is unitary if and only if :
(Let us recall that $f(c) = \sum f(c',c).c'$ where c' ranges over C')

$$\forall c' \in C', \forall c \in C \quad f(c',c) = 0 \text{ or } f(c',c) = 1$$

In a model, the functions are almost always unitary as projections, orthonormal functions and identities are currently used. The quasi injectivity is a significant property for the unfolding of the coloured net and is fulfilled by a wide class of standard colour functions.

Definition 8 Let f be a function from Bag(C) to Bag(C') and g be a function from Bag(C') to Bag(C"). Then the composition of f and g is a function g o f from Bag(C) to Bag(C") defined by :

$$g \circ f(c) = g(f(c)) = \sum \left(\sum g(c'',c').f(c',c) \right) . c''$$

where c' ranges over C' and c" ranges over C".

We now define the unfolded Petri net of a coloured net as in [Jen81b]. This net is the low level representation of the coloured net. It has exactly the same behaviour as the coloured net and then properties of the ordinary net are properties of the coloured net. This equivalence is fundamental since it is the basis of the theory of high level nets.

Definition 9 Let R be a coloured net then R' the unfolded Petri net of R is defined by :

- $P' = \cup (p,c)$ where the union is over $p \in P$ and $c \in C(p)$
- $T' = \cup (t,c)$ where the union is over $t \in T$ and $c \in C(t)$
- I'^- and I'^+ the backward and the forward matrices are defined by :
  $I'^-(p,c)(t,c') = I^-(p,t)(c,c')$ and $I'^+(p,c)(t,c') = I^+(p,t)(c,c')$
- $M'(p,c) = M(p)(c)$

For the implicit place transformation, we need the definition of the flows of a coloured net. There are different ways of defining them. Here we choose a simple definition : the flows of a coloured net are the flows of the corresponding unfolded Petri net.

Definition 10 The incidence matrix I of a Petri net is defined by : $I = I^+ - I^-$. Then I(p,t) belongs to $\mathbf{Z}$

Definition 11 A flow v of a Petri net is a vector of $\mathbf{Z}^P$ (where P is the set of places) that verifies : $I^t.v = 0$. The support of a flow is the subset P' of P defined by :

$$p \in P' \Longleftrightarrow v_p \neq 0$$

Definition 12 A flow v of a coloured net R is a flow of the unfolded Petri net of R.

Remark Depending on the authors, the flows are sometimes called invariants. We prefer our notation since invariants are not necessarily linear and even if they are linear they do not necessarily verify the flow equation (See definition 11).

At last , all the reductions we will define here preserve the main properties of a net. So we collect them in a set.

Definition 13 The set of the main properties of a Petri net is defined by :
{boundedness, safeness, invariant covering, normal end, home state, unavoidable state, liveness, pseudo-liveness, quasi-liveness, abstraction properties}

Remark For more information about these properties the reader may refer to [Bra83] and especially for the abstraction properties to [And81]

3 PRINCIPLE OF THE EXTENSION OF A REDUCTION

The schema below summarizes the process we have followed to specify and to validate a coloured extension of an ordinary reduction.

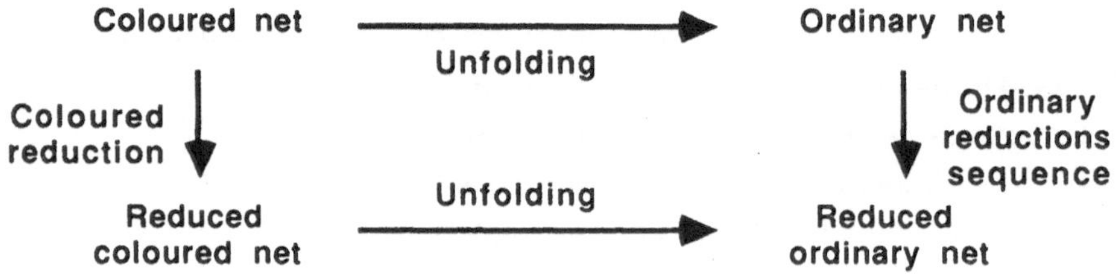

a. Specifying a coloured reduction

As the corresponding ordinary reduction, the coloured reduction is defined by two specifications, the application conditions and the transformation rule.

When specifying a coloured reduction, the conditions have to be decomposed in two parts :

- Structural reductions which (if possible) must be the same ones as the ordinary reduction. Example : a transition does not share its input places. (see the ordinary pre-agglomeration)
- Functional conditions which can not be predefined but are just those necessary to ensure a "good" unfolding. Example : a quasi-injective function valuating an arc from a place to a transition ensures that each unfolded transition does not share its input places. (see the coloured pre-agglomeration)

The transformation rule must verify these two principles :

- It may not "increase" the colour domain of the transitions of the net. (this is an imperative condition in order to have practical use of a coloured reduction)
- It only allows the composition and inverse of coloured functions in order to build new valuations since these are the only significant operations over coloured functions.

For instance we forbid the use of generalized inverses since a function may have several generalized inverses and moreover their signification (for the behaviour of the net) is unclear.

b. Validating the coloured reduction

Once the reduction is defined, we still have to prove that the previous diagram commutes. That is to say , each of the three steps (unfolding of the original net, reductions sequence, unfolding of the reduced net) is implicitly done in the specification of the coloured reduction. Let us detail it :

Unfolding of the original coloured net

One must recognize in the unfolded net the application conditions of one or more ordinary reductions as consequences of the application conditions in the coloured net.

Reductions sequence

One must verify that once an ordinary reduction is applied in the unfolded net, the conditions of reductions not yet applied are still true.

Unfolding of the reduced coloured net

Once all the ordinary reductions have been done, one must prove that the reduced net is the unfolded net of the coloured reduced net.

4 IMPLICIT PLACE SIMPLIFICATION

4.1 Ordinary implicit place simplification [Ber83]

From the original definition of an implicit place, we have excluded the case of multiple initial markings and unbounded implicit places which can be found by the covering graph and which generally represent a significant mistake in the modeling. As this definition is a restricted definition, all the results remain true. We have also suppressed a condition from Berthelot's definition, since this condition is never used in the proofs and thus, in our opinion, is not necessary.

Definition 1 **Implicit place**

Let (R,Mo) be a marked Petri net, a place p of R is implicit related to a subset of places P' if and only if :

(1) There is a flow f the support of which is {p} U P' :

$$f = a_p.p - \sum_{q \in P'} a_q.q \text{ with } a_p, a_q \in \mathbb{N}$$

(2) $\forall t \in T,\ a_p.I^-(p,t) - \sum_{q \in P'} a_q.I^-(q,t) \leq a_p.Mo(p) - \sum_{q \in P'} a_q.Mo(q)$

Interpretation An implicit place will never disable the firing of a transition since it does not initially disable it because of (2), and this condition is reproducible for all the reachable markings because of (1).

Definition 2 **Implicit place simplification**

The reduced net (R_r,Mo_r) obtained from the net (R,Mo) by simplification of the implicit place p is defined by :

- $P_r = P - \{p\}$
- $T_r = T$
- $\forall t \in T_r, \forall p' \in P_r, I_r^-(p',t) = I^-(p',t)$ and $I_r^+(p',t) = I^+(p',t)$
- $\forall p' \in P_r, Mo_r(p') = Mo(p')$

Interpretation One deletes the implicit place (arcs and marking included)

Theorem Let (R_r,Mo_r) be a reduced net obtained from the net (R,Mo) by simplification of the implicit place, π a main property different from safeness. Then :

(R,Mo) verifies π <=> (R_r,Mo_r) verifies π

(R,Mo) is safe => (R_r,Mo_r) is safe

Proof in [Ber83]

4.2 Coloured implicit place simplification

In contrast to the other reductions that we will present here, the implicit place is based on an algebraic property (existence of a particular flow). Then the generalization of this reduction implies the existence of a flow computation for coloured nets (See for instance [Had86] , [Had87b] or [Sil85]). Since the computation of flows in coloured nets is much more complex than in ordinary Petri nets, this reduction, which could be done by hand in Petri nets, now requires a good flow computation. We have not chosen a behavioural definition of an implicit place (as it is done in [Col86]) since the verification of the conditions of this definition requires the examination of all the reachable markings !

Definition 1 **Coloured implicit place**

Let (R,Mo) be a coloured net. A place p is implicit if and only if :

(1) $\forall c \in C(p)$, There is a flow f_c the support of which is $\{(p,c)\} \cup P'$

where $P' = \{ (q_1,c_1),\ldots,(q_k,c_k) \}$

$$f_c = a_{pc}.(p,c) - \sum_{i=1\ldots k} a_i.(q_i,c_i) \text{ with } a_{pc},a_i \in \mathbf{N} \text{ and } \forall c' \in C(p), (p,c') \notin P'$$

(2) $\forall t \in T, \forall c_t \in C(t)$

$$a_{pc}.I^-(p,t)(c,c_t) - \sum_{i=1\ldots k} a_i.I^-(q_i,t)(c_i,c_t) \leq a_{pc}.Mo(p,c) - \sum_{i=1\ldots k} a_i.Mo(q_i,c_i)$$

Example $C(t) = C(t') = C(p) = C(q) = \{c_1,\ldots,c_n\}$, $Mo(p) = Mo(q) = 0$

$X(c_i) = c_i$ and $(S-X)(c_i) = \sum_{i \neq j} c_j$

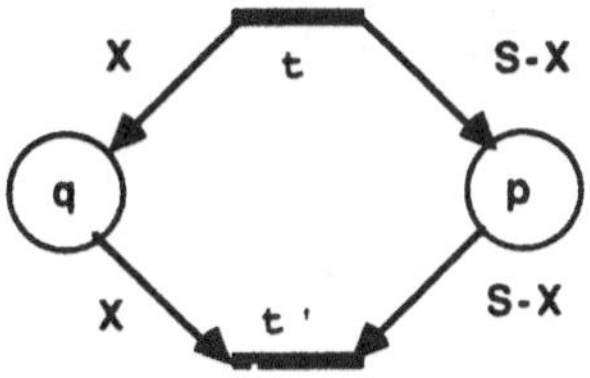

Then $f_i = (p,c_i) - \sum_{i \neq j} (q,c_j)$ and p is implicit.

Notice that even on two places, the computation of this flow is not obvious and that an inattentive reader may believe q is the implicit place ! Let us prove that q is not implicit (n=3) :

Mo[t(c1).t(c2)>M1 with M1(q)=c1+c2 and M1(p)=c1+c2+2.c3.

Then t'(c3) is not enabled for M1 because M1(q)(c3)= 0, while M1(p) ≥ c1+c2

As in ordinary Petri nets, the transformation deletes the implicit place and the related arcs.

Definition 2 **Implicit coloured place simplification**

The reduced net (R_r,Mo_r) obtained from the net (R,Mo) by simplification of the implicit place p is defined by :

- $P_r = P - \{p\}$
- $T_r = T$
- $\forall t \in T_r, \forall p' \in P_r, C_r(p') = C(p')$ and $C_r(t) = C(t)$
- $\forall t \in T_r, \forall p' \in P_r, I_r^-(p',t) = I^-(p',t)$ and $I_r^+(p',t) = I^+(p',t)$
- $\forall p' \in P_r, Mo_r(p') = Mo(p')$

Example (continued)

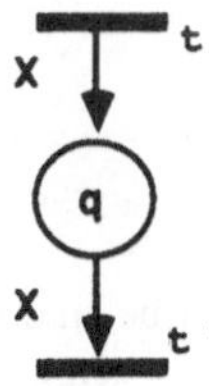

Theorem Let (R,Mo) be a coloured net and (R_r,Mo_r) be the net reduced by simplification of an implicit place. The unfolded net of (R_r,Mo_r) is obtained by a sequence of simplification of implicit places in the unfolded net of (R,Mo).

Proof

Step 1

Let (R',Mo') be the unfolded net of (R,Mo). In this net each place (p,c) has the corresponding invariant: $a_{pc}.(p,c) - \sum_{i=1...k} a_i.(q_i,c_i)$ with $a_{pc},a_i \in \mathbf{N}$ and $\forall c', (p,c') \notin P'$

with $\forall t \in T, \forall c_t \in C(t)$

$$a_p.I'^-[(p,c)(t,c_t)] - \sum_{i=1...k} a_i.I'^-((q_i,c_i)(c_i,c_t)) \leq a_p.Mo'(p,c) - \sum_{i=1...k} a_i.Mo'(q_i,c_i)$$

Hence every place (p,c) fulfills the conditions of an implicit place.

Step 2

Moreover the suppression of a place (p,c) does not change the conditions of the other places, since (p,c) **does not belong** to the support of any flow $f_{c'}$. So one can successively apply the suppression of an implicit place to every (p,c).

Step 3

Then the reduced net is exactly the unfolded net of (R_r,Mo_r) since the reductions have neither changed the initial marking, nor the incidences of the other places. ◊◊◊

Corollary Let (R_r,Mo_r) be a reduced net obtained from the net (R,Mo) by coloured simplification of the implicit place, π a main property different from safeness. Then :

(R,Mo) verifies π <=> (R_r,Mo_r) verifies π and (R,Mo) is safe => (R_r,Mo_r) is safe

5 ORTHONORMALIZATION

The reduction we present now is not a generalization of an ordinary reduction. However it is a very useful one, since it allows one to extend the application conditions of the other reductions. In fact this reduction can be considered as an equivalent transformation [Gen88] and can be proved by the rules presented in this paper. For the sake of simplicity, we have chosen to prove directly its correctness. The principle of this reduction is the following : a renaming of the colours of a transition induced by an orthonormal function of the colour domain of this transition.

Definition **Orthonormalization of a transition**

Let (R,Mo) be a marked coloured net, t be a transition of R and f be an orthonormal function of C(t). Then the reduced net (R_r,Mo_r) obtained from the net (R,Mo) by the f-orthonormalization of t is defined by :

- Pr = P, Tr = T
- $\forall t \in Tr, \forall p \in Pr, C_r(t) = C(t)$ and $C_r(p) = C(p)$
- $\forall t' \in Tr- \{t\}, \forall p \in Pr, I_r^+(p,t') = I^+(p,t'), I_r^-(p,t') = I^-(p,t')$
- $\forall p \in P_r, I_r^-(p,t) = I^-(p,t) \circ f, I_r^+(p,t) = I^+(p,t) \circ f$
- $\forall p' \in P_r, Mo_r(p') = Mo(p')$

Example

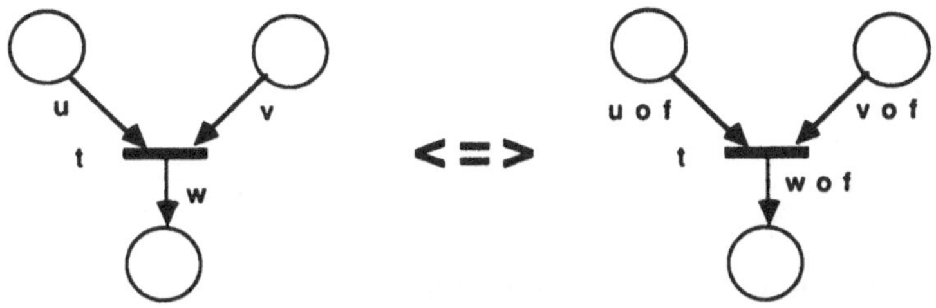

Theorem Let (R,Mo) be a marked coloured net and (R_r,Mo_r) be the reduced net obtained from the net (R,Mo) by the f-orthonormalization of t . Then the unfolded nets of these two nets are identical within an isomorphism which is the identity for the coloured places and the coloured transitions different from (t,c) and which maps (t,c) on $(t, s^{-1}(c))$ where s is the substitution associated to f.

Proof

Let us denote I'^+ (I'^-) the forward (backward) incidence matrix of the unfolded net of (R,Mo) and $I'_r{}^+$ ($I'_r{}^+$) the forward (backward) incidence matrix of the unfolded net of (R_r,Mo_r) . Then we only have to verify that :

$I'_r{}^-((p,c),(t, s^{-1}(c'))) = I'^-((p,c),(t,c'))$ and $I'_r{}^+((p,c),(t, s^{-1}(c'))) = I'^+ (p,c),(t,c'))$

$I'_r{}^-((p,c),(t, s^{-1}(c'))) = I_r{}^-(p,t)(c, s^{-1}(c')) = I^-(p,t) \text{ o } f(c, s^{-1}(c'))$

$= \sum I^-(p,t)(c,c'').f(c'', s^{-1}(c'))$ where c" ranges over C(t)

$= I^-(p,t)(c,c') = I'^-((p,c),(t,c'))$

The proof of the second identity is similar ◊◊◊

Corollary Let (R_r,Mo_r) be a reduced net obtained from the net (R,Mo) by orthonormalization of a transition, π a main property. Then :

(R,Mo) verifies π <=> (R_r,Mo_r) verifies π

6 PRE-AGGLOMERATION

6.1 Ordinary pre-agglomeration [Ber83]

Definition 1 **Pre-agglomerable transitions**

Let (R,Mo) be a marked Petri net, a subset of transitions F is pre-agglomerable if and only if there is a place p and a transition $h \notin F$ such that the following conditions are fulfilled :

(1) $I^+(p,h) = 1$ and $\forall\, t \neq h$, $I^+(p,t) = 0$

$\forall\, f \in F$, $I^-(p,f) = 1$ and $\forall\, t \notin F$, $I^-(p,t) = 0$

$Mo(p) = 0$

{ The single input transition of p is h and the output transitions of p are F }

{ All the arcs related to p are valuated by 1 }

{ p is unmarked }

(2) $\forall\, p' \neq p$, $I^+(p',h) = 0$ { The single output place of h is p }

(3) $\exists\, p' \in P$,such that $I^-(p',h) \neq 0$ { h has an input place }

(4) $\forall\, p' \in P$, $\forall\, t \in T- \{h\}$, $I^-(p',h) \neq 0 \Rightarrow I^-(p',t) = 0$

{ h does not share its input places }

Interpretation

p is an intermediate state accessed by the firing of h and left by the firing of any transition of F. The principle of the pre-agglomeration is the following : in every sequence of firings with an occurrence of h followed later by an occurrence of a transition f of F, one can postpone the firing of h and "merge" it with the firing of f.

Definition 2 **Pre-agglomeration of transitions**

The reduced net (R_r,Mo_r) obtained from the net (R,Mo) by pre-agglomeration of h and F is defined by :

- $P_r = P - \{p\}$
- $T_r = T - \{h\}$
- $\forall\, t \in Tr / F$, $\forall\, p' \in Pr$, $I_r^-(p',t) = I^-(p',t)$ and $I_r^+(p',t) = I^+(p',t)$
- $\forall\, f \in F$, $\forall\, p' \in P_r$, $I_r^-(p',f) = I^-(p',f) + I^-(p',h)$ and $I_r^+(p',f) = I^+(p',f)$
- $\forall\, p' \in P_r$, $Mo_r(p') = Mo(p')$

Interpretation The transition h disappears since in the reduced net it is merged with each transition of F. The reduced incidence matrices take this merging into account.

Theorem Let (R_r,Mo_r) be a reduced net obtained from the net (R,Mo) by pre-agglomeration of transitions, π a main property. Then :

$$(R,Mo) \text{ verifies } \pi \iff (R_r,Mo_r) \text{ verifies } \pi$$

Proof in [Ber83]

6.2 Coloured pre-agglomeration

In order to define the conditions of a coloured pre-agglomeration, there must be, as in ordinary Petri nets a place p, a transition t and a set of transitions F verifying the structural conditions of the ordinary pre-agglomeration. We are going to explain (before the proof) the additional functional conditions :

- The valuation of the arc between h and p must be an orthonormal function (u) since it implies that in the unfolded net, each place (p,c) has only one input transition with valuation 1 namely $(h,u^{-1}(c))$.

- The valuation of an arc between p and any transition of F must be an unitary function since it implies that in the unfolded net, each arc between (p,c) and (f,c') with $f \in F$ has valuation 1.

- The valuation of an input arc of h must be a quasi injective function since it implies that in the unfolded net, each transition (h,c) does not share its input places.

Definition 1 **Pre-agglomerable transitions**

Let (R,Mo) be a marked coloured net. A subset of transitions F is pre-agglomerable if and only if there is a place p and a transition $h \notin F$ such that the following conditions are fulfilled :

(1) $\forall t \neq h$, $I^+(p,t) = 0$ and $\forall t \notin F$, $I^-(p,t) = 0$

$C(p) = C(h)$ and $I^+(p,h)$ is an orthonormal function

$\forall f \in F$, $I^-(p,f) \neq 0$ and $I^-(p,f)$ is an unitary function

$Mo(p) = 0$

(2) $\forall p' \neq p$, $I^+(p',h) = 0$

(3) $\exists p' \in P$, such that $I^-(p',h) \neq 0$

(4) $\forall p' \in P$, $\forall t \in T-\{h\}$,

$I^-(p',h) \neq 0 \Rightarrow I^-(p',t) = 0$ and $I^-(p',h)$ is a quasi injective function

Comparison If we compare our reduction rule with the reduction rule n° 8 given in [Col86], we can observe that our rule extends the rule n° 8 :

- In our rule, p may have severals ouput transitions (the subset F) whereas in the rule n° 8 , only a single transition is possible.
- In our rule, the coloured functions from p to F are unitary whereas they are orthonormal in the rule n° 8 (an orthonormal function is always an unitary function).

However in our rule, there are conditions (quasi-injectivity) on the coloured functions valuating arcs going to h whereas there are none in the rule n° 8 . But then it can be proved with a simple counter-example that without these supplementary conditions, the rule n° 8 does not ensure the equivalence of liveness for the two nets.

Example

$C(p') = C1 \times C2 \times C3$ $C(h) = C(p) = C(f1) = C(f2) = C1 \times C2$

The colour functions are defined as usual, for instance :

$<X_1,X_2,S_3>(c_1,c_2) = \sum_{c \in C3} (c_1,c_2,c)$ and $<S_1-X_1,X_2>(c_1,c_2) = \sum_{c \in C1,\ c \neq c1} (c,c_2)$

The reader may verify that $<X_1,X_2,S_3>$ is quasi-injective and $<X_1,S_2>$ and $<S_1-X_1,X_2>$ are unitary.

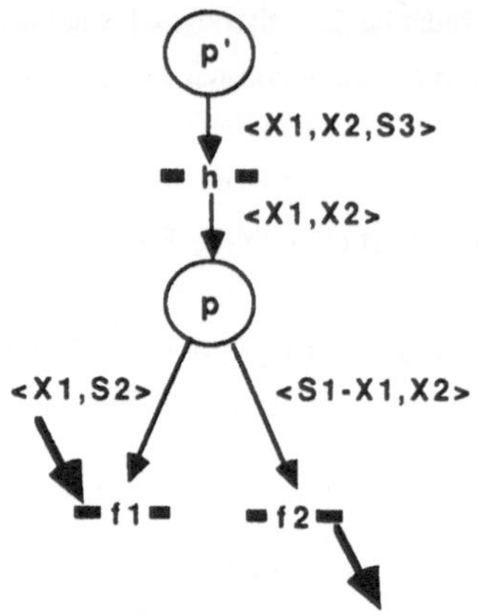

As in the ordinary pre-agglomeration the place p and the transition h disappear. The input arcs of h now become input arcs for each transition f of F. The functions valuating these arcs are obtained by the composition of the initial valuation with the inverse of the function valuating the arc between h and p and with the function valuating the arc between p and f.

Definition 2 **Pre-agglomeration of transitions**

The reduced net (R_r,Mo_r) obtained from the net (R,Mo) by a coloured pre-agglomeration of h and F is defined by :

- $Pr = P - \{p\}$
- $Tr = T - \{h\}$
- $\forall t \in Tr, \forall p' \in Pr, C_r(t) = C(t)$ and $C_r(p') = C(p')$
- $\forall t \in Tr, \forall p' \in Pr, I_r^+(p',t) = I^+(p',t)$

Let $P_h = \{ p' \in Pr / I^-(p',h) \neq 0 \}$

- $\forall t \in T, \forall p' \notin P_h, I_r^-(p',t) = I^-(p',t)$
- $\forall f \in F, \forall p' \in P_h, I_r^-(p',f) = I^-(p',h) \circ I^+(p,h)^{-1} \circ I^-(p,f)$
- $\forall p' \in P_r, Mo_r(p') = Mo(p')$

Example (continued)

For this kind of functions the composition can be symbolically done by substitution (See [Had87b]). In our example :

$<S_1-X_1,X_2,S_3> = <X_1,X_2,S_3> \circ <S_1-X_1,X_2>$

$<X_1,S_2,S_3> = <X_1,X_2,S_3> \circ <X_1,S_2>$

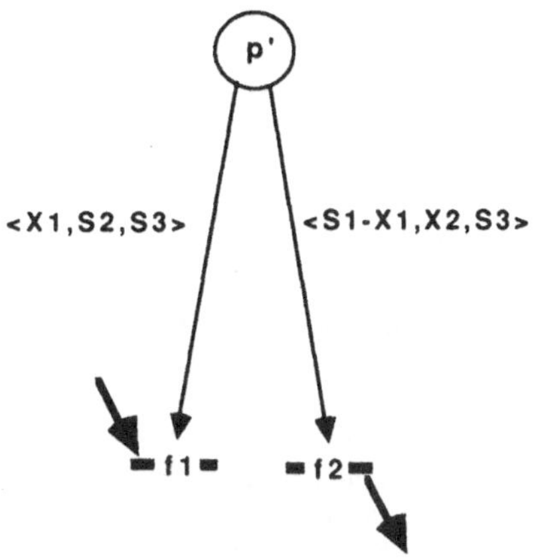

Theorem Let (R,Mo) be a coloured net and (R_r,Mo_r) be the net reduced by a pre-agglomeration of transitions. Then the unfolded net of (R_r,Mo_r) is obtained by a sequence of pre-agglomerations in the unfolded net of (R,Mo).

Proof

The proof organizes in two parts. First we prove the theorem in the case where the orthonormal functions are identities. Then we prove that the general case can be reduced to the particular case.

Part A $I^+(p,h)$ is an identity function

Step 1 of part A

Let us verify that the transition (h,c) is pre-agglomerable in the unfolded net :

- Its single output place is (p,c) which has only (h,c) for input transition. This place is unmarked. The valuation of the arc between these two nodes is 1 (because of the identity function valuating the arc h -> p and the structural conditions).

- The output transitions of (p,c) are the transitions (f,c') which verify the conditions (a) and (b)

(a) $f \in F$

(b) $I^-(p,f)$ (c,c') $\neq 0$ and then equal to 1 (since $I^-(p,f)$ is unitary)

- The input places of (h,c) are elements of the set {(p',c') where p' is any input place of h in the coloured net } . Since the p' has h for single output transition and since $I^-(p',h)$ is a quasi injective function, if (p',c') is an input place of (h,c) then (h,c) is the single output transition of (p',c').

Step 2 of part A

Let us verify that the reduction applied to (h,c) does not change the conditions of the pre-agglomeration of any (h,c'). It is sufficient to show that (h,c) and (h,c') do not share a "neighbouring" place :

- (p,c) cannot be an input place of (h,c') since p is not an input place of h in the coloured net and (p,c) cannot be an output place of (h,c') since its single input transition is (h,c).

- Let (p',c') be an input place of (h,c). Then (p',c') cannot be an output place of (h,c') since in the coloured net p' is not an output place of h and (p',c') cannot be an input place of (h,c') since (h,c) does not share its input places. Hence one can successively apply all the pre-agglomerations.

Step 3 of part A

Let us have a look at the reduced net .

- All the transitions (h,c) have disappeared
- All the places (p,c) have disappeared
- Let q≠p such that q is not an input place of h in the coloured net. Then for each c ∈ C(q), all the arcs of (q,c) are unchanged.
- Let p' be an input place of h in the coloured net, then for every place (p',c) all its input arcs are unchanged and it has an output arc for each (f,c) such that

a) ∃ c" such that $I^-(p',h)(c,c") \neq 0$

(this c" is unique because of the quasi injectivity)

b) f ∈ F

c) $I^-(p,f)\ (c",c') = 1$

The valuation of this arc is $I^-(p',h)(c,c")$ and since c" is unique, it can be rewritten as :

$$\sum_{d \in C(h)} I^-(p',h)(c,d)\ .\ I^-(p,f)\ (d,c') = I^-(p',h) \circ I^-(p,f)\ (c,c')$$

Then this reduced net is clearly the unfolded net of the coloured reduced net.

Part B

First we reduce the coloured net by an orthonormalization of h where the orthonormal function is $I^+(p,h)^{-1}$. Then the reduced net verifies the conditions of the part A and it is easy to see that the final net obtained by the reductions in the part A is the reduced coloured net ◊◊◊

Corollary Let (R_r, Mo_r) be a reduced net obtained from the net (R,Mo) by a coloured pre-agglomeration, π a main property. Then :

(R,Mo) verifies π <=> (R_r, Mo_r) verifies π

7 POST-AGGLOMERATION

7.1 Ordinary post-agglomeration [Ber83]

From the original definition of post-agglomeration, we have excluded the case of heterogeneous valuations since they do not appear in practice and they lead to technical complications. As this definition is a restricted definition, all the results remain true.

Definition 1 **Post-agglomerable transitions**

Let (R,Mo) be a marked Petri net, a subset of transitions F is post-agglomerable if and only if there is a place p and a subset of transitions H with H ∩ F = Ø such that the following conditions are fulfilled :

(1) $\forall\ h \in P$, $I^+(p,h) = 1$ and $\forall\ t \notin H$, $I^+(p,t) = 0$

$\forall\ f \in F$, $I^-(p,f) = 1$ and $\forall\ t \notin F$, $I^-(p,t) = 0$

$Mo(p) = 0$

{ The input transitions set of p is H and the ouput transitions set of p is F }

{ All the arcs related to p are valuated by 1 }

{ p is unmarked }

(2) $\exists\ f \in F$, $\exists\ p' \in P$, such that $I^+(p',f) \neq 0$

{ There is a transition of F which has an output place }

(3) $\forall\ f \in F$, $\forall\ p' \neq p$, $I^-(p',f) = 0$

{ The single input place of every transition of F is p }

Interpretation

p is an intermediate state accessed by the firing of any transition of H and left by the firing of any transition of F. The principle of post-agglomeration is the following : in every sequence of firings with an occurrence of a transition h of H followed later by an occurrence of a transition f of F, one can fire f immediately after the firing of h.

Definition 2 **Post-agglomeration of transitions**

The reduced net (R_r,Mo_r) obtained from the net (R,Mo) by post-agglomeration of H and F is defined by :

- $P_r = P - \{p\}$, $T_r = T \cup (H \times F) / (H \cup F)$
- $\forall f \in F, \forall h \in H$, one denotes hf the transition (h,f) de H x F
- $\forall t \in T_r - (H \times F)$, $\forall p' \in Pr$, $I_r^-(p',t) = I^-(p',t)$ and $I_r^+(p',t) = I^+(p',t)$
- $\forall h \in H, \forall f \in F$,
- $\forall p' \in P_r$, $I_r^-(p',hf) = I^-(p',h)$ and $I_r^+(p',hf) = I^+(p',h) + I^+(p',f)$
- $\forall p' \in P_r$, $Mo_r(p') = Mo(p')$

Interpretation The transitions of H and F disappear since they are merged by the cartesian product in the reduced net. The reduced incidence matrices take into account this product.

Theorem Let (R_r,Mo_r) be a reduced net obtained from the net (R,Mo) by post-agglomeration of transitions, π a main property. Then :

(R,Mo) verifies π <=> (R_r,Mo_r) verifies π

Proof in [Ber83]

7.2 Post-agglomeration with multiple outputs

In order to define the conditions of a coloured post-agglomeration, there must be, as in ordinary Petri nets a place p, a set of transitions H and a set of transitions F verifying the structural conditions of the ordinary post-agglomeration. We are going to explain (before the proof) the additional functional conditions :

- The colour domain of each transition of F must be the same as the colour domain of p and moreover the colour domain of p must be a projection of the colour domain of each transition of H.
- The coloured function valuating each arc from any transition of H to p must be the composition of the projection function and an orthonormal function (u_h) and the coloured function valuating each arc from p to any transition of F must be an orthonormal function (v_f).
- So when a transition (h,<c,c'>) is fired it gives a single token d (where d is such that <d,d'> = u_h(<c,c'>)) in p which can be used only by the firing of some transition (f, v_f^{-1} (d)) where f ranges over F. Then the cartesian product H x F will be well defined in the unfolded net.

Definition 1 **Post-agglomerable transitions - with multiple outputs -**

Let (R,Mo) be a coloured Petri net, a subset of transitions F is post-agglomerable if and only if there is a place p and a subset of transitions H with $H \cap F = \emptyset$ such that the following conditions are fulfilled :

(1) $\forall t \notin H, I^+(p,t) = 0$ and $\forall t \notin F, I^-(p,t) = 0$

$\forall h \in H, \exists C_h$ such that $C(h) = C(p) \times C_h$

and $I^+(p,h)$ is the composition of u_h , an orthonormal function of C(h), and the projection of C(h) over C(p)

$\forall f \in F, C(f) = C(p)$ and $I^-(p,f)$ is an orthonormal function we call it v_f.

$Mo(p) = 0$

(2) $\forall c \in C(p), \exists f \in F, \exists p' \in P$, such that $I^+(p',f)(c) \neq 0$

(3) $\forall f \in F, \forall p' \neq p,\ I^-(p',f) = 0$

Remark In the second condition $I^+(p',f)(c)$ is an item of Bag(p') (see the notations)

Comparison If we compare our reduction rule with the reduction rule n° 2 given in [Col86], we can observe that our rule extends the rule n° 2 :

- In our rule, p may have severals ouput transitions (the subset F) and several input transitions (the subset H) whereas in the rule n° 2, only a single output transition and a single input transition are possible.
- In our rule, the colour functions from H to p are compositions of an orthonormal function and a projection whereas in the rule n° 2 the functions are orthonormal (as the identity is a special case of projection, our category of functions is larger than the one of the rule n°2).

Example

C(h1) = C1 , C(h2) = C1 x C3 , C(f1) = C(f2) = C(p) = C(s) = C1 , C(q) = C2 , C(r) = C1 x C3

The colour functions are defined as usual. Notice that the symbol X1 valuating the arc h2->p denotes a projection since C(h2) = C(p) x C2 while the same symbol X1 valuating the arc p->f1 denotes identity since C(f2) = C(p) (i.e. the meaning of the function denotations in coloured nets depends on the domain of the transitions).

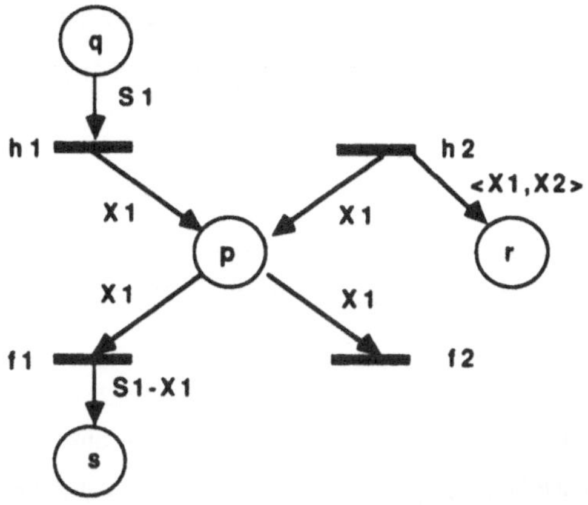

In the post-agglomeration with multiple outputs, the place p disappears and one substitutes the "product" transitions of H x F to the transitions of H and F. The arcs related to these transitions are obtained by the union of the arcs of H and F. The valuation of the output arcs of F are obtained by composition of the initial valuation with the inverse of the function valuating the arc between p and f and with the function valuating the arc between h and p. The multiple outputs denote the set F.

Definition 2 **Post-agglomeration of transitions with multiple outputs**

The reduced net (R_r, Mo_r) obtained from the net (R,Mo) by a coloured post-agglomeration of H and F is defined by :

- $P_r = P - \{p\}$, $T_r = T \cup (H \times F) / (H \cup F)$

  $\forall f \in F, \forall h \in H$, one denotes hf the transition (h,f) of H x F
- $\forall t \in T_r / (H \times F), \forall p' \in Pr, C_r(t) = C(t)$ and $C_r(p') = C(p')$

  $\forall f \in F, \forall h \in H, C_r(hf) = C(h)$
- $\forall t \in T_r / (H \times F), \forall p' \in P_r, I_r^-(p',t) = I^-(p',t)$ and $I_r^+(p',t) = I^+(p',t)$
- $\forall h \in H, \forall f \in F,$

  $\forall p' \in P_r, I_r^-(p',hf) = I^-(p',h)$ and $I_r^+(p',hf) = I^+(p',h) + I^+(p',f) \circ I^-(p,f)^{-1} \circ I^+(p,h)$
- $\forall p' \in P_r, Mo_r(p') = Mo(p')$

Example (continued)

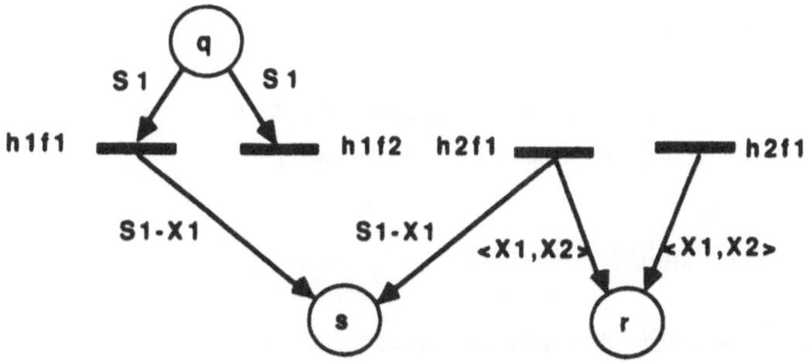

Theorem Let (R,Mo) be a coloured net and (R_r,Mo_r) be the reduced net by a post-agglomeration with multiple outputs. Then the unfolded net of (R_r,Mo_r) is obtained by a sequence of pre-agglomerations in the unfolded net of (R,Mo).

Proof

The proof organizes in two parts. First we prove the theorem in the case where the orthonormal functions are identities. Then we prove that the general case can be reduced to the particular case.

Part A All u_h and v_f are identies.

Step 1 of part A

Let us verify that the transitions (f,c) with $f \in F$ are post-agglomerable :

- Their single input place is (p,c) and conversely the only output transitions of (p,c) are (f,c). This place is unmarked. The valuation of the arc between (p,c) and any (f,c) is 1.
- The input transitions of (p,c) are (h,<c,c'>) with $h \in H$ and $c' \in C_h$
- The valuation of the arc between (p,c) is 1.
- There is a transition f such that (f,c) has an output place.

Step 2 of part A

Let us verify that the reduction applied to $\{(f,c) \,/\, f \in F\}$ does not change the conditions of the post-agglomeration of any $\{(f,c') \,/\, f \in F\}$. It is sufficient to show that (p,c) and (p,c') do not share their neighbouring transitions and this is clear :

$$[\ \{(f,c) \,/\, f \in F\} \cup \{(h,<c,c''>) \,/\, h \in H \text{ and } c'' \in C_h\}\ ] \cap$$

$$[\ \{(f,c') \,/\, f \in F\} \cup \{(h,<c',c''>) \,/\, h \in H \text{ and } c'' \in C_h\}\ ] = \emptyset$$

Hence one can successively apply all the post-agglomerations.

Step 3 of part A

Let us have a look at the reduced net .

- All the transitions (h,<c,c'>) and (f,c) have disappeared
- All the places (p,c) have disappeared
- New transitions (h.f,<c,c'>) have appeared
- Let $q \neq p$ such that q is not connected to $H \cup F$ in the coloured net. Then for each $c \in C(q)$, all the arcs of (q,c) are unchanged.
- The input places of any (h.f,<c,c'>) are exactly the same as the input places of (h,<c,c'>) and have the same valuation.
- The valuation of the arc from (h.f,<c,c'>) to an output place (p',c") is :

$$I^+(p',h)\ (c'',<c,c'>) + I^+(p',f)\ (c'',c)$$

and since $I^+(p,h)$ is the projection function

($d \neq c \Rightarrow I^+(p,h)\ (d,<c,c'>) = 0$ and $I^+(p,h)\ (c,<c,c'>) = 1$), this valuation can be rewritten :

$$I^+(p',h)\ (c'',<c,c'>) + \sum_{d \in C(p)} I^+(p',f)\ (c'',d)\ .\ I^+(p,h)\ (d,<c,c'>) =$$

$$[\ I^+(p',h) + I^+(p',f) \circ I^+(p,h)\ ]\ (c'',<c,c'>)$$

Then this reduced net is clearly the unfolded net of the coloured reduced net.

Part B

First we reduce the coloured net by the successive u_h^{-1} orthonormalization of h where h ranges over H and v_f^{-1} orthonormalization of f where f ranges over F. Then the reduced net verifies the conditions of the part A. Once we have reduced this net by the reductions of part A, we apply the u_h orthonormalization of hf again where hf ranges over HxF and it is easy to see that the final net is the reduced coloured net ◊◊◊

Corollary Let (R_r,Mo_r) be a reduced net obtained from the net (R,Mo) by a coloured post-agglomeration with multiple outputs, π a main property. Then :

$$(R,Mo) \text{ verifies } \pi \iff (R_r,Mo_r) \text{ verifies } \pi$$

7.3 Post-agglomeration with a single output

In contrast to the post-agglomeration with multiple outputs, here F is reduced to a single transition. Then the coloured function which valuates an arc from a transition of H to the place p is less constrained : it must be an unitary function (a very weak condition). There are no more constraints on the colour domains of the transitions of H. The other conditions are the same as those of the post-agglomeration with multiple outputs.

Definition 1 **Post-agglomerable transitions - with a single output -**

Let (R,Mo) be a coloured Petri net, a transitions f is post-agglomerable if and only if there is a place p and a subset of transitions H with $H \cap \{f\} = \emptyset$ such that the following conditions are fulfilled :

(1) $\forall t \notin H$, $I^+(p,t) = 0$ and $\forall t \neq f$, $I^-(p,t) = 0$

$I^+(p,h) \neq 0$ and $I^+(p,h)$ is an unitary function

$C(f) = C(p)$ and $I^-(p,f)$ is an orthonormal function we call it v_f

$Mo(p) = 0$

(2) $\forall c \in C(p)$, $\exists p' \in P$, such that $I^+(p',f)(c) \neq 0$

(3) $\forall p' \neq p$, $I^-(p',f) = 0$

Remark In the second condition $I^+(p',f)(c)$ is an item of Bag(p') (see the notations)

Comparison If we compare our reduction rule with the reduction rule n° 2 given in [Col86], we can observe that our rule extends the rule n° 2 :

- In our rule, p may have several input transitions (the subset H) while in the other rule, only a single input transition is possible.
- In our rule, the coloured functions from H to p are unitary while in the other rule they are orthonormal. (orthonormal functions are unitary)

Example

$C(h1) = C2$, $C(h2) = C(p) = C(f) = C1 \times C2$, $C(q) = C1$

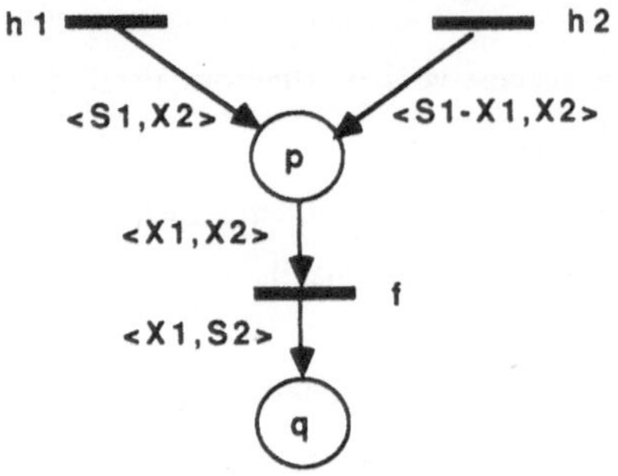

In the post-agglomeration with a single output, the place p and the transition f disappear and one substitutes the "product" transitions of H x {f} to the transitions of H. The arcs related to these transitions are obtained by the union of the arcs of H and {f}. The valuation of the output arcs of f are obtained by composition of the initila valuation with the inverse of the function valuating the arc between p and f and with the function valuating the arc between h and p.

Definition 2 **Post-agglomeration of transitions with a single output**

The reduced net (R_r,Mo_r) obtained from the net (R,Mo) by a coloured post-agglomeration of H and f is defined by :

- $P_r = P / \{p\}$
- $T_r = T / \{f\}$
- $\forall t \in T_r / H, \forall p' \in Pr, C_r(t) = C(t)$ and $C_r(p') = C(p')$
- $\forall t \in T_r / H, \forall p' \in P_r, I_r^-(p',t) = I^-(p',t)$ and $I_r^+(p',t) = I^+(p',t)$
- $\forall h \in H, \forall p' \in P_r$,
  $I_r^-(p',h) = I^-(p',h)$ and $I_r^+(p',h) = I^+(p',h) + I^+(p',f) \circ I^-(p,f)^{-1} \circ I^+(p,h)$
- $\forall p' \in P_r, Mo_r(p') = Mo(p')$

Example (continued)

Notice that here, the composition of the functions is obtained by a symbolic substitution [Had87b] :

$<S_1,S_2> = <X_1,S_2> \circ <S_1,X_2>$ and $<S_1 - X_1,S_2> = <S_1 - X_1,S_2> \circ <S_1,X_2>$

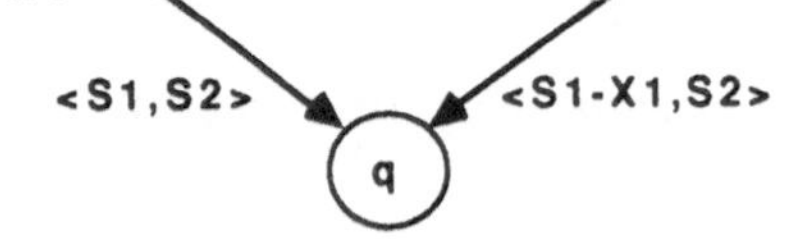

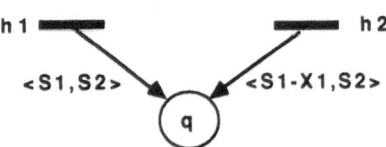

Theorem Let (R,Mo) be a coloured net and (R_r,Mo_r) be the net reduced by a post-agglomeration with a single output. The unfolded net of (R_r,Mo_r) is obtained by a sequence of post agglomerations in the unfolded net of (R,Mo).

Proof

The proof organizes in two parts. First we prove the theorem in the case where the orthonormal functions are identities. Then we prove that the general case can be reduced to the particular case.

Part A v_f is an identity function.

Step 1 of part A

Let us verify that the transition (f,c) is post-agglomerable :

- Its single input place is (p,c) and conversely the only output transition of (p,c) is (f,c). This place is unmarked. The valuation of the arc between (p,c) and any (f,c) is 1.
- The input transitions of (p,c) are (h,c') with $h \in H$ and $I^+(p,h)(c,c') \neq 0$
- The valuation of the arc between (p,c) and (h,c') is 1since $I^+(p,h)$ is unitary
- The condition (2) implies that (f,c) has an output place.

Step 2 of part A

Let us verify that the reduction applied to (f,c) does not change the conditions of the post-agglomeration of any (f,c'). Since there is a single transition in the set F' the new transitions of H' x F' may be identified with the transitions of H'.The places (p,c) and (p,c') may share their input transitions (some subset of {(h,c")}). But when the reduction is applied since (f,c) has not (p,c) for an output place, the valuation of the arc between a transition (h,c") and the place (p,c') is unchanged.

Hence one can successively apply all the post-agglomerations.

Step 3 of part A

Let us have a look at the reduced net .

- All the transitions (f,c) have disappeared
- All the places (p,c) have disappeared
- Let $q \neq p$ such that q is not connected to H U {f} in the coloured net. Then for each $c \in C(q)$, all the arcs of (q,c) are unchanged.
- The input places of any (h,c) are unchanged.
- The valuation of the arc from (h,c) to an output place (p',c') is exactly:

$$I^+(p',h)\,(c',c) + \sum_{d \in C(p)} I^+(p',f)\,(c'',d)\;.\;I^+(p,h)\,(d,c) =$$

$$[\, I^+(p',h) + I^+(p',f) \circ I^+(p,h)\,]\,(c',c)$$

The sum over d is obtained by the successive reductions.

Then this reduced net is clearly the unfolded net of the coloured reduced net.

Part B

First we reduce the coloured net by the v_f^{-1} orthonormalization of f. Then the reduced net verifies the conditions of the part A. Once we have reduced this net by the reductions of part A, it is easy to see that the final net is the reduced coloured net ◊◊◊

Corollary Let (R_r,Mo_r) be a reduced net obtained from the net (R,Mo) by a coloured post-agglomeration with a single ouput, π a main property. Then :

(R,Mo) verifies $\pi \iff (R_r,Mo_r)$ verifies π

8 APPLICATION TO THE DATA BASE MANAGEMENT MODEL

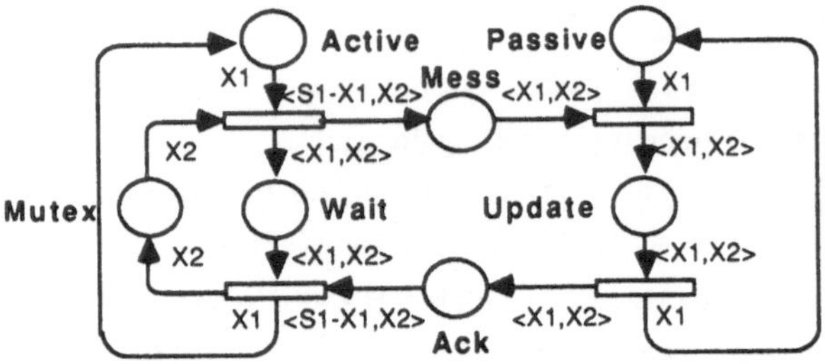

Post agglomeration with a single output around Update

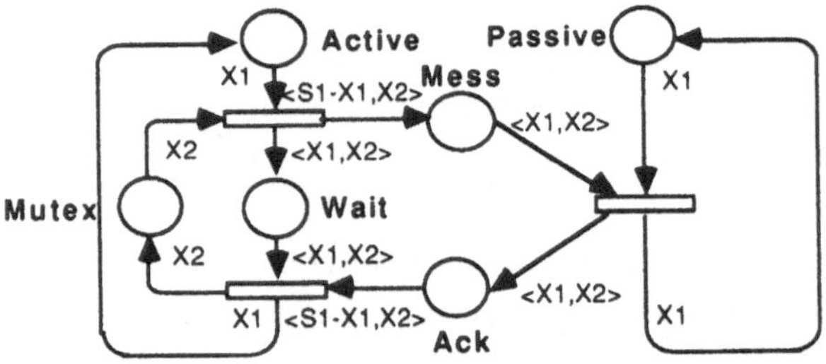

Simplification of the implicit place Passive

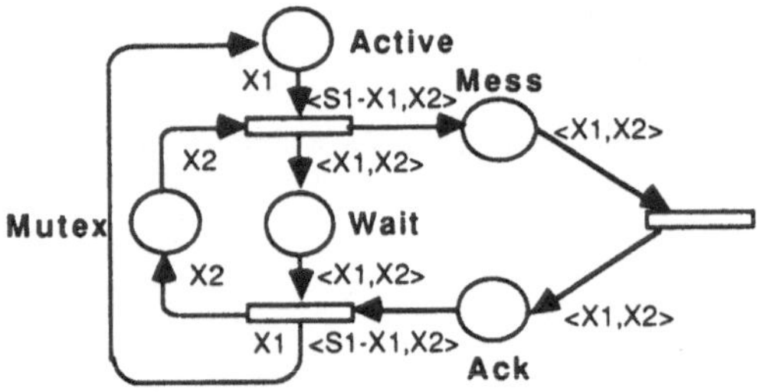

Pre agglomeration around Ack

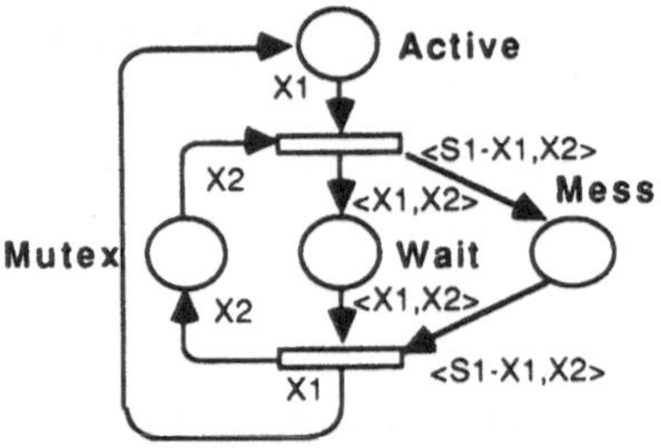

Simplification of the implicit place Mess

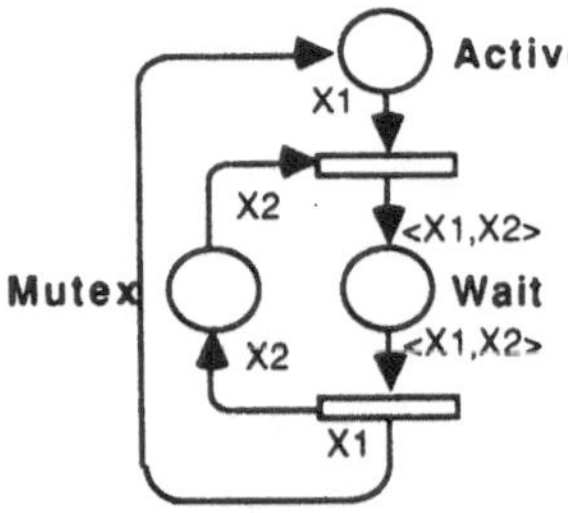

Post agglomeration
with a single output around Wait

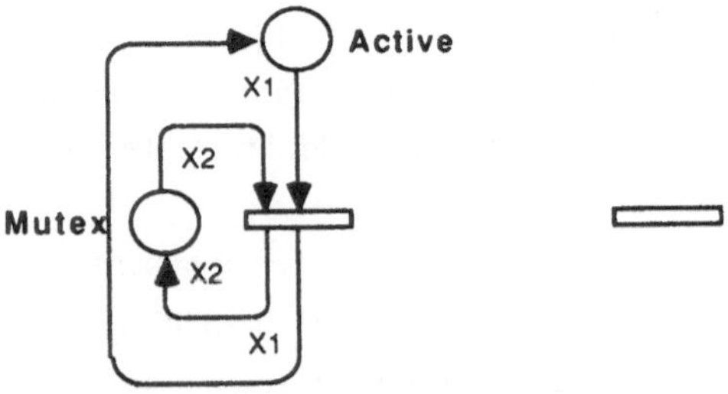

Simplification of the implicit
pl aces Mutex and Active

We present now the model of a data base management with multiple copies. This model is an improved version of those of [Jen81a]. Each site has two processes, an active one and a passive one. The access grant to a file of the data base is centralized and submitted to the mutual exclusion. In order to modify a file the active process of a site must get its grant and once it has modified the file, it sends messages to the other sites with the updated file. Then the passive processes update their own data base and send an acknowledgment. Once the active process has received all the acknowledgments, it releases the grant. Simultaneous accesses to different files are allowed.

In the net, an active process must get in Mutex the single token coloured by the file it wants to access. The messages are composed by the name of the receiver followed by the name of the file. The acknowledgments are composed by the name of the sender followed by the name of the file. Accessing and modifying a file is modelled by a single transition (indivisible step) while the updating of the passive process is modelled by a place (divisible step). Initially there is a token per site in Active and Passive and a token per file in Mutex.

C1 = { Sites } , C2 = { Files }
C(Active) = C(Passive) = C1 , C(Mutex) = C2
C(Wait) = C(Update) = C(Mess) = C(Ack) = C1 x C2
For every transition t , C(t) = C1 x C2
The coloured functions are defined as in the previous examples.

In the final net (a single transition) all the main properties are verified. Thus the original net also verifies the main properties (boundedness, liveness, ...).

9 CONCLUSION

Here we have presented a methodology to generalize reductions to coloured nets. This methodology is based on two principles :

- Not define, if possible, additional structural conditions for the extended reduction rules.
- Only define the functional conditions necessary to ensure the equivalence between the reduced net and the original net:

In order to illustrate this methodology we have extended Berthelot's most frequently used rules.

The advantages of our reductions are twofold :

- on the one hand, they are strictly equivalent to the reductions defined by Berthelot and they then have the numerous properties proved by him;
- on the other hand the functional conditions are not predefined but are as weak as possible to obtain this equivalence in each case and then the reductions have a large field of application.

With the coloured reductions, we have completely reduced an improved version [Had87b] of the data base management [Jen81a] with multiple copies. As we discussed at the beginning, reduction rules can still be improved. One way to do so, is to combine our reductions with the equivalence transformations of Genrich as we have already done with the orthonormalization of transitions improving our previous reductions [Had88]. Another way is to extend the fusions defined in [Ber86] , but it must be noticed that these reductions are based on behavioural conditions and then the problem of their extension is quite different.

ACKNOWLEDGMENTS

I would like to acknowledge the four anonymous referees who helped me to improve my paper and Claude Dutheillet for checking of the final version.

REFERENCES

[And81] **C. ANDRE** : Systèmes à évolutions parallèles : Modélisation par réseaux de Petri à capacités et analyse par abstraction. Thèse d'état. Université de Nice (1981)

[Bae81] **J.L. BAER, C.GIRAULT, G. GARDARIN, G. ROUCAIROL** : The two step comitment protocol : modeling specification and proof methodology. Fifth international conference on software enginering IEEE/ACM San diego pp 363-373 (1981)

[Ber83] **G. BERTHELOT** : Transformation et analyse de réseaux de Petri, applications aux protocoles. Thèse d'état. Université P. et M. Curie. Paris (1983)

[Ber85] **G. BERTHELOT** : Checking properties of nets using transformations. LNCS 222. Advances in Petri nets 85, G. Rozenberg ed, Springer-Verlag 1986.

[Ber86] **G. BERTHELOT** : Transformations and decompositions of nets. LNCS 254. Advances in Petri nets 86, G. Rozenberg ed, Springer-Verlag 1987.

[Bra83] **G.W. BRAMS** : Réseaux de Petri. Théorie et pratique. Masson editeur, Paris (1983)

[Col86] **J M COLOM, J MARTINEZ, M SILVA** : Packages for validating discrete production systems modeled with Petri nets. IMACS-IFAC Symposium, Lille, France (1986)

[Gen81] **H.J. GENRICH, K. LAUTENBACH:** System modelling with high-level Petri nets. Theoretical computer science 13,1981, pp 103-136.

[Gen88] **H.J. GENRICH** : Equivalence transformations of PrT-nets.Ninth european workshop on applications and theory of Petri nets. Venise, Italie (1988).

[Had86] **S. HADDAD, C. GIRAULT:** Algebraic structure of flows of a regular net. Seventh european workshop on applications and theory of Petri nets, Oxford England , june 1986, in "Advances in Petri nets 87", L.N.C.S. n° 266, G.Rozenberg ed., Springer Verlag, 1987, pp 73-88.

[Had87a]S. **HADDAD:** Un calcul d'une base de flots pour les réseaux colorés. Deuxième colloque C3 Angoulême, France,1987.

[Had87b] **S. HADDAD** : Une catégorie régulière de réseau de Petri de haut niveau: définition, propriétés et réductions. Application à la validation de systèmes distribués.Thèse de l'Universite Pierre et Marie Curie. Paris (1987)

[Had88] **S. HADDAD** : Generalization of reduction theory to coloured nets. Ninth european workshop on applications and theory of Petri nets. Venise, Italie (1988).

[Jen81a] **K. JENSEN** : How to find invariants for coloured Petri nets. 10 th symposium on Mathematical foundations of computer science 1981, L.N.C.S. vol 118, Springer-Verlag,1981, pp 327-338.

[Jen81b] **K. JENSEN** : Coloured Petri nets and the Invariant method. TCS 14 (1981)

[Jen82] **K. JENSEN** : High-level Petri nets. Third european workshop on applications and theory of Petri nets, Varenna Italy, september1982, pp 261-276.

[Kro89] **P. KROHN, M. RAUHAMAA, A. YLIKOSKI** : Reduction transformations of PrT_nets. Submitted to the sixth symposium on theoretical aspects of computer science. Paderborn.

[Lau85] **K. LAUTENBACH, A. PAGNONI** : Invariance and duality in predicate transition nets and in coloured nets. Arbeitspapiere der G.M.D, n°132.

[Sil85] **M. SILVA, J. MARTINEZ, P. LADET, H. ALLA** : Generalized inverses and the calculation of symbolic invariants for coloured Petri nets. Technique et science informatique, Vol.4 n°1, 1985, pp 113-126.

15.
Equivalence Transformations of PrT-Nets

H.J. Genrich

G. Rozenberg (ed.): Advances in Petri nets 1989. Lecture Notes in Computer Science, vol. 424.
Springer, Berlin Heidelberg New York 1990, pp. 179-208

Abstract: There is a net theoretical system model called ***Predicate Transition Nets (PrT-Nets)*** that describes distributed systems as first-order ***dynamic*** structures. A single PrT-net can present a whole family of related systems in the same way as a set of logical formulae may describe a whole family of (static) systems – the class of its set-theoretical models.
In this note two PrT-nets are called *equivalent* if they describe the same family of systems. A list of formal transformations (re-writing rules) for PrT-nets that are *consistent* with equivalence is presented. For a non-trivial subclass of PrT-nets this set of rules is shown to characterize equivalence; it is also *complete*.

Key Words: Higher-level Petri nets; predicate transition nets; semantical equivalence; rewriting rules.

Contents

1 Introduction

Predicate transition nets (PrT-nets) have been introduced in [3] as higher-level Petri nets that describe distributed systems as first-order *dynamic* structures.

A PrT-net presents a distributed system in terms of a first-order language consisting of function symbols and relation symbols where some of the relation symbols have a *dynamic* interpretation (relative to a *fixed* domain of individuals). The places of the underlying net represent the relations whose extensions change dynamically. They are consequently marked by tokens consisting of tuples of individuals to denote the 'current' interpretations of the associated relation symbols. These current interpretations capture in a natural way the distributed states of the system.

The transitions of the underlying net correspond to *schemes* of elementary changes (events). An occurrence of an event changes the 'old' interpretations of the relation symbols that are in its reach into 'new' interpretations. This effect is expressed through *token patterns* annotating the flow arcs impinging upon a transition. An event proper is generated by binding the individual

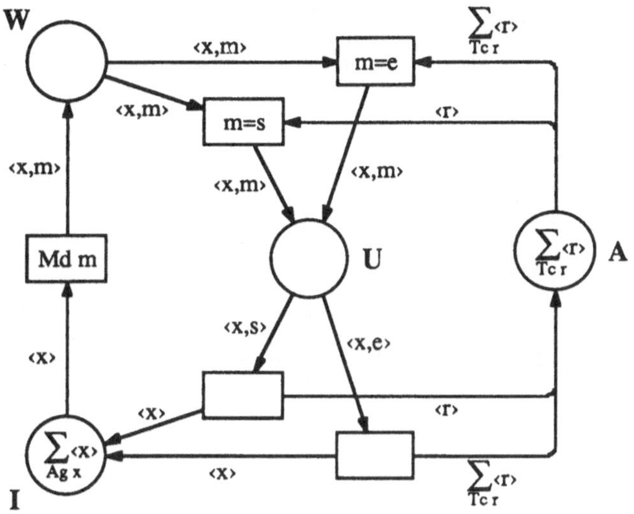

Vocabulary

Constants

s: shared access mode; **e**: exclusive access mode.

Static relations

$Ag\,x$: x is an agent; $Md\,m$: m is an access mode; $Tc\,r$: r is a ticket.

Dynamic relations

$I\,x$: agent x is idle; $W\,xm$: x wants access in mode m; $U\,xm$: x is accessing in mode m;
$A\,r$: ticket r is available.

Representative Constellation

All agents are idle and all tickets are available.

Figure 1: Example of a predicate transition net

variables in the neighbouring token patterns so that a 'guard' associated with the transition is satisfied. Such guards are composed out of function symbols and relation symbols that have a *static* (i.e. fixed) interpretation over the domain of individuals.

As an example we present in figure 1 a simple ticket mechanism that regulates the usages of a single commodity by a group of agents. There are N agents that can have access to the commodity in two modes: either exclusively (mode **e**), or shared (mode **s**) - up to L agents at the same time. The access which is not depicted itself is coordinated by means of L tickets; an agent needs one ticket to get shared access and all of them to get exclusive access.

A revised version of the PrT-net model was presented in [5]. It maintains the original approach but adds a few features that increase the modelling power and flexibility of PrT-nets in mainly two respects.

- The revised PrT-nets fully support the power of using the interpretation of the static symbols as a parameter, allowing to represent a whole family of related systems as a single *abstract* PrT-net.

- The same family of systems may be represented at *various degrees of detail.* Places and transitions may be split or merged locally thus trading the simplicity of a part of the underlying net for the simplicty of some annotating expressions, and vice versa.

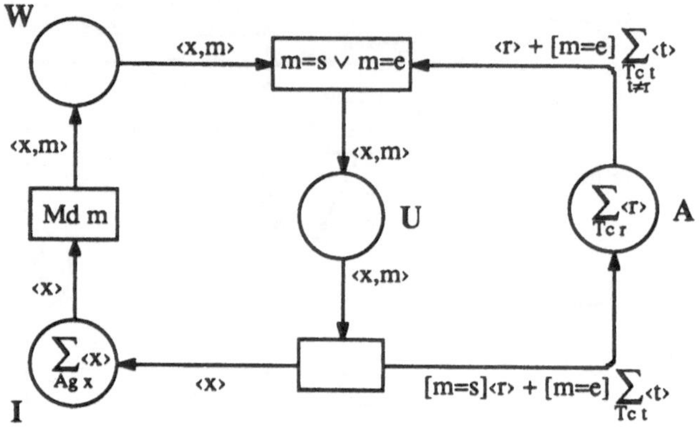

Figure 2: Boolean coefficients and non-uniform transitions

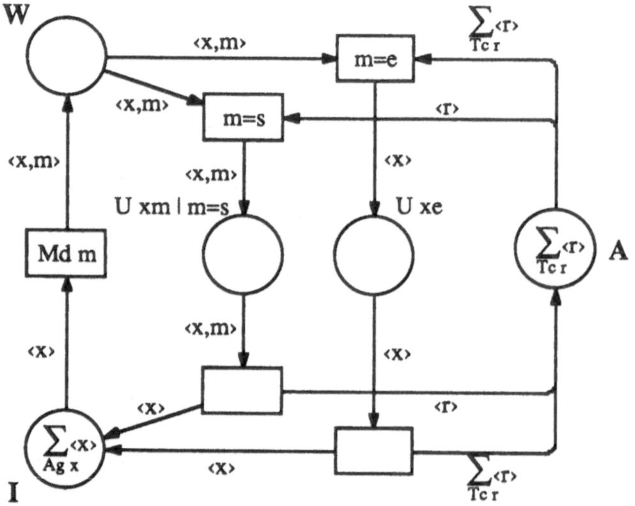

Figure 3: A dynamic relation spread over more than one place

It is the modelling *flexibility* of PrT-nets in particular that this note is concerned with. We shall present and study a collection of formal transformations – a kind of re-write rules – that leave the class of systems represented by a PrT-net unchanged. An instance of such an equivalence is already indicated in figure 1. The same effect that the guards $m{=}\mathbf{s}$ respectively $m{=}\mathbf{e}$ have in the upper part of the net is achieved in the lower part by **s** and **e** occurring in the respective token patterns.

We shall now use the same example to demonstrate how the idea of equivalence transformation has influenced the enhancement of the original, basic PrT-net model. We begin with observing that the two transitions representing the access to the commodity look so similar that one may wish to merge them into a single one. The same holds for the two transitions representing the return from using the commodity. The difficulty is that the result of merging would no longer be *uniform* [6]. Merging yields an arc whose annotation does not evaluate to the same number of tokens in all occurrence modes. To allow non-uniform arc annotations, the use of the transition

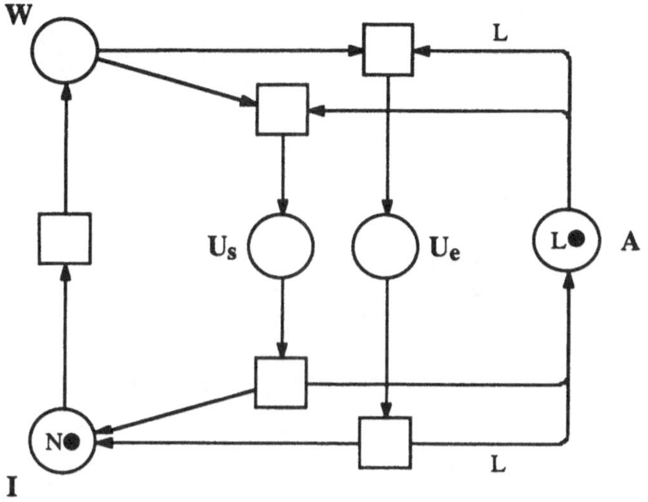

Figure 4: A merely quantitative view: multiple tokens and weighted arcs

guards was generalized. The symbolic sums of tuples annotating the arcs were allowed to contain 'guards' as well. Every summand may be multiplied by a *boolean coefficient*, or *conditional* $[\varphi]$ that evaluates to either 1 or 0 depending on whether the formula φ evaluates to *true* ($\top$) or *false* ($\bot$).

The result of merging the two pairs of related transitions is shown in figure 2 in two different forms. It is easy to see that the two arc annotations

$$\langle r\rangle + [m{=}\mathbf{e}]\sum_{Tc\,t,\,t\neq r}\langle t\rangle \quad\text{and}\quad [m{=}\mathbf{s}]\langle r\rangle + [m{=}\mathbf{e}]\sum_{Tc\,t}\langle t\rangle$$

have the same value for all interpretations of the vocabulary and all values of the variables r, m. In full generality, however, arc annotations containing $\sum$ (whose maximal size may therefore depend on the static interpretation) are difficult to handle – to the effect that the completeness result we shall present later only applies to PrT-nets without $\sum$ occurring at arcs.

Analogously to transitions one may also want to operate on the places of the net. In figure 3 we have divided the relation U into its two aspects *shared use* and *exclusive use* – once more in two slightly different yet equivalent ways. This representation can then be used, for example, to verify the ticket mechanism by means of standard integer S-invariant techniques. To this end it can be simplified, by a *non-equivalence* transformation, into a place transition net – a conventional Petri net with arc weights – as shown in figure 4.

To allow partitioning of dynamic relations in general, a place in the enhanced PrT-net model can be annotated by an atomic formula that is derived from a dynamic predicate and functions as a *condition scheme*. It can be followed by yet another kind of 'guard' which selects a subset of all the possible instances of the condition scheme and hence is called a *place selector*. Different places must be strictly *disjoint* in such that for no interpretation of the static vocabulary it is possible to replace the variables at two different places in such a way that the condition schemes become identical and the selectors both hold.

Note how the number of different variables occurring in a condition scheme determines the *degree* of the place it annotates, i.e. the length of tuples occurring at the arcs impinging upon it. Annotating a place by just a dynamic predicate can now be viewed as a shorthand for a condition

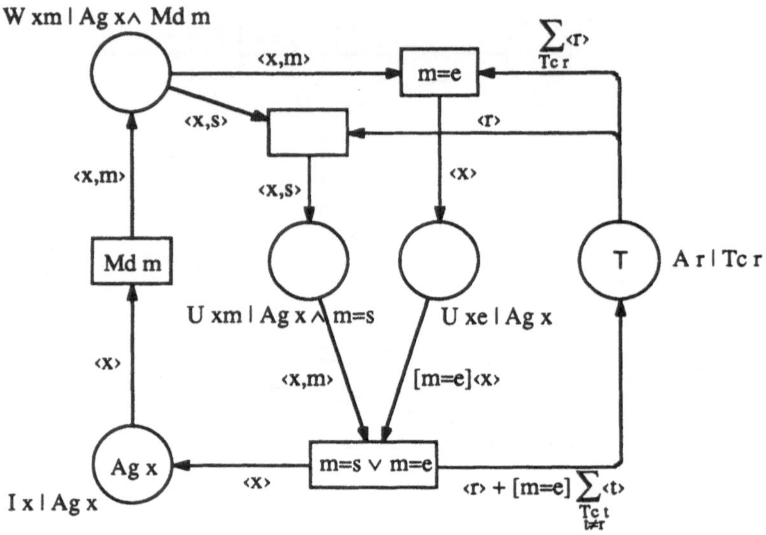

Vocabulary

Constants

s: shared access mode; **e**: exclusive access mode.

Static relations

$Ag\,x$: x is an agent; $Md\,m$: m is an access mode; $Tc\,r$: r is a ticket.

Dynamic relations

$I\,x$: agent x is idle; $W\,xm$: x wants access in mode m; $U\,xm$: x is accessing in mode m; $A\,r$: ticket r is available.

Properties of the Static Support

Md consists of the two different access modes **s** and **e**.

$$\mathbf{s} \neq \mathbf{e} \quad \wedge \quad (Md\,m \leftrightarrow (m = \mathbf{s} \vee m = \mathbf{e}))$$

Representative Constellation

All agents are idle and all tickets are available.

$$I^0\,x \leftrightarrow Ag\,x\ ;\quad A^0\,r \leftrightarrow Tc\,r\ ;\quad W^0\,xm \vee U^0\,xm \;\rightarrow\; \bot$$

Figure 5: Fully parameterised (abstract) PrT-net

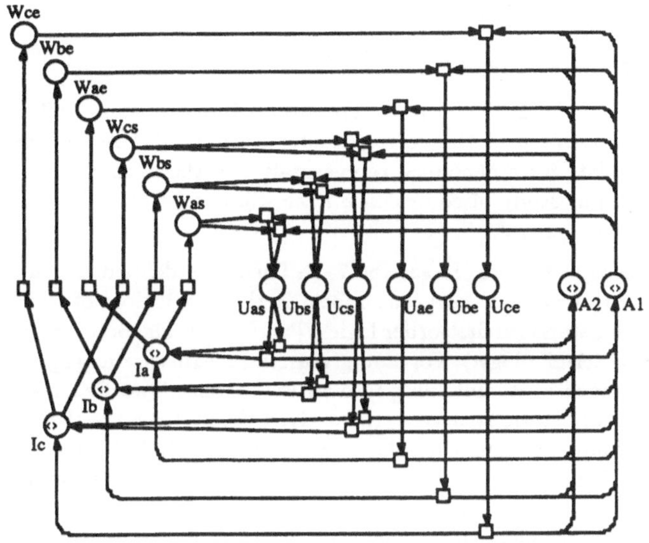

Static Support: $Ag = \{\mathbf{a},\mathbf{b},\mathbf{c}\}$; $Tc = \{1,2\}$; $Md = \{\mathbf{s},\mathbf{e}\}$

Figure 6: Complete unfolding of a concrete PrT-net (3 agents, 2 tickets)

scheme where different variables occur at all positions – the degree of the place being the same as the arity of the predicate.

The generalized form of annotating places also allows to express their initial marking in a way that directly corresponds to the formal specification of the representative constellations for each member of the family of systems presented as an abstract PrT-net.

An example of such an abstract presentation of the ticket mechanism is shown in figure 5. Place selectors are used here also to specify the type of each dynamic relation in terms of the three types of individuals, *agents*, *tickets*, and *modes*. The irregularity of the presentation – analogous parts are presented in different rather than analogous ways – is meant again to hint at possible equivalence transformations.

A particular member of the family of systems represented by an abstract PrT-net is chosen by providing a static ***support***, i.e a structure for the static vocabulary. The interpretation of the static symbols can be restricted by providing a set of formulae expressing some required properties of the supports – a kind of 'super-guard' that applies to the whole PrT-net. A specific finite support can be provided by assigning proper names to all individuals in a one-to-one fashion and listing the interpretations of the static function and relation symbols.

For any given support, the represented system is an *Elementary Net System (EN system)* [8] that can be derived by totally unfolding the abstract PrT-net for this support. If the support is finite, the unfolding can be done by a sequence of equivalence transformations that step by step split places and transitions into its constituent conditions respectively events. For N=3 agents a, b, c and L=2 tickets $1, 2$ the resulting EN system is shown in figure 6.

Note that the ***unit*** $\langle\rangle$ is nothing but the ***token*** of the EN system model. Since *all arcs* of the total unfolding are annotated by $\langle\rangle$, the arc annotations have been omitted in figure 6. The interpretation of concrete PrT-nets as EN systems reflects the fact that in our approach the places of PrT-nets are marked by sets, not by multi-sets. Higher-level Petri net models with multi-sets as place markings as well as conventional Petri nets like the one in figure 4 can be treated as a special case (rather than a generalization) within this *strict*, set-theoretic approach. [4]

The EN system of figure 6 seen as a special PrT-net is equivalent to the PrT-net of figure 5 to whose support specification we add

$$Ag\,x \leftrightarrow (x = \mathbf{a} \vee x = \mathbf{b} \vee x = \mathbf{c})\,; \quad Tc\,r \leftrightarrow (r = 1 \vee r = 2)$$

assuming that the individual constants are always interpreted by different individuals. Since it provides the denotation of the corresponding concrete PrT-net we have, for finite concrete PrT-nets, a candidate for a canonical representative of the class of all equivalent PrT-nets. An important part of our study of equivalence later will consist of finding such a representative for abstract PrT-nets in general.

We now proceed as follows. In section 2, PrT-nets are defined as syntactical objects. Syntax and evaluation of those expressions that appear in the annotations of PrT-nets are defined. (This is, to a large extent, standard first-order logic [7] and a little algebra; it mainly serves as a reference for notations and terminology.) For any given static structure as support, the denotation of a PrT-net is determined as its unfolding into an EN system. Two PrT-nets are then defined to be equivalent if for every static structure, the derived EN systems are the same.

The sections 3 and 4 contain the substance of this note. In 3 the list of equivalence transformations is formulated and proved to be sound. Section 4 develops, in several steps, the completeness result.

In the conclusion, some open problems and issues of future work are discussed.

2 The Formalism of PrT-Nets

In this section (which is taken – with minor changes – from [5]), PrT-nets are introduced as formal objects that can be manipulated like logical formulae or algebraic expressions. We define the syntax and semantics of the expressions that we derive from a given vocabulary of function and relation symbols and use for annotating the PrT-nets. This is, to a large extent, standard first-order logic [7] and a little algebra. It helps to make this note technically self-contained and mainly serves as a reference for notations and terminology.

For any given interpretation of the static part of the vocabulary, the denotation of a PrT-net is determined unfolding it into an EN system. Two PrT-nets are then defined to be equivalent if for every static structure, the derived EN systems are the same.

The language for annotating PrT-nets consists of individual terms and formulae of *first-order predicate logic* plus a class of simple algebraic expressions being symbolic sums of tuples of individual terms.

Definition 2.1 For each index $n \geq 0$, let $\Omega^{(n)}$ be a set of n-ary operators and $\Pi^{(n)}$ a set of n-ary predicates and let $\Omega = \bigcup_{n \in \mathbb{N}} \Omega^{(n)}$ and $\Pi = \bigcup_{n \in \mathbb{N}} \Pi^{(n)}$ be finite. In addition, there is an infinite set of symbols, V, disjoint from Ω and Π whose elements serve as (individual) *variables*, and unsigned integers are denoted in the usual way. Operators and predicates form the *vocabulary* of a language $\mathbf{L} = \mathbf{L}(\Omega, \Pi)$ that consist of three kinds of expressions: $\mathbf{L}$ is $\mathbf{T} \cup \mathbf{F} \cup \mathbf{S}$ where $\mathbf{T}$ is the set of *individual terms*, $\mathbf{F}$ is the set of *first-order logical formulae*, and $\mathbf{S} = \bigcup_{k \in \mathbb{N}} \mathbf{S}^{(k)}$ is the set of *symbolic sums*. Terms in $\mathbf{T}$, formulae in $\mathbf{F}$ and symbolic sums in each $\mathbf{S}^{(k)}$ are built in the following way.

1. *Terms*:
   (a) A variable is in $\mathbf{T}$.
   (b) If $f^{(n)}$ is a n-ary operator and $v_1, \ldots, v_n$ are terms then $f(v_1, \ldots, v_n)$ is in $\mathbf{T}$. (Note that 0-ary operators are terms; they are used, as usual, as proper names of distinct individuals.)

(c) No other expression is in **T**.

2. *Formulae*:

   (a) The constants $\top$ and $\bot$ are ***atomic*** formulae in **F**.

   (b) If v_1 and v_2 are terms then $v_1{=}v_2$ is an atomic formula in **F**.

   (c) If $P^{(n)}$ is a n-ary predicate and $v_1, \ldots, v_n$ are terms then $Pv_1 \ldots v_n$ is an atomic formula in **F**. (Note that 0-ary predicates are atomic formulae; they are unstructured propositions, the propositional variables of propositional logic.)

   (d) If p_1 and p_2 are in **F** then $\neg p_1$ and $(p_1 \vee p_2)$ are in **F**.

   (e) If x is a variable and p is in **F** then $(\exists x)p$ is in **F**.

   (f) No other expression is in **F**.

   Remark: The connectors $\wedge$, $\rightarrow$, $\leftrightarrow$ and $\forall$ are derived from $\neg$, $\vee$ and $\exists$ in the usual way.

3. *Symbolic Sums*:

   (a) The constant **0** is in $\mathbf{S}^{(k)}$.

   (b) If $v_1, \ldots, v_k$ are terms then the k-tuple $\langle v_1, \ldots, v_k \rangle$ is in $\mathbf{S}^{(k)}$.

   (c) If l_1, l_2 are in $\mathbf{S}^{(k)}$ then $(l_1 + l_2)$ is in $\mathbf{S}^{(k)}$.

   (d) If l is in $\mathbf{S}^{(k)}$ and z denotes a non-negative integer then zl is in $\mathbf{S}^{(k)}$; z is called an ***integer coefficient***.

   (e) If l is in $\mathbf{S}^{(k)}$ and p is a formula then $[p]l$ is in $\mathbf{S}^{(k)}$; $[p]$ is called a ***boolean coefficient*** or ***conditional***.

   (f) If l is in $\mathbf{S}^{(k)}$ and c_1, c_2 are (integer or boolean) coefficients then $(c_1{+}c_2)l$ is in $\mathbf{S}^{(k)}$.

   (g) If x is a variable and l is in $\mathbf{S}^{(k)}$ then $\sum_x l$ is in $\mathbf{S}^{(k)}$.

   (h) No other expression is in $\mathbf{S}^{(k)}$. □

An occurrence of a variable x in an expression E is called a *free occurrence* if it is not in the range of a $(\exists x)$, $(\forall x)$ or $\sum_x$. The set of variables that have a free occurrence in E is called the *degree* of E. An expression is called ***open*** iff all its occurrences of variables are free, and it is called ***closed*** iff no occurrence is free. Closed terms are compound names of individuals; closed formulae are propositions.

Free occurrences of a variable in an expression serve as place holders for individuals; they can be replaced by terms.

Definition 2.2 Let E be an expression of **L**, $x_1, \ldots, x_n$ different variables and $v_1, \ldots, v_n$ terms in **T**. Then $E{:}\{x_1{\leftarrow}v_1, \ldots, x_n{\leftarrow}v_n\}$ designates the result of substituting v_i for each free occurrence of x_i, simultaneously for $1 \leq i \leq n$.
$\alpha = \{x_1{\leftarrow}v_1, \ldots, x_n{\leftarrow}v_n\}$ is called a ***substitution***, and $E{:}\alpha$ is called the ***α-instance*** of E. Note that $E{:}\alpha$ is a term, formula or symbolic sum iff E is. □

Closed expressions are evaluated with respect to a given structure $\mathcal{R}$. A closed term denotes an individual and a closed formula denotes a truth value, as usual. The purpose of closed symbolic sums is to denote ***multi-sets*** (***bags***). We define multi-sets as non-negative linear combinations.

Definition 2.3 Let D be a set. A mapping $\lambda{:}\, D{\rightarrow}\mathbf{Z}$ is called a ***linear combination in D with integer coefficients***. The set of all linear combinations in D is denoted by $\mathcal{L}(D)$, $\mathcal{L}(D) = [D{\rightarrow}\mathbf{Z}]$. For $D = D_1 \times \cdots \times D_n$ we write $\mathcal{L}(D_1, \ldots, D_n)$. (Note that $\mathcal{L}()$ is isomorphic to **Z**.)
The set of non-negative linear combinations in D is denoted by $\mathcal{L}^+(D)$; its elements are ***multi-sets*** over D. (Note that for every subset of D, its characteristic function belongs to $\mathcal{L}^+(D)$).

The combinations whose coefficients are all 0 or all 1 are denoted by **0** and **1**, respectively. (**0** corresponds to the empty set and **1** corresponds to D.)
Our notation for a single linear combination is such that for $D = \{a,b,c,d\}$, $2\langle a\rangle - 3\langle b\rangle + \langle d\rangle$ denotes $\{a \mapsto 2,\ b \mapsto -3,\ c \mapsto 0,\ d \mapsto 1\}$. □

In $\mathcal{L}(D)$, componentwise addition, multiplication, multiplication by scalars, and comparison are defined in the straightforward way. (Note that for characteristic vectors of subsets of D, componentwise multiplication corresponds to intersection.)

Definition 2.4 Let $\lambda, \lambda_1, \lambda_2$ be in $\mathcal{L}(D)$ and z an integer. Then

1. $(\lambda_1 + \lambda_2)(x) := \lambda_1(x) + \lambda_2(x) \quad (x \in D)$
2. $(\lambda_1 \bullet \lambda_2)(x) := \lambda_1(x) \cdot \lambda_2(x) \quad (x \in D)$
3. $(z\lambda)(x) := z\lambda(x) \quad (x \in D)$
4. $\lambda_1 \leq \lambda_2 :\Longleftrightarrow \forall x \in D : \lambda_1(x) \leq \lambda_2(x)$ □

For evaluating the closed expressions of **L**, every operator of **L** must be interpreted by a function and every predicate of **L** must be interpreted by a relation of the respective arity.

Definition 2.5 For $\mathbf{L} = \mathbf{L}(\Omega, \Pi)$, a structure $\mathcal{R} = (D; F_1, \ldots, F_k; R_1, \ldots, R_n)$ is called a **L**-*structure* if every operator f of $\Omega^{(m)}$ denotes a m-ary function of $\mathcal{R}$ designated by $f_\mathcal{R}$, $f_\mathcal{R} : D^m \rightarrow D$, and every predicate P of $\Pi^{(m)}$ denotes a m-ary relation of $\mathcal{R}$ designated by $P_\mathcal{R}$, $P_\mathcal{R} \subseteq D^m$.
Note that $D^m = [m \rightarrow D]$ where $0 = \emptyset$ and $m{+}1 = m \cup \{m\}$ in the set-theoretic model of the natural numbers. Hence $D^0 = \{\emptyset\} = 1$, $D^1 = \{\emptyset, x \mid x \in D\} \simeq D$, and for every $f \in \Omega^{(0)}$, $f_\mathcal{R} = \{(\emptyset, x)\} \simeq x$ for some $x \in D$. □

To ensure that each individual in the domain of $\mathcal{R}$ can be named in an expression, we now add to the vocabulary of **L** a new set, $U_\mathcal{R}$, of *constants* denoting the individuals of $\mathcal{R}$ in a one-to-one fashion. For our purposes we assume that $\Omega^{(0)}$ is contained in $U_\mathcal{R}$ and consequently, different proper names in $\Omega^{(0)}$ denote different individuals. The individual denoted by a constant u is designated as $u_\mathcal{R}$. Every constant is defined to be a term. The language derived from this augmented vocabulary is designated as $\mathbf{L}_\mathcal{R}$.

A **L**-structure $\mathcal{R}$ assigns to every closed term, v, of $\mathbf{L}_\mathcal{R}$ an individual of $\mathcal{R}$, designated by $\mathcal{R}(v)$, to every closed formula (proposition), p, of $\mathbf{L}_\mathcal{R}$ the truth value $\top$ (*true*) or $\bot$ (*false*), designated by $\mathcal{R}(p)$, and to every symbolic sum l of $\mathbf{L}_\mathcal{R}$ a non-negative linear combination (a multi-set) of tuples of individuals of $\mathcal{R}$, designated by$\mathcal{R}(l)$.

For symbolic sums l containing $\sum$, we define $\mathcal{R}(l)$ first for finite $\mathcal{R}$. This is extended to the infinite case if l is *bounded*, i.e. iff there is a $k \geq 0$ such that for all finite $\mathcal{R}$, $\mathcal{R}(l) \leq k\,\mathbf{1}$. For example, $\sum_x \langle x\rangle$ is bounded but $\sum_x \langle a\rangle$ is not. It is easy to see that a symbolic sum without $\sum$ is bounded and that it diverges for some infinite $\mathcal{R}$ if it is unbounded.

Definition 2.6 Let $\mathcal{R}$ be a **L**-structure and $E \in \mathbf{L}_\mathcal{R}$ be a closed expression. $\mathcal{R}(E)$ is defined recursively on the syntactic structure of E.

1. (a) If v is a constant u, $\mathcal{R}(v)$ is the individual denoted by u, $u_\mathcal{R}$.
   (b) If v is $f^{(n)}(v_1, \ldots, v_n)$ then $\mathcal{R}(v) = f_\mathcal{R}(\mathcal{R}(v_1), \ldots, \mathcal{R}(v_n))$.
2. (a) $\mathcal{R}(\top) = \top$ and $\mathcal{R}(\bot) = \bot$.
   (b) If p is $v_1 = v_2$ then $\mathcal{R}(p) = \top$ iff $\mathcal{R}(v_1)$ and $\mathcal{R}(v_2)$ are the same individual.
   (c) If p is $P^{(n)} v_1 \ldots v_n$ then $\mathcal{R}(p) = \top$ iff $\langle \mathcal{R}(v_1), \ldots, \mathcal{R}(v_n)\rangle \in P_\mathcal{R}$.
   (d) If p is $\neg q$ then $\mathcal{R}(p) = \top$ iff $\mathcal{R}(q) = \bot$.

(e) If p is $(p_1 \vee p_2)$ then $\mathcal{R}(p) = \top$ iff $\mathcal{R}(p_1) = \top$ or $\mathcal{R}(p_2) = \top$.

(f) If p is $(\exists x)q$ then $\mathcal{R}(p) = \top$ iff there is a constant d such that $\mathcal{R}(q{:}\{x \leftarrow d\}) = \top$.

3. (a) $\mathcal{R}(\mathbf{0})$ is $\mathbf{0}$.

   (b) If l is $\langle v_1, \ldots, v_n \rangle$ then $\mathcal{R}(l){:}\, D^n \to \mathbf{Z}$ is such that $\mathcal{R}(l)(d_1, \ldots, d_n) = 1$ iff $d_i = \mathcal{R}(v_i)$ for $1 \leq i \leq n$, and $\mathcal{R}(l)(d_1, \ldots, d_n) = 0$ otherwise, for $\langle d_1, \ldots, d_n \rangle \in D^n$.

   (c) If l is $(l_1 + l_2)$, $\mathcal{R}(l)$ is $\mathcal{R}(l_1) + \mathcal{R}(l_2)$.

   (d) If l is zl', $\mathcal{R}(l)$ is $z\mathcal{R}(l')$.

   (e) If l is $[p]l'$, $\mathcal{R}(l)$ is $\mathcal{R}(l')$ iff $\mathcal{R}(p) = \top$, and $\mathbf{0}$ otherwise.

   (f) If l is $(c_1{+}c_2)l'$, $\mathcal{R}(l)$ is $\mathcal{R}(c_1 l') + \mathcal{R}(c_2 l')$.

   (g) If l is $\sum_x l'$ and D is $\{d_1, \ldots, d_k\}$, $\mathcal{R}(l)$ is $\sum_{i=1}^{k} \mathcal{R}(l'{:}\{x \leftarrow u_i\})$.

   (h) If l is $\sum_x l'$, l' is bounded and $U_{\mathcal{R}}$ is $\{u_i\}_{i \in \mathbf{N}}$, $\mathcal{R}(l)$ is $\sum_{i=0}^{\infty} \mathcal{R}(l'{:}\{x \leftarrow u_i\})$.

□

The following notions are based on the evaluation of expressions.

Definition 2.7 Let $\mathbf{A}$, $\mathbf{B}$ be finite sets of formulae of $\mathbf{L}$, p, q formulae of $\mathbf{L}$, l, l' bounded symbolic sums of $\mathbf{L}$, $\mathcal{R}$ a $\mathbf{L}$-structure and r a formula of $\mathbf{L}_{\mathcal{R}}$. Then

1. r is said to be *valid* in $\mathcal{R}$ iff each closed instance of r is true in $\mathcal{R}$. Notation: $\mathcal{R} \models r$.
2. p is said to be *valid* iff it is valid in every $\mathbf{L}$-structure. Notation: $\models p$.
3. q is said to be implied by p iff $(p \rightarrow q)$ is valid. Notation: $p \Rightarrow q$.
4. $\mathcal{R}$ is called a *model* of $\mathbf{A}$ iff every formula of $\mathbf{A}$ is valid in $\mathcal{R}$. Notation: $\mathcal{R} \models \mathbf{A}$
5. p is called a *logical consequence* of $\mathbf{A}$ iff p is valid in every model of $\mathbf{A}$. Notation: $\mathbf{A} \models p$
6. $\mathbf{A}$ and $\mathbf{B}$ are called *logically equivalent* iff for all structures $\mathcal{R}$, $\mathcal{R} \models \mathbf{A}$ iff $\mathcal{R} \models \mathbf{B}$.
7. l and l' are called equivalent iff for all finite structures $\mathcal{R}$ and all substitutions α, $\mathcal{R}(l{:}\alpha) = \mathcal{R}(l'{:}\alpha)$. Notation: $l = l'$.
8. $[p]$ and $[q]$ are called equivalent iff for all symbolic sums l, $[p]l = [q]l$. Notation: $[p] = [q]$. □

From these definitions we get immediately the following laws (equivalence transformations) for boolean coefficients and symbolic sums.

Proposition 2.8 Let p, q be formulae, l_1, l_2, l_3 bounded symbolic sums and x, y variables. Then

1. $[\top] = 1$, $[\bot] = 0$
2. $[p] = [q]$ iff $p \Leftrightarrow q$.
3. $[p] + [\neg p] = 1$
4. $[p][q] = [p \wedge q]$
5. $[p] + [q] = [p \vee q] + [p \wedge q]$
6. $(l_1 + (l_2 + l_3)) = ((l_1 + l_2) + l_3)$
7. $(l_1 + l_2) = (l_2 + l_1)$
8. $[p]([q]l) = ([p][q])l$
9. $([p]l_1 + [p]l_2) = [p](l_1 + l_2)$
10. $([p]l + [q]l) = ([p] + [q])l$
11. $\sum_x [p]l = [p] \sum_x l$ if x does not occur freely in p.

12. $\sum_x(l_1+l_2) = \sum_x l_1 + \sum_x l_2$

13. $\sum_x \sum_y l = \sum_y \sum_x l$

Proof: The verification of these laws is straightforward. □

In a usual first-order structure $\mathcal{R}$, all functions and relations are static. The representation of dynamic systems as PrT-nets, however, requires that we distinguish between predicates denoting static relations and predicates denoting dynamic relations (we do not consider dynamic functions). Consequently we divide the set of predicates, Π, into a set of static predicates, Π_{st}, and a set of dynamic predicates, Π_{dy}.

We designate the sublanguage $\mathbf{L}(\Omega, \Pi_{st})$ of $\mathbf{L}$ as $\mathbf{L}_{st}$. Dynamic predicates are used in this note only to build *condition schemes*. (In general they will appear in statements that may hold at some or all states of the modelled systems. [5]) Condition schemes are atomic formulae of the form $Pw_1 \ldots w_m$ where P is in $\Pi_{dy}^{(m)}$ and the w_i are individual constants and variables (no compound terms). We designate the class of condition schemes as $\mathbf{C}_{dy}$.

We are now prepared to define three classes of PrT-nets with annotations in $\mathbf{L} = \mathbf{L}(\Omega, \Pi)$ where $\Pi = \Pi_{st} \cup \Pi_{dy}$ with $\Pi_{st} \cap \Pi_{dy} = \emptyset$.

Definition 2.9 $\mathcal{A}(\mathbf{L})$ is the class of *abstract* PrT-nets. It consists of marked annotated nets, $\mathcal{N} = (N, A, M^0)$, where N is the underlying directed net, A is its *annotation*, and M^0 is its *initial marking*.

1. N is a directed pre-net, $N = (S, T; F)$, with $S \cap T = \emptyset$ and $F \subseteq (S \times T) \cup (T \times S)$ (we do not exclude isolated elements in N because they may appear – temporarily – in the transformation processes we define later);
2. A is the annotation of N, $A = (A_N, A_S, A_T, A_F)$ where
   (a) $A_N \subseteq \mathbf{F}_{st}$ is a finite set of static formulae that specifies the permissible supports of $\mathcal{N}$.
   (b) $A_S{:}S \to \mathbf{C}_{dy} \times \mathbf{F}_{st}$ is a mapping that assigns to every place s a pair of formulae (π, p) – written as $\pi|p$ – such that
      - π is an atomic formula $Pu_1 \ldots u_m$ built from a dynamic predicate $P \in \Pi_{dy}^{(m)}$ and individual constants and variables (no compound terms) which is called the *condition scheme* of s;
      - p is an *open, static* formula called the *selector* of s;
      - for every place $s' \neq s$ annotated by $\pi'|p'$, the set of conditions derivable from π that satisfy the selector p is disjoint from the set of conditions derivable from π' that satisfy p', for all supports of $\mathcal{N}$ (if π and π' are built from the *same* dynamic predicate P and for some $\mathcal{R} \models A_N$, there are substitutions $\alpha, \alpha'{:}V \to U_{\mathcal{R}}$ such that $(Pu_1 \ldots u_m){:}\alpha$ and $(Pu'_1 \ldots u'_m){:}\alpha'$ are identical expressions in $\mathbf{L}_{\mathcal{R}}$, then $\mathcal{R}(p{:}\alpha) = \bot$) or $\mathcal{R}(p'{:}\alpha') = \bot$);
   (c) $A_T{:}T \to \mathbf{F}_{st}$ is a mapping that assigns to every transition $t \in T$ an *open, static* formula q called the *selector* of t;
   (d) $A_F{:}F \to \mathbf{S}_{st}$ is a mapping that assigns to every arc $(x, y) \in F$ a symbolic sum l such that the boolean coefficients of l, called *arc selectors*, are formed by *open, static* formulae and l is in $\mathbf{S}_{st}^{(n)}$ if at the adjacent place s ($x = s$ or $y = s$), n is the number of different variables occurring in the condition scheme of s.
3. $M^0{:}S \to \mathbf{F}_{st}$ is a mapping that assigns to each place $s \in S$ an *open, static* formula m called the *initial marking* of s. □

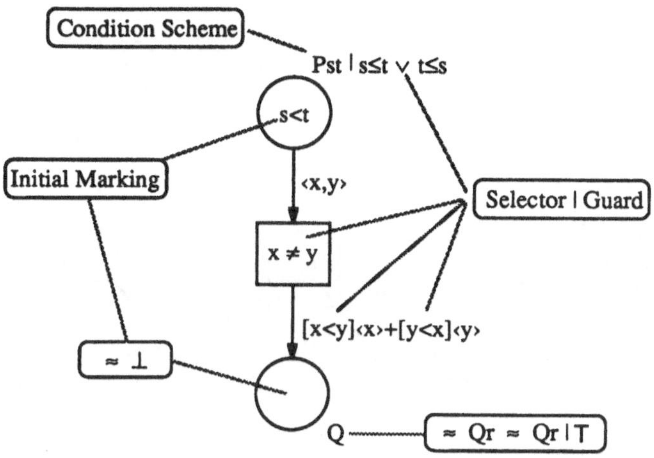

Vocabulary:

$\Omega = \emptyset$; $\Pi_{st} = \{<^{(2)}\}$; $\Pi_{dy} = \{P^{(2)}, Q^{(1)}\}$

Support Axioms:

$\neg x < x$; $(x < y \wedge y < z) \rightarrow x < z$

Representative Constellation:

$P^0\, st \leftrightarrow (s < t \vee t < s \vee s = t)$; $Q^0\, r \rightarrow \perp$

Figure 7: Example of an abstract PrT-net

There is a special subclass of $\mathcal{A}(\mathbf{L})$ that will later play a central role in proving the completeness result. We call it the *basic form* of PrT-nets. It is very much like the original PrT-nets that do not have boolean coefficients at the arcs or dynamic relations split into several places. In addition, they do not contain several transitions of the same 'type'. As we shall see, the basic form is a kind of normal form that allows to compare PrT-nets that are supposed to be equivalent.

Definition 2.10 A PrT-net $\mathcal{N} = (N, A, M^0)$ in $\mathcal{A}(\mathbf{L})$ is said to be in *basic form* iff

- N is a net, i.e. there are no isolated elements in N;
- the annotations of all places have the form $Px_1 \ldots x_m|\mathsf{T}$ such that the x_i are all different variables;
- the annotations of all arcs are free of (boolean or integer) coefficients; they do not contain terms other than individual constants and variables;
- there are no two different transitions with the same preset and the same postset of places such that at each pair of corresponding arcs, the annotations consist of the same number of tuples (transitions of the same *type* are identical).

The class of PrT-nets in basic form is designated by $\mathcal{B}(\mathbf{L})$. □

The third class of PrT-nets we want to identify is the class of *concrete* PrT-nets. Here an abstract PrT-net and a $\mathbf{L}_{st}$-structure that satisfies its support specification are chosen to denote a specific system rather than a whole family of them.

Definition 2.11 A pair $(\mathcal{N}, \mathcal{R})$ of a PrT-net $\mathcal{N} = (N, A, M^0) \in \mathcal{A}(\mathbf{L})$ and a $\mathbf{L}_{st}$-structure $\mathcal{R}$ satisfying the support specification $A_{\mathcal{N}}$ of $\mathcal{N}$ is called a *concrete* PrT-net. The class of concrete PrT-nets is designated by $\mathcal{C}(\mathbf{L})$. □

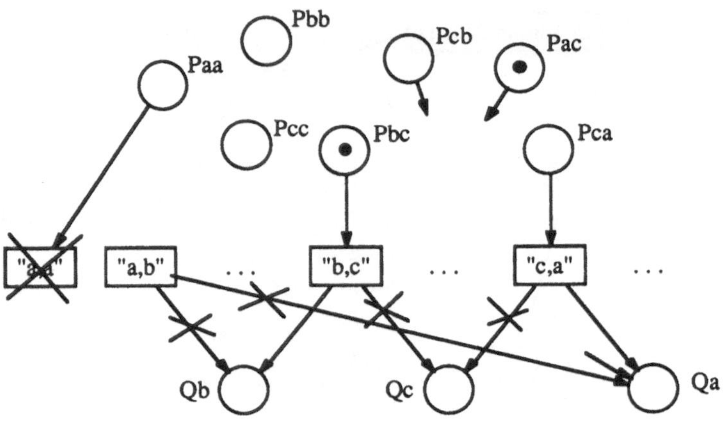

Support: $\mathcal{R} = (\{a, b, c\}; \{(a, c), (b, c)\})$

Figure 8: Unfolding of a concrete PrT-net

Figure 7 shows a very small and rather meaningless example of an abstract PrT-net. Of all the equivalent PrT-nets shown in figure 9 only the upper right one is in basic form. If we associate with the abstract PrT-net in figure 7 some ordered set $(D; <)$, for example the set $D = \{a, b, c\}$ ordered by $< = \{(a, c), (b, c)\}$), we have a concrete PrT-net.

Figure 7 explains the different kinds of annotations of abstract PrT-nets and introduces some notational conventions (indicated by $\approx$). The vocabulary, the support specification and the representative constellation are presented in a kind of legend but are meant as formal as all other annotations.

> The support is specified as a (partially) ordered set. The upper place represents that portion of a dynamic binary relation P that contains only pairs of individuals that are ordered or identical. Its initial marking consists of all pairs (s, t) with s less than t.
> The lower place represents a dynamic unary relation Q without restriction. Its initial marking is empty.
> The transition takes pairs of different individuals from the ordered portion of P and adds the respective less one to the relation Q.

If we have an abstract PrT-net $\mathcal{N} = (N, A, M^0)$ in $\mathcal{A}(\mathbf{L})$ and a $\mathbf{L}_{st}$-structure $\mathcal{R}$ satisfying the support specification A_N, the concrete PrT-net $(\mathcal{N}, \mathcal{R})$ denotes a single dynamic system that we designate as $\mathcal{R}(\mathcal{N})$. $\mathcal{R}(\mathcal{N})$ is now determined by unfolding $\mathcal{N}$ into an elementary net system [8], $\mathcal{R}(\mathcal{N}) = (B, E; \widehat{F}, c^0)$.

Definition 2.12 An *elementary net system* is a tuple $\Sigma = (B, E; F, c^0)$ where

- B is the set of *conditions* and E is the set of *events* such that $B \cap E = \emptyset$ and $B \cup E \neq \emptyset$.
- F is the *flow* relation between conditions and events, $F \subseteq (B \times E) \cup (E \times B)$ such that no condition or event is isolated.
- c^0 is a set of conditions serving as the representative of some suitable reachability class of cases under the strict transition rule, for example the forward reachabilty class, $[c^0\rangle$. □

In $\mathcal{R}(\mathcal{N})$ the conditions are closed atomic formulae built from the dynamic predicates, the events are instances of the transitions of $\mathcal{N}$. Figure 8 shows the result of unfolding the PrT-net in figure 7 for the set $\{a, b, c\}$ ordered by the relation $< = \{(a,c),(b,c)\}$. (Those conditions that are shown in the diagram as legal instances of a place but are not connected to any legal instance of a transition will be eliminated during the unfolding process.)

For the precise presentation of the unfolding procedure, we need some auxiliary notions.

Definition 2.13

1. Let s be a place of $\mathcal{N}$ annotated by $Pw_1 \ldots w_m|p$. The list $\langle x_1, \ldots, x_n \rangle$ of different variables occurring in $(w_1, \ldots, w_m)$ and being ordered by their first occurrences is called the *degree* of s.
2. Let t be a transition of $\mathcal{N}$. The set $\{y_1, \ldots, y_k\}$ of those variables that occur at some arc impinging upon t is called the *degree* of t.
3. The *scope* of a variable occurring freely *at a place* s are the annotation and the marking of s.
4. The *scope* of a variable occurring freely *at a transition* t are the annotations of t and its adjacent arcs.
5. Those variables that occur at a node (place or transition) n but do not belong to the degree of n are called *dangling* variables.
6. If n is a node and q a formula, we designate by q/n the result of binding all variables that do not belong to the degree of n by $\exists$. □

For the operation of restricting the degree of an open formula, it is good to know the following facts.

Proposition 2.14 Let q, q_1, q_2 be formulae and $I = \{x_1, \ldots, x_n\}$ a set of variables. Let q/I designate the result of binding all variables that have a free occurrence in q but do not belong to I by $\exists$. Then

1. $q \Rightarrow q/I$,
2. $(q_1/I \vee q_2/I) \Longleftrightarrow (q_1 \vee q_2)/I$,
3. there is a q_1' with $q_1/I \Longleftrightarrow q_1'/I$ such that $(q_1/I \wedge q_2/I) \Longleftrightarrow (q_1' \wedge q_2)/I$.

Proof: 1 and 2 are obvious. In 3, q_1' is the result of consistently renaming the variables to be bound in q_1 such that they differ from all the variables to be bound in q_2. □

In the sequel let $\mathcal{N} = (N, A, M^0)$ be an abstract PrT-net in $\mathcal{A}(\mathbf{L})$, let $\mathcal{R}$ be a $\mathbf{L}_{st}$-structure satisfying the support specification of $\mathcal{N}$, A_N, and let t be a transition of $\mathcal{N}$ that is annotated by the selector q and whose degree is $\{y_1, \ldots, y_k\}$.

Definition 2.15 Let α be a substitution that replaces all variables occurring at t by constants of $U_{\mathcal{R}}$. α is called *$\mathcal{R}$-feasible* at t iff

1. α satisfies the selector q, $\mathcal{R}(q{:}\alpha) = \top$;
2. α does not create multiple arcs; for every symbolic sum l at an adjacent arc, $\mathbf{0} \le \mathcal{R}(l{:}\alpha) \le \mathbf{1}$;
3. α does not generate an impurity; for every pair of opposite adjacent arcs annotated by symbolic sums l_1 and l_2, respectively, $\mathcal{R}(l_1{:}\alpha) \bullet \mathcal{R}(l_2{:}\alpha) = \mathbf{0}$. □

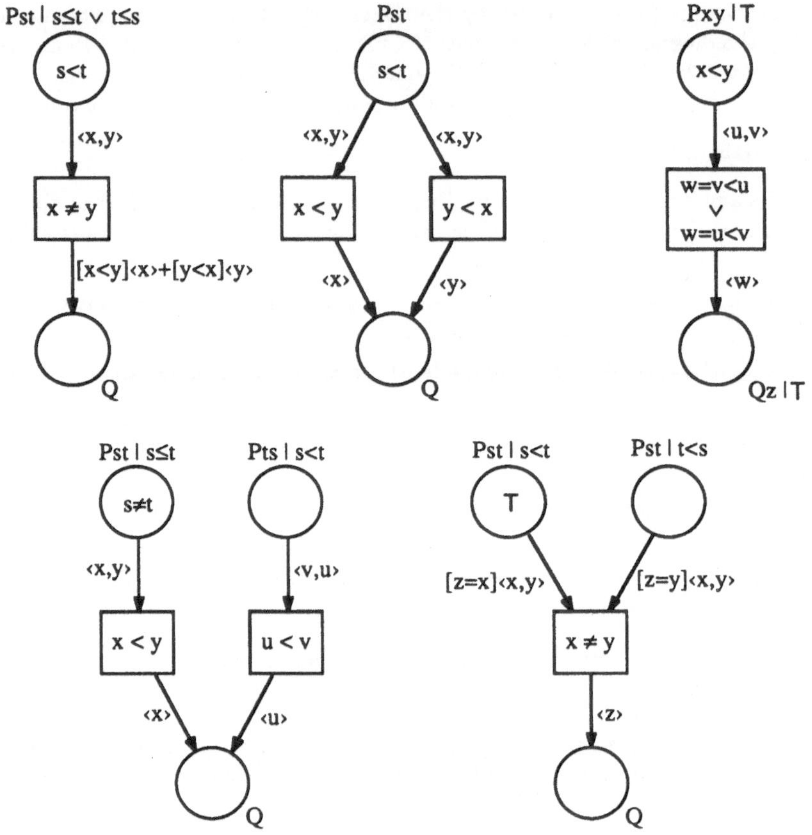

Figure 9: Equivalent PrT-nets

Definition 2.16 Let α be a $\mathcal{R}$-feasible substitution for t. The *α-instance* of t, designated by $t{:}\alpha$, is a (non-trivial) pair of sets of conditions, a preset ${}^\bullet(t{:}\alpha)$ and a postset $(t{:}\alpha)^\bullet$ with ${}^\bullet(t{:}\alpha) \cup (t{:}\alpha)^\bullet \neq \emptyset$, such that the following holds.

A condition $Pu_1 \ldots u_m$ with $P \in \Pi_{dy}^{(m)}$ and $u_1, \ldots, u_m \in U_{\mathcal{R}}$ belongs to ${}^\bullet(t{:}\alpha)$, respectively $(t{:}\alpha)^\bullet$, iff

1. there is a place s of $\mathcal{N}$ annotated by $Pw_1 \ldots w_m | p$ with degree $\langle x_1, \ldots, x_n \rangle$ and a substitution $\tilde{\alpha} = \{x_1 \leftarrow u_{i1}, \ldots, x_n \leftarrow u_{in}\}$ such that
   - $Pu_1 \ldots u_m$ is generated by $\tilde{\alpha}$, i.e. $\mathcal{R}(w_j{:}\tilde{\alpha}) = u_{j\mathcal{R}}$ for $1 \leq j \leq m$, and
   - $\tilde{\alpha}$ satisfies the selector p of s, i.e. there is a substitution β for the dangling variables of p such that $\mathcal{R}((p{:}\tilde{\alpha}){:}\beta) = \top$;
2. there is an arc leading from s to t, respectively from t to s, annotated by a symbolic sum l such that the coefficient of $\langle u_{1\mathcal{R}}, \ldots, u_{n\mathcal{R}} \rangle$ in $\mathcal{R}(l{:}\alpha)$ is 1. □

The feasible instances of the transitions of $\mathcal{N}$ are now put together to form the elementary net system $\mathcal{R}(\mathcal{N})$.

Definition 2.17 The denotation of the concrete PrT-net $(\mathcal{N}, \mathcal{R})$ is the elementary net system $\mathcal{R}(\mathcal{N}) = (B, E; \widehat{F}, c^0)$ where

- the set E of events is the set of feasible instances of transitions of $\mathcal{N}$;
- the set B of conditions is the union of all pre-sets and post-sets of events, $B = \bigcup_{e\in E}({}^\bullet e \cup e^\bullet)$;
- the flow relation $\widehat{F}$ is derived from the pre/post relation, $\widehat{F} = \bigcup_{e\in E}({}^\bullet e \times \{e\} \cup \{e\} \times e^\bullet)$;
- the representative case, $c^0 \subseteq B$, is the set of those conditions that satisfy the marking formula of their place in $\mathcal{N}$: $Pd_1 \dots d_m \in B$ belongs to c^0 iff there is a place s annotated by $Pu_1 \dots u_m|p$ with degree $\langle x_1, \dots, x_n\rangle$ and marked by the formula m such that the substitution $\tilde{\alpha} = \{x_1 \leftarrow d_{i1}, \dots, x_n \leftarrow d_{in}\}$ that generates $Pd_1 \dots d_m$ also satisfies m; there is a substitution β binding the dangling variables of $p \wedge m$ such that $\mathcal{R}(((p \wedge m){:}\tilde{\alpha}){:}\beta) = \top$. □

Figure 9 shows several abstract PrT-nets for which it is easy to see that for every ordered set $\mathcal{R} = (D; <)$ as support they all unfold into the same elementary net system. This finally leads to the following notion of equivalence of abstract PrT-nets.

Definition 2.18 Two abstract PrT-nets $\mathcal{N}_1$ and $\mathcal{N}_2$ in $\mathcal{A}(\mathbf{L})$ are called *(semantically) equivalent* – $\mathcal{N}_1 \cong \mathcal{N}_2$ – iff

- a relational $\mathbf{L}_{st}$-structure $\mathcal{R}$ is a support of $\mathcal{N}_1$ iff it is a support of $\mathcal{N}_2$;
- for every support $\mathcal{R}$, $\mathcal{R}(\mathcal{N}_1) = \mathcal{R}(\mathcal{N}_2)$. □

The remaining and main part of this paper is now devoted to a list of equivalence transformations for abstract PrT-nets that characterise $\cong$.

3 A List of Equivalence Transformations

We will now formulate a list of transformations of PrT-nets that preserve semantical equivalence $\cong$ in $\mathcal{A}(\mathbf{L})$. The list is coarsely divided into five groups indicating scope or purpose of the transformations. Some transformations come in pairs such that one is the reverse of the other; all other transformations are supposed to be self-reverse.

The transformations are formulated in terms of changing a given PrT-net $\mathcal{N} = (N; A; M^0)$. s is a generic place annotated by $Pu_1 \dots u_m|p$ with degree $\langle x_1, \dots, x_n\rangle$ and marked by m. t is a generic transition with degree $\{y_1, \dots, y_k\}$ annotated by q. A generic arc connects s and t and is annotated by the symbolic sum $l \in \mathbf{S}^{(n)}$.

Transformations at the Net

1. *Rewriting the support specification*

   Replace A_N by a set of formulae that is logically equivalent to A_N.

2. *Rewriting a static formula*

   Replace an open static formula r occurring somewhere at the net except in A_N by an open static formula r' such that $A_N \models (r \leftrightarrow r')$.

3. *Local renaming*

   Replace, consistently within the scope of each node, a variable by a new one.

Transformations at an Arc

4. *Rewriting with the transition selector*

   Replace the arc annotation l by a symbolic sum $l' \in \mathbf{S}^{(n)}$ such that with the adjacent transition selector q, $[q]l = [q]l'$.

5. *Rewriting with the place selector*

   Replace the formula r in some $[r]\langle v_1, \ldots, v_n\rangle$ occurring in l by an open static formula r' such that with the adjacent place selector p,
   $(p/s):\{x_1 \leftarrow v_1, \ldots, x_n \leftarrow v_n\} \Rightarrow (r \leftrightarrow r')$.

6. *Removing/inserting a void arc*

   (a) Remove an arc annotated by $\mathbf{0}$.
   (b) Insert an arc annotated by $\mathbf{0}$ between some place s and some transition t not yet connected by an equally directed arc.

Transformations at a Place

7. *Rewriting the place selector*

   Replace the selector p by an open static formula p' such that $p/s \Longleftrightarrow p'/s$.

8. *Rewriting with arc selectors*

   Replace the selector p by an open static formula p' such that the place is still disjoint from all other places and for every summand $[r]\langle v_1, \ldots, v_n\rangle$ occurring at some adjacent arc,
   $r \Rightarrow ((p/s):\{x_1 \leftarrow v_1, \ldots, x_n \leftarrow v_n\} \leftrightarrow (p'/s):\{x_1 \leftarrow v_1, \ldots, x_n \leftarrow v_n\})$.

9. *Rewriting the place marking*

   Replace the marking formula m by an open static formula m' such that $(p \wedge m)/s \Longleftrightarrow (p \wedge m')/s$.

10. *Increasing/decreasing the degree*

    (a) Let for some number i, u_i in $Pu_1 \ldots u_m$ be a constant b, let j be the number of different variables occurring before u_i, and let z be a variable new at s. Replace u_i by z, append the equation $z{=}u_i$ to the selector p, and insert b in all tuples at all adjacent arcs after the j-*th* position.
    (b) Let for some number i $(1 \leq i \leq n)$ and some constant b, the selector p be $p' \wedge x_i{=}b$ and let in all tuples at the adjacent arcs, the i-*th* element be b. Replace all occurrences of x_i at s by b and remove the i-*th* element of all tuples at the adjacent arc.

11. *Removing/adding an isolated place*

    (a) Remove an isolated place.
    (b) Add an isolated place annotated by $Px_1 \ldots x_m | \perp$ and marked by $\perp$ for some dynamic predicate $P \in \Pi_{dy}^{(m)}$ and arbitrary variables $x_1, \ldots, x_m$.

Transformations at a Transition

12. *Rewriting the selector with feasibility requirements*

    Let at some arc, the annotation be $\langle v_1, \ldots, v_n\rangle + \langle v_1', \ldots, v_n'\rangle + l'$, or let at some pair of opposite adjacent arcs, the annotations be $\langle v_1, \ldots, v_n\rangle + l_1'$ and $\langle v_1', \ldots, v_n'\rangle + l_2'$, respectively. Replace the selector q by an open static formula q' such that $\langle v_1, \ldots, v_n\rangle \neq \langle v_1', \ldots, v_n'\rangle \Rightarrow (q/t \leftrightarrow q'/t)$.

13. *Rewriting the selector with sums*

    Let an adjacent arc of t be annotated by a sum of the form $\langle z_1, \ldots, z_n \rangle + \langle z'_1, \ldots, z'_n \rangle + l'$ such that the z_i and z'_j are all different variables that occur nowhere else at an adjacent arc of t (but possibly in the selector). Let $\gamma = \{z_1 \leftrightarrow z'_1, \ldots, z_n \leftrightarrow z'_n\}$ be the substitution that replaces simultaneously each z_i by z'_i and each z'_i by z_i. Replace q by an open static formula q' such that $(q \vee q{:}\gamma)/t \iff (q' \vee q'{:}\gamma)/t$.

14. *Rewriting the selector with common arc selectors*

    Replace the selector q by a open, static formula q' such that for all arc selectors $[r]$ at the adjacent arcs, $r \Rightarrow (q \leftrightarrow q')$.

15. *Increasing/decreasing the degree*

    (a) If v is an individual term occurring in some tuple at some adjacent arc and z is a variable that is new at t, replace v by z and the selector q by $q \wedge z{=}v$.
    (b) If q is $q' \wedge z{=}v$ where v is a term not containing z, replace all occurrences of z at t by v.

16. *Removing/adding an isolated transition*

    (a) Remove an isolated transition.
    (b) Add an isolated transition annotated by $\perp$.

Refinement and Abstraction

17. *Splitting/merging of nodes*

    (a) Let at some node n, the selector r be $(r_1 \vee r_2)$ such that $A_N \models \neg(r_1 \wedge r_2)$. Replace the selector of n by r_1 and add to the net a copy of n with its adjacent arcs whose selector is r_2.
    (b) Let n_1, n_2 be nodes that except for their selectors r_1 and r_2 are copies of each other, including the adjacent arcs. Remove n_2 and replace the selector of n_1 by $(r_1 \vee r_2)$.

Here ends our list of transformations. Before we actually prove that they are really consistent with equivalence we give a few examples for the kind of use we are going to make of them.

Some of these rules ask for a pictorial presentation. As an example, we show a distributive law for transition and arc selectors (figure 10) and the splitting and merging of places (figure 11).

Proposition 3.1 The selectors of a transition and its adjacent arcs satisfy the distributive law as expressed in pictorial form in figure 10.

Proof: Both directions are a combination of the rules for rewriting an arc annotation (rule 4: $[q \wedge r]l = [q \wedge r][r]l$) and rewriting the transition selector with the arc selectors (rule 14: $r \Rightarrow (q \leftrightarrow (q \wedge r))$). □

Another example of the interplay of several rules concerning transitions is the following

Proposition 3.2 If $<$ denotes a strict total order, the three nets shown in figure 12 are equivalent.

Proof: Since $\langle x \rangle + \langle y \rangle = \langle y \rangle + \langle x \rangle$, the selector $x<y$ can be weakened to $x<y \vee y<x$ (rule 13) which in a strict total order is equivalent to $x \neq y$ (rule 2). $\top$ can be strengthened to $x \neq y$ because of the feasibility constraints for events (rule 12). □

Our last example of combining some rules to meet specific needs is the following

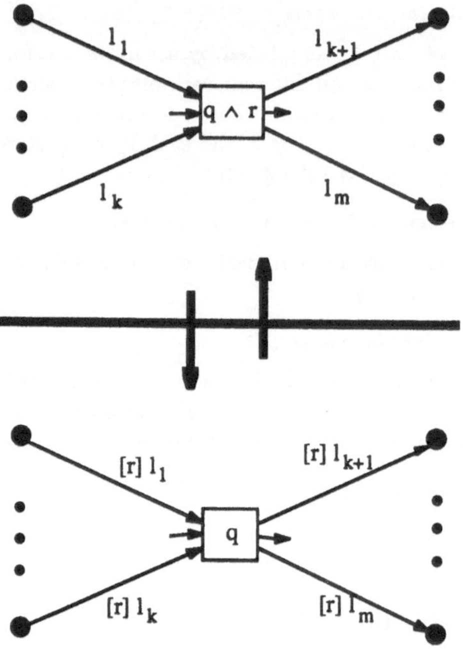

Figure 10: Distributive law for transition and arc selectors

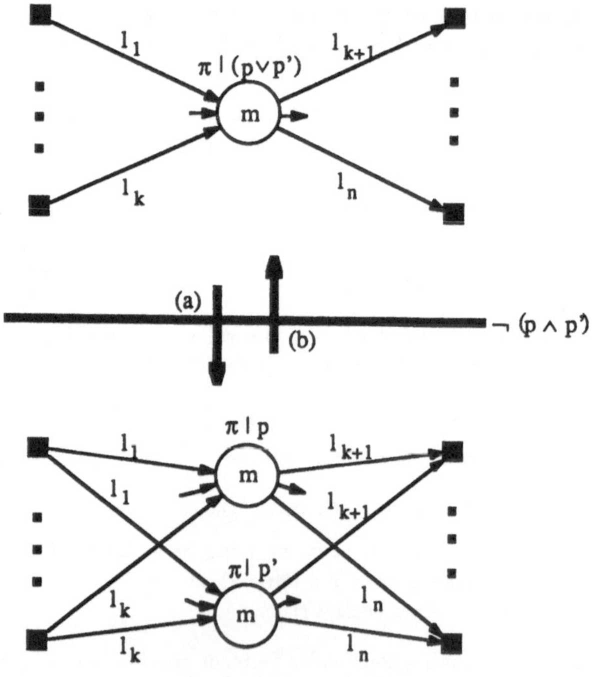

Figure 11: Splitting/merging of places in pictorial form

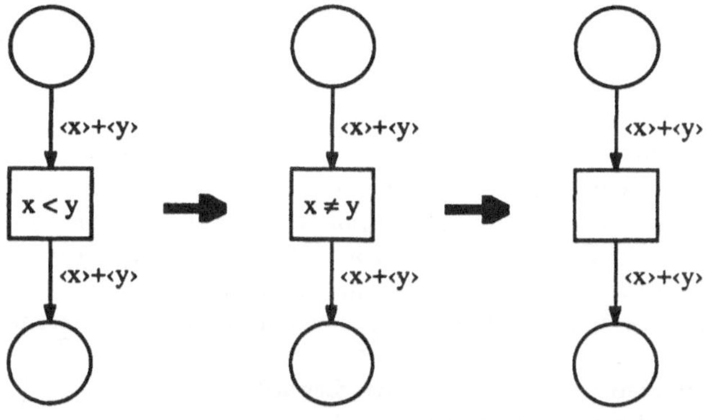

Figure 12: Commutativity of + and feasibilty constraints

Proposition 3.3 If the selector of a node contradicts the support specification, that node can be eliminated.

Proof: The selector is replaced by $\perp$ (rule 2) and distributed to all adjacent arcs (rule 4 or 5). The annotations of these arcs are replaced by **0** (rule 4) and the arcs removed (rule 6). The isolated node is removed (rule 11 or 16). □

In the next section, when proving the completeness result, we shall see all the rules being employed. Our task now is to verify that the rules formulated above are really equivalence transformations, i.e. are consistent with semantical equivalence of PrT-nets as defined in definition 2.18.

Theorem 3.4 Let $\mathcal{N}_1$ be a PrT-net in $\mathcal{A}(\mathbf{L})$ and let $\mathcal{N}_2$ be the result of applying one of the given transformations to $\mathcal{N}_1$. Then $\mathcal{N}_2$ is a PrT-net in $\mathcal{A}(\mathbf{L})$ that is semantically equivalent to $\mathcal{N}_1$.

Proof: It is easy to see that $\mathcal{N}_2$ is again a PrT-net in $\mathcal{A}(\mathbf{L})$. All transformations observe the syntactical constraints. All selectors remain open, static formulae. The disjointness of places is observed by rules 8 and 17. Transformation 10 affecting the degree of places observes the constraints concerning the length of tuples at arcs.

Let us now assume that $\mathcal{N}_2$ is the result of applying some transformation to $\mathcal{N}_1$. Let $\mathcal{R}$ be an arbitrary **L**-structure. We have to show that

- $\mathcal{R}$ is a support of $\mathcal{N}_1$ iff it is a support of $\mathcal{N}_2$,
- if $\mathcal{R}$ is a support then $\mathcal{R}(\mathcal{N}_1) = \mathcal{R}(\mathcal{N}_2)$.

The only rule affecting the support specification of $\mathcal{N}$ is rule 1. By definition of logical equivalence, it does not alter the family of supports. Hence we can concentrate on $\mathcal{R}(\mathcal{N}_1) = \mathcal{R}(\mathcal{N}_2)$. This means in essence to show that the sets of feasible instances of transitions are the same for $\mathcal{R}(\mathcal{N}_1)$ and $\mathcal{R}(\mathcal{N}_2)$.

Transformations at the Net

1. ***Rewriting the support specification***: Obvious.
2. ***Rewriting a static formula***: Obvious; by definition of $\models$, $\mathcal{R}(r) = \mathcal{R}(r')$.
3. ***Local renaming***: Obvious.

Transformations at an Arc

4. ***Rewriting with the transition selector***: Obvious; l and l' are evaluated for the same feasible substitutions α satisfying q.

5. *Rewriting with the place selector*: A feasible substitution α at the adjacent t creates an arc corresponding to $\langle v_1, \ldots, v_n\rangle{:}\alpha$ only if $\mathcal{R}((p/s){:}\{x_1 \leftarrow (v_1{:}\alpha), \ldots, x_n \leftarrow (v_n{:}\alpha)\}) = \top$ for the selector p of the adjacent s. Then, $\mathcal{R}([r{:}\alpha]) = \mathcal{R}([r'{:}\alpha])$.
6. *Removing/inserting a void arc*: Obvious.

Transformations at a Place

7. *Rewriting the place selector*: For a condition $(Pu_1 \ldots u_m){:}\tilde{\alpha}$ to satisfy the selector p, the dangling variables of p can be bound by any substitution β. After the transformation, for all such $\tilde{\alpha}$ there is some β with $\mathcal{R}((p{:}\tilde{\alpha}){:}\beta) = \top$ iff there is some β' with $\mathcal{R}((p'{:}\tilde{\alpha}){:}\beta') = \top$.
8. *Rewriting with arc selectors*: Similar to rule 5 (converse).
9. *Rewriting the place marking*: Same as 7; a condition in the representative marking satisfies $p \wedge m$.
10. *Increasing/decreasing the degree*: Straightforward.
11. *Removing an isolated place*: Trivial.

Transformations at a Transition

12. *Rewriting the selector with feasibility requirements*: Every feasible substitution at t satisfies both the transition selector and avoids multiple arcs and impurities.
13. *Rewriting the selector with sums*: Since the two tuples are not related to any other tuples by common variables, interchanging the role of the two tuples does not change the set of feasible instances of the transition.
14. *Rewriting the selector with common arc selectors*: Similar to rule 4 (converse).
15. *Increasing/decreasing the degree*: Straightforward.
16. *Removing an isolated transition*: Trivial.

Refinement and Abstraction

17. *Splitting/merging of nodes*: Straightforward for transitions. For a place note that for every adjacent transition, the disjoint place selectors select exactly one of the two arc copies. □

For the completeness proof that we present in the following section, it is essential that every application of a transformation rule can be reversed. Hence we state explicitly that

Theorem 3.5 Each transformation can be reversed by either applying the same rule or, if it belongs to a pair, by the respective other item, possibly in combination with other rules.

Proof: It is easy to check that all rules that do not come in pairs are symmetrical with respect to before/after. For the paired rules we get:

6. *Removing/inserting a void arc*: One item is the reverse of the other.
10. *Increasing/decreasing the degree of a place*: One item is the reverse of the other.
11. *Removing/adding an isolated place*: The annotation of any isolated place can be easily brought into the form assumed by (b).
15. *Increasing/decreasing the degree of a transition*: One item is the reverse of the other.
16. *Removing/adding an isolated transition*: The annotation of any isolated transition can be easily brought into the form assumed by (b).
17. *Splitting/merging of nodes*: One item is the reverse of the other. □

4 A Completeness Result

Finally we are going to prove that our equivalence transformations for PrT-nets are – up to a certain point – complete. In other words, we identify a non-trivial subclass of abstract PrT-nets such that any two equivalent PrT-nets in this class can be translated into each other through a finite sequence of transformations.

The limitations for completeness are due to the generalised +-operator $\sum$ that we have to use whenever the maximal size of an arc annotation is not independent of the static support. In our introductory PrT-net, for example (see figure 1), $\sum$ is needed when *all* tickets are moved in one event. The free use of $\sum$ can lead to expressions that we don't know how to translate into the kind of normal form that we need in our completeness proof.

Note that the difficulty with $\sum$ encountered here does not affect the use of the equivalence transformations. It only limits the scope of the completeness result. Only for the purpose of this section we assume that

The PrT-nets of this section are $\sum$-free.

The 'technicality' that depends on the absence of $\sum$ is the following

Proposition 4.1 Every (remember, $\sum$-free) arc annotation can be translated into the form $[r_1]\langle v_1^1, \ldots, v_n^1\rangle + \ldots + [r_k]\langle v_1^k, \ldots, v_n^k\rangle$.

Proof: Follows easily from lemma 2.8. Apply rule 4 for rewriting the arc annotation. □

The completess result is now developed in two major steps. First we show that every PrT-net $\mathcal{N} = (N, A, M^0) \in \mathcal{A}(\mathbf{L})$ can be brought into an equivalent basic form $\widetilde{\mathcal{N}} \in \mathcal{B}(\mathbf{L})$. Remember (definition 2.10) that in a PrT-net in basic form, each dynamic relation is represented by only one place, there are no (integer or boolean) coefficients contained in the arc annotations, and all transitions are of different types.

Since the *type* of a transition will play an essential role in what follows, let us define it explicitly.

Definition 4.2 Let t be a transition such that all surrounding arc annotations are plain sums of tuples (there are no $\sum$ and no integer or symbolic coefficients at its adjacent arcs). A pair (I_t, O_t) of functions $I_t, O_t{:}\Pi_{dy}{\rightarrow}\mathbf{N}$ is called the *type* of t iff for every predicate $P \in \Pi_{dy}$, there is an arc from place P into t annotated by i tuples iff $I_t(P) = i > 0$, and there is an arc from t into P annotated by i tuples iff $O_t(P) = i > 0$. (Hence $i = 0$ means no arc.) □

In the second step we show for two equivalent PrT-nets $\widetilde{\mathcal{N}_1}$ and $\widetilde{\mathcal{N}_2}$ in basic form how to translate one into the other. From the definition 2.18 of equivalence it is clear that two PrT-nets $\mathcal{N}_1$ and $\mathcal{N}_2$ are equivalent iff for all static supports, their transitions create the same sets of feasible instances. In the basic form there is for each type only one transition that represents all events of this type. This will allow us to reduce the equivalence of $\widetilde{\mathcal{N}_1}$ and $\widetilde{\mathcal{N}_2}$ to the logical equivalence of the guards of corresponding transitions. Since all transformations are reversible (theorem 3.5), we have at the end a translation between $\mathcal{N}_1$ and $\mathcal{N}_2$.

We are now going to develop the completeness result we are after in a series of propositions that reflect the different stages of translating two equivalent PrT-nets into the same target PrT-net.

The first step is to bring the places into basic form. An example for the following procedure can be found in figure 13.

Proposition 4.3 The places of an abstract PrT-net $\mathcal{N}$ can be translated into basic form.

Proof:
1. The degree of each place is made maximal, i.e. equal to the arity of the annotating dynamic predicate, by means of rule 10.
2. The annotation of each arc is transformed into the form $[r_1]\langle v_1^1, \ldots, v_n^1\rangle + \ldots + [r_k]\langle v_1^k, \ldots, v_n^k\rangle$ (see proposition 4.1).
3. The selector p of each place is merged into all arc selectors at its adjacent arcs; each summand

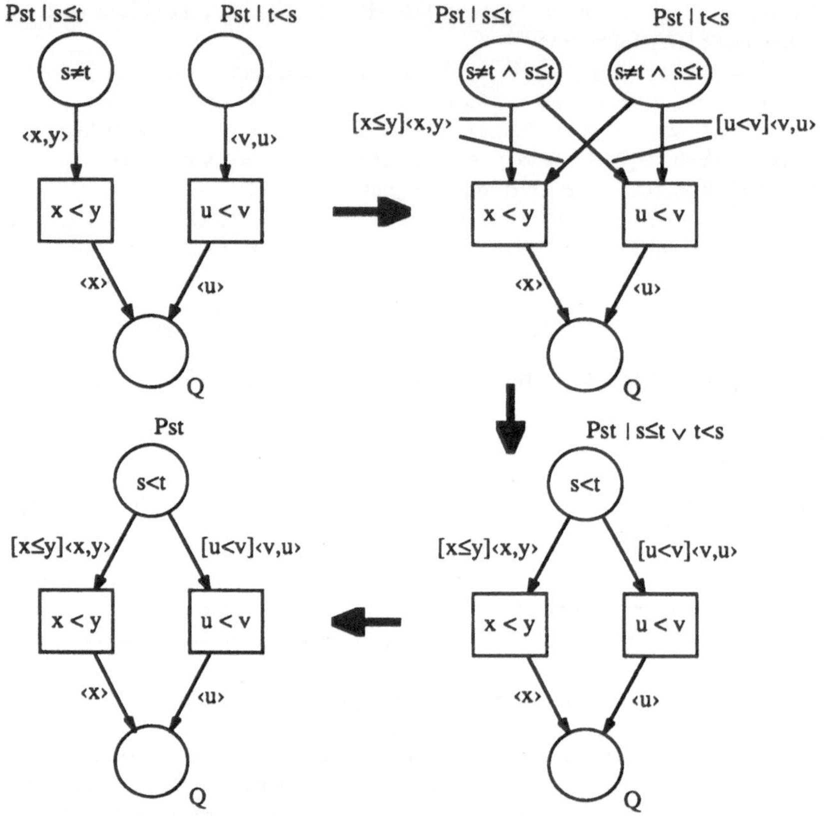

1. Copy place selectors into arcs; add void arcs; add-up annotations of corresponding arcs; unify condition schemes; add-up markings.

2. Merge identical places.

3. Remove place selector.

Figure 13: Translating of places into basic form

$[r]\langle v_1, \ldots, v_n\rangle$ is replaced by $[r \wedge (\tilde{p}{:}\{x_1 \leftarrow v_1, \ldots, x_n \leftarrow v_n\}]\langle v_1, \ldots, v_n\rangle$ by rule 5 where $\tilde{p}$ is the result of consistently renaming the dangling variables of p such that they do not occur at the adjacent transition.

Let now P be a dynamic predicate and $s_1, \ldots, s_k$ the places whose condition schemes are built from P.

4. The condition schemes of the $s_1, \ldots, s_k$ are made identical by consistently renaming the variables (rule 3).

Since the places are pairwise disjoint, we have for the (revised) place selectors, $A_N \models \neg(p_i \wedge p_j)$ for $i \neq j$.

5. The pre-sets and post-sets of all $s_1, \ldots, s_k$ are made identical by inserting, if necessary, arcs annotated by **0**.

6. For every transition t_j in the pre-set, respectively the post-set, of some (hence all) s_i, replace

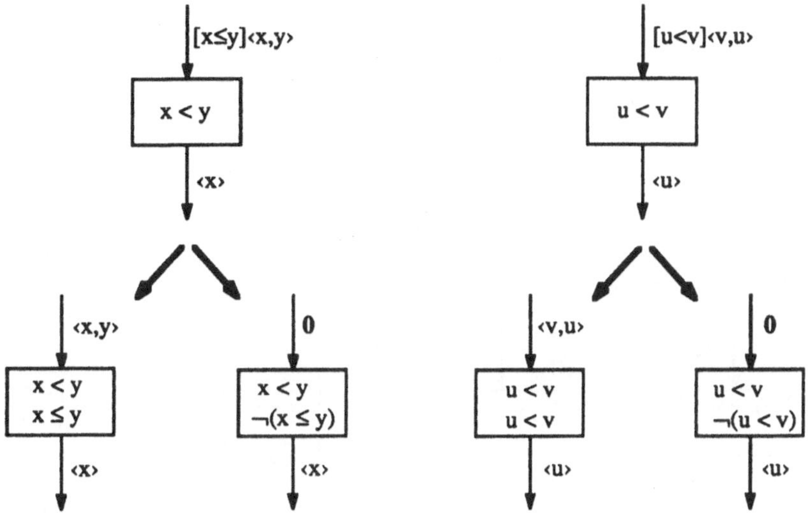

Figure 14: Eliminating arc selectors

the annotation l_i^j of the corresponding arcs by the sum of all these annotations, $l_1^j + \ldots + l_k^j$. This means an repeated application of rule 4 replacing **0** by some $[\bot]\langle v_1, \ldots, v_n\rangle$ and rule 5 replacing $[\bot]$ by some $[r]$ under the protection of the *disjoint* place selectors.

7. Replace the markings m_i by the equivalent (under the protection of the disjoint selectors) $(m_1 \wedge p_1) \vee \ldots \vee (m_k \wedge p_k)$.

8. The $s_1, \ldots, s_k$ have become identical except for their selectors. They are merged (rule 17(b)) into a single place whose selector is $(p_1 \vee \ldots \vee p_k)$. This selector is now redundant; it can be replaced by $\top$ since the constituent selectors have been copied into all adjacent arcs (rule 8).

As a result, the places of the new equivalent net are in basic form. □

Next we show how to eliminate the boolean coefficients that may occur at some arcs. An example is shown in figure 14.

Proposition 4.4 Boolean coefficients that occur in an arc annotation can be eliminated.

Proof: Let l be a symbolic sum annotating some arc that contains a boolean coefficient. Then l is transformed into $[r]l_1 + l_2$ where l_2 may be **0** (see proposition 4.1). The selector, q, of the adjacent transition is replaced by the logically equivalent formula $(q \wedge r) \vee (q \wedge \neg r)$ (rule 2) and the transition is split using rule 17(a). At the r-copy of t, the annotation of the copy of the arc concerned is replaced by $l_1 + l_2$ (rule 4: $[q \wedge r]([r]l_1 + l_2) = [q \wedge r](l_1 + l_2)$). At the $\neg r$-copy, the annotation of the copy of the arc concerned is replaced by l_2 in the same way. If a resulting arc is annotated by **0**, it is removed (rule 6) as well as transitions that become isolated (rule 16). □

The last step of the translation into basic form is the merging of transitions of the same type. See figure 15.

Proposition 4.5 Transitions without boolean coefficients at their arcs that are of the same type can be merged into a single one.

Proof: First, at all transitions the degree is made maximal by replacing all non-variables at the arcs by new variables (rule 15(a)). For transitions of the same type, the annotations of corresponding

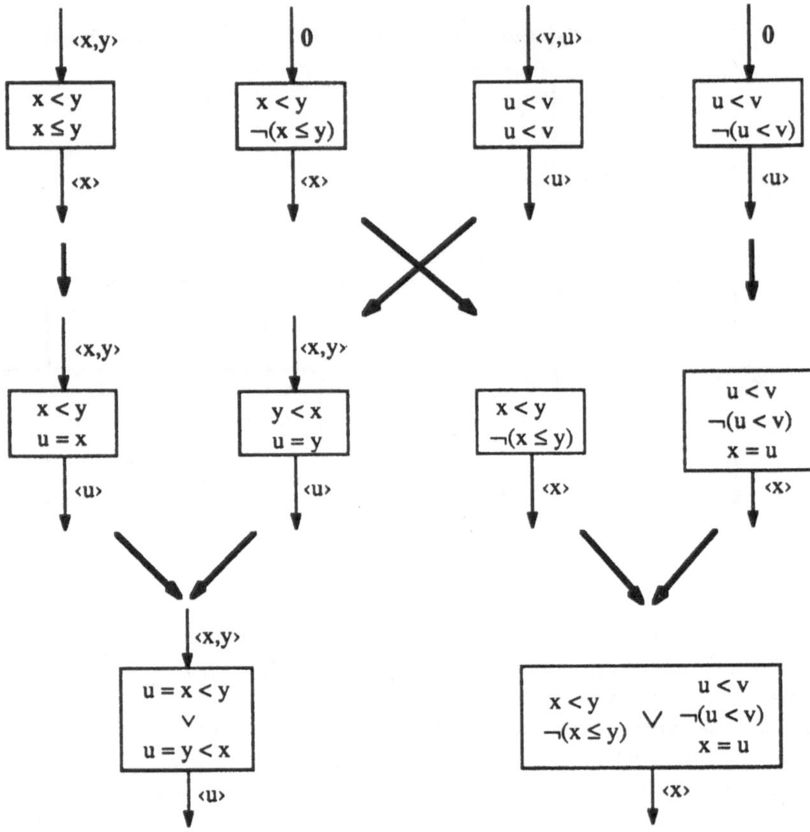

1. Eliminate void arcs; unify corresponding arcs of transitions of the same type by consistent renaming.

2. Merge transitions with identical surroundings.

Figure 15: Merging of transitions of the same type

arcs (they consist of the same number of tuples of equal length) are made identical by consistent renaming (rule 3), and finally the transitions are merged into a single one (rule 17(b)). □

At this point we have completed the first part of completeness proof. We summarize the result in the following

Theorem 4.6 Every PrT-net $\mathcal{N} = (N, A, M^0) \in \mathcal{A}(\mathbf{L})$ can be transformed, through a finite sequence of equivalence transformations, into a PrT-net $\widetilde{\mathcal{N}}$ that is in basic form.

Proof: We have shown the three major steps:
1. Merging of places belonging to the same dynamic predicate and elimination of place selectors (proposition 4.3).
2. Elimination of boolean coefficients at arcs (proposition 4.4).
3. Merging of transitions of the same type (proposition 4.5). □

The final step now consists of showing that once two equivalent PrT-nets have been translated into basic form, it is possible to make them identical. Our example is continued in figure 16. It takes as the second, equivalent net $\mathcal{N}_2$ the one in the right upper corner of figure 9 that is already in basic form.

Theorem 4.7 Let $\mathcal{N}_1$ and $\mathcal{N}_2$ be two equivalent PrT-nets. $\mathcal{N}_1$ can be transformed into $\mathcal{N}_2$ through a finite sequence of equivalence transformations.

Proof: Let $\widetilde{\mathcal{N}_1}$ and $\widetilde{\mathcal{N}_2}$ be PrT-nets in basic form that are equivalent to $\mathcal{N}_1$ respectively $\mathcal{N}_2$. They exist according to theorem 4.6 and are equivalent since $\mathcal{N}_1$ and $\mathcal{N}_2$ are equivalent. We show how to make them identical.

1. Since the two nets are equivalent, the support specifications A_{N_1} and A_{N_2} have the same models; they are logically equivalent. A_{N_2} is replaced by A_{N_1} (rule 1).

2. At every pair of places belonging to the same dynamic predicate, the variables are consistently renamed so that the condition schemes become identical (rule 3).

3. If for some transition type (I, O) a transition t exists in one net but no transition of the same type exists in the other net, t has no feasible instances for any support $\mathcal{R}$. Hence its selector contradicts the support specification (there are no arc or place selectors anymore). t is eliminated (see proposition 3.3).

4. At every transition, all variables occurring at the adjacent arcs are made different (the degree of the transition is made maximal using rule 15(a)). For two transitions t_1, t_2 of the same type (in different nets), all variables are consistently renamed (rule 3) such that corresponding arcs are annotated identically.

5. The selector q of every transition is made insensible against interchanging tuples at the arcs (rule 13); for every pair of different tuples $\langle z_1, \ldots, z_n \rangle$, $\langle z'_1, \ldots, z'_n \rangle$ at an adjacent arc, q is repeatedly replaced by $q \vee q{:}\{z_1 \leftrightarrow z'_1, \ldots, z_n \leftrightarrow z'_n\}$.

6. The inequalities excluding multiple arcs and impurities are $\wedge$-ded to the selector (rule 12).
(For step 5 and 6 see example 3.2.)

7. Every pair of transitions t_1, t_2 of the same type I, O generate the same set of feasible instances for every support $\mathcal{R}$. Since their selectors q_1, q_2 are now closed under permutations of tuples at the arcs and contain the feasibility requirements, this implies for every substitution $\alpha{:}V \rightarrow U_{\mathcal{R}}$ of the variables, $\mathcal{R}((q_1/t_1){:}\alpha) = \mathcal{R}((q_2/t_2){:}\alpha)$. Hence the selectors of the transitions in $\widetilde{\mathcal{N}_1}$ can replace the selectors of the corresponding transitions in $\widetilde{\mathcal{N}_2}$ according to rule 13.

8. Finally the marking formula of every place s in both nets is restricted to those conditions of s that can actually occur in some feasible instance of some adjacent transition. By means of rules 4, 7 and 9, the selector p is replaced by $p \wedge \bigvee_j q_j{:}\{z_1^j \leftarrow x_1, \ldots, z_n^j \leftarrow x_n\}$ where every q_j is the selector of an adjacent transition t_j and $\langle z_1, \ldots, z_n \rangle$ is some tuple at the corresponding arc (no matter which one). Let now $(m_1 \wedge p_1), (m_2 \wedge p_2)$ be the restricted marking formulae of corresponding places in $\widetilde{\mathcal{N}_1}$ and $\widetilde{\mathcal{N}_2}$. For every relational structure $\mathcal{R}$ and for all valuations α, $\mathcal{R}(m_1{:}\alpha) = \mathcal{R}(m_2{:}\alpha)$, hence $A_N \models (m_1 \leftrightarrow m_2)$. Consequently the marking of every place in $\widetilde{\mathcal{N}_2}$ is replaced by the marking of the corresponding place in $\widetilde{\mathcal{N}_1}$ (rule 2).

As a result, $\widetilde{\mathcal{N}_1}$ and $\widetilde{\mathcal{N}_2}$ have become identical. The same PrT-net was derived from both $\mathcal{N}_1$ and $\mathcal{N}_2$ in a finite sequence of equivalence transformations. Since all transformations can be reversed, $\mathcal{N}_1$ and $\mathcal{N}_2$ can be derived from each other in a finite translation process. □

As an immediate consequence of this result we note that under certain circumstances, the procedure for translating equivalent nets into each other can be modified for testing the equivalence of nets. It is, in fact, the undecidability of first-order predicate logic that is responsible for the undecidability of the theory of abstract PrT-nets in general.

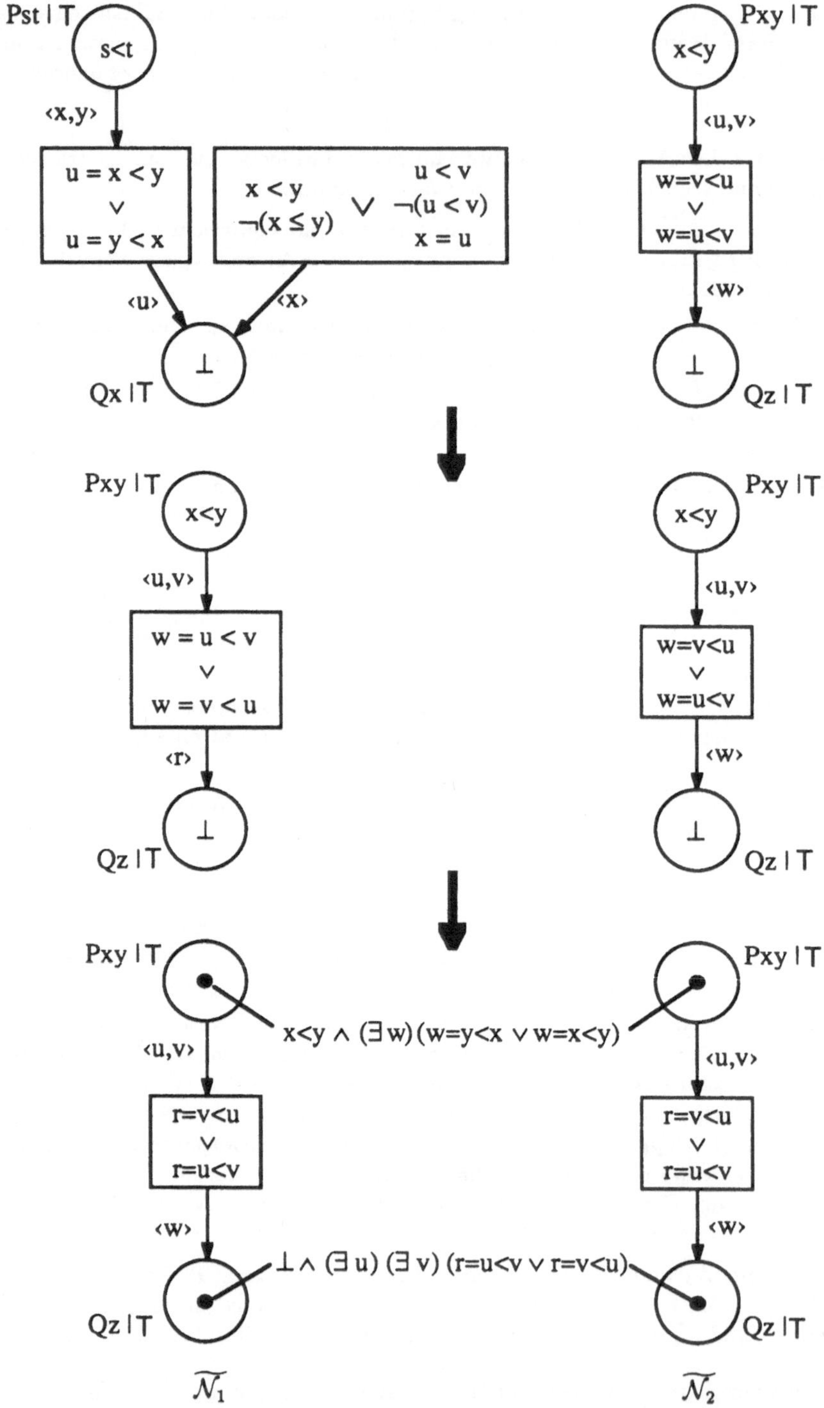

Figure 16: Unifying two equivalent nets in basic form

Lemma 4.8 Within a given decidable theory of static supports, the equivalence of PrT-nets is decidable.

Proof: If there is an algorithm to check whether, within a given family of supports, two static formulae are equivalent, the procedure for translating equivalent PrT-nets into each other can also be used to test whether two PrT-nets are equivalent. At those points in the procedure where the equivalence of formulae (namely transition selectors and place markings) is concluded from the equivalence of nets, it must be verified. □

5 Conclusion

We have presented the semantics of a subclass of *abstract PrT-nets* – those without the generalised +-operator $\sum$ appearing at the arcs – in two complementary ways. In the primal semantics we have defined what can be represented by an abstract PrT-net: the meaning of a PrT-net is the family of elementary net systems that is generated by unfolding it for all static supports that satisfy the support specification. In the complementary semantics we have defined what can be *done* with an abstract PrT-net: the meaning of a PrT-net is its equivalence class generated by a set of formal transformations. The soundness and completeness theorems establish the consistency of the two approaches. (Note the analogy to assigning, in turn, a meaning to a marked elementary net either by unfolding it into the set of its processes or by defining the transformation rule for its markings.)

The style of this note was rather theoretical. Origin and purpose, however, of investigating the modelling power of PrT-nets, and of studying transformations in particular, are rooted in the author's practical interest in supporting system design with PrT-nets by computer tools. Hence it is hoped that equivalence transformations will become an integral part of software packages that support higher-level Petri nets. To this end more work, both theoretical and practical one, has to be done. Some issues shall end this note.

1. More general notions of equivalence:

A major result reported in this note is the completeness theorem 4.7. To get it, a restricted class of abstract PrT-nets and the most restrictive notion of equivalence between them was needed. Too restrictive, in fact, for many practical purposes. Three examples:

- In actual system modelling – as in our introductory example –, the size of an arc annotation (the number of constituent tuples) very often depends on the cardinality of the set of individuals. One needs symbolic sums at the arcs that contain the generalised plus operator $\sum$. In such cases, an equivalent basic form cannot exist in general. Hence the completeness proof doesn't work. A different kind of standard form for transitions and of comparing transitions of the same type would be necessary.

- One would like to have the possibility of altering the type of static supports of a PrT-net $\mathcal{N}$ without affecting the family of elementary net systems denoted by $\mathcal{N}$. This would require the following kind of equivalence notion: two PrT-nets $\mathcal{N}_1, \mathcal{N}_2$ are called *extensionally equivalent* if for every support $\mathcal{R}_1 \models A_{\mathcal{N}_1}$ of $\mathcal{N}_1$ there is a support $\mathcal{R}_2 \models A_{\mathcal{N}_2}$ of $\mathcal{N}_2$, and for every support $\mathcal{R}_2 \models A_{\mathcal{N}_2}$ of $\mathcal{N}_2$ there is a support $\mathcal{R}_1 \models A_{\mathcal{N}_1}$ of $\mathcal{N}_1$, such that $\mathcal{R}_1(\mathcal{N}_1) = \mathcal{R}_2(\mathcal{N}_2)$.

- One would also like to be able to rename the conditions of the underlying elementary net systems according to a scheme like

$$\begin{array}{ccc} Px_1 \dots x_n a^1 & \Longleftrightarrow & Q^1 x_1 \dots x_n \\ \cdot & \cdot & \cdot \\ \cdot & \cdot & \cdot \\ \cdot & \cdot & \cdot \\ Px_1 \dots x_n a^k & \Longleftrightarrow & Q^k x_1 \dots x_n \end{array}$$

for $a^1, \dots, a^k \in \Omega^{(0)}$, $P \in \Pi_{dy}^{(n+1)}$, and $Q^1, \dots, Q^k \in \Pi_{dy}^{(n)}$.

This would require yet another equivalence notion: two PrT-nets $\mathcal{N}_1, \mathcal{N}_2$ are called *weakly equivalent* if for every support $\mathcal{R}_1 \models A_{N_1}$ of $\mathcal{N}_1$ there is a support $\mathcal{R}_2 \models A_{N_2}$ of $\mathcal{N}_2$, and for every support $\mathcal{R}_2 \models A_{N_2}$ of $\mathcal{N}_2$ there is a support $\mathcal{R}_1 \models A_{N_1}$ of $\mathcal{N}_1$, such that $\mathcal{R}_1(\mathcal{N}_1)$ and $\mathcal{R}_2(\mathcal{N}_2)$ are *isomorphic* under some bijection of the set of conditions.

For all three cases, a complete characterisation would be impossible. And much more general notions of *behavioural equivalence*, like case graph isomorphy, bisimulation, observational equivalence, interface equivalence etc., are being studied for net systems and other models of concurrent systems to support structured system design.

Hence for practical purposes, the emphasis should be on the existence of (sound!) equivalence transformations rather then on their completeness.

2. Relative complexity of deciding equivalence:
We do not claim that the procedure for translating equivalent PrT-nets into each other is clever in terms of efficiency. Assume that the theory of the static supports is decidable with known complexity. It would be good to know how much of complexity is added by the pure PrT-net part.

3. Other kinds of formalisms:
Other formalisms for dealing with structures are needed. The first-order logical approach was chosen mainly for reasons of foundation. In software design, abstract data types are being used in combination with PrT-nets. Other areas like database design or protocol design may develop their own net theoretical models of distributed systems. Other formalisms for higher-level Petri nets may allow similar ways of investigating equivalence notions. They may profit from the PrT-net approach, or show a better way to do it.

4. Software support:
Software tools for PrT-nets require a combination of graphical system design, computer algebra, symbolic logic, algebraic specification, and much more. The ultimate goal is to integrate them into a single uniform environment for net based system design like the recently released *Design/CPN*. [1]

Acknowledgement: I gratefully acknowledge the help I got from Wil Dekkers during the process of collecting, properly phrasing and testing of equivalence transformations. I would also like to thank the anonymous referee of *Advances in Petri Nets* whose comments and suggestions helped improve the presentation of these results.

References

[1] Albert, K.; Jensen, K.; Shapiro, R.: *DESIGN/CPN - A Tool Package Supporting the Use of Coloured Petri Nets.* Meta Software Corporation. Camridge, MA (1989)

[2] Brauer, W.; Reisig, W.; Rozenberg, G. (eds.): *Petri Nets: Central Models and Their Properties.* Advances in Petri Nets 1986, Part I. Lecture Notes in Computer Science 254. Springer 1987

[3] Genrich, H.J.; Lautenbach, K.: *System Modelling with High-Level Petri Nets.* Theoretical Computer Science 13 (1981) 109–136

[4] Genrich, H.J.: *Projections of CE Systems.* In: Advances in Petri Nets 1985, Lecture Notes in Computer Science 222. Springer 1986

[5] Genrich, H.J.: *Predicate/Transition Nets.* In [2]

[6] Jensen, K.: *Coloured Petri Nets.* In [2]

[7] Shoenfield, J.R.: *Mathematical Logic.* Addison-Wesley 1967

[8] Thiagarajan, P.S.: *Elementary Net Systems.* In [2].

Section G
Analysis of Stochastic Nets

Temporal concepts were introduced in Petri net models in the mid 1970s. This initiated a hot debate—about the suitability of such an extension, and about the most appropriate ways to make it. Both Petri nets with timed places and timed transitions were proposed, and in the latter case there was either a time delay (before the atomic occurrence of a transition) or a time interval (between the consumption of input tokens and the generation of output tokens). Whether deterministic or stochastic timing should be used was also an issue of the discussion.

Stochastic Petri nets were proposed, simultaneously by S. Natkin & G. Florin, and by M.K. Molloy. Similar ideas can be found in the work of F.J.W Symons. The new net class paved the way to the utilization of Petri nets for performance evaluation, and pushed the choice of the modeling paradigm towards timed transitions with atomic occurrences. Later generalizations, e.g. those proposed by M. Ajmone-Marsan, G. Balbo and G. Conte, increased the usability of stochastic nets—maintaining a good balance between generality of description and possibility of obtaining analytic results.

The increased complexity of the modelled systems and the introduction of high-level Petri nets naturally led to the development of high-level versions of stochastic Petri nets. The first results in this area were obtained by A. Zenie and by the authors of the first paper in this section. The great flexibility of the first classes of high-level stochastic nets made it difficult to provide a concise specification of the time concepts. Moreover, the flexibility made it possible to construct models where it was difficult to detect symmetry properties. This meant that it was difficult to keep down the calculation complexity of analytic solutions. Newer versions of high-level stochastic nets avoid this problem by being more restrictive. Regular stochastic Petri nets was the first of these net classes, and it is presented in the second paper of this section.

The third and fourth paper illustrate techniques for the automated generation of compound Markovian chains (reduced state spaces). The third paper deals with a general net model and thus the generation of compound markings is complex. The fourth paper deals with a more restricted class of net models (which is a successor of regular Petri nets) and thus the generation of compound markings is simpler than in the previous case.

16.

Stochastic High-level Petri Nets and Applications

C. Lin and D.C. Marinescu

IEEE Transactions on Computers *37*, *7* (1988) 815-825

Abstract—**A new class of stochastic Petri nets is proposed in this paper. The stochastic high-level Petri nets (SHLPN's) are high-level Petri nets augmented with exponentially distributed firing times. The main advantage of modeling homogeneous systems using SHLPN's is that the resulting models are simpler, more intuitive, and have a smaller number of states as our examples show.**

Index Terms—**Continuous time Markov chains, high-level Petri nets, multiprocessor systems, performance evaluation, stochastic high-level Petri nets, stochastic Petri nets.**

I. Introduction

RAMAMOORTHY [19], Sifakis [21], and others have used Petri nets for the performance evaluation of concurrent systems. The introduction of stochastic Petri nets (SPN's) proposed by Molloy [16] and by Natkin [18] has triggered a number of interesting developments in the area of timed Petri nets.

Ajmone Marsan and his coworkers have introduced GSPN's, generalized stochastic Petri nets [13], and have examined the impact of different execution policies upon the semantic of the model [14]. Meyer and his group have defined stochastic activity networks, suited to modeling of performability [15]. Dugan, Trivedi, and the group from Duke University have defined ESPN's, extended stochastic Petri nets, which allow arbitrary probability distributions for transition firing time [3]. Efforts to design automated tools for performance analysis using GSPN's and ESPN's have been reported in the literature [1], [2].

Molloy has proved that there is an isomorphism between k-bounded Petri nets with exponentially distributed transition rates and finite Markov processes. Two stochastic systems are isomorphic if

1) there are one-to-one mappings between the state spaces of the two systems, and between the set of state transitions of the two systems,

2) the probability of a transition from one state to another in one system equals the probability of a transition between the corresponding states of the other system.

The importance of this result is that the methodology used to find the transient and the steady-state solutions for a Markov chain can be used for an SPN system in which each marking corresponds to a Markov state.

Manuscript received July 21, 1986; revised June 20, 1987. D. C. Marinescu was supported in part by ARO Grant DAAL03-86-K-0106 and by NSF Grant NCR-8702115.

C. Lin is with the State Planning Committee, Beijing, China.

D. C. Marinescu is with the Department of Computer Sciences, Purdue University, West Lafayette, IN 47907.

IEEE Log Number 8717763.

Although the stochastic Petri nets do not provide more modeling power than Markov processes, they can be used as a convenient description of the system being modeled. Since the size of the state space of a stochastic Petri net is equal to the size of the Markov process space, the complexity of solving an SPN model is the same as in the case of the model based upon the Markov process.

To reduce the complexity of solving an SPN model, Marsan and his coworkers have proposed to define two types of transitions, timed and immediate, and to separate the state space of an SPN into two subsets, one containing *vanishing states,* which enable immediate transitions, and one containing *tangible states,* which enable timed transitions. The time spent by the system in vanishing states is zero. The existence of vanishing states increases the computational effort by enlarging the size of the transition matrix. The effort to increase the modeling power of GSPN's has lead to the introduction of stochastic activity networks and of extended stochastic Petri Nets. In the case of stochastic activity networks, transitions (activities) have associated cases and gates which describe how a transition is enabled and how its firing affects the next state of the model. ESPN's allow the definition of logical conditioning functions for the enabling of transitions as Boolean expressions involving relational and numerical operators and the number of tokens in various places.

This paper takes a different approach, it defines stochastic high-level Petri nets based upon high-level Petri nets augmented with exponentially distributed transition rates. High-level Petri nets are extensions of regular Petri nets. An example of high-level Petri nets are the predicate transition nets [6] in which individual (distinguishable) tokens are allowed [20], and predicates may be associated with some or all transitions. The term "token" is used throughout this paper but "item" (see, for example, [6]) is also adequate to describe the different entities flowing through the system.

As a general rule, high-level Petri nets lead to simpler models, with a more readable graph than the corresponding regular Petri nets. The size of the state space of stochastic high-level Petri nets models can be further simplified due to the introduction of the *compound marking* technique described in the following sections. Zenie has proposed colored stochastic Petri nets [23]. The differences among high-level Petri nets, colored Petri nets, and predicate transition nets are outlined in [10].

In fact, we have concentrated our attention upon *homogeneous systems*. Informally, we define a homogeneous system as one consisting of identical processing elements which carry out identical tasks. A more detailed discussion of this topic can be found in Section IV-A in connection with the modeling assumptions for a multiprocessor system. When modeled using stochastic high-level Petri nets, these systems have subsets of *equivalent* states. Such states can be grouped together in such a way that the stochastic high-level Petri net model of the system with compound markings contains only one *compound state* for each group of *individual states* in the original stochastic high-level Petri net model. In this case, an *equivalence relation* exists among the stochastic high-level Petri net model with compound markings and the original stochastic high-level Petri net model.

Since the SPN are isomorphic to continuous time Markov chains, the stochastic high-level Petri nets are also isomorphic to continuous time Markov chains. The compound marking concept represents a grouping of markings in the Petri net domain and it corresponds to state grouping in the Markov domain. It can be shown that stochastic high-level Petri nets with compound markings are homomorphic with continuous time Markov chains with a grouping operation.

Two systems are then modeled using the technique described above. We start with an example based upon the problem of philosophers who share dinnerware, and conclude with another example, similar to the one treated by Marsan [13]: the analysis of the performance of a multiprocessor system.

The main advantages of introducing SHLPN's with compound markings are: the model of a system has a considerably lower number of states than the same model constructed using SPN's or SHLPN's, the graph associated with the model is simpler, easier to read, often it is invariant to system size, and the analysis methods for Petri net models can still be applied to SHLPN models. Since the SHLPN model with compound markings is isomorphic with a Markov chain, its steady-state probabilities can be determined using techniques well developed by stochastic analysis.

The extended stochastic Petri nets, as well as the stochastic activity networks, allow predicates to be associated with transitions. The main difference is that SHLPN's allow tokens with multiple attributes, and that the predicates associated with transitions may be expressed in terms of the attributes of the tokens present in the input places of the transition.

This paper is focused on steady-state analysis of systems using SHLPN's.

II. Informal Introduction to Stochastic High-Level Petri Nets

In order to introduce stochastic high-level Petri nets, the definition of stochastic Petri nets will be reviewed. Then stochastic high-level Petri nets will be introduced informally and an example illustrating the use and the advantages of SHLPN's will be given.

The SPN's are obtained by associating with each transition in a Petri net a possibly marking dependent, transition rate for the exponentially distributed firing time. A formal definition of SPN is thus the following.

$$SPN = (P, T, A, M, \lambda)$$

1) P is the set of places
2) T is the set of transitions
3) $P \cap T = \emptyset$, $P \cup T \neq \emptyset$
4) A is the set of input and output arcs; $A \subseteq (P \times T) \cup (T \times P)$
5) M is the initial marking
6) λ is the set of transition rates.

A marking of a Petri net, or of a stochastic Petri net, is a distribution of tokens on its places. A stochastic Petri net is said to be k-bounded if there exists a finite nonnegative integer k such that for every marking M in the reachability set, and for every place P_i, $M(P_i) < k$. A marking may be viewed as a mapping from the set of places P to the natural numbers N. We can associate with each marking a state of the system and in the following the terms *state* and *marking* are used with essentially the same meaning.

The SPN's are isomorphic to continuous time Markov chains due to the memoryless property of the exponential distribution of firing times. The SPN markings correspond to the states of the corresponding Markov chain so that the SPN model allows the calculation of the steady-state and transient system behavior.

In SPN analysis, as in Markov analysis, ergodic (irreducible) systems are of special interest. For ergodic SPN systems, the steady-state probability of the system being in any state always exists and is independent of the initial state. If the firing rates do not depend upon time, a stationary (homogeneous) Markov chain is obtained. In particular, k-bounded SPN's are isomorphic to finite Markov chains. In this paper, we consider only ergodic, stationary, and k-bounded SPN's (or SHLPN) and Markov chains.

As an example, consider a system consisting of five hungry philosophers who spend some time thinking between copious meals. There are only five forks on a circular table and there is a fork between two philosophers. To eat, each philosopher needs the two adjacent forks. When they become free, the philosopher hesitates for a random time, exponentially distributed with average $1/\lambda_1$, and then moves from the thinking phase to the eating phase where spends an exponentially distributed time with average $1/\lambda_2$. This system is described by the stochastic Petri net in Fig. 1. The model has 15 places and ten transitions, all indexed on variable i,

$i \in$ [1, 5] in the following description.

T_i is the "thinking" place. If T_i holds a token, the ith philosopher is pretending to think while waiting for forks.

E_i is the "eating" place. If E_i holds a token, the ith philosopher is eating.

F_i is the "free fork" place. If F_i holds a token, the ith fork is free.

G_i is the "getting forks" transition. This transition is enabled when the hungry philosopher can get the two adjacent forks. The transition firing time is associated with $1/\lambda_1$ and it is related to the time the philosopher hesitates before taking the two forks and starting to eat.

R_i is the "releasing forks" transition. A philosopher releases the forks and returns to the thinking stage,

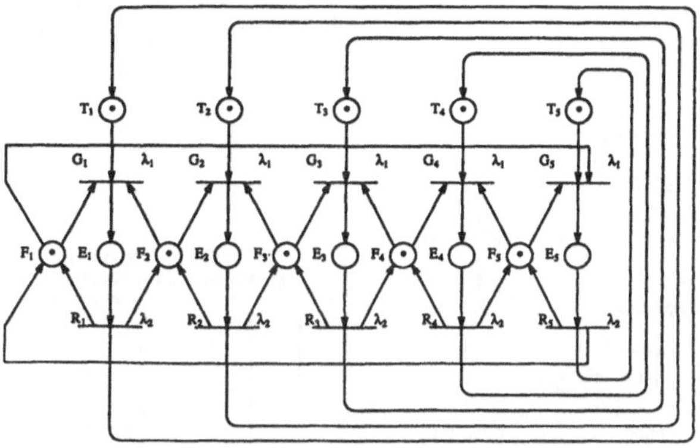

Fig. 1. Modeling of the philosopher system using stochastic Petri nets.

TABLE I
THE MARKINGS OF THE PHILOSOPHER SYSTEM

	T_1	T_2	T_3	T_4	T_5	E_1	E_2	E_3	E_4	E_5	F_1	F_2	F_3	F_4	F_5
M_0	1	1	1	1	1						1	1	1	1	1
M_1		1	1	1	1	1							1	1	1
M_2	1		1	1	1		1				1			1	1
M_3	1	1		1	1			1			1	1			1
M_4	1	1	1		1				1		1	1	1		
M_5	1	1	1	1						1		1	1	1	
M_6		1		1	1	1		1							1
M_7		1	1		1	1			1				1		
M_8	1		1		1		1		1		1				
M_9	1		1	1			1			1				1	
M_{10}	1	1		1				1		1		1			

after the eating time exponentially distributed with average $1/\lambda_2$.

The stochastic Petri net model of the philosopher system has a state space size of 11 and its states (markings) are presented in Table I. The state transition diagram of the corresponding Markov chain is shown in Fig. 2. Using the methods mentioned above (see, for example, [22]), the steady-state probabilities that the system is in state i, p_i, can be obtained:

$$p_i=\begin{cases} \dfrac{\lambda_2^2}{5\lambda_1(\lambda_1+\lambda_2)+\lambda_2^2} & i=0 \\ \dfrac{\lambda_1\lambda_2}{5\lambda_1(\lambda_1+\lambda_2)+\lambda_2^2} & i=1,\ 2,\ 3,\ 4,\ 5 \\ \dfrac{\lambda_1^2}{5\lambda_1(\lambda_1+\lambda_2)+\lambda_2^2} & i=6,\ 7,\ 8,\ 9,\ 10. \end{cases}$$

Our objective is to model the same system but using a representation which leads to a model with a smaller number of states. High-level Petri nets provide a more compact representation of complex systems, hence, they represent a natural choice for an alternative representation if the time

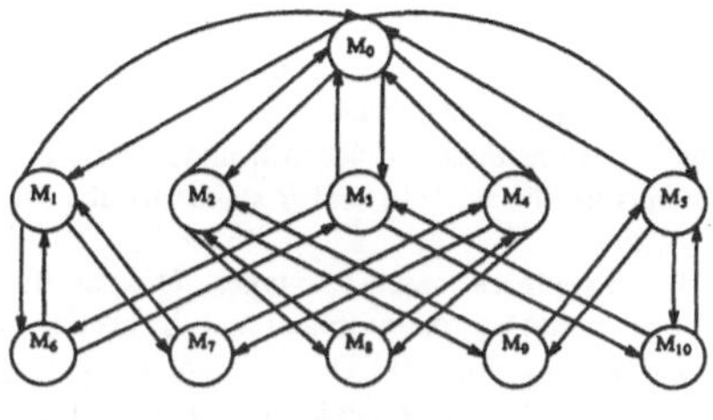

Fig. 2. The state transition diagram of the philosopher system.

concept can be embedded into them. Different types of high-level Petri nets (HLPN's) have been proposed, for example predicate transition nets [5], [6], colored Petri nets [9], relation nets [20], but all of them are conceptually similar. Moreover, the model of a system constructed using one type of HLPN can be informally translated into any other type of HLPN [10], [20]. The predicate transition nets, for example, support intuition and modeling elegance since they allow variables which represent arc labels or token attributes to appear in the conditions associated with the firing of a transition. The stochastic high-level Petri nets are extensions of high-level Petri nets in which each transition has an exponentially distributed firing time associated with it.

The following notation is used throughout this paper: $a \oplus b \pmod{p}$ stands for addition modulo p. $|\{\ \}|$ denotes the cardinality of a set. The relations between the element and the set, $\in$ and $\notin$, are often used in the predicates.

The SHLPN's will be introduced by means of an example which illustrates the fact that an SHLPN model is a scaled down version of an SPN model, it has a smaller number of places, transitions, and states than the original SPN model. Fig. 3 presents the SHLPN model of the same philosopher system described in Fig. 1 using an SPN. In the SHLPN

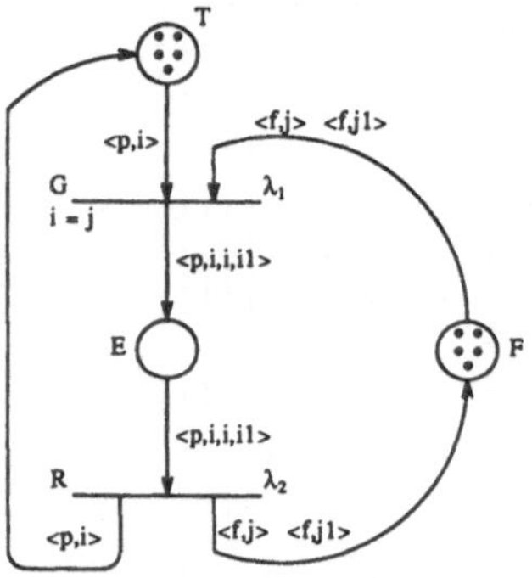

Fig. 3. Modeling the philosopher system using SHLPN.

model, each place and each transition stands for a set of places or transitions in the SPN model. The number of places is reduced from 15 to 3, the place T stands for the set $\{T_i\}$, E stands for $\{E_i\}$, and F stands for $\{F_i\}$, for $i \in [1, 5]$. The number of transitions is reduced from ten to two; the transition G stands for the set $\{G_i\}$ and R stands for the set $\{R_i\}$ with $i \in [1, 5]$.

The three places contain two types of tokens, the first type is associated with the philosophers and the second is associated with forks (see Fig. 3). The arcs are labeled by the token variables. A token has a number of attributes, the first attribute being its type and the second attribute being its *identity, id.* The tokens residing in the place E, the eating place, have four attributes; the last two attributes are the ids of the forks currently used by the philosopher. The transition G is associated with the predicate which specifies the correct relation between a philosopher and the two forks used by him. The predicate inscribed on transition G (Fig. 3) as $i = j$ is a concise form of expressing that the second attribute of a $\langle p, i\rangle$ token should be equal to the second attribute of the two tokens representing the forks. This means that a philosopher can eat only when the two adjacent forks are free; for example, the forks $\langle f, 3\rangle$ and $\langle f, 4\rangle$ must be free in order to allow the philosopher $\langle p, 3\rangle$ to move to the eating place. Note that a predicate expresses an imperative condition which must be met in order for a transition to fire. A predicate should not be used to express the results associated with the firing of a transition. There is no predicate associated with transition R in Fig. 3, although there is a well-defined relationship between the attributes of the tokens released when R fires.

In an SHLPN model, the transition rate associated with every transition is related to the markings which enable that particular transition. To simplify the design of the model, only the transition rate of the individual markings is shown in the graph, instead of the transition rate of the corresponding compound markings. For example, in Fig. 3, the transition rates are written as λ_1 for the transition G and λ_2 for the transition R. As shown in Fig. 4, the system has three states, S_i with $i \in [0, 1, 2]$ representing the number of philosophers in the eating place. The actual transition rates corresponding to the case when the transition G fires are $5 \times \lambda_1$ and $2 \times \lambda_1$ depending upon the state of the system when the transition G fires. If the system is in state S_0, then there are five different philosophers which can go to the eating place; hence, the actual transition rate is $5 \times \lambda_1$.

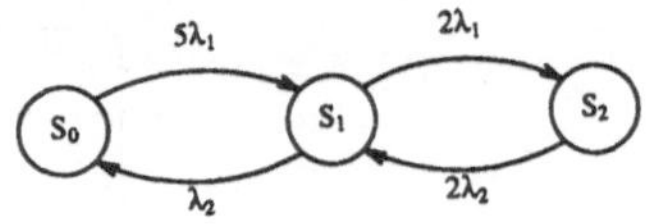

Fig. 4. The state transition diagram of the philosopher system with compound markings.

TABLE II
THE STATE TABLE OF THE PHILOSOPHER SYSTEM WITH INDIVIDUAL MARKINGS. EACH STATE (MARKING) IS DESCRIBED BY THE DISPOSITION OF ALL TOKENS IN EVERY PLACE

State	Place index		
	T	E	F
0	<p,1>, <p,2>, <p,3>, <p,4>, <p,5>	0	<f,1>, <f,2>, <f,3>, <f,4>, <f,5>
1	<p,2>, <p,3>, <p,4>, <p,5>	<p,1,1,2>	<f,3>, <f,4>, <f,5>
2	<p,1>, <p,3>, <p,4>, <p,5>	<p,2,2,3>	<f,1>, <f,4>, <f,5>
3	<p,1>, <p,2>, <p,4>, <p,5>	<p,3,3,4>	<f,1>, <f,2>, <f,5>
4	<p,1>, <p,2>, <p,3>, <p,5>	<p,4,4,5>	<f,1>, <f,2>, <f,3>
5	<p,1>, <p,2>, <p,3>, <p,4>	<p,5,5,1>	<f,2>, <f,3>, <f,4>
6	<p,2>, <p,4>, <p,5>	<p,1,1,2>, <p,3,3,4>	<f,5>
7	<p,2>, <p,3>, <p,5>	<p,1,1,2>, <p,4,4,5>	<f,3>
8	<p,1>, <p,3>, <p,5>	<p,2,2,3>, <p,4,4,5>	<f,1>
9	<p,1>, <p,3>, <p,4>	<p,2,2,3>, <p,5,5,1>	<f,4>
10	<p,1>, <p,2>, <p,4>	<p,3,3,4>, <p,5,5,1>	<f,2>

The problem of determining the compound markings and the transition rates among them is discussed in the following. The markings (states) of the philosopher system based upon HLPN are given in Table II. The initial population of different places is five tokens in T, five tokens in F, and no token in E. When one or more philosophers are eating, E contains one or more tokens.

In many systems, a number of different processes have a similar structure and behavior. To simplify the system model, it is desirable to treat similar processes in a uniform and succinct way. In the HLPN models, a token type may be associated with the process type and the number of tokens with the same type attribute may be associated with the the number of identical processes [11]. A process description, a subnet, can specify the behavior of a type of process and defines variables unique to each process of that type. Each process is a particular and independent instance of an execution of a process description (subnet).

The tokens present in SHLPN's have several attributes: type, identity, environment, etc. In order to introduce compound markings, such attributes are represented by variables with a domain covering the set of values of the attribute.

In the philosopher system, we can use a variable i to replace the *identity* attribute of the philosopher and the *environment variable* attribute representing fork tokens to each philosopher process. The domain set of the variable i is $[1, 5]$, i.e., the $\langle p, i\rangle$ represents anyone among $\langle p, 1\rangle$, $\langle p, 2\rangle$, $\langle p, 3\rangle$, $\langle p, 4\rangle$, $\langle p, 5\rangle$, and the $\langle f, i\rangle$, represents anyone among $\langle f, 1\rangle$, $\langle f, 2\rangle$, $\langle f, 3\rangle$, $\langle f, 4\rangle$, $\langle f, 5\rangle$. The compound marking (state) table of the philosopher system is shown in Table III. The size of the state space is reduced compared to the previous case. The equivalent marking concept was also proposed for reachability trees

TABLE III
THE STATE TABLE OF THE PHILOSOPHER SYSTEM WITH COMPOUND MARKINGS

State	Place index		
	T	E	F
0	<p,i>,<p,i1>,<p,i2>,<p,i3>,<p,i4>	0	<f,i>,<f,i1>,<f,i2>,<f,i3>,<f,i4>
1	<p,i1>,<p,i2>,<p,i3>,<p,i4>	<p,i,i,i1>	<f,i2>,<f,i3>,<f,i4>
2	<p,i1>,<p,i3>,<p,i4>	<p,i,i,i1>,<p,i2,i2,i3>	<f,i4>

of the HLPN in [7]. Our compound marking concept is convenient for computing the reachability set and for understanding the behavior of the system modeled.

The markings of Table III correspond to the Markov chain states shown in Fig. 4 and are obtained by grouping the states from Fig. 2. The transition rates between the grouped states (compound markings) can be obtained after determining the number of possible transitions from one individual marking in each compound marking to any individual marking in another compound marking. In our case, there is one possible transition from only one individual marking of the compound marking S_0 to each individual marking of the compound marking S_1 with the same rate. So, the transition rate from S_0 to S_1 is $5\lambda_1$. Using a similar argument, we can obtain the transition rate from S_1 to S_2 as $2\lambda_1$, from S_2 to S_1 as $2\lambda_2$, and from S_1 to S_0 as λ_2. The steady-state probabilities of each compound marking (grouped Markov state) can be obtained as

$$p_0 = \frac{\lambda_2^2}{5\lambda_1(\lambda_1+\lambda_2)+\lambda_2^2}$$

$$p_1 = \frac{5\lambda_1\lambda_2}{5\lambda_1(\lambda_1+\lambda_2)+\lambda_2^2}$$

$$p_2 = \frac{5\lambda_1^2}{5\lambda_1(\lambda_1+\lambda_2)+\lambda_2^2}.$$

The probability of every individual marking of a compound marking is the same and can be easily obtained since the number of individual markings in each compound marking is known.

III. Stochastic High-Level Petri Nets

The previous example has presented the advantage of using high-level Petri nets augmented with exponentially distributed firing times called, in this paper, stochastic high-level Petri nets, for modeling of concurrent systems.

This section is organized as follows: first we present a formal definition of stochastic high-level Petri nets and discuss the isomorphism of SLPNS with continuous time Markov chains. Then the concept of *compound marking* is discussed.

A. Formal Definition of SHLPN's

A high-level Petri net consists of the following elements.

1) A directed graph (P, T, A) where

- P is the set of places
- T is the set of transitions
- A is the set of arcs; $A \subseteq (P \times T) \cup (T \times P)$.

2) A structure set Σ consisting of some types of individual tokens (u_i) together with some operations (op_i) and relations (r_i), i.e., $\Sigma = (u_1, \cdots, u_n; op_1, \cdots, op_m; r_1, \cdots, r_k)$.

3) A labeling of arcs with a formal sum of n attributes of token variables (including the zero-attributes indicating a no-argument token).

4) An inscription on some transitions being a logical formula constructed from the operation and relations of the structure Σ and variables occurring at the surrounding arcs.

5) A marking of the places of P with n attributes of individual tokens.

6) A natural number K which assigns to the places an upper bound for the number of copies of the same token.

7) *Firing rule:* Each element of T represents a class of possible changes of markings. Such a change, also called *transition firing*, consists of removing tokens from a subset of places and adding them to other subsets according to the expressions labeling the arcs. A transition is enabled whenever, given an assignment of individual tokens to the variables which satisfies the predicate associated with the transition, all input places carry enough copies of proper tokens, and the capacity K of all output places will not be exceeded by adding the respective copies of tokens. The *state space of the system* consists of the set of all markings connected to the initial marking through such occurrences of firing.

Definition 3.1: A continuous time stochastic high-level Petri Net is an HLPN extended with the set of markings related, transition rates, $\lambda = \{\lambda_1, \lambda_2, \cdots, \lambda_R\}$. The value of R is determined by the cardinality of the reachability set of the net.

In order to have an equivalence between a timed Petri net and the stochastic model of the system represented by the net, the following two elements need to be specified:

- the rules for choosing from the set of enabled transitions, the one that fires, and
- the conditioning upon the past history.

A race model for the execution policy of SHLPN's has been selected. Since we have restricted the distribution of firing times to the negative exponential, the memoryless property of this distribution makes the distinction between different race models (with resampling $R-R$, with age memory $R-A$, and with enabling memory $R-E$) immaterial [14]. The sojourn time in any state is given by the minimum among the exponential random variables associated with the transitions enabled by that particular state.

The SHLPN's introduced in this paper do not have immediate transitions. The predicate associated with a transition performs the selection function using the attributes of the tokens in the input places of the transition. On the other hand, although actions do not occur instantaneous in real systems, the need to incorporate immediate transitions into SHLPN's might occur in the future in order to reduce the state space of the model. Then, preselection must be introduced into race models by defining probability distributions associated with immediate transitions.

A one-to-one correspondence between each marking of a stochastic high-level Petri net and a state of a Markov chain representing the same system can be established. Following the arguments presented in [16], the following can be stated.

Theorem 3.1: Any finite place, finite transition, stochastic high-level Petri net is isomorphic to a one-dimensional, continuous time, finite Markov chain.

As in the case of SPN's, this isomorphism is based upon the marking sequence and not upon the transition sequence. Any number of transitions between the same two markings is indistinguishable.

B. The Compound Marking of an SHLPN

The compound marking concept is based on the fact that a number of entities processed by the system exhibit an identical behavior and they have a single subnet in the SHLPN model. The only distinction between such entities is the *identity* attribute of the token carried by the entity. If, in addition, the system consists of identical processing elements distinguished only by the identity attribute of the corresponding tokens, it is possible to lump together a number of markings in order to obtain a more compact SHLPN model of the system. Clearly, the model can be used to determine the global system performance in case of homogeneous systems when individual elements are indistinguishable.

Definition 3.2: A compound marking of an SHLPN is the result of partitioning an individual SHLPN marking into a number of disjoint sets such that

- the individual markings in a given compound marking have the same distribution of tokens in places, except for the identity attribute of tokens of the same type,
- all individual markings in the same compound marking have the same transition rates to all other compound markings.

These ideas can be clearly followed in the previous example. If we now consider the example presented in Section IV, (Fig. 6 shows the SHLPN model of a multiprocessor system), we see that the compound marking indicated as state 2 in Table IV, corresponds to 15 individual markings as shown in Table V.

Let us now consider a few properties of the compound marking.

P1. A compound marking enables all transitions enabled by all individual markings lumped into it.

P2. If the individual reachability set of an SHLPN is finite, its compound reachability set is finite.

P3. If the initial individual marking is reachable with a nonzero probability from any individual marking in the individual reachability set, the SHLPN initial compound marking is reachable with a nonzero probability from any compound marking in the compound reachability set.

We denote by p_{ij} the probability of a transition from the compound marking i to the compound marking j and by $p_{i_nj_k}$ as the probability of a transition from the individual marking i_n to the individual marking j_k, where $i_n \in i$ and $j_k \in j$. The relation between the transition probability of compound markings and the transition probability of individual markings is

$$p_{ij}=\sum_k p_{i_nj_k}. \tag{3.1}$$

The relation between the transition rate of compound markings and the transition rate of individual markings is

$$q_j(t)=\frac{d\left(\sum_i p_{ji}\right)}{dt}=\frac{\sum_i d\left(\sum_k p_{j_ni_k}\right)}{dt} \tag{3.2}$$

$$q_{ij}(t)=\frac{dp_{ij}}{dt}=\frac{\sum_k d(p_{i_nj_k})}{dt}. \tag{3.3}$$

If the system is ergodic, then the sojourn time in each compound marking is an exponentially distributed random variable with average

$$\left[\sum_{i\in H}(q_{jk})_i\right]^{-1} \tag{3.4}$$

where H is the set of transitions that are enabled by the compound marking and q_{jk} is the transition rate associated with the transition i firing on the current compound marking j.

Since there is an isomorphism between stochastic high-level Petri nets and Markov chains, any compound markings of an SHLPN correspond to grouping, or lumping of states in the Markov domain.

In order to be useful, a compound marking must induce a correct grouping in the Markov domain corresponding to the original SHLPN. Otherwise, the methodology known from Markov analysis, used to establish whether the system is stable and to determine the steady-state probabilities of each compound marking, cannot be applied. The problem of state grouping for a Markov process is treated in [4] and [8]. While emphasis is placed upon discrete time Markov chain, [8] also gives the groupability conditions for a continuous time Markov process as well. The compound marking of an SHLPN induces a partition of the Markov state space which satisfies the conditions for grouping. Without proof, we can state the following theorem.

Theorem 3.2: A stochastic high-level Petri net with a compound marking operation is homomorphic with a continuous time Markov chain with a grouping operation.

As a conclusion, given the SHLPN model of a system, after constructing a compound marking as defined in the previous section, we can study the behavior of the system applying Markov techniques. In particular, we can determine the steady-state probabilities for each compound marking (grouped state) if the system is ergodic.

Rather than giving additional rules for constructing a compound marking, we describe in detail an example of modeling with SHLPN's augmented with compound markings.

TABLE IV
STATE TABLE OF THE MULTIPROCESSOR SYSTEM MODEL

Marking	Place index				
(State)	P	Q	M	B	A
1	5	0	i, j, k	2	0
2	4	i	i, j, k	2	0
3	3	i, j	i, j, k	2	0
4	3	i, i	i, j, k	2	0
5	2	i, j, k	i, j, k	2	0
6	2	i, i, j	i, j, k	2	0
7	2	i, i, i	i, j, k	2	0
8	1	i, i, j, k	i, j, k	2	0
9	1	i, i, i, j	i, j, k	2	0
10	1	i, i, j, j	i, j, k	2	0
11	1	i, i, i, i	i, j, k	2	0
12	0	i, i, i, j, k	i, j, k	2	0
13	0	i, i, j, j, k	i, j, k	2	0
14	0	i, i, i, i, j	i, j, k	2	0
15	0	i, i, i, j, j	i, j, k	2	0
16	0	i, i, i, i, i	i, j, k	2	0
17	4	0	j, k	1	i
18	3	j	j, k	1	i
19	3	i	j, k	1	i
20	2	j, k	j, k	1	i
21	2	i, j	j, k	1	i
22	2	i, i	i, k	1	j
23	2	i, i	j, k	1	i
24	1	i, j, k	j, k	1	i
25	1	i, i, j	i, k	1	j
26	1	i, i, j	j, k	1	i
27	1	i, i, i	i, k	1	j
28	1	i, i, i	j, k	1	i
29	0	i, i, j, k	j, k	1	i
30	0	i, i, i, k	i, k	1	j
31	0	i, i, j, j	i, j	1	k
32	0	i, j, j, k	j, k	1	i
33	0	i, i, i, i	i, k	1	j
34	0	i, i, i, j	j, k	1	i
35	0	i, i, j, j	j, k	1	i
36	0	i, i, i, j	i, k	1	j
37	0	i, i, i, i	j, k	1	i
38	3	0	k	0	i, j
39	2	k	k	0	i, j
40	2	i	k	0	i, j
41	1	i, i, k	i, k	1	j
42	1	i, k	k	0	i, j
43	1	i, j	k	0	i, j
44	1	i, i	k	0	i, j
45	0	i, i, k	k	0	i, j
46	0	i, i, i	i	0	j, k
47	0	i, j, j	j	0	i, k
48	0	i, j, k	k	0	i, j
49	0	i, i, i	k	0	i, j
50	0	i, i, j	k	0	i, j
51	1	k, k	k	0	i, j

TABLE V
THE 15 INDIVIDUAL MARKINGS (STATES) FOR PLACES P AND Q, CORRESPONDING TO THE COMPOUND MARKING DEFINED AS MACROSTATE 2 IN TABLE IV.

P				Q
<p,2>	<p,3>	<p,4>	<p,5>	<p,1,1>
				<p,1,2>
				<p,1,3>
<p,1>	<p,3>	<p,4>	<p,5>	<p,2,1>
				<p,2,2>
				<p,2,3>
<p,1>	<p,2>	<p,4>	<p,5>	<p,3,1>
				<p,3,2>
				<p,3,3>
<p,1>	<p,2>	<p,3>	<p,5>	<p,4,1>
				<p,4,2>
				<p,4,3>
<p,1>	<p,2>	<p,3>	<p,4>	<p,5,1>
				<p,5,2>
				<p,5,3>

IV. MODELING AND PERFORMANCE ANALYSIS OF A MULTIPROCESSOR SYSTEM USING SHLPN'S

To assess the modeling power of stochastic high-level Petri nets, we consider now a multiprocessor system as shown in Fig. 5. Clearly, the performance of a multiprocessor system depends upon the level of contention for the interconnection network and for the common memory modules.

There are two basic paradigms for interprocessor communication, determined by the architecture of the system, namely, message passing and communication through shared memory. The analysis carried out in this section is designed for shared memory communication but it can be extended to accommodate message passing systems. To model the system, we assume that each processor executes in a number of domains and that the execution speed of a given processor is a function of the execution domain. The model assumes that a random time is needed for the transition from one domain to another.

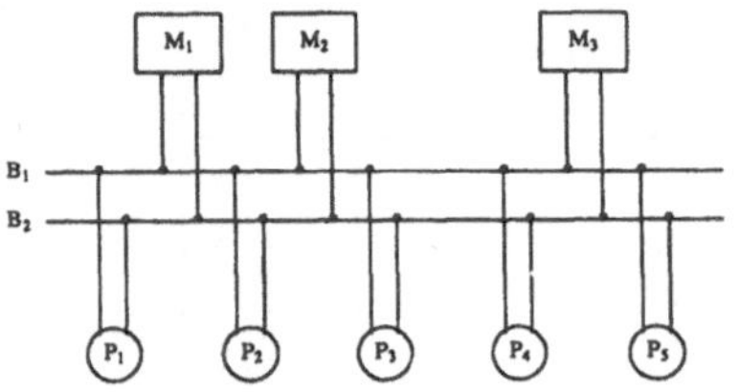

Fig. 5. The configuration of the multiprocessor system used in SHLPN modeling.

This section is organized as follows: first we describe the basic architecture of a multiprocessor system and the assumptions necessary for system modeling. Then the SHLPN model of the system is presented. The methodology to construct a model with a minimal state space is presented and the equilibrium equations of the system are solved using Markov chain techniques. Based upon the steady-state probabilities associated with system states, the performance analysis is carried out.

A. System Description and Modeling Assumptions

As shown in Fig. 5, a multiprocessor system consists of a set of n processors $\boldsymbol{P} = \{P_1, P_2, \cdots\cdots P_n\}$ interconnected by means of an interconnection network to a set of q common memory modules $\boldsymbol{M} = \{M_1, M_2, \cdots, M_q\}$. The simplest topology of the interconnection network is a set of r buses $\boldsymbol{B} = \{B_1, B_2, \cdots\cdots B_r\}$. Each processor is usually connected also to a private memory module through a private bus.

As a general rule, the time to perform a given operation depends on whether the operands are in local memory or in the common one. When more than one processor is active in common memory, the time for a common memory reference will increase due to contention for buses and common memory modules characterized the load factor ρ. A possible definition for this load, adopted in this paper, is the ratio between the time spent in an execution region located in the common domain and the time spent in an execution region located in the private domain.

A common measure of the multiprocessor system performance is the processing power of a system with n identical processors expressed as a fraction of the maximum processing power (n times the processing power of a single processor executing in its private memory). Considering an application which is decomposed into n identical processes, the actual processing power of the system depends upon the ratio between local memory references and common memory ones.

The purpose of our study is to determine the resource utilization when the load factor increases. The basic assumptions made for our model are the following.

A1: All processors exhibit identical behavior for the class of applications considered. It is assumed that the computations performed by all processors are similar and they have the same pattern of memory references. More precisely, it is assumed that each processor spends an exponentially distributed random time with mean $1/\lambda_1$, while executing in its private domain and then an exponentially distributed random time

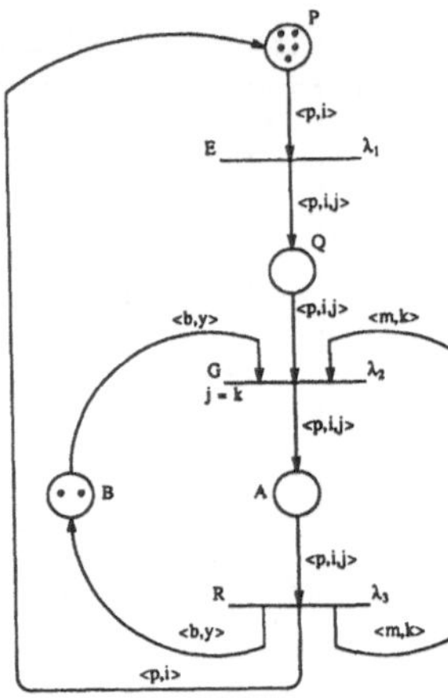

Place and transition significance

- *P* is the "private memory" place. When the place *P* holds tokens, the corresponding processors are active in their own memory.
- *Q* is the "queuing" place. When the place *Q* holds tokens, the corresponding processors are queued for common memory.
- *A* is the "accessing" place. When the place *A* holds tokens, the corresponding processors are accessing common memory.
- *M* is the "idle common memory" place. When the place *M* holds tokens, the corresponding common memories are idle.
- *B* is the "available bus" place. When the place *B* holds tokens, the corresponding buses are available.
- *E* is the "end of activity" in minute memory transition. When it fires, a processor ends its activity in its private memory.
- *G* is the "getting common memory" transition. The transition is enabled when the common memory module and a bus are free. The transition firing time is related to $1/\lambda_2$.
- *R* is the "releasing common memory" transition. After accessing common memory, the processor releases the common memory and the bus and returns to the private memory execution.

Fig. 6. Modeling of the multiprocessor system with an SHLPN.

with mean $1/\lambda_3$ while executing in a common domain. To model that common memory references are evenly spread into the set of available common memory modules, it is assumed that after finishing an execution sequence in private memory, each processor draws a random number k, uniformly distributed into the set $[1, q]$, which determines the module where its next common memory reference will be.

A2: The access time to common memory modules has the same distribution for all modules and there is no difference in access time when different buses are used.

A3: When a processor acquires a bus and starts its execution sequence in the common memory, then it releases the bus only after completing its execution sequence in the common domain.

The first assumption is justified since common mapping algorithms tend to decompose a given parallel problem into a number of identical processes, one for every processor available in the system. The second and the third assumptions are clearly realistic due to hardware considerations.

B. Model Description

Fig. 6 presents a stochastic high-level Petri net model of a multiprocessor system. Although the graph representing the model is invariant to the system size, the state space of the SHLPN clearly depends upon the actual number of processors n, common memory modules q, and buses r. For our example, $n = 5$, $q = 3$, and $r = 2$.

The graph consists of five places and three transitions. Each place contains tokens whose type may be different. The general format of a token is ⟨type, id, $i, j, \cdots$⟩. A token has a number of attributes, the first attribute being its type. We recognize three different types: p—processor, m—common memory, b—bus. The second attribute of a token is its *identity, id.* The id attribute is a positive integer with values depending upon the number of objects of a given type. In our example, when type $= p$, the id takes values in the set [1, 5]. The tokens residing in place Q have a third attribute: the id of the common memory module they are going to refer next.

The meaning of different places and the tokens they contain are presented in Fig. 6. The notation used should be interpreted in the following way: the place P contains the set of tokens of type processor with two attributes $\langle p, i\rangle$, with $i \in [1, 5]$. The maximum capacity of place P is equal to the number of processors. The transition E corresponds to an end of execution in the private domain and it occurs with a transition rate exponentially distributed with mean λ_1. As a result of this transition, the token moves into place Q where it selects the next common memory reference. A token in place Q has three attributes $\langle p, i, j\rangle$ with the first two as before and the third attribute describing the common memory module $j \in [1, 3]$ to be accessed by processor i. The processor could wait to access the common memory module when either no bus is available or the memory module is busy. Transition G occurs when a processor switches to execution in common domain, and when the predicate $j = k$ (see Fig. 6) is satisfied. This is a concise representation of the condition that the memory module referenced by the processor i is free. Another way of expressing this condition is: the third attribute of token $\langle p, i, j\rangle$ is equal to the second attribute of token $\langle m, k\rangle$. The place B contains tokens representing free buses and the place M contains tokens representing free memory modules. The maximum capacities of these places are equal to the number of buses and memory modules. The rate of transition G is λ_2 and it is related to the exponentially distributed communication delay involved in a common memory access. The place A contains tokens representing processes executing in the common domain. The maximum capacity of the places in our graph are

$$\text{Capacity } (P) = n$$

$$\text{Capacity } (Q) = n$$

$$\text{Capacity } (M) = q$$

$$\text{Capacity } (B) = r$$

$$\text{Capacity } (A) = \min (n, q, r). \tag{4.1}$$

The compound markings of the system are presented in Table IV. In order to simplify this table, the following convention is used: whenever the attributes of the tokens do not have any effect upon the compound marking, only the number of the tokens present in a given place is shown. When an attribute of a token is present in a predicate, only that attribute is shown in the corresponding place if no confusion about the token type is possible.

For example, the marking corresponding to state 2 has 4 tokens in place P (the token type is p according to model description), two tokens in place B (type $= b$), 0 tokens in place A. Only the third attribute i of the token present in place Q (the id of the memory module of next reference) is indicated. Also shown are the ids of the tokens present in place M, namely i, j, and k.

As a general rule, it is necessary to specify in the marking the attributes of the tokens referred to by any predicate which may be present in the SHLPN. In our case, we have to specify the third attribute of tokens in Q and the second attribute of the tokens in M since they appear in the predicate associated with transition G.

Table IV shows the state transition table of the system. For example, state 2 can be reached from the following states: state 1 with the rate $15 \times \lambda_1$, state 18 with the rate λ_3, and state 19 with the transition rate equal to λ_3. From state 2, the system goes either to state 3 with the transition rate equal to $8 \times \lambda_1$, to state 4 with rate $4 \times \lambda_1$, or to state 17 with rate λ_2.

State 2 corresponds to the situation when any four processors execute in the private domain and the fifth has selected the memory module of its next common domain reference to be module i. It should be pointed out that state 2 is a *macrostate* obtained due to the use of the compound marking concept and it corresponds to 15 atomic states. These 15 states are distinguished only by the identity attributes of the tokens in two places, P and Q, as shown in Table V. The transition rate from the compound marking, denoted as state 1 in Table IV, to the one denoted by state 2 is $15 \times \lambda_1$ since there are 15 individual transitions from one individual marking of state 1 to the 15 individual markings in the compound marking corresponding to state 2.

C. Performance Analysis

To determine the average utilization of different system resources, it is necessary to solve the equilibrium equations and then identify the states when each resource is idle and the occupancy of that state, the number of units of that resource which are idle. The following notation is used: size $[B]_i$ is the occupancy of place B when the system is in state i and p_i is the probability of the system being in state i. Then the average utilization of a processor η_p, a common memory module η_m, and a bus η_b are defined as

$$\eta_p = 1 - \sum_{i \in S} \frac{p_i \times \text{size } [Q]_i}{n} \tag{4.2}$$

$$\eta_m = 1 - \sum_{i \in S} \frac{p_i \times \text{size } [M]_i}{q} \tag{4.3}$$

$$\eta_b = 1 - \sum_{i \in S} \frac{p_i \times \text{size } [B]_i}{l}. \tag{4.4}$$

We assume a fixed ratio between the transition rate related transition G and that related transition E, $\lambda_1/\lambda_2 = 10^{-3}$. This expresses the fact that the communication delay is much smaller than the time spent in executing in the private domain.

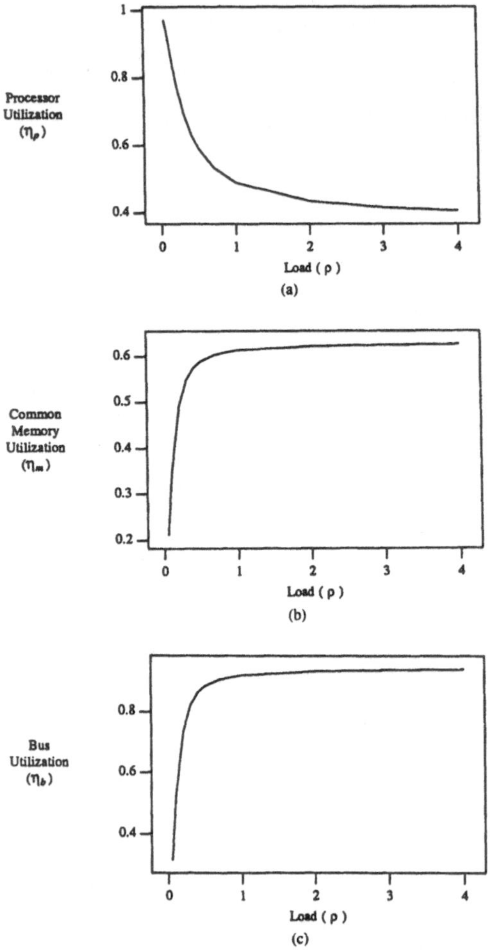

Fig. 7. Resource utilization for the multiprocessor system.

The load for common resources is defined as

$$\rho = \frac{\lambda_1}{\lambda_3}. \tag{4.5}$$

We study the resource utilization for $5 \times 10^{-2} \leq \rho \leq 4$.

Fig. 7(a) shows the processor utilization when the load placed upon shared system resources increases. Even when the load is close to 0.7, the actual speed up factor of the system is only about 50 percent of its maximum value. Bus utilization increases [Fig. 7(c)] rapidly even for lower values of the load and indicates that the system is communication bound and no significant performance improvement can be obtained by adding more memory modules. This corroborates with the information provided by the memory utilization curve shown in Fig. 7(b) which shows a maximum common memory utilization of about 60 percent.

The number of original states is very high, larger than 500, and we have reduced the model to only 51 states. As mentioned earlier, the same conceptual model can be used to model a message passing system. In such a case, λ_2 will be related to the time necessary to pass a message from one processor to another, including the processor communication overhead at the sender and at the receiving site as well as the transmission time dependent upon the message size and the communication delay. In case of a synchronous message passing system, λ_3 will be related to the average blocking time in order to generate a reply.

The technique described here can be used for the performance analysis of considerably larger systems. The analysis of a multiflex machine consisting of eight interconnected flex systems, each with nine processors, is reported in [12]. Each processor can access local, locally shared, and global memory. Two models for the execution of PDE splitting algorithms have been developed, one leading to a system with 145 states and one which embeds the synchronization among processors, and leads to a system with 1505 states. In both cases, we have taken advantage of the compound marking, the original state space of the models is considerably larger than the one obtained with SHLPN's.

V. Conclusions

The stochastic high-level Petri nets introduced in this paper represent a powerful and convenient tool for performance analysis of communication and computing systems. They are extensions of stochastic Petri nets but, generally, lead to models with a lower size of the state space. The compound marking concept allows a considerable reduction of the number of states and it induces a correct grouping of states in the Markov domain. SHLPN models of multiprocessor systems have been developed. The current research effort is directed toward the design of automated tools for SHLPN's.

Acknowledgment

The authors express their thanks to the anonymous referees for their careful review and for their suggestions which have improved the quality of this paper.

References

[1] G. Chiola, "A software package for the analysis of generalized stochastic Petri net models," in *Proc. Int. Workshop Timed Petri Nets*, Torino, Italy, July 1985, pp. 136–144.

[2] J. B. Dugan, G. Ciardo, A. Bobbio, and K. S. Trivedi, "The design of a unified package for the solution of stochastic Petri net models," in *Proc. Int. Workshop Timed Petri Nets*, Torino, Italy, July 1985, pp. 6–13.

[3] J. B. Dugan, K. S. Trivedi, R. M. Geist, and V. F. Nicola, "Extended stochastic Petri nets: Application and analysis," in *Performance 84*, Paris, France, Dec. 1984, pp. 507–519.

[4] E. Gelenbe and I. Mitrani, *Analysis and Synthesis of Computer Systems*. New York: Academic, 1980.

[5] J. H. Genrich and K. Lautenbach, "The analysis of distributed systems by means of predicate transition-nets," *Lecture Notes Comput. Sci.*, vol. 70, pp. 123–146, 1979.

[6] J. H. Genrich and K. Lautenbach, "System modeling with high-level Petri nets," *Theoret. Comput. Sci.*, vol. 13, pp. 109–136, 1981.

[7] P. Huber, A. M. Jensen, L. O. Jepsen, and K. Jensen, "Towards

reachability trees for high-level Petri nets," *Lecture Notes Comput. Sci.*, vol. 188, pp. 215-233, 1985.

[8] M. Iosifescu, *Finite Markov Processes and their Applications.* New York: Wiley, 1980.

[9] K. Jensen, "Coloured Petri nets and the invariant-method," *Theoret. Comput. Sci.*, vol. 14, pp. 317-336, 1981.

[10] K. Jensen, "High-level Petri nets," *Informatik-Fachberichte*, vol. 66, pp. 166-180, 1983.

[11] C. Lin and D. C. Marinescu, "Application of modified predicate transition nets to modeling and simulation of communication protocols," Comput. Sci. Dep., Purdue Univ., Tech. Rep. CSD-TR-599, May 1986.

[12] D. C. Marinescu and J. R. Rice, "Domain oriented analysis of PDE splitting algorithms," *J. Inform. Sci.*, vol. 43, pp. 3-24, 1987.

[13] M. A. Marsan, G. Conte, and G. Balbo, "A class of generalized stochastic Petri nets for the performance evaluation of multiprocessor systems," *ACM Trans. Comput. Syst.*, vol. 2, pp. 93-122, May 1984.

[14] M. A. Marsan, G. Balbo, A. Bobbio, G. Chiola, G. Conte, and A. C. Cumani, "On Petri nets with stochastic timing," in *Proc. Int. Workshop Timed Petri Nets*, Torino, Italy, July 1985, pp. 80-87.

[15] J. F. Meyer, A. Movaghar, and W. H. Sanders, "Stochastic activity networks: Structure, behavior and applications," in *Proc. Int. Workshop Timed Petri Nets*, Torino, Italy, July 1985, pp. 106-115.

[16] M. K. Molloy, "Performance analysis using stochastic Petri nets," *IEEE Trans. Comput.*, vol. C-31, pp. 913-917, Sept. 1982.

[17] ——, "Discrete time stochastic Petri nets," *IEEE Trans. Software Eng.*, vol. SE-11, pp. 417-423, 1985.

[18] S. Natkin, "Les reseaux de Petri stochastique et leur application a l' evaluation des systems informatiques," These de Docteur Ingegneur, CNAM, Paris, 1980.

[19] C. V. Ramamoorthy and G. S. Ho, "Performance evaluation of asynchronous concurrent systems using Petri nets," *IEEE Trans. Software Eng.*, vol. SE-6, pp. 440-449, Sept. 1980.

[20] W. Reisig, "Petri nets with individual tokens," *Informatik-Fachberichte*, vol. 66, pp. 229-249, 1983.

[21] J. Sifakis, "Petri nets for performance evaluation," in *Measuring, Modeling, Evaluating Comput. Syst., Proc. 3rd Int. Symp. IFIP Working Group 7.3*, H. Beilner and E. Gelenbe, Eds., 1977, pp. 75-93.

[22] K. S. Trivedi, *Probability and Statistics with Reliability, Queueing and Computer Science Applications.* Englewood Cliffs, NJ: Prentice-Hall, 1982.

[23] A. Zenie, "Coloured stochastic Petri nets," in *Proc. Int. Workshop Timed Petri Nets*, Torino, Italy, July 1985, pp. 262-271.

17.
Regular Stochastic Petris Nets

C. Dutheillet and S. Haddad

G. Rozenberg (ed.): Advances in Petri nets 1990. Lecture Notes in Computer Science, vol. 483. Springer, Berlin Heidelberg New York 1990, pp. 186-210

ABSTRACT : An extension of regular nets, a class of colored nets, to a stochastic model is proposed. We show that the symmetries in this class of nets make it possible to develop a performance evaluation by constructing only a graph of symbolic markings, which vertices are classes of states, instead of the whole reachability graph. Using algebraic techniques, we prove that all the states in a class have the same probability, and that the coefficients of the linear system describing the lumped Markov process can be calculated directly from the graph of symbolic markings.

Keywords : Higher-level net models, stochastic nets.

CONTENTS

0. INTRODUCTION

As they are a function of the number and the complexity of the processes to be represented, Petri net models of distributed computer systems must quickly face problems due to the huge size of the reachability graph.
The use of higher level models (e.g., colored nets, Pr/T nets) can make the task of the modeler easier. However, the existing analysis methods often make it necessary to go back to an equivalent ordinary net, so that there is no improvement in the complexity.
In order to face the exponential increase in the number of states, Jensen [Jen 81a] has introduced in colored Petri nets some equivalence relations that take into account the symmetries of the model. He showed that those relations were directly depending on the type of the color functions. He also defined analysis methods associated with the use of simple functions [Jen 81a, Hub 84].

Research has been done to extend the symmetries in colored nets to the stochastic domain [Zen 85, Lin 87, Chi 88]. Some attempts to reuse the symmetry properties have been proposed. One approach [Zen 85] has been to extend the reduced graph developed by Jensen to a stochastic model. However, when using this technique, the partition of states in classes can lead to a non Markovian process. In [Lin 87], the reduced process is actually Markovian, but no method is provided to build it automatically. The approach in [Chi 88] was to build the reachability graph, to group the states in classes according to some symmetry relation, and then to partition these classes in subclasses until the resulting process is Markovian. In fact, as soon as the color functions are general, no existing method avoids the development of the whole reachability graph. Indeed the simplification due to the symmetries can be used only once the reachability graph has been constructed, even sometimes requiring an expensive preliminary analysis.

This paper aims at showing that in the case of colored nets that do not use general functions, the results obtained during the structural analysis can be extended to the performance evaluation. In order to optimize the analysis, our study is based on a particular class of colored nets, the regular nets [Had 87]. In a regular net, the objects in a class have similar behaviors. The color domains of places and transitions are Cartesian products of object classes. The color functions are Cartesian products and linear combinations of two basic functions, one selecting a unique object in a class, the other synchronizing all the objects of the class. For this class of nets, a condensed representation of the reachability graph, the symbolic reachability graph, can be defined. The symmetry properties of this graph are used to simplify the quantitative analysis of the model.

The paper is organized as follows. In the next section, we give the definition and an example of regular net. Section 3 will presents the construction of the graph of symbolic markings. The stochastic model derived from a regular net will be presented in Section 4. In Section 5, we propose an algorithm for computing the state probabilities from the graph of symbolic markings. Our algorithm maps the symbolic reachability graph on a lumped Markov chain and uses the labels of the symbolic arcs to compute the transition rates. In the last section we prove the correctness of the algorithm. The proof organizes into three steps. First we prove that the developed Markov chain always has a solution such that all the ordinary markings within a symbolic marking have the same probability. Then we show that the lumping condition is verified by the Markov chain, and as a consequence, that the linear system to solve can be reduced. Finally we prove that this reduced system is the one obtained with our algorithm. Moreover, as the number of ordinary markings within a symbolic marking is computed by the algorithm, the ordinary probabilities can be derived without additional operations.

1. REGULAR NETS

Even if they do not have the same expression power as general colored nets, regular nets allow one to model a large class of systems. They have been the starting point for developing important theoretical results, such as reductions and computation of linear place-invariants. They have also allowed to formalize the parametrization which leads to a validation of the system that does not depend on the values of some parameters such as the number of sites, or the number of processes.

1.1 Definition :

A regular net RN = <P, T, C, I^-, I^+> is defined by :

P the set of places,

T the set of transitions,

C the set of object classes : $C = \{C_1, \ldots, C_n\}$, with $C_i \cap C_j = \emptyset$,

I^+ and I^- the input and output matrices defined on P x T, which elements I^+, $I^-(p, t)$ are standard color functions of p (defined below).

The color domains C(p) for a place, and C(t) for a transition, are defined as follows : a color domain is made either of the neutral color, or of a Cartesian product of object classes such that all the elements in the product are distinct.

<u>Definition 1.1</u> : A normalized RN is an RN in which all places and transitions have the same color domain $\boldsymbol{C} = C_1 \times \ldots \times C_n$.

As every RN can be transformed in a normalized RN without modifying its structural behavior, we will limit our study to that specific class of regular nets. However, the case of non-normalized regular nets and the transformation rule are presented in [Dut 89].

<u>Definition 1.2</u> : Marking of an object.

A marking is a function $m : P \times \boldsymbol{C} \rightarrow \mathbb{N}$, such that m(p, c) is the number of marks of color c in p. The marking of an object c_i can be defined by the function below :

$m : C_i \rightarrow [P \times C_1 \times \ldots \times C_{i-1} \times C_{i+1} \times \ldots \times C_n \rightarrow \mathbb{N}]$

$m(c_i)(p, c_1, \ldots, c_{i-1}, c_{i+1}, \ldots, c_n) = m(p, c)$ where $c = (c_1, \ldots, c_n)$.

Definition 1.3 : Marking of a color.
The marking of a color c can be defined by the function below :

$$m : C \to [P \to \mathbb{N}]$$

$$m(c)(p) = m(p, c)$$

As the three definitions are equivalent, we use the same letter for all the marking functions.

The standard color functions of a regular net are defined from two basic functions :

$$X_i : C_i \times C \to \mathbb{N} \quad \text{such that } X_i(c_i', (c_1, \ldots, c_n)) = Id(c_i, c_i') = (\text{If } c_i = c_i' \text{ then } 1 \text{ else } 0).$$

An arc labeled X_i distinguishes exactly one object in the class C_i. The behavior of this object will be independent of the behavior of the other objects of the class when firing the transition.

$$S_i : C_i \times C \to \mathbb{N} \quad \text{such that } S_i(c_i', (c_1, \ldots, c_n)) = 1.$$

An arc labeled S_i means that all the objects in the class C_i play a similar part. It corresponds to a synchronization of all the objects of the class if it labels an input arc. It represents a diffusion to all the objects of the class if it labels an output arc.

Those two basic functions can be combined by :

$$\mathbf{a_i . S_i + b_i . X_i} : C_i \times C \to \mathbb{N} \quad \text{such that } a_i.S_i + b_i.X_i(c_i', (c_1, \ldots, c_n)) = (\text{If } c_i = c_i' \text{ then } (a_i+b_i) \text{ else } a_i).$$

As a consequence, we have $a_i \geq 0$ and $(a_i+b_i) \geq 0$.

Definition 1.4 : A standard color function $< a_1.S_1+b_1.X_1, \ldots, a_n.S_n + b_n.X_n>$ of a regular net, also denoted by $\prod_{i=1}^{n} <a_i.S_i + b_i.X_i>$, is defined by :

$$\prod_{i=1}^{n} <a_i.S_i + b_i.X_i> : C \times C \to \mathbb{N},$$

$$\prod_{i=1}^{n} <a_i.S_i + b_i.X_i> ((c'_1, \ldots, c'_n), (c_1, \ldots, c_n)) = \prod_{i=1}^{n} <a_i.S_i + b_i.X_i> (c'_i, (c_1, \ldots, c_n))$$

The initial marking of a regular net, denoted by m_0, must be symmetric, i.e., it must verify the following property :

$$\forall p \in P, \forall c, c' \in C(p), m_0(p, c) = m_0(p, c').$$

1.2 Example :

We consider the net in Figure 1, which models the behavior of a distributed database. This model is derived from a model presented in [Jen 81b], and has also been studied in [Had 87]. A site is made of an active and a passive part.The active part of a site can modify a file, whereas the passive part only takes into account the modifications performed by the other sites. When a site modifies a file, its active part sends a message to all the other sites so that they can take the modification into account, whereas it waits for the acknowledgements (transition T1). The passive part of a site that receives a message modifies its copy of the file (transition T3). If there are several messages for a site, the modifications are done one at a time. Once its copy is modified, a site sends back a message to the one that originated the modification (transition T4). Once all the sites have acknowledged the modification, the waiting site resumes its activity and another site can in turn work on the modified file (transition T2).

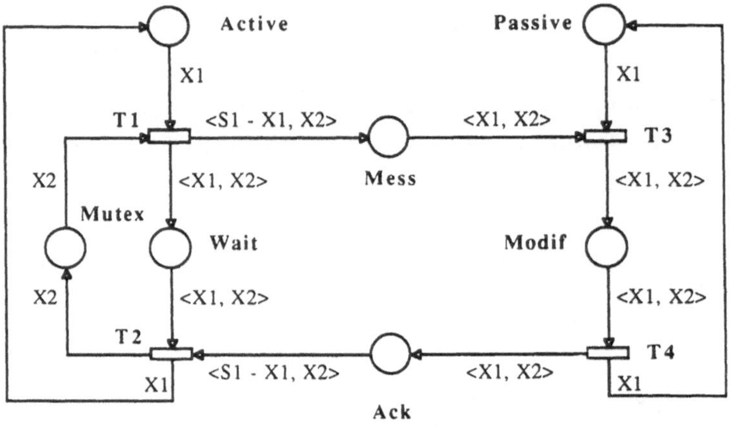

Figure 1 : Regular net model of a database

We consider a distributed database with three sites and two files. There are two color classes in the net: the site class SIT = {s1, s2, s3}, and the file class FIL = {f1, f2}. The initial marking of the net is the following : Active = Passive = SIT, and Mutex = FIL.

In the next section, we show that it is possible to develop a reduced reachability graph for regular nets, the symbolic reachbility graph.

2. SYMBOLIC REACHABILITY GRAPH

In this section, we first present some properties of regular nets that are used in the construction of the symbolic reachability graph (SRG). Unlike Jensen's reachability tree [Hub 84] that develops only once equivalent subtrees, the SRG is obtained by grouping states a priori, thus avoiding to develop any reachable state of the unfolded net. The method for constructing the SRG is then developed. We define the representation of a symbolic marking, and we present a symbolic firing rule that can be used directly on the symbolic markings.

2.1 Definitions and properties :

The properties given here result from the definition of the color domains. Also due to the specificity of the color functions, a basic symmetry property of the model is presented.

Definition 2.1 : let s_i a permutation of C_i. A permutation $s = <s_1, \ldots, s_n>$ of $(C_1 \times \ldots \times C_n)$ is defined by

$$s(<c_1, \ldots, c_n>) = <s_1(c_1), \ldots, s_n(c_n)>.$$

The group of the permutation of the type $<s_1, \ldots, s_n>$ will be denoted by S.

Definition 2.2 : let m a marking, $s \in S$ a permutation. Then s.m is a marking defined by :

$$\forall p \in P, \quad s.m(p, c) = m(p, s(c)).$$

Proposition 2.1 : s.m defines an operation of the group S on the marking set MS, i.e.,

$$\forall s, s' \in S, \forall m \in MS, \ (s \circ s').m = s.(s'.m)$$
$$id.m = m.$$

Definition 2.3 : the orbit of m, reg(m), is defined by reg(m) = $\{s.m, s \in S\}$.

The three following corollaries are standard properties of the operation of a group on a set [Lan 77].

Corollary 2.1 : the orbits reg(m) define a partition of MS that induces an equivalence relation **R** :

$$m \mathbf{R} m' \Leftrightarrow reg(m) = reg(m').$$

The equivalence classes of the relation **R** are called symbolic markings, denoted by M.

Corollary 2.2 : Let s a permutation, and M a symbolic marking. Let $f_s : M \rightarrow M$, defined by

$$\forall m \in M, \quad f_s(m) = s.m.$$

Then f_s is a bijection.

Corollary 2.3 : Let m, m' $\in$ M. $|\{s \in S, s.m = m'\}| = |S| / |M|$.

We finally give the basic theorem [Had 87], which is a consequence of the type of the color functions :

Theorem 2.1 : $\forall m, m' \in M, \forall t \in T, \forall c \in (C_1 \times \ldots \times C_n), \forall s \in S,$

$$m[t(c)>m' \Leftrightarrow s.m[t(s(c))>s.m'.$$

2.2 Construction of the SRG :

The construction of the SRG requires that we define the representation of a symbolic marking, and the firing rule that can be applied directly on these symbolic markings. We also present some properties of the SRG. These properties are used to prove that the SRG is relevant for performance evaluation.

2.2.1 Optimal representation of a symbolic marking :

The basic principle for representing a symbolic marking consists in grouping in a subclass all the objects of a class that have the same marking. Therefore, permuting two objects within a subclass will not modify the marking. The identity of the objects in a subclass is then forgotten and only the number of objects is taken into account for each subclass. In this goal, each object class C_i is partitioned in a number of subclasses, $C_i = \{C_{i,1}, \ldots, C_{i,si}\}$, such that all the objects in a subclass have the same marking. The cardinalities of each subclass, which values are in $\mathbb{N}^+$, verify $\sum_{j=1,si} |C_{i,j}| = |C_i|$. The marking of each place is then similar to an ordinary marking where the subclasses are considered as objects. Thus we can define the marking of a subclass, the marking of a product of subclasses and the permutation on subclasses in the same way as they are defined for objects in Definitions 1.2, 1.3 and 2.1.

Moreover, the grouping must be maximal, i.e., two different subclasses must have different markings. The representation of symbolic markings as defined above is unique within a permutation $<s_1, \ldots, s_n>$ of the set of subclasses. However, it is possible to define and to calculate a canonical representation for each symbolic marking by an adequate ordering based on the marking of the products of subclasses [Had 87].

Notice that the decomposition in subclasses is local to a symbolic marking. Thus, the subclass $C_{i,j}$ appearing in a symbolic marking M and the same subclass $C_{i,j}$ appearing in a symbolic marking M' may not have related meanings.

2.2.2 Example :

Figure 2 represents one reachable ordinary marking of the net in Figure 1, in which the site s2 has sent a message to the two other sites s1 and s3. The net in Figure 3 gives the symbolic representation of the marking in Figure 2. In this representation, the subclasses have the following cardinalities :

|SIT1| = 2 |FIL1| = 1
|SIT2| = 1 |FIL2| = 1

But note that ther is not a one to one correspondence between the markings in Figure 2 and Figure 3. In fact, the symbolic marking in Figure 3 represents any ordinary reachable marking in which one site has sent a message to both others.

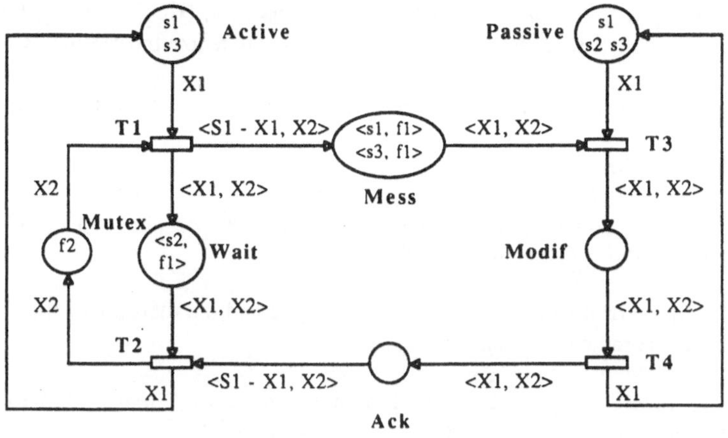

Figure 2 : A reachable ordinary marking of the database model.

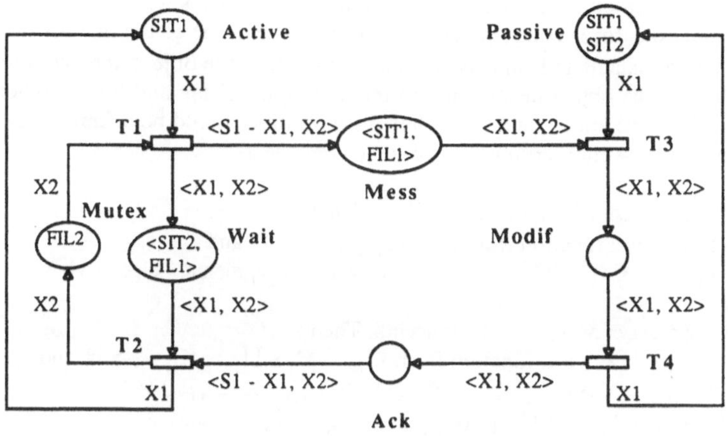

Figure 3 : Symbolic representation of the marking in Figure 2.

<u>2.2.3 Symbolic firing rule</u> :

In order to build the SRG, we first define a symbolic firing rule on the symbolic markings which must be sound i.e., an ordinary marking enables a colored transition if and only if its symbolic marking enables an equivalent symbolic firing and the ordinary marking obtained by the firing belongs to the symbolic marking obtained by the symbolic firing.

The first effect of the symbolic firing will be to split each instantiated subclass in two subclasses, one with the object instantiated in the underlying firing and the other with the remaining objects of the subclass. Thus we formally define this splitting.

Notation : In M, we will denote the partition of the class C_i in $\{C_{i,1}, \ldots, C_{i,si}\}$ by $\mathbb{C}_i = \{C_{i,1}, \ldots, C_{i,si}\}$. If confusion may arise, $|C_{i,j}|_M$ will denote the cardinality of $C_{i,j}$ in M.

Definition 2.4 : Let M a symbolic marking. Then $M[C_{1,u1}, \ldots, C_{n,un}]$ is a symbolic marking defined by :

- If $|C_{i,ui}| > 1$ then
  $\mathbb{C}_i = \{C_{i,1}, \ldots, C_{i,si}, C_{i,si+1}\}$ with
  $|C_{i,si+1}| = |C_{i,ui}|_M - 1$, $|C_{i,ui}| = 1$, $|C_{i,j}| = |C_{i,j}|_M$ for any $j \neq u_i$ and $j \neq s_i+1$
  Else the partition of C_i is unchanged.

- The marking of the old subclasses is unchanged and the marking of the new subclass $C_{i,si+1}$ is the same as the one of $C_{i,ui}$.

Notice that in $M[C_{1,u1}, \ldots, C_{n,un}]$ the grouping is not always maximal and that even if the grouping is maximal the representation may not be canonical. But it does not matter since this symbolic marking is just an intermediate marking and it will not appear in the SRG.

The instantiation of a transition in a symbolic firing will be made by choosing a subclass per class instead of an object per class in an ordinary firing. Thus we must define the value of the colored functions for subclasses. This definition is the same as the one for the objects. In the case where an instantiated subclass contains more than one object, the symbolic firing should be enabled for the object instantiated in the underlying firing and for the other objects of the subclass which are not instantiated. Thus the definition should be different but since we apply our definitions on split markings, this case never appears.

Definition 2.5 : Let M a symbolic marking. Then :
$\langle a_i.S_i + b_i.X_i \rangle$ is a function from $\mathbb{C}_i \times \Pi_{j=1,n} \mathbb{C}_j \rightarrow \mathbb{N}$ and
$\langle a_i.S_i + b_i.X_i \rangle (C_{i,vi} , (C_{1,u1}, \ldots, C_{n,un}))$ = If $u_i \neq v_i$ then a_i else (a_i+b_i)

Definition 2.6 : Let M a symbolic marking. Then :
$\Pi_{j=1,n} \langle a_i.S_i + b_iX_i \rangle$ is a function from $\Pi_{i=1,n} \mathbb{C}_i \times \Pi_{i=1,n} \mathbb{C}_i \dashrightarrow \mathbb{N}$ and
$\Pi_{j=1,n} \langle a_i.S_i + b_iX_i \rangle ((C_{1,v1}, \ldots, C_{n,vn}) , (C_{1,u1}, \ldots, C_{n,un})) =$
$\Pi_{j=1,n} (\langle a_i.S_i + b_iX_i \rangle (C_{i,vi} , (C_{1,u1}, \ldots, C_{n,un})))$

Let M_j a symbolic marking, t a transition, and $(C_{1,u1}, \ldots, C_{n,un})$ a tuple of subclasses such that $C_{i,ui} \in \mathbb{C}_i$. $C_{i,ui}$ is the distinguished subclass of C_i for the firing of t.

Definition 2.7 : t is enabled from M for $(C_{1,u1}, \ldots, C_{n,un})$ iff :
$\forall p \in P$,
$M[C_{1,u1}, \ldots, C_{n,un}] (p, C_{1,v1}, \ldots, C_{n,vn}) \geq I^-(p, t) ((C_{1,v1}, \ldots, C_{n,vn}), (C_{1,u1}, \ldots, C_{n,un}))$

The symbolic marking M' obtained by firing $t(C_{1,u1}, \ldots, C_{n,un})$ is calculated with the three following steps :

Step 1 : We apply the incidence functions on $M[C_{1,u1}, \ldots, C_{n,un}]$ giving a new symbolic marking M_1

$\forall\, p \in P$ (we denote $I = I^+ - I^-$),

$$M_1(p, C_{1,v1}, \ldots, C_{n,vn}) = M[C_{1,u1}, \ldots, C_{n,un}]\,(p, C_{1,v1}, \ldots, C_{n,vn}) + I(p,t)\,((C_{1,v1}, \ldots, C_{n,vn}), (C_{1,u1}, \ldots, C_{n,un}))$$

Step 2 : as the grouping of states may not be maximal in M_1, it consists in grouping all the subclasses that have same markings giving a new marking M_2. In fact, only the splitted subclasses may be equivalent to previously existing ones.

Step 3 : calculation for M_2 of the canonical representative marking M'.

2.2.4 Example :

We apply the technique to our example. From the initial marking, we show the possible transition firings in the reachability graph (RG), and we represent the same step in the SRG. We will consider the marking of the different places in the following order :

(Active) (Passive) (Mess) (Mutex) (Wait) (Modif) (Ack).

The possible transition firings in the RG are represented in Figure 4 :

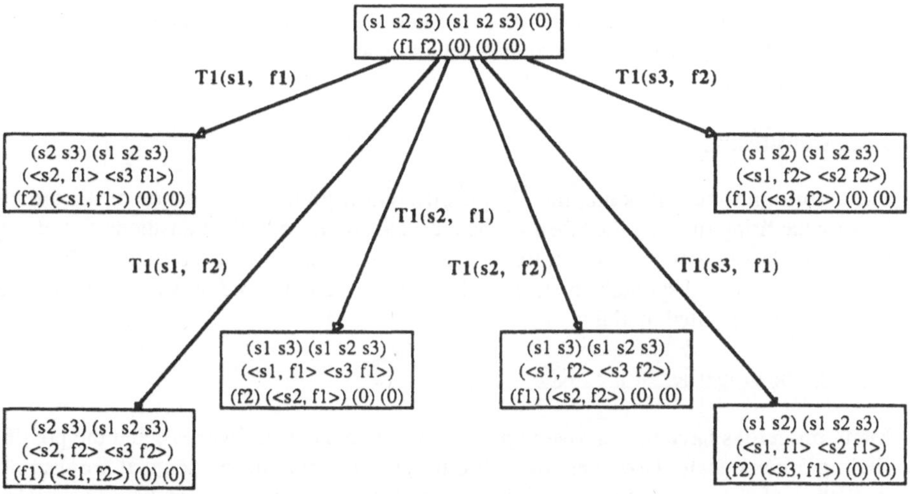

Figure 4 : A set of ordinary firings in the database model.

The SRG corresponding to the same step is the following one : At the beginning, places Active and Passive are marked with a subclass SIT1 of SIT, which is equal to SIT, and Mutex is marked with FIL1 which is a subclass of FIL, FIL1 = FIL. The firing of transition T1 for any couple of objects in SIT1 x FIL1 leads to a unique symbolic marking. This marking is obtained by splitting SIT1 in two subclasses SIT1 and SIT2, and FIL1 in two subclasses FIL1 and

FIL2. The splitting is necessary because the couple of objects for which T1 has been fired now have a marking different from that of the other objects of their former subclasses. This SRG is represented in Figure 5 :

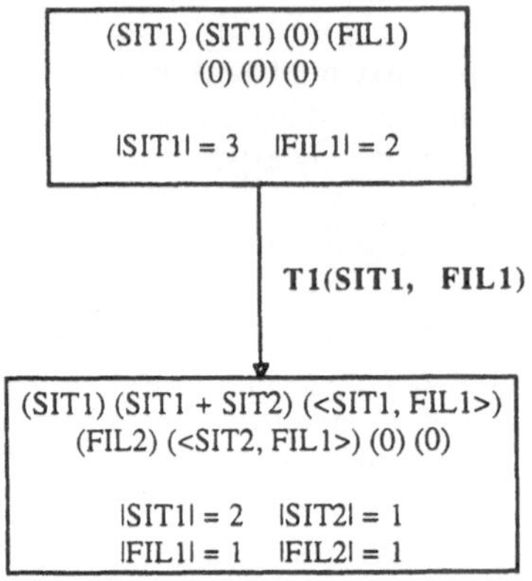

Figure 5 : The SRG associated with Figure 4.

The structure of the complete SRG in case of two sites and two filesis represented in Figure 6. Each node of the graph can be interpreted without knowing the identity of the file or the site that marks any place.

2.2.5 Construction :

The algorithm for constructing the SRG is different from the construction of the ordinary RG only by the firing rule and the labels of the arcs that are made of the transitions and the tuples of firing subclasses, whereas in the RG we have the transition and the tuple of firing objects. Notice that the initial symbolic marking only contains the initial ordinary marking because of the symmetry of the initial marking.

2.2.6 Some properties of the SRG :

Many properties have been proved on the SRG, such as quasi-liveness, and the possibility of finding a home state. However, we will only present some properties that are useful for the derivation of a stochastic model. All the proofs of the propositions given here are in [Had 87].

Proposition 2.2 : The reachability property is equivalent for the ordinary and the symbolic markings :

$$m \in RG \Leftrightarrow reg(m) \in SRG.$$

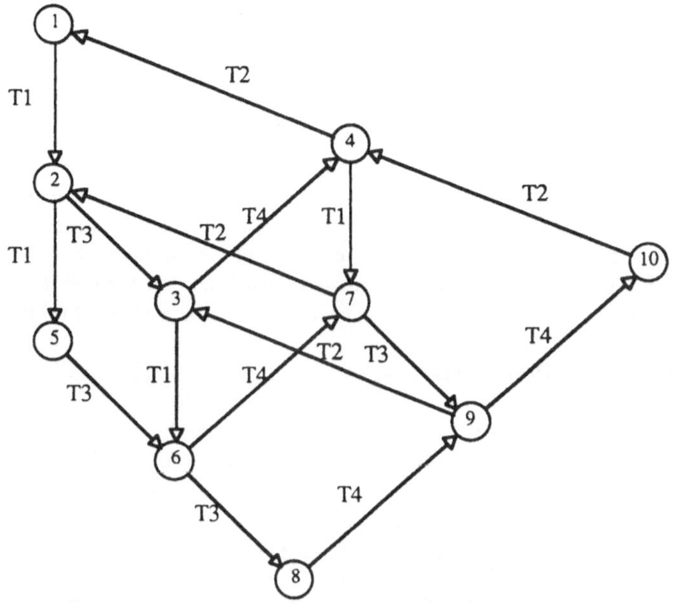

1 : no file being modified
2 : 1 file being modified, a message has been sent
3 : 1 file being modified, the second site is performing the modification
4 : 1 file being modified, the second site has sent an acknowledgement
5 : 2 files being modified, 2 messages have been sent
6 : 2 files being modified, one site is performing the modification
7 : 2 files being modified, one site has sent an acknowledgement
8 : 2 files being modified, both sites are performing a modification
9 : 2 files being modified, one site has sent an acknowledgement, the other is performing a modification
10 : 2 files being modified, 2 acknowledgements have been sent

Figure 6 : The complete SRG of the database model.

Proposition 2.3 : There is an exact relation between the arcs of the RG and the SRG.
Let M = reg(m) and M' two symbolic markings of the SRG. Let $\mathcal{A}_{m,M'}$ the set of arcs going out of m to any m' $\in$ M', $\mathcal{A}_{M,M'}$ the set of symbolic arcs leading from M to M'. Then there is an application mapping $\mathcal{A}_{m,M'}$ on $\mathcal{A}_{M,M'}$ such that the reciprocal image of a symbolic arc labelled by $t(C_{1,u1}, \ldots, C_{n,un})$ is a set of arcs labelled by some $t(c_1, \ldots, c_n)$. The cardinality of this set is $\Pi_{j=1}^{n} | C_{j,uj} |$.

3. REGULAR STOCHASTIC PETRI NETS

Stochastic Petri nets (SPN's) [Mol 81, Flo 85] are a tool well adapted to the modeling and the performance evaluation of distributed computer systems. In an SPN, a firing delay is associated with each transition. Firing delays are instances of random variables that have a negative exponential probability distribution. The probability that two timed transitions sample exactly the same delay time is zero, so that priority zero transitions are assumed to fire one at a time. The selection of the transition instance to fire among the set of the enabled ones follows a "race" policy [Ajm 85] (the transition that has drawn the least delay is the one that fires). Under these assumptions, the reachability graph is isomorphic to a Markov chain [Mol 81].

Regular Stochastic Petri Nets :

The association of a timing semantics to colored nets, and the extension of some notions such as conflict or confusion, have been clearly defined in [Chi 88]. However, it has also been shown that it was almost impossible to avoid the construction of the reachability graph of the unfolded net when trying to analyze the model. This is the reason why we decided to build our stochastic model only on a subclass of colored nets

As regular nets are a class of colored nets, the rules defined in [Chi 88] apply to Regular Stochastic Petri Nets (RSPN's). A RSPN is then defined as a couple (RN, λ), where RN defines a Regular Net, and λ is a weight function $\lambda : T \times \mathcal{C} \rightarrow \mathbb{R}^+$ associating a positive real number with each transition. The value of $\lambda_{t(c)}$ is the mean delay between the moment when (t, c) is enabled and the moment when it actually fires.

Restrictions on the model :

We will neither consider the case of immediate transitions, nor marking-dependent weights, but these extensions have been developed in [Dut 89]. We will further impose that all the color instances of a transition have the same firing rate, so that λ can be redefined as a function $T \rightarrow \mathbb{R}^+$. As all the objects in a class behave the same way, it is not a heavy restriction to believe that they have the same timing constraints.

4. COMPUTATION OF THE STATE PROBABILITIES

All the performance measures of a specific system can be obtained from the computation of the steady-state probabilities. However, complex systems can generate large reachability graphs, thus making the probability computation impossible. In this section, we present an algorithm for computing the steady state probabilities without developping the whole reachability graph. This algorithm is based on a stochastic aggregation of states that is directly derived from the symbolic reachability graph. We first recall the algorithm used in a general case, then we present an algorithm that can be used in the case of regular nets.

4. 1 Usual algorithm

The probability vector P, where p(i) is the steady probability of state i, is usually obtained from the following algorithm :

(1) Unfolding of the net.

(2) Construction of the reachability graph of the unfolded net (the unicity of P is ensured iff there is only one absorbing strongly connected component in the graph), and valuation of the arcs with the rate of the associated transition.

(3) Computation of the square matrix A defined by :
the dimension of A is the number N of reachable markings,

$$\text{if } i \neq j, \quad a_{i,j} = \sum_{t(c),\ \text{the firing of } t(c) \text{ leads from } i \text{ to } j} \lambda_{t(c)} ,$$

$$a_{i,i} = - \sum_{j \neq i} a_{i,j}$$

(4) Resolution of the system

$$\begin{cases} P.\,A = 0 \\ \| P \| = \sum_{i=1}^{N} p(i) = 1 \end{cases}$$

This algorithm is very general and does not use the particular structure of the Markov chain derived from a regular net. We now present an algorithm which is valid only in the case of regular nets.

4.2 Improved algorithm

The algorithm presented in this section uses some properties of regular nets in order to reduce the complexity of the resolution. In the sequel, a symbolic marking of the SRG will be denoted M_I, I = 1, ..., N.

(1) Construction of the SRG.
The number of ordinary markings within a symbolic marking is computed during this step with the following formula :

$$|M| = \frac{1}{K} \cdot \prod_{i=1}^{n} \frac{|C_i|!}{|D_{i_1}|! \times \ldots \times |D_{i_{n_i}}|!}$$

where n_i is the number of subclasses of C_i in the symbolic marking M, and K is the cardinality of the set of the permutations that leave M unchanged when applied to the subclasses of object classes :

$$K = | \{\langle s_1, \ldots, s_n\rangle \in S, \langle s_1, \ldots, s_n\rangle.M = M\} |.$$

(2) Computation of the square matrix A^* defined by :

the dimension of A^* is the number N' of reachable symbolic markings,

$$\text{if } I \neq J, \quad a^*_{I,J} = \sum_{t(D_1, \ldots, D_n) \in \mathcal{A}_{M_I,M_J}} \lambda_t . | D_1 | . \ldots . | D_n |$$

$$a^*_{I,I} = - \sum_{J \neq I} a^*_{I,J}$$

(3) Resolution of the system

$$\begin{cases} Q^*. A^* = 0 \\ \| Q^* \| = \sum_{I=1}^{N'} q^*(M_I) = 1 \end{cases}$$

(4) Computation of the state probability vector Q defined by $q(i) = q^*(M_I) / | M_I |$ for any marking i belonging to the symbolic marking M_I.

The fourth step is optional and depends on the will of the user to get either the probabilities of the classes of states, or the probabilities of the individual states.

N' is generally much less than N, and as the whole reachability graph is never built, the algorithm provides a real improvement in the place necessary for storing the data. Moreover, the improvement is still greater when looking at the complexity of the computation, as the exact resolution of the matricial system has a complexity $O(n^3)$. As for approximate methods, the complexity depends both on the number of states and the number of arcs, our algorithm is still an improvement. It is therefore possible to obtain exact results for large reachability graphs. The results obtained are also often more significant for the user, because the symbolic markings can be easily interpreted. In the next section, we will prove the correctness of the improved algorithm by showing that vectors P and Q are equal.

5. PROOF OF THE ALGORITHM

In this section, we detail the different steps that make the transformation of the algorithm correct. The process associated with the unfolded net is supposed to be ergodic. We first give a sketch of the method we will use to prove the correctness of our improved algorithm. The proof organizes into three steps.

(1) All the markings within a symbolic marking have the same probability.

(2) We prove that the Markovian lumping condition is fulfilled, and the linear system corresponding to the ordinary matricial equation can be reduced to a system with fewer variables, but which is no longer probabilistic (the solution is not a probability vector). However, the reduced system can be transformed into a probabilistic system by a simple change of variables.

(3) This probabilistic system is exactly the one computed by our algorithm.

Notation : According to the context, the index of a symbolic marking will be denoted with a capital or a lower case letter. An ordinary marking of the RG will be denoted (i, j) where M_I is the associated symbolic marking, and j is an order number within the symbolic marking. Let $s \in S$ a permutation. Then s.(i, j) will be denoted (i, s.j), where s.j is the order number of s.(i, j) within M_I.

5.1 Equiprobability of the markings :

In this section, we show that as the linear system associated to the Markov chain has a unique solution, the solution is a probability vector such that all the markings in a symbolic marking have the same probability.

We will denote by Eq(i, k) the equation of the system Eq defined by P.A = 0, corresponding to the ordinary marking number k of the symbolic marking number i:

$$\text{Eq}(i, k) : \sum_{J=1}^{N} \sum_{(j, q) \in M_J} p_{(j,q)} \cdot a_{(j, q)(i, k)} = 0.$$

Proposition 5.1 : The transition rate between a state (j, q) and a state (i, k) is the same as the transition rate between (j, s.q) and (i, s.k).

$$\forall (j, q), \forall (i, k), \forall s \in S, \quad a_{(j, q)(i, k)} = a_{(j, s.q)(i, s.k)}.$$

Proof : Two cases must be considered.
case 1 : $(j, q) \neq (i, k)$: this is a straightforward consequence of the basic Theorem 2.1.
case 2 : $(j, q) = (i, k)$: $a_{(i, s.k)(i, s.k)}$ is written as

$$a_{(i, s.k)(i, s.k)} = - \left[\sum_{\substack{J=1 \\ J \neq I}}^{N} \sum_{(j, q) \in M_J} a_{(i, s.k)(j, q)} + \sum_{\substack{(i, q) \in M_I \\ q \neq s.k}} a_{(i, s.k)(i, q)} \right]$$

From Corollary 2.2, we know that a permutation s defines a bijection on a symbolic marking M_I. Therefore, it is equivalent to sum on (i, q) or (i, s.q).

$$a_{(i, s.k)(i, s.k)} = - \left[\sum_{\substack{J=1 \\ J \neq I}}^{N} \sum_{(j, q) \in M_J} a_{(i, s.k)(j, s.q)} + \sum_{\substack{(i, q) \in M_I \\ q \neq k}} a_{(i, s.k)(i, s.q)} \right]$$

Applying case 1, we obtain :

$$a_{(i,\ s.k)(i,\ s.k)} = -\left[\sum_{\substack{J=1\\ J\neq I}}^{N}\ \sum_{(j,\ q)\in M_J} a_{(i,\ k)(j,\ q)} + \sum_{\substack{(i,\ q)\in M_I\\ q\neq k}} a_{(i,\ k)(i,\ q)}\right] = a_{(i,\ k)(i,\ k)} \qquad \Delta\Delta\Delta$$

Proposition 5.2 : $\{(p_{(i,\ s.k)})_{(i,\ k)}\}$ is a solution of the same equation as $\{(p_{(i,\ q)})_{(i,\ k)}\}$.

$$\forall\ (i,\ k),\quad \sum_{J=1}^{N}\ \sum_{(j,\ q)\in M_J} p_{(j,\ s.q)} \cdot a_{(j,\ q)(i,\ k)} = 0.$$

Proof : we write Eq(i, s.k) :

$$\sum_{J=1}^{N}\ \sum_{(j,\ q)\in M_J} p_{(j,\ q)} \cdot a_{(j,\ q)(i,\ s.k)} = 0.$$

and applying Corollary 2.2, we change the index (j, q) by (j, s.q)

$$\sum_{J=1}^{N}\ \sum_{(j,\ q)\in M_J} p_{(j,\ s.q)} \cdot a_{(j,\ s.q)(i,\ s.k)} = 0.$$

Applying Proposition 5.1, we get

$$\sum_{J=1}^{N}\ \sum_{(j,\ q)\in M_J} p_{(j,s.q)} \cdot a_{(j,\ q)(i,\ k)} = 0. \qquad \Delta\Delta\Delta$$

As we assume that the system is ergodic, the solution is unique. So $p_{(i,\ k)} = p_{(i,\ s.k)}$, and all the markings in a symbolic marking have the same probability.
However, we are going to show that in case the system has multiple solutions, there is always one such that all the markings in a symbolic marking have the same probability, and this solution is computed by our algorithm. If the reader is only interested in ergodic systems, he can jump to Section 5.2.

Proposition 5.3 : $\left\{ \left(\frac{1}{|M_I|} \cdot \sum_{(i,\,k')\in M_I} p_{(i,\,k')} \right)_{(i,\,k)} \right\}$ is a solution of Eq, i.e.,

$$\sum_{J=1}^{N} \sum_{(j,\,q)\in M_J} \left(\frac{1}{|M_J|} \cdot \sum_{(j,\,q')\in M_J} p_{(j,\,q')} \right) \cdot a_{(j,\,q)(i,\,k)} = 0.$$

Proof : we apply Proposition 5.2, and we sum on all the possible permutations.

$$\sum_{s\in S} \sum_{J=1}^{N} \sum_{(j,\,q)\in M_J} p_{(j,s.q)} \cdot a_{(j,\,q)(i,\,k)} = 0.$$

$$\sum_{J=1}^{N} \sum_{(j,\,q)\in M_J} \left(\sum_{s\in S} p_{(j,\,s.q)} \right) \cdot a_{(j,\,q)(i,\,k)} = 0.$$

The application of Corollary 2.3 gives :

$$\sum_{J=1}^{N} \sum_{(j,\,q)\in M_J} \left(\frac{|S|}{|M_J|} \cdot \sum_{(j,\,q')\in M_J} p_{(j,\,q')} \right) \cdot a_{(j,\,q)(i,\,k)} = 0.$$

Dividing by $|S|$, we get $\sum_{J=1}^{N} \sum_{(j,\,q)\in M_J} \left(\frac{1}{|M_J|} \cdot \sum_{(j,\,q')\in M_J} p_{(j,\,q')} \right) \cdot a_{(j,\,q)(i,\,k)} = 0.$ ΔΔΔ

5.2 Reduction of the linear system :

We prove here that the steady-state probability of being in a symbolic marking can be calculated from a system with a reduced number of variables.

Proposition 5.4 : The transition rate out of a symbolic marking M_J to an ordinary marking in M_I has the same value for every marking in M_I. Conversely, the transition rate out of an ordinary marking in M_I to a symbolic marking M_J has the same value for every marking in M_I.
$\forall$ J, $\forall$ (i, k), $\forall$ s,

$$\sum_{(j,\,q)\in M_J} a_{(j,\,q)(i,\,k)} = \sum_{(j,\,q)\in M_J} a_{(j,\,q)(i,\,s.k)}$$

$$\sum_{(j,\,q)\in M_J} a_{(i,\,k)(j,\,q)} = \sum_{(j,\,q)\in M_J} a_{(i,\,s.k)(j,\,q)}$$

Note that the second equation of Proposition 2 is equivalent to the strong Markovian lumping condition [Kem 60], which ensures that the aggregation of states will preserve the Markovian property of the process.

Proof : The result is obtained by changing the indexes of the sums and applying Proposition 6.1. ΔΔΔ

Let $\bar{p}_I = \frac{1}{|M_I|} \cdot \sum_{(i,k)\in M_I} p_{(i,k)}$ the probability of any marking in M_I for the equiprobable solution,

and $\hat{p}_I = |M_I| \cdot \bar{p}_I$ the probability of being in the symbolic marking M_I.

Let $\overline{Eq}(i, k)$ the following equation : $\sum_{J=1}^{N} \sum_{(j,q)\in M_J} \bar{p}_J \cdot a_{(j,q)(i,k)} = 0.$

Proposition 6.5 : The system $\overline{Eq}$ is such that the equations associated with all the markings in a symbolic marking are the same, i.e.,

$$\forall\ (i, k),\quad \overline{Eq}(i, k) = \overline{Eq}(i, s.k)$$

Proof : $\overline{Eq}(i, k)$: $\sum_{J=1}^{N} \sum_{(j,q)\in M_J} \bar{p}_J \cdot a_{(j,q)(i,k)} = 0.$

$$\sum_{J=1}^{N} \bar{p}_J \cdot \sum_{(j,q)\in M_J} a_{(j,q)(i,k)} = 0.$$

Applying Proposition 5.4, $\sum_{J=1}^{N} \bar{p}_J \cdot \sum_{(j,q)\in M_J} a_{(j,q)(i,s.k)} = 0.$ ΔΔΔ

As a consequence, we can consider only one equation by symbolic marking. The system we have to solve is then :

$$\begin{cases} \sum_{J=1}^{N} \bar{p}_J \cdot \sum_{(j,q)\in M_J} a_{(j,q)(i,k)} = 0 \\ \sum_{I=1}^{N} |M_I| \cdot \bar{p}_I = 1. \end{cases}$$

Let $\bar{a}_{I,J} = |M_J| \cdot \sum_{(i,k)\in M_I} a_{(i,k)(j,q)}$

Proposition 5.6 : $\sum_{J=1}^{N} \bar{a}_{I,J} = 0.$

Proof : $\sum_{J=1}^{N} \bar{a}_{I,J} = \sum_{j=1}^{N} |M_J| . \sum_{(i,k)\in M_I} a_{(i,k)(j,q)}$

Applying Proposition 6.4,

$$= \sum_{J=1}^{N} \sum_{(j,q)\in M_J} \sum_{(i,k)\in M_I} a_{(i,k)(j,q)}$$

$$= \sum_{(i,k)\in M_I} \left(\sum_{J=1}^{N} \sum_{(j,q)\in M_J} a_{(i,k)(j,q)} \right) = 0. \quad \Delta\Delta\Delta$$

Multiplying the first equation by $|M_I|$, we can transform our system in

$$\begin{cases} \sum_{J=1}^{N} \bar{p}_J . \bar{a}_{J,I} = 0 \\ \sum_{I=1}^{N} |M_I| . \bar{p}_I = 1. \end{cases}$$

Introducing the new notation $\hat{a}_{I,J} = \frac{\bar{a}_{I,J}}{|M_I|}$, we obtain the final system :

$$\begin{cases} \sum_{J=1}^{N} \hat{p}_J . \hat{a}_{J,I} = 0 \\ \sum_{I=1}^{N} \hat{p}_I = 1. \end{cases}$$

where $\hat{a}_{I,J} = \frac{1}{|M_I|} . \sum_{(i,k)\in M_I} \sum_{(j,q)\in M_J} a_{(i,k)(j,q)} \left(= \sum_{(j,q)\in M_J} a_{(i,k)(j,q)} \right)$, with $\sum_{J=1}^{N} \hat{a}_{I,J} = 0.$

The system we have to solve is a stochastic system, and the usual techniques for calculating the steady state probabilities can then be used. The same system would have been obtained using Markovian lumping techniques. And in that case too, it would have been necessary to develop an additional demonstration to prove that all the markings within a symbolic marking have the same probability.

Applied to our example, the above technique allows us to transform a system with 487 variables for the ordinary markings in a system with 46 variables for the symbolic markings.

The gain considerably increases with the cardinalities of the object classes. Moreover, the probabilities of the symbolic markings are often more significant for the modeler than the probabilities of the ordinary markings, making it all the more useful to compute directly the values for the lumped states.

In the next part, we will show that the coefficients $\hat{a}_{I,J}$ can be derived directly from the SRG, and that it is possible to calculate the number of ordinary markings in a symbolic marking, thus allowing us to derive the ordinary probabilities from the probabilities of the symbolic markings.

5.3 Computation of the coefficients of the reduced linear system :

The properties we are going to use to compute the coefficients of the linear system are directly linked to Proposition 2.3.

Let φ the function mapping $\mathcal{A}_{m,M'}$, the set of arcs leading from an ordinary marking m to any marking in the symbolic marking M', on $\mathcal{A}_{M,M'}$, the set of symbolic arcs leading from M, with $m \in M$, to M'. Then we have the following properties :

(1) φ is surjective,

(2) the same transition labels an arc and its image by φ :

$$\varphi[t'(c_1, \ldots, c_n)] = t(D_1, \ldots, D_n) \Rightarrow t' = t,$$

(3) $| \varphi^{-1} [t(D_1, \ldots, D_n)] | = \Pi_{i=1}^{n} | D_i |$

We denote λ_t the rate associated with a transition t, t(A) the transition labeling the symbolic arc A, and D_i^A the subclass of C_i instantiating t(A). So, if t(a) is the transition labeling an ordinary arc a, then t(a) = t(A) for any $a \in \varphi^{-1}(A)$.

Then the coefficients of the reduced linear system can be directly computed with the following formula :

Proposition 6.7 : $$\hat{a}_{I,J} = \sum_{A \in \mathcal{A}_{M_I,M_J}} \lambda_{t(A)} . | D_1^A | . \ldots . | D_n^A |$$

Proof : From the definition of $\hat{a}_{I,J}$, we have for any $(i, k) \in M_I$:

$$\hat{a}_{I,J} = \sum_{(j,q) \in M_J} a_{(i,k)(j,q)} = \sum_{a \text{ leading from } (i,k) \text{ to } (j,q) \in M_J} \lambda_{t(a)}$$

The set of arcs leading from (i, k) to $(j, q) \in M_J$ can be partitioned according to their image by φ. Thus we get :

$$\hat{a}_{I,J} = \sum_{A \in \mathcal{A}_{M_I,M_J}} \sum_{a \in \varphi^{-1}(A)} \lambda_{t(a)}$$

As φ preserves the transition names, this can be also written

$$\hat{a}_{I,J} = \sum_{A \in \mathcal{A}_{M_I,M_J}} \sum_{a \in \varphi^{-1}(A)} \lambda_{t(A)}$$

which is equal to

$$\hat{a}_{I,J} = \sum_{A \in \mathcal{A}_{M_I,M_J}} \left|\varphi^{-1}(A)\right| . \lambda_{t(A)}$$

Applying Proposition 2.3, we obtain

$$\hat{a}_{I,J} = \sum_{A \in \mathcal{A}_{M_I,M_J}} \prod_{i=1}^{n} \left|D_i^A\right| . \lambda_{t(A)} \qquad \Delta\Delta\Delta.$$

Those values can therefore be calculated directly from the SRG, by giving to an arc a weight depending on the cardinalities of the subclasses of its label. The values of the coefficients for I = J are derived of the nullity of the sum.

6. CONCLUSION

The symmetry properties of regular nets have been used to prove many structural results. In this paper, we have shown that the symmetries can be used also in the case of a quantitative analysis.
We have proved that all the states in a symbolic marking have the same probability, and therefore, that the probabilities of all the ordinary markings can be derived from the resolution of a system which size depends only on the number of symbolic markings. We have given an algorithm for calculating the coefficients of the reduced system directly from the graph of symbolic markings.

The advantages of our method are twofold . On the one hand, the user can choose if he wants the probabilities of the ordinary markings, or if he is only interested in the probabilities of the symbolic markings which are often more significant. On the other hand, the reduction of the system will bring a dramatic improvement in the memory space and the CPU time required to solve large models, and will increase the class of models that can be analytically solved.

Our research directions will be to develop a software tool that will automatically construct the graph of symbolic markings. This tool could be later interfaced with powerful softwares for stochastic Petri nets, such as GreatSPN [Chi 87], or RDPS [Flo 86]. The introduction of immediate transitions and marking-dependent weights has been presented in [Dut 89]. We now intend to analyze less restricted classes of nets that still have symmetry properties. These extended nets include successor functions [Had 88], a non-symmetric initial marking, or color domains with several occurrences of the same object class.

Acknowledgements :

We gratefully acknowledge the three anonymous referees whose pertinent remarks allowed us to improve the quality of this paper.

References

[Ajm 85] M. Ajmone Marsan, G. Balbo, A. Bobbio, G. Chiola, G. Conte, and A. Cumani, "On Petri Nets with Stochastic Timing", in proc. International Workshop on Timed Petri Nets, pp 80-87, IEEE-CS Press, Torino, Italy (July 1985).

[Ajm 87] M. Ajmone Marsan, G. Balbo, G. Chiola, and G. Conte, "Generalized Stochastic Petri Nets Revisited : Random Switches and Priorities", in proc. International Workshop on Petri Nets and Performance Models, pp 44-53, IEEE-CS Press, Madison, WI, USA (August 1987).

[Chi 87] G. Chiola, "A Graphical Petri Net Tool for Performance Analysis", 3rd International Workshop on Modeling Techniques and Performance Evaluation, AFCET, Paris (March 1987).

[Chi 88] G. Chiola, G. Bruno, and T. Demaria, "Introducing a Color Formalism into Generalized Stochastic Petri Nets", in proc. 9th European Workshop on Application and Theory of Petri Nets, pp 202-215, Venezia, Italy (June 1988).

[Dut 89] C. Dutheillet, S. Haddad, "Aggregation of States in Colored Stochastic Petri Nets. Application to a Multiprocessor Architecture", in proc. PNPM 89, pp 40-49, IEEE-CS Press, Kyoto, Japan (December 1989).

[Flo 85] G. Florin, S. Natkin, "Les Réseaux de Petri Stochastiques", AFCET TSI, vol. 4, no. 1, pp 143-160 (January 1985).

[Flo 86] G. Florin, "RDPS, a Software Package for the Validation and the Evaluation of Dependable Computer Systems", 3rd Proc. IFAC SAFECOMP Workshop, Sarlat, France (October 1986).

[Had 87] S. Haddad, "Une Catégorie Régulière de Réseaux de Petri de Haut Niveau : Définition, Propriétés et Réductions", Thèse de Doctorat, RR87/197, Laboratoire MASI, Université Paris VI, Paris, France (October 1987).

[Had 88] S. Haddad and J.M. Couvreur, "Towards a General and Powerful Computation of Flows for Parametrized Coloured Nets", in proc. 9th European Workshop on Application and Theory of Petri Nets, pp 202-215, Venezia, Italy (June 1988).

[Hub 84] P. Huber, A.M. Jensen, L.O. Jepsen, and K. Jensen, "Towards Reachability Trees for High Level Petri Nets", 5th European Workshop on Application and Theory of Petri Nets, Aarhus, Denmark (1984).

[Jen 81a] K. Jensen, "Coloured Petri Nets and the Invariant Method", Theoretical Computer Science 14, pp 317-336 (1981).

[Jen 81b] K. Jensen,
"How to Find Invariants for Coloured Petri Nets", Mathematical Foundations of Computer Science, LNCS 118, Springer - Verlag (1981).

[Kem 60] J. G. Kemeny and J. L. Snell,
"Finite Markov Chains", Van Nostrand, Princeton, NJ (1960).

[Lan 77] S. Lang,
"Algebra", Addison - Wesley (1977).

[Lin 87] Chuang Lin and D. Marinescu,
"On Stochastic High Level Petri Nets", in proc. International Workshop on Petri Nets and Performance Models, pp 44-53, IEEE-CS Press, Madison, WI, USA (August 1987).

[Mol 81] M. K. Molloy,
"On the Integration of Delay and Throughput Measures in Distributed in Distributed Processing Models", Ph.D. Dissertation, University of California, Los Angeles, CA, USA (September 1981).

[Zen 85] A. Zenie,
"Colored Stochastic Petri Nets", in proc. International Workshop on Timed Petri Nets, pp 262-271, IEEE-CS Press, Torino, Italy (July 1985).

18.
Automated Construction of Compound Markov Chains from Generalized Stochastic High-level Petri Nets

J.A. Carrasco

Proc. 3rd Int. Workshop on Petri Nets and Performance Models.
IEEE Computer Society Press, Kyoto 1989, pp. 93-102

Abstract

A new type of Petri nets: Generalized Stochastic High-Level Petri nets (GSHLPN's), collecting the qualities of GSPN's and SHLPN's, is presented. The automated construction of compound continuous-time Markov chains (CTMC's) from GSHLPN's is also considered. A formalism for the description of compound markings allowing a symbolic firing of the net to obtain a compound CTMC with correct state grouping is derived. The construction of the compound CTMC requires an algorithm to test the equivalence of compound markings. It is shown that, in the general case and for bounded number of rotation groups, the problem is polynomially equivalent to GRAPH ISOMORPHISM, a problem whose classification in the NP world is currently open.

1. INTRODUCTION

The success of Petri nets (PN's) to represent and analyze the qualitative behaviour of complex systems with synchronization and concurrency has motivated the introduction of stochastic Petri nets (SPN's) for the quantitative analysis of such systems [1], [2]. SPN's are PN's with exponentially distributed firing times associated to transitions. Bounded SPN's are isomorphic to finite continuous-time Markov chains (CTMC's). Therefore, SPN's do not provide more modeling power than CTMC's, but they are indeed an appropriate specification methodology for Markov models having CTMC's of large size. Since the introduction of SPN's, there have been a number of interesting developments. Some of them extend the modeling power of SPN's. Among these, we have the extended stochastic Petri nets [3] and the stochastic activity networks [4], both of which allow arbitrary firing time distributions. Other efforts are oriented to the development of more expressive types of Petri nets with the same modeling power as SPN's.

Generalized stochastic Petri nets (GSPN's) [5] extend SPN's by introducing inhibitor arcs and immediate transitions. This class of nets has recently be reextended [6] by allowing probabilities and priorities to be added to immediate transitions. GSPN's not only are more expressive than SPN's, but also support the specification of models with smaller CTMC's when some activities are much faster than others so that the first can be modeled by immediate transitions firing in zero time. Efficient algorithms for the generation of CTMC's from GSPN's have recently been presented [7].

Some systems are composed of sets of processing elements with identical behaviour. These systems have been called homogeneous. Since the size of the CTMC is the limiting factor for the numerical analysis of the model, it is highly desirable to take advantage of the symmetries exhibited by those systems to reduce the size of the CTMC. Although SPN's provide some support, they fail to exploit the symmetries in the behaviour of the system when, as it is often the case, processes have relationships. Since the tokens in a SPN do not convey any information, except the place where they are, the relationships can only be expressed by individualizing the processes. This not only yields CTMC's of undue size, but also enlarges the Petri net, making more arduous the specification of the model. In order to overcome this problem Lin and Marinescu [8], [9] have introduced stochastic high-level Petri nets (SHLPN's). SHLPN's are high-level Petri nets (HLPN's) with constant, but possibly marking dependent, firing rates attached to transitions. HLPN's are an extension of standard PN's in which tokens have colours, arcs have expressions with variables yielding bags of colours, and transitions have predicates (guards) on the variables used in the expressions of the surrounding arcs. HLPN's were proposed by Jensen [10] to summarize the qualities of predicate transition nets and coloured Petri nets. Similar to SHLPN's, stochastic coloured Petri nets, were proposed by Zenie [11]. SHLPN's have been shown to be a powerful methodology for the specification of models of homogeneous systems [8], [9], [12]. They allow the specification of complex homogeneous systems in a more succinct and readable way than standard PN's and yield CTMC's of much smaller size.

Although some work has been done towards the automated analysis of HLPN's [13], it is apparent that some problems have not yet received a satisfactory solution. These problems are identified for SHLPN's in [8], [9]: derivation of compound markings so that compound CTMC's have a correct state grouping, computation of transition rates between compound markings, and efficient testing of the equivalence of compound markings.

The contents of this paper is as follows. Section 2 introduces a new type of stochastic Petri nets: Generalized Stochastic High-Level Petri Nets (GSHLPN's), combining the qualities of SHLPN's and GSPN's. Section 3 presents a methodology to derive a symbolism for the description of compound markings which allows the symbolic firing of the net to obtain compound CTMC's with correct state grouping. It is apparent that the determination of the equivalence of compound markings is a difficult combinatorial problem. In section 4 we analyze the complexity of the problem from the point of view of the NP-completeness theory. Finally, conclusions and suggested directions for future work are presented in Section 5.

2. GSHLPN's

Generalized stochastic high-level Petri nets are obtained from SHLPN's by the introduction of immediate transitions with priorities, inhibitor arcs, and cases. This results in a better framework for the modeling of conflict and contention than SHLPN's. We first give a formal definition of GSHLPN's. The modeling power of the nets is illustrated next with an example.

2.1. Definitions

A generalized stochastic high-level Petri net is a 9-tuple:

$$GSHLPN = (S, P, T, I(), TO(), CO(), H(), M_0),$$

where:

S is the set of token types. Each token type $s \in S$ has a list of attributes with finite domains which, without lost of generality, are assumed to be of the form $D = \{1, ..., |D|\}$. A token of type s is denoted by $<s, a_1, ..., a_k>$, where a_i is the value of the ith attribute of s.

P is the set of places.

T is the set of transitions. As in GSPN's, there are two types of transitions: timed transitions and immediate transitions. Each timed transition t has associated a, possibly marking dependent*, firing rate $\lambda(t)$ and each immediate transition t a priority level $\pi(t)$. A transition t may have associated a predicate $P(t)$, i.e., a logical expression over the free variable set of t (defined in the next paragraph). A transition may have cases or not. The set of cases associated to a transition t is denoted by $C(t)$. Each case $c \in C(t)$ has a, possibly marking dependent*, probability $q(c)$.

* this dependency is restricted to be expressed in terms of the number of tokens of given type in each place, i.e., ignoring token attributes.

$I()$ is the input function which assigns sets of places to transitions. Each pair (p, t) with $p \in I(t)$ is called an input arc and p is said to be an input place of t. Each input arc (p, t) has associated a set of symbolic input tokens $s_I(p, t)$. The values of the attributes of the symbolic input tokens of a transition are specified by independent free variables constituting the *free variable set* of the transition.

$TO()$ is the transition output function which assigns sets of places to transitions without cases. Each pair (t, p) with $p \in TO(t)$ is called a transition output arc and p is said to be an output place of t.

$CO()$ is the case output function which assigns sets of places to cases. Each pair (c, p) with $p \in CO(c)$ is called a case output arc and p is said to be an output place of c.

$H()$ is the inhibitor function which assigns sets of places to transitions. Each pair (t, p) with $p \in H(t)$ is called an inhibitor arc and p is said to be an inhibitor place of t.

M_0 is the initial compound marking.

Transition output arcs, case output arcs and inhibitor arcs have associated bags of symbolic tokens $b_{TO}(t, p)$, $b_{CO}(c, p)$, and $b_H(t, p)$, respectively. The attributes of the tokens contained in these bags are either specified by expressions on the free variable set of t ($c \in C(t)$ for $b_{CO}(c, p)$) or left unspecified.

An *atomic token* is a token with all its attributes bound to values of their respective domains. An *atomic marking* m of the net is an assignment of bags of atomic tokens to places of the net. A *compound marking* M is an assignment of bags of symbolic tokens to places, where the attributes of the tokens are specified by expressions over a set of free variables taking values in the respective attribute domains. A compound marking M represents the subset of atomic markings $S(M)$ obtained by binding the free variables with values of the attribute domains. Two compound markings are *equivalent* if they represent the same subset of atomic markings.

In the graphical representation of the net, places are represented by circles, timed transitions by rectangles, immediate transitions by thin bars, and cases by ovals. Input arcs have an arrow in the transition side, transition and case output arcs have an arrow in the place side, and inhibitor arcs have a small circle in the transition side. Cases are connected to their transitions by lines.

The facts that in GSHLPN's tokens have attributes and transitions have predicates make necessary a more detailed definition of the enabling and firing rules than for GSPN's. The rules will be defined in terms of atomic markings. A *firing set* F of a transition t in an atomic marking m is a selection for each $p \in I(t)$ of a subset of tokens $F(p) \subset m(p)$ binding $s_I(p, t)$ satisfying the predicate $P(t)$ and such that for no $p \in H(t)$ there exists a subset of $m(p)$ binding $b_H(p, t)$ with the values imposed by F to the variables in the free variable set of t. Note that $F(p)$ is a subset, and replicated elements in $m(p)$ are seen as different for the determination of the firing sets. Consider, for instance, the transition and atomic marking shown in Figure 1. The firing sets of the transition are:

$$\begin{array}{ll} F_1: & F_1(P2)=\{\langle B,1,2,3\rangle_1\},\ F_1(P3)=\{\langle C,2\rangle,\langle C,3\rangle\} \\ F_2: & F_2(P2)=\{\langle B,1,2,3\rangle_2\},\ F_2(P3)=\{\langle C,2\rangle,\langle C,3\rangle\} \\ F_3: & F_3(P2)=\{\langle B,1,3,4\rangle\},\ F_3(P3)=\{\langle C,3\rangle,\langle C,4\rangle\} \\ F_4: & F_4(P2)=\{\langle B,2,4,5\rangle\},\ F_4(P3)=\{\langle C,4\rangle,\langle C,5\rangle\} \end{array} \quad (1)$$

If the atomic token in place $P1$ were <A, 1> instead of <A, 6>, the only firing set of the transition would be F_4. Two firing sets F_i, F_j of a transition t are *conflicting* if for some $p \in I(t)$, $F_i(p) \cap F_j(p) \neq \emptyset$. A *firing distribution* is a maximum cardinality set of non-conflicting firing sets. The firing distributions for the transition $T1$ of Figure 1 are:

$$\begin{array}{l} FD_1 = \{F_1, F_4\} \\ FD_2 = \{F_2, F_4\} \end{array} \quad (2)$$

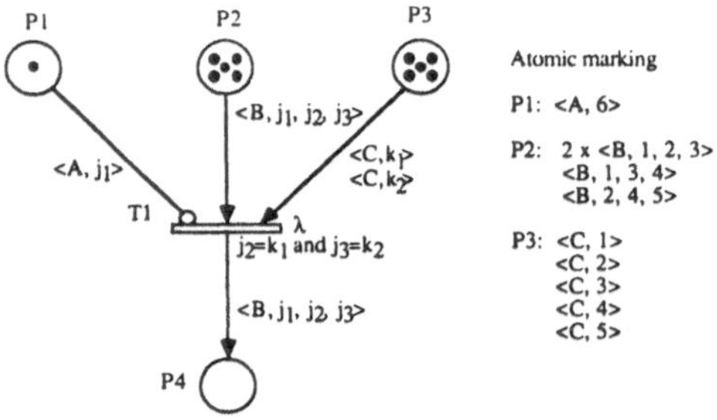

Figure 1. Example to illustrate enabling and firing rules in GSHLPN's.

A transition t (immediate or timed) is *firable* in an atomic marking m if t has some firing set in m. Firable timed transitions are *enabled* only if no immediate transition is firable. When there are firable immediate transitions only those with the highest priority are enabled. Only enabled transitions actually fire. Immediate transitions fire instantaneously in each firing set with the same probability. Timed transitions fire in each firing set belonging to some firing distribution with rates which are computed by assigning the same probability to each firing distribution and a rate $\lambda(t)$ to each firing set of the distribution. The firing of a transition t in a firing set F removes from each input place p the set of tokens $F(p)$. Cases have the same semantics as in stochastic activity networks [4]: if t has cases, one is sampled according to the probabilities $q(c)$. Output tokens are added to each transition (case) output place p according with the bags of symbolic tokens $b_{TO}(t,p)$ $(b_{CO}(c,p))$. Specified attributes are computed using the expressions defining them with the values for the free variables bound by F. For unspecified attributes, a value from the domain of the attribute is taken at random. For instance, the transition $T1$ in Figure 1 has the firing distributions (2) and a probability 1/2 is assigned to each of them. Given a firing distribution, the transition fires in each firing set of the distribution with rate λ. Then, $T1$ fires in F_1 and F_2 with rate $\lambda/2$ and in F_4 with rate λ. If $T1$ were immediate it would fire in each firing set (1) with probability 1/4. The firing of $T1$ in the firing set F_4 removes <B, 2, 4, 5> from $P2$, <C, 4> and <C, 5> from $P3$, and adds <B, 2, 4, 5> to $P4$.

The original definition of GSPN's [5] allowed nets with reachable markings enabling several immediate transitions and left to the user the responsability of defining the random switches required to disambiguate the model. Recognizing the practical problems of the approach, GSPN's were redefined in [6]. The new definition imposes some structural restrictions to the nets and allows priorities for immediate transitions. Weights are associated to immediate transitions as required in a computer-aided process to disambiguate the net. In GSHLPN's probabilistic selection is modelled by cases and the net has to be defined so that no reachable marking enables immediate transitions in the same immediate subnet of the corresponding priority. Immediate subnets of a given priority are [7] the connected components obtained by removing in the original net timed transitions and immediate transitions of lower priority. This approach has the advantage of allowing a highly efficient construction of reachability graphs with reduced number of vanishing states [7].

The firing policies adopted for immediate and timed transitions have been chosen to provide a useful default. Immediate transitions are commonly used as contention points and the firing rule adopted for this type of transitions assumes a random policy for the resolution of conflicts among contending processes. Timed transitions are commonly used as resource usage points and the firing rule adopted for this type of transitions implements a random policy within the constraint of optimal use of resources. The modeling of more complex contention and resource usage policies would require the use of subnets with combinations of immediate and timed transitions.

2.2. An example

In order to illustrate the expressive power of GSHLPN's we give an example in Figure 2. In this figure ⊕ represents the addition module the size of the domain of the involved attributes. The model is a slightly modified version of one described in [8], [9] and describes a shared memory multiprocessor system in which a set of identical processors with their own local memory access a set of common memory modules through a set of buses. A token-passing protocol is used in each bus for access control. The token is circulated among processors and its reception gives the processor the right to use the bus for accessing a common memory module. The execution of a processor alternates between the private and the common domains. In the private domain the

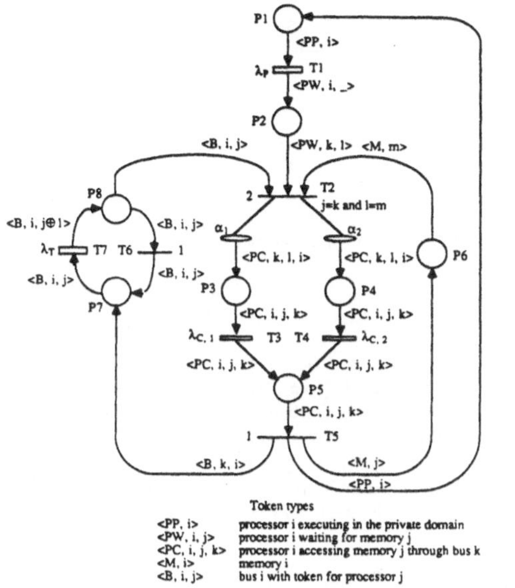

Figure 2. GSHLPN for a shared memory multiprocessor system with a token passing protocol.

processor accesses its local memory. In the common domain, it accesses a given common memory module taken at random from the set. The duration of an execution in the private domain is exponentially distributed with mean $1/\lambda_P$. The duration of an execution in the common domain has an hyperexponential distribution with parameters α_1, $\lambda_{C,1}$, and α_2, $\lambda_{C,2}$. The token pass delay has an exponential distribution with mean $1/\lambda_T$.

We follow the formal definition and give to each attribute of the sets of input tokens associated to a given transition a unique free variable, even if they are required to be equal by the transition predicate (see transition *T2*). An alternative would be to use the same free variable for the attributes required to be equal by the predicate. Unspecified attributes of output and inhibitor tokens are represented by an underscore. In order to model the contention for tokens and memory modules, we use a immediate transition *T2*, which is given priority over the immediate transition *T6* modeling the start of a token pass. Thus, if a processor waiting for an access to a memory module receives a token, it captures the token and starts the execution in the common domain if the memory module is free (transition *T2*), and sends the token to the next processor if the memory module is busy (transition *T6*). The execution in the common domain is modeled by introducing cases in the immediate transition *T2* with probabilities equal to the selection parameters of the hyperexponential distribution, and followed by places and timed transitions with the rates associated to each mode of the distribution.

GSHLPN's have over GSPN's the same advantages than SHLPN's have over SPN's: compactness and reduced size of the state-level model if a compound CTMC is constructed instead of the atomic CTMC. The introduction of immediate transitions makes also possible an economical modeling of processes which, compared to others, complete in a negligeable amount of time. For instance, using SHLPN's instead of GSHLPN's, the previous example would have required the use of timed transitions with very high rates for the modeling of the contention of processors for busses and memories, and the decission about whether to pass the token to the next processor. This not only would increase the size of the compound CTMC but also would stress numerical methods during its solution.

3. A SYMBOLISM FOR COMPOUND MARKINGS

Assume that the atomic reachability graph obtained by firing the GSHLPN from a given atomic marking of $S(M_0)$ enabling only timed transitions is finite, strongly connected, and contains all the atomic markings in $S(M_0)$. Under these conditions an ergodic atomic CTMC $(\Omega, X(t))$ isomorphic to the GSHLPN can be obtained in two steps as follows. In the first step, the atomic reachability graph G is constructed by firing the net from any atomic marking $m_0 \in S(M_0)$, using the rules described in the previous section. G has two types of nodes: tangible markings (enabling only timed transitions) and vanishing markings (enabling only immediate transitions), and associates rates to arcs coming out of tangible markings and probabilities to arcs coming out of vanishing markings. In the second step, vanishing markings are reduced as shown in [5] to obtain $(\Omega, X(t))$.

We are however interested in the construction of a compound CTMC $(\Omega', X'(t))$ whose states be tangible compound markings. The compound CTMC has to be a correct grouping of $(\Omega, X(t))$, i.e., has to satisfy the following requisites [14]:

1) the sets of atomic markings associated to the compound markings in Ω' are a partition of Ω

2) for all $M_1, M_2 \in \Omega'$, $M_1 \neq M_2$ and all $m_1 \in M_1$:

$$\lambda_{m_1, M_2} \equiv \sum_{m_2 \in S(M_2)} \lambda_{m_1, m_2} = \lambda'_{M_1, M_2}$$

The construction of the compound CTMC can be done in analogy with the construction of the atomic CTMC, substituting the atomic firing of transitions by a symbolic firing working with compound markings, and merging equivalent compound markings instead of identical atomic markings. The requisites 1) and 2) are satisfied if the symbolism for the representation of compound markings and the symbolic firing are defined such that:

Property 1 *two compound markings* M_1, M_2 *are either disjoint* $(S(M_1) \cap S(M_2) = \emptyset)$ *or equivalent* $(S(M_1) = S(M_2))$.

Property 2 *all the atomic markings represented by a compound marking enable the same transitions.*

Property 3 *symbolic firings are isomorphic to atomic firings. More explicitely, the symbolic firing of a transition t in a compound marking M yields compound markings $M_1, M_2, \ldots, M_n$ (some could be equivalent) with rates λ_{M, M_i} if t is timed, or probabilities p_{M, M_i} if t is immediate, such that the firing of t in any $m \in S(M)$ yields atomic markings $m_1, m_2, \ldots, m_n$ with $m_i \in S(M_i)$ and with rates $\lambda_{m, m_i} = \lambda_{M, M_i}$ if t is timed, or probabilities $p_{m, m_i} = p_{M, M_i}$ if t is immediate.*

If properties 1, 2, and 3 are satisfied then the compound CTMC is a correct grouping of the atomic CTMC. A formal proof can easily be derived by induction on the elementary processes involved in the construction of the compound CTMC from the GSHLPN, however it is rather long and is not given here. In this section we define a symbolism for the description of compound markings of a GSHLPN and a symbolic firing satisfying properties 1, 2, and 3, and thus yielding compound CTMC's with a correct state grouping. The symbolism is based on a classification of the attributes of the token types of the net into groups. The classification also considers the place where token types can be held.

A *located token type* of a net is a pair *(s, p)*, where *s* is a token type and *p* is a place of the net which can hold tokens of type *s*. Let *PH(s)* be the set of places which, assumming that the GSHLPN is not overspecified, can hold tokens of type *s*. *PH(s)* can be found by examination of the arc labels: $p \in PH(s)$ if some input or inhibitor arc *(p, t)* has a token of type *s*, or some transition (case) output arc *(t, p)* (*(c, p)*) has a token of type *s*. The net of Figure 2 has the following located token types:

(PP, P1),
(PW, P2),
(PC, P3), *(PC, P4)*, *(PC, P5)*,
(M, P6),
(B, P7), *(B, P8)*

Although in the net of the example each place can only hold tokens of one type, it is clear that in general a place may hold tokens of several types. An attribute of a located token type will be called *located attribute* and the notation *att(s, p, i)* will be used to indicate the ith attribute of the located token type *(s, p)*. Before defining the classification of located attributes, it is necessary to select a particular syntax for predicates and expressions used in inhibitor and output token attributes. The syntax chosen is appropriate to capture permutation and cyclic symmetries in the values of the token attributes.

It is assumed that the predicate of a transition *t* is a logical expression with clauses of the types:

$$i = j \tag{3}$$
$$i \neq j \tag{4}$$
$$A_1 i_1 \oplus \ldots \oplus A_n i_n = K \tag{5}$$
$$A_1 i_1 \oplus \ldots \oplus A_n i_n \neq K \tag{6}$$

and that expressions for attributes of inhibitor and output tokens of a transition *t* have one of the forms:

$$i \tag{7}$$
$$A_1 i_1 \oplus \ldots \oplus A_n i_n \oplus K \tag{8}$$

where, i, j, and i_k belong to the free variable set of t, the attributes of the free variables in the same clause, and, for expressions, the attributes of the free variables and the attribute receiving the expression, have the same domain D, K and A_k are non-negative integers smaller to $|D|$, and the summation is taken modulus $|D|$. Note that $i = j \oplus K$, $i \neq j \oplus K$ are clauses of, respectively, the types (5), (6), and $i \oplus K$ is an expression of type (8).

Two located attributes are *connected* if the value of one of them in some inhibitor or output token is defined by an expression using a free variable of the other. For instance, *att(PW, P2, 1)* and *att(PP, P1, 1)* of the net of Figure 2 are connected. Consider the equivalence relation *link* defined as the reflexive and transitive (connect is symmetric by definition) closure of connect. We define the following relation from the equivalence relation link: two located attributes *a, b* are *related* if they have free variables appearing in a clause of some predicate or are linked to located attributes satisfying that condition. Note that the relation is symmetric and transitive, but not necessarily reflexive. A located attribute is *irrelevant* if it is not related to any other located attribute. Since the relation is symmetric and transitive, related attributes are classified into subsets, with all the attributes in a subset being related to each other. Such subsets are called *groups*. A group *G* is a *rotation group* if free variables for several located attributes in *G* appear together in a clause of types (5), (6) or an expression of type (8). A *permutation group* is a group which is not a rotation group.

Irrelevant attributes indicate overspecification of token types. They can be omitted in the corresponding located token types and we will assume they are. A net with irrelevant located attributes removed will be called *normalized*. For the net of Figure 2 it is found that all the located attributes corresponding to the first attribute of the token type *B* and the third attribute of the token type *PC* are irrelevant. In this case, the normalization of the net is accomplished by simply removing those attributes from those token types and the fact that these attributes can be removed indicates that the identity of the busses is irrelevant for the modeling of the multiprocessor system. In the general case, it may be necessary to define new token types, since token types have by definition a fixed numbers of token attributes and the same attribute could be irrelevant in some places and relevant in others. The relevant attributes of the net of Figure 2 are classified into one rotation group R_1 and one permutation group P_1:

$$R_1 = \{att(PP, P1, 1), att(PW, P2, 1), att(PC, P3, 1), att(PC, P4, 1), att(PC, P5, 1), att(B, P7, 2), att(B, P8, 2)\}$$

$$P_1 = \{att(PW, P2, 2), att(PC, P3, 2), att(PC, P4,2), att(PC, P5, 2), att(M, P6, 1)\}$$

The located attributes of R_1 are processor identifiers and the located attributes of P_1 are memory identifiers. Even if attributes in the same group have different semantics, the domain requirements imposed for clauses and expressions guarantee that all the attributes in a group will have the same domain.

Let R_i and P_i be, respectively, the rotation and permutation groups of the net and denote by RD_i and PD_i the domains of the located attributes in the respective groups. We associate to each rotation group R_i a free variable r_i taking values in RD_i , and to each permutation group P_i , a set of distinguished free variables p_{ij} , $j = 1, ..., |PD_i|$, taking *different* values in PD_i. A *compound marking* M is an assignment of bags of symbolic tokens to places of the net where the located attributes of each group R_i are specified by expressions $r_i \oplus K$ (K constant, $0 \le K \le |RD_i|-1$), and the located attributes of each group P_i by distinguished free variables p_{ij}, $j = 1, ..., U_i(M)$, $U_i(M) \le |PD_i|$. The set $S(M)$ of atomic markings represented by M is obtained by assigning values to the free variables from their respective domains (different values for the distinguished free variables associated to the same permutation group).

Assume that the multiprocessor system taken as example has 4 processors, 2 memories and 2 busses. Then $|RD_1|=4$ and $|PD_1|=2$. Let r be the free variable associated to R_1 and p_1, p_2, p_3, p_4 the distinguished free variables associated to P_1. A reachable compound marking of the normalized net is:

$$\begin{array}{ll} P1: & \langle PP, r\rangle, \langle PP, r\oplus 1\rangle \\ P2: & \langle PW, r\oplus 2, p_1\rangle \\ P3: & \langle PC, r\oplus 3, p_1\rangle \\ P6: & \langle M, p_2\rangle \\ P7: & \langle B, r\rangle , \end{array} \qquad (9)$$

which describes the state of the system in which two consecutive (according to the cyclic ordering defined by the token passing protocol) processors are executing in the private domain, the next processor is waiting for a memory currently used, the remaining processor is executing in the common domain in the mode with mean duration $1/\lambda_{C,1}$, accesing the memory for which the latter is waiting, and the token of the free bus is being passed to the second processor executing in the private domain. We show next that the compound markings thus defined satisfy Property 1.

Proposition 1 *Two compound markings M_1, M_2 of a GSHLPN are either disjoint or equivalent.*

Proof

From the definition of compound markings it is clear that the elements of $S(M)$ can be generated from any atomic marking $m \in S(M)$ by combinations of rotations $\rho_i(x) = x \oplus K$ ($0 \le K \le |RD_i| - 1$) of the values of located attributes in each rotation group R_i and substitutions of the $U_i(M)$ different values of the located attributes of each permutation group P_i by different values of PD_i. Assume $m \in S(M_1) \cap S(M_2)$. This imposes $U_i(M_1) = U_i(M_2)$ for all the permutation groups P_i. Then, all the elements in $S(M_1)$ and $S(M_2)$ can be obtained by combinations of the same mappings of the attribute values of m and $S(M_1) = S(M_2)$ $\therefore$

The symbolic firing of a GSHLPN is defined in complete isomorphism with its atomic firing but using the free variables supporting the description of compound markings, instead of values of the attribute domains, for the determination of the firing sets and firing distributions, and for the evaluation of the predicates and specified attributes of output and inhibitor tokens. For unspecified attributes, each of the expressions r_i, $r_i \oplus 1$, ..., $r_i \oplus (|RD_i|-1)$ is taken with the same probability if the located attribute belongs to the rotation group R_i, and each of the variables p_{ij}, $j = 1, ..., |PD_i|$ is taken with the same probability if the located attribute belongs to the permutation group P_i *.

For instance, the transition *T1* of the GSHLPN of Figure 2 is enabled in the compound marking (9) and has one firing distribution comprising the firing sets $\{\langle PP, r\rangle\}$ and $\{\langle PP, r\oplus 1\rangle\}$. The transition fires in each firing set with rate λ_P. Since the second attribute of the output token of the transition is unspecified, there are two modes for each firing set, each associated to a memory module of the system. Each mode has a probability 1/2. Thus, the firing of *T1* yields 4 compound markings, each with rate $\lambda_P/2$. The compound markings are:

$$\begin{array}{lll} M_1 & P1: & \langle PP, r\rangle \\ & P2: & \langle PW, r\oplus 1, p_1\rangle, \langle PW, r\oplus 2, p_1\rangle \\ & P3: & \langle PC, r\oplus 3, p_1\rangle \\ & P6: & \langle M, p_2\rangle \\ & P7: & \langle B, r\rangle \end{array}$$

* Although this is not the most efficient approach, it shows clearly the isomorphism (Property 3). A more efficient and equivalent method is to take any of the U_i free variables of the permutation group P_i remaining in the compound marking with probability $1/|PD_i|$, and, if $U_i < |PD_i|$, a different free variable of the group with probability $(|PD_i| - U_i)/|PD_i|$.

M_2 $P1$: $\langle PP, r\rangle$
$P2$: $\langle PW, r\oplus 1, p_2\rangle, \langle PW, r\oplus 2, p_1\rangle$
$P3$: $\langle PC, r\oplus 3, p_1\rangle$
$P6$: $\langle M, p_2\rangle$
$P7$: $\langle B, r\rangle$

M_3 $P1$: $\langle PP, r\oplus 1\rangle$
$P2$: $\langle PW, r, p_1\rangle, \langle PW, r\oplus 2, p_1\rangle$
$P3$: $\langle PC, r\oplus 3, p_1\rangle$
$P6$: $\langle M, p_2\rangle$
$P7$: $\langle B, r\rangle$

M_4 $P1$: $\langle PP, r\oplus 1\rangle$
$P2$: $\langle PW, r, p_2\rangle, \langle PW, r\oplus 2, p_1\rangle$
$P3$: $\langle PC, r\oplus 3, p_1\rangle$
$P6$: $\langle M, p_2\rangle$
$P7$: $\langle B, r\rangle$

Given the definition of the symbolic firing of a transition, it should be clear that properties 2 and 3 are satisfied if the symbolism for compound markings is coherent with the predicates and specified attributes of output and inhibitor tokens. This is stated by the following proposition:

Proposition 2 *The symbolism for compound markings satisfies the following properties:*

a) located attributes with free variables in the same clause belong to the same group G, and if the clause is of type (5) or (6) G is a rotation group.

b) the located attributes with free variable in an expression (7) or (8) and the located attribute whose value is defined by the expression belong to the same group G, and if the expression is of type (8) G is a rotation group.

Proof

To prove a), consider first clauses of types (3), (4), let *a*, *b* be the located attributes of, respectively, *i*, *j*, and assume $a \neq b$ ($a=b$ is a trivial case). Then, by definition, *a* and *b* are related and belong to the same group. Parallel arguments show that the located attributes with free variable in a clause of type (5) or (6) are related and belong to the same group *G*. In addition, since *G* has located attributes with free variables appearing together in a clause of type (5) or (6), *G* is a rotation group.

To prove b), consider an expression of type (7), let *a* be the located attribute of *i*, *b* the located attribute whose value is defined by the expression, and assume $a \neq b$. By definition, *a* and *b* are connected and, since link is the closure of the connect relation, linked. Since *a* is not irrelevant (irrelevant attributes are suppressed), *a* is related to some other located attribute *c*. But, being *a* and *b* linked, *b* is also related to *c*. Therefore, *a* and *b* belong to the same group. It can be shown similarly that the located attributes with free variable in an expression of type (8) and the located attribute whose values is defined by the expression belong to the same group *G*. In addition, since *G* has located attributes with free variables appearing together in an expression of type (8), *G* is a rotation group ∴

Therefore, we can state the result looked for in this section:

Theorem 1 *The compound CTMC obtained by firing symbolically a GSHLPN with the compound markings as defined in this section is a correct state grouping of the atomic CTMC of the GSHLPN.*

To end this section we justify why located attributes and not just attributes of token types have been used to define the symbolism for compound markings. The reason is that disregarding the place would produce in some cases less compact compound markings. Consider, for instance, the GSHLPN shown in Figure 3, where all the attributes have the same domain *D*, with $|D|=2$. Disregarding the place would force compound markings with all the attributes in a unique rotation group. Considering, as proposed, located attributes two rotation groups are obtained:

$$R_1 = \{att(B, P2, 1), att(C, P3, 1)\}$$
$$R_2 = \{att(B, P4, 1), att(D, P5, 1)\}$$

This makes that, in general, each of the last compound marking contain several of the former. For instance, the compound marking:

$P2$: $\langle B, r_1\rangle$
$P3$: $\langle C, r_1\oplus 1\rangle$
$P4$: $\langle B, r_2\rangle$
$P5$: $\langle B, r_2\oplus 1\rangle$

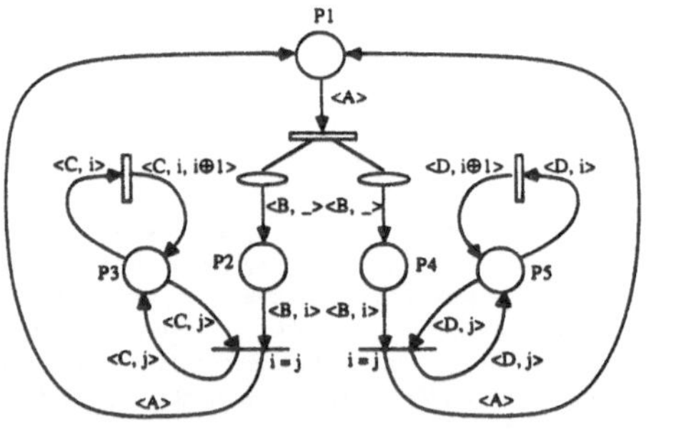

Figure 3. A GSHLPN to illustrate the advantages of using located attributes.

represents the same set of atomic markings that the following compound markings together:

$P2$: $\langle B, r\rangle$	$P2$: $\langle B, r\rangle$
$P3$: $\langle C, r \oplus 1\rangle$	$P3$: $\langle C, r \oplus 1\rangle$
$P4$: $\langle B, r\rangle$	$P4$: $\langle B, r \oplus 1\rangle$
$P5$: $\langle D, r \oplus 1\rangle$	$P5$: $\langle D, r\rangle$

4. COMPOUND MARKING EQUIVALENCE

The construction of the compound CTMC requires the determination of the equivalence of compound markings. Two compound markings M_1, M_2 are equivalent if the subsets $S(M_1), S(M_2)$ coincide. By renaming free variables of permutation groups we obviously get equivalent compound markings. Therefore, without lost of generality, it will be assummed that the set of distinguished free variables of each permutation group used in any compound marking M has the form $\{p_{ij}, 1 \le j \le U_i(M)\}$. If $U_i(M_1) \neq U_i(M_2)$ for some P_i, the compound markings are clearly not equivalent. Assuming $U_i(M_1) = U_i(M_2)$ for all P_i, the markings are equivalent if and only if there exists a rotation $\rho_i(x) = x \oplus K_i$ for the located attributes of each rotation group and a permutation of each set of distinguished variables p_{ij} associated to a permutation group P_i mapping M_1 into M_2. Since the construction of large compound CTMC's will involve a large number of equivalence tests, an efficient algorithm should be found to carry them out. However, although heuristics which will work well in most cases are easily found, our attempts to derive an algorithm with a polynomial complexity were unsuccessful. Therefore, we decided to study the complexity of the problem from the point of view of the NP-completeness theory (see, for instance [15]). We present in this sections the results of the analysis.

Let RG be the number of rotation groups and PG the number of permutation groups of the GSHLPN. We first show that for bounded RG, the problem is polynomially equivalent to GRAPH ISOMORPHISM, a well-known problem whose classification in the NP world is still unknown. We then show that if token types have at most one permutation attribute the problem is in P (a polynomial deterministic algorithm exists).

The problem is obviously in NP. In addition it is known [15] that DIRECTED GRAPH ISOMORPHISM is polynomially equivalent to GRAPH ISOMORPHISM. Therefore, it is enough to find Karnap reductions, in both directions, between COMPOUND MARKING EQUIVALENCE and DIRECTED GRAPH ISOMORPHISM. Before describing the reductions we give some additional definitions. A *classification* of a compound marking M is a partitioning of the tokens in M in classes holding tokens in the same place, of the same token type, with the same multiplicity, and with the same expressions for each located attribute belonging to a rotation group. A *grouping* of a compound marking M is a partitioning of the classes of M into groups holding classes associated to the same place, token type and multiplicity.

Reduction from COMPOUND MARKING EQUIVALENCE to DIRECTED GRAPH ISOMORPHISM

Compute the classes and groups of M_1 and M_2. If it is not possible to define a one-to-one mapping between corresponding groups (groups with the same place, token type and multiplicity), or the numbers of classes in some pair of corresponding groups are different, the markings are not equivalent. Note that the mapping of groups is unique if exists. Take classes $C_i^1, 1 \le i \le k$ of M_1 from different groups $G_i^1, 1 \le i \le k$ such that all the variables r_i instantiated in M_1 appear in some C_i^1 (note that this can always be done with $k \le RG$). Consider the groups G_i^2 mapped to G_i^1, and all the selections of classes $C_i^2, 1 \le i \le k$ of M_2 with $C_i^2 \in G_i^2$ and such that $|C_i^2| = |C_i^1|$. If there not exists any selection with $|C_i^2| = |C_i^1|, 1 \le i \le k$, the markings are not equivalent. For a given selection, each pair (C_i^1, C_i^2) requires a rotation ρ_j for each located attribute of R_j in the token type of the classes. For instance, if the tokens in C_i^1 are of the form $\langle s, ..., r_1, r_2 \oplus 1\rangle$ and those in C_i^2 are of the form $\langle s, ..., r_1 \oplus 1, r_2 \oplus 3\rangle$, the rotations $\rho_1(x) = x \oplus 1$ and $\rho_2(x) = x \oplus 2$ are required. The required rotations may be incompatible. If the required rotations are incompatible for all the selections, the markings are not equivalent.

Consider the selections requiring compatible rotations. By construction, there is a rotation for each rotation group with located attributes instantiated in the compound markings, mapping the rotation attributes of the tokens in C_i^1 into those of C_i^2, but this may not be the case for the remaining classes. The sets of rotations mapping rotation attributes for all the classes constitute an *arrangement of compatible rotations*. If no arrangement of compatible rotations exists for any selection $C_i^2, 1 \le i \le k$, the markings are not equivalent. Since tokens in classes have unique expressions for their rotation attributes, an arrangement defines a one-to-one mapping of the classes of the compound markings. The arrangement maps M_1 into M_2 except for the variables p_{ij}. It is enough to check whether a permutation exists for each set of variables $p_{ij}, 1 \le j \le U_i$ mapping each class of M_1 into the corresponding class of M_2. We assume any ordering of the classes of M_1 and consider the ordering of M_2 induced by the one-to-one mapping between classes. Then, for each compound marking, we construct a directed graph "writing" the compound marking except for permutations of the variables p_{ij} as follows.

Let M be the compound marking, CS the set of classes, with ordering $C_1, ..., C_m$, and VS the set of variables p_{ij} instantiated in M. The directed graph D associated to M contains a node r, and the subsets CS and VS. D also contains the arcs required to build the circuit $(r, C_1, ..., C_m)$, and rooted at each node C_i a tree constructed to represent the set of tokens in C_i, with each node of the tree receiving an arc from some node in VS. The construction is illustrated in Figure 4. It should be clear that there

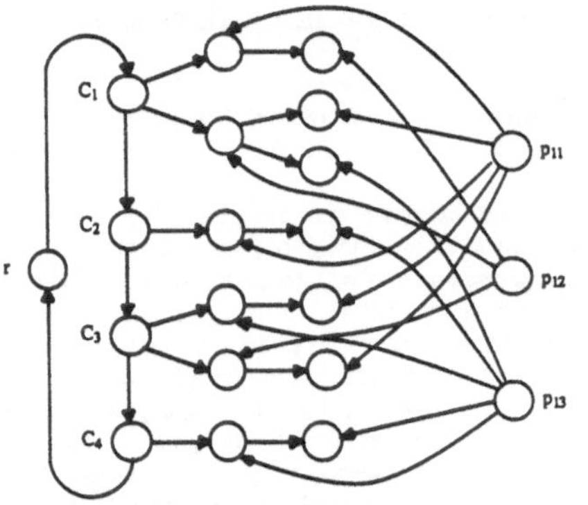

Class	Tokens	Multiplicity
C_1	$\langle \cdots, p_{11}, p_{12} \rangle$	2
	$\langle \cdots, p_{12}, p_{11} \rangle$	2
	$\langle \cdots, p_{12}, p_{13} \rangle$	2
C_2	$\langle \cdots, p_{11}, p_{13} \rangle$	1
C_3	$\langle \cdots, p_{13}, p_{11} \rangle$	1
	$\langle \cdots, p_{12}, p_{11} \rangle$	1
C_4	$\langle \cdots, p_{13}, p_{13} \rangle$	1

Figure 4. Example illustrating the construction of directed graphs in the reduction of COMPOUND MARKING EQUIVALENCE to DIRECTED GRAPH ISOMORPHISM.

is a one-to-one mapping between the set of leaves of the tree rooted at C_i and the set of tokens C_i. If the directed graphs D_1 and D_2 obtained for some arrangement of compatible rotations are isomorphic, the markings are equivalent. Otherwise, they are not ∴

It can easily be seen that the reduction is correct. Note that $(r, C_1, ..., C_m)$ is the only cycle of the directed graphs, r the only node in the cycle with output degree 1, and p_{ij} the only nodes without input arcs. Therefore, r, $C_1, ..., C_m$, the trees rooted at nodes C_i, and the set of nodes $\{p_{ij}\}$ are identified in the directed graphs and any permutation of nodes mapping the graphs has to be restricted to $\{p_{ij}\}$. Let N be the number of tokens in the compound markings. Since at most RG classes C_i^1 will have to be taken and the number of tokens in a class is bounded by N, the number of sets of rotations is bounded by N^{RG} and so is the number of arrangements of compatible rotations. Therefore, the number of times that DIRECTED GRAPH ISOMORPHISM is invoked is bounded by N^{RG}. By construction, the size of the graph is linearly related with the size of the compound markings. It is a trivial task to check that all the other operations involved are polynomial in the size of the compound markings. Therefore, the reduction is polynomial for bounded RG.

Reduction from DIRECTED GRAPH ISOMORPHISM to COMPOUND MARKING EQUIVALENCE

This reduction is trivial. For each directed graph D a compound marking M is built having tokens of the same type, in the same place and with multiplicity 1. The marking uses distinguished free variables of one permutation group. Tokens have two attributes and a distinguished variable i is used for each node of the directed graph. There is a token $\langle A, i, j \rangle$ for each arc (i,j) of D. The compound markings thus obtained are checked for equivalence. The directed graphs are isomorphic if the compound markings are equivalent ∴

The last reduction shows that is enough to allow token types with more than one permutation attribute to have a polynomial equivalence between the problems. In the case in which token types have only one permutation attribute, the invocations of the DIRECTED GRAPH ISOMORPHISM PROBLEM done in the first reduction are trivial and can be solved in polynomial time. Therefore, in this case, COMPOUND MARKING EQUIVALENCE is polynomial. This allows to state the following theorems:

Theorem 2 *COMPOUND MARKING EQUIVALENCE with bounded number of rotation groups is polynomially equivalent to GRAPH ISOMORPHISM if the compound markings have token types with more than one permutation attribute.*

Theorem 3 *COMPOUND MARKING EQUIVALENCE with bounded number of rotation groups is polynomial if compound markings have token types with at most one permutation attribute.*

5. CONCLUSIONS

We have presented generalized stochastic high-level Petri nets, a new type of nets obtained by the combination of the qualities of stochastic high-level Petri nets and generalized stochastic Petri nets and have discussed their semantics, paying special attention to the firing policies, which become more complicated than for standard stochastic Petri nets due to the existence of token types with attributes. After illustrating the expressive power of the nets with an example we have considered the automated construction of compound CTMC's from GSHLPN's.

We have given properties of symbolisms for compound markings which guarantee that the compound CTMC obtained by symbolically firing the net will have a correct state grouping. For a particular, yet highly expressive, syntax, a symbolism for compound markings with those properties has been derived. The procedure is amenable of automation. The construction of the compound CTMC requires an algorithm to test the equivalence of compound markings. We have shown that, in the general case and for bounded number of rotation groups, the EQUIVALENCE COMPOUND MARKING problem is polynomially equivalent to the well-known GRAPH ISOMORPHISM problem. The classification of the last problem is currently open, but there is strong evidence that it cannot be NP-complete (see, [16] for details). Although the existence of a polynomial algorithm to test the equivalence of compound markings would constitute an important asset for GSHLPN's (and SHLPN's as well), its inexistence does not rule out their application, since it seems apparent that heuristics which will be efficient in most cases can be developed. We are currently working in that direction.

REFERENCES

[1] M. K. Molloy, *Performance Analysis using Stochastic Petri Nets*, IEEE Trans. on Computers, vol. C-31, no. 9, pp. 913-917, September 1982.

[2] S. Natkin, *Les reseaux de Petri Stochastiques et leur application a l'evaluation des systèmes informatiques*, Thèse de Docteur Ingénieur, CNAM, Paris, 1980.

[3] J. B. Dugan, G. Ciardo, A. Bobbio, and K. S. Trivedi, *The design of a unified package for the solution of stochastic Petri net models*, in Proc. Int. Workshop on Timed Petri Nets, Torino, Italy, pp. 6-13, July 1985.

[4] J. F. Meyer, A. Movaghar, and W. H. Sanders, *Stochastic activity networks: structure, behaviour and applications*, in Proc. Int. Workshop on Timed Petri Nets, Torino, Italy, pp. 80-87, July 1985.

[5] M. A. Marsan, G. Conte, and G. Balbo, *A class of generalized stochastic Petri nets for the performance evaluation of multiprocessor systems*, ACM Trans. on Computer Systems, vol. 2, no. 1, pp. 93-122, May 1984.

[6] M. A. Marsan, G. Balbo, G. Chiola, and G. Conte, *Generalized Stochastic Petri nets revisited: Random switches and priorities*, in Proc. Int. Workshop on Petri Nets and Performance Models, 1987, pp. 44-53.

[7] G. Balbo, G. Chiola, G. Franceschinis, and G. Molinar Roet, *On the efficient construction of the tangible reachability graph of generalized stochastic Petri nets*, in Proc. Int. Workshop on Petri Nets and Performance Models, 1987, pp. 136-145.

[8] Ch. Lin and D. C. Marinescu, *On Stochastic High-Level Petri Nets*, in Proc. Int. Workshop on Petri Nets and Performance Models, 1987, pp. 34-43.

[9] Ch. Lin and D. C. Marinescu, *Stochastic High-Level Petri Nets and Applications*, IEEE Trans. on Computers, vol. 37, no. 7, pp. 815-825, July 1988.

[10] K. Jensen, *High-level Petri Nets*, Informatik-Fachberichte, vol. 66, pp. 166-180, 1983.

[11] A. Zenie, *Coloured stochastic Petri nets*, in Proc. Int. Workshop on Timed Petri Nets, Torino, Italy, pp. 262-271, July 1985.

[12] D. C. Marinescu, *Domain oriented analysis of PDE splitting algorithms*, Journal of Information Science, vol. 43, pp. 3-24, 1987.

[13] P. Huber, A. M. Jensen, L. O. Jepsen, and K. Jensen, *Towards Reachability trees for high-level Petri nets*, Lecture Notes in Computer Science, vol. 188, pp. 215-133, Springer-Verlag, 1985.

[14] I. A. Papazoglou, E. P. Gyftopoulos, *Markov Processes for Reliability Analysis of Large Systems*, IEEE Trans. on Reliability, vol. R-26, no. 3, pp. 232-237, August 1977.

[15] M. R. Garey and D. S. Johnson, *Computers and Intractability: A Guide to the Theory of NP-Completeness*, W. H. Freeman and Co., New York, 1983.

[16] D. S. Johnson, *The NP-Completeness Column: An Outgoing Guide*, Journal of Algorithms, vol. 9, no. 3, Sept. 1988, pp. 426-444.

CORRECTIONS

The definitions of the *connect* and *relate* relations and *irrelevant* attributes given in Section 3 have to be revised as follows:

Two located attributes are *connected* if the value of one of them in some output token is defined by an expression using a free variable of the other.

A located attribute is *related* to itself if there is some input token of multiplicity greater than 1 with that attribute. Two (non necessarily different) located attributes are related if different free variables associated to them appear in a clause of some predicate or if the value of one of them in some inhibitor token is defined by an expression using a free variable of the other. Finally, located attributes linked or related to related located attributes are also related.

A located attribute is *irrelevant* if it is not related to any located attribute (including itself).

19.
Stochastic Well-Formed Coloured Nets and Multiprocessor Modelling Applications

G. Chiola*, C. Dutheillet, G. Franceschinis and S. Haddad

Submitted to IEEE Transactions on Computers

Abstract

The new class of Stochastic Well-formed coloured Nets (SWN) is defined as a syntactic restriction of Stochastic High-Level Nets. The interest of the introduction of restrictions in the model definition is the possibility of exploiting the Symbolic Reachability Graph (SRG) to reduce the complexity of Markovian performance evaluation with respect to classical Petri net techniques. It turns out that SWNs allow the representation of any colour function in a structured form, so that any unconstraint high-level net can be transformed into a well formed net. Moreover, most constructs useful for the modelling of distributed computer systems and architectures directly match the "well form" restriction, without any need of transformation. A non trivial example of the usefulness of the technique in the performance modelling and evaluation of multiprocessor architectures is included.

Key words: High-Level Petri nets, model symmetry, symbolic reachability graph, performance evaluation, computational complexity, Markov chains, lumpability condition, multiprocessor systems, bus and memory contention.

1 Introduction

Stochastic Petri nets have been been proposed in the literature as a good modelling tool for the study and performance evaluation of multiprocessor computer architectures [1,2,3,4]. The technique is much easier to use than more classical Markovian modelling techniques, but still leads to models whose size is too large when modelling realistic systems. Techniques for the development of "compact" or "folded" models [5,6] have been proposed, but they had not been used by many researchers outside the restricted group that developed them. This is due to the deep knowledge required by the modeller of the behavioural characteristics and symmetries of the system under study. High level Petri net models have been proposed from time to time as a more adequate tool (see, e.g., [7,8]) for the modelling of complex multiprocessor systems. Although easier to use even by inexperienced modellers because of their higher level of abstraction, classical high-level Petri nets do not alleviate the need for a thorough understanding of the symmetries of the system in order to allow the performance evaluation of reasonably large systems.

Recently, stochastic models based on a restricted subclass of coloured nets called Regular Stochastic Petri Nets (RSPN) have been introduced for performance evaluation purposes [9,10]. The steady-state performance of an RSPN model can be obtained by numerically solving a Markov chain (MC) isomorphic to the *symbolic reachability graph* (SRG) generated by the RSPN. The complexity in the computation of this solution is polynomial in the number of symbolic markings of the SRG, which can be much less than the number of ordinary markings generated by the Place/Transition net (P/T) resulting from the unfolding of the RSPN.

Some kinds of symbolic marking representations to exploit the symmetries of the reachability graph (RG) have already been proposed for general CPNs [11,12,8,13] as well, but in all these cases heuristics

*This work has been done while G. Chiola was visiting researcher at the Lab. MASI of the University of Paris 6, with the financial support of a NATO-CNR annual research grant.

were needed to decide the type of aggregations. These heuristics were based on an explicit knowledge by the modeller of the symmetries present in a particular model. Thus none of these methods could be implemented in a general algorithmic form. From this point of view, the superiority of RSPNs was due to the availability of a generic ***symbolic firing rule*** which allows the computation of the SRG without any actual instantiation of colours and without explicit knowledge of the symmetry of the model.

This use of the SRG for performance evaluation purposes is the main motivation that led us to the proposal of the extended SRG computation for more general classes of CPNs called "well-formed" coloured nets (WN) [14]. RSPNs can be used to model many interesting systems, but unfortunately the strong restrictions imposed on the definition of the arc labelling functions, the object colour sets, and the initial marking, prevent their use in general cases.

In this paper we define the new class of stochastic well-formed coloured nets (SWN) as extensions of RSPNs. The Markovian analysis of SWNs can be substantially improved by using the SRG instead of the usual RG, exactly as in the case of RSPNs. In particular, the Markov chain obtained from the SRG is shown to be an exact lumping [15] of the Markov chain that would be obtained from the RG. Moreover, the lumped Markov chain is obtained directly, i.e., without first generating the complete Markov chain and then applying usual aggregation methods. The lumping implicitly performed by using the SRG instead of the RG exploits the symmetry properties of models with respect to different colours belonging to same basic colour classes.

From the modelling power point of view, the SWN formalism allows a natural representation of complex distributed systems as the unconstrained coloured net formalism. In principle any coloured net can be translated into a well-formed net by re-writing the colour functions and the composition of colour classes using our predefined formalism. Moreover, in practical modelling this formalism translation is hardly needed: most (if not all) CPN models published in the literature can be directly represented as WNs, even without exploiting the power of predicate guards on the arc labelling functions. In order to support our claim, we present a complete modelling example taken from the literature on multiprocessor computer architecture. The example shows how the technique can be exploited even by non expert modellers to obtain good performance models of complex systems. The symmetries of the system are automatically taken into account by the SRG construction algorithm without any explicit intervention of the modeller. The time and space complexity of the analysis is however equal to the one of the best model devised by an expert modeller that explicitly used the Markovian lumping technique.

The balance of the paper is as follows. Sections 2 presents the definition of WNs, and Section 3 presents the SRG, a sketch of the SRG construction algorithm with some examples, and some relevant mathematical properties of the graph. Section 4 contains the definition of SWNs, and the performance evaluation technique. Section 5 presents a non trivial example of the application of SWNs to the performance evaluation of a well known multiprocessor computer architecture. Finally, Section 6 contains some concluding remarks and perspectives of this work. The formal notation has been somewhat sacrificed, and the presentation has been based mainly on examples to provide easier comprehension for the non expert. A more rigorous notation and the formal proofs of many results can be found in [14] or in [16].

2 Well-Formed Coloured Nets

We start by giving a short outline of the notation used in the paper.
Let $(F_i)_{i\in I}$ be a family of sets. The Cartesian product of this family is denoted by $\bigotimes_{i\in I} F_i$. An element of $\bigotimes_{i\in I} F_i$ is denoted $\bigotimes_{i\in I} c_i$, with $c_i \in F_i$. Any n-tuple $(c_1,\ldots,c_n)$ is denoted $\bigotimes_{i=1}^{n} c_i$ as well.
Let E be a set, F_i be a family of sets, and $f_i : E \to F_i$ be a family of functions. Then $\bigotimes_{i\in I} f_i$ is a function defined by : $\bigotimes_{i\in I} f_i : E \to \bigotimes_{i\in I} F_i$

$$\left(\bigotimes_{i\in I} f_i\right)(c) = \bigotimes_{i\in I} f_i(c)$$

A multi-set a over a non-empty set A is a mapping $a \in [A \rightarrow I\!N]$. Intuitively, a multi-set is a set that can contain several occurrences of the same element. It can be represented by a formal sum:

$$a = \sum_{x \in A} a(x).x$$

in which the non-negative integer $a(x)$ denotes the number of occurrences of the element x in the multi-set a. We denote $Bag(A)$ the set of finite multi-sets over A.
Let $A = \{x_1, \ldots, x_n\}$. An element a of $Bag(A)$ will be also denoted as a n-tuple (vector) $(a(x_1), \ldots, a(x_n))$. In case an order relation is defined among the elements of A, then this is reflected also in the ordering of the components in the vector representation of $Bag(A)$.
In any case an arbitrary order relation can be defined among Bags based on the lexicographic order of their vector representation. Moreover, the usual vector algebraic operations apply to $Bag(A)$.

2.1 Coloured Nets : an Introduction

Because of the large size of the models obtained when representing complex systems by P/T nets, coloured Petri nets have been introduced as an abbreviation [17]. In a coloured net, subnets representing similar processes are folded one upon another. This folding consists in representing all the processes by a single subnet and identifying different resources or processes by different colours.

A colour domain is then attached to each place and transition. The colour domain of a place determines the different colours of tokens that are allowed in this place, whereas the colour instances of a transition represent the different possibilities for firing this transition. An arc between a place and a transition is labelled with a function mapping the colour domain of the transition on the bag associated with the colour domain of the place. When firing a transition, one colour is chosen in the transition domain, and the functions labelling the input arcs of the transition are evaluated for this colour. This evaluation gives the number of tokens required for each colour in the input places of the transition. If the current marking satisfies the firing condition, coloured tokens are removed from the input places according to the values of the functions, and they are put in the output places according to the values of the functions labelling the output arcs.

Examples of coloured nets. The first example presents an operating system that supports both batch and interactive processes. When a process enters the system, it remains in state "ready" while the central unit is busy. As soon as the CPU is available, the process becomes "active". We assume that interactive processes can be interrupted before the end of their activity, whereas batch processes cannot. All the nodes of the model except 'Unit' have a colour domain C which is the set of processes. Two subsets of process colours are defined: interactive processes (D_1) and batch processes (D_2). The function X labelling the arcs selects one object in C for each transition firing. The transition 'Interrupt' is guarded by a predicate restricting its colour domain to D_1 i.e., only objects in D_1 can be chosen to fire the transition. As the central unit is unique there is no need to attach a specific colour to it. Thus it is represented by a neutral colour.

The second net models the "dining philosophers" problem. A group of philosophers, each with one chopstick, are sitting around a table. Thinking makes them hungry, so that from time to time they decide to eat. To do so, they need both their own chopstick and that of their left-hand neighbour. If one of these chopsticks is not available, they have to wait until it is released. At that time, they eat their rice until they are full, then they release their chopsticks and resume thinking. In this particular case, the ("successor") function $\oplus X$ is used to select the chopstick of the left-hand neighbour.

The third example represents a communication system in which a site sends a message to a distant site. This message is split into three packets and each packet is transmitted separately on the network. Once the three packets have been received by the distant site, it sends an acknowledgement to the site that originated the communication. On reception of this acknowledgement, the communication ends and the sender becomes idle again.

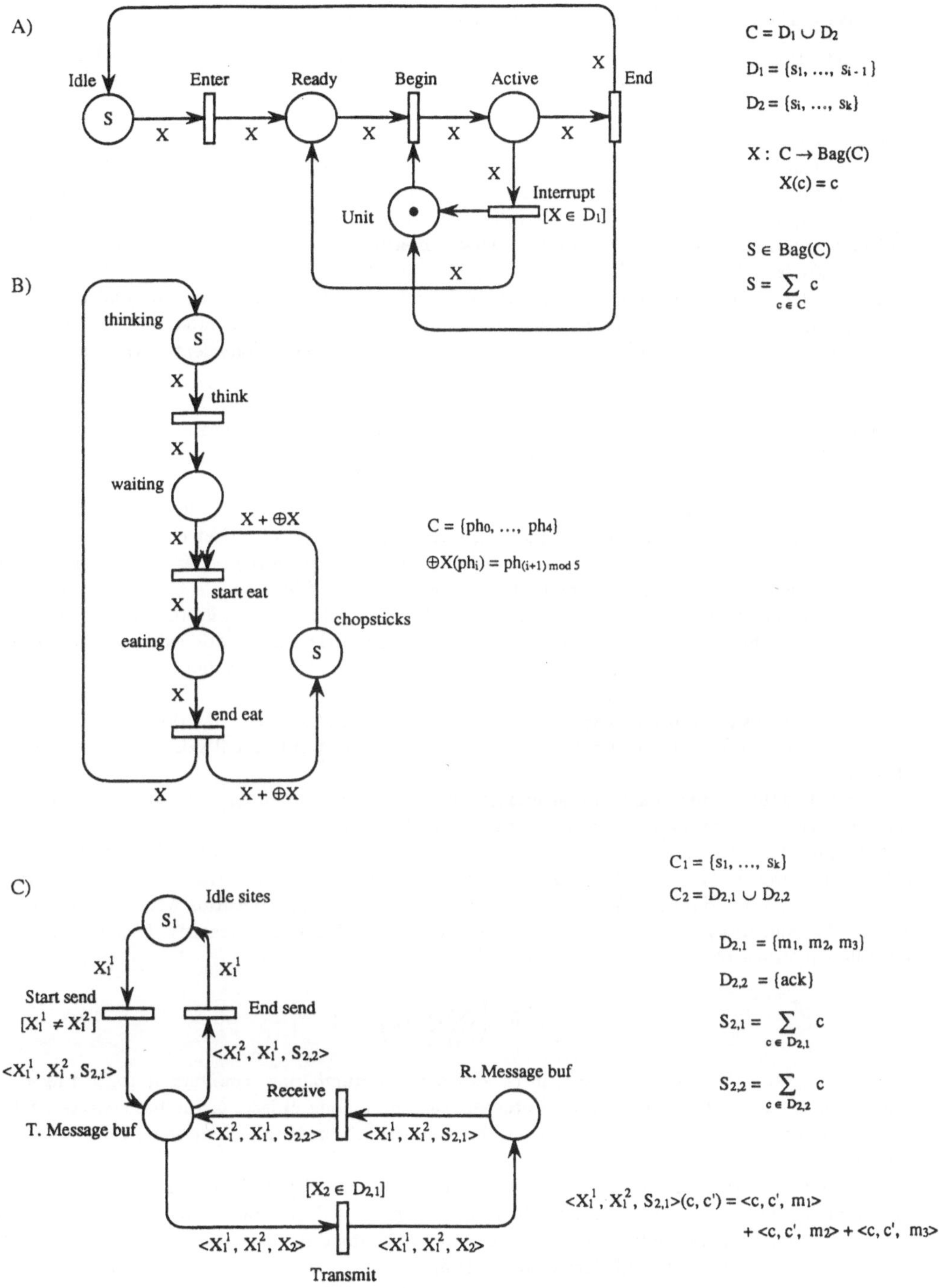

Figure 1 : Three examples of coloured nets

2.2 Well-Formed Coloured Nets : an Introduction

In the definition of coloured nets, a colour function is defined by an integer matrix, whose lines (resp. columns) are indexed by the colours of places (resp. transitions). However, in practice, one usually substitutes this extensive description with an intensive description having intuitively evident semantics. Thus, for instance, in Figure 1 the symbol X stands for the identity matrix.

This substitution is possible thanks to the intrinsic structuring of the models studied. But the structuring is irremediably lost as soon as the colour functions are described in matrix form. Therefore, the use of coloured nets leads to heuristic validation methods where the modeller has to intervene giving additional information (e.g., an algebraic property on a function, or a behavioural symmetry) in order for the analysis to be completed.

Only a syntactic description preserves the information on the model structure and can be used to derive validation algorithms. To the best of our knowledge, well-formed coloured net is the first coloured net formalism including a syntax that has a modelling power strictly equivalent to that of coloured nets defined using matrix representations.

We now introduce the basic principles for building a well-formed net. The first one concerns the construction of colour domains. The colours representing resources of the same type, such as a set of files, a set of identical processors, or a set of memory modules, are grouped in a class. Colour classes of a net will be denoted C_i, with i ranging from 1 to n.

If we call objects the basic components of the system, such as files or memories, we can notice that both place and transition domains are made of Cartesian products of object classes. These classes can be ordered if such a relation exists between the objects they contain (e.g. sites on a ring). Colours will denote the elements of colour domains that are actually tuples of objects.

When objects belonging to the same class have different behaviours, it is convenient to partition the class in static subclasses, so that all the objects belonging to the same static subclass behave the same way. Thus, the model obtained respects the intrinsic symmetries of the system. The static subclasses of C_i are denoted $D_{i,q}$, $q = 1, \ldots n_i$.

In the previous examples we can also notice that the number of colour functions used in these models is limited. This is due to the fact that they are used to represent only usual synchronization mechanisms of parallel systems.

We can actually distinguish three main families of colour functions. All these functions can be composed by Cartesian product and/or linear combinations.

The projection functions, denoted by X_i^j allow one to select one particular object in a class C_i. The behaviour of the selected object is independent of the behaviour of the other objects of the class when firing the transition. If several functions X_i^j appear around the same transition with the same superscript j, then the same object must be selected to instantiate the different occurrences of the function, whereas for different values of j, the objects can be different.

$$X_i^j \left(\bigotimes_{k=1}^{n} \bigotimes_{l=1}^{e_i} c_k^l \right) = c_i^j$$

Successor functions are also used, in case of message numbering, counters or token ring protocols. These functions denoted $\oplus X_i^j$ imply an ordering relation among objects of the basic classes. Of course, $\oplus X_i^j$ makes sense only if X_i^j appears near the same transition, and it represents the successor of the object selected by X_i^j.

The diffusion functions S_i allow one to synchronize all the objects belonging to the same class C_i if they label an input arc, or to diffuse to all the objects of C_i if they label an output arc. In some cases, a diffusion function applies only to objects belonging to the same static subclass. This function is denoted $S_{i,q}$, where $D_{i,q}$ is the static subclass to which it applies.

$$\forall c \in C(t), \quad S_{i,q}(c) = \sum_{c_i \in D_{i,q}} 1.c_i$$

Note that the value of the diffusion function $S_{i,q}$ does not depend on the objects chosen to instantiate the transition and thus C_i may or may not appear in the transition domain. In fact, $S_{i,q}$ is a constant function. In particular, it can be used to denote a constant object if $|D_{i,q}| = 1$.

It is sometimes necessary to restrict the firing possibilities of a transition. When several objects are selected in the same class, we can require that these instances be different. This is the case for the sender and the receiver of a phone call. In this goal, transitions may be guarded by a predicate. Predicates can also apply to colour functions, in order to "cancel" the arc between a place and a transition for some colour instantiations. Allowed forms of (standard) predicates are any logic combination of $X_i^j \in D_{i,k}$, or $X_i^j = X_i^k$ or $\oplus X_i^j = X_i^k$.

The basic Petri net structure is extended by including inhibitor arcs and different transition priority levels. These features increase the modelling power when concurrency and/or model parametrization are taken into account [18,19].

We call Well-formed coloured Nets (WNs), the class of nets including the features we have just described. A formal definition of WNs can be found in [14].

2.3 Basic Properties of WN

A major interest of WN is that they provide a model in which symmetries appear naturally and automatically, while preserving the modelling power of coloured nets. The two next propositions of WN illustrate this situation.

2.3.1 Modelling power

Proposition 2.1 *Any CPN can be transformed into a WN with the same basic structure, same colour domains (possibly partitioned in static subclasses), equivalent arc labelling.*

Sketch of proof: Let f be a function $C_{J(t)} \to Bag(C_{J(p)})$. If f is not a WN standard colour function, or if either $C_{J(t)}$ or $C_{J(p)}$ is not a standard colour domain, then we use the following transformation. The domain of each node is partitioned in as many static subclasses as the number of colours it contains.

$$C(p) = \bigcup_{c \in C(p)} D_{i,c} \quad \text{with } D_{i,c} = \{c\}$$

$$C(t) = \bigcup_{c' \in C(t)} D_{j,c'} \quad \text{with } D_{j,c'} = \{c'\}$$

Then the standard colour function $g \in [C_{J(t)} \to Bag(C_{J(p)})]$ defined by :

$$g = \sum_{c' \in C(t)} \sum_{c \in C(p)} f(c')(c).[X \in D_{j,c'}].S_{i,c}$$

is equal to f. Actually, $\forall c_0 \in C(t)$,

$$g(c_0) = \sum_{c' \in C(t)} \sum_{c \in C(p)} f(c')(c).[X \in D_{j,c'}].S_{i,c}(c_0) = \sum_{c \in C(p)} f(c_0)(c).c = f(c_0)$$

2.3.2 Symmetry and Permutation Equivalence Classes

Definition 2.1 (Colour permutation) *Let $\xi = \{ s = \langle s_1, \ldots, s_h, s_{h+1}, \ldots, s_n \rangle \}$ (where, as usual, s will be also denoted $\bigotimes_{i=1}^n s_i$) be a subgroup of the permutations on $\bigotimes_{i=1}^n C_i$ such that:*

- *$\forall 0 < i \leq h$ s_i is a permutation on C_i such that $\forall D_{i,q}$, $s_i(D_{i,q}) = D_{i,q}$;*
- *$\forall h < i \leq n$ s_i is a rotation on C_i such that $\forall D_{i,q}$, $s_i(D_{i,q}) = D_{i,q}$. Note that this condition implies that if $n_i > 1$ then the only allowed rotation s_i is the identity.*

Let $c \in C_J$, $s \in \xi$. Then $s(c)$ is defined by : $s\left(\bigotimes_{i=1}^n \bigotimes_{j=1}^{e_i} c_i^j\right) = \bigotimes_{i=1}^n \bigotimes_{j=1}^{e_i} s_i\left(c_i^j\right)$

Definition 2.2 (Marking permutation) *Let M be a marking, and $s \in \xi$ a permutation. Then $M' = s.M$ is a marking defined by:*

$$\forall p \in P,\ \forall c \in C(p), \quad s.M(p, s(c)) = M(p, c)$$

For instance, if $C(p) = C_J$, then $M' = s.M$ is defined on p by:

$$\forall c \in C_J, \quad M'\left(p, \bigotimes_{i=1}^{n} \bigotimes_{j=1}^{e_i} s_i(c_i^j)\right) = M\left(p, \bigotimes_{i=1}^{n} \bigotimes_{j=1}^{e_i} c_i^j\right)$$

Proposition 2.2 *The firing property is preserved by applying a permutation both on the markings and the transition instantiation. $\forall M$ ordinary marking, $\forall t \in T$, $\forall c \in C(t)$, $\forall s \in \xi$,*

$$M[t, c\rangle M' \iff s.M[t, s(c)\rangle s.M'$$

2.3.3 Symbolic Marking

Definition 2.3 (Symbolic marking) *Let Eq be the equivalence relation defined by:*

$$M\ Eq\ M' \iff \exists s \in \xi,\ M' = s.M$$

An equivalence class of Eq is called a symbolic marking, denoted with $\mathcal{M}$.

Rather than considering a single initial marking M_0, we allow a WN to be initially marked by a symbolic marking $\mathcal{M}_0$. In this case, the WN no longer represents a single net, but a set of nets, each being initially marked by one of the markings contained in the equivalence class $\mathcal{M}_0$.

3 SRG

The symbolic reachability graph of a well-formed net is based on the idea of symmetry of objects of the basic colour classes. In WNs it is possible to identify two basic kinds of symmetry: *rotation*, and *general permutation inside subsets.* Combinations of the two basic symmetry kinds can be found in actual models. The idea is to represent in a symbolic way the symmetry properties that hold in a model state rather than the actual state itself. A symbolic marking thus represents an equivalence class on the state space of the WN model, and the equivalence is in terms of the possible basic colour permutations that yield the same behaviour. An efficient algorithm for the computation of the SRG has been proposed in [14], whose application does not require any a-priori knowledge on the symmetries of the system modelled by the WN (i.e., the algorithm is based only on the syntax of the WN representation, and not on any sort of semantics). Here we recall the SRG formalism by means of an example, then we analyze the properties of the SRG that are relevant for the Markovian analysis of performance models based on SWNs.

3.1 SRG Computation

We start by defining the *representation of symbolic markings* that, together with a *symbolic firing rule*, allows the construction of the symbolic reachability graph in an efficient algorithmic way. The choice of a good representation of the data is always the first problem in the definition of an algorithm. Let us present informally the idea of symbolic marking by considering the example of the message communication system depicted in Figure 1.c in the case of three sites ($C_1 = \{s_1, s_2, s_3\}$).

In the initial marking one token for each colour in C_1 is present in place 'Idle sites'. Hence the symbolic representation of the initial marking $\mathcal{M}_0$ uniquely identifies one initial marking M_0. The firing of transition 'Start send' with colour $\langle s_1, s_3\rangle$ from marking M_0 yields marking M_1 such that M_1(Idle sites) = s_2 and M_1(T.Message buf) = $\sum_{c \in D_{2,1}} \langle s_1, s_3, c\rangle$. A complete symmetry exists in this case among the colours in C_1, i.e., any permutation of basic colours yields a valid reachable marking that has the same characteristics of M_1. For example, s_2 and s_3 can be exchanged, thus yielding marking M_2 such that M_2(Idle sites) $= s_3$ and M_2(T.Message buf) $= \sum_{c \in D_{2,1}} \langle s_1, s_2, c \rangle$.

Markings M_1 and M_2 have in common the fact of having *one* token (of some colour belonging to C_1) in place 'Idle sites', and three tokens in place 'T.Message buf', differing from each other for the third component of the triple. In general all markings equivalent to M_1 and M_2 can be characterized by a single *symbolic marking* $\mathcal{M}_1$ such that $\mathcal{M}_1$(Idle sites) $= Z_1^1$ and $\mathcal{M}_1$(T.Message buf) $= \langle Z_1^2, Z_1^3, Z_2^1 \rangle$. Z_1^1, Z_1^2, and Z_1^3 are *dynamic subclasses* of cardinality 1, that represent any arbitrary partition of C_1 in three subsets containing one element each. Z_2^1 is a dynamic subclass of cardinality 3 corresponding to the subset $D_{2,1}$. This symbolic representation identifies all markings that are equivalent to M_1 from the point of view of the strong lumpability condition for Markov chains, and that can be thus grouped in a single aggregate state.

From the above informal reasoning it appears convenient to directly represent, with an appropriate data structure, the equivalence classes of markings based on the permutation and symmetry property of WNs rather then the individual markings as usual in Petri nets. We then introduce the concept of *dynamic subclasses*, representing sets of colours that are not specified explicitly, but that are known to be permutable one with the other producing markings that belong to the same equivalence class. Informally, a dynamic subclass is characterized by its cardinality (i.e., the number of different basic colours represented by the dynamic subclass), and by the static subclass to which the represented basic colours belong. Two different basic colours can be represented by the same dynamic subclass only if they have the same token distribution in all places, and if they belong to the same static subclass. In case of ordered basic classes, only contiguous objects can be represented by the same dynamic subclass and the ordering relation among basic objects is reflected by the ordering of the indexes of the dynamic subclasses.

More formally, the representation of a symbolic marking can be defined as the composition of a syntax and a semantics. From a syntactical point of view, a representation $\mathcal{R}$ of a symbolic marking $\mathcal{M}$ is defined as a 4-tuple $\mathcal{R} = \langle m, card, d, mark \rangle$, where

- $m : I \rightarrow I\!N^+$, such that $m(i)$ (which will be also denoted m_i) is the number of dynamic subclasses of C_i in $\mathcal{M}$.
- The set of dynamic subclasses of C_i is denoted: $\hat{C}_i = \{Z_i^j \mid 0 < j \leq m_i\}$
- $card : \left(\bigcup_{i \in I} \hat{C}_i\right) \rightarrow I\!N$, such that $\forall i$, $\sum_{j=1}^{m_i} card(Z_i^j) = |C_i|$
- $d : \left(\bigcup_{i \in I} \hat{C}_i\right) \rightarrow I\!N$ such that $\forall Z_i^j$, $d(Z_i^j) = q \in [1..n_i]$ and $\forall 0 < i \leq n$, $\forall 0 < j < k \leq m_i$, $d(Z_i^j) \leq d(Z_i^k)$
- $\forall p \in P$, $mark(p) : \bigotimes_{i \in I} \left(\hat{C}_i\right)^{e_i} \rightarrow I\!N$

The associated semantics can be stated as follows:

- $\hat{C}_i$ is a non-instantiated partition of the basic subclass C_i
- Z_i^j represents any subset $C_{i,j}$ of C_i such that $|C_{i,j}| = card(Z_i^j)$
  in case C_i is ordered ($i > h$), the elements of $C_{i,j}$ are contiguous
- $\forall M \in \mathcal{R}$, $\forall i \in I$, $\exists \eta_i : C_i \rightarrow \hat{C}_i$
- $\eta_i : C_i \rightarrow \hat{C}_i$ preserves
  - static subclass partitioning (function d)
  - cardinality (function $card$)
  - marking (function $mark$)
  - ordering relation in case of ordered classes (i.e. if $i > h$)

- $\eta_i^{-1}(Z_i^j) = C_{i,j}$ is a possible instantiation of a dynamic subclass that defines one marking $M \in \mathcal{R}$

When no ambiguity can arise, we denote the components of a representation simply by m, $card$, d, and $mark$. When referring to more than one representation, we remove the ambiguity by prefixing a proper representation identifier (e.g. $\mathcal{R}_1.m$).

For example, the symbolic marking $\mathcal{M}_1$ considered above can be represented as follows:

$$\begin{aligned}\mathcal{R}_1: \quad & n_1 = 1, \quad m_1 = 3, \quad \forall 0 < j \leq 3\ d(Z_1^i) = 1, \quad card(Z_1^1) = card(Z_1^2) = card(Z_1^3) = 1 \\ & n_1 = 2, \quad m_1 = 2, \quad d(Z_2^1) = 1, \quad d(Z_2^2) = 2 \quad card(Z_2^1) = 3, \quad card(Z_2^2) = 1 \\ & mark(\text{Idle sites}) = Z_1^3, \quad mark(\text{T.Message buf}) = \langle Z_1^2, Z_1^1, Z_2^1 \rangle\end{aligned}$$

Note that using the above representation without further constraints, one can find many representations for a given symbolic marking $\mathcal{M}$. For example, for the same marking considered above we can find the two following alternative representations as well:

$$\begin{aligned}\mathcal{R}_2: \quad & n_1 = 1, \quad m_1 = 3, \quad \forall 0 < j \leq 3\ d(Z_1^i) = 1, \quad card(Z_1^1) = card(Z_1^2) = card(Z_1^3) = 1 \\ & n_1 = 2, \quad m_1 = 2, \quad d(Z_2^1) = 1, \quad d(Z_2^2) = 2 \quad card(Z_2^1) = 3, \quad card(Z_2^2) = 1 \\ & mark(\text{Idle sites}) = Z_1^2, \quad mark(\text{T.Message buf}) = \langle Z_1^3, Z_1^1, Z_2^1 \rangle\end{aligned}$$

$$\begin{aligned}\mathcal{R}_3: \quad & n_1 = 1, \quad m_1 = 3, \quad \forall 0 < j \leq 3\ d(Z_1^i) = 1, \quad card(Z_1^1) = card(Z_1^2) = card(Z_1^3) = 1 \\ & n_1 = 2, \quad m_1 = 3, \quad d(Z_2^1) = d(Z_2^2) = 1, \quad d(Z_2^3) = 2 \quad card(Z_2^1) = 2, \quad card(Z_2^2) = 1, \quad card(Z_2^3) = 1 \\ & mark(\text{Idle sites}) = Z_1^3, \quad mark(\text{T.Message buf}) = \langle Z_1^2, Z_1^1, Z_2^1 \rangle + \langle Z_1^2, Z_1^1, Z_2^2 \rangle\end{aligned}$$

The first step in devising an efficient algorithm for the enumeration of the symbolic reachability graph of a WN is the definition of a *unique* representation for each symbolic marking $\mathcal{M}$. To obtain a *canonical form* we exploit two properties defined on the type of representation that we have devised, called *minimality* and *ordering* of dynamic subclasses [14]. Minimality avoids representations of the type of $\mathcal{R}_3$, where more than one dynamic subclass (in this case Z_2^1 and Z_2^2) is used to represent basic objects belonging to the same static subclass and having identical markings in each place. The minimal representation of a symbolic marking is unique within a permutation of the indexes of dynamic subclasses. A lexicographic order is also imposed in order to select among minimal representations (such as e.g. $\mathcal{R}_1$ and $\mathcal{R}_2$) the one which is minimal from the point of view of lexicographic order of the names of the dynamic subclasses. For instance, given that place 'Idle sites' is selected to precede place 'T.Message buf' in the extensive representation of the '*mark*' function, the symbolic representation $\mathcal{R}_1$ is chosen as the canonical representation of $\mathcal{M}_1$ (since it is lexicographically minimal with respect to any arbitrary permutation of the names of the dynamic subclasses).

Proposition 3.1 *Let $\mathcal{M}$ be a symbolic marking. Then there is one and only one minimal and ordered representation of $\mathcal{M}$, which is called canonical.*

In the following, we will use $\mathcal{M}$ to denote both a symbolic marking and its canonical representation, unless an explicit distinction is needed. In our example above the representation $\mathcal{R}_1$ is ordered as well as minimal, so that it is the canonical representation of the considered symbolic marking $\mathcal{M}_1$.

In order to build the SRG directly starting from a symbolic initial marking $\mathcal{M}_0$ (i.e., without building the RG and then grouping markings into equivalence classes which would be much easier but too costly), we first defined a symbolic firing rule on the symbolic marking representations. In a symbolic firing, dynamic subclasses are instantiated instead of basic colours. Thus we need an extension of the evaluation of colour functions and predicates in order to operate on dynamic subclasses.

Definition 3.1 (Symbolic instantiation) *Let t be a transition with $C(t) = \bigotimes_{i=1}^{n} \bigotimes_{j=1}^{e_i} C_i$. Let $\mathcal{R}$ be a symbolic representation. Then $[\bigotimes_{i=1}^{n} \bigotimes_{j=1}^{e_i} Z_i^{\lambda_i(j),\mu_i(j)}]$, also denoted $[\lambda, \mu]$ for short, is an instantiation of $\hat{C}(t)$ for $\mathcal{R}$ defined by:*

$\lambda = \{\lambda_i : \{1, \ldots, e_i\} \rightarrow I\!N^+\}, \qquad \mu = \{\mu_i : \{1, \ldots, e_i\} \rightarrow I\!N^+\},$

such that $\forall i \in I$, $\forall 0 < x \leq e_i$,

- $\lambda_i(x) \le \mathcal{R}.m(i)$,
- $\mu_i(x) \le \mathcal{R}.card(\mathcal{R}.Z_i^{\lambda_i(x)})$
- *if* $i \le h$ *then* $\forall 0 < k < \mu_i(x)$, $\exists x' < x$ *such that* $\lambda_i(x') = \lambda_i(x) \wedge \mu_i(x') = k$.

$\lambda_i(x)$ is used to choose the subclasses of $\hat{C}_i$ to be instantiated. In case of non ordered classes C_i, $\mu_i(x)$ is used to distinguish already instantiated elements from the other elements within the subclass selected by $\lambda_i(x)$. The additional conditions on $\mu_i(x)$ guarantee that the functions λ and μ define a partition of the ordinary arcs of the reachability graph in symbolic arcs (an ordinary arc cannot satisfy two different pairs $[\lambda, \mu]$ simultaneously).
We denote $\mu_i^j = \sup(\mu_i(x) \mid \lambda_i(x) = j)$ the number of different instantiations in the dynamic subclass Z_i^j. For any Z_i^j that is not instantiated (i.e.: $\nexists x$ such that $\lambda_i(x) = j$), then by definition $\mu_i^j = 0$.

In order to separate the objects that will be selected for the firing from the other objects represented by a same dynamic subclass we introduce the notion of *split symbolic marking.*

Definition 3.2 (Splitting) *Let $\mathcal{R}$ be a representation of a symbolic marking $\mathcal{M}$. Then $\mathcal{R}_s = \mathcal{R}[\lambda, \mu]$ is defined by the following transformations on $\mathcal{R}$:* $\mathcal{R}_s.\hat{C}_i = \{Z_i^{j,k}\}$ *with $Z_i^{j,k}$ such that:*

1. $Z_i^j \in \mathcal{R}.\hat{C}_i$
2. **if** $i \le h$ *(C_i is not ordered),* **then**
   either $\exists x$ *such that* $\langle \lambda_i(x), \mu_i(x) \rangle = \langle j, k \rangle$ *(subclasses of instantiated objects)*
   or $k = 0 \wedge \mu_i^j < \mathcal{R}.card(Z_i^j)$ *(subclass of remaining objects)*
3. **if** $i > h$ *(C_i is ordered),* **then** $0 < k \le \mathcal{R}.card(Z_i^j)$
4. $\mathcal{R}_s.card(Z_i^{j,k}) = (\textit{ if } (k = 0) \textit{ then } (\mathcal{R}.card(Z_i^j) - \mu_i^j) \textit{ else } 1\ )$.
5. $\forall u, v, \quad \mathcal{R}_s.mark\left(p, \otimes_{i=1}^{n} \otimes_{e=1}^{e_i} Z_i^{u(i,e),v(i,e)}\right) = \mathcal{R}.mark\left(p, \otimes_{i=1}^{n} \otimes_{e=1}^{e_i} Z_i^{u(i,e)}\right)$
6. $\mathcal{R}_s.d(Z_i^{j,k}) = \mathcal{R}.d(Z_i^j)$

Proposition 3.2 *Let $\mathcal{R}$ be a representation of the symbolic marking $\mathcal{M}$. Then $\mathcal{R}_s = \mathcal{R}[\lambda, \mu]$ is another representation of $\mathcal{M}$.*

With such a definition of the split marking, subclasses can substitute objects in the transition firing.

A *symbolic firing rule* is formally defined in [14] which allows the construction of the SRG directly, without even considering the actual markings of a well-formed net. The symbolic firing is composed of four sequential steps: splitting, actual firing, grouping, and ordering. The first step identifies the instantiation of the transition that is about to fire in terms of dynamic subclasses instead of basic colours. The second step is the actual transformation of the symbolic marking by removing the input tokens and adding the output tokens. The third and fourth steps are the computation of the canonical form of the reached symbolic marking by first obtaining a minimal representation and then transforming it into an ordered one. A complete example of symbolic firing is depicetd in Figure 2. The WN considered in this example is artificial, and has no particular meaning. The top-left portion of the figure depicts the canonical representation of an arbitrary marking $\mathcal{M}$. The other portions represent the four steps of the symbolic firing of transition t for $\langle Z_1^2, Z_2^1 \rangle$ (which is enabled in $\mathcal{M}$, as the reader can check). The details of the algorithms can be found in [14].

The Symbolic firing rule can be easily cast into the usual algorithm structure for the computation of the Reachability Graph of a Petri net (provided that the Symbolic canonical representation is used to store the markings of the reachability set). The resulting general algorithm for the computation of the SRG can thus be outlined as follows:

Given $\mathcal{R}_0$, a symbolic representation of the initial marking
  compute $\mathcal{M}_0$; *(* canonical representation *)*
  $SRS := \{\mathcal{M}_0\}$;

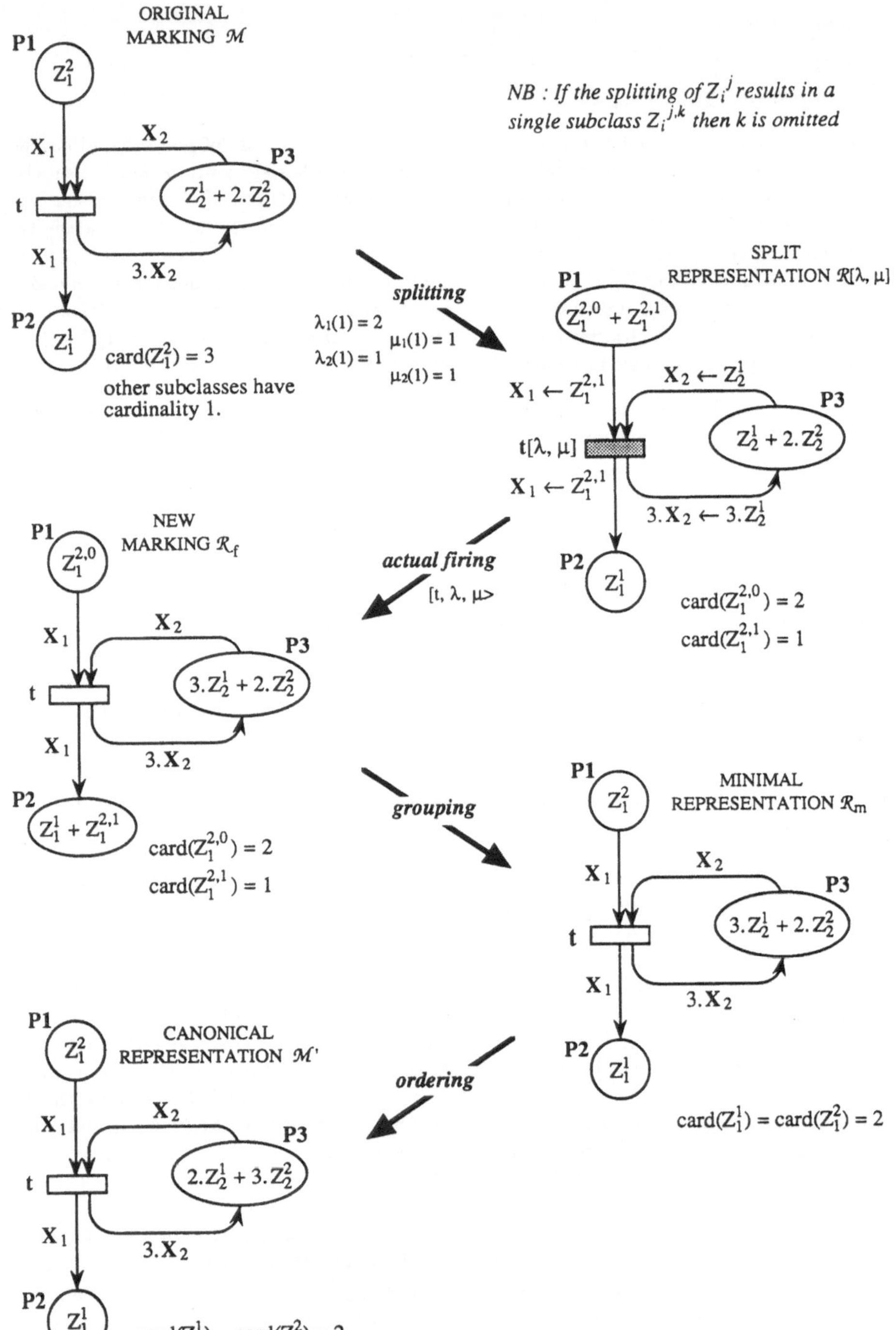

Figure 2 : An example of symbolic firing

```
new_mark := {M_0};
SRG := ∅;
while  new_mark≠ ∅ do
    extract M from new_mark;
    foreach  t ∈ T do
        foreach  [λ,μ] do   (* try symbolic firing *)
            compute R_s = M[λ,μ];   (* splitting *)
            if [t,λ,μ] is enabled in R_s
            then   (* do symbolic firing *)
                foreach  p ∈ P do
                    R_f.mark(p) := R_s.mark.(p) + Ŵ+(p,t)(λ,μ) − Ŵ−(p,t)(λ,μ);
                od;
                R_m := minimal(R_f);
                M' := ordered(R_m);
                if M' ∉ SRS
                then   (* add to SRS *)
                    SRS := SRS ∪ {M'};
                    new_mark := new_mark ∪{M'};
                fi;
                SRG := SRG ∪ {M[t,λ,μ⟩M'}   (* add to SRG *)
            fi
        od
    od
od.
```

3.2 SRG relevant properties

We present now the most interesting properties of the symbolic reachability graph that can be exploited for a performance evaluation of WN models. Other interesting properties can be shown concerning the qualitative behaviour of WN models based on the analysis of the SRG, but they are not reported here for the sake of conciseness.

The first two properties that we consider establish the equivalence between the RG and the SRG from the point of view of the reachability of markings.

Property 3.1 *Equivalence between symbolic and ordinary reachability.*
Let $\mathcal{M}_0$ be a symbolic marking, and $[\mathcal{M}_0\rangle$ be the set of symbolic markings reachable from $\mathcal{M}_0$. Then

$$\bigcup_{M_0 \in \mathcal{M}_0} [M_0\rangle \quad = \quad [\mathcal{M}_0\rangle$$

Property 3.2 *Cardinality of a symbolic marking.*
Let $\mathcal{M}$ be a symbolic marking and $|\mathcal{M}|$ be the number of ordinary markings belonging to the equivalence class of $\mathcal{M}$.
Let $K(\mathcal{M})$ be the number of permutations computed during the ordering phase of the symbolic firing algorithm. Then

$$|\mathcal{M}| \quad = \quad \frac{1}{K(\mathcal{M})} \left(\prod_{i=1}^{h} \prod_{q=1}^{n_i} \frac{|D_{i,q}|!}{\prod_{d(\mathcal{M}.Z_i^j)=q} card(\mathcal{M}.Z_i^j)!} \right) \prod_{i=h+1}^{n} \nu(i)$$

where $\nu(i) =$ if $(\mathcal{M}.m(i) > 1 \wedge n_i = 1)$ then $|C_i|$ else 1.

The above properties ensure that no information on reachability is lost by analyzing the SRG instead of the RG. However, in order to derive an improved technique for performance evaluation based on the

SRG instead of the RG, we still need to know how to test the ergodicity of the Markov chain and how to compute its transition rates. These additional requirements lead to the formulation of the following additional propositions.

Property 3.3 *Necessary condition for ergodicity.*
Strong connection of RG $\Longrightarrow$ Strong connection of SRG, but not vice-versa.

In [14] an example is given in which the SRG is strongly connected while the underlying RG is not.

Property 3.4 *A sufficient condition for ergodicity.*

Strong connection of SRG
$\wedge$ $\forall 0 < i \leq h$, $\exists \mathcal{M}' \in$ *SRG such that* $(\mathcal{M}'.m(i) = n_i)$
$\wedge$ $\forall h < i \leq n$, $((n_i > 1) \vee (\exists \mathcal{M}' \in$ *SRG such that* $(\mathcal{M}'.m(i) = 1)\ )$
$\Longrightarrow$ *Strong connection of RG*

The above property is an extension of a property already proven in [20] for the restricted case of Regular nets. Indeed, the additional condition on the dynamic subclass partition is always satisfied by the (symmetric) initial marking in an RN. On the other hand, the condition is not necessary in order for the RG to be strongly connected, even in the simpler case of RNs. Indeed, in [14] an example of WN is given in which the SRG is strongly connected, dynamic subclasses are never completely grouped, and yet the RG is strongly connected.

Property 3.5 *Equivalence between symbolic firing and ordinary firing.*
Let $\mathcal{M}$ and $\mathcal{M}'$ be two symbolic markings of the SRG, and $M \in \mathcal{M}$ be an ordinary marking of the RG. Let $A_{M,\mathcal{M}'}$ be the set of arcs of the RG connecting M to any marking $M' \in \mathcal{M}'$, and $A_{\mathcal{M},\mathcal{M}'}$ be the set of symbolic arcs of the SRG connecting $\mathcal{M}$ to $\mathcal{M}'$.
Then there exists a mapping ω from $A_{M,\mathcal{M}'}$ onto $A_{\mathcal{M},\mathcal{M}'}$ such that:

- *if the label of an arc $a \in A_{M,\mathcal{M}'}$ is $[t, \bigotimes_{i=1}^{n} \bigotimes_{j=1}^{e_i} c_i^j]$ then the label of $\omega(a)$ is $[t, \lambda, \mu]$ with $c_i^j \in D_{i,q} \Longleftrightarrow \mathcal{M}.d(Z_i^{\lambda_i(j)}) = q$*
- *if the label of a symbolic arc $a \in A_{\mathcal{M},\mathcal{M}'}$ is $[t, \lambda, \mu]$ then the cardinality of the reciprocal image of a denoted $|\omega^{-1}(a)|$ is*

$$\prod_{i=1}^{h} \prod_{j=1}^{m_i} \frac{card(Z_i^j)!}{(card(Z_i^j) - \mu_i^j)!}$$

where $\mu_i^k = \sup_{x\ :\ \lambda_i(x)=k} \mu_i(x)$

4 Derivation of a lumped Markov chain

Stochastic Regular nets where already defined in [9]. The basic principle for timing a coloured Petri net is to associate a function from the markings to positive real numbers to each arc of the reachability graph. In this way a discrete-state semi-Markov process is defined whose state space is isomorphic to the reachability set of the coloured net. In this section, after a formal definition of stochastic well-formed coloured nets (SWN), we recall the usual Markovian solution technique based on the RG, and then prove that an aggregate Markovian process can be defined based on the SRG in order to compute the same performance estimates with a lower computational cost.

4.1 Stochastic well-formed nets

In WNs a priority structure is defined on transitions. This priority is reflected in the timing semantics of SWNs in the same way as was originally defined for GSPNs [21]. Transitions with priority level 0 are called *timed* transitions, and they fire at the instant of the elapsing of a delay from the instant of the transition enabling. The delay is determined for each instantiation of the enabling of a transition according to a random process with negative exponential probability distribution. Transitions with priority level greater than 0 are called *immediate* transitions, and they fire in zero time at the instant of their enabling. In case of conflicting immediate transitions a firing probability is assigned to each conflicting transition, proportionally to a weight. The probability is computed by normalizing the weights of all conflicting transitions enabled in the same marking.

In order to guarantee the presence of symmetry not only from a logical but also from a stochastic point of view, we restrict the possibility of marking dependency for the mean values of transition firing delays. The average firing rate of a transition can be a function of the static subclass to which the firing basic colour belongs, and not the colour itself. In this way all basic colours of a given static subclass determine the same transition firing delay. This can be formalized by introducing the notation

$$\tilde{C}_i = \{\ D_{i,1}, \ldots, D_{i,q}\ \}$$

In analogy with the notation introduced for the representation of symbolic markings and dynamic subclasses, given a transition t with colour domain $C(t) = C_{J(t)}$ we define

$$\tilde{C}(t) \;=\; \left\{ \bigotimes_{i=1}^{n} \bigotimes_{j=1}^{e_i} D_{i,u(i,j)} \;\mid\; 0 < u(i,j) \leq n_i \right\}$$

For any $c = \bigotimes_{i=1}^{n} \bigotimes_{j=1}^{e_i} c_i^j \in C(t)$ we also define $\tilde{c} = \bigotimes_{i=1}^{n} \bigotimes_{j=1}^{e_i} \tilde{c}_i^j \in \tilde{C}(t)$ such that $\tilde{c}_i^j = D_{i,q}$ iff $c_i^j \in D_{i,q}$. Finally we define the *static partition of a marking M* denoted $\tilde{M}(p) \in Bag(\tilde{C}(p))$ as follows:

$$\tilde{M}(p)(\tilde{c}) \;=\; \sum_{c' \,:\, \tilde{c}'=\tilde{c}} M(p)(c')$$

The static partition of a marking represents, for each place and for each Cartesian product of basic static subclasses, the number of tuples in the place that belong to the same Cartesian product of static subclasses.

Property 4.1 *Static partition of symbolic markings.*

$$\forall M,\ M' \in \mathcal{M},\quad \forall p \in P,\quad \tilde{M}(p) = \tilde{M}'(p)$$

Hence we can define the static partition of a symbolic marking as well and denote it $\tilde{\mathcal{M}}$.

Definition 4.1 (SWN) *A stochastic well-formed coloured net is a pair* $SWN = \langle WN, \theta \rangle$ *such that*

WN is a well-formed coloured Petri net

θ is a function defined on the set of transitions T such that

$$\theta(t) \;:\; \tilde{C}(t) \times \bigotimes_{p \in P} Bag(\tilde{C}(p)) \longrightarrow I\!R^{+}$$

For any timed transition t, the function $\theta(t)(\tilde{c}, \tilde{M})$ represents the average firing rate for any instance of transition $[t, c]$ enabled in marking M. In case of immediate transitions, the same function is interpreted as the weight to be normalized within a conflict set in order to obtain the firing probability:

$$\frac{\theta(t)(\tilde{c}, \tilde{M})}{\sum_{M[t',c'\rangle} \theta(t')(\tilde{c}', \tilde{M})}$$

4.2 RG based Markovian technique

Stochastic Petri nets (without distinction between ordinary or coloured ones) can be used as a convenient modelling tool for the definition of a semi-Markov stochastic process based on the reachability graph of the net. The steady-state probability distribution of this process can be numerically computed by deriving the embedded Markov chain and solving the associated eigenvector equations. We report here a brief algorithmic outline of the technique in the case of coloured Petri nets. Note that in this case the colour structure is not taken into account for the Markovian analysis. Note also that the normal RG accounting for the actual basic colours is considered here, not the SRG.

The embedded chain is defined on the instants of transition firings that determine a change of marking in the net. We assume that all transitions (if any) from a state to itself (corresponding to the case $\forall p \in P$, $W^+(p,t)(c) - W^-(p,t)(c) = 0$) have been previously eliminated from the RG representation. The probability distribution vector $\psi = [\psi_1, \ldots, \psi_k]$, where ψ_l denotes the probability of the l-th state (or marking), is obtained as follows:

1. Construction of the state space description
   - construction of the RG of the net
   - elimination of self-loop arcs
   - for each arc of the RG, computation of $\theta(t)(\tilde{c}, \tilde{M})$
2. Construction of the square matrix $Q = [q_{i,j}]$ of dimension $N = |RS|$

$$\text{if } i \neq j, \quad q_{i,j} = \frac{\sum_{[t,c]\,:\,M_i[t,c\rangle M_j} \theta(t)(\tilde{c}, \tilde{M}_i)}{\sum_{[t,c]\,:\,M_i[t,c\rangle} \theta(t)(\tilde{c}, \tilde{M}_i)}$$

   $q_{i,i} = -\sum_{j \neq i} q_{i,j}$

3. Numerical solution of the system of linear equations

$$\begin{cases} Y.Q = 0 \\ \|Y\| = \sum_{i=1}^{N} Y_i = 1 \end{cases}$$

   The solution is unique (*egodic* model) if the RG contains a unique strongly connected component

4. Computation of the steady-state probability distribution for tangible markings:

$$\psi_i = Y_i . \frac{\frac{1}{\sum_{[t,c]\,:\,M_i[t,c\rangle} \theta(t)(\tilde{c}, \tilde{M}_i)}}{\sum_j \frac{Y_j}{\sum_{[t',c']\,:\,M_j[t',c'\rangle} \theta(t')(\tilde{c}', \tilde{M}_j)}}$$

4.3 SRG based lumped process

We present now an algorithm that exploits the partition in equivalence classes of markings implicitly determined by the computation of the SRG of an SWN in order to reduce the cost of the numerical Markovian analysis.

We denote by SRS the set of all reachable symbolic markings $\mathcal{M}_i$ of the WN. An actual marking of the equivalence class defined by $\mathcal{M}_i$ is denoted here with a double index $M_{i,k}$, where k represents an internal ordering within the equivalence class. The weight function θ is extended to symbolic subclasses as follows: $\forall \mathcal{M}$, $\forall [\lambda, \mu]$, $\forall M \in \mathcal{M}$, $\forall t \in T$, $\forall c \in C(t)$ such that $\eta_i(c_i^j) = Z_i^{\lambda_i(j)}$, $\Theta(t)(\lambda, \mu, \tilde{\mathcal{M}}) = \theta(t)(\tilde{c}, \tilde{M})$. The performance evaluation algorithm can thus be outlined as follows:

1. Construction of the aggregate state space description
   - construction of the SRG of the WN

- elimination of self-loop arcs $(\hat{W}^+(p,t)(\lambda,\mu) - \hat{W}^-(p,t)(\lambda,\mu) = 0)$
- for each arc of the SRG, computation of $\Theta(t)(\lambda,\mu,\tilde{\mathcal{M}})$

2. Construction of the square matrix $\mathcal{Q} = [q_{i,j}]$ of dimension $N' = |SRS|$

$$\text{if } i \neq j, \quad q_{i,j} = \frac{\sum_{[t,\lambda,\mu]\,:\,\mathcal{M}_i[t,\lambda,\mu\rangle\mathcal{M}_j} \Theta(t)(\lambda,\mu,\tilde{\mathcal{M}}_i).|\eta^{-1}(t,\lambda,\mu)|}{\sum_{[t',\lambda',\mu']\,:\,\mathcal{M}_i[t',\lambda',\mu'\rangle} \Theta(t')(\lambda',\mu',\tilde{\mathcal{M}}_i).|\eta^{-1}(t',\lambda',\mu')|}$$

$q_{i,i} = -\sum_{j\neq i} q_{i,j}$

3. Numerical solution of the system of linear equations

$$\begin{cases} \mathcal{Y}.Q = 0 \\ ||\mathcal{Y}|| = \sum_{i=1}^{N} \mathcal{Y}_i = 1 \end{cases}$$

The solution is unique if the SRG contains a single strongly connected component

4. Computation of the steady-state probability distribution for tangible symbolic markings:

$$\Psi_i = \mathcal{Y}_i.\frac{\frac{1}{\sum_{[t,\lambda,\mu]\,:\,\mathcal{M}_i[t,\lambda,\mu\rangle} \Theta(t)(\lambda,\mu,\tilde{\mathcal{M}}_i).|\eta^{-1}(t,\lambda,\mu)|}}{\sum_j \frac{\mathcal{Y}_j}{\sum_{[t',\lambda',\mu']\,:\,\mathcal{M}_j[t',\lambda',\mu'\rangle} \Theta(t')(\lambda',\mu',\tilde{\mathcal{M}}_j).|\eta^{-1}(t',\lambda',\mu')|}}$$

From the probability distribution Ψ of the tangible symbolic markings it is always possible, if needed, to compute the probability distribution ψ of the tangible actual markings of the SWN as:

$$\forall M_j \in \mathcal{M}_i, \quad \psi_j = \frac{\Psi_i}{|\mathcal{M}_i|}$$

Note that usually the actual marking distribution is not needed for the computation of performance indexes defined at the net level and that in any case the complexity of the above outlined numerical analysis is polynomial in the cardinality of the SRG instead of the cardinality of the RG. Thus, the proposed techniques exploits at its greatest extent the colour structure of the model in order to reduce the size of the state space of the stochastic process to be analyzed.

4.4 Proof of lumpability

Now we prove that the analysis of the stochastic process defined on the SRG yields the same steady-state solution that can be computed from the general technique based on the RG. The proof is divided in three steps.

1. The embedded Markov chain is strongly lumpable [15]. [1] Hence it is possible to derive a reduced linear system whose variables are the symbolic marking probabilities.
2. The coefficients of the reduced linear system can be computed directly from the SRG.
3. All actual markings within a symbolic marking have the same probability.

Note that properties 3.3 and 3.4 give two partial characterizations of the ergodicity of the model in terms of the strong connection of the SRG. In particular, they provide respectively a necessary and a sufficient ergodicity condition that can be checked directly on the SRG without construction of the RG.

[1] Remember that the strong lumpability condition for exact aggregation of states of a Markov chain requires that for any two states s_1, s_2 in every aggregate A_1, the sum of the transition rates from s_1 to any given aggregate A_2 is equal to the sum of the transition rates from s_2 to the same aggregate A_2.

4.4.1 Lumpability

Because of the symmetry of the weighting function θ and the fact that the firing relation is preserved by applying a permutation both to the markings and to the transition instantiation, we can show the following property.

Proposition 4.1 *The probability of going from $M_{i,k}$ to $M_{j,q}$ is the same as the probability of going from $M_{i,s.k}$ to $M_{j,s.q}$, i.e.:*

$$q_{(i,k)(j,q)} = q_{(i,s.k)(j,s.q)}$$

As a consequence, it is easy to show that the lumping condition is satisfied.

Proposition 4.2 *The embedded Markov chain verifies the lumping condition [15], i.e.:*

$$\forall \mathcal{M}_j, \ \ \forall M_{i,k}, \ \ \forall s \in \xi, \qquad \sum_{M_{j,q} \in \mathcal{M}_j} q_{(i,k)(j,q)} = \sum_{M_{j,q} \in \mathcal{M}_j} q_{(i,s.k)(j,q)}$$

From the lumping theory we know that a reduced system can be computed whose variables are the symbolic marking probabilities and whose coefficients are: $\sum_{M_{j,q} \in \mathcal{M}_j} q_{(i,k)(j,q)}$. We show now that these coefficients can be computed using the information contained in the SRG alone.

4.4.2 Computation of the coefficients

Using the values given by the RG based algorithm one obtains:

$$\sum_{M_{j,q} \in \mathcal{M}_j} q_{(i,k)(j,q)} = \frac{\sum_{M_{j,q} \in \mathcal{M}_j} \sum_{[t,c] \ : \ M_{i,k}[t,c\rangle M_{j,q}} \theta(t)(\tilde{c}, \tilde{M}_{i,k})}{\sum_{[t,c] \ : \ M_{i,k}[t,c\rangle} \theta(t)(\tilde{c}, \tilde{M}_{i,k})}$$

We now state two propositions showing that this expression can be computed from the SRG information.

Proposition 4.3

$$\sum_{M_{j,q} \in \mathcal{M}_j} \ \sum_{[t,c] \ : \ M_{i,k}[t,c\rangle M_{j,q}} \theta(t)(\tilde{c}, \tilde{M}_{i,k}) = \sum_{[t,\lambda,\mu] \ : \ \mathcal{M}_i[t,\lambda,\mu\rangle \mathcal{M}_j} \Theta(t)(\lambda, \mu, \tilde{\mathcal{M}}_i).|\eta^{-1}(t, \lambda, \mu)|$$

Proposition 4.4

$$\sum_{[t,c] \ : \ M_{i,k}[t,c\rangle} \theta(t)(\tilde{c}, \tilde{M}_{i,k}) = \sum_{[t,\lambda,\mu] \ : \ \mathcal{M}_i[t,\lambda,\mu\rangle} \Theta(t)(\lambda, \mu, \tilde{\mathcal{M}}_i).|\eta^{-1}(t, \lambda, \mu)|$$

The proof of the two propositions is similar. As stated in Property 3.5 in Section 3.2, the ordinary arcs between $M_{i,k}$ and any marking of the equivalence class $\mathcal{M}_j$ can be partitioned according to their image by η. Hence we have

$$\sum_{M_{j,q} \in \mathcal{M}_j} \ \sum_{[t,c] \ : \ M_{i,k}[t,c\rangle M_{j,q}} \theta(t)(\tilde{c}, \tilde{M}_{i,k}) = \sum_{[t,\lambda,\mu] \ : \ \mathcal{M}_i[t,\lambda,\mu\rangle} \ \sum_{\eta^{-1}(\lambda,\mu)=[t,c]} \Theta(t)(\lambda, \mu, \tilde{\mathcal{M}}_i)$$

$$= \sum_{[t,\lambda,\mu] \ : \ \mathcal{M}_i[t,\lambda,\mu\rangle \mathcal{M}_j} \Theta(t)(\lambda, \mu, \tilde{\mathcal{M}}_i).|\eta^{-1}(t, \lambda, \mu)|$$

The coefficients of the system computed using the SRG based algorithm are the same as the coefficients of the system obtained when lumping the Markov chain built on the RG. The SRG based algorithm allows one to derive the symbolic marking probabilities. With the additional property that all the markings within a symbolic marking have the same probability it is immediate to derive the ordinary marking probabilities.

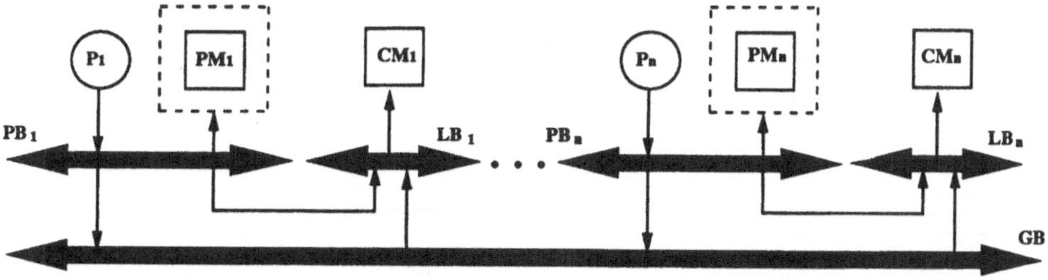

Figure 3: A multiprocessor architecture

4.4.3 Equiprobability

Proposition 4.5

$$\forall \mathcal{M}_i, \quad \forall M_{j,q}, \quad \forall s \in \xi, \qquad \sum_{M_{i,k} \in \mathcal{M}_i} q_{(i,k)(j,q)} = \sum_{M_{i,k} \in \mathcal{M}_i} q_{(i,k)(j,s.q)}$$

This proposition, together with Proposition 4.1, proves that all the actual markings within a symbolic marking have the same input rate from any symbolic marking and the same outptut rate to any symbolic marking. It is then easy to prove that they have the same probability.

5 A Complex Application to Multiprocessor Architectures

In this section we apply the SWN formalism to the modelling and performance evaluation of a multiprocessor system already studied in [22,2,6]. The reasons for reconsidering this almost classic architecture are many. First, it is intrinsically more complex and more realistic than other systems already studied by coloured or high-level Petri nets, such as, e.g., [8,10]. Second, the solutions based on Markov or Stochastic Petri net techniques presented in [22,2] are not satisfactory due to their inherent complexity in terms of increasing of the number of states as a function of the number of processors considered in the system, that limited the availability of results to very few processors (≤ 3). Third, the solution based on folded GSPN model presented in [6], although much better from the computational complexity point of view was still not satisfactory from a modelling point of view due to the complexity of the GSPN model itself, whose construction required a great deal of ingenuity and a deep understanding of the behavioural symmetries of the system.

The aim of this example is twofold. On one hand we show the considerable difference in size of the (aggregated) Markov chain directly obtained from the SRG with respect to the size of the Markov chain obtained from the detailed RG, which is equivalent to those presented in [22,2]. On the other hand we compare the aggregations automatically performed in the SRG generation phase with those devised in [6] at the price of a thorough (human) analysis of the system behaviour and find that the same level of lumping is now achieved without any effort on the part of the modeller. Indeed the gain in space enlarges the class of numerically solvable models while the automatic aggregation procedure could itself suggest to the modeller the interesting behavioural symmetry properties of the system [23].

The multiprocessor architecture analyzed in this example is depicted in Figure 3. Each processor p_i is associated with a local memory composed of two sections, a private one (PM_i) and a common one (CM_i). Private memory areas (PM) can be accessed only by the corresponding processor through its private bus (PB). Common memory areas (CM) are accessible to all the processors in the system. Accesses to the local module of the CM are performed through the private bus plus the local bus (LB), while accesses to non local CM modules are performed using the global bus (GB) plus the local bus of the destination CM module.

Contention arises in the use of GB as well as of the local busses (and the CM modules). A processor is delayed when some of the resources it is trying to access are busy. We assume however that external access requests to CM modules have priority over the local CM accesses and cause their preemption.

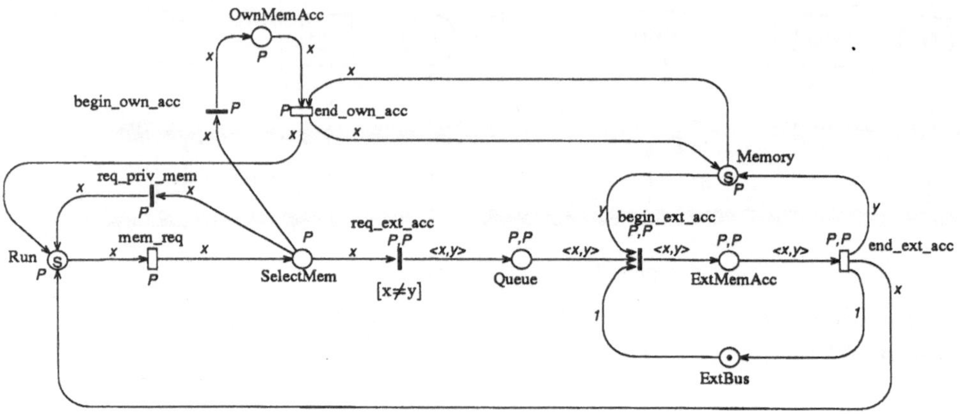

a) An intuitive WN model.

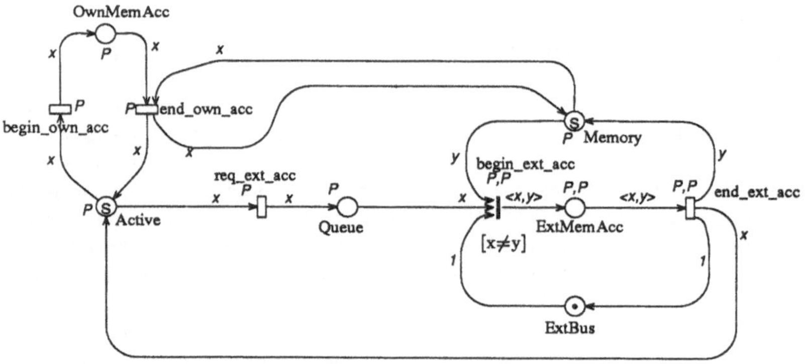

b) A more compact WN model.

Figure 4: Two WN models of the multiprocessor architecture.

The overall behaviour of the system can be described as follows: processors alternate periods of processing requiring only access to the private memory (we call these periods CPU bursts), with periods of CM modules accesses. For simplicity we assume that the system is made up of n identical processor-memory modules.

In order to build the WN model of the system it can be useful to classify the possible states of a processor as follows:

Active Processor executing in its private memory;

Accessing local CM Processor performing a local CM module access;

Accessing remote CM Processor performing an external CM module access;

Queued Waiting for the GB to become available;

Blocked Waiting to continue a local CM access preempted by an external access.

The behaviour of the system is described in a straightforward manner by the WN model in Figure 4.a). Note that in Figure 4.a) we have used a simplified notation with respect to the formal definition of WNs given in Section 2, where various instances of the projection function near the same transition are denoted x, y, z, ... instead of X_i^j. Moreover, since there is only one basic colour set in this model representing processor identities, we called it 'P' instead of C_1; its Cartesian square is denoted 'P, P' instead of $C_1 \otimes C_1$.

Places represent the possible states of each processor as follows. 'Run' contains tokens whose color represent the identity of processors in the "Active" state. Similarly, places 'ExtMemAcc' and 'Queue' represent processors in the "remote access" and in the "Queued" states, respectively. Place 'OwnMemAcc' can represent either state "local access" or "Blocked", depending on the fact that a token of the same colour is or is not present in place 'Memory'. A probabilistic choice among private, local, or external memory access is modelled by the three conflicting immediate transitions 'req_priv_mem', 'begin_own_acc', and 'req_ext_acc'. In the latter case, the choice of the external memory is represented by the choice of the second component (y) of the colour domain of transition 'req_ext_mem' (which is enabled for any $y \neq x$).

This WN representation of the system, although correct and quite easy to understand, is not the simplest one can draw. For example, the three conflicting immediate transitions can be "agglomerated" into their preceeding timed transition 'mem_req' by applying a well known net reduction rule that preserve all behavioural properties of the net [24] as well as the timing semantics of the model. Moreover, the timed transition resulting from the fusion of transitions 'mem_req' and 'req_priv_mem' in the WN in Figure 4.a) can be deleted since its firing determines no change in the marking of the net. Finally, the WN representation can be further simplified by "delaying" the choice of the external memory to be accessed until the global bus is available: indeed, the information about the identity of the memory module to be accessed is useless before the GB access is granted to the processor. Notice that this further simplifications implies a change of the colour domain of place 'Queue', which becomes of type 'P' instead of 'P, P'.

By applying all the above discussed simplifications, one can finally draw the more compact WN model depicted in Figure 4.b). The possible process states are represented in the WN model in Figure 4.b) by the places:

Active this place of domain P contains tokens representing the active processors;

Queue this place of domain P contains token representing the processors in *Queued* state;

ExtMemAcc this place, of domain $P \otimes P$, can contain pairs $\langle processor, CMmodule \rangle$ representing the external accesses currently going on (of course, if there is only one GB, there cannot exist more than one token at a time in this place);

OwnMemAcc this place, of domain P, contains both processors in *Blocked* and in *Access local CM* states, the former (latter) being characterized by the absence (presence) of the corresponding identifier in place *Memory*;

Memory this place, of domain P, contains tokens corresponding to the CM modules not used by any external processor;

ExtBus this place, of neutral domain, describes the GB state: empty for GB busy, filled with one token for GB idle.

The transitions represent the possible state changes; here is a list of the transitions semantics. Transitions *begin_own_acc*, *end_own_acc*, and *req_ext_acc*, have domain P. Transition *begin_ext_acc* has domain P, P and has an associated predicate.[2] The changes in the state of processors modelled by the different transition firings are resumed in the following list:

[2]The reason we used the predicate was that this transition represents the beginning of an *external* access of processor x to the CM module y. Because of the predicate only firing instances with $x \neq y$ are enabled. Indeed $x = y$ would represent a *local* access.

begin_own_acc(Proc1)

$Active_{(f)} \rightarrow AccessinglocalCM$
$Active_{(b)} \rightarrow Blocked$

end_own_acc(Proc1)

$AccessinglocalCM \rightarrow Active_{(f)}$
$Blocked \rightarrow Active_{(b)}$

req_ext_acc(Proc1)

$Active_{(i)} \rightarrow Queued(i)$

begin_ext_acc(Proc1,Proc2)

Proc1:

$Queued_{(i)} \rightarrow AccessingremoteCM$

Proc2:

$Active_{(f)} \rightarrow Active_{(b)}$
$Queued_{(f)} \rightarrow Queued_{(b)}$
$AccessinglocalCM \rightarrow Blocked$

end_ext_acc(Proc1,Proc2)

Proc1:

$AccessingremoteCM \rightarrow Active_{(f)}$

Proc2:

$Active_{(b)} \rightarrow Active_{(f)}$
$Queued_{(b)} \rightarrow Queued_{(f)}$
$Blocked \rightarrow AccessinglocalCM$

Note that in the above list the *Active* and *Queued* states are displayed with a subscript. This is to make a distinction between processors in the *Active* (or *Queued*) state whose local CM module is not being accessed (free, with subscript "f") and those whose local CM module is being accessed (busy, with subscript "b"). Note also that the list is introduced here only for the sake of a further explanation of the model and that it adds nothing to the information already contained in the WN model in Figure 4.b).

Transition *begin_ext_acc* is immediate because we assume that both arbitration and release time of busses are negligible (and thus set to zero). Furthermore we have assumed that both CPU bursts and CM access periods are independent, exponentially distributed, random variables. Finally, external access requests from each processor are directed to any non local CM module with probability $\frac{1}{n-1}$; this is represented in the model by stating that all the possible (conflicting) instantiations of transition *begin_ext_acc* have equal probability. The parameters of our model are the number n of processor-memory modules, the average CPU burst length $(1/\lambda)$, and the average external accesses duration $(1/\mu)$. As in [22,2,6], we define the system load factor ρ as the ratio λ/μ. Some performance figures that can be obtained from our model are the average number of active processors ($E[AP]$) and the GB utilization ($U[GB]$). These performance figures can be computed from the steady state probability of the symbolic markings using the following formulas:

- $E[AP] \;=\; \sum_{\mathcal{M} \in tang(SRG)} \delta(\mathcal{M}) \; \#Active$
  where $\delta(\mathcal{M})$ is the steady state probability of the symbolic marking $\mathcal{M}$, while $\#Active$ is the number of elements in place 'Active' defined as

$$\#Active = \sum_{<Z_1^j> \in \mathcal{M}(Active)} card(Z_1^j)$$

- $U[GB] \;=\; \sum_{\mathcal{M} \in tang(SRG): \mathcal{M}(GB)=0} \delta(\mathcal{M})$

n	ρ	# tang. SRG	# tang. RG	$E(AP)/n$	$U(GB)$
2	0.2	6	10	0.6752411	0.27009645
2	0.5			0.4285714	0.42857145
2	1.0			0.2608695	0.52173917
3	0.2	13	62	0.6600941	0.39605642
3	0.5			0.4045638	0.60684595
3	1.0			0.2399574	0.71987262
4	0.2	23	340	0.642929	0.51434340
4	0.5			0.3765453	0.75309119
4	1.0			0.2149914	0.85996634
5	0.2	36	1652	0.6227463	0.62274684
5	0.5			0.3444203	0.86105182
5	1.0			0.1882871	0.94143486
6	0.2	52	7354	0.5991253	0.71895120
6	0.5			0.31032	0.93095914
6	1.0			0.1632581	0.97954751
7	0.2	71	30746	0.5719982	0.80079870
7	0.5			0.2771187	0.96991400
7	1.0			0.1419933	0.99395176
8	0.2	93	122728	0.5417373	0.86678098
8	0.5			0.2471198	0.98847777
8	1.0			0.1248069	0.99845428
9	0.2	118	472904	0.5092124	0.91658070
9	0.5			0.2213536	0.99608951
9	1.0			0.1110726	0.99965196
10	0.2	146	1772494	0.475696106	0.95139024
10	0.5			0.199762593	0.99881109
10	1.0			0.099993149	0.99992986

Table 1: Results

In Table 1 the performance figures computed for different values of the number of processors and of the load factor are shown; the mean number of active processors is here divided by n in order to obtain a normalized version of this figure. These results are also displayed in graphic form in Figure 5, the leftmost and rightmost graphics display the values of $U[GB]$ and $E[AP]/n$ respectively, as a function of n for three values of ρ.

In Table 1 the size of the aggregate and complete Markov chains are also reported for different values of n, to give a flavour of the considerable gain that can be achieved with this method.

Let us compare the automatic aggregations performed by the SRG generation algorithm with those devised in [6]. In that paper the uncoloured GSPN model depicted in Figure 6 was built already having in mind the aggregated state information required. That is, the redundant information of the original (much simpler and easy to build) model was discovered by carefully studying the system behaviour and eliminated by changing the kind of state encoding at the net level. It is interesting to observe that the same redundancies are automatically detected on the simpler WN model by the SRG generation algorithm as can be observed comparing the tangible SRG of the WN model and the tangible RG of the uncoloured model shown in Figure 7. Furthermore notice that the uncoloured model in Figure 6 could be interpreted as the *semi-unfolding* of the WN model based on the possible partition of the basic colour classes (in this case P) in subclasses. This possible partition is obtained from the SRG and in general depends on the cardinality of the colour classes. In this case, however, the partition is invariant for values of $|P|$ greater than two.

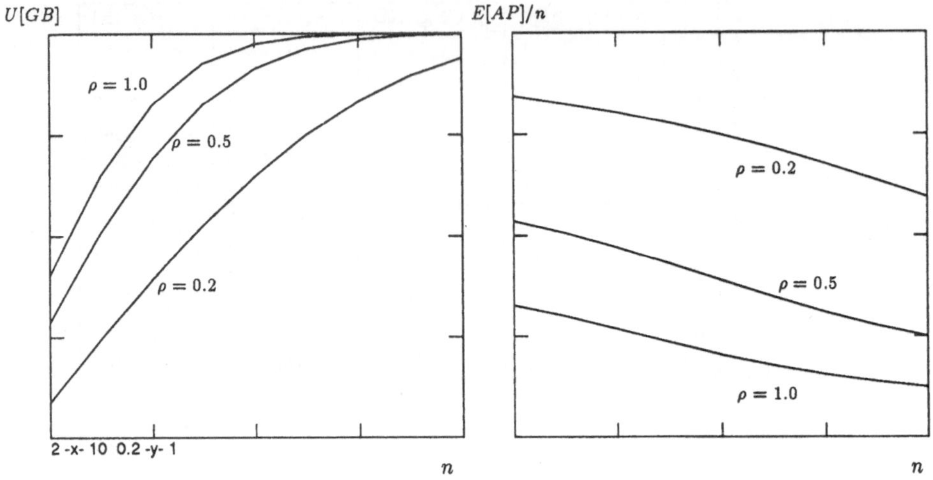

Figure 5: Performance Figures Curves.

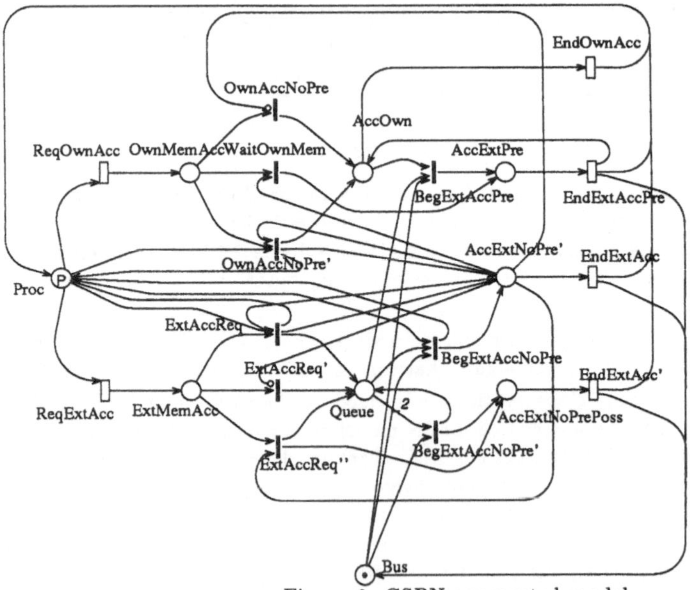

Figure 6: GSPN aggregated model

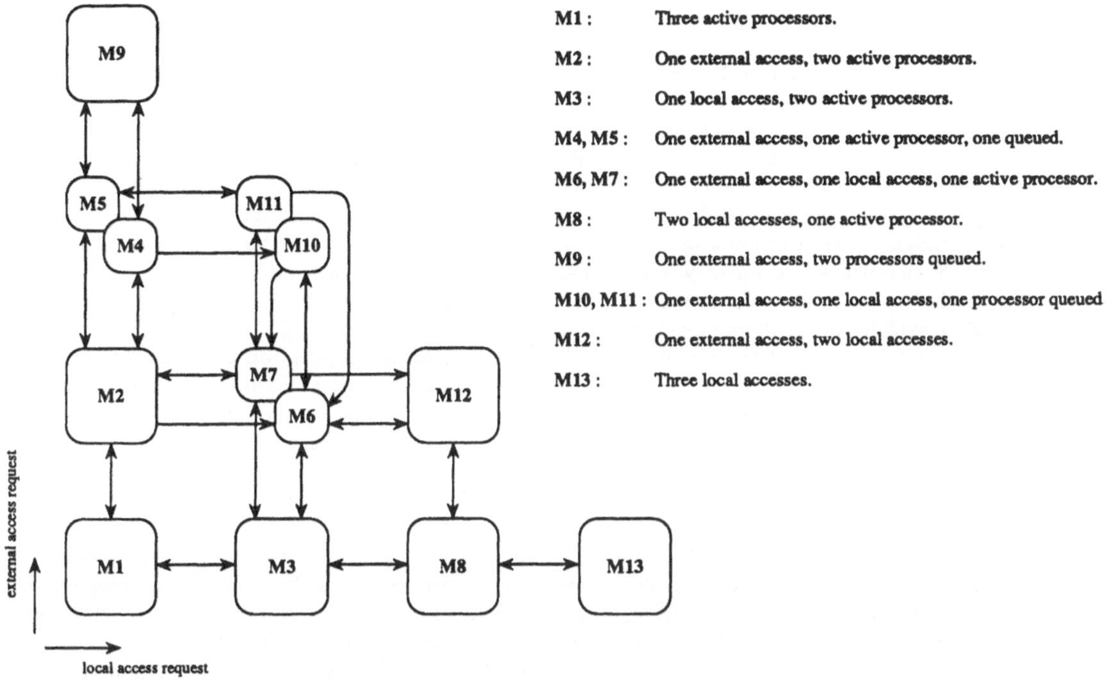

Figure 7: The WN SRG and the GSPN RG

Let us explain better what we mean by "semi-unfolding". First we analyze the SRG in Figure 7 in order to discover which (symbolic) tokens can be observed in each place on the whole reachability set and which (symbolic) instances of each transition can be observed on the SRG arcs. Then we try to find a relation between pairs $\langle place, token(symbolic)colour \rangle$ and the corresponding "unfolded" place (or set of places) in the uncoloured net and between transition firing instances and the corresponding "unfolded" transitions (or transition sequences) in the uncoloured net.

For example, transition *begin_own_acc* can fire for at most two different subclasses. Actually place *Active*, which represents the set of processors that have not yet issued any access requests, can contain two subclasses representing respectively the processors whose local CM module is currently being accessed through the external bus and those whose local CM module is not in use. When the former instance of *begin_own_acc* fires, it represents the beginning of a local memory access which is immediately preempted. This corresponds to the firing of transition *ReqOwnAcc* immediately followed by *WaitOwnMem* in the net in Figure 6. When the latter instance of *begin_own_acc* fires, it represents the beginning of a non-preempted local access. In the net in Figure 6 this corresponds to the firing of transition *ReqOwnAcc* immediately followed either by *OwnAccNoPre* or by *OwnAccNoPre'*. Note that *OwnAccNoPre* is never in conflict with *WaitOwnMem* or *OwnAccNoPre'*, and its enabling conditions correspond to the presence of the latter subclass alone in place *Active* of the net in Figure 6. All other transitions of the GSPN model can be justified with similar arguments in terms of the WN and its SRG.

The relation between the places of the WN model and the uncoloured one is more complex. Place *ExtMemAcc* may contain tokens of three kinds, and thus it is unfolded into the three places *AccExtPre*, *AccExtNoPre'*, and *AccExtNoPrePoss*. Instead, Places *Active* and *Queue* have a one-to-one correspondence with the omonimous places in the uncoloured model even if they can contain two kinds of tokens. The reason is that the information on the partition into two different kinds of the tokens contained in the uncoloured places *Active* and *Queue* is in the marking of places *AccExtNoPre'* and *AccExtNoPrePoss*. Thus we could say that *Active* is unfolded into *Active* and *AccExtNoPre'* while *Queue* is unfolded into *Queue* and *AccExtNoPrePoss*.

6 Conclusions

Stochastic Petri nets were introduced some time ago as a good modelling tool for the performance evaluation of multiprocessor computer systems. They offer a much higher level tool than previously used Markov chains, but they suffer from the same large dimension problem that often hampers the possibility of obtaining results for realistic size systems. High-level Petri net formalisms have already been employed to further reduce the cost of the model definition and validation, and sometimes also to perform lumpings of the underlying Markovian models in order to reduce the cost of the performance evaluation for larger system configurations. Until now, however, it was the total responsibility of the modeller to identify such system characteristics as symmetry in behaviour, in order to exploit the characteristics to obtain models that are less complex to analyze. For the first time a completely algorithmic approach is proposed that allows the exploitation of these model characteristics for the construction of lumped Markov chains. The user need not be aware of the lumping technique, as long as he specifies the behaviour of the system in terms of well-formed coloured nets. These intrinsic symmetries are automatically detected and used at their greatest extent by means of the construction of the *symbolic reachability graph*. The technique presented in this paper is a direct extension of that already proposed for a restricted class of "regular nets". The novelty of the result presented here is that, with the extension of the SRG technique to well-formed coloured nets, the technique is now applicable to any coloured model.

We have therefore presented an example of use of stochastic well-formed nets for the study of a non trivial multiprocessor architecture. The example shows how the technique overcomes the difficulties found in the use of other performance modelling techniques. On the one hand the model is very easy to devise in terms of the WN formalism, and fairly easy to understand even by non Petri net experts. On the other hand, the exploitation of the SRG construction algorithm allows an evaluation whose cost is comparable to the analysis cost of thorougly designed GSPN models where all crucial behavioural symmetries of the system have been fully identified by the modeller himself.

Finally, notice that the method described in this paper allows to reduce the number of states in the RG, and thus in the corresponding Markov Chain, exploiting the intrinsic symmetries of the model that produce many different portions of a graph representing essentially the same behaviour, while it does not attack another cause of state space grouth, namely the representation of parallelism in terms of all possible interleavings that can be observed on the global state of the model. This problem is inherent to the state-space representation, and in our opinion can be avoided only by using non state-space based analysis techniques. The initial simplification of the WN model in Figure 4.a) in order to obtain the one in Figure 4.b) could be an example of use of formal net structural reduction techniques for the reduction of the cardinality of the state space. Work is currently in progress in order to formalize these reduction techniques in the performance evaluation framework provided by SWNs.

References

[1] M. K. Molloy. Performance analysis using stochastic Petri nets. *IEEE Transaction on Computers*, 31(9):913–917, September 1982.

[2] M. Ajmone Marsan, G. Balbo, G. Conte, and F. Gregoretti. Modeling bus contention and memory interference in a multiprocessor system. *IEEE Transactions on Computers*, 32(1):60–72, January 1983.

[3] M. Ajmone Marsan, G. Balbo, and G. Conte. A class of generalized stochastic Petri nets for the performance analysis of multiprocessor systems. *ACM Transactions on Computer Systems*, 2(1), May 1984.

[4] M. Ajmone Marsan, G. Chiola, and G. Conte. Generalized stochastic Petri net models of multiprocessors with cache memories. In *Proc. 1st Int. Conf. on Supercomputing Systems*, IEEE, St. Petersburg, FL, December 1985.

[5] M. Ajmone Marsan, G. Balbo, and G. Conte. *Performance Models of Multiprocessor Systems.* MIT Press, Cambridge, USA, 1986.

[6] M. Ajmone Marsan and G. Chiola. Construction of generalized stochastic Petri net models of bus oriented multiprocessor systems by stepwise refinements. In *Proc.* 2nd *Int. Conf. on Modeling Techniques and Tools for Performance Analysis*, ACM, Sophia Antipolis, France, June 1985.

[7] M. Ajmone Marsan, G. Balbo, G. Chiola, and G. Conte. Modeling the software architecture of a prototype parallel machine. In *Proc. 1987 SIGMETRICS Conference*, ACM, Banf, Alberta, Canada, May 1987.

[8] Chuang Lin and C. Marinescu. Stochastic high level Petri nets and applications. *IEEE Transactions on Computers*, 37(7):815–825, July 1988.

[9] C. Dutheillet and S. Haddad. Regular stochastic Petri nets. In *Proc. 10th Intern. Conf. Application and Theory of Petri Nets*, Bonn, Germany, June 1989.

[10] C. Dutheillet and S. Haddad. Aggregation and disaggregation of states in colored stochastic Petri nets: application to a multiprocessor architecture. In *Proc. 3rd Intern. Workshop on Petri Nets and Performance Models*, IEEE-CS Press, Kyoto, Japan, December 1989.

[11] P. Huber, A.M. Jensen, L.O. Jepsen, and K. Jensen. Towards reachability trees for high-level Petri nets. In G. Rozenberg, editor, *Advances on Petri Nets '84*, pages 215–233, Springer Verlag, 1984.

[12] A. Zenie. Colored stochastic Petri nets. In *Proc. Int. Workshop on Timed Petri Nets*, pages 262–271, IEEE-CS Press, Torino, Italy, July 1985.

[13] J. A. Carrasco. Automated construction of compound Markov chains from generalized stochastic high-level Petri nets. In *Proc. 3rd Intern. Workshop on Petri Nets and Performance Models*, pages 93–102, IEEE-CS Press, Kyoto, Japan, December 1989.

[14] G. Chiola, C. Dutheillet, G. Franceschinis, and S. Haddad. On well-formed coloured nets and their symbolic reachability graph. In *Proc.* 11th *International Conference on Application and Theory of Petri Nets*, Paris, France, June 1990.

[15] J.G. Kemeni and J.L. Snell. *Finite Markov Chains.* Van Nostrand, Princeton, NJ, 1960.

[16] Claude Dutheillet. *Symétries dans les réseaux colorés: Définition, analyse et application à l'évaluation de performances.* PhD thesis, Laboratoire MASI, Université Paris 6, France, January 1991. thèse de l'Université P. et M. Curie (in French).

[17] K. Jensen. Coloured Petri nets and the invariant method. *Theoretical Computer Science*, 14:317–336, 1981.

[18] G. Chiola, S. Donatelli, and G. Franceschinis. *Priorities, Inhibitor Arcs, and Concurrency in P/T nets.* Technical Report, Dipartimento di Informatica, Università di Torino, Corso Svizzera 185, 10149 Torino, Italy, January 1991. (submitted to ATPN'91 Aarhus).

[19] G. Chiola, S. Donatelli, and G. Franceschinis. *On Parametric P/T nets and their Modelling Power.* Technical Report, Dipartimento di Informatica, Università di Torino, Corso Svizzera 185, 10149 Torino, Italy, January 1991. (submitted to ATPN'91 Aarhus).

[20] S. Haddad. *Une Categorie Regulier de Reseau de Petri de Haut Niveau: Definition, Proprietes et Reductions.* PhD thesis, Lab. MASI, Universite P. et M. Curie (Paris 6), Paris, France, Oct 1987. These de Doctorat, RR87/197 (in French).

[21] M. Ajmone Marsan, G. Balbo, G. Chiola, and G. Conte. Generalized stochastic Petri nets revisited: random switches and priorities. In *Proc. Int. Workshop on Petri Nets and Performance Models*, pages 44–53, IEEE-CS Press, Madison, WI, USA, August 1987.

[22] M. Ajmone Marsan, G. Balbo, and G. Conte. Comparative performance analysis of single bus multiprocessor architectures. *IEEE Transactions on Computers*, 31(12), December 1982.

[23] G. Chiola and G. Franceschinis. Colored GSPN models and automatic symmetry detection. In *Proc. 3rd Intern. Workshop on Petri Nets and Performance Models*, IEEE-CS Press, Kyoto, Japan, December 1989.

[24] S. Haddad. Generalization of reduction theory to coloured nets. In *Proc.* 9th *Europ. Workshop on Application and Theory of Petri Nets*, Venezia, Italy, June 1988.

Section H
Application of High-level Nets

High-level Petri nets have been used in a large number of practical projects—many of these in an industrial setting. Unfortunately, only a small number of these practical experiences have been documented in the Petri net literature.

This section contains eight papers describing different applications of high-level nets. The first paper deals with a concurrent algorithm for mutual exclusion. The next three papers deal with different kinds of protocols. The fifth paper deals with flexible manufacturing systems, while the sixth paper describes a development methodology which can be used for a large range of different systems. Finally the last two papers show how hierarchical high-level nets can be used to model and analyse an electronic funds transfer system and a VLSI chip. The eight papers are totally independent and they can be read in any order.

The number of industrial applications of high-level Petri nets is rapidly growing—this goes together with the development of the theory, and with the development of suitable computer tools supporting the use of high-level Petri nets for large scale modelling and analysis. It is very desirable that the Petri net community learns about applications, and thus we encourage participants of practical projects to report their experiences. This can be done by means of journal and conference papers, but also through the poster sessions of the Petri net conferences, and through technical notes published in the Petri Net Newsletter (cf. Appendix).

20.
An Example of Modelling and Evaluation of a Concurrent Program Using Coloured Stochastic Petri Nets: Lamport's Fast Mutual Exclusion Algorithm

G. Balbo, S.C. Bruell, P. Chen and G. Chiola

Submitted to IEEE Transactions on Parallel and Distributed Systems

Abstract

We use a coloured generalized stochastic Petri net (CGSPN) model to study both the correctness and performance of Lamport's concurrent algorithm to solve the mutual exclusion problem on machines lacking an atomic "test and set" instruction. The aim of the paper is to illustrate the potential of the GSPN modelling technique on an academic but non trivial example of distributed system.

Index Terms: stochastic Petri nets, well-formed coloured nets, concurrent algorithms, modelling, correctness proof, performance evaluation, mutual exclusion, liveness.

1 Introduction

Many computational problems are partitionable into a set of concurrently executing sequential processes. The resulting concurrent algorithms (programs) are often difficult to analyze with respect to two important criteria: correctness and performance. Correctness can (sometimes) be established by the construction of a mathematical proof. Performance characteristics can (sometimes) be established by carefully monitoring the execution of the concurrent processes. The two types of studies usually exploit different tools and techniques and are performed separately.

This paper describes a different approach to the study of concurrent algorithms. It uses Generalized Stochastic Petri Nets (GSPNs) [1,2] and their extension to Coloured Generalized Stochastic Petri nets (CGSPNs) [3,4] as a means for constructing realistic models of a concurrent algorithm. A detailed coloured GSPN model is derived directly from the code of the algorithm by applying straightforward statement translation rules. The structural properties of the resulting CGSPN model are used as a first step in proving correctness. The (exact) Markovian analysis of a bounded CGSPN model provides insight into the actual performance of the concurrent algorithm. We use Lamport's fast mutual exclusion algorithm [5] as an illustration of the proposed technique.

The main contribution of this paper is that the correctness **and** the performance of a concurrent algorithm can be studied hand-in-hand by analyzing appropriate GSPN models. The importance of this technique is not that of obviating the need for a formal mathematical proof of correctness. Instead the technique shows that a single model can be used to both make objective statements about the correctness of a concurrent algorithm and to derive indications about its efficiency, with the certainty that the performance figures are obtained from a model that captures the main features of the algorithm. The technique can also be applied when the proof of correctness of an algorithm is difficult (if not impossible) to derive mathematically, as well as when a performance model is difficult to devise and to validate with classical queueing theoretic approaches. Indeed, models obtained with this technique that are not only bounded, but also scaled down in terms of number of basic components, can always be analyzed exactly by straightforward enumeration of the state space, both from a qualitative and quantitative point of view. In addition, the GSPN model (analyzed with an appropriate tool such as that described in [6]) can aid the user in visualizing the many possible execution sequences resulting from statement interleaving and concurrent execution of statements. Being able to actually see the *dynamic* behaviour of a concurrent algorithm provides the implementor with a powerful debugging tool. Finally, the complexity of the model analysis used to study the qualitative behaviour of the algorithm can be kept under control by starting a (formal) reduction process that yields the construction of more compact models that are easier to analyze also from a quantitative point of view, while maintaining a clear relationship with the original "exact" model.

Prior work in the area is divided along the two separate tracks of correctness and performance. Correctness of concurrent programs has been studied formally as in [7,8] and by computer-assisted methods as in [9,10]. Performance has been treated by mathematical analysis (see for example [11]) and by Petri net models as in [12,13]. To the best of our knowledge, no unified studies of correctness and performance of concurrent algorithms have been reported in the literature so far.

The balance of this paper is outlined as follows. Section 2 provides a brief description of Lamport's fast mutual exclusion algorithm. Section 3 then describes the derivation of a well-formed coloured GSPN model from the source code of Lamport's algorithm. The model represents most of the important features of Lamport's algorithm. In Section 4 the original net is reduced to obtain a behaviourally equivalent well-formed CGSPN model that is used for validation purposes. Although the coloured net obtained by direct translation of Lamport's algorithm looks quite simple, its performance evaluation (i.e., the construction of its underlying stochastic process) requires that it be *unfolded.* The size of the unfolded net grows larger as the number of processes vying for the critical section increases and its state space accordingly grows exponentially; the model quickly becomes numerically intractable. Section 5 describes the model evaluation process showing how a sequence of models can be obtained from the original net by first exploiting the symmetries of the problem and then replacing deterministic choices with probabilistic ones. We also summarize our preliminary observations on the performance of Lamport's algorithm in this section. Section 6 shows how different implementation details can be included in the basic model to study the efficiency of the mapping of the algorithm on a given architecture, as well as the effect of minor implementation choices. Finally, Section 7 provides some concluding remarks and directions for future research. The reader interested in the formal definition of well-formed coloured nets may refer to the original papers [4,14] which are reprinted in this book.

2 Description of the Example

Lamport has recently developed a new, efficient mutual exclusion algorithm for shared-memory multiprocessor computers that lack atomic test-and-set instructions [5]. The shared memory is assumed to support atomic read and write operations. Previous algorithms required that each of N processes vying for entry into a critical section inspect N shared variables even when there was no contention [15]. Lamport's algorithm is based on the observation that *if* contention for a critical section occurs rarely, then there is no need for a process to inspect the state of the other processes. The algorithm requires only five writes and two reads to the shared memory *in the absence of contention*, independent of the number of processes in the system. When contention is detected, the algorithm requires the inspection of the state of all N-processes. Figure 1 depicts the code that process i executes under Lamport's algorithm; angle brackets delimit atomic operations and the **await** statement represents a "busy wait".

The code in Figure 1 looks deceptively simple.[1] The complexity (and some of its obscurity) is due to

[1] In the figure we adopted the unusual indenting convention of alligning the **fi** under the corresponding **then** keyword.

```
start: ⟨b[i] := true⟩;
       ⟨x := i⟩;
       if     ⟨y ≠ 0⟩                      then
              ⟨b[i] := false⟩;
              await ⟨y = 0⟩;
              goto start                   fi;
       ⟨y := i⟩;
       if     ⟨x ≠ i⟩                      then
              ⟨b[i] := false⟩;
              for j := 1 to N /* forall j ∈ 1..N */
                     do await ⟨¬b[j]⟩ od;
              if     ⟨y ≠ i⟩                      then
                     await ⟨y = 0⟩;
                     goto start                   fi;
/*            ⟨b[i] := true⟩; redundant */ fi
       Critical Section
       ⟨y := 0⟩;
       ⟨b[i] := false⟩
```

Figure 1: Lamport's Fast Mutual Exclusion Algorithm (Process i's Program); from the original paper. Commented statements are additions to the algorithm presented in the original paper that do not change its behaviour and that are included for convenience.

---

the fact that the code must work no matter how the operations on the various processors are executed. The code uses $N+2$ shared variables: x, y, and $b[1..N]$. The i-th process sets $b[i]$ to *true* to indicate its intention to enter the critical section. The sharing of variables x and y in the sequence "write-x", "read-y", "write-y", "read-x" provides a clever method of implementing (via software) the effect of an atomic "test and set" instruction. Because of the clever manner in which these variables are used, it is difficult to ascribe specific meanings to them. For instance, from a cursory inspection of the code the identity of the process in the critical section seems to be contained in the variable x. Indeed, a process can enter into the critical section if it was the "last process" to have completed writing its process-id into x. Unfortunately, while the process that was the "last" to set x is in the critical section another process can overwrite x. We can thus conclude that x represents only the identity of the last process that has shown its intention of entering the critical section. Similarly, y does not contain the identity of the process in the critical section as the reader can verify (consider what happens if multiple processes pass the **if** $\langle y \neq 0\rangle$ test, prior to any of them setting $\langle y := i\rangle$).

3 Model Construction and Description

Lamport's algorithm is well-suited to be described by a coloured Petri net. We use the formalism of "Well-formed coloured Nets" (WN) proposed in [14] to represent our models. This formalism can be viewed as a particularization of the coloured Petri nets as defined by Jensen in [16], by addition of some syntactic constraints for the specification of colour domains, arc functions, and predicates.

The algorithm makes use of its shared variables in the following manner: the $b[i]$'s can take on the boolean values *true* or *false*; x is either uninitialized or set to a process identifier; and y is initialized to

This is to clarify the structure of the conditional statements. We avoided having the **fi** keyword on a separate line in order to maintain the original convention of having one line per each executable statement of the program.

zero, set to a process identifier, and then reset to zero. Since these shared variables can only assume a finite number of values, we use basic colours to represent these values.

In our model the time for executing statements is represented by timed transitions. When all the input places of a transition have a token, the statement commences execution. Note that several processes may simultaneously be executing the same statement; this is conveniently represented by multiple tokens in the input place of the statement and by associating an "infinite server" policy[2] with the timed transition. The identity of the processor(s) executing the statement is dictated by the colour(s) of the token(s) in the input place. We use "P" to describe the set of colours used to represent processors.

3.1 The Process of Translating Statements into Coloured Net Constructs

Figure 2 contains a semi-formal description of how the statements of Lamport's algorithm are represented in a coloured net. We adopt the usual GSPN notation of drawing timed transitions as rectangular boxes, immediate transitions as bars, and places as circles. Each section refers to a generic statement which is mapped in the corresponding net. An example of instantiation of each generic statement is also provided. What follows briefly describes the contents of this figure.

The program counter positions of processors are represented by places containing tokens of colour "P". Since the assignment statements (:=) of Lamport's algorithm always store some value into shared variables, the net representation of an assignment statement associates distinct places with such variables. In addition, associated with each shared variable place is a colour set "Q" that represents the possible values that the variable can acquire. The execution of an assignment statement simultaneously "consumes" the old value and "deposits" a new value into the shared variable place; hence, an "assignment" transition has a two-dimensional colour set "P,Q", defined by the Cartesian product P×Q. Since the old value is really immaterial for the execution of the assignment statement, the function y that tags the input arc from the shared variable place to the transition is a projection function that matches any token colour in the input place with the second component of the transition colour set. The function x labeling both an input and an output arc of the assignment statement represents the fact that a specific processor has executed the statement. This function associates the colour of any token found in the input place with the first component of the transition colour set. The "v" function denotes the value to be assigned to the shared variable, which in Lamport's algorithm is always either a constant (denoted by an S function in the WN formalism) or the processor identity matched by the function x. Z_q denotes the initial value of the shared variable, and represents one element of the set 'Q' (without specifying which one).

The **if** statement evaluates a boolean expression by inspecting a shared variable place. The type of arc emanating from the shared variable place depends on what condition is actually labeling the arc. The immediate transition labeled 'true' is either enabled or inhibited by the tokens in the shared variable place depending on whether the condition or its negation holds. The colour set of place 'A' represents the domain of the condition function. Z_q represents the initial value of the shared variable. The time for performing the test is accounted for separately on both branches of the conditional statement (because only one of the branches will actually be followed). The decision about which path to follow is made using the value of the shared variable at the *beginning* of the statement execution, so that a subsequent modification of that value has no effect on the execution of the conditional statement.

The **await** statement waits for a certain condition to occur. The condition is tested by evaluating a function applied to the shared variable. Inhibitor arcs are sometimes used to simplify the definition of

[2]The infinite server policy of a transition is easily explained in the case of only one input place for the transition itself. This policy implies that as soon as a token enters the input place, it begins the activity ideally associated with the transition itself. When n tokens are in the input place of a transition this is equivalent to activating the enabling of the transition n times concurrently.

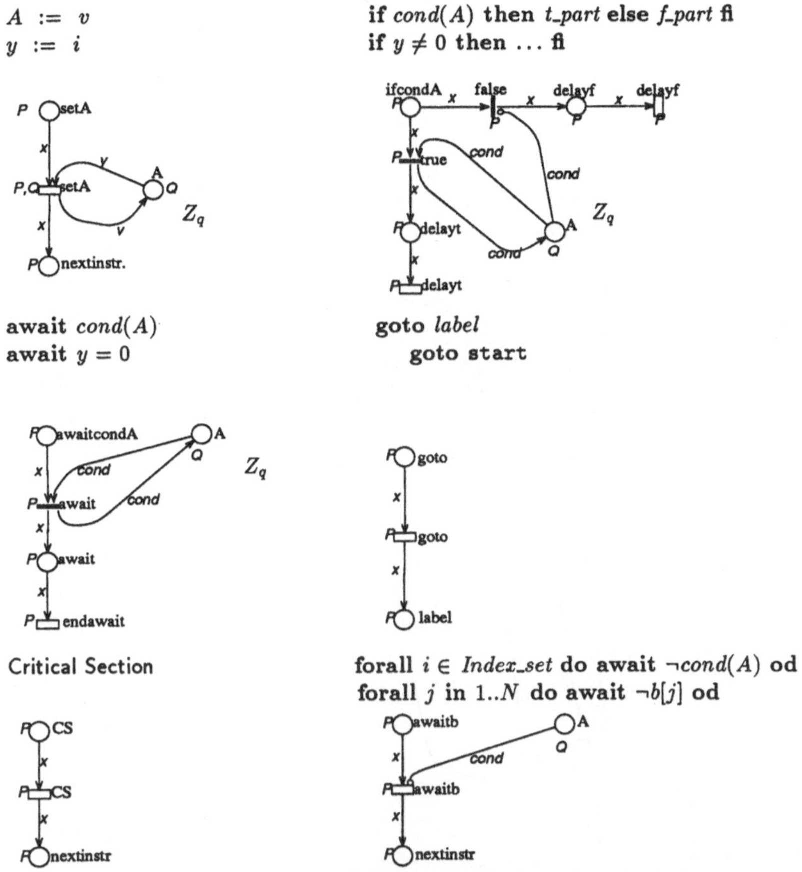

Figure 2: Translation of Statements into Coloured Net Constructs.

the function. The timed transition following the occurrence of the condition only accounts for one access to the shared variable (the one in which the condition becomes satisfied). The use of colour sets, label functions, and initial values, are very similar to those explained for the **if** statement.

The **goto** statement simply transfers control to a specific statement (place). As usual this transfer of control consumes some time.

The **Critical Section** statement is modelled by an input place preceding a timed transition that is used to represent the time a process spends inside it. Exiting the critical section deposits a token in the output place of this transition.

The **for** statement is modelled as a simple conditional delay that ends when all components of the colour set associated with place 'B' are removed from place 'B' itself. The function S_P labelling the inhibitor arc represents the combined condition that holds true whenever place 'B' is not empty. More detailed implementations of the **for** statement together with their impact on the performance of the algorithm are discussed in Section 6.2.

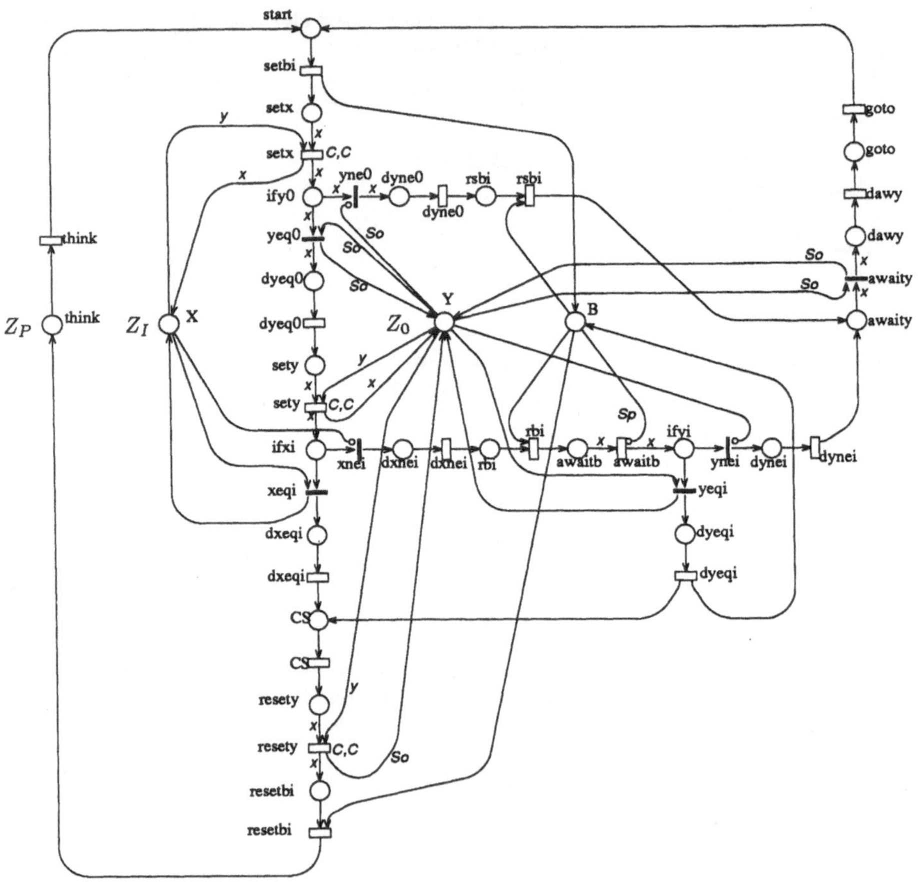

$P = \{ < p_i > \mid 1 \leq i \leq N \}$, $\quad P_0 = \{ < p_0 > \}$, $\quad C = P \cup P_0$, $\quad C,C = C \times C$;
$Z_0 = P_0$, $\quad Z_P = P$, $\quad Z_I \subset P \; : \; |Z_I| = 1$;
$\forall c \quad S_P(c) = \sum_P < p_i >$, $\quad S_0(c) = < p_0 >$

In order to simplify the drawing, C is assumed as the colour set for places and transitions, and x as the default function labelling the arcs, unless explicitly marked otherwise.

Figure 3: The CGSPN Model of Lamport's Algorithm.

3.2 The CGSPN Model of Lamport's Algorithm

Using the procedures outlined in the previous section, we constructed the CGSPN model of Lamport's algorithm that is depicted in Figure 3. The WN formalism [14] used for its representation allows the use of a large number of efficient analysis algorithms to compute qualitative and quantitative characteristics of the model. One of the peculiar features of this formalism is the definition of the initial marking of place 'X' in terms of the "dynamic subclass" Z_I; the meaning of this notation is that place 'X' in the initial marking contains one token of colour $p_i \in P$, but we do not care which one. Instead the initial marking is completely defined for all other places: empty for all places except 'think', that contains one token for each colour in P (initial value Z_P), and 'Y', that contains the $\langle p_0 \rangle$ token (initial value Z_0).

The structure of this net closely follows the structure of the code shown in Figure 1. To simplify the net somewhat, we have chosen not to represent the explicit execution of the **for** loop. Instead we simply wait for all of the $b[i]$'s to become false before proceeding to the next statement. This corresponds to a replacement in the original algorithm of the statement **for** j := 1 **to** N **do await** $\langle\neg b[j]\rangle$ **od** with either the logically equivalent statement: **await** $\langle\neg b[1] \wedge \neg b[2] \wedge \ldots \wedge \neg b[N]\rangle$, or the more abstract statement **forall** j **in** $1..N$ **do await** $\langle\neg b[j]\rangle$ **od** (as shown in the comment).

The presence of the $< p_i >$ token in place 'B' denotes that $b[i] = true$, while its absence signifies $b[i] = false$. Instead of using the translation procedure outlined before for the assignment to numerical variables, the values to the boolean variables $b[i]$ are updated in a simplified manner by neglecting the withdrawal from the corresponding place of the old value, and by assuming that new values of the variables $b[i]$ are always the negations of the previous ones. Of course this in not true in general, but in the case of our algorithm the assignments to the boolean variables $b[i]$ are represented correctly by adding an arc from transition 'dyeqi' to place 'B', which simulates an additional $b[i] := true$ statement in the "else" part[3] of the **if** $\langle y \neq i\rangle$ statement. The added statement (not present in the original paper [5]) is functionally redundant in Lamport's algorithm, in the sense that its absence does not affect the correct execution of the algorithm, but it clarifies the role of the $b[i]$ boolean variable as an indication of the desire for processor i to enter the critical section. On the other hand, having modelled the statement implicitly with a single arc does not change the timed behaviour of the model with respect to the "optimized" implementation without the redundant statement.

As in Lamport's algorithm, the access to the critical section in the *absence of contention* is represented by a straight walk from setting $\langle b[i] := true\rangle$ (place 'start') to setting $\langle b[i] := false\rangle$ (transition 'resetbi'). The labels of the places in the vertical straight line flow path in the CGSPN refer to the seven instructions that are used by Lamport's algorithm ('setbi', 'setx', 'ify0', 'sety', 'ifxi', 'resety', 'resetbi'), except for the upper one, labelled 'start'. The top horizontal flow path is followed by processes that find that the critical section is potentially busy (i.e., other processes have stated their intention to occupy the critical section and a decision has either been made or will be made). The second horizontal flow path is followed by all processes that stated their intention to occupy the critical section before y is set non-zero, but were not the last to set x.

The net in Figure 3 provides a compact high-level representation of Lamport's algorithm shown in Figure 1 and modified as explained before. The colour set C for places and transitions, and the identity function x for arcs are assumed by default and not explicitly shown in the picture (unless needed) to simplify its appearance. This net captures all of the qualitative behaviour of the algorithm because it can reproduce all possible execution sequences. Therefore, this net can be used as an alternative way of studying the formal properties of Lamport's algorithm.

The formal notation used in Figure 3 is an extension of that originally introduced by Haddad in [17], for the definition of "regular nets", a restricted subclass of coloured nets. The advantage of working on a restricted class of coloured nets such as regular nets was that complete and efficient algorithms were proposed for their structural analysis [18] and performance evaluation [3], while for more general cases of coloured nets a preliminary transformation into a place/transition net (unfolding) was required. Very recently, with the introduction of WNs [14], the modelling power has been extended with respect to regular nets, without losing the efficient analysis algorithms initially devised for regular nets. In summary, using the WN formalism it is possible to model any system that can be modelled by general coloured nets or by Predicate/Transition nets (using the same number of places, transitions, and arcs), and still be able to compute generative families of invariants [19] and the "symbolic reachability graph" [14] for the automatic detection of model symmetries.

[3]The use of the **goto** statement implies that the statements after the **if** ... **then** ...; **goto** *label* **fi** are executed only if the expression is false, so that they can be seen as the else part of the conditional statement.

4 Model Validation

The correctness of the CGSPN representation of Lamport's mutual exclusion algorithm can be formally proven by considerations based on the structure of the CGSPN, its initial marking, and the linear invariant relations that can be computed using the place invariant (P-invariant) technique [20,21,18]. The same result can also be obtained from the construction of the reachability graph of the Petri net representation of the algorithm, computed for a given value of the number N of processes using the algorithm. The advantage of a proof based on structural properties over the (not less formal) one based on reachability analysis is that the former is parametric with respect to variations of N, and that the latter is feasible only for very small values of N (in practice $N < 4$), as the reachability graph becomes extremely large even in these small models.

The proof of correctness considers two different properties: liveness and mutual exclusion in the access to the critical section. To prove the mutual exclusion property, Lamport used a generalization of assertional reasoning based on the Owicki and Gries method [7], since the ordinary assertional method cannot prove the correctness of an algorithm with timing considerations. To prove the freedom from deadlock property, Lamport used an informal sketch of the method of Owicki and Lamport [8].

We use two classical Petri net analysis techniques to prove the correctness of our net model of Lamport's algorithm: behaviour-preserving structural reduction, and computation and use of P-invariants. Haddad has proposed an extension to coloured nets of the well-known technique of structural reduction [22]. Preliminary results in this sense were presented by Colom et al. in [23]. The idea is to simplify the net description by applying formal reduction rules that are guaranteed to preserve the properties of the model that one wants to prove. The proof is carried out on the simplified net, and the result is guaranteed to hold also on the original net due to the properties of the reduction rules. Figure 4 depicts the result of the application of some net simplification transformations that are guaranteed to preserve the liveness and the "home space" properties with respect to the original net model in Figure 3. The simplification rules employed in order to derive the model in Figure 4 are mainly "transition agglomerations" which, in principle, can be performed automatically, or at least supported by a proper Petri net editing tool.

Note that the net in Figure 4 is not the simplest representation of the model that could be obtained by applying reduction rules to the model of Figure 3; for instance, transition 'think' could be combined into transition 'endCS', thus erasing also place 'think'. However, we did not complete the reduction procedure for the sake of convenience and clarity of the proof.

From the coloured net of Figure 4, one can compute four different (parametric[4]) P-invariants that are labeled I1–I4 below.

I1 there is always exactly one token in place 'X' ($|M(X)| = 1$);

I2 there is always exactly one token in place 'Y' ($|M(Y)| = 1$);

I3 $\forall i \in [1..N]$, exactly one token of colour $< p_i >$ is always present in one of the places 'think', 'start', 'ify0', 'awaity', 'sety', 'ifxi', 'awaitb', 'ifyi', and 'CS';

I4 $\forall i \in [1..N]$, exactly one token of colour $< p_i >$ is always present in one of the places 'think', 'start', 'B', 'awaity', 'awaitb', and 'ifyi'.

Before proceding with the actual proofs of the properties of the net, let us remark that all of them are based on the structure of the coloured net model, its initial marking, and the linear invariant relations I1–I4. No information on model semantics is required in order to carry out the proof. Semantic considerations

[4] Parametric in the sense that the formulation of the P-invariants is independent of the cardinality of the basic colour class P. These four P-invariants are a generative set of all P-invariants of the coloured net, hence they contain all the information concerning the linear invariant relations of the model. Using these parametric P-invariants it is possible to set up a proof that is valid for any number N of processes running Lamport's algorithm.

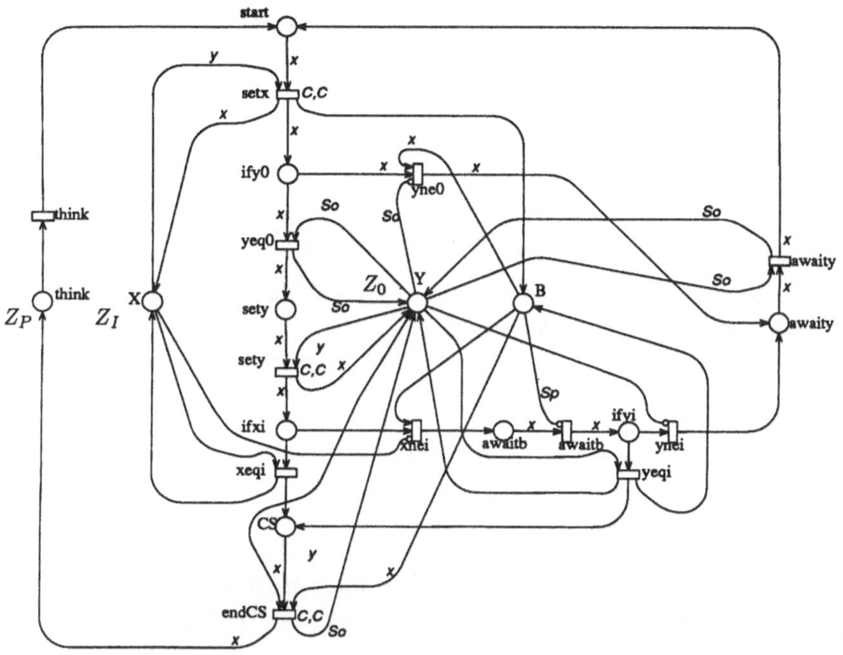

Figure 4: Behaviourally equivalent reduced model.

are introduced only to give an interpretation of the formal results obtained in terms of the original problem formulation. Moreover, the semantic information needed to interpret the validation results is the same contained in the mapping shown in Figure 2.

4.1 Proof of Liveness

The liveness proof verifies (1) that at all times *at least* one transition is enabled to fire (absence of deadlock in the net) for each basic colour p_i, (2) that the initial marking is a home state, which in this particular case implies the liveness of transition 'think' and the fact that each process can always request and obtain access to the critical section, and (3) that place 'CS' is inevitable, meaning that if a request is issued for the access to the critical section, then the access is always granted to some process as soon as possible (efficiency of the algorithm).

4.1.1 Absence of deadlock in the net

We start by proving the following auxiliary result based on the knowledge of I3 and I4:

Lemma 1. For any $1 \leq i \leq N$, one of the places in the set

$$L1 \quad = \quad \{ \text{'ify0'}, \text{'sety'}, \text{'ifxi'}, \text{'CS'} \}$$

contains a token $< p_i >$, if and only if place 'B' contains a token of colour p_i.

Proof :

Consider the expression I5=I3−I4, i.e.

I5: $\forall 1 \leq i \leq N,\ M(\text{B}, < p_i >) = M(\text{ify0}, < p_i >) + M(\text{sety}, < p_i >) + M(\text{ifxi}, < p_i >) + M(\text{CS}, < p_i >)$

Since I3 and I4 are invariant relations, also their linear combination I5 is an invariant relation, that directly proves the statement.

Q.E.D.

Now we prove directly the absence of deadlock from a subset of the places of the net:

Lemma 2. Let $L2$ be the following set of places:

$$L2 = \{ \text{think}, \text{start}, \text{ifyi}, \text{ify0}, \text{sety}, \text{ifxi}, \text{CS} \}$$

For any $1 \leq i \leq N$, a token of colour $< p_i >$ in one place $p \in L2$ implies that (one of) the output transition(s) of that place is enabled for that colour.

Proof :

p =**'think'** transition 'think' that has only one input arc labelled x from place 'think' is enabled.

p =**'start'** transition 'setx' is enabled since the other input arc to the same transition labelled y, independently matches any possible token colour in place 'X' (places 'X' always contains a token due to the first P-invariant I1).

p =**'sety'** transition 'sety' is enabled since the other input arc to the same transition labelled y, independently matches any possible token colour in place 'Y' (which is marked due to the P-invariant I2).

p =**'ifyi'** There are two mutually exclusive output transitions from this place with complementary enabling conditions on the marking of place 'Y', so that one of them can always fire.

p =**'ify0'** *or* **'ifxi'** As for place 'ifyi' there are two similar complementary pairs of output transitions originating from these two places; in addition, by Lemma 1, the input arc connected to place 'B' can never prevent the enabling of transitions 'yne0' or 'xnei'.

p =**'CS'** Transition 'endCS' is enabled because a token of colour p_i must be in place 'B' (by Lemma 1) and one token capable of binding the function y is in place 'Y' (by I2).

Q.E.D.

Finally, we prove an auxiliary result concerning the non reachability of some dangerous states:

Lemma 3. For any reachable marking the following condition holds

$$M(\text{awaity}) + M(\text{think}) + M(\text{start}) = P \implies M(\text{Y}) = S_0$$

Proof :

By contradition. The condition

$$NOT3: \; M(\text{awaity}) + M(\text{think}) + M(\text{start}) = P \wedge M(\text{Y}) \neq S_0$$

is not reachable because:

- $NOT3$ is false in the initial marking ($M_0(\text{think}) = P \wedge M_0(\text{Y}) = S_0$);

- If $M(\mathrm{Y}) = S_0$ then only the firing of 'sety' may change the marking of 'Y', by inspection of the arcs connected to 'Y';
  If $M(\mathrm{Y}) \neq S_0$ then only the firing of 'endCS' may put S_0 into 'Y', by inspection of the arcs connected to 'Y' (neither 'yeq0' nor 'awaity' is enabled due to the input arcs labelled S_0, and 'sety' cannot fire with x bound to $< p_0 >$ due to I3);
  Hence, from the initial marking it would be necessary to fire 'sety' without firing 'endCS' in order to render condition $NOT3$ true;
- The firing of 'sety' with x bound to some colour $< p_i >$ puts a token of this colour in place 'ifyi' (by inspection of the output arc). Due to I3 a token of the same colour cannot be in 'awaity', 'think', or 'start' at the time of the firing of 'sety', hence $NOT3$ is not satisfied at that time;
- Consider the state machine obtained by projecting the net on the set of places corresponding to I3: by inspection of the state transition arcs of this state machine it is clear that the only way of moving from state 'ifxi' to states 'think' or 'start' is through the firing of 'endCS'. Hence, in order to satisfy $NOT3$ it is necessary to move all colours from 'ifxi' to 'awaity';
- From the same state machine structure it appears that in order to move from 'ifxi' to 'awaity' it is necessary to fire 'ynei' with the same colour $< p_i >$ for which transition 'sety' fired. This is impossible for the last colour for which 'sety' fired since a token of this same colour is added in 'Y' by this firing. Indeed, if by hypothesis 'endCS' has not fired after the last firing of 'sety', then the token $< p_i >$ remains in place 'Y' thus representing an inhibition condition for 'ynei'.

Q.E.D.

Now we are ready to prove the absence of deadlock.

Theorem 1 The net has no deadlock state, i.e., for any reachable marking there is always at least one transition enabled in that state.

Proof :

By contradiction, let us assume that a marking M_d exists in which no transition is enabled.

if $M_d(\mathbf{B}) \neq 0$ Due to Lemma 1 this implies that the set $L1$ is not empty. Since $L2 \supset L1$, also $L2$ must contain at least one token of some colour $< p_i >$. Hence by Lemma 2 at least one transition is enabled.

else if $M_d(\mathbf{B}) = 0$ then three cases should be considered:

if $|M_d(\mathbf{awaitb})| > 0$ then $L1$ must be empty by Lemma 1, so that transition 'awaitb' is enabled;

if $\exists p \in L2 \ : \ |M_d(p)| > 0$ then at least one transition is enabled by Lemma 2;

if $\forall p \in L2 \cup \{\mathbf{awaitb}\}$, $|M_d(p)| = 0$ then by I3 we must have $M_d(\mathrm{awaity}) = P$. Hence by Lemma 3 $M_d(\mathrm{Y}) = S_0$, so that 'awaity' is enabled.

No other alternatives are possible.

Q.E.D.

4.1.2 The initial marking is a Home State

An additional result that can be shown is that the initial marking is a home state. Since for each colour $p_i \in P$ transition 'think' is enabled in the initial marking, this directly implies that transition 'think' is live. Moreover, if the initial marking is a home state, then all reachable states are home state (the reachability graph is strongly connected). Hence, after issueing a request, a generic process p_i can always return into a state in which it can issue another access request for the critical section. Due to the structure of the net (transition 'think' is the only output and 'endCS' is the only input of place 'think'), this implies also the liveness of transition 'endCS', which means that each process p_i can always access the critical section indefinitely.

Lemma 4 The net is in its initial symbolic marking if and only if the following condition holds:

$$\text{Cond0}: \quad M(\text{think}) = P$$

Proof :

- By P-invariant I4, Cond0 implies that places 'start', 'B', 'awaitb', 'ifyi', and 'awaity' are all empty;
- By P-invariant I3, Cond0 implies that places 'ify0', 'sety', 'ifxi', and 'CS' are all empty as well.
- By I3 and Lemma 3, Cond0$\Longrightarrow$ $M(\text{Y}) = S_0$;
- By P-invariant I1, place 'X' contains one token. The actual colour of the token (not specified in the initial symbolic marking given in terms of the dynamic subclass Z_P) is immaterial since:
  1. any colour p_j can be left there depending only on the actual sequence of firings of transition 'setx';
  2. the colour of the token in place 'X' does not affect the enabling condition of any transition since it is just overwritten by the next firing of 'setx'.

  Indeed a set of actual markings differing only for the actual colour in 'X' corresponds to the symbolic initial marking.

Q.E.D.

Due to the above Lemma 4, we can concentrate on the position of the tokens of colour $< p_i >$ among the places of P-invariant I3 in order to study the possibility of returning to M_0.

Theorem 2 The initial symbolic marking is a home state, i.e., from any reachable marking it is always possible to reach M_0.

Proof :

- From any reachable marking it is possible to move all tokens $< p_i > \in P$ of P-invariant I3 into the three places 'think', 'awaitb', and 'awaity'. This is proven by repeated application of Lemma 2 to all tokens that are possibly present in places of the set $L3 = L2 - \{\text{think}\}$;
- Since $L3 \supset L1$, by Lemma 1 place 'B' must be empty in these reachable markings so that all tokens that are possibly present in place 'awaitb' can be moved to place 'ifyi' by firing transition 'awaitb'. Subsequently all tokens in 'ifyi' can be moved either to 'think' or to 'awaity' by repeated application of Lemma 2. Hence markings with $M(\text{think}) + M(\text{awaity}) = P$ can always be reached from any reachable marking;
- From the above markings two alternatives are possible:

M(awaity) $= 0$ In this case M(think) $= P$, so that by Lemma 4 this marking is M_0;

M(awaity) $\neq 0$ By Lemma 3 transition 'awaity' must be enabled so that all the tokens that are present in place 'awaity' can be moved into place 'start'. Each token present in place 'start' can afterwards enable the sequence of transitions 'setx' (Lemma 2), 'yeq0', (Lemma 3 plus the fact that the marking of place 'Y' is not changed by the firing of 'awaity' and 'setx'), 'sety' (Lemma 2), 'xeqi' (because of the previous firing of 'setx'), 'endCS' (Lemma 2). Hence each token can move from 'awaity' to 'think'. At that point Lemma 3 is again applicable with one more token in 'think' and one less token in 'start'. Hence another similar sequence can fire until 'start' becomes empty and a marking satisfying the first alternative is reached.

Q.E.D.

This completes the proof that the initial marking is a home state, and that the two transitions 'think' and 'endCS' are live for each colour $p_i \in P$.

4.1.3 Inevitability of 'CS'

Until now we have shown not only that the model is deadlock free, but also that any access request has always a chance to be satisfied, and that each process can repeat the access to the critical section for an infinite number of times. This means that its behaviour *can be* the expected one, i.e. that the algorithm never blocks and that access requests can always be granted an access permission. However, this is not enough to guarantee the efficiency of the algorithm. In principle, one could raise the question whether access permission is always promptly granted to some of the processes that have requested it. We thus complete the "liveness proof" by showing that from any marking with one or more tokens in place 'start' it is not possible to reach a marking in which all requests are refused, i.e., that all these tokens reach place 'start' again without encountering an intermediate marking in which one of them is present in 'CS'.

Theorem 3 Starting from a marking M_s such that

$$M_s(\text{think}) + M_s(\text{start}) = P \quad \wedge \quad M_s(\text{start}) \neq 0$$

and firing any non empty sequence of transitions it is not possible to return to a marking M_e such that

$$M_e(\text{think}) = M_s(\text{think}) \;\wedge\; M_e(\text{start}) = M_s(\text{start})$$

avoiding an intermediate marking M_a such that $M_a(\text{CS}) \neq 0$.

Proof :

- By Lemma 3, $M_s(\text{Y}) = M_e(\text{Y}) = S_0$.
- If a transition sequence contains 'endCS', then marking M_a is reached in the sequence (by inspection of the input of 'endCS').
- Due to I3 the only transitions enabled in M_s are 'setx' and possibly 'think' (the latter only in case $M_s(\text{think}) \neq 0$). Hence any non empty transition sequence σ should start with one of these two transitions.

  σ starting with 'think': Transition 'endCS' must be in σ because it is the only input of place 'think' that can re-establish the condition $M_e(\text{think}) = M_s(\text{think})$ after the firing of 'think'. Hence M_a is traversed.

σ **starting with 'setx':** Since $M_s(\mathrm{Y}) = S_0$ and the firing of 'setx' does not change the marking of 'Y' transition 'yeq0' is enabled and transition 'yne0' is not enabled. By the structure of the state machine projection of the net on the places in I3 the token put in state 'ify0' by the firing of 'setx' cannot return to state 'start' without firing either 'yne0' or 'yeq0'. Hence the sequence must contain 'yeq0' in order to lead from M_s to M_e.
Due to the same state machine structure, σ must contain transition 'sety' as well. The firing of 'sety' results in a marking M' such that $M'(\mathrm{Y}) \neq S_0$ (due to the binding of the output function x to a colour $p_i \notin S_0$).
The only transition that can recover the condition $M_e(\mathrm{Y}) = S_0$ is 'endCS' (by inspection of the arcs).
Hence also in this case M_a is traversed.

Q.E.D.

The efficiency of the algorithm modelled by the net in Figure 4 (and thus of that originally modelled by the net in Figure 3 from which the former is derived by behaviour preserving transformations) follows from Theorem 1, Theorem 3, and from the interpretation of the marking implicit in the rules depicted in Figure 2.

Corollary 1 Starting from a marking M_s such that

$$M_s(\mathrm{think}) + M_s(\mathrm{start}) = P \quad \wedge \quad M_s(\mathrm{start}) \neq 0$$

it is always possible to fire a non empty sequence of transitions such that all intermediate markings M' satisfy the condition

$$M'(\mathrm{think}) \geq M_s(\mathrm{think})$$

and lead back to the starting marking M_s. Such sequence always contains an intermediate marking M_a such that

$$M_s(\mathrm{start}) \geq M_a(\mathrm{CS}) \neq 0$$

Proof :

The proofs of Theorems 1, 2, and 3 are independent of the cardinality of the set P of basic colours. Hence, if any result holds for a given model for P, it is also true for a model obtained by considering a subset of colours $P' \subset P$ (provided that $|P'| > 0$).
Construct the auxiliary model obtained by substituting $P' = M_s(\mathrm{start})$ into the original model, and consider the starting marking M'_s such that

$$M'_s(\mathrm{think}) = 0 \quad \wedge \quad M'_s(\mathrm{start}) = P'$$

Theorem 2 applies to this auxiliary model, implying that a sequence exists that leads from M_s to itself. Theorem 3 also applies, showing the unavoidability of M'_a such that $M'_a(\mathrm{CS}) \neq 0$. Moreover, $M'_a(\mathrm{CS}) \in P'$ by P-invariant I3. Finally, the same transition sequences with the same colour instantiations can be found in the two models if by hypothesis the tokens in 'think' remain freezed in the original model.

Q.E.D.

Corollary 1 is a Petri net formalization of the intuitive concept of efficiency of Lamport's mutual exclusion algorithm. Theorem 3 is comparable with the result provided by Lamport in the original paper under the name "absence of deadlock" [5]. Our liveness results are instead not easily comparable to the original proof of the algorithm, and in this sense could be considered as an additional new result.

4.2 Sketch of Mutual Exclusion proof

The mutual exclusion proof shows that no more than one token at a time can ever be in the place 'CS' (i.e., that $|M(CS)| \leq 1$). We provide only an informal sketch instead of the formal proof for the sake of brevity.

The condition is satisfied in the initial marking. From I3 we know that if several tokens are in place 'CS', then they must be of different colours. By inspection of the input and output arcs of place 'CS' it is evident that the only way for two different tokens $< p_i >$ and $< p_j >$ to simultaneously occupy place 'CS' is that both fire either transition 'xeqi' or 'yeqi' (for colours p_i and p_j), without firing transition 'endCS' (for these two colours).

We need only consider the portion of the net below the transition 'yeq0'. Indeed if several colours have fired 'sety' without firing 'endCS', the only token present in place 'Y' (by I2) cannot be $< p_0 >$, so that transition 'yeq0' has no more chance to fire until the firing of 'endCS' (as already argued before). Hence only the tokens already contained in places 'sety', 'ifxi', 'awaitb', and 'ifyi' might have a chance to enter place 'CS' by firing either 'xeqi' or 'yeqi'. At most one of the tokens that have already passed 'yeq0' can enjoy the enabling condition of transition 'xeqi' (due to I1 and the fact that neither $< p_i >$ nor $< p_j >$ can be added into place 'X' by any transition); in case one token does fire 'xeqi', then all the others will queue up in place 'awaitb' waiting for the firing of transition 'endCS' to remove $< p_i >$ from 'B' (that must be there due to Lemma 1). If no token enters 'CS' by firing 'xeqi' (i.e., if all tokens flow through 'awaitb'), then the inhibitor arc to transition 'awaitb' forces all tokens to wait for the withdrawal of all other tokens in 'B' by the firing of 'xnei'. When 'awaitb' eventually becomes enabled, place 'sety' must be empty (by Lemma 1), so that the value of the token in 'Y' remains unchanged, and no more than one of the tokens that enters place 'ifyi' can enjoy the enabling condition of transition 'yeqi'. This token enters place 'CS' while all the others are moved back to the 'start' place. This leads to the exclusive access of the critical section.

5 Model Evaluation

This section describes how coloured nets can be used to study performance aspects of a concurrent algorithm. We model the execution of statements with timed transitions that fire with unit rate. To keep the model analytically solvable, all timed transitions are exponential and are of the infinite server variety.

In the specific case of Lamport's algorithm, the interesting performance measures include the total time required for a process to use the critical section (response time = R), the indication of the efficient use of the critical section (utilization = U), and the rate at which processes exit the algorithm (throughput = X). Considering that all these quantities are functions of

(1) the number N of processes accessing the critical section,

(2) the arrival rate λ of processes entering the algorithm at the 'start', and

(3) the rate μ at which a process in the critical section departs from the same,

the following relations exist:

$$X(N,\lambda,\mu) = U(N,\lambda,\mu)\,\mu$$

$$R(N,\lambda,\mu) = \frac{N}{X(N,\lambda,\mu)} - \frac{1}{\lambda}$$

We can hence focus our attention on one or the other of these, say X, to study its behaviour as a function of the parameters λ, μ, and N. This study is conducted using a succession of GSPN models derived from the CGSPN model of Figure 3 (that was used for validation). All numerical results were obtained using the *GreatSPN* [24,25] software package.

5.1 Evaluating Coloured Nets

(C)GSPN models can be easily translated into Markovian models that are useful for performance evaluation purposes [1,2,3]. The compactness of the CGSPN models of Figure 3 hides the enormous complexity of the underlying stochastic process. This represents the advantage of using CGSPN models for studying problems that would otherwise be practically impossible to analyze with direct approaches (the direct construction of their underlying stochastic processes would just be too costly and error prone). However, because of this complexity, the numerical evaluation of the model is practically limited to cases in which there are few processes (< 4) vying for the critical section. Nonetheless, these nets can be automatically translated into simulation models that can be used to study the performance of the algorithm for any number of processes [6].

The numerical evaluation of a coloured net is accomplished via an intermediate step. The description of the net together with the incidence functions are translated into an uncoloured net in which the identity of each colour is explicitly accounted for by the replication of places, transitions, and arcs. This unfolding process thus produces one copy of the original net *per colour*, together with the interactions between the subnets.

The resulting unfolded net is the one that is used to construct the underlying Markovian model. The solution of this Markovian model is then used to compute performance measures defined on the coloured net.

5.2 Exploiting the Marking Symmetries of the CGSPN Model

When all processes behave in the same manner (execute the same algorithm), it is possible to exploit the inherent symmetry of this situation to obtain a more compact model (with respect to the size of the state space). This is accomplished by focusing attention on the execution of certain *distinguished processes* and considering all *other* processes lumped together.

We call the process that last sets x and passes the **if** $\langle y \neq 0 \rangle$ statement, the *x-distinguished process.* Similarly, the process that last sets y and can take the 'then' branch of the **if** $\langle x \neq i \rangle$ test is called the *y-distinguished process.* The process that is x-distinguished and has not yet started the **if** $\langle x \neq i \rangle$ test will become *not x-distinguished* if another process sets x in the meantime. Likewise, the process that is y-distinguished and has not yet started the **if** $\langle y \neq i \rangle$ test will become *not y-distinguished* if another process either sets y or resets $\langle y := 0 \rangle$ in the meantime. Note that the x-distinguished process is the process that can enter the critical section directly (through the $x = i$ branch). The y-distinguished process is the process that can enter the critical section through the indirect path caused by contention (i.e., entrance into the critical section is deferred due to the requirement of testing the state of all the other processes).

Using these ideas, we constructed the net of Figure 5.a that represents the basic structure of Lamport's algorithm and which differs from the net of Figure 3 in the following manner. The immediate transition 'xisset' is introduced in order to capture the displacement of the x-distinguished process into the *not x*-distinguished class. Similarly, the immediate transitions 'yisset' and 'yisreset' capture the displacement of the y-distinguished process into the *not y*-distinguished class. In addition, immediate transitions 'yne0' and 'xeqi' of the original model have been split into two in order to more easily handle the tokens of different colours flowing through the original transitions. We use four different colours to identify the four categories of processes described above. These four colours are pertinent only for a subset of places. Aside from this subset of places the identity of a process is immaterial.

The subnet depicted in Figure 5.b illustrates the effect that the setting of x may have on the x-distinguished process. A token in place 'nox-d' signals the nonexistence of an x-distinguished process. A process that is the first to set x becomes the x-distinguished process in place 'ify0' and empties

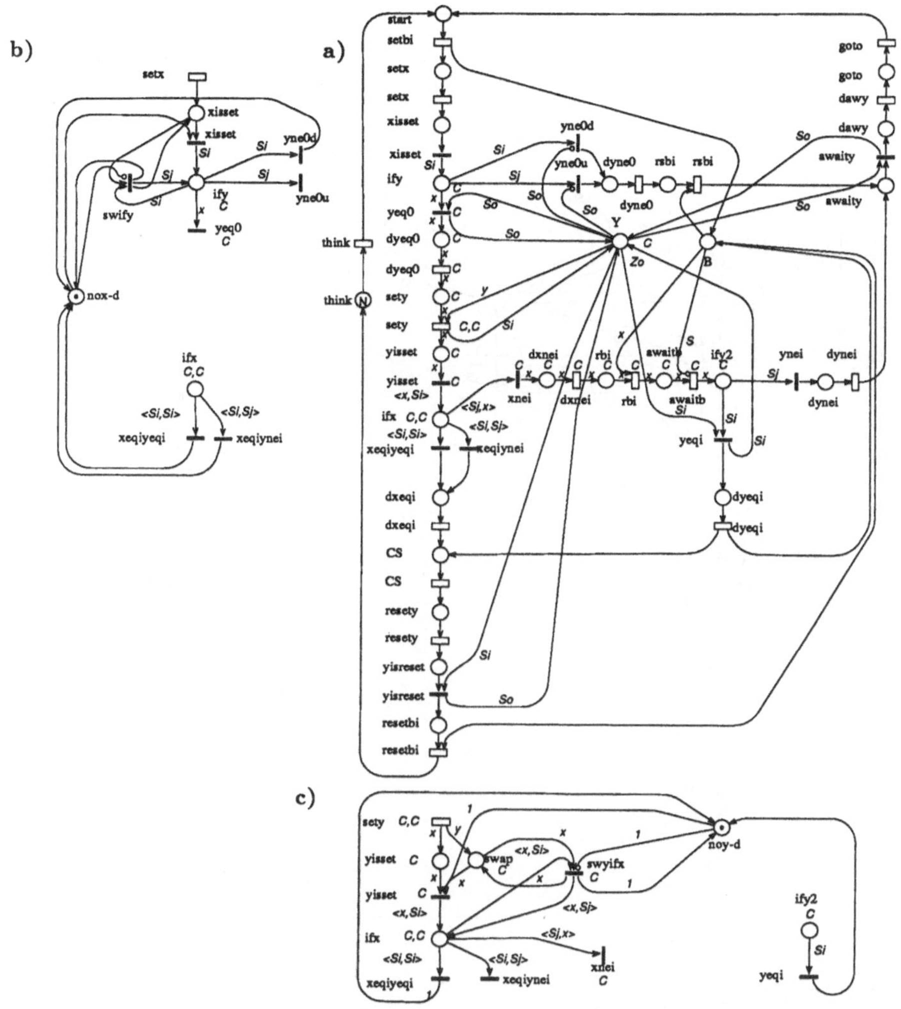

$P_i = \{< pi >\}, \quad P_j = \{< pj >\}, \quad P_0 = \{< p_0 >\}, \quad C = P_i \cup P_j \cup P_0, \quad C, C = C \times C$

$Z_0 = P_0$

All places and transitions not labelled are not coloured. All arcs not labelled have default multiplicity 1.

Figure 5: Building Blocks of Compact Coloured Net.

the token from place 'nox-d' to signify that an x-distinguished process now exists. A process that is not the first to set x may find the x-distinguished process in place 'ify0'. This new process, before becoming the x-distinguished one, changes the categorization of the current x-distinguished process to *not* x-distinguished. This is accomplished by transition 'swify' that tests whether the x-distinguished process is waiting in the place 'ify0' and changes it to *not* x-distinguished. When a process completes the **if** $\langle y \neq 0 \rangle$ test and finds y to be nonzero, it will follow the horizontal path. If that process happens to be the x-distinguished process, it loses its identity and signals that there is no current x-distinguished process by adding a token to place 'nox-d'. Figure 5.c illustrates the effect that setting y has on the y-distinguished process and is quite similar to Figure 5.b as the reader may observe.

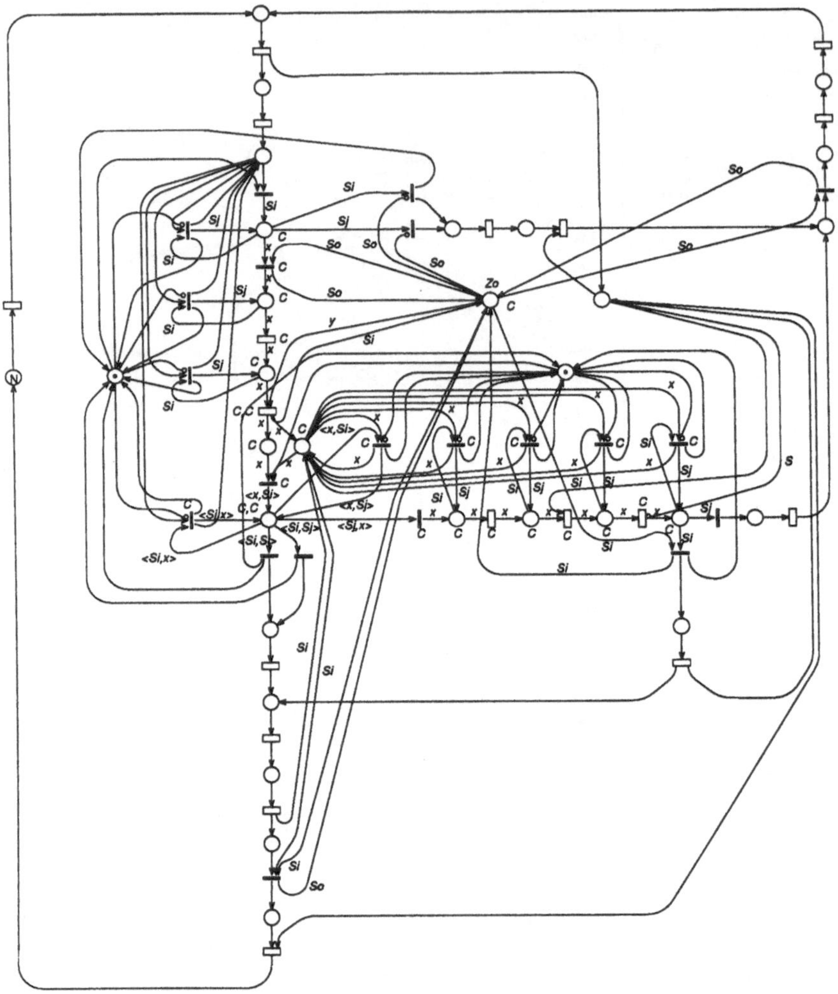

Figure 6: Full Compact Coloured Net.

Figure 6 depicts the full model obtained by creating multiple instances of the swapping mechanism depicted in Figures 5.b and 5.c. The net of Figure 6 is equivalent to that of Figure 3 from a performance evaluation point of view because its underlying Markov chain is an exact aggregation (lumping) of the Markov chain of the detailed coloured net. The equivalence can also be shown by computing the Symbolic Reachability Graph (SRG) of the WN of Figure 3 using the algorithm described in [14] and checking that it is isomorphic to the Reachability Graph of the net in Figure 6. The aggregation technique based on the SRG analysis of a Stochastic Well-formed Net (SWN) model rather than of the RG has been recently proven in [4]. Indeed, the model depicted in Figure 6 can be obtained directly from the WN in Figure 3 and from the knowledge of its SRG using a technique proposed in [26].

The net in Figure 6 is thus the *exact model* that will be used for evaluating the performance of Lamport's algorithm and that exploits the symmetries to reduce the computational complexity of the evaluation; we call this model "E". Unfortunately, this exact model, although reasonably compact with

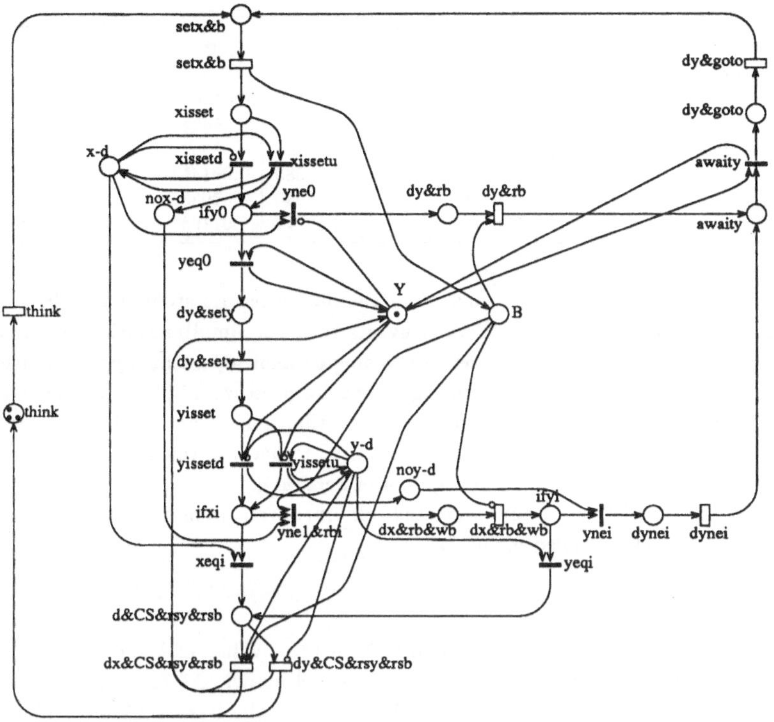

Figure 7: Model "A".

respect to the size of its unfolded version (being independent of N), still has a state space that grows quite rapidly with increasing N (this model could be used only for $N < 5$). We thus attempted to construct approximate models capable of reproducing the general behaviour of the exact net (and thus of the algorithm) with a smaller state space.

5.3 Probabilistic Approximations

The swapping mechanism of Figure 6 keeps explicit track of the positions that the last processes to set x and y have in the set of places $S1$ ={'ifyO', 'dyeqO', 'sety', 'ifxi'} and of places $S2$ ={'ifxi', 'dxnei', 'rbi', 'awaitb', 'ifyi'}, respectively. This allows the model to make a *deterministic* choice at the **if** $\langle x \neq i \rangle$ and **if** $\langle y \neq i \rangle$ tests. The state space can be compacted by replacing the two deterministic choices by *probabilistic* ones. This is accomplished by replacing the swapping mechanisms by two places that signal the existence of an x-distinguished and a y-distinguished process, respectively. Then at the **if** $\langle x \neq i \rangle$ test, the probability that a process has of entering the critical section is computed by dividing the number of tokens in the x-distinguished signal place (either 0 or 1 token) by the total number of tokens found in the set of places $S1$. A similar argument is used in calculating the **if** $\langle y \neq i \rangle$ probability. Making these modifications, we obtained another model that provides a slight reduction in the size of the state space. This new model over-estimates the throughput because it allows a process to enter the critical section earlier than it normally would. In the subsequent section we refer to it as model "P", whose net description is omitted for the sake of brevity.

In order to further reduce the size of the state space, we combined successive timed transitions into a single one whose firing time is the sum of the average firing time of the individual transitions. The

N	Throughput for $\lambda = 0.2$ E	P	A	Throughput for $\lambda = 1.0$ E	P	A	Throughput for $\lambda = \infty$ E	P	A
1	0.05882	0.05882	0.05882	0.07692	0.07692	0.07692	0.08333	0.08333	0.08333
2	0.07500	0.07543	0.07369	0.08022	0.08080	0.08069	0.08465	0.08514	0.08593
3	0.07879	0.07984	0.07764	0.08074	0.08202	0.08183	0.08434	0.08545	0.08589
4	0.07929	0.08104	0.07920	0.07989	0.08193	0.08168	0.08307	0.08491	0.08526
5	0.07811*		0.07929	0.07810*		0.08080	0.08145*		0.08397
6	0.07735*		0.07857	0.07658*		0.07938	0.07935*		0.08224
7	0.07680*		0.07739	0.07657*		0.07777	0.07763*		0.08029

Table 1: Throughput for Models "E", "P", and "A"; (simulation results are indicated by a star '*')

impact of this modification is that we force distinct events to occur simultaneously. For example, the $\langle b[i] := true\rangle$ and $\langle x := i\rangle$ statements are combined into a single atomic operation by this transformation. Using this modification, the mean values of the firing delays are preserved, but the higher moments are not. This leads to modifications of the state probability distributions computed from the reduced net and thus to approximations. Formally, the reduction rules used to perform this transformation are of the same kind of those used to obtain the model in Figure 4 in Section 4.

Figure 7 shows the final net where we have indicated combined transitions by names with ampersands (&). In the next section this net is called model "A".

5.4 Numerical Results

We solved each of the three models "E", "P", and "A" for three values of λ and various values of N. Models "E" and "P" could be solved exactly for $1 \leq N \leq 4$ and model "A" for $1 \leq N \leq 7$. The exact model ("E") was also simulated for $5 \leq N \leq 7$. All simulation results have a ±2% confidence interval at the 99% confidence level. It should be noted that the event-driven simulations were obtained by a direct and automatic translation of the net description into a C code [6]. Obviously, no limits exist for the value of N when the models are simulated. In all models the average time in the critical section was kept constant and equivalent to five instruction executions ($\mu = 0.2$), in agreement with the fact that critical sections should be short [5].

Figure 8 displays the results obtained from these evaluations; Table 1 contains the numerical values. The throughput of the exact model has an interesting "trashing" shape (see Figure 8.b): it is not a monotonic function of N since it first increases to a maximum and then decreases. This behaviour can be explained as follows. When few processes are vying for entry into the critical section, the likelihood that one of them will enter directly is high. But as the number of actively competing processes increases, the likelihood of (directly) entering the critical section decreases. As a consequence, access to the critical section will more often be preceded by a delay that depends on the number of processes executing the algorithm. This overhead not only increases with N but also reduces the throughput for increasing N.

As Figure 8.b indicates, the maximum of the curves (at $N^*(\lambda)$) shifts to the right as the arrival rate decreases. Lamport's algorithm loses some of its efficiency if it operates in regions where N exceeds the value of $N^*(\lambda)$ for a given λ. The "A" model, although approximate, is able to reproduce the interesting behaviour of the exact curves, and thus provides useful information about the location of $N^*(\lambda)$ (see Figure 8.a).

In practice, of course, one would like to evaluate these models for larger values of N. This can be done by simulation. Figure 8.b shows the results for model "E" where simulation values are indicated by starred points. Observe that the general behaviour predicted by model "A" in Figure 8.a is confirmed by the simulation results for model "E". However, due to the inherent variability of the simulation results it is difficult to confirm conjectures about what the actual shape of the curves will really be.

Throughput versus N ($1 \leq N \leq 7$) for $\lambda = 0.2,\ 1.0,\ \infty$

a) b)

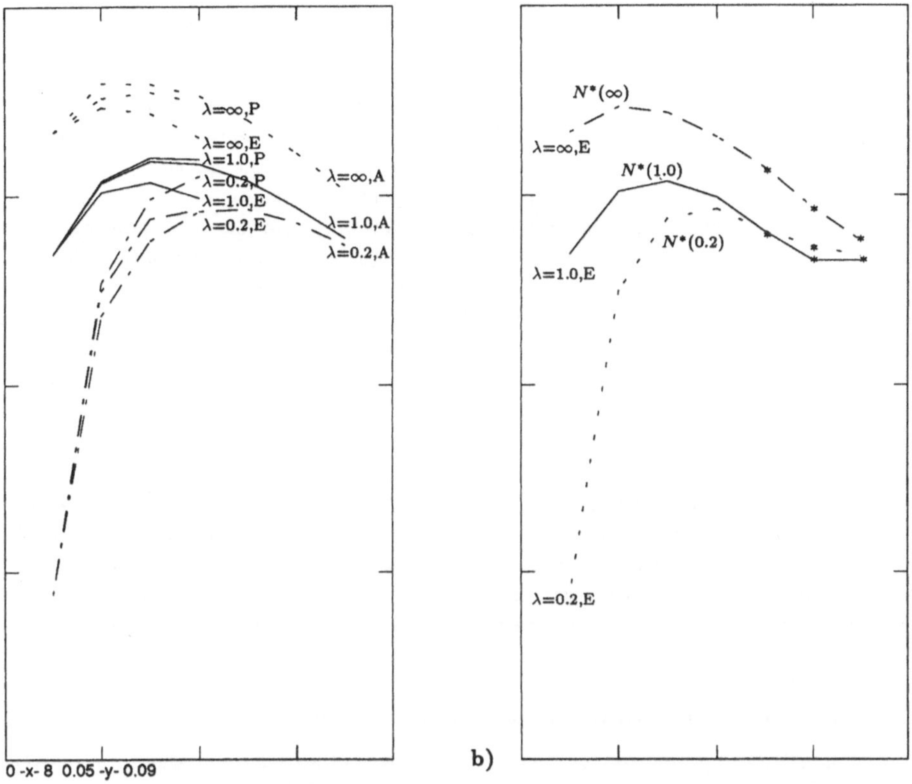

Figure 8: Results of Model Evaluations.

6 Model Variations

6.1 Modeling Contention for Shared Memory

Although we believe that the models of the previous sections represent the qualitative behaviour of Lamport's algorithm, they may not tell the whole story of the performance of the algorithm. For example, these models do not account for the contention for memory that occurs when multiple processors attempt to simultaneously access a shared variable. In real systems, memory contention is resolved by hardware arbitration, which allows one processor at a time to successfully complete an access. This hardware feature is not accounted for in the previous models in that they allow concurrent execution of, for example, an assignment to the same shared variable. Moreover, these models also allow multiple processes to concurrently execute *different statements* using the same shared variables. The first inaccuracy results from the use of an infinite server policy at each timed transition. The second inaccuracy is due to the fact that different statements are represented by different timed transitions that, if not explicitly synchronized, act independently. In the previous models we completely neglected the memory access time; by so doing, these models provide optimistic performance figures.

A conservative approach, instead, assumes that the time for executing a statement is dominated by its memory access time (a more precise representation of the real system would be a compromise between this approach and the previous one). In this case since the memory hardware serializes references to the same shared variable, all statements referring to the same shared variable consequently become serialized. Assuming that memory is interleaved, we can certainly permit the concurrent execution of statements that access different shared variables. All of these hardware features can be included in our models by explicitly representing the "availability" of each shared variable. This can be accomplished by adding as many places (containing a single token) as there are shared variables and by introducing an immediate transition to permit execution of a statement only if the shared variable referenced by the statement is not currently in use. Figure 9 shows the transformation rules that are applied to encode Lamport's algorithm when memory contention is considered. Note that also these transformation rules are behaviour-preserving, so that their application does not affect the validity of the validation performed on the original model [22]. Figure 10 compares the results of the exact solution of model "E" to those obtained from the exact solution of the memory contention model ("M"), for $N = 3$. (Including more immediate transitions and more places makes the state space grow unwieldy for N larger than 3.)

One can clearly see that accounting for memory contention has a profound effect on the throughput (between a 20 and 30% variation). The two sets of curves depicted in Figure 10 present a different behaviour as a function of λ. In the non-memory contention model, the throughput increases as λ goes from 0.2 to ∞; in the memory contention model the throughput has a non-monotonic shape that presents the possibility for an optimal value for λ. This is due to the fact that the presence of memory contention emphasizes the thrashing phenomenon observed before. For $\lambda = 0.2$ the thrashing effect due to congestion is moved drastically to the left, so that a maximum for the throughput curve appears for $N = 2$ instead of $N = 4$ as it was observed for the non-memory contention model. For large values of λ the throughput drastically decreases as the number of processes increases (cf. results for $\lambda = 1.0$ and ∞), and there is no longer the possibility of seeing the maximum of the curves.

6.2 Variations in the Implementation of the Loop

CGSPN models can be used to study an implementation of a concurrent algorithm in even finer detail than shown before. One issue that can affect the efficiency of the execution of Lamport's algorithm on actual machines is the implementation of the **for** loop, which guarantees that a process refrains from trying to enter the critical section until a safe decision can be made.

In the detailed CGSPN model of Figure 3, we represent only the intended effect of the loop rather than its possible implementation on an actual processor. An alternative representation is to model the **for** loop by explicitly accounting for its component statements (the initial assignment $\langle j := 1\rangle$, the test **if** $\langle j \leq N\rangle$, and so on). A second alternative could be to try to optimize the test in various ways: avoiding the **await** on the i-th element (which will certainly succeed) or starting the loop from different points in order to reduce contention for the same array elements from different processes. A possible optimized version of the loop is the following:

$j := next(i)$;
while $j \neq i$ **do await** $\langle \neg b[j] \rangle$; $j := next(j)$ **od**

Both the explicit representations of the **for** and the **while** loop are trivial to map into well-formed CGSPN models, as depicted in Figure 11. Again it is possible to formally show that both models can be reduced to the original one depicted in Figure 3, so that also these variations do not affect the validation performed on the first model.

To understand these nets, consider the one representing the **for** loop. When transition 'rbi' fires, it deposits a token of colour $< p_i, p_1 >$ in place 'for'. (The identity of the process performing the **for** loop is p_i and the initial value in the loop is p_1.) Transition 'doawait' waits for the condition $b[1]$ to

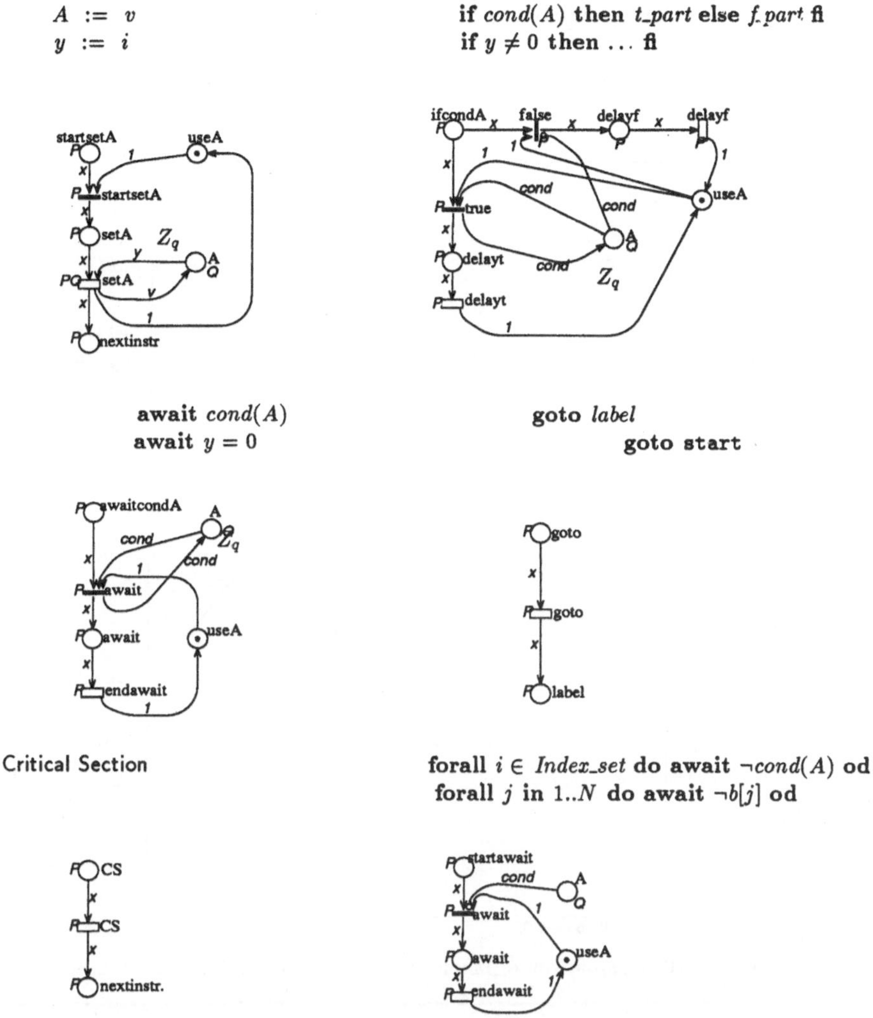

Figure 9: Translation of Statements (Considering Memory Contention).

become false and then deposits a token of colour $< p_i, p_2 >$ in place 'for' (removing the one of colour $< p_i, p_1 >$). When the colour of the token in place 'for' finally assumes the colour $< p_i, p_N >$, transition 'odawait' becomes enabled and upon its firing the **for** loop is completed. The net for the **while** loop can be explained using similar reasoning.

Table 2 shows the results that were obtained from these two modified CGSPN models for $N = 3$ and $\lambda = 1$. It is interesting to note that in the case of Lamport's proposed **for** loop implementation, the three processors behave differently from one another. In particular, processor 1 (the one from which the **for** loop starts) obtains access to the critical section more quickly than the others. About an 8% difference can be observed in the throughput of processes 1 and 3. An ordering is established because of the particular succession of the variables tested in the loop. On the other hand, the **while** loop is fair in granting access to the different processors.

Throughput versus N $(1 \leq N \leq 3)$ for $\lambda = 0.2,\ 1.0,\ \infty$

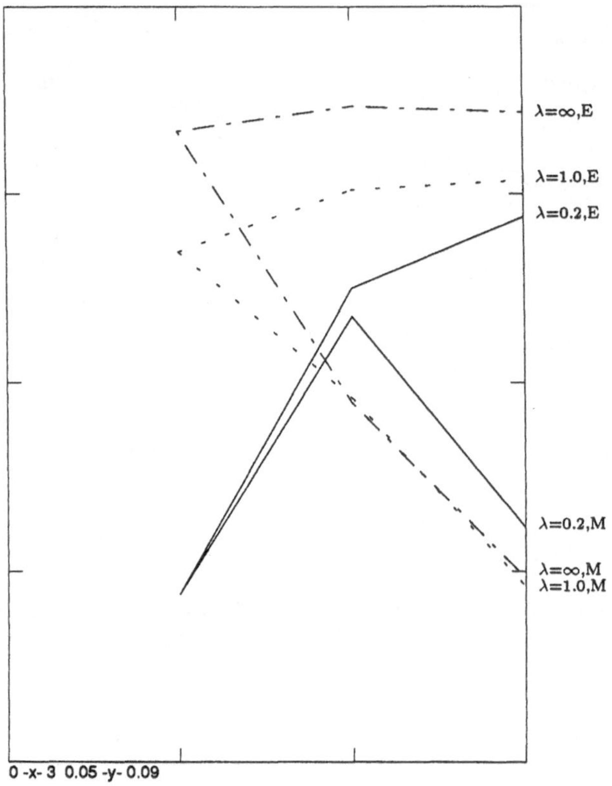

Figure 10: Comparison of Memory vs. Non-memory Contention Models.

Model	Process 1	Process 2	Process 3
for	.0271266	.0266812	.0249734
while	.0262779	.0262779	.0262779

Table 2: Throughput for the **for** and **while** Loop Models.

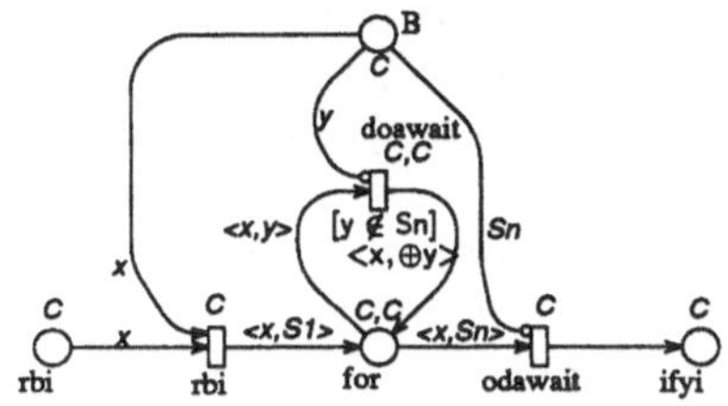

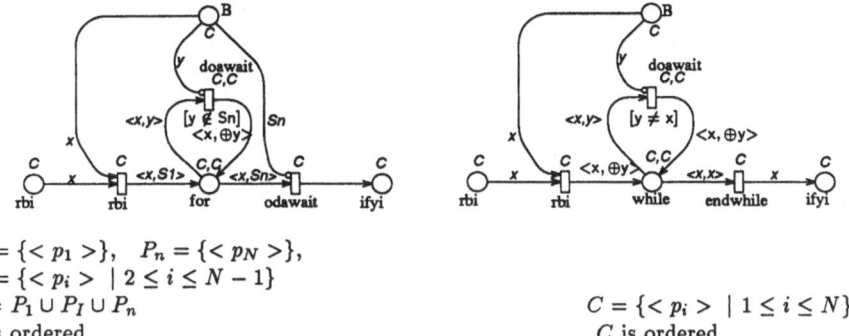

$P_1 = \{< p_1 >\}, \quad P_n = \{< p_N >\},$
$P_I = \{< p_i > \mid 2 \leq i \leq N-1\}$
$C = P_1 \cup P_I \cup P_n$
C is ordered

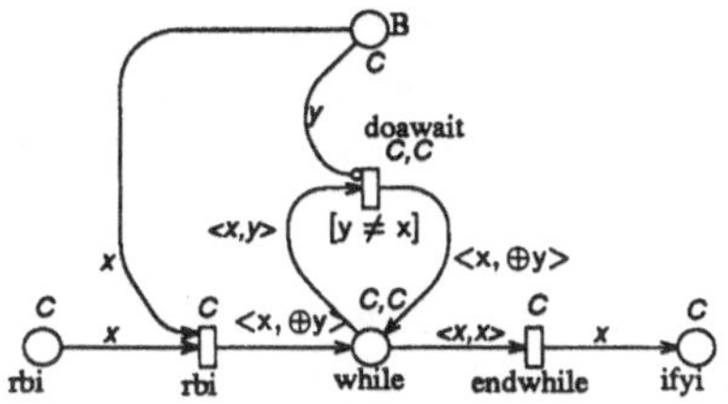

$C = \{< p_i > \mid 1 \leq i \leq N\}$
C is ordered

Figure 11: Coloured Net Representation of **for** and **while** Loop.

7 Conclusions

Looking back at the code for Lamport's fast mutual exclusion algorithm in Figure 1, one cannot overlook how compact it is. Nonetheless, even in this short amount of code, there exist *many* possible flow paths or interleavings. It is a futile exercise to try to trace through all such paths in order to establish correctness. This highlights the importance of formal mathematical proofs of correctness such as the one Lamport provides in [5]. But when an algorithm cannot be formally proven correct, it still behooves the designer to ascertain properties of the new concurrent algorithm.

We have demonstrated one tool that can aid the designer in this step: the construction of a representative well-formed coloured GSPN model. We have shown how formal properties can be derived from this model. In addition, the *same* model is used as a basis for analyzing the performance characteristics of the algorithm. Unfortunately, this model may be too complex to analyze exactly for large numbers of basic colours because of its level of detail. Nevertheless it provides a very good starting point for developing less costly approximate solution models. By successive refinements it is possible to derive abstract representations of the original algorithm that still capture the most important performance characteristics of the algorithm. Of course, simulation can always be used, but the results obtained with this technique may be difficult to interpret due to their variability.

The concurrent program chosen for this example allows also to show that when the processes vying for the critical section have identical behaviours, the symmetries of the algorithm can be exploited to obtain an abstract representation of the original problem that still captures the most important performance characteristics of the algorithm while its state space is greatly reduced. The abstract model is subsequently used as the starting point for the development of an approximate net that replaces deterministic choices with probabilistic ones and that fuses certain transitions to further reduce its state space.

Finally, we have shown that the technique is flexible enough to allow an analyst to successively include and/or neglect implementation/hardware aspects of the algorithm in order to investigate their impact on the overall performance of the algorithm.

More work still needs to be done to make the technique described in this paper practical for the analysis of more complex concurrent algorithms. One step in this direction could be the automatic generation of the coloured net from the specification of the concurrent algorithm. The level of detail of the coloured net could be a parameter to the generator; conceivably one could select between a net that closely represents the hardware under consideration or a less detailed net that more closely represents the structure of the code. Once the detailed model has been constructed, guidelines must be provided to help the user in the refinement process that leads to the definition of simpler and more compact nets useful both for the qualitative and quantitative analysis of the algorithm. The example discussed in this paper shows that well known behaviour preserving reduction rules can be applied to simplify the model. Even the probabilistic modifications that were employed to construct a highly compact model useful for an approximate evaluation of the efficiency of the algorithm could be formalized in order to make possible their automatic application. However, we must point out that the difficult step in the refinement process presented in the paper is the construction of the compact coloured net of Figure 6 that exploits the symmetries of the algorithm. Recent results on the analysis of well-formed coloured nets show that an automatic implicit identification of symmetries is possible in order to construct the so called "symbolic reachability graph" directly [14]. The restriction on the syntax of colour functions and domains posed by WNs with respect to general coloured nets [16] is essential in order to render the symbolic reachability graph construction algorithm effective. This newly developed technique can be used for performance evaluation based on Markovian exact aggregation [4] as well as for qualitative behavioural analysis. Finally, an important future direction of work is the formalization of techniques for deriving correctness proofs.

References

[1] M. Ajmone Marsan, G. Balbo, and G. Conte. A class of generalized stochastic Petri nets for the performance analysis of multiprocessor systems. *ACM Transactions on Computer Systems*, 2(1), May 1984.

[2] M. Ajmone Marsan, G. Balbo, G. Chiola, and G. Conte. Generalized stochastic Petri nets revisited: random switches and priorities. In *Proc. Int. Workshop on Petri Nets and Performance Models*, pages 44–53, IEEE-CS Press, Madison, WI, USA, August 1987.

[3] C. Dutheillet and S. Haddad. Aggregation and disaggregation of states in colored stochastic Petri nets: application to a multiprocessor architecture. In *Proc. 3rd Intern. Workshop on Petri Nets and Performance Models*, IEEE-CS Press, Kyoto, Japan, December 1989.

[4] G. Chiola, C. Dutheillet, G. Franceschinis, and S. Haddad. *Stochastic Well-Formed Coloured Nets and Multiprocessor Modelling Applications.* Technical Report 90/41, Université Paris 6, 4 Place Jussieu, 75252 Paris Cedex 05, France, October 1990. IBP Tech. Report, (submitted for publication to IEEE–TC).

[5] L. Lamport. A fast mutual exclusion algorithm. *ACM Transactions on Computer Systems*, 5(1):1–11, February 1987.

[6] G. Balbo and G. Chiola. Stochastic Petri net simulation. In *Proc. 1989 Winter Simulation Conference*, Washington D.C., December 1989.

[7] S. Owicki and D. Gries. An axiomatic proof technique for parallel programs. *Acta Informatica*, 6(4):319–340, 1976.

[8] S. Owicki and L. Lamport. Proving liveness properties of concurrent programs. *ACM Transactions on Programming Languages and Systems*, 4(3):455–495, July 1982.

[9] D. Good. Principles of proving concurrent programs in gypsy. In *Proceedings Sixth ACM Symposium on Principles of Programming Languages*, San Antonio, Texas, January 1979.

[10] J. Nagle and S. Johnson. Practical program verification: automatic program proving for real-time embedded software. In *Proc. ACM 10th Symp. on Principles of Programming Languages*, pages 48–58, January 1983.

[11] M. Abrams and A. K. Agrawala. *Exact Performance Analysis of Two Distributed Processes with One Synchronization Point.* Technical Report, University of Maryland, 1988.

[12] C. Chou. Performance evaluation of concurrent programs modeled by timed pq-net. In *IEEE COMPSAC'87*, pages 465–473, October 1987.

[13] D. Peng and K. G. Shin. Modeling of concurrent task execution in a distributed system for real-time control. *IEEE Transactions on Computers*, 36(4):500–516, April 1987.

[14] G. Chiola, C. Dutheillet, G. Franceschinis, and S. Haddad. On well-formed coloured nets and their symbolic reachability graph. In *Proc.* 11th *International Conference on Application and Theory of Petri Nets*, Paris, France, June 1990.

[15] E.W. Dijkstra. Cooperating sequential processes. In F. Genvys, editor, *Programming languages and systems*, pages 43–112, Academic Press, New York, 1968.

[16] K. Jensen. Coloured Petri nets: a high level language for system design and analysis. In G. Rozenberg, editor, *Advances on Petri Nets '90*, Springer Verlag, 1991. (to appear).

[17] S. Haddad. *Une Categorie Regulier de Reseau de Petri de Haut Niveau: Definition, Proprietes et Reductions.* PhD thesis, Lab. MASI, Universite P. et M. Curie (Paris 6), Paris, France, Oct 1987. These de Doctorat, RR87/197 (in French).

[18] S. Haddad and J.M. Couvreur. Towards a general and powerful computation of flows for parametrized coloured nets. In *Proc.* 9th *Europ. Workshop on Application and Theory of Petri Nets*, Venezia, Italy, June 1988.

[19] J.M. Couvreur. The general computation of flows for coloured nets. In *Proc.* 11th *International Conference on Application and Theory of Petri Nets*, Paris, France, June 1990.

[20] K. Jensen. Coloured Petri nets and the invariant method. *Theoretical Computer Science*, 14:317–336, 1981.

[21] M. Silva, J. Martinez, P. Ladet, and H. Alla. Generalized inverses and the calculation of symbolic invariants for coloured Petri nets. *Technique et Science Informatiques*, 4:113–126, 1985.

[22] S. Haddad. Generalization of reduction theory to coloured nets. In *Proc.* 9th *Europ. Workshop on Application and Theory of Petri Nets*, Venezia, Italy, June 1988.

[23] J.M. Colom, J. Martinez, and M. Silva. Packages for validating discrete production systems modeled with Petri nets. In P. Borne and S. Tzafestas, editors, *Applied Modelling and Simulation of Technological Systems*, pages 529–536, Elsevier Science Publ. (North Holland), 1987.

[24] G. Chiola. A graphical Petri net tool for performance analysis. In *Proc.* 3rd *Int. Workshop on Modeling Techniques and Performance Evaluation*, AFCET, Paris, France, March 1987.

[25] G. Chiola. *GreatSPN* 1.5 software architecture. In *Proc.* 5th *Int. Conf. Modeling Techniques and Tools for Computer Performance Evaluation*, Torino, Italy, February 1991.

[26] G. Chiola and G. Franceschinis. Colored GSPN models and automatic symmetry detection. In *Proc. 3rd Intern. Workshop on Petri Nets and Performance Models*, IEEE-CS Press, Kyoto, Japan, December 1989.

21.
PROTEAN: A High-level Petri Net Tool for the Specification and Verification of Communication Protocols

J. Billington, G.R. Wheeler and M.C. Wilbur-Ham

IEEE Transactions on Software Engineering, Special Issue on Tools for Computer Communication Systems, SE-*14, 3* (1988) 301-316

***Abstract*—A computer aid for the specification and analysis of computer communication protocols has been developed over a period of 7 years by the Telecom Australia Research Laboratories. It is based on a formal specification technique called Numerical Petri Nets. The computer aid, known as PROTEAN (PROTocol Emulation and Analysis), provides both graphical (color) and textual interfaces to the protocol designer. Numerical Petri Net (NPN) specifications may be created, stored, appended to other NPNs, structured, edited, listed, displayed, and analyzed.**

Interactive simulation, exhaustive reachability analysis, and several directed graph analysis facilities are provided. Reachability graphs can be automatically laid out and displayed. PROTEAN determines liveness (dead code, deadlocks, and livelocks) from the reachability graph and its strongly connected components. Language analysis, involving the automatic reduction of reachability graphs to language graphs, can be used to study sequences of key system events. This allows a protocol to be compared with its service specification. Elementary cycles of graphs can be generated, allowing interesting cycles to be highlighted on reachability and language graphs. Facilities are provided for debugging the specification, once a problem with the protocol has been discovered. They allow sequences of events, which lead to the undesired behavior, to be traced.

The paper commences with a comparison of specification languages, concentrating on extended finite state machines and high-level Petri nets. NPNs and PROTEAN's facilities are then described and illustrated with a simple example. The application of PROTEAN to complex examples is mentioned briefly. A discussion of the approach, its limitations and future work is presented in the context of other developments reported in the literature. Work towards a comprehensive Protocol Engineering Workstation is also discussed.

***Index Terms*—Automated verification, computer graphics, Petri nets, protocols, software CAD, specification animation.**

I. Introduction

COMPUTER communication networks and the services that they provide are becoming more sophisticated, leading to the proliferation of complex communication protocols. Protocol architectures, services, procedures, and formats are being standardized internationally [1] paving the way for the world-wide provision of advanced information services using heterogeneous computing equipment. Protocols are not only required for accessing networks and interworking between networks and services, but also for their efficient operation, including advanced network management strategies.

It is accepted [2] that there is a growing need for rigorous methods for the design of computer communication systems. This implies the development of a methodology based on mathematics for the rigorous design of protocols, supported by a comprehensive set of computer aided design (CAD) tools. CAD tools are required for: specification creation and maintenance; specification execution/animation; logical analysis; performance analysis; assisted or automatic implementation; implementation validation (conformance); and protocol maintenance. (For example see the proceedings of the annual IFIP WG 6.1 Workshop/Symposium on Protocol Specification, Testing and Verification [3] and [4]–[6].) Ideally these tools would be integrated into a Protocol Engineering Workstation (PEW), with an advanced color graphics user interface.

The main purpose of this paper is to describe a CAD Tool for the management, animation, and logical analysis of protocol specifications as a step towards the development of a comprehensive PEW. The tool is known as PROTEAN as it is a versatile tool for PROTocol Emulation and ANalysis. It uses a *high level* inhibitor Petri net, called Numerical Petri Nets (NPNs) for the formal specification of protocols. The paper briefly describes NPNs but places emphasis on the description of PROTEAN. Protocol design issues and the verification methodology are not addressed, but can be found in [7].

Our approach to protocol specification and analysis is similar to that being undertaken by research groups at GMD Darmstadt [8], the Helsinki University of Technology [9], the LAAS, Toulouse [10], and the University of California, Irvine [11]. All groups use a form of high-level Petri net for specification and have developed CAD tools for analyzing the specification. A detailed comparison of these systems and others based on different specification techniques is beyond the scope of this paper. An overview of Petri net tool requirements [12] and a brief summary and comparison table of 19 net tools (including PROTEAN) [13] have recently been published. A survey

Manuscript received November 17, 1986; revised September 1, 1987. J. Billington was supported by a Telecom Australia Postgraduate Scholarship and an ORS Award.
J. Billington is with the University of Cambridge Computer Laboratory, Corn Exchange Street, Cambridge, CB2 3QG, England.
G. Wheeler and M. Wilbur-Ham are with the Telecom Australia Research Laboratories, P.O. Box 249, Clayton, Victoria, 3168, Australia.
IEEE Log Number 8718691.

of protocol development tools was undertaken in 1985/1986 for the Department of Communications of Canada and some brief results are reported in [14], which also contains a tools bibliography. Further information can be found in [3].

II. Specification Languages

A. Requirements

In order to specify protocols unambiguously and to be able to analyze such specifications, it is essential that specification languages have a firm mathematical foundation. It is also important that the technique is able to express specification ideas at many levels of detail (abstraction) to facilitate refinement; is able to support implementation independence via nondeterminism and concurrency; and is able to manage complexity by structuring a specification. Finally, it would be desirable for a design and implementation team to be able to learn the specification language easily.

B. Experience with Extended Finite State Machines

Initial work concentrated on using an "extended finite state machine," CCITT's Specification and Description Language (SDL) [15], for specification of protocols. Our experience with SDL revealed the following difficulties [16]:

- A lack of formal semantics. Although some work has been done on the formal definition of SDL using VDM, it is our understanding that this work is far from complete. It has been our experience that SDL has been interpreted in many different ways.
- A lack of analysis techniques.
- Structuring is tied to the level of concurrency. The static structure is defined at the level of sequential EFSMs called processes. We wanted to structure our specifications according to functions, each of which can behave with differing amounts of concurrency. EFSMs require a different structure for each level of concurrency as the amount of concurrency depends on the number of processes. For implementation independence this means that we would require a set of specifications with different structures to express each different amount of concurrency. This is clearly undesirable. A consequence of this is the need to prove equivalence between the variations of the specification, a rather difficult task with imprecise semantics.
- An inflexible communication mechanism. EFSM techniques assume a particular synchronization mechanism between processes. For example, SDL assumes an infinite FIFO queue while others assume synchronous (rendez-vous) communication. In the case of SDL this led to a number of problems in synchronizing processes and the development of much more complicated specifications in which extra signals had to be introduced. It was clear that a more abstract description was needed.
- Although the graphical form of SDL is good for showing flow of control, it is totally inadequate for showing the flow of data.
- There is inadequate support for concurrency.
- There is effectively no support for nondeterminism. The SDL EFSMs are deterministic automata. Some nondeterminism is introduced via the communication mechanism but this is not sufficient.
- SDL is difficult to modify. A specification using SDL was developed for CCITT's X.25 LAP protocol. This was then taken as a basis for developing a specification of the LAPB protocol; however, it become obvious that a complete restructuring was necessary to avoid an explosion in the size of the specification. Thus a relatively minor modification to the protocol necessitated a complete overhaul of the specification.

C. Experience with Petri Nets

1) Suitability as a Specification Technique: In parallel with the work on SDL, experience was being gained with the use of Petri net theory [17]–[20], a theory originated in 1962 by C.A. Petri, which now has a bibliography of over 2000 articles [21]. Net theory is founded upon a notion of concurrency, it expresses nondeterminism simply and can be used to express system concepts at different levels of abstraction. Nets can be structured in a number of ways. For example, simple *Channel/Agency* nets [22], can be used to indicate a system structure. These nets can then be refined in various steps to define the dynamic behavior of the system [16]. High-level nets also provide ways of grouping system features to indicate system structure in another way.

Nets are usually presented in a simple graphical form which is easy to relate to physical systems. This makes learning and understanding the language relatively easy.

Perhaps the greatest advantages of Petri nets are a solid mathematical foundation and the large number of techniques being developed for their analysis. These include:

1) reachability analysis
2) invariants analysis (a technique using linear algebra)
3) transformations (including reductions) preserving desired properties
4) structure theory
5) formal language theory
6) synchronic distance
7) decomposition and equivalence of nets.

The formal basis of nets allows them to be related to other models of concurrency that may be useful for the specification and analysis of distributed systems. For the latest information the reader is referred to [19], [20].

2) High-Level Nets: Early work with Petri nets in specifying communication protocols [23] recognized that place/transition nets [24] (also called Petri nets [25]) were too primitive to model complex protocols conveniently as this lead to a proliferation of net elements. The simple nets were also very inconvenient when representing information in message headers and compound state information (control state, sequence number, address and multiplexing data, timeout limits etc.). This led to the ideas behind Numerical Petri Nets (NPNs) [23]: tokens were generalized to tuples of variables; a set of data variables was added and the net was inscribed to express the more complex enabling and firing rules. Further development is reported in [26], [27].

In the initial work emphasis was placed on representing rather than analyzing systems. (It can be shown that a number of the extensions were equivalent to the addition of the inhibitor arc to Petri nets, thus raising its power to that of a Turing machine.) A few years later, papers were published on other "high-level" nets [28], [29] (i.e., nets where tokens are tuples of variables and arcs are inscribed by expressions or functions) where emphasis was placed on preserving the analytical properties of Petri nets. Influenced by this work and that of others [30], [31], Wheeler developed NPNs to include self-modifying nets and predicate/transition nets [28] as subclasses and formally specified the language [32]. A guide to the full language was also developed [33] to exemplify the various constructs.

Since NPNs have Turing machine modeling power, they could be used to model any distributed system. For systems requiring less modeling power various subclasses can be used to take advantage of the greater decision power of these simple nets. In practice, only a small subset of the NPN notation and extensions defined in [32] are used for modeling protocols and services; these correspond to predicate/transition nets with some extensions. This is illustrated in [34].

It must be emphasised that high-level nets are NOT a hybrid technique. They are based on Petri nets and use the concepts of place and transition folding to obtain abbreviated descriptions. (For the latest developments of these nets see [19].)

The aim in developing NPNs has been to maximize modeling convenience to provide compact specifications, while providing various classes of modeling power to suit different applications, and to take advantage of developments in the analysis of nets. Care has also been taken to conform to the syntax used for other nets, particularly predicate/transition nets. Formal definition of syntax and semantics has accompanied the development. We feel confident that the problems with EFSMs (SDL) that were listed in the previous section have been largely overcome.

D. Other Techniques

There are a large number of other techniques used for protocol engineering, including state transition techniques, abstract data types, grammars, temporal logic, high-level programming languages [3], set theory (e.g., Z [35]), and process algebras (TCSP, CCS) [36], [37], as well as hybrid techniques (e.g., the ISO languages ESTELLE and LOTOS [38]). A few brief comments on some of these techniques follows.

Most extended state transition techniques suffer from several of the problems listed above for SDL, particularly points 1 and 2. However some state machine techniques have been developed with a sound mathematical basis. For example, the *selection/resolution* model [39], [40] being developed at Bell Laboratories claims to provide a formal framework which has: mathematically precise semantics; a formal algebraic product for the composition of state machines; and methods of reduction in an attempt to solve the state explosion problem of reachability analysis and to provide for stepwise refinement. In comparison to nets, possible difficulties are that: the model is apparently synchronous, with all processes clocked by the same global clock [40] (not the general case for distributed systems); flow of data is not represented; the synchronization mechanism is fixed (shared variable); structuring is by (sequential) process rather than by (possibly concurrent) function; and the level of abstraction tends to be at too low a level (see [41] for example), leading to comparatively complex models and hence possibly to difficulties in using reachability analysis for complex systems.

Of the programming languages which support concurrency, few support nondeterminism, most are relatively complex and difficult to analyze and are not abstract enough for implementation independence. The obvious advantage is that the specifications can be directly implemented so long as compilers are available.

Process algebras are promising as they have been developed with composition in mind to manage complexity and have well developed theories of equivalence. Links with net theory [20] may provide a well structured model with a well developed set of analysis techniques. More abstract approaches (temporal logic and Z) have the advantage of more compact specifications, but do not provide any support for the intuition in specifying a complete set of properties. Analyzing even moderately complex protocols appears to be difficult.

A comparison of all these techniques would be a vast undertaking and is well beyond the scope of this paper. Unfortunately there are very few comparative studies of the different techniques. Early surveys may be found in [42] and more recent work in [43]. Other useful compendiums are [3], [44], [47].

III. Terminology

The terminology used within this paper is based on Petri net terminology. We assume a familiarity with Petri nets as discussed in [25], [18]. Perhaps the most confusing of the terminology is the concept of *liveness* because of overloading of the term in other contexts. Other possibly confusing terms are *safeness* and *safety*.

Liveness in a Petri net was studied extensively in [48]. It is unfortunate that the term *liveness* was also used 5 years later in the context of parallel program verification using inductive assertions [49] to categorize a set of properties where *something good must eventually happen*. At the same time the concept of *safety* was introduced to categorize the set of properties where *something bad never happens*. Recently some attempts at formalizing these notions have been made in [44], [50] in terms of temporal logic formulas.

In this paper we use the concept of liveness as defined in [18], [24]. (The concept of safeness is not used in the rest of this paper. It is the notion that the number of tokens in any place of a net never exceeds unity. For further discussion of safeness and boundedness in Petri nets see [25].)

The idea of liveness in a Petri net is that every transition in the net is potentially firable in every marking of the reachability set. A transition t, is potentially firable in a particular marking M, if there exists a firing sequence from M to a new marking M' in which t is enabled. Proving that a net is live ensures that the modeled system is free from deadlock, livelock, and dead code (transitions which are never enabled). A formal definition may be found in [24] and further discussion in [18], [25].

To avoid confusion, where the word liveness is used, it is meant in the sense of Petri net liveness. Where the word is used in the sense of progress properties it will be placed in inverted commas ("liveness") and similarly for "safety" properties.

IV. Numerical Petri Nets

This section provides an informal introduction to NPNs, and a subset of notation sufficient for understanding the example of Section V. (For a mathematical treatment and the complete definition, see [32].) The reader is assumed to have a knowledge of Petri nets [25], [18] or place/transition systems as defined in [24], [51].

A. Extensions

Numerical Petri Nets are Place/Transition (P/T) Systems with the following extensions.

- Tokens have been generalized to tuples of variables (cf. predicate/transition (PrT) nets [28]). This allows the convenient modeling of the fields in protocol messages, as well as other protocol entity features such as compound states, counters and look-up tables.
- A set of data variables is associated with the net. Only very simple types (integer, modulo, Boolean, enumerated, strings) are presently considered so that precise semantics can be obtained in terms of the net model. (A data variable can always be represented by a place, with appropriate input and output arcs and a token carrying its present value.) It is introduced purely for modeling convenience for objects such as counters.
- There are two different types of place capacity. The first, K, sets a bound on the number of tokens of a particular value that can be resident in a place (this is the same as for PrT systems). The second, K^*, sets a bound on the total number of tokens allowed in a place.
- Three inscriptions are associated with the arcs of the underlying net.

1) An Input Condition (IC) is inscribed to the left of a transition's input arc, as seen by an observer at the transition. It defines a condition which may be satisfied by a collection of tokens in the associated input place.
2) The Destroyed Tokens (DT) are inscribed to the right of each input arc, as seen by our observer. It defines a bag (multiset) of tokens, which is removed from the associated input place (by bag subtraction), when the transition fires.
3) The Created Tokens (CT) are inscribed next to each output arc of a transition. It defines a bag of tokens, which is deposited into the associated output place, when the transition fires.

- There are two optional inscriptions associated with each transition.

1) A Transition Condition (TC), delimited by square brackets, and written next to or inside the associated transition. It defines a condition on net data variables and the variables associated with tokens residing in the transition's input places.
2) A Transition Operation (TO), written next to or inside its transition. It defines an operation on the data variables.

- An initial Marking (Mo), defining the initial allocation of tokens to each of the places in the net, and the initial value of the data variables.

B. A Generic Example

An example which illustrates the extensions is shown in Fig. 1. Places and transitions are given names which are strings of alphanumeric characters commencing with a letter. Places are represented by ellipses (usually circles) and transitions by rectangles or bars. Place capacities are represented by integers written next to the place (e.g., 2 or 2*). Note that the underlying transitions, places and arcs constitute a directed net as in P/T systems.

C. Marking

An NPN marking is the net marking (the bags of tokens associated with each place) together with a vector consisting of the value of each of the data variables. The net marking is restricted by the capacities of the places, as is the vector by the type of each variable. The NPN marking may be thought of as the global state of a distributed system.

D. Enabling

A transition is enabled when

- the Input Condition of each of its input places is true
- its Transition Condition is true
- if fired, the capacities of its output places will not be exceeded.

E. Transition Firing Rule

When a transition fires (occurs), the following events occur indivisibly and concurrently

- for each input place, its Destroyed Tokens are removed (bag subtraction)
- for each output place, its Created Tokens are added (bag addition)
- the Transition Operation is performed.

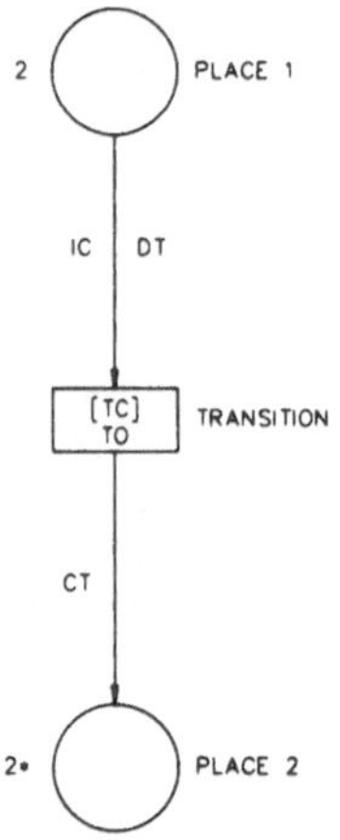

Fig. 1. A generic Numerical Petri Net.

F. Net Execution and Reachability Analysis

When an NPN is defined, it is endowed with an initial marking. For this initial marking a set (possibly empty) of transitions will be enabled. (Note that a single transition may be enabled a number of times, if a number of different values of variables satisfy the IC's and TC.) An arbitrary choice is made as to which transition occurs (and which set of allowed variable values is bound). This occurrence generates a new marking, with a new set of enabled transitions. The net can be executed further, generating a set of reachable markings. The complete set of markings that can be generated this way is known as the Reachability Set. These markings are related to one another by a set of transition firing (occurrence) sequences. A directed graph which relates the set of markings (nodes of the graph) to transition occurrences (edges of the graph) is called a Reachability Graph. (Note that we have assumed interleaving semantics for simplicity. In general, partial order semantics allowing the expression of "true concurrency," could be used.)

1) Binding Variables: Free variables (those associated with tokens) may be part of the IC's, DT's, CT's, TC, and TO specifications associated with a transition. The scope of these variables is restricted to the transition concerned. When the transition fires, the variables are bound to a particular value via consistent substitution.

G. Notation

A commonly used subset of NPN notation is presented in this section, the full notation being given in [32]. In the following: m and n are positive integers; *tok*, *tok*1, and *tok*2 are tokens; $M(p)$ is the bag of tokens in place p; and $mult(x, M(p))$ is the multiplicity of token x in the bag of tokens $M(p)$.

1) Tokens: In general tokens are tuples of variables/constants separated by commas and enclosed in angular brackets. Variable names are strings of alphanumeric characters commencing with a letter. Some examples are $<7>$, $<red, yellow>$, and $<x, y, z>$. Alphanumeric strings may be variables or constants, the context making it clear.

2) Input Conditions:

Notation	Condition on Input Place Marking, $M(p)$
tok	$tok \in M(p)$
mtok	$mult(tok, M(p)) \geq m$
*mtok*1 + *ntok*2	$mult(tok1, M(p)) \geq m$ and $mult(tok2, M(p)) \geq n$
0	$M(p) = \phi$
#	$M(p) = M(p)$ (always true)
$o = tok$	"oldest" token in input place is *tok*
$\sim tok$	$tok \notin M(p)$
*tok*1& $\sim$ *tok*2	$tok1 \in M(p)$ but $tok2 \notin M(p)$

3) Destroyed Tokens:

Notation	Destroyed Tokens Bag $D(p)$
tok	$\{tok\}$
mtok	$mult(tok, D(p)) = m$ and $\vert D(p) \vert = m$
*mtok*1 + *ntok*2	comprises *mtok*1 tokens and *ntok*2 tokens
0	ϕ
#	$M(p)$
"blank" or:	the "enabling" tokens
o:	"oldest" token singleton bag

4) Created Tokens:

Notation	Created Tokens Bag $C(p)$
tok	$\{tok\}$
mtok	$mult(tok, C(p)) = m$ and $\vert C(p) \vert = m$
*mtok*1 + *ntok*2	comprises *mtok*1 tokens and *ntok*2 tokens

5) Transition Condition: Transition conditions comprise Boolean expressions with the usual logical connectives ("&" and **and**; "|" for **inclusive or**, and "$\sim$" for **not**) and two place predicates ($<$, $\leq$, $=$, $\geq$, $>$, and $<>$ for $\neq$). They may involve variables, natural number expressions, and string expressions.

6) Transition Operation: The transition operation consists of the sequential composition of assignments of data variables to natural number, Boolean, or string of character expressions.

H. Analysis

The previous section listed a number of analysis techniques that could be applied to nets. At this stage of development of PROTEAN, we have only implemented analysis techniques related to reachability analysis. (Future developments are discussed later.) The techniques that can be applied to analyzing the reachability graph are thus similar to those developed in EFSM analysis.

V. PROTEAN

PROTEAN is a computer aid developed for PROTocol Emulation and ANalysis. Development of an early prototype began in 1980 and has been reported in [52]. Graphical facilities were added to PROTEAN during 1983 and a number of other analysis features over the last few years. A more mature PROTEAN is reported in [7], [53].

PROTEAN and NPNs were developed concurrently. A major revision (known as PROTEAN 2) occurred in early 1986, aligning PROTEAN with the current NPN definition. It also provided for some enhancements to the analysis, simulation and graphical facilities.

A. Requirements

Specification Management: A major goal in the development of PROTEAN was for it to provide for the development and management of complex protocol specifications, particularly those being developed for standardization by the CCITT, such as many of the X-series, Q-series, and I-series Recommendations.

Specification Analysis: A set of analysis capabilities, to allow for the checking of liveness (i.e., absence of deadlock, livelock, and dead code) and boundedness (resource limits) as well as checking the protocol against its specific requirements (service specification) were seen to be necessary. User controlled testing of a protocol specification, preferably with animation, was seen as desirable to allow for checking of the input specification and to give increased confidence in the quality of the specification.

Support for Refinement: It is seen as desirable that PROTEAN supports stepwise refinement and abstraction facilities, to allow for top-down designs with iteration.

Friendly User Interface: It is important that the tool is readily available and easy to use by the specifiers of telecommunication and information services.

Maintainability: It was recognized that PROTEAN would go through a number of design phases and that ease of maintenance and enhancement, in the face of changing personnel and requirements, would be important.

Commercial Standard: An increasingly important requirement has been for PROTEAN to be of a commercial standard, for marketing to telecommunication administrations and switching manufacturers.

Portability: A desirable but nonessential feature. Portability is becoming more important for marketing and release to Australian research and teaching establishments under nondisclosure agreements.

Efficiency: It was appreciated that the complexity of the analysis algorithms is considerable and that efficient use of memory and CPU would be critical to PROTEAN's success.

B. Facilities

1) General: PROTEAN facilities have been developed in a hierarchy. At each level the user is prompted by a menu of commands. Two of these commands are always present: EXIT and HELP. The user may always exit to the next highest level (or exit from PROTEAN), or may ask for help at the level (s)he is at. Care has been taken to provide helpful error messages for exception handling.

PROTEAN may be operated in interactive or batch mode. All preliminary specification and analysis will be handled interactively, but sometimes it is convenient to submit large analysis jobs in batch mode.

PROTEAN facilities cover four main areas: net creation and manipulation, interactive net simulation, reachability analysis, and directed graph analysis. These are summarized in the following subsections. A detailed description of each facility is given in the User's Manual [54].

Before describing the four main areas mentioned above, it is necessary to describe the facilities offered by the graphical editor.

Graphical Editor: The graphical editor (GE) is a high level editor in that it is aware of the connectivity of the graphs upon which it operates. Thus movement of a node will automatically ensure that the connecting arcs are also moved appropriately. The GE has standard zoom and pan facilities and will allow movements of any graphical element. It also allows nodes (e.g., places or markings) of the same type to be swapped. Arcs are drawn as straight lines, but intermediate points between nodes are allowed (i.e., an arc may be represented by a set of line segments). Labels are automatically placed near graph elements, if there is room for them. If no labels are shown, they can be revealed by zooming in. With nets, the marking is also indicated, and details can be found by zooming.

2) Net Creation and Manipulation: A subset of the present Numerical Petri Net language constructs has been implemented. Nets are input textually using a linear form of NPNs [54] via a simple editor. Components are entered and checked for syntax errors, one at a time.

A layout facility allows a graphical representation to be created. PROTEAN prompts for each place and transition to be positioned on the screen via the cross-hairs. Places are represented by circles, transitions by bars and arcs by straight line segments. Default values are used for place radius and transition bar length in the initial layout. These may be changed with the graphical editor. Any component may be skipped. Net arcs are drawn automatically with transitions. The graphical editor can then be used to refine the layout.

Three different forms of net representation are provided for different purposes. A raw textual form for rapid input; a verbose textual form, which also gives a summary of net components, useful for checking net input; and the graphical form for greatest visual impact and understanding.

The net may be stored, retrieved and structured into a number of subnets or modules [27], [55]. These subnets may be merged for simulation or analysis.

3) Interactive Net Simulation: Once the net has been created and checked, the user may wish to increase her/his understanding of the net (or the system being modeled by the net), by observing its operation. The net may be executed interactively in SINGLESTEP mode by the user selecting which event is to occur at each marking.

The net is given an initial marking in its definition. PROTEAN will display the transitions that are enabled at this marking, and wait for the user to choose one. PROTEAN will fire the chosen transition, generating a new marking and will display the new set of transitions that are enabled. Ths user can continue to explore the behavior of the net in this fashion. At any stage a previous marking can be selected, and the net will be reset to this marking. If a terminal marking is reached (i.e., one in which no transitions can occur), then a previous marking may be selected. At each stage the user may access details of the current marking or exit from SINGLESTEP.

Interactive net simulation may be performed either graphically or textually. The graphics mode provides animation, where at each step, the distribution of tokens in the net is displayed. Enabled transitions are shown in red. This allows the user to observe the flow of control (e.g., a local system state) and data (e.g., a message in a protocol) in the specification simultaneously.

Interactive simulation is extremely useful in the early stages of protocol specification as it allows the designer to test each part of the specification as it is completed, and can provide considerable insight into the system's behavior. It may also be used for debugging and learning about protocols.

4) Reachability Analysis: Once the designer is happy with the specification, an exhaustive analysis of the protocol may be required. PROTEAN performs reachability analysis, generating the reachability set and the reachability graph for the net of interest. PROTEAN finds all the deadlocked markings (as leaf nodes of the RG) and also gives a list of all transitions that have not occurred (e.g., dead code). Our present algorithm only deals with finite reachability graphs.

In order to manage the complexity of practical protocols, reachability analysis is performed stepwise on a set of NPN modules representing different functions of the protocol. Details can be found in [55], [56].

Analysis of the reachability graph (RG) and reachability set allows a number of other protocol properties (event sequences, livelocks, resource bounds) to be investigated. These aspects are discussed below.

5) Directed Graph Analysis: A finite reachability graph can be viewed as a directed graph (digraph) or as finite state automaton (FSA). It may therefore be analyzed by algorithms developed for the analysis of diagraphs or FSAs.

Strongly Connected Components: A strongly connected component (SCC) is the maximal set of nodes of a digraph in which, for each node in the component, a path from one node can be found to all of the other nodes. PROTEAN's LIVENESS facility generates the strongly connected components of the reachability graph, together with their associated graph, the *liveness graph*, showing how the strongly connected components are related. Leaf nodes of this graph are either deadlocks (1 marking) or livelocks (2 or more markings). For example, if the RG of a data transfer protocol contains only one SCC, then the protocol is free of deadlocks and livelocks.

Language Analysis: The directed graph is treated as a finite state automaton. A subset of the transitions may be chosen as ones of interest. We are then interested in the language (set of sequences of these transitions) accepted by the automaton. After a set of final states (markings) have been entered the minimum state deterministic FSA, equivalent to the nondeterministic FSA with transitions on empty inputs (the original digraph with selected transitions) is then computed [57]. This is a digraph which contains only the subset of selected transitions as edges.

This graph is called a *language graph* in PROTEAN and the facility for generating the language graph is called LANGUAGE. LANGUAGE may be used on reachability and language graphs to greatly reduce the size of the original graph, allowing only sequences of key transitions (important events) to be considered. It is useful for comparing protocol requirements, expressed in service specifications [58], with the protocol [10], [7], [53].

Scenario Generation: A PROTEAN facility, SCENARIO, finds paths in a digraph that are consistent with a given input sequence of transitions or nodes. Particular transitions or nodes may be excluded from the sequence. The number of paths generated may be set by the user, who may select all paths, all shortest paths or just a finite number. This facility is useful for debugging specifications once a problem has been found, especially when using language analysis [53]. (It enables you to answer the question: *How could this behavior possibly occur?*)

Elementary Cycles: PROTEAN's CYCLE facility generates all elementary cycles of a digraph and identifies the largest cycle. This may be used for a loop layout of the graph (see below). CYCLE may be usefully run on reachability and language graphs to examine cyclic protocol behavior.

Directed Graph Display: This facility takes as input a directed graph and will provide automatic layout, either based on a tree or a cycle, the cycle having been identified in earlier analysis. The graph is then displayed and may be edited with the graphical editor. Strongly connected components, elementary cycles and scenarios associated with the graph can be highlighted on the graph. Deadlocked or terminal nodes (leaf nodes) are always highlighted. Final states of a language graph are also highlighted.

C. Implementation Environment

PROTEAN has been developed under a VMS® environment running on DEC® VAX® computers. The programs have been written in Pascal for portability. PROTEAN has been ported to a number of other VAX or micro-VAX® installations without difficulty. A UNIX® version is presently under investigation.

Without graphics, PROTEAN can be run using DEC compatible terminals. Terminals compatible with the ReGIS® Graphic command language are required for

®VMS, DEC, VAX, microVAX, and ReGIS are registered trademarks of Digital Equipment Corporation.

®UNIX is a registered trademark of AT&T Bell Laboratories.

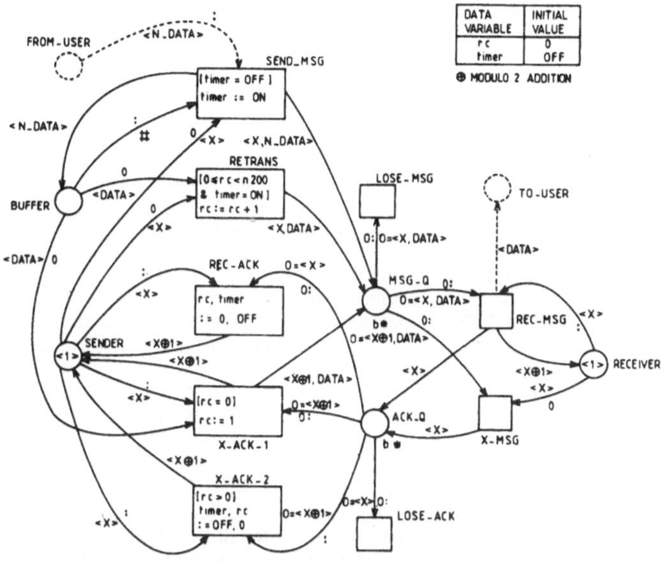

Fig. 2. NPN specification of the single frame procedures.

(color) graphics. (It is also possible to use personal computers as terminals, with the appropriate emulation.)

Hardcopy is presently obtained using a dot matrix printer (black and white graphics) or an ink jet printer (color graphics). The use of a laser printer for high quality B&W hardcopy is also possible.

VI. An Example

The example of a particular version of the alternating bit protocol has been chosen because of its relative simplicity and to illustrate most of the features of PROTEAN. It was the first version of the single frame procedures of the D-Channel Link Level Protocol for Basic Access to the Integrated Serviees Digital Network contained in draft Recommendation Q921 [59]. The work presented here is a slightly modified and expanded version of that given in [53].

A. Protocol Description

The protocol provides for two way data transfer, using a modified version of the standard alternating bit protocol [60], to recover from spontaneous loss of synchronization (due to power fail for example). The protocol operates over a medium which may lose messages, but does not duplicate, reorder, or otherwise mutilate the data. The protocol provides a single frame transfer service, the user having to wait for an acknowledgment before sending further frames. Loss of a frame (or acknowledgment) is recovered by retransmissions. Duplicates are detected by the addition of a single bit sequence number to the frame. When a new message is sent the sequence number is complemented. This allows the receiver to detect duplicates and discard frames with an unexpected sequence number. Some extra procedures have been added in an attempt to recover from spontaneous desynchronizations (details are given in the next section).

It is assumed that a connection has been established for the transfer of data and that recovery in the case of reaching a predetermined retransmission limit will be taken by an unspecified management entity.

B. NPN Protocol Specification

The two directions of information flow are independent. Thus we only consider one way data transfer, between a *sender* and *receiver*. The NPN specification is given in Fig. 2.

The left side represents the sender. Three state variables and five actions are associated with it. The three state variables are: timer $\in \{on, off\}$; rc (retransmission counter) $\in \{0, 1 \cdots n200\}$, with $n200$ a nonnegative integer representing the maximum number of retransmissions allowed for any frame; and ssn (sender sequence number) $\in \{0, 1\}$. The three state variables could be represented by NPN variables or by places containing tokens. In Fig. 2, we have represented both the timer and retransmission counter as (housekeeping) variables, and the sender's sequence number we maintain in a place. This was chosen as we wished to highlight the changes in sequence number value, through the visual imagery (see next section) of values of tokens changing in the place labeled SENDER. (Of course other possibilities could be chosen, for example to highlight the value of the timer if this was required.)

The five actions are: sending a new message; retransmission on timeout; receiving a *good* acknowledgment (one containing the expected sequence number); and two actions on receipt of a *bad* acknowledgment (one containing an unexpected sequence number) depending on whether or not a retransmission has occurred; represented by the transitions: SEND_MSG, RETRANS, REC_ACK, X_ACK_1, and X_ACK_2, respectively.

The sender maintains a buffer (place BUFFER) which contains the current message. The sender interacts with its user via a data store (FROM_USER), and with the transmission medium (MSG_Q and ACK_Q).

With the retransmission counter set to zero ($rc = 0$), the timer off ($timer = off$) and the send sequence number set to one ($< 1 >$ token in place SENDER), a new message from the user may be sent (occurrence of SEND_MSG) with a one bit appended. When it is sent, the timer is switched on ($timer = on$), the buffer is overwritten with the contents of the new message and the sender waits for an acknowledgment. If an acknowledgment has not been received ($timer = on$), the sender may retransmit up to $n200$ times (by firing RETRANS).

If an acknowledgment is received with the correct sequence number, the retransmission counter and timer are reinitialized ($rc = 0$, $timer = off$) and the send sequence number is complemented ($M(\text{SENDER}) = < 0 >$). A new message may now be sent with the zero bit appended. If a *bad* acknowledgment is received and no restransmissions have occurred ($rc = 0$), it is assumed that the receiver was not properly synchronized. To regain synchronization, the sender complements its sequence number and retransmits the message (X_ACK_1). If the message has been retransmitted ($rc > 0$) and an incorrect acknowledgment is received, it is assumed that the first acknowledgment was lost but the message correctly received. The retransmission counter and timer are reinitialized and the sequence number is complemented (X_ACK_2), allowing a new message to be transmitted.

The transmission medium is modeled by two bounded lossy FIFO queues of capacity b, one for each direction of information flow. The queues are represented by places MSG_Q and ACK_Q and loss of messages (acks) is represented by transition LOSE_MSG (LOSE_ACK). The FIFO discipline is maintained by using the "oldest" rule for the IC and DT inscriptions associated with arcs leaving the two queue places.

The receiver is shown on the right side of Fig. 2. It has one state variable ESN (expected sequence number) $\in \{0, 1\}$, represented by the place RECEIVER. It may receive *good* messages (those with the expected sequence number) via transition REC_MSG or *bad* messages (those with an unexpected sequence number) via transition X_MSG. On receipt of a good message it passes the message to its user (TO_USER) and sends an acknowledgment to the sender via the transmission medium. Bad messages are discarded and the expected sequence number returned in the acknowledgment.

The following sections illustrate how some of PROTEAN's facilities can be used to investigate the operation of the protocol. Figs. 3-7 illustrate PROTEAN's graphics. They are redrafted photographs of PROTEAN's display for crisp reproduction in black and white. In the figures, **bolding** is used for highlighting corresponding to *red* on the display. The display also uses different colors for displaying the net and text. This has not been shown in the figures.

C. Interactive Simulation

To obtain a good understanding of the dynamics of the protocol, interactive simulation can be used. An example of the use of interactive simulation using PROTEAN's SINGLESTEP is shown in Fig. 3, for the case when three characters "a", "b", and "c" are to be sent from sender to receiver. The figures provide a graphical view of the marking of the total system for the expected sequence of events (the transition sequence: SEND_MSG, REC_MSG, REC_ACK), which has successfully transferred "a" from the sender to the receiver. The sender is now ready to transmit "b". In the figures, enabled transitions are shown in bold. Note that many other event sequences could be explored by selecting different transitions to fire.

Due to resolution limitations of our color monitors, the NPN inscriptions (IC, DT, CT, TC, TO, place capacities) are not shown in the diagrams. This information can be obtained by the user quite simply during the simulation if required, as can the details of the net marking in textual form.

D. Reachability Analysis

In the above, message values were included in the simulation to increase clarity. When performing reachability analysis it is important to reduce the global state as much as possible. To do this, we take advantage of the fact that the protocol behavior is not dependent on the content of messages. Thus we no longer consider the contents of messages and assume that messages are always available to be sent.

With the above simplification in force, we can investigate the specification under a variety of conditions, varying the retransmission limit ($n200$) and the queue bounds (b). The simplest situation is to consider $b = 1$ and $n200 = 0$. The reachability graph generated by PROTEAN using the loop layout option is given in Fig. 4. It consists of a single loop of six markings, denoting the correct sending and receipt of messages, together with four deadlocked markings (shown in bold). The deadlocks are expected as the system will always deadlock if the number of lost messages or acknowledgments exceeds the retransmission limit $n200$. In this case it is left to a management entity to take appropriate recovery action. This has not been included in the model (as it was not included in Q921).

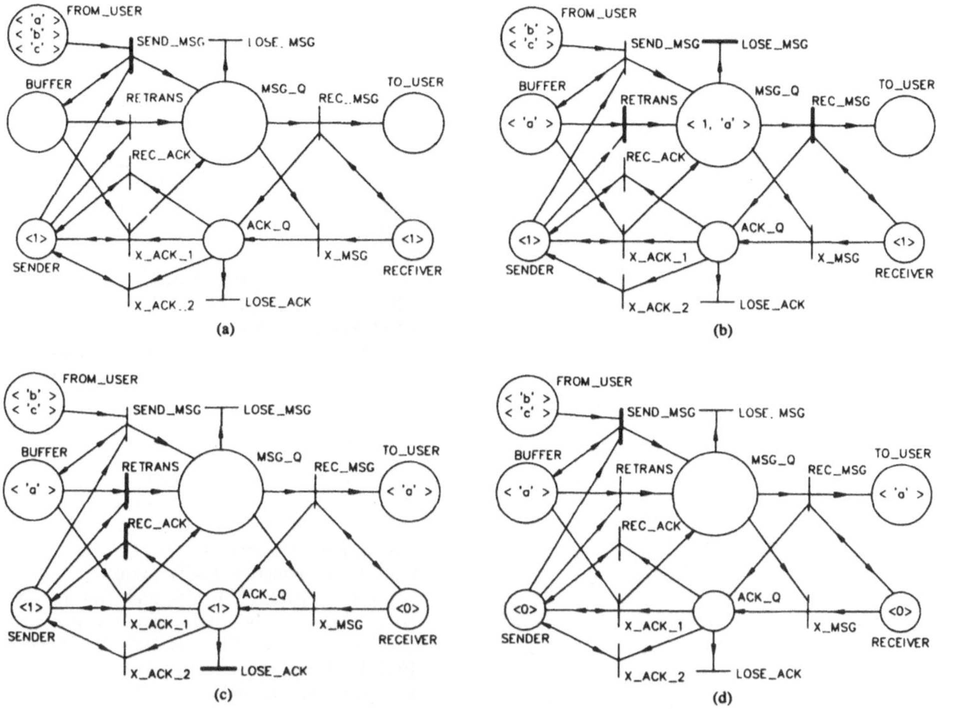

Fig. 3. SINGLESTEP simulation of single frame procedures: expected event sequence from initial marking. (a) Marking 1: PROTEAN display of NPN with initial marking (SEND_MSG enabled). (b) Marking 2: after firing SEND_MSG from marking 1 (RETRANS, LOSE_MSG, REC_MSG enabled). (c) Marking 3: after firing REC_MSG from marking 2 (RETRANS, REC_ACK, LOSE_ACK enabled). (d) Marking 4: after firing REC_ACK from marking 3 (SEND_MSG enabled).

Further reachability analysis can be performed for different values of $n200$ and b. For example, for $n200 = 1$ and $b = 2$, the reachability graph contains 348 markings. In this case graphical display is not very useful.

E. Strongly Connected Components

The strongly connected components of the reachability graph for the case of $b = 2$, $n200 = 1$, is shown in Fig. 5. The leaf nodes are the same four deadlocks as mentioned above.

F. Language Analysis

We now turn our attention to whether or not the protocol provides its service—the delivery of data without loss or duplication. The correct operation is for there to be alternating send and receive events, corresponding to the alternating occurrences of transitions SEND_MSG and REC_MSG. The reachability graph for $b = 2$, $n200 = 1$, is reduced by PROTEAN to find the language of sends and receives and the language graph of Fig. 6 is produced. It can be seen that there are four loops: one corresponding to the desired behavior of alternating send/receive; another corresonding to loss of messages where there can be an unlimited number of sends without a receive; and two other loops where duplication occurs, as there are two receives for every send.

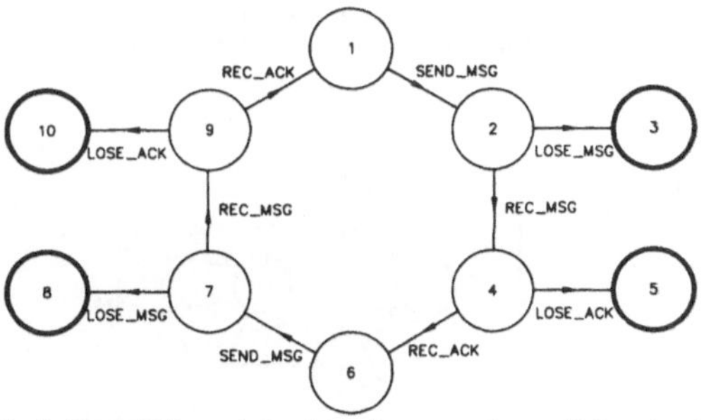

Fig. 4. Reachability graph for single frame procedures with $b = 1$ and $n200 = 0$.

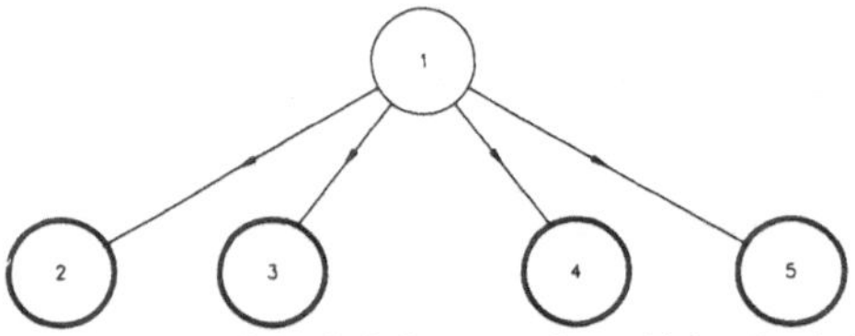

Fig. 5. Liveness graph for single frame procedures with $b = 2$ and $n200 = 1$.

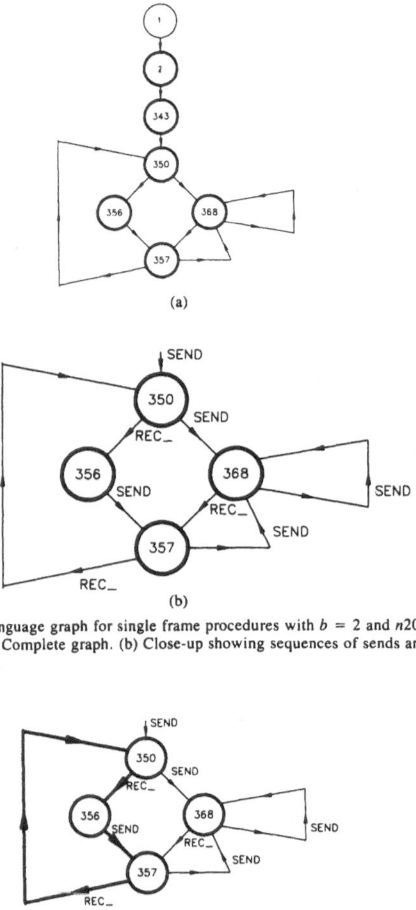

(a)

(b)

Fig. 6. Language graph for single frame procedures with $b = 2$ and $n200 = 1$. (a) Complete graph. (b) Close-up showing sequences of sends and receives.

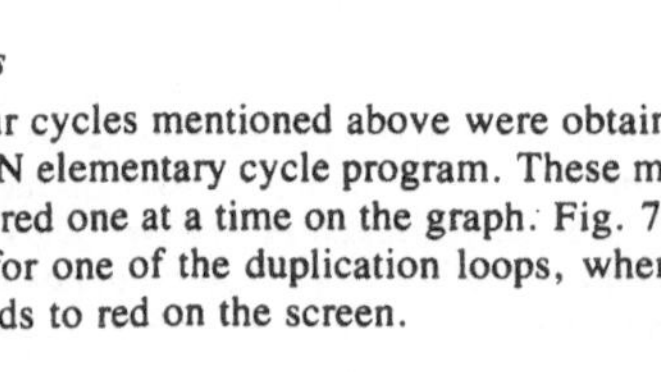

Fig. 7. A highlighted cycle demonstrating a duplication lock-up mode.

G. Cycles

The four cycles mentioned above were obtained by the PROTEAN elementary cycle program. These may be displayed in red one at a time on the graph. Fig. 7 shows an example for one of the duplication loops, where bolding corresponds to red on the screen.

H. Scenario

Once maloperations have been discovered, PROTEAN's SCENARIO facility can be used to "debug" the specification. For example, the case of loss of a message without loss by the medium is shown to occur by the following sequence, which returns to the initial marking: "**SEND_MSG**, **REC_MSG**, RETRANS, REC_ACK, X_MSG, REC_ACK, **SEND_MSG**, X_MSG, RETRANS, X_ACK_2, X_MSG, **SEND_MSG**, REC _ACK, **REC_MSG**, X_ACK_1, X_MSG, X_ACK_2". Another example showing duplication is the sequence: "**SEND_MSG**, **REC_MSG**, RETRANS, X-MSG, REC_ACK, **SEND_MSG**, REC_ACK, **SEND_MSG**, **REC_MSG**, **REC_MSG**, X_ACK_1, X_ACK_2, **SEND_MSG**, **REC_MSG**, **REC_MSG**". These scenarios can be investigated further using PROTEAN's SINGLESTEP.

VII. Other Applications

PROTEAN has also been applied to the analysis of several other protocols and signaling systems. These include: Telecom Australia Specifications of line signaling Systems T6 and P1 based on CCITT's R2D; the Open Systems Interconnection (OSI) Transport Protocols Class 0 and 2 [55], [61], [62], [56] and Transport Service [26], [27], [7]; and the SDL specifications of level 3 of CCITT's Common Channel Signalling System Number 7 [63]. Each of these investigations revealed problems with the specifications, mainly deadlocks. The references may be consulted for details, with a brief summary in [53].

The Level 3 procedures of the D-channel for ISDN Primary Rate Access have been specified using NPNs [64] and explored using PROTEAN [65]. This has revealed several problems with the definition of the Data Link Service [66], [67] and the Layer 3 procedures.

Finally, as part of collaborative research with the Commonwealth Scientific and Industrial Research Organization's Division of Information Technology, it is planned to specify the OSI File Transfer, Access and Management (FTAM) protocol in NPNs and to verify the specification on PROTEAN.

VIII. Discussion of Approach, Limitations, and Future Work

A. High-Level Net Specification Languages

It has been our experience that high-level nets provide a good model for the specification of protocols and services. Control and data flow can be expressed in the same model and analyzed together as illustrated in the example. The foundation of high-level nets in Condition/Event systems provides a firm semantic base and opens the way to a number of other analysis techniques. Perhaps the greatest limitation of the approach has been the lack of structuring techniques. We now discuss this and other issues.

1) Structure: Net models were not developed with structuring in mind. Structuring has been addressed to a certain extent by the introduction of high-level nets and channel-agency nets. Channel-agency nets allow a static structure to be refined to a point where the dynamic behavior can be described using high-level nets.

Good software engineering practice needs to be enforced with the specification of complex systems to avoid the "spaghetti net" phenomenon. The service or protocol is decomposed by function and phase and specified on separate diagrams. Different aspects, such as normal operation and various exception conditions can then be specified separately (see [27], [55] for one particular approach and [68] for another).

2) Synchronization Mechanism: An integral part of most structuring schemes is an assumption about the means of synchronization between the decomposed parts (modules). For example, often synchronous communication is assumed. With nets, no synchronization mechanism is assumed as a primitive for communication (e.g., synchronous, shared variable, queues, or more detailed mechanism). This has the disadvantage that every synchronization has to be specified explicitly but the advantage of flexibility in that the most appropriate synchronization mechanism can be chosen at the particular level of abstraction. We presently believe this to be important for providing an appropriate level of abstraction. For example, this allows service primitives to be represented as synchronous events in a service specification and then refined to more complex interactions at the protocol level.

3) Specification Methodology: A detailed methodology needs to be devised for the specification of services and protocols that provides guidelines for structuring and the net constructs to be used for particular functions. Once this is achieved, a specification "store" can be developed which houses generic net models of service or protocol functions. These net modules could then be used to construct new protocols. A method of composition would be of considerable value here (see below).

4) Abstraction: Nets include the notion of state and event as integral parts of the model. Hence they cannot be used to specify at a level of detail where all state information has been abstracted. For example if we wish to have a purely black box view of protocol services, we need only specify the allowed sequences of service primitives at each of the service access points and specify how they are related. This may be done using some form of grammar or temporal logic or perhaps Z. Whether or not this is an important limitation is yet to be determined. The required sequences may be *generated* from a gray box specification using nets (i.e., the corresponding Petri net language) [7].

Nets allow services to be specified from an understanding of the sequences allowed at the set of service access points and the queueing discipline of an abstract queue which provides the communication mechanism. Our present view is that this approach gives the designer more confidence that the specification of the service is what is wanted. This is because it provides a model which is much closer to the physical reality of the communication service.

It appears that nets provide a good compromise in abstraction. They can represent systems at many different levels of detail, but will always contain some state information. They appear to provide a model which aids the intuition of a designer of distributed systems to specify his requirements completely. This does not appear to be true with the more abstract approaches (e.g., temporal logic). The advantage of more abstract approaches is that specifications can be expressed more concisely. This raises the hope that they will be simpler to analyze, but so far just the opposite appears to be the case.

5) Expressiveness: Numerical Petri Nets are still evolving and further syntatic extensions are envisaged. For example, token attributes are variables which currently have simple types (e.g., integer or Boolean) and simple operators (e.g., arithmetic operators). More complex types and operators are being considered for the convenient representation of queues and operations on data which are relevant in applications protocols.

6) Stepwise Refinement: Ideally, a protocol specifier would like to be able to design a protocol from the specifications of the underlying service and the service to be provided, in a series of refinements. Mappings would be required to ensure that the constructed protocol retained the desired properties of the service to be provided. No general method exists for this type of top-down design. General net theory may provide some answers with net morphisms. Berthelot's work on the transformation and decomposition of Petri nets [69] is a start in this direction but further research is required for the results to be generalized for high-level nets and then applied to protocols. (A state machine based top/down design methodology is mentioned in [70], but the details are as yet unpublished in the open literature.)

B. PROTEAN

There are a number of areas where PROTEAN can be improved, including: the use of advanced color graphics hardware; additional user facilities; and improved performance. These are discussed below.

1) Graphics Workstation: Although the present system provides a relatively friendly user interface, advanced graphics workstations with multiple windows and providing direct graphical input via a mouse or tablet would considerably increase its functionality. A number of options are being considered.

2) Additional User Facilities:

Query Language: PROTEAN does not provide a flexible query language for asking questions about protocol properties on the reachability set and graph. For example, "what is the bound on a resource?"; "is a mutual exclusion property satisfied?"; or specific progress ("liveness") properties such as "will a message eventually reach its destination?" and questions of fairness. Logic-based query languages are being developed to interrogate the RG ("model checking") in a flexible way [11], [71]-[73], [40]. It is planned to incorporate these developments into PROTEAN.

Infinite Reachability Sets: Our present reachability

analysis will not terminate if the reachability set is infinite. This is indicative of the protocol being modeled requiring unbounded resources. It is thus a property which is of interest to system designers. We would like to implement an algorithm for the coverability graph (which is finite) based on the Karp and Miller algorithm [25], but extended for Numerical Petri Nets.

Language Comparison: We would like to compare sets of sequences of events, one given at a high level of abstraction defining our requirements, and the other generated by a protocol specification at a lower level, to see if they are consistent. We plan to automate such a facility.

Refinement: The present system does not give any support for stepwise refinement. It is planned to enhance PROTEAN to provide facilities for displaying net morphisms and managing specifications at many different levels. Facilities need to be provided to allow for the modification of higher levels when a change has been made at a lower level, in order to maintain a consistent set of specifications.

Simulation: The addition of backwards simulation (executing the net backwards) and automatic simulation, where enabled transitions can be selected to occur automatically according to some probability distribution, would also add to PROTEAN's functionality.

3) Other Analysis Techniques:

Improved Reachability Analysis: The major problem with reachability analysis is that the algorithm is of high complexity, requiring large amounts of CPU cycles and memory. Therefore methods to reduce the size of the generated reachability graph are important. The first step will be to use the notion of equivalence classes of markings, similar to that in [74]. Other techniques are induction over protocol parameters [75], [6] and truncation of the reachability graph on detection of errors.

Invariants Analysis: Structural analysis of nets has been developed for some time [76]–[79], [29]. This analysis is performed by standard linear algebraic techniques and does not require enumeration of the complete state space. It may be used to prove invariants of the system which are always true ("safety" properties). Unfortunately, our experience with invariants analysis, leads us to the conclusion that quite often it does not provide the desired results [80].

Despite these limitations, the linear algebraic approach may be able to provide answers where reachability analysis fails, particularly if the set of desired invariants is stated as part of the service specification. In this case one may just check that the invariant exists, rather than generating all the invariants of the system [12].

Net Reduction: As mentioned previously rules exist which allow Petri nets (PT systems) to be reduced while maintaining desirable properties. Work is progressing on applying these results to high-level nets. The reduced net will hopefully have a smaller reachability set and graph and will be easier to analyze for invariants. It may also provide a mechanism for formal refinement of specifications in a top down development methodology.

Compositional Approach: The global protocol system is decomposed into a number of processes which are analyzed individually. The total system is then composed according to rules which guarantee that the desired properties are retained. Some initial work on Petri nets has been achieved in [81], [69]. Further work is required to extend this approach to high-level nets and to apply it to practical protocols.

4) Performance: New algorithms for analysis (see [82] for example) and graphical layout are being monitored and will be implemented in a new version of PROTEAN.

5) Future: A specification of a future PROTEAN system is presently being written. It is planned to incorporate most of the features mentioned above.

IX. A Protocol Engineering Workstation

PROTEAN allows for the management of protocol specifications and for their verification and conformance testing as well as protocol maintenance.

Automatic or semiautomatic implementation [5] leads to considerable productivity gains in protocol prototyping and maintenance as new versions of protocols are ratified by standards bodies such as ISO and CCITT. Under contract to Telecom Australia, an NPN interpreter has been developed by a local software house and applied to the implementation of the OSI Transport Protocols Classes 0–3.

We are also interested in rigorous methods of conformance testing based on NPN specifications. Testing sequences can be generated from the reachability graph using the SCENARIO facility. They could also be generated interactively using the interactive simulation facility, SINGLESTEP. For complex protocols it is likely that the full reachability graph will not be able to be generated due to resource limitations. However, the SINGLESTEP facility could always be used as a controller to generate test sequences and to compare them to the expected or desired behavior. It is planned to develop conformance testing facilities based on NPN specifications.

Performance specification and analysis are also very important and not addressed by our present tools. A number of performance analysis tools have been developed based on timed and stochastic Petri nets [83], [84]. Work is also progressing on introducing time into high-level nets [83], [85], a necessary development for the modeling of practical systems. This work is being assessed with a view to the development of tools for performance specification and evaluation.

It is hoped that within the next few years an integrated set of tools will be developed for the specification, verification, performance evaluation, implementation, testing, and maintenance of protocols to provide a comprehensive protocol engineering workstation [6].

X. Conclusions

A first generation protocol specification and verification computer aid, PROTEAN, has been developed and applied successfully in an industrial environment to protocols of moderate complexity, such as the OSI Transport Protocols. Most of the requirements set out in Section V-A have been achieved.

The high-level Petri net approach compares favorably with other techniques as it provides most of the requirements desired of a specification language and opens the door to a powerful set of analysis techniques (reachability being the only one considered here).

Although several other Petri net tools exist, it is believed that it is the first such tool to combine:

- facilities for the input and analysis of several classes of nets including a high level inhibitor net, known as Numerical Petri Nets;
- color graphics for the display of nets and the analysis results (with automatic layout);
- interactive simulation (with color graphics);
- reachability analysis;
- directed graph analysis to allow:
  - —liveness and cyclic behavior to be investigated,
  - —verification against a service specification (language analysis), and
  - —specification debugging;
- on-line help at every stage; and
- detailed user and program documentation.

PROTEAN is being continually enhanced to improve user facilities and real-time performance. For example a powerful query language on the reachability set and graph will be developed in the near future. PROTEAN is limited by its reliance on reachability analysis, and breakthoughs in analysis techniques are required to overcome the state explosion problem. Current research on invariants analysis, net transformations, and reachability graph reduction may prove useful in this regard.

Although PROTEAN has mainly been used for analyzing protocols, it can be used for general systems design, particularly distributed systems.

A first step has been taken toward the design of a comprehensive protocol engineering workstation. There is still much to be done!

Acknowledgment

The authors would like to thank a number of their colleagues and fellow researchers for valuable discussions during the development of PROTEAN. In particular, the following people require special thanks: F. Symons who initiated the work on Numerical Petri Nets, protocol verification, and the development of PROTEAN within the Telecom Australia Research Laboratories; the late N. Gaylard who developed the early prototypes of PROTEAN; J. Gilmour for major development, enhancements, testing, and maintenance of PROTEAN; D. Hopkins and A. Borley of UNICO Computer for recent enhancements and maintenance work; S. Young and S. McPherson, of Software Contracts, for developing PROTEAN's graphical facilities; M. Bearman, of Canberra College of Advanced Education, for discussions on PROTEAN facilities and collaborative work on the OSI Transport Protocols; T. Batten for investigating invariants analysis and possible enhancements to PROTEAN; R. Fone and H. Everitt for support in preparing PROTEAN's User Manual; and finally to R. Fone and K. Parker of CSIRO Division of Information Technology, for feedback on PROTEAN while working on Complex OSI and ISDN protocols.

The permission of the Director Research, Telecom Australia, to publish this paper is hereby acknowledged.

References

[1] CCITT, *Red Book*, vols. III, VI, VII, VIII. Geneva, Switzerland: ITU, 1985.

[2] L. A. Jackson, "Special issue on systems design engineering," *British Telecom Technol. J.*, vol. 4, July 1986.

[3] *Proc. IFIP WG6.1 Int. Workshops Protocol Specification, Testing and Verification*, vols. II-VII. Amsterdam, The Netherlands: North-Holland, 1982-1987.

[4] T. F. Piatkowski, "Protocol engineering," in *Proc. ICC*, June 1983, pp. 1328-1332.

[5] T. P. Blumer and D. P. Sidhu, "Mechanical verification and automatic implementation of communication protocols," *IEEE Trans. Software Eng.*, vol. SE-12, pp. 827-843, Aug. 1986.

[6] J. Billington, "Protocol engineering and nets," in *Proc. Eighth European Workshop Application and Theory of Petri Nets*, Zaragoza, Spain, June 1987, pp. 137-156.

[7] J. Billington, M. C. Wilbur-Ham, and M. Y. Bearman, "Automated protocol verification," in *Protocol Specification, Testing and Verification, V*, M. Diaz, Ed. Amsterdam, The Netherlands: Elsevier, 1986, pp. 59-70.

[8] C. Paule and H. Eckert, "The Net Simulation SYstem NESSY, summary and example," GMD, Arbeitspapiere 182, Nov. 1985.

[9] M. Antilla, H. Eriksson, J. Ikonen, R. Kujansuu, L. Ojala, and H. Tuominen, "Tools and studies of formal techniques—Petri nets and temporal logic," in *Protocol Specification, Testing and Verification, III*, H. Rudin and C. H. West, Eds. Amsterdam, The Netherlands: Elsevier, 1983, pp. 138-148.

[10] J. P. Courtiat, J. M. Ayache, and B. Algayres, "Petri nets are good for protocols," in *ACM SIGCOMM '84 Symp. Communications Architectures and Protocols*, Montreal, Canada, June 1984, pp. 66-74.

[11] E. T. Morgan and R. R. Razouk, "Computer aided analysis of concurrent systems," in *Protocol Specification, Testing and Verification, V*, M. Diaz, Ed. Amsterdam, The Netherlands: Elsevier, 1986, pp. 49-58.

[12] K. Jensen, "Computer tools for construction, modification and analysis of Petri nets," in *Petri Nets: Applications and Relationships to Other Models of Concurrency* (LNCS, vol. 255), W. Brauer, W. Reisig, and G. Rozenberg, Eds. Berlin: Springer-Verlag, Feb. 1987, pp. 4-19.

[13] F. Feldbrugge and K. Jensen, "Petri net tool overview 1986," in *Petri Nets: Applications and Relationships to Other Models of Concurrency* (LNCS, vol. 255), W. Brauer, W. Reisig, and G. Rozenberg, Eds. Berlin: Springer-Verlag, Feb. 1987, pp. 20-61.

[14] G. v. Bochmann, "Usage of protocol development tools: The results of a survey," in *IFIP TC6, Protocol Specification, Testing and Verification: VII An Int. Symp.—Participant's Proceedings*, Zurich, May 5-8, 1987.

[15] CCITT, "Recommendations Z.100 to Z.104: Specification and Description Language," in *Red Book*. Geneva; Switzerland: ITU, 1985.

[16] Australia, "Reply to questionnaire on question 2/X," CCITT Study Group X, Question 2/X, Com X—No. 16, Oct. 1985.

[17] W. Brauer, Ed., *Net Theory and Applications* (LNCS, vol. 84). Berlin: Springer-Verlag, 1980.

[18] W. Reisig, *Petri Nets, An Introduction* (EATCS Monographs on Theoretical Computer Science, vol. 4). Berlin: Springer-Verlag, 1985.

[19] W. Brauer, W. Reisig, and G. Rozenberg, Eds., *Petri Nets: Central Models and Their Properties* (Lecture Notes in Computer Science, vol. 254). Berlin: Springer-Verlag, 1987.

[20] —, *Petri Nets: Applications and Relationships to other Models of Concurrency* (Lecture Notes in Computer Science, vol. 255). Berlin: Springer-Verlag, 1987.

[21] S. Drees, D. Gomm, H. Plünnecke, W. Reisig, and R. Walter, "Bibliography of net theory," in *Advances in Petri Nets 1987*, G. Rozenberg, Ed. Berlin: Springer-Verlag, Apr. 1987, pp. 309-451.

[22] W. Reisig, "Petri nets in software engineering," in *Petri Nets: Applications and Relationships to Other Models of Concurrency*, W.

Brauer, W. Reisig, and G. Rozenberg, Eds. Berlin: Springer-Verlag, 1987, pp. 63-96.

[23] F. J. W. Symons, "Modelling and analysis of communication protocols using Numerical Petri Nets," Ph.D. dissertation, Dep. Elec. Eng. Sci., Univ. Essex, Telecommun. Syst. Group Rep. 152, May 1978.

[24] E. Best and C. Fernandez, "Notations and terminology on Petri net theory," GMD, Arbeitspapiere 195, Jan. 1986.

[25] J. L. Peterson, *Petri Net Theory and the Modeling of Systems*. Englewood Cliffs, NJ: Prentice-Hall, 1981.

[26] J. Billington, "Specification of the Transport Service using Numerical Petri Nets," in *Protocol Specification, Testing and Verification*, C. Sunshine, Ed. Amsterdam, The Netherlands: North Holland, 1982, pp. 77-100.

[27] —, "Abstract specification of the ISO Transport Service definition using labelled Numerical Petri Nets," in *Protocol Specification, Testing and Verification, III*, H. Rudin and C. H. West, Eds. Amsterdam, The Netherlands: Elsevier, 1983, pp. 173-185.

[28] H. J. Genrich and K. Lautenbach, "System modelling with high-level Petri nets," *Theoret. Comput. Sci.*, vol. 13, pp. 109-136, 1981.

[29] K. Jensen, "Coloured Petri nets and the invariant-method," *Theoret. Comput. Sci.*, vol. 14, pp. 317-336, 1981.

[30] R. Valk, "Self-modifying nets, a natural extension of Petri nets," in *Automata, Languages and Programming, Udine* (Lecture Notes in Computer Science, vol. 62). Berlin: Springer-Verlag, 1978, pp. 464-476.

[31] E. Dijkstra, *A Discipline of Programming*. Englewood Cliffs, NJ: Prentice-Hall, 1976.

[32] G. R. Wheeler, "Numerical Petri Nets—A definition," Telecom Australia, Res. Labs. Rep. 7780, May 1985.

[33] M. C. Wilbur-Ham, "Numerical Petri Nets—A guide," Telecom Australia, Res. Labs. Rep. 7791, Sept. 1985.

[34] —, "Numerical Petri Nets: A guide—version 2," Telecom Australia Res. Labs., Switching and Signalling Branch Paper 111, Mar. 1987.

[35] I. Hayes, Eds., *Specification Case Studies* (International Series in Computer Science). Englewood Cliffs, NJ: Prentice-Hall, 1987.

[36] C. Hoare, *Communicating Sequential Processes* (International Series in Computer Science). Englewood Cliffs, NJ: Prentice-Hall, 1985.

[37] R. Milner, *A Calculus of Communicating Systems* (Lecture Notes in Computer Science, vol. 92). Berlin: Springer-Verlag, 1980.

[38] H. Rudin and C. H. West, Eds., *Tutorial Notes*, IBM Zurich Res. Lab. May 1987 (Lecture Notes from the Tutorial preceding the Seventh Int. Meeting on Protocol Specification, Testing and Verification).

[39] S. Aggarwal, R. Kurshan, and K. Sabnani, "A calculus for protocol specification and validation," in *Protocol Specification, Testing and Verification, III*, H. Rudin and C. H. West, Eds. Amsterdam, The Netherlands: Elsevier, 1983, pp. 19-34.

[40] S. Aggarwal, D. Barbará, and K. Meth, "SPANNER: A tool for the specification, analysis and evaluation of protocols," *IEEE Trans. Software Eng.*, vol. SE-13, Dec. 1987.

[41] S. Aggarwal, D. Barbará, and K. Meth, "Specifying and analyzing protocols with SPANNER," in *IEEE Int. Conf. Communications*, Toronto, Canada, 1986, pp. 556-562.

[42] C. A. Sunshine, Ed., *Communication Protocol Modeling*. Dedham, MA: Artech House, 1981.

[43] *Formal Methods Applied to Protocols: Literature Survey*, Rep. 1 to Alvey Directorate for FORMAP Project (includes reviews of 270 papers), July 1985.

[44] M. Paul and H. Siegert, Eds., *Distributed Systems: Methods and Tools for Specification*. (Lecture Notes in Computer Science, vol. 190). Berlin: Springer-Verlag, 1985.

[45] W. Bibel and K. Jantke, Eds., *Mathematical Methods of Specification and Synthesis of Software Systems 1985* (Lecture Notes in Computer Science, vol. 215). Berlin: Springer-Verlag, 1986.

[46] J. de Bakker, W. de Roever, and G. Rozenberg, Eds., *Current Trends in Concurrency* (Lecture Notes in Computer Science, vol. 224). Berlin: Springer-Verlag, 1986.

[47] G. Müller and R. Blanc, Eds., *Networking in Open Systems* (Lecture Notes in Computer Science, vol. 248). Berlin: Springer-Verlag, 1987.

[48] F. Commoner, "Deadlocks in Petri nets," Massachusetts Computer Associates, Wakefield, Rep. CA-7206-2311, June 1972.

[49] L. Lamport, "Proving the correctness of multiprocess programs," *IEEE Trans. Software Eng.*, vol. SE-3, pp. 125-143, Mar. 1977.

[50] A. Pnueli, "Applications of temporal logic to the specification and verification of reactive systems: A survey of current trends," in *Current Trends in Concurrency*, J. de Bakker, W. de Roever, and G. Rozenberg, Eds. Berlin: Springer-Verlag, Mar. 1986, pp. 510-584.

[51] W. Reisig, "Place/transition systems," in *Petri Nets: Central Models and Their Properties*, W. Brauer, W. Reisig, and G. Rozenberg, Eds. Berlin: Springer-Verlag, Feb. 1987, pp. 117-141.

[52] M. C. Wilbur-Ham and J. Billington, "A protocol emulation and analysis tool," in *IREECON Int.*, Sydney, pp. 30-32, Sept. 1983.

[53] G. R. Wheeler, M. C. Wilbur-Ham, J. Billington, and J. A. Gilmour, "Protocol analysis using Numerical Petri Nets," *Advances in Petri Nets 1985*, (Lecture Notes in Computer Science, vol. 222). Berlin: Springer-Verlag, 1986, pp. 435-452.

[54] M. C. Wilbur-Ham, *PROTEAN V2.1 User's Manual*. Telecom Australia Res. Labs., Clayton, Victoria, Australia, Apr. 1987.

[55] M. Y. Bearman, M. C. Wilbur-Ham, and J. Billington, "Some results of verifying the OSI class 0 Transport Protocol," in *Proc. ICCC*, Sydney, Nov. 1984, pp. 597-602.

[56] M. Y. Bearman, "Formal specification of the Open Systems Interconnection Transport Protocol class 2 using NPNs," CSIRONET, Tech. Rep. 25, 1986.

[57] W. A. Barrett and J. D. Couch, *Compiler Construction: Theory and Practice*, SRA, 1979.

[58] C. A. Vissers and L. Logrippo, "The importance of the service concept in the design of communication protocols," in *Protocol Specification, Testing and Verification, V*, M. Diaz, Ed. Amsterdam, The Netherlands: Elsevier, 1986, pp. 3-17.

[59] CCITT, "ISDN user-network interface data link layer specification," Draft Recommendation Q.921, Working Party XI/6, Issue 7, Jan. 1984.

[60] K. A. Bartlett, R. A. Scantlebury, and P. T. Wilkinson, "A note on reliable full-duplex transmission over half-duplex lines," *Commun. ACM*, vol. 12, pp. 260-261, May 1969.

[61] M. Y. Bearman, M. C. Wilbur-Ham, and J. Billington, "A formal specification of the OSI Class 0 Transport Protocol using NPNs," Telecom Australia, Victoria, Australia, Res. Lab. Rep. 7736, Oct. 1984.

[62] —, "Analysis of Open Systems Interconnection Transport Protocol Standard," *Electron. Lett.*, vol. 21, pp. 659-661, July 1985.

[63] N. Kim, "Protocol analysis of signalling system no. 7 level 3," Telecom Australia, Victoria, Australia, Res. Lab. Rep. 7652, 1985.

[64] R. J. Fone, "Specification of level three of Australian ISDN primary rate access interface using Numerical Petri Nets," Draft Telecom Australia Res. Lab. Rep., July 1987.

[65] —, "Verification of level three of the Australian ISDN primary rate access interface using PROTEAN," Draft Telecom Australia Res. Lab. Rep., July 1987.

[66] —, "Inconsistencies between X.212 and Q.920-Q.921," submitted to CCITT WP VII/5, Q.43, May 1986.

[67] —, "Another sequence of primitives in X.212," submitted to CCITT WP VII/5, Q.43, Sept. 1986.

[68] B. Baumgarten, P. Ochsenschläger, and R. Prinoth, "Building blocks for distributed system design," in *Protocol Specification, Testing and Verification, V*, M. Diaz, Ed. Amsterdam, The Netherlands: Elsevier, 1986, pp. 19-38.

[69] G. Berthelot, "Transformations and decompositions of nets," in *Petri Nets: Central Models and Their Properties*, W. Brauer, W. Reisig, and G. Rozenberg, Eds. Berlin: Springer-Verlag, 1987, pp. 359-376.

[70] J. Katzenelson and R. Kurshan, "S/R: A language for specifying protocols and other coordination processes," in *Proc. Fifth Annu. Phoenix Conf. Computers and Communications '86*, Washington, Mar. 1986, pp. 286-292.

[71] H. Tuominen, "Temporal logic as a query language for Petri net reachability graphs," in *Seventh European Workshop Application and Theory of Petri Nets*, July 1986.

[72] E. Clarke, E. Emerson, and A. Sistla, "Automatic verification of finite state concurrent systems using temporal logic specifications," *ACM Trans. Program. Lang. Syst.*, vol. 8, pp. 244-263, July 1986.

[73] M. C. Browne, "An improved algorithm for the automatic verification of finite state systems using temporal logic," in *Proc. 1986 Conf. Logic Computer Science*, Cambridge, MA, June 1986, pp. 260-267.

[74] K. Jensen, "Coloured Petri nets," in *Petri Nets: Central Models and Their Properties*, W. Brauer, W. Reisig, and G. Rozenberg, Eds. Berlin: Springer-Verlag, Feb. 1987, pp. 248-299.

[75] E. Clarke, O. Grümberg, and M. C. Browne, "Reasoning about networks with many identical finite-state processes," in *Proc. Fifth Annu. ACM Symp. Principles of Distributed Computing*, Aug. 1986, pp. 240-248.

[76] K. Lautenbach and H. Schmid, "Use of Petri nets for proving correctness of concurrent process systems," in *Proc. IFIP Congress*. Amsterdam, The Netherlands: North-Holland, 1974, pp. 187-191.

[77] G. Memmi and G. Roucairol, "Linear algebra in net theory," in *Net Theory and Applications*, W. Brauer, Ed. Berlin: Springer-Verlag, 1980.
[78] G. Memmi and J. Vautherin, "Analyzing nets by the invariant method," in *Petri Nets: Central Models and Their Properties*, W. Brauer, W. Reisig, and G. Rozenberg, Eds. Berlin: Springer-Verlag, 1987, pp. 300-336.
[79] H. J. Genrich and K. Lautenbach, "The analysis of distributed systems by means of predicate/transition-nets," *Lecture Notes in Computer Science*, vol. 70. Berlin: Springer-Verlag, 1979, pp. 123-146.
[80] T. J. Batten, "An extension to invariants analysis techniques applied to Petri net models of protocols," Sept. 1986, *Australian Telecommun. Res.*, to be published.
[81] G. Winskel, "Petri nets, algebras and morphisms," Univ. Cambridge Comput. Lab., Tech. Rep. 79, Sept. 1985.
[82] G. J. Holzmann, "On limits and possibilities of automated protocol analysis," in *IFIP TC6, Protocol Specification, Testing and Verification: VII An international Symp.—Participant's Proceedings*, Zurich, May 5-8, 1987.
[83] *International Workshop on Timed Petri Nets*, IEEE Comput. Soc., Torino, Italy, July 1985.
[84] R. Razouk and C. Phelps, "Performance analysis using timed Petri nets," in *Protocol Specification, Testing and Verification, IV*, Y. Yemini, R. Strom, and S. Yemini, Eds. Amsterdam, The Netherlands: Elsevier, 1985.
[85] *International Workshop on Petri Nets and Performance Models*, IEEE Comput. Soc., Madison, WI, Aug. 24-26, 1987.

packet switching techniques, the specification of OSI services and protocols and has headed a protocol engineering group developing specification and verification techniques and computer aids for communication protocols. In July 1986 he took up a Telecom Australia Postgraduate Research Award to study for a doctorate in the area of protocol engineering and distributed computing systems at the University of Cambridge Computer Laboratory.

CORRECTIONS

In Section IX, A Protocol Engineering Workstation, the first paragraph is in error due to a misprint and should be replaced by:

PROTEAN allows for the management of protocol specifications and for their verification. Areas not addressed by PROTEAN include automatic implementation of specifications, conformance testing and protocol maintenance.

22.
Specification and Properties of a Cache Coherence Protocol Model

C. Girault, C. Chatelain and S. Haddad

G. Rozenberg (ed.): Advances in Petri nets 1987. Lecture Notes in Computer Science, vol. 266. Springer, Berlin Heidelberg New York 1987, pp. 1-20

Abstract
This paper describes a cache coherence protocol for an architecture composed of several processors, each with their own local cache, connected via a switching structure to a shared memory itself split into several modules managed by independent controllers. The protocol prevents processors from simultaneously modifying their respective copies and always provides a processor requiring a copy of a memory location with the most up-to-date version. A top down description and modeling of the protocol is given using Predicate/Transition nets. This modeling allows to formally describe the complex synchronizations of this protocol. Then invariants are directly obtained without unfolding the Predicate/Transition net. They are the basis for studying behavioral properties.

Keywords
Petri nets, protocols, specification, verification, cache coherence, memory hierarchy, multiprocessors.

1 Introduction

Caches are small memories that match the processor speed. They are inserted between the processor and the primary memory to speed up the memory hierarchy. The caches and the primary memory are respectively partitioned into small **blocks** and **lines** of fixed size that are automatically managed by hardware just as the "page on demand" feature is managed by an operating system. For multiprocessors, the use of private caches instead of a common one is essential to reduce memory conflicts. We consider architectures composed of several processors, each with their own local cache, connected via a switching structure to a shared main memory.

The use of local caches in a multiprocessor gives rise to the **"cache coherence problem"**. Several caches can contain a copy of a particular main memory line but it is required that all programs continue to run exactly like for a unique cache without having to introduce extra synchronizations or data reorganizations. It is necessary to automatically prevent processors from simultaneously modifying their respective copies of a same line and a processor must not read a line while another is writing it. At last, a processor requiring a copy of a memory location must always receive the most up-to-date version, otherwise inconsistencies would arise. These requirements are the same as for the classical "readers and writers" problem but here there is a high level of parallelism, the sequence of accesses by a processor may be interrupted by requests from other processors, the line replacements must be managed and the speed is very crucial.

Formal models are needed to understand and to study such complex synchronization protocols. Our main purpose is to give a top down description of the model of one of the best cache coherence protocol [Archibald 84]. This model is based on Predicate / Transition nets [Genrich 79, 86, Brams 82, Jensen 86]. Our second purpose is to underline some key behavioral points and explain how colored invariants are a basis for further behavioral studies, such as correctness, fairness, and bounds for the hardware resources required.

The Petri nets have been designed to keep the main invariants associated with processors, caches and controllers. Other invariants concerning data structures and messages have been directly constructed together with the net itself. The nets have been especially drawn to hightlight these invariants as early as the design stage. The size of the model, its complex tokens, and its arrays of FIFO queues would make it difficult to find meaningful invariants by unfolding the colored net into an uncolored one or by using the methods and tools that are now appearing for colored nets [Haddad 86a, 86b, Jensen 86, Memmi 86].

This paper is organized as follows. In the next section we describe the hardware components and their physical and logical interconnections. Then in section 3 we describe in detail a particular coherence protocol for which we give in section 4 the structure of our Predicate/Transition model. Sections 5 and 6 detail the cache and controller models. Section 7 gives the basic ideas for validation and finally Section 8 concludes on possible extensions and other applications of our model.

2 Multiprocessor architecture

A cache is an high speed small memory C which is inserted between the processor P and the primary memory M (cf. Figure 1). The effective memory access time is reduced because most accesses concern data already in the cache and are performed at the cache speed, others need to first load a line into some block of the cache. The access time ratio between primary memory and cache is 10 versus over 10000 in the secondary-primary memory case and the line transfers are speeded up by organizing the memory into interleaved banks. The cache small size induces frequent line faults (miss ratio around 10^{-2} instead of 10^{-5} for pages) that must be very quickly managed by hardware. Cache organizations, management and performances are extensively covered in [Smith 82, 85, Hwang 84].

For large numbers of processors, sharing a unique common cache, even with several buses, becomes inefficient because of cache access conflicts. Some projects [Gadjski 83, Gottlieb 83, Kuehn 85] envision several hundreds of processors. Thus local caches are needed. Additional management is needed when a processor requires a line that has been modified in another cache.

For the "**write-through** strategy" the primary memory is updated each time a processor performs a write access. Thus any cache may obtain up-to-date information either directly from the memory or by continuously listening to bus exchanges (snooping caches). This elementary strategy does not apply for large numbers of processors. The more efficient "**write-back** strategy" lightens the bus congestion by avoiding these systematic updates: instead a complete block update is only required when a modified block must be replaced or accessed by another cache [Censier 78, Archibald 84, Katz 85, Yen 85].

Again, the bus and memory bottlenecks may be avoided by splitting the memory into parallel modules M managed by parallel controllers K and by using multibuses or networks allowing parallel exchanges [Papamarcos84, Archibald 85]. (cf. Figure 1). All these types of parallelism complicate the coherence problem. Hypotheses on the network and careful serializations must be introduced to deal with the asynchronism of caches and controllers.

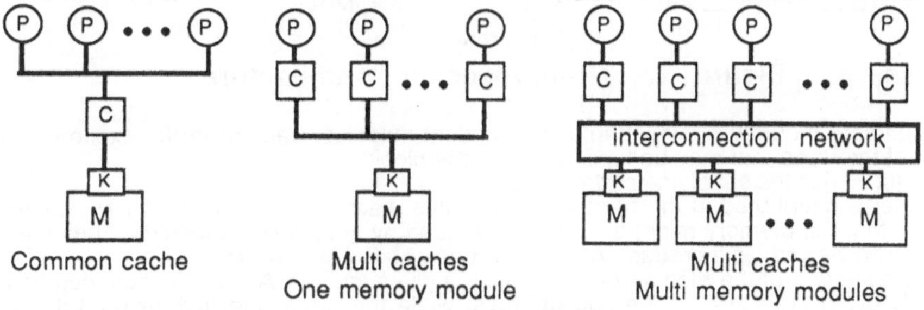

Figure 1. Cache Organizations

The multiprocessor architecture (cf. Figure 2.) consists of:

- Processor-cache pairs Pk-Ck: **N**=[0..N-1] is the set of their indices,
- Primary memory modules Mx and associated controllers Kx: **M**=[0..M-1] is the set of their indices,
- Two independent networks: the first one from caches towards modules and the second one from modules towards caches.

A processor Pk, while executing instructions produces elementary **access orders** for the cache and consumes the delivered results. Each processor Pk is provided with its own cache memory element Ck and communicates only with it by a private interface. FIFO buffers may allow the processor to prepare several access orders in advance but here this pipelining will not be considered.

A cache Ck manages a set of **blocks** which may contain copies of primary memory **lines**.

• **B**=[0..B-1] is the set of block addresses in a cache. Let $b \in$ **B**.
The cache consists of:
• An associative **directory** to recognize if a given line a is present in some block b of the cache.
Moreover one **state bit s[b]** per block allows to know whether block b is **valid** (i.e. it contains an up to date copy of the line a) or is **invalid**. A **modified bit m[b]** indicates whether the cache may write and modify the block b (the line is said dirty) or may only read it (the line is said clean).
• A set of data values corresponding to the contents of the respective blocks.
• FIFO queues for incoming and outgoing messages.

On the one hand a cache has a local part which responds to the orders of the corresponding processor and a global part which responds to the messages from all the controllers. These parts share common data structures needing partial exclusion. On the other hand a cache is composed of two separate units: a control unit and a memory unit. They may be organized into pipelined stages: buffer management, associative search, replacement, order and message analyses, and data transfers [Baer 85].

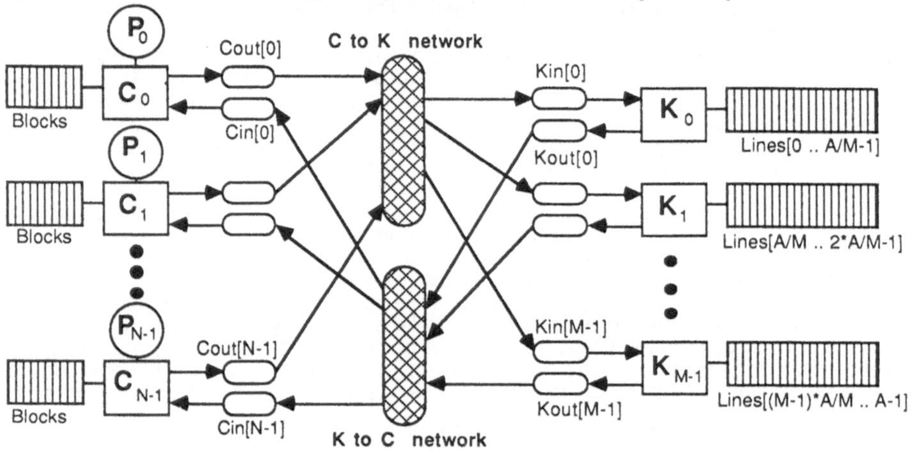

Figure 2. Multiprocessor Architecture

The primary memory is divided into independent modules. Each **module** Mx stores a fixed subset of the **lines** that may be loaded into cache blocks.
• **A**= [0..A-1] is the set of line addresses.
This set is distributed in the M memory modules, each one containing A / M consecutive lines, so that the memory might be simply upgraded by adding new modules. Thus the set **A** may also be viewed as a product **A**= [0..M-1, 0..A / M-1] = **M** × **A** / M.
• The number x of the module Mx containing a given line $a \in$ **A** is easily obtained from the **high order bits x=â** of the line address, while the remaining **low order bits ã** give the relative line number within Mx.
• This address decomposition is denoted by: **a = (â,ã)** with $\hat{a} \in$ **M** and $\tilde{a} \in$ **A** / M.

The **controller Kx** of a module Mx has to manage all the lines of Mx. It contains :
• a bit map to encode the **global state g[x,ã]** of each line a=(x,ã) of Mx: access rights given by Kx to some caches and information about owner caches,
• a set of data values that are the contents of the respective lines of Mx,
• FIFO queues for incoming and outgoing messages.

Each controller is divided into a control unit and a memory unit. The first one manages informations on all the module lines, receives requests and returns from the caches, and broadcasts queries. The memory unit manages the line contents and sends grants together with data to the caches. Moreover the FIFO queues and the network interface are managed in parallel.

The interconnection facility is a crucial point for large multiprocessors. Any cache may be connected with any controller because protocols imply the exchanges of control and data messages between all the caches and all the controllers. To avoid path conflicts between messages in opposite directions, this simultaneous transmission of messages between several partners is distributed on **two interconnection networks** [Auguin 84, Baer 85b, Chi Yuan 84, Siegel 85]. The network from controllers to caches must allow any controller to broadcast messages to all caches.

Each network may transfer different types of messages of various lengths, distinguished by appropriate headers and cut into consecutive packets. Packets may contain any combination of orders, addresses, status bits and data. An **header flag** (F= L or S) is used to distinguish either short messages (one packet) for only control information or long ones (several packets) that include also data for line contents. Emitter and receiver units will recognize and manage these packets [Baer 85b].

For verification purposes the model must be independent of the network design: hence we assume as little as possible about the message tranfers. Messages between the same partners must be delivered in the emission order; entries in the input-output queues of the controllers and buffering must avoid any loss of messages and so all the ordered transfers are assumed to be performed. Also the message broadcasting is the responsability of the network layer. All broadcast messages are assumed to be synchronously delivered.

3 Coherence protocol

With a common cache several processes would be allowed to write or read objects stored into a same line provided they do not interfere or they are conveniently synchronized. They are only synchronized for sharing of logical objects and do not care how objects are stored into lines and how lines are loaded into caches. Thus, with multicaches, the Bernstein's conditions apply to lines instead of objects and must be automatically enforced. A line cannot be simultaneously loaded for write into several caches and cannot be loaded for read into one cache and for write into another one. When a line has been modified into one cache it must be stored again in the primary memory before being loaded in another one so that a processor requiring a copy of a memory location always receives the most up-to-date version.

Various solutions have been proposed according to the amount of information distributed among caches and controllers [Censier 78, Archibald 84, Yen 85]. A global state of N+1 bits per line may allow a controller to exactly know the state and all the owners of each line. But as the lines are small and thus numerous the amount of information for all the lines would become costly when the number N of processors increases. This paper deals with a solution of J.Archibald and J-L.Baer, that uses only two bits per line, allowing an easy expansion of the number of processor-cache pairs. For each line the global state distinguishes three cases: "**Absent**" (the line is in no cache), "**PresentW**" (only one cache has been granted to write the line that is considered as modified), and "**PresentR**" (any number of caches have been granted to only read the line which remains not modified).

Since the information given by these states does not include the location of the owner caches, a memory controller might need to broadcast queries to all caches (e.g. for writing-back or invalidating a given line). This feature may be efficiently implemented by the network. The number of messages and the involved delaysare also a crucial point. Waiting positive or negative answers from all the caches would induce congestion and long delays according to the most loaded answering caches. Thus when a query is broadcast the modeled protocol requires only an answer if one of the caches must write-back a line. Moreover line contents are always transferred together with command messages to avoid extra synchronizations.

This solution has been already modeled to study the architectural design and its performance [Baer 85a, 85b]. Here we present a simplified variant: it deletes some states and messages useful for performance considerations but is more suitable to study the protocol properties.

Four cases are considered, depending on a read or write order, and on a **hit** (valid copy of the desired line already present in the cache) or a **miss**.

- ***Read hit*** : In case of immediate access, no special action is needed. As this is the case for most of the accesses, its detection must be efficient.
- ***Write hit*** : An immediate access is possible if the line has been already modified in this cache. Otherwise Ck sends a REQUEST to the controller Kx monitoring it. This REQUEST might entail a broadcast QUERY to **invalidate** the line if it is present in some other caches. Further the GRANT will be given by Kx.

- ***Read miss*** : A REQUEST for a load on read is submitted to the controller, possibly after making room by the replacement of another line. If the wanted line is not modified in some other cache, Kx can immediately give the GRANT. Otherwise Kx first sends a QUERY to force the owner cache to **update** the line by a write-back. Finally the GRANT and the data are sent to the requiring cache.
- ***Write miss*** : Kx must broadcast a QUERY to **invalidate** the line if it is owned by some caches only for read or to enforce a **purge** if the line is owned for writing in one cache. A purge implies both an invalidation and a write-back. Finally, the GRANT and the data are sent to the requiring cache.

Additional messages are implied by line replacements:

- ***Replacement*** : When Ck needs to make room for a new line, it must choose a block and replace its line content. This line is called the **old line**. If it is valid and has been modified a write-back is needed: the controller must be informed (for a global state change) by a RETURN command and moreover data must be written back. Otherwise the Ck may use freely the block containing the old line.

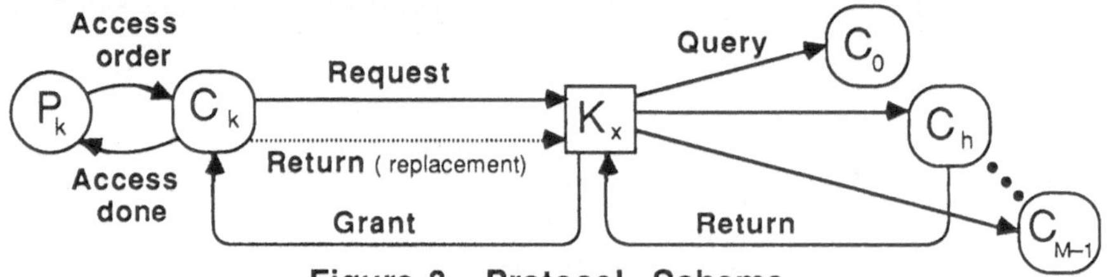

Figure 3. Protocol Schema

Figure 3. shows the messages sent by the caches (**C→K**) and the controllers (**C←K**).

• **C→K.** For any access such that a cache owns a valid copy of a line and does not need a higher grant, the cache Ck may immediately access its line copy: this is the case of a "read hit" or the case of a "write hit "for a modified line. But if Ck owns a line "a" only for read while Pk wants to modify it (case write hit for an unmodified line) or if it does not own such a valid copy (case read miss or case write miss), Ck must send a command REQUEST(k,a,t) to the controller x=â, where t=r or t=w indicates the desired GRANT for this line.

• **C→K.** Before getting a line "a", Ck may also need to make room for it by replacing another old line "o". If the copy of this old line is still valid and has been modified, the cache warns the controller Ky (with y=ô) by sending a command RETURN(k,a,**C**) together with the line content **C**, in order to purge (i.e write-back and invalidate) the line. If the copy is invalid or unmodified, Ck has only to free the associated block without warning Ky.

• **C←K.** When a line requested by a cache Ck is owned for write by some other cache (state "PresentW"), the controller Kx must broadcast a QUERY message to ensure that the unknown owner writes back the line. Hence Kx waits for one RETURN command with the line content before sending a GRANT to the requesting cache. In case of a READ REQUEST the owned line must be updated but may remain valid and shared. In case of a WRITE REQUEST the line must be purged by the owner cache since no more than one cache may be allowed to write a given line.

When the requested line is absent or only owned for reading, the controller does not need to obtain a RETURN for writing-back and thus may immediately send a GRANT(k,â,ã,g,**C**) to Ck, where g= r or w tells whether read or write is allowed, and **C** is the line content. In case of a WRITE REQUEST and if Kx suspects that there is at least one owner (state "PresentR"), it broadcasts a QUERY forcing the invalidation by all owners. To avoid traffic overload Kx does not wait for any acknowledgement from the caches. So the coherence protocol needs to be sure that all invalidations will be made before the granted access. In case of a READ REQUEST or if Kx is sure that no cache owns the line (state "Absent"), no broadcast is needed.

A message QUERY (h,â,ã,u) is broadcast to the set of all caches excepting the requesting one (h belongs to the set (**N**-k) of the destination cache numbers). The parameter u tells the new line-use i.e. whether the line a=â,ã must remain valid or not.

• **C→K.** Whenever a cache Ck receives a QUERY about a line, it determines its behavior according to its directory. If the line is not present or is invalid, the cache has nothing to do. A valid queried line which is unmodified has only to be invalidated without sending a RETURN whereas if it is modified the cache must purge or update the line according to u and send a RETURN.

Each time a line is returned by a cache for purge, update or replacement, the controller has to manage its global state and to copy the line content into the module.

• **C←K.** Any return for writing-back a requested line is called an **awaited** return. Such an awaited return may be an answer to a query but may also have been sent for a replacement initiated by the cache during a query broadcast for this same line. In both cases, it allows the controller Kx to send a GRANT with the new line content to the requesting cache. Conversely, the other returns (that concern only replacements) are called **unwaited** returns and do not imply to send grants.

4. Petri net structure

The structure of our model (cf. Figure 4.) reflects the underlying hardware structure (cf. Figure 2.). It consists of four nets working in parallel and interconnected by three interfaces modeling common information structures or buffers. These nets are:

- the P-net for processor actions,
- the B-interface for the private bus of a processor-cache pair,
- the L-net for local cache-processor interaction,
- the C-interface for the cache common data structures shared by L and G,
- the G-net for the global cache actions upon receiving messages from the controller,
- the N-interface for the network communications between L, G and K,
- the K-net for the controller flow of actions.

All the place and transition identifiers are prefixed according to this structure.

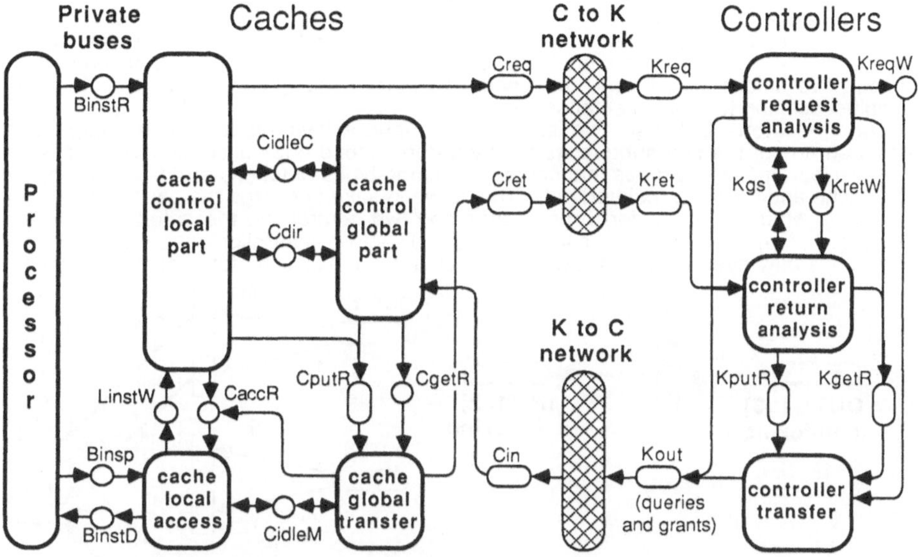

Figure 4. Complete Model Structure

4.1 Token domains

Two main sets **N**=[0..N-1] and **M**=[0..M-1] of colored tokens are associated with caches and controllers. Let h, k, l ∈ **N** and x ∈ **M**.

A =[0..A-1] is the set of line addresses. Let a, o ∈ **A**.

B=[0..B-1] is the set of block addresses in a cache. Let b ∈ **B**.

Other sets of colors are used for the values of control variables and message parameters.

t ∈ (r, w) is an access type: either read or write.

m ∈ (r, w) is a block modified bit: either read, or write.

s ∈ (v, i) is a block state bit: either valid or invalid.

u ∈ (v, i) is new block use queried by a controller: either valid or invalid.

g ∈ (z,r,w) is a line global state: respectively absent, presentR or presentW.

F ∈ (S,L) is an header flag to distinguish either pure commands (requests and queries) or commands together with data (returns and grants).

Note that h, k, l, x, a, o, b, t, m, s, u, g, F always denote parameters while r, w, z, v, i, S, L always denote constants.

4.2 Queue arrays

All places contain colored tokens, most of these tokens are compound ones with several fields. These places are private to caches or controllers and one of the token fields is always a cache or controller number. According to the architecture, they are equivalent to place arrays having **N** or **M** for range.

There are two types of places: normal ones (drawn as circles) and queues (drawn as ovals). Normal places may only contain unordered sets of tokens. Queues must contain ordered tokens in order to insure serialization and fairness conditions. However as these queues model private buffers belonging to caches or controllers, it is essential not to mix all the tokens. Thus these places are arrays of queues, with **distinct orders** for each cache or for each controller. Any queue-array place in a cache net is indexed by the cache number modeling a private independent buffer for each cache. A token is directed into the subqueue corresponding to its cache number field. Similarly, any queue-array place in a controller net is indexed by the controller number modeling a private independent queue for each controller.

Ordering of each queue is FIFO within each cache or each controller. However, lines are independent, thus serialization conditions apply separately for each line: if an inhibitor arc apply to a transition needing a token in a subqueue, the first authorized token from the subqueue (if any) may be extracted. If an inhibitor arc is issued from a queue, any token, whatever its place in the subqueue may prevent to fire the transition.

4.3. Network interface

All the cache subnets communicate with all the controller subnets through two sets of arrays of FIFO places that model the streams of the common interconnection networks. Indeed two unidirectional networks, used for control messages and data transfers, are the simplest solution for implementation.

The transitions between two associated queue-arrays model the message transfer between any of the first subqueues to any of the second ones according to the emitter and receiver parameters. Queues that contain the same types of tokens are differently indexed in the cache and controller colored nets. The networks take in charge the packet routing. Figure 5. shows what would be, for three caches and two controllers, the unfolding of a network moving (k,x,m) tokens from a cache queue OUT into a controller queue IN. The unfolded transitions may even work synchronously, provided the exclusion for destination queues

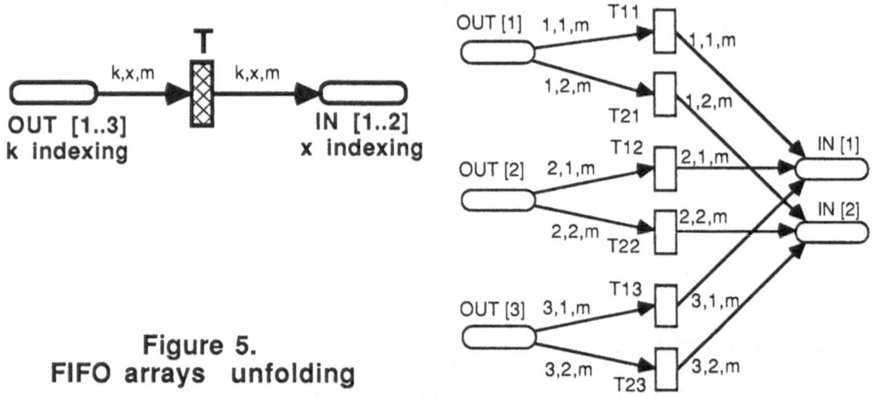

Figure 5.
FIFO arrays unfolding

Due to space constraints, we do not detail how supplementary network interface units would allow multiplexing and demultiplexing packets (cf. [Baer 85b]). The header flag (F= L or S) in the first packet of a message is used to distinguish either short messages (one packet for requests or queries) for only control information or long ones (several packets for returns and grants) that include also data. This implies that interface places may contain two types of tokens. Besides this, the two networks are not symetrical.

For the **C to K network** the requests and returns sent by a cache for a given block generally concern different lines and do not need serialization and when a serialization is needed it is well managed by the cache. Queues are exploited for fairness rather than for serialization, so some freedom may be left to introduce some transfer priorities. Thus Figure 4. deviates from Figure 2. by allowing two parallel input queues and output queues for the C to K network.

- Places Creq (F=S) or Cret (F=L) are used for the tokens modeling commands sent by the caches to the controllers.
- An auxiliary place Kin (not shown in Figure 4.) is used to receive both tokens. Then they are directed into places Kreq (L=S) or Kret (F=L) to be independently managed by the parallel request unit or return unit of each controller.

The **K to C network** must insure serialization of grants and queries that are mixed into places Kout and Cin. Moreover it implements synchronous broadcast.

- Place Kout is used for the messages sent by the controllers to the caches. The FIFO ordering of mixed queries (F=S) and grants (F=L) is essential from each controller towards each cache.
- Place Cin is used for all the messages received by the caches from controllers (either F=S or F=L). The FIFO ordering of queries and grants for each line is essential within each cache.

4.4 Processors and bus interface

A sequential **processor net** is sketched (cf. Figure 6.). When it is ready (k token in Pr) it may submit a new access order to its cache (transition Pinst) and the processor becomes waiting (k token in Pw) for the access to complete. After the instruction execution (transition Pexec) it becomes ready again.

At the initialization place Pr contains a complete set of k tokens for all k in **N**.

The **bus interface** is composed of 3 places (cf. Figures 4. and 6.).

- Place BinstR is used by the processor to transmit the desired line address a and access type t . Thereby a (k,a,t) token is put in this place and the cache analyzes only this information to perform all needed coherence actions before allowing the access.
- Place Binstp may contain a (k,f,d,c) token used later to perform the access itself: f specifies the access function, d is the displacement within the line a and c is the new value of the content of the word (a,d) in case of a write or is a dummy value in case of a read.
- Place BinstD may contain a (k,c') token modeling the result delivered to the processor: this new value c' is the previous content of the word (a,d) in case of a read or a dummy value in case of a write. Functions for atomic exchanges or incrementations are thus allowed.

These places as well as places Pr and Pw may be FIFO ones, allowing (in a further model) the processor to give several access orders in advance.

5. Caches

A cache is composed of a local part managing its processor orders and a global part managing the messages coming from all the controllers. These two parts share common data structures containing information on all the blocks owned by this cache (cf. Figure 4.).

5.1 Common data structure Interface

- Place Cdir models the cache directory: it always contains a set of (k,b,a,s,m) tokens for all possible values of k and b, allowing the associative search of the block "b" that may contain a given line "a". To achieve high speed this search is implemented by hardware. Indeed a given line may be only placed in a very small subset of blocks (set associative cache [Smith 82]), but this restriction would have little effect on the model.

  The s component gives the state of the line copy contained in the block "b": it indicates whether the content of this block "b" is valid or not. The m component gives the mode m (r=read or w=write) of the block "b": it indicates what accesses are allowed for this block and thus whether this block "b" has been modified or not.

  At the initialization, the s components of all the (k,b,a,s,m) tokens must be invalid, the "a" components are thus meaningless, and for convenience all the m components are set to r.

Each cache is split into a control unit and a memory unit. Both units perform either local or global work and need to communicate.

- Place CidleC contains a k token whenever the control unit of Ck is idle. It insures a mutual exclusion between instruction analysis in the local part and query or grant analysis in the global part. At the initialization it contains a complete set **N** of tokens.
- Place CidleM contains ak token whenever the memory unit of Ck is idle. It insures a mutual exclusion between word access in the local part and block transfers in the global part. At the initialization it contains a complete set **N** of tokens.
- Places CgetR and CputR may contain the tokens denoting the blocks that are to be

loaded or written-back by the cache memory unit. Place CputR is a FIFO one, allowing some transfers to remain waiting. Place CgetR is 1-bounded if the processor orders are not pipelined and must be a FIFO one otherwise.

- Place CaccR indicates that the access wanted by the processor may be performed.

5.2 Cache local part

The cache local part waits for the processor orders. It has to analyse whether the accesses are allowed or requests must be sent to controllers, to manage the block contents and to finally perform the ordered accesses (cf. Figure 6.).

Presence analysis

The cache Ck has to accept and analyze a new access order (k,a,t) produced by its processor. Place LinstW serves to limit the number of orders that may be looked ahead. Here its initialization by one k token per cache allows each local part to manage only one order at a time.

- Transition LAinst: It must be idle and waiting for a new order. Moreover an access must be avoided when an invalidation, an update or a purge is required: an inhibitor arc from Cin prevents this acceptance as soon as a query for the line a is received. Transition LAinst moves the token (k,a,t) from BinstR to LinstA.

Then the cache checks if a copy of the line a is present in some block b. This is known by the presence or the absence of a (k,b,a,s,m) token in the associative directory Cdir. The state of a copy present in a block b is indicated by the s component with either s=v (valid copy) or s=i (invalid copy).

- Transition LvalhitA: The line "a" is present in the block "b" of Ck and still **valid** (s=v): thus no replacement nor line getting are needed. Moreover if (t=r) or (m=w) no request is needed and the local access may be immediately allowed. Therefore the (k,a,t) token in LinstA becomes a (k,b) token in CaccR while the directory is left unchanged and the cache control unit becomes idle again (a k token is put into CidleC).
- Transition LvalhitM: If the line a is present and **valid** (s=v), but if (t=w) and (m=r), a request to the controller is needed because the processor wants to write an unmodified line. Therefore the (k,a,t) token in LinstA is moved to Creq while the directory is unchanged and the control unit becomes idle.
- Transition LinvhitRW: The line "a" has been present in block "b" but is now **invalid** because either a purge or an invalidation has occurred. As the block b is free no replacement is needed. Therefore the (k,a,t) token in LinstA is moved to Creq denoting a READ or a WRITE REQUEST.
- Transition Lmiss: The line a does not appear in any cache block. This condition is modeled by an inhibitor arc from Cdir requiring the absence of all tokens (k,b,a,s,m) for b belonging to the set **B**. The token from LinstA is moved to Lrep for the replacement algorithm.

Replacement algorithm

The three next transitions choose a block "b" for the line "a", move the token from LinstA to Creq for a READ or WRITE REQUEST and check the old line "o" that was previously in "b". The (k,b,o,s,m) token allows to distinguish three cases.

- Transition LinvrepN: If the line "o" is invalid (s=i) in the block "b" nothing is needed.
- Transition LvalrepF: If the line "o" is valid but unmodified (s=v and m=r) it may be freed without any warning because no writing-back is needed and the controller does not keep track of the owners of a given line.
- Transition LvalrepP: If the line "o" is valid and modified a purge is needed therefore a (k,b,o) token is placed into CputR ordering the cache memory unit to write-back the block content into the memory module.

" The token in Cdir is changed into a (k,b,a,i,r) token denoting that from now on the block "b" is associated with the new line a but remains invalid and by convention unmodified until the line "a" is obtained.

Cache access

The access may either immediately occur in case of a valid hit without request or on the contrary after a grant and a line transfer. In any case the access is allowed as soon as the information is available, that is when a (k,b) token appears in CaccR.

- Transition Lacc: This transition performs the access using the parameters (k,f,d,c) found in Binstp: d is the displacement within the line a, f denotes the access function (load, store or exchange) and c is the new value of the content of the word (a,d) in case of a write or

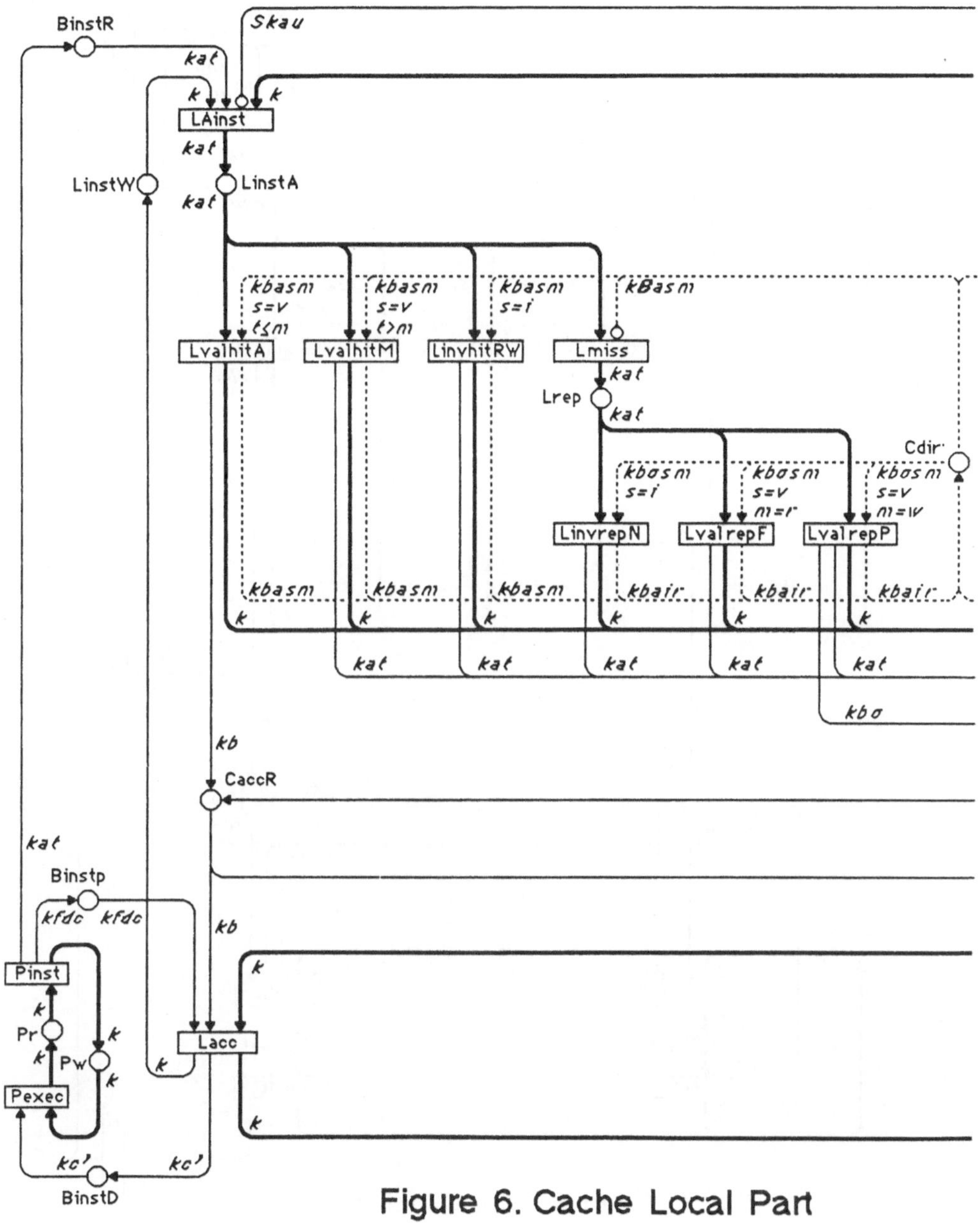

Figure 6. Cache Local Part

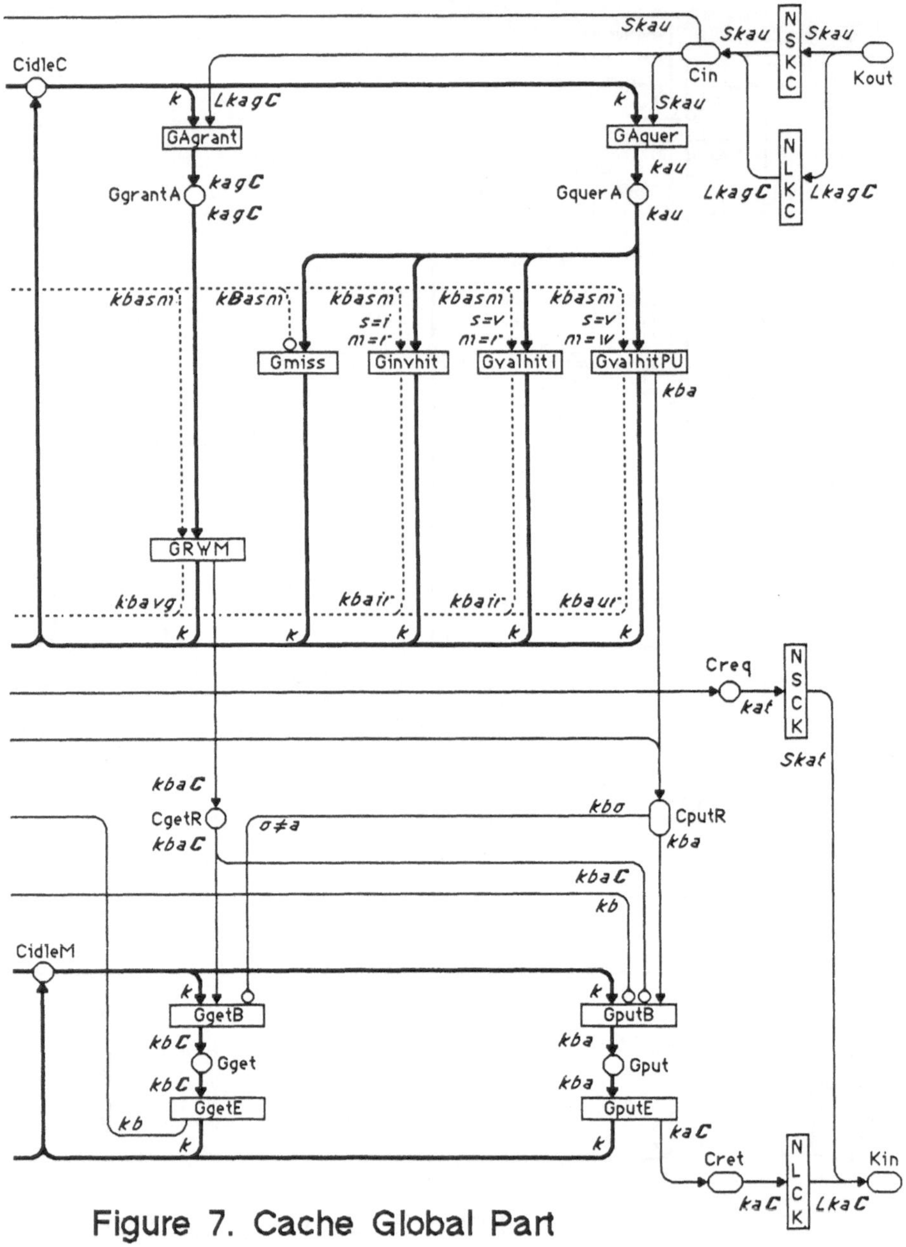

Figure 7. Cache Global Part

exchange or is a dummy value in case of a read. It delivers a (k,c') token to the processor, via place BinstD where c' is the previous content of the word (a,d) or is dummy in case of a write. Atomic exchange accesses may be performed for process synchronizations.

An inhibitor arc from place CaccR to transition GputB plays a very important role in the cache global part: in case of a concurrent query for the same line it prevents any writing-back of the block b (and thus any modification by another cache) before the wanted local access has been performed. This also insures fairness of the processor access, because the block content may not be purged or updated before the access. A parallel invalidation is possible but it does not matter because this is the last read access allowed for Pk. Another processor Ph may perform a write but, as the processes of Ph and Pk are not suspended by a logical synchronization, the accesses involve independent data in the same line. The invalidation will immediately prevent further accesses: if, after some synchronization with Ph, Pk tries to read the line it will be forced to request the new line version.

5.3 Cache global part

The cache global part has to analyse the queries broadcast by the controllers, to wait for the grants and to transfer the block contents (cf. Figure 7.).

Queries and grants are kept in order in a unique FIFO place Cin, and are distinguished by the flag F: either F=S for queries or F=L for grants that also convey the line content. This ordering is essential: when conflicts occur between two caches, it happens that the queue contains for the same line an invalidation query, followed by a grant because in parallel the cache has requested a write, and again followed by a purge because another cache has also requested a write. The corresponding actions must be performed strictly in this order. Of course independent messages for different lines may be merged in the queue and the FIFO is also useful for fairness.

The controllers communicate only in terms of line addresses within the primary memory and ignore their placement into blocks. Thus when receiving a query or a grant for a line "a", an associative search in the directory Cdir is needed. This search determines if a copy of the line "a" is associated with some block "b". In this case, it cheks whether this copy is valid or not and whether it has been modified or not.

Query analysis

Queries must be extracted from Cin and then analyzed.

- Transition GAquer: By firing of this transition the cache Ck accepts a message QUERY(S,k,a,u) broadcast from a controller where u is the new use of the line queried by the controller (u=i for an invalidation or a purge and u=v for an update). Removing the k token from CidleC insures exclusion for the cache control unit. A (k,a,u) token is put into GquerA to be analyzed by one of the three next transitions.
- Transition GvalhitPU: If a valid modified copy of the line "a" is present in some block "b" then a write-back is needed. The token in GquerA is completed by "b" to give a (k,b,a) token in the FIFO place CputR, thereupon ordering a RETURN to the memory unit. The directory Cdir is immediately updated according to the u component of the query: if u=v the cache is still allowed to read the block.
- Transition GvalhitI: If a valid unmodified copy of the line "a" is in block "b" no write-back is needed but only an invalidation. Thus a (k,b,a,i,r) token is put in Cdir where "a" becomes meaningless.
- Transition Ginvhit: If a copy of the line "a" has been present in some block "b" but has become invalid the cache control unit ignores this query.
- Transition Gmiss: If there does not exist any block b containing line "a" the cache control unit is again not concerned.

Grant analysis

The grants inform the cache about the transfers of line contents performed by the controllers: the global part of the cache gets the transferred data into the associated blocks.

- Transition GAgrant accepts the message GRANT(L,k,a,g,**C**) sent by a controller where g is the new type of granted accesses. The controller sets g=r for a READ GRANT and g=w for a WRITE or MODIFY GRANT.

Here to simplify, the line content **C** is always transferred together with the GRANT message. In case of a MODIFY GRANT this may be superfluous. Avoiding this is quite complicated to manage because a line may be invalidated after a cache has sent a MODIFY REQUEST and the controllers do not know which cache owns the line [Baer 85a, 85b].

- Transition GRWM updates the directory Cdir: the (k,b,a,v,g) token specifies the new access rights to the block b. A (k,b,a,**C**) token is put into CgetR to notify the cache memory unit that it must copy the transferred data into the block b before performing the access itself. It would be possible to avoid the associative search by storing the b-a association at the request time.

Data transfers

The cache memory unit has either to **get** into a block the copy of a line transferred from a module or to transfer a block content that will be **put** into the associated line of a module. Sufficient output buffers are needed for these block contents: these buffers are modeled by the FIFO place Cret. The memory operations are not atomic, so we explicit their begins and ends between which the cache memory unit remains busy.
Several serialization conditions must be insured before starting the data transfers notified by the tokens in places CputR and CgetR.

- Transition GgetB: A (k,b,a,**C**) token in CgetR indicates that the content **C** of the line "a" must be got into the block "b". But if a replacement is needed for "b", it must be done before the get or, at least, the old block content must already be put into the output buffer Cret. Thus there is an inhibitor arc applying to any (k,b,o) token in the queue CputR block "b" of Ck but for an old line "o" distinct from the new line "a".
- Transition GputB : A (k,b,a) token in CputR indicates that the content of the block b must be extracted from the cache memory, put into the FIFO output buffer Cret and transferred.
  However because of the request of another cache, a query for a purge or an update of the line a may be received soon after this line has been granted; but the corresponding return must only occur after the line has been got and also after the granted access has been performed (possibly modifying the line). Thus an inhibitor arc must delay this return as long as there is a (k,b,a,**C**) token in GgetR for the same line a.
  The second inhibitor arc prevents the return while there is a (k,b) token in CaccR for any line (it might have been restricted to the same line a but it is better to give a complete priority to all accesses). It is possible to have at the same time a (k,b,o) token in CPutR with o≠a, a (k,b,a,**C**) token in CgetR and a (k,b,a) token in CputR; however no deadlock will occur because the replacement, the block getting, the access and the write-back will occur in this order. It would be possible to avoid these inhibitor arcs by using a unique queue instead of CgetR and CputR. But, with the presented solution, the access to the requested line may be fastened by giving priority to CgetR.

The last two transitions finish the put or get operations and release the cache memory unit.

- Transition GgetE: A (k,b,**C**) token remains in Gget while the line content **C** is being got into the block b. As soon as this is done, a (k,b) token is placed into CaccR to allow the access by the cache local part.
- Transition GputE: A (k,b,a) remains in Gput while the block b content is extracted. Then a (k,a,**C**) token where **C** is the whole block content is placed into the FIFO place Cret that models the cache output buffers. It will then be transferred and put into the corresponding primary memory module.

5.4 Cache interface with the networks

Requests and returns, even sent sent by the same cache for the same block, concern generally different controllers. It is not necessary to serialize them. The well ordering of replacements and transfers is managed by the cache alone.

- Transition NSCK removes a (k,a,t) token from Creq to transfer an (S,k,a,t) token into Kin.
- Transition NLCK removes a (k,a,**C**) token from Cret to transfer an (L,k,a,**C**) one into Kin.

To insure coherence queries and grants for a same line must be carefully serialized: the emission order in the controller must be kept as well in the K to C network as in the cache queue management and processing.

- Transition NSKC moves a (S,k,a,u) token from Kout to Cin for a short query transfer. Doing so for each cache, it implements the broadcast utility.
- Transition NLKC moves an (L,k,a,g,**C**) token from Kout to Cin for a long grant transfer including the line content **C**: thus the order of grants and queries is kept.

6. Memory controllers

A controller Kx deals with all the REQUESTS and RETURNS from all caches but concerns only the lines contained in the module Mx. A request from one cache may need the controller to broadcast a set of QUERIES to all other caches. In the case of a write-back (either after a query or for a replacement) a RETURN together with the line content is received from the owner cache. When a controller has the most recent version of a requested line it sends a GRANT with the line content to the requesting cache (cf. Figure 8.).

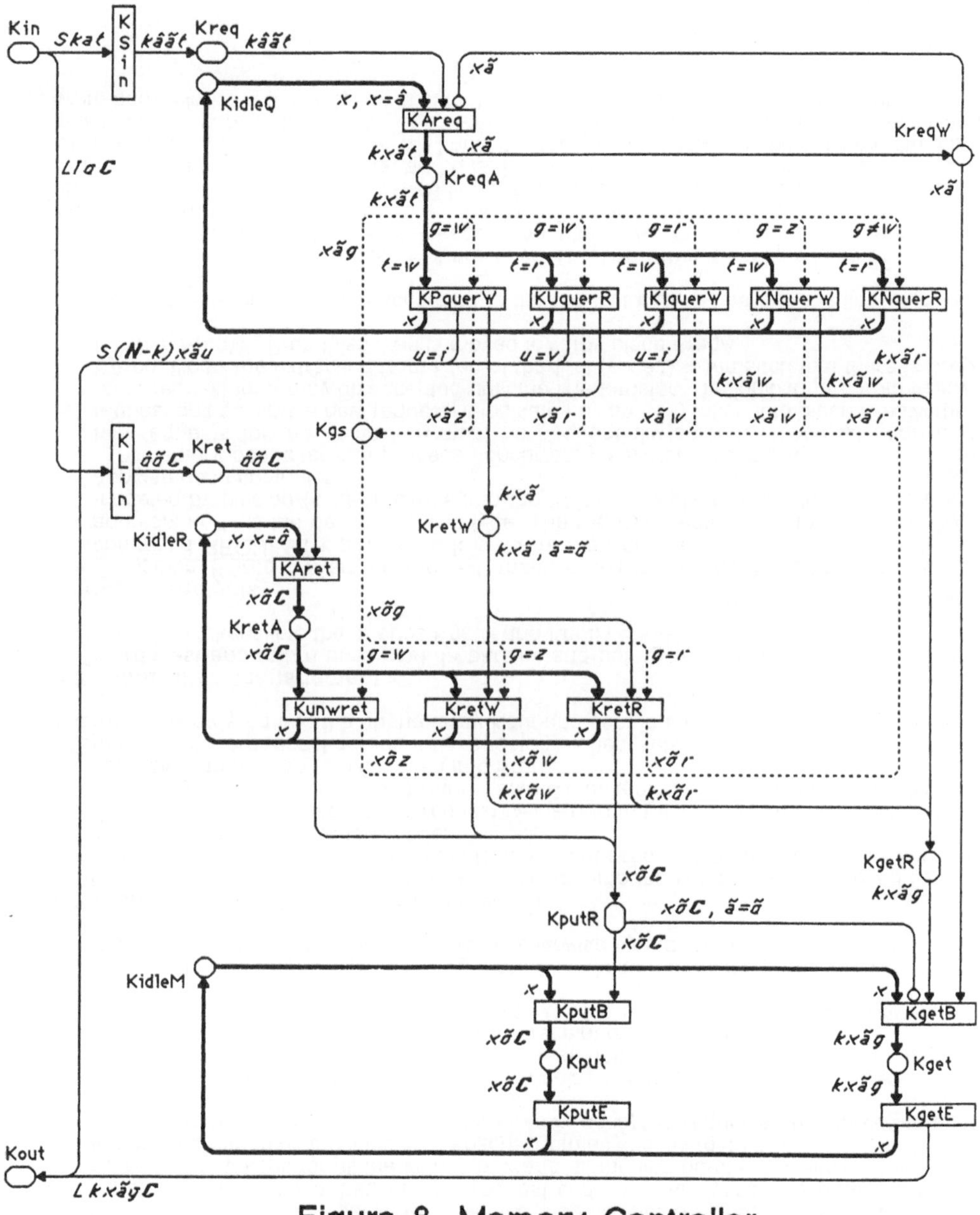

Figure 8. Memory Controller

The cache and network must only deal will line addresses. Thus each line address a is also expressed as a=(â,ã) where the high order bits â determine the controller Kx such that x=â and the low order bits ã give the relative line address inside Kx. Conversely a line address a=(â,ã) may be constructed from â=x and ã.

A controller is composed of two separate units: the control unit and the memory unit. We have split each control unit into a request analysis unit and a return analysis unit. These two control subunits may work in parallel but share information on all the lines owned by the module.

6.1 Common data structure

- Places KidleQ, KidleR and KidleM contain an x token respectively in when the request, return and memory units of Kx are available.
- Place Kgs keeps the global states of all the module lines and insures a mutual exclusion, separately for each line, between the request and return analyses. Kgs always contains a set of (x,ã,g) tokens for all the values of x and all the line numbers ã within each module. At the initialization all the lines are "Absent" from all caches (g=z).
- Place KreqW remembers what are the already accepted requests: this serves to accept only one request for a given line ã at a time.
- Place KretW is used whenever a request is delayed until the end of a write-back from the cache holding the wanted line. The request unit places an (k,x,ã) token into it to remember the request parameters until the receipt of the awaited return.
- The FIFO-array KputR contains the (x,õ,**C**) tokens placed by the return unit to warn the memory unit for the write-backs.
- The FIFO-array KgetR contains the get orders indicated by (k,x,ã,g) tokens placed by the request unit if no write-back is needed or otherwise by the return unit.

6.2 Controller interface with the networks

As soon as a command from a cache is received into the subqueue Kin indexed by x, it is directed into one subqueue of the arrays Kreq or Kret according to its flag and the address is decomposed into upper and lower bits.

- Transition KSin removes a short request (S,k,a,t) from Kin and puts a (k,â,ã,t) into Kreq.
- Transition KLin removes a long return (L,l,o,**C**), deletes the now useless returning cache number l and puts an (ô,õ,**C**) token into Kret.

Short tokens broadcast for queries and long ones selectively sent for grants will be appropriately ordered into the queue array Kout and sent in the same order.

6.3 Controller management

The request and return units work in parallel and both give tranfer orders to the memory unit by puting tokens into the queues KgetR and KputR.

Request analysis

- Transition KAreq: When the request analysis unit is idle this transition may accept a command REQUEST(k,a,t) where k is the requesting cache number, a=â,ã is the line address, and t is the desired access type. The requests received by the controller Kx are modeled by placing (k,x,ã,t) tokens into the FIFO-array place Kreq where distinct orders apply to each controller.

  A memory controller can process concurrently several requests for different lines but two requests for the same line must be sequentially served. Hence as soon as the request unit accepts a new request, it indicates by an (x,ã) token into place KreqW that no other request for ã may be accepted until the line transfer. The inhibitor arc from KreqW to KAreq allows the extraction from Kreq of the first (k,x,ã,t) token such that there is no (x,ã) token in KreqW. Then this token is moved towards place KreqA.

Five cases are distinguished according to the request parameters and the line global state given by the (x,ã,g) token in Kgs.

We first consider the two cases where g=w. The line a has been modified so the owner cache Cl must write-back the corresponding block. As this owner Cl is unknown, the controller broadcasts a query to all caches Ch for h belonging to the set **N**-k. This "broadquery" is denoted by a set (S,**N**-k,â,ã,u) of tokens in the FIFO place Kout where S indicates a short query message, and u is the future valid bit imposed by Kx (u=i for a purge or u=v for an update). The owner will know by inspecting the modified bit of its directory that it must send its copy of the line a. The controller will send a grant to Ck only after receiving the write-back. Hence a (k,x,ã) token is put into KretW to wait for the corresponding return.

- Transition KPquerW : If the REQUEST is a WRITE (t=w) the requesting cache must obtain the line a for it alone. Thus Kx sets u=i. The new global state in Kgs becomes "Absent" (x,ã,z) until the purge return will be received.
- Transition KUquerR : If the REQUEST is an UPDATE (t=r) the requesting cache may only obtain the line a in a sharing mode. Thus Kx sets u=v allowing the owner to keep the line. The new global state becomes "PresentR" (x,ã,r).

The three remaining cases concern a line that is not used for writing (g≠w). No write-back is needed because an updated copy of the line "a" still remains in the module. Thus a GRANT may be immediately sent to Ck and a (k,x,ã,g) token in KgetR notifies the memory unit to promptly send the GRANT with the line a to Ck. The controller must check whether an invalidation is needed or not.

- Transition KIquerW: In case of g=r and a REQUEST with t=w from Ck, the controller broadcasts, perhaps uselessly, a QUERY for invalidation to all the caches of the set **N**-k. This query is denoted by a set (S,**N**-k,x,ã,i) of tokens in Kquer. All the owner caches will find a modified bit set to r and will only reset the associated valid bit to i without sending any return. This avoids congesting the network with useless answers. The new global state Kgs becomes (x,ã,w) for "PresentW" since Ck will be the unique owner of the line a.

  When g=r the grant may be in fact overvalued (perhaps the line a is not used at all and so g=z would be better): because of the two-bits encoding the global state g the controller is unable to exactly count the number of owners of a line. This cost only some superfluous invalidation requests.

  After an invalidation Kx does not wait for an acknowledgement: the grant message may be sent immediately but remains serialized after the invalidation query in the FIFO-array places Kout and Cin. The broadcast will keep the whole interconnection network busy: thus any grant towards any cache as well as for this request or for a logical synchronization (implying a write) will be postponed.

The controller avoids broadcasting a QUERY for invalidation in the two important cases of a READ REQUEST or of a line certainly absent.

- Transition KNquerW: When g=z the line a is absent, thus the controller has no query to broadcast even for a WRITE request (t=w). The new global state in Kgs becomes "PresentW" (x,ã,w).
- Transition KNquerR: In case of a READ REQUEST (t=r) provided g≠w, there is again no query to broadcast. Ck will become a new reader so the global state is always set to "PresentR" . An (x,ã,r) token is put into Kgs.

Return analysis

When the return analysis unit of Kx is idle a command RETURN (ô,õ,**C**) may be accepted: as usual x=ô, õ is the addresss of the line, **C** is the line content. The number of the sending cache does not need to be transferred since the line content **C** itself comes together with the command.

- Transition KAret accepts the return modeled by the first token (ô,õ,**C**) of the FIFO-array place Kret. The token is moved to KretA.

Some returns are **awaited** answers to previously broadcast queries in which case there is a (k,x,ã) token in KreqW precisely for the same line ã=õ. Most of these awaited returns have been sent by the cache global parts because of an update or purge query but some, originated by the cache local parts for replacements, must be considered as an anticipated answer to a query sent in parallel. This unification of returns whatever their cause is very important to avoid deadlocks and in any cases the cache directory has been coherently set. Other returns for replacements remain **unwaited**.

The analysis distinguishes three cases according to the global state (x,õ,g) of the returned line x,õ. In the three cases an (x,õ,**C**) token is put into KputR for ordering the memory unit to store the line content.

For an awaited return, it does not matter if a replacement purge occurs when only an update was awaited because the cache owner count is not managed. The (k,x,ã) token in KretW with ã=õ remembers the requesting cache. A (k,x,ã,g) token is put into KgetR for ordering the memory unit to send the GRANT and the line content to the requesting cache. This unit will correctly send the grant only after the writing-back.

- Transition KretR: If g=r, the return is a purge or update awaited after a READ request. As the requesting cache becomes either the unique or a supplementary owner the global state always becomes "PresentR". So, an (x,õ,r) token is put into Kgs. As g=r, a (k,x,ã,r) token is put into KgetR.

- Transition KretW: If g=z, this is an awaited purge return after a WRITE request. The requesting cache Ck becomes the unique owner of the line o: the state becomes "PresentW". So, an (x,ō,w) token is put into Kgs. A (k,x,ã,w) token is put into KgetR.
- Transition Kunwret: If g=w this may onky be an unwaited purge return for a replacement at the initiative of Cl while an update was waited for by Kx as an answer to its query. It would be superfluous to check by an inhibitor arc that the return is unwaited because g=w only in this case. As the sending cache was the unique owner of the line, the line state now becomes"Absent". So, an (x,ō,z) token is put into Kgs.

Data transfers

The last two steps concern the supervision of transfers by the controller memory unit either to **put** into the module a line content written-back by some cache or to send to a cache a line copy that it will **get** into the associated block. These operations are not atomic, thus we have explicitly shown their beginings and ends between which the controller memory unit becomes busy (an x token is removed from KidleM).

- Transition KgetB: A (k,x,ã,g) token in KgetR orders to send the content of the line a=x,ã to the requesting cache Ck. However the transfer must not begin if a write-back remains to be done for the same line. Such a write-back is indicated by a (x,ō,**C**) token anywhere in the FIFO place KputR: so this condition is checked by an inhibitor arc issuing from KputR with ã=ō. This inhibition also applies when, because of the memory load, a return remains waiting in KputR. Further, the request unit orders directly a grant in KgetR.
- Transition KputB: A returned line o=ô,ō is modeled by a (x,ō,**C**) token in KputR where x=ö, and **C** denotes the line content to put into the module. This operation is always possible and will allow further line sendings by KgetB. Thus no deadlock may occur.

It would be possible to avoid the inhibitor arc by merging the FIFO arrays KputR and KgetR but this would give less possibilities to speed up the grants.

The last two transitions terminate the data management.

- Transition KgetE: A (k,x,ã,g) token remains in Kget while the line content is extracted. Subsequently the GRANT g, the line address a and its content **C** must be transferred by the network to the requesting cache Ck. For that, a (L,k,x,ã,g,**C**) token for a long message is placed into the FIFO ouput buffer Kout.
- Transition KputE: An (x,ō,**C**) token remains in Kput while the contents **C** of the line o=(x,ō) is written-back into the memory module.

7 Validation

The Predicate/Transition net has been designed for keeping elementary invariants corresponding to the work of each unit, to the correct structuring of each block and line state, and to the management of each request.

To simply express complex invariants we need concise notations.

- Let P be a place of color domain D and c a set of colors belonging to this domain D, then P \ c will denote the number of tokens in P of colors belonging to c.
  This may be extended if D is a product of component domains and c is a set of compound colors.

For the sake of simplicity we assume that all elementary domains may be distinguished by adequate renamings.

- Let D be a product of distincts component domains and D' a subdomain of D i.e. a product of only some of the components of D. Moreover these components may be differently ordered in D' than in D. Then proj(P / D') will denote the bag of tokens of P, restricted to the D' components and reordered with respect to D'.
  For instance place CaccR may contain (k,b) tokens for $k \in \mathbf{N}$ and $b \in \mathbf{B}$, CaccR \ k denotes the number of tokens (k,*) concerning the cache Ck whatever the block b may be.
- Let c be a set of colors in D', and P \ c = proj(P / D') \ c. Then (P + Q) \ c is an abbreviation for P \ c + Q \ c.

With these notations partial token information diversely combined in various places may be used to construct invariants.The following invariants take into account the conventional initial markings without processor pipelining.

For each processor $k \in \mathbf{N}$:

• (Pr + Pw) \ k =1 ;
There is one unique token of color k either in Pr or in Pw.
• (Pr + Binstp + BinstD) \ k =1 ;
There is one token of color k in Pr or one (k,*,*,*) token in Binstp or one (k,*) token in BinsD. Thus place Pw, useful to explain that the processor is waiting for its result, is in fact redundant.
• LinstW \ k = (BinstD + Pr + BinstR) \ k
This indicates that place LinstW is redundant according to the initialization. With pipe-lining the place would serve to bound the number of orders accepted by a cache independently of the look ahead possibilities of the processor.
For each cache k ∈ **N** :
• (CiddleC + LinstA + Lrep + GquerA + GgrantA) \ k =1 ;
There is one token of color k in CidleC or one compound token with a k component in one of the places of the cache control unit.
• (CidleM + Gput + Gget) \ k =1 ;
There is one k token or component in the places of the cache memory unit.
For each cache k ∈ **N** and each block b ∈ **B**:
• Cdir \ (k,b) =1;
There is one (k,b,*,*) token for each block of each cache.

For each controller x ∈ **M** :
• (KidleQ + KreqA) \ x =1 ;
• (KidleR + KretA) \ x =1 ;
• (KidleM + Kput + Kget) \ x =1;
For each controller x ∈ **M** and each relative line address ã ∈ **A** / M :
• Kgs \ (x,ã) =1 ;
There is one (x,ã,*) token for each line of each module.
• KreqW \ (x,ã) = (KreqA + KretW + KgetR) \ (x,ã)
This shows that place KreqW is redundant. However its removal would need three inhibitor arcs.

More interesting invariants describe the request and return treatments involving communications between caches and controllers. For instance we obtain for the requests:
For any cache and processor k ∈ **N**:
• (Pr + BinstR + LinstA + Lrep + Creq) \ k + Kin \ (S,k) +
(Kreq + KreqA + KretW + KgetR + Kget) \ k + (Kout + Cin) \ (L,k) +
(GgrantA + CgetR + Gget + CaccR + BinstD) \ k =1;
There is either one k token in Pr or one (k,*,*) in BinstR, LinstA, Lrep, Creq or one (S,k,*,*) in Kin or one (k,*,*,*) in Kreq, KreqA, KgetR, Kget or one (k,*,*) in KretW or one (L,k,*,*,*,*) in Kout or one (L,k,*,*,*) in Cin, or one (k,*,*,*) in CgrantA, CgetR or one (k,*,*) in Gget or one (k,*) in CaccR, BinstD. This means that the processor is idle or its order is managed by its own cache or by some controller. Note that for place Kin only the (S,k,*,*) request tokens are considered thus transition KLin does not interfere with this invariant. For places Kout and Cin only the (L,k,*,*,*,*) and (L,k,*,*,*) grant tokens are considered thus transitions NSKC and GAquer do not interfere.
• Pr \ k + (BinstR + LinstA + Lrep + Creq) \ (k,a) + Kin \ (S,k,â,ã) +
(Kreq + KreqA + KretW + KgetR + Kget) \ (k,â,ã) + Kout \ (L,k,â,ã) +
Cin \ (L,k,a) + (GgrantA + CgetR) \ (k,a) + (Gget + CaccR + BinstD) \ k =1;
This expresses the fact that when processor Pk is not idle, a request for a given line a is being managed. Moreover transition LvalhitA keeps the invariant by removing a (k,a,*) token in LinstA and puting a (k,b) one in CaccR but checks that there is a (k,b,a,*,*) token in Cdir. Transitions LvalhitM, LinvhitRW, LinvrepN, LvalrepF, LvalrepP move a (k,a,*) token towards Creq but insure also that there is a (k,b,a,*,*) token in Cdir. Transition GRWM will not change this token but put a (k,b,*,*) token in CgetR that is moved first in Gget and then in CaccR without any change of Cdir so that b is exactly the block associated with a.
A refinement would allow to check that the access type is correctly transmitted. The crucial point concerns transitions KPquerW and KUquerR that remove a (k,â,ã,t) token in KreqA for putting an incomplete (k,ä,ã) token in KretW and an (â,ã,g) token in Kgs with respectively g=z or r for t=w or r. Subsequently transitions KretW and KretR restore the matching. Thus the g component in KgetR, Kget, Kout, Cin, GgrantA and then Cdir always matches with the ordered t.
The next invariant is very important because it insures that a line may be owned for write by

at most one cache.

For any module a = (x,ã) ∈ **A** :

- Kgs \ (w,x,ã) + KretW \ (x,ã) =
(KgetR + Kget) \ (x,ã,w) + Kout \ (L,x,ã,w) + Cin \ (L,a,w) + (GgrantA + Cdir) \ (a,w) + (CputR + Cput +Cret) \ a + Kin \ (L,a) + (Kret + KretA) \ (x,ã)
Only tokens for line "a" with a w field are involved. Transitions KPquerW and KUquerR remove one token from Kgs for puting one into KretW while other query transitions are not concerned because g=w. Transition Kunwret removes a token from both Kgs and KretA; transitions KretR and KretW remove one token from both KretW and KretA but KretA also puts one token both into Kgs and KgetR. The directory place Cdir is involved for this invariant just for the relevant fact concerning the right for writing a. Transition GRWM moves a token from GgrantA to Cdir, while transitions LvalrepP and GvalhitPU move one token from Cdir to CputR respectively for replacement or for queries.

To study returns it will be necessary to distinguish invalidations from other queries, blocks replaced without notification as well as awaited returns from unwaited ones. Such properties need too much behavioral study to be detailled here. Redundant service places and tokens have been introduced to take more advantage of the invariants as well as colored net transformations to reduce the net size.

8 Perspectives and conclusion

The Petri net modeling of the protocols is very valuable for understanding the difficulties that can arise from the synchronization and exchanges of messages and for obtaining behavioral properties.

Such a modeling allow also to easily experiment with variants or to increase the parallelism inside the proposed architectural design [Baer 85b]. Slight modifications of the underlying Petri net model may allow the study of a spectrum of solutions. The current model assumes that messages always arrive at their destination but it could be expanded to include acknowledgements and time-out mechanisms in order to extend the protocol to distributed multiprocessors. The model could be used as an entry point for a simulation program, a Stochastic Petri net [Florin 85, Ajmone Marsan 85] or a Timed Petri net [Chretienne 84] to obtain performance estimates.

Acknowledgements

The authors thankfully acknowledge the four referees for having suggested very helpful improvements as well for the paper structure as for numerous details.

References

Ajmone Marsan,M., Chiola,G. and Conte,G. " Generalized stochastic Petri net models of multiprocessors with cache memories" Proc of 1rst Int. Conf. on Supercomputing Systems, St Petersbourg Florida (December 1985), IEEE 85CH2216-0, pp690-696.

Archibald,J. and Baer,J-L. "An economical solution to the cache coherence problem." Proc of 11th Int. Symp. on Computer Architecture, IEEE, 1984, pp. 355-362.

Archibald,J. and Baer,J-L. "Cache coherence protocols: evaluation using a multiprocessor simulation model" *ACM TOCS*, 4(4), november 1986, pp. 273-298.

Auguin,M. and Boeri,F. "Etude comparative de reseaux d'interconnection dans une architecture MIMD". Congres sur les nouvelles architectures pour les communications, Paris (sept 1984)

Baer,J-L. and Girault,C. "A Petri net solution for the cache coherence problem." Proc of 1rst Int. Conf. on Supercomputing Systems, St Petersbourg Florida (December 1985), IEEE 85CH2216-0, pp. 680-689.

Baer,J-L. and Girault,C. "Design of a parallel architecture for a solution to the cache coherence problem." Parallel computing, Berlin,September 1985, North Holland.

Berthelot,G. "Analyse de processus parallèles par transformation de réseaux de Petri." AFCET, *T.S.I.* ,vol4 n1, Janvier 1985, pp 73-82.

Brams,G.W. "Réseaux de Petri : théorie et pratique." Masson ed. vol 1 and 2, Paris 1982 and 1983.

Carlier,J., Chretienne,Ph. and Girault,C. "Modeling scheduling problems with timed Petri nets" 4th. European workshop on application and theory of Petri nets, Toulouse (September 1983).

Censier,L.M. and Feautrier,P. "A new solution to coherence problems in multicache systems." *IEEE TC C-27*,12 (Dec 1978), pp. 1112-1118.

Chi Yuan Chin and Kai Hwang. "Connection principles for multipat packet switching

networks" Proc of 11th Int. Symp. on Computer Architecture, IEEE, June1984, pp. 99-108.
Chretienne,P. "Exécutions controlées des réseaux de Petri temporisés" AFCET *T.S.I.*, vol3 n1, Janvier 1984, pp 23-31.
Diaz,M. "Petri net based models for the specification and validation of protocols". 5th. European workshop on application and theory of Petri nets, Aarhus (June 1984).
Finkel,A. and Memmi,G. "FIFO nets: a new model of parallel computation", 6th G.I. conference on theoretical computing, Dortmund (January 1983).
Florin,G. and Natkin,S. "Les réseaux de Petri stochastiques." AFCET, *T.S.I.* ,vol4 n1, Janvier 1985, pp 143-160.
Gajski,D., Kuck,D., Lawrie,D. and Sameh,A. "CEDAR: a large multiprocessor." *Computer Architecture News 11, 1* (March 1983), pp. 7-11.
Genrich,H.J. and Lautenbach,K. "The analysis of distributed systems by means of Predicate / Transition nets" semantics of concurrent computation, Lecture Notes in Computer Science n°70, Springer Verlag 1979.
Genrich,H.J. " Predicate / Transition nets" Advanced course on Petri nets, Bad Honnef, September 1986.
Gottlieb,A., Grishman,R., Kruskal,C.P., Mc Auliffe,K.P., Rudolph,L. and Snir,M. "The NYU ultra computer: Designing an MIMD shared memory parallel computer." *IEEE TC C-32,2* (Feb. 1983), pp. 175-189.
Haddad,S. and Bernard,J.M. "ARP a sotware for specification and validation of protocols and distributed applications", 3rd Conference-Exhibition on Software engineering, AFCET, Versailles, May 1986.
Haddad,S and Girault.C. "Algebraic structure of flows of a regular Colored Petri Nets". 7th. European workshop on application and theory of Petri nets, Oxford (June 1986).
Huber,P., Jensen,A.M., Jensen,L.O. and Jensen,K. "Towards Reachability Tree for high-level Petri Nets". 5th. European workshop on application and theory of Petri nets, Aarhus (June 1984).
Jensen,K. "Coloured Petri nets and the invariant method" T.C.S. 14, n 3, North Holland pub., (June 1981).
Jensen,K. "Coloured Petri nets" Advanced course on Petri nets, Bad Honnef, September 1986.
Hwang,K. and Briggs,F.A. "Computer architecture and parallel processing", 1984, McGraw-Hill Books Company.
Katz,R.H., Eggers,S.J., Wood,D.A., Perkins,C.L. and Sheldon,R.G. "Implementing a cache consistency protocol" Proc of 12th Int. Symp. on Computer Architecture, IEEE, Boston, June 1985, pp. 276-283.
Kuehn,J.T.,Schwederski,T. and Siegel,H.J. "Design of a 1024-Processor PASM system". Proc of 1rst Int. Conf. on Supercomputing Systems, St Petersbourg Florida (December 1985), IEEE 85CH2216-0, pp 603-612.
Kujansuu,R. and Lindqvist,M. "Efficient algorithms for computing S-invariants for predicate/transition nets" 5th. European workshop on application and theory of Petri nets, Aarhus, (June 1984).
Memmi,G. "Méthodes d'analyse de réseaux de Petri, réseaux à files et applications aux systèmes en temps réel." ,Thèse d'état, Université Paris 6, June 1983.
Memmi,G. and Vautherin,J. "Advanced Algebraic Techniques" Advanced course on Petri nets, Bad Honnef, September 1986.
Papamarcos,M. and Patel,J. "A Low Overhead Coherence Solution for multiprocessors with Private Cache Memories". Proc of 11th Int. Symp. on Computer Architecture, IEEE, June1984, pp. 348-354.
Rudolf,L. and Segall,Z. "Dynamic Decentralized Cache Schemes for MIMD Parallel Processors". Proc of 11th Int. Symp. on Computer Architecture, IEEE, 1984, pp. 340-347.
Siegel,H.J. "Interconnection networks for large-scale parallel processing." Lexington Books, 1985.
Smith,A.J. "Cache memories." *Computing Surveys 14,3* (Sept. 1982), pp. 473-530.
Smith,A.J. "Cache evaluation and the impact of workload choice" Proc of 12th Int. Symp. on Computer Architecture, IEEE, Boston, June 1985, pp. 276-283.
Vautherin,J. and Memmi,G. "Computation of flows for unary Predicate / Transition nets" 5th. European workshop on application and theory of Petri nets, Aarhus (June 1984).
Vautherin,J. "Non linear invariants for safe coloured Petri nets and application to the proof of parallel programs." 6th. European workshop on application and theory of Petri nets, Espoo, Finland (June 1985).
Yen,W.C., Yen,D.W.L. and King-Sun Fu "Data Coherence Problem in a Multicache System". *IEEE TC C-34,1* (Jan 1985), pp. 56-65.

23.
Use of Reentrant Nets in Modular Analysis of Colored Nets

G. Chehaibar

11th Int. Conference on Applications and Theory of Petri Nets, Paris 1990

Abstract This paper deals with the properties of modular constructs such as replacement and composition for a restricted net class called reentrant nets. A reentrant net has a distinguished place subset (interface) and satisfies some behavioral properties like the existence of home spaces. A notion of interface equivalence (I-eq) is defined over this class. Then a new equivalence notion is defined, observational home equivalence (OH-eq), which is a bisimulational relation analogous to the observational one labelling places and transitions, but different from previous notions in that we only consider sequences yielding markings in a given home space. A replacement theorem is established: if we replace a reentrant subnet whose border is formed by its interface places with another I-equivalent reentrant net, the resulting net is OH-eq to the original one. Finally, we define composition operations on reentrant nets, which allow a hierarchical and modular design of reentrant nets and cyclic protocols: choice, sequential and ring composition. All these notions are defined for colored nets.

Keywords Colored Petri Nets, Modular Analysis, Reentrant Net, Observational Home Equivalence, Replacement, Composition.

Acknowledgements This paper is part of Bull's contribution to DEMON (ESPRIT BRA 3148) and is a revised version of the paper *Validation of Phase-Executed Protocols Modelled with Colored Petri Nets* (Eleventh International Conference on Application and Theory of Petri Nets, Paris 90) prepared when the author was given a grant of the *Ecole Nationale des Ponts et Chaussées,* and supervised by *Claude GIRAULT* who has suggested several corrections and improvements.

1 Introduction

In this paper, we study some modular constructs applied on a restricted net class called *reentrant nets.* These operations are the replacement of a place-bordered reentrant subnet by another reentrant net and the composition of reentrant nets by merging places. A replacement operation can be a refinement or an abstraction depending on wether the replacement net is more or less detailed than the subnet replaced; composition operation builds large systems from small ones easier to design and analyse.

Net analysis is modular if, in the case of replacement, a relation between the original net and the resulting one is determined only by comparing the subnet replaced with its replacement net, and in the case of composition, the properties of a composed net are determined by those of the components. Since our aim is to obtain such results on a practically useful class net we consider colored nets: another work in progress studies this problem basically and more generally.

This paper is organized as follows. In the second section we give the basic definitions and notations, the definition of a reentrant net and that of an interface equivalence.

The intuitive interpretation of a reentrant net is a protocol phase which may correspond to the execution of a request by a server or to a procedure call: its execution causes objects to go from an initial state s_1 to a final state s_2: this is is modelled with a net having a place interface that represents the states s_1 and s_2. The interface of a reentrant net consists of two place subsets: the initial places and the final ones. There exists a linear place invariant (flow) whose support contains the initial places and the final ones, and for any distribution of tokens in the initial places, the state where all the tokens of the flow support are in the final places is a home space. There are other constraints that standardize the behavior of a reentrant net enough to allow the definінition of a simple interface equivalence which only deals with the number of interface places.

Reentrant nets are a generalization of clean termination nets [Brams 83]. The main extensions to the classical termination are reentrance—several concurrent executions of a reentrant net are possible— and memory— a reentrant net may not return to its initial state when all executions are completed. In addition, since a phase may be iterated the final places of a reentrant net representing the final states may have outgoing arcs.

In the third section we consider reentrant nets that are subnets whose borders are formed by their interface places and we establish a replacement theorem (Theorem 4.2): when we replace in a net R_1 a reentrant subnet by an interface-equivalent reentrant net resulting in R_2, R_1 and R_2 are OH-equivalent. Observational Home equivalence (OH-eq) is a new bisimulational equivalence notion labelling places and transitions, analogous to the classical observational one, but its distinguishing feature is that we only consider sequences yielding markings in a given home space. The corollary of this theorem is preservation of properties like liveness and home space.

This approach is complementary to the approaches of [André 82], [Bourguet-Rouger 88] and [Baumgarten 88] who have studied equivalence notions and subnet substitutions: but these subnets have a *transition* border. By putting the accent on place role (through flows and home spaces) and considering place-bordered subnets and composing nets by merging places we are closer to [Pomello&Simone 89] and [Lausen 88].

This section is concluded with a comparison between transition-bordered and place-bordered subnets, and a discussion about related equivalence notions and their connections to this work.

In the fourth section we define three composition operations on reentrant nets. The two first ones are general choice and sequential composition and they hierarchically build reentrant nets: from two reentrant nets we construct a third one by appropriately merging the interface places of the two nets. The third operation consists in forming a ring of sequentially composed reentrant nets wich is deadlock-free.

2 Reentrant Nets

Throughout this paper, we consider colored nets, and results are proved straight on colored nets, although usually results are obtained on PT-nets, then colored net classes are defined, which satisfy the requirements of the basic results, when unfolded. This is motivated by our wish to apply the results to examples which are not only toy examples, and by the fact that linear place invariants (flows), which can be symbolically computed for colored nets, play an important role here. It slightly makes notations—not reasonning—more complex.

The following definition recalls the basic notions about colored nets and the notations used throughout the paper. Please be careful of the overloading of some symbols and the different notations defined for convenience on multisets.

Definition 1 (Basic Notions and Notations)

Multiset A multiset over X is a function $f: X \to \mathbf{N}$ ($\mathbf{N}$ is the set of natural integers); the support of f is $Y \subseteq X$ such that $Y = \{x \in X, f(x) \neq 0\}$, and f is represented by the formal sum $\Sigma_{x \in X} f(x)x$. The set of multisets over X is denoted $\mathbf{N}^X$.

Sum: The sum of $f_1 : X_1 \to \mathbf{N}$ and $f_2 : X_2 \to \mathbf{N}$, denoted $f_1 + f_2$, is $f : X_1 \cup X_2 \to \mathbf{N}$, defined by $f(x) = f_i(x)$ if $x \in X_i - X_j$, and $f(x) = f_1(x) + f_2(x)$ if $x \in X_1 \cap X_2$.

Comparison: If $f \in \mathbf{N}^X$ and $g \in \mathbf{N}^Y$ where $X \subseteq Y$, then $f \leq g$ iff $\forall x \in X, f(x) \leq g(x)$.

Subtraction: If $f \in \mathbf{N}^X$ and $g \in \mathbf{N}^Y, f - g = h \in \mathbf{N}^X$ is defined by

$$h(x) = f(x) \quad \text{if } x \in X - Y$$
$$= max\ (0, f(x) - g(x)) \quad \text{if } x \in X \cap Y.$$

Restriction: If $f \in \mathbf{N}^X$ and $Y \subseteq X$, then $f/Y = g \in \mathbf{N}^Y$ such that $\forall y \in Y,\ g(y) = f(y)$.

If $f: X \to \mathbf{N}^Y$, then f is represented by the formal sum $\Sigma_{x \in X} f(x)x$ and it will be sometimes identified with $g \in \mathbf{N}^{X \times Y}$ such that $g(x,y) = f(x)(y)$.

Colored Net A colored net is denoted $N = (P,T;C;W)$ where P and T are finite sets (the set of places, and the set of transitions), C is the color function: for all x in $P \cup T$, $C(x)$ is a finite set called the (color) domain of x; and W is the weight function: if (p,t) is in $P \times T$, $W(p,t)$ and $W(t,p)$ are functions from $C(t) \times C(p)$ to $\mathbf{N}$. For convenience, we denote $W(p,t)((ct,cp))$ by $W(p,t,ct,cp)$, and $W(p,t,ct)$ is the function $g : C(p) \to \mathbf{N}$ such that $g(cp) = W(p,t,ct,cp)$. The directed graph associated to a net is defined by $G = (X,U)$ where $X = P \cup T$ is the set of nodes, and $U = \{(x,y) \in (P \times T) \cup (T \times P);\ W(x,y) \neq 0\}$ is the set of arcs. The postset of $x \in P \cup T$ is $x^\bullet = \{y \in P \cup T;\ W(x,y) \neq 0\}$ and $Y^\bullet = \cup_{y \in Y} y^\bullet$. The preset of $x \in P \cup T$ is ${}^\bullet x = \{y \in P \cup T;\ W(y,x) \neq 0\}$ and ${}^\bullet Y = \cup_{y \in Y} {}^\bullet y$.

Flow We consider a restricted definition of a flow: A flow whose support is $B \subseteq P$, is a multiset f over P whose support is B and $\forall t \in T, \Sigma_{p \in B} f(p)[W(t,p)-W(p,t)] = 0$.

Marking The set of all potential markings of a net is denoted $POT(P)$:

$POT(P) = \{M : P \to \cup_{p \in P} \mathbf{N}^{C(p)};\ \forall p \in P,\ M(p) \in \mathbf{N}^{C(p)}\}$

Sum of markings. If $M_i \in POT(P_i)$ where $i=1,2$, then the sum of M_1 and M_2 is

$M_1 + M_2 \in POT(P_1 \cup P_2)$ defined by $(M_1 + M_2)(p) = M_1(p) + M_2(p)$

where $M_1(p) + M_2(p)$ is the multiset sum defined above.

A marked colored net is a tuple $(N;M_0)$ where N is a colored net and M_0 a marking of N called the initial marking.

Firing Sequences A transition t is enabled under a marking M for the color $ct \in C(t)$ (or (t,ct) is enabled under M) iff $\forall p \in P, M(p) \geq W(p,t,ct)$, and it is denoted $M\ ((t,ct)>$. If (t,ct) is enabled under M, the firing of (t,ct) yields the marking M' and is denoted $M\ ((t,ct)> M'$. M' is defined by $M'(p) = M(p)+W(t,p,ct)-W(p,t,ct)$. The enabling and the firing of a sequence are defined by induction.

In order to simplify notations and since all sequences are formed by tuples (t,ct), if $A \subseteq T$, we denote $A^* = \cup_{n \geq 0}\{(t,ct);\ t \in A,\ ct \in C(t)\}^n$ and λ is the empty word.

Let s be a sequence in T^*, then $/s/$ indicates the length of s, $Oc(s,(t,ct))$ is the number of occurrences of (t,ct) in s, and if $A \subseteq T$, s/A is the sequence formed from s by omitting all tuples (t,ct) such that $t \notin A$.

Reachability The reachability (or accessibility) set of a marked net is denoted

$ACC(N;M_0) = \{M \in POT(P);\ \exists s \in T^*,\ M_0\ (s> M\}$. The language of a marked net is

$L(N;M_0) = \{s \in T^*;\ M_0\ (s>\}$

$E \subseteq POT(P)$ is a T'-home space of $(N;M_0)$, where T' is a subset of T, iff

$\forall M \in ACC(R),\ \exists s \in T'^*,\ M\ (s> M' \in E$. The difference with the usual definition of home space is the possibility to choose a sequence in T'^* to reach the home space; but other sequences are not excluded..

Liveness Let $t \in T$, $R = (N;M_0)$, $M \in POT(P)$ and $A \subseteq T$.

t is quasi-live under $M \Leftrightarrow \forall ct \in C(t),\ \exists s \in T^*,\ M\ (s.(t,ct)>$

t is live in $R \Leftrightarrow \forall M \in ACC(R)$, t is quasi-live under M

R is live $\Leftrightarrow \forall t \in T\ t$ is live

A is pseudo-live $\Leftrightarrow \forall M \in ACC(R),\ \exists t \in A,\ \exists ct \in C(t),\ \exists s \in T^*,\ M\ (s.(t,ct)>$

R is pseudo-live $\Leftrightarrow T$ is pseudo-live

◊

The replacement and composition operations studied in this paper are defined on a restricted net class called reentrant nets. The intuitive interpretation of this restricted net class is a service execution or a phase (cf. introduction). Some of the requirements a reentrant net must fulfill are structural—graphical constraints, flows, and some other are behavioral—home spaces. Immediately below the definition, each requirement is informally commented, and the comments link up with the intuitive interpretation of a reentrant net.

Definition 2 (Reentrant Net) A reentrant net is a four-tuple $RN = (N;INI;FIN;M_0)$ where $N = (P,T;C;W)$ is a colored net and INI and FIN two place subsets such that $INI \cap FIN = \varnothing$. INI is the set of initial places, and FIN the set of final places; $ITF = INI \cup FIN$ is the reentrant net interface.

This net observes the following conditions:

H1 The preset of INI is empty: $^\bullet INI = \varnothing$.

H2 All places of ITF have the same domain C_{ITF}: $\forall p \in ITF,\ C(p) = C_{ITF}$.

H3 There exists a flow whose support FS contains ITF; the domain of all places of FS is C_{ITF}. The coefficients of places of ITF are equal:

$$\forall t \in T, \Sigma_{p \in ITF}[W(t,p)-W(p,t)] + \Sigma_{p \in FS-ITF}a(p)[W(t,p)-W(p,t)] = 0$$

where $a(p)$ is a positive rational.

H4 There is no path between any two places of *FIN*.

H5 Initially, the places of *FS* are empty: $M_0/FS = 0$.

H6 $COM = \{M \in POT(P);\ M/(FS-FIN) = 0\}$. For any marking $M_{INI} \in POT(INI)$, *COM* is a $(T-FIN^\bullet)$-home space of $(N; M_0+M_{INI})$ (cf.. Definition 1).

H6' $PAR = \{M \in POT(P);\ M/(FS-ITF) = 0\}$. For any marking $M_{INI} \in POT(INI)$, *PAR* is a $(T-ITF^\bullet)$-home space of $(N; M_0+M_{INI})$.

H7 Let $M_{INI} \in POT(INI)$. $\forall M \in ACC(N; M_0+M_{INI}) \cap PAR$, $\forall c \in C_{ITF}$, $\forall pini \in INI$, $\forall pfin \in FIN$, if $M(pini,c)>0$, then

$$\exists s \in (T-FIN^\bullet)^*, M\ (s> M' \in PAR\ , M'/ITF = (M/ITF - (pini,c)) + (pfin,c)$$

◊

(H1) This is justified by properties 1 and 2 stated below. There is no restriction on the preset and the postset of *FIN*.

(H3) This flow represents the data path: *INI* represents the waiting states, *FS – ITF* the execution states an *FIN* the termination states. It may not be unique, but if there exists another one with support *FS'* and corresponding sets *PAR'* and *COM'* , it is easy to show that these sets satisfy H6, H6' and H7. The domain of a place $p \in FS-ITF$ can have the form $C_{ITF} \times C'(p)$: then, H3 must be satisfied with the places (p,c) where $c \in C'(p)$.

(H4) It prevents a token from moving from a final place to another one: this explains why outgoing arcs from final places are allowed. This condition is to standardize the reentrant net behavior, and then to make interface comparision easier, since we intend to define reentrant net equivalence and to perform replacements.

(H5) This means that there are no executions in progress $(M_0/(FS-ITF) = 0)$, or completed executions $(M_0/FIN = 0)$, or waiting requests $(M_0/INI = 0)$.

(H6) Reentrance property: complete termination state is reached for any initial marking of the initial places. *COM* is the set of complete termination states: there are no executions in progress $(M/(FS-ITF) = 0)$ or waiting requests $(M/INI = 0)$. Executions in progress and waiting requests can be completed (*COM* is a home space) regardless of previous results (without firing transitions in $FIN^\bullet$).

(H6') Partial termination state is reached for any initial marking of the initial places. *PAR* is the set of partial termination states: there are no executions in progress, but there may be waiting requests in the initial states (M/INI may not be null). Executions in progress can be completed (to empty places of $FS-ITF$) whatever the previous results (without firing transitions in $FIN^\bullet$) and whatever the waiting requests (without firing transitions in $INI^\bullet$).

(H7) For any initial state, every final state is possible and the choice between the final states is indeterminist.

The most important requirement is the existence of a home space reachable without firing output transitions of interface places (*PAR* in H6'). The other requirements (like H1, H3, H4 and H7) are specially usefull for simplifying the comparison between two reentrant nets (and the proofs) and making the verification of a reentrant net practicable.

The following property is an immediate consequence of H1: there is no synchronization between the tokens of an initial place and the executions in progress are independeant of the waiting requests.

Property 1 Let RN be a reentrant net, and let $M_{INI} \in POT(INI)$. If M_0+M_{INI} $(s>$ $M+m_{INI}$ such that $M/INI = 0$ and $m_{INI} \in POT(INI)$, then $M_0+(M_{INI}-m_{INI})$ $(s>$ M.

Property 2 In the reentrant net definition the constraints are redundant, that is

$$(H1, H2, H3, H4, H5, H6, H7) \Leftrightarrow (H1, H2, H3, H4, H5, H6', H7)$$

Proof

$\Leftarrow$) (H6', H7) $\Rightarrow$ H6 is trivially verified.

$\Rightarrow$) We will show (H1, H6) $\Rightarrow$ H6'. We consider a marking M such that M_0+M_{INI} $(s>$ M.

Let us show that we can reach PAR from M.

We define the marking MI_s by: $MI_s/(P-INI) = 0$ and

$$\forall p \in INI, MI_s(p) = \Sigma_{(t,ct)\in s} Oc(s,(t,ct))W(p,t,ct).$$

Therefore, M_0+MI_s $(s>$ M_s where $M_s/(P-INI) = M/(P-INI)$ and $M_s/INI = 0$, due to Property 1 (which is a consequence of H1).

By H6, there exists a sequence w such that M_s $(w>$ M', $w/FIN^\bullet = \lambda$ and $M' \in COM$. From $M_s/INI = 0$ and $^\bullet INI = \varnothing$, we deduce $w/INI^\bullet = \lambda$, then $w/ITF^\bullet = \lambda$. However, $M_s/(P-INI) = M/(P-INI)$ and $M/INI \geq M_s/INI$; consequently, M $(w>$ M'', with $M''/(P-INI) = M'/(P-INI)$, and particularly $M''/(FS-ITF) = 0$: $M'' \in PAR$. ♦

The requirements H1 to H7 strongly standardize the behavior of a reentrant net making the comparison of two reentrant nets very simple: two reentrants nets are said to be interface equivalent iff their initial (resp. final) places are equal (up to a bijection). Then any reentrant net has a straightforward equivalent one, which clearly shows how a reentrant net schematically behaves. Notice that even if the original net has outgoing arcs from FIN, its canonical equivalent has not such arcs (because of H4).

Definition 3 Interface Equivalence

Two reentrant nets RN_1 and RN_2 are (interface) equivalent iff $INI_1 = INI_2$, $FIN_1 = FIN_2$ and $C_{ITF1} = C_{ITF2}$. This equivalence is denoted $\equiv_I$.

Proposition 1 (Canonical Interface Equivalent Net) Each reentrant net has a canonical interface equivalent reentrant net constructed as shown in Figure 1.

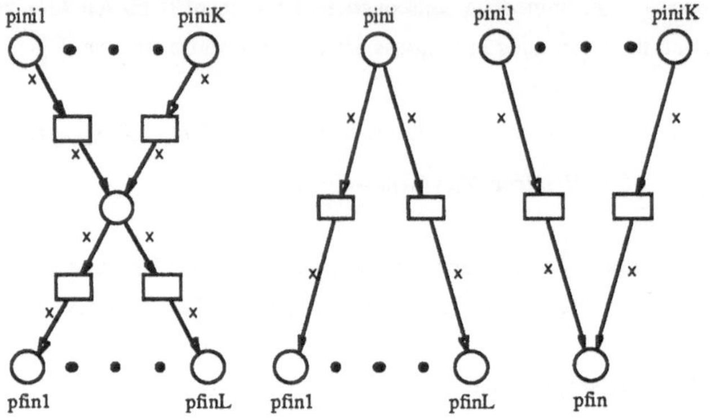

a. General case b. |INI|=1 c. |FIN|=1

Figure 1: Canonical interface equivalent reentrant net

Example 1 : Operating System

Our first example is Figure 2 which models a simplified operating system managing loading, saving and execution of tasks. It is not a protocol, but the evolution of a task in an OS can be decomposed in phases.

The task execution phase is modelled by reentrant subnet RN_1 : the places W and Q compose the subnet interface. Informally, we can say that subnet RN_1 moves tokens from the initial place W to the final place Q. A token put in W can be brought to Q (termination property) and this is true for any initial number of colored tokens put in W (reentrance property).

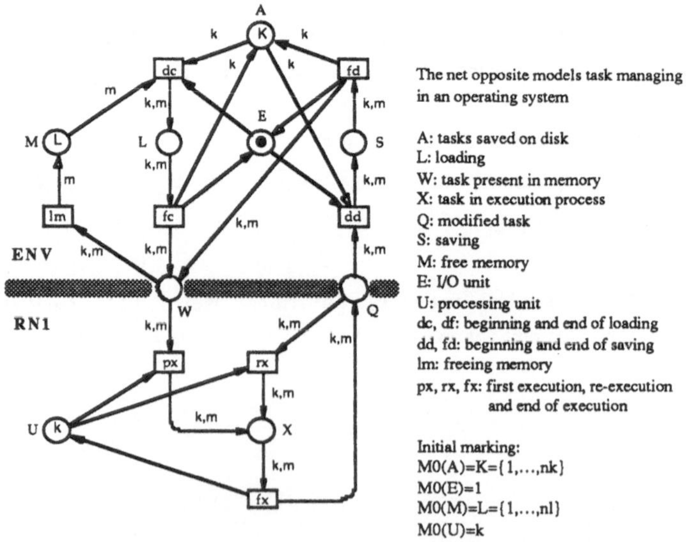

Figure 2: Operating system. RN1 is a reentrant net modelling the execution phase

The interface sets of RN_1 are $INI = \{P\}$ and $FIN = \{Q\}$; its flow is $W+X+Q$: its places are initially empty. Here H6 means that the places P and X can be emptied without firing the transition rx, H6' that X can be emptied without firing px or rx.

Though tokens in Q can go back in the subnet, Q remains the final place when we empty the intermediate place X of the flow between W and Q.

Example 2 : Client-Server

Our second example is a client-server model (Figure 3). The server is connected to a set of n clients, and is supplied with a buffer for waiting requests. After sending a request, a client waits for an acknowledgement that can be positive (executed request) or negative (discarded request). When a request is received, it is immediately served if the server is idle; if the server is busy and the buffer not full, the request is put in the buffer; if the server is busy and the buffer is full, the request is discarded and a negative acknowledgement is sent; if the request is erroneous, it is discarded whatever the states of the server and the buffer. In our model the negative acknowledgement is indeterminist because we do not model the case of erroneous requests.

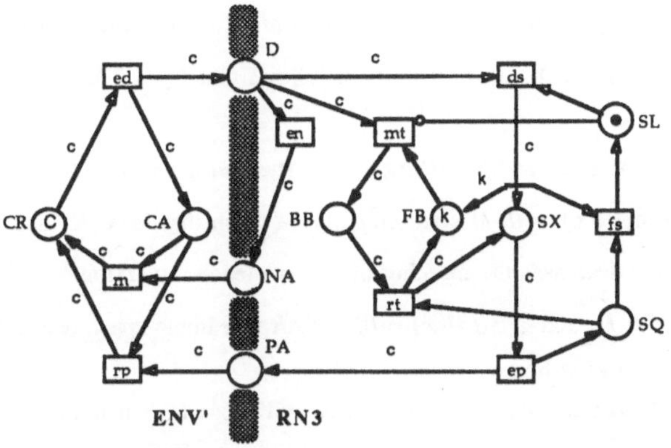

CR: idle clients
CA: waiting clients

D: request
NA: negative ack
PA: positive ack

BB: buffered requests
FB: number of free buffer positions

SL: idle server
SX: server executing a request
SQ: server cheking the buffer

ed: sending a request
rn: receiving neg. ack
rp: receiving pos. ack

mt: puts a request in the buffer
rt: gets a request from the buffer
en: sending a negative ack

ds: beginning of a service exec.
ep: sending a positive ack
fs: end of a service exec.

Initial marking: M0(CR)=C={1,...,n} M0(FB)=k M0(SL)=1

The inhibitor arc is an abbreviation since it is connected to a bounded place.

Figure 3: Client-Server. RN3 is a reentrant net modelling the service execution phase

The service execution phase is modelled by reentrant subnet RN_3 with interface sets $INI = \{D\}$ and $FIN = \{PA,NA\}$; its flow is $D+NA+BB+SX+PA$. H6 means that the places D, BB and SX can be emptied, and H6' that the places BB and SX can be emptied without firing transitions in $\{mt,ds,en\}$. H7 means that a token in D can be freely moved to PA or to NA when BB and SX are empty.

Remark 1 (Verification of H6, H6' and H7)

H6' is easier to verify than **H6**: since these two conditions are redundant, **H6'** will be chosen to verify that a net is reentrant. Though this property is a behavioral one and it has to be cheked for an infinite number of initial markings, its proof can be made easier by using structural properties like flows. In fact, the use of place-invariants (linear or non-linear ones) helps in proving reachability properties with parameterized markings. We follow here the home space refinement method of [Memmi&Vautherin 87]. We have to show that we can empty some places (which are in the support of a flow) without firing some transitions. Here is an example of such a proof for the reentrant subnet RN_3 (Figure 3).

• Proof of **H6'**. We wish to prove that for any initial marking of D, we can empty the places BB and SX without firing any transition in $\{mt,ds,en\}$. First we compute the invariants of the net: there are two linear invariants deduced from the flows of the net and a non-linear one that can easily be verified by induction:

$$M(SL)+ \Sigma_c M(SX,c)+M(SQ) = 1 \qquad (1)$$

$$\Sigma_c M(BB,c)+M(FB) = k \qquad (2)$$

$$\forall c,\ M(BB,c).M(SL) = 0 \qquad (3)$$

We define $E_0 = \{M$ satisfying $(1), (2)$ and $(3)\}$ which is a home space.

If $\Sigma_c M(SX,c) \neq 0$, by (1) there exists c such that $M(SX,c) = 1$ and for every $c' \neq c$, $M(SX,c') = 0$: then M $((ep,c)> M' \in E_0$ where $\Sigma_c M(SX,c) = 0$. It follows that $E_1 = \{ M \in E_0$ and $\Sigma_c M(SX,c) = 0\}$ is a home space.

Now assume that $M \in E_1$ and $\Sigma_c M(BB,c) \neq 0$. Then $M(SL) = 0$ by (3), and $M(SQ) = 1$ ($M \in E_1$ and (1)). For c such that $M(BB,c) > 0$, M $((rt,c).(ep,c)> M'$ such that $M' \in E_1$ and $M'(BB) < M(BB)$. By iterating this operation that decreases the marking of BB, we reach a marking $M \in E_1$ such that $\Sigma_c M(BB,c) = 0$.

Hence $E_2 = \{ M \in E_1$ and $\Sigma_c M(BB,c) = 0\} = PAR$ is a home space reachable without firing mt or ds or en: PAR is a $(T-\{mt,ds,en\})$-home space.

• Proof of **H7**. A token c in D can go in NA by firing en. When a marking M is in PAR, $M(SQ)+M(SL)$ $= 1$ by the invariant (1) and $\Sigma_c M(BB,c) = 0$: it is sufficient to fire the sequence $fs.(ds,c).(ep,c)$ (or $(ds,c).(ep,c)$) to verify **H7**.

3 Replacement of Reentrant Subnets and OH-Equivalence

In this section we study the replacement of a reentrant subnet by an interface-equivalent one, where the reentrant subnet is connected to its environment via its interface places. In this case, the net can be regarded as the composition of two nets (the reentrant net and its environment) by place merging.

Definition 4 (Composition on a Set of Places) If $N_i = (P_i,T_i;C_i;W_i)$, $i=1,2$, are two colored nets, the composition of N_1 and N_2 on S, defined if $P_1 \cap P_2 \subseteq S$, $\forall p \in P_1 \cap P_2\ C_1(p) = C_2(p)$, and $T_1 \cap T_2 = \emptyset$, is the net denoted $N_1 \otimes_S N_2 = (P,T;C;W)$ such that:

- $P = P_1 \cup P_2$
- $T = T_1 \cup T_2$
- $\forall p \in P,\ C(p) = C_i(p)$ if $p \in P_i$
- $W(x,y) = W_i(x,y)$ if $(x,y) \in (P_i \times T_i) \cup (T_i \times P_i)$

  $= 0$ if $(x,y) \in ((P_i - P_j) \times T_j) \cup (T_j \times (P_i - P_j))$

Definition 5 (Reentrant Subnet) A reentrant net $RN = (N_r;INI;FIN;M_{0r})$ is a subnet of $(N';M_0')$ if there exists $ENV = (N;M_0) = (P,T;C;W;M_0)$ such that:

(i) $N' = N \otimes_{ITF} N_r$

(ii) $\forall t \in T, \forall pfin \in FIN,\ W(t,pfin) = 0$

(iii) $M_0/FIN = 0$

(iv) $M_0' = M_0 + M_{0r}$

ENV is called the environment of *RN* in $(N';M_0')$, and $ENV \times RN$ indicates the net $(N';M_0')$.

◊

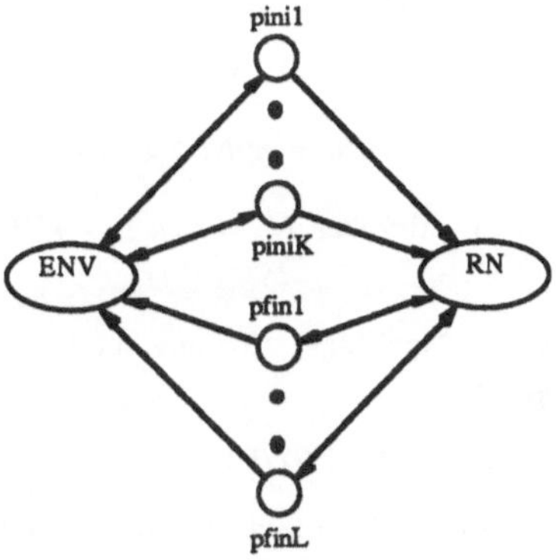

The initial places of the reentrant net RN may have input and output transitions in ENV, the final places only output transitions.

Figure 4: The net ENV x RN.

(i) The structure of the net is the composition on *ITF* of the structure of the environment with that of the reentrant net.
(ii) The final places of *RN* have no input transition in *ENV* in order to avoid the entering of tokens in *RN* through the final places.
(iii) The final places are initially empty for the same reason as above.

Notice that $\otimes_S$ is the composotion of two unmarked colored nets by merging their common places, where x is the composition of two marked nets such that, one of them is reentrant, the composition is by sharing the interface places and the conditions (i), (ii), (iii) and (iv) are satisfied.

First, we show that some properties are transferred from RN to $ENV \times RN$: the names indexed by r refer to RN and the plain names refer to ENV. The following lemma states the locality of the properties of a reentrant subnet: the restriction of a reachable marking of $ENV \times RN$ can be reached in RN by a "local" sequence from some initial marking of INI. It implies the "extension" of H6, H6' and H7 to $ENV \times RN$ (Cf. Lemma 2). This lemma will often be used implicitly.

Lemma 1 For any sequence $s \in L(ENV \times RN)$ such that $M_0\ (s>\ M$ and $s/T_r = y$, there exists a marking $M_s \in POT(INI)$, $M_{0r}+M_s\ (y>_r M'$, $M'/(P_r - FIN) = M/(P_r - FIN)$ and $M'/FIN \geq M/FIN$.

Proof By induction on $n = /s/$.

-) For $n=0$, define M_s by $M_s/INI = M_0/INI$.

-) Induction step. $s = u.(t,ct)$ and $M_0\ (u>\ m\ ((t,ct)>\ M$.

By induction hypothesis, there exists a marking $m_u \in POT(P_r)$ such that

$m_u/(P_r - INI) = 0$, $M_{0r}+m_u\ (v>_r m'$, where $u/T_r = v$,

$m'/(P_r - FIN) = m/(P_r - FIN)$ and $m'/FIN \geq m/FIN$

• If $t \in T$, $s/T_r = v$. The marking

$$M_s = m_u + \Sigma_{p \in INI}(W(t,p,ct) - W(p,t,ct))p$$

is suitable: t can only decrease the marking of FIN and does not change the marking of $P_r - ITF$, and the enabling of v under $M_{0r}+M_s$ follows from $m'/INI = m/INI \geq \Sigma_{p \in INI}W(p,t,ct)p$ and Property 3.1.

• If $t \in T_r$, $s/T_r = v.(t,ct)$. It is easy to verify that $M_s = m_u+\Sigma_{p \in INI}W(p,t,ct)p$ is suitable. ♦

Notation 1 (Extension of Sets of Markings) To any set H of markings of RN corresponds a set XH of markings of $ENV \times RN$ defined by

$$XH = \{M \in POT(P \cup P_r);\ M/P_r \in H\}.$$

XH is called the extension of H to $P \cup P_r$. Then in the following, $XCOM$ and $XPAR$ denote the extension of COM and PAR.

The extensions of the home spaces of a reentrant net RN are preserved when it is composed with an environment: this is due to the constraint on firing sequences.

Lemma 2 $XCOM$ (resp. $XPAR$) is a $(T_r - FIN^\bullet)$-home space of $ENV \times RN$ (resp. $(T_r - ITF^\bullet)$-home space of $ENV \times RN$). Moreover if $M \in ACC(ENV \times RN)$ such that $M \in XPAR$ then H7 is true for M/P_r.

Proof If $M \in ACC(ENV \times RN)$, by Lemma 1 there exist $M_{INI} \in POT(P_r)$ and M' such that $M_{INI}/(P_r - INI) = 0$, $M' \in ACC(RN;M_{INI})$ and $M'/(P_r - FIN) = M/(P_r - FIN)$. Therefore, the conclusion follows from Condition H6 (concerning $XCOM$) and Condition H6' (concerning $XPAR$). ♦

The following lemma is an immediate consequence of H3, the existence of a flow whose support contains the interface places: when no transition is fired in the environment, the weighted sum of tokens in the places of the flow support is not modified.

Lemma 3 If $M \in ACC(ENV \times RN)$ and $w \in (T_r)^*$, such that $M\ (w> M'$, then

$$\Sigma_{p\in ITF}M'(p) + \Sigma_{p\in FS-ITF}a(p)M'(p) = \Sigma_{p\in ITF}M(p) + \Sigma_{p\in FS-ITF}a(p)M(p)$$

Definition 6 (Replacement of a Reentrant Subnet) Let R_1 be a colored net such that $R_1 = ENV \times RN_1$ where RN_1 is a reentrant net. Let RN_2 be an interface-equivalent reentrant net to RN_1: the replacement of RN_1 by RN_2 results in the net $R_2 = ENV \times RN_2$.

If we replace reentrant subnet RN_1 (Figure 2) by its canonical equivalent RN_2, we obtain Figure 5. The analysis of this net shows that the transitions of ENV are live, and that the set of markings $\{M;\ M(L)=M(S)=M(W)=M(Q)=0\}$ is a home space. The results we prove below state that these properties are verified by $ENV \times RN_1$. We will even be able to say that $\{M;\ M(L)=M(S)=M(W)=M(Q)=M(X)=0\}$, which is reduced to the initial marking (due to place invariants), is a home state.

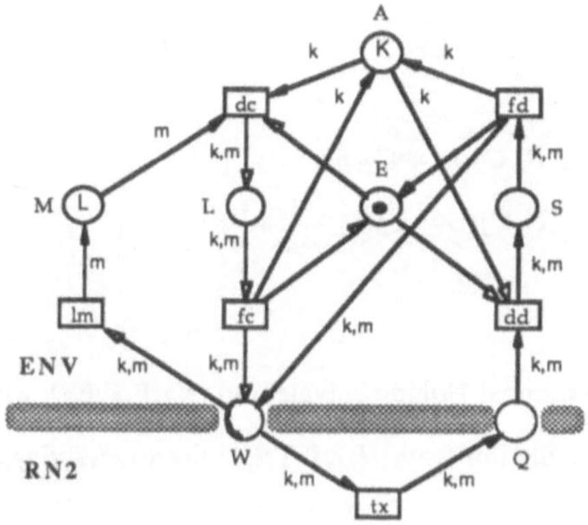

Figure 5: It is obtained by replacing RN1 with RN2

Figure 6 is obtained by replacing RN_3 with RN_4 in $ENV' \times RN_3$ (Figure 3). We will show that the analysis of $ENV' \times RN_4$ allows us to deduce that the initial marking is a home state of $ENV' \times RN_3$.

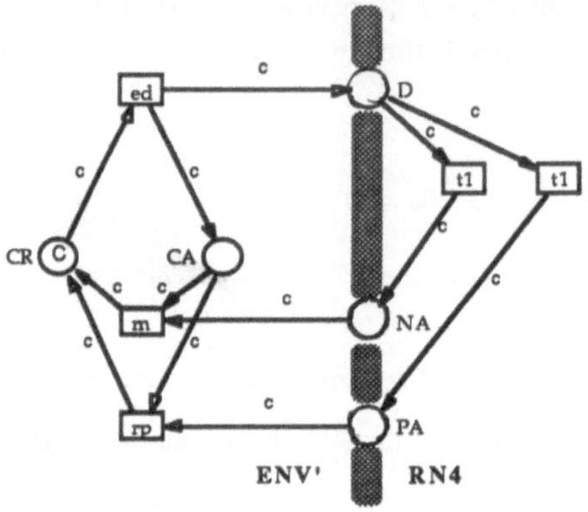

Figure 6: It is obtained by replacing RN3 with RN4.

Before we give the replacement theorem, we define an equivalence notion, OH-equivalence (Observational Home equivalence), which will relate R_1 to R_2 where R_2 is obtained by replacing a reentrant subnet of R_1 with an interface-equivalent one, and we state the properties preserved by OH-equivalence.

Definition 7 (OH-Equivalence) Let $R_i = (P_i,T_i;C_i;W_i;M_{0i}), i=1,2$, two marked colored nets such that

- $P_1 \cap P_2 = P$, and $\forall p \in P, C_1(p) = C_2(p)$
- $T_1 \cap T_2 = T$, and $\forall t \in T, C_1(t) = C_2(t)$
- E_i is a home space of R_i, $i = 1,2$.

Then $(R_1;P,T;E_1)$ is Observational Home equivalent to $(R_2;P,T;E_2)$, written $(R_1;P,T;E_1) \equiv_{OH} (R_2;P,T;E_2)$, iff there exists an OH-bisimulation from $(R_1;P,T;E_1)$ to $(R_2;P,T;E_2)$, that is a relation $\Re_{OH} \subseteq E_1 \times E_2$ such that:

(1) $M_{01}\ \Re_{OH}\ M_{02}$

(2) $M_1\ \Re_{OH}\ M_2 \Rightarrow$

(i) $M_1/P = M_2/P$

(ii) $\exists s_1 \in T_1^*, M_1(s_1> M_1' \in E_1 \Rightarrow \exists s_2 \in T_2^*, M_2(s_2> M_2' \in E_2$

$s_i/T = s_j/T, M_1'\ \Re_{OH}\ M_2'$

(iii) as (ii) but with the role of 1 and 2 reversed.

Actually, OH-equivalence is inspired from Observational equivalence (O-eq) defined in [Milner 80], and adapted to Petri nets in [Pomello 86] by only labelling transitions, and in [Voss 87] by labelling transitions and places: then OH-eq is closer to the latter. The significant difference between O-equivalence and

OH-equivalence is the requirement $M_i \in E_i$: we only observe sequences that yield markings in a given home space; in fact, O-eq can be considered a particular case where $E_i = ACC(R_i)$. There is a brief comparison of OH-eq with other equivalence notions at the end of this section.

When we stated the above definition we mainly aimed at liveness preservation; however, beside liveness and quasi-liveness, other properties concerning language, markings and home space are preserved as is shown in the following proposition (the proof is straightforward).

Notation 2

• Let $R=(P,T;C;W;M_0)$ be a marked net and E a subset of $POT(P)$, then the set of tuples (s,M) such that s is a sequence yielding the marking M in E is denoted:

$LM(R,E)=\{(s,M) \in T^* \times POT(P);\ M_0\ (s{>}\ M,\ M \in E\}$

and the restriction of $LM(R,E)$ to (T',P') where T' is a transition subset and P' a place subset, is

$LM(R,E)/(T',P') = \{(s',M');\ \exists(s,M) \in LM(R,E),\ s' = s/T',\ M' = M/P'\}$

Proposition 2 (Property Preservation) Let R_1 and R_2 be two marked nets such that $(R_1;P,T;E_1) \equiv_{OH} (R_2;P,T;E_2)$:

a. (i) $LM(R_1,E_1)/(T,P) = LM(R_2,E_2)/(T,P)$

(ii) $L(R_1)/T = L(R_2)/T$

(iii) $(ACC(R_1) \cap E_1)/P = (ACC(R_2) \cap E_2)/P$

b.(i) $M_1\ \Re_{OH}\ M_2 \Rightarrow\ \forall t \in T$, t is quasi-live under $M_1 \Leftrightarrow t$ is quasi-live under M_2

(ii) T is pseudo-live in $R_1 \Leftrightarrow T$ is pseudo-live in R_2

(iii) Let $t \in T$: t is live in $R_1 \Leftrightarrow t$ is live in R_2

c. Let $H \subseteq POT(P)$ and $H_i = \{M \in POT(P_i);\ M/P \in H\}$, $i = 1,2$

$H_1 \cap E_1$ is a home space of $R_1 \Leftrightarrow H_2 \cap E_2$ is a home space of R_2

Theorem 1 (Replacement Theorem)

1) $XPAR_i$ is a $(T_{ri}-ITF^\bullet)$-home space of $ENV \times RN_i$. (Lemma 2).

2) $RN_1 \equiv_I RN_2 \Rightarrow (ENV \times RN_1;P,T;XPAR_1) \equiv_{OH} (ENV \times RN_2;P,T;XPAR_2)$

This replacement theorem is the most important result of this paper: if you replace a reentrant subnet by an equivalent one you obtain an OH-equivalent net to the original one. The common places and transitions of $ENV \times RN_1$ and $ENV \times RN_2$ are those of ENV. It is worth noting that the existence of home spaces in RN_i and the extension of this property to $ENV \times RN_i$ play a central role in this theorem.

Proof Define $\Re_{OH}$ by $M_1 \ \Re_{OH} \ M_2$ iff $M_i \in ACC(ENV \times RN_i) \cap XPAR_i$ and $M_1/P = M_2/P$. We prove by induction on $n=/y_1/$ the following assertion **A1**, which implies the required result.

Notation: $CF_i(M) = \Sigma_{p\in ITF}M(p) + \Sigma_{p\in FSi-ITF}a_i(p)M(p)$, where M is a marking. $P_i = P_{ri} \cup P$.

A1 M_1 and M_2 being reachable markings satisfying $M_i \in XPAR_i$ and $M_1/P = M_2/P$, if $M_1 \ (y_1> m_1$ then there exists a sequence $y_2 \in (T_2)^*$ such that:

$$y_2/T = y_1/T \quad (1)$$

$$M_2 \ (y_2> m_2 \in XPAR_2 \quad (2)$$

$$m_2/(P-ITF) = m_1/(P-ITF) \text{ and } m_2/INI \geq m_1/INI \quad (3)$$

$$CF_1(m_1) = CF_2(m_2) \quad (4)$$

$$\forall w_1 \in (T_{r1})^*, \ \forall m_{INI} \in POT(P_1) \text{ such that } m_{INI}/(P_1-INI) = 0$$

$$m_1+m_{INI} \ (w_1> m_{11} \in XPAR_1 \Rightarrow m_{11}/FIN \geq m_2/FIN \quad (5)$$

Base step. For $n=0$, $y_2 = y_1 = \lambda$, $m_1 = M_1$ and $m_2 = M_2$.

In order to show (5) we need the following property which can be proved by induction on the length of the sequence:

If a place p is in the support of a flow FS and $M \ (s> M'$ such that $M'(p) < M(p)$ then there exists $p' \neq p$ in FS such that $M'(p') > 0$ and there exists a path from p to p'.

Then, if there exist $pfin$ and c such that $m_{11}(pfin,c) < M_2(pfin,c) = M_1(pfin,c)$, since $^{\bullet}INI = \varnothing$ (by **H1**) and $m_{11} \in XPAR_1$, it follows that there exists a path from $pfin$ to $pfin'$, which contradicts Condition **H4**.

Induction step. $y_1 = u_1.(t,ct)$, $M_1 \ (u_1> m_{u1} \ ((t,ct)> m_1$, $M_2 \ (u_2> m_{u2} \in XPAR_2$.

- If $t \in (T-FIN^{\bullet})$, then (t,ct) is enabled under m_{u2} (due to induction hypothesis (3)). Therefore $m_{u2} \ ((t,ct)> m_2$ and $y_2 = u_2.(t,ct)$.

We still have to verify (5) with m_1 and m_2: it follows from

$$m_1/P_{r1} = (m_{u1} + \Sigma_{p\in INI}(W(t,p,ct)-W(p,t,ct))p)/P_{r1}$$

$$m_2/FIN = m_{u2}/FIN \ \ (\text{since } t \in (T-FIN^{\bullet}) \text{ and } T \cap {}^{\bullet}FIN = \varnothing)$$

- If $t \in T_{r1}$, $y_2 = u_2$ and $m_2 = m_{u2}$. Condition (5) with m_1, m_2 and w_1 is implied by induction hypothesis (5) with m_{u1}, m_{u2} and $(t,ct).w_1$.

- If $t \in (T \cap FIN^{\bullet})$. There exists a sequence $v_1 \in (T_{r1}-ITF^{\bullet})^*$ such that $m_{u1} \ (v_1> m_{u11} \in XPAR_1$ (by **H6'**) and:

$$CF_1(m_{u11}) = CF_1(m_{u1}) = CF_2(m_{u2}) \ \ (\text{by Lemma 3 and ind. hyp.(4)})$$

$$CF_1(m_{u11}) = \Sigma_{p\in ITF}m_{u11}(p) \text{ and } CF_2(m_{u2}) = \Sigma_{p\in ITF}m_{u2}(p) \ \ (\text{for } m_x \in XPAR_i)$$

$$m_{u11}/FIN \geq m_{u2}/FIN \ (\text{by ind. hyp. (5)}) \text{ and } m_{u11}/FIN \geq m_{u1}/FIN \ \ (v_1/FIN^{\bullet} = \lambda)$$

$$m_{u11}/INI = m_{u1}/INI \leq m_{u2}/INI \ \ (\text{ind. hyp. (3)})$$

Then by applying **H7** we can move the missing tokens from *INI* to *FIN* in $ENV \times RN_2$, to enable (t,ct): there exists a sequence $v_2 \in (T_{r2}-FIN^{\bullet})^*$ such that

$$m_{u2}\ (v_2>\ m_{u22} \in XPAR_2 \ \text{and}\ m_{u22}/INI \geq m_{u11}/INI = m_{u1}/INI$$

and for every $pfin \in FIN$

$$m_{u22}(pfin) = W(pfin,t,ct) \ \text{if}\ m_{u2}(pfin) < W(pfin,t,ct)$$

$$m_{u22}(pfin) = m_{u2}(pfin) \ \text{if}\ m_{u2}(pfin) \geq W(pfin,t,ct)$$

Therefore $m_{u2}\ (v_2.(t,ct)>\ m_2$ and $y_2 = u_2.v_2.(t,ct)$.

Verification of (5). It follows from these facts:

$$m_{u1} = m_1+\Sigma_{pfin\in FIN}W(pfin,t,ct)pfin$$

$$m_1+m_{INI}\ (w>\ m_{11} \in XPAR_1 \Rightarrow m_{u1}+m_{INI}\ (w>\ mr_{11} \in XPAR_1$$

$$mr_{11} = m_{11}+\Sigma_{pfin\in FIN}W(pfin,t,ct)pfin$$

$$mr_{11}/FIN \geq m_{u2}/FIN \ \text{(by ind. hyp. (5))}$$

$$m_2(pfin) = m_{u22}(pfin)-W(pfin,t,ct) = \sup(m_{u2}(pfin)-W(pfin,t,ct),\ 0)$$

then $m_{11}/FIN \geq m_2/FIN$.

We apply **A1** with $y_1 = s_1$ and $m_1 = M_1'$. Then by **H7**, there exists a sequence $w_2 \in (T_{r2})^*$ such that m_2 $(w_2>\ mr_2$, $mr_2/ITF = m_1/ITF$ and $mr_2 \in XPAR_2$.

Hence $s_2 = y_2.w_2$ and $M_2' = mr_2$.

When we replace RN_1 by RN_2 in Figure 2, we obtain Figure 5. Here $P=\{A,E,M,L,S,W,Q\}$ and $T=\{lm,dc,fc,fd,dd\}$. In $ENV \times RN_2$ the transitions *dc*, *fc*, *fd* and *lm* are live: then they are live in $ENV \times RN_1$. The set $H = \{m;\ m(A) = m_0(A),\ m(M) = m_0(M),\ m(E) = m_0(E)\}\}$ is a home space of $ENV \times RN_2$: then the set $\{m;\ m \in H,\ m(X) = 0\}$ is a home space of $ENV \times RN_1$ (Proposition 2(c)): but it coincides with the initial marking. Then to show that $ENV \times RN_1$ is live it is sufficient to prove that the transitions of RN_1 are quasi-live under the initial marking.

When we replace RN_3 by RN_4 in Figure 3 we obtain Figure 6: $P=\{CR,CA,D,NA,PA\}$ and $T=\{ed,rn,rp\}$. In $ENV' \times RN_4$ the transitions *ed*, *rn*, and *rp* are live: then they are live in $ENV' \times RN_3$. The set $H = \{m;\ m(CR) = m_0(CR)\}$ is a home space of $ENV' \times RN_4$: then the set

$$\{m;\ m(CR) = m_0(CR),\ m(BB) = m(SX) = 0\} = \{m;\ m(CR) = m_0(CR),\ m(SQ)+m(SL) = 1\}$$

is a home space of $ENV' \times RN_3$: but if $m(SQ) = 1$, it is sufficient to fire *fs* to reach the initial marking. The liveness of the transitions of RN_3 can be deduced from their quasi-liveness under m_0.

Discussion

Transition-border vs place-border. A subnet with a transition-border (resp. place-border) is a subnet generated by a place subset (resp. transition subset); any net can be viewed, in several ways, as the composition of two subnets with either transition borders or place borders (Fig.4.2).

The composition by transition fusion corresponds to the synchronized parallel composition defined in [Hoare 85]. The behavior of a transition-bordered subnet is characterized by language properties of the border transitions: then its use fits equivalence notions defined on transition-labelled systems (where event sequences are considered) [Pomello 86].

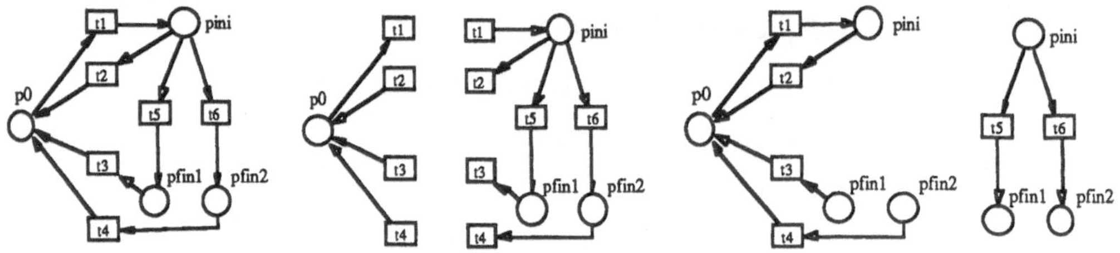

N=TB1xTB2=PB1xPB2 | TB1 gen.by p0 | TB2 gen. by {pini,pfin1,pfin2} | PB1 gen. by {t1,t2,t3,t4} | PB2 gen. by {t5,t6}

Figure 7: Transition border vs place border

The composition by place fusion corresponds to sequential composition, to the choice operator and to resource sharing (Cf. [Kotov 78]). The behavior of a place-bordered subnet is characterized by marking properties of the border places: then its use fits equivalence notions defined on place-labelled systems (where we consider state transformations) [Pomello&Simone 89].

Comparison with other equivalence notions. First, we compare reentrant nets with "S-observable systems" and EF-equivalence defined in [Pomello&Simone 89]. The interface places of a reentrant net are analogous to the observable places of an S-observable system and the observable markings—where unobservable places are not marked—are analogous to *PAR* —markings where internal places of the flow (≠ interface places) are not marked. But reentrant nets can have loops containing at most one interface place while S-observable systems cannot (Cf. subnet RN2 in Figure 8).

OH-equivalence is less restrictive then other equivalence notions that preserve deadlock/deadlock-freeness situations. If only transitions are labelled, OH-equivalence may be compared to Behavior-equivalence (B-eq), Observation-equivalence (O-eq), Exhibited-Behavior-equivalence (EB-eq) and Failure-equivalence (F-eq) discussed by [Pomello 86], who has proved that B-eq ⇒ O-eq ⇒ EB-eq ⇒ F-eq. For any X-eq, XH-eq can be defined by adding the requirement that sequences yield markings in a given home space *E:* then it is easy to prove that X-eq ⇒ XH-eq by taking *E* equal to the reachability space. Figure 8 shows two nets that are OH-equivalent for a suitable choice of home spaces, while they are not even F-equivalent (notice that the home spaces are not the same when places are labelled or not).

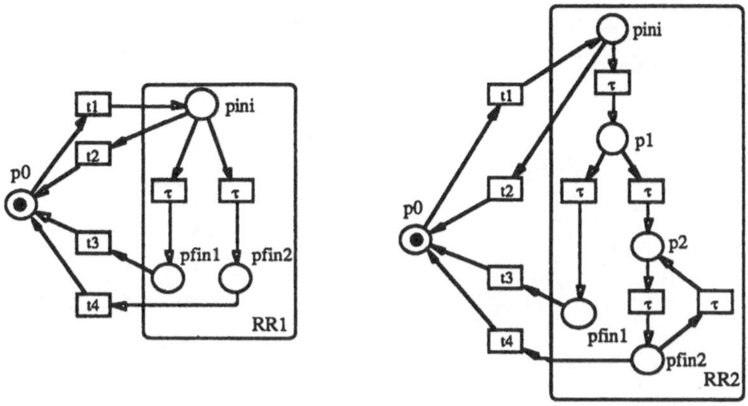

Consider nets R1 (leftmost) and R2 (rightmost). RN1 and RN2 are I-equivalent reentrant subnets with ITF={pini,pfin1,pfin2}, PAR1={m∈POT(ITF)} and PAR2={m; m(p1)=m(p2)=0}. Then, for T={t1,t2,t3,t4}, P={p0, pini,pfin1,pfin2}, E1=ACC(R1) and E2={m; m(p1)=m(p2)=0}, (R1;P,T;E1) and (R2;P,T;E2) are OH-eq. If we wish to only label transitions, we take T={t1,t2,t3,t4}, P=∅, E1={m; m(pini)=0} and E2={m(pini)=m(p1)=m(p2)=0}: then (R1;P,T;E1) and (R2;P,T;E2) are OH-eq but (R1;T) and (R2;T) are not failure equivalent (consider the sequence p0(t1>p1 in R2).

Figure 8: Comparison with other equivalences.

4 Composition of Reentrant Nets

The reentrance proof being difficult for it is a behavioral property, it is dersirable to have at one's disposal composition operators on reentrant nets to hierarchically construct large reentrant nets from small ones easy to verify. In this section, we define two operations consisting in composing reentrant nets by sharing interface places. Another operation is defined consisting in composing reentrant nets in a ring: the resulting net is not reentrant butis deadlock-free.

The results of this section only concern two subclasses of reentrant nets: the loopless and the memoryless reentrant nets. The final places of a loopless reentrant net have no output transitions: tokens that have reached the final places cannot go back. A memoryless reentrant net can return to its initial state when there are no executions in progress (but its final places may have output transitions). These reentrant net subclasses have a common property: the entering in their final places of tokens that have not passed through the initial ones does not put out of order their behavior (ie, H6, H6' and H7 remain true even if $M_{INI} \in POT(ITF)$ instead of $M_{INI} \in POT(INI)$).

Definition 8 (Loopless Reentrant Net) A reentrant net is *loopless* if $FIN^{\bullet} = \varnothing$. The set of loopless reentrant nets is denoted *LLS*.

Definition 9 (Memoryless Rentrant Net) A reentrant net *RN* is *memoryless* if, $\forall M_{INI} \in POT(INI), \forall M \in ACC(N;M_0+M_{INI}) \cap PAR$, there exists M' such that $M(s>M'$ and $M'/(P-ITF) = M_0/(P-ITF)$. The set of memoryless reentrant nets is denoted *MLS*.

Reentrant nets RN_1 (Figure 2) and RN_3 (Figure 3) are memoryless, and RN_3 is loopless. Figure 9 shows reentrant nets with a memory.

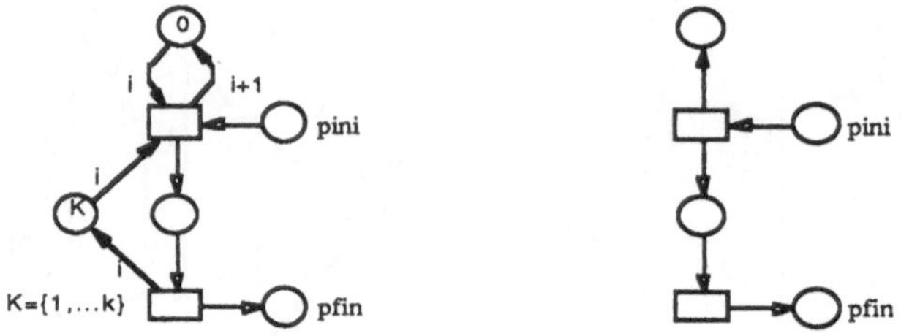

a. Bounded b. Unbounded

Figure 9: Reentrant nets with a memory

We define two composition operators on reentrant nets: a choice and a sequential composition. The choice composition superposes the initial places and the final places of a reentrant net, respectively on the initial places and the final places of another one. The sequential composition superposes the initial places of a reentrant net on the final places of another one. These operations are not free since the net interfaces must satisfy some conditions: the following definition states the precise requirements.

Definition 10 (Choice and Sequential Composition of Reentrant Nets) Let RN_1 and RN_2 be two reentrant nets:

(i) If $INI_1 = INI_2$ or $FIN_1 = FIN_2$ or ($INI_i \subset INI_j$ and $FIN_i \subset FIN_j$) where $\{i,j\}=\{1,2\}$, and $N_1 \otimes_S N_2$ is defined where $S = (INI_1 \cap INI_2) \cup (FIN_1 \cap FIN_2)$, then the choice composition of RN_1 and RN_2 is defined by (Figure 10)

$$RN_1 + RN_2 = (N_1 \otimes_S N_2; M_{01} + M_{02})$$

(ii) If $INI_2 \supseteq FIN_1$ or $FIN_1 \supseteq INI_2$ and $N_1 \otimes_S N_2$ is defined where $S = (FIN_1 \cap INI_2)$, then the sequential composition of RN_1 and RN_2 is defined by (Figure 11)

$$RN_1 \circ RN_2 = (N_1 \otimes_S N_2; M_{01} + M_{02})$$

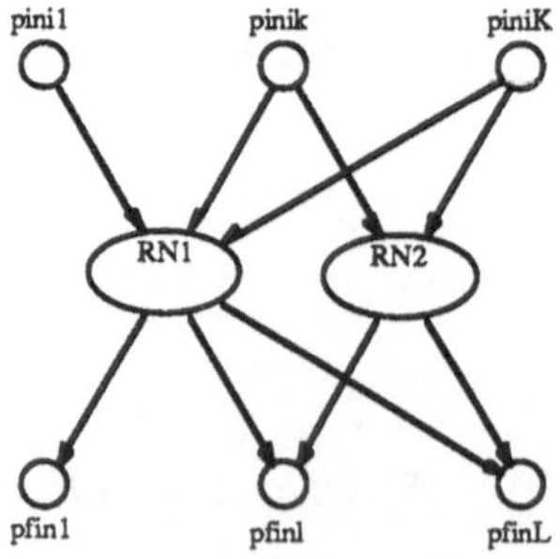

Figure 10: RN1+RN2

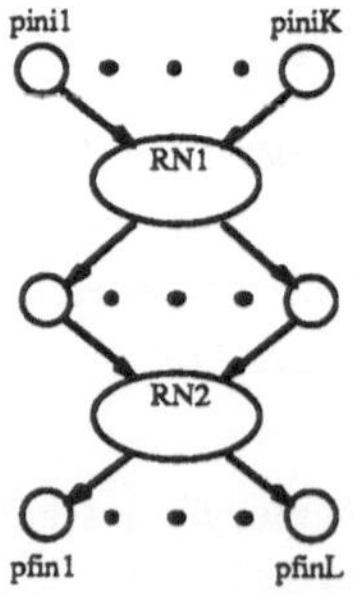

Figure 11: RN1∘RN2

The operator + is a generalized choice operator. Figure 10 is a particular case where $INI_2 \supseteq INI_1$: notice that in the general case the interfaces can verify $INI_1 \cap INI_2 = \varnothing$ and $FIN_1 = FIN_2$.

The operator $\circ$ is the sequential composition: RN_1 is followed by RN_2. Figure 11 is a particular case where $INI_2 = FIN_1$. More generally the interfaces can verify $INI_2 \supset FIN_1$ (or $FIN_1 \supset INI_2$) and $INI_2 \neq FIN_1$.

Proposition 3 Let RN_1 and RN_2 be two reentrant nets.

(i) If RN_1 and RN_2 are in $LLS \cup MLS$ and RN_1+RN_2 is defined, then $RN = RN_1+RN_2$ is a reentrant net whose interface $ITF = INI \cup FIN$ satifies

$$INI = INI_1 \cup INI_2$$

$$FIN = FIN_1 \cup FIN_2$$

(ii) If $RN_1 \circ RN_2$ is defined, then $RN = RN_1 \circ RN_2$ is a reentrant net whose interface $ITF = INI \cup FIN$ satisfies

$$INI = INI_1 \cup (INI_2 - FIN_1)$$

$$FIN = FIN_2 \cup (FIN_1 - INI_2)$$

Proof The proofs of (i) for nets in *LLS*, and of (ii), are obvious. The only tricky case is the + -composition with a reentrant net in *MLS*: we have to show that the final places of a memoryless reentrant net RN_1 can have input transitions in RN_2, i.e., RN_1 can receive in its final places tokens that have not passed through its initial places. This is true, as it was pointed out when the memoryless reentrant nets were defined, because, informally, a memoryless reentrant net doesn't keep signs of the flow of tokens. ♦

The *RING*-operation defined below, "closes" a reentrant net *RN* by superposing its initial places on its final places. When *RN* itself is a sequential composition of reentrant nets, $RING(RN_1 \circ \ldots \circ RN_n, f)$ is an arrangement of reentrant nets in a ring (Figure 12).

Definition 11 (Ring Operation) Let $RN = (P,T;C;W;M_0)$ be a reentrant net such that $|INI| = |FIN|$ and f be a bijection $INI \rightarrow FIN$. We define the net $RING(RN,f) = (P',T';C';W';M_0')$ by

- $P' = P - FIN$
- $T' = T$
- $C' = C/(P' \cup T)$
- $W'(x,y) = W(x,y)$ if (x,y) or (y,x) is in $(P'-INI) \times T'$
  $= W(x,y)+W(f(x),y)$ if $x \in INI$
  $= W(x,f(y))$ if $y \in INI$.
- $M_0' = M_0$

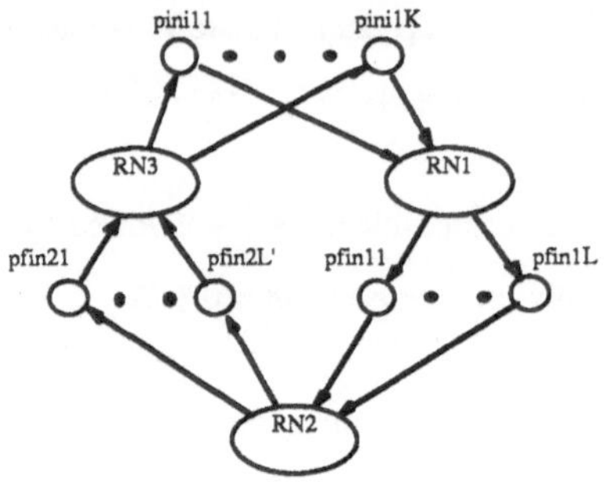

Figure 12: Reentrant nets arranged in a ring: RING(RN1∘RN2∘RN3,f)

Proposition 4 Let $RN_1,\ldots,RN_n$ be reentrant nets such that $RN_n \in LLS \cup MLS$, $FIN_i = INI_{i+1}$, $|FIN_n| = |INI_1|$, and $f : INI_1 \rightarrow FIN_n$ be a bijection. Then $RN_1 \circ\ldots\circ RN_n$ is defined, its interface subsets are INI_1 and FIN_n, and $RING(RN_1 \circ\ldots\circ RN_n,f)$ is defined and verifies:

$\forall MI \in POT(INI_1), \forall i \in 1..n,\ E_i = \{M;\ \Sigma_{p\in FIN_i}M(p) = \Sigma_{p\in INI_1}MI(p)\}$ is a home space of $R = (RING\ RN_1 \circ\ldots\circ RN_n,f);MI)$.

Proof It is based on the same idea of the proof of Proposition 3: this justifies the requirement $RN_n \in LLS \cup MLS$.

This operation is suited to the designing of cyclic protocols that are a sequence of phases. The set of markings E_i, which represents the completed-execution states of the ith phase, is reachable for each phase. Then each phase can be executed and completed an arbitrary number of times and therefore the system cannot be blocked.

5 Conclusion

Some results have been shown about the use of reentrant nets in modular analysis of colored nets. The most important one is the replacement theorem: if you replace a place-bordered reentrant subnet by an I-equivalent reentrant net, the resulting net is OH-equivalent to the original one. Reentrant nets seem to be useful in practice, specially for modelling phase-decomposable protocols (each phase is modelled by a reentrant net).

Future work will be focused on the deepening of reentrant net class study. The application of reentrant nets to large examples is in progress and it leads us to relax some constraints of the definition and to consider other composition operations. Also, it is possible to study the sharing of a reentrant net by two nets, similarly to what is done in [Souissi&Memmi 89].

Another work in progrees considers the problems of replacement of place-bordered subnet and composition of nets by merging places in a more therotical and general setting.

References

[André 82] C.André: *Use of the Behavior Equivalence in Place-Transition Net Analysis.* Applications and Theory of Petri Nets, Informatik-Fachberichte 52, C.Girault and W.Reisig editors, Springer Verlag, 1982, pp 241-250.

[Baumgarten 88] B.Baumgarten: *On Internal and External Characterizations of PT-nets Building Block Behavior.* Advances in Petri Nets 87, LNCS 340, Springer Verlag, 1988, pp 44-61.

[Bourguet-Rouger 88] A.Bourguet-Rouger: *External Behavior Equivalence between two Petri Nets.* Concurrency 88, LNCS 335, F.A.Vogt editor, Springer Verlag, 1988, pp 237-256.

[Brams 83] G.W. Brams: *Réseaux de Petri: théorie et pratique.* Masson, vol 1 and 2, Paris 1982 and 1983.

[Hoare 85] C.A.R. Hoare: *Communicating Sequential Processes.* Prentice Hall International, 1985.

[Kotov 78] V. E. Kotov: *An Algebra for Parallelism Based on Petri Nets.* Proc. of MFCS 78, LNCS 64, Springer Verlag, 1978, pp 39-55.

[Lausen 88] G. Lausen: *Modelling and Analysis of the Behavior of Information Systems.* IEEE Transactions on Software Engineering, Vol. 14, N° 11, November 1988, pp 1610-1620.

[Memmi&Vautherin 87] G. Memmi and J. Vautherin: *Analysing Nets by the Invariant Method.* Petri Nets: Central Models and their Properties, LNCS 254, W.Brauer W.Reisig G.Rozenberg editors, Springer Verlag 1987, pp 300-337.

[Milner 80] R. Milner: *A Calculus for Communicating Systems.* LNCS 92, Springer Verlag, 1980.

[Pomello 86] L. Pomello: *Some Equivalence Notions for Concurrent Systems: an Overview.* Advances in Petri Nets 85, LNCS 222, G.Rozenberg editor, Springer Verlag, 1986, pp 381-400.

[DDPS 88] F. De Cindio, G. De Michelis, L. Pomello, C. Simone: *A State Transformation Equivalence for Concurrent Systems: Exhibited Functionality-Equivalence.* Concurrency 88, LNCS 335, F.A.Vogt editor, Springer Verlag, 1988, pp 222-236

[Pomello&Simone 89] L. Pomello and C. Simone: *A State Transformation Preorder Over a Class of EN-Systems.* Tenth International Conference on Application and Theory of Petri Nets, Bonn, June 1989.

[Souissi&Memmi 89] Y. Souissi and G. Memmi: *Composition of Nets via a Communication Medium.* Tenth International Conference on Application and Theory of Petri Nets, Bonn, June 1989.

[Voss 87] K. Voss: *Interface as a Basi- Concept for System Specification and Verification.* Concurrency and Nets, K.Voss, H.Genrich, G.Rozenberg editors, Springer Verlag 1987, pp 585-604.

24.

Modeling, Validation and Software Implementation of Production Systems Using High Level Petri Nets

J. Martínez, P. Muro and M. Silva

Proc. IEEE Int. Conference on Robotics and Automation, Raleigh (USA) 1987, pp. 1180-1185

Abstract: This paper discusses the design of the coordination subsystem of a Flexible Manufacturing System (FMS) and its integration into the integrated control system of the FMS. The approach adopted is based on the use of high level Petri nets (colored Petri nets) for the subsystem modeling, its qualitative and quantitative analysis and its software implementation. The design of the real time scheduling subsystem as an expert system is proposed. The interaction between the coordination subsystem and the local controllers and the scheduling subsystem is done by using a terminology based on the colors of the coordination model.

1) INTRODUCTION

An FMS consists, basically, of a set of workstations (NC-machines, assembly stations, quality control systems, etc.), product and tool stores connected by an automatic transport system , with the whole operating as a *computer- controlled integrated system.*

The production systems's degree of flexibility will not only be conditioned by the flexibility of its elements (workstations, storage, handling and transport system, etc.) but will also depend fundamentally on the integrated control system. This can be seen as a hierarchical system in which the following levels, among others, can be distinguished: 1) specific controllers for elements of the plant, such as machines, robots, product and tool storage systems, transport systems, etc., 2) subsystem for coordinating the above, 3) real time scheduling subsystem, 4) overall scheduling subsystem, and 5) production planning subsystem.

This paper centres on the design of FMS's coordination subsystem. The principal characteristic of this subsystem is that it is highly concurrent. For this reason our approach is based on the use of tools from the family of Petri nets for the modeling, analysis and realization of the subsystem. Another aspect of great importance which will be considered here is the integration of the coordination subsystem into the FMS integrated control system and its connection with the levels which are strictly below (specific controllers) and above (task scheduling and real time resource assignment subsystem) it.

2) MODELING PRODUCTION SYSTEMS USING PETRI NETS

Generalized Petri nets, Colored Petri nets and Predicate/Transition nets are members of the family of tools for modeling concurrent systems which is known by the generic name of Petri nets (PN). They have been used in such widely varied fields as logic controllers[1], computer networks, communication protocols[2], operating systems[3], as well as production systems[4,5].

From a practical point of view, Generalized PNs , Colored PNs and Predicate/Transition Nets have similar descriptive power. Generalized PNs can be considered as low-level tools (assembler-like nets), while Colored. PNs and Predicate/Transition Nets permit the realization of descriptions at a higher level of abstraction, thus giving more concise models. For this reason these last two types are classed as high level Petri nets (HLPN). The introduction of time and stochastic hypotheses leads to timed and stochastic Petri nets, which are of use in performance evaluation[6].

Last but not least among the advantages of Petri nets is the fact that *a single family of tools* can serve at different levels from design to implementation.

The main difficulties involved in modeling the coordination subsystem of an FMS do not normally lie in the complexity of the interactions between concurrent tasks, but in the large number of elements which are involved in its functioning. Thus, although Generalized PNs permit the description of these subsystems[5], the resulting models are huge, with a corresponding drastic reduction in their legibility and manageability.

High level Petri nets will be used here to model the FMS. With regard to notation, at times that of Colored PNs will be used, while at others that of Predicate/Transition Nets will be used. Nevertheless, the transformation of colored models into Predicate/Transition models and vice versa is straightforward.

Section 3 introduces some basic ideas and concepts on how to model, analyze and implement the coordination subsystem of an FMS. They will be introduced in the course of an example. A formalization of concepts can be found, for instance, in references[7,5].

3) DESIGN OF THE COORDINATION SUBSYSTEM OF AN FMS

3.1) Description of the functioning of an FMS

The manufacturing system scheme given in figure 1 has been discussed before[4], but a simpler, more intuitive model is given here. The system consists of *n* workstations and a unidirectional transport system made up of *n* roller tables.

Each station has a workpost, *P*, and two tables, *L* and *U*, which act as entry and exit buffers for the station, respectively. Table *L* temporarily stores each part arriving at the workstation until *P* is free, at which point the part is loaded onto the station. Table *U* stores each part unloaded from *P* until it is taken away by the transport system. Both the workpost *P* and the loading and unloading tables of each station have the capacity to store a single part. That is, at any given moment the maximum theoretical number of parts at a station is three (one on*L*, another on *U* and a third on the workpost *P*).

Each table in the transport system can contain a maximum of one part. The basic tasks the system must carry out are: 1.- loading table T_1 from the exterior, 2.- transferring a part from table T_i to table T_{i+1}, 3.- loading the station S_i via table T_i, 4.- unloading station S_i via table T_i and, 5.- unloading table T_n to the exterior.

To load and unload the station S_i, the corresponding table T_i can pivot, as shown in figure 1.

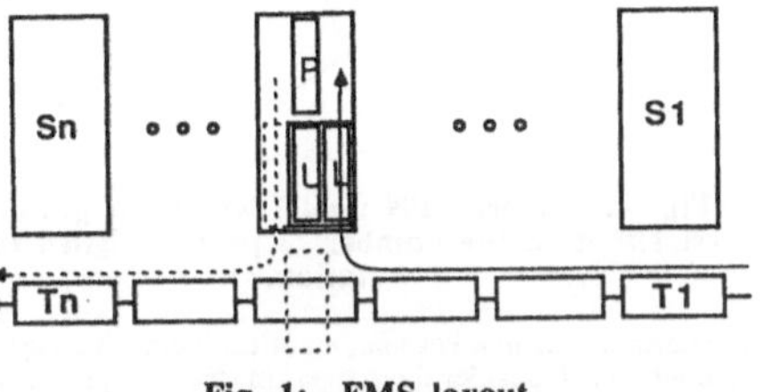

Fig. 1: FMS layout.
Load path: continuous line. Unload path: dotted line

The parts enter the production system via table T_1 and leave via table T_n after visiting one of the stations. In principle, no restrictive hypotheses are made regarding the nature of the parts or the way of assigning a station to each of them.

3.2) Modeling the coordination subsystem

To construct a model of the coordinating system's behavior, it is necessary to make an abstraction of its functional aspects. In the problem posed, the following general comments can be made:

1) All the stations have identical behavior.
2) A part which arrives at a station goes through the following sequence of states: waiting at table L, processing on the workpost P, and waiting for removal on table U.
3) The transport system has a FIFO-like behavior, with the possibility of unloading any part situated on any table onto the corresponding station, and of loading it again once the station has finished working on it.
4) The tables belonging to the transport system and the workstations have capacity for a single part, as is the case with the workposts of each station.
5) The transformation of a single part is irrelevant to the system's global behavior (it belongs to a hierarchically inferior level).

The above points naturally lead to a model which is constructed with a colored Petri net, as shown in figure 2.

This model is a faithful reflection of the above considerations. In effect, the stations have been modeled using the same scheme (sequence of three places, *L*, *P* and *U*, together with the restriction on the capacity of places, *FL* and *FU*, and of the workstation, *FP*). Stations will be distinguished by the set of colors {s[i], 1<=i<=n} involved in the model.

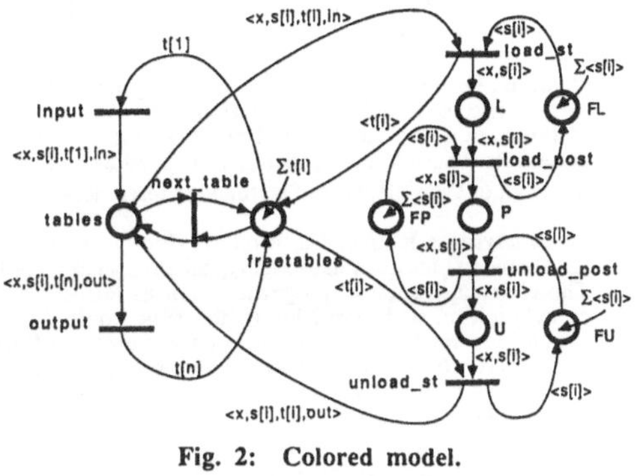

Fig. 2: Colored model.

The transport system is modeled by a subnet representing a FIFO queue having a special interaction with the subnet of stations.

Although the coordinator behavior is independent of the type of parts which reach the system, this information has been taken into consideration in the model by means of the set of colors PARTS.

The transitions which appear in the model correspond to elementary tasks to be carried out by the FMS. These transitions are described below, together with the subset of colors associated with firing them.

input {(part,s[i],t[1],in), 1<=i<=n}: load table T_1 from the exterior

output {(part,s[i],t[n],out), 1<=i<=n}: unload table T_n to the exterior

load_station {(part,s[i],t[i],in), 1<=i<=n}: load one of the stations (table L) from the transport system

load_post {(part,s[i]), 1<=i<=n}: load a workpost from the corresponding table L

unload_post {(part,s[i]), 1<=i<=n}: unload a workpost onto the corresponding table U.

unload_station {(part,s[i],t[i],out), 1<=i<=n}: unload one of the stations (table U) onto the transport system.

nextable {(part,s[i],t[j],dir), (1<=i<=n) AND (1<=j<=n-1) AND ((dir=in) AND (i<j))) OR ((dir=out) AND (i>=j))}: transfer a part from table T_j to table T_{j+1}

3.3) Analysis of the model

We can distinguish two categories of model analysis: *qualitative* and *quantitative* analysis. The first concentrates on studying properties such as liveness, absence of deadlocks, boundedness, mutual exclusions, conflicts, etc. In a quantitative analysis, on the other hand, the aim is to evaluate the modeled system's performance (occupation rates for machines and transport system, average size of waiting queues, average waiting times, etc.).

There are three classic techniques in qualitatively analyzing

Generalized PN's: analysis by enumeration, structural analysis and analysis by reduction[1]. The extension of these methods to the area of high level Petri nets has been under development in recent years. A representative subset of references is given below: analysis by enumeration[8], structural analysis[7,9,10], analysis by reduction[11].

Quantitative model analysis has been approached both by formal methods and by simulation. Among the formal methods, those based on timed and stochastic models built around Petri nets deserve a mention. A good idea of the state of the art of these models can be found in reference[6].

The analysis techniques based on *synchrony theory* of nets falls between the quantitative and qualitative analysis of PN's. The characterization of properties such as fairness and the calculation of such variables as the synchronic distances, synchronic leads or deviations in the number of firings between two transitions or between two groups of transitions can be used to partially characterize the dynamic behavior of a model[12,13].

By focussing on the model in figure 2, and analyzing the net, important information for completing the design task can be obtained. From the qualitative analysis of the model it is also possible to deduce that the system *can deadlock* if, at a given instant, more than four parts are assigned to any one station. A marking, *m*, which verifies:

m(L) >= (*,s[i])
m(P) >= (*,s[i])
m(U) >= (*,s[i])
m(tables) >= (*,s[i],t[i],in)

leads, unavoidably, to the deadlock situation shown in figure 3a.

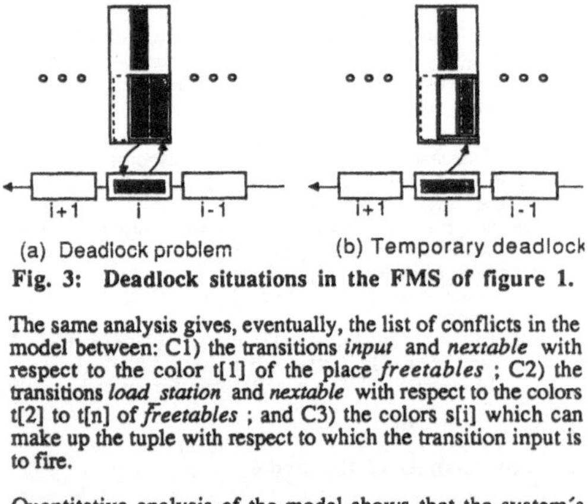

(a) Deadlock problem (b) Temporary deadlock

Fig. 3: Deadlock situations in the FMS of figure 1.

The same analysis gives, eventually, the list of conflicts in the model between: C1) the transitions *input* and *nextable* with respect to the color t[1] of the place *freetables* ; C2) the transitions *load_station* and *nextable* with respect to the colors t[2] to t[n] of *freetables* ; and C3) the colors s[i] which can make up the tuple with respect to which the transition input is to fire.

Quantitative analysis of the model shows that the system's performance drops considerably if the number of parts assigned simultaneously to a station is more than two. This is due to the fact that temporary deadlocks of the transport system are produced, for example that described in figure 3b ,which prevents loading and unloading the other stations. The decision to limit to two the maximum number of parts which can be assigned to a station at a given instant can be incorporated into the model (figure 4) or can be considered in designing the decision system (see section 4).

3.4) Decision problems

It has been pointed that in model analysis there are certain aspects of the coordination subsystem's functioning which are not specified in the colored Petri net in figure 2. Some of these indeterminacies can be solved by modifying the model. This is the case with figure 4, where the number of parts assigned simultaneously to a station is limited to two. However, these and other decisions can be taken at a higher level in the hierarchy of the FMS's integrated control system.

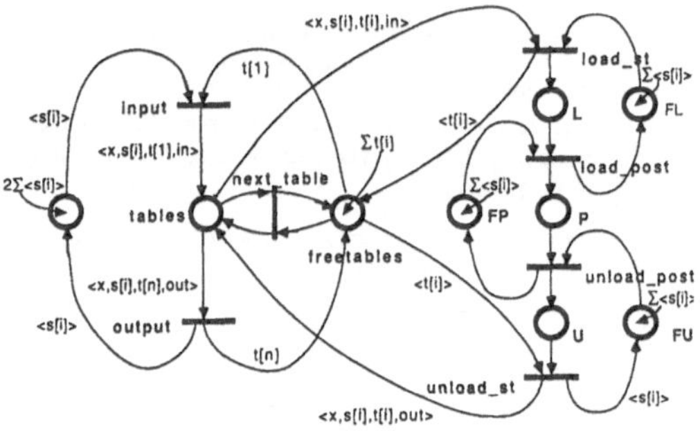

Fig. 4: Colored PN model which integrates the restriction on the number of parts assigned to each station.

Indeterminacies in a Petri net manifest themselves, essentially, as conflicts. Consider three types of decision problems which arise in the model in figure 2 and which correspond to each of the types of conflict detected in the net analysis:

C1) A decision must be made regarding who to assign T_1 to when it is requested simultaneously by a part entering the system and by another which wishes to abandon the first station.

C2) A decision must be made regarding who to assign table T_{j+1} (with j>=1) to when it is requested simultaneously by a part which is on T_j and wishes to advance along the transport system, and by another which wishes to abandon the j+1-th station (figure 5).

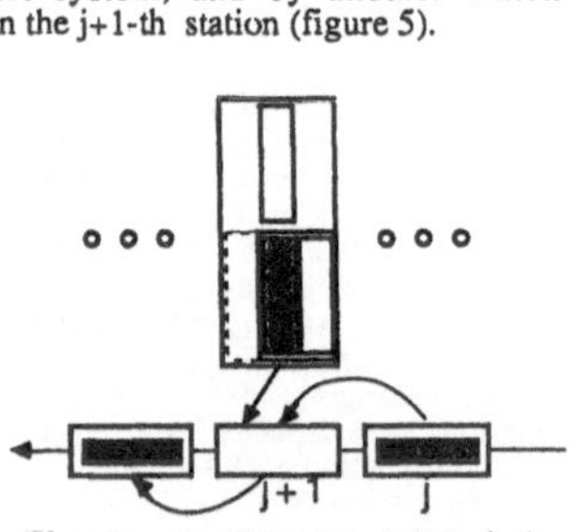

Fig. 5: Conflict for table j+1.

C3) Every time a new part enters the system, the station where it is to be processed must be decided. The number of parts assigned to the same station at any given instant will be limited to two in order to avoid deadlocks and performance losses, as was deduced in the model analysis.

In this case, all the decisions to be made refer to the *policy of resource* (workstations and transport system tables) *assignment* to a series of concurrent processes (parts which visit the production system). The result of these decisions is:

* a series of tasks, ordered in time, which are given to the transport system and to each workstation, and

* a route defined for each part.

These decisions constitute the FMS's scheduler. In section 4, the design of the decision system as an expert system is proposed.

3.5) Realization

The coordination subsystem of an FMS interacts in real time with the low level (plant level) *local controllers* , and with the *real time scheduling* subsystem. It is worth pointing out that, depending on the computer architecture of the integrated control system, communication between subsystems can require the use of a local network or may consist of simple communication of concurrent tasks being executed on the same machine (computer).

Firing a transition with respect to a color (simple or tuple of colors) is translated into orders to certain local controllers for the execution of tasks by plant elements (machines, manipulators, transports, etc.) controlled by them. The message to be sent to the local controllers is coded by the transition and color which define each firing. The coordination subsystem freezes the colored tokens which must be withdrawn from the transition's input places, leaving them unavailable. Once the local controller reports the conclusion of the task, the transition will be effectively fired (evolution of the model's marking).

The coordination subsystem appeals to the decision system to solve conflicts in transition firings and to report breakdowns and alarms detected at its level or in lower levels of the control system. In communicating a conflict, information on the transitions and colors which produce it must be included. The decision system will respond with the transition-color couple chosen.

Programmed realization of the coordination system of an FMS described by a (Colored) Petri net can be approached using various techniques[14,15,16,17]. First, it is possible to consider a *centralized* realization on a computer, or a *decentralized* one on a machine with distributed architecture. Within a centralized realization it is possible to choose an *interpreted* scheme (a program interprets the net, represented by a suitable data structure) or a *compiled* scheme (from the description of the net, code is generated in a given computer language; its execution simulates the behavior of the original net). The basic idea behind a decentralized realization is the decomposition of the net into a set of *sequential processes* (which can be assigned to different processors in a distributed machine) which *communicate* among themselves. For efficient interpretation/simulation of Petri nets there are various concepts and techniques, and the reader is referred to the above-mentioned references.

4) SCHEDULING IN AN FMS

The problem of scheduling consists, basically, of the search for an "optimal" behavior of the system (according to some predetermined criteria) under a set of structural and functional restrictions. The flexibility of an FMS leads to a substantial increase in the scheduling possibilities with respect to the classic manufacturing structures.

Unfortunately, the scheduling problems which arise in an FMS are of such computational complexity that exact calculation methods are not viable. Approximate methods have been developed and applied[18].

Below is a brief resume of some of the characteristic features which are most relevant to the scheduling problem in an FMS:

* Its computational complexity, mentioned above.
* The frequent ill definition of the problem and its objectives
* The existence of conflicts between the different criteria with respect to which optimization is to be carried out.
* The existence of extensive knowledge of the problem, in the hands of human experts.
* The need to incorporate the experience acquired while operating the FMS
* The need to work under unforessen functioning conditions.

This type of problem outlines a field of application which is very suitable for searching for solutions by applying artificial intelligence techniques[19]. For this reason, it is proposed that the decision system be designed as a *knowledge-based system* , in particular as an *expert system* .

An expert decision system of an FMS consists of: 1) A *facts base* which includes the state of the control system (the state of the coordination subsystem is determined by the *marking* of the Petri net which models it), as well as the static and dynamic characteristics associated with the physical components of the FMS and with the manufacturing orders (physical restrictions, functioning rates, operation times, etc.). 2) A rule-based *knowledge base.* These rules allow the materialization of different policy decisions obtained from scientific analyses and/or contributions from human experts in scheduling. These rules can be based on such principles as : priorities, first-in first-out, due date, cost minimization, use rates, etc. 3) An *inference engine* whose job is to manage the knowledge by activating and chaining rules, based on information from the facts base.

```
(IF      ((s[i] is_in free_load)
  AND (((s[i] /is_in free_post)
           AND ((time_end_process (x  s[i])) < low))
        OR (s[i] is_in free_post)))
THEN (free_for_work[i] := T))
(IF (conf_i_1s
  AND ((y  s[1]) is_in unload)
  AND (((y  s[1]) is_in load) OR ((y  s[1] is_in post))
        OR (unload.y.priority >= medium)
        OR (unload.y.wait_time > 50))
THEN (y.priority := maximum)
ELSE (increment_priority col_inp lineal))
```

TABLE I : Some rules of a Real-Time Scheduling Subsistem.

Returning to the manufacturing system studied in section 3, its decision system can be designed as an expert system. Table I gives, as an example, a simple knowledge base which is capable of solving the decision problems posed above. It is worth pointing out that communication between the coordination and scheduling subsystems is done by means of a code based on the *colors* defined in the coordination model.

Certain behavior specifications for the FMS adopted at the coordination level can be transferred to the scheduling subsystem. For example, the restriction on the number of parts assigned simultaneously to a station, introduced in figure 4, can be expressed as a rule of the knowledge base of the scheduling expert system.

The pre- and post-conditions of some rules of the knowledge base can be expressed exclusively in terms of the coordinator model's marking. Such rules can be modeled in the form of Petri net transitions, and be incorporated into the coordinator model. The new (colored) Petri net thus obtained allows a more exact analysis of the FMS functioning (include a part of the scheduler).

So far we have worked under the hypothesis that an FMS always evolves according to certain specifications which do not consider breakdowns or faults. Obviously, this is utopic and such contingencies must be provided for in designing the integrated control system.

Our approach is based on faults and breakdowns being detected by the control subsystem corresponding to their physical origin (local controller or coordination system). If the design of the corresponding detecting system does not contain specifications as to how to proceed, the fault or breakdown will be reported to the level immediately above. Eventually, these reports reach the decision system which, finally, carries out a diagnosis and adopts the appropriate action. Some of these possible actions are given below, as an example:

* Replacement of the model (Petri net) of the coordination system's *normal* functioning with another model (Petri net) of a functioning which is "better adapted" to the system's new situation.
* Freezing of some tokens of, in general, the external transformation of the marking, in order to confine the fault by avoiding the use of certain resources.
* Carrying out exceptional manoeuvers (e.g. manual operations) and updating the coordination model's marking in accordance with the manoeuvers carried out, once these have finished.

In any case, the experience acquired from operating the FMS in both normal and exceptional conditions should serve as a basis for improving the decision system's knowledge. Designing this as an expert system permits its knowledge to be updated simply.

5) CONCLUSION

It has been briefly shown that it is possible to describe, analyze and realize the coordination subsystem of an FMS using colored Petri nets as the modeling tool. This approach facilitates the integration of the coordination subsystem with the local controllers and with the decision subsystem (real time scheduler). Communication between them is achieved via the information contained on the colors defined on the coordination model.

Despite the conceptual simplicity of Petri-net-based models, they may be unfamiliar to FMS designers. For this reason, the development of CAD systems to help describe, analyze and realize FMSs is very important. One line of research being followed in our department is aimed at designing CAD systems for the design of concurrent systems and, as one field of special interest, FMSs[11].

With regard to the scheduling subsystem, it is proposed to design it as an expert system. Its facts base will include the coordination system's marking and its knowledge base will be made up of rules which can be formulated using the terminology of the coordinator model's colors.

It has been shown that certain behavior specifications for an FMS can be either be handled within the coordinator model or can be phrased as rules of the scheduling system.

The isolated analysis of the coordination model does not guarantee an optimum behavior of the FMS since its performance depends, to a great extent, on the scheduling adopted. Nevertheless, properties verified on the coordinator model, such as the absence of deadlocks, mutual exclusions between activities or the boundedness of queue are preserved, independently of the scheduling.

Finally, it is worth mentioning recent research into establishing a bridge between Petri-net-based models and logic[20,21]. This opens up new possibilities for the integration of expert systems and systems designed with Petri nets and for their joint analysis.

ACKNOWLEDGEMENTS

This work has been realised under Grants: CAICYT (Comisión Asesora para la Investigación Científica y Técnica) No. 3359-83 and DGA (Diputación General de Aragón) No. IT-2-86.

REFERENCES

1 M.Silva: *Las redes en la Automática y la Informatica*, Ed. AC, Madrid. An English translation is to appear in 1987, published by Kluwer Eds.

2 M.Diaz: Modelling and Analysis of Communication and Cooperation Protocols using Petri net based models, *Computer Networks*, Vo.6, No.6, December 1982, pp. 419-441.

3 J.L.Peterson: Petri nets theory and the modelling of systems, Prentice-Hall, 1981.

4 J.Martínez, M.Silva: A langage for the description of Concurrent Systems Modelled by Colored Petri nets: Application to the Control of Flexible Manufacturing Systems, Chapter 8 in *Languages for Automation,* Plenum Publishing Co., New York, 1985, pp. 369-388.

5 J.Martínez, H.Alla, M.Silva: Petri nets for the specification of FMSs, in [Kusiak 86], 1986, pp. 389-406.

6 M.Ajmone Marsan, G.Balbo, K.Trivedi (Eds.): *IEEE/ACM Pocs. International Workshop on Timed Petri Nets*, Torino (Italy), July 1985.

7 K.Jensen: Coloured Petri nets and the invariant method, *Theoretical Computer Science 14,* North-Holland, Amsterdam, 1981, pp. 317-336.

8 P.Huber, A.M.Jensen, L.O.Jepsen, K.Jensen: Towards reachability trees for high level Petri nets, *Advances in Petri Nets 1984*, Lecture Notes in Computer Science 188, Springer-Verlag, Berlin, pp. 215-233.

9 H.J.Genrich, K.Lautenbach: System Modelling with High-Level Petri Nets,*Theoretical Computer Science*, 13, 1981,pp.109-136.

10 M.Silva, J.Martínez, H.Alla, P.Ladet: Generalized inverses and the calculation of symbolic invariants for coloured Petri nets, *Technique et Science Informatique*, Vol. 4, No.1, 1985, pp.113-126.

11 J.M.Colom, J.Martínez, M.Silva: Packages for validating discrete production systems modeled with Petri nets, Procs. of IMACS-IFAC Symposium on *Modelling and Simulation for Control of Lumped and Distributed Parameter Systems,* Villeneuve d'Ascq (France), June 1986, pp. 457-462.
12 M.Silva, T.Murata: B-fairness and Structural B-fairness in Petri nets models of concurrent systems, *Internal report*, Dpto. Automática, University of Zaragoza, June 1986.
13 M.Silva: Towards a synchronic theory in Petri nets, *Internal report*, Dpto. Automática, University of Zaragoza, September 1986.
14 M.Silva, S.Velilla: Programmable Logic Controllers and Petri Nets: A Comparative Study, *Proc. of Software for Computer Control, SOCOCO'82*, Pergamon Press, 1982, pp. 83-88.
15 R.Nelson, L.Haibt, P.Sheridan: Casting Petri Nets into Programs, *IEEE Trans. on Software Engineering,* Vol. SE-9, No.5, 1983, pp. 590-602.
16 J.M.Colom, M.Silva, J.L.Villarroel: On software implementation of Petri nets and colored Petri nets using high-level concurrent languages, 7th European Workshop on Application and Theory of Petri nets, Oxford, July 1986, pp. 207-241.
17 G.Bruno, G.Marchetto: Process-Translatable Petri Nets for the Rapid Prototyping of Process Control Systems, *IEEE Trans. on Software Engineering,* Vol. SE-12, No.2, February 1986, pp. 346-357.
18 A.Kusiak (ed.): *Modelling and Design of Flexible Manufacturing Systems,* 1986, Elsevier, Amsterdam.
19 M.S.Fox, S.F.Smith: ISIS, A Knowledge-Based System for Factory Scheduling, *Expert Systems, the Int'l J. Knowledge Engineering*, Vol.1, July 1984, pp. 25-49.
20 K.Lautenbach: On Logical and Linear Dependencies, GMD Report No.147, 1985, Sankt Augustin, Germany.
21 T.Murata, D.Zhang: A High-Level Petri Net Model for Parallel Interpretation of Logic Programs, *Procs. of the 1986 Inter. Conf. on Computer Languages*, IEEE Comp. So., October 1986.

25.
PROTOB: An Object Oriented Methodology for Developing Discrete Event Dynamic Systems

M. Baldassari and G. Bruno

Computer Languages *16, 1* (1991) 39-63

(*Received* 22 *January* 1990)

Abstract—This paper presents PROTOB, an object oriented language and methodology based on PROT nets, and the CASE environment that supports it. PROT nets integrate extended dataflows and Petri nets into an object oriented formalism. The CASE environment consists of several tools supporting specification, modelling and prototyping activities using the PROTOB language within the framework of the operational software life cycle paradigm. As its major application area it addresses distributed systems, such as real-time embedded systems, communication protocols and manufacturing control systems. The CASE environment automatically generates the distributed and object oriented implementation code in Ada or C providing advanced features such as multitasking and system distribution over a LAN. A model of a Flexible Manufacturing System and its production control software are analysed as a case study.

Automatic code generation Dataflows Distributed systems Executable specification Object oriented design Object-oriented specification Operational software life cycle Petri nets PROT nets Rapid prototyping Structured analysis

1. INTRODUCTION

This paper addresses the problem of developing software for discrete event dynamic systems. Such systems react to external non-continuous events according to their internal state, generating events and controlling physical activities. Real-time embedded systems, manufacturing control systems, communication protocols, avionic systems, automotive systems and distributed systems in general are all examples of discrete event dynamic systems.

Designing software for such systems is a difficult undertaking because of their inherent complexity both at the conceptual and implementation level.

A global software life cycle approach is needed to support the early phases of specification and high level design. A representation of the system is required to be unambiguous, suitable for behavioral prototyping and performance evaluation. Furthermore, a strong link as much automated as possible must exist between the model and the actual code generation.

In particular we focus the attention on the following issues:

Conceptual representation of the system—Complex discrete event dynamic systems require a formalism capable of capturing a reality consisting of elements which operate in parallel and asynchronously. Elements interact with each other through synchronization and information passing mechanisms. They respond to external events with timing constraints according to their internal state, which depends on time and on their interactions with other components.

Such representation must be used to build as requirements model which formalizes the behavior of the system. It must be the reference for end user, software developers and subcontractors.

The specification formalism should be object oriented [7, 9], executable and powerful yet simple.

It should be object oriented so that company knowledge may be formalized in building blocks and reused. Previously defined specification blocks that have been validated may be composed to specify new systems.

It should be directly executable (operational) to allow a behavioral analysis of the model to take place at all stages of its refinement.

It should be powerful enough to capture all the details and aspects of such complex systems. Yet it has to be the simplest possible and well organized to be handled and understood by non-experts.

Analysis of the model—Because of its complexity, a static description of the system does not provide sufficient insight into its real behavior. Model simulation is necessary to introduce a dynamic dimension, allowing both the validation of functional aspects and the performance evaluation of quantitative aspects to be carried out. Such analysis of the model, quite similar to a discrete event simulation, is to be performed on the requirements model itself instead of requiring an *ad hoc* simulation model to be built.

Code generation—The gap between the requirements model and the actual software implementation is a critical aspect which accounts for the difficulty of developing and maintaining a complex software system, perhaps consisting of several concurrent tasks distributed over a network.

The object oriented approach is receiving an increasing consensus owing to reusability and information hiding issues. At the moment it only addresses the architectural design and coding phases. On the contrary, current specification techniques are still bound to the functional decomposition approach, so there is a methodological gap between specification and design. Specification formalisms should be object oriented too.

The model of the software system should be detailed enough to enable the automatic code generation in the framework of the operational software life cycle. Software maintenance may then be performed on the model rather than on code. When systems are very large coding is not the problem. It is by far more important to build a correct and accurate model of the total system. Code optimization is a local aspect. It is of little use if the system specification and design are incorrect or have not been optimized.

Recent research in software engineering, such as **Executable Specification** [6], **Rapid Prototyping** [12], **Object Oriented Paradigm** [7] and **Visual Programming** [11], provide the necessary background to stimulate the development of more powerful integrated approaches.

The designer should work with executable models which allow qualitative and quantitative analysis. Executable models improve the understanding and reuse of the work done by others at all levels: specification, design, implementation. Furthermore they enable system testing to be performed incrementally, on-line with the system development. Logical errors may be detected and corrected immediately. Correcting a logical error detected at the end of the implementation phase requires the correct understanding and modification of code that was perhaps thought of and written by others. A correction on code is extremely costly and would often require an architectural design modification.

When developing large systems, each subsystem may work well on its own but fail to respond to all the prescribed events when connected to others. This is often the case when subsystem composition is tested only at the end of the development phase. Prototypes should be generated automatically from models in order to test the software architecture, perhaps on a distributed hardware. Refining the model and adding suitable interfaces to the devices of the target application should have automatic code generation produce the deliverable system as final result.

Models should be object oriented to permit the system to be decomposed in building blocks that may be developed by different people at different times. Company knowledge may be captured in formal and executable models which become components to be reused. Knowledge is transferred to other people that understand models and learn how to modify and reuse them. This should happen during both specification and design phases.

Concurrency and synchronization aspects are well perceived visually. If events may be originated by many causes producing several effects, the correlation of causes to effects may be described in a more understandable way by using graphical notations. Dependencies among components of a very complex system may also be captured visually, using the well known property of parallel processing performed by the eye.

2. USING PROTOB FOR THE PRODUCTION OF SOFTWARE

On the basis of the previous considerations the **PROTOB** [1] methodology and an object oriented CASE tool that supports it have been developed aiming at modelling, prototyping and implementing distributed systems according to a development cycle which consists schematically of three major phases:

(1) Modelling—An executable model of the system is built according to the object oriented paradigm, where each object has a graphical representation in terms of a formalism called **PROT net** [2]. PROT nets are extended Petri nets [16] which can be considered executable extended dataflows [13–15] where dataflows, control flows and state transition diagrams are integrated into a uniform notation.

Objects, representing classes of software components, can be built on top of other objects in a hierarchical architecture. They communicate with each other by sending and receiving messages. The overall model is obtained by generating and interconnecting instances of PROTOB objects graphically.

The model can be executed in terms of discrete event simulation providing a graphical animation which allows the interactive validation of the system behavior to be carried out. During the simulation, statistical data about the system performance can be gathered thus assisting the designer in the proper dimensioning of the system.

Time is modelled by associating a delay to the execution of each elementary action. During simulation, time flow is accelerated and it is just a reference to measure the distance in time between events.

(2) Emulation—While the model of the system is actually constructed and tested in a design environment where the high level logic is of concern, in this phase the model is refined by taking into account implementation details and a new architecture of interconnected instances of PROTOB objects may be obtained.

The aim of this phase is to build an emulator of the final system that has all the requested functionalities, that runs on the target architecture, but that has no connection with the physical devices. The latter are emulated by PROTOB objects so that system testing may be performed.

Such emulator is produced automatically from the PROTOB model by specifying the allocation of the instances of PROTOB objects to processors and tasks of the target distributed system. Since the target system is well known, the communication and synchronization mechanisms are generated automatically. Moreover timing constraints present in PROT nets are now real constraints to be managed by the underlying operating system.

(3) Application generation—Exercising the emulator can result in the adjustment of the allocation and implementation details. In this phase the link with the target devices is performed. This implies the replacement of PROTOB objects emulating the devices with appropriate interfaces which can be generated automatically since the interaction with the devices is well known. The interface generation results from the definition of its model consisting of data and commands to be transferred, and also from the definition of the events to be perceived.

This paper illustrates **PROTOB** as an environment for modelling, prototyping and implementing a distributed system consisting of a set of distinct entities that communicate with each other by message passing. Each entity may stand for a single process and may reside on any node of the network that will host the distributed system. In the modelling phase the designer should not bother with the problem of implementing the communication between processes but should focus instead on the conceptual decomposition of the system into distinct entities and the job each entity is to carry out. PROTOB directly implements the model of the distributed software system in a way that is totally transparent to the user. It is in fact a fully integrated CASE package which supports the editing and documentation of models of concurrent and distributed software systems and their automatic translation into a simulation code which is executed and animated on a graphical terminal. Models can also be translated into a distributed architecture on VAX/VMS computers with the automatic generation of processes and communication mechanisms.

The strength of PROTOB is its modelling language which has a graphical representation and procedural semantics. The graphical representation allows the end user to easily understand the dynamic behavior of the system. Owing to its procedural semantics, it can be associated with code written in a standard programming language, thus allowing the automatic translation into a running program to be performed by exploiting compiling technology.

The three phases of modelling, emulating and application generation are carried out in the same environment and in the same language through the cyclic refinement of the model.

PROTOB is a specification tool which supports hierarchical object-oriented modelling and represents the behavior of objects using a graphical representation based on PROT nets.

PROTOB is a simulation tool which allows a discrete event dynamic model to be built using an innovative approach based on the object oriented paradigm. The model is then automatically translated into simulation code and executed with graphical animation and data collection for statistical purposes.

PROTOB is a prototyping tool producing code in different languages (Pascal, C and Ada) which can be directly incorporated into target applications.

Practical experiences with PROTOB involved modelling and simulation activities of real systems, in particular manufacturing lines, flexible manufacturing systems, monitoring systems and communication protocols. The complexity of such models ranged from 10 to 50 different objects made up of 1–15 views. Some types of modelled and prototyped systems are:

Communication protocol systems—The first protocol to be modelled with PROTOB was a three-layer system. Each layer is modeled by an object: **User, Stop-and-Wait, Aloha, Channel**. Each user requires the services of stop-and-wait layer which, in turn, uses the services of an aloha layer. All alohas are connected to the channel (the physical layer) and require its services to send packets to other alohas.

Using PROTOB to model communication protocols is quite straightforward. Communication layers are commonly defined by Petri nets and it is natural to consider each layer an object. ISO/OSI standards give an object-oriented definition of layers, regardless of their implementation but focusing on their interaction with adjacent layers. Objects having the same interface may implement the same layer in different ways. Communication protocols are modelled through the composition of different implementations of layers and simulation is used to test each implementation and choose the best [3].

Automation systems—The Object-Oriented properties of PROTOB and the PROT net descriptions of the synchronization of concurrent events are well suited to model automation systems. This article describes the model of a simple Flexible Manufacturing System composed of several objects: **Scheduler, Cart and Cart Manager, Machine, Buffer, Warehouse.** Each object models an entity of the FMS. Different objects may give different implementations of the same entity.

The building blocks of the system are first modelled with PROT nets and then connected and simulated. Simulation is the sole means of deciding on the efficiency of the system layout and of the interactions among the system entities. Different layouts and components may be tested in a comparably short time so that the system may be tuned to fully satisfy its requirements.

Real time systems—The classical Ward–Mellor's Cruise Control example [14 (Appendix A)] of structured analysis has been remodelled using PROTOB [4]. The example points out how the PROTOB language models STDs, dataflows and control flows with a single smooth notation and how the structured analysis methodology is conveniently applied to construct PROTOB models. The model is obtained through a one-to-one translation followed by its refinement and simplification which proves the conceptual and expressive power of PROTOB.

Distributed systems—The entire life cycle must be the shortest possible to allow the experimentation of new algorithms to be carried out before a comprehensive implementation of the system. The use of the PROTOB environment has been very time effective to test the correctness and improve the performance of the proposed algorithms.

A case study was proposed by experienced chief programmers working at Olivetti at basic software development. It illustrates how PROTOB may be used to model, simulate and implement a real-time distributed file system [5].

3. CHARACTERIZATION OF PROT NETS

Like a Petri net, a PROT net models a system's behavior in terms of **States** and **State Transitions**. Possible states are represented by circles called **Places**. The elementary event or action that causes a state transition is a **Transition** and it is represented by a rectangle. A state transition is **Caused** by certain states and has the **Effect** of activating other states. This is represented graphically with directed arcs going from places to the transition they cause and from the transition to the places which model its effects. A PROT net statically describes in a visual way the relationships of cause and effect among states (places) and actions (transitions). Names of places should be nouns that describe a state while names of transitions should be verbs describing an action.

A transition of a PROT net, together with its cause and effect places, may be considered a graphical if-then rule. Requirements naturally come in rules. Associating data types with places and operations with transitions completes the formalism making it suitable for the specification of the system behavior.

Marking the active states by putting a **Token** in the corresponding places makes a PROT net a dynamic model. During the animation of the execution of the net, the place contains the number of tokens it containes. This is quite similar to having a cursor pointing to the next instruction of a flow-chart. A transition will then activate its effect-places by moving the tokens from the cause-places to the effect-places when all the cause-places are active. It corresponds to moving the cursor on to the next instruction during the execution of a flow-chart. In this sense PROT nets are executable models. Figure 1 is a simple Petri net modelling the actions of opening and closing a tap in order to fill a reservoir. When the water level is low the closed tap will be open. The tap being open will fill the reservoir getting the water level from low to high. When the water level is high the open tap will be closed. Note how events are considered to be atomic or discrete. The number inside places is the number of tokens. The first identifier near a place is the place name, the second, after a comma, is the name of the type of tokens it may contain. The number of tokens in each place at the beginning of execution is written in the figure after the place name and type and defines the initial marking of the net. The label in the bottom-left corner PROT.PROT.MAINVIEW is formed by the names of the instance, of the object and of the view. The graph of a large object may be conceptually decomposed in views as it will be explained later.

PROT nets can be executed and their execution can be supplemented by such activities as traceability and data collection, so as to lead to a real simulation of the system. This is a remarkable

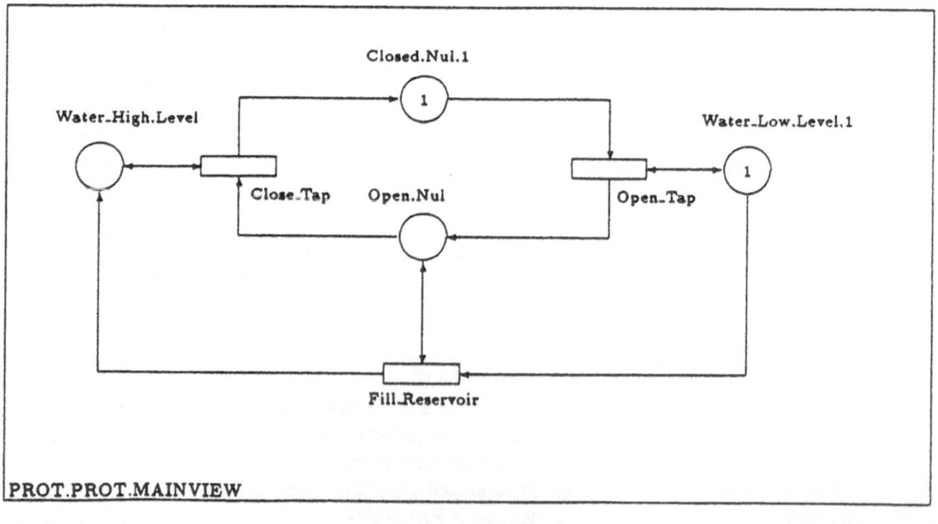

Fig. 1. A simple PROT net.

aspect because specification and simulation activities are usually carried out in unrelated environments precluding the rapid analysis of the consequences of a change in a specification.

PROT net support process oriented [25], transaction oriented [26] and rule oriented [27] programming paradigms. Depending on the paradigm, states can represent particular situations in the process life cycle or particular occurrences in a database or knowledge base.

PROT nets are a visual programming language. In fact pieces of code written in the target language can be associated with the transitions of the net. Transitions may therefore carry out well-defined actions while the overall control structure is visually established by the PROT net. The final program results from a translation phase which assembles these actions into the appropriate framework—i.e. tasks, transactions or rules.

PROT nets and structured analysis

PROT nets can be regarded as a kind of extended dataflow model which is executable and object-oriented. Extended dataflows are the most popular technique for specifying discrete event dynamic systems. They add a conceptual level to the traditional dataflow paradigm [13] in order to capture control and timing information. In fact, a key issue of a dynamic system is the dependency of its behavior, that is, its response to external stimuli, upon its internal state. The state is a mode of operation determining which activities are appropriate and which are not. The mechanism by which a subsystem changes state is referred to as an event. An event can be a message received from another subsystem or the notification that a certain time interval has elapsed. A state-based behavior cannot be expressed using the traditional dataflow technique which accounts for data driven computations. For these reasons several extensions of Structured Analysis were proposed, such as Hatley–Pirbhai's [15] and Ward–Mellor's [14]. The basic idea of such extensions is to add to the usual dataflow specification a control specification based on a state transition diagram or STD. An STD describes the evolution of a subsystem's behavior through states and event-driven state transitions. At each state transition the proper dataflow transformations are enabled while the others are disabled. However such extensions have two major drawbacks:

- There is no possibility of executing the specification model according to the **Operational Software Life Cycle** [6]. Furthermore the specification model has to be substantially reworked to show the introduction of implementation details and to obtain the software architectural design consisting of processors, tasks and communication mechanisms.
- There is no real support of the **Object Oriented Paradigm** [7] thus preventing the user from building a collection of reusable building blocks.

PROT nets are an evolution of extended dataflows. The main advantages provided by PROT nets in software development are:

- The ability of graphically modelling a system at a conceptual level by means of the intuitive notions of:
  —dataflows, controls, states;
  —data transformations, control transformations, state transitions based on time and mutual interaction.
- The capability of executing or simulating the model in order to produce behaviors of the intended system.

PROT nets define with a single, smooth notation an extended dataflow model derived from the Hatley–Pirbhai's [15] and Ward–Mellor's [14] formalism by taking the following steps:

(1) Give **operational semantics** to data transformations. For the model to become executable dataflows and data transformations must be defined by operational specifications that may also include implementation details. This involves two major assumptions:

(a) Dataflows are viewed as flows of *tokens* (i.e. information packets) through *places* which are temporary holders of tokens and act as data-stores. The type of the token (that is, of the data) must be unambiguously and formally defined.

(b) The data dictionary is given a formal content because both data structure and P-spec associated with data transformations (bubbles) are written in a *standard programming language* like Pascal, C or Ada rather than in a pseudo-code.

(2) Provide for **multiple threads of control** by extending state transition diagrams. State transition diagrams can be modified so as to have the *same semantics* as that defined for data transformations. In fact this can be achieved by materializing as places both the *events* and the *states* of the automation, thus allowing data structures to be associated with events and states of the automation.

A limit of extended dataflows is that multiple threads of control must be represented by *separate automata* interacting with control flows. This complicates the specification of *synchronization* issues. However, interacting automata can be *merged* making the model simpler, by relying on the same semantics which has been previously introduced. The theoretical foundation for this new model is provided by Petri nets [16]. In fact the formalism that has been introduced is similar to high level Petri nets where tokens are moving data. PROT nets are "engineered" high level Petri nets which can be translated into actual programs.

(3) Unify the model by merging data transformations and control transformations. Using extended dataflows, because of separate data and control transformations, one level of the model can span two, three or four kinds of representation, thus making the model rather intricate.

Using PROT nets dataflow and controlflow transformation have the same semantics so there is no reason for keeeping them separate; they can be merged in a single kind of representation which consists of two types of elements:

Transitions, representing *state transitions* or *data transformations.*
Places, representing *states, datastores, dataflows* or *control flows.*

The actual role of such elements can be pointed out by suitable names or even by a different colour in the graphical scheme.

Let us have a look back at the PROT net of Fig. 1. This is what an extended dataflow looks like after the application of steps (1)–(3). The two transitions. *Close_tap* and *Open_tap* are actual state transitions of a STD which has two states: *Open* and *Closed.* The transition to state *Open* is triggered by event *Water_low*. State *Open* is also a control. In a dataflow diagram it would be represented by a dashed line that enables the activity of transition. *Fill_reservoir* that would be a data transformation (bubble). *Water_low* and *Water_high* are dataflows and events at the same time. The type of the dataflow is level. The data transformation reads the low level value and calculates the high level value. When the high level has been calculated an event is generated that makes the STD switch to *Closed.* The equivalent extended dataflow diagram would look like Fig. 2.

(4) Continuing our demonstration of transforming a dataflow into a PROT net, the next step is: providing **horizontal decomposition into views**. Using views, hierarchy is not artificially introduced when a large model does not fit in a single page. Views allow some places to appear in more than one view. Transitions on the other hand cannot be duplicated. The places to which a transition is connected must all be on the same view. This is so because transitions are like graphical if–then rules which connect causes to effects. Places, which may be duplicated, are the only elements that two views may have in common.

(5) A further improvement can be obtained if models based on PROT nets are structured within an **object-oriented framework**, where each object is represented by an autonomous net exchanging messages (i.e. tokens) with the other objects of the system. In particular, it is their object-oriented structure that distinguishes PROT nets from other realizations of high level Petri nets and from the classical dataflow model.

Object orientedness stimulates the design of reusable software components characterized by a graphical structure. A new element, the **Gate**, was added to the PROT net to represent the sending of a token to some other PROT net. A token reaching a gate is sent to a place of another net. Sending tokens makes the object-oriented connection of PROT nets possible. Connecting PROT nets by simply merging places might cause transitions of

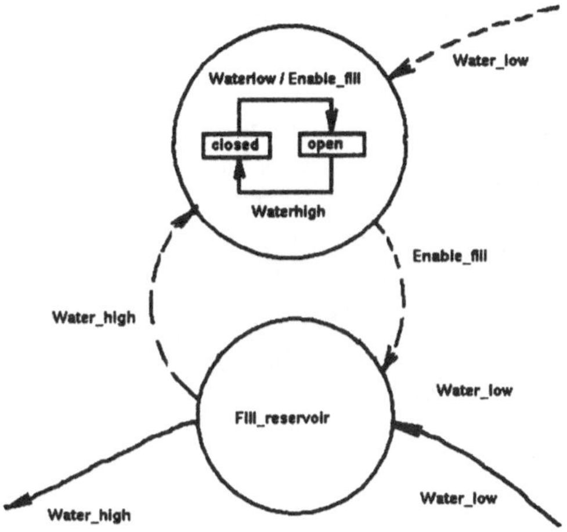

Fig. 2. An extended dataflow equivalent to the previous PROT net.

different PROT nets to be in conflict and thus alter the original behavior of the nets going against the object oriented paradigm.

4. OBJECT-ORIENTED PARADIGM WITH PROTOB

PROTOB was designed to support the construction of models according to an object-oriented methodology which divides the system in a hierarchy of objects to improve its comprehensibility and to simplify its modification and the reuse of system elements.

An object is a model of a real world entity which groups data and operations on that data. Objects are defined by their external properties only, whereas the internal structure is hidden to the user, thus giving a view of how it interacts with other objects. An object is an abstraction of a set of real world things having the same characteristics and conforming to the same rules (class).

The decomposition of a system into a set of objects implies two major issues:

(1) Specify *objects*: their I/O ports, parameters and internal behavior. Objects are identified as proposed in [7] and [9]. The interface of an object is here determined by I/O ports. An Input port is the entry point to an object. A service provided by an object is requested by sending a token to the corresponding Input place. The input parameters are the fields of its data type. The internal behavior of an object is defined by a PROT net. PROT nets respond asynchronously by sending back a message from an Output port.

(2) Define *structuring* and *interconnection* mechanisms of objects. Once the object interface has been defined (its I/O ports), objects may be connected by defining with lines which output port of an object should be connected to which input port of another object. Objects are generally composed of objects or PROT nets.

The **Object** is the building block of a PROTOB model. An object captures the notion of autonomous entity or activity characterized by a lifecycle depending on time and on the interactions with other objects. Examples of objects are: a machining center, a production control module, a protocol level.

An object is **Active** if its response to external stimuli may be delayed, otherwise it is **Passive**. An active object receives an asynchronous trigger/controlflow and reaction to this request/stimulus may be delayed according to the internal state of the object. In a passive object the required service is performed immediately. A PROT net is an active object since the processing of the input place tokens depends on its state. A stack is an example of a passive object.

An object represent a class of individuals all having the same structure but separate existence. Individuals of a class are called instances. Instances are graphically represented by squares within

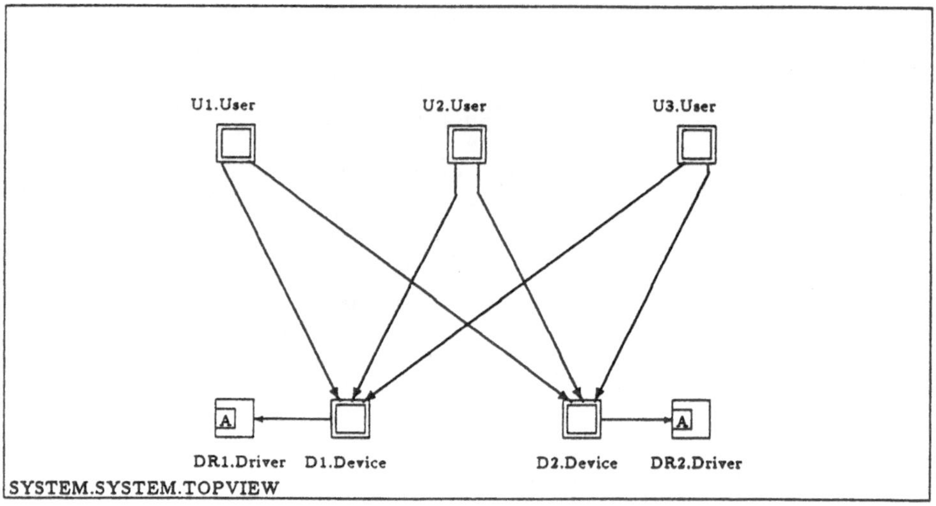

Fig. 3. Users use devices that use drivers.

squares. Objects may have parameters which allow instances to be properly characterized. Defining a class and then instantiating any number of individuals conceptually simplies both the definition and modification of the system.

Objects can be composed to form objects at a higher conceptual level thus creating a hierarchy of objects. The overall system is modelled by the object at the highest level in the hierarchy. Objects can be structured and interconnected in two types of hierarchies:

Usage (senior/junior) relationship—Senior objects may use services provided by junior objects. The usage relationship is represented by drawing a directed line from the senior to the junior object. Active objects may use other objects freely, but passive objects may use only passive objects.

Composition (parent/child) relationship—Parent objects may be decomposed into several child objects. The composition relationship is represented by drawing the children inside the parent.

Note that there is no need to include an object as a child in order to use it or, in other words, to require its services. The difference between **Composition** and **Usage** should be clear in Fig. 3. It shows users requesting services from devices. *System* is an object **composed** of two *Device*, two *Driver* and three *User* instances. The first identifier is the name of the instance. The second is the name of the object (class). Each *User* object uses two *Device* objects. Each *Device* uses a *Driver*, which is an active PACKAGE object. *Device* and *User* are PROT objects. *DR1* and *DR2* are instances of the *Driver* PACKAGE object. They are represented by a small square on the left side of a large square. Their internal behavior is not defined by a PROT net but by plain code structured using the pure HOOD methodology. the "A" stands for Active.

Note that in PROTOB an object does not know which object will provide the services it requires unless its junior objects are also its children. The user is free to connect a senior object to any junior object that provides the required services, without having to change the definition of the senior object. The usage relationships are specified in the definition of the parent object that includes the senior and junior objects.

PROTOB has two kinds of objects: **PROT objects**, containing PROT nets, are always active; **PACKAGE** objects may be either active or passive and are defined textually like an Ada package. Package objects may be modelled in greater detail using the HOOD methodology. PACKAGE objects may be composed of other PACKAGE Objects but not of PROT objects whereas PROT objects may be composed of objects of both kinds.

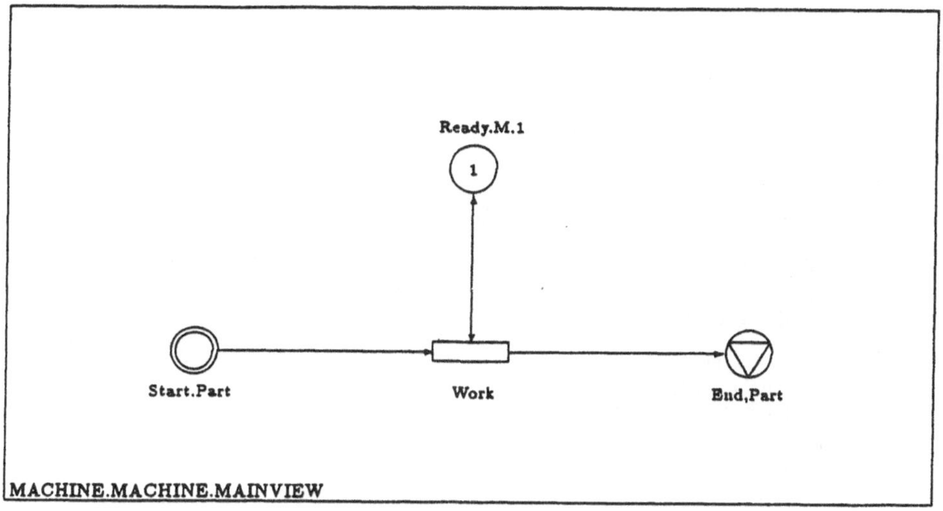

Fig. 4. Graph of a simple object that models a machine with a PROT net.

Objects are defined by their external properties only, whereas the internal structure is hidden to the user, thus giving a view of "how it appears to other objects" [7]. From the outside, an object is known to a higher conceptual level by its name and its interactions with the other objects. The object interaction is defined by the object's **Interface** that, in a PROT object, consists of the object's set of Input and Output ports, which conceptually correspond to a list of provided and required services or messages to be received and sent (representing events, dataflows or controlflows).

The interface of a PACKAGE object is the list of provided and required services (i.e. entry points or routines).

Hidden inside the object we find its **Body** at a conceptually lower but more detailed level that implements the behavior of the object. The body may be defined by other objects interacting with an optional PROT net in a **Compound Object** or by a PROT net or PACKAGE definition alone in a **Simple Object**. Simple objects are the objects at the lowest level in the hierarchy while compound objects permit the hierarchical decomposition of systems into subsystems that are still modeled with objects. A **Subobject** is a child and component object of the parent object that includes it. An object definition consists of a graphical **(Graph)** and textual **(Script)** description. They will be explained in detail in Section 5.

Figure 4 is the graph of a simple object that models a machine with a PROT net. The PROT object interface to the outside world is defined by the two I/O ports *Start* and *End*. The object receives from the outside a token of type *part*, modelling the part the machine is to work, in input place *start*. Transition *work* takes the token from *start* when a token is in place *ready*, i.e. when the machine is ready. Initially place *ready* will contain a token but the transition will fire and put the token back only after a certain delay, modelling the device service time, has expired. It will also put the *part* token in output port *end* and this will cause the token to be sent outside to some other connected port.

4.1. *Using PACKAGE objects*

A transition may need to use in its action a service provided by a PACKAGE object. Below is a fragment of two script files that show how a PROT object, *Example*, uses a PACKAGE object. The script file completes the graph defining the types of the tokens and the actions each transition is to perform when it fires.

```
OBJECT example IS
  [...]
  TRANSITION store IS
    ACTION %% push(accepted.info);
              push(coded.info); %%;
  ENDTR store;
  TRANSITION send IS
    action: %%out.data: = process(pop,pop);
           %%;
  ENDTR send:
  [...]
  END example.

PACKAGE stack IS
  PROCEDURE push (x: item);
  FUNCTION pop RETURN item;
End stack;
PACKAGE BODY stack IS
  [...]
  END stack.
```

Transition *Store* uses the *Push* and transition *Send* the *Pop* operations to be provided by a PACKAGE object. *Stack* is a PACKAGE object that provides procedure *Push* and function *Pop*. Which object is actually to provide the required operations will be defined in the graph of the parent object with a directed line from the senior to the junior object. This would correspond to the WITH and USE pragmas in an Ada implementation of the senior object. Figure 3 defines that the services required by each *device* object will be provided by a *driver* active PACKAGE object.

Of course, a PACKAGE Object too may use the operations provided by another PACKAGE object.

4.2. Connecting PROT objects

The PROT net is the essence of a PROT object body, being subobjects a plain mean of system decomposition. PROT objects **Use** other PROT objects by requesting their services. This may be defined graphically by connecting objects with lines with an arrow pointing to the used object like in Fig. 3. A PROT net can request a service by sending a token to the input port of the junior object that is associated with that service. The junior object may then reply by sending back another token.

The connection between PROT objects is clearly between the objects' PROT nets. PROT nets are connectable by setting **Links** from gates to places. For their common nature of being connectable by links, places and gates are generically called **Ports**. Places are ports from which no links start. A gate is a port from which links start and arcs end. The token that reaches a gate will leave its PROT net and enter another PROT net queueing up in the place connected by a link to that gate. Considering the token as a message, the link may be said to be a transmission line a gate may use to send the connected place a message. PROT nets thus communicate by message passing. The object *Line* of Fig. 5 simply groups three machines into a cascade connection. The body of each machine is defined by the PROT net of Fig. 4. Input port *In* accepts tokens from the outside and passes them on to port *Start* of the first machine while output node *Out* sends outside the tokens it receives from port *End* of the last machine. Ports are connected with links. If a connected port is inside a subobject its name will be displayed by the link, before or after an arrow depending on whether the link starts or ends on that port.

It sometimes may be useful to connect a gate to a place of the same PROT net but links usually connect gates to place of different PROT nets. One of the two connected PROT nets must be in a subobject while the other one may be in another subobject or be part of the body.

The information hiding mechanism needed to ensure the reusability of objects prevents us from being able to reach any port of a subobject. Only **I/O ports** are visible from the outside of a subobject and may be connected to any port of the parent object. An Input port accepts an

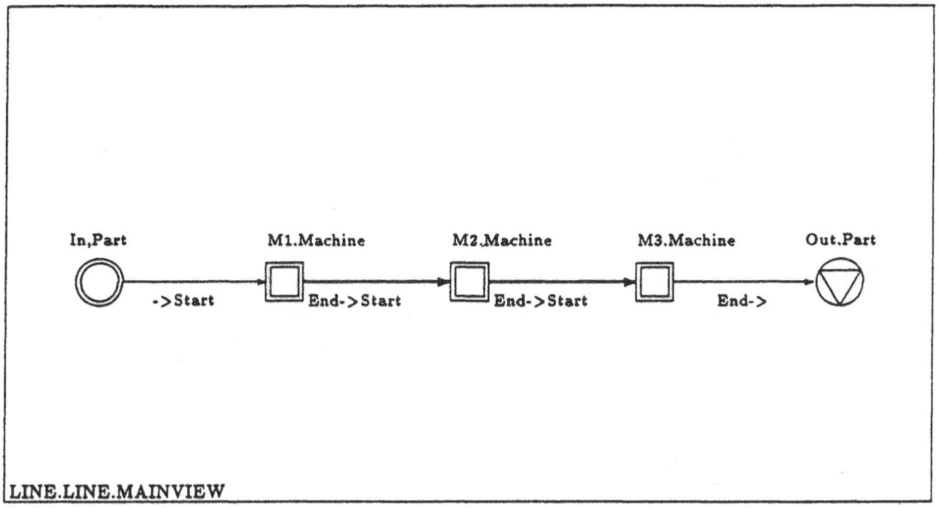

Fig. 5. Object LINE is a compound object with no PROT net.

incoming link from the outside while an outgoing link is allowed to start only from an Output port. Places may only be Input ports while gates may only be Output ports. No port may be both of input and output.

I/O ports define the **Interface** of an object. Connecting a link to a non-I/O port would be the same as changing the definition of the object. For this reason subobjects are not visible from outside the parent object. Links may not be directly connected from outside the parent object to the I/O ports of a subobject. This is so because reaching a subobject directly with a link would modify the definition of the parent and this is forbidden by the object oriented paradigm. However, an object can be defined so that a subobject may be connected to the outside through a special I/O port called **Node**. Nodes are I/O ports, visible from the outside, that are connected by links to subobjects. Reaching a sub-subobject is then possible by reaching with links the nodes of the subobject that includes it. The node simply passes the token on to the port of the next hierarchical level; upwards if output node, downwards if input node. Nodes may not be linked to ports of the same object as this does not make sense. The communication line between a gate and a place thorugh the hierarchy of objects has a structure that is well captured by the expression $G \rightarrow \{N \rightarrow\} P$ where $\rightarrow$ stands for the link.

Two ports connected by a link *must be of the same type* since they handle the same tokens. A port which is or will be connected by a link is therefore called a **Communication Port** and the types of communication ports are called **Communication Types** that must be defined uniquely for all PROT nets using them. Nodes and gates are always communication ports so as any place to which a link arrives, even if they are not I/O ports.

All I/O ports are of course communication ports but of a special kind as they define the interface of the object. The rest of the communication ports are connected to ports of subobjects or to other ports of the same PROT net. The object that has no I/O ports is a **Closed Object** that may not be included in the definition of other objects because it has no interface. The closed object models the system at the top level. Figure 6 shows a closed compound object that contains a subobject and a **PROT** net. Note that gate *Raw* and place *Finished* are communication ports but are not I/O ports as they interface the PROT net to a subobject.

4.3. Selective routing and broadcasting

It may occur that a gate need to send messages to more than one place. Consider for example the case of a device that provides a service to several users. A user would request the service sending

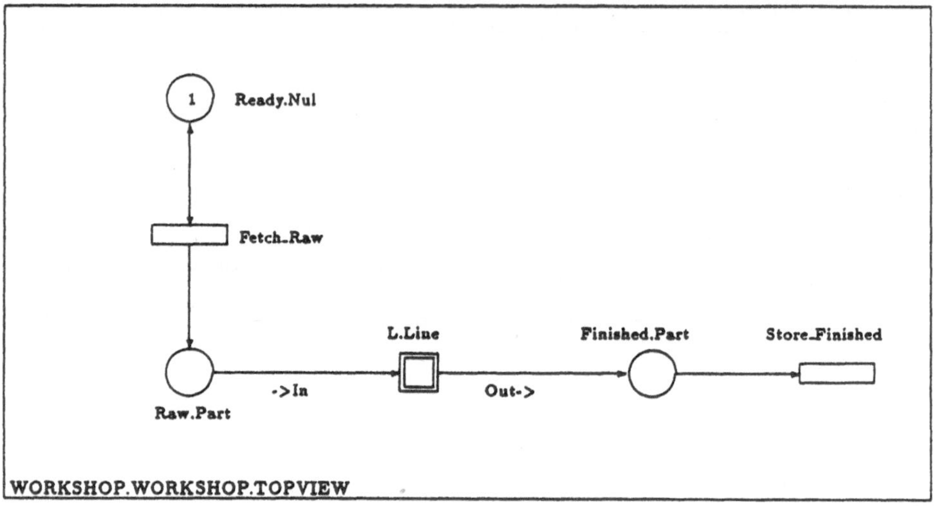

Fig. 6. Graph of a closed compound object, i.e. with no I/O ports.

a request-token to the request-accepting input place of the device which, upon completion, will answer back sending a reply-token to the user reply-receiving place. The gate of the device is therefore connected to the reply-receiving place of all the users. How is the device supposed to select the user to reply to? Note that the device is built to serve any number of users implemented in any way as long as they have the same interface.

Let us now suppose that more than one device is available, say two. We are in the case of Fig. 3. Each user may partition its service requests on the two devices. The user is built to receive service from any number of however implemented devices. At the beginning of operation each device must send its identity to all the connected users to let them know of its existence. Each user will therefore receive an identity-token from each available device to which it may send service requests. Figures

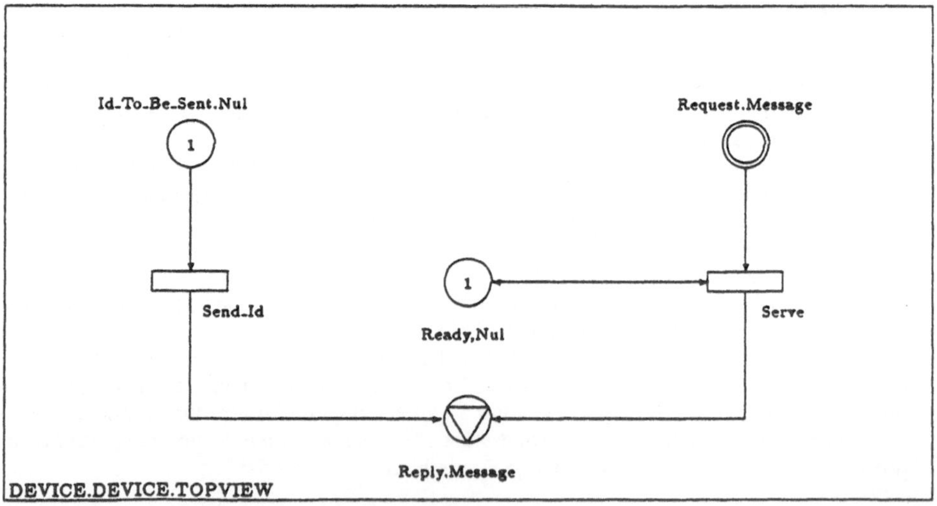

Fig. 7. Device object.

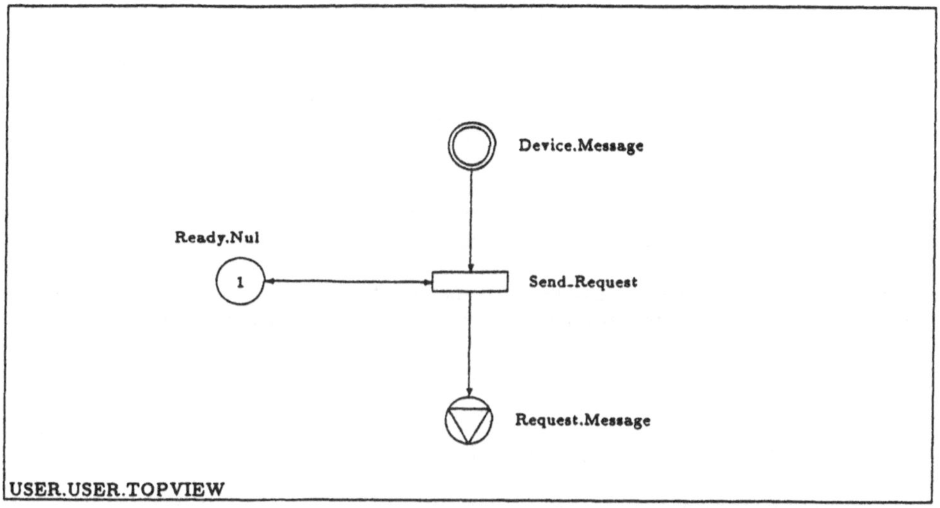

Fig. 8. User object.

7 and 8 show the PROT nets that define the body of the *User* and *Device* objects of Fig. 3.

If a gate has more than one link starting from it, it is connected to more than one place. This is also true if more than one link starts from some node in the communication line. The forking of the communication line arises the question of which way the token should take. To maintain the object-oriented structure an object must be defined independently from the number and type of objects to which it will eventually be connected. Thus the object may not select itself the instance to which the token is to be actually sent. We find the same problem in distributed systems where each node may not know beforehand which nodes of the network are actually up. To solve this problem the most general protocol requires each instance to send its identity to all the other instances to which it is connected. Each instance receives the identity of all the reachable instances. It will then be able to send tokens selectively by mentioning the identity of the destination object. PROTOB provides two ways of sending tokens:

Selective routing—Each token upon leaving the PROT net through a gate may be stamped with the address of the destination instance. This may be accomplished by using primitive xx_setaddress(). An instance may know its own address by reading it with function xx_myaddress() and store it in a variable of type XX_ADDRESS. An instance may send its address to some other instance by writing it in a field of typed XX_ADDRESS of a token.

Broadcasting—When a gate is connected to more than one place and the token has not been stamped with a destination address, the token is copied and sent to each connected place. Broadcasting is the sole protocol an object may use when the identities of the destination instances are unknown.

Linking a gate to more than one place requires special care to design the objects involved and their communications. Ports that allow more than one link to start from them are special and said **Derouting Ports**.

A superlink stands for a bundle of links and is represented like a link with no name. The superlink of Fig. 3 that connects users to devices is made up of two links: *Reply*$\rightarrow$*Device* and *request*$\leftarrow$*Request*. Each device initially broadcasts its address to all users with a token from gate *Reply*. Each user keeps the identities of the reachable devices in place *Devices*. As soon as the user is ready, a service request message, modelled by the token of type *Message*, is sent selectively to a device. When the device is ready it serves the request and when it has finished it sends a reply message to the user. This time the reply is not broadcast but is routed selectively back to the user that had sent the request.

5. THE PROTOB LANGUAGE

PROTOB is a hierarchical object-oriented graphical programming language. Its main features are the definition of objects, the structure of PROT nets and the interconnection mechanism.

5.1. The graph

Most information on the object is conveyed visually by a formal drawing called **Graph** which defines at a conceptual level the PROT net and the interconnections between subobjects.

The PROT Object Graph. A **PROT net** is the primitive definition of a PROT object behavior in terms of states and state transitions. Its graph is composed of the following elements:

Places—They contain **Tokens**, which are structured mobile data, similar to a Pascal record. Each place can contain more than one token, but they all have to be of the same type. Tokens are queued in FIFO order but may also be retrieved according to a particular policy defined by the user. Places model **States**, **Data** and **Events**.

Places are represented graphically by a circle which is labelled with the name of the place, followed by the name of the type of tokens to be contained and optionally by the number of tokens present at the beginning of the execution (the initial marking of the net). During the animation of the PROT net, the current number of contained tokens is displayed at the center of the place.

Gates—A gate is similar to a place because it accepts tokens of a certain type but it does not keep them. A token that reaches a gate is sent to the place to which the gate is connected. Gates model **States**, **Data** and **Events** that are messages to be sent.

Gates have the same graphical representation of places because there is no semantic difference between them.

I/O ports—Places and gates are generically called **Ports**. Special ports are the object's input and output ports which allow objects to send and receive tokens to and from other external objects. I/O ports define the object's **Interface** to the outside. Places do not send tokens and so may not be output ports. Gates do not contain tokens and so may not be Input ports.

Input ports are represented by a smaller circle inscribed in the outer circle that characterizes all ports. Output ports have a triangle inscribed upsidedown.

Arcs—An arc is a graphical means of defining the input and output ports of a transition. It is an oriented segment going from a transition to one of its output ports or from one of its input places to a transition. No arc may go from a gate to a transition (i.e. no gate may be of input to a transition) since gates do not contain tokens.

Transitions—They model actions caused by events that modify states or data and cause events. A transition may fire when all its input places contain at least one token and its predicate, an explicit firing conditions, is satisfied. When a transition fires, it moves a token from each input place to the corresponding output port and performs an action, which is a sequence of instructions. Each input place is in fact implicitly mapped onto the output place of the same type. Consequently, a transition cannot have more than one input place or more than one output port of the same type, except for the *nul* type. Tokens of the *nul* type are like a record that has no fields and so they are undistinguishable. Tokens removed from an input place that has no output place that matches its type are destroyed. A token is created instead for every output place that has no type matching input place. When two or more transitions may fire at the same time but the firing of one disables the others, they are said to be in **Conflict**. Two transitions are in conflict if they have a place in common and a token in that place satisfies the predicate of both. When transitions are in conflict the one with the highest priority is chosen to fire.

A transition is graphically represented by a rectangle which is labelled with the name of the transition optionally followed by a number indicating its priority. During the animation of the PROT net, the number of times the transition has fired since the start of execution is displayed at the center of the transition.

In addition to the elements of a PROT net a PROT object can contain **Subobjects** that are connected to one another and to the PROT net by means of **Links**, **Superlinks** and **Uselinks**:

Subobjects—A subobject is an object which is part of another object. The use of subobjects makes the definition of an object hierarchical by allowing instances of other objects to be components of instances of more complex objects. PROT subobjects exchange tokens with each other and with their parent's PROT net. PACKAGE subobjects are used by other subobjects and by their parent's PROT net that request their services. When an object is instantiated each of its subobjects is instantiated too and its instance is connected to the rest of the instances of which the object's instance is composed.

PROT subobjects are graphically represented by an icon consisting of a square framed by another square, while PACKAGE subobjects by a square with a smaller square inside aligned on its left side. Each icon is labelled with the name of the subobject, followed by the name of the object of which it is an instance.

Two standard variables are always associated with a PROT subobject: **State** (an integer value) and **Colour**. When the model is executed the editor displays in the middle of the PROT subobject icon the state value in the associated color. The user can take advantage of these variables to show some information on the internal condition of the PROT subobject. During the animation of the model, the state of each PROT subobject is displayed in the selected colour at the center of the framed square which represents the subobject.

Links and superlinks—A link connects a gate to a place. The token leaving a PROT net instance through a gate will queue up in the place of another PROT instance following the link that connects the two ports. Being the token a message, a link can be considered a transmission line which an instance can use to send messages to the places of another instance. Ports that are connected by a link are called **Communication Ports** and they must be of the same type as they obviously accept the same tokens. Their types are called **Communication Types** and they must be defined in a unique and global way for all the objects that use them. All I/O ports are communication ports. All the ports that are connected to the I/O ports of a subobject are also communication ports even if they are not I/O ports themselves.

Links, like arcs, are represented by directed segments. They go from a gate to a PROT subobject, from a PROT subobject to a place or from a PROT subobject to another PROT subobject. A link starting from a subobject is labelled with the name of the subobject gate from which it logically starts, followed by an arrow. If the link goes to a subobject the arrow is followed by the name of the subobject place the link logically reaches.

Superlinks are sets of links connecting ports of two subobjects. They are defined in a textual file simplifying the connection of two subobjects. Superlinks are graphically represented like links and arcs by oriented segments. A superlink goes from a subobject to another subobject. The orientation of a superlink has no practical effect but it may be given a semantic meaning.

Nodes—They are special I/O ports that are connected to subobjects with links. No arc may be connected to nodes. They permit the communication of subobjects with the outside of the object of which they are part. A token that reaches a subobject through an incoming link is sent on to the port to which the outgoing link is connected. Nodes may not be connected to other ports of their object. Nodes are represented like I/O ports but are connected to subobjects by links.

Uselinks—Uselinks connect PROT or PACKAGE objects to PACKAGE objects. PACKAGE objects have no I/O ports; their interface is the list of provided and required services. A uselink specifies that a set of services an object requires is provided by a PACKAGE object. A uselink is like the combination of the Ada WITH and USE pragmas inserted in the senior object to declare the junior one is being used. Uselinks are represented by directed segments that go from the senior to the junior object that is being used. The mapping between required operations at senior level and the operations provided at junior level is represented graphically at a lower conceptual level associated to the uselink.

When a PROTOB graph is so complex that it does not fit on the screen, it can be logically decomposed into several **Views**. Views point out the relationships among some logically related transitions and subobjects. The same place is allowed to appear in more than one view. Transitions and arcs cannot be duplicated, because the context of a transition, i.e. its input and output places, must be defined completely in the same view. Subobjects and links may not be duplicated because the interface to a subobject, i.e. the ports and subobject to which it is linked, must be all in one view.

The PACKAGE object graph. A PACKAGE object contains no PROPT net and so it consists of subobjects and uselinks only.

5.2. The script

The **Script** of an object is a textual file which completes the graph defining in detail the entities that in the graph are known at a more conceptual level only by name. There are two kinds of scripts, the PROT script and the PACKAGE script, since there are two kinds of objects.

The PROT script. It contains the definition of the token types, of the local variables and a detailed description of the transitions' predicates and actions. Sections defining data and sequential operations are written in the standard target programming language (Ada, Pascal and C versions exist at present) external routines written in other languages may be called if needed. The script file therefore contains segments of embedded target language. It is generally composed of the following sections:

(1) Definition of **Token Types**. Tokens are *structured moving data*, similar to Pascal records. If a *scalar* field is preceded by a tilde, it is possible to examine its value during the execution of the model. Communication types, that is the types of input/output ports, are defined in a separate file common to all the objects in the model. In the script only the names of the communication types involved in the corresponding object are mentioned after the keyword COMMUNICATION. The *nul* type is a standard predefined type which has a null data structure: null tokens are only simple flags.

(2) Declaration of the object **Parameters**. *These must be scalars.* If the parameter name is preceded by a tilde, it is possible to examine its value during execution. The values of all parameters are read from the parameter file at the start of execution.

(3) Definition of the **Local Variables**, which are visible only to the PROT net they belong to. They can be of any type. It is clear that the values of the local variables of each instance will change separately from those of other instances. It is also possible to initialize the local variables by writing into the script a sequence of instructions, the *initial action*, which will be executed at the beginning of the program. There is a standard variable (*xx_tracefile*) denoting the output file, where the trace of the execution is written and where the actions of transitions can write.

(4) Definition of **Transitions** in terms of the following attributes:

   (a) Optional **Predicate** in *embedded target language*. It is a condition that must be satisfied for the transition to be enabled. The implicit condition set by the PROT net is that a transition may fire if all of its input places have at least one token in them. The predicate is an additional explicit condition, specified in the script by a boolean expression, which can be used to select the tokens from the input places according to a particular policy. The predicate may depend on the values of local variables and parameters, but these do not influence the synchronization of transitions, which is only determined by the flow of tokens. In fact a change in a local variable is not allowed to bring about the firing of a transition as a side effect: a transition can be enabled only when a new token arrives at one of its input places, and only then will its predicate be evaluated.

   (b) Optional **Action** in *embedded target language*. An action is a sequence of operations that a transition must carry out when it fires. External procedure calls are permitted, but the external code must be declared as such in the script. The action may involve reading and modifying the value of local variables and parameters, of the output tokens of the transition and of the input tokens that will be destroyed. To refer to a particular token involved in the action of a transition, the name of the output place where it will be moved, or of the input one if the token will be destroyed, must be used. Since the token is a Pascal-like record, its fields are identified with the usual dotted notation, e.g. place-name.field-name.

(5) Optional **Initial Action**. It is used mainly to give an initial value to local variables.
(6) Optional **Final Action**. The final action is a sequence of instructions similar to the initial action. It is carried out at the end of the execution of the model: its main purpose is to allow the user to compute and present in the desired format performance statistics and other calculations.

Transitions can be timed to remove tokens from its input places without adding them to their output places until a time delay has expired. The delay value must be set during the execution by using the primitive *xx_setdelay*.

The PACKAGE script. It is a target language definition of a program module or unit. It declares the provided services or access points to the package and it defines their implementation. Using Ada as the target language the script has the usual structure:

```
PACKAGE obj_name IS
  [definition of the provided services and types]
END
PACKAGE BODY obj_name IS
  [Implementation of the provided services]
END obj_name.
```

6. THE PROTOB CASE ENVIRONMENT

The environment is composed of four separate but strongly interconnected tools.

Editor/animator—It carries out a twofold function: as an editor it allows the graphical representation of PROTOB objects to be built and modified while performing immediate consistency checks, as an animator it is able to animate the model during its simulation. In the latter case the editor controls the simulator by sending commands to one mailbox and waiting for replies from a second mailbox. The editor displays on the screen the current state of the elements as it is received from the simulator. Simulation commands are given by the user to the editor that forwards them to the simulator.

Translator—It translates the PROTOB model into an executable program written in the target language: Pascal, C or Ada.

In general each object is composed of a PROT net and of subobjects. For efficiency purposes the execution mechanism handles instances of PROT nets only, objects and subobjects being considered as a structuring facility. For this reason the translator, starting from the closed object, must open iteratively each subobject in order to find all the instances of PROT nets involved in the model. At the end of this process the translator generates and compiles for each PROT net a program module, an Abstract Data Type (ADT) which encapsulates the predicates and the actions of transitions taking them from the related script.

The translator then compiles and generates one more module—the *mailer*—which glues together the modules implementing the ADT of each PROT net, handles inter-object communication and interfaces the system to the editor during the animation of the simulation.

Finally it links all these modules with the execution kernel, thus generating the executable code for the PROTOB closed object.

The translator presents the important feature of **Program Generation**. It automatically produces the software that implements the modelled system. The translator is able to produce simulation code which is a single executable program that simulates the flow of time. Once the model has been validated it is translated in the implementation code that can be distributed program hosted over a local computer network. The parallel execution of subobjects is obtained by generating several processes (tasks) to be loaded on the nodes of the LAN. The distributed program is also installed automatically by the translator according to the system configuration requirements. Task to task communication on the same node and over DECNET is handled automatically.

Simulator/emulator—It executes the PROTOB closed object and consists of two parts: one is automatically generated by the translator, the other is invariant. The invariant part is the *inferential engine* which iteratively selects the transition to fire and returns control to the modules generated by the translator for the test of the predicate and for the execution of the action. The mailer module is the interface which makes it possible to connect a variable number of objects of different kinds to form the final system.

Report generator—It assists the user in preparing the documentation of the model and produces a high quality report combining both graphical and textual descriptions. The user is required to add comments to explain the meaning of each name used in the graph and in the script. The output format is handled automatically and may be customized.

Even if the above-mentioned programs can be used independently, the editor is the natural access point to the environment, as it provides a transparent use of all the tools.

6.1. State of the art of the PROTOB CASE environment

The PROTOB environment is no longer a prototype. ARTIS is the CASE company that has implemented a sound and userfriendly environment called ARTIFEX that supports the PROTOB methodology. It is available on several platforms and operating systems and has a user interface based on X-WINDOWS.

7. MODEL OF A PRODUCTION CONTROL SYSTEM

This case study clearly illustrates how the principles of structured analysis and object-oriented programming are helpful to build a better structured and more comprehensive PROTOB model. On the other hand it demonstrates the power of the PROTOB language modelling distributed systems composed of several entities with an intuitive and user-friendly graphical notation. Each entity is modelled using STDs, data flows and data transformations, control flows and control transformations represented with a single, smooth notation: the PROT net.

The production control system is a software package that controls the machines of an automated production plant. Its purpose is to manage the production plant at its best in order to accomplish the production of lots according to the production plans formulated by the scheduler. The available resources are to be used so as to maximize the throughput of the system by controlling the flow of parts entering and leaving the plant and their internal routing. The control system must also guarantee the plant to be fault-tolerant and capable to continue production with a lower performance if some machine breaks down by routing parts to alternative machines or continuing with the production of alternative parts.

In order to test the behavior of the production control system a model of the entire production plant is also necessary. In fact the performance of the production plant is due to the production control system but it is strongly influenced by the nature and availability of the controlled resources. So the model of an entire FMS (Flexible Manufacturing System) was defined and tested. It consists of three workstations W1, W2, W3; two intermediate buffers B1, B2; a load/unload station L; and an AGV C which routes the parts round the six stations of the plant. The actul production control system is modelled by the dispatcher D. The FMS model is shown in Fig. 9.

7.1. The workstation

The workstation consists of an input buffer, a machining place and an output buffer. Buffers contain one part at most and are used as a temporary storage before and after machining in order to reduce the lead time due to loading and unloading activities. The machine will have a null waiting time if the input buffer is always full and the output buffer is always empty when the machining of a part is over.

The sections of the workstation, the two I/O buffers and the working place, are managed by a control mechanism which is responsible for synchronizing the activities of each section with those

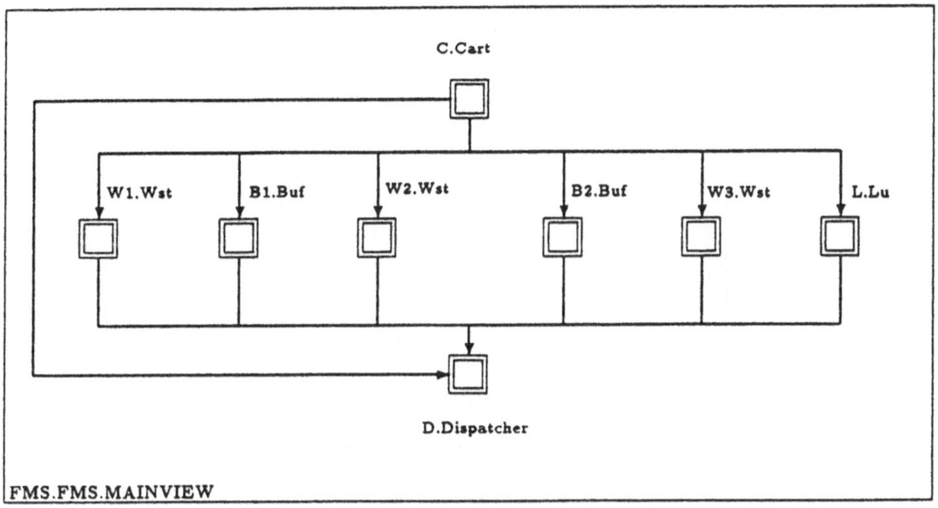

Fig. 9. The model of FMS.

of the other sections. The mechanism is defined in Fig. 10. In fact a part may not be moved from the input buffer to the machining place if the latter is not empty. If the output buffer is not empty a part may not be moved to it from the machining place. The control mechanism of the workstation consists of three interacting state transition diagrams.

Wait_I indicates that the input buffer is empty. It enables transition *Loading* that shuttes a part from the cart into the input buffer. An acknowledgment signal is sent to the cart through the output gate *Load_Done* and *Full_I* indicates the input buffer is full.

Transition *I_To_W* fires when the input buffer is full and the machining place is empty (a token is in place *Wait_W*) and puts a token in place *Ready_W* enabling transition *Working* to fire. The input buffer is now empty and thus a token is put in place *Wait_I* and a request for a new part

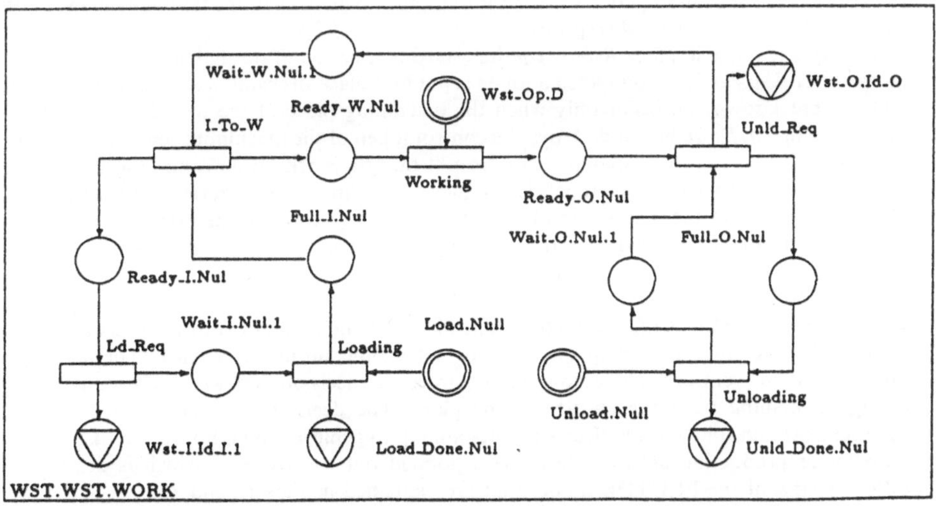

Fig. 10. The workstation, view WORK.

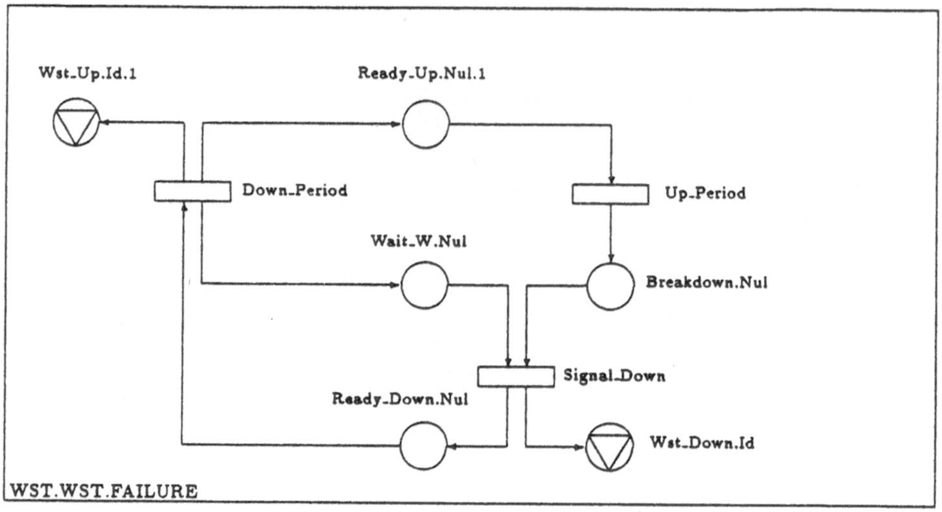

Fig. 11. The workstation, view FAILURE.

is sent to the dispatcher through gate *Wst_I*. Transition *Working* represents the actual machining of the part and, for simulation purposes, the operation to perform and the time it requires are received from the dispatcher at the input place *Wst_Op*. When the operation time is over, transition *Working* puts a token in place *Ready_O* indicating that a part is ready for the output buffer.

When a token is both in places *Ready_O* and *Wait_O*—meaning that the output buffer is empty—transition *Unld_Req* fires modelling the movement of the worked part from the machining place to the output buffer by moving the token from *Wait_O* to *Full_O*. The transition also sends an output request to the dispatcher through the output gate *Wst_O*. Then, when the cart is ready to unload the output buffer, an interaction similar to that described for the input buffer takes place between the cart and the output buffer through ports *Unload* and *Unld_Done*.

The second view, in Fig. 11, models the breakdown of workstations. We assume that a workstation has alternate periods of operation and repair, which are represented by transitions *Up_Period* and *Down_Period* respectively.

Initially a token is in place *Ready_Up* thus starting a period of normal activity. When the firing of transition *Up_Period* is completed a token is put into place *Breakdown*. For simplicity we assume that the breakdown can occur only when the machining place of the workstation is idle. When transition *Signal_Down* fires it destroys the control token of the machining centre in place *Wait_W* and a signal is sent to the dispatcher to inform it that the workstation is no longer available, and a repair period is started. When repair is completed, transition *Down_Period* notifies the dispatcher of the availability of the workstation by sending a message through the output port *Wst_Up* and puts a token into place *Ready_Up* thus starting a new period of normal activity.

7.2. The dispatcher

The DISPATCHER is the production control system model that controls the six stations of the plant and routes the parts from one station to the other using the CART. The activity of the dispatcher is an event-driven decision-making process as decisions are taken in response to requests or signals coming from the entities of the plant. The decisions are based on the production requirements and follow some heuristic rules aiming to achieve a high utilization of the resources. Two major production activities have been pointed out, each one of which is dealt with by a different view of the PROTOB model that are explained in Figs 12 and 13.

Introduction of Pallets into the Plant—Fig. 12 shows two graphical if–then rules. Places in this case are used as datastores. The scheduler assigns to each lot of parts to be produced a given

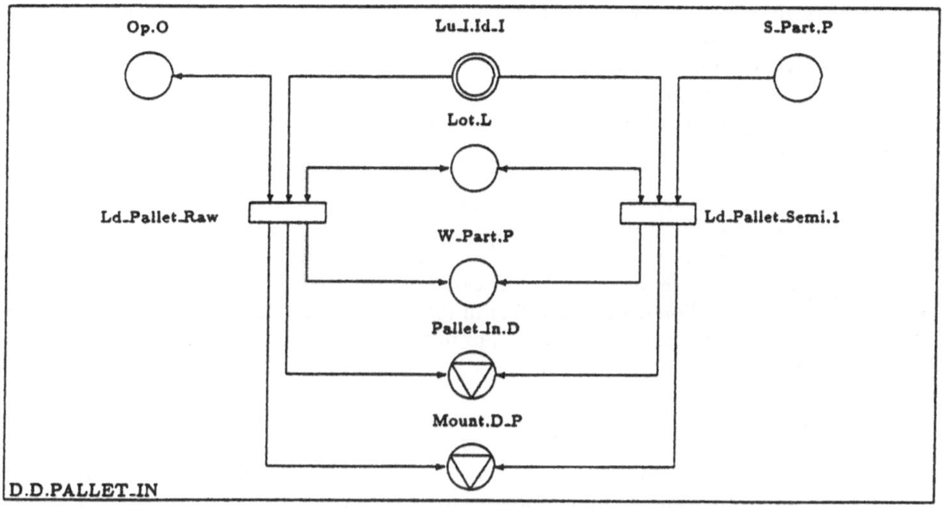

Fig. 12. Dispatcher, view dealing with the introduction of pallets into the plant.

number of pallets which must circulate in the plant simultaneously in order to obtain the desired throughput. Usually a pallet contains one or more parts that a workstation works at the same time. When the number of circulating pallets of a lot is less than the scheduled amount, the dispatcher must introduce a new pallet in the plant. If a semiworked part may be reintroduced in the plant it will be by transition *Ld_Pallet_Semi*, otherwise a raw part will be introduced by transition *Ld_Pallet_Raw* that has a lower priority. The number following the transition's name is in fact its priority that is omitted when zero. Place *Lot* is a datastore containing tokens of type *L* that describe the lots to be produced and contain the scheduled and actual number of pallets along with other information. Input place *Lu_I* contains the identifier of the load/unload stations that are ready to perform a new operation. *W_Part* and *S_Part* contain tokens of type *P* which describe the parts

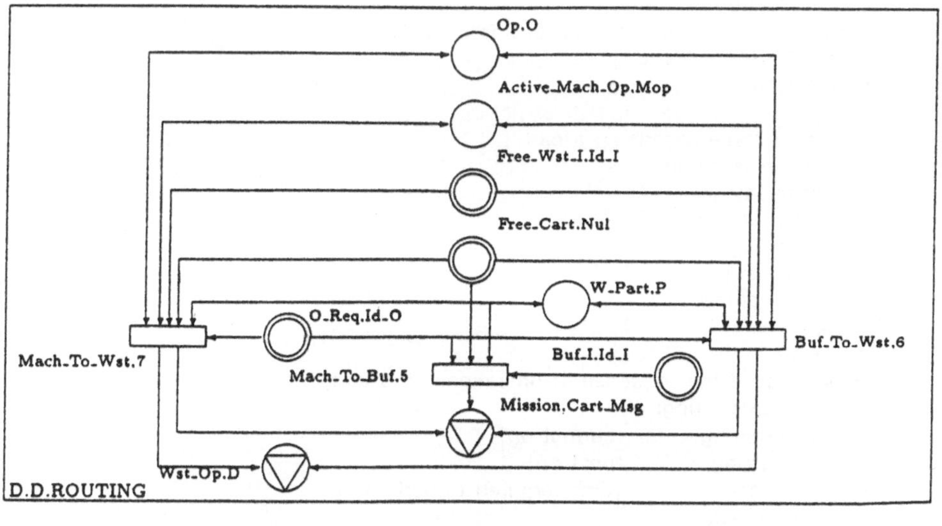

Fig. 13. View dealing with the routing parts.

inside the plant. When a raw part is introduced into the plant a new descriptor for the part is created and stored in place *W-Part* along with the other parts which are being worked. When a part is not finished yet but for the time being it may not be worked upon, it is suspended, removed from the plant and its token is moved from *W_Part* to *S_part*. When transition *Ld_Pallet_Semi* resumes the suspended part the token is moved back to *W_Part*. *Pallet_in* and *Mount* are two output gates which model the signals sent to the load/unload station to prepare a new pallet and mount a part on it.

Routing of parts—The routing of parts is controlled by the dispatcher according to the scheduled production plans and the available resources. It is actually performed by the cart which carries the part from one station of the plant to another. Figure 13 shows the view with the routing rules. *Mission* is the output gate that controls the cart. Its tokens hold the identity of the station where the part is to be loaded on the cart and that of the station where the part is to be unloaded. There are three kinds of missions with decreasing priority that are commanded by the three transitions. For any mission to take place the cart must be free as signalled by a token in place *Free_Cart*.

Mach_to_Wst moves a part from a machine—a load/unload station or a workstation—to a free workstation. For such event to take place a request from a machine for its output buffer to be emptied must have been received in input place *O_Req*. Also a free workstation must exist that may perform the next operation the part is to undertake. Place *Op* is a datastore containing tokens—triples—reporting the next operation to be performed on a part knowing the lot and the current operation. *W_Part* records the current operation. *Active_Mach_Op* is a datastore containing the tokens describing the operations each workstation may perform. *Free_Wst_I* contains the identifiers of the workstations which are free.

Buf_to_Wst is the same as *Mach_to_Wst* but takes the part from a full buffer instead of a machine. Note that it has lower priority because it is more important to empty a machine output buffer first to ensure that the machine will not stop working.

Mach_to_Buf moves a part from a machine to a buffer if there is no workstation that may perform the next operation and there is an empty buffer signalled in input place *Buf_I*.

The script of the dispatcher is quite long and therefore large cuts have made in it to reduce it to a reasonable size. However it still is a good example of a PROTOB script in Pascal. Scripts in C or Ada are organized in the same way. The syntax of the embedded C or Ada obviously differs from that of Pascal.

```
OBJECT dispatcher IS
CMN fms; {fms is a file containing the definition of the communication types}
TOKEN_TYPE
{communication types defined in fms.cmn}
COMMUNICATION id_i;
COMMUNICATION id_o;
[...]
p = record {type describing a part}
     lot_name: string;
     code: integer;
     start_time: real;
     op: string;
     entity: string;
     name: string;
  end;
o = record
{descriptor of the sequence of operations on a part}
     lot_name: string;
     name: string;          {current op}
     next: string;          {next op}
     time: real;            {time required to perform it}
     avail: integer;
  end;
```

```
[...]
LOCAL  lot_f:text;  wst_f:text; [...]
TRANSITION mach_to_wst IS
    PRIORITY 7;
    PREDICATE |
      (w_part.lot_name = op.lot_name) and
      (w_part.op = op.name) and
      (w_part.op = active_mach_op.op) and
      (active_mach_op.mach = wst_i.name) and
      (op.name <> 'dismount') and
      (op.avail > 0) and
      (o_req.entity = w_part.entity) and
      (w_part.entity <> 'buffer') and
      (o_req.name = w_part.name); |
    ACTION |
      w_part.op: = op.next;
      w_part.entity: = wst_i.entity;
      w_part.name: = wst_i.name;
      wst_op.lot_name: = op.lot_name;
      wst_op.op_delay: = op.time;
      mission.s_pos: = o_req.pos;
      mission.d_pos: = wst_i.pos; |
ENDTR mach_to_wst;
[...]
INITIALIZE | {initial action}
    open(lot_f,'lot.txt',old);
    reset(lot_f);
    open(wst_f,'wst.txt',old);
    reset(wst_f);
    n_tot_lot: = 0;
    sim$setcolour(2); |
FINAL | sim$stat; | {final action --- writes final statistics}
ENDOBJ dispatcher.
```

8. CONCLUSIONS

Practical experiences with PROTOB concerned mainly modelling and simulation activities of real systems, in particular manufacturing lines, flexible manufacturing systems, monitoring systems and communication protocols. The complexity of such models ranged from 10 to 50 different objects made up of 1–15 views. Users unfamiliar with Petri nets, dataflows and object oriented design readily understood the modelling language and methodology and learned quickly how to use the environment to obtain new and complex models. Thanks to rapid prototyping the models were tested and enhanced in a very short time to outstanding quality standards. We would like to thank our users for their precious hints and feedbacks that were used to improve the environment.

The application of the methodology supported by the ARTIFEX CASE environment has brought about tremendous productivity gains. A system which took ten people two years to specify using traditional approaches was remodelled in eight months by two people who had not taken part in the first modelling activity.

The automatically generated code stands in a size ratio of 15 against the implementation model files. This means that the environment has reduced the size of the files to be handled roughly by a factor of fifteen. The improvement is even greater if we consider that handling code is harder than working on its model. Furthermore the system maintenance may be performed on the model rather than on code.

Acknowledgement—This work has been partly supported by the C.N.R., Progetto Finalizzato Robotica, obiettivo ALPI, contract number 89.00495.67.

REFERENCES

1. Baldassari, M. and Bruno, G. An environment for object-oriented conceptual programming based on PROT nets. In *Advances in Petri Nets. Lecture Notes in Computer Science No. 340*, pp. 1–19. Berlin: Springer; 1988.
2. Bruno, G. and Marchetto, G. Process-translatable Petri nets for the rapid prototyping of process control systems. *IEEE Trans. Software Engng* **SE-12**: 356–357; 1986.
3. Ajmone, M., Barbetta, L. and Neri, F. A Petri net simulation model of HDLC. In *TENCON 89*, Bombay; November 1989.
4. Baldassari, M. and Bruno, G. Operational software engineering with PROTOB. In *CASE 89*, London; July 1989.
5. Baldassari, M., Bruno, G., Russi, V. and Zompi, R. PROTOB, a Hierarchical Object-Oriented CASE Tool for Distributed Systems. In *European Software Engineering Conference 1989*, Coventry; September 1989.
6. Zave, P. The operational versus the conventional approach to software development. *Commun. ACM* **27**: 104–118; 1984.
7. Booch, G. Object oriented development. *IEEE Trans. Software Engng* **SE-12**: 211–221; 1986.
8. Booch, G. *Software Engineering with Ada*. New York: Benjamin Cummings.
9. HOOD Working Group. *HOOD Reference Manual*, Issue 3.0. The Netherlands: ESA, Noordwijk; September 1989.
10. HOOD Working Group. *HOOD User Manual*, Issue 3.0. The Netherlands: ESA, Noordwijk; December 1989.
11. *IEEE Computer*. Special issue on visual progamming; August 1985.
12. Balzer, R., Cheatham, T. E. and Green, C. Software technology in the 1990's: Using a new paradigm. *Computer* **16**: 39–45; November 1983.
13. Marco, De T. *Structured Analysis and System Specification*. Englewood Cliffs, NJ: Prentice–Hall; 1979.
14. Ward, P. T. and Mellor, S. J. *Structured Development of Real-Time Systems*, Vols 1–30. Yourdon Press; 1985.
15. Hatley, D. and Pirbhai, I. *Strategies for Real-Time System Specification*. Dorset House; 1987.
16. Reisig, W. Petri nets for software engineering. In *Petri Nets: Applications and Relations to Other Models of Concurrency*, pp. 63–96. Berlin: Springer; 1986.
17. Zave, P. An operational approach to requirement specification for embedded systems. *IEEE Trans. Software Engng* **SE-8**: 250–269; 1982.
18. Berzins, V. and Gray, M. Analysis and design in MSG.84: Formalizing functional specifications. *IEEE Trans. Software Engng* **SE-11**: 657–670; 1985.
19. *IEEE Software*. Special issue on CASE; March 1988.
20. Jensen, K. Coloured Petri nets and the invariant-method. *Theoret. Comput. Sci.* **14**: 317–336; 1981.
21. Genrich, H. J. and Lautenbach, K. System modelling with high level Petri nets. *Theoret. Comput. Sci.* **13**: 109–136; 1981.
22. Wheeler, G. R., Wilbur-Ham, M. C., Billington, J. and Gilmour, J. A. Protocol analysis using numerical Petri nets. In *Advances in Petri nets 1985*, pp. 435–452. Berlin: Springer; 1986.
23. Colom, J. M., Silva, M. and Villarroel, J. L. On software implementation of Petri nets and coloured Petri nets using high-level concurrent languages. In *Proc. Application and Theory of Petri Nets*, pp. 207–241, Oxford; June 1986.
24. Nelson, R. A., Haibt, L. M. and Sheridan, P. B. Casting Petri nets into programs. *IEEE Trans. Software Engng* **SE-9**: 590–602; 1983.
25. Bruno, G. and Balsamo, A. Petri net-based object-oriented modelling of distributed systems. In *ACM Conf. on Object-Oriented Programming*, pp. 284–293, Portland, Oregon; October 1986.
26. Bruno, G. and Elia, A. Extending the entity-relationship approach for dinamic modelling purposes. In *5th International Conference on Entity–Relationship Approach*, pp. 327–339, Dijon, France; November 1986.
27. Bruno, G. and Elia, A. Operational specification of process control systems: execution of PROT nets using OPS5. In *10th World IFIP Congress*, pp. 35–40, Dublin; September 1986.
28. Brownston, L., Furrell, R. and Kant, E. *Programming Expert Systems in OPS5*. Reading, MA: Addison–Wesley; 1985.
29. Valette, R., Courvoisier, M. and Mayeux, D. Control of flexible production systems and Petri nets. In *Informatik-Fachberichte 66: Application and Theory of Petri Nets*, pp. 264–277. Berlin: Springer; 1983.
30. Martinez, J., Muro, P. and Silva, M. Modelling, validation and software implementation of production systems using high level Petri nets. In *IEEE Int. Conf. on Robotics and Automation*, pp. 1180–1185, Raleigh, NC; March 1987.

About the Author—MARCO BALDASSARI received his M.Sc. in electronics in 1987 from the Politecnico di Torino, where he is now studying for his Ph.D. in computer science. His current research in software engineering focuses on specification and design methodologies and the CASE tools that support them aiming at an automated code generation. With Giorgio Bruno, he is co-founder of the PROTOB environment and methodology.

About the Author—GIORGIO BRUNO received the degree in electronic engineering in 1977 from Politecnico di Torino, Italy, where he is currently an Associate Professor of computer science with the Dipartimento di Automatica e Informatica. His current research interests include programming languages and environments, distributed systems, discrete event simulation and software engineering especially addressed to industrial automation. He has authered nearly eighty papers on those topics. He has been involved in several Italian national research projects, ESPRIT projects and ESA projects. He has also been a consultant to several major companies on simulation, special-purpose programming languages and software engineering issues. He has developed an original approach based on extended Petri nets for modelling and prototyping discrete event dynamic systems, such as computer-integrated manufacturing systems, communication protocols and real-time systems. He is a member of the IEEE Computer Society and the ACM.

26.
An Integrated Software Development Methodology Based on Hierarchical Colored Petri Nets

V. Pinci and R.M. Shapiro

11th Int. Conference on Applications and Theory of Petri Nets, Paris 1990

Abstract
This paper presents a new integrated software development methodology that uses SADT diagrams, Hierarchical CP-nets, Standard ML, and automatic mechanisms for translating SADT diagrams into Hierarchical CP-nets and for converting Hierarchical CP-nets into Standard ML executable code.

A case study describes the development of a software application called the Debt Manager's Assistant, which implements a non-FIFO strategy for electronic funds transfer. This application was built for the Marine Midland Bank of New York and Societé Générale. During application development, the system specification and requirements analysis were done with the aid of the Structured Analysis and Design Technique. The system design and verification used Hierarchical Colored Petri Nets and the implementation was supported with the automatic production of executable Standard Meta Language code.

0. Introduction
In this paper we describe an integrated methodology for software development in which the system specification and requirements analysis are done with the aid of Structured Analysis and Design Technique (SADT) [1]. The prototype and simulation use Hierarchical Colored Petri net models (HCP-net) [2,3] and the implementation is supported with the automatic production of executable Standard Meta Language (Standard ML) code [4,5]. The different development phases are connected via automatic bridges.

The methodology is abstracted from our own experience in developing a software application, called the Debt Manager's Assistant (DMA), which implements a non-FIFO strategy for electronic funds transfer. This application was built for the Marine Midland Bank of New York and Societé Générale.

The paper starts (1) by giving a short overview of the components of the methodology and of their interconnections. Then (2) the case study is presented. Then there are five sections followed by some concluding remarks (8). The first of these five sections (3) describes the specification phase, the second one (4) is dedicated to the translation of SADT models into Hierarchical CP-nets, the third one addresses (5) the prototype and simulation phase and the fourth one (6) discusses implementation. The last of these sections (7) explains code generation. For each of these five sections we explain what the main objectives are, the method used, the tool support and examples from the application.

1. An Integrated Software Development Methodology

The software development methodology that is presented has the following components (fig.1):

- SADT is used in the early design phase to specify the functional decomposition of the system and the data flow dependencies between components
- Hierarchical CP-nets are used to add data descriptions and dynamic aspects to the SADT specification and study the behavior of the system through simulation and formal validation
- Standard ML is used to implement complex algorithms and as the target language for the automatic production of executable code

These components are integrated in such a way that:

- SADT specifications are translated into HCP-nets
- Hierarchical CP-nets are used to generate Standard ML executable code.

The software tools that are used to support the different phases are also represented in figure 1.

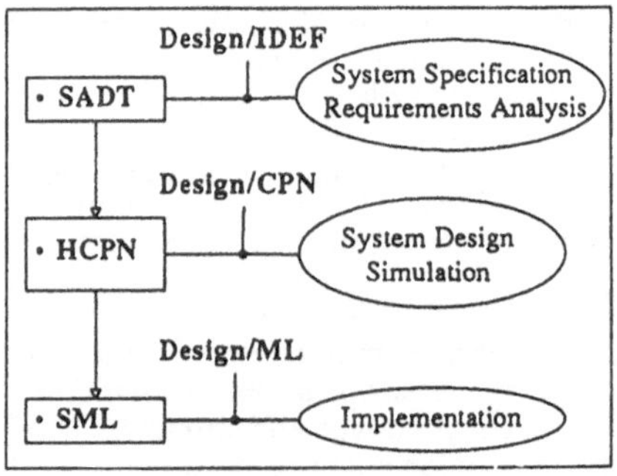

Fig. 1: Integrated Software Development Methodology and Software Tools

Some of the most obvious benefits of this integrated methodology are the following:

- SADT's loose notation, its simple graphic lay-out conventions, its strong decomposition primitives enable designers to become familiar with it very quickly and to generate representations of systems that are easy to understand and that in many cases are well-structured and depict accurately the functionality of the system under study.
- By automatically translating SADT specifications into Hierarchical CP-net models it is possible to step into the next phase of software development, typically design and prototype, in the framework of the original SADT representation of the system, therefore ensuring consistency in the development process. Alternative approaches that use SADT for the initial specification and then use different methods for the detailed design phase in a non integrated fashion, suffer from being error-prone, time consuming and from not being able to ensure consistency between the different development phases.
- Hierarchical CP-nets enable a designer to rapidly prototype the system under study, by adding to the original SADT specification a sound representation of the behavior using HCP-net inscriptions.
- By simulating and formally analyzing the Hierarchical CP-net model, it is possible to validate the initial SADT specification early in the development process.
- The availability of a powerful programming language in the framework of a Hierarchical CP-net model, makes it easy to extend the initial system prototype to a more detailed implementation.
- Finally, code that encapsulates the functionality of the system is generated automatically from the Hierarchical CP-net model. Such code can be used for many different purposes:

to study performance issues, to undertake statistical analysis of the results, and ultimately as a stand-alone application to be used in production environments.

A more general use of Hierarchical CP-nets in software development is stressed in figure 2, where different "front-ends" for system specification are hypothesized together with different "back-ends" for code generation:

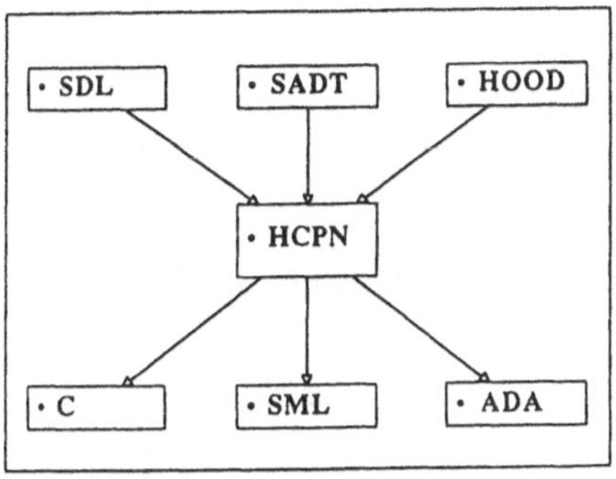

Fig. 2: The role of Hierarchical CP-nets in Software Development

2. Case Study: Electronic Fund Transfer & Debt Utilization

Electronic bank-to-bank funds transfer is the large electronic payment process that effects the movement of funds between banks, primarily through the use of payment networks [6]. The combined average daily value of electronic funds transfer (CHIPS[1] and FEDWIRE[2] networks) amounts to trillions of dollars and is constantly increasing.

The widespread use of bank-to-bank electronic funds transfer has created the potential for major financial disruption that can be triggered by the failure of a single banking house. The speed with which a transaction is processed allows banks to execute secondary and tertiary payments based upon expected incoming funds: a potential disaster if the payor fails to deliver the cash.

Still things may go wrong. For example, when a computer failure at Mellon Bank required an extension of the business day to correct their systems, the Federal Reserve granted the extension but failed to give notice of the situation. Many banks ended the day in debt since they expected payments which that day never arrived at their destinations. Consequently, many banks had to buy settlement funds at the end of the day to cover their debts, driving the cost of overnight borrowing far above the norm and producing losses for several institutions.

Risk management has been introduced to help control this situation. Payments that would cause debits to exceed preset limits are delayed. This, in turn, reduces the rate of flow of money. Currently, transaction processing is managed by FIFO rules (fig. 3). However, by analyzing a queue of pending transactions and reordering them, it is possible to reduce or eliminate the need for excessive debt utilization, thereby increasing the rate of flow of money without additional risk (fig.3).

The developed application, called the Debt Manager's Assistant, implements a number of different strategies to reorder the queue of pending transactions and quantify the dollar impact of transaction reordering on debt utilization.

---

1 Clearing House Interbank Payment System, built and supported by approximately 140 banks and used primarily for international dollar payments.

2 Built and supported by the US Federal Reserve Bank and used primarily for U.S domestic payments.

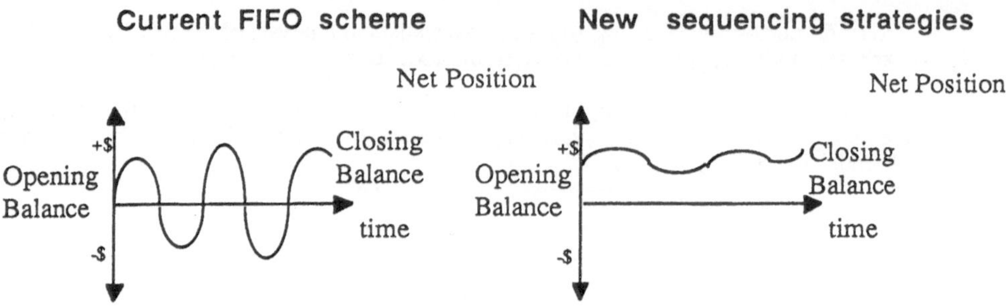

Fig. 3: Simplified Payment Flow

At the same time the application incorporates all the limits and self-regulations that the market and the federal institutions have imposed.

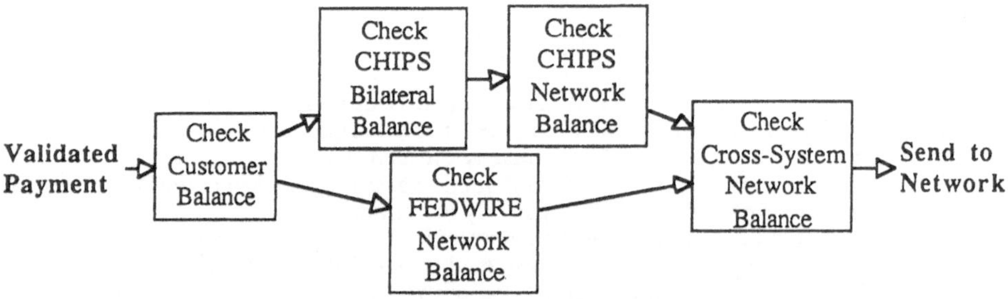

Fig. 4: Risk Management Hurdles

The sequence of limits that a payment must be checked against before it can be send to the network (fig. 4) are:

- the customer balance, which limits the amount of funds that may be withdrawn from the account of a customer of the bank
- the CHIPS bilateral balance, which limits the debt that each bank will accept from another with respect to bilateral agreements
- the network balances, which limits the overall debt any institution may have within the CHIPS and the FEDWIRE networks, alone or cross-checked [6].

Since the number of payments that can be executed every hour is limited by the network capacity, the rate of flow of money can be increased by grouping together transactions with similar characteristics, such as the debit and credit parties. Also the net result of incoming and outgoing payments with similar characteristics may be calculated, thus reducing the amount of money that travels through the networks.

The application is used in two ways to process the day's business in advance: first, to predict the debt positions as the day evolves and second to determine in what sequence the payments should be executed.

3. System Specification & Requirement Analysis

The objective of this phase is to create the initial specification of the software application.

The SADT methodology is chosen, at this stage, because of its very simple semantics and its "loose" syntax; both easy to learn by the bank managers responsible for the creation of the initial model.

SADT (Structured Analysis and Design Technique) is a graphic notation for system specification and requirements analysis[1]. In SADT, a system is represented by a collection of boxes and arrows. Boxes are used to represent the activities that compose the system and are accompanied by text in natural language describing the activity they represent. Arrows represent Input, Control, Output and Mechanism (ICOM) for an activity box. The underlying systems principle is that activities transform Inputs into Outputs, Controls are applied to determine when the transformation may or should occur, and Mechanisms are used to perform the transformation. Arrows have attached labels in natural language describing what they represent.

Arrows, then show how the activities inter-connect, exchange data, and exercise control on each other. Outputs of one activity can be Inputs or Controls or Mechanisms for another one.

Boxes can be decomposed. A decomposition represents a detailed description of the activity box which it represents. This detailed description is again made of inter-connected activity boxes, some of which may be further decomposed. The decomposition forms the boundaries between a box and its detailed representation. The arrows that touch a box define precisely the interface between the box and its decomposition.

SADT models are composed of set of diagrams, connected to form a hierarchy. Each diagram has, typically, three to six boxes and represents a detailed description of an activity box on the next higher level of the hierarchy.

Arrows can branch or join. A branch can represent fan-out (multiple identical copies) or splitting (different parts of the same thing). A join can represent fan-in or merging.

SADT is not considered adequate for direct representation of executable models[2]. The arrow structure is intentionally ambiguous in respect to the representation of such concepts as concurrency and choice, these being regarded as implementation details to be left unspecified until needed. The activation rules by which the detailed behavior of undecomposed activities is specified are too primitive and a formal theory of behavior based on these rules has not been elaborated. As a result SADT models cannot be directly validated.

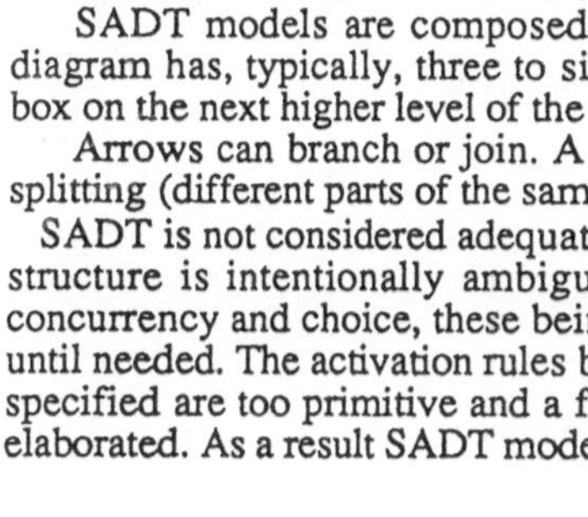

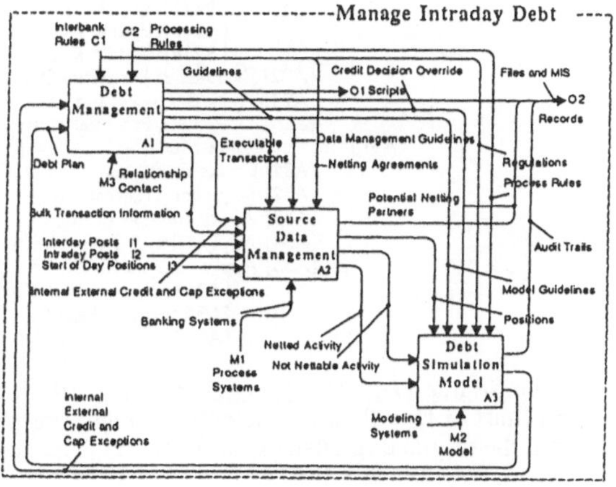

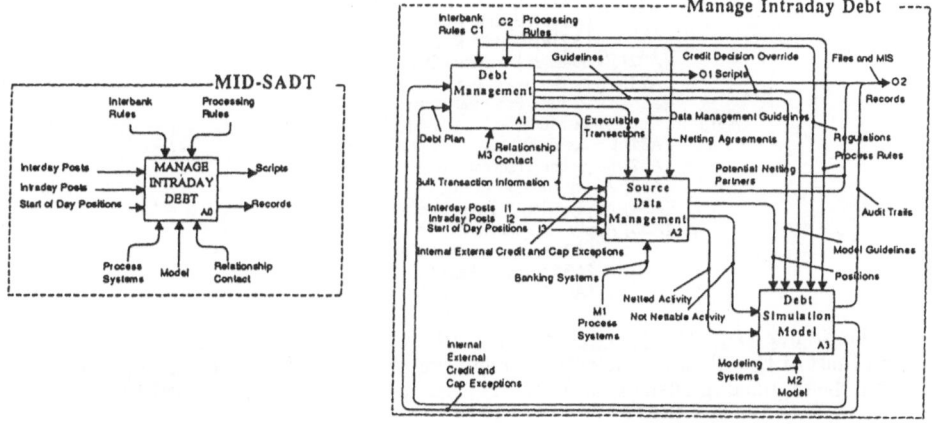

Fig. 5: MID-SADT top-level diagram Fig. 6: MID-SADT Manage Debt diagram

The software tool that was chosen to support the development of the SADT model is Design/IDEF [8].

[1] The Integrated Computer-Aided Manufacturing Program of the US Department of Defense recognized the usefulness of SADT by standardizing a subset of it called IDEF0 [1].

[2] An executable model is a generic term for a program. It may be textual, graphical, or a combination of both.

The SADT specification, called the MID-SADT model, portraying the new method of managing intraday debt, was built by a team of four people, representing different job functions, at Marine Midland Bank of New York in a total of five man weeks.

The top-level diagram of this model (fig. 5) shows the activity of managing intraday debt as a single box with labelled arrows representing the inputs, the outputs, the mechanisms and the controls of the model.

The inputs for the model are bank-to-bank transactions (i.e., Intraday_Posts and Interday_Posts) and the debt position at the beginning of the day. The outputs are Scripts containing the order in which the transactions should be executed and Records of the transactions.

The decomposition of this diagram (fig. 6) shows the main activities of the model: Debt Management (A1) , Source Data Management (A2) and Debt Simulation Model (A3) and their mutual dependencies.

The complete activity report[1] for the model is the following:

- [A1] Debt Management
  - [A11] Process Decision
    - [A111] Review
    - [A112] Determine Possible Adjustments
    - [A113] Select Adjustments
  - [A12] XCAP Decision
  - [A13] CHIPCAP FEDCAP Decision
  - [A14] Bilat Credit Decision
  - [A15] Internal Credit Decision
- [A2] Source Data Management
  - [A21] Data Collection
    - [A211] Load Inputs
      - [A2111] Load Positions
      - [A2112] Load Interday Posts
      - [A2113] Load Intraday Posts
      - [A2114] Load Netting Agreements
    - [A212] Split Transactions Expected and Orders
    - [A213] Update Actual Positions
    - [A214] Update Expected Positions
  - [A22] Evaluate & Distribute Data
  - [A23] Track & Process Futures
  - [A24] Determine Potential Nettables
- [A3] Debt Simulation Model
  - [A31] Create Debt Plan
  - [A32] Sequence Transactions
  - [A33] Apply Internal Limits
  - [A34] Apply CHIPS_Bilat Limits
  - [A35] Apply External CAP
    - [A351] Apply CHIP Cap
    - [A352] Apply FED Cap
    - [A353] Apply XsysCap

In the Source Data Management activity, all necessary data (Interday_Posts, Intraday_Posts and Start of Day Position) is collected and pre-processed (A21). The payments are collected from the originating business areas (A211) and split into two subsets (A212):

- the actuals set, composed of transactions that have already been executed through the networks and that update the current debt position (A213)
- the futures set, composed of transactions that still have to be processed.

The customer balance, the bilateral balance, and the network balance (see section 1) are also loaded in the system (A2111).

The futures set can be evaluated to determine the transactions for which the expiration date is close in which case they will be processed immediately and the set of transactions that have not reached their expiration dates (A22).

---

[1] This is a file created by the Design/IDEF tool that contains all activities of the model and their decomposition.

Transactions with similar characteristics are grouped together and the net result of incoming and outgoing payments is calculated (Netted_Activity) reducing the amount of money that travels through the networks (A24). The outputs of the A2 activity are therefore Netted or Non Nettable transactions.

In the Debt Simulation Model activity (fig. 7) the transactions are sequenced (A32) with the objective of reducing the average debt as shown in fig. 2. Then the transactions are checked against the three limits: customer (A33), bilateral (A34) and network (A35). Different payments need to be checked against different limits as shown in fig. 1. When all payments have been checked a debt plan is produced (A31).

In the Debt Management activity the payments that exceed the limits (Internal External Credit and Cap Exceptions) are processed. These payments generate exceptions that can be overridden on the basis, for example, of an expected receivable. If an exception is overridden, then it is re-inserted into the list of payments that are being processed (Credit Decision Override). If the exception is not overridden then the payment that generated it will not be executed at that time. The debt plan is reviewed and proposed as the script for executing transfers or, alternatively, some of the parameters that control the method may be changed (Data Management Guidelines), such as the sequencing scheme, so that a different debt plan will be produced.

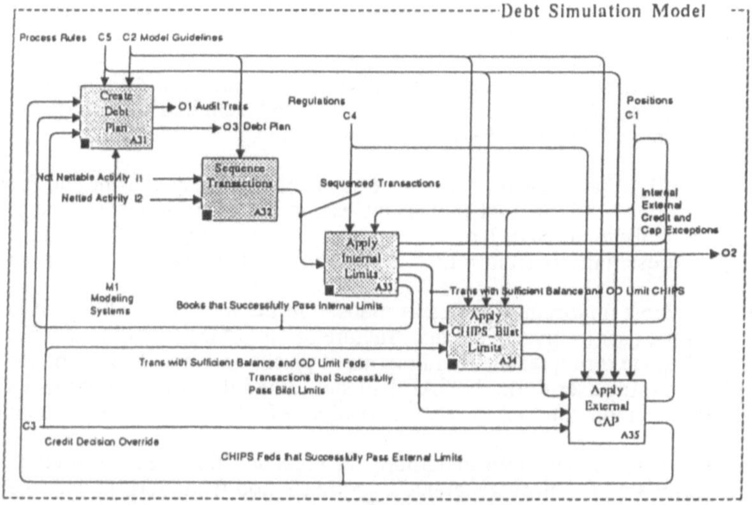

Fig. 7: MID-SADT Debt Simulation Model diagram

4. Automatic translation of SADT diagrams into Hierarchical CP-net models

The objective of this step is to be able to generate a representation of SADT models that is suitable for validation. This representation must convey all the information that is contained in the SADT models plus it must enable a designer to address those issue such as concurrency and choice that in the SADT models are left unspecified.

To achieve this purpose it was decided to translate SADT models into Hierarchical CP-net models where detailed behavioral analysis is possible.

SADT to HCP-net translation: Hierarchical CP-nets are used to give formal semantics to the informal specification notation of SADT [7]. While it is beyond the scope of this paper to present all the details of such a translation process, we will present the general paradigm and some related issues.

SADT activity boxes are interpreted as CP-net transitions and SADT arrows as a combination of CP-net places and arcs.

If an SADT activity box has a decomposition then it is interpreted as a substitution transition (see section 5 for a description of substitution transitions).

In order to represent SADT diagrams as Hierarchical CP-net models, a minor syntactic restriction is imposed on the SADT arrow structure [7]. SADT models that obey this restriction can be translated into equivalent HCP-net models with similar graphical layout, where detailed behavioral modelling is possible. Problems related to concurrency and choice, unspecified in the SADT model, can now be addressed properly with inscriptions by a system designer.

Some observations on the translation process are necessary. Information in a SADT specification is captured in two ways: through graphics and text. As we have seen so far, the translation from SADT to HCP-net is such that the graphic structure of SADT models is mapped into a correspondent HCP-net graphic description.

Furthermore, the textual information in SADT is expressed mainly through labels attached to arrows. This information is translated into HCP-net in the form of a color set name. The reason for doing this is that a label in SADT is not a sufficient description of data that is "transmitted" between activities. A more appropriate specification of that data can surely be done via a color set declaration (i.e. a data type description). By mapping SADT arrow labels into color set names, the process of moving from an informal specification to a formal one is easier to understand.

At the same time it should be clear that in the translation process nothing is artificially added to the SADT specification. From an ambiguous SADT model a ambiguous HCP-net model can be generated. Nevertheless this skeleton HCP-net model can now be used to address the behavioral aspects of a system that go beyond the possibilities of SADT.

In addition it should be noticed that conceptually it makes no difference to add the HCP-net inscriptions to the SADT model or to its HCP-net representation. As a result, SADT models with HCP-net inscriptions are executable and may also be studied with formal methods, since the behavior is defined by the underlying HCP-net model.

The Design/IDEF tool has been extended to enable its use as a front-end for Hierarchical CP-nets. On the one hand, the tool enables the user to add inscriptions, such as guards and arc expressions, directly in the SADT model. On the other hand, the tool translates SADT models into textual representations from which correspondent Hierarchical CP-net models can be generated.

One of the most important features introduced in the tool is the CPN syntax check. This feature allows the user to investigate the structure of the SADT model to discover if its arrow structure is adequate for representation in terms of places and arcs. If it is not, the CPN syntax check will prompt the user to fix the SADT model. When the structure of the SADT model is adequate for translation, the tool shows the positions of all places before the actual translation occurs.

The tool also supports the user in defining the color sets that are associated with places. By default, the color set of a place generated by a translation from SADT, uses the SADT label describing the flow along the arrows from which the existence of the place was inferred. The tool individuates all the SADT labels (i.e., color set names) that must be associated with data types definitions (i.e., color set declarations).

In our case study, some interaction was necessary between SADT-trained users at META and experts of the banking community before the model could be successfully translated into a Hierarchical CP-net model. During this interaction some changes were made to the arrow structure of the MID-SADT model by experts at META.

The MID-CPN model, generated by means of an automatic translation from the MID-SADT model, is hierarchical with 4 levels of decomposition. There is a total of 9 pages. There are 24 non-decomposed transitions (i.e., transitions that do not have a subpage that replaces them). A diagram (called the hierarchy page) shows the hierarchical relationships between the pages of this model (fig. 10). This diagram is related to the hierarchical structure of the SADT activity report (see section 3) in such a way that all decomposed activities in the SADT model represent a node on the hierarchy page in the HCP-net.

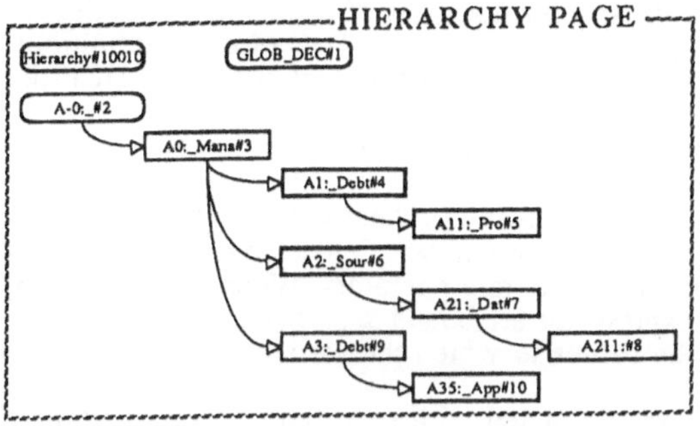

Fig. 8: MID-CPN Hierarchy Page

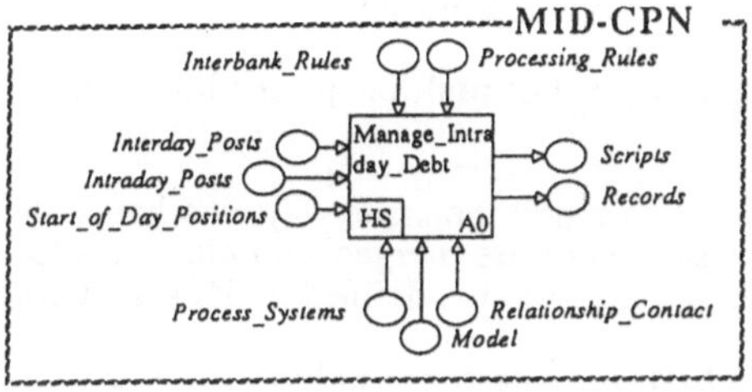

Fig. 9: MID-CPN top-level page

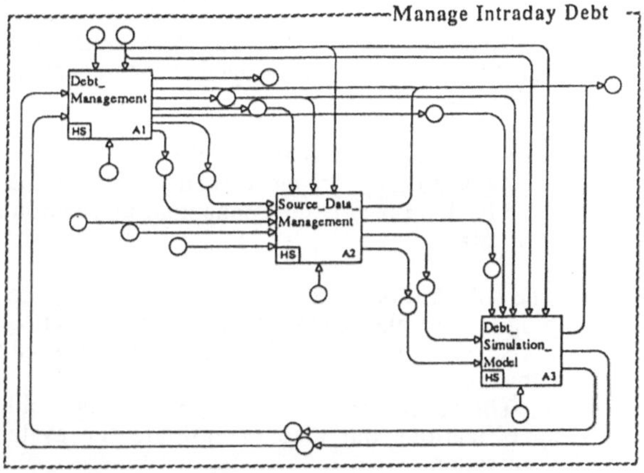

Fig. 10: MID-CPN Manage Intraday Debt sub-page

The top page of this model (fig. 9) shows the activity of managing intraday debt as a substitution transition. The labelled arrows have become places inscribed with color set names. The subpage that replaces it (fig. 10)[1] shows the activities A1, A2 and A3 of the corresponding SADT diagram also interpreted as substitution transitions. Places have been automatically generated during the translation process according to the rules specified in [7]. The color set names have been hidden on this last page to emphasize the net structure resulting from the translation.

5. Prototype & Simulation

The objective of this phase is to rapidly generate a prototype of the application to validate the correctness of the initial SADT specification. It is important to be able to initiate such validation process as early as possible in order to reduce the impact of serious design errors detected after final implementation. Hierarchical CP-nets are used for this purpose.

Hierarchical CP-nets are a high-level graphical formalism for the design and analysis of complex concurrent systems [2]. The behavior of a system, using Hierarchical CP-nets, is specified in terms of a graphical net structure and of formal net inscriptions such as arc expressions and guards. Therefore HCP-nets combine the power of formal mathematical expressions with the clarity of graphical models to provide a concise representation of complex systems.

In Design/CPN [9], HCP-net formal inscriptions include the following:

---

[1] Color set names for this and the following CPN sub-pages have been hidden since they can be read from the MID-SADT diagrams.

- Color set definitions specify the types of tokens in the model. Tokens can be complex data objects (e.g., tuples, records, lists). The color set definitions are analogous to type declarations in programming languages. Each place in a HCP-net has a color set which constrains the tokens allowed to occupy it.
- Markings specify the set of tokens occupying each place.
- Arc Expressions specify the multi-set[1] of tokens that are removed from input places and added to output places as a result of the occurrence of a transition. Arc expressions may contain variables of a given color set, pre-declared or user-declared functions[2], boolean selectors and ordinary algebraic operators. When a transition occurs, variables in the surrounding arc expressions are bound to particular values.
- Guards restrict the values to which the variables occurring in arc expressions can be bound.
- Code Segments are associated with transitions and may be used to perform complex calculations of token values, as well as to create well-defined side-effects (e.g., read input files, generate reports).

HCP-nets have hierarchies [3] to structure large HCP-nets as a set of subnets (called pages) related in such a way that together they constitute a single model. The relationships include the following:

- Substitution transition is the replacement of a transition (called supernode) with a page (called subpage) that contains its detailed description.
- Fusion place is the conceptual unification (i.e., folding) of multiple places into a single place.

The substitution of a transition with a subpage works in much the same way as a hardware plug-in. The places (called sockets) surrounding the supernode are merged with places (called ports) on the subpage. The semantics of the substituted transition is defined by its associated subpage.

The Hierarchical CP-net methodology, unlike SADT, is used not only to build system specifications but also to validate the systems design through simulation and formal verification.

However, HCP-nets are not considered adequate for directly executing the final version of the system under development, because their high-level interactive graphical environment limits the efficiency of their execution.

The software tool used to support this phase is Design/CPN [9].

This tool is composed of:

- a graphical Editor with which it is possible to draw HCP-net models and add all the inscriptions that regulate the model's behavior
- a graphical interactive Simulator with which all the testing of these inscriptions can be done.

A special feature of this tool translates the text representation of an SADT model automatically generated by the Design/IDEF tool into a HCP-net model. Another enhancement of this tool enables the automatic generation of stand-alone SML applications that exhibit the same behavior as the HCP-net model from which they originate.

During the development of the first prototype of our case study, the basic data structures of the model were defined in terms of color sets and the basic operations to be performed on this data were specified in terms of arc inscriptions, guards and code segments. Since the purpose of this initial phase was to validate the logic of the net structure of the model, not too much effort was put into the process of specifying all the details of the color sets and of the net

[1] A multi-set is a set that may have multiple occurrences of the same element.

[2] The full generality of SML can be used in writing any net inscription.

inscriptions. Nevertheless all of the most important functions of the model were written at the level of detail considered to be relevant. This model was then validated in a single meeting with the bank managers, in which the graphical properties of the simulation tool were used to display the model's behavior in terms of the flow and content of the markings. The potential of this new approach for solving financial problems was recognized by the higher management in the bank and it was decided to continue with the development process.

In the sequel we discuss in detail how HCP-net inscriptions were used to model the behavior of the system.

Color Sets

Color sets play a central role in the construction of a HCP-net model and can be interpreted as being equivalent to data types in ordinary programming languages.

A Color set can be a primitive SML type (e.g., a unit, a boolean, an integer, a real, a string of characters) or it can be an enumerative set, an indexed set, a product set, a record set (which is similar to a product color set where the elements of a tuple are accessed by a labelled field instead of the position in the tuple), a list set, a subset set, a duplicate set (equal to a previously defined color set) as well as any combination of them.

It must be remembered that in this methodology, all the required color sets to be defined are specified in the SADT model as labels on arrow structures and are transported from the SADT specification to the Hierarchical CP-net model via the automatic translation.

In our case study the input data for the model (e.g. Intraday_Posts as specified in the initial SADT model) can be described as a list of bank-to-bank transactions. Transactions are characterized by the identification number, the method of payment, the amount, the debit and credit parties.

The set of bank-to-bank transactions is defined as a list of individual payments:

color Intraday_Posts = list Trans ;

The payments are defined as records with labelled fields:

*color Trans = record iD:String * PaymeT:Paymet * AmounT:Real *
DebtiD:String * CrediD:String ;*

The transactions method (i.e., which network the payment is going to be sent through) and the type (credit or debit) are defined as elements of an enumerative set:

color Paymet = with BOOK | CHIPOUT | FEDOUT | CHIPIN | FEDIN ;

In the HCP-net model many color sets are conceptually the same data type but have different names. This happens because the color set names are inherited from the labels of the SADT model, where, for example, in different parts of the model lists of transactions are referred to as Transactions Within Measurement Period or Transactions Outside Measurement Period or Executed Transactions. Many of the color sets are therefore duplicate color sets, i.e., defined in terms of a previously defined color set, as in the following case:

color Executed_Transactions = Intraday_Posts ;

Variables of a given color set, to be used in arc expressions, guards and code segments, are declared in a programming language fashion, as in the following example:

var interlist, intralist : Intraday_Posts ;

Using Design/CPN the color set declarations, the variable declaration and the user-defined SML functions that are used in the inscriptions of the HCP-net, are placed in special objects, called declaration nodes.

Arc Expressions

Arc expressions are of two types:

- input arc expressions determine which and how many tokens are removed from the input places
- output arc expressions determine which and how many tokens are created and put in the output places.

Input arc expressions may contain variables of a given color set and output arc expressions may contain SML operators and functions applied to the input variables or variables whose value is calculated within a guard or a code segment (warning: some familiarity with the notation and syntax of Standard ML is required to understand the details of the function definitions). Let us examine the following transition[1], as it was modelled during the prototype phase:

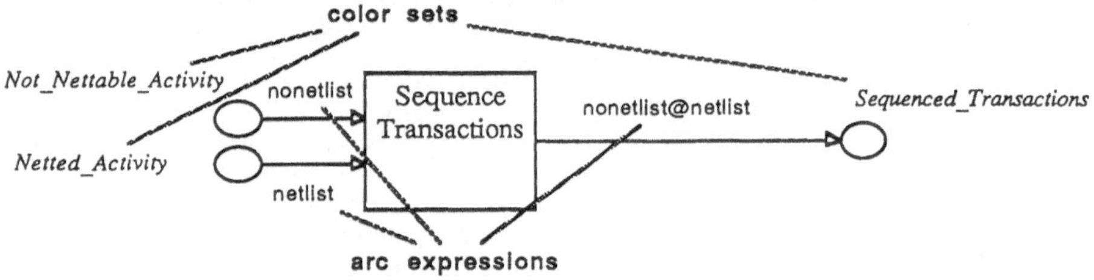

Fig. 11: MID-CPN Debt Simulation Model page

In this case we are modelling the sequencing of the transactions. At one point in the model's execution, two tokens (holding different lists of transactions) will be present on each one of the input places, thus enabling the transition to occur. The value for the token that is produced from the occurrence of the transition is calculated by the output arc expression. This consists of the concatenation of the input list values.

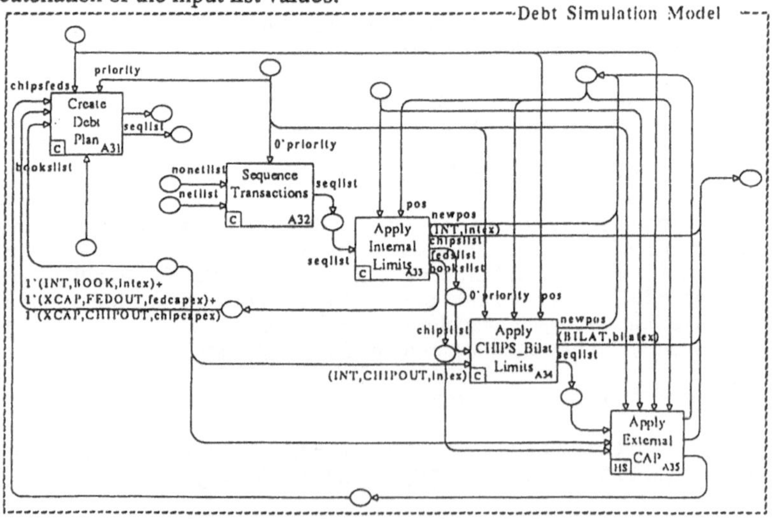

Fig. 12: MID-CPN Debt Simulation Model page

[1] In all figures, color sets are displayed in italic, arc expressions in plain text, initial markings in underlined bold and finally code segments are in dotted line boxes.

In the early prototype of the application it is not necessary to define the algorithms that reorder the transactions. Our focus is on the architectural description. We are interested in validating the interaction between the different components of the system, making sure that all data flows have been specified accurately, and that all the information that is necessary to perform the activities is been modelled explicitly in the SADT specification.

The diagram in fig. 12 example shows the Debt Simulation Model with all the arc expression that define its behavior.

Interactive graphical simulation: manual and automatic modes:

Interactive graphical simulation of the MID-CPN model started with the testing of the individual transitions in a manual fashion. Several bindings for the transitions were inspected at this stage and the occurrence sets (the sets of transitions to be fired) were selected manually. The feature of Design/CPN that permits one to change markings during a simulation session was used extensively to allow transitions to be tested.

After the correctness of the individual transitions was checked, the model was tested in automatic mode. In this simulation mode, random choices are made by the Design/CPN Simulator of which enabled transitions should occur.

Using Design/CPN, it is possible for the user to decide what subset of the model should be included in the simulation. If the simulator is asked to run in automatic mode, it may be instructed to display the occurring transitions by opening the appropriate pages, or it may be asked to run in non observable mode. Also the simulator may be asked to show the occurrence of concurrent transitions in a sequential mode.

The focus of the simulation was then to find out whether the logic of the model satisfied the modeler's expectations, in terms of resource sharing, synchronization, and so on. The main interest is therefore to observe the token flow within the model. As input data for the simulation we used few transactions (maximum 50-100) and in some cases none (i.e., empty lists), since the occurrence sequence of this model does not change by using a large or small number of transactions as input data.

6. Implementation & Testing

The objective of this phase is to implement the application. This is done by refining the initial data descriptions and the algorithms that perform transformations on data values. The programming language Standard ML is chosen since it is the language that was used for the implementation of the Design/CPN simulation capabilities and therefore is available for the implementation of algorithms of any desired functionality. When the code is written it can be plugged in the model in the form of arc expressions, guards and code segments and tested in the framework of an existing HCP-net model.

Standard ML (SML) is a textual programming language [4,5] which does not use graphics. SML provides the syntax of HCP-net inscriptions in Design/CPN and is used both as the execution engine for the Design/CPN simulator and to build production versions of HCP-net models. Here are some of the key aspects of SML:

- SML is a functional programming language, where functions can be passed as arguments, returned as results, and stored in variables. The principal control mechanism in SML is recursive function application.
- SML is strongly typed. Every legal expression has a type which is determined automatically by the compiler so that no type errors may be generated during run-time.
- SML has a polymorphic type system. Each phrase has a uniquely-determined general typing that specifies the context in which that phrase may be legally used.
- SML supports abstract data types.
- SML is statically scoped. Identifier references are resolved at compile time.
- SML supports exceptions to handle unusual conditions.

The software tool used to execute the production version of HCP-net models is Design/ML [10], a development environment that integrates the SML compiler with the ability to manipulate graphical objects. Using this tool, it is possible to build customized interfaces that rely on the logic of the HCP-net model.

At this stage of our case study, more detailed analysis of the data that flows within the model is required, together with a more appropriate description of the operations to be performed on this data. The color sets then become a real description of all the data that is needed in order for the model to execute properly. Several databases are modelled with color sets to represent customers balances, network balances and agreements on netting opportunities. Since the purpose of this phase is no longer the validation of the logic of the model but the validation of the computational results, most of the effort is put into the writing and testing of the SML code (i.e., of the inscriptions of the HCP-net model). The validation of this model was done by the bank's technical staff in a week by means of extensive simulation. During this phase a serious design error in the SADT model was discovered. The discovery of this error required changes in both the MID-SADT and MID-CPN models. However, this took only three days, thanks to the flexibility and locality properties of the models used.

Standard ML plays a major role in the development of the MID-CPN model since most of the detailed behavior associated with the occurrence of transitions is implemented as SML code.

Let us consider the case in which an algorithm that performs the reordering of the transactions must be implemented.

In SML this can be done be defining a function, such as the following:

fun Sequence (x : Netted_Activity) = x ;

where we have specified that the function Sequence has a single argument of type Netted_Activity and returns its argument unchanged as a result of its application, which corresponds to a FIFO ordering scheme.

When this is done it is possible to use such a function in arc expressions, as in the following case:

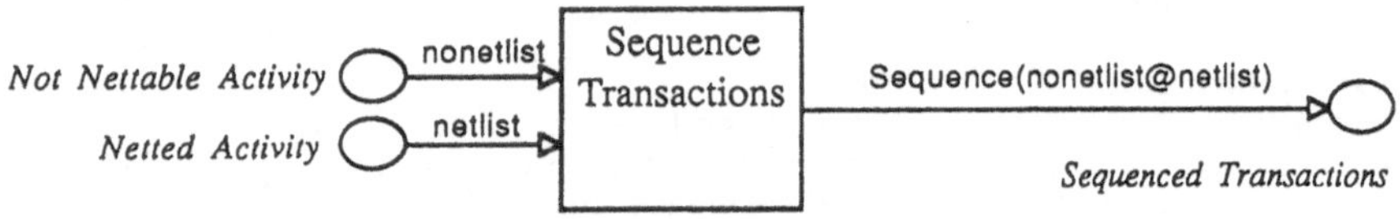

Fig. 13: Sequence Transactions transition

The Sequence function can be redefined so that it performs a complex ordering scheme based on type of transaction and amount without making any changes to the HCP-net model. Furthermore, many different sequencing schemes can be implemented. As a result the functionality of the Sequence function may change by adding a parameter to determine which sequencing strategy will be performed.

This change in functionality can be reflected in the HCP-net model by adding one arc and one place with proper inscriptions, as in the following way:

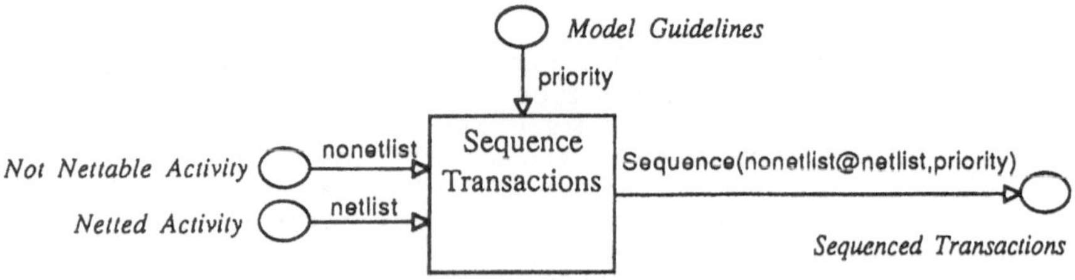

Fig. 14

During the development of the MID-CPN model, progressive extensions made to the SML code have produced nearly 1300 lines of SML function declarations, while the size of the model in terms of number of pages (9-10) and number of transitions (20-24) has been relatively stable during the whole modelling process.

It should be clear that the major difference between the prototype and implementation phases of the methodology is mainly in the volume and accuracy of the SML code written. This introduces the idea of progressively shifting interest, during the evolution of a software development effort, from a more graphical interaction with the model, in which the behavior is displayed in terms of the occurrence sequences, to a less graphical and more batch-oriented interaction, in which the interest is in the results of the occurrence of the transitions. These results become far too complex to be conveniently inspected in the form of token contents so that it is necessary to display them in a more adequate fashion. It is then easy to print them on output files that can subsequently be analyzed and/or displayed in dialog boxes with the desired format.

7. Code Generation

The objective of this last phase is to test the application on large sets of data and to generate SML code that runs very efficiently to be eventually used in the production environment.

Two distinct steps have been taken in our case study to demonstrate the value of this new approach to software development. Initially the objective was to demonstrate that the SML code written to implement the detailed algorithms, could process large sets of data in a useful time frame. To achieve this purpose we decided to manually embed the functionality of the HCP-net model into a SML program. When it appeared clear that the performance of the code that was written was sufficiently fast to calculate the desired results within the required time-frame, a technique that automatically generates the SML code from the existing HCP-net model was implemented.

Manual Code Generation

All the SML code written and tested in the final MID-CPN model is structured in such a way that its execution reproduces the same behavior as the MID-CPN model. This process is very straightforward since, for the MID-CPN model, a fixed sequence of transitions that correspond to a desired behavior can be found. Therefore the SML code that is executed within this sequence can be easily stripped out from its HCP-net environment. The applicability of this method is limited, since it depends on the complexity of the model to be manually converted.

The tool used to support the manual production of this SML application is Design/ML.

In our case study, the speed of the transaction processing in the simulator was too slow to make it useful in real time. The number of transactions that are processed every day in the real world is between 10,000 and 30,000 depending on the particular institution, and for an application to be of practical use it must process them at least once in less then 15 minutes.

It was time to build what we call a production version of the HCP-net model. This application, a Standard ML program, was created in less than a week by a single person. Most of the time was spent doing tests and the application itself was built in less than a day.

This SML program, the MID-ML, was based on a fixed transition occurrence sequence recorded during simulation. Since all the events in the MID-CPN model occurred in this sequence, it was possible to strip out all the enabling calculation and occurrence set calculation by writing appropriate SML code.

Every transition of the MID-CPN model was then manually converted into a sequence of SML instructions.

For example, the transition that sequences the transactions (fig. 14) can be written using SML syntax in the following way:

val seq_trans = Sequence (nonetlist@netlist,priority);

This process of associating every transition with SML code leads to the creation of an equivalent SML application of approximately an extra 100 lines, many of which deal with input and output of data while the core of the application consists of 30 lines.

With this compact code we were able to execute a very large number of transactions. The results shown in the fig. 15 diagram were obtained running 11,698 transactions with the New Jersey SML compiler on a dedicated SUN SPARC server with a 32MB RAM memory.

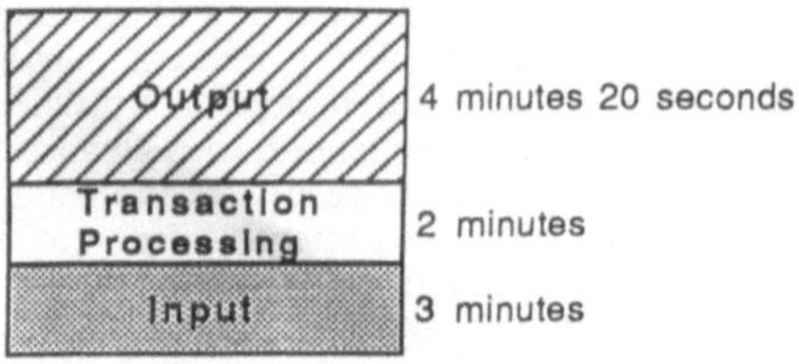

Fig. 15: Performance Components of compiled SML application

It is interesting to note that of the 9 minutes and 20 seconds that are spent in executing these transactions, 3 minutes are consumed by the input operation of loading them, 2 minutes are spent to process them and 4 minutes and 20 seconds are spent writing output files. These contain all the information concerning which transactions were netted together, how the payments were sequenced, which exceptions were generated and much more. Of this total output time, 50 seconds were spent writing the proposed script for the next transfer and writing the updated position database.

Automatic generation of SML code from HCP-net models

The model is separated from its concrete graphical representation and encapsulated in a single SML function. This function is built automatically by the system, re-using simulation functions previously defined for each transition (e.g., functions that determine if a transition is enabled, and that specify the tokens that should be removed from the input places and added on the output places when the transition occurs). This function attempts to execute every transition in the model in a random or a fixed occurrence sequence, depending on the user's instructions. Then, this function is exported as executable object code using an SML feature.

The tool used to support the automatic production of this SML application is Design/CPN, while Design/ML is used to execute it.

Using the automatic code generator, the production version of a HCP-net model can be built from the Design/CPN tool in a few minutes. This is particularly useful in those cases in which it would be very complicated to manually produce the code, as when the HCP-net model has several hundred transitions.

This is made possible by the existence of a number of SML functions that are built by Design/CPN for every transition in a given model. These functions remove tokens, execute code actions, generate new tokens and update the current markings.

In the Design/CPN simulator the execution of these functions is driven by the graphical interface which, in order to provide user feed-back on the enabled transitions, the markings etc, executes them in a very fragmented way. This means that their execution is interrupted continuously to send results to the graphical interface that displays them in the form of tokens being removed from the input places and committed to a binding, and tokens added to the output places, and so on. On the other hand, the execution of these functions isolated from this interface produces performance results that are identical with the results from the manually produced code.

Therefore, it is possible to generate a SML application which executes all these functions without being interrupted by the requests of the graphical interface. This SML application can be executed both as an independent application or still within the Design/CPN tool. The process generating such an application can be viewed as a Design/CPN compilation facility in which models and sub-models can be compiled into executable code.

When this code is executed, no tokens are seen being removed or added to places. Not even places or transitions can be seen. Yet all the components of the HCP-net model (e.g., places, transitions, and initial markings) are encapsulated in the SML function that executes the model. Since no tokens are shown during the execution of this SML function, it is impossible to know the state of the model. For this reason a user interface should be built on top of the model. This user interface can be in the form of report generation (e.g., statistics of the execution of the model) or be interactive (e.g., animation, dialogues) or both.

When compiling a model, the user may choose an occurrence sequence. The occurrence sequence determines the order in which the compiled model will attempt to execute its transitions. If the user does not choose the order of execution, then conflicts are resolved randomly at run-time.

If the user specifies the occurrence sequence, the generated SML code will try to execute every transition in the compiled net once in every step according to the order specified by the user. Otherwise, the generated SML code will try to execute every transition in the compiled net according to a probability defined by the user.

If a binding cannot be found for any of the transitions the execution quits, since there are no more enabled transitions. In any case, the user can specify the number of compiled steps he wants to execute.

8. Conclusions

The most significant objectives of this project are:

- the demonstration that the complete software life cycle could be supported using an integrated modeling approach, based on SADT, HCP-nets and Standard ML.
- the creation of a HCP-net model which implements a new strategy for electronic funds transfer

For the accomplishment of these objectives, automatic mechanisms have been implemented for translating SADT models into HCP-net models and for converting HCP-net models into SML applications.

While the benefits of formal validation techniques have yet not been utilized, due to the current lack of tool support, the advantages of this new approach to software development are evident:

- costs of development and maintenance of software applications are greatly reduced due to the creation of compact and executable documentation. Changes can be tested quickly and at the desired level of detail. Enhanced applications can be generated within days. The documentation itself is easily transferred among many developers.
- gaps in different software phases are filled with automatic translation mechanisms that enforce reliability between final software and initial design. Close interaction between high-level software designers and developers becomes natural.

The system under development has now been specified, prototyped, simulated and implemented within the framework of integrated tools with the possibility of making changes at any stage of the development process without traumatic consequences.

The future plan for this project involves extending the current model to incorporate the credit side of the transactions and a 15-minute interval for cross-system checking. The model could then be approved as a standard for federal regulations and adopted by more banks that heavily participate in electronic funds transfer.

A model to evaluate the benefits of this new method for debt management has also been scheduled, so that it will be possible to quantify the results of the new strategy from a financial point of view.

The application of this integrated software development methodology in different areas of the bank operations is under evaluation.

The current tools will be extended by the implementation of a more sophisticated conversion process from HCP-net models into SML applications to support random occurrence sequences as well as pre-recorded ones. The possibility of building graphical user interfaces for SML applications within Design/CPN and Design/ML will also be supported. In this way the graphical interface may be easily prototyped using Design/ML; it can then be inserted in the final version of the HCP-net model.

Acknowledgments
We would like to thank Kurt Jensen and Peter Huber for reviewing the first version of this paper and for providing us useful comments.
A special thanks to Fiorella De Cindio for the contribution given to improve the quality and the significance of this work.
We also would like to thank the anonymous referees for their valuable comments and suggestions.

References

[1] D. A. Marca, C. L. McGowan: "SADT". McGraw-Hill, New York, 1988.
[2] K. Jensen: "Coloured Petri Nets: A High Level Language for System Design and Analysis". To appear in: G. Rozenberg (eds.): Advances in Petri Nets 1990, Lecture Notes in Computer Science, Springer-Verlag.
[3] P. Huber, K. Jensen, R. M. Shapiro: "Hierarchies in Colored Petri Nets". K. Jensen: "Coloured Petri Nets: A High Level Language for System Design and Analysis". To appear in: G. Rozenberg (eds.): Advances in Petri Nets 1990, Lecture Notes in Computer Science, Springer-Verlag..
[4] R. Harper: "Introduction to Standard ML". LFCS, Dep. Comp. Science, University of Edinburgh, 1986.
[5] Åke Wikström: "Functional Programming Using Standard ML". Prentice Hall, New York, 1987
[6] J. F. McDermott, R. S. Cottrell: "Application of IDEF to Banking: Building an IDEF Model to Manage Bank-to-Bank Intraday Debt". Presented at IDEF User Group Conference, Los Angeles, U.S.A., October, 1989.
[7] R. M. Shapiro, V. O. Pinci, R. Mameli: "Modeling a NORAD Command Post using SADT and Colored Petri Nets". Proceedings of the IDEF User Group, Washington D.C., U.S.A, May, 1990 .
[8] Design/IDEF User's Manual. Meta Software Corporation. Cambridge, 1989.
[9] Design/CPN User's Manual. Meta Software Corporation. Cambridge, 1989.
[10] Design/ML User's Manual. Meta Software Corporation. Cambridge, 1989.

27.
Validation of a VLSI Chip Using Hierarchical Colored Petri Nets

R.M. Shapiro

Microelectronics and Reliability, Special Issue on Petri Nets, Pergamon Press 1991

Abstract

In this paper we focus on the task of modelling and validating the behavior of a VLSI chip using hierarchical Colored Petri Nets (CP-Nets or CPN's). We discuss current practice in hardware design at the register transfer level. We describe an approach whereby engineering block diagrams, supplemented with suitable formal inscriptions, can be mapped directly to a CPN model. We show in detail a CPN model of an actual digital filter chip from a super-computer. We discuss the possibility of using this model to validate the logic of the design. We describe the potential of using formal analysis methods and propose a simplification technique for reducing the combinatorics involved in Occurrence Graph Analysis. We discuss performance issues and propose an extension to Colored Petri Nets that incorporates the concept of time.

1. Introduction & Overview

Hardware designers often take a graphical approach toward specifying a new device. Each level of the specification consists of an interconnected set of blocks. Each block is characterized by data flow in and out and internal transformation of the data. The transformation is the behavioral description of the block. Typically, in a top down system design the behavior of the initial high level blocks is crudely specified. As the design is refined each such block is thought of as a new block diagram in which the detailed behavior can be elaborated. Eventually, at the lowest (most detailed) level, all the blocks have a precise behavioral description.

Having created such a specification, the designer wants to validate it. Validation requires two different kinds of analysis. Firstly, the logic of the design must be correct. Secondly, performance and timing issues must be examined. In this paper our main focus will be in the area of logical validation. In current practice this usually involves manual translation of the design into an executable model. The executable model is typically written in one of the many simulation languages available for such purposes. Once translated, the properties of the design may be tested by extensive simulation.

The design and validation process would be more efficient if the graphical design could serve directly as an executable model. In this paper we describe the use of hierarchical Colored Petri Nets to accomplish this. We present in detail the design of a filter chip in a signal processing computer. We discuss the potential value of formal analysis as a supplement to simulation in validating the correctness of the design.

Overview of the Project

The project originated in response to a request from a manufacturer of supercomputers. The manufacturer was interested in speeding up the design phase of the production cycle for new hardware. As part of the design phase, all register transfer level designs (RTL) were standardly validated by writing a behavioral description of the design in a special dialect of the 'C' programming language. A custom-built pre-compiler and operating system then allowed translation and execution of the behavioral description under user control. The debugged 'C' program would be applied to a large number of test vectors (input values) and the outputs analyzed to detect malfunctions in the design logic.

The manufacturer was interested in eliminating the programming step which was error-prone and time-consuming. It was hoped that the block diagrams drawn by the design engineer could serve as the basis for automated generation of an executable model. It was understood that the graphics of the block diagramming technique would have to be formalized and some inscriptions added to the graphics to provide a complete behavioral description. We proposed the use of hierarchical high level Petri nets as a starting graphical representation. Later on a graphical front end tailored to the needs and conventions of the hardware designer would be provided to facilitate creating the design.

To test this approach, the manufacturer provided us with a register transfer level design of a VLSI chip used in their most recent super computer. The design included a block diagram of the chip, supplemented by some number of pages of textual description. Additionally, we were provided with a computer program, written in a dialect of 'C', that simulated the behavior of the design. The ultimate objective of the project was to develop a design approach whereby the graphical description, with suitable inscriptions, would suffice both as a design medium and as a complete directly executable specification of the behavioral properties.

Our objective from the outset was to establish whether or not hierarchical high level Petri nets were useful in specifying and studying the behavioral aspects of this kind of hardware design. We were not concerned with the details of a graphical front end tuned to the needs of a hardware designer. A suitable front end would be specified only after establishing the adequacy of the net representation. This is discussed more in the sequel.

Restated, the main objective was to construct a hierarchical Colored Petri Net that satisfactorily modeled the behavior of the register transfer level design. To accomplish this we used the software tool Design/CPN [1].

The original design made use of two hierarchical levels of graphic description. The simulation program followed this architecture. We found that a computer tool that facilitated hierarchical modeling led to significantly more use of hierarchies. This had the effect of showing graphically major architectural features of the design that had been previously very difficult to see. It also reduced the complexity of the most complicated submodels. Note that further use of hierarchies could have been made. We believe that this would happen naturally in a new design as a consequence of the availability of computer support for working this way.

The hierarchical net model is presented in detail in what follows.

Overview of the Filter Chip

An array of filter chips perform the role of a digital filter in generating an image on a graphical display. If you are familiar with digital filters read the following two paragraphs which provide a brief description of the filter function. Otherwise skip them since what follows can be understood independently.

The filter chip performs the final function necessary to calculate half of the filtered pixel intensity for one pixel of a video display and to provide data for an adjacent pixel's value. Each filter chip receives eight sample point intensity values for a single pixel from a video buffer and applies the filter function to the eight samples. The chip adds the eight filtered sample values together and sums the result with an adjacent pixel to calculate a left or right half pixel output value.

Each sample intensity is multiplied by left and right weighting factors which are stored in an on chip register file. The sum of the eight right-weighted samples is output to an adjacent chip. The sum of the eight left-weighted samples is added to the input value from an adjacent chip and this result is output to the video card.

The filter chip design permits a great deal of concurrency. A set of eight samples is treated as a single 'wave' of inputs. The chip's pipelined architecture processes six waves of inputs concurrently, each wave at a different stage in the pipeline. Before proceeding we will describe briefly what a pipelined architecture is.

In the diagram on the right, with a suitable initial marking the transformations in stage one and stage two can execute concurrently without violating safety criteria. Thus if the transformations have long duration relative to the other transitions, the throughput of the system approaches the time of the worst-case transformation.

Note the dependency of stage two on the results of stage one. The concurrency is achieved by allowing stage two to work on data supplied by the input generator at time i-1, while stage one works on data supplied at time i.

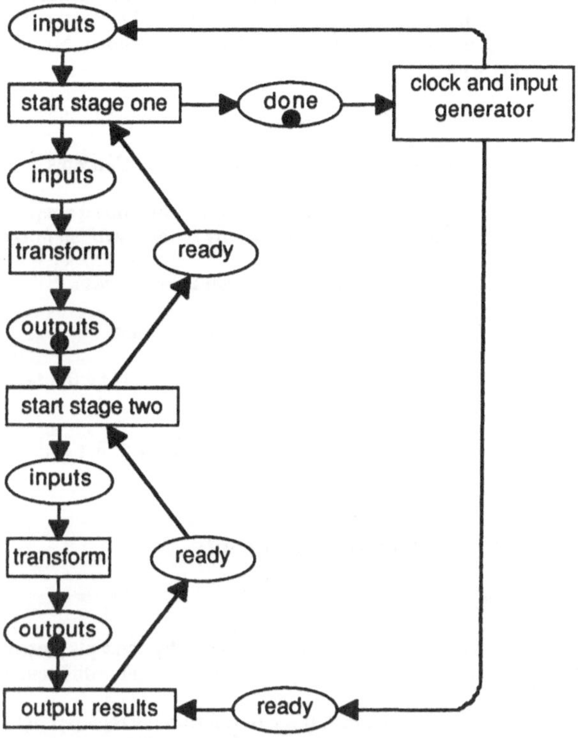

Refer to FIGURE THREE for a representation of the filter chip in six stages. Here we summarize the functionality of each stage, including the behavioral detail from several levels of description.

Stage 1: fetch the eight data samples and the weighting factor address. Read the appropriate left and right weighting factors for each sample from their individual weight factor files and perform 16 concurrent multiplications.

Stage 2: sum the left and right products for two groups of four samples.

Stage 3: sum the final left and right results; use any intermediate overflows to force the maximum result and limit both the left and right results to the high order 12 bits. When the next clock signal arrives, output the right result to the adjacent chip on the right.

Stage 4: fetch the result from the adjacent chip on the left.

Stage 5: sum the left result with the result from the adjacent chip on the left and limit this result to 12 bits. When the next clock signal arrives, output the result to the graphics display.

Stage 6: generate the parity bit. When the next clock signal arrives, output the parity bit.

This pipelined architecture allows a large number of filter chips to be connected together (via left and right adjacency). They can then execute concurrently.

In this paper we assume the reader is familiar with high level Petri nets [9,10]. The hierarchy concepts are described in 'Hierarchies in Colored Petri Nets [2] and we review them briefly in what follows. The inscription language is based on the functional language Standard ML [3] and is described in [1]. It is not essential that you be familiar with it in order to understand the paper.

The remainder of the paper is organized as follows:

- In Part 2 we present Basic Modeling Concepts.
- In Part 3 the details of the filter chip model are presented.
  This part may be skipped by readers with no interest in the design itself.
- In Part 4 we discuss the validation process.
- In Part 5, we discuss the timing issues.
- In Part 6, we discuss appropriate graphical front-ends for register transfer level models.
- In Part 7, we present a discussion of future work and our conclusions.

2. Basic Modeling Concepts

Hierarchy Concepts

Here we briefly review the hierarchy concepts presented in [2]. A hierarchical CP-net contains a number of *interrelated subnets* — called *pages*. There are three different ways to relate these pages to each other:

1) *Substitution*. A place or transition is *replaced* by a page — which often gives a more detailed description of the corresponding state/activity. This works in a similar way as hardware plug-ins.

2) *Invocation*. The occurrence of a transition may *trigger* the creation and execution of a new *page instance*. This works in a similar way as subroutines.

3) *Fusion*. A set of places (or a set of transitions) may be *unified* — i.e. folded into a single conceptual node. The nodes may reside on the same page or on different pages.

For substitution and invocation, we call the related node a *supernode* and its page a *superpage*. The related page is called a *subpage*. In the sequel we restrict ourselves to the use of transition substitution.

Components are *reusable* — in the sense that several nodes can be related to the *same* page. This means that a page — during execution — may have several *instances*, each with their own marking and each related to a single node.

There is a well defined *interface* between each subpage and the corresponding superpage. This interface relates *socket nodes* (surrounding the supernode) to *port nodes* (at the subpage). This is similar to the wiring of hardware plug-ins. The interface tells how the superpage *interacts* with the subpage.

Substitution of transitions works in a similar way as the refinement primitives found in many system description languages — e.g. IDEF or data flow diagrams.

The semantics of a substitution transition can be understood by replacing it with the corresponding subpage. Each socket place is merged into its assigned port. The substitution transition and its arcs are ignored (i.e. the guard and arc expressions are ignored).

Note that in hierarchical models, it is often useful to provide inscriptions for a substitution transition and its adjacent arcs. This allows the model to be executed without needing or making use of the lower level detail. Thus at various times during the development and testing of the model, a simulation can be carried out without the detail. It is not necessary that the higher level inscriptions be equivalent to the lower level detail. Often, the higher level inscriptions provide a simpler characterization of the behavior.

Pipelines, Registers and Clocks

The pipelined architecture depends upon the use of registers. When a register is 'clocked' the values present on the register's inputs are transferred to the register's outputs. These new outputs then induce changes which propagate through the circuit. The propagation is 'blocked' by arriving at the input side(s) of one or more registers. Thus the registers partition the circuit into a sequenced set of stages.

After a register is clocked, it takes some time for the changes induced to 'settle' to a stable value in the next stage. This time is usually referred to as the 'settling time.' If a register is clocked before its inputs have settled, potentially indeterminate results will be transferred to the next stage. In performance optimization of a synchronous design, the clock frequency is limited by the worst-case settling time across all the stages of the pipeline.

In current hardware design approaches, the validation of logic is often separated totally from settling time issues. Thus a register transfer level simulation will typically assume all stages have settled before the next clock pulse. With this assumption, the clocking of registers can be assumed to be a synchronous event triggered instantly everywhere by a clock tick.

In our filter chip design we have modeled the clock explicitly. The model represents the effect of a clock tick in a purely causal and logical manner. No assumption is made about the speed of propagation of the clock pulse or in fact about the speed of any of the data transformations taking place in the various stages of the circuit. This approach is very similar to that taken in the design of 'self-timed' circuits and data-flow computers. For the filter chip, a synchronous design will have to replace the causal logic provided by the clock subnet with an explicit set of timing assumptions about signal propagation and settling that guaranteed the same outcome. More about this in the sequel.

Five Different Kinds of Transition

Blocks in the register transfer level design correspond to transitions in the hierarchical Colored Petri Net. Thus transitions represent the **transformations** in the system. Transmission lines, busses and memory correspond to places and their associated arcs. Tokens at places represent the **data** in the system.

The filter chip model contains five types of transitions:

1) Data transforming transitions where the same transformation is needed in more than one location or is sufficiently complex. The behavior is then specified in a subpage. Arc expressions are needed on the arcs adjacent to such transitions only if we wish to execute the model without using the detailed behavior specified in the subpage. In the sequel we use all the subpages. All type 1 transitions are referred to as **substitution** transitions.

2) Data transforming transitions where the transformation is used only once and is relatively simple. The transformation is specified by the inscriptions associated with the transition. These include all the input and output **arc expressions** as well as the **guard**.

3) Transitions which act as registers. All of these transitions have one or more data inputs and clock input(s). The data is passed through unaltered. The clock is passed through if it is needed for further coordination purposes. Simple arc expressions suffice for specifying this kind of transition.

4) Clock transitions. These are used to either fan out the clock signal or synchronize the arrival of two or more clock signals. All arcs in and out have the same arc expression, **c**, standing for the clock pulse.

5) Fan-out transitions. These are used to generate multiple copies of data for subsequent concurrent processing. Fanout can be represented in transition types 2, 3 and 4 described above, but it is sometimes useful to represent fanout by itself. In hardware fanout is often associated with busses.

The notational conventions are derived from user-settable defaults within the **Design/CPN** software. Place names are text within the places. Color set names for places are indicated in script adjacent to the places and initial markings are underlined. Arc expressions are positioned on the appropriate arcs. Note the absence of arc expressions on arcs associated with transitions that have a refinement (e.g. a subpage giving detailed description). Guards appear in left and right brackets.

Declarations for the color sets and variables are all given in a **global declaration node** which is for convenience on another page of the model.

FIGURE FIFTEEN shows the declaration node. This is discussed in more detail in the sequel.

3. The Detailed Model

Hierarchical Structure of the Filter Chip Model

A diagram showing the relationships between the various levels of the model is shown in **FIGURE ONE.** Each node represents a page, and the shape of each such page node tells what kinds of supernodes the page can have. Ellipse shape indicates that all supernodes must be places, box shape that they must be transitions and rounded box shape that there is no restriction. Each arc represents a hierarchical relationship between two pages. Inscriptions on the arcs list the names of all the supernodes that refer to the same subpage. All diagrams in this paper were created with Design/CPN [1].

In the executable model we show three filter chips on the highest level of the hierarchical design. This could obviously be extended as need be.

The behavior of a single filter chip is described on the second level. The pipelined architecture is visible at this level. Each stage of the pipeline is then detailed on the third level. Most of the blocks on this level are detailed at a fourth level of detail. These blocks include:

1: the filter multiplier cells
2: the adders
3: the various 'or' blocks.
4: the 'limit' blocks that transform a 16 bit result with overflow indicator into the most significant 12 bits.

The filter multiplier cell fetches weighting factors from a loadable memory device. The behavior of this memory device is detailed at a fifth level of detail.

The Filter Chip Array

The diagrams at the end of this paper are hard copy printouts from which the multi-page interrelationships are not as easy to follow as it would be on computer. It is necessary for the reader to go from one subpage to another manually in order to establish the connections.

To aid in reading the diagram, we have intentionally omitted color set information from some number of places. In addition, we have included arc inscriptions in some of the higher level models, even where not necessary.

The following description refers to the page **fcarray (FIGURE TWO)**.

At the highest level of the model the page shows three filter chips wired up via their adjacent inputs and outputs. The page also includes a test environment which supplies random values and clock pulses to one of the filter chips. In the testing phase, once the behavior of a single chip is validated, the test environment can be extended to include the complete set of chips. The test environment can easily be altered to supply test vectors from a file of such vectors, or interactively from another program.

The **fload** input, when true, forces the data present on the data sample inputs **s1** through **s8** to be used as new weighting factors. This is done by storing them in the weighting factor files (WFfile, c level 5) at an address specified by **ed, ul, s8x,** and the high order bit of the data sample itself. In any case, the sample data is summed using the appropriate weights. The details are provided by the lower levels of the model.

The Filter Chip (FIGURE THREE)

The filter chip submodel exhibits the six stages of the pipeline architecture. Refer to the overview for a description of the six stages. Briefly stated these are:

1) fetch data and calculate left and right products using weights
2) sum left and right products
3) sum left and right results and limit them to high order 12 bits.
4) fetch adjacent input
5) sum and limit result to generate pixel for display
6) generate parity bit.

The filter chip page generates a command on the basis of the **fload, ed, ul** and **s8** inputs (see transition at the top of the page). This command is clocked into the components within **Stage 1** after the appropriate sample data have been loaded.

Stage 1 (FIGURE FOUR)

The fanout transition on the upper left sends a copy of the weighting factor command (including address) to eight instances of the filter multiplier. The results of the left and right multiplications generated by the filter multiplier subpage are passed to the next stage.
The two clock transitions on the right synchronize a pair of clock signals and then fanout the clock signal to four instances of the filter multiplier. The clock transition on the upper left synchronizes eight clock signals, one from each instance of the filter multiplier.

Notice that there are no registers visible at this level of stage 1. However, the **Filter multiplier cell** contains registers. This is essential for the staged pipeline architecture.

Stage 2 (FIGURE FIVE)

Stage 2 sums the left and right products for two groups of four samples, simultaneously detecting overflows and setting an overflow indicator for the pair of left results and the pair of right results. Stage 2 contains 4 registers: SUM1L, SUM1R, SUM2L, SUM2R, which hold the 4 partial sums.

Stage 3 (FIGURE SIX)

Stage 3 consists of two disjoint subnets. Each subnet takes a pair of sums and performs a final sum (left or right). An input overflow flag is combined with the flag generated by the add and used by the limiter to generate an appropriately clipped 12 bit result. Stage 3 contains 2 registers: SUM3L, which holds the final left sum for the 8 sample data and ADDOUT which contains the final right sum for the sample data.

When **ADDOUT** occurs it outputs a result to be used by the adjacent chip on the right.

Stage 4 (FIGURE SEVEN)

Stage 4 reads in an input from the adjacent chip (**ADIN**) and also delays the computed left result (**DELAY**) in order to synchronize the wave of data appropriately for the next stage.

Stage 5 (FIGURE EIGHT)

First the left result and adjacent value are summed. Then transition **LIMIT** computes a 12 bit value on the basis of a 16 bit input and a prior overflow indication.

When **PIXOUT** occurs it outputs a pixel for the display generator.

Stage 6 (FIGURE NINE)

The **parity generator** transition calculates a parity bit using a predefined ML function.
When **POUT** occurs it outputs a parity bit for the pixel generated in Stage 5.

The Filter Multiplier Cell (fmcell FIGURE TEN)

The filter multiplier cell uses the subpage **WFfile** to obtain left and right weighting factors. Having obtained these, it executes two concurrent multiplications of the low order 10 bits of the sample data with the weighting factors. The two 16 bit products are generated as outputs of **fmcell.**

The transitions **SDATA, MULTL** and **MULTR** represent registers. They partition this subpage into two stages.When **SDATA** is clocked it supplies a new sample to be multiplied by the appropriate weighting factors. When **MULTL** and **MULTR** are clocked, the left and right products waiting at **LP** and **RP** respectively, are allowed to pass into the next stage and thence out of the subpage.

Please note the 'nondestructive reads' indicated graphically by the double hollow-headed arrows. Since **WFfile** is a substitution transition the input arc from **Sout** to **WFfile** needs no arc inscription. Refer to the subpage **WFfile** for details. The left and right multiplications of the sample data execute nondestructive reads of the contents of **Sout** .

The Weight File Register (WFfile FIGURE ELEVEN)

A single bit in the **WFADD** input to the **WFfile** subpage determines whether a new weight should be stored in the writable memory. If a new weight should be stored (w= true), the address of the appropriate weighting factor is constructed and the new value replaces the old one. In any case, the address of the left and right weighting factor pair is passed to the next transition which then reads the appropriate left and right values and puts them in the output places.

The Adder (adder FIGURE 12)

The adder adds two sixteen bit integers. The low order 16 bits of the sum is placed in the result. If the sum exceeds 2**16 -1 the overflow indicator is set to true; otherwise it is set to false.

The OR Gates (or FIGURE 13)

All of the **or** subpages generate a true on the output if there is a true on one or more of the inputs; otherwise the result is false. We show only the 4-input subpage.

The 12-bit Limiter (limiter FIGURE 14)

The limit subpage generates a maximal 12-bit value (ie 2**12 -1) if the **overflow** input is true. Otherwise, it generates a 12-bit value obtained from the high order 12 bits of the **arg** input.

ML Functions Used in the Model

There are several **ML** functions used in the filter chip model. These are all defined in the global declaration node (see **FIGURE FIFTEEN**). The **bits** function is used in a number of inscriptions and its purpose must be understood to comprehend the model. The **bits** function allows an integer to be viewed as a binary number and extracts from it contiguous bits starting at some designated position, returning the result once again as an integer. Thus bit(x,p,n) extracts n bits from x starting at high bit position p, where the least significant bit of x is position 0.

4. The Validation Process

It is natural to validate sections of a design before validating or even specifying the whole design. Validation employs a number of different techniques. These include:

1) Simplification: Reduce the complexity of the model in order to validate specific aspects of it. This is employed in conjunction with the other techniques.

2) Partial Execution: simulate parts of the model under user control.

3) Extensive Execution: simulate the model on a large set of input vectors and verify that the results produced are as intended.

4) Formal analysis, including Occurrence Graph Analysis [4, 9] and Invariance Analysis [5, 9].

During these phases it is important to be able to work with a single model. Once the designer is forced to create separate models for the different stages of testing, it is difficult to guarantee that all the models are consistent and remain so. It is important to be able to work with different views of the same complete model. Generation of the different views should be as automatic as possible, thereby guaranteeing consistency.

Working from the bottom up, we tested each lowest level subpage by itself. To do this values are manually inserted in the inputs to a subpage and then the subpage executed in *manual* or *automatic* mode. In manual mode, the set of bindings for each individual transition could be examined. Specific outcomes could be forced in order to explore a particular set of possibilities. In automatic mode the simulator would make random choices of occurrence sets. In this way each of the subpages used by a higher level model was first 'validated' before testing the higher level model. Please note that this technique cannot totally validate a model. In general, real-world models often deal with such a large set of input combinations and ordering choices that only a small fraction of them are explored during testing.

It also is of great use to be able to study and validate a higher level page without executing lower level subpages used by it. One reason for this may simply be that the subpage has not yet been specified. Alternatively, the detailed model may involve complexity that is irrelevant to the validation at a higher level and would simply slow down the debugging process. Higher level inscriptions for a decomposed transition (the arc inscriptions, guard and code segments) are very useful in this context. These are normally ignored, but the user may request that the transition's behavior be defined by them rather than the subpage.

5. Timing Issues

Earlier in the paper we briefly discussed settling time issues in the context of pipelines, registers and clocks. The model we have presented incorporates *time* only in the sense that the causal logic to be *guaranteed* by clock pulses in conjunction with worst-case propagation delays is explicitly represented in the net. This could have been done in many different ways. At one extreme we could have threaded up all of the registers into one sequence. Alternatively, the clock logic could have been fanned out as much as possible, allowing a larger number of possible sequences of occurrences. The possibilities are dictated by the partial ordering requirements of the individual calculations and by the way in which the registers partition the entire model.

The question we raise now is as follows: If we wish to study timing issues, because for instance we are concerned with performance analysis, how would we do this ?

Here we outline two approaches. The first can be done within Hierarchical Colored Petri Nets as they are now. The second is basically an extension to Colored Petri Nets to allow timing issues to be represented directly. The basic approach of the the first method is to extend the color set definitions and the inscriptions to explicitly represent time in the tokens themselves.

1. Extend the color set definitions to include 'time'.

    a: Clock signals would be represented by integers or reals instead of units (colorless tokens)
    b: All other color sets would be augmented by another element of type integer or real. Thus all of these sets would become two element tuples.

2. Rewrite the arc expressions and guards so that the time component of any output token is set equal to the maximum of all the time components on input tokens plus, for transitions that represent transformations, a suitable time delay for the transformation taking place.

3. For each register transition add a test to ascertain whether the time component of the clock input is greater than the time component of all the other inputs. If this test fails then the clock rate is too fast in comparison with the settling time arising from the time delays introduced in step 2.

A second and more general approach is an extension to directly deal with performance analysis. Here we present an informal introduction to this approach. The basic idea behind performance analysis is to be able to define duration times for the individual state changes - and in this way make it possible to describe and analyze the performance of the modelled system. The performance analysis can be done either by means of simulation or by solving a set of Markovian equations [8]. Here we discuss only simulation. The ideas are based on [15] and [16].

In a timed simulation, each system state exists at a given *model time*. The model time is represented by a global variable of type integer/real. This variable is monotonically increased throughout the simulation. There are two kinds of tokens: those with a time stamp and those without. Each color set declaration determines whether time stamps are present or not — and this means that the tokens of a given color set either are all with time stamps or all without. An occurrence element (i.e. a pair consisting of a transition and one of its bindings) is said to be *color enabled* if and only if it is enabled in the usual sense (ie. there are enough input tokens of the correct colors and the guard is satisfied). For each color enabled occurrence element the *enabling time* is defined to be the maximum of the current model time and those time stamps which are present at the corresponding input tokens. A color enabled occurrence element is said to be *time enabled* if and only if there are no color enabled occurrence elements which have a smaller enabling time. A set of occurrence elements can only be concurrently time enabled if they are concurrently color enabled — and have the same enabling time. If a transition has no timed inputs it is time enabled if and only if it is color enabled. Such transitions are said to be immediate.

In a simulation without time, it is demanded that all elements of the occurrence sets must be color enabled. In a timed simulation, all elements of the occurrence sets must be time enabled (which is a stronger requirement). Stated in another way, this means that the time mechanism demands that certain occurrence elements must be chosen before others — and this means that the possible occurrence sequences in the timed model are a subset of the possible occurrence sequences in the non-timed model. Whenever a simulation step has been executed, the enabling time for all color enabled occurrence elements are calculated — and the model time is advanced to match the minimum of these enabling times. This works very much in the same way as the event queue in a traditional simulator.

When a time enabled transition occurs, time stamps are generated for those of its output tokens which are in a "timed" color set. These time stamps are equal to the current model time plus a *transition delay* plus an *arc delay*.

This approach to timing has several advantages:

1. It is more automatic, thereby eliminating a number of potential errors.
2. It produces simpler models.
3. It separates the timing issues from the color issues making some forms of analysis easier.

6. An Appropriate Front-End for Register Transfer Level Models

Please note that in this paper we are not proposing hierarchical Colored Petri Nets be used directly by logic design engineers. There are special properties of these diagrams that allow a reduction in the graphical complexity without loss of meaning. There are also existing conventions in drawing logic design diagrams that are useful notational devices. We discuss each of these briefly in the sequel. Our objective has been merely to demonstrate the adequacy of CPN as a formal basis for the behavioral specification. An appropriate graphical front end would allow the design engineer to work in a notation already familiar supplemented by the extensions necessary to completely specify the behavioral information.

In register transfer level diagrams (and circuit diagrams more generally) there is no explicit representation of local 'state' or value. A line between two transforming elements, often with an inscription that designates the data width of the line, suffices to designate the output arc from the first transformation, the intermediate place and the input arc to the second transformation:

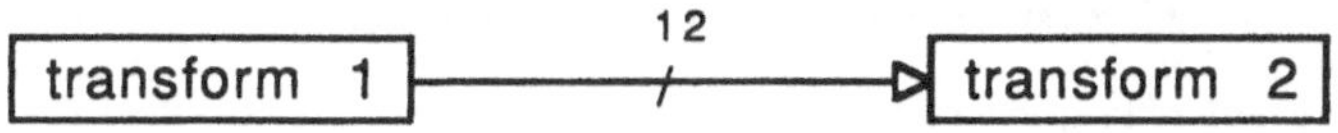

instead of:

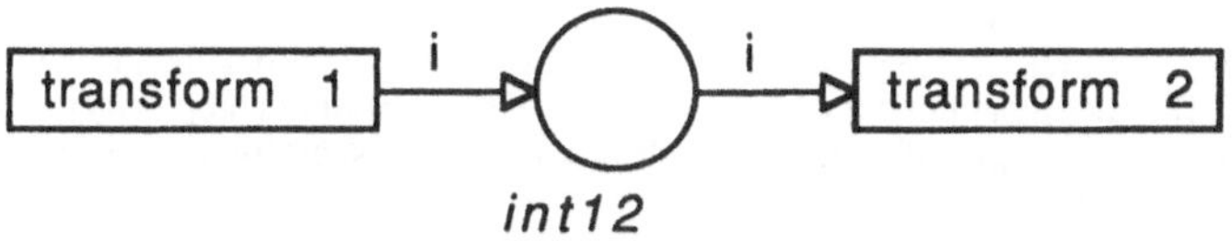

In point of fact, the graphics in such diagrams often consists of nodes and connections, where the nodes are of a single type, rather than the bipartite nature of Petri nets. In this way they resemble **Marked Graphs** [7]. In the circuit diagrams, with very few exceptions all of the transformations are applied independent of the particular data values. The representation of choice is contained within the inscriptions associated with the transitions. Under these circumstances places need not be drawn explicitly and a single simple arc expression suffices. Thus an appropriate default convention for such diagrams would be to allow places to be omitted.

Circuit diagrams also make use of a number of icons understood by the engineering community to designate common transformations. These include standard representation of 'or' gates, 'and' gates and so forth. These icons can be used directly as transitions without having to explicitly write inscriptions to define their behavior.

There already exists one example of a graphical front end for CPN, though not intended for VLSI design. IDEF/CPN [14] is an extension to IDEF0 (also known as SADT) that allows an analyst to create an IDEF model, supplement the model with behavioral inscriptions and automatically translate the model into a hierarchical Colored Petri Net.

The use of front ends to simulation always raises the following question: How easy is it for the user to understand the simulation in terms of the original front-end model? When the graphics of the front-end model and the simulation model are close to identical, this problem is greatly facilitated.

7. Future Work & Conclusions

During the course of the project we did not have an implementation of the Formal Analysis Techniques. The project did raise some important questions in regards to the use of such techniques for large models, and this led to the development of some simplification ideas. The issue here is simply that some of the color sets in the model contain a very large number of colors (e.g. 16 bit integers). Furthermore, known techniques for generation of occurrence graphs (based on symmetries or equivalence classes) are not effective in this situation. Thus it would at first appear that Occurrence Graph Generation is out of the question because of the number of states required.

The key here is simplification. Looking at the filter chip model, a sensible first question might be: Is the synchronization structure correct? Suppose we wash out the distinctions introduced by complex tokens and focus only on questions of coordination. Choices made on the basis of token content would now have to be made randomly. This is almost equivalent to regarding the net as a classical Place Transition net [6]. Now all the inscriptions can be ignored and what's more, the size of the Occurrence Graph has been reduced by many orders of magnitude.

There is an error in the CPN model of the filter chip. It is related to the details of the causal logic representing the role of the clock. It is an error in the synchronization structure that is detectable in the simplified model. This class of errors can be formulated and found using search procedures in the Occurrence Graph. Alternatively, a place invariant can characterize this class of errors and Invariant Analysis carried out to validate or invalidate the model. We leave it to the reader as an exercise finding the error!!

A less extreme simplification would be to ignore the different token colors for all the input data samples, but retain the distinctions represented by the command and addressing logic so that most of the logic of the chip could be tested, but none of the details of the actual computations.

The next step could be as follows. Rather than simply converting a complex color set to a single 'colorless' token, reducing the set to a few representative values and redefining the operations on these values in a suitable way would result in a model that retains most of the logic and computations without the combinatorial explosion caused by large color sets.

The simplification techniques are now being specified. We expect to implement them by mid-1990 and incorporate them in Design/CPN.

We have in the course of the paper mentioned some technical ideas that were stimulated in part by the filter chip project. These lie principally in the area of validation by formal analysis techniques. Our concern here is to be able to apply Occurrence Graph Analysis and Invariance Analysis to real problems. The use of simplification techniques would appear to be critical in order to avoid the combinatorial explosion caused by large color sets. A related phenomenon can be observed when time is represented explicitly in the model.

We have developed a preliminary specification for the use of both simplification techniques [12] and the representation of time [13]. The filter chip model (and other models of of similar complexity) will be rewritten and tested using these extensions to Design/CPN. We expect these concepts to be useful in model construction and simulation, as well as Formal Analysis.

Acknowledgements

I would like to thank Hartmann Genrich for his suggestions based on reading a very early version of this paper and many discussions on the topic of simplification. Kurt Jensen provided an extremely detailed critique of the paper and, with Peter Huber, developed the initial specifications for both the simplification techniques and the representation of time.

FIGURE ONE The Hierarchical Structure of the Model

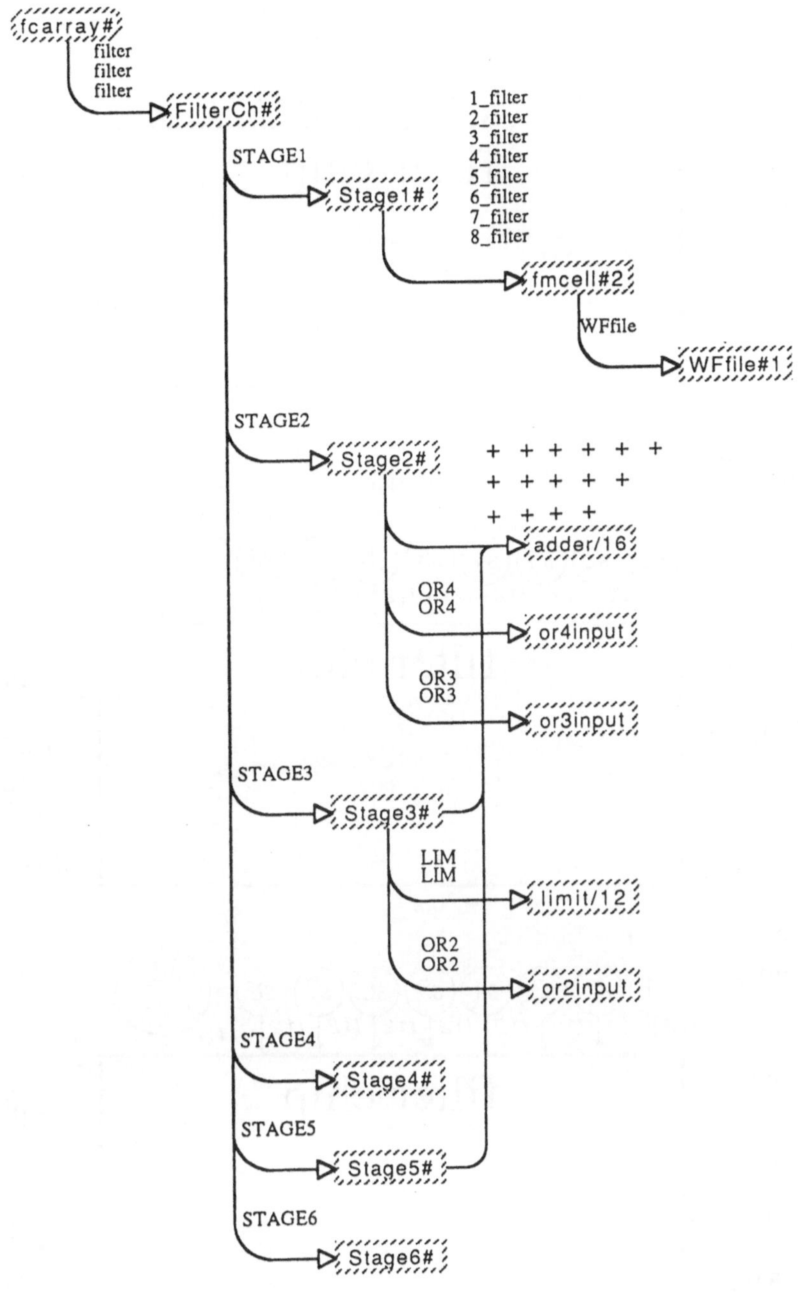

FIGURE TWO The Filter Chip Array

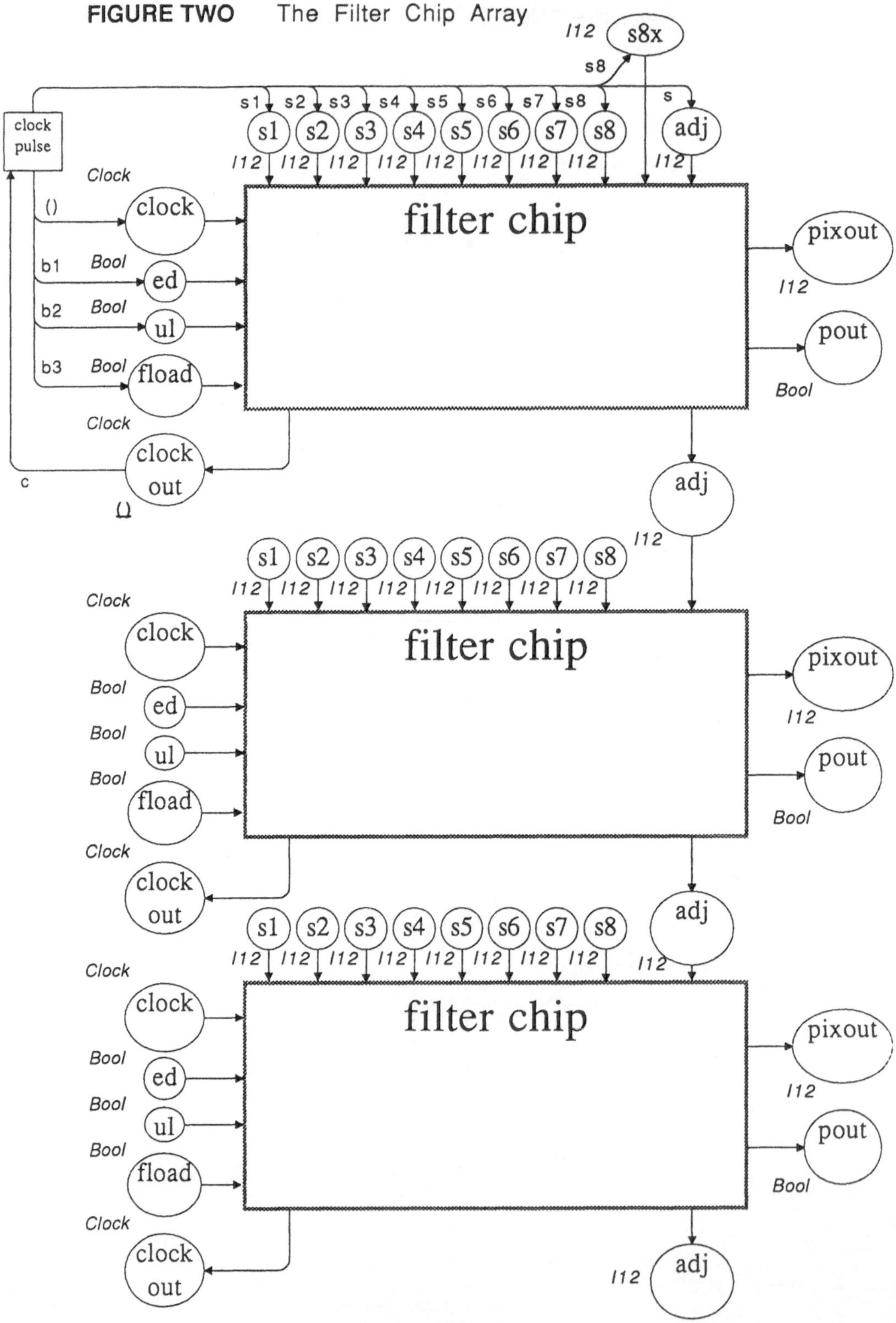

FIGURE THREE The Filter Chip

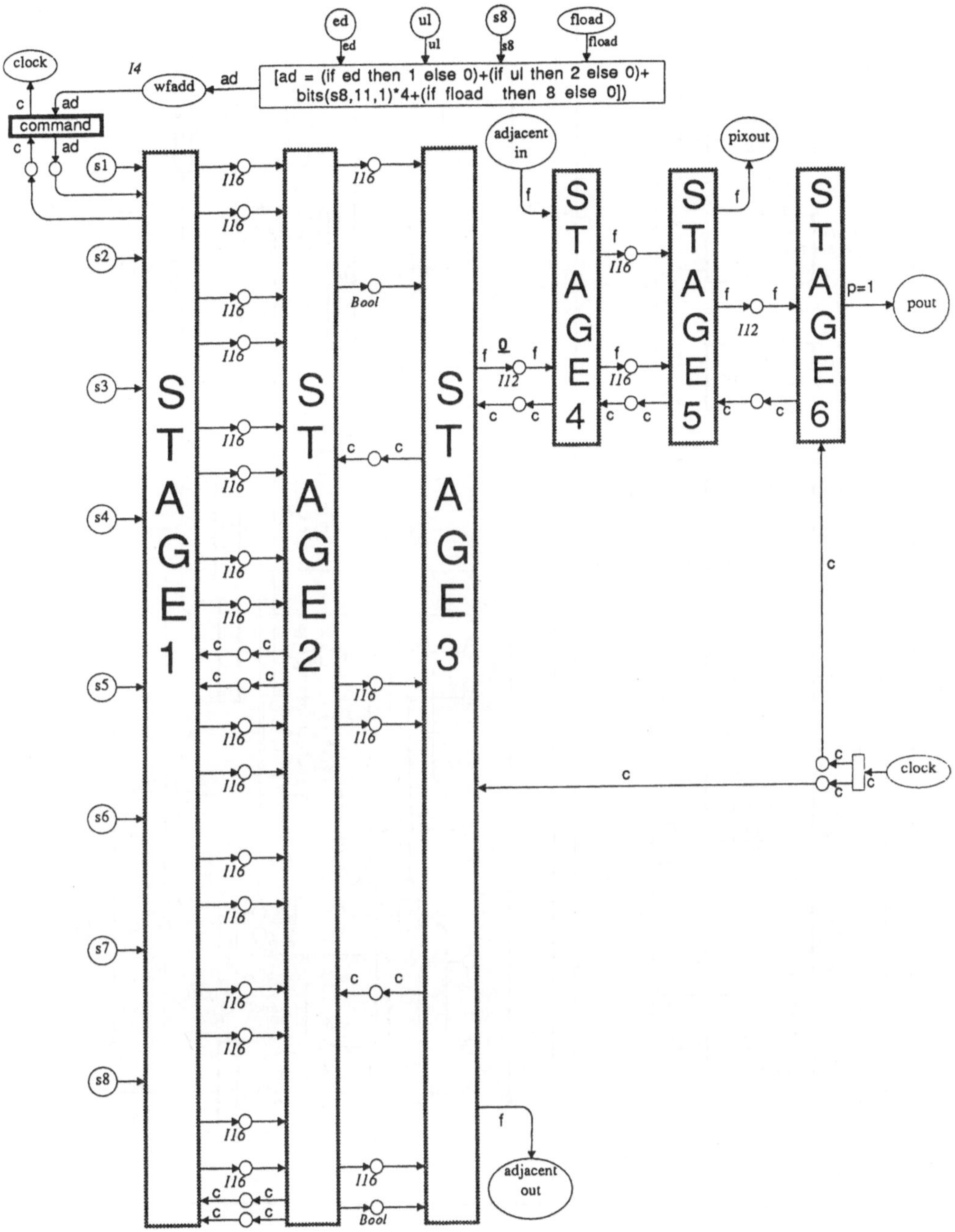

FIGURE FOUR Stage 1

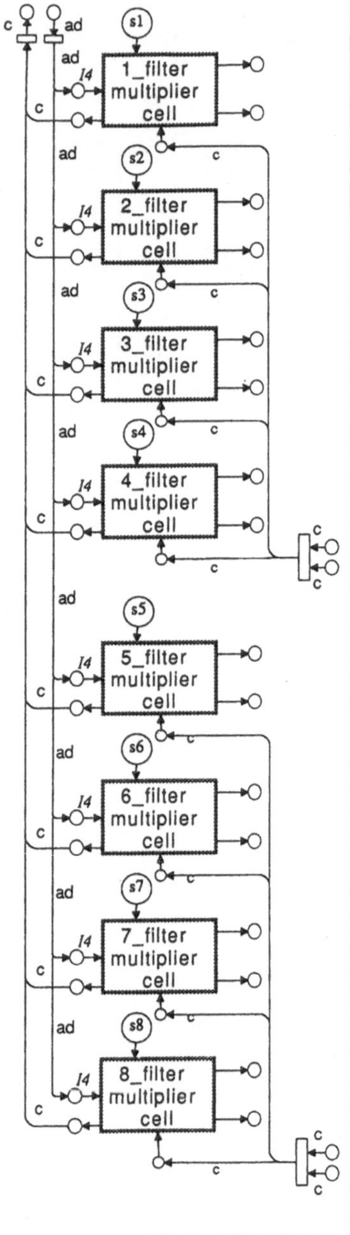

FIGURE FIVE Stage 2

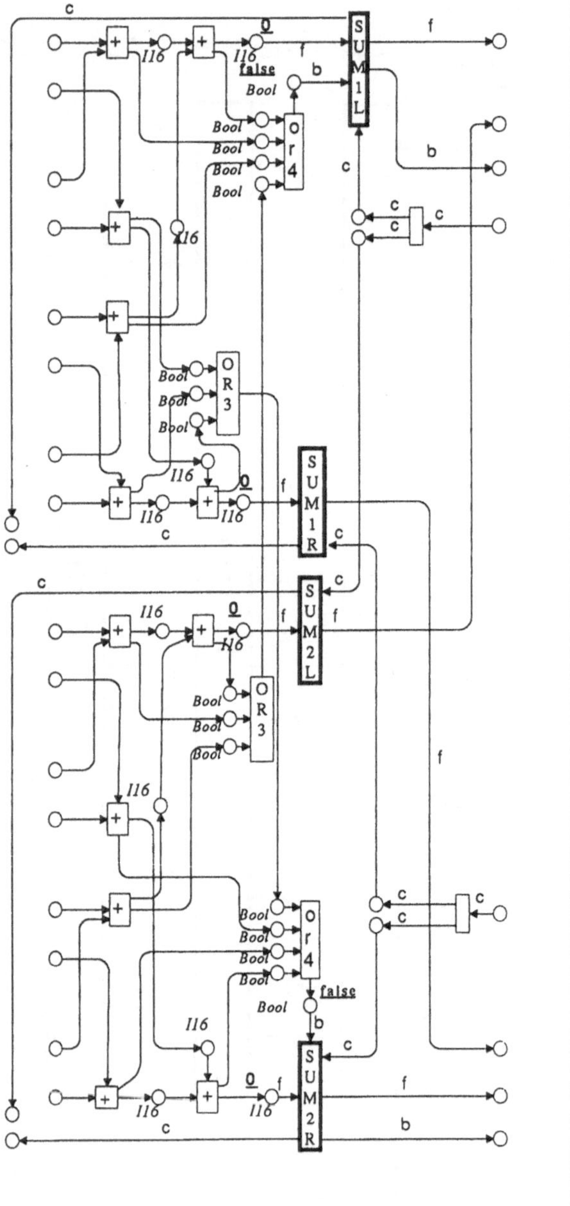

FIGURE SIX Stage 3

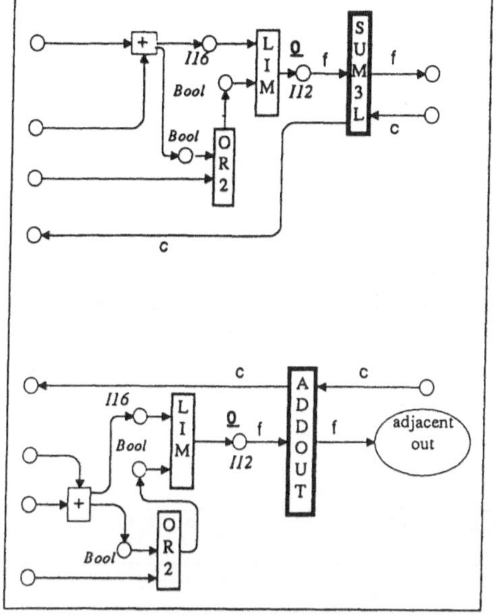

FIGURE SEVEN Stage 4

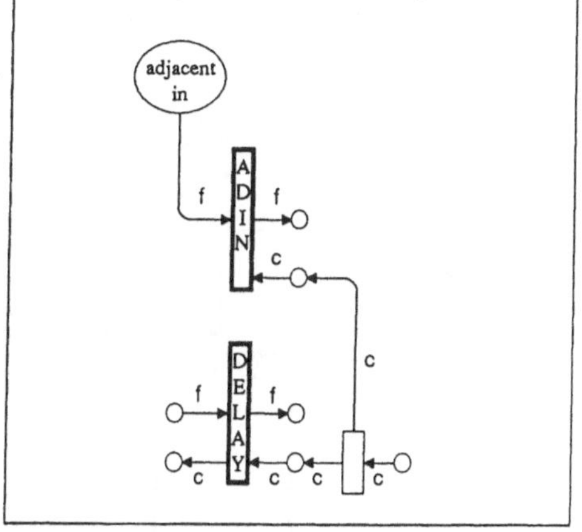

FIGURE NINE Stage 6

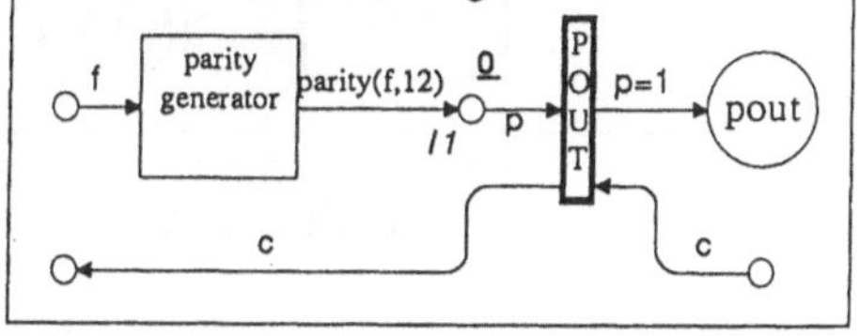

FIGURE EIGHT Stage 5

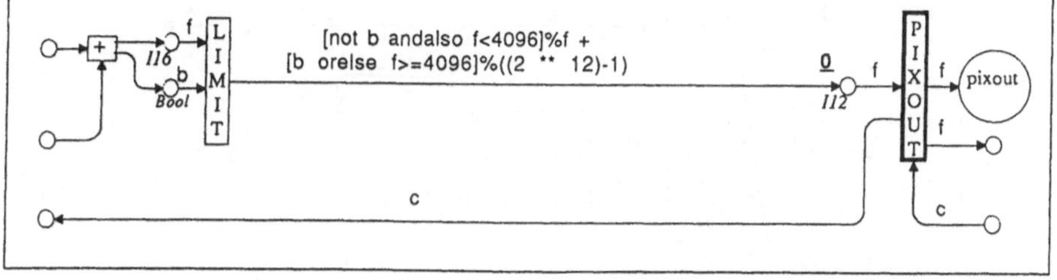

FIGURE TEN The Filter Multiplier Cell

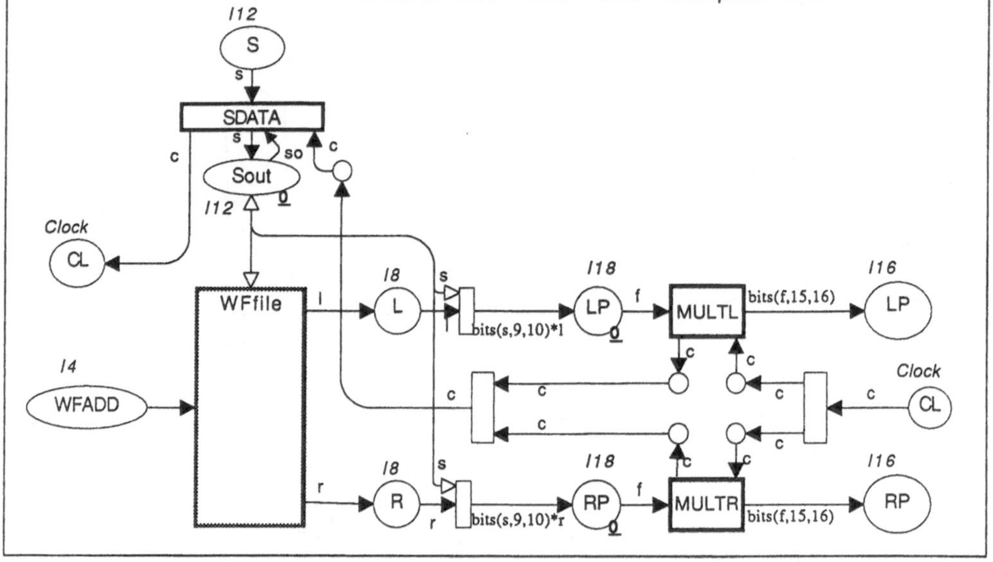

FIGURE ELEVEN WFfile: On-Chip Memory

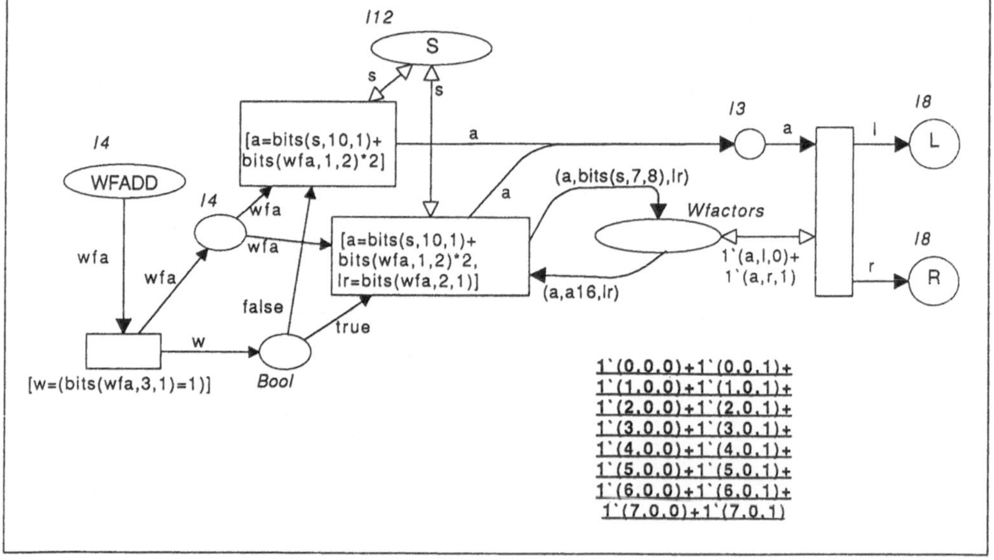

FIGURE TWELVE 16-Bit Adder

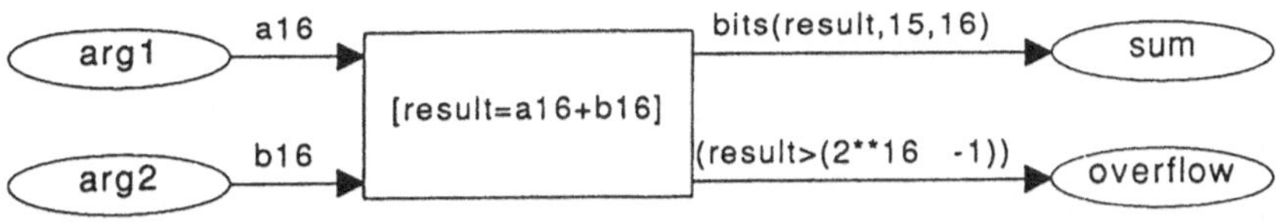

FIGURE THIRTEEN OR Gate

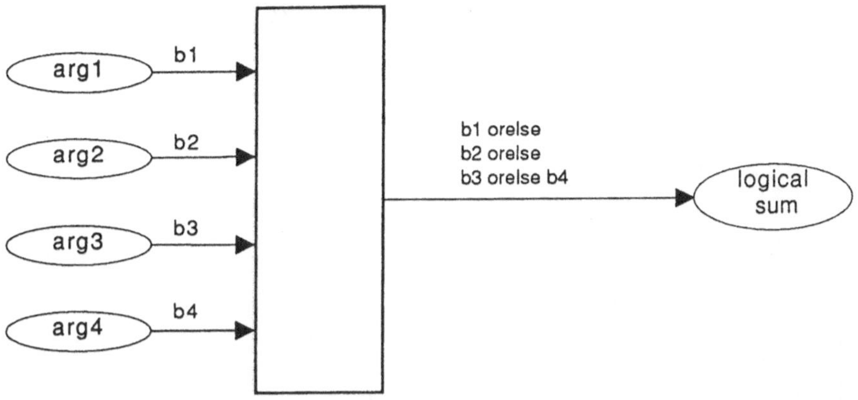

FIGURE FOURTEEN 12-bit Limiter

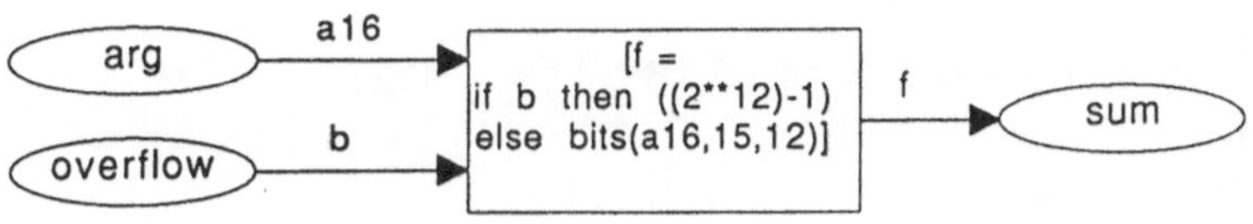

FIGURE FIFTEEN The Declaration Node

```
fun parity(x:int,y:int) = 1;

infix 8 **;
fun ((x:int) ** 0) = 1
  | (x ** y) = x * (x ** (y-1)) ;

fun bits (x:int,y:int,z:int) =
 (x mod (2**(y+1))) div (2 ** (y-z+1));


color Clock = unit;
color Bool = bool;
color I1 = int;
color I3 = int;
color I4 = int;
color I8 = int;
color I16 = int;
color I12 = int with 0 .. 4095;
color I18 = int;
color Wfactors = product I3 * I8 * I1;

var c,c1 : Clock;
var b,b1,b2,b3,b4,w,ed,ul,fload : Bool;
var a16,b16 : I16;
var result : I18;
var l,r : I8;
var lr,p : I1;

var s,so,s1,s2,s3,s4,s5,s6,s7,s8 : I12;
var f : I18;
var a : I3;
var ad,wfa : I4;
```

REFERENCE LIST

[1] **Design/CPN User's Manual**, Version 1.21: Meta Software Corporation, Cambridge, Massachusetts, USA, October 1989.

[2] P. Huber, K. Jensen, R. Shapiro: **Hierarchies in Colored Petri Nets.** Meta Software Corporation, Cambridge, Massachusetts, USA, 1989. Presented at the 10th International Conference on Application and Theory of Petri Nets, Bonn, June 1989.

[3] Robert Harper, Robin Milner, and Mads Tofte: **The Definition of Standard ML, Version 2.** Technical Report ECS-LFCS-88-62, University of Edinburgh, LFCS, Department of Computer Science, University of Edinburgh, The King's Buidings, Edinburgh EH9 3JZ, August 1988.

[4] P. Huber, A.M. Jensen, L.O. Jepsen and K. Jensen: **Reachability Trees for High-Level Petri Nets.** Theoretical Computer Science 45 (1986), 261-292.

[5] K. Jensen: **Coloured Petri Nets and the Invariant Method.** Theoretical Computer Science 14 (1981), 317-336.

[6] M. Jantzen and R. Valk: **Formal Properties of Place/Transition Nets.** In: W. Brauer, G. Goos and J. Hartmanis (eds.): Net Theory and Applications: Proceedings of the Advanced Course on General Net Theory of Processes and Systems, Hamburg, 1979, Lecture Notes in Computer Science, Vol. 84, Springer-Verlag 1979, 165-21.

[7] F. Commoner, A.W. Holt, S. Even. A. Pnueli: **Marked Directed Graphs**, Computer and System Sc. 5, 511-523 (1971).

[8] Ajmone Marsan M., Balbo G., Conte G., **A Class of Generalized Stochastic Petri Nets for the Performance Evaluation of Multi-Processors Systems;** ACM Trans. on Computer Systems, Vol. 2, N. 2, May 1984

[9] K. Jensen: **Coloured Petri Nets** In: W. Brauer, W. Reisig and G. Rozenberg (eds.): Applications and Relationships to Other Models of Concurrency, Advanced in Petri Nets 1986 - Part I, Lecture Notes of Computer Science, vol. 254, Springer-Verlag 1987, 248-299.

[10] H. J. Genrich, K. Lautenbach, P. S. Thiagarajan: **Elements of General Net Theory.** In: W. Brauer, G. Goos and J. Hartmanis (eds.): Net Theory and Applications: Proceedings of the Advanced Course on General Net Theory of Processes and Systems, Hamburg, 1979, Lecture Notes in Computer Science, Vol. 84, Springer-Verlag 1980, 76-92.

[11] D. A. Marca and C. L. McGowan: **SADT**, McGraw-Hill, New York, 1988.

[12] K. Jensen and P. Huber: **Colour Set Restrictions**, Specification paper for the CPN Palette - Part 1a. Version 1.0. Meta Software Corporation, Cambridge, Massachusetts, USA, 1989.

[13] K. Jensen and P. Huber: **Performance Analysis**, Specification paper for the CPN Palette - Part 4a, Version 1.0. Meta Software Corporation, Cambridge, Massachusetts, USA, 1989.

[14] **Design/IDEF User's Manual**, Version 1.5. Meta Software Corporation, Cambridge, Massachusetts, USA, 1989.

[15] Carlo Ghezzi, Dino Mandrioli, Sandro Morasca, Mauro Pezze: **A General Way To Put Time In Petri Nets**, ACM, January 1989 (ACM 0-89791-305-1/89/0500/0060).

[16] K.M. van Hee, L.J. Somers and M. Voorhoeve: **Executable Specifications For Distributed Information Systems.** In: Information System Concepts: An In-Depth Analysis, E.D. Falkenberg, Paul Lindgreen (eds.), Elsevier Science Publishers B.V., IFIP, 1989.

Section I
Computer Tools for High-level Nets

As for all other description techniques, the practical use of high-level Petri nets is strongly dependent upon the existence of adequate computer tools—helping the user to handle all the details of a large and complex description. For high-level Petri nets one needs editors (supporting construction, syntax check and modification) as well as analysis programs (supporting the full range of different analysis methods). The recent development of fast and cheap raster graphics provides an opportunity to work directly with the graphical representations of high-level nets.

The most important advantage of using computerized Petri net tools is the possibility to obtain better results. A graphical Petri net editor provides the user with a precision and drawing quality, which by far exceeds the normal manual capabilities of humans beings. Analogously, computer support for complex analysis methods (e.g. reachability graphs) makes it possible to obtain results, which could not have been achieved manually (since the calculations are too error-prone).

A second advantage is the possibility to create faster results. A Petri net editor often multiplies the speed by which minor modifications can be made—it is easy to change the size, form, position and text of the individual net elements without having to redraw the entire net. It is also possible to construct new parts of a net by copying and modifying existing subnets. Analogously, analysis methods may be fully or partially automated. The manual construction of a reachability graph is an extremely slow process—but a computer can do it in a few minutes/hours (even when there are several hundred thousand nodes).

A third advantage is the possibility to make interactive presentations of the analysis results. A Petri net simulator makes it easy to trace the different occurrence sequences in a high-level Petri net. Between each step the enabled transitions can be seen, and conflicts can be resolved (manually or automatically). Analogously, some tools make it possible to make an interactive investigation of complex reachability graphs.

A fourth advantage is the possibility of "hiding" technical aspects of net theory inside the tools. This allows the users to apply complicated analysis methods without having a detailed knowledge of the underlying mathematics.

For industrial applications the possibility of producing fast results of good quality—without requiring too deep knowledge of net theory—is a necessary prerequisite for the use of Petri nets. Furthermore it is important to be able to use high-level Petri nets together with other specification/implementation languages. Many groups all over the world are working on the development of computer tools for high-level Petri nets. The paper in this section lists available tools. It is, however, obvious that such a list becomes outdated rather fast, and hence the reader should consult the updated tool lists regularly published in the Petri Net Newsletter and Advances in Petri Nets. These tool lists are compiled by F. Feldbrugge, and they constitute a very useful source of information. Tool exhibitions at the Petri net conferences (cf. Appendix) usually present the most recent developments.

28.
Computer Tools for High-level Petri Nets

F. Feldbrugge and K. Jensen

Updated version of the tool list by F. Feldbrugge in: G. Rozenberg (ed.): Advances in Petri nets 1989. Lecture Notes in Computer Science, vol. 424. Springer, Berlin Heidelberg New York 1990. The updated list contains only tools that support high-level nets.

Abstract

This paper provides an overview of computer tools for high-level Petri nets. The information about the individual tools are provided by the developers. The list contains all currently available high-level net tools which we are aware of.

Introduction

This paper contains a description of 10 computer tools supporting the use of high-level Petri nets. Some of the tools have been described in [1], while others are new. The list has been obtained by asking each developer to make a description of his own tool. Such a request has been made for all high-level net tools which we were aware of.

Updated tool lists—for high-level net tools and all other kinds of Petri net tools—will be published in the Petri Net Newsletter and the Advances in Petri Nets, with regular intervals. Developers of new tools are invited to send a description of their tool to the first author. Analogously, a description should be sent when a tool is significantly changed. The descriptions must follow the format in this paper, and an e-mail with a machine readable version is appreciated. Only tools which are generally available will be included in the lists.

[1] F. Feldbrugge: **Petri net tool overview 1989.** In: G. Rozenberg (ed.): Advances in Petri Nets 1989. Lecture Notes in Computer Science vol. 424, Springer-Verlag 1990, 151-178.

1. AMI

Version number:	1.2
Year of completion:	1990
Language of the interface:	French, English, German, Spanish, Italian
Language of the tools:	French, English
Language of documentation:	French

Environment

Computer system:	net of SUN/3 and/or SUN/4
Operating system:	SUN OS 4.0 and Mac OS 6.0.2 or later
Graphics system:	Macintosh with Mac OS
Open/closed system:	Open
Source program language:	C
Extensibility:	Interface library for new tools (C, Ada, Pascal, Lisp)

Availability

Price:	Educational institutions: nominal fee Not Commercialized yet. User license after negotiation.
Documentation:	User's Manual
Maintenance:	No guarantee; reported errors will be fixed in due time.

Contact person:

Name:	Dr. Jean-Marc Bernard or Dr. Jean-Luc Mounier
Institute:	University Pierre & Marie Curie MASI Laboratory
Address:	4, Place Jussieu (Tour 56-66) 75252 PARIS Cedex 05 France
E-mail:	mars@masi.ibp.fr
Telephone:	(+33) 1 44 27 61 89
Telefax:	(+33) 1 44 27 62 86

Functionality

- AMI is an highly interactive software environment based on typed graphs with attributes. It is a sufficiently general model to allow easy extensions to new kinds of graphs or Petri nets.

  The AMI software environment groups together an interactive toolkit for creation, manipulation, transformation, analysis and checking of graphs. It is basically designed to be extended to further domains.

- A graphic editor, MACAO, allows multiple net entry and results display on a dedicated workstation, either in connection with the workshop or alone.

- An inference engine MIAMI, based on constraints, is integrated to allow logic programming into expert system rules.
- AMI is an open workshop which makes easy the integration of new applications. At this time, it integrates a lot of applications which work in majority with Petri nets: Combag [Trèves], coverability graph [Finkel], Structural properties algorithms from GreatSPN [Chiola], OCCAM prototyping [Bréant], ADA prototyping [Kordon].

Future plans

- X-Window version of the user interface
- Hierarchical modelling techniques assisted by the graphical interface.
- Tools for Well Formed Nets
- Performance evaluation tools from GreatSPN
- Synchronous timed Petri nets

References

Bernard J-M, Mounier J-L, Beldiceanu N, Haddad S: "*AMI: An extensible Petri net Interactive Workshop*", Proceedings of the 9th European Workshop on Application and Theory of Petri nets, Venice, Italy, pp 101-117, June 1988.

Beldiceanu N, Souissi Y: "*Deterministics systems of sequential processes: theory and tools*", Proceedings of the International Conference on Super Computers, Hamburg, RFA, October 1988.

Bernard J-M, Mounier J-L: "*L'atelier logiciel AMI, un environnement multi-utilisateurs, multi-sessions pour une architecture distribuée*", Séminaire Franco-Brésilien sur les systèmes informatiques répartis, Florianopolis, Brésil, pp 200-206, 14 Septembre 1989.

Kordon F, Estraillier P, Card R: "*Rapid ADA Prototyping: principles and example of a complex application*", Proceedings of the International Phoenix Conference on Computer and Communications, April 1990.

Chiola G, Dutheillet C, Franceschinis G, Haddad S: "*On Well-Formed Coloured Nets and their Symbolic Reachability Graph*", Proceedings of the 11th International Conference on Application and Theory of Petri Nets, Paris, France, pp 387-410, June 1990.

Bréant F: "*Tapioca: OCCAM Rapid Prototyping from Petri Nets*", Proceedings of the Vth Jerusalem Conference on Information Technology, Jerusalem, Israel, October 1990.

Bernard J-M, Mounier J-L: "*Conception et Mise en Oeuvre d'un environnement système pour la modélisation, l'analyse et la réalisation de systèmes informatiques*", university Thesis, PARIS VI, 1991.

2. CPNA – Coloured Petri Net Analyser

Version number:	1.3 or later
Year of completion:	First shipped 1989
Language of tool:	English
Language of documentation:	German (English in preparation)

Environment

Computer system:	DEC-Workstation, VAX SUN-Workstation IBM-PC and compatibles
Operating system:	VMS UNIX MS/DOS
Graphic system:	—
Open/closed system:	Open
Source program language:	MODULA-2
Extensibility:	Communication with the program and between its parts is done by ASCII-files, the user may interfere with own programs.

Availability

All versions are available.

Price:	DM 15000,- (Workstations) DM 10000,- (IBM-PC) Universities 50% discount DM 1000,- (PC-Version for Education only)
Documentation:	Users Reference Guide
Maintenance:	One year maintenance free. Extensions to be negotiated. Consulting services available.

Contact person:

Name:	Prof. Dr. P. Starke
Institute:	Humboldt-Universität, FB Informatik
Address:	PSF 1297 0-1086 Berlin, Germany
E-mail:	starke@hubinf.uucp

Functionality

CPNA is a tool package supporting the analysis of Coloured Petri nets (CPN).

A CPN is a net where with every place there is adjoined a finite set of (token) colours, with every transition there is adjoined a finite set of firing modes (called transition colours) and where every arc is inscribed with a function. If an arc connects the place p with the transition t, then its function is a mapping from the colourset of the transtion t to the set of all multisets (bags) of colours of the place

p. Markings of CPN adjoin with every place a multiset of appropritate token colours.

Coloured Petri Nets form the semantic of all kinds of High Level Nets.

CPNA combines the following:
- a textual editor for CPN
- a by-hand simulation part
- an analysis part to compute
    - structural information
    - place invariants, transition invariants
    - reachability and coverability graphs,

which then are used to verify properties like boundedness, liveness, reversability, etc.

The editor gives the possibility to construct and to edit CPN. One can use any graphical editor too, if one knows the structure of its database file by writing a program which converts it to a CPNA-file. Using the CPNA-editor editing the arc functions is supported by a lot of standards.

The by-hand simulation part allows starting at a given marking to fire forward or backward single transition colours or maximal steps, in this way the user can traverse parts of the reachability graph. The analysis part contains a small expert system which draws all the consequences from the known properties of the actual net. The structural informations computed include

- conflicts (static, dynamic) and their structure (e.g. free choice),
- deadlocks and traps (deadlock-trap-property),
- state machine decomposition and covering.

For certain subclasses of CPN these properties can be used to deduce dynamic properties.

Invariant analysis can be done by computing generator sets of all place/transition invariants and of all non negative invariants. Vectors can be tested for invariant properties.

For checking boundedness and coverability, the coverability graph of the actual net can be computed and analysed. For bounded nets the reachability graph can be computed and analysed for liveness, reversability, dynamic conflicts, realisable transition invariants, livelocks etc. The symmetries of the actual net can be computed and used to reduce the size of the reachability graph. The reachability graph can be printed either as a tree or as a (adjacency) matrix. Many options help to avoid printing unnecessary informations.

Except the educational version there are no restrictions on the size of the net other than the RAM-capacity of the computer used. During a session all the commands and options are saved to a command file to that the session can be repeated automatically with a different net.

3. Design/CPN

Version number:	1.5 or later
Year of completion:	First shipped 1989
Language of tool:	English
Language of documentation:	English

Environment

Computer system:	We support three different possibilities: • Apple Macintosh SE, II, IICX, IICI , IIFX (8MB RAM) • SUN3, SUN4, SPARC Workstations (16MB+ RAM) HP 9000 Workstations (16MB+ RAM) • Macintosh front-end + simulation engine on Workstation
Operating system:	Apple OS SunOS 4.0.3 or later HP-UX 7.0
Graphic system:	Macintosh X-Windows version 11, release 4
Open/closed system:	Open
Source program language:	C and Standard ML.
Extensions:	C part programmed by means of Design/OA (a library of C routines for the manipulation of graphical objects). ML part uses Design/ML (ML interpreter with graphical interface, based on the University of Edinburgh's SML interpreter).

Availability

Price:	$20,000 (Universities 50% discount)
Documentation:	Users Reference Guide
Maintenance:	$4000 per year includes product upgrades. Local maintenance is also available from distributors. CPN modeling and consulting services available from Meta.

Contact person:

Name:	Robert Seltzer
Institute:	Meta Software Corporation
Address:	150 CambridgePark Drive, Cambridge, MA 02140, USA
E-mail:	seltzer%metasoft@bbn.com
Telephone:	(+1) 617 576 6920
Telefax:	(+1) 617 576 0519

Local distributors for:

France:	IGL Technology; Philip Kelley; tel. (+33) 1-4773 1100
Germany:	CIT GmbH; Gert Scheschonk; tel (+49) 30-434 6561
Denmark:	Elektronikcentralen; John Mølgaard; tel. (+45) 4286 7722
Italy:	currently being selected.

Functionality

Design/CPN is a tool package supporting Coloured Petri Nets (CP-nets). Design/CPN combines the following:

- Colored Petri Nets, with which the size of models is kept much smaller than with ordinary Petri Nets, while at the same time retaining the rigorous formality of the simple Petri Nets.
- Hierarchical Models, with which it is possible to construct large, complex, system models using a sound notion of hierarchy that is consistent with CPN simulation and formal analysis techniques.
- Computer Tool Support, with which it is possible to create an executable hierarchical CP-net, in a user-extensible environment that includes an evaluation engine for code segments associated with the CPN model.
- Open Architecture, with which it is possible to extend the system to handle several types of Petri Nets, provide new analysis tools, or interface with other modeling techniques.

Design/CPN allows the user to construct and edit hierarchical CP-nets. This means that the CP-net may contain a set of subnets, which are related to each other in a formal way. The subnets are called pages and they can be related to each other in three different ways: substitution, invocation and fusion. Each of these relationships constitute a powerful structuring method, and together they allow the user to construct large CP-nets which are modular and manageable. Hierarchical CP-nets are described in a paper presented at the 10th International Conference on Application and Theory of Petri Nets, Bonn, 1989.

The user works directly with the CP-net graph, at a graphical workstation with a high-resolution screen and a mouse. Each editing operation is immediately reflected on the screen, which always contains an exact image of the current CP-net (except for the inevitable degradation caused by the fact that the screen has less resolution than the final laser writer output).

The Design/CPN Editor recognizes the structure of CP-nets. This means that it will refuse to perform operations that have no meaning for CP-nets (e.g., drawing an arc between two transitions). When the user moves or reshapes a place or a transition, all the corresponding arcs are automatically updated. When the user deletes a place or transition, all its arcs will be deleted. With these facilities, editing of CP-nets can be performed much more quickly and smoothly than with ordinary drawing tools (where places, transitions and arcs are unrelated objects, many of which must be manually adjusted each time one of them is changed). The Design/CPN Editor also has a set of "power commands", which allow the user to shape and position arcs in a very precise, general and effective way.

The Design/CPN Editor contains a set of default attributes (size, form, shading, font, color, etc.) for each of the different kinds of objects which can be contained in a CP-net (places, transitions, arcs, initial markings, color sets, guards, arc expressions, etc.). The user can change all defaults, and can overwrite them (i.e. set the attributes for each individual object). The Design/CPN Editor also has a large number of optional syntax checks. These checks play the same role as type checking in a programming language. This means that the system is able to detect potentially unsafe structures—and report them to the user.

The user can construct hierarchies top-down, by creating new pages (giving a more detailed description of existing nodes), or bottom-up, by moving existing subnets to new pages (replacing each subnet by a single node). Moreover, there is great freedom in the layout of the CP-net, and the user can apply all the fancy forms of graphics that are available in modern drawing systems.

Design/CPN also allows the user to simulate hierarchical CP-nets. Each page in a hierarchical CP-net is shown in its own window. The enabled transitions are highlighted by the system, and the occurrence of a simulation step is animated, allowing the user to see input tokens (moving along input arcs, towards the transition) and output tokens (moving along output arcs, away from the transition). The Design/CPN Simulator contains a large number of options, which allow the user to perform and observe the simulation process in many different ways.

The simulation can be manual, in the sense that each simulation step is defined by the user. Alternatively, the simulation can be automatic, in the sense that each simulation step is generated by the system (making random choices when necessary). In both cases, it is the Design/CPN Simulator which calculates the effect of the simulation steps.

In manual mode, the user has complete control over the simulation steps (defining how the variables in the arc expressions and guard are bound). It is, however, still possible to get a lot of help from the system. For example, the user can ask the Design/CPN Simulator to calculate all possible bindings for a given transition (or those bindings that are extensions of a given partial binding). Such system facilities allow the user to concentrate on the interpretation of the simulation, instead of using energy on the calculation of possible simulation steps.

In addition to manual and automatic simulation modes, a so-called super-automatic mode is also provided. In this mode there is no user-interaction (for the selection of bindings) and there is no feedback during the simulation. In this way the simulation runs much faster since it is performed by a SML program alone (while an ordinary simulation is performed by a SML program and a C program with intense intercommunication). This SML program can be used independently from the Design/CPN tool and thus be used as a stand-alone application.

Standard ML (SML) is a functional programming language developed at the University of Edinburgh by a design group headed by Robin Milner. Design/CPN uses an extension of SML—called CPN ML—for the following purposes:

- The user specifies the inscriptions of a CP-net (the color sets, initial markings, arc expressions and guards) in CPN ML.
- The internal algorithms that calculate the enabling and occurrence of transitions are written in CPN ML.
- Each transition can have an attached a code segment, which is executed each time the transition occurs. The user writes the code segments in CPN ML. In particular, the code segments can calculate values for output tokens, but are often used for input/output and animation.

It is important to notice that the first two uses of CPN ML are entirely independent of the last one. This means that in later versions of Design/CPN, the user will be able to write code segments in another programming language and still use CPN ML for the first two purposes.

The Design/CPN Editor and the Design/CPN Simulator are integrated into a single program, which has an editing mode and a simulation mode. The user can perform the simulation upon the CP-net constructed by means of the editor, without having to make any modifications or translations. Simulation parameters can be set up in the editor, and it is easy to switch between editing and simulation (e.g., to correct minor modelling errors discovered during the simulation).

Design/CPN may be used also in conjunction with another tool built by Meta called Design/IDEF; this tool supports primarily the creation of SADT models. The tools are integrated in such a way that it is possible to automatically generate a CP-net from an SADT model for simulation and for formal analysis (when it will be available).

The routines of Design/OA for the manipulation of graphical objects are available to the user of Design/CPN via a SML interface. In this way models built with Design/CPN can be animated with Design/OA routines in code segments attached to transitions. Stand-alone applications may also be generated automatically with sophisticated user graphical interfaces.

Additional functionalities to the Design/CPN tool may be built by the users themselves writing SML programs. For example users may write a program that parses a text file and generates a CPN diagram (SML equivalents of the Unix tools Lex and Yacc are also available); this approach is used to integrate Design/CPN with other tools via text interfaces (a textual specification language for hierarchical Colored Petri Nets is provided); following this approach Design/CPN is currently being integrated with a tool called ADA/Nice that supports the HOOD methodology. SML programs may be written also to implement tools for the analysis of CP-net models.

Future Plans

Version 2.0 of Design/CPN, to be released during the first quarter of 1991, will support timed CP-nets (an extension to ordinary CP-nets that make it easier to describe systems that are time driven). This means that the same CP-net model can be used to analyze both the logic correctness and the performance of a system. Version 2.0 will also support reporting facilities that will allow easier visualization of the simulation results (e.g. bar and pie charts). Later versions of Design/CPN will support occurrence graph analysis (year-end 1991) and place invariants (1992).

References

A list of publications concerning Design/CPN can be obtained via the above address.

Other Remarks

Meta Software has invested more than 60 man-years over the past decade in developing and enhancing its software to support large hierarchical and executable graphic models. Design/CPN is the product of a close cooperation with Kurt Jensen, Aarhus University. Hartmann Genrich, GMD, has also extensively influenced the development of the Design/CPN tool.

4. ExSpect

Version number:	2.2
Year of completion:	1990 (of version 2.2)
Language of the tools:	English
Language of documentation:	English

Environment

Computer system:	SUN3 and SUN4
Operating system:	SUNOS 4.1 or higher
Graphics system:	Sunview
Open/closed system:	Open
Source program language:	C
Extensibility:	Simulator is extensible by means of user defined processes, for example coded in C

Availability

Price:	Educational institutions: nominal fee Not Commercialized yet. User license after negotiation.
Documentation:	User's Manual
Maintenance:	No guarantee; reported errors will be fixed in due time.

Contact person:

Name:	L.Somers
Institute:	Dept. of Math. and Computing Science, Eindhoven University of Technology
Address:	P.O. box 513 5600 MB Eindhoven The Netherlands
E-mail:	wsinlou@win.tue.nl or wsinfsys@win.tue.nl
Telephone:	(+31) 40 472 733 or (+31) 40 472 805
Telefax:	(+31) 40 436 685

Functionality

ExSpect is an acronym for executable specification tool. It is a software tool for designing and simulating hierarchical colored timed petri nets.

Once defined, subnets can be instantiated several times. The subnets can be heavily parametrized with encourages the reuse of specifications. The colors are based on a typed functional language. Tokens may have a time stamp denoting the earliest time they may be consumed.

The ExSpect package consists of several parts:

- A shell controlling all other components.

- A graphical editor for designing a specification. The graphical editor allows the user to build his nets and functions top-down by means of separate subnets.
- A type checker for checking the consistency of the instantiations and definitions of the subnets and the typing rules of the functions that describe the transitions.
- A simulator for interpreting the specification and simulating the dynamics of the specified system.
- A runtime user interface to interact with a running simulation (inspecting places, adding tokens, reading from and writing to files or other applications).
- Applications to analyze the results statistically, like an interactive bar graph.

Future plans

Migration to Xwindows.
Integration with object oriented data modeling.
Time analysis methods independent of simulation.
Course development.

References

A list of publications concerning ExSpect can be obtained via the above address.

5. FORSEE

FORSEE, a formal systems engineering environment, is a collection of tools that provides computer-based assistance in the development of concurrent systems. Eventually it is hoped that it will include tools for specification, verification, performance analysis, simulation/animation, compilation and testing. Currently, there are three components of this environment: Design/CPN, TORAS and PROMPT. Design/CPN is the tool package from Meta Software Corporation that provides a graphical front end and a simulator for CPNs; it is described elsewhere in this list of tools. TORAS and PROMPT are a verifier and a high-level net compiler, respectively. They are described below:

TORAS component

Version number:	0.x
Year of completion:	not yet complete but prototype exists
Language of tool:	English
Language of documentation:	English

Environment

Computer system:	Sun workstations
Operating System:	Unix
Graphics system:	—
Open/Closed system:	Open
Source program language:	Eiffel, C
Extensibility:	Designed so that code depending on the various formal description techniques (including various net types) can be easily interfaced to the reachability analysis code. Additional analysis software can also be added.

Availability

Not yet available.

Contact Persons:

Name:	J.Billington and G.Wheeler
Institute:	Telecom Australia Research Laboratories
Address:	P.O. Box 249, Clayton, Victoria 3168, Australia.
E-mail:	j.billington@trl.oz.au, g.wheeler@trl.oz.au
Telephone:	(+61) 3 541 6416 (+61) 3 541 6415
Telefax:	(+61) 3 544 2362

Functionality

TORAS [1]TORAS [1] is a tool for analysing Petri nets (and other techniques) using reachability analysis. To combat the well-known state explosion problem faced by reachability analysis it uses the stubborn set method to remove redundant interleavings [2].

As well as a simple algorithm which generates the complete state space, three variations of the stubborn set method are available. The basic algorithm guarantees that deadlocks and fact occurrence are preserved in the reduced state space. The other variations will, in addition, preserve livelocks, traces, failures and testing equivalence.

Another available option is Holzmann's reachability analysis technique [3]. This technique, unlike the stubborn set method, does not guarantee preservation of deadlocks and other properties but it can allow the exploration of more of a large state space than complete reachability analysis. It may be combined with the stubborn set algorithms under some circumstances.

Currently, TORAS accepts a text form of place/transitions nets as input to the analyser.

Future Plans

- Interface to CPNs, P-nets [4] and possibly LOTOS.
- The addition of various other reachability analysis algorithm techniques to provide practical analysis of diverse systems.
- Graphical interface based on Design/CPN.
- Sophisticated pre- and post-generation analysis.

References

[1] G. Wheeler, A. Valmari, J.Billington: "Baby TORAS Eats Philosophers but Thinks about Solitaire", Proceedings of the 5th Australian Software Engineering Conference, Sydney, pages 283-288, 1990.

[2] A. Valmari: "Stubborn Sets for Reduced State Space Generation", Proceedings of the 10th International Conference on Application and Theory of Petri Nets, Bonn, II, pages 1-22, 1989.

[3] G.J. Holzmann: "An Improved Protocol Reachability Analysis Technique", Software Practice and Experience, 18(2), pages 137-161, February 1988.

[4] J. Billington: "Many-sorted High-level Nets", Proceedings of the Third International Workshop on Petri Nets and Performance Models, Kyoto, 11-13 December 1989, IEEE CS Press.

[5] J. Billington, G. Wheeler, M. Wilbur-Ham: "PROTEAN: A High-level Petri Net Tool for the Specification and Verification of Communication Protocols", IEEE Transactions on Software Engineering, vol. 14, no. 3, March 1988, pages 301-316.

Other remarks

TORAS is the successor of PROTEAN [5] which was also developed in Telecom Australia Research Laboratories.

PROMPT component

Version number:	2.1 or later
Year of completion:	1990 (Version 1 in 1989)
Language of tool:	English
Language of documentation:	English

Environment

Computer system:	Sun Workstations, Unix machines, Vax/VMS, IBM PC AT, 386 and compatibles
Terminal types:	VT100 for Unix and VMS systems
Operating System:	Unix, VMS MS-DOS (limited version)
Graphics system:	Sunview, Curses (X-Windows under development)
Open/Closed system:	Open
Source program language:	C
Extensibility:	Uses standard C, user-written C code can be incorporated

Availability

Versions for all machines are available.

Price:	Negotiable on application Research licence available to Universities
Documentation:	Users' reference manual (versions for Sunview and Curses)
Maintenance:	One year maintenance included in purchase price. (Support not guaranteed for research licence.) Consulting services available from Unico Computer Systems Pty Ltd.

Contact Persons:

Name: Michael Palmer
Institute: Unico Computer Systems Pty Ltd
Address: 43 Linacre Rd, Hampton VIC 3188 AUSTRALIA
E-mail: unico@trl.oz.au
Telephone: (+61)-3-8991101
Telefax: (+61)-3-8991019

Name: Ken Parker
Institute: Telecom Research Laboratories
Address: 770 Blackburn Rd, Clayton VIC 3168 AUSTRALIA
E-mail: k.parker@trl.oz.au
Telephone: (+61)-3-5416797
Telefax: (+61)-3-5442362

Functionality

PROMPT is a tool for implementation of protocols and other applications from a high level Petri net language into C code. It provides the following components:

- A compiler which translates into C code nets which have been expressed in XNL, the eXtended Net Language. XNL caters for a rich set of data types, as well as for the logic of nets. It describes a net as a set of variables, whose values define the current state of the net and a set of transitions which defines how states of the net may change. XNL allows enabling conditions and transition operations to be expressed easily. XNL also supports sub modules, which appear to the calling net as a single transition, but which are in fact made up of multiple transitions at the lower level.
- A full screen symbolic debugger, XDB, which allows debugging to be performed on the protocol specification (expressed in XNL) rather than on the resultant C code. This allows the designer to concentrate on the protocol specification and not on C implementation considerations. Apart from the usual symbolic debugger features, XDB also allows the user to fire one transition at a time or to step through several transitions; to choose transition firing order; to set traces on transitions and variables; and to change variable values dynamically.
- An interface for studying and controlling net execution, called SCI (System Control Interface). It allows up to 16 nets to run as separate processes simultaneously, and a user can load a new process, or suspend, resume or abort existing processes. SCI displays on the screen various statistics on the processes as they run. It also provides control of a trace logger facility allowing the user to specify various logging details such as what types of events for each process are to be logged.
- A standalone utility, LFA (Log File Analyser) for reading and formatting log files created during SCI runs. The user may specify selection criteria on the records in the log files so that only records of interest are displayed. The LFA performs some symbolic data translation so that, for example, transitions are printed out under their names rather than as numbers.
- A set of libraries which provide non net functions as sub modules. These functions provide support in the following areas:
  - timers and delays, including means to advance time in a controlled manner when debugging;
  - file, socket and queue input/output, with the data involved encoded in ASCII, ASN.1, the PROMPT system's own encoding scheme, or a user-supplied encoding scheme;
  - gathering of timing statistics on the running of the net.
- A capability for a user to incorporate his own C-coded non-net functions into an XNL net.

Future Plans

- Porting of PROMPT to X-windows.
- Addition of a run-time error checking capability.
- Division of the documentation into a tutorial manual and a separate reference manual.
- Provision of a capability to allow a user to test a protocol specification under a variety of automatically generated situations.

References

[1] Unico Computer Systems, "PROMPT System User Documentation", Unico Computer Systems and Telecom Australia, December 1990.

[2] Billington, J., Wheeler, G.R. and Wilbur-Ham, M.C. "PROTEAN: A High-level Petri Net Tool for the Specification and Verification of Communication Protocols", IEEE Trans. Soft. Eng., Vol. 14, No. 3, March 1988, pp. 301-16.

[3] Billington, J. and Kinny, D. "Computer Aided Protocol Engineering", in Proc. I.E. Aust. Conf. on New Business Applications of Information Technology (Melbourne, Australia, April 1989), pp. 69-73.

[4] Illing, G.C. "Automatic Petri Net based Protocol Implementation", in IREECON International (Melbourne, Australia, September 1989), pp. 358-61.

[5] Parker, K.R. "The PROMPT Automatic Implementation Tool - Initial Impressions", in Proc. Third Int. Conf. on Formal Description Techniques (FORTE '90) (Madrid, Spain, November 1990), pp. 701-7.

Other remarks

PROMPT has evolved from the PROTEAN Tool (see Reference (2)), which includes the capability to analyse NPNs with interactive simulation, by displaying enabled transitions at each marking, and having the user select which is to fire next. PROMPT has been developed by UNICO Computer Systems Pty Ltd under a contract from Telecom Australia Research Laboratories. It is jointly owned by Unico and Telecom Australia.

6. GRASPIN

Version number:	2.0
Year of completion:	1990
Language of the tools:	English
Language of documentation:	English

Environment

Computer system:	Symbolics LM, Sun with UX400 board MacII with Ivory board
Operating system:	Genera 7.2, Genera 8.0
Graphics system:	Genera
Open/closed system:	Open
Source program language:	Common Lisp
Extensibility:	User obtains source code, Menus can be changed. New net types and tools may be added.

Availability

Price:	Licence under reasonable conditions (cost for tape)
Documentation:	User Guide, Reference Manuals
Maintenance:	None, but major bugs will be fixed when reported

Contact person:

Name:	Hans Nieters
Institute:	Institut für Systemtechnick Fachbereich Konstruktiontechnik
Address:	Gesellschaft für Mathematik und Datenverarbeitung Postfach 1240 D-5205 St. Augustin 1, Germany
E-mail:	nit@gmdzi.gmd.de
Telephone:	(+49) 2241-14-2444

Functionality

Interactive graphic design of diagrams, graphs and nets, including the usual layout operations (selection, alignment, adjustment and profile options, buffer and file operations).

Net classes: Supported are Pr/T systems with PT, PR/E and EN as subclasses.

The operations and variables use din inscriptions are defined in an algebraic specification language SEGRAS, written with the host text editor (ZMACS). SEGRAS specifications are checked for consistency, completeness and termination properties, as equations are transformed to rewrite rules which are then compiled into Common Lisp functions.

A simulator for flat (non-hierarchical) Pr/T systems with lots of options and a dedicated simulator for EN-systems are available. Analysis tools for graphs and EN-systems. Coarsening and refinement. For presentation and other purposes during simulation, transitions may be associated with hand-coded Common Lisp functions. They are executed after each transition firing.

References

A list of publications concerning GRASPIN can be obtained via the above address.

7. PACE

Version number:	2.1
Year of completion:	1988
Language of the tools:	English
Language of documentation:	English

Environment

Computer system:	80386 abd 486 wutg MS-DOS and Windows 3.0 Apple Macintosh II Atari DECstation 3100 Sun-3, Sun-4, Sun386i, SPARCstation Apollo Series 2500, 3500, 4500 Hewlett-Packard HP-9000 IBM RISC System/6000 PCS Cadmus and Cadmus RCU
Graphics system:	—
Open/closed system:	open
Source program language:	Smalltalk-80
Extensibility:	Extension of tool functionality, customization of user interface

Availability

Price:	available on request
Documentation:	User's Manual
Maintenance:	Maintenance contract with update service available

Contact person:

in Germany:	GPP mbH, D-8024 Oberhaching, tel. 089 61 304 1
in Austria:	C.S.E. GmbH, A-1090 Wien, tel. 0222 48 64 72
in Switzerland:	Zühkle Engineering AG, CH-8952 Schlieren-Zürich, tel. 01 730 70 55

Name:	Jacques Dähler,
Address:	Neptunstrasse 16 CH-8280 Kreuzlingen
E-mail:	
Telephone:	+41 72 72 50 50
Telefax:	+41 72 72 16 57

Functionality

Interactive user interface based on multiple windows, context sensitive menus, mouse, etc.

Fully integrated graphical editor and interactive simulator/animator. Other functions include: statistical analysis for performance evaluation, automatic program generation, PostScript documentation, on-line help.

Customization of user interface and dynamic animation through user defined icons for nodes and tokens. Icon editor and import/export for standard paint tools.

High level Petri Nets combined with object-oriented data modelling in Smalltalk. Smalltalk-80 standard classes available for token attributes.

Hierarchical nets, refinement of t-elements and s- elements, partial net views; Timed Petri Nets; Stochastic Petri Nets.

Reusable subnet libraries.

Interactive syntax consistency analysis.

Automatic, step mode, background, forward and backward execution; debugging facilities.

Animation of any selection of net behaviour and statistical diagrams.

Automatic C program generation for fast simulation or final target implementation (e.g. for process control systems, communication protocols).

References

Application reports (in German) and list of selected installations available.

8. PAPETRI

Version number:	1.0
Year of completion:	1990
Language of the tools:	English
Language of documentation:	French and English

Environment

Computer system:	SUN
Terminal types:	graphical workstations
Operating system:	SUNOS 3.5 and upper
Graphics system:	X Windows
Open/closed system:	open
Source program language:	C, Pascal
Extensibility:	user may obtain source code new tools can be added

Availability

Price:	Educational and non-profit institutions: free
Documentation:	User's Guide
Maintenance:	No guarantee, reported errors will be fixed in due time

Contact person:

Name:	Laure Petrucci Gérard Berthelot
Institute:	Institut d'Informatique d'Enterprise
Address:	18, allée Jean Rostand BP 77 F-91002 Evry cedex
E-mail:	berthe@cnam.cnam.fr
Telephone:	(+33) 1-60 77 97 40
Telefax:	(+33) 1-60 77 96 99

or

Name:	Colette Johnen
Institute:	Laboratoire de Recherche en Informatique
Address:	Université Paris-Sud Bât. 490 F-91405 Orsay cedex
E-mail:	colette@lri.lri.fr
Telephone:	(+33) 1-69 41 64 53
Telefax:	(+33) 1-69 41 65 86

Functionality

Highly interactive tool for construction, modification, validation of P/T nets, coloured nets and algebraic nets. Manipulation of algebraic nets uses

ASSPEGIQUE, an environment for algebraic specification developed at the LRI. Ordinary nets can also be validated using rewriting techniques by means of REVE software, developed by the CRIN-Nancy and the MIT.

• Graphical editor for SUN Workstations: object manipulation with an Apple Macintosh style of interaction.

• Model Validation:
  - interactive token game (for P/T and algebraic nets)
  - composition of nets
  - computation of invariants (for P/T and coloured nets)
  - covering and minimal covering graph (for P/T and coloured nets)
  - checking classical properties using covering graph
    - (un)boundedness of every place
    - (non) quasi-liveness of every transition
    - (non) liveness of every transition
    - termination
    - fireable sequence from initial state
    - home states
    - mutual exclusion
  - analysis using rewriting techniques. The net is transformed into a set of oriented equations representing its behaviour. Knuth-Bendix's completion transforms these equations into a rewriting system. The following properties can be checked independently of the initial marking:
    - invariants
    - quasi-liveness
    - reachability
    - termination
  - validation of algebraic nets:
    - syntactical and type verification of abstract data terms
    - generation of skeleton and normed net
    - verification of termination
    - verification of P-invariants

Future plans

• decomposition of nets in order to check a given property
• computation of P-invariants for normed nets
• reachability graph for algebraic nets
• transformations (reductions) of algebraic nets.

References

C. Choppy: *ASSPEGIQUE user's manual.* Research report LRI, no. 452, Orsay, 1988.

R. Forgaard, J.V. Guttag: *REVE: a term rewriting system generator with failure resistant Knuth-Bendix.* Proc. of an NSF Workshop on the rewrite rule laboratory, 1983.

G. Berthelot, C. Johnen, L. Petrucci: PAPETRI: Environment for the Analysis of PETRI nets. Proc. of the 2nd Computer-Aided Verification Workshop, to appear in ACM/AMS DIMACS and LNCS, 1990.

9. PSItool NET

Version number:	4.0
Years of development:	1980-91
Language of the tools:	English and German
Language of documentation:	English and German

Environment

Computer system:	VAX-GPX and VAX Stations >=2000 running VMS or Ultrix. Other Unix platforms on request. An older version on DEC-VAX or Micro VAX running VMS is no longer supported.
Graphics system:	UIS for the VMS version. X-Windows/Motif for the Unix version.
Open/closed system:	
Source program language:	VMS-Pascal for the VMS version and C++ for the UNIX version.
Extensibility:	

Availability

Price:	Licenses can be bought or rented. Price upon request. Special conditions for universities and research centres. Six month's guarantee. Future updates free available within a maintenance contract.
Developer:	PSI Gesellschaft für Prozeßsteurerungs- und Informationssysteme mbH
Contact person:	

Name:	Frank Itter
Institute:	c/o PSI, Division of Simulation
Address:	Kurfürstendamm 67 D-1000 Berlin 15, Germany
E-mail:	itter@psise.uucp
Telephone:	(+49) 30-88423-0
Telefax:	(+49) 30-8824256

Functionality

With NET the user may construct, analyse, execute and animate so-called NET nets. NET nets are a modified version of predicate/transition nets. There are additional features which can be characterized by time and stochastic Petri nets. NET nets can be structured hierarchically to facilitate reading and understanding.

Aim of this set of definitions is to have a tool for practical simulation purpose. It is then on the responsibility of the user to delimit some features while modelling in order to be able to apply theoretical analysis methods.

The NET System consists of four integrated components: an Editor, a Simulator, an Analyser and an animation component.

The Editor combines syntax-sensitive editing and documentary functions and makes it possible to create hierarchical NET nets on a graphical and interactive basis.

For the net analysis features of the Petri-Net-Machine (see references) have been integrated in the tool. They consider NET nets as place/transition nets and therefore ignore node inscriptions. So the result of the analysis has to be interpreted in terms of inscribed NET nets.

During the execution a series of firing events and markings are calculated by the Simulator. It can be used in online and in offline mode.

In online mode a stepwise execution of a net is possible. Between steps the user may inspect or interactively intervene in the execution process by modifying current markings. Online statistics and firing protocols and monitoring can be used to observe an execution process. Monitoring gives a graphic representation of the qualitative development of the modelled systems dynamics. Several predefined monitors, such as bar charts and line diagrams are available.

In offline mode the Simulator makes it possible to conduct simulations in batch mode. It also enables series of experiments to be carried out for simulation investigations. Therefore all information which is necessary and desired to execute a NET net is stored in a simulation control file. It is thus possible to vary experiments without amending the net.

Simulation results are written into separate files in the form of result protocols and statistics for each simulation run. Statistical data about all markings and firings may be reported.

The animation component allows to visualize the behaviour of the execution process in a net independent and application specific graphical form on the screen.

Nets of up to 5000 nodes have been edited and executed. Depending on the size of a net, the complexity of the inscriptions and the number of current tokens, about one second of CPU-time is consumed by up to 50 firings.

The NET system needs 15 MByte diskspace. There are about 50 installations in industrial and research environments. The application areas are mainly in the field of production system simulation.

Future plans

There are plans to develop the PSItool NET further. Presently, dedicated net-pattern-libraries for various production systems are under construction and soon additional features for performance evaluation will be available. The Ultrix/-Motif version written in C++ will be ported to VMS/Motif in order to substitute the older VMS/UIS version.

References

Net User's Manual and other internal reports available upon request. Published application reports in English are:

F. Itter, C. Relewicz: Computer Supported Design of Kanban Controlled Production-Integrated System Analysis and Simulation with higher Order Petri Nets; in G. Rozenberg (ed.): Advances in Petri Nets 1990; Lecture Notes in Computer Science, Springer Verlag.

F. Itter, M. Lang: Simulation of a Chemical Production System Using Petri Nets; in R. Eckermann (ed.): European Symposium on Computer Application in the Chemical Industry, Erlangen, Germany, Dechema, vol. 116, 1989.

A separate decription of the analysis features used are given in:

P.H. Starke: Validation of Petri Net Models by Petri-Net-Machine; in A. Tal (ed.), Oxford, England, Pergamon, 1986.

10. SPECS

Version number:	1.0
Year of completion:	1991
Language of the tools:	English
Language of documentation:	English

Environment

Computer system:	Apple Macintosh II 386-based PC with MS-DOS, Windows 3.0 SUN workstation other workstations with Objectworks 2.5 Transputer board (needed for very fast simulation only)
Operating system:	Objectworks 2.5 (in future Objectworks 4.0)
Graphics system:	Objectworks 2.5 (in future Objectworks 4.0)
Open/closed system:	open
Source program language:	Objectworks for Smalltalk-80, OCCAM-2
Extensibility:	source code available

Availability

Price:	educational institutions: nominal fee (non disclosure contract) others: to be announced
Documentation:	introductory user manual user reference manual SpecsLingua report annotated examples
Maintenance:	reported errors will be fixed

Contact person:

Name:	Rob Esser Rudi Mattmann Heinz Oswald
Institute:	Corporate Reserch & Development - 4001 Landis & Gyr Betriebs Corp.
Address:	CH-6301 Zug Switzerland
E-mail:	
Telephone:	(+41) 42 24 11 24
Telefax:	(+41) 42 24 46 93

Functionality

high level Petri nets

inscription language for transitions, places, and arcs: SpecsLingua (Pascal like, functional, strong type checking) graphical editor for nets, textual editor for inscriptions

integrated simulator/animator

hierarchical structuring of nets

subnet libraries, extendible by user, any subnet can be made reusable

IO elements for entering data and displaying results during simulation

hypertext like extensions for documenting nets

transparent very fast simulation on Transputer based parallel computer system

integrated tools for
- accessing, naming, and managing several nets and subnet libraries
- definition of reusable constants, types, and functions
- constructing icons replacing the standard form of net elements
- constructing IO elements
- net analysis for validating the net decomposition
- generating documents in PostScript format (including EPSF)

longtime saving and restoring of nets

Future plans

Support for Objectworks 4.0 (April 1991)
further analysis methods

References

Bütler B., Esser R., Mattmann R. *"A Distributed Simulator for High Order Petri Nets"*, Tenth International Conference on Application and Theory of Petri Nets, Bonn, 1989.

Oswald H., Esser R., Mattmann R. "An Environment for Specifying and Executing Hierarchical Petri Nets", IEEE 12th Intern. Conference on Software Engineering, Nice, IEEE 1990.

Bütler B., Esser R., Mattmann R. *"A Distributed Simulator for High Order Petri Nets"*, in G. Rozenberg (ed.) Advances in Petri Nets 1990, Springer Lecture Notes in Computer Science.

Other remarks

SPECS is the successor of the original SPECS tool which has been developed at ETH Zürich.

Appendix
Petri Net Activities

This appendix describes some of the on-going activities in the area of Petri nets. Details about the conferences, publications, tutorials and advanced courses can be found in the Petri Net Newsletter—described below.

International Conference on Application and Theory of Petri Nets

This conference was formerly called the European Workshop on Applications and Theory of Petri Nets and the meetings have taken place annually since 1980:

1980	Strasbourg, France	1987	Zaragoza, Spain
1981	Bad Honnef, Germany	1988	Venice, Italy
1982	Varenna, Italy	1989	Bonn, Germany
1983	Toulouse, France	1990	Paris, France
1984	Aarhus, Denmark	1991	Aarhus, Denmark
1985	Espoo, Finland	1992	Sheffield, UK
1986	Oxford, UK		

The conference has approximately 200 participants—from industry, universities and research institutions. The conference takes place in the last week of June.

International Workshop on Petri Nets and Performance Models

This conference was formerly called the International Workshop on Timed Petri Nets and the meetings have taken place each second year since 1985:

1985	Torino, Italy	1989	Kyoto, Japan
1987	Madison, USA	1991	Melbourne, Australia

The conference has approximately 100 participants—from industry, universities and research institutions. The conference deals primarily with timed and stochastic Petri nets.

Petri Net Newsletter

The newsletter is published by the Special Interest Group on Petri Nets and Related System Models of the Gesellschaft für Informatik, three times a year. The newsletter contains short research articles, announcements of meetings, abstracts of recent publications, etc. The newsletter is edited by O. Herzog, W. Reisig and R. Valk and can be ordered from the following address:

Prof. W. Reisig	Phone:	+49 89 2105 2405
Institut für Informatik, TU München	Telefax:	+49 89 2105 8207
Postfach 202420	E-mail:	reisig@lan.informatik.
D-8000 München 2		tu-muenchen.dbp.de
Germany		

Advances in Petri Nets

This series is published by the Springer-Verlag (within the Lecture Notes in Computer Science series). The volumes are edited by G. Rozenberg. The intention of this series is to present the most significant recent results in the application and theory of Petri Nets to the broad computer science community. The majority of papers appearing in the Advances of Petri Nets are selected from those presented at the conferences. These papers appear in the Advances in a revised and improved form. Research papers and survey articles for the Advances can also be submitted directly to:

Prof. G. Rozenberg	Phone:	+31 71 27 70 55
Dep. of Mathematics & Computer Science	Telefax:	+31 71 27 58 19
University of Leiden, Niels Bohr Weg 1	E-mail:	rozenber@hleru15.bitnet
P.O. Box 9512, NL-2300 CA Leiden		
The Netherlands		

Tutorials and Advanced Courses

Tutorials on Petri nets are offered in connection with each of the two conferences mentioned above. The length of the tutorials is one or two days. Moreover advanced courses are organized periodically, in order to present the progress in Petri nets in a systematic way. Until now there have been two such courses: Hamburg 1979 and Bad Honnef 1986. The material from these courses has also been published by Springer-Verlag (within the Lecture Notes in Computer Science series).

Contributing Authors

Hassane Alla
Laboratoire d'Automatique de Grenoble
Institut National Polytechnique de Grenoble
Domaine Universitaire, BP. 46
F-38402 Saint-Martin-d'Hères Cedex
France
Phone: +33 76 82 62 44
Telefax: +33 76 82 63 88

Gianfranco Balbo
Dipartimento di Informatica
Università degli Studi di Torino
Corso Svizzera 185
I-10149 Torino
Italy
Phone: +39 11 771 2002
Telefax: +39 11 751 603
E-mail: balbo@di.unito.it

Marco Baldassari
Dipt. di Automatica e Informatica
Politecnico di Torino
Corsa Duca degli Abruzzi, 24
I-10129 Torino
Italy
Phone: +39 11 564 7003
Telefax: +39 11 564 7099

Eugenio Battiston
Dipt. di Scienze dell'Informazione
Università degli Studi di Milano
via Moretto da Brescia 9
I-20131 Milano
Italy
Phone: +39 2 7575. 266
Telefax: +39 2 7611 0556
E-mail: bernardi@imiucca.unimi.it

Jonathan Billington
Telecom Australia
Research Laboratories
770 Blackburn Road
Clayton North, Vic., 3168
Australia
Phone: +61 3 541 6416
Telefax: +61 3 544 2362
E-mail: j.billington@trl.oz.au

Steven C. Bruell
Computer Science Department
University of Iowa
Iowa City, IA 52242
USA
Phone: +1 319 335 0734
Telefax: +1 319 335 0627
E-mail: bruell@cs.uiowa.edu

Giorgio Bruno
Dipt. di Automatica e Informatica
Politecnico di Torino
Corso Duca degli Abruzzi, 24
I-10129 Torino
Italy
Phone: +39 11 564 7003
Telefax: +39 11 564 7099
E-mail: bruno@itopoli.bitnet

Juan A. Carrasco
Departament d'Enginyeria Electrònica
UPC (Univ. Politécnica de Catalunya)
Diagonal 647, plta. 9
E-08028 Barcelona
Spain
Phone: +34 3 401 6652
Telefax: +34 3 401 6600
E-mail: carrasco@eel.upc.es

Christian Chatelain
Gec-Alsthom
Tour Neptune, Cedex 20
F-92086 Paris la Défense Cedex
France
Phone: +33 1 47 44 98 89
Telefax: +33 1 47 44 90 15

Ghassan Chehaibar
BULL Corporate Research Center
Bldg F3.2G.28
Rue Jean Jaures
F-78340 Les Clayes-Sous-Bois
France
Phone: +33 1 30 80 69 15
Telefax: +33 1 30 80 69 53

Po-zung Chen
Computer Science Department
Tamkang University
Tamsui, Taipei Hsien
Taiwan 25137
Republic of China

Giovanni Chiola
Dipartimento di Informatica
Università degli Studi di Torino
Corso Svizzera 185
I-10149 Torino
Italy
Phone: +39 11 771 2002
Telefax: +39 11 751 603
E-mail: chiola@di.unito.it

Fiorella De Cindio
Dipt. di Scienze dell'Informazione
Università degli Studi di Milano
via Moretto da Brescia 9
I-20131 Milano
Italy
Phone: +39 2 7575. 220
Telefax: +39 2 7611 0556
E-mail: decindio@imiucca.unimi.it

Jean M. Couvreur
Laboratoire MASI
Université Pierre et Marie Curie
4, Place Jussieu
F-75252 Paris Cedex 05
France
Phone: +33 1 44 27 43 63
Telefax: +33 1 44 27 62 86
E-mail: couvreur@masi.ibp.fr

Claude Dutheillet
Laboratoire MASI
Université Paris 6
4, Place Jussieu
F-75252 Paris Cedex 05
France
Phone: +33 1 44 27 31 92
Telefax: +33 1 44 27 62 86
E-mail: dutheillet@masi.ibp.fr

Frits Feldbrugge
Philips TDS
P.O. Box 245
7300 AE Appeldorn
The Netherlands
Phone: +31 55 43 24 87
Telefax: +31 55 43 24 87
E-mail: frits@idca.tds.uucp

Giuliana Franceschinis
Dipartimento di Informatica
Università degli Studi di Torino
Corso Svizzera 185
I-10149 Torino
Italy
Phone: +39 11 771 2002
Telefax: +39 11 751 603
E-mail: giuliana@di.unito.it

Hartmann J. Genrich
Institut für Methodische Grundlagen
GMD
Schloss Birlinghoven
D-5205 Sankt Augustin 1
Germany
Phone: +49 2241 14 0
Telefax: +49 2241 14 2889
E-mail: genrich@gmdzi.gmd.de

Claude Girault
Laboratoire MASI
Université PARIS 6
4, Place Jussieu
F-75252 Paris Cedex 05
France
Phone: +33 1 44 27 43 63
Telefax: +33 1 44 27 62 86
E-mail: girault@masi.ibp.fr

Serge Haddad
Laboratoire MASI
Université PARIS 6
4, Place Jussieu
F-75252 Paris Cedex 05
France
Phone: +33 1 44 27 39 21
Telefax: +33 1 44 27 62 86
E-mail: haddad@masi.ibp.fr

Peter Huber
Meta Software Corporation
150 Cambridge Park Drive
Cambridge, MA 02140
USA
Phone: +1 617 576 6920
Telefax: +1 617 576 0519
E-mail: huber%metasoft@bbn.com

Arne M. Jensen
Mølleparken 8
DK-8320 Mårslet
Denmark
Phone: +45 86 29 56 12

Kurt Jensen
Computer Science Department
Aarhus University
Ny Munkegade, Bldg. 540
DK-8000 Aarhus C
Denmark
Phone: +45 86 12 71 88
Telefax: +45 86 13 57 25
E-mail: kjensen@daimi.aau.dk

Leif O. Jepsen
RC International
Klamsagervej 19
DK-8230 Aabyhøj
Denmark
Phone: +45 86 25 04 11
Telefax: +45 86 25 09 92

Bernd Krämer
Institut für Systemtechnik
GMD
Schloss Birlinghoven
D-5205 Sankt Augustin 1
Germany
Phone: +49 2241 14 2448
Telefax: +49 2241 14 2618
E-mail: kraemer@gmdzi.gmd.de

Pierre Ladet
Laboratoire d'Automatique de Grenoble
Institut National Polytechnique de Grenoble
Domaine Universitaire, BP. 46
F-38402 Saint-Martin-d'Hères Cedex
France
Phone: +33 76 82 62 44
Telefax: +33 76 82 63 88

Chuang Lin
Information Science Institute
State Information Center
Beijing
Republic of China

Markus Lindqvist
Nokia Telecommunications
P.O.Box 33
SF-02601 Espoo
Finland
Phone: +358 0 511 61
Telefax: +358 0 511 5615
E-mail: lindqvist@tele.nokia.fi

Dan C. Marinescu
Department of Computer Science
Purdue University
West Lafayette, IN 47907
USA
Phone: +1 317 494 6018
Telefax: +1 317 494 0739
E-mail: dcm@cs.purdue.edu

Javier Martínez
Departamento de Ingeniería Eléc. e Informática
Universidad de Zaragoza
María de Luna 3
E-50015 Zaragoza
Spain
Phone: +34 76 51 72 74
Telefax: +34 76 51 29 32
E-mail: javier@etsii.unizar.es

Giancarlo Mauri
Dipartimento di Scienze dell'Informazione
Università degli Studi di Milano
via Moretto da Brescia 9
I-20131 Milano
Italy
Phone: +39 2 7575. 231
Telefax: +39 2 76110556
E-mail: decindio@hermes.unimi.it

Gérard Memmi
MA30/821A
300 Concord Road
Billerica, MA 01821
USA
Phone: +1 508 294 2617
Telefax: +1 508 294 2617
E-mail: memmi@pws.bull.com

Pedro R. Muro
Departamento de Ingeniería Eléc. e Informática
Universidad de Zaragoza
María de Luna 3
E-50015 Zaragoza
Spain
Phone: +34 76 51 72 74
Telefax: +34 76 51 29 32
E-mail: muro@etsii.unizar.es

Valerio Pinci
Meta Software Corporation
150 Cambridge Park Drive
Cambridge, MA 02140
USA
Phone: +1 617 576 6920
Telefax: +1 617 576 0519
E-mail: pinci%metasoft@bbn.com

Wolfgang Reisig
Computer Science Department
Technical University of Munich
Arcisstrasse 21
D-8000 Munich 2
Germany
Phone: +49 89 2105 2405
Telefax: +49 89 2105 8207
E-mail: reisig@lan.informatik.tu-muenchen.dbp.de

Heinz-Wilhelm Schmidt
International Computer Science Inst.
1947 Center Street, Suite 600
Berkeley, CA 94704-1105
USA
Phone: +1 415 643 9153
Telefax: +1 415 643 7684
E-mail: hws@icsi.berkeley.edu

Robert M. Shapiro
Meta Software Corporation
150 Cambridge Park Drive
Cambridge, MA 02140
USA
Phone: +1 617 576 6920
Telefax: +1 617 576 0519
E-mail: robert%metasoft@bbn.com

Manuel Silva
Departamento de Ingeniería Eléc. e Informática
Informática
Universidad de Zaragoza
María de Luna 3
E-50015 Zaragoza
Spain
Phone: +34 76 51 72 74
Telefax: +34 76 51 29 32
E-mail: silva@etsii.unizar.es

Jacques Vautherin
Lab. de Recherche en Informatique
Université Paris Sud, Bat. 490
F-91405 Orsay Cedex
France

Geoffrey Wheeler
Telecom Australia
Research Laboratories
770 Blackburn Road
Clayton North, Vic., 3168
Australia
Phone: +61 3 541 6415
Telefax: +61 3 544 2362
E-mail: g.wheeler@trl.oz.au

Michael Wilbur-Ham
Telecom Australia
Corporate Customer Division
11/172 William Street
Melbourne, Vic., 3000
Australia
Phone: +61 3 606 6932
Telefax: +61 3 670 2562

oležal, Dampferzeugung

erichtigungen

eite	Ort	falsch	richtig
19	8.Z.v.o.	um 13 %	bis zu 13 %
20	5.Z.v.o.	...Überhitzung...	...Zwischenüberhitzung...
46	Gl.(4.38)	$r_b \cdot \gamma_{H_2OB}$	$r_b \cdot \mu_{H_2OB}$
11	Gl.(13.2)u.f.	dt_K	dt
16	6.Z.v.o.	10^2	10^{-2}
16	1.Z.v.u.	(4.9)	(4.5)
81	6.Z.v.o.	Bild 22.10	Bild 22.9
84	3.Z.v.o.	Bild 22.16	Bild 22.15
92	2.Z.v.u.	...Rohrwasser...	...Rohwasser...
15	1.Z.v.o.	(22.11)	(22.21)
17	11.Z.v.o.	$\Delta p_{AU} = \kappa \Delta p_w$	$\Delta p_{AU} = \Delta p_w$
19	1.Z.v.o.	Abschnitt 22.5	Abschnitt 22.4
25	5.Z.v.o.	(25.18)	(25.14)
27	10.Z.v.o.	Bild 33.14	Bild 33.11
45	1.Z.v.o.	Abschn. 22.2	Abschn. 22.1.2
58	Bild 32.10	kJ/kg	kJ/kgK
63	4.Z.v.o.	Abschn. 32.3.1	Abschnitt 32.3.2
64	11.Z.v.o.	(22.24)	(22.25)
64	Gleichungen für A_{SR} und Z_{SR} miteinander vertauschen		
65	10.Z.v.o.	(33.2)	(33.3)
69	1.Z.v.o.	(25.18)	(22.14)
69	Bild 33.5	Δp innerhalb des Diagramms entfällt	
75	3.Z.v.u.	(33.22)	(33.32)
78	19.Z.v.o.	Bild 40.1	Bild 39.3
78	5.Z.v.u.	Abschnitt 39.1	Abschnitt 38.1
82	7.Z.v.u.	...38.18 die Trommel...	...26.11 die Trommel mit Innenmantel...
96	2.Z.v.u.	...$\dot{M}_D$ aus. Der an der Stelle 1 in...	...$\dot{M}'$ aus. Der in den Dampf...
98	unterste Gl.	$\chi = \frac{1}{v} \frac{\partial v}{\partial p_T}$	$\chi = - \frac{1}{v} \left(\frac{\partial v}{\partial p} \right)_T$
18	Bild 38.4	untere und obere Bildbeschriftung vertauschen	